STRUCTURAL MECHANICS COMPUTER PROGRAMS
Surveys, Assessments, and Availability

Structural Mechanics Computer Programs

Surveys, Assessments, and Availability

Edited by

W. Pilkey
K. Saczalski
H. Schaeffer

University Press of Virginia
Charlottesville

THE UNIVERSITY PRESS OF VIRGINIA

First published 1974

Library of Congress Cataloging in Publication Data

Pilkey, Walter D
Structural mechanics computer programs.

Papers presented at a symposium held at the University of Maryland, June 1974.
1. Structural engineering--Computer programs--Congresses. I. Saczalski, K., joint author. II. Schaeffer, Harry G., joint author. III. Title.
TA641.P54 624'.17'0285425 74-8300 ISBN 0-8139-0566-4

Printed in the United States of America

PREFACE

The growth and extension of computational capabilities which have taken place during the last decade have resulted in a quantum leap forward in the complex problems in many technological disciplines, including structural mechanics. Unfortunately, these capabilities have not been achieved without considerable growing pains. Concomitant with the greatly increased computational capabilities for the designer, the analyst, and the technician is the likelihood that future progress in the computing technology field will be accompanied by many other severe problems.

One particular discipline where problems are manifold is the field of structural mechanics software development. Inaccessibility of software, lack of user confidence in software developed by others, inadequate documentation, and lack of unified theoretical and algorithmic approaches to the solution of complex structural problems in many functional and disciplinary areas have led to much duplication of effort. As a result, a paradoxical situation arises wherein considerable effort and funds are expended annually for software development, but the engineers and scientists for whom this software has presumably been generated cannot fully absorb and use this technology.

As a result of the concern for the many problems associated with structural mechanics software development and dissemination, a unique symposium was conceived and held at the University of Maryland in June 1974. In contrast to traditional meetings emphasizing theoretical aspects and/or numerical algorithms this symposium highlighted the computer program, the program developer, and the program user.

The symposium sessions were categorized into three broad areas: (1) user response to structural mechanics software and interaction between program developers and the user community; (2) reviews and summaries of available computer programs, relying heavily upon responses to questionnaires sent to program developers and users as well as descriptions of code capabilities provided by developers; and (3) future considerations and possible means of facilitating the development and dissemination of analysis and design programs.

In addition, representative programs of the current state-of-the-art in structural mechanics software were modified by the addition of preprocessors and postprocessors to permit execution in an interactive mode during and prior to the symposium. The assistance of several program developers who modified their own programs and Y. H. Chen, A. Jay, G. Johnson, C. Thasanatorn, and D. D. Lee of the University of Virginia in modifying programs provided by other developers is gratefully appreciated.

The quality of the papers included in this volume is a tribute to the effort expended by each of the authors. Several of the authors were directly supported by the National Science Foundation or by the Office of Naval Research. The editors of this volume would like to take this opportunity to recognize, in a collective sense, all of these contributions and to express our appreciation. The encouragement, support, and advice of Dr. Nicholas Perrone of the Office of Naval Research and Dr. Michael Gaus of the National Science Foundation has been a major factor leading to the symposium and the publication of this volume.

The symposium, from which the papers in this volume were taken, was sponsored jointly by the National Science Foundation and the Office of Naval Research and was presented in cooperation with the University of Maryland and the University of Virginia. Special note is also made of the cooperation and support received from the American Institute of Aeronautics and Astronautics, The American Society of Civil Engineers, The American Society of Mechanical Engineers, The American Society of Engineering Education, the Society of Naval and Marine Architects, and the Society of Automotive Engineers.

Special thanks are due to A. Jay and S. DeMasters of the University of Virginia, who assisted in the collection of information, including the distribution of questionnaires, on behalf of many of the authors.

We wish to acknowledge the major editorial assistance of B. Pilkey. To a great extent, the uniformity of appearance of the papers is due to the editorial, typing, and artistic efforts of S. DeMasters, S. Wood, R. Turk, D. Bibb, J. Hawkins, M. Rovin, and C. Lepine.

March 1974

W. Pilkey
K. Saczalski
H. Schaeffer

CONTRIBUTORS

R. J. Allwood
The Genesys Center
University of Technology
Loughborough Leicestershire
Leics LE11, 3TU England

B. O. Almroth
Lockheed Palo Alto Research Labs
Palo Alto, California 94304

H. A. Armen, Jr.
Applied Mechanics Group
The Grumman Aircraft Corp.
Bethpage, Long Island, New York 11714

M. L. Baron
Weidlinger & Associates
110 East 59th Street
New York, New York 10022

K. J. Bathe
Dept. of Civil Engineering
University of California
Berkeley, California 94720

T. Belytschko
Dept. of Engineering Materials
University of Illinois at Chicago Ctr.
Chicago, Illinois 60680

S. E. Benzley
Analytical Development Div. 1541
Sandia Laboratories
Albuquerque, New Mexico 87115

L. J. D. Bernier
Air Force Flight Dynamics Lab.
Wright-Patterson Air Force Base
Ohio 45433

D. Bushnell
Dept. 52-53, Building 205
Lockheed Palo Alto Research Ctr.
Palo Alto, California 94304

H. N. Christiansen
Dept. of Civil Engineering
Brigham Young University
Provo, Utah 84601

D. E. Cooley
Vehicle Dynamics Division
Air Force Flight Dynamics Lab.
Wright-Patterson Air Force Base
Ohio 45433

I. W. Cotton
Institute for Computer
Sciences and Technology
National Bureau of Standards
Washington, D. C. 20234

G. E. Everstine
Code 1844
Naval Ship Research
Development Center
Bethesda, Maryland 20034

R. H. Ewald
Computer Center
University of Colorado
Boulder, Colorado 80302

L. H. Feeser
Computing Center
University of Colorado
Boulder, Colorado 80302

C. A. Felippa
Lockheed Palo Alto Research Labs.
Palo Alto, California 94304

S. J. Fenves
Dept. of Civil Engineering
Carnegie-Mellon University
Pittsburgh, Pa. 15213

E. I. Field
Universal Analytics, Inc.
7151 West Manchester Avenue
Los Angeles, California 90045

G. Gaggero
Commission of the European
Communities
Joint Research Center
Ispra Establishment
21020 Ispra, Italy

D. Gouirand
Dept. of Civil Engineering
Carnegie-Mellon University
Pittsburgh, Pa. 15213

K. K. Gupta
Jet Propulsion Laboratory
4800 Oak Grove Drive
Pasadena, California 91103

W. E. Haisler
Dept. of Aerospace Engineering
Texas A & M University
College Station, Texas 77843

J. K. Haviland
Dept. of Engineering Science and Systems
University of Virginia
Charlottesville, Virginia 22901

E. Heer
Jet Propulsion Laboratory
4800 Oak Grove Drive
Pasadena, California 91103

E. D. Herness
Boeing Computer Services
P. O. Box 24346
Seattle, Washington 98124

A. Jay
Dept. of Engineering Science and Systems
University of Virginia
Charlottesville, Virginia 22901

S. E. Johnson
Universal Analytics, Inc.
7151 West Manchester Avenue
Los Angeles, California 90045

R. F. Jones, Jr.
Code 1745, Structures Dept.
Naval Ship Research & Development Ctr.
Bethesda, Maryland 20034

H. A. Kamel
Dept. of Aerospace Engineering
The University of Arizona
Tucson, Arizona 85721

N. S. Khot
Air Force Flight Dynamics Labs.
Wright-Patterson Air Force Base
Ohio 45433

J. L. Korf
Virginia Highway Research Council
Box 3817, University Station
Charlottesville, Virginia 22901

W. E. Lorensen
Watervliet Arsenal
Computer Science Office
Watervliet, New York 12189

P. V. Marcal
Division of Engineering
Brown University
Providence, Rhode Island 02912

M. W. McCabe
Dept. of Aerospace Engineering
The University of Arizona
Tucson, Arizona 85721

J. M. McCormick
Weidlinger & Associates
110 East 59th Street
New York, New York 10022

J. M. McKee
Code 1844
Naval Ship Research Development Ctr.
Bethesda, Maryland 20034

W. T. McKeel
Virginia Highway Research Council
Box 3817, University Station
Charlottesville, Virginia 22901

J. S. Mescall
Army Materials and Mechanics Research Center
Watertown, Massachusetts 02172

C. Mongini-Tamagnini
Commission of the European Communities
Joint Research Centre
Ispra Establishment
21020 Ispra, Italy

R. Monti
Instituto Di Aerodinamica
Universita Degli Studi di Napoli
P. Le V. Tuchio, 80 Napoli, Italy

P. Murio
Instituto Di Aerodinamica
Universita Degli Studi di Napoli
P. Le V. Tuchio, 80 Napoli, Italy

L. G. Napolitano
Instituto Di Aerodinamica
Universita Degli Studi di Napoli
P. Le V. Tuchio, 80 Napoli, Italy

M. Newman
Computer Aided Design Group
Israel Aircraft Industries, Ltd.
LOD Airport
Israel

R. E. Nickell
Analytical Development Div. 1541
Sandia Corporation
Albuquerque, New Mexico 87115

D. M. Parks
Brown University
Providence, Rhode Island 02912

W. D. Pilkey
Dept. of Engineering Science and Systems
University of Virginia
Charlottesville, Virginia 22901

J. Potts
Code 1745, Structures Dept.
Naval Ship Research & Development Ctr.
Bethesda, Maryland 20034

D. M. Purdy
Structural Mechanics Subdiv.
Douglas Aircraft Company
McDonnell Douglas Corporation
Long Beach, California 90801

D. R. Rehak
Dept. of Civil Engineering
Carnegie-Mellon University
Pittsburgh, Pa. 15213

N. F. Rieger
Rochester Institute of Tech.
Dept. of Mechanical Engr.
Rochester, New York 14603

G. Ruoff
Lehrstuhl für Baumechanik
Technische Hochschule
3000 Hannover
Welfengarten 1, Germany

R. S. Schiffman
Dept. of Civil Engineering
University of Colorado
Boulder, Colorado 80302

E. Schrem
Institute Für Statik und Dynamic Der Luft-und Raumfahrtkon
Universitat Stuttgart
7 Stuttgart 80,
Pfaffenwaldring 27, Germany

E. Schweid
Computer Aided Design Group
Israel Aircraft Industries, Ltd.
LOD Airport
Israel

J. Shomrat
Computer Aided Design Group
Israel Aircraft Industries, Ltd.
LOD Airport
Israel

E. Stein
Lehrstuhl für Baumechanik
Technische Hochschule
3000 Hannover
Welfengarten 1, Germany

R. B. Stillman
Systems & Software Division
National Bureau of Standards
Washington, D. C. 20234

H. Stralberg
Universal Analytics, Inc.
7151 West Manchester Avenue
Los Angeles, California 90045

J. A. Stricklin
Dept. of Aerospace Engineering
Texas A & M University
College Station, Texas 77843

J. A. Swanson
Swanson Analysis Systems, Inc.
Elizabeth, Pa. 15037

V. A. Tischler
Air Force Flight Dynamics Labs.
Wright-Patterson Air Force Base
Ohio 45433

J. L. Tocher
Boeing Computer Services
P. O. Box 24346
Seattle, Washington 98124

D. W. Vannoy
Dept. of Civil Engineering
University of Virginia
Charlottesville, Virginia 22901

V. B. Venkayya
Air Force Flight Dynamics Lab.
Wright-Patterson Air Force Base
Ohio 45433

W. A. Von Riesemann
Analytical Development Div. 1541
Sandia Corporation
Albuquerque, New Mexico 87115

E. L. Wilson
Dept. of Civil Engineering
University of California
Berkeley, California 94720

W. B. Wright
A. D. Little, Inc.
25 Acorn Park
Cambridge, Mass. 02139

CONTENTS

I. REVIEWS AND SUMMARIES OF PROGRAMS

NONLINEAR CONTINUA

W. A. Von Riesemann
Sandia Laboratories
Albuquerque, New Mexico

J. A. Stricklin and W. E. Haisler
Texas A&M University
College Station, Texas

ABSTRACT

The software available for nonlinear continua is surveyed, and where possible assessments are made. Primarily, the nonlinearities considered are geometric (large displacement), material, and large strain. The main emphasis of the paper is on the actual software and not on the methods of analysis, though they are briefly discussed. The programs surveyed fall into two categories: the so-called hydrodynamic or wave propagation codes, which are primarily finite difference, and structural response codes,which primarily use the finite element method for solids and both finite difference/element methods for shells. Some guidelines and suggestions for potential users are discussed. The survey includes programs which will be available in the near future.

INTRODUCTION

This paper is concerned with nonlinear structural computer codes primarily for continua. Nonlinearities [1] in a structural system may be due to: geometric nonlinearities (the so-called large displacement problem), material nonlinearities, large strains, nonlinear boundary conditions (such as opening and closing of gaps, sliding boundaries, ablation), thermal effects, nonconservative loading, cracking, buckling and instability, and certainly other effects. The main emphasis in this paper will be on the first three items; the fourth and fifth items will only be briefly covered, while the others are the subject matter of other papers in this volume.

The analyst (potential user of a program) is thus faced with knowing a priori which nonlinearities are present in his problem; however, in most cases he must examine several of the above items to ascertain which ones are, in fact, present. As an aside, it should be noted that unlike linear analyses, a nonlinear analysis may require several attempts with a given program in order to obtain a correct solution.

The field of nonlinear analysis has progressed rapidly over the last few years, and the user of a program must be cognizant of the current state-of-the-art. In addition, one must be aware of the fact that there is no single method of solution for nonlinear problems. For example, what might be recommended for geometric nonlinearities will not be appropriate for material nonlinearities.

Furthermore, for geometric nonlinearities and for large strain problems, either a Lagrangian, an Eulerian, or a modified formulation may be used. In conjunction with these formulations different stress and strain measures are required. This also requires caution in the exact definition of the required input material definition to the code (e.g., engineering vs. true stress and strain).

For static problems most programs use a Newton-Raphson method or a simple modification to solve the nonlinear equilibrium equations. A few use a simple iterative scheme, but caution must be exercised in this case as the method might not converge.

In the case of dynamic problems the transient response is usually obtained by either an explicit or an implicit time integrator. Both methods have merit and the "best" method depends on such factors as the geometry of the structure, frequency content of the loading, and the frequency response of interest to the user. Both consistent and lumped mass matrices are used. Recently, mode superposition [2] has been applied for nonlinear problems; however, the method is too new to comment on.

It should also be noted that in the finite element method the nonlinearities may be incorporated in two different ways. Roughly speaking, in one approach, the so-called tangent stiffness method, the element stiffness matrices are recomputed at each increment; in the other approach (either the initial strain or stress methods) the nonlinearities are treated as pseudo loads and the stiffness matrix remains unchanged. Again, both approaches have merit.

In regard to material nonlinearities many of the programs include material models only for plastic behavior of metals. That is, soils, rocks, foams, and such materials are only beginning to be considered. For all of these materials, including metals, there is much yet to be learned about their nonlinear behavior and characterization. See, for example, [3] which examines in detail the various hardening rules for metals.

Finally, the degree and type of nonlinearity play a role in what method or methods are practicable.

It is not the intent of this paper to discuss all of the ramifications of nonlinear structural analysis. Rather the reader is directed to the ever-growing literature in the field. See, for example, References [1,4,5,6].

Again it must be emphasized that to evaluate properly and to use a program the user must know the methods of nonlinear analysis, since all of the points raised above, in addition to others, play a role in the performance (idiosyncrasies) of a program. There are no clear-cut answers (and perhaps never will be) as to which method is the best. In fact, due to varying requirements (e.g., design, analysis, degree of accuracy required) several programs (i.e., methods) may be desirable even for a given class of problems.

A listing of nonlinear programs for wave propagation, two-dimensional solids, general purpose, shell of revolution and lumped parameters follows. In many cases the information was taken directly from user's manuals, abstracts, or developers' responses. After the listing of codes, some brief comments are made regarding selection of program. The sources for the programs are given in Appendix A.

WAVE PROPAGATION CODES

In this section we shall discuss the category of codes which have been variously titled as hydro or hydrodynamic, or shock wave, or wave propagation codes. In general, they are finite difference codes, one exception being SWAP, a method of characteristics code, involving the response of materials subjected to severe loadings. They are not concerned with structural response but rather material response; however, the distinction in meaning between these two categories is not clear cut. An example of a problem which these codes exclude is the flexural response of a beam. However, there is an area of overlap with the two- and three-dimensional codes (primarily finite element) which will be discussed below. In general, the wave propagation codes are concerned with the first few microseconds (possibly milliseconds) of response of a structure subjected to a severe environment. Most of them contain elaborate equations of state, that is, material models. In order to run the codes the material characteristics must be fairly well known, and thus one tends to see a close relationship between experimental groups and code developers and users. For many problems the codes are long running; several hours on a CDC 6600 is not too uncommon. Also, most have been continually updated and maintained over several years, and therefore it is important to check with the developers of the codes to make sure the latest version is in fact being used. For example, rezoning in one of the codes has reduced the execution time by a factor of 5 to 20. The question of Lagrangian vs. Eulerian coordinates will not be discussed herein; the interested reader should see, e.g., [7].

Due to space limitations, only a few of the available codes will be reported herein. The ones included here were chosen primarily on the basis of their availability and the fact that they have been maintained. As an indication of the number of codes available, a limited survey prepared in 1970 [8] lists 38 codes.

One-Dimensional Codes: Lagrangian Method

CHART D [9,10] is a code for computing coupled hydrodynamic motion and radiation diffusion. Lagrangian finite difference equations of motion including energy transport terms are solved in one-dimensional rectangular, cylindrical, or spherical coordinates. Elastic-plastic (p-α model) and high explosive material models are available. Thermal and electron conduction, material spall, and later rejoin calculations are provided. In addition, voids or gaps may be modeled.

WONDY-IV [11] is the latest version of a series of one-dimensional finite difference wave propagation programs called WAVE and WONDY. The program solves the finite difference equations for one-dimensional motion using a Lagrangian coordinate system. Geometries may be either spherical or cylindrical with motion in the radial direction only or rectangular with motion along one axis only (uniaxial strain). For problems involving the propagation of pressure discontinuities or shock waves, the method of artificial viscosity is used. Equations of state (material models) are available for a number of different materials, including hydrodynamic, elastic-plastic, ideal gasses, detonating high explosives, and porous (p-α). The elastic-plastic model is based on the von Mises yield criterion in which the yield strength can be infinite, constant, or variable. The program is continually being updated with additional equations of state. There is a restart capability, and plotting using the SD-4020 is available. The program is written in FORTRAN IV for the CDC 6600 and requires about 61,500 storage locations. The program runs at a rate of between 4.5 and 10 million zone-cycles (number of mass elements times the number of time increments) per hour. The code is available from the Argonne Code Center.

PISCES-1DL [12] is a one-dimensional explicit Lagrangian finite difference shock wave code developed by Physics International Company, San Leandro, Calif. The code calculates the nonlinear (material, strain, and geometry) response due to such occurrences as impact, explosions, projectile penetration, and ricochet. The code contains the following options: (a) static solutions, (b) extending rezone, (c) microzone, (d) heat flow and restart. The program is available on the CDC CYBERNET system.

One-Dimensional Code: Method of Characteristics

SWAP (Stress Wave Analyzing Program) [13] analyzes problems involving shock waves in one-dimensional strain configurations in an elastic-plastic solid. The method of characteristics approach is used, which represents all wave shapes by a series of shock waves. The problems which can be solved include those involving both hydrostatic and elastic-plastic materials, as well as several interfaces, shocks, free surfaces, impacts, and spalls. Work hardening, changes in elastic constants, and yield strength with pressure and internal energy and spall at a given tensile stress are features incorporated in the programs. The materials are assumed to be strain rate independent. The program is written in FORTRAN IV, operational on a CDC 6600, and available from the Argonne Code Center.

Two-Dimensional Codes: Lagrangian Method

CSQ [14] is a two-dimensional energy flow hydrodynamic code and is similar in many respects to the one-dimensional counterpart, CHART D. The current version includes an elastic-plastic model, a simple porous material model, internal energy sources, a high explosive model, and a gravity calculation. Radiation flow, plasma, and thermal conduction are treated in the diffusion approximation. The program is coded in FORTRAN IV and is operational on CDC 6600 computers; it should be available in the near future.

HEMP (Hydrodynamic, Elastic, Magnets, and Plastic) [15] is a two-dimensional Lagrangian finite difference code for the solution of dynamic elastic-plastic flow. It, like TOODY, is based on the difference equations which were developed by Wilkins [16]. The program is written in FORTRAN IV and is operational on CDC 6600 and 7600 computers. It is the original two-dimensional wave propagation code.

PISCES-2DL [17], like PISCES-1DL, is a finite difference Lagrangian code developed by Physics International Co. and available on the CDC CYBERNET system. It is a two-dimensional code, and the available options include (a) automatic coordinate generator, (b) untangling rezone, (c) static solution, (d) slideline, (e) linear elastic end condition (used for nonreflecting boundaries), and (f) restart.

REXCO-H [18] is a time-dependent, hydrodynamic, two-dimensional wave propagation code, primarily written for calculating the response of a primary reactor containment system to a high energy excursion. The hydrodynamic equations and the equations of state are expressed in Lagrangian form. Equations of state for porous materials are included. Cylindrical symmetry is assumed. Shock discontinuities are eliminated by the use of the von Neumann-Richtmyer pseudoviscosity. A rezoning option is included. A modified version, called REXCO-HEP, is under development. It will include shell equations which will be used to determine the dynamic response of the reactor containment.

TOODY IIA [19], like HEMP, is a two-dimensional wave propagation code based on the conventional finite difference analogs to the Lagrangian equations of motion. The equations of state include the Mie-Grueneisen equation, elastic-plastic response, high explosive, vapor, perfect gas, and a distended material (p-α) model. The code includes a sliding interface option which is useful when the motion of two adjacent meshes is very different. The program is currently being tested with a rezone capability. The code is written in FORTRAN IV for the CDC 6600 and is available from the Argonne Code Center.

Two-Dimensional Code: Eulerian Method

TOIL [20] is a two-dimensional, continuous Eulerian hydrodynamic code which can treat the inviscid, nonheat-conducting, compressible fluid flow of two different materials in the presence of shock. The code was developed by Johnson [21] as a modification of his OIL (a one material) code. The geometry is limited to cylindrical coordinate systems. A basic difference between TOIL and other Eulerian codes is the technique of following interfaces. A continuum mass transport scheme is used, whereas Particle-in-Cell (PIC) codes use discrete particle transport across the fixed cells. Consequently TOIL smooths oscillatory results caused by the discrete mass transport and decreases computer usage time. The code is written in FORTRAN IV and is operational on CDC 6600 computers.

Two-Dimensional Codes: Other Methods

In addition to the codes mentioned above, codes have been developed which allow the user to specify any coordinate motion. Examples include AFTON and ADAM, which are discussed in [7].

Three-Dimensional Codes

As far as is known there are no available three-dimensional wave propagation production codes. Current developments include the work by D. L. Hicks and H. S. Lauson of Sandia Laboratories on the code THREEDY, and L. D. Buxton from the same organization on the code called TRIHYD, which uses the Eulerian method.

Additional information on the wave propagation codes listed above and on others can be found in [7,8,22,23].

TWO-DIMENSIONAL SOLIDS

In this section we will discuss some of the codes available for the nonlinear analysis of two-dimensional solids. The codes are all based on the finite element method, and those that perform dynamic analysis use either implicit or explicit time integrators. As mentioned previously, the dynamic codes in this section will, in some cases, do the same class of problems as those listed in the Wave Propagation section. However, those in this section are primarily structural response codes.

Static Codes

EPAD (Elastic-Plastic Analysis of Domes) [24] is a finite element program developed for the Air Force Weapons Laboratory, Kirtland Air Force Base, New Mexico. The original version was written for an IBM 360/65 and was later modified at KAFB to run on an CDC 6600 in conjunction with an extended core storage (ECS) of 50,000 words. The documentation and program are subject to special export controls. The program is based on SAAS-II (see below) and uses a quadrilateral ring element which is composed of four triangular elements with an assumption of a linear displacement field within the element. The program determines the static elastoplastic deformations, stresses, strains, and plastic strains for solids of revolution subjected to asymmetric mechanical loading (i.e., no thermal loading is allowed). The initial strain method is used. Incremental stress-strain relationships are derived from the Prandtl-Reuss constitutive equations and the von Mises yield criterion. The unsymmetric loading (axisymmetric about the 0°-180° axis) is handled by a Fourier decomposition in the circumferential direction. The equilibrium equations which are coupled through the nonlinearities and which treat the nonlinearities as pseudo loads are solved in an iterative fashion for each increment of load. (Note - a simple iterative solution is used. No reports on convergence difficulties have been received.) The stress-strain curve for the material(s) is input as a series of data points which describe a maximum of 10 linear segments. Though anisotropic values may be used for the elastic analysis, the inelastic portion assumes that the material is isotropic. A maximum of 700 nodal points or 500 elements and 4 materials and 10 terms in the Fourier expansion and a maximum difference between any nodal point number within an element of 22 are allowed.

EPIC-II [25] is a finite element program for the static elastic-plastic-creep analysis of two-dimensional solids (axisymmetric or plane). The program contains three quadrilateral elements (user option): Q4, QM5, and Q12. An incremental technique using the tangent modulus method is used, and geometric stiffness matrices may be formed. The program uses the Prandtl-Reuss flow law and the Soderberg relationship for creep. The program is a production code at Lockheed but is proprietary. It is being updated.

H326 [26] is a finite element program for the elastic-plastic static analysis of axisymmetric solids subjected to axisymmetric mechanical and/or thermal loading. The program is a modification of Wilson's SAAS code and uses a quadrilateral ring element consisting of four triangular elements with linear displacement fields. The plasticity considers only transversely isotropic materials (elastic solutions are valid for orthotropic materials) whose properties can be a function of temperature. The assumptions include: (a) uniaxial stress-strain curves (nondimensional) for a material along principal material axes should be nearly identical, (b) plastic flow does not cause any volume change. The yield stress is based on Hill's quadratic function and requires knowledge of the yield stress and strain in the R-Z plane and in the θ direction. The equations of equilibrium are solved in an iterative fashion through updating of the stiffness matrix. An accelerating procedure can be applied to a monotonically increasing stress-strain curve to speed convergence. The capacity of the program is 650 nodal points or 600 elements; it operates on a CDC 6600 and can be used on a UNIVAC 1108 with minor modifications. The reported run times for some sample problems are on the order of, e.g., 70 seconds for 6 cycles of iterations for a problem with 265 equations.

OASIS (Orthotropic Axisymmetric Solution of Inelastic Solids) [27] is a finite element code for the static anisotropic elastic-plastic-thermal analysis of bodies of revolution and plane stress and plane bodies. An incremental anisotropic theory is used with an extension of the Jensen, Falby, and Prince anisotropic yield criterion. The method of solution is the tangent stiffness method and an iterative procedure is used at each increment of load. In addition to the thermal loading, the materials may be temperature dependent. The input for the material consists of the stress-strain curve in the principal material direction.

The code is apparently proprietary. Information regarding the performance of the code was unavailable.

PLAST2 [28,29] is a program within the FESS (Finite Element Solution System) family of codes developed at University College of Swansea, Great Britain, to obtain the static solution for two-dimensional solids (plane and axisymmetric bodies) having nonlinear materials undergoing small strains. The available material models are elastic, perfectly plastic, or elastic with work-hardening plasticity. Four different yield conditions are included: von Mises, Tresca, Drucker-Prager, and Beltrami. The materials are assumed to be isotropic. Three different algorithms for solving the nonlinear equations are available: (1) constant stiffness initial stress method, (2) a two-step process where the stiffness matrix is updated on the second iteration of an otherwise initial stress process, (3) a regular tangent stiffness method. Convergence is checked by evaluation of the force residuals at any load increment. A frontal method is used for the equation solution.

Isoparametric quadrilateral elements are used with a user choice of linear, quadratic, or cubic displacement fields. The number of Gauss points for numerical integration of the stiffness matrices is a user option. The program is operational on several machines. The program's many features make it ideal for research activities and for checking other programs.

SAAS III [30] and ASAAS [31] are the latest versions in a series of programs which are based on the original work by E. L. Wilson. Both programs use a quadrilateral ring element which is composed of four triangular elements with linear displacement fields. SAAS III is restricted to axisymmetric solids with axisymmetric mechanical and thermal loading. While in ASAAS the initial geometry of the body must be axisymmetric, the inclusion of asymmetric thermal loading and temperature-dependent material properties allows for asymmetric solids. Note, however, that since a Fourier expansion is used in the circumferential direction and only a limited number of harmonics are allowed (they are coupled), the asymmetric loading must be fairly smooth. SAAS III also includes an option for plane stress/strain bodies. Both programs allow for orthotropic temperature-dependent materials (nine independent material constants for ASAAS and seven [due to symmetry] for SAAS III). Both programs contain a mesh generator and plotting subroutines. However, while the programs which were originally written for an IBM 360 are easily converted to other machines, such as the CDC 6600, the plotting package is machine dependent. The mesh generator does include an option for circular regions. Earlier versions did not, and the Laplacian smoothing could cause nodal points to fall outside the body. SAAS III includes an option for inputing a bilinear stress-strain curve with a different yield point for each principal material direction; however, the ratio of the slope of the second segment to the slope of the first segment must be the same in all directions. A simple iterative scheme is used to obtain a solution. If the slope of the second segment is fairly flat, many iterations will be required to obtain convergence. The program is available from the Aerospace Corp.

Dynamic Codes

DYNS [32,33,34], also known as NONDYN and as GOLLY at Sandia Laboratories, is a two-dimensional finite element dynamic nonlinear program developed at the University of California at Berkeley and modified at the U.S. Army Engineer Waterways Experiment Station [35] and at Sandia Laboratories. It is an extension of an earlier linear program by E. L. Wilson for elastic dynamic response of axisymmetric structures.

The element used in the program is either a so-called Q4M or Q5M quadrilateral, depending on the version of the code. The Q4M is a four-point quadrilateral (bilinear displacement field) with a modification in the method of evaluating the shear strain. Likewise the Q5M is a five-point (the fifth node is at the centroid) with the same modification. The modification consists of using a one-point Gaussian integration method which will in effect assume that the shear is constant within the element. Further discussion of these elements and further modifications to these elements, such as the addition of incompatible displacement modes, can be found in the literature. In brief, the proper choice of element(s) is still being discussed.

The program treats nonlinearities by updating of the stiffness matrix (frequency of which is a user option). A lumped mass matrix is used, and the equations of motion are integrated through use of an implicit Wilson-θ method. For a linear problem the method is unconditionally stable for a specified range of θ values.

For a highly nonlinear problem the program will operate slowly due to the need of constantly updating the stiffness matrix and decomposing the equations (the integrator is implicit).

The program contains a geometric stiffness matrix but does not apparently use large strain theory. For some problems this would appear to be in conflict.

The nonlinear material model is for a soillike material in which the behavior is specified by the bulk behavior (hydrostatic pressure vs. volumetric strain), the unloading bulk modulus, and the shear modulus. The material is assumed to have infinite strength in shear. One of the attractive features of the programming is the way the nonlinear material model is implemented as a self-contained subroutine; consequently additional material models may be added without much difficulty. The program uses dynamic dimensioning in which a vector of adjustable length contains most of the key variables.

HONDO [36,37] is a finite element program to compute the time-dependent displacements, velocities, accelerations, and stresses within elastic or inelastic, two-dimensional or axisymmetric bodies of arbitrary shapes and materials. The material models are elastic-plastic, strain hardening, strain rate behavior, soil, crushable foam, and viscoelastic and rubber elasticity. Only mechanical loading is permitted. A quadrilateral isoparametric element with bilinear displacements is used. Based on the changes over a time step the strain is updated. Using the new strain and/or the current strain rate, a new stress is calculated. A numerical divergence of the stress field is taken to obtain the nodal forces which are accumulated to provide the new accelerations. The accelerations lead to new velocities and new positions, and the whole process is repeated. (See also the program SAMSON.) This procedure saves time and storage requirements.

The simultaneous equations of motion are integrated by a central difference expression for velocity and acceleration. A lumped mass matrix is used, and hence the scheme is explicit and fast per time step. The integration is conditionally stable with respect to time step size; however, the program has a built-in control on the magnitude of the step size.

SAMSON [38] is a two-dimensional finite element code for the dynamic nonlinear stress analysis of plane and axisymmetric bodies subjected to mechanical loading. The code is a modification of the SLAM code [39] and includes a new solution procedure and the addition of nonlinear constitutive models, including a cap model for soils based on the work of DiMaggio and Sandler and a cracking model for representing the behavior of reinforced concrete. If the reinforcing bars are not included in this latter model, the model is suitable for rock behavior. There is no elastic-plastic model for metals. The program contains the following elements: a triangular plane and axisymmetric element with a linear displacement field, a rectangular plane and axisymmetric element with a bilinear displacement field, a two-node bar element with constant cross-sectional area and axial stresses only, a one-node bar element (circumferential bars) for axisymmetric problems only, and a membrane linear element.

A direct method of solution is used without explicit formulation of the stiffness matrix. This saves computer storage and time. A Newmark beta method with $\beta = 0$ (equivalent to the central difference method) is used to integrate the equations of motion. For linear problems the time step must be less than 0.318 of the shortest period of the mesh. For nonlinear problems no comment is given on time step.

Two types of artificial damping are available: linear, which is proportional to a percentage of critical damping, and nonlinear. Sliding and debonding interfaces with specified restrictions are allowed. Static solutions may be obtained through the use of dynamic relaxation.

The code is relatively new and has not seen much use; therefore, no comments can be made in regard to its performance. Initial distribution of the documentation is limited to U.S. Government agencies only. Others may request the document from the Air Force Weapons Laboratory, Kirtland AFB, New Mexico. The program is written in FORTRAN IV for operation on a CDC 6600 machine as a three-part structure using an overlay feature.

Other Two-Dimensional Codes

In the area of finite deformations of elastic solids using the finite element method the work of Oden and his colleagues [5] is noteworthy. Most of the work is understandably research oriented with the end goal not being a production computer program but rather understanding the physical situation.

Table 1 Two-dimensional Structural Codes

		EPAD	EPIC-II	H326	OASIS	PLAST2	SAAS III	ASAAS	DYNS	HONDO	SAMSON
Static		x	x	x	x	x	x	x			x
Dynamic									x	x	x
Thermal Loading			x	x	x		x	x			
Temperature Dependent Materials			x	x	x		x	x			
Axisymmetric Solids	Axisymmetric Loading	x	x	x	x	x	x	x	x	x	x
	Asymmetric Loading	x						x			
Geometric Nonlinearities			x						x	x	
Large Strains										x	
Material Model	Metal Plasticity	x	x	x	x	x	x	x		x	x
	Soils/Rocks					x			x	x	x
	Crushable Foams									x	
	Rubber Elasticity									x	

Costantino and his colleagues (e.g., [40]) have developed a series of dynamic codes for nonlinear media.

BERSAFE [41,42] is a system of finite element programs being developed by the Central Electricity Generating Board (CEGB) in the United Kingdom. Within the package is a program (TESS) for two-dimensional creep and plasticity problems.

WESTES [43] is a finite element code for the static analysis with soillike nonlinear materials. The code is a modification of PLAXYM (developed at Agbabian Associates).

EPSOLA [44] is a finite element code for static stress analysis of axisymmetric solids with both geometric and material (small strains) nonlinearities. The program uses the Q4M element. A dynamic version is also available [45,46].

STRAW [47] (Structural Transient Response of Assembly Wrappers) is a finite element code for nonlinear, large displacement, small strain, dynamic problems. The program contains an Euler beam element and a linear displacement triangular element for plane bodies. The authors use a so-called convected coordinate system which does not deform with the elements but does rotate. A sliding-debonding interface option is included for fluid-solid interfaces. An explicit (hence a lumped mass matrix is used) central difference time integrator is used. The program is currently being modified to include a membrane element with large strains (three-dimensional). This version is known as SADCAT.

BOPACE [48] is a finite element program for the static analysis of plane bodies considering thermal, elastic, plastic, and creep strains. A triangular element with a linear displacement field is used. There are two versions: one for 300 D.OF and the other for 1000 D.OF. The program is operational on IBM 360 and UNIVAC 1108.

GAP [49] is a modification of an early version of SAAS for inclusion of nonlinear effects due to elastic contacts such as would occur in multicomponent structures. Included in the program are slippage at boundaries and either the separation or closing of contact points.

CREEP-PLAST, EPACA, and PLACRE are programs which perform elastic-plastic-creep analysis and are discussed in the paper on creep analysis.

GENERAL PURPOSE CODES

There are several general purpose finite element codes which do nonlinear analyses, though one of these, NASTRAN, has only a limited capability and then only for a few of the elements within the library. No special purpose codes which only have a three-dimensional element are included in this report; however, all of the programs in this section do have a three-dimensional element. The following programs are not included herein because they lack the capability of directly performing nonlinear analyses: STARDYNE, EASE, MAGIC, ELAS75, and SAP-IV.

ANSYS (Engineering ANalysis SYStem) [50] is a general purpose finite element computer program developed by and maintained by Swanson Analysis Systems, Inc. The program is also available from SRDC (Structural Dynamics Research Corporation), Control Data's CYBERNET system, USS Engineers and Consultants, Inc., UCC (University Computing Company), Westinghouse Tele-Computer Systems Corporation, and Basic Technology, Inc. It has been the topic of several educational seminars. A brief description of its capabilities follows--more detail may be found in the publications of the organizations listed above and in [41,51]. The program is operational on CDC, IBM, and UNIVAC machines.

The program contains more than 60 finite elements (such as beam, plane stress, axisymmetric solid and shell, general shell, and three-dimensional solid) of which 17 have nonlinear formulations and includes the capability of performing static or dynamic analyses, fluid flow, and heat transfer. The nonlinear capabilities include plasticity (small strain), creep (thermal and irradiation induced), irradiation induced swelling, large deflection, and buckling. It therefore has found wide application, in particular, in the analysis of nuclear power reactors. The dynamic capabilities include eigenvalue-eigenvector, steady state harmonic response, and linear and nonlinear transient response. The materials may be either isotropic or anisotropic and may include nonlinear temperature dependency.

The static analysis employs the wavefront method of solution, and hence there is no restriction on the bandwidth and no limit (except machine time) on the number of elements. Two-dimensional problems with more than 8,000 nodes and three-dimensional problems with more than 3,000 nodes have been solved. The dynamic analysis employs a consistent mass matrix and an explicit quadratic integration routine.

Extensive plotting capabilities exist, including geometry, stresses, displacements, and temperatures.

The ASKA (Automatic System for Kinematic Analysis) family of general purpose finite element programs have been developed at the University of Stuttgart by Argyris and his colleagues over the last 15 years. The current version of the program is written in FORTRAN IV and is operational on CDC 6600, UNIVAC 1108, and IBM 360/50, 65, 75, 85, and 370/155 computers. Since the program is well known and well documented only few comments will be given here. The linear versions of the program are described in [52], while some of the elastic-plastic capabilities are described in [53]. A users' review of the program is given in [54]; however, the comments are limited to the linear version only. Material nonlinearities, which include creep and plasticity, are contained in the program ASKA Part III-1 [55,56]. The program is restricted to small deformations and isotropic materials but does allow for temperature-dependent material properties. A limited number of elements are available for one-, two-, and three-dimensional problems. The initial strain method is used. Geometric nonlinearities are treated in the separate program ASKA Part III-2.

MARC [57] is a general purpose static and dynamic finite element computer program developed by Pedro V. Marcal (MARC Analysis Research Corp., Providence, Rhode Island) and his colleagues. It is a proprietary program and is available from the developers or may be used on Control Data's CYBERNET system. It is primarily available for use on CDC 6000/7000 machines. MARC is a series of six separate programs of which two are mesh generations, two are postprocessor plotting programs, and the remaining two are a transient heat conduction program and the main program (i.e., stress analysis).

The main program contains some 19 elements including an eight-node and a twenty-node isoparametric three-dimensional hexahedrons. The nonlinear capabilities include elastic-plastic behavior with large displacements, creep analysis, and buckling behavior. Elastic-plastic behavior is based on isotropic materials with temperature-dependent elastic properties, a von Mises yield stress criterion, and either isotropic or kinematic strain hardening. The creep analysis is based on a von Mises flow criterion and isotropic behavior described by a user-supplied equivalent creep rate law. In addition, three material models based on a hydrostatic yield dependence (Mohr-Coulomb) for soil- and rocklike materials are available. (Finite strains are included through use of a Lagrangian formulation.) The equations of equilibrium are solved by the Crout reduction method. Out-of-core block over relaxation is also available. The dynamic solution is obtained through integration of the equations of motion by (a) the Newmark method ($\beta = 1/4$, $\gamma = 1/2$), (b) the Houbolt method, or (c) central differences. Mode superposition is also available. Additional descriptions of the program are available in [41,51,58].

The input and output features of the program could be improved. For example, there is only one material which may have temperature-dependent properties, and the stresses are given at integration points within the element without specification of the coordinates of these points. In addition, the requirement to specify materials by element blocks appears to be somewhat awkward.

Since NASTRAN [59] (NAsa STRuctural ANalysis) has been the subject matter of many papers, seminars and colloquia, and a newsletter, only the pertinent details regarding nonlinear analyses and usage will be discussed here. Currently NASTRAN is being managed by the NASTRAN Systems Management Office at NASA Langley Research Center, and the latest official version, Level 15.5, is available from COSMIC. A level 16 is tentatively due for release in the fall of 1974. Regarding nonlinear capability, the only apparent addition in Level 16 will be a new differential stiffness rigid format (SOL 4). NASTRAN is also available on the CDC/CYBERNET system, from McDonnell Douglas Automation Company, St. Louis, Missouri; Westinghouse Tele-Computer Systems Corporation, Pittsburgh, Pa.; USS Engineers and Consultants, Inc. (UEC), Pittsburgh, Pa.; Naval Ship Research and Development Center (NSRDC), Bethesda, Md.; SRDC, Cincinnati, Ohio; and MacNeal-Schwendler Corp., Los Angeles, Calif. The versions of NASTRAN available from these sources are not identical to the one available from COSMIC.

Regarding Level 15.5, there is only a limited nonlinear capability available as one of the rigid formats, and this includes a

Table 2 General Purpose Codes

		ANSYS	ASKA III-1	MARC	NASTRAN	NEPSAP	NONSAP
Static		x	x	x	x	x	x
Dynamic		x	x	x	x		x
Elements	1-D	x	x	x	x	x	x
	2-D	x	x	x	x	x	x
	3-D	x	x	x		x	x
Shells	Shells of Revolution	x		x			
	Arbitrary	x		x		x	
Thermal Loading		x	x	x		x	
Temp. Dependent Material Properties		x	x			x	
Geometric Nonlinearities		x		x	x	x	x
Large Strains				x		x	x
Material Model	Metal Plasticity	x	x	x		x	
	Soils/Rocks			x			

which are, however, not discussed herein. Also, it must be noted that several of the general purpose codes contain nonlinear shell elements; these are not repeated in this section.

Geometric Nonlinearities

SATANS [79,80] is a finite difference code developed by R. E. Ball for the nonlinear (geometric) static and dynamic analysis of shells of revolution. The static code uses the method of successive substitution to solve the nonlinear algebraic equation, which limits the degree of nonlinearity that can be treated. The same procedure is used in the dynamic analysis, but in this case it is not limited to moderate nonlinearities. The program can be used to analyze any shell of revolution for which the following conditions hold:

1 The geometric and material properties of the shell are axisymmetric but may vary along the shell meridian.

2 The applied pressure and temperature distributions are symmetric about, but may vary along, a meridian.

3 The shell material is isotropic, but the modulus of elasticity may vary through the thickness. Poisson's ratio is constant.

4 The boundaries of the shell may be closed, free, fixed, or elastically restrained.

The governing partial differential equations are based upon Sander's nonlinear thin shell theory for the condition of small strains and moderately small rotations. The in-plane and normal inertial forces are accounted for, but the rotary inertia terms are neglected. The set of governing nonlinear partial differential equations is reduced to a finite number of sets of four second-order differential equations in the meridional and time coordinates by expanding all dependent variables in a sine or cosine series in terms of the circumferential coordinate. The sets are uncoupled by utilizing appropriate trigonometric identities and by treating the nonlinear coupling terms as pseudo loads. The meridional derivatives are replaced by the conventional central finite difference approximations, and the displacement accelerations are approximated by the implicit Houbolt backward differencing scheme. This leads to sets of algebraic equations in terms of the dependent variables and the Fourier index. At each load or time step, an estimate of the solution is obtained by extrapolation from the solutions at the previous load or time steps. The sets of algebraic equations are repeatedly solved using Potter's form of Gaussian elimination, and the pseudo loads are recomputed, until the solution converges.

Basically, there are four fundamental features of the method of solution: (1) circumferential series, (2) meridional finite differences, (3) pseudo load concept, and (4) the Houbolt timewise differencing scheme.

SOR Codes. A family of compatible computer codes for the analysis of the shell of revolution (SOR) structures has been developed by researchers at Texas A&M University. These analyses employ the matrix displacement method of structural analysis utilizing a curved shell element. Geometrically nonlinear static and dynamic analyses can be

restart the program at a specified time without having to expend the computer time necessary to regenerate the prior response.

The SOR codes are available from COSMIC.

Material and Geometric Nonlinearities

The SHORE code [84] developed by Philip Underwood computes the transient response of shells with midsurfaces defined by a surface of revolution. The equations solved by two-dimensional finite difference techniques are those derived by Sanders and include the nonlinear geometric effects. Time integration is by the central difference method. Inelastic material behavior based on Hill's anisotropic yield function is valid for initially orthotropic materials for both perfect plasticity and linear strain hardening. A fracture criterion also is available. The loadings considered are initial radial velocity, initial step change in temperature, and surface pressure histories; all of these can be specified for each mesh point. The shell can have either one or two different material layers obeying the Kirchhoff assumptions with additional weight and coordinate direction springs at each mesh point. The boundary conditions include the classic simple and clamped supports plus conditions of symmetry and antisymmetry. The code is checked out for cylinders and cones for elastic, inelastic, linear, and nonlinear kinematic response. Other shapes have been analyzed, but no real check cases are available.

One of the most impressive aspects of the SHORE code is its speed of solution. These computer run times have in certain cases been lower than the computer time for codes which solve the shell of revolution using a Fourier series.

DYNAPLAS. The large deflection, elastic-plastic, dynamic analysis of ring-stiffened, thin shells of revolution subjected to arbitrary mechanical loadings may be achieved through the application of the SAMMSOR IV and DYNAPLAS II computer codes. These two codes represent an extension of the SAMMSOR II and DYNASOR II codes which are part of the SOR series of codes operational since 1970. The documentation for these codes is presented in three reports: user's guide for SAMMSOR IV program [85], user's guide for DYNAPLAS II program [86], and detailed theoretical formulation of the analysis procedure [87].

As in the SOR series of codes the dynamic analysis is conducted by first executing the SAMMSOR IV code to obtain an output tape containing the stiffness and mass matrices for the particular geometry being studied. DYNAPLAS II is then executed to solve the dynamic equations for a specific set of initial conditions, boundary conditions and loading history. The equations of motion are integrated by either the Houbolt or the central difference method. A subsequent analysis of the same problem with, for example, a different loading history requires only the execution of DYNAPLAS II.

The geometry input is handled entirely by the SAMMSOR IV code which utilizes a highly refined, meridionally curved, axisymmetric, thin shell of revolution element in addition to beam-type ring stiffeners. The shell element utilizes cubic displacement functions for the normal and in-plane displacements and, through static condensation, a basic eight degree of freedom element stiffness matrix is generated. Each element is assumed to have constant thickness and material properties although discontinuities are allowed at element boundaries. The shell material may be either isotropic or orthotropic. The ring stiffeners are assumed to have zero products of inertia but may be eccentrically located and may be formed by as many as three rectangular flange members.

In addition to the stiffness and mass matrices generated and stored on tape by SAMMSOR IV, DYNAPLAS II requires the uniaxial stress-strain data for the shell and stiffeners, the boundary conditions, the initial displacement and velocity conditions, the applied load as a function of time, and a number of other control constants. The applied load history may be described by specifying either a pressure distribution over the element or the Fourier coefficients of the distribution at discrete time intervals with a linear variation being assumed between the specified times. For a particular element, the pressure distribution is assumed to be constant in the meridional direction and to vary as a step function in the circumferential direction. In addition, the code is capable of accepting concentrated ring loads at each node. The uniaxial stress-strain data is described by inputing a piecewise linear curve with as many as five stress-strain data points. In the plastic region, either isotropic hardening or the mechanical sublayer model may be used [3]. In addition, strain rate effects (Cowper, Symonds) model may be included. The code may of course be run with plasticity and/or geometric nonlinear effects omitted. In addition, the code allows springs to be attached at various points on the shell.

The codes will be available from the Argonne Code Center.

TWORNG [88] is a finite difference computer code for calculating the free in-plane response of two bonded concentric rings to the combined effects of impulse, internal stress, and thermal heating. It is an extension of the Ring Bond [89] code. The rings are assumed to be narrow and thin and to experience small deformations, whereas the bond is massless and shear deformable. The single, one-layer ring is also treated. The loadings may have arbitrary distributions around the rings and through the thickness. Although the rings are assumed to be thin, the through-the-thickness profiles of strain, stress, and temperature are calculated in order to account for the nonlinear and thermal effects in the materials. For strain there are contributions from both the dynamic response and the thermal relaxation. For stress, temperature-dependent strength properties of the materials as well as extensional and flexural strain rate damping stresses are included. The relaxation of the temperature gradients is calculated simultaneously with the dynamic response calculations. User-specified time scaling is also performed to allow the entire response from the initial dynamic state through the long-term quasi-static thermal relaxation to be calculated in minimum computation time.

The code is written in FORTRAN IV and can be accommodated in 41,000 core storage. The complete data set includes:

1. Description of the geometry, finite difference mesh, and time scaling for damping and conduction
2. Basic material data for density, damping, and fracture strain
3. Piecewise linear approximations of the temperature-dependent materials data for the elastic and plastic moduli, the yield stress, thermal expansion, specific heat, and thermal conductivity
4. Impulse loading conditions
5. Initial internal stress distributions
6. Initial internal energy distributions

For problems which do not utilize the full potential of the code, subsets of the above data set are required. The code is due to be available shortly; however, dissemination will be limited.

SMERSH [90] is a finite difference computer code for calculating the dynamic response of two bonded concentric shells to impulse and internal stress loading. This version is an update of a previous release [91]. The shells are assumed to be thin, to have rotational symmetry but arbitrary meridional geometry, and may contain integral ring stiffeners. The bond is massless and shear deformable. The single, one-layer shell is also treated. Various standard boundary conditions may be applied. The deformations may be finite as long as the strains are small and the rotations moderate. The material constitutive behavior is elastic/plastic and uses the von Mises yield criterion and Prandtl-Reuss plastic flow. The loadings may have arbitrary distributions over the shell and through the thickness. Although the shells are assumed to be thin, the through-the-thickness profiles of strain and stress are calculated in order to account for the nonlinear effects of material behavior.

The code is written in FORTRAN IV. Minimum core storage is 22,600 with actual total storage allocated dynamically upon specification of the problem. Large problems may require 65,000 or more total. The complete data set includes:

1. Descriptions of the geometry, finite difference mesh, and boundary conditions
2. Material data of density, elastic/plastic parameters, and fracture strains
3. Impulse loading conditions
4. Initial internal stress distributions

The code will be available soon. The dissemination will be limited.

JET 3 [92,93] is a finite element computer code developed by Wu and Witmer of MIT and is capable of the following large deflection elastic plastic dynamic analysis.

1. A structural ring, complete or partial, whose geometrical shape can be circular or arbitrarily curved with variable thickness
2. A structural ring, with various support conditions, subjected to arbitrarily distributed elastic restraints
3. A structural ring subjected to arbitrary initial velocity distribution
4. A structural ring subjected to transient mechanical loads which vary arbitrarily in both space and time

In analysis, the transient structural responses of the ring are assumed to consist of planar (two-dimensional) deformations. Also, the Bernoulli-Euler (or Kirchhoff) hypothesis is employed; that is, transverse shear deformation is excluded.

The displacement behavior within each finite element is represented by a cubic polynomial for circumferential displacement and the normal displacement w. For application to arbitrarily curved, variable thickness, ring structures, the finite elements are described by reading in at each nodal station the two coordinates Y and Z, the slope, and the thickness of the discretized structure, where X, Y, Z represent global reference Cartesian coordinates. Within each element, the slope is approximated by a quadratic function and the thickness is approximated as being piecewise linear between nodes. For application to a circular, uniform thickness ring structure, and in view of the computer storage and operation considerations, the structure is modeled by uniform mesh, uniform thickness, circular ring elements. The local reference coordinate system of each element is arranged to take advantage of the symmetry of the element geometry.

As for the support conditions of the structure, the JET 3 program includes three types of prescribed nodal displacement conditions, (1) symmetry, (2) ideally clamped, (3) smoothly hinged, and two types of elastic restrains:

1. Point elastically restrained (elastic restoring spring) at given locations

2. Distributed elastically restrained (elastic foundation) over a given number of elements

A global effective stiffness matrix supplied by the elastic foundation and/or the restoring springs is evaluated in the program from the virtual work statement for the case in which the structure is subjected to one or both of these two types of elastic restraints.

The resulting equations of motion are solved by applying an appropriate timewise finite difference operator whereby one obtains a recurrence equation which provides a solution step-by-step in finite time increments.

Two options of solution procedures for this code are provided: (1) the explicit three-point central difference operator or (2) the implicit Houbolt operator (four-point backward difference).

The work at the University of Alabama in Huntsville by T. J. Chung and R. L. Eidson should also be noted. They have developed a series of programs, the latest being a finite element code for the visco-elastoplastic response of an anisotropic shell under dynamic loads [94]. The generalized Maxwell model is incorporated into the von Mises anisotropic yield function.

SABOR/DRASTIC 6A [95] is a modification of the well-known SABOR series to include material and geometric nonlinearities. The program uses the curved shell element of the SABOR series and Fourier harmonics to represent the asymmetric variation in loads and shell properties. The nonlinearities are treated as pseudo loads in the equations of motion. These equations are integrated by the Newmark method (beta = 1/4). A consistent mass matrix is used, and viscous damping is allowed. The program does not have provision for thermal loading. There also is a capability of performing a nonlinear static analysis, however, the technique is unknown. A user's manual is in preparation.

LUMPED MASS CODES

In addition to the codes discussed in the previous sections there are the so-called lumped mass codes, some of which have nonlinear capabilities and are useful for performing preliminary analyses for either design calculations or to obtain some indication of the nonlinearity of the problem. In many situations another type of code would be used after these preliminary calculations were made.

One such code is SHOCK [96], which calculates the transient response of lumped mass systems using the Newmark-Beta method of integration. The program allows for 100 masses and 200 springs which may be nonlinear. Viscous damping is allowed. Only axial, lateral, or torsional models of beam-type elements are allowed. Furthermore, the combined axial and lateral problems are not allowed. A modification to this code, under development, called SHELL-SHOCK, will eliminate this restriction and also will add shell of revolution and axisymmetric solid elements.

One difficulty in using lumped mass models is the proper modeling of a structure. In most cases the user must have considerable experience in this area.

GUIDELINES FOR SOFTWARE SELECTION AND IMPLEMENTATION

Since so many variables enter into the subject matter of choosing software one cannot definitively give guidelines on this subject matter. In addition to the usual comments, personal experience has shown it useful to attack the question of which program to pick as an engineering problem in itself. One must use <u>all</u> the available resources regarding pertinent information about the program. Also, the intended uses of the program should be known.

Once a program arrives, it must be implemented onto <u>your</u> system and checked (verified and qualified) with a series of benchmark problems, preferably in the area of your interest. Surprisingly, in many cases simple problems are capable of producing erroneous results. The benchmark problems should be documented for user's use and for checking other programs, or for use whenever the compiler or operating system is changed or modified. One point should also be realized, and that is, that plotting packages are not only machine dependent but also in many cases installation dependent.

There should be adequate training classes on the program; that is, the user must also become qualified to run the program. Obviously, here a good user's manual is required.

One must also realize that not every program that is implemented should be carried through all the steps; that is, some programs will be found inadequate and should be deleted. In addition, a good index and line of communications on the status of programs within an organization must be established.

For a further discussion in this area see [97].

FUTURE TRENDS

To a large extent, the technology of the numerical methods has exceeded the ability to characterize materials. In addition, the material models in many cases do not capture the basic behavior of the material. Consequently, one can see an effort in the future for all three groups (numerical analysts, experimentalists, and developers of models) to work together. In sharing their problems and experiences one would also hope to see software that would obtain reasonable answers using less machine time.

ACKNOWLEDGMENT

The material in this paper was, in part, obtained from user's manuals, from the developers, and from the documents from software centers. The work by W. A. Von Riesemann was supported by the United States Atomic Energy Commission.

REFERENCES

1 Felippa, C. A., and Sharifi, P., "Computer Implementation of Nonlinear Finite Element Analysis," in Numerical Solution of Nonlinear Structural Problems, Hartung, R. F., ed., AMD-Vol. 6, The American Society of Mechanical Engineers, N.Y., 1973, pp. 31-49.

2 Nickell, R. E., "Nonlinear Dynamics by Mode Superpositon," to be presented at AIAA/ASME/SAE 15th Structures, Structural Dynamics, and Materials Conference, Las Vegas, Nevada, April 17-19, 1974.

3 Vaughan, D. K., "A Comparison of Current Work-Hardening Models Used in the Analysis of Plastic Deformations," M.S. Thesis, Aerospace Engineering Dept., Texas A&M University, College Station, Texas, 1973.

4 Oden, J. T., Finite Elements of Nonlinear Continua, McGraw-Hill Book Co., Inc., New York, 1972.

5 Oden, J. T., "Annual Report on Basic Research in Methods of Approximation in the Nonlinear Mechanics of Solids and Structures," The University of Alabama in Huntsville, Huntsville, Alabama, AFOSR-TR-72-0782, (AD 743278), Jan. 1972.

6 Stricklin, J. A., Von Riesemann, W. A., Tillerson, J. R. and Haisler, W. E., "Static Geometric and Material Nonlinear Analysis," Advances in Computational Methods in Structural Mechanics and Design, Oden, J. T., Clough, R. W. and Yamamoto, Y., editors, UAH Press, The University of Alabama in Huntsville, Huntsville, Alabama, 1972, pp. 301-324.

7 Walsh, R. T., "Finite Difference Methods," Dynamic Response of Materials to Intense Impulsive Loading, Chou, P. C. and Hopkins, A. K., eds., Air Force Materials Laboratory, Wright-Patterson AFB, Ohio, 1973, pp. 363-403.

8 Bade, W. L., Davis, G. J., and Oston, S. G., "Potential Applications of Multidimensional Material-Response Codes to Reentry Vehicle Vulnerability Problems," Defense Atomic Support Agency, DASA 2538, July 1970, Washington, D. C. 20305.

9 Thompson, S. L., and Lauson, H. S., "Improvements in the CHART D Radiation Hydrodynamic Code II: A Revised Program," Sandia Laboratories, SC-RR-71-0713, Feb. 1972, Albuquerque, New Mexico.

10 Thompson, S. L., and Lauson, H. S., "Improvements in the CHART D Radiation Hydrodynamic Code III: Revised Analytical Equations of State," Sandia Laboratories, SC-RR-71-0714, Mar. 1972, Albuquerque, New Mexico.

11 Lawrence, R. J., and Mason, D. S., "WONDY IV - A Computer Program for One-Dimensional Wave Propagation with Rezoning," Sandia Laboratories, SC-RR-71-0284, Aug. 1971, Albuquerque, New Mexico.

12 "PISCES 1DL, Manual A - General Description and Finite-Difference Equations," Publication No. 86617100, Control Data Corp., Minneapolis, Minn., June 1, 1973.

13 Barker, L. M., "SWAP-9: An Improved Stress Wave Analyzing Program," Sandia Laboratories, SC-RR-69-233, August 1969, Albuquerque, New Mexico.

14 Thompson, S. L., "Tentative Input Instructions for CSQ," Sandia Laboratories, Albuquerque, New Mexico, 1973 (unpublished report).

15 Giroux, E. D., "HEMP Users Manual," Lawrence Livermore Laboratory, UCRL-51079, June 24, 1971, Livermore, California.

16 Wilkins, M. L., "Calculation of Elastic-Plastic Flow," Methods in Computational Physics, Vol. 3, Alder, B., Fernbach, S., and Rotenburg, M., ed., Academic Press, New York, 1964.

17 "PISCES 2DL, Manual A - General Description and Finite-Difference Equations," Publication No. 86617500, Control Data Corp., Minneapolis, Minn., Feb. 15, 1973.

18 Chang, Y. W., Guildys, J. and Fistedis, S. H., "Two Dimensional Hydrodynamics Analysis for Primary Containment," Argonne National Laboratory, ANL-7498, Nov. 1969, Argonne, Illinois.

19 Benzley, S. E., Bertholf, L. D. and Clark, G. E., "TOODY II-A, A Computer Program for Two Dimensional Wave Propagation - CDC 6600 Version," Sandia Laboratories, SC-DR-69-516, Nov. 1969, Albuquerque, New Mexico.

20 Hill, L. R., "Users' Manual for the TOIL Code at Sandia Laboratories," Sandia Laboratories, SC-DR-70-61, March 1970, Albuquerque, New Mexico.

21 Johnson, W. E., "TOIL (A Two-Material Version of the OIL Code," Gulf General Atomic, GAMD-8073, Addendum, AD834058, Nov. 30, 1967, San Diego, California.

22 Hicks, D., "Hydrocode Test Problems," Air Force Weapons Laboratory, AFWL-TR-67-127, Feb. 1968, Kirtland AFB, New Mexico

23 Herrmann, W. and Hicks, D. L., "Numerical Analysis Methods," Metallurgical Effects at High Strain Rates, Rohde, R. W., Butcher, B. M., Holland, J. R., and Karnes, C. H., eds., Plenum Press, New York, 1973, pp. 57-91.

24 Cappelli, A. P., Agrawal, G. L., Romero, V. E., "EPAD - A Computer Program for the Elastic Plastic Analysis of Dome Structures and Arbitrary Solids of Revolution Subjected to Unsymmetrical Loadings," Air Force Weapons Laboratory, AFWL-TR-69-146, Feb. 1970, Kirtland AFB, New Mexico.

25 Cyr, N. A., and Teter, R. D., "Finite Element Elastic-Plastic Creep Analysis of Two Dimensional Continuum with Temperature Dependent

Material Properties," Computers & Structures, Vol. 3, No. 4, July 1973, pp. 849-863.

26 Chao, H., and Hauner, K. S., "An Elastic-Plastic Analysis of Axisymmetric Solid of Orthotropic Solids," McDonnell Douglas Astronautics Co., DAC-63159, March 1969, Huntington Beach, California.

27 Weiler, F. C., and Rodriguez, D. A., "Advanced Methods of Anisotropic, Elastic-Plastic-Thermal Stress Analysis for Bodies of Revolution," Graphitic Materials for Advanced Re-Entry Vehicles, Forney, D. M., Jr., ed., Air Force Materials Laboratory, AFML-TR-70-133, Part 1, 1970, Wright-Patterson AFB, Ohio, pp. 367-452.

28 Nayak, G. C., "Plasticity and Large Deformation Problems by the Finite Element Method," Ph.D. Thesis, University College of Swansea, Great Britain, Dec. 1971.

29 Nayak, G. C., and Zienkiewicz, O. C., "Elasto-Plastic Stress Analysis with Curved Isoparametric Elements for Various Constitutive Relations," to be published in International Journal for Numerical Methods in Engineering.

30 Crose, J. G. and Jones, R. M., "SAAS III, Finite Element Stress Analysis of Axisymmetric and Plane and Solids with Different Orthotropic, Temperature-Dependent Material Properties in Tension and Compression," The Aerospace Corp., SAMSO-TR-71-103, June 22, 1971, Los Angeles, California.

31 Crose, J. G., "ASAAS, Asymmetric Stress Analysis of Axisymmetric Solids with Temperature Dependent Material Properties that can Vary Circumferentially," The Aerospace Corp., SAMSO-TR-71-297, Dec. 29, 1971, Los Angeles, California.

32 Farhoomand, I. and Wilson, E., "A Nonlinear Finite Element Code for Analyzing the Blast Response of Underground Structures," U. S. Army Engineer Waterways Experiment Station, Contract Report N-70-1, Jan. 1970, Vicksburg, Miss.

33 Farhoomand, I., "Nonlinear Dynamic Stress Analysis of Two-Dimensional Solids," Ph.D. Dissertation, University of California, Berkeley, California, 1970.

34 Wilson, E. L., Farhoomand, I., Bathe, K. J., "Non-Linear Dynamic Analysis of Complex Structure," Earthquake Engineering and Structural Dynamics, Vol. 1, 1973, pp. 241-252.

35 Kirkland, J. L., and Walker, R. E., "Fundamental Studies of Medium Structure Interaction, Report 1, Finite Element Analysis of Buried Cylinders," U. S. Army Engineer Waterways Experiment Station, Technical Report N-72-7, June 1972, Vicksburg, Miss.

36 Key, S. W., "HONDO - A Finite Element Computer Program for the Large Deformation Dynamic Response of Axisymmetric Solids," Sandia Laboratories, SLA-73-1009, Dec. 1973, Albuquerque, New Mexico.

37 Key, S. W., "A Finite Element Procedure for the Large Deformation Dynamic Response of Axisymmetric Solids," to be presented at ASME-PVP, Materials, and Nuclear Conference, Miami Beach, Florida, June, 1974.

38 Belytschko, T. and Chiapetta, R. L., "A Computer Code for Dynamic Stress Analysis of Media-Structure Problems with Nonlinearities (SAMSON), Vol. 1, Theoretical, Manual," Air Force Weapons Laboratory, AFWL-TR-72-104, Vol. 1, Feb. 1973, Kirtland AFB, New Mexico.

39 Costantino, C. J., et al., "Stress Waves in Layered Arbitrary Media - SLAM Code Free Field Study," Vols. 1-4, IIT Research Institute, SAMSO-TR-68-181, July 1968, Chicago, Illinois.

40 Heifitz, J. H., and Costantino, C. J., "Dynamic Response of Nonlinear Media at Large Strains," Journal of the Engineering Mechanics Division, ASCE, Vol. 98, No. EM6, Dec. 1972, pp. 1511-1528.

41 Cruse, T. A., and Griffin, D. S., ed., "Three-Dimensional Continuum Computer Programs for Structural Analysis," The American Society of Mechanical Engineers, New York, 1972.

42 Jaeger, T. A., compiler, "Summaries of Contributed Papers and Invited Lectures," 2nd International Conference on Structural Mechanics in Reactor Mechanics, Berlin, Germany, Sept. 10-14, 1973.

43 Radhakrishnan, N., "Analysis of Stress and Strain Distributions, in Triaxial Tests Using the Method of Finite Elements," U. S. Army Engineer Waterways Experiment Station, Technical Report S-73-4, May 1973, Vicksburg, Miss.

44 Hartzman, M., "Static Stress Analysis of Axisymmetric Solids with Material and Geometric Nonlinearities by the Finite Element Method," Lawrence Livermore Laboratory, UCRL-51390, Jan. 19, 1973, Livermore, California.

45 Hartzman, M., "Nonlinear Dynamic Analysis of Axisymmetric Solids by the Finite Element Method," Lawrence Livermore Laboratory, UCRL 74978, Aug. 1973, Livermore, California.

46 Hartzman, M. and Hutchinson, J. R., "Nonlinear Dynamics of Solids by the Finite Element Methods," Computer and Structures, Vol. 2, Feb. 1972, pp. 47-77.

47 Belytschko, T., and Kennedy, J. M., "Response of Reactor Core Subassemblies to Impulsive Loading," Paper No. 73-DET-93, Presented at ASME Design Engineering Technical Conference, Cincinnati, Ohio, Sept. 9-12, 1973.

48 Vos, R. G., "SSME Structural Computer Program Development, Volume 1: Bopace Theoretical Manual," Boeing Aerospace Co., NASA CR-124390, July 31, 1973, Huntsville, Alabama.

49 Nosseir, S. B., Takahashi, S. K., and Crawford, J. E., "Stress Analysis of Multicomponent Structures," Naval Civil Engineering Laboratory, Technical Report R743, Oct. 1971, Port Hueneme, California.

50 Swanson, J. A., "ANSYS - Engineering ANalysis SYStem Users' Manual," Swanson Analysis Systems, Inc., Elizabeth, Pa.

51 Nickell, R. E., and Yeung, S. F., ed., "Thermal Structural Analysis Programs," The American Society of Mechanical Engineers, New York, 1972.

52 Argyris, J. H., "ASKA - Automatic System for Kinematic Analysis," Nuclear Engineering and Design (Holland), Vol. 10, 1969, pp. 441-455.

53 Argyris, J. H., and Scharpf, D. W., "Methods of Elasto-plastic Analysis," Journal of Applied Mathematics and Physics (ZAMP), Vol. 23, July 25, 1972, pp. 517-552.

54 Meijers, P., "Review of the ASKA Program," Numerical and Computer Methods in Structural Mechanics, Fenves, S. J., Perrone, N., Robinson, A. R., and Schnobrich, W. C., eds., Academic Press, New York, 1973, pp. 123-149.

55 Balmer, H. A., and Doltsinis, J. St., Institut für Statik und Dynamik der Luft-und Raumfahrtkonstruktionen, University of Stuttgart,

ISD-Report No. 132, 1972, Stuttgart, Germany.

56 Balmer, H. A., Doltsinis, J. St., and Koenig, M., "Elasto-plastic and Creep Analysis with the ASKA Program System," Computer Methods in Applied Mechanics and Engineering, Vol. 3, 1974, pp. 87-104.

57 "MARC-CDC, Non-Linear Finite Element Analysis Program, User Information Manual," Publication No. 17309500, Control Data Corp., Minneapolis, Minn.

58 Ayres, D. J., "Elastic-Plastic and Creep Analysis via the MARC Finite-Element Computer Program" Numerical and Computer Methods in Structural Mechanics, Fenves, S. J., Perrone, n., Robinson, A. R., and Schnobrich, W. C., eds., Academic Press, New York, 1973, pp. 247-263.

59 MacNeal, R. H., ed., "The NASTRAN Theoretical Manual," NASA SP-221(01), National Aeronautics and Space Administration, April 1972.

60 Tocher, J. L., and Herness, E. D., "A Critical View of NASTRAN," Numerical and Computer Methods in Structural Mechanics, Fenves, S. J., Perrone, N., Robinson, A. R., and Schnobrich, W. C., eds., Academic Press, New York, 1973, pp. 151-174.

61 Tocher, J. L., et al., "A Technical Evaluation of the NASTRAN Computer Program," Boeing Computer Services, Inc., Seattle, Wash., Feb. 1971.

62 "Proceedings of the Fourth Navy - NASTRAN Colloquium," Naval Ship Research and Development Center, Washington, D. C., AD 764508, March 27, 1973.

63 Sharifi, P. and Yates, P., "Nonlinear Thermo-Elastic-Plastic and Creep Analysis by the Finite Element Method," AIAA Paper No. 73-358 presented at the AIAA/ASME/SAE 14th Structures, Structural Dynamics, and Materials Conference, Williamsburg, Va., March 20-22, 1973.

64 Bathe, K. J., and Wilson, E. L., "NONSAP - A General Finite Element Program for Nonlinear Dynamic Analysis of Complex Structures," Paper M3/1, Proceedings 2nd International Conference on Structural Mechanics in Reactor Technology, Berlin, Sept. 1973.

65 Bathe, K. J., Ramm, E., and Wilson, E. L., "Finite Element Formulations for Large Displacement and Large Strain Analysis," Dept. of Civil Engineering, University of California, Report No. UC SESM 73-14, Sept. 1973, Berkeley, California.

66 Bathe, K. J., Wilson, E. L., and Peterson, F. E., "SAP IV, A Structural Analysis Program for Static and Dynamic Response of Linear Systems," Earthquake Engineering Research Center, Univ. of California, Report No. EERC 73-11, June 1973, Berkeley, California.

67 Bushnell, D., "Computer Analysis of Shell Structures," Paper No. 69-WA/PVP-13, presented at ASME Winter Annual Meeting, Nov. 16-20, 1969, Los Angeles, California.

68 Hartung, R. F., "An Assessment of Current Capability for Computer Analysis of Shell Structures," Air Force Flight Dynamics Laboratory, AFFDL-TR-71-54, April 1971, Wright-Patterson AFB, Ohio.

69 Hartung, R. F., and Ball, R. E., "A Comparison of Several Computer Solutions to Three Structural Shell Analysis Problems," Air Force Flight Dynamics Laboratory, AFFDL-TR-73-15, April 1973, Wright-Patterson AFB, Ohio.

70 Hartung, R. F., "Numerical Solution of Nonlinear Structural Problems," AMD-Vol. 6, The American Society of Mechanical Engineers, New York, 1973.

71 Hartung, R. F., ed., "Computer Oriented Analysis of Shell Structures," Air Force Flight Dynamics Laboratory, AFFDL-TR-71-79, June 1971, Wright-Patterson AFB, Ohio.

72 Fulton, R. E., "Numerical Analysis of Shells of Revolution," Proceedings of IUTAM Conference of High Speed Computing in Elastic Structure, University of Liege, August 1970.

73 Bushnell, D., "Stress, Stability, and Vibration of Complex Shells of Revolution: Analysis and User's Manual for BOSOR3," Lockheed Missiles and Space Company, N-SJ-69-1, Sept. 6, 1969, Sunnyvale, California.

74 Balmer, H. A., "Improved Computer Programs--DEPROSS 1, 2 and 3--To Calculate the Dynamic Elastic-Plastic Two-Dimensional Responses of Impulsively-Loaded Beams, Rings, Plates, and Shells of Revolution," Aeroelastic and Structures Research Laboratory, MIT, ASRL TR 128-3, Aug. 1965, Cambridge, Mass.

75 Krieg, R. D., and Duffey, T. A., "UNIVALVE II, A Code to Calculate the Large Deflection Dynamic Response of Beams, Rings, Plates, and Cylinders," Sandia Laboratories, SC-RR-68-303, Oct. 1968, Albuquerque, New Mexico.

76 Almroth, B. O., Brogan, F. A., Meller, E., Zele, F., and Petersen, H. T., "Collapse Analysis for Shells of General Shape, Vol. II, User's Manual for the STAGS - A Computer Code," Air Force Flight Dynamics Laboratory, AFFDL-TR-71-8, March 1973, Wright-Patterson AFB, Ohio.

77 Sobel, L., Silsby, W., and Wrenn, B. G., "User's Manual for the STAR (Shell Transient Asymmetric Response) Code," Lockheed Missiles and Space Co., LMSC-D006673, April 1970, Palo Alto, California.

78 Santiago, J. M., Wisniewski, H., and Huffington, Jr., N. J., "A User's Manual for the REPSIL Code," U. S. Army Ballistic Research Laboratories, Aberdeen Proving Grounds, Maryland, (in preparation).

79 Ball, R. E., "A Geometrically Nonlinear Analysis of Arbitrarily Loaded Shells of Revolution," NASA CR-909, Jan. 1968.

80 Ball, R. E., "A Program for the Nonlinear Static and Dynamic Analysis of Arbitrarily Loaded Shells of Revolution," Proceedings of Computer Oriented Analysis of Shell Structures, Air Force Flight Dynamics Laboratory, AFFDL-TR-71-79, Wright-Patterson AFB, Ohio, and Computers and Structures, Vol. 2, 1972, pp. 141-162.

81 Tillerson, J. R., and Haisler, W. E., "SAMMSOR II - Finite Element Program for the Stiffness and Mass Matrices of Shells of Revolution," Sandia Laboratories, SC-CR-70-6168, Albuquerque, New Mexico (also Aerospace Engineering Department, Report 70-18, Texas A&M University), November 1971.

82 Haisler, W. E., and Stricklin, J. A., "SNASOR II - A Finite Element Program for the Static Nonlinear Analysis of Shells of Revolution," Sandia Laboratories, SC-CR-71-5155, Albuquerque, New Mexico (also Aerospace Engineering Department, Report 70-20, Texas A&M University), November 1971.

83 Tillerson, J. R., and Haisler, W. E., "DYNASOR II - A Finite Element Program for the Dynamic Nonlinear Analysis of Shells of Revolution," Sandia Laboratories, SC-CR-70-6169, Albuquerque,

New Mexico (Aerospace Engineering Department, Report 70-19, Texas A&M University), October 1971.

84 Underwood, P., "User's Guide to the SHORE Code," Lockheed Palo Alto Research Laboratory, LMSC-D244589, Palo Alto, California.

85 Haisler, W. E., and Stricklin, J. A., "SAMMSOR IV - A Finite Element Program to Determine Stiffness and Mass Matrices of Ring-Stiffened Shells of Revolution," Sandia Laboratories, SLA-73-1105, Albuquerque, New Mexico (Aerospace Engineering Department, TEES 2926-73-1, Texas A&M University, College Station, Texas), October 1973.

86 Haisler, W. E., and Vaughan, D. K., "DYNAPLAS II - A Finite Element Program for the Dynamic, Large Deflection, Elastic-Plastic, Analysis of Stiffened Shells of Revolution," Sandia Laboratories, SLA-73-1107, October 1973, Albuquerque, New Mexico.

87 Stricklin, J. A., Haisler, W. E., Von Riesemann, W. A., Leick, R., Hunsaker, B., and Saczalski, K., "Large Deflection, Elastic-Plastic, Dynamic Response of Stiffened Shells of Revolution," Sandia Laboratories, SLA-73-0128, Albuquerque, New Mexico (Aerospace Engineering Department, Report 72-25, Texas A&M University, College Station, Texas), December 1972.

88 TWORNG, Personal communication, Paul A. Wieselmann, Kaman Sciences Corp., to Walter A. Von Riesemann, Dec. 19, 1973.

89 Franke, R., "Ring Bond, 'A Computer Program for Multiple Layered Rings,'" Kaman Nuclear, KN 65-397, Nov. 23, 1965, Colorado Springs, Colorado.

90 SMERSH, Personal communication, Paul A. Wieselmann, Kaman Sciences Corp., to Walter A. Von Riesemann, Dec. 19, 1973.

91 Hubka, W. F., Windholz, W. M., and Karlsson, T., "A Calculation Method for the Finite Deflection, and Elastic Dynamic Response of Shells of Revolution," Kaman Nuclear, KN 69-660(R), Jan. 30, 1970, Colorado Springs, Colorado.

92 Wu, R. W.-H., and Witmer, E. A., "Computer Program--JET 3--To Calculate the Large Elastic-Plastic Dynamically-Induced Deformations of Free or Restrained, Partial and/or Complete Structural Rings," Aeroelastic and Structural Research Laboratory, MIT, ASRL TR 154-7 (also NASA CR-120993), Cambridge, Mass.

93 Wu, R. W.-H., and Witmer, E. A., "Finite-Element Analysis of Large Transient Elastic-Plastic Deformations of Simple Structures, with Application to the Engine Rotor Fragment Containment Deflection Problem," Aeroelastic and Structural Research Laboratory, MIT, ASRL TR 154-4 (also NASA CR-120886), Jan. 1972, Cambridge, Mass.

94 Chung, T. J. and Eidson, R. L., "Dynamic Analysis of Viscoelastoplastic Anisotropic Shells," Computers and Structures, Vol. 3, No. 3, May 1973, pp. 483-496.

95 Klein, S., "The Elastic-Plastic Dynamic Analysis of Shells of Revolution by the Finite Element Method," Stanford Research Institute, Poulter Lab. TR 002-73, Jan. 1973, Menlo Park, California.

96 Gabrielson, V. K., and Reese, R. T., "SHOCK Code User's Manual, A Computer Program to Solve the Dynamic Response of Lumped-Mass Systems," Sandia Laboratories, SCL-DR-69-98, Nov. 1969, Livermore, California.

97 "Use of the Computer in Pressure Vessel Analysis," Papers presented at ASME Computer Seminar, Dallas, Texas, Sept. 20, 1968, The American Society of Mechanical Engineers, New York, 1969.

APPENDIX A

Source for Programs

1 CHART D:	Dr. Samuel L. Thompson Division 5166 Sandia Laboratories Albuquerque, NM 87115
2 WONDY IV:	Argonne Code Center Argonne National Laboratory 9700 South Cass Avenue Argonne, IL 60439
3 PISCES-1DL:	CDC CYBERNET System
4 SWAP:	See no. 2
5 CSQ:	See no. 1
6 HEMP:	Dr. Mark L. Wilkins, L-24 Lawrence Livermore Laboratory University of California P. O. Box 808 Livermore, CA 94550
7 PISCES-2DL:	CDC CYBERNET System
8 REXCO-H:	See no. 2
9 TOODY IIA:	See no. 2
10 TOIL:	Dr. Leslie R. Hill Division 8314 Sandia Laboratories P. O. Box 969 Livermore, CA 94550
11 EPAD:	Maj. Richard A. Mirth AFWL (WLCP) Kirtland AFB, NM 87117
12 EPIC-II:	Dr. Norman A. Cyr Lockheed Missiles & Space Co. P. O. Box 504 Sunnyvale, CA 94088

13 H326: H. Chao
Structural Research Branch
McDonnell Douglas Astronautics Co.
5301 Bolsa Ave.
Huntington Beach, CA 92647

14 OASIS: Dr. Frank C. Weiler
Weiler Research Inc.
744 Holbrook Place
Sunnyvale, CA 94087

15 PLAST2: Prof. O. C. Zienkiewicz
Dept. of Civil Engineering
University College of Swansea
Singleton Park
Swansea SA2-8PP
United Kingdom

16 SAAS II and ASAAS: Aerospace Corporation
P. O. Box 95085
Los Angeles, CA 90045

17 DYNS: Prof. Edward L. Wilson
Dept. of Civil Engineering
University of California
Berkeley, CA 94720

18 HONDO: Will be available from source given in no. 2

19 SAMSON: Capt. Harvey D. Bartel
AFWL (DEV)
Kirtland AFB, NM 87117

20 ANSYS: CDC CYBERNET System

or Dr. John A. Swanson
Swanson Analysis Systems, Inc.
870 Pine View Drive
Elizabeth, PA 15037

21 ASKA: Prof. John H. Argyris
Institut für Statik und Dynamik der Luft und Raumfahrtkonstruktionen
Technische Hochschule Stuttgart
7 Stuttgart - Vaihingen
Robert Leichtstr. 225
Germany

22 MARC: CDC CYBERNET System

or Dr. Pedro Marcal
MARC Analysis Corp.
105 Medway St.
Providence, RI 02906

23 NASTRAN: CDC CYBERNET System

or COSMIC
Barrow Hall
University of Georgia
Athens, GA 30601

or Source stated in text

24 NEPSAP: Will be available at the
Control Data Center
Palo Alto, CA

25 NONSAP: See no. 17

26 SATANS: Prof. Robert E. Ball
Code 57Bp
Naval Postgraduate School
Monterey, CA 93940

27 SOR Codes: Prof. Walter E. Haisler
Dept. of Aerospace Engineering
Texas A&M University
College Station, TX 77843

or COSMIC
Barrow Hall
University of Georgia
Athens, GA 30601

28 SHORE: Mr. Philip Underwood
Lockheed Palo Alto Research Lab.
3251 Hanover St.
Palo Alto, CA 94304

29 DYNAPLAS: See no. 27

30 TWORNG: Dr. Paul A. Wieselmann
Kaman Sciences Corp.
P. O. Box 7463
Colorado Springs, CO 80933

31 SMERSH: See no. 30

32 JET 3:	Prof. Emmett A. Witmer Dept. of Aeronautics and Astronautics Massachusetts Institute of Technology Cambridge, MA 02139
33 SABOR/ DRASTIC 6A:	Dr. Stanley Klein Philo Ford Corp. Ford Road Newport Beach, CA 92663

PLASTIC ANALYSIS

H. Armen, Jr.
Grumman Aerospace Corporation
Bethpage, New York

ABSTRACT

This paper provides an overview and evaluation of available general purpose plasticity programs and also presents some theoretical considerations necessary for the treatment of plasticity within the framework of a comprehensive structural analysis program. Definitions of several basic concepts are included to provide a user of these programs with the necessary ingredients for an understanding of the limitations of the applied numerical procedures. Accordingly, this information should be of value in an assessment of currently available software. Descriptions of the capabilities (e.g., element library, plasticity theories, solution scheme, etc.) of several comprehensive programs are presented. This list will facilitate a preliminary assessment of the available options open to a potential user. It is further hoped, however, that a familiarization with the basic concepts and the consequences of choosing various idealizations associated with plastic behavior will place the user in a more favorable position to make a decision concerning a particular need than would otherwise be possible.

NOMENCLATURE

- A - Cross-sectional area
- C - An array relating increments of plastic strain to stress, Eq. (3)
- D - A differential operator in matrix form, Eq. (10)
- E - Young's modulus
- E - Matrix of elastic constants
- E_T - An array relating increments of stress to total strain, Eq. (6)
- e - Total strain
- K - Element stiffness influence coefficient matrix, Eq. (15)
- K_I - Initial strain stiffness matrix, Eq. (15)
- K_S - Initial stress stiffness matrix, Eq. (16)
- K_T - Tangent modulus stiffness matrix, Eq. (13)
- $\tilde{N}$ - Matrix of shape functions used to describe displacement fields, Eq. (9)
- P - Vector of applied external loads, Eq. (11)
- Q - Effective plastic loads, Eq. (14a)
- R - Equilibrium imbalance forces, Eqs. (18), (19)
- u - Displacements
- u^* - Generalized nodal displacements

V - Volume of finite element
α - Displacement of yield surface in stress space
ϵ - Plastic strains
Δ - Used to denote incremental quantities
$d\lambda$ - Nonnegative proportionality factor, Eq. (1)
σ - Stress

INTRODUCTION

The increased demands from governmental and industrial organizations associated with aerospace, naval, and nuclear reactor technology for determining accurate stress, strain, and displacement fields have been a motivating force behind the advances in the development of general purpose programs for the plastic analysis of structures. As in the case of linear elastic analyses, the progress associated with treating material nonlinearities has been the direct result of advances in several interdisciplinary areas that include: structural mechanics, numerical analysis, and computer systems engineering. Specifically, in the area of structural mechanics, significant advances have been those associated with the finite-element method. The contribution from numerical analysis methods notably has been the development of efficient algorithms to solve, on a repetitive basis, large systems of equations. The ease with which the theoretical considerations and numerical algorithms are efficiently processed and translated into (hopefully) meaningful answers is attributable to the ingenuity and talents of computer systems engineers who develop the necessary hardware and, in some cases, software to satisfy what must seem to them the insatiable appetite of the developers of large scale programs.

The advances in the abovementioned areas have resulted in the development of many software systems to treat plasticity. These range from simple instructional programs to general comprehensive computer systems capable of treating the entire visible spectrum of two- and three-dimensional problems. Following the example of Schrem [1] an attempt to classify the programs into a tabular form is presented in Fig. 1.

Class	Organization	Element library	Applications	Limitation due to core storage
A	In-core	Limited	Specialized to a particular class	Serious 100-200 DOF
B	Sequential	Moderate	General with some limitations	Moderate 500-1000 DOF
C	Direct access	Large < 10	General for present and forseeable needs	None

Fig. 1 Classification of programs

This review will not consider those programs that fall into class A. This does not, in any manner, intend to diminish the efforts associated with developing special purpose programs. In fact, it is reasonable to assume that many, if not all, of the programs that fall into classes B and C represent an outgrowth of some special purpose development.

Prior to a discussion of the highlights of individual programs, a review of some pertinent theoretical considerations will be presented. Finally, some conclusions and thoughts are drawn concerning the present state of developments and the options open for future efforts.

THEORETICAL CONSIDERATIONS

Plasticity Theories

"The tensile test [is] very easily and quickly performed but it is not possible to do much with its results, because one does not know what they really mean. They are the outcome of a number of very complicated physical processes. . . . The extension of a piece of metal [is] in a sense more complicated than the working of a pocket watch and to hope to derive information about its mechanism from two or three data derived from measurement during the tensile test [is] perhaps as optimistic as would be an attempt to learn about the working of a pocket watch by determining its compressive strength."

E. Orowan, F.R.S., Proc. Instn. Mech. Engrs., Vol. 151, p. 133, 1944.

The above quote, which serves as the lead-in to the introduction in Ref. [2], is deemed to be equally applicable here. It is appropriate that a discussion concerned with the nature of this article should begin with a discussion of those theories that attempt to describe, in meaningful terms and on a rational basis, that complex phenomenological process known as plastic deformation. There is not, in this author's view, any "best" plasticity theory to describe all the known facets of nonlinear material behavior. There are, however, some theories better suited for particular needs than others. It is therefore desirable to be sufficiently familiar with those theories currently available and to be in a position to choose the theory that combines mathematical and computational simplicity with a proper representation of experimentally observed behavior.

In the following, some of the available plasticity theories that have been incorporated in the various plastic analysis computer programs are presented and briefly discussed. A more thorough examination of these theories is to be found in the reference cited.

As indicated by Ziegler [3], the plastic behavior of a material in a multiaxial stress state can be described by specifying the following:

1. An initial yield condition defining the elastic limit of the material in a multiaxial stress state

2. A flow rule to serve as the constitutive relations between plastic strains and stresses
3. A hardening rule, used to establish conditions for subsequent yield from a plastic state

The theories associated with a multiaxial yield criterion, in general, are the extension of the single point representation of a yield stress (or strain) in a uniaxial state to an n-dimensional yield surface that separates elastic and plastic regions in an n-dimensional stress space. The two most popular criteria considered are those attributed to von Mises and to Tresca. The former is equivalent to the condition that yielding at an arbitrary point in a material will begin whenever the combination of stresses is such that the strain energy of distortion per unit volume absorbed at the point is equal to the corresponding energy developed in a bar uniaxially stressed to the elastic limit, as occurs in a simple tension (or compression) test. For a state of two dimensional principal stresses σ_1 and σ_2, the von Mises yield criterion is represented by the ellipse shown in Fig. 2a.

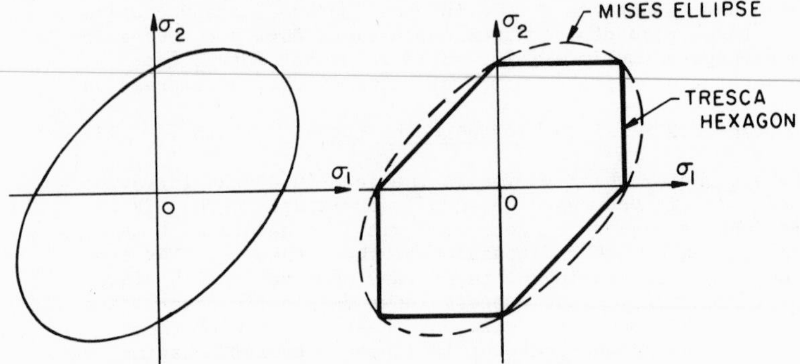

(a) von Mises yield ellipse (b) Tresca yield hexagon

Fig. 2 Yield surfaces in $\sigma_1 - \sigma_2$ plane

The Tresca condition states that inelastic action at any point in a body begins only when the maximum shearing stress on some plane through the point reaches a value equal to the maximum shearing stress in a tension specimen when yielding starts. A piecewise linear curve in the $\sigma_1 - \sigma_2$ principal stress space, as shown in Fig. 2b, illustrates this yield criterion.

A flow rule that is generally used to describe elastic-ideally plastic behavior is the Prandtl-Reuss relation, which is a generalization of the Levy-Mises equations. The Prandtl-Reuss assumption is that the plastic strain increments, $d\epsilon_{ij}$, is proportional to the corresponding stress deviation, σ'_{ij}, i.e.,

$$d\epsilon_{ij} = \sigma'_{ij} d\lambda \tag{1}$$

where the instantaneous nonnegative value of $d\lambda$ is left to the inventiveness of the user of these equations. The concept of an effective stress or effective plastic strain is usually introduced to reduce the complexity of a multiaxial situation to one that can be related to uniaxial behavior. Thus, the proportionality parameter, $d\lambda$, can be the ratio of the effective plastic strain increment to the prevailing effective stress.

A more general approach to determining a flow rule is the use of the concept of a plastic potential. The assumption is made that there exists a scalar function of stress, say $f(\sigma_{ij})$, from which the components of plastic strain increments may be determined from the partial derivative of $f(\sigma_{ij})$ with respect to σ_{ij}. In a form similar to Eq. (1), we may express this condition as

$$d\epsilon_{ij} = \frac{\partial f(\sigma_{ij})}{\partial \sigma_{ij}} d\lambda \tag{2}$$

where once again $d\lambda$ is a nonnegative constant of proportionality. If the von Mises yield condition is used as the scalar plastic potential $f(\sigma_{ij})$, then $\partial f/\partial \sigma_{ij} = \sigma'_{ij}$, and Eqs. (1) and (2) are identical.

If $f(\sigma_{ij})$ represents a yield surface in stress space, then Eq. (2) also represents a result of Drucker's postulate [4], which states that the work done by an external agency during a complete cycle of loading and unloading must be nonnegative.

Equations (1) or (2) lead to an incremental, or flow theory of plastic behavior, in which there is a path dependence of a final state of stress and strain as reached from some previous state. Flow theory is, in general, distinct from the deformation theory of plasticity, in which the total plastic strains are related to the final stress state. According to this latter theory, a relationship between final states of stress and strain exists for any given loading process, unloading being specified by a separate law. In the case of proportional loading, in which the stress vector remains fixed in direction, flow and deformation theories coincide.

Assuming Drucker's postulate to constitute a criterion for physical soundness, Budianski [5] has shown deformation theories to be consistent with this postulate only for loading paths in the vicinity of proportional loading. Since the purpose of a comprehensive program is to treat problems with general loading paths, including reversed loading, the use of deformation theory is usually not considered.

Having selected a yield condition and a flow rule, there remains the choice of describing the subsequent response of the model beyond initial yielding, i.e., a hardening rule. The subsequent yield surface, termed the "loading surface," will represent a convenient mathematical idealization of some macroscopically observed behavior. It should have the ability to describe quantitatively hardening behavior as determined from experimental results for a particular material.

There have been several hardening rules proposed for use in the plastic analysis of structures. The choice of a specific hardening rule will depend upon the ease with which it can be applied in the method of analysis to be used, in addition to its capability of representing the actual hardening behavior of structural metals. These requirements, together with the necessity of maintaining mathematical consistency with the yield function, constitute the criteria for the final choice of the hardening rule. An appraisal of the various hardening rules currently available has been previously presented in [6] and in a more extensive investigation recently completed and reported in [7]. The following appraisal represents a distillation and excerption from these studies.

Isotropic Hardening

This theory assumes that during plastic flow the loading surface expands uniformly about the origin in stress space, maintaining the same shape, center, and orientation as the yield surface. Figure 3 illustrates, on the basis of a simplification to a two-dimensional plot, the yield and loading surfaces when the stress state shifts from point 1 to 2. Unloading and subsequent reloading in the reverse direction will result in yielding at the stress state represented by point 3. The path 2-3 will be elastic, and 0-2 is equal to 0-3.

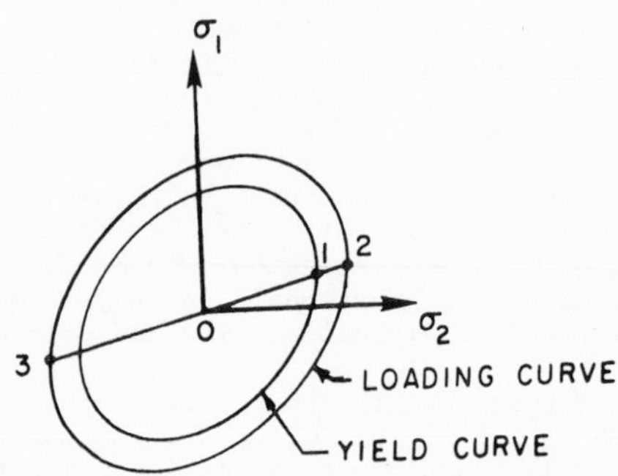

Fig. 3 Isotropic hardening

It is seen that the isotropic representation of work hardening does not account for the Bauschinger effect exhibited by most structural materials. In fact, this theory provides that, due to work hardening, the material will exhibit an increase in the compressive yield stress equal to the increase in the tensile yield stress. Furthermore, since plastic deformation is an anisotropic process, it cannot be expected that a theory that predicts isotropy in the plastic range will lead to realistic results when complex loading paths, involving changes in direction of the stress vector in stress space (not necessarily completely reversed), are considered. This conclusion has been indicated experimentally in [8-11].

Slip Theory

Utilizing the physical concept of slip surfaces in crystals, Batdorf and Budiansky [12] have developed a theory that describes a loading surface that is distorted relative to the yield surface and previous loading surfaces. This theory predicts the formation of corners at the instantaneous stress state on the loading surface during plastic deformation. A representation of the growth of the yield function in going from a stress state at the origin 0 to the final state represented by point 3 is given in Fig. 4. In this figure, the unshaded region is that enclosed by the yield surface, and the various shaded regions indicate the stages in the formation of the loading surfaces in going from 0 to 3.

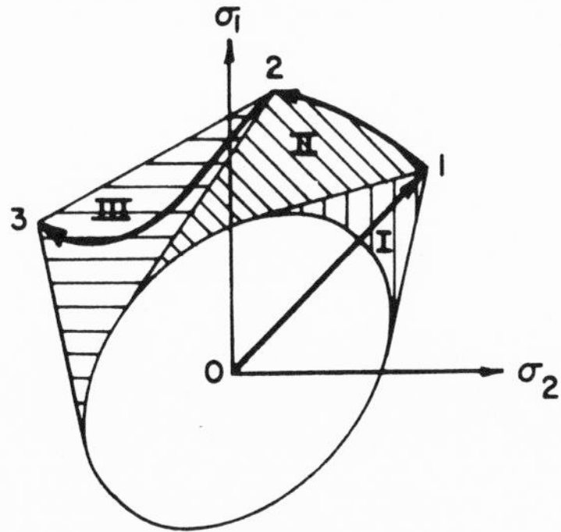

Fig. 4 Slip theory hardening

Since the stress state is almost always in a corner, the resulting constitutive relation between stresses and strains becomes quite complex. For this reason this theory is usually rejected for application. Although there are some experimental results indicating the formation of corners and the distortion of the loading surface [9, 10, 13], these results and those of [14] do not fully substantiate the behavior represented by slip theory. Furthermore, the Bauschinger effect is not taken into account.

Piecewise Linear Plasticity

In this representation, the yield surface consists of a finite number of plane surfaces, whose intersections constitute corners.

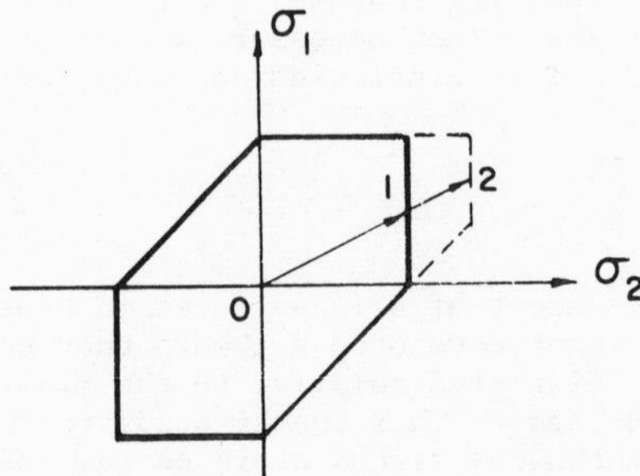

(a) Independent loading surfaces

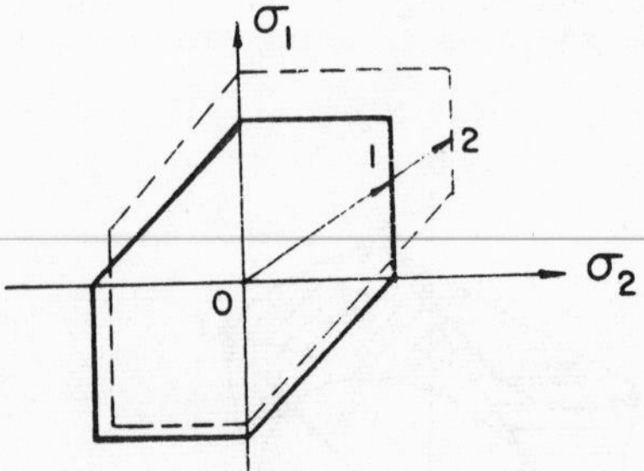

(b) Interdependent loading surfaces

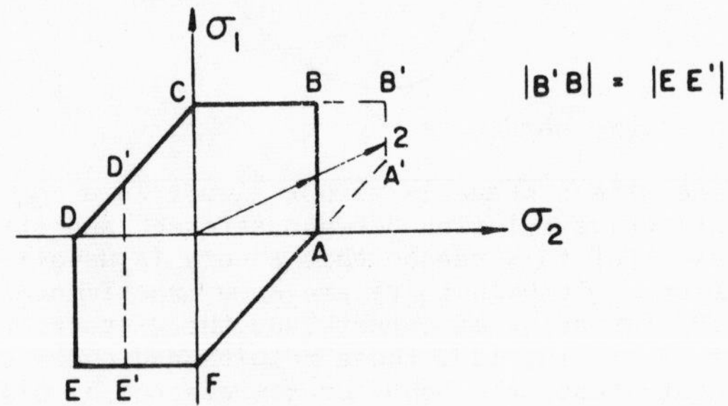

(c) Special case of (b)

Fig. 5 Piecewise linear hardening

The oldest and most widely used piecewise linear yield surface is that associated with the Tresca yield condition. The loading surface is assumed also to consist of plane surfaces, and the subsequent hardening behavior can be classified as

1. The hardening rule of independent plane loading surfaces. One of the earliest discussions of this representation of the hardening behavior is given in [15] and is illustrated in Fig. 5a. As seen from this figure, in which σ_1 and σ_2 are the only nonzero stress components, a loading path, 0-2, in any quadrant of the stress plane does not affect the loading surface in the remaining quadrants. Thus, this hardening rule does not take the Bauschinger effect into account.

2. The hardening rule of interdependent loading surfaces. This type of hardening rule, originally proposed by Hodge [16], is a generalization of the hardening rule described in the previous paragraph. By specifying a dependence between the planes that comprise the loading surface, a loading path intersecting any one plane of this surface may effect changes in each of the remaining planes. As illustrated in Fig. 5b, this hardening rule can be used to specify any piecewise linear loading surface and is capable of taking the Bauschinger effect into account.

A special case of the interdependent loading surfaces is considered in [17]. It is assumed that plastic strain is due to slipping along three independent slip planes, along any one of which the shear is a maximum. Piecewise linear stress-strain relations are written in terms of coefficients representing the hardening behavior of the material. These coefficients are functions of stress and are dependent upon a linear strain-hardening rule employed in the analysis. By specifying the correspondence between various segments of the yield surface and the slip planes, total plastic strains for any loading are computed as the sum of the contributions from the three independent sets of slip planes. It is further assumed that the corresponding segments of the yield surface must maintain a constant elastic range from positive to negative yielding. An illustration of the subsequent loading surfaces determined in this way is shown in Fig. 5c. It is seen from this figure that the Bauschinger effect can be taken into account.

Kinematic Hardening

The hardening behavior postulated in this theory assumes that during plastic deformation the loading surface translates as a rigid body in stress space, maintaining the size, shape, and orientation of the yield surface. The primary aim of this theory, due to Prager [18] and [19], is to provide a means of accounting for the Bauschinger effect. For piecewise linear yield surfaces, kinematic hardening may be considered to be a special case of the hardening rule of interdependent loading surfaces. However, it is not limited to piecewise linear yield surfaces.

An illustration of kinematic hardening, as applied in conjunction with the von Mises yield curve in the σ_1, σ_2 plane, is provided in Fig. 6. The yield surface and loading surface are shown in this figure for a shift of the stress state from point 1 to point 2. The translation of the center of the yield surface is denoted by α_{ij}.

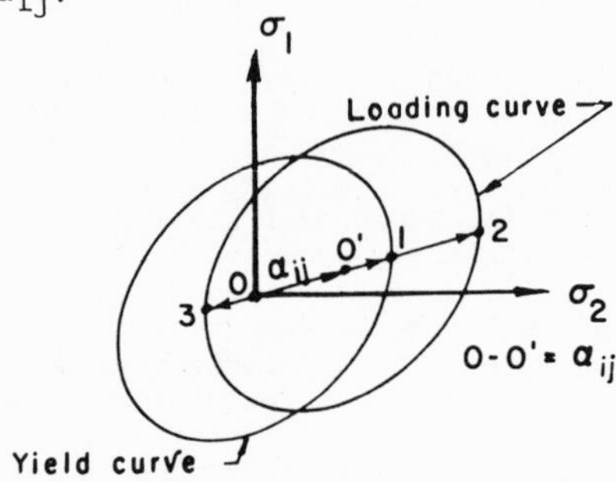

Fig. 6 Kinematic hardening

As a consequence of assuming a rigid translation of the loading surface, kinematic hardening predicts an ideal Bauschinger effect for completely reversed loading conditions. That is, the magnitude of the increase of yield stress in one direction will result in a decrease of yield stress of the same magnitude in the reversed direction. Kadashevitch and Novozhilov [20] have concurrently developed a hardening rule identical to Prager's kinematic hardening rule. In their theory, the total translation of the yield surface is regarded as being associated with "internal microstresses" that remain in the body upon unloading. It is these internal microstresses that are considered to be responsible for the Bauschinger effect.

This hardening theory, as set forth by Prager, predicts that the increments of translation of the loading surface in 9-dimensional stress space occur in the direction of the exterior normal to the surface at the instantaneous stress state. However, as indicated in [21-23], inconsistencies arise when the theory is applied in various subspaces of stress, that is, when the symmetry of the stress tensor is taken into account or when there are zero stress components.

These inconsistencies mentioned in connection with Prager's rule produce the result that the loading surface will not, in general, translate in the direction of the exterior normal in a subspace of stress when it is made to do so in the full 9-dimensional stress space. Reference [22] specifies stress conditions under which a linear transformation of variables enables the loading surface, in the transformed subspace, to translate in the direction of the exterior normal. It is also indicated in [22] that the application of Prager's rule to the Tresca yield condition, in the case in which more than one normal stress is zero, results in a deformed yield locus.

In order to avoid the difficulties associated with the use of complete kinematic hardening, Ziegler [2] has proposed a modification of Prager's rule that deals with the expression for the increment of translation of the yield surface. With this modification there is no inconsistency between the behavior in any subspace and in the full 9-dimensional stress space, and the loading surface will translate without distortion in the former when it is presumed to do so in the latter.

Comparisons between Prager's rule and the modification of Ziegler indicate that, in general, the two rules do not coincide. However, as shown by Nagdhi [24], the application of Prager's rule results in a translation of the loading surface coincident with that predicted by Ziegler's modification in the case of plane stress with the following additional conditions: (1) the von Mises yield condition is used, (2) the parameter characterizing the hardening behavior of the material of an instantaneous stress state is assumed to be constant.

Fields of Work-hardening Moduli (Mróz model)

In proposing a work-hardening model to account for the behavior of metals under cyclic loading conditions, Mróz [25] has introduced the notion of a field of work-hardening moduli and the variation of this field during the course of plastic deformation. In this proposed model, a stress-strain curve of an initially isotropic material is represented by n linear segments of constant tangent plastic moduli, as shown in Fig. 7. In stress space, this approximation can be represented by n hypersurfaces $f_0, f_1, \ldots, f_n$, where f_0 is the initial yield surface, and f_1 to f_n define regions of constant work-hardening moduli. Figure 8 illustrates these hypersurfaces in the $\sigma_1 - \sigma_2$ plane for an initially isotropic material. As seen in this figure, the surfaces $f_0, f_1, \ldots f_n$ are similar and concentric, and for simplicity are schematically represented by a family of circles. If we consider proportional loading in the σ_2 direction, corresponding to σ in Fig. 7, and if we assume that the surfaces can experience a rigid translation without experiencing a change of size or orientation, then when the stress state reaches point A on Fig. 8a, the surface f_0 will translate until it reaches the circle f_1 at the stress corresponding to point B. The circles f_0 and f_1 translate together until point C is reached, where now f_0, f_1, and f_2 are attached at a common point of contact. For unloading and subsequent reversed loading, when the stress reaches a point corresponding to point E (Fig. 8b), reverse plastic flow occurs and the surface f_0 translates downward along the σ_2 axis until it reaches the surface f_1 at F. Mróz further proposes that the curve of reverse loading in Fig. 7 joins the curve $OA'B'G$ that is obtained by symmetry with respect to the origin

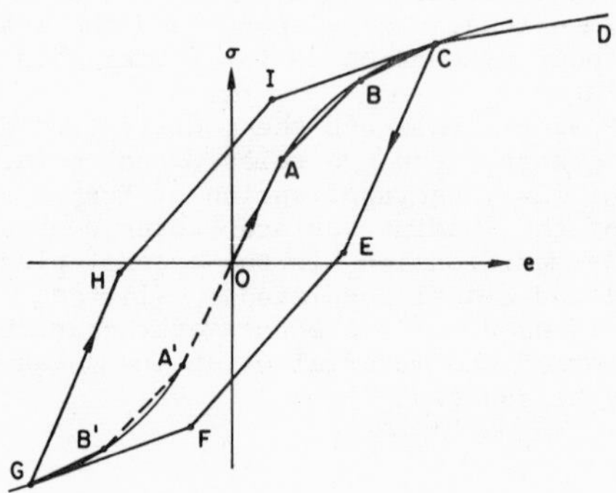

Fig. 7 Representation of typical cyclic stress-strain curve by constant tangent moduli

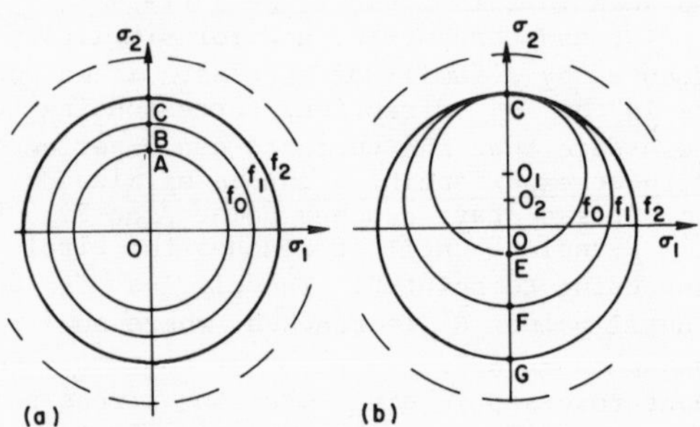

Fig. 8 Representation of hypersurfaces of constant work-hardening moduli

from OABC. Thus, the curve of reverse loading EFG is uniquely defined by the curve of primary loading, represented by an equation of the form $\sigma = f(e)$. If a new coordinate system $(\overline{\sigma},\overline{e})$ with origin at C is used, we have for the curve CEFG

$$\frac{1}{2}\,\overline{\sigma} = f(\frac{1}{2}\,\overline{e})$$

In the generalization of this model to nonproportional loading, it is assumed that during translation of the hypersurfaces the individual surfaces do not intersect but consecutively contact and push each other.

It should be noted that when f_1 tends to infinity, so that the work-hardening modulus is constant (the work-hardening curve being represented by a straight line), the theory proposed by Mróz is identical to Prager's kinematic hardening model.

The further generalization of the theory of work-hardening moduli is associated with an expansion or contraction of the surfaces $f_0, f_1, \ldots, f_n$, so that transitory phenomena (work-stiffening, work-softening, or nonisothermal conditions) can be treated. Thus the hypersurfaces f_k are not constants but functions of a monotonically increasing scalar parameter during plastic flow. One suggestion for this scalar is presented in [26].

Mechanical Sublayer Model

A technique to model the arbitrary nonlinear mechanical behavior of a solid by means of a parallel assemblage of elastic ideally plastic solids can be traced to Duwez [27], with extensions by White [28] and Besseling [29]. This modeling concept equates the integrated effect of a network of ideally plastic solids to the actual behavior. An illustration is shown in Fig. 9 for the case of a simple bar. Here the sublayer model is represented by n parallel bars, each having different yield stresses and cross-sectional areas. The individual areas and yield stresses are adjusted so as to represent the actual material behavior. The actual stress-strain curve is in turn idealized by n linear segments as shown in Fig. 10. Although not a necessary condition, the bars are assumed to have the same elastic modulus E_1.

With respect to a multiaxial stress condition, each sublayer in a typical $\sigma_1 - \sigma_2$ stress space has a corresponding yield surface, as shown in Fig. 11. Although there is assumed to be no expansion or translation of the individual sublayer yield surfaces with respect to their origin, there is an effective translation of the sublayer surfaces with respect to each other as shown in Fig. 12. The net effect of this representation is the desired result that the stress state in an individual sublayer in the plastic range remains on the yield surface (there is no penetration or shift of the surface with respect to its own axis) and at a value measured with respect to its own axis. The relative degree of plastic deformation (i.e., plastic strain) of any sublayer with

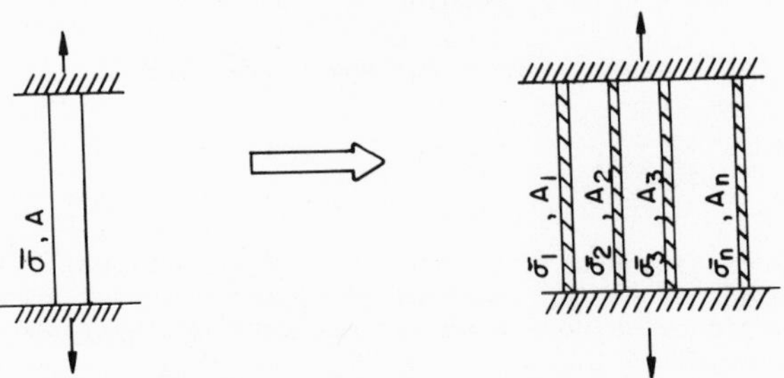

Fig. 9 Idealization of simple bar for mechanical sublayer model

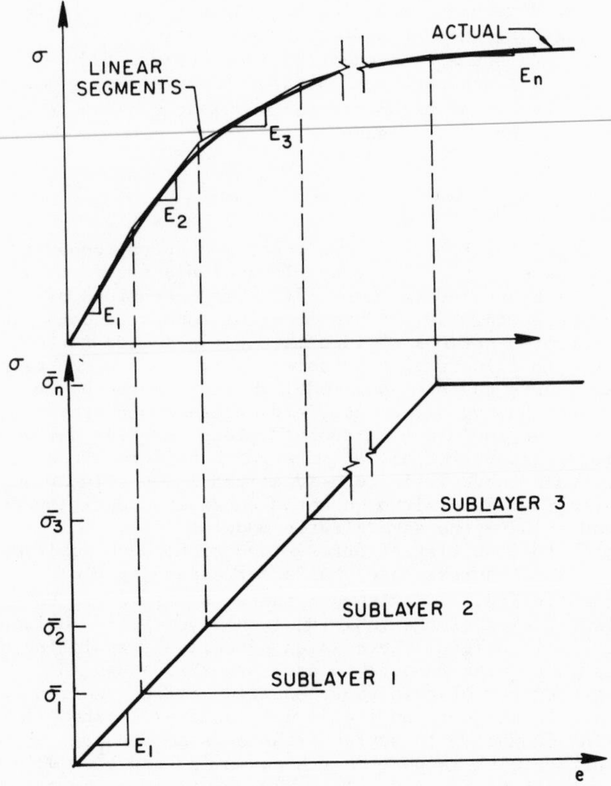

Fig. 10 Idealization of stress-strain curve for mechanical sublayer model

respect to another is represented by the relative shift of axes between the sublayers. This aspect of the model is depicted in Fig. 12,where it is also evident that the treatment of the Baushinger effect is within the capabilities of this model.

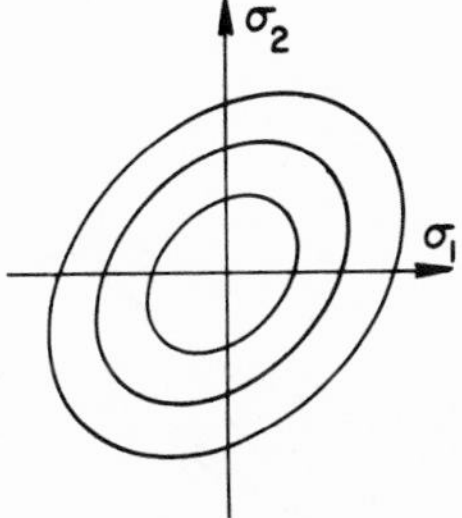

Fig. 11 Representation of sublayer yield surfaces for mechanical sublayer model

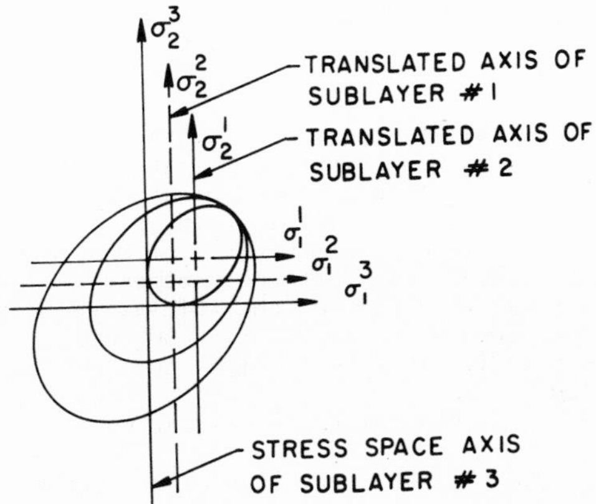

Fig. 12 Representation of sublayer loading surfaces for mechanical sublayer model

A summary of the plasticity theories presented here is presented by means of the hypothetical cyclic stress-strain curve of Fig. 13 and Table 1. The stress-strain curve illustrates the fact that all of the theories are capable of treating the monotonic loading situation. For cyclic loading, including reversed yielding, kinematic and isotropic hardening represent the limits to the actual behavior, whereas the remaining theories are capable of falling anywhere within these limits.

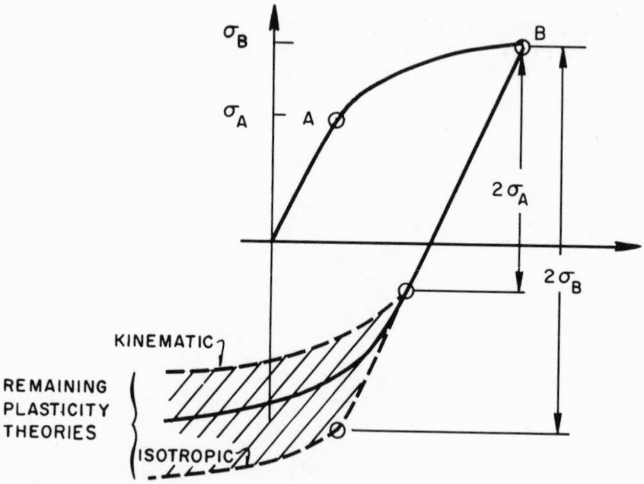

Fig. 13 Typical stress-strain curve for reversed yielding

The plasticity theories discussed thus far are used to describe the hardening behavior of materials subsequent to initial yielding. They attempt to describe process of hardening graphically as an expansion, translation, distortion, etc. of the initial yield surface. For the case of an elastic-ideally plastic material, that is one that does not exhibit a "knee" in the uniaxial stress-strain curve, the initial yield surface is generally assumed to remain unchanged. In [30] the treatment of multiaxial ideally plastic behavior requires that the following conditions be satisfied:

1. The stress increment vector must be tangent to the loading surface
2. The plastic strain increment vector must be normal to the loading surface

The first condition provides a linear relationship among the components of the stress increment, and the second condition provides a linear relation among the various components of the plastic strain increment. The combination of independent increments of stress and plastic strain are subsequently determined as outlined in [30].

Table 1. Comparison of Various Theories of Plasticity

Theory	1	2	3	4
• Isotropic	E	ROL	L	G
• Slip Surfaces	D	ROL	G	L
• Piecewise Linear	M	L	G	M
• Kinematic	M	ROL	G	G
• Mróz	M	L	G	L
• Mechanical Sublayer	E	L	G	G

Column 1: Ease of Use

E - relatively simple
M - moderate
D - difficult

Column 2: Required information

L - linear segmented uniaxial stress-strain curve
RO - Ramberg-Osgood representation of uniaxial stress-strain curve (or any nonlinear functional representation)
ROL - either of the above

Column 3: Generality

L - limited loading situations
G - general

Column 4: Usage

L - limited to special purpose programs
M - moderate usage
G - general usage

Governing Equations

For flow theories of plasticity, the combination of the flow and hardening rules will provide a relationship (explicit or implicit) between increments of plastic strain and stress in the following matrix form

$$\{\Delta\epsilon\} = [C]\{\Delta\sigma\} \tag{3}$$

where $[C]$ is an array whose terms reflect the instantaneous states of stress and the hardening state of the material. Thus the terms of $[C]$ are dependent upon the particular plasticity theory chosen for use.

Additional relations are developed by considering the strain generated during the application of an increment of external load to be composed of the sum of elastic and plastic components, as

$$\{\Delta e^T\} = \{\Delta e\} + \{\Delta\epsilon\} . \tag{4}$$

The elastic component of the strain increment $\{\Delta e\}$ is related to the corresponding stress increments as

$$\left\{\Delta e\right\} = [E]^{-1} \left\{\Delta\sigma\right\} \tag{5}$$

where the elements of $[E]^{-1}$ contain the usual elastic constants.

The combined use of Eqs. (3) through (5) result in the following incremental relations between stress and total strain

$$\left\{\Delta\sigma\right\} = [E_T]^{-1} \left\{\Delta e^T\right\} \tag{6}$$

where $[E_T] = [C] + [E]^{-1}$. The corresponding relation between increments of plastic strain and total strain can be written as

$$\left\{\Delta\epsilon\right\} = [C][E_T]^{-1} \left\{\Delta e^T\right\} \tag{7}$$

Equations (6) and (7) are particularly significant since they indicate that the states of stress and plastic strains can be determined from total strains, a quantity that is, in turn, determined from displacement fields by means of kinematic relations that are independent of material behavior.

The appropriate next step is the incorporation of the inelastic constitutive equations into a set of equations governing the response of the system to a prescribed set of conditions.

Due to familiarity, an approach that is particularly popular is the treatment of plasticity as a modification to a linear elastic analysis. The development of linear elastic finite element analysis is, of course, prevalent for virtually the entire spectrum of problem classifications. The corresponding development in the case of nonlinear analyses is beginning to receive the attention [30-40] that is necessary to provide a better understanding of the underlying principles and limitations of such techniques. The goal here is to provide a succinct development of the governing equations so that the relationships between the various techniques will become evident.

Following a conventional linear finite element formulation (or any Ritz method, for that matter), we assume a displacement field u in terms of discrete variables u^* (i.e., generalized nodal displacements, for finite element approach) in the following form

$$u = \sum_{i=1}^{n} u_i^* f_i(x,y,z) \tag{8}$$

where n is the number of variables used to describe the displacement field in a prescribed domain, and f_i are appropriate functions chosen to satisfy kinematic compatibility conditions, constant strain states, and rigid body motions, so as to achieve convergence of the solution upon grid refinement. It is customary, although not necessary, to take f_i as a complete polynomial of the coordinates x, y, and z. In matrix form Eq. (8) becomes

$$\{u\} = [N]\{u^*\} \tag{9}$$

where [N] consists of the functions f_i for each patch of the domain.

The strains are determined from Eq. (9) by applying the appropriate differential operators on {u} as,

$$\{e^T\} = [D]\{u\} = [B]\{u^*\} \tag{10}$$

where [B] = [D][N] and [D] is the differential operator in matrix form.

The principle of virtual work is employed to yield an approximation on the equilibrium state as

$$\sum\left(\int_V [B]'\{\sigma\}\,dV - \{P\}\right) = 0 \tag{11}$$

where {P} is the vector of external forces, V is the volume of the individual patches (or elements), and the summation is over all the patches used to model the actual structure.

Using an Euler forward difference approximation on stress strain, displacement, and loads, Eq. (9) can be written in incremental form as

$$\int_V [B]'\{\Delta\sigma\}\,dV - \{\Delta P\} = 0 \tag{12}$$

For convenience, the summation symbol is abandoned.

It is at this point in the development that one of two paths are generally taken. The first path introduces Eqs. (6) and (10) into Eq. (12) to produce the following equation:

$$[K_T]\{\Delta u^*\} = \{\Delta P\} \tag{13}$$

where $[K_T] = \int_V [B]'[E_T]^{-1} [B]dV.$

Casting the equation as shown above is, in a broad sense, referred to as the tangent modulus method [31-33]. It is seen that for linear elastic behavior the $[E_T]$ array reduces to the usual elastic coefficients $[E]$.

The alternate path is associated with incorporating Eqs. (3-5) into Eq. (12) to produce

$$[K] \{\Delta u^*\} = \{\Delta P\} + [K_I] \{\Delta \epsilon\} \qquad (14)$$

where

$$[K] = \int_V [B]'[E][B]dV \qquad (15)$$

and

$$[K_I] = \int_V [B]'[E]dV$$

The $[K]$ array is the elastic stiffness matrix, and $[K_I]$ is termed the initial strain matrix. Note that in Eq. (14) the plastic strains (interpreted as initial strains) need not be constant quantities (at nodes or centroids of elements). If a distribution of plastic strains within an element is assumed, its functional form must be considered in the integration for $[K_I]$. The solution procedure that incorporates Eq. (14) is generally termed the "initial strain method." The commonality of this approach with the tangent modulus method is linked to Eq. (12).

An alternate form of Eq. (14) replaces $\{\Delta\epsilon\}$ with $\{\Delta\sigma\}$ from Eq. (3), so as to write

$$[K] \{\Delta u^*\} = \{\Delta P\} + [K_S] \{\Delta \sigma\} \qquad (16)$$

where $[K_S] = [K_I][C]$ and is termed the "initial stress matrix." Appropriately, the use of Eq. (16) is referred to as the "initial stress method," [36].

An equivalent form for Eq. (13) that utilizes total quantities is

$$[K_T] \{u^*\} = \{P\} \tag{13a}$$

and correspondingly for Eq. (14),

$$[K] \{u^*\} = \{P\} + \{Q\} \tag{14a}$$

where $\{Q\} = [K_I]\{\epsilon\}$ and is termed the "effective plastic load."

The use of either the total or incremental forms of the equations requires the loading to be applied in increments, resulting in a piecewise linearization of the nonlinear response.

The linearization process induces the solution to "drift" from the actual exact equilibrium state. In general, the drifting associated with the tangent modulus approach is less than that with the initial strain method for equal load increments. This added accuracy is at the expense of reformulating the $[K_T]$ array and processing it through a matrix solution scheme. The initial strain approach is, in general, solved by approximating the nonlinear effective plastic load at any step by its value at the previous step, so that for the kth step Eq. (14a) becomes [30,35]

$$[K] \{u^*\}^k = \{P\}^k + \{Q\}^{k-1} \tag{17}$$

The drifting of the solution using the initial strain approach is more pronounced than that associated with the tangent modulus method due to the use of an estimated value of the effective plastic load. The equilibrium imbalance is further aggravated with the use of large load increments. The obvious advantage to offset the requirement of having to use relatively small increments of load is that the coefficient matrix $[K]$ is unchanged throughout the entire solution. Thus, the subsequent cost associated with the incremental solutions is considerably reduced if some form of decomposition scheme is employed to process the stiffness matrix into the product of upper and lower triangular arrays.

The obvious question as to which of the two approaches, i.e., tangent modulus or initial strain, is best suited for nonlinear analysis is perhaps best answered (or circumvented) by stating that, in this author's opinion, a hybrid approach is perhaps the optimum answer. The initial strain method can be used with moderate size increments until the structure develops regions of extensive plastic deformation. It would then be desirable to switch the solution to the tangent modulus approach. This was accomplished with some success in [6].

A measure of the degree of drifting at any stage in the loading is determined analytically by inspecting the equilibrium imbalance vector [37-40] defined for the tangent modulus approach as

$$\{R\}^k = - [K_T]^k \{u^*\}^k + \{P\}^k \tag{18}$$

and for the initial strain approach as

$$\{R\}^k = - [K] \{u^*\}^k + \{P\}^k + \{Q\}^k \tag{19}$$

This force imbalance is to be added to the governing equations as follows

$$[K_T] \{\Delta u^*\} = \{\Delta P\} + \{R\} \tag{13b}$$

and

$$[K] \{\Delta u^*\} = \{\Delta P\} + \{\Delta Q\} + \{R\} \tag{14b}$$

and substantially reduces drifting.

A very thorough discussion of the role of the equilibrium imbalance force and suggestions for obtaining significant improvements in accuracy and stability of the incremental procedures for both material and geometrically nonlinear analyses is presented in [34, 37, and 39].

Additional Considerations

The following represents some further thoughts concerning additional techniques or procedures that should be considered in the development of a comprehensive plasticity capability. These procedures, although equally applicable to a linear elastic analysis, become particularly significant in terms of cost-effectiveness when one considers the repetitive or incremental nature associated with a numerical method for nonlinear analysis. No particular significance is assigned to the order of the following items. Each item, in my opinion, contributes to the overall effectiveness.

Substructuring. For two- and particularly three-dimensional problems where plastic flow is judged (by the experience of the analyst) to be contained within a localized region, the use of a substructuring scheme can result in a significant reduction in cost. The cost associated with the additional computations of substructuring (or developing "superelements" to represent large regions of the structure) is offset by performing the bulk of the computations for only that portion of the structure that is of interest.

Restart capability. In many situations where the final result cannot be predicted with any confidence, it is judicious to examine the results up to some prescribed value in the load time history. A decision to continue or cease operations can be made after this examination. This process eliminates further computations that may be unnecessary or meaningless.

Pre- and postprocessors. Satellite programs that can be used to automatically generate finite element meshes are, of course, equally significant for elastic and plastic analyses. Since plasticity is a path-dependent phenomenon the amount of information generated for a complete time history can be quite substantial. Thus, postprocessers that can digest the information determined from the program and graphically produce a desired set of results become a very desirable feature.

Choosing an equation solver. An excellent review of the linear equation solvers currently available is presented in [1]. The appropriate choice from among these, to be used in a plastic analysis, depends very much on whether the tangent modulus or initial strain approach is chosen as the solution technique. For the tangent modulus approach an iterative procedure using over-relaxation appears to be quite efficient. For the initial strain approach, the obvious choice is some form of the Cholesky decomposition or Gaussian elimination.

Are assumptions being violated? In many situations the effect of plasticity precipitates the development of large deformations, or strains. Subsequent results become meaningless if the treatment of such behavior is not accounted for. A further example is associated with developing increments of stress or strain that may be so large as to violate the assumptions of the flow theory of plasticity used in the analysis. These examples are but two of the many possible violations of assumptions that illustrate the need to include self-monitoring logic to check the validity of the underlying assumptions.

Is linear behavior being taken advantage of? In situations involving complex loading paths, including reversed loading, the entire structure may be behaving as an elastic body during some period of the prescribed load history. This behavior eliminates the need to apply the load in small increments. Therefore, it is recommended that a provision be included to determine the limits of the elastic range for initial loading and subsequent reloadings.

AVAILABLE SOFTWARE

The following items are used to describe the individual programs that fall under class B or C of the introduction:

1. Structural modeling capability
   - element library
2. Treatment of material nonlinearity
   - plasticity theory(ies)
   - material description
3. Solution technique
   - linearization scheme (initial strain, tangent, modulus, etc.)
   - linear equation solver

A specific goal of this paper is to provide a critical review of the available programs. Therefore, this section concludes with an evaluation of these programs on the basis of those features considered to be relevant to this author. Individual ratings of each feature are given for each program according to a scale subsequently discussed. They include the following items:

1. Element library
2. Ease of use/documentation
3. Ability to treat a variety of material behavior
4. Special features
   - substructuring
   - restart
   - self-monitoring assumption checks
   - equilibrium corrections
   - selective I/O
   - pre- and postprocessors
   - specifying multipoint constraint conditions (tying nodes)
   - accounting for linear behavior

Program Description

The table of element information originally presented in [41] is used here to describe the element libraries associated with each of the finite element programs. An explanation of the code used in this table follows.

In column B: Beams

Subcolumn a) Position of nodal point in cross section fixed
b) Arbitrary position of nodal point in cross section

In columns E and F: Plates and Shells

Subcolumn a) Kirchhoff: transverse shear neglected
b) Reissner: transverse shear included

In line 4: Model

K = Equilibrium model
V = Displacement model
H = Hybrid model
G = Mixed model

In line 5: Types of elements

L = Line shaped
D = Triangle
R = Rectangle
V = Arbitrary quadrilateral
T = Tetrahedron
P = Pentahedron
Q = Cube
H = Arbitrary hexagon

With elements L, D, R, V one may also specify:

r = body of revolution
s = sector of body of revolution

i.e., L_r = shell of revolution
D_s = sector with triangular meridional cross section

In lines 6 and 7: Element shape boundary

G = Rectilinear
K = Curvilinear
E = Plane
S = Shallow surface
B = Arbitrarily curved surface

For ring and sector types of elements these variables refer to the meridional cross section.

In lines 8 and 9: Cross-sectional data
Material data

Properties:

Q = Area of cross section or thickness
T = Moment of inertia
I = Isotropic
O = Orthotropic
A = Arbitrary anisotropy
K = constant
V = variable

For example: QK, QV, IV, AK

In line 10: External loading

P = concentrated load, L = line load, F = surface load,
V = Volume load

In line 11: Initial strains

N = Not implemented
T = Isotropic thermal strains only
B = Arbitrary initial strains

A blank field indicates that the information is not applicable. An X indicates that the information was not obtained.

Program Name: ANSYS

Developer. Swanson Analysis Systems, Inc.
870 Pine View Drive
Elizabeth, Pennsylvania 15037

Documentation. [42]

Method of analysis. Finite element analysis (FEA)-Displacement method

State of development. Fully operational

Element information. (For plastic analysis)

ANSYS	A	B		C	D	E		F		G
	Rod	Beam a)	b)	Plane membrane	Membrane in space	Plate a) Kirchhoff	b) Reissner	Shell a) Kirchhoff	b) Reissner	Three dimensional continuum
1. No. of elements available	2	1		2		1		1		3
2. No. of nodes per element	2,2	2		3,4		3		3		4,6,8
3. No. of freedoms per element	4,6	6		6,8		3		18		12,18,24
4. Model	V	V		V		V		V		V
5. Element configuration	L	L		D,V		D		D		T
6. Element shape				E, D_r, V_r		E		E		
7. Boundary				G		G		G		G
8. Cross-sectional data	QK	QTK		QK		QK		QK		
9. Material data				OK		OK		OK		OK
10. External loading	P	P.L		P,L		P.L,F		P,L,F		P,E,F,V
11. Initial strains	B	B		B		B		B		B

Comments. Program capabilities include many other elements than those listed above (i.e., special pipe and elbow elements for nonlinear piping analysis). In addition to an impressive static and dynamic structural analysis capability (including elastic-plastic, creep, and swelling, small and large deflections), the program provides for an extensive heat transfer analysis (steady-state and transient; conduction, convection, and radiation).

Program Name: ASAS

Developer. Atkins Research and Development
Ashley Road
Epsom, Surrey, England

Documentation. [X]

Method of analysis. (FEA)-Displacement and equilibrium

State of development. Fully operational

Element information.

ASAS	A	B		C	D	E		F		G
	Rod	Beam a)	b)	Plane membrane	Membrane in space	Plate a) Kirchhoff	b) Reissner	Shell a) Kirchhoff	b) Reissner	Three dimensional continuum
1. No. of elements available	2	2		3	4	1	1	3		3
2. No. of nodes per element	3	2		6,4,6	6,4,6,8	3	8	2,3,5		8,20,32
3. No. of freedoms per element	9,7	12		12,8,12	18,12,12,16	18	48	8,27,54		24,60,96
4. Model	V,K	V,K		V	V,V,K,K	V	V	V		V
5. Element configuration	L	L,K		D,V,D	D,V,D,V	D	V	L_T,D_S		H
6. Element shape				E,E,D_T	E,E,K,K	E	E	K,E		
7. Boundary				G	G	G	G	G		G
8. Cross-sectional data	QK	QTK		QK	QK	QK	QK	QK		
9. Material data				AK	AK	AK	AK	AK		AK
10. External loading	P	P,L		P,L	P,L,F	P,L,F	P,L,F	P,L,F		P,L,F,V
11. Initial strains	B	B		B	B	B	B	B		B

Comments. Two additional elements not included above are

1. A 6-cornered, 30-node, 90-degree-of-freedom, 3-D element
2. A quadrilateral crack-tip element having 11 nodes and 22 degrees of freedom

This reviewer was previously unaware of the existence of this program.

Program Name: ASKA

Developer. ASKA-Group
Pfaffenwaldring 27
Stuttgart-80, West Germany

Documentation. [43]

Method of analysis. (FEA)-Displacement

State of development. Fully operational

Element information.

ASKA	A	B		C	D	E		F		G
	Rod	Beam a)	b)	Plane membrane	Membrane in space	Plate a) Kirchhoff	b) Reissner	Shell a) Kirchhoff	b) Reissner	Three dimensional continuum
1. No. of elements available	2			2	2					4
2. No. of nodes per element	2,3			3,6	3,6					4,10,3,6
3. No. of freedoms per element	6,9			6,12	9,18					12,30,6,12
4. Model	V			V	V					V
5. Element configuration	L			D	D					T,Dr
6. Element shape				E	E					
7. Boundary				G	G					G
8. Cross-sectional data	QV			QV	QV					
9. Material data				I	I					I
10. External loading	P,V			P,L	P,L					P,F,V
11. Initial strains	B			B	B					B

Comments. An extensive system that has recently concentrated its efforts toward building a nonlinear analysis capability.

Program Name: BERSAFE

Developer. CEGB
Berkeley Nuclear Laboratory
Berkeley, Glos., U.K.

Documentation. [44]

Method of analysis. (FEA)-Displacement

State of development. Fully operational

Element information.

BERSAFE	A	B		C	D	E		F		G
	Rod	Beam a)	b)	Plane membrane	Membrane in space	Plate a) Kirchhoff	b) Reissner	Shell a) Kirchhoff	b) Reissner	Three dimensional continuum
1. No. of elements available		1		4		1				3
2. No. of nodes per element		2		12,8,12,8		3				32,20,15
3. No. of freedoms per element		12		24,16,24,16		12				96,60,45
4. Model		V		V		V				V
5. Element configuration		L		V,V,V_r,V_r		D				H,H,P
6. Element shape				E		E				
7. Boundary				G		G				G
8. Cross-sectional data		QTK		QK		QK				
9. Material data				IK		IK				IK
10. External loading		P,L		P,L		P,L,F				P,L,F,V
11. Initial strains		B		B		B				B

Comments. This reviewer was previously unaware of the existence of this program.

Program Name: EPACA

Developer. Mechanical and Nuclear Engineering Department
Franklin Institute Research Laboratories
The Franklin Institute
Philadelphia, Pennsylvania 19103

Documentation. [45]

Method of analysis. (FEA)-Displacement

State of development. Operational

Element information.

EPACA	A	B		C	D	E		F		G
	Rod	Beam a)	b)	Plane membrane	Membrane in space	Plate a) Kirchhoff	b) Reissner	Shell a) Kirchhoff	b) Reissner	Three dimensional continuum
1. No. of elements available				2	6	2	2		4	
2. No. of nodes per element				3,4	3,4,6,8 10,12	3,4	3,4		6,8,10,12	
3. No. of freedoms per element				6,8	9,12,18 24,30,36	15,20	15,20		30,40,50,60	
4. Model				V	V	V	V		V	
5. Element configuration				D,V	D,V,D,V,D,V	D,V	D,V		D,V,D,V	
6. Element shape				E	E,E,K,K, K,K	E	E		K	
7. Boundary				G	G	G	G		G	
8. Cross-sectional data				QV	QV	QV	QV		QV	
9. Material data				IV	IV	IV	IV		IV	
10. External loading				P,L,F,V	P,L,F,V	P,L,F,V	P,L,F,V		P,L,F,V	
11. Initial strains				B	B	B	B		B	

Comments. A program designed for elastic-plastic-creep analysis of thin or thick three-dimensional shells made up of curved or flat shell elements. Additional capabilities to perform transient and steady-state heat transfer analysis and stationary creep rate computation are planned for completion in the future. Continuous or discrete elastic foundations effects are also included in present version of program.

Program Name: MARC-CDC

Developer. Marc Analysis Research Corporation
105 Medway Street
Providence, Rhode Island 02906

Documentation. [46]

Method of analysis. (FEA)-Displacement

State of development. Fully operational

Element information.

MARC-CDC	A	B		C	D	E		F		G
	Rod	Beam a)	b)	Plane membrane	Membrane in space	Plate a) Kirchhoff	b) Reissner	Shell a) Kirchhoff	b) Reissner	Three dimensional continuum
1. No. of elements available	1	2		3	2	1		3		1
2. No. of nodes per element	2	2,2		4,3,3	4,6	6		2,2,3		8
3. No. of freedoms per element	3	6,12		8,6,6	12,10	24		6,8,27		24
4. Model	V	V		V	V	V		V		V
5. Element configuration	L	L		V,D,D_T	V	D		L_T,L_T,D		H
6. Element shape		K		E	B,E	E		K		
7. Boundary				G	G	G		G		G
8. Cross-sectional data	QK	QTK		QK	QK	QK		QK,QV,QV		
9. Material data				AK	AK	AK		AK		AK
10. External loading	P,L	P,L		P,L	P,L,F	P,L,F		P,L,F		P,L,V
11. Initial strains	B	B,B		B	B,B	B		B,B,B		B

Comments. A program designed, developed, and built specifically for nonlinear analysis (plasticity and large deflection). Elements, in addition to those listed above include many thin-walled open and closed section beams, and pipe bend (elbow) elements. Additional capabilities are included for problems associated with soil mechanics.

This program is possibly the best known and most widely used program for nonlinear analysis.

Program Name: NEPSAP

Developer. Lockheed Corporation
Dept. 81-12, Bldg. 154
P.O. Box 504
Sunnyvale, California 94088

Documentation. [51]

Method of analysis. (FEA)-Displacement

State of development. (Fully operational)

Element information.

NEPSAP	A	B		C	D	E		F		G
	Rod	Beam a)	Beam b)	Plane membrane	Membrane in space	Plate a) Kirchhoff	Plate b) Reissner	Shell a) Kirchhoff	Shell b) Reissner	Three dimensional continuum
1. No. of elements available		1		2	1	1		1	1	1
2. No. of nodes per element		2		4,4	4	4		5,9	16	8
3. No. of freedoms per element		12		8,8	12	24		24,24	48	24
4. Model		V		V	V	V		V	V	V
5. Element configuration		L		V, V_r	V	V		V,V	H	H
6. Element shape				E	E	E		S	B	
7. Boundary				G	G	G		G	K	G
8. Cross-sectional data		QTK		QK	QK	QV		QV	QV	QV
9. Material data				AK	AK	AK		AK	AK	AK
10. External loading		P,L		P,L	P,L	P,L,F		P,L,F	P,L,F	P,L,F,V
11. Initial strains		B		B	B	B		B	B	B

Comments. An improved, extended version of this program to treat transient thermal analysis and nonlinear dynamics is now being prepared and will be completed by mid-1974.

Program Name: PAFEC 70+

Developer. Mechanical Engineering Department
Nottingham University
University Park
Nottingham N67, 2RD, U.K.

Documentation. [47]

Method of analysis. (FEA)-Displacement and hybrid

State of development. (see comment below)

Element information.

PAFEC	A	B		C	D	E		F		G
	Rod	Beam a)	b)	Plane membrane	Membrane in space	Plate a) Kirchhoff	b) Reissner	Shell a) Kirchhoff	b) Reissner	Three dimensional continuum
1. No. of elements available	1	2		4	2	4	1	3	1	3
2. No. of nodes per element	2	2		10,17,10,17	10,17	6,12,6,4	16	4	16	20,32,15
3. No. of freedoms per element	12	12		20,34,20,34	20,34	18,36,18,12	48	20	48	60,96,45
4. Model	V	V		V	V	V,V,H,H	V	V	V	V
5. Element configuration	L	L		D,V,D_r,V_r	D,V	D,V,D,V	X	V,D,V	X	H,H,P
6. Element shape		K		E	K	E	E	S,S,B	E	
7. Boundary				G	G	G	X	G	X	G
8. Cross-sectional data	QK	QTK		QK	QK	QK	QK	QK	QK	QK
9. Material data	AK	AK		AK	AK	AK	AK	AK	AK	AK
10. External loading	P,L	P,L		P,L	P,L,F	P,L,F	P,L,F	P,L,F	P,L,F	P,L,F,V
11. Initial strains	B	B		B	B	B	B	B	B	B

Comments. The developers of this program indicate that those features of the program specifically concerned with plastic analysis are added on an ad hoc basis from various special purpose programs not included in basic package for general release.

This reviewer was previously unaware of the existence of this program.

Program Name: PLANS

Developer. Applied Mechanics Group
Research Department
Grumman Aerospace Corporation
Bethpage, New York 11714

Documentation. [48]

Method of analysis. FEA-Displacement

State of development. (see comment below)

Element information.

PLANS	A	B		C	D	E		F		G
	Rod	Beam a)	b)	Plane membrane	Membrane in space	Plate a) Kirchhoff	b) Reissner	Shell a) Kirchhoff	b) Reissner	Three dimensional continuum
1. No. of elements available	2		2	5	4	1	–	2		1
2. No. of nodes per element	2,3		2,1	3,4,5,6,3	3,4,5,6	3		2,3		8-20
3. No. of freedoms per element	4,6		12,3	6,8,10,12,9	9,12,15,18	36		12,36		24-60
4. Model	V		V,V	V	V	V		V		V
5. Element configuration	L		L,L_r	D,D,D,D,D_r	D,D,D,D	D		L_r,D		H
6. Element shape					E	E		K,B		
7. Boundary				G	G	G		G		G
8. Cross-sectional data	QK		QTK	QK	QK	QK		QK		
9. Material data				OK	OK	OK		OK		OK
10. External loading	P		P,L	P,L,F	P,L	P,L,F		P,L,F		P,L,F,V
11. Initial strains	B		B	B	B	B		B		B

Comments. A program designed, developed,and built specifically for nonlinear analysis (plasticity and large deflection). Elements in addition to those listed above include many thin-walled open and closed section beams, and special crack tip elements (combined FEA and conformal mapping techniques). Portions of this program are operational at NASA-Langley Research Center. General distribution is expected by mid-1974.

The author of this review is a codeveloper of PLANS.

Analysis Information

The additional analytic features of the programs are presented in the following table.

Table 2. Analysis Features of Programs

Program	Plasticity theory (1)	Material property (2)	Linearization technique (3)	Linear equation solver (4)
ANSYS	I,K,RO	P,L,ML	R	W
ASAS	I	P,L,ML RO	R	F,PT,W
ASKA	I,K	P,L	R	PT
BERSAFE	X	P,L,ML	X	W
EPACA	I,K	P,L,RO	TM	G
MARC-CDC	I,K	P,L,ML, RO	TM	G
NEPSAP	I,K	P,L,ML	TM	S
PAFEC 70+	I	P,RO	R	F,B,W
PLANS	K	P,L,RO	R	PT

The code used in the above table is as follows:

In column (1): Plasticity Theory
I = Isotropic hardening
K = Kinematic hardening

In column (2): Material Property
P = Elastic-perfectly plastic
L = Elastic-linear hardening
ML = Elastic-multilinear segments
RO = Ramberg-Osgood or any other power-law representation

In column (3): Linearization Technique
R = Right-hand side (initial strain, or stress)
TM = Tangent modulus

In column (4): Linear Equation Solver
W = Wavefront
F = Full matrix
PT = Partitioning
IT = Iterative
G = Gaussian elimination
S = Skyline
B = Constant bandwidth

Comparison of Available Features

The following table is intended to serve as a means of evaluating the individual programs in terms of some select items chosen by this reviewer. By and large, the items listed are those discussed in the previous section.

Table 3. Comparison and Rating of Available Features of Programs

Program	Element Library (1)	Material Treatment (2)	Ease of Use/ Documentation (3)	Special Features (4) a	b	c	d	e	f	g	h
ANSYS	3	3	3	1	1	1	1	0	X	1	X
ASAS	3	1	2	1	1	1	1	1	X	1	X
ASKA	3	2	3	1	1	0	1	1	X	1	X
BERSAFE	2	X	X	0	0	1	1	1	X	1	X
EPACA	2	3	3	1	0	0	1	1	0	0	0
MARC-CDC	3	3	3	1	0	1	1	1	1	1	1
NEPSAP	3	3	2	1	0	1	1	1	1	1	1
PAFEC 70+	3	1	X	1	1	1	1	0	X	1	X
PLANS	3	3	2	0	0	1	1	1	0	1	1

The code used in the above table is as follows:

In column (1): Element library
1 = Limited to a special class of problems
2 = Moderately general
3 = General

In column (2): Material treatment
1 = Limited
2 = Moderately general
3 = General

In column (3): Ease of use/documentation

This rating is based on either of the above items and includes some user reaction

1 = Difficult/no documentation
2 = Moderate/limited documentation
3 = Easy/extensive documentation

In column (4): Special features

Subcolumn (a) Restart
(b) Substructuring
(c) Multipoint constraints (tying nodes)

(d) Pre- and postprocessors
(e) Equilibrium checks
(f) Assumption checks
(g) Selective I/O
(h) Taking advantage of linear range

The rating for the special features is based on the following code.

0 = Not available
1 = Available

CONCLUDING REMARKS

The computational capability available for the plastic analysis of structures has experienced a tremendous growth during the past 10 years. Indeed, the level of structural analysis capability that has been achieved has outstripped our ability to describe accurately complex material behavior such as cyclic, time, and temperature dependent plasticity. Prior to the development of the programs now available, the designer or analyst confronted with a problem involving material nonlinearities was left with a choice of using his engineering judgment alone or in conjunction with potentially expensive laboratory tests. He now has the further option of performing numerical analysis to gain insight into the behavior of the structure.

Most nonlinear analysis programs, with the exception of a few, have been developed as a spin-off of existing programs that were originally designed for linear structural analysis. Although this development is a natural one, as indicated in [49], this added dimension of generality has placed great responsibilities on the user of such programs. Perhaps the greatest asset of these programs, i.e., their ability to solve sophisticated problems also represents a potential liability, i.e., they always produce numbers. The user must still exercise engineering judgment in order to interpret the results meaningfully. Hopefully, the analytic results will confirm these feelings and provide him with additional insight. However, he now has the luxury of having his intuition fail him without suffering the consequences of a catastrophic failure or an overdesigned system.

This review has restricted its attention to the static, time, and temperature independent plastic analysis of structures. A natural extension of this type of capability is to include creep, (time and temperature dependent materials). Since creep can be considered as time and temperature dependent plasticity,its implementation within the framework of a comprehensive structural analysis program is analogous to the implementation of plasticity alone.

Investigations of creep and thermoplasticity are already under way [50-55] and are particularly significant for nuclear reactor technology [56].

Another area of investigation currently generating a great deal of interest is associated with nonlinear (material and geometric) dynamic analysis of structures [57-60]. The applica- tion of such a capability to the area of crashworthiness evalua- tion of vehicular systems represents a potential payoff.

Two recommendations concerning the economics of nonlinear analysis systems are in order. The first is concerned with com- puting costs, a subject appropriately discussed in [49]. Fig- ure 14 is reproduced from [49] and schematically illustrates that total system time increases with increasing computer speeds. No attempt is made to show relative computing costs as determined from some algorithm involving core, CPU, and I/O times. The figure does, however, show that an advantage in cost is obtained by solving large problems on a time-shared machine. A further

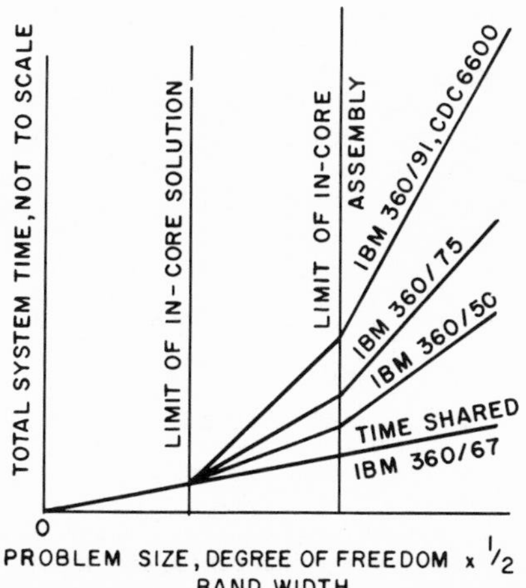

Fig. 14 Computing time — problem size. Reprinted by permission from H. D. Hibbitt, H. J. Levy, P. V. Marcal, "On General Purpose Programs for Nonlinear Finite Element Analysis," Special ASME publication, <u>On General Purpose Finite Element Computer Programs</u>, edited by P. V. Marcal, November 1970, © 1970, ASME.

possibility that can significantly reduce computing costs may be linked to the use of minicomputers for large-scale problems. A study concerned with the feasibility of such a proposal is pre- sented in [61].

The second recommendation worth considering is associated with the decision to buy and extend or develop a nonlinear analysis capability tailored to a specific need. Most of the programs mentioned in this review are available at a nominal or modest cost (in comparison with the development cost). Further- more, they are constantly undergoing extensions and modifications to prolong their useful half-life. Therefore, the obvious sug- gestion is made that costs and effort associated with extending an ongoing capability should be determined before a decision is made to develop a specialized program from scratch.

A final comment is concerned with the fact that the author of this review is a codeveloper of the PLANS system. It is earnestly hoped that, in the writing of this review, the natural inclination to overstate one's achievements has been sufficiently tempered by the sense of modesty felt when confronted with the impressive accomplishments of others whose work is reviewed here.

ACKNOWLEDGMENTS

The author acknowledges a debt of gratitude to his two colleagues A. Pifko and H. Levine for their suggestions concerning the nature of this review. Thanks also go to the many developers and users of the programs cited here for filling out the questionnaires and to those who were gracious enough to allow telephone interviews. Final thanks go to Mrs. M. Sudwischer for typing.

REFERENCES

1 Schrem, E., "Computer Implementation of the Finite-Element Procedure," Numerical and Computer Methods in Structural Mechanics, ed. S. J. Fenves, N. Perrone, A. R. Robinson, and W. C. Schnobrich, Academic Press, New York, 1973, pp. 79-121.

2 Johnson, W., and Mellor, P. B., Plasticity for Mechanical Engineers, D. Van Nostrand Co., London, 1962, p. 1.

3 Ziegler, H., "A Modification of Prager's Hardening Rule," Quart. Appl. Math., Vol. 17, No. 1, 1959, p. 55.

4 Drucker, D. C., "A More Fundamental Approach to Plastic Stress-Strain Relations," Proc. 1st U.S. Natl. Congr. Appl. Mech. (Chicago, 1951), New York, 1952, p. 487.

5 Budiansky, B., "A Reassessment of Deformation Theories of Plasticity," J. Appl. Mech., Vol. 26, June 1959, p. 259.

6 Isakson, G., Armen, H., Jr., and Pifko, A., Discrete-Element Methods for the Plastic Analysis of Structures, NASA CR-803, 1967.

7 Hunsaker, B., Vaughan, D. K., Stricklin, J. A., and Haisler, W. E., "A Comparison of Current Work-Hardening Models Used in the Analysis of Plastic Deformations" (to appear as an ONR Research Report).

8 Phillips, A., and Kaechele, L., "Combined Stress Tests in Plasticity," J. Appl. Mech., Vol. 23, 1956, p. 43.

9 Marin, J., and Hu. L. W., "Biaxial Plastic Stress-Strain Relations of a Mild Steel for Variable Stress Ratios," Trans. ASME, April 1956, p. 499.

10 Hu. L. W., and Bratt, J. F., "Effect of Tensile Plastic Deformation on the Yield Condition," J. Appl. Mech., Vol. 25, September 1958, p. 111.

11 Marin, J., and Hu. L. W., "On the Validity of Assumptions Made in Theories of Plastic Flow for Metals," Trans. ASME, Vol. 75, No. 6, 1953, p. 1181.

12 Batdorf, S. B., and Budiansky, B., "A Mathematical Theory of Plasticity Based on the Concept of Slip," NACA TN 1871, 1949.

13 Parker, J., and Basset, M. B., "Plastic Stress-Strain Relationships - Some Experiments to Derive a Subsequent Yield Surface," J. Appl. Mech., Vol. 31, December 1964, p. 676.

14 Peters, R. W., Dow, N. F., and Batdorf, S. B., "Preliminary Experiments for Testing Basic Assumptions of Plasticity Theories," Proc. Soc. Exp. Stress Anal., Vol. 7, 1950, p. 127.

15 Sanders, J. L., Jr., "Plastic Stress-Strain Relations Based on Linear Loading Functions," Proc. 2nd U.S. Nat. Congr. Appl. Mech., 1954, p. 455.

16 Hodge, P. G., Jr., "Piecewise Linear Plasticity," Proc. 9th Intern. Congr. Appl. Mech. (Brussels, 1956), Vol. 8, p. 65, 1957.

17 Stricklin, J. A., "Large Elastic, Plastic, and Creep Deflections of Curved Beams and Axisymmetric Shells," AIAA Jour., Vol. 2, No. 9, September 1964, p. 1613.

18 Prager, W., "The Theory of Plasticity: A Survey of Recent Achievements," (James Clayton Lecture), Proc. Instn. Mech. Engrs., Vol. 169, 1955, p. 41.

19 Prager, W., "A New Method of Analyzing Stress and Strains in Work-Hardening Plastic Solids," J. Appl. Mech., Vol. 23, 1956, p. 493.

20 Kadashevitch, Yu, I., and Novozhilov, B. V., "The Theory of Plasticity Taking Residual Micro-Stresses into Consideration," Prikl. Mat. Mekh., Vol. 22, No. 1, 1955, p. 78.

21 Budiansky, B., Discussion of Reference 19, J. Appl. Mech., Vol. 24, 1957, p. 481.

22 Shield, R. T., and Ziegler, H., "On Prager's Hardening Rule," ZAMP, Vol. 9a, 1958, p. 260.

23 Perrone, N., and Hodge, P. G., Jr., "Applications of a Consistent Theory for Strain-Hardening Plastic Solids," PIBAL Report 403, September 1957.

24 Naghdi, P. M., "Stress-Strain Relations in Plasticity and Thermoplasticity," Proc. 2nd Symp. Naval Structural Mechanics, 1960, p. 121.

25 Mróz, Z., "On the Description of Anisotropic Workhardening," J. Mech. Phys. Solids, Vol. 15, 1967, pp. 163-175.

26 Mróz, Z., "An Attempt to Describe the Behavior of Metals under Cyclic Loads Using a More General Workhardening Model," Acta Mechanica, Vol. 7, 1969, pp. 199-212.

27 Duwez, P., "On the Plasticity of Crystals," Physical Review, Vol. 457, 1935, p. 494.

28 White, G. N., Jr., "Application of the Theory of Perfectly Plastic Solids to Stress Analysis of Strain Hardening Solids," Brown University Tech. Rept. 51, 1950.

29 Besseling, J. F., "A Theory of Plastic Flow for Anisotropic Hardening in Plastic Deformation of an Initially Isotropic Material," Rept. S410, National Aeronautical Research Institute, Amsterdam, The Netherlands, 1953.

30 Armen, H., Jr., Pifko, A., and Levine, H., Finite Element Analysis of Structures in the Plastic Range, NASA CR-1649, February 1971.

31 Marçal, P., "A Stiffness Method for Elastic-Plastic Problems," Int. J. Mech. Sci., Vol. 7, 1965.

32 Marçal, P., "A Comparative Study of Numerical Methods of Elastic-Plastic Analysis," AIAA Journal, Vol. 6, No. 1, January 1968, p. 157.

33 Mallett, R., and Marçal, P., "Finite Element Analysis of Nonlinear Structures," Proceedings of ASCE, J. Struc. Div., Vol. 94, No. ST9, September 1968.

34 Stricklin, J., Haisler, W., and Von Riesemann, W., Formulation, Computation, and Solution Procedures for Material and/or Geometric Nonlinear Structural Analysis by the Finite Element Method, Sandia Corporation Report SC-CR-72-3102, Albuquerque, New Mexico, January 1972.

35 Armen, H., Jr., Levine, H., Pifko, A., and Levy, A., "Nonlinear Analysis of Structures," Grumman Research Department Report RE-454, May 1973 (also to appear as NASA Contractors Report).

36 Zienkiewicz, O., Valliappan, S., and King, J., "Elasto-Plastic Solutions of Engineering Problems 'Initial Stress,' Finite Element Approach," International Journal for Numerical Methods in Engineering, Vol. 1, No. 1, January-March 1969, pp. 75-100.

37 Haisler, W., Stricklin, J., and Stebbins, F., "Development and Evaluation of Solution Procedures for Geometrically Nonlinear Structural Analysis," AIAA Journal, Vol. 10, No. 3, March 1972, pp. 264-272.

38 Hofmeister, L., Greenbaum, G., and Evensen, D., "Large Strain, Elasto-Plastic Finite Element Analysis," AIAA Journal, Vol. 9, No. 7, 1971, p. 1248.

39 Tillerson, J. R., Stricklin, J. A., and Haisler, W. E., "Numerical Methods for the Solution of Nonlinear Problems in Structural Analysis," in Numerical Solution of Nonlinear Problems, edited by R. F. Hartung, ASME Special Publication AMD-Vol. 6, November 1973.

40 Levine, H., Armen, H., Jr., Winter, R., and Pifko, A., "Nonlinear Behavior of Shells of Revolution under Cyclic Loading," Computers and Structures, Vol. 3, pp. 584-617, 1973.

41 Buck, Scherem, and Stein, "Einige allgemeine Programmsysteme fur finite Elemente," from Finite Elemente in der Statik, Wilhelm Ernst and Sokne, 1973.

42 De Salvo, G. J., and Swanson, J. A., "ANSYS Engineering Analysis Systems Manual," Swanson Analysis Systems, Inc., Elizabeth, Pennsylvania, October 1972.

43 Schrem, E., Roy, J. R., "An Automated System for Kinematic Analysis," Institut für Statik und Dynamik der Luft — und Raumfahrtkonstructionen, University of Stuttgart, ISD Report No. 98, 1971.

44 Hellen, T. K., and Protheroe, S. J., "The BERSAFE Finite Element System," to appear in Computer Aided Design, January 1974.

45 Zudans, Z., et al., "Theory and User's Manual for EPACA," Franklin Institute Report F-C-3038, June 1972.

46 "MARC-CDC Nonlinear Finite Element Analysis Program," User Information Manual, Vol. 1, Control Data Corp., 1971.

47 Henshell, R. D., "PAFEL 70+ Users Manual," Mechanical Engineering Department, Nottingham University, Nottingham, U.K., 1972.

48 Pifko, A., Levine, H., and Armen, H., Jr., "PLANS — A Finite Element Program for Nonlinear Analysis of Structures," Grumman Research Department Report (to be published).

49 Hibbitt, H. D., Levy, H. J., and Marçal, P. V., "On General Purpose Programs for Nonlinear Finite Element Analysis," Special ASME publication, On General Purpose Finite Element Computer Programs, edited by P. V. Marçal, November 1970.

50 Zudans, Z., et al., "Elastic-Plastic-Creep Analysis of High Temperature Nuclear Reactor Components," 2nd International Conference on Structural Mechanics in Reactor Technology, Berlin, September 1973.

51 Sharifi, P., and Yates, D. N., "Nonlinear Thermo-Elastic-Plastic and Creep Analysis by the Finite Element Method," AIAA paper No. 73-358 presented at 14th Structures Conference, Williamsburg, Va., March 1973.

52 Hibbitt, H. D., and Marçal, P. V., "A Numerical, Thermo-Mechanical Model for Welding and Subsequent Loading of a Fabricated Structure," presented at National Symposium on Computerized Structural Analysis and Design, Washington, D.C., March 1972.

53 Oden, J. T., Chung, T. J., and Key, J. E., "Analysis of Nonlinear Thermoelastic and Thermoplastic Behavior of Solids of Revolution by the Finite Element Method," First International Conference on Struct. Mech. Reactor Technology, Paper No. MS/6, Berlin, September 1971.

54 Ueda, Y., and Yamakawa, T., "Thermal Nonlinear Behavior of Structures," Advances in Computational Methods in Structural Mechanics and Design, edited by J. T. Oden, R. W. Clough, and Y. Yamamoto, UAH Press, Huntsville, Alabama, 1972.

55 Cyr, N. A., and Teter, R. D., "Finite-Element Elastic-Plastic-Creep Analysis of Two-Dimensional Continuum with Temperature Dependent Material Properties," Computer and Structures, Vol. 3, No. 4, July 1973.

56 Greenstreet, W. L., Corum, J. M., and Pugh, C. E., "High-Temperature Structural Design Methods for LMFBR Components," Oak Ridge National Laboratory Report ORNL-TM-3736, December 1971.

57 Chung, T. J., et al., "Dynamic Elasto Plastic Response of Geometrically Nonlinear Arbitrary Shells of Revolution under Impulsive Mechanical and Thermal Loadings," Shock and Vibration Bulletin, Vol. 41, No. 7, December 1971.

58 Marçal, P. V., and McNamara, J. F., "Incremental Stiffness Method for Finite Element Analysis of the Nonlinear Dynamic Problem," Int. Symposium on Numerical and Computer Methods in Structural Mechanics, Urbana, Illinois, September 1971.

59 Stricklin, J. A., et al., "Large Deflection Elastic-Plastic Dynamic Response of Stiffened Shells of Revolution," Texas A & M University, Texas Engineering Experiment Station, TEE'S-RPT-72-25, December 1972.

60 Wu, R., and Witmer, E. A., Finite Element Analysis of Large Transient Elastic-Plastic Deformation of Simple Structures with Application to the Engine Rotor Fragment Containment/Deflection Problem," NASA CR-120886, MIT ASRL, January 1972.

61 Kottler, C. F., Jr., and McGill, R., "The Feasibility of Using Minicomputers for Reducing Large Problem Solving Costs," Grumman Research Report RE-421, February 1972.

FRACTURE MECHANICS

S. E. Benzley
Sandia Laboratories
Albuquerque, New Mexico

D. M. Parks
Brown University
Providence, Rhode Island

ABSTRACT

Some fracture mechanics computer programs currently being used are surveyed. Most of this software represents very recent developments in computational fracture mechanics. Included are brief abstracts of computer programs that treat the characteristic near field singularity that is present at the tip of a crack.

A critical review of computational fracture mechanics is offered, and some suggestions are given as to the proper selection of fracture mechanics software for a particular problem. In addition, a discussion of future trends in this field is presented.

COMPUTATIONAL FRACTURE MECHANICS

Fracture mechanics analysis is concerned with the proper characterization of the stress and displacement fields in the region near the tip of a crack and, consequently, the ability to predict any catastrophic failure that may occur. Figure 1 shows a typical crack geometry and graphically represents the three basic deformations associated with a flaw of this type.

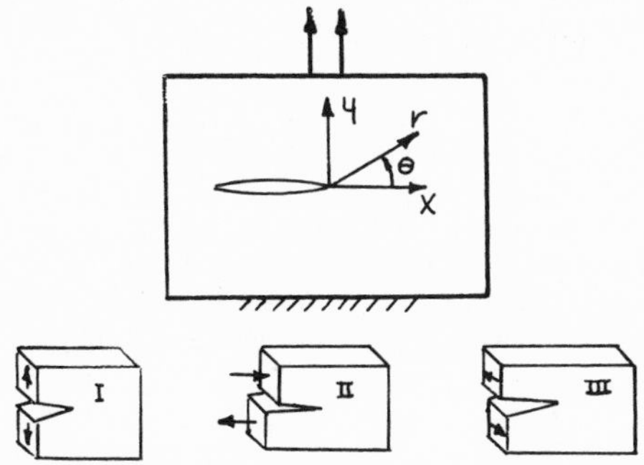

Fig. 1 Crack tip coordinate system and three basic modes of near tip deformations

Let f_{ij}, g_{ij}, h_{ij} and $\hat{f}_i$, $\hat{g}_i$, $\hat{h}_i$ be functions that define the θ dependence of crack tip stresses and displacements, respectively. Assuming linear elastic conditions at the crack tip, the dominate singular stresses are defined as in [1].

$$\sigma_{ij} = \frac{1}{(2\pi r)^{\frac{1}{2}}} [K_I f_{ij}(\theta) + K_{II} g_{ij}(\theta) + K_{III} h_{ij}(\theta)] + \ldots \quad (1)$$

and the near field displacements are

$$u_i = \frac{1}{2G}\left(\frac{r}{2\pi}\right)^{\frac{1}{2}} [K_I \hat{f}_i(\theta) + K_{II} \hat{g}_i(\theta) + K_{III} \hat{h}_i(\theta)] + \ldots \quad (2)$$

The parameters, K_I, K_{II}, and K_{III}, are called stress intensity factors. These quantities measure the strength of the singularities of the corresponding three types of stress and displacement fields. Stress intensity factors are related to a material property called the fracture toughness in a manner analogous to the way that stress is related to strength. For example, the Mode I plane strain crack toughness, K_{Ic}, is defined as the value of K_I corresponding to the load at which unstable crack extension is observed. Therefore, a linear elastic fracture mechanics analysis predicts failure when the computed value of stress intensity equals or exceeds the fracture toughness of the material being analyzed.

Linear elastic fracture mechanics is valid only for applications where the inelastic deformation which actually does occur is confined to a region about the crack tip that is small in comparison to other relevant dimensions of the problem. In this small-scale yielding regime, the elastic stress intensity factors provide asymptotic boundary conditions which characterize the near tip fields. Thus, the stress intensity factor provides a convenient measure of the load sensed in the crack tip region. Among the applications typified by small-scale yielding conditions are the brittle fracture of cracked bodies of high-strength materials and high-cycle, low-amplititude fatigue crack propagation.

An energy relationship which is useful for the determination of stress intensity factors is that the potential energy, P, of a loaded cracked body decreases with an increment of crack growth, $d\ell$, under constant load. For plane strain and unit thickness, the potential energy decrease per unit crack advance is given by [1]

$$-\left.\frac{\partial P}{\partial \ell}\right|_{load} = \mathcal{G} = \frac{1-\nu^2}{E}(K_I^2 + K_{II}^2) \quad (3)$$

Let $\underset{\sim}{T}$ be a traction vector according to the outward normal along Γ; $\underset{\sim}{u}$ be a displacement vector; U be a strain energy density. An alternative computation of the energy release rate for two-dimensional planar configurations may be based on the path-independent J-integral [2]:

$$J = \int_\Gamma U dy - \underset{\sim}{T} \cdot \frac{\partial \underset{\sim}{u}}{\partial x} ds \quad (4)$$

The physical significance of J is as the potential energy release rate, $\mathcal{G}$ [1]. Γ is an arbitrary path surrounding a crack tip.

Many analytical techniques have been applied to the elastic analysis of cracks [3,4,5]. Most of these methods treat only highly idealized crack configurations with the most simple loading conditions. The elastic analysis of many realistic fracture mechanics problems is complicated by arbitrary crack configurations in two and three dimensions and arbitrary loadings. In addition, the arbitrary loadings of the near crack tip region is complicated by the nonlinear material behavior occasioned by crack tip plasticity [2,6,7,8], nonlinear geometric behavior associated with the finite geometry changes due to crack tip blunting [9], and nonlinear strain conditions.

This paper includes abstracts that describe various computer programs currently being used to analyze fracture mechanics problems. A brief description of the specific computational fracture mechanics method used by each program precedes the set of computer program abstracts that implement that particular method. Information concerning the computer for which the program was written, the total number of cards in the source deck, and the principal developer is listed after each abstract. If not otherwise noted, the code may be obtained directly from the developer. A critical assessment of the programs described is given, and some guidelines for selecting the proper method or code for a particular application are offered. Finally, the shortcomings of present numerical schemes are discussed, and some suggestions as to what developments are still needed to treat adequately fracture mechanics problems on a continuum scale are offered.

Finite Element Codes

Recently, much activity has centered on applying the finite element method [10,11] to the fracture problem. Several excellent survey papers [12,13,14] have reviewed the approaches that may be used by finite element schemes to determine stress intensity factors. Briefly, these schemes are identified as: (1) direct methods where extremely small sized conventional elements are used to define the stresses and strains in the vicinity of the cracks, (2) superposition methods where conventional elements are used in conjunction with a classical singular solution, (3) energy release methods that determine the change in potential energy as a crack changes position, thus defining the magnitude of K, and (4) singularity element formulations that embed the correct singularity in the element assumptions to determine the stress intensity factors as a part of the total solution.

Direct Method

This method may be used with conventional continuum finite element programs. It is incorporated by defining the vicinity of the crack tip with a large number of very small elements. The stress intensity factors may then be estimated from the near field finite element solution by use of Eqs. (1) and (2). The most direct procedure is to match a calculated stress or displacement measure in the immediate crack tip region with the dominant term of the Williams expansion, thus giving the magnitude of the stress intensity factor, K. Another implementation of the direct technique involves extrapolation of a field parameter on a given ray from the crack tip [15]. For example, Eq. (1) can be rearranged, for $\theta = 0$, to give

$$K_I = \lim_{r \to 0} \{[\sigma_{yy}(r)]\sqrt{r2\pi}\} \quad (5)$$

Thus the product of $r^{\frac{1}{2}}$ with a stress measure or $r^{-\frac{1}{2}}$ with a displacement can be plotted as a function of r for a particular ray. A tangent extrapolation of this product to the crack tip, $r = 0$, gives an estimate of the stress intensity factor. It should be noted that the extrapolation procedure, which has no theoretical justification, will obtain a result which may be significantly different from the matching procedure described earlier.

In general, it is probably best to base direct estimations of the stress intensity factor on calculated displacements when using displacement method finite element programs, since stresses are obtained from differentiating displacement fields and are therefore likely to be less accurate than displacements.

Superposition Method

The superposition of classical and finite element schemes allows a coarse finite element idealization. The concept is applied by using a classical solution for a problem that resembles the problem being analyzed. The classical solution will not satisfy all conditions for the actual problem, but the disparities can be interpreted as body forces and edge loadings. These body forces and edge loadings are applied to the finite element analysis in reversed direction, and the results of the finite element and classical solutions are superimposed.

CRACK ANALYSIS [16]. In this program the stress distribution near the crack tip is approximated by Westergaard's solution for the opening and sliding modes. The technique used consists of superimposing the analytical solution and a finite element solution. This superposition is applied only to a small region that contains a crack tip (i.e., a crack substructure). The method is applicable to bodies with multiple cracks.

IBM-360, UNIVAC 1108

3,500 cards

K. Ando
Ship Strength Research Laboratory
Nagasaki Technical Institute Technical Headquarters
Mitsubishi Heavy Industries, Ltd.
Fukaburi 5-717-1 Nagasaki 851-03, Japan

SIF-PLANE [17] and SIF-AXISYM [18]. These codes calculate stress intensity factors by superimposing analytical solutions with finite element solutions. SIF-PLANE treats plate structures and uses constant strain triangular elements, and SIF-AXISYM treats axisymmetric bodies. Satisfactory results can be obtained with rather coarse mesh subdivisions.

HITAC 5020E, 8700, 8800

1,000-1,500 cards

Y. Yamamoto, N. Tokuda, Y. Sumi
Department of Naval Architecture
University of Tokyo
Bunkyo-ku, Tokyo 113, Japan

Energy Release Methods

Energy methods can generally be easily incorporated in conjunction with conventional continuum finite element codes for the analysis of cracks. The great advantage of energy methods over direct methods is that comparable accuracy in the determination of stress intensity factors is generally obtainable with considerably smaller formulations. However, for problems with general loading, the magnitudes of the individual Mode I and II stress intensity factors cannot be separately determined.

The simplest finite element formulation of energy difference methods consists of solving two problems of slightly differing crack lengths, ℓ and $\ell + \Delta\ell$ [19,20]. For each solution, a potential energy is obtained, and the difference in potential energies $\Delta P = P_{\ell+\Delta\ell} - P_\ell$ is also defined by

$$-\Delta P = \int_{\text{Crack Front}} \mathcal{G}(s)\Delta\ell(s)ds \qquad (6)$$

For two-dimensional cases, where the energy release rate and crack length difference are constants all along the crack front, Eq. (6) can be substitutied into Eq. (3) to give

$$\frac{-\Delta P}{\Delta\ell \int_{\text{Crack Front}} ds} = \mathcal{G} = \frac{(1-\nu^2)}{E}(K_I^2 + K_{II}^2) \qquad (7)$$

for problems of in-plane loading.

A more efficient implementation of this procedure, which requires only one finite element solution, is obtained by expressing $-\Delta P$ in terms of the change in the master stiffness matrix, $\Delta[K]$, and the computed nodal displacements $\{u\}$. Here $\Delta[K]$ is associated with a virtual crack extension accomplished by incrementing the coordinate of the crack tip node by an amount $\Delta\ell$ [21]. In this case,

$$-\Delta P = -\frac{1}{2}\{u\}^T\Delta[K]\{u\} \qquad (8)$$

which can then be substituted into Eq. (6) or Eq. (7). This method offers great promise in providing acceptably accurate and inexpensive three-dimensional crack analyses [21]. It should be noted here, however, that the energy release Eq. (6) applies only for crack advance $\Delta\ell(s)$ parallel to the existing crack face surfaces. To treat adequately the problem of direction of crack propagation, one must use a formulation as reported by Hussian et al. [22].

An alternative energy difference procedure using only one finite element solution may be based, in two-dimensional planar configurations, on the J-integral. In the numerical evaluation of the integral in terms of the computed stresses, strains, and displacements, it is generally advisable to choose the contour of integration removed from the immediate crack tip region. This is recommended because the near tip region is where the field parameters, comprising the integrand of J, as computed in the conventional finite element formulation, are likely to be in greatest error.

When conventional finite elements are used to model the crack tip region, for either direct or energy difference calculations of stress intensity factor, it is often advantageous to use a mesh which joins several elements to the crack tip node. That is, for the same order of crack tip element size, better accuracy is obtained for formulations which concentrate more elements in the crack tip region.

AXISOL [20]. This code is a modification of a digital computer program developed by Wilson [23]. It may be used for the analysis of elastic plane stress (or plane strain) structures or axisymmetric structures subjected to axisymmetric loads. To treat the crack tip problem, the stored strain energy is computed for two slightly different crack lengths. These results are then numerically differentiated to produce the strain energy release rate, $\mathcal{G}$, which is directly related to the stress intensity factor, K.

UNIVAC 1108

2,000 cards

W. E. Anderson
Battelle Memorial Institute
Richland, Washington

BERSAFE [24,25,26]. The BERkeley Stress Analysis by Finite Elements system is a set of integrated computer programs designed to perform the general stress analysis of arbitrary structures subjected to pressures, point loads, body forces, rotational forces, and temperature distributions. The finite element method, with displacements as unknowns, is used to calculate displacements and stresses throughout the structure. A virtual crack extension method has been incorporated in BERSAFE whereby many crack tip displacements (or virtual extensions) are permitted within one computer run with only small increases in computer cost. The method works in both two and three dimensions and, by using special elements containing the $r^{\frac{1}{2}}$ displacement function, gives excellent results with only moderate meshes. Peripheral programs include automatic mesh generation, output plotting, and thermal transients. A nonlinear J-integral calculation is currently being added to the existing plasticity version of BERSAFE.

IBM 370

20,000 cards

T. K. Hellen
Berkeley Nuclear Laboratories
Central Electricity Generating Board
Berkeley, Gloucestershire, United Kingdom

NV344 [27,28,29]. The computer program is based on the multilevel superelement technique (multilevel substructuring) and thus can only be used in combination with the large-scale program system SESAM-69. The fracture analysis procedure is based on the strain energy release rate. Fatigue propagation analysis is carried out based on computed stress intensity factor. Two- and three-dimensional geometries can be analyzed.

UNIVAC 1108

25,000 cards

Bjarne Aamodt
The Norwegian Insitutute of Technology
The University of Trondhein
Trondheim, Norway

SASL [30,23]. The code titled Stress Analysis at Sandia Laboratories is the result of extensive revisions and additions to the SAAS II code developed by the Aerospace Corporation for the Air Force Systems Command. It is a finite element, user-oriented code for the static analysis of axisymmetric and plane solids subjected to various axisymmetric and plane loadings. Major differences from SAAS II include a revised input data specification, the addition of a contour plotting capability, a revised finite element model, a new equation solver, and the calculation of energy release rates for fracture mechanics analysis.

CDC 6600

6,000 cards

M. L. Callabresi, S. T. Heidelberg
Sandia Laboratories
Livermore, California

Singularity Element Formulations

Finite element computer programs that incorporate a special singular element provide a very attractive analysis method for fracture mechanics problems. Many different singular elements have been proposed. These elements can be categorized according to the basic assumptions used to generate the governing equations. Those that use assumed displacement fields to generate element stiffness matrices are some of the most popular schemes. This approach as applied to crack tip singular elements can be further categorized as formulations that (1) enrich the conventional assumed displacement field with the proper near field crack tip displacement equations (i.e., Eq. (2))[31], (2) define a circular core region around the crack tip that contains the appropriate equations [32], (3) use a centered fan element representation at the crack tip that has a $\sqrt{r}$ displacement assumption in the radial direction [33], and (4) incorporate Muskhelishvili's [34] equations in a displacement sense. Singularity elements can also be generated by assuming the singular stress field at the element level. Here Muskhelishvili's stress functions may be used. A hybrid element can be formulated by using both stress and displacement assumptions [35]. Table 1 categorizes the singular element codes surveyed in this paper according to the formulation of the crack tip element.

Special convergence considerations must be applied to singular finite element solutions. As is the case with conventional elements, monotone convergence is preserved only when interelement compatibility is maintained. However, in contrast to conventional elements, Tong and Pian [36] have shown that decreasing the singular element size does not produce convergent results. Their study demonstrates that the region defined by singular elements must remain finite.

Table 1 Singularity Element Codes Categorized According to Crack Tip Element Formulation

ASAS CAPS CHILES CRAK1 SING	CAPS JODAV	TF0747	FEMWC MAGIC III	CRACK-1 CRACK-2	TEXGAP UD1 PINTO
Enriched	Core	Fan	Muskhelishvili	Muskhelishvili	
DISPLACEMENT				STRESS	HYBRID

ASAS. The Atkins Stress Analysis System is a general purpose finite element program. Fracture mechanics calculations are treated with a four-cornered, 11-node displacement-formulated crack tip element. Atkins is a computer bureau currently serving twenty ASAS clients.

IBM 360, 370, CDC 6600, UNIVAC 1108, 1106

50,000 source cards

Atkins Research and Development
Ashley Road
Epsom, Surrey, England

CAPS. This computer program performs a finite element analysis for general two-dimensional problems. A 12-node isoparametric element with an assumed cubic displacement function is incorporated in the code. Computations for K_I and K_{II} are accomplished by the introduction of a singular circular element at the crack tip, or by surrounding the crack tip (or tips) with isoparametric elements "enriched" with the known singular solution. K_I and K_{II} are computed directly by the program, as well as the displacements and stresses in the structures. The program philosophy is "out-of-core," accounting for the capability of solving large systems with only 20,000 decimal words of core.

CDC 6400, 6600, 6700

3,000 cards

L. Nash Gifford
Oles Lomacky
Submarine Division (Code 172) Structures Department
Naval Ship Research and Development Center
Bethesda, Maryland 20034

CHILES [37]. This finite element computer program calculates the strength of singularities in linear elastic bodies. A generalized quadrilateral finite element that includes a singular point at a corner node is incorporated in the code. The displacement formulation is used, and interelement compatibility is maintained so that monotone convergence is preserved. Plane stress, plane strain, and axisymmetric conditions are treated. Crack tip singularity problems are solved by this version of the code, but any integrable singularity may be properly modeled by modifying selected subroutines in the program.

CDC 6600, FORTRAN IV

1,500 cards

S. E. Benzley
Sandia Laboratories
Albuquerque, New Mexico

Available from:
Argonne Code Center
Argonne National Laboratory
9700 South Cass Avenue
Argonne, Illinois

CRAK1 [38]. Incorporated in this computer program is a rectangular cracked element that uses many of the higher symmetric modes of deformation permitted by the Williams series and thus can accurately represent a comparatively large fraction of a structure. A single precision-banded Cholesky decomposition is used for the solution procedure. Geometry and element generators are included, and a choice of output options is provided.

UNIVAC 1108

2,000 cards

J. M. Anderson
Georgia Institute of Technology
Engineering Science and Mechanics Dept.
Atlanta, Georgia

J. A. Aberson
Lockheed-Georgia Company
Dept. 72-26
Zone 459
Marietta, Georgia

SING. This finite element computer program is based on the SAAS III code [23]. Arbitrary quadrilaterals with singular nodal points are included. Plane strain, plane stress, and axisymmetric geometries are treated. The singular terms are carried as global functions so that interelement compatibility is maintained. Mixed mode problems can be run.

CDC 6600

2,500 cards

A. F. Emery
Dept. of Mechanical Engineering
University of Washington
Seattle, Washington

JODAV [39]. This program uses a circular singular crack tip element of radius typically 2 percent of crack length. The two stress intensity factors are used as generalized nodal displacements. The singular crack tip element is "tied" at discrete nodal points to regular four-node isoparametric elements, which are used to model the rest of the body. Plane strain, plane stress, and axisymmetric configurations are treated.

CDC 6400

700 cards

Prof. Peter D. Hilton
Dept. of Mech. Engr.
Lehigh University
Bethlehem, Pennsylvania 18015

TF07471S, TF07475S, TF07441 [40]. This series of programs incorporates displacements that vary as $a + b\sqrt{\zeta} + c\zeta$ with distance ζ from the crack tip. The J-integral is evaluated. The codes have been applied to various stress analysis problems that occur in turbogenerators.

IBM 360/370

W. S. Blackburn
C. A. Parsons and Co.
Newcastle upon Tyne
United Kingdom

FEMWC, FEMWC.ANIS [41]. Incorporated here is a special cracked element with a square root stress singularity. The "normal" elements are constant strain triangles. Automatic mesh generation is included for rectangular plates with edge cracks.

IBM 370/165

750 cards

Esben Byskov
Structural Research Laboratory
Technical University of Denmark, Build 118
DK-2800 Lyngby, Denmark

MAGIC III [42]. This code is capable of determining stress distribution and strain fields in the vicinity of cracks from which stress intensity factors are directly determined. The numerical techniques are formulated to accommodate two dimensional linear elastic fracture mechanics theory considering both plane stress and plane strain. Through the thickness, embedded and edge cracks are treated.

Two distinct methods of modeling cracks are included. Both approaches are based on a boundary point least-squares technique incorporating a series representation of Muskhelishvili's [34] stress functions. One approach, called the "appended capability," is essentially a finite element superposition technique incorporating a boundary collocation procedure on a crack substructure. The other approach leads to the formulation of an "equilibrium compatible" singular element. The stiffness matrix of this element is generated by a boundary point least-squares method.

IBM 360/65, CDC 6400, 6600

55,000 source cards

Stephan Jordan, Chief
Advanced Structural Analysis
Bell Aerospace Co.
P. O. Box 1
Buffalo, New York

Also available from:
Air Force Flight Dynamics Lab.
Structures Division, FBR
Wright-Patterson AFB, Ohio 45433

CRACK-1 and CRACK-2 [43]. These two subroutines generate necessary element matrices for a generally loaded cracked finite element and a symmetrically loaded cracked finite element, respectively. They are coupled to the Lockheed research structural analysis program and Lockheed's NASTRAN system. Two singular stress functions are used to represent the singular behavior of the stress fields at the crack tip. The coefficients for the leading terms of each function correspond to the mode I and mode II stress intensity factors, respectively.

IBM 360/91

350 cards - CRACK-1, CRACK-2
3,500 cards - Lockheed Research Structural Analysis Program
16,000 cards - NASTRAN

Lockheed - California Company
P. O. Box 551
Burbank, California 91503

PINTO [44]. This hybrid finite element computer program utilizes a superelement at the crack tip. The stress field within the superelement is expressed in terms of a truncated power series representation of Muskhelishvili's stress functions. Because of the higher order terms of the stress field, the size of the crack tip element can be relatively large compared to crack length; consequently, the total number of degrees of freedom necessary to represent the entire body is smaller than "conventional" singular element formulations. Both mode I and II stress intensity factors can be obtained for planar configurations.

IBM 370/155

500 cards

Prof. Pin Tong
33-305
Mass. Inst. of Tech.
Cambridge, Mass. 02139

TEXGAP [45,46]. The TEXas Institute for Computational Mechanics Grain Analysis Program is a finite element code for the analysis of two-dimensional, linearly elastic plane or axisymmetric bodies. For axisymmetric bodies, loadings in the circumferential coordinate may be analyzed using a Fourier series. The primary intent is to provide for highly accurate determination of stress and deformation fields in solid propellant rocket motors, expecially in the critical areas near case-insulation-liner-propellant interfaces. TEXGAP is FORTRAN coded with a few easily replaced routines in CDC assembly language. This program has a special hybrid fracture mechanics element that permits calculation of Mode I and II stress intensity factors at the tip of plane and axisymmetric cracks.

CDC 6600, 6400, 7400, FORTRAN IV

4,200 cards

R. S. Dunham, E. B. Becker
The Texas Institute for Computational Mechanics
University of Texas
Austin, Texas

UD1 [35,44]. This finite element computer program incorporates a four-node hybrid model super element to determine the stress intensity at the tip of a crack. Results from the code have shown good agreement with theoretical solutions and test results.

IBM 360/55

2,000 cards

K. S. Chu
General Dynamics Corporation
Convair Aerospace Division
Fort Worth, Texas

Other Finite Element Schemes

Some finite element schemes do not fall in the general categories previously outlined. One of these methods couples the conformal mapping procedure with conventional finite elements. This code is described below.

FAST/PLANS [47]. This computer program is based on the finite element displacement method. Linear elastic crack tip singularities are treated by a coupled finite element conformal mapping technique. The basic philosophy of this method is to transform the actual problem containing the crack and its associated singular stress field into an imaginary plane by means of conformal transformation functions. These functions transform the crack boundary into a unit circle (or any other convenient geometric configuration). The problem is then discretized and solved in the transformed region. The results are then determined for the actual problem by means of appropriate inverse transformations.

IBM 370/168, CDC 6600

3,000 source cards

H. A. Armen, Jr.
Grumman Aerospace Corporation
Plt. 35, Dept. 584
Bethpage, New York 11714

Boundary Collocation Method

Boundary collocation procedures [48,49] were some of the first numerical methods used to gain fracture mechanics solutions. Generally, this method is applied to a series representation of Muskhelishvili's stress functions. The coefficients of the series are chosen to match the imposed stress conditions at a discrete number of external boundary points. Generally an excessive number of collocation points are chosen which results in an overdetermined system. These equations can then be solved in a least-squares sense by minimization of total error over the discrete points. This procedure has been used by many investigators [50,51,52]; however, most of the actual software is either proprietary or not suited for general applications.

Boundary Integral Method

The solution of continuum elasticity problems may be given in the form of an integral over the boundary of a fundamental singular solution times an unknown weighting function. This method has been used by Cruse and co-workers [53,54,55,56,57] in the analysis of plane two-dimensional and three-dimensional elastic crack problems.

After solving the algebraic equations for the unknown discretized surface data, the interior stresses and displacements at any point can be obtained by direct quadrature of the surface data. The stress intensity factor can then be determined by the "direct" matching or extrapolation techniques discussed earlier. Alternatively, the boundary integral equation method can be used in conjunction with the potential energy difference procedures mentioned previously.

The great advantage of the boundary integral equation method lies in its reduction of the dimensionality of region to be modeled. This greatly reduces the total number of degrees of freedom to be incorporated. However, the matrices of the resulting equations are not banded, as in finite elements formulations, but are, in general, full. Thus, the most readily applicable method of energy difference techniques is probably the potential energy difference of two solutions, with slightly different crack lengths.

BIE3D2 [55]. This code is a boundary integral equation computer program for the elastic analysis of arbitrary three-dimensional structures. The Kelvin solution is used as the kernel. Cracks are modeled as open cavities with an aspect ratio (length/opening) of order 25. Stress intensity factors may be inferred from calculated interior stress measures or from energy differences between solutions of slightly different crack lengths.

CRX2D1 [56]. This code is also a boundary integral equation program capable of analysis of cracked plane, anisotropic bodies. The Green's function used is the exact two-dimensional model of a single crack, so that the only boundary which need be analyzed is other than the crack surface. Both in-plane stress intensity factors are obtained for either internally or externally cracked bodies.

UNIVAC 1108, CDC 6600, IBM 370

2,000 cards - BIE3D2
1,200 cards - CRX2D1

Dr. Thomas A. Cruse
Pratt and Whitney Aircraft
Engineering Dept. EB3S1
East Hartford, Connecticut 06108

Other Fracture Mechanics Software

Some fracture mechanics software does not fall within the general categories previously outlined. An example of this type of analysis is given here.

CRACKS [58]. This computer program analyzes crack propagation in cyclic loaded structures. The program has the option of using relationships derived by Forman or by Paris for crack growth. Provisions are made for both surface flaws and through cracks as well as the transition from the former to the latter. The program utilizes a block loading concept wherein the load is applied for a given number of cycles rather than applied from one cycle number to another cycle number. Additional features of the program are: variable print interval, variable integration interval, and optional formats for loads input.

IBM 360, 370, 7094, CDC 6400, 6600

1,500 cards

R. M. Engle
AFFDL/FBR
Wright-Patterson AFB, Ohio

Software Comparison

Table 2 is a matrix categorizing the fracture mechanics computer programs reviewed here. The codes are listed vertically according to the general numerical method used and horizontally according to the complexity of the problem solved. No attempt has been made to establish the relative accuracy, efficiency, usability, or reliability of this software.

Guidelines for Code Selection and Use

The basis for selecting fracture mechanics software has many variables. Some of the things that should be considered are the frequency fracture mechanics calculations are made, the software directly available, the fracture mechanics background of the user, and the problem that must be solved. It is imperative that a user have adequate knowledge of fracture mechanics principles before any attempt is made to use a fracture mechanics computer code as a design or analysis tool. Linear elastic fracture mechanics has many pitfalls for the untrained stress analyst, and disastrous results may be obtained if improper use of fracture calculations is made. Specifically, one must know whether or not linear elastic fracture mechanics principles adequately represent the problem being analyzed. If they do, the linear codes listed in this paper may be properly applied to the problem. Care must also be used in defining the proper stress state (i.e., plane stress or plane strain) that exists. Problems that involve significant crack tip plasticity are not yet properly treated by available fracture mechanics software.

The direct finite element approach to linear elastic fracture mechanics is available to anyone who has access to a two-dimensional continuum finite element computer program. This technique is probably best suited to an installation that has relatively few fracture mechanics problems to solve and thus can use an available conventional finite element code. The main drawback to this method is the large number of elements needed to define the crack tip region, thus requiring long computational times. Also, mixed mode problems are not treated.

Finite element superposition methods offer good accuracy but are somewhat cumbersome in application although CRACK-ANALYSIS appears to be a user-oriented code.

Energy release methods can be directly incorporated in existing continuum codes with only slight modifications such as was done in SASL and AXISOL. Improvements of the method as shown by Parks [21] and incorporated in BERSAFE offer definite advantages in solving three-dimensional problems. Substructuring as done in NV344 is also applicable to three dimensions. The energy release method requires more direct involvement between user and code than some of the other techniques. Mixed mode problems are not treated by this scheme. Computational times may be reduced when using energy release methods by eliminating stress computations in that the potential energy P may be calculated directly from the applied loads and displacements.

Table 2 Matrix of Available Fracture Mechanics Software: Methods and Applications

Method		Two-dimensional: Single mode	Two-dimensional: Mixed mode	Two-dimensional: Fatigue	Two-dimensional: Nonlinear	Three-dimensional: Single mode	Three-dimensional: Mixed mode	Three-dimensional: Fatigue	Three-dimensional: Nonlinear
Other				CRACKS					
Boundary integral		CRX2D1				BIE3D2			
Finite element	Singular elements	ASAS CAPS CHILES CRACK-1,2 CRAK-1 FEMWC JODAV MAGIC PINTO SING TEXGAP TF0747 UD1	ASAS CAPS CHILES CRACK-1,2 CRAK-1 FEMWC JODAV MAGIC PINTO SING TEXGAP TF0747 UD1						
Finite element	Other	FAST/PLANS							
Finite element	Energy release	AXISOL BERSAFE NV344 SASL		NV344		BERSAFE NV344		NV344	
Finite element	Super-position	CRACK ANALYSIS SIF-PLANE SIF-AXISYM	CRACK ANALYSIS SIF-PLANE SIF-AXISYM						
Finite element	Direct	Conventional 2-D programs			Conventional Nonlinear 2-D programs	Conventional 3-D programs			Conventional Nonlinear 3-D programs
		Application							

Finite element codes incorporating singular elements are becoming increasingly popular. Codes such as MAGIC, TEXGAP, ASAS, and Lockheed's version of NASTRAN incorporating CRACK-1 or CRACK-2 are used for a broad range of problems besides fracture mechanics. TEXGAP also properly treats the incompressible problem associated with solid rocket propellants. CHILES, SING, and CRAK-1 are smaller codes written primarily to address the fracture problem. This method has the advantage of being easy to use but, in a similar vain, suffers from an equal ease of being misused.

Boundary collocation codes are often not as generally applicable as some of the finite element codes but, more often than not, are used by very expert fracture mechanicians.

Boundary integral formulations have the advantage of eliminating one dimension of the problem. This technique therefore appears attractive for three-dimensional problems. This advantage is lost, however, if more than a relatively few specific stress and/or displacement locations are desired because a complete boundary numerical integration must be done for each stress location output.

FUTURE TRENDS IN COMPUTATIONAL FRACTURE MECHANICS

Table 2 graphically shows the gaps that exist in fracture mechanics software. As previously mentioned, problems involving crack tip plasticity are not yet state-of-the-art. Work is progressing in this area, but many problems such as ductile failure criteria [59], near field singular definitions [60], effect of finite strains, to mention a few, remain unresolved. Most of the efforts to this date to numerically treat the plastic problem have been by the direct approach [61,62].

Much is left to do in the analysis of three-dimensional problems. Recently, however, Tracey [63] has formulated a wedge-shaped three-dimensional version of his two-dimensional [33] element. Some of the difficulties that arise in three-dimensional analyses are out-of-plane stress intensity variation, automatic mesh generation, and reasonable output.

Running cracks offer challenges that can possibly be met by numerical methods. Direction of propagation, unloading, and dynamic effects are a few of the problems that need to be treated.

Little work has been done on the analysis of fracture in visco-elastic materials. Here again, numerical methods could be incorporated to give the analyst a more complete description of the problem.

In summary, computational fracture mechanics is essentially in its infancy. Most of the fracture mechanics programs reviewed here have been available for only a short period of time. Linear, plane (or axisymmetric) problems including mixed mode conditions are readily treated by this software. Three-dimensional, nonlinear (i.e., geometric and material), running crack and viscoelastic conditions offer a great deal of opportunity for further work in both the theoretical development of fracture mechanics models and the application of these models in user-oriented computer programs.

ACKNOWLEDGMENT

This work was supported by the United States Atomic Energy Commission.

REFERENCES

1 Rice, J. R., "Mathematical Analysis in the Mechanics of Fracture," Fracture An Advanced Treatise, ed. H. Liebowitz, Academic Press, New York, 1968, pp. 191-311.

2 Rice, J. R., "A Path Independent Integral and the Approximate Analysis of Strain Concentration by Notches and Cracks," Journal of Applied Mechanics, Vol. 35, June 1968, pp. 379-386.

3 Sneddon, I. N., and Lowengrub, M., Crack Problems in the Classical Theory of Elasticity, Wiley, New York, 1969.

4 Sih, G. C., ed., Methods of Analysis and Solutions of Crack Problems, Noordhoff, Leyden, The Netherlands, 1973.

5 Tada, H., The Stress Analysis of Cracks Handbook, Del Research Corp., Hellertown, Pa., 1973.

6 Hutchinson, J. W., "Singular Behavior at the End of a Tensile Crack in a Hardening Material," Journal of the Mechanics and Physics of Solids, Vol. 16, 1968, pp. 13-31.

7 Hutchinson, J. W., "Plastic Stress and Strain Fields at a Crack Tip," Journal of the Mechanics and Physics of Solids, Vol. 16, 1968, pp. 337-347.

8 Rice, J. R., and Rosengren, G. F., "Plane Strain Deformation Near a Crack Tip in a Power Law Hardening Material," Journal of the Mechanics and Physics of Solids, Vol. 16, 1968, pp. 1-12.

9 Rice, J. R., and Johnson, M. A., "The Role of Large Crack Tip Geometry Changes in Plane Strain Fracture," Inelastic Behavior of Solids, ed., M. F. Kanninen, McGraw-Hill, New York, 1970, pp. 641-672.

10 Zienkiewicz, O. C., The Finite Element Method in Engineering Science, McGraw-Hill, London, 1971.

11 Rice, J. R., and Tracey, D. M., "Computational Fracture Mechanics," Numerical and Computer Methods in Structural Mechanics, Academic Press, New York, 1973, pp. 585-623.

12 Gallagher, R. H., "Survey and Evaluation of the Finite Element Method in Fracture Mechanics Analysis," First International Conference in Reactor Technology, Berlin, Sept., 1971.

13 Jeram, K., and Hellen, T. K., "Finite Element Techniques in Fracture Mechanics," International Conference on Welding Research Related to Power Plants, Southampton, 1972.

14 Oglesby, J. J., and Lomacky, O., "An Evaluation of Finite Element Methods for the Computation of Elastic Stress Intensity Factors," Journal of Engineering for Industry, Vol. 95, No. 1, Feb. 1973, pp. 177-185.

15 Chan, S. K., Tuba, I. S., and Wilson, W. K., "On the Finite Element Method in Linear Fracture Mechanics," Engineering Fracture Mechanics, Vol. 2, No. 1, Feb. 1970, pp. 1-17.

16 Ando, K., "Crack Analysis of a Ship's Structure - Two Dimensional Elastic In-Plane Problems," Mitsubishi Heavy Industry Journal, Vol. 10, No. 3, 1973, pp. 327-335. (In Japanese)

17 Yamamoto, Y., Tokuda, N., "Determination of Stress Intensity Factors in Cracked Plates by the Finite Element Method," International Journal for Numerical Methods in Engineering, Vol. 6, No. 3, 1973, pp. 427-439.

18 Yamamoto, Y., Tokuda, N., Sumi, Y., "Finite Element Treatment of Singularities of Boundary Value Problems and its Application to Analysis of Stress Intensity Factors," Theory and Practice in Finite Element Structural Analysis, University of Tokyo Press, Tokyo, 1973.

19 Anderson, G. P., Ruggles, V. L., and Stibor, G. S., "Use of Finite Element Computer Programs in Fracture Mechanics," International Journal of Fracture Mechanics, Vol. 7, No. 1, March 1971, pp. 63-76.

20 Watwood, V. B., "Finite Element Method for Prediction of Crack Behavior," Nuclear Engineering and Design, Vol. 11, No. 2, March 1970, pp. 323-332.

21 Parks, D. M., "A Stiffness Derivative Finite Element Technique for Determination of Elastic Crack Tip Stress Intensity Factors," NASA NGL 40-002-080/13, May 1973.

22 Hussain, M. A., Pu, S. L., and Underwood, J., "Strain Energy Release Rate for a Crack Under Combined Mode I and II," presented at Seventh National Symposium on Fracture Mechanics, American Society for Testing and Materials, College Park, Maryland, 1973.

23 Wilson, E. L., "Structural Analysis of Axisymmetric Solids," AIAA Journal, Vol. 3, No. 12, Dec. 1965, pp. 2269-2274.

24 Hellen, T. K., "The Calculation of Stress Intensity Factors Using Refined Finite Element Techniques," RD/B/N2583, March 1973, Central Electricity Generating Board, Berkeley Nuclear Laboratories, Berkeley, Gloucestershire, United Kingdom.

25 Hellen, T. K., "BERSAFE (Phase 1) A Computer System for Stress Analysis, Part 1: User's Guide," RD/B/N1761, Central Electricity Generating Board, Berkeley Nuclear Laboratories, Berkeley, Gloucestershire, United Kingdom.

26 Hellen, T. K., "BERSAFE (Phase 1) A Computer System for Stress Analysis, Part 2: Advice and Sample Problems," RD/B/N1813, Central Electricity Generating Board, Berkeley Nuclear Laboratories, Berkeley, Gloucestershire, United Kingdom.

27 Aamodt, B., Bergan, P. G., and Klem, H. F., "Calculation of Stress Intensity Factors and Fatigue Crack Propagation of Semi-Elliptical Part Through Surface Cracks," Proceedings of the Second International Conference on Pressure Vessel Technology, San Antonio, Texas, October 1973.

28 Bergan, P. G. and Aamodt, B., "Finite Element Analysis of Crack Propagation in Three-Dimensional Solids Under Cyclic Loading," Proceedings of the Second International Conference on Structural Mechanics in Reactor Technology, Vol. 3, Berlin, September 1973.

29 Bergan, P. G., and Aamodt, B., "NV344, SESAM-69 Users Manual," Cp Report No. 71-12, AIS Computas, P. O. Box 167, Okern, Oslo 5, Norway.

30 Callabresi, M. L., and Heidelberg, S. T., "SASL: A Finite Element Code for the Static Analysis of Plane Solids Subjected to Axisymmetric and Plane Loadings," SC-DR-72 0061, Dec. 1972, Sandia Laboratories, Livermore, California.

31 Strang, G., and Fix, G. J., An Analysis of the Finite Element Method, Prentice-Hall, Englewood Cliffs, 1973, pp. 257-280.

32 Hilton, P. D., and Sih, G. C., "Applications of the Finite

Element Method to the Calculations of Stress Intensity Factors," Methods of Analysis and Solutions of Crack Problems, ed., G. C. Sih, Noordhoff, Leyden, The Netherlands, 1973, pp. 426-483.

33 Tracey, D. M., "Finite Elements for Determination of Crack Tip Elastic Stress Intensity Factors," Engineering Fracture Mechanics, Vol. 3, No. 3, Sept. 1971, pp. 255-265.

34 Muskhelishvili, N. I., Some Basic Problems of the Mathematical Theory of Elasticity, P. Noordhoff, Groningen, Holland, 1953.

35 Pian, T. H. H., Tong, P., and Luk, C. H., "Elastic Crack Analysis by a Finite Element Hybrid Method," Proceedings of the Third Conference on Matrix Methods in Structural Mechanics, Wright-Patterson AFB, Ohio, 1971.

36 Tong, P., and Pian, T. H. H., "On the Convergence of the Finite Element Method for Problems with Singularity," International Journal for Solids and Structures, Vol. 9, No. 3, March 1973, pp. 313-322.

37 Benzley, S. E., "CHILES - A Finite Element Computer Program that Calculates the Intensities of Linear Elastic Singularities," SLA-73-0894, Sept. 1973, Sandia Laboratories, Albuquerque, N.M.

38 King, W. W., Anderson, J. M., and Morgan, J. D., "Dynamics of Cracked Structures Using Finite Elements," Proceedings of Thirteenth Annual Symposium on Fracture and Flaws, American Society of Mechanical Engineers, Albuquerque, N.M., 1973.

39 Hilton, P., and Hutchinson, J. W., "Plastic Intensity Factors for Cracked Plates," Engineering Fracture Mechanics, Vol. 3, No. 4, Dec. 1971, pp. 435-451.

40 Blackburn, W. S., "Calculation of Stress Intensity Factors at Crack Tips Using Special Finite Elements," The Mathematics of Finite Elements and Its Applications, Brunel University, Academic Press, 1972, pp. 327-336.

41 Byskov, E., "The Calculation of Stress Intensity Factors Using the Finite Element Method," International Journal of Fracture Mechanics, Vol. 6, No. 2, June 1970, pp. 159-167.

42 Jordan, S., Padlog, J., Hopper, A. T., Rybicki, E. F., Hulbert, L. E., and Kanninen, M. F., "Development and Application of Improved Analytical Techniques for Fracture Analysis Using MAGIC III," AFFDL-TR-73-61, June 1973, Air Force Flight Dynamics Laboratory, Wright-Patterson AFB, Ohio.

43 , "Formulation of Crack Finite Elements," LR25939, Sept. 1973, Lockheed--California Company, Burbank, California.

44 Tong, P., Pian, T. H. H., and Lasry, S., "A Hybrid Element Approach to Crack Problems in Plane Elasticity," International Journal for Numerical Methods in Engineering, Vol. 7, No. 3, 1973, pp. 297-308.

45 Dunham, R. S., and Becker, E. B., "TEXGAP - The Texas Grain Analysis Program," TICOM Report 73-1, August 1973, University of Texas, Austin, Texas.

46 Koteras, J. R., "A Hybrid Finite Element for Determination of Stress Intensity Factors at a Crack Tip," TICOM Report 73-2, August 1973, University of Texas, Austin, Texas.

47 Armen, H., Saleme, E., Pifko, A., and Levine, H. S., "Nonlinear Crack Analysis with Finite Elements," Numerical Solution of Nonlinear Structural Problems, ed., R. F. Hartung, ASME, AMD - Vol. 6, 1973.

48 Gross, B., and Srawley, J. E., "Stress Intensity Factors for Three Point Bend Specimens by Boundary Collocation," NASA Technical Note D-3092, 1965.

49 Gross, B., and Srawley, J. E., "Stress Intensity Factors for Single Edge Notched Specimens in Bending or Combined Bending and Tension," NASA Technical Note D-2603, 1965.

50 Bowie, O. L., and Neal, D. M., "A Modified Mapping Collocation Technique for Accurate Calculation of Stress Intensity Factors," International Journal of Fracture Mechanics, Vol. 6, No. 2, June 1970, pp. 199-206.

51 Kobayashi, A. S., Cherepy, R. B., and Kinsel, W. C., "A Numerical Procedure for Estimating the Stress Intensity Factor of a Crack in a Finite Plate," Journal of Basic Engineering, Vol. 86, No. 4, Dec. 1964, pp. 681-684.

52 Hussain, M. A., Lorensen, W. E., Kendall, D. P., and Pu, S. L., "A Modified Collocation Method for C-Shaped Specimens," R-WV-T-X-6-73, Feb. 1973, Benet Weapons Laboratory, Watervliet Arsenal, Watervliet, N.Y.

53 Cruse, T. A., "Application of the Boundary-Integral Method to Three-Dimensional Stress Analysis," Computers and Structures, Vol. 3, No. 3, May 1973, pp. 509-527.

54 Cruse, T. A., and Van Buren, W., "Three Dimensional Elastic Stress Analysis of a Fracture Specimen with an Edge Crack," International Journal of Fracture Mechanics, Vol. 7, No. 1, March 1971, pp. 1-15.

55 Cruse, T. A., "An Improved Boundary Integral Equation Solution Method for Three-Dimensional Elastic Stress Analysis," Report SM-73-19, Department of Mechanical Engineering, Carnegie-Mellon University, Pittsburgh, Pa.

56 Snyder, M. D., and Cruse, T. A., "Crack Tip Stress Intensity Factors in Finite Anisotropic Plates,' AFML-TR73-209, Air Force Materials Laboratory, Wright-Patterson AFB, Ohio.

57 Cruse, T. A., "Lateral Constraint in a Cracked Three-Dimensional Body," International Journal of Fracture Mechanics, Vol. 6, No. 3, Sept. 1970, pp. 326-328.

58 Engle, R. M., "CRACKS, A Fortran IV Digital Computer Program for Crack Propagation Analysis," AFFDL-TR-70-107, Oct. 1970, Air Force Flight Dynamics Laboratory, Wright-Patterson AFB, Ohio.

59 Landes, J. D., and Begley, J. A., "Test Results from J Integral Studies," presented at Seventh National Symposium on Fracture Mechanics, American Society for Testing and Materials, College Park, Maryland, 1973.

60 Shih, C. F., "Small Scale Yielding Analysis of Mixed Mode Plane-Strain Crack Problems," presented at Seventh National Symposium on Fracture Mechanics, American Society for Testing and Materials, College Park, Maryland, 1973.

61 Nair, P., and Reifsnider, K. L., "UNIMOD: An Application Oriented Finite Element Scheme for Fracture Mechanics," presented at Seventh National Symposium on Fracture Mechanics, American Society for Testing and Materials, College Park, Maryland, 1973.

62 Anderson, H., "Finite Element Analysis of a Fracture Toughness Test Specimen in the Nonlinear Range," Journal of the Mechanics and Physics of Solids, Vol. 20, 1972, pp. 33-51.

63 Tracey, D. M., "Finite Elements for Three-Dimensional Elastic Crack Analysis," to appear in Nuclear Engineering and Design.

THERMAL STRESS AND CREEP

R. E. Nickell
Sandia Laboratories
Albuquerque, New Mexico

ABSTRACT

Computer programs designed to analyze problems of time-dependent flow in solids (creep and/or recovery) are surveyed. Special attention is given to the manner in which the temperature field is supplied to the stress analysis module and to data input procedures for temperature-dependent flow parameters. The computer programs are divided into four classes, based primarily on geometrical capability: (1) historical, (2) one-dimensional, (3) two-dimensional, and (4) general purpose. Algorithmic details that are particularly germane to the creep problem--such as automatic increment control, iteration within the increment based on error tolerances, tangent modulus vs. initial strain approaches, flow and hardening rules, and equations of state--are discussed. Program sources, documentation, availability, and maintenance organizations are listed.

NOMENCLATURE

s_{ij} = components of deviatoric stress tensor

σ_{ij} = components of stress tensor

$\bar{\sigma}$ = effective stress

ε_{ij} = components of total strain tensor

e_{ij} = components of deviatoric strain tensor

ε_{ij}^{c} = components of creep strain tensor

$\bar{\varepsilon}^{c}$ = effective creep strain

ε_{ij}^{E} = components of elastic strain tensor

$\dot{\varepsilon}_{ij}^{c}$ = components of creep strain-rate tensor

t = time

T = temperature

INTRODUCTION

The need to design structures that are subject to sustained or cyclic loads while operating at elevated temperatures has led to a proliferation of computer programs that incorporate time-dependent flow. In solid materials such behavior is referred to as rate-dependent plasticity, viscoplasticity, creep, and relaxation, or often by the single generic term "creep." This survey is aimed specifically at the class of software that deals with creep, but in addition, some attention will be given to the manner in which the elevated temperature field is supplied to the stress analysis program, in terms of both temperature-dependent material properties (e.g., yield stress, elastic moduli, creep parameters) and induced thermal stresses.

A distinction will be made between this class of computer programs and those that deal with time-independent flow (see the survey on plasticity programs by Armen) and viscoelasticity (see the survey by Gupta), although overlap is inevitable. In spite of this distinction, a number of similar terms will occasionally be encountered during the discussion of particular programs, such as "flow rule" and "hereditary theory." For the most part, however, the program descriptions will be limited to those features that are unique to the creep problem.

TIME-DEPENDENT FLOW CONCEPTS

Incorporating time-dependent flow into a computer program first involves three theoretical considerations [1]:

1. Given that a set of one-dimensional, isothermal creep results have been experimentally obtained for various constant stress levels and temperatures, the analyst usually attempts to fit the data with some convenient functional form. This functional fit is referred to as a "creep law" or, in some of the literature, as an "equation of state." The latter term is undesirable since it implies some physical meaning to the process. During the data fitting procedure, it is often convenient to distinguish between "primary creep" (decreasing creep rate), "secondary creep" (minimum constant creep rate), and "tertiary creep" (increasing creep rate). Figure 1 illustrates these regions. The uniaxial characterization of time-dependent flow could also have been obtained from isothermal "relaxation" experiments (constant total strain and temperature) or from some other series of tests; historically, however, creep experiments have been the rule because of their simplicity.

2. Given that a uniaxial creep law has been determined, a method of generalizing the result to multiaxial states of stress and strain is needed. This generalization is referred to as a "flow rule," in accord with similar terminology in time-independent flow. Most flow rules in current use are based on proportionality between creep strain rate components and deviatoric stresses, just as in the Prandtl-Reuss flow rule in plasticity. The usual form is

$$\dot{\varepsilon}^{c}_{ij} = \frac{3}{2}\,\frac{\dot{\bar{\varepsilon}}^{c}(\bar{\sigma},t,T)}{\bar{\sigma}}\,s_{ij} \qquad (1)$$

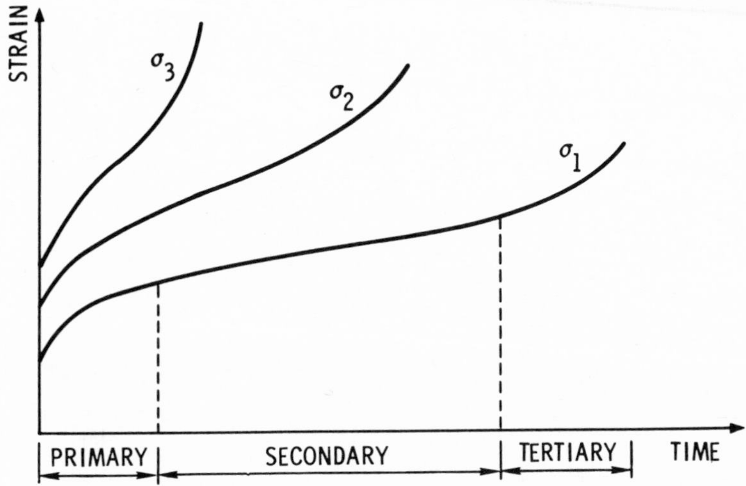

Fig. 1 Creep rate zones

where

$$s_{ij} = \sigma_{ij} - \frac{1}{3}\sigma_{kk}\delta_{ij} \tag{2}$$

$$\bar{\sigma}^2 = \frac{3}{2} s_{ij}s_{ij} \tag{3}$$

and

$$\bar{\varepsilon}^{c^2} = \frac{2}{3} \varepsilon^c_{ij}\varepsilon^c_{ij} \tag{4}$$

Implied in this form is the incompressibility of creep strains. An argument could be made that, at many temperature and stress levels, the material retains some elastic strength and that, therefore, incompressibility of creep strains represents a severe assumption. Further, this elastic strength would seem to be related, in some way, to the delayed recovery that would occur upon removal of the load in a creep test [2] (Fig. 2). However, there have been no successful attempts, to my knowledge, to correlate either transient creep or delayed recovery with creep strain compressibility. When volumetric creep, such as that encountered in hot pressing, is included [3], a nonassociated flow rule of the type used in soil mechanics can be introduced. The creep strains are then related to both deviatoric stresses and mean pressure.

3. Given that a flow rule has been established, some means for generalizing the theory to time-varying stress levels is, in general, required. The rules that are postulated in this regard are referred to as "hardening laws." An excellent discussion of various hardening laws is contained in [1] and will be summarized here. "Time-hardening"

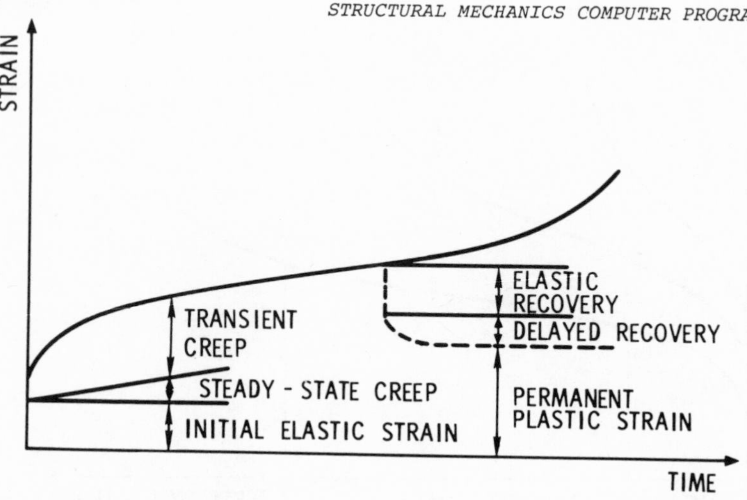

Fig. 2 Creep and recovery

theory is an attempt to use an analogy with viscous flow and is, therefore, very effective when secondary creep dominates (e.g., at very high temperatures). The creep strain rate at any time and stress level is assumed not to depend upon the current value of creep strain (Fig. 3). If the stress is changed from σ_1 to σ_2 at time t_1, the creep rate is determined at point B. Accordingly, the creep rate is found from a formula of the type

$$F(\bar{\sigma},\dot{\bar{\varepsilon}}^c,t,T) = 0 \tag{5}$$

Strain-hardening theory, on the other hand, is an attempt to use an analogy with work-hardening plasticity theories. The creep strain rate at any time and stress level is assumed to depend upon the total (or, often, only the primary) creep strain (Fig. 3). If the stress is changed from σ_1 to σ_2 at time t_1, the creep rate is determined at point C. Accordingly, the creep rate is found from a formula of the type

$$F(\bar{\sigma},\dot{\bar{\varepsilon}}^c,\bar{\varepsilon}^c,T) = 0 \tag{6}$$

Other hardening rules have been proposed. The most notable have been the "hereditary law" proposed by Rashid [4] and the attempts to tinker with the strain-hardening law in order to obtain a more meaningful theory [5]. These latter modifications are often referred to as "auxiliary rules for reversed loading."

The hereditary law is an extension of the conventional creep compliance form of the one-dimensional constitutive equation of linear viscoelasticity to nonlinear dependence upon stress. Superposition is assumed to remain valid. Then

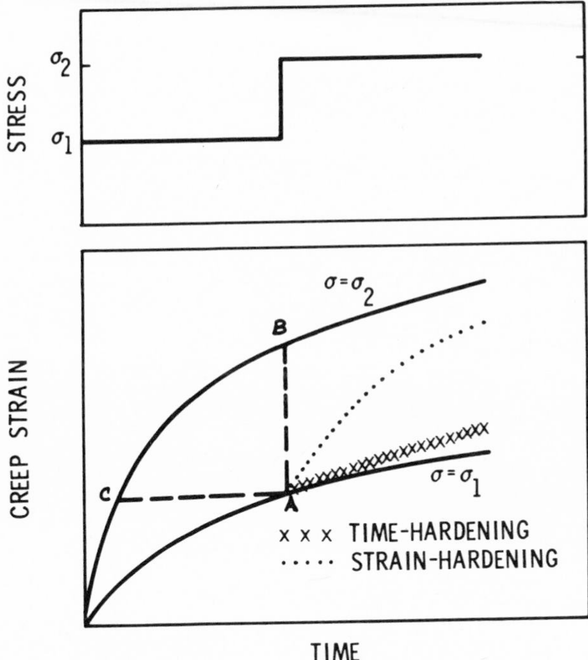

Fig. 3 Time and strain hardening

$$\overline{\varepsilon}^c(t) = \int_{-\infty}^{t} J(\overline{\sigma}, t-\tau) \frac{d\overline{\sigma}(\tau)}{d\tau} d\tau \tag{7}$$

where $J(\overline{\sigma},t)$ is the creep compliance at a stress level $\overline{\sigma}$ or, conversely, the derivative, with respect to effective stress, of the usual uni-axial creep law. This law also accepts the notion of time-temperature equivalence [4], where the arguments of the creep compliance are replaced by a "pseudo-time" that depends upon the history of stress and temperature.

The auxiliary hardening rules are meant to correct the deficiencies in the strain-hardening theory that are evidenced when unloading occurs (Fig. 4). As the stress is changed from σ_1 to $-\sigma_1$ at time t_1, the instantaneous creep strain rate would be expected to change sign while maintaining very nearly the same magnitude. As time progresses and the effective creep strain decreases, however, the magnitude of the creep strain rate will begin to increase, precisely the opposite of experimental results. In order to force primary creep behavior upon unloading, a new effective creep strain origin can be defined at that time. The current total equivalent creep strain must be accumulated so that the appropriate value is available if the loading is again reversed (note the procedure indicated in Fig. 4).

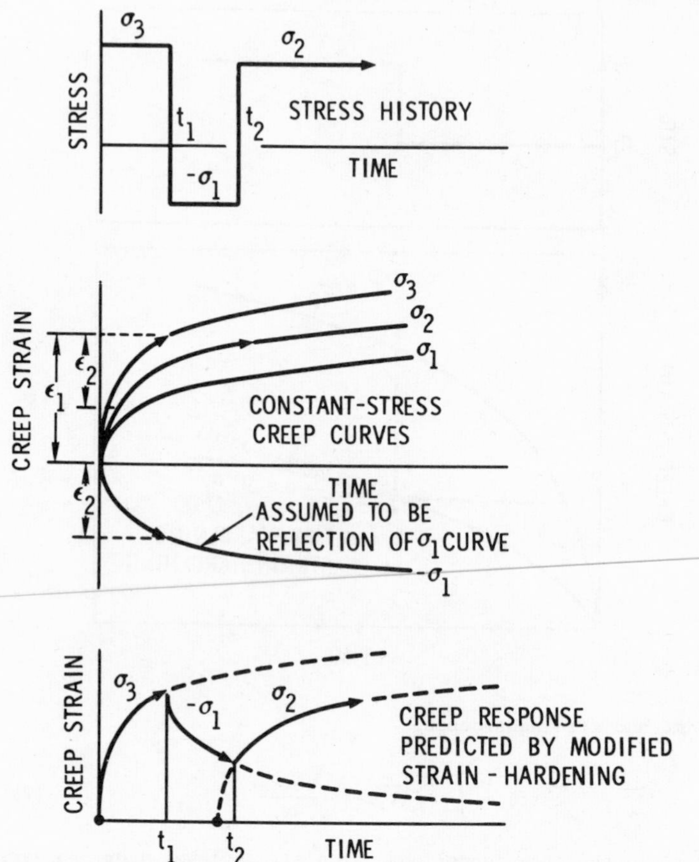

Fig. 4 Auxiliary rules

COMPUTATIONAL CONCEPTS

The original applications of numerical methods to somewhat general (more than one dimensional) creep problems were based upon the similarity to elastic thermal stress analysis [6]. There, a set of "initial strains," depending upon temperature and the coefficient of thermal expansion, can be calculated and converted to "initial stresses" and, hence, to "initial loads" through the use of elastic and geometric properties. The problem is linear since none of the "initial" quantities depend on the deformation or stress level in the structure. Creep strains are dependent upon stress level, at a minimum; therefore, the initial strains are only an approximation to the actual creep strains. The linearization and iteration algorithm for dealing with this nonlinear problem is described in [7]. The primary advantage of this procedure is its ease of incorporation into an existing analysis package. Another apparent advantage is that the governing system of equations need be factored but once (if a direct solution algorithm is employed), although several hundred right-hand side reductions and back-substitutions may be required. The disadvantage of the initial strain method is the small time steps that are required in order to ensure proper satisfaction of the creep law and achieve convergence. Several iterations within each time increment may be needed, as well, to satisfy the creep law.

As a means for alleviating this difficulty, some computer programs are based upon variable stiffness. This requires that the creep strain rate at the current time depend, to some extent, upon the current (unknown) stress level, in addition to any dependence upon previous (known) stress levels. As pointed out in [4], the hereditary integral form leads to a variable stiffness if a Simpson's rule integration technique is used to evaluate the integral. When conventional hardening rules are used, the creep strain rate components are integrated with respect to time to yield creep strain increments. Rather than have the arguments in the integrand (equivalent stress, deviatoric stress components, total equivalent creep strain, etc.) constant during the time increment, however, some simple time variation is assumed. If this variation depends upon unknown values, a variable stiffness results [8].

Another approach has been suggested that purportedly leads to an improved approximation for the creep strain increments without the variable stiffness and the resulting refactorizations of the stiffness matrix [9]. This approach is also an initial strain method, but during the iterations within the time increment, a weighted average of initial strains from previous iterations is calculated. Time steps that are one or two orders of magnitude larger than those for the conventional initial strain method are claimed, using three to five iterations per time increment.

Another computational feature of many creep programs is a built-in alteration of the time increment, based on preset tolerances of some type. The user of the program may supply an increment that is either too large for convergence to take place or so small that insufficient creep occurs during the increment. In either case, the program computes a new step size and attempts to continue the analysis. The most common internal measures are the ratio of incremental equivalent creep strain to equivalent elastic strain (initially) and the ratio of two successive equivalent creep strain increments (for later times). Some programs monitor the stress increments, as well, so that relaxation problems are treated in an accurate manner.

Many initial strain programs iterate only to establish a suitable time increment, without ensuring that the stresses at the end of the increment are compatible with the creep strains calculated at the beginning of the increment. These programs often compute the unbalanced forces that result from a lack of satisfaction of the stress-strain law and add these forces to the incremental forces for the next interval. This procedure is referred to as "equilibrium load correction." Variable stiffness programs that do not iterate within the increment often use this procedure, as well. One measure of convergence for the nonlinear creep problem is, in fact, the norm of this residual load vector as a function of the iterative cycle.

There are a number of other computational features that will be discussed within the context of individual programs. Most of these features (e.g., finite element libraries) have a well-established terminology and will not be introduced here.

HISTORICAL PROGRAMS

CRAB. The earliest treatment of creep for any general class of geometries was the work by Padlog, Huff, and Holloway [10], who used the method of initial strains. The first identifiable computer program, however, was the CRAB (CReep in Axisymmetric Bodies) Code [11]. The theory was later described in [7]. The constant-strain triangular ring represented the element library, a time-hardening law with an associated flow rule was proposed, and the method of initial strains was used. Temperatures interior to the body were obtained by linear interpolation of card input inside and outside nodal values, similar to automatic mesh generation procedure. An average centroidal temperature is the basis for material property evaluation. The automatic increment control feature employed a 1/25 ratio of initial effective creep strain to maximum effective elastic strain; subsequent time steps were chosen on the basis of the ratio of incremental effective stress to total effective stress. Since the uniaxial creep law was of the form

$$\bar{\varepsilon}^c = A_1 \bar{\sigma}^m t^n \tag{8}$$

where

$$A_1 = B_1 e^{(B_2 + B_3 T + B_4 T^2)} \tag{9}$$

then

$$\Delta t_1 = \left[\frac{A_1 \bar{\sigma}^{(1-m)}}{25E} \right]^{1/n} \tag{10}$$

and

$$\frac{\Delta t_{i+1}}{\Delta t_i} = \frac{0.03}{(\Delta\bar{\sigma}/\bar{\sigma})_{max}} \tag{11}$$

A further limitation that provides an internal convergence test is the requirement that the incremental effective creep strain must always be less than the maximum elastic strain. No plasticity is available in the program, and Gaussian elimination is the solution technique for the linearized equations.

AXICRP. The next computer program to receive some general use was AXICRP, which, contrary to its name, was designed to solve plane stress and strain problems, in addition to axisymmetric solids [12,13]. The library consisted of either triangular or quadrilateral elements--the quadrilateral being composed of four constant-strain triangles with condensed centroidal unknowns. Initial strain methods, time- or strain-hardening, an associated flow rule, and an incremental formulation of the equilibrium equations were programmed. Isotropic properties and small deformation were assumed, and no plasticity was included. Similar automatic increment controls to that in CRAB were specified. The initial incremental effective creep strain was required to be from 1/10 to 1/25 of the maximum effective elastic strain, and the fractional change in effective stress in consecutive intervals was required to be between 3 and 10 percent. At the same time, the increase in time step size was restricted to be less than 100 percent. Internal tests were designed to find steady-state creep conditions automatically, thus enabling an extrapolation to the final time. No mention was made of iterating within the increment or using equilibrium load correction.

ONE-DIMENSIONAL PROGRAMS

Over the past few years, the great concern over predicting the creep collapse of nuclear fuel cladding has spawned a large number of computer programs aimed at analyzing one-dimensional geometries [14-21]. Primarily, these programs solve axisymmetric, plane strain, elastic-plastic and creep problems. It is likely that every company involved in reactor fuels and most of the AEC national laboratories have their own programs of this type. As an aside, one wonders why such software has received disproportionate attention, since the cladding collapse problem is at best two dimensional (plane-strain ring, deforming nonaxisymmetrically) and at worst three dimensional (finite length cylinder, deforming nonaxisymmetrically). The incipient cracking of the fuel elements themselves could conceivably be predicted, but such analysis is of little value. Postcracking behavior, axial densification, and ratcheting phenomena are most pressing concerns; none can be treated in an adequate manner by a one-dimensional program.

Nevertheless, such software can serve a useful purpose in providing benchmark calculations for more complex programs and are, in themselves, useful educational tools for the training of analysts. Two of these programs are typical of the remainder and will be discussed in some detail.

CRASH. (CReep Analysis in a fuel pin SHeath) was written by M. Guyette of Belgonucleaire in Brussels, Belgium, and is also maintained at the same location. The program solves for the radial, steady-state temperature distribution subject to volumetric heat sources and boundary data. The boundary data are not general (e.g., the heat transfer coefficient at the outer wall is not allowed to be a function of time and/or temperature). With the temperature available, a quasi-closed-form solution is sought for the equilibrium equations, using Newton-Raphson iteration to converge at each time step. Plastic deformation is computed prior to beginning creep calculations. Provisions are made to calculate intermittent contact stresses between fuel and sheath, should contact occur. Several axial stations may be computed, but no axial coupling between the sections is allowed. Several alternative uniaxial creep laws are available: (a) a steady-state power law (Norton's law); (b) a primary creep power law; (c) a primary creep hyperbolic law; and (d) a creep swelling law that includes both primary and secondary effects. Either time or strain hardening may be flagged. Automatic increment control is built into the program to assure convergence. The time steps provided by the user may be subdivided so that the incremental equivalent creep strains are less than or equal to the total equivalent creep strain. The initial comparison is with equivalent elastic strain. These convergence criteria are much more relaxed than those required for initial strain methods.

LIFE. The LIFE program was developed by the Argonne National Laboratory and has been modified extensively by that organization, as well as by numerous other organizations that have acquired it from ANL. The program is difficult to discuss, not only because so many different versions exist, but also because a number of algorithmic "tricks" are tried in order to achieve an asserted capability that seems questionable. One of these tricks is the use of axisymmetric, generalized plane strain theory (one-dimensional), meanwhile leaving the impression with the user that axial variations are being accounted for. Another ambiguous development is the treatment of fuel cracking. The only cracks that could possibly be treated by such a theory are circumferential, while the impression is left with the user that radial cracks (that destroy the axisymmetry) and horizontal cracks (that cause severe axial gradients) are accounted for. Most of the other program features are similar to CRASH--steady-state, radial temperature distribution; treatment of intermittent contact between fuel and cladding; an associative flow rule, etc. Apparently, however, the program does not deal with time-independent flow at all; also, the only uniaxial creep law appears to be for steady creep, and only time-hardening theory is used. Dilatational creep is included for clad and fuel swelling. Automatic increment control is available for convergence control with the Newton-Raphson solution technique. All in all, the program is a very special purpose, one-dimensional software package with severe limitations (e.g., only three radial zones are allowed in the fuel). An alternative to LIFE would be a more general, one-dimensional program such as that described in [5], developed by Foster Wheeler Corporation. This program, called R-1045, provides solutions for combined elastic-plastic and creep analysis of axisymmetric bodies under conditions of generalized plane strain. The material properties can be temperature dependent, but the temperature distribution is not calculated within the program, being input by cards or through a subroutine.

TWO-DIMENSIONAL PROGRAMS

The packages CRAB and AXICRP were both limited to two-dimensional problems. Since their development in the mid to late sixties, however, a number of other two-dimensional programs have been written--primarily to extend the capability to large deformation and finite strain, as well as to include plastic deformation. Most of these programs are company-proprietary, but whether generally available or not, they represent the state-of-the-art and will be discussed accordingly.

<u>CREEP-PLAST</u>. This program was written by Y. R. Rashid at General Electric [4,22] under contract to the Oak Ridge National Laboratory (ORNL) [23] and is being maintained and distributed by

Computer Librarian
Room A228, Building 4500-N
Oak Ridge National Laboratory
P. O. Box X
Oak Ridge, Tennessee 37830

A small service charge is required. The program is designed to analyze plane stress, plane strain, or axisymmetric structures. The most recent version at ORNL includes an overlaid transient heat conduction capability. Other versions depend upon a set of total nodal point temperatures being read from disk or tape. There is no automatic subdivision or combining of time intervals from the temperature data. The elastic-plastic-creep problem is formulated incrementally, and the linearized solutions are found by Gaussian elimination. The element library contains the constant-strain triangle and a uniaxial bar. All volume integration is done in closed form. Uniaxial creep laws are supplied by the user in special subroutines; an associated flow rule is built in; and two forms of strain hardening are available--one based on total creep strain and the other on primary creep strain only. The recommended method for treating time-varying stress is a memory, or hereditary, theory. The theory is based on a modified superposition principle and attempts to overcome the logical inconsistencies of the hardening theories. The auxiliary ORNL rules for treating creep under load reversal [5] are included for strain hardening. Time-hardening theory is not permitted. CREEP-PLAST is a modified-stiffness program when the hereditary theory is flagged, but it is an initial strain program for the strain-hardening laws. The user specifies the number of iterations within a time increment (one iteration is recommended), and each iteration requires a total matrix solution when the hereditary theory is used. Time increments are normally specified by the user, but a sophisticated set of internal controls can be flagged, if desired, in order to optimize time step selection. These controls are similar to those discussed previously.

A General Electric company-proprietary version, called SAFE-2D, has been extended to treat creep swelling, in addition to deviatoric creep, by the hereditary theory. A flow rule based on a dilatational-dependent flow surface is used [3]. Fuel cracking is modeled with anisotropic material behavior and fuel-clad interaction is treated through intermittent constraints.

EPIC-II (Elastic-Plastic Incremental Code) is a company-proprietary program [24] developed and maintained at Lockheed Missiles and Space Corporation. Inquiries may be directed to:

Roger D. Teter
Group Engineer - Reentry Structures
Lockheed Missiles and Space Company
Department 81-12, Building 154
P. O. Box 504
Sunnyvale, California 94088

The program is designed for plane stress, plane strain, or axisymmetric structures undergoing finite strain. The element library has a standard, first-order isoparametric quadrilateral (Q4), a quadrilateral modified for bending (QM5), and an incompatible mode quadrilateral (Q6). The uniaxial creep law is of the steady-state, power law type; the flow rule is associative; and time hardening is used. A modified stiffness approach enables moderately large time increments to be chosen. The user supplies the time increments (no automatic control) at which the linearized system is solved by Gaussian elimination and equilibrium load correction can be flagged. The program is coupled to heat transfer codes by magnetic tape. Extensive pre- and post-processing modules enable the program to be highly user oriented. EPIC has not been extensively applied to creep problems; therefore, little effort has been made to optimize time step selection. An advanced version with more refined elements is under development. Combined creep and plasticity and/or creep buckling are possible analytical choices.

GOLIA was developed at the EURATOM Joint Nuclear Research Center, Ispra, Italy, and is designed to solve creep (no time-independent flow) problems in plane stress, plane strain, generalized plane strain, and axisymmetric structures [25]. The generalized plane strain model allows translation in the out-of-plane direction but no rotation. The element library consists of a constant strain triangle. Strains are assumed infinitesimal. The method of initial strains and an incremental formulation are used to linearize the equations. A fourth-order Runge-Kutta integration technique is used to allow larger time steps. Automatic increment controls are similar to those in CRAB and AXICRP, however, and the tolerances do not appear to be significantly improved over those imposed by the Euler integration technique commonly used in initial strain programs. The initial creep strain is limited to about 15 to 20 percent of the elastic strain, and the maximum change in effective stress allowed in successive increments is on the order of 20 to 30 percent. The time increment is always limited to a 200 percent change, and the incremental effective creep strain is never allowed to exceed the elastic strain. The program does not appear to be state-of-the-art.

Wilson Program. This program was apparently developed by E. A. Wilson [9] at Honeywell Information Systems, Phoenix, Arizona, and is designed to solve plane stress, plane strain, or axisymmetric elastic-plastic-creep problems by an incremental, initial strain method. The element library consists of an incompatible quadrilateral that should be viewed with suspicion. Steady-state creep is the only form considered, and the flow rule is associative. An iterative procedure based on "weighted" initial strains is used to enlarge the convergent time step. The effective stress in the uniaxial creep law is thus taken to be a "weighted" average of the effective stress at the end of the previous increment, the effective stress used to form the initial strain at the last iteration, and the effective stress found from the last iteration. This program does not appear to be state-of-the-art.

TESS was developed by the Central Electricity Generating Board (CEGB), Berkeley Nuclear Laboratories, Berkeley, United Kingdom, and is a proprietary package. The transient heat conduction program HETRAN provides the temperature distribution. The element library contains only the constant strain triangle. TESS appears to be quite similar to CREEP-PLAST except that the initial strain method, rather than the modified stiffness method, is used. Apparently, both time and strain hardening are options; some thought has even been given to hereditary formulas [26], but such a generalization has not yet been coded. An averaging of effective stresses similar to that described in [9] is used to calculate initial strains. The purpose of averaging the effective stresses is to allow larger convergent time increments in the analysis. In lieu of calculating the equivalent creep strain increment based on the effective stress at the beginning of the time interval, that effective stress is averaged with the effective stress derived from the most recent iteration. In addition, averaged values of time, temperature, and total equivalent creep strain are used.

BOPACE (BOeing Plastic Analyis Capability for Engines) [27] has an element library consisting of the constant strain triangle. Both plasticity and creep are treated for plane stress, plane strain, and generalized out-of-plane behavior. Axisymmetry is not an option. The uniaxial creep law must apparently be supplied by the user in functional form. The flow rule is associative. Three hardening rules are allowed--time hardening, strain hardening, or work hardening (the creep rate is a function of the integral of effective stress and equivalent creep strain). A load reversal scheme, using a creep-hardening parameter, is available that is similar to the ORNL auxiliary rules. When complete load reversal occurs, the parameter is set to zero, and the initial point on the required creep curve is taken as a new starting point. Complete reversal is defined as an incremental creep strain vector of exactly opposite sense to the previous incremental creep strain vector. For imcomplete reversal, the parameter is multiplied by $(1 + \alpha)/2$, where α is the cosine of the angle between successive incremental creep strain vectors and is summed with the previous value of the parameter. Contrast this with the ORNL auxiliary rules [5].

BOPACE uses the initial strain method with equilibrium load correction. Iteration may be flagged in order to reduce the norm of the load correction vector. No details on automatic increment control and temperature field input are available.

GENERAL PURPOSE PROGRAMS

In the past few years most of the innovative code development has been concerned with the general purpose programming concept. The thesis behind this is the recognition that good algorithms for solving nonlinear structural mechanics problems should not be wasted on a special geometric subset but, instead, should be embedded in a code framework that allows the user wide modeling flexibility. The additional cost (due to the logical switching within the general purpose program) and increased storage requirements (due to the size of the element library) are felt to be a reasonable compromise with the flexibility. There are now about a dozen of these programs under active development today that address themselves to problems of time-dependent flow. In all cases, the programs also treat time-independent flow, as well.

As one might expect, the cost of using or acquiring these programs reflects the respective developers' investments. Only one of them--EPACA (Elastic-Plastic And Creep Analysis)--can be considered nonproprietary, since the initial development was paid for by ORNL [28]. The others range in price from about $6000 for Nottingham University's PAFEC-70+ System up to about $250,000 for MARC or ANSYS. In the discussion that follows, the programs will not be treated separately, since many of their characteristics are the same; instead, they will be discussed by topic (e.g., hardening rules). The programs to be discussed are:

1. ANSYS (ANalysis SYStem) [5]
   Swanson Analysis Systems, Inc. (c/o J. A. Swanson)
   870 Pine View Drive
   Elizabeth, Pennsylvania 15037
   (412)751-1940 or (412)872-9555

2. ASAS (Atkins Stress Analysis System)
   Atkins Research and Development (c/o R. K. Henrywood)
   Ashley Road
   Epsom, Surrey, United Kingdom

3. ASKA (Automatic System for Kinematic Analysis) [29]
   Institut fur Statik und Dynamik der Luft-Und-Raumfahrtkonstruktionen (c/o J. H. Argyris)
   Pfaffenwaldring 27
   Stuttgart, Germany

4. BERSAFE (BERkeley Structural Analysis by Finite Elements) [26]
   Central Electricity Generating Board (c/o T. K. Hellen)
   Berkeley Nuclear Laboratories
   Berkeley, Glos., United Kingdom

5. EPACA (Elastic-Plastic And Creep Analysis) [28]
   Computer Librarian
   Room A228, Building 4500-N
   Oak Ridge National Laboratory
   P. O. Box X
   Oak Ridge, Tennessee 37830

6. FESS-FINESSE (Finite Element Structural System) [29]
   Department of Civil Engineering (c/o O. C. Zienkiewicz)
   University of Wales, Swansea
   Singleton Park, Swansea SA2 8PP, United Kingdom

7. MARC (MARCal) [5,30]
   MARC Analysis Research Corporation (c/o P. V. Marcal)
   105 Medway Street
   Providence, Rhode Island 02906
   (401)751-9120

8. NEPSAP (Nonlinear Elastic-Plastic Structural Analysis Program) [8]
   Lockheed Missiles and Space Co. (c/o D. N. Yates)
   Department 81-12, Building 154
   P. O. Box 504
   Sunnyvale, California 94088
   (408)742-1397

9. PAFEC (Acronym meaning unknown) [31]
   Mechanical Engineering Department (c/o R. D. Henshell)
   University of Nottingham
   University Park
   Nottingham, NG7 2RD, United Kingdom
   Nottingham 56101, x-2641

Formulation and Solution

All of the programs are based on the finite element displacement method, although PAFEC includes some hybrid plate-bending elements and ASAS has some equilibrium elements in their respective libraries. In all cases, the nonlinear problem is incrementally solved with equilibrium load correction at least an optional choice for the user (most of the codes automatically assemble the corrective load vector). Because of the variety of elements in their respective libraries, most of the packages use a frontal solution scheme--all, however, use some form of elimination. Large deformations are included in ANSYS, ASAS, EPACA, MARC, and NEPSAP. ASKA and FESS modules for large deformation are available but not, apparently, for creep and plasticity options. ASAS, ANSYS, and EPACA use an updated Lagrangian coordinate frame for geometrically nonlinear problems, while MARC and NEPSAP are

Lagrangian codes (MARC has an inactive updated Lagrangian capability, as well). All the programs treat elastic-plastic problems by a modified stiffness procedure; many have iteration available within the load increment for convergence to yield criteria and stress-strain behavior. The iteration in every case appears to be based on successive load vector substitution (modified Newton-Raphson). Specifically with regard to creep, all the programs are based on initial strain except NEPSAP, which modifies the stiffness for both creep and plastic strain.

Uniaxial Creep Laws; Hardening and Flow Rules

Detailed information on some of the programs is not yet available; however, it would appear that ASKA, FESS, PAFEC, and BERSAFE have research capabilities in creep analysis that have not yet been translated into firm programming decisions (i.e., the potential user could bargain with the developer for particular creep laws or flow rules, depending upon the amount one wishes to pay). BERSAFE does, apparently, depend upon tabular uniaxial data from which the required input to the hardening rule is interpolated or extrapolated. Documentation on ASAS is sufficient to determine only that primary creep is available, probably with an associated flow rule. NEPSAP reflects its aerospace origins, including a steady-state creep law, an associated flow rule, and a time-hardening rule for varying stress states. EPACA was programmed under specific requirements for ORNL and thus has the recommended ORNL creep law

$$\bar{\varepsilon}^c(\bar{\sigma},t,T) = f(\bar{\sigma},T)[1-e^{-r(\bar{\sigma},T)t}] + g(\bar{\sigma},T)T \tag{12}$$

where f, g, and r are experimentally determined functions of effective stress and temperature. This law has both a primary and a steady-state creep term. Strain hardening based on either primary or total creep strain is available, the flow rule is associative, and the auxiliary ORNL rules for creep strain addition under load reversal are built in. ANSYS has a library of primary and secondary creep laws built in, including the ORNL law, includes both strain and time hardening, uses an associated flow rule, and has an optional choice of the auxiliary rules. Irradiation swelling is treated as a pseudo-thermal strain and is not stress dependent. Not every element in the library can be used for creep analysis. MARC provides a user-supplied subroutine that describes both the uniaxial creep law and the hardening rule. The default capability is steady-state creep with time hardening. Creep swelling is included; therefore, either an associated or a volumetric flow rule may be employed. It is unclear whether the program has the option of auxiliary summation rules.

Convergence and Automatic Increment Control

Again, data are scarce for those programs that are still in the research phase for creep analysis. BERSAFE uses a variant of the Wilson [9] method, calculating a mean stress over the increment for each iteration in an attempt to increase the convergent time step. Most of the experience with this procedure appears to have been gained with the TESS program, however, rather than with BERSAFE. PAFEC appears to use conventional techniques for shortening or lengthening the time increment. MARC has an automatic time increment selection scheme that is based upon user-supplied tolerances on stress change and creep strain increment per elastic strain, again a convential procedure. ANSYS uses either stress change or creep strain increment per elastic strain, depending upon the user's ability to recognize creep or relaxation. No automatic control is available in NEPSAP; the user specifies the time increments by card input. EPACA computes the time step internally based on creep strain increment per elastic strain.

Temperature Distribution

Both ANSYS and MARC have stand-alone transient heat conduction modules that communicate directly to the stress analysis package by magnetic tape or disk. EPACA does not yet have such capability, although development is being considered. ASKA and FESS have modules that can be interfaced directly with stress analysis modules, although documentation on temperature input to the creep packages is not available. BERSAFE has an associated package called FLHE that can generate compatible thermal fields. Information on ASAS and PAFEC is unavailable. Preprocessors are used to prepare the temperature input for NEPSAP from output obtained by aerothermodynamic programs. Insufficient information is available to determine whether interpolation and/or extrapolation of temperature input is possible (since the time steps for thermal diffusion are radically different from creep time steps, in general, this would seem to be an important consideration).

SUMMARY

The trend with regard to creep computation is clear. Because the field is so dominated by high-temperature design of components for nuclear steam supply systems, almost all of the relevant software packages have catered to the user by providing the uniaxial creep law recommended by ORNL (see Eq. (12)), an associated flow rule, strain hardening based on either primary or total equivalent creep strain, and auxiliary hardening rules for the case of load reversal. There is no apparent trend, as yet, to abandon the initial strain method entirely--in favor of hereditary theories or other modified stiffness approaches. Rather, it seems, the initial strain calculations are no longer being based on values of time, temperature, effective stress, and equivalent creep strain at the beginning of the current increment; instead, various averaging (integration) techniques are being applied to the creep rate equation in order to allow larger convergent time steps to be used. If these integration techniques depended upon unknown values, the modified stiffness method would result. The tendency is to use predictor-corrector kinds of algorithms instead, thus preserving the initial strain method.

General purpose programs seem to be gaining in popularity among users and will continue to improve their position. Many are extremely user-oriented, especially in terms of pre- and postprocessing graphics. Notable, in this regard are ANSYS and BERSAFE. A useful adjunct is the compatible heat conduction program with convenient temperature field data transmission to the stress analysis program. Most of the general purpose systems are well equipped in this area.

Two troubling trends can be discerned, however. One is that almost all of the software is proprietary and essentially unavailable to the user, who must depend upon the largesse of the developer when problems occur (e.g., modeling or "real" bugs). On this point, it seems clear that a state-of-the-art general purpose program which is freely available to the user would be universally accepted within the user community. The second problem concerns the indiscriminate comparison of programs, especially for creep, that use radically different algorithms. Modified stiffness programs are often condemned by users because of long running times, whereas initial strain programs are usually praised. The solution to this difficulty is to provide a user option in the program so that, for most problems, initial strain could be flagged. When sensitive results are needed, however, modified stiffness would be available.

ACKNOWLEDGMENT

This work was supported by the United States Atomic Energy Commission.

REFERENCES

1 Boresi, A. P., and Sidebottom, O. M., "Creep of Metals Under Multiaxial States of Stress," Nuclear Engineering and Design, Vol. 18, 1972, pp. 415-456.

2 Finnie, I., "Stress Analysis for Creep and Creep-Rupture," in: Applied Mechanics Surveys, ed. by Abramson, et al, Spartan Books, Washington, D. C., 1966, pp. 373-387.

3 Rashid, Y. R., Tang, H. T., and Johansson, E. B., "Mathematical Treatment of the Mechanical Densification of Reactor Fuel," Proceedings of the 2nd International Conference on Structural Mechanics in Reactor Technology, Berlin, Germany, September 10-14, 1973.

4 Rashid, Y. R., "Part I, Theory Report for CREEP-PLAST Computer Program: Analysis of Two-Dimensional Problems Under Simultaneous Creep and Plasticity," GEAP-10546, AEC Research and Development Report, January 1972.

5 Anon., "Requirements for Design of Nuclear System Components at Elevated Temperatures (Supplement to ASME Code Case 1331)," Volume I, RDT Standard F 9-4, Draft 3, October 1973.

6 Zienkiewicz, O. C., The Finite Element Method in Engineering Science, McGraw-Hill, London, 1971, pp. 395-404.

7 Greenbaum, G. A., and Rubinstein, M. F., "Creep Analysis of Axisymmetric Bodies Using Finite Elements," Nuclear Engineering and Design, Vol. 7, 1968, pp. 379-397.

8 Sharifi, P., and Yates, D. N., "Nonlinear Thermo-Elastic-Plastic and Creep Analysis by the Finite Element Method," AIAA Paper 73-358, AIAA/ASME/SAE 14th Structures, Structural Dynamics, and Materials Conference, Williamsburg, Virginia, 1973.

9 Wilson, E. A., "A Design Oriented Approach to Creep and Plasticity in Finite Element Programs," ASME Paper No. 70-WA/DE-4, Winter Annual Meeting, 1970.

10 Padlog, J., Huff, R. D., and Holloway, G. F., "Unelastic Behavior of Structures Subjected to Cyclic, Thermal and Mechanical Stressing Conditions," Report WPADD-TR-60-271, Bell Aerosystems, Co., December 1960.

11 Greenbaum, G. A., "Radioisotope Propulsion Technology Program (POODLE). Volume IIB - Analytical Program for Creep in Axisymmetric Bodies (CRAB)," Final Report STL-517-0049, TRW Systems, October 1966.

12 Sutherland, W. H., "A Finite Element Computer Code (AXICRP) for Creep Analysis," Report BNWL-1142, Battelle Memorial Institute, October 1969.

13 Sutherland, W. H., "AXICRP - Finite Element Computer Code for Creep Analysis of Plane Stress, Plane Strain and Axisymmetric Bodies," Nuclear Engineering and Design, Vol. 11, 1970, pp. 269-285.

14 Chang, L. K., "Study of the LIFE-I Code," Report ANL-EBR-64, Argonne National Laboratory, January, 1973.

15 Jankus, V. Z., and Weeks, R. W., "LIFE-II - A Computer Analysis of Fast-Reactor Fuel-Element Behavior as a Function of Reactor Operating History," Nuclear Engineering and Design, Vol. 18, 1972, pp. 83-96.

16 Friedrich, C. M., "CYGRO - Stress Analysis of the Growth of Concentric Cylinders," Report WAPD-TM-514, Westinghouse Bettis Atomic Power Laboratory, September, 1965. "CYGRO - I Stress Analysis of the Growth of Concentric Cylinders," Report WAPD-TM-514, Addendum No. 1, Westinghouse Bettis Atomic Power Laboratory, July, 1966.

17 Bard, F. E., Jr., "PECT - I. A Fortran IV Computer Program to Determine the Plastic-Elastic Creep and Thermal Deformations in Thick-Walled Cylinders," Report BNWL-1171, Battelle Memorial Institute, December, 1969.

18 Stillman, W. E., "TRANS - A One-Dimensional Fortran IV Program for Computing the Time Response of Fuel Rods to a Loading-Unloading Environment," Report ORNL-TM-3293, Oak Ridge National Laboratory, May 1971.

19 Puthoff, R. L., "A Digital Computer Program for Determining the Elastic-Plastic Deformation and Creep Strains in Cylindrical Rods, Tubes and Vessels," Report NASA-TM-X-1723, Lewis Research Center, January, 1969.

20 Guyette, M., "CRASH - A Computer Program for the Analysis of Creep and Plasticity in Fuel Pin Sheaths," Report KFK-1050, Institut fur Angewandte Reaktorphysik, August, 1969.

21 Guyette, M., "Cladding-Strength Analysis Under the Combined Effect of Creep and Plasticity in Fast-Reactor Environments," Nuclear Engineering and Design, Vol. 18, 1972, pp. 53-68.

22 Rashid, Y. R., "Part II: User's Manual for CREEP-PLAST Computer Program," GEAP-13262-1, AEC Research and Development Report,

March, 1973.

23 Clinard, J. A., and Crowell, J. S., "ORNL User's Manual for CREEP-PLAST Computer Program," USAEC Report ORNL-TM-4062, Oak Ridge National Laboratory, November, 1973.

24 Cyr, N. A., and Teter, R. D., "Finite Element Elastic-Plastic-Creep Analysis of Two-Dimensional Continuum with Temperature Dependent Material Properties," Computers and Structures, Vol. 3, 1973, pp. 849-863.

25 Donea, J., and Giuliani, S., "Creep Analysis of Transversely Isotropic Bodies Subjected to Time-Dependent Loading," Nuclear Engineering and Design, Vol. 24, 1973, pp. 410-419.

26 Lewis, D. J., and Hellen, T. K., "Analysis Techniques for Elevated Temperature Applications," Paper C233/73, International Conference on Creep and Fatigue in Elevated Temperature Applications, Philadelphia, Pa., September 1973.

27 Vos, R. G., "SSME Structural Computer Program Development, Volume 1: BOPACE Theoretical Manual," NASA-CR-124390, Boeing Aerospace Company, Huntsville, Alabama, July 31, 1973.

28 Zudans, Z., et al, "General Purpose Elastic-Plastic-Creep Finite Element Analysis Program for Three-Dimensional Thick Shell Structures. Theory and Users Manual for EPACA," Final Report F-C3038 to Oak Ridge National Laboratory, Franklin Institute Research Laboratories, Philadelphia, Pa., June 30, 1972.

29 Anon., Thermal Structural Analysis Programs: A Survey and Evaluation, ASME, New York, 1972.

30 Anon., "MARC-CDC Nonlinear Finite Element Analysis Program. Volume I. User Information Manual," Control Data Corporation, Minneapolis, Minnesota (Revised), 1973.

31 Parkes, D. A. C., and Webster, J. J., "Finite Element Solutions for Two Transient Creep Problems," Paper C157/73, International Conference on Creep and Fatigue in Elevated Temperature Applications, Philadelphia, Pa., September, 1973.

THICK SHELLS

K. J. Bathe and E. L. Wilson
Department of Civil Engineering
University of California
Berkeley, California

ABSTRACT

The current solution techniques for general thick shell analysis are surveyed, and available computer programs are evaluated. Desirable future developments in solution techniques and software are discussed.

INTRODUCTION

The analysis of thick shell structures is of considerable interest in various areas of structural engineering. Arch dams, nuclear containment vessels, and certain components of turbines can be idealized as thick shell structures.

Until the advent of the electronic computer and the development of powerful analysis programs, only limited results as to the behavior of shell structures could be obtained. However, the need for accurate structural representation was recognized early [10, 28].

A definition of what may be classed as a thick shell is rather arbitrary. Commonly, a thick shell is considered to be a shell structure which, because of its thickness and radii of curvature, violates some of the assumptions used in thin shell theory. Depending on the severity with which these assumptions are not fulfilled, a thin shell analysis may give good approximate results or predict a rather unrealistic shell response [17, 26].

The assumptions used in classical shell theory are those concerning the geometric approximation of the shell surface, the kinematics on deformation behavior, and the existence of certain shell strains and stresses. In essence, these assumptions are utilized to reduce an intractable three-dimensional problem to a problem which can be formulated in terms of a few governing differential equations, the solution of which represents an approximation to the actual structural response. Realizing that a thick shell may actually behave much like a three-dimensional continuum, the governing differential equations become very complex, and resort to numerical methods for solution is necessary. Today, almost all practical large scale shell analyses are conducted using numerical procedures, and the computer programs available are designed corresponding to the techniques used.

The purpose of this paper is to present a survey of the current software available for thick shell analysis. The analysis of general three-dimensional thick shells is considered, with axisymmetric analysis being a special case. The paper is divided into four parts. First, the general analysis procedures used are briefly reviewed. Next, the desirable requirements of an analysis technique and, therefore, of the corresponding computer program are discussed. This leads to the presentation of the computer programs currently available for general thick shell analysis. In this context, the results of a questionnaire distributed to assess the current software capabilities are presented. Finally, desirable requirements for future developments in the area of shell analysis procedures and corresponding software are summarized.

ANALYSIS PROCEDURES

Essentially three different procedures are currently used for the analysis of shell structures, namely, the conventional finite difference method, the finite difference energy method, and the finite element method. The purpose in this section is to summarize briefly the essential features of each of these analysis techniques with emphasis on their current practical usage in computer programs.

Finite Difference Methods

Prior to the development of the electronic computer much attention was directed to the use of finite difference solutions of the governing differential equations of equilibrium [12,15,18,26]. The main advantage of such analysis, which, in general, is referred to as conventional finite difference method, is the ease with which the governing difference equations can be set up once the appropriate finite difference molecules have been derived. The solution of the equations requires the main effort, and only with the introduction of digital computers could large and complex structures be analyzed [25,29]. The main limitations of the conventional finite difference analysis are the difficulties associated with the derivation of the required finite difference molecules for general boundary conditions, material variations,and applied loads. In addition, numerical difficulties can arise in the solution of the governing sets of difference equations which are not necessarily positive definite. Because of the lack of generality and solution difficulties, the conventional finite difference method has practically been abandoned for general structural analysis.

During the last years a quite different and very promising finite difference method in which the energy functional is approximated by finite differences has obtained new impetus [7,15]. This procedure, referred to as finite difference energy method, has been applied successfully to the analysis of thin shells. The main advantage of the finite difference energy method is that difference expressions of much lower degree can be used for approximating the expressions in both the energy functional and the boundary conditions. Currently, however, only a restricted number of programs are based on this analysis technique.

Finite Element Method

Concurrent with the advent of the electronic computer a very versatile numerical procedure, namely, the finite element method, was developed. In this technique, the actual physical structure is idealized by an assemblage of finite elements, interconnected at nodal points. The generality of structural shapes, displacements, material properties, and boundary conditions that can be appropriately idealized in a finite element analysis made the procedure very attractive, and currently most general purpose structural analyses computer programs are based on this technique. The basic steps followed in a finite element analysis are well documented [3,14,28], and regarding thick shell analysis, attention need only be focused on the characteristics of the specific element used.

Various finite elements have been proposed for thin shell analysis which may account to some degree for the thickness of the shell if a shear deformation mechanism is introduced [9,11,19]. Displacement, mixed, and hybrid models have been developed, but mainly displacement models are used in practical analysis [23,28].

An effective way of analyzing thick shells is to employ isoparametric elements or one of their degenerate forms. The concept of isoparametric elements was originally introduced by Irons and successfully applied and extended by various researchers (for a list of references see the book by Zienkiewicz [28]). The basis of isoparametric element analysis is the interpolation of coordinates and displacements by the same interpolation functions, written in terms of the natural coordinates of the element. The generality and flexibility of isoparametric finite elements as regards computer implementation and usage is best realized by briefly focusing attention on the construction of the interpolation functions used in three-dimensional and thick shell analysis.

Figure 1 shows a general, three-dimensional element with its natural coordinate system r,s,t. The coordinate and displacement interpolations for this element are, respectively,

$$\left.\begin{aligned} X_i &= \sum_{k=1}^{N} h_k\,(r,s,t)\,X_i^{\,k} \\ u_i &= \sum_{k=1}^{N} h_k\,(r,s,t)\,u_i^{\,k} \end{aligned}\right\}\quad i = 1,2,3 \tag{1}$$

in which h_k = k'th interpolation function, $X_i^{\,k}$, $u_i^{\,k}$ = coordinate and displacement corresponding to direction i of nodal point k. The element may have any number of nodes between 8 and 20, with the corresponding interpolation functions defined by

$$h_1 = g_1 - (g_9 + g_{13} + g_{19})/2$$

$$h_2 = g_2 - (g_{10} + g_{13} + g_{18})/2$$

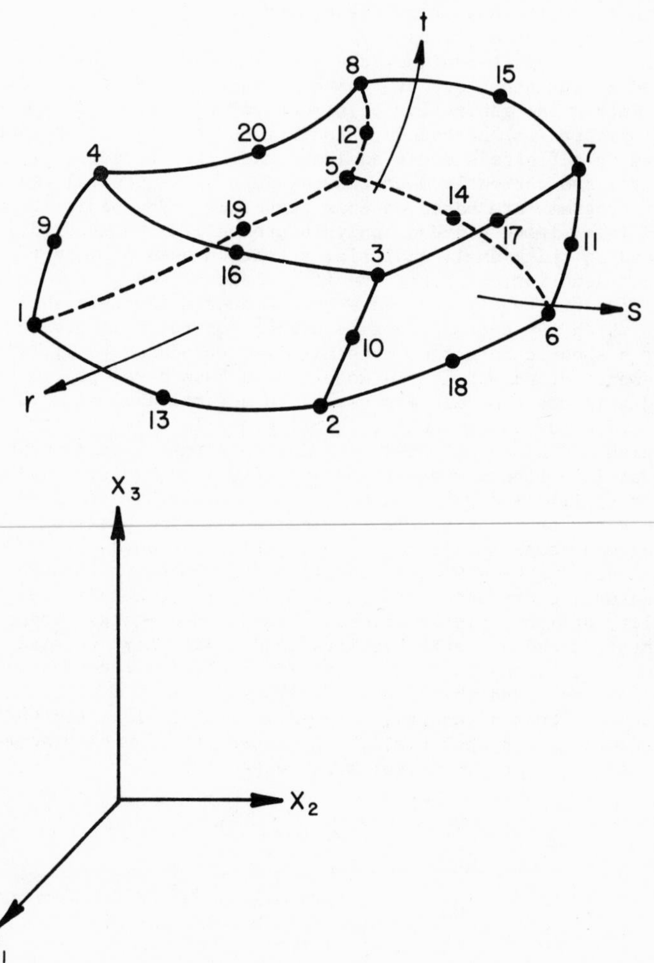

Fig. 1 Eight to twenty variable-number-nodes three-dimensional element

$$\left.\begin{aligned}
h_3 &= g_3 - (g_{10} + g_{16} + g_{17})/2 \\
h_4 &= g_4 - (g_9 + g_{16} + g_{20})/2 \\
h_5 &= g_5 - (g_{12} + g_{14} + g_{19})/2 \\
h_6 &= g_6 - (g_{11} + g_{14} + g_{18})/2 \\
h_7 &= g_7 - (g_{11} + g_{15} + g_{17})/2 \\
h_8 &= g_8 - (g_{12} + g_{15} + g_{20})/2 \\
h_j &= g_j \quad \text{for } j = 9, 10, 11, \ldots, 20
\end{aligned}\right\} \tag{2}$$

where

$$g_i = G(r, r^i)\, G(s, s^i)\, G(t, t^i) \tag{3}$$

and

$$\begin{aligned}
G(\beta, \beta_i) &= 1/2\,(1 + \beta_i\, \beta) \quad \text{for } \beta_i = \pm 1 \\
G(\beta, \beta_i) &= (1 - \beta^2) \qquad\quad \text{for } \beta_i = 0
\end{aligned} \tag{4}$$

If any of the nodes 9 to 20 are omitted, the corresponding value of g is zero.

In analogy to the procedure above for writing the required interpolation functions, elements with practically any number of nodes can be defined in essentially the same computer program. However, in practice, usage is still restricted to lower-order elements, simply because the computer time required for the formation of higher-order isoparametric elements is still relatively large.

The 16- or 20-node element derived from the general variable-number-nodes element in Fig. 1 is commonly used in three-dimensional and thick shell analysis. On examining the deformation patterns of these elements, it is observed that constant bending moments can be represented, but that for a linearly varying moment excessive shear strain energy is stored. In order to improve the element behavior, selective or reduced order Gauss integration can be used, or incompatible modes may be employed [2,21,22,27,30]. Selective integration has found limited application in general analysis programs, and incompatible modes need to be used with care, but can be very effective [5,10]. In most cases the element is simply used with a lower-order integration rule, in order to avoid integrating the excess shear strain energy.

To obtain improved results, a cubic isoparametric element can be used in three-dimensional and thick shell analysis. Figure 2 shows the general cubic element, which can again be employed as a variable-number-nodes element; i.e., a reduced number of nodes would reasonably

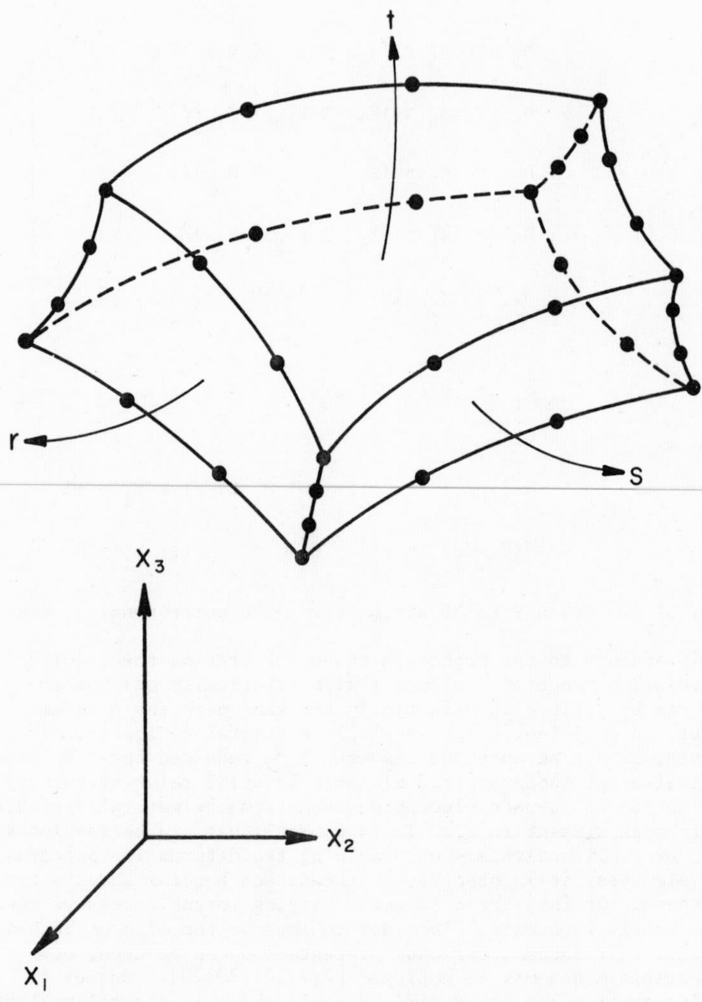

Fig. 2 Cubic finite element for three-dimensional and thick shell analysis

be used through the thickness of the shell element. This element has the capability of representing linearly varying moments. The main objection to employing a cubic element in general three-dimensional analysis is the relatively large element formation time required. Furthermore, if the element is relatively thin; i.e., the element is used in thick to thin shell analysis, the stiffness coefficients corresponding to the transverse displacement degrees of freedom are considerably larger than those corresponding to the longitudinal displacements, which results in numerical ill-conditioning of the assembled equations. This can be circumvented by using relative displacement degrees of freedom through the thickness of the element as indicated in Fig. 3. However, since the relative displacements are small and do not contribute significantly to the overall response of the shell, it is more effective to employ a super-parametric element for shell analysis [1,2,21,28]. The basic assumptions in the development of the degenerate element from the isoparametric element in Fig. 3, as originally proposed by Ahmad et al., are that the normals to the middle surface are taken as inextensible and straight and that the elastic modulus in the direction of the normals is taken as zero, in order not to include normal stresses [1]. Since the surface normals can rotate relative to the middle surface of the shell, shear deformations are approximately included. It must be realized that degenerate isoparametric elements can only be used in essentially thin shell analysis with moderate shear deformations. For this reason and because of the flexibility to account for thick shell and three-dimensional behavior using the same element, in most thick shell analyses the 16- or 20-node elements shown in Fig. 1 are employed.

DESIRABLE PROGRAM CHARACTERISTICS

Before actually presenting the current programs available it is of value to summarize the desirable characteristics of a thick shell analysis program.

Geometry and Materials

An important requirement of a program is that the geometry, material properties, boundary conditions, and loading on the shell can be appropriately represented. This condition requires a considerable flexibility of the program since the shell may behave almost like a thin shell in one part of the structure and like a fully three-dimensional continuum in another part. In addition, orthotropy of material and material variation through the thickness and along the surface of the shell may need to be considered.

Linear and Nonlinear Analyses

In most cases a linear static or dynamic analysis is required. In dynamic analysis, low mode response is usually predominant, and mode

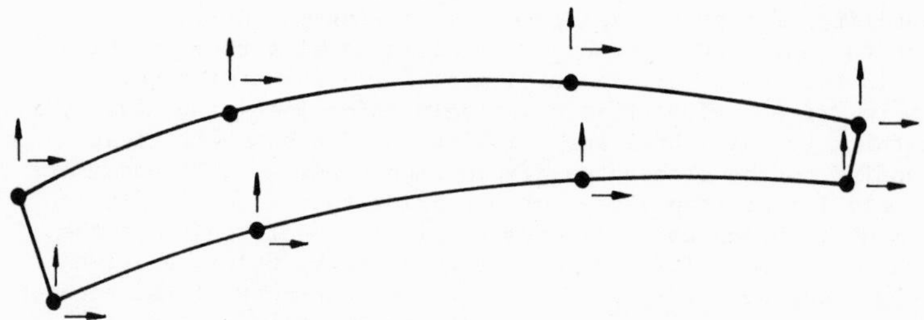

Original isoparametric element

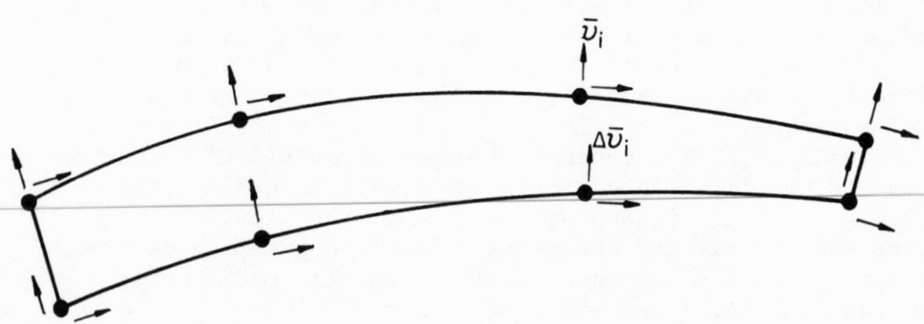

Isoparametric element using relative displacement degrees of freedom

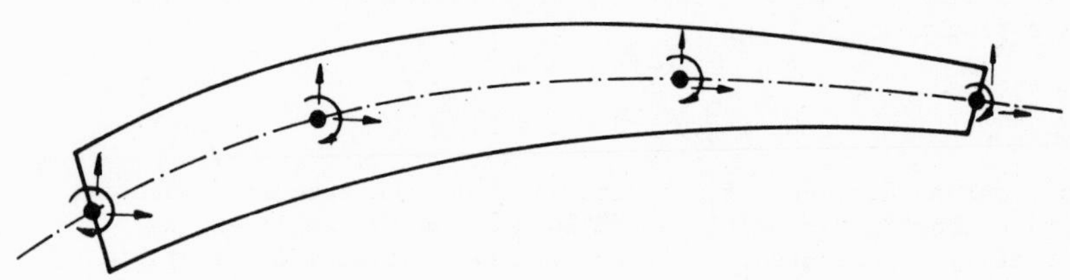

Superparametric Element

Fig. 3 Displacement degrees of freedom definitions for thick and thin shell analysis

superposition using effective numerical techniques for the calculation of frequencies and mode shapes is important [8]. Direct integration could be effective if a short duration response need be predicted [8].

Considering nonlinear behavior of thick shell structures, the primary nonlinearities are due to nonlinear material behavior. Geometric nonlinearities can be important when the shell becomes thin. Therefore, buckling analyses of thick shell structures are, in practice, not very frequent. However, it may be important to have various nonlinear material models available to account for plastic and creep effects. In most cases only nonlinear static analyses are considered, since, firstly, nonlinear dynamic thick shell analyses are still prohibitively expensive and, secondly, the results are often questionable.

Preprocessing and Postprocessing

An important aspect for program usage is the capability for pre- and postprocessing. Effective handling of output is particularly important in nonlinear analysis, since a large amount of data is calculated. In addition, the definition of nodal point and element data without generation capability can be tedious because the nodal points and elements are lying in three-dimensional space. Included in the preprocessing phase should be extensive data checking with meaningful error messages.

Program Stability

A most disappointing aspect is often the lack of program stability. A variety of options may exist; however, considerable care may be required in their usage. Good program stability results from stable numerical techniques and effective program implementation. Also, care must be taken in the installation of a program on machines other than the one on which the program was originally developed. In this context, the need for double precision arithmetic on IBM, UNIVAC, and other machines with relatively few (say, less than ten) characters per word need be realized. Thick shell problems can be ill-conditioned, and truncation errors due to not using enough digits can severely affect the results [5,24].

CURRENT PROGRAMS AVAILABLE

In the course of this survey a questionnaire was sent out in order to establish the current program capabilities for thick shell analysis. The questions posed were based on the considerations discussed in the preceding section.

Tables 1 to 3 summarize the response obtained on the questionnaire. The general program characteristics are presented in Table 1, whereas Tables 2 and 3 summarize the program capabilities in the

Table 1 Summary of Thick Shell Analysis Programs

Program name	Geometry and boundary conditions		Loading			Analysis type			
	Axisymmetric	3-D general	Axisymmetric	General	Temperature	Linear static	Nonlinear static	Linear dynamic	Nonlinear dynamic
ADAP	No	√	No	√	√	√	No	√	No
ANSYS	√	√	√	√	√	√	√	√	√
ASEF & DYNAM	√	√	√	√	√	√	√	√	No
ASKA	√	√	√	√	√	√	No	√	No
BERSAFE	√	√	√	√	√	√	No	No	No
DANUTA	√	√	√	√	√	√	√	√	No
ELAS	√	√	√	√	√	√	√	No	No
MAGIC III	√	√	√	√	√	√	√	√	No
MARC	√	√	√	√	√	√	√	√	√
NASTRAN	√	√	√	√	√	√	√	√	√
SAP IV	√	√	√	√	√	√	No	√	No
SESAM-69	√	√	√	√	√	√	No	√	No

(√ = Yes)

Table 1 (cont.)

Program name	Material variation				Associated heat transfer program available
	Isotropic	Orthotropic	Variation through thickness	Variation along surface	
ADAP	√	√	√	√	No
ANSYS	√	√	√	√	√
ASEF & DYNAM	√	√	√	√	No
ASKA	√	No	√	√	No
BERSAFE	√	No	√	√	√
DANUTA	√	No	√	√	No
ELAS	√	√	√	√	No
MAGIC III	√	√	√	√	No
MARC	√	√	√	√	√
NASTRAN	√	√	√	√	√
SAP IV	√	No	√	√	No
SESAM-69	√	No	√	√	√

(√ ≡ Yes)

Table 2 Program Nonlinear Analysis Capabilities

Program name	Linearized static buckling analysis	Post-buckling analysis	Large displacement effects	Large strain effects	Material nonlinearities		
					Isothermal plasticity	Nonisothermal plasticity	Creep
ADAP	No	No	No	No	No	No	No
ANSYS	No	No	√	√	√	√	√
ASEF & DYNAM	√	No	No	No	No	No	No
ASKA	No	No	No	No	No	No	No
BERSAFE	No	No	No	No	No	No	No
DANUTA	No	No	No	No	√	No	No
ELAS	No	No	No	No	√	√	√
MAGIC III	√	No	No	No	No	No	No
MARC	√	√	√	√	√	√	√
NASTRAN	√	No	No	No	√	√	No
SAP IV	No	No	No	No	No	No	No
SESAM-69	No	No	No	No	No	No	No

(√ ≡ Yes)

Table 3 Linear Dynamic Analysis Capabilities

Program name	Frequency analysis	Response spectrum analysis	History analysis by mode superposition	History analysis by direct integration
ADAP	✓	✓	✓	No
ANSYS	✓	✓	No	✓
ASEF & DYNAM	✓	No	✓	✓
ASKA	✓	No	✓	✓
BERSAFE	No	No	No	No
DANUTA	✓	✓	✓	No
ELAS	No	No	No	No
MAGIC III	✓	No	No	No
MARC	✓	No	✓	✓
NASTRAN	✓	✓	✓	✓
SAP IV	✓	✓	✓	✓
SESAM-69	✓	✓	✓	No

(✓ ≡ Yes)

specific areas of nonlinear analysis and linear dynamic analysis, respectively. It is noted that only general purpose programs currently available are listed in the tables. The programs have special axisymmetric analysis capabilities. In addition, a large number of special purpose programs for axisymmetric analysis only are available [6,13,20] (see Appendix). The addresses at which additional and detailed information about the programs in Tables 1 to 3 can be obtained are given in the Appendix.

It should be mentioned that all programs for which the questionnaire was returned employ the displacement-based finite element method. This enables the use of the programs for relatively complicated shell geometries and material variations, although the shell element itself may be simple; i.e., by using a sufficient number of elements along and through the surface of a shell, supposedly arbitrary material variations can be approximated. However, element compatibility requirements need be considered, and it should be realized that the cost of analysis can become very large as the number of elements is increased [28].

Tables 1 to 3 seem to indicate that practically any thick shell structure can currently be analyzed without difficulties. In fact, however, all programs must be used with care, and in some analyses a great deal of experience with the program and the analysis techniques employed is necessary.

It may require considerable experience with a program, even for linear analysis of a specific shell structure, before the analyst can establish, a priori, a cost-effective finite element idealization [10]. In particular, a great deal of experience is required for nonlinear shell analyses, principally because certain solutions are not yet possible on a routine basis [4,19]. It is, therefore, important to understand the theoretical basis of a program and to apply the program only under the conditions and assumptions for which it was developed.

DESIRABLE FUTURE DEVELOPMENTS

Important improvements in the computer programs currently available for thick shell analysis are to be envisaged concerning various aspects.

Ideally, one program can be used for the analysis of thin and thick shells and three-dimensional continua. Recalling that thin and thick shells are merely defined in order to reduce the analysis effort required by treating the shell as a three-dimensional continuum, it follows that by improving three-dimensional analysis capabilities, general shell analysis is also enhanced. This is particularly the case in nonlinear analysis, because appropriate kinematic and static assumptions on the behavior of the shell may be difficult or impossible to make. Reference should here be made to the comparative study of nonlinear shell solutions reported by Hartung and Ball [19], in which the nonlinear results obtained using various programs showed virtually no agreement. A general variable-number-nodes finite element can currently be used for general shell and three-dimensional analysis; however, the use of higher order elements, particularly in nonlinear analysis, is still very costly.

Considering analysis capabilities for nonlinear shell behavior, a great deal more research is required. Assuming that the material can be identified appropriately, the main problems lie in obtaining stable and economical solution algorithms. At the present time, much experimenting must often be performed by the analyst in order to obtain a reliable solution, which can be expensive. This experimenting is largely due to inadequate numerical techniques, since in most cases a physical solution is well known to exist. The numerical algorithms must be based on a consistent nonlinear formulation, stable constitutive relations, and appropriate computer implementation. It is this interaction between the disciplines of continuum mechanics, numerical analysis, and computer hardware usage that makes the development of effective nonlinear analysis programs a great deal more difficult than may be realized.

CONCLUSIONS

The current solution techniques and available general programs for thick shell analysis have been surveyed. The programs use the displacement-based finite element method.

A relatively large number of computer programs is available for linear analysis. Linear static solutions can be obtained without much difficulty, but a considerable amount of experience may still be required to perform an effective linear dynamic analysis.

Programs are also available for nonlinear thick shell analysis. However, a great deal of experience with the structure under consideration, the program used, and its theoretical basis may be required in order to obtain a valid solution. Since increasing emphasis is placed on nonlinear analysis, it is primarily in this area that a considerable development is necessary.

REFERENCES

1 Ahmad, S., Irons, B. M., and Zienkiewicz, O. C., "Curved Thick Shell and Membrane Elements with Particular Reference to Axisymmetric Problems," Proceedings Second Conference on Matrix Methods in Structural Mechanics, Wright Patterson AFB, Ohio, 1968.

2 Ahmad, S., Irons, B. M., and Zienkiewicz, O. C., "Analysis of Thick and Thin Shell Structures by Curved Elements," International Journal for Numerical Methods in Engineering, Vol. 2, 1970, pp. 419-451.

3 Argyris, J. H., "Continua and Discontinua," Proceedings Conference on Matrix Methods in Structural Mechanics, Wright Patterson AFB, Ohio, 1965.

4 Bathe, K. J., Ozdemir, H., and Wilson, E. L., "Static and Dynamic Geometric and Material Nonlinear Analysis," Report No. SESM 74-4, University of California, Department of Civil Engineering, Berkeley, February 1974.

5 Bathe, K. J., Wilson, E. L., and Peterson, F. E., "SAP IV - A Structural Analysis Program for Static and Dynamic Response of Linear Systems," Report No. EERC 73-11, College of Engineering, University of California, Berkeley, June 1973.

6 Benzley, S. E., "TACOS - A Finite Element Computer Program for the Transient Analysis of Cylindrical Obese Shells," Report No. SC-RR-72 0454, Analytical Development Division, Sandia Laboratories, Albuquerque, 1972.

7 Bushnell, D., Almroth, B. O., and Brogan, F. A., "Finite Difference Energy Methods for Nonlinear Shell Analysis," Journal Computers and Structures, Vol. 1, 1972, pp. 361-387.

8 Clough, R. W., and Bathe, K. J., "Finite Element Analysis of Dynamic Response," Proceedings 2nd U.S.-Japan Symposium on Recent Advances in Computational Methods of Structural Analysis and Design, Berkeley, California, 1972.

9 Clough, R. W., and Felippa, C. A., "A Refined Quadrilateral Element for Analysis of Plate Bending," Proceedings Conference on Matrix Methods in Structural Mechanics, Wright Patterson AFB, Ohio, 1968.

10 Clough, R. W., Raphael, J. M., Mojtahedi, S., "ADAP - A Computer Program for Static and Dynamic Analysis of Arch Dams," Report No. EERC 73-14, College of Engineering, University of California, Berkeley, June 1973.

11 Clough, R. W., and Wilson, E. L., "Dynamic Finite Element Analysis of Arbitrary Thin Shells," Journal Computers and Structures, Vol. 1, No. 1, 1971, pp. 33-56.

12 Collatz, L., The Numerical Treatment of Differential Equations, Springer-Verlag, Berlin, 1967.

13 Crose, J. A., and Jones, R. M., "SAAS III - Finite Element Stress Analysis of Axisymmetric and Plane Solids with Different Orthotropic, Temperature-Dependent Material Properties in Tension and Compression," Air Force Report No. SAMSO-TR-71-103, San Bernardino Operations, The Aerospace Corporation, San Bernardino, California.

14 Felippa, C. A., "Refined Finite Element Analysis of Linear and Nonlinear Two-Dimensional Structures," Report No. SESM 66-22, Department of Civil Engineering, University of California, Berkeley, 1966.

15 Felippa, C. A., "Finite Element and Finite Difference Energy Technique for the Numerical Solution of Partial Differential Equations," Proceedings Computer Simulation Conference, Montreal, July 1973.

16 Felippa, C. A. and Sharifi, P., "Computer Implementation of Nonlinear Finite Element Analysis," Proceedings ASME Symposium, Detroit, November 1973.

17 Flügge, W., Stresses in Shells, Springer-Verlag, Berlin, 1960.

18 Forsythe, G. E., and Wasow, W. R., Finite Difference Methods for Partial Differential Equations, John Wiley and Sons, New York, 1960.

19 Hartung, R. F., and Ball, R. E., "A Comparison of Several Computer Solutions to Three Structural Shell Analysis Problems," Report No. AFFDL-TR-73-15, Wright Patterson AFB, Ohio, 1973.

20 Klein, S., "The SABOR Manual," Aerospace Report No. ATR-72(S9990)-2, San Bernardino Operations, The Aerospace Corporation, San Bernardino, California.

21 Pawsey, S. F., "The Analysis of Moderately Thick to Thin Shells by the Finite Element Method," Report No. SESM 70-12, Department of Civil Engineering, University of California, Berkeley, August 1970.

22 Pawsey, S. F., and Clough, R. W., "Improved Numerical Integration of Thick Shell Finite Elements," International Journal for Numerical Methods in Engineering, Vol. 3, 1971, pp. 575-586.

23 Pian, T. H. H., "Formulations of Finite Element Methods for Solid Continua," Proceedings 1st U.S.-Japan Symposium on Recent Advances in Matrix Methods of Structural Analysis, Tokyo, Japan, 1968.

24 Roy, J. R., "Numerical Errors in Structural Solutions," ASCE Journal of the Structural Division, April 1971.

25 Szilard, R., "A Matrix and Computer Solution of Cylindrical Shells of Arbitrary Shape," Proceedings International Symposium on Shell Structures, Budapest, Hungary, September 1965.

26 Timoshenko, S. P., and Woinowski-Krieger, S., Theory of Plates and Shells, McGraw Hill, London, 1959.

27 Wilson, E. L., Taylor, R. L., Doherty, W., and Ghaboussi, J., "Incompatible Displacement Models," Proceedings ONR Symposium on Numerical Methods in Engineering, University of Illinois, Urbana, 1971.

28 Zienkiewicz, O. C., The Finite Element Method in Engineering Science, McGraw Hill, London, 1971.

29 Zienkiewicz, O. C., and Cheung, Y. K., "Finite Element Method of Analysis for Arch Dam Shells and Comparison with Finite Difference Procedures," Proceedings Symposium on Theory of Arch Dams, Southampton University, 1964.

30 Zienkiewicz, O. C., Taylor, R. L., and Too, J. M., "Reduced Integration Technique in General Analysis of Plates and Shells," International Journal for Numerical Methods in Engineering, Vol. 3, 1971, pp. 275-290.

APPENDIX

Addresses for Obtaining Information on the Programs Listed in Tables 1 to 3

Program ADAP Cost: $250.00	NISEE, 720 Davis Hall University of California Berkeley, California 94720 Attn: Ken Wong
Program ANSYS Cost: negotiable	John A. Swanson Swanson Analysis Systems, Inc. 870 Pine View Crive Elizabeth, Pa. 15037
Programs ASEF and DYNAM Cost: $50,000.00	G. J. Sander Laboratoire d'Aeronautique Universite de Liege 75 Rue du Val Benoit Liege 4000, BELGIUM
Program ASKA Cost: negotiable	Horst Parisch Institut für Statik und Dynamik der Luft-und Raumfahrtkonstruktionen Universitat Stuttgart 7 Stuttgart 80 Pfaffenwaldring 27 WEST GERMANY
Program BERSAFE Cost: ₤ 7000	Computing Bureau Central Electricity Generating Board 85 Park Street London SE1, UNITED KINGDOM
Program DANUTA Cost: negotiable	S. A. Chacour Allis-Chalmers Corp. Hydro Turbine Division Box 712 York, Pa. 17405
Program ELAS Cost: $1,000.00 (ELAS65)	Computer Structural Analysis Fund Duke University Civil Engineering Department Durham, N.C. 27706

Program MAGIC III Cost: negotiable	Bell Aerospace Company or U.S.A.F. Flight Dynamics Lab. Structures Division FBR, WPAFB, Ohio 45433
Program MARC Cost: negotiable	H. D. Hibbitt MARC Analysis Research Corp. 105 Medway Street Providence, R.I. 02906
Program NASTRAN Cost: $1700.00	Richard H. MacNeal The MacNeal-Schwendler Corp. 7442 N. Figueroa Street Los Angeles, California 90041 Also: COSMIC INFORMATION SERVICES 112 Barrows Hall The University of Georgia Athens, Georgia 30602 Attn: E. C. Martin
Program SAP IV Cost: $300.00	NISEE 720 Davis Hall University of California Berkeley, California 94720 Attn: Ken Wong
Program SESAM-69 Cost: negotiable	GEOCOM 2122 Governors Circle Houston, Texas

Further requests on the availability of a required program may be placed to:

(1) COSMIC INFORMATION SERVICES
112 Barrows Hall
The University of Georgia
Athens, Georgia 30602

Attn: E. C. Martin

(2) NISEE
720 Davis Hall
University of California
Berkeley, California 94720

Attn: Ken Wong

PIPING SYSTEMS

W. B. Wright
Arthur D. Little, Inc.
Cambridge, Massachusetts

ABSTRACT

This article surveys computer programs used for piping analysis and presents detailed information on the basis of analytical capability to aid the user in selecting programs. To perform this function the article emphasizes program capability by describing types of problems that may be solved and input and output, which are normally strong points of user interest. This emphasis automatically relegates to a lesser position of importance the numerical techniques of solution, the mathematical formulation, and the amenities of the computer system used. The programs discussed here range from the very special purpose to the very general purpose. The special purpose programs are those where the pre- and postprocessing are an integral part of the main program, usually to the point of being indistinguishable. The general purpose programs' analytical capabilities with regard to piping analysis are highlighted here. These programs are characterized by their ability to solve a broad class of problems which includes piping analysis as one capability. More detailed attention of the analyst to the pre- and postprocessing is required because of this general capability. Other capabilities of these general purpose programs are reported elsewhere in this book.

INTRODUCTION

There are several steps in evaluating a computer program for acceptance and use. The first step in answering the rather fundamental question, What program(s) is available to solve my problem? is the subject of the body of this article. To aid in answering this question, information is presented in three sections. The first section discusses a standard format used for tabular comparison of programs and then presents the program capability using this tabular form. The second section lists the names and addresses of those who have supplied program data and who may be contacted for information not included in the first section's standard format. The third section briefly discusses some further steps that are appropriate to program evaluation by the user.

STANDARD COMPARISON OF PROGRAMS

The basic comparison of programs is provided in Table 1 by eight main categories with additional descriptive subcategories that are of immediate interest to the person evaluating a program. These main cate-

Table 1 Piping Program Capability Check List

PROGRAM NAME	ADLPIPE	ANSYS	DYNAL	FESAP	ISOPAR SHL	MARC	MEC 21 1	MEC 21 2	MEC 21 3	MEL 21 1	MEL 21 2	MEL 40	MSC NASTRAN	PIPDYN II	PIPESD	PIRAX I	SAP IV	SAP V
TYPES OF LOADING:																		
STATIC: Thermal	X	X		X		X	X	X	X	X	X	X	X	X	X	X	X	X
Deadweight	X	X		X	X	X	X	X	X	X	X	X	X	X	X	X	X	X
Externally Applied Loads	X	X		X	X	X	X	X	X	X	X	X	X	X	X	X	X	X
DYNAMIC: Response Spectra	X	X	X	X		X							X	X	X		X	X
Time History-Linear	X	X	X			X							X	X			X	X
Time History-Non-linear:																		
Elastic		X		X		X							X					X
Elastic-Plastic		X				X												
TRANSIENT THERMAL: Linear Response		X			X	X	X						X	X		X		X
Non-Linear Response		X				X							X			X		
Plasticity/Creep Response		X				X										X		
MODELING CAPABILITY:																		
STRAIGHT PIPE as a: Beam	X	X	X	X		X	X	X	X	X	X	X	X	X	X	X	X	X
Shell		X		X	X	X											X	
ELBOW as a: Modified Curved Beam	X	X	X			X	X	X	X	X	X	X		X	X	X	X	X
Shell		X		X	X	X												
TEES as a: Beam Intersection		X	X			X	X	X	X		X				X	X		
Modified Beam Intersection	X	X		X		X								X				X
Intersection of Two Shells		X		X	X	X												
SPECIAL STIFFNESS ELEMENTS: 6 x 6 Stiffness Matrix:																		
To Ground	X	X	X			X	X			X	X		X	X		X	X	X
Between Elements		X	X			X	X			X	X		X					X
Spring Hangers:																		
To Ground	X	X	X			X	X	X	X	X	X	X	X	X	X	X		X
Between Elements	X	X	X			X	X		X	X	X		X					X
ANCHOR (GROUND) RESTRAINTS: Partial Restraint	X	X	X	X		X	X	X	X	X	X	X	X	X	X	X	X	X
Full Restraint	X	X	X	X	X	X	X		X	X	X	X	X	X	X	X		X
SKEW (GROUND) RESTRAINT: Guided	X	X	X	X	X	X	X	X	X	X	X		X	X	X	X	X	X
PRE-PROCESSING:																		
Input Data Preparation: Specifically for Piping	X	X				X	X	X	X			X		X	X	X	X	X
General		X	X	X	X	X			X					X			X	X
Input Data Error Diagnostics: Non-fatal	X	X	X				X				X	X	X		X		X	X
Fatal	X	X	X		X	X	X	X	X	X	X	X	X	X	X		X	X
Plotting: Dimensioned Isometrics	X	X		X		X	X	X	X	X	X	X		X			X	X
Dimensioned Orthographics	X	X		X										X	X		X	X
Stress Isometrics	X			X			X										X	
INTERMEDIATE DATA:																		
Input data printed in interpreted form	X	X	X	X	X		X	X	X	X	X	X		X	X	X	X	X
Mathematical manipulations may be called for evaluation	X	X	X			X	X			X	X	X	X			X	X	X
Forces and moments are printed: Static	X	X		X	X	X	X	X	X	X	X	X	X	X	X	X	X	X
Time History	X	X	X	X		X							X	X		X	X	X
Deflections and rotations are printed: Static	X	X		X	X	X	X	X	X	X	X	X	X	X	X	X	X	X
Time History	X	X	X	X		X							X	X		X	X	X
Non-linear accumulated strains/deflections are printed		X														X		

NOTE: (X) mark denotes capability.

Table 1 (cont.)

PROGRAM NAME	ADLPIPE	ANSYS	DYNAL	FESAP	ISOPAR SHL	MARC	MEC 21			MEL 21		MEL 40	MSC NASTRAN	PIPDYN II	PIPESD	PIRAX I	SAP IV	SAP V
POST-PROCESSING:																		
STRESSES are computed to meet:																		
B31.1.0 Power Piping	X	X							X	X	X	X		X	X			X
B31.2 Fuel Gas Piping																		
B31.3 Petroleum Refinery Piping							X					X						
B31.5 L. P. Transportation Piping												X						
B31.6 Chemical Process Systems																		
B31.7 Nuclear Power Piping	X	X								X				X	X			X
B31.8 Gas Transmission & Distribution Piping																		
ASME Section III, Nuclear Components Code, Class 1	X	X												X	X			X
Class 2	X	X												X	X			X
Class 3	X	X																X
STRESSES are computed as a tensor					X	X												
PLOTTING: Static Deformation		X		X		X	X		X	X			X	X				X
Dynamic: Mode Shapes	X	X	X	X		X							X	X				X
Displacement Time History		X	X	X		X							X				X	
Strain Time History		X				X												
Stress Time History	X		X		X								X					
Force Time History		X	X			X							X				X	
GENERAL INFORMATION:																		
TOTAL PROGRAM CAPABILITY: Reported Here	X	X			X	X			X		X			X		X	X	
Reported Elsewhere in Book		X	X	X		X	X								X		X	
DOCUMENTATION LEVEL:																		
Data Preparation Manual	X	X	X	X	X	X	X	X	X	X	X	X	X	X	X	X	X	X
Programmer's Manual				X		X				X	X		X	X	X	X	X	
Mathematical Formulation	X	X	X			X	X		X	X	X	X	X	X	X	X	X	X
Numerical Techniques	X		X			X	X						X			X	X	X
Comparative Sample Problem Solutions	X	X	X			X	X	X	X	X	X	X	X	X	X	X	X	X
Listing Available	X			X				X	X	X	X			X		X	X	
OPERATIONAL ON:																		
IBM	X	X	X	X		X	X	X	X	X		X	X	X				
CDC	X	X				X			X	X	X		X	X	X	X	X	X
UNIVAC	X	X			X	X		X	X				X	X				
BURROUGHS																		X
GE																		
Other	X	X							X									
PROGRAM DEVELOPMENT:																		
The program is under active development	X	X	X		X	X	X		X						X	X	X	X
The program is actively maintained	X	X	X	X		X	X		X	X			X	X			X	X
PROGRAM AVAILABILITY:																		
Federal Software Centers: COSMIC								X	X			X						
ARGONNE																		
Commercially Available: Purchase						X		X						X			X	
License	X	X	X			X							X	X			X	
Utility Network (use basis)	X	X	X			X		X	X					X	X			
The program is not available outside reporting company				X	X		X		X									X

NOTE: **(X)** mark denotes capability.

gories are described briefly in the following paragraph.

The "Types of Loading" category describes the time-independent (static) loadings and the time-dependent (dynamic, thermal) loadings of usual interest to the analyst. The "Modeling Capability" describes the structural elements (pipe elements, elbows, and so on) that are available to describe a piping system and the types of constraints (partial, skew, and so on) that may be placed on the system. The "Preprocessing" is of practical importance, particularly where large numbers of analyses are being performed; and this section is intended to give an indication of preprocessing capability. To cover this important consideration adequately the analyst is probably well advised to obtain a detailed input description of the several candidate programs. "Intermediate Data," coupled with preprocessor error diagnostics, provides one of the best methods for evaluating input data and intermediate mathematical manipulations and computations. Normally, the intermediate data may be selectively requested by control cards so the analyst can retrieve selected meaningful data for further scrutiny. "Postprocessing" normally contains the purposeful information of the analysis which is used to determine the adequacy of the design by evaluating load, deflection, and stress values against established criteria. It is difficult to overemphasize the value of plotting select data so that a visual image can be formed. "Documentation Level" is important in understanding the program's established and documented level of reliability. This subject is discussed in detail in the third section and reflects some of the author's personal convictions. "Operational-On" indicates the computer system's generic name by hardware manufacturers and does not attempt to define the machine versions on which the program has been implemented. This information does provide an indication of the program's machine dependence or independence. "Program Development" and "Program Availability" listings provide a quick assessment of the program's current activity and the method(s) of procurement.

PROGRAM DEVELOPERS

The nuances and latest development of a computer program are best described by one who is intimately involved in its development and use. This section lists names and addresses of the developers/users who can be contacted for further specific information.

PROGRAM	CONTACT FOR FURTHER INFORMATION
ADLPIPE	I. W. Dingwell (617)864-5770 Arthur D. Little, Inc. 20 Acorn Park Cambridge, Massachusetts 02140
ANSYS	J. A. Swanson (412)751-1940 Swanson Analysis Systems, Inc. 870 Pine View Drive Elizabeth, Pennsylvania 15037

PROGRAM	CONTACT FOR FURTHER INFORMATION	
DYNAL	A. Y. Cheung Ontario Hydro 620 University Avenue Toronto 2, Ontario Canada	(416)368-6767
FESAP	D. Van Fossen Babcock & Wilcox Company 1562 Beeson Street Alliance, Ohio 44601	(216)821-9110
ISOPAR-SHL	A. Gupta Sargent & Lundy 140 S. Dearborn Chicago, Illinois 60603	(312)346-7600
MARC	Dr. H. D. Hibbitt MARC Analysis Research Corporation 105 Medway Street Providence, Rhode Island 02906	(401)751-9120
MEC 21 (1)	P. Baker Computing Department Phillips Petroleum Company 7th Floor, Adams Building Bartlesville, Oklahoma 74004	(918)661-6187
MEC 21 (2)	R. V. Cramer PFACS 5 Bonita Avenue Napa, California 94558	(707)224-0092
MEC 21 (3)	D. B. Mitchell Code 244.5, Stop 060 Mare Island Naval Shipyard Vallejo, California 94592	(707)646-2444
MEL 21 (1)	B. J. Round Combustion Engineering, Inc. 1000 Prospect Hill Road Windsor, Connecticut 06095	(203)688-1911
MEL 21 (2)	W. C. Kroenke Babcock & Wilcox Company 1570 S. Hawkins Avenue Akron, Ohio 44320	
MEL 40	L. Kaldor Naval Ship R & D Center Annapolis, Maryland 21402	(301)267-2447
MSC/NASTRAN	C. W. McCormick The Macneal-Schwendler Corporation 7442 N. Figueroa Street Los Angeles, California 90041	(213)254-3456

PROGRAM	CONTACT FOR FURTHER INFORMATION	
PIPDYN II	Yung-Lo Lin Franklin Institute Research Labs 20th Street and Parkway Philadelphia, Pennsylvania	(215)448-1595
PIPESD	D. Mann Control Data Corporation Box 0 Minneapolis, Minnesota 55440	(612)853-3188
PIRAX I	Dr. G. Workman Battelle Columbus 505 King Avenue Columbus, Ohio 43201	(614)299-3151
SAP IV	J. P. Scott Fluor Corporation 2500 S. Atlantic Boulevard Commerce, California	(213)262-6111
SAP V	C. S. Parker General Electric Corporation 175 Curtner Avenue San Jose, California 95114	(408)297-3000

FURTHER PROGRAM EVALUATION AND COMMENT

This section briefly discusses three important items which are adjuncts to the previous section's description of piping programs.

First, it is recognized that the list of available programs is not complete. There are two reasons for this. Some companies indicated that their programs are proprietary, used in-house, and little is gained by making data available to the public. At the time this article was written, several companies and developers had not responded with current information on their programs.

Second, in piping analysis, where many similar system analyses are performed, there are three highly interdependent time considerations: data preparation time, program execution time, and results interpretation time. With the interdependency of these three considerations on total dollar cost of analyses, it is felt that reporting "typical program running times" may be more misleading than helpful, and therefore these have not been included here.

A third general item applicable to all computer programs is verification. In the particular area of structural analyses one method of verification has been introduced and implemented in the literature, verification by benchmark calculation [1].

ACKNOWLEDGMENT

Sincere thanks are offered to all who have participated in developing this information exchange venture with me, particularly to Joan Byington for the help with many details and typing.

REFERENCES

1. Tuba, I. S. and Wright, W. B., ed., Pressure Vessel and Piping/1972 Computer Programs Verification, The American Society of Mechanical Engineers, New York, New York.

SHIP STRUCTURES

R. F. Jones, Jr.
Naval Ship Research and Development Center
Bethesda, Maryland

ABSTRACT

The results of a survey conducted among engineers and naval architects working in the naval structural field are presented. The primary purpose of the survey was to determine the computer programs being used in the field. Guidelines for selecting computer software and recommendations for future research and development are also presented.

INTRODUCTION

Many aspects in the design and analysis of ship structures are very similar to those of other structures. Ship structures here will include both submersible and surface type craft. The results of a questionaire sent to 225 engineers and naval architects in the United States, Europe, and Japan indicated that the computer programs being used in this field are in many cases the same programs being applied to various types of other structures.

In the following, the computer programs are separated into two different categories. The first group consists of general application type programs. They are being used primarily in this country and Europe, and since they are applied to a wide range of structures they are capable of solving many types of problems. Their development was, in most cases, based on the intent that they be exercised by a large number of users, and thus more effort has gone into their documentation. The second group of programs consists of the special purpose type which have been developed primarily for ship structure application. A significant portion of these programs has been developed in Japan. Since load prediction and evaluation is a large part of the task in designing surface type vehicles, many of the programs have been developed for this purpose.

The descriptions of the programs provided are intended to make potential users aware of the many programs being used in this field and can possibly be used as a starting point for broadening an organization's analysis and design capabilities. Since a large number are presented, space does not allow an evaluation of these programs. Furthermore, it would be difficult to provide objective evaluations even for a few since such evaluations tend to be biased by each user's own needs and capabilities. However, general suggestions are given for program selection.

The results of the survey indicated that there presently exist many programs to perform the most sophisticated analyses, even those which include both nonlinear material and geometric behavior. It appears that a major portion of the near term research and development resources that are spent on numerical methods should be in making better use of the capability that we now have. This would be primarily in making the programs easier to use through automatic data generation methods and also in determining the capabilities of the programs through systematic evaluations. The evaluations should be thoroughly documented so that they may be used as guidelines for designers and analysts in the optimum use of the programs for subsequent work.

SHIP STRUCTURES SOFTWARE

The following provide program descriptions that were obtained in the survey mentioned earlier. The programs that are listed here are only those that are in active use and are operational and well tested. For each program a brief description of the capability, the operating hardware, and person or organization to contact on its availability is provided. Also, reference for each program in the open literature is provided if available.

General Application Programs

In order to describe the capabilities of the general application programs, the following code will be used to indicate the various type of analyses, that is, options available with each program.

1. Small displacement
2. Large displacement
3. Incremental plasticity
4. Creep
5. Temperature dependent material
6. Natural frequencies, mode shapes
7. Transient response
8. Data generation
9. Graphic displays
10. Multielement library
11. Thermal effects
12. Bifurcation buckling

1. ANSYS
   Options: 1 through 12
   Systems: IBM 360, 370, CDC 6000, UNIVAC 1100
   Program is proprietary and is available on lease or royalty basis.
   Swanson Analysis Systems, Inc.
   870 Pine View Drive
   Elizabeth, Pennsylvania 15037

2. ASKA [1]
   Options: 1 through 4, 6 through 12
   Systems: IBM 360, 370, CDC 6000, UNIVAC 1100
   Program is proprietary and is available for sale or for use on in-house basis.
   IKO Software Service GmbH
   7 Stuttgart 80
   Vaihingerstr. 49
   Germany

3. BOSOR4 [2] General Axisymmetric Stiffened Shells
   Options: 1, 2, 6, 9, 11, 12
   Systems: IBM 360, 370, CDC 6000, UNIVAC 1100
   Available thru developer, charge is $300.00.
   Dr. David Bushnell
   Department 5233
   Lockheed Missiles and Space Company
   3251 Hanover Street
   Palo Alto, California

4. DAISY [3]
   Options: 1, 8, 9, 10, 11
   Systems: IBM 360, 370, CDC 6000
   Availability restricted to agreement with American Bureau of Shipping.
   Professor H. A. Kamel
   Aerospace Engineering
   University of Arizona
   Tucson, Arizona

5. EASE2
   Options: 1, 5, 8, 9, 10, 11
   System: CDC 6000
   Program is proprietary and is available for use thru Control Data Corporation.
   Engineering Analysis Corporation
   1611 South Pacific Coast Highway
   Redondo Beach, California 90277

6. GIFTS [4]
Options: 1, 8, 9
System: CDC 6000
Program available after official release.
Office of Naval Research
Washington, D.C.

or

Professor H. A. Kamel
Aerospace Engineering
University of Arizona
Tucson, Arizona

7. MARC-CDC [5]
Options: 1 through 12
Systems: IBM 360, 370, CDC 6000
Program is proprietary and is available for use thru Control Data Corporation or program developer.
MARC Analysis Corporation
105 Medway Street
Providence, Rhode Island 02906

8. NASTRAN [6]
Options: 1, 6, 7, 9, 10, 11, 12
Systems: IBM 360, 370, CDC 6000, UNIVAC 1100
There are two versions of this program, one distributed thru COSMIC[1] and one available thru data centers throughout the world. Information may be obtained thru the program developer.
MacNeal - Schwendler Corporation
7442 N. Figueroa Street
Los Angeles, California 90041

9. SAP-IV [7]
Options: 1, 5, 6, 7, 8, 10, 11
Systems: IBM 360, 370, CDC 6000, UNIVAC 1100, GE Telefunken, Siemens
Cost of program is $250 and can be obtained thru the program developer.
Dr. K. J. Bathe
410 Davis Hall
University of California, Berkeley

10. STAGS B [8] Shell Application
Options: 1 through 5, 7 through 12
Systems: CDC 6000, UNIVAC 1100
Available through developer; charge is $1000.
Mr. Bo Almroth
Department 5233
Lockheed Missiles and Space Company
3251 Hanover Street
Palo Alto, California

---

[1]Computer Software Management and Information Center, Barrow Hall, University of Georgia, Athens, Georgia 30601

11. STARDYNE [9]
Options: 1, 6, 7, 8, 9, 10, 11
System: CDC 6000
Program is proprietary and is available for use thru Control Data Corporation. For information contact program developer.

Dr. Richard Rosen
Mechanics Research Inc.
9841 Airport Boulevard
Los Angeles, California 90045

12. STRESS
Options: 1
Systems: IBM 1130, UNIVAC 1100
Available through developer.

Professor R. D. Logcher
Massachusetts Institute of Technology
Cambridge, Massachusetts

13. STRUDL II [10]
Options: 1, 2, 6, 7, 8, 9, 10, 11
Systems: IBM 360, 370, UNIVAC 1100
Program is available on batch or time sharing basis and information on availability may be obtained through the developer.

ICES USER'S GROUP INC.
P. O. Box 8243
Cranston, Rhode Island 02920

Special Application Programs

A brief paragraph is given on each of the special application programs. Since a large portion of the ship structural analyst effort must be spent on load prediction and evaluation, a significant number of the programs are concerned with this task.

14. ARCL8

The program calculates block loading for a ship in a rigid drydock. It is based on the finite element method. Up to 100 blocks on the center line may be included. The program generates much of the data used for the finite element solution. Operation is on the CDC 6000 system. Running time is approximately 30 seconds. Information may be obtained through:

Commander, Mare Island Naval Shipyard
Attention: Mr. Ron Munden
Code 244.5, Stop 060
Vallejo, California 94592

15. BCSTAP [11]

The program performs structural analysis of hold parts of bulk carriers. The finite element method is used with beam and stiffened plate type elements. The program can be applied to most conventional type bulk carriers. Stress distributions can be obtained either through diagrams by line printers or plotters. The computer system used is the IBM 360 series. Running time is approximately 3 minutes (CP) on the IBM 360/195. The user's manual is in Japanese. Information may be obtained through:

Professor Yoshiyuki Yamamoto
Department of Naval Architecture
University of Tokyo
Bunkyo-ku, Tokyo, Japan

16. CONDESS (CONtainer Ship DESign System)

The program determines structural arrangement and member scantling. The three-dimensional structural analysis is based on frame theory. The program operates on the IBM 370 or UNIVAC 1100 systems. The cost of the program will be determined at time of inquiry. Information may be obtained through:

Mr. Ryojiro Naito
Mitsubishi Heavy Industries, Ltd.
5-1, Marunouchi 2 Chome
Chiyoda-Ku
Tokyo, Japan

17. FINEL [12]

The program performs a linear stress analysis of a general three-dimensional structure by the finite element method. Bar membrane and plate elements are used in the analysis. The program operates on the CDC 6000 series machine. It is available to U.S. Government agencies and its contractors. For information contact:

Dr. John Adamchak
Code 173
Naval Ship Research and Development Center
Bethesda, Maryland 20034

18. Hydrodynamic Pressures on Ship Hull [13]

The hydrodynamic pressures induced on the hull surface are calculated theoretically by using the solution for ship motions. The analysis is based on the assumption that the ship's section can be represented by the Lewis formulation. The program operates on the IBM 360 and UNIVAC 1100 systems. For the person to contact see Program No. 16.

19. ISTRAN/PL

A three-dimensional stress analysis is performed by the finite element method on structures that are made up of planar components. The program has data generation and plotting capability. It operates on UNIVAC 1100, IBM 370, and TOSBAC 5600 systems. Typical running time is approximately 75 minutes (CP) for a case which has 26,000 elements. No indication was given on which machine this is based. For person to contact, see Program No. 15.

20. Lateral Wave Loads [14]

Lateral wave loads, i.e., shear force, bending moment, and torsional moment acting on a ship hull, are calculated in oblique regular waves. The calculations are based on the strip theory. The program operates on the IBM 360 and UNIVAC 1100 systems. Approximately 2 seconds are required for each run. For person to contact see Program No. 16.

21. Longitudinal Strength

The program computes bending moments and shear forces for both wave and still water conditions by integrating the differences in weight and buoyancy curves. It is restricted to a trapezoidal weight distribution and no more than 50 concentrated weights. The program provides an output plot of the shear and bending moment diagrams. The program operates on the CDC 6000 system. Typical running time is 8 seconds for each condition. Information on the availability of the program can be obtained through:

Computer Analysis Group
Maritime Administration
Room 4069A
14th and E St. N. W.
Washington, D. C. 20235

22. MEC21

Flexibility analyses are performed on piping systems and beam-type structural networks. A maximum of 999 nodes is allowed. The program will handle bends, flexible joints, rigid elements, anchors, restraints, and structural type elements. It also considers weight effects, thermal expansion, internal pressure, corrosion allowance, and stress intensifiers. Optional Calcomp plotting is available. The program operates on the IBM 7094 or the IBM 360, UNIVAC 1100, CDC 6000, and Honeywell 6060 systems. The program is available thru COSMIC, University of Georgia, or by contacting:

Commander, Mare Island Naval Shipyard
Attention: David B. Mitchell
Code 244.5, Stop 060
Vallejo, California 94592

23. MIDAS [15]

The program designs the longitudinal scantlings of a midship section. Any practical combination of decks, platforms, and longitudinal bulkheads for the midship section configuration may be used. Options to include an inner bottom structure and to perform a nuclear air blast analysis of shell and upper strength deck structure are provided. An iterative process is used in the design. The design criteria is that established by the U.S. Navy. The program operates on the CDC 6000 system. An S-C 4020 plotting option is available. For information on program availability contact:

Mr. N. S. Nappi
Code 173
Naval Ship Research and Development Center
Bethesda, Maryland 20034

24. MIDSEC

The program determines the geometry, scantlings, and weight per foot of the midship section. Linear programming optimization techniques are used. The program does not design transverse webs but can handle mixed framing. It is stated that the program is extremely efficient for parametric studies. The program is written for the CDC 6000 system. For information on program availability contact:

Mr. Otto Jons
Hydronautics, Inc.
Pindell School Road
Laurel, Maryland 20810

25. PASSAGE (Program for Analysis of Ship Structures with Automatically Generated Elements) [16]

The program performs a linear elastic membrane analysis of a total ship hull structure. A multilevel substructure method is used. An interpolation approach is used in addition to the standard method for inner point elimination. There are a maximum of 1000 degrees of freedom allowed in each of the maximum 1000 substructures. The program is oriented toward the ship designer with automatic input generation and graphical output. The program operates on the CDC 6000 or CYBER 74 systems. Typical CP time for a 30,000 degree of freedom problem is 4700 seconds. The availability of the program has not been decided. For information contact:

Mr. Ken Takata
The Shipbuilding Research Association of Japan
Senpakushinko Building
35 Shiba-Kotohiracho
Minato-ku
Tokyo, Japan

26. SASP (Structural Analysis of Stiffened Plate)

The stress analysis of a reinforced rectangular plate under lateral loads is performed. The differential equation for the plate is solved using Fourier series. The program is limited to simply supported rectangular plates with one-directional stiffeners parallel to one side. It is stated that the program has easy input and short computing time. The program operates on the IBM 370 system. For information on the availability of the program contact:

Dr. M. Higuchi
Ship Research Section
Technical Research Center
Nippon Kokan K. K.
Keihin Works
Kawasaki, Japan

27. Section Modulus Calculation (MSD/NA/1)

The program calculates the ship's section modulus at the midship section. It is programmed for the IBM 1130 with 8K word storage, the IBM 360 series, and the General Automation 18/30. Typical running time is one minute. For information on availability of program contact:

Mr. S. G. Kinkaid
Manager - Technical Computer Center
Maryland Shipbuilding and Drydock Company
P. O. Box 537
Baltimore, Maryland 21203

28. Section Moduli of Local Strength Members (MSD/NA/9)

The program computes the sectional properties of flanged plates, angle bars, and T-sections such as webs and girders. It operates on the IBM 1130, IBM 360 system, and the General Automation 18/30 machine. For person to contact on the availability, see Program No. 27.

29. SESAM-69 [17]

The program has the capability for performing many types of analyses and is based on the multilevel substructuring method. This approach enables the program to solve problems which are modeled with a very large number of degrees of freedom. Structures modeled with 240,000 degrees of freedom have been reported in the literature. Although the program was developed primarily for ship applications, it has capabilities that are comparable to the programs listed previously in the general purpose category. The program operates on all major computer systems and it is proprietory. For information on its use and availability, contact:

Mr. P. O. Araldsen
Det norske Veritas
P.O. Box 6060
Etterstad, Oslo 6, Norway

30. Ship Hull Characteristics Program (SHCP)

The hull form parameters and longitudinal bending moments are computed. The approach used is to balance the ship on waves by iteration. The weight curves that are input are used to calculate shear and bending moment curves. A maximum of 40 stations are allowed along the hull. Plots of weight, buoyancy, load, shear, and bending moment curves are drawn. The program operates on the IBM 1130. The typical running time is five minutes for still water, hogging, and sagging conditions. For information on availability contact:

Mr. Mike Aughey
Code 6102C
Naval Ship Engineering Center
Prince Georges Center
Hyattsville, Maryland 20782

31. Still Water Shear Force and Bending Moments (MSD/NA/3)

The program produces the still water shear force and bending moment diagram ordinates as required by the American Bureau of of Shipping. The integral factors method is used. There is no limit on the number of stations per condition. Also, there is no limit on number of loading arrangements or number of ships. The program operates on the IBM 1130, IBM 360 system, and the General Automation 18/30 machine. For person to contact on the availability, see Program 27.

32. Structural Analysis of Longitudinally Framed Ships [18]

The program considers both the longitudinal and transverse strength of longitudinally framed Ships. Both finite element and transfer matrix methods are used in the analysis with the bulkheads modeled by finite elements. The program is designed to solve large problems. For information on the availability, contact:

Dr. Pin-Yu Chang
George Sharp Inc.
100 Church Street
New York, N. Y. 10007

33. Three-Dimensional Strength Calculation of Bulk Carrier [19]

The program performs the strength analysis of double bottom and transverse ring of bulk carriers. Space frame analysis is used in the program. The ship's side is single hull. The program operates on the CDC 6000 system and requires approximately 400 seconds (CP) per load. For person to contact on the availability, see Program No. 16.

34. Three-Dimensional Strength Calculation of the Horizontal Girder [20]

A stress analysis of the primary supporting members on the transverse bulkhead and the transverse rings adjacent to the transverse bulkhead is performed. The slope deflection method is used in the program. The maximum number of horizontal girders is 4. Approximate running time is three minutes per load. The types of machines on which the program operates are not indicated. For person to contact on the availability, see Program No. 16.

35. Three-Dimensional Strength Calculation of Oil Tanker [21]

A stress analysis of the longitudinal strength member and the transverse rings of oil tankers for all tank parts is performed. Slope deflection and transfer matrix methods are used in the analysis. The maximum number of tanks and transverse rings in one tank is six. Upright and heeled conditions can be calculated. The program operates on the IBM 360 and UNIVAC 1100 systems. Approximate running time is sixty seconds per load. For person to contact on the availability, see Program No. 16.

36. Three-Dimensional Strength Calculation of Ore Carrier [22]

A stress analysis of the longitudinal strength members, the double bottom,and the transverse rings of an ore carrier is performed. The transfer matrix method is used for the longitudinal members. The maximum number of holds is eleven, and the maximum number of transverse rings in any one hold is sixteen. The program operates on the UNIVAC 1100 system. Approximate running time per load is two minutes. For person to contact on the availability, see program No. 16.

37. TITO 11

The program calculates the hydrostatic properties including volume, displacement, LCB, block coefficient, prismatic coefficient, midship coefficient, waterplane coefficient, TPI, BMT, BML, KB, KMT, and KML for each water line considered. The program is based on matrix methods and Simpson's rule. It operates on the IBM 1130 machine. Typical running time is 10 minutes. For information on availability, contact:

Mr. T. Penn Johnson
Chief Design Engineer
Levingston Shipbuilding Company
Orange, Texas 77630

38. TITO 14

The program computes cross curves of stability, Bonjean, displacements, G2, and moments for each ten degrees of heeling up to 60 degrees. It is stated that with a small change, the program could be used to calculate values for every 5 degrees up to 45 degrees or for every 15 degrees up to 40 degrees. The program operates on the IBM 1130 machine. Typical running time is 10 minutes. For person to contact on availability, see Program No. 37.

39. TITO 17

The program calculates the still water bending moment. Typical running time is 15 minutes on the IBM 1130 machine. For person to contact on availability, see Program No. 37.

40. TSHULGDR [23]

The program calculates the fluctuating normal stresses in the hull girder caused by the ship motions and dynamic wave loads in the seaway for a given probability of occurrence. The hull girder is simulated as a continuous beam with variable cross sections. The general bending theory for thin-walled beams is implemented by the finite element method. The calculation of the ship motion, wave loads, and inertia forces is based on strip theory. Automatic data generation is used with the program. Approximately three hours is required for an analysis on the FACOM 230-25 machine. For person to contact on availability, See Program No. 15.

41. VASP

A vibration analysis of a reinforced rectangular plate subjected to in-plane loads is performed. Direct energy minimization by nonlinear programming is used to obtain the solution in terms of Fourier or power series. The program is limited to rectangular plates with one-directional stiffeners parallel to one side. Various boundary and load conditions can be analyzed. It is stated that the program has easy-to-prepare input. Approximate running time is one minute on the IBM 370/155 machine. For person to contact on availability, see Program No. 26.

42. Vertical Ship Motions and Wave Loads [24]

The program computes the heaving and pitching motions of ships in regular waves by the strip method. Longitudinal wave bending moment and shearing forces in regular waves are calculated at midship and at various points along the ship length. The cases of wave length, wave-to-course angles, and ship's speed must be less than 30, 10, and 20 respectively. The program operates on the IBM 360 and UNIVAC 1100 systems. Approximate running time is 1/2 second for a single wave length, wave-to-course angle, and ship speed. For person to contact on availability, see Program No. 16.

43. ZPLATE [25]

The program performs stress analysis of large three-dimensional structures which can be modeled by plain or orthotropic plates. It is based on the finite element stiffness method. It is stated that there is practically no restriction on the number of elements or the bandwidth of the equations to be solved. The program can be run on the UNIVAC 1108 or IBM 370/155. An example case is given with 5239 nodes, 5808 elements, and six load cases. The running time was just over 82 minutes (CP). This included processing time for pictorial displays. It is not indicated on which of the above two machines the problem was run. For information on the availability, contact:

Dr. I. Neki
Ishikawajima-HARIMA
Heavy Industries
Tokyo Office 2-16
3-Chome Toyosu
Koto-ku, Tokyo, 135
Tokyo, Japan

44. X1Z058 [26]

A frame analysis of tankers is performed by the stiffness and transfer matrix methods. Only the main hold part of the tanker is analyzed. The program runs on the UNIVAC 1108 machine. The approximate running time is six minutes (CP) for one loading condition. For person to contact on availability, see Program No. 15.

45. X1Z072

The bending and torsional strength of ships having large hatch openings is determined. The program is based on thin-wall beam theory. A maximum of 29 ship sections is allowed with a maximum of 80 joints in each section. The program runs on the UNIVAC 1108. For person to contact on availability, see Program No. 15.

46. ZUNIT [27]

The program performs the stress analysis of frame structures. It is based on the matrix stiffness method. The maximum number of joints is 800 while the maximum number of members is 3200. Up to six loading cases can be analyzed in a single run. The program operates on the UNIVAC 1108, IBM 370, and TOSBAC 5600 machines. Five minutes running time is required for each 100 members. It is not stated for which machine this applies. For person to contact on availability, see Program No. 15.

47. ZVIBRA [28]

The program determines the vibration response of frame structures. It is based on the matrix stiffness method. The maximum number of joints is 800 and the maximum number of members is 1600. Six loading cases can be analyzed in one run. The program has automatic plotting capability. It runs on the UNIVAC 1108, IBM 370, and TOSBAC 5600 machines. Five minutes running time is required for each 100 members. It is not stated for which machine this applies. For person to contact on availability, see Program No. 15.

SELECTING SOFTWARE

The selection of computer software to be used in the day-to-day solution of structural problems deserves a great deal of research and planning. Significant amounts of labor costs and computing money will be consumed in using the programs,and it is important that proper choices are made.

Besides the obvious question of the accuracy of a program, there are many other factors that have to be considered in making a choice. One of these factors is the way in which the program will be used. Each program will have an optimum way in which it can be exercised or may, in fact, be restricted in the way it may be used. The selection of a program usually infers that calculations will be performed at a central data center or a remote teletype or batch terminal may have to be leased to use the program.

Another factor to be considered is the frequency of turnaround that one can expect to achieve with the program. This is particularly true of programs that operate in the batch mode. The more central core a program requires, the less frequent one can expect to get output. This is an important consideration since productivity will be adversely affected if it takes a long time to get the computed results. Some problems, because of their size, will have to be run overnight or on weekends, but there are many instances when it is necessary for the designer to interact with the machine. If it takes him 24 hours to find out that he has a control card error or a piece of data is punched in the wrong field, a lot of time can be wasted.

The above are just two examples of the many characteristics that should be considered in choosing a program. Examples of others are the program's documentation, efficiency, ease of input preparation, ease of output interpretation, options in types of loading,and on and on. Obviously, there is much information that should be known about a candidate program before an attempt is made to make the first run.

Since no one organization at the present has the responsibility of endorsing or certifying computer programs, and thus being a source of information for program selection, the only way to plan and research an approach to extend an organization's numerical analysis capability is the informal one of reading the appropriate literature or talking to people who are already knowledgeable in the field. A lot of free information is available. Of course, any information obtained has to be weighed in terms of one's own needs and also the capability of their personnel.

Thus the brief descriptions of the 47 programs given here can only serve as a starting point for those who want to extend their computing capability. For each of the programs, an attempt has been made to give at least one reference in the literature, if available, and the developer or the person responsible for the maintenance of the program or that has knowledge of the program's availability. Additional names of persons who have used a particular program can usually be obtained from the person listed for that program.

Information on the first fourteen programs will be easier to obtain since they are intended for a wide range of users and thus have, in most cases, been used more than the programs in the second category. These programs will probably have more thorough documentation and also will have been run more extensively than those in the second group and thus have the higher probability of being more "bug-free." It does not follow that the programs in the first group will be easier to use because they will in general have a greater capability. With this increased capability comes more options on the type of input that has to be prepared. Also, since they can do more, they demand more of the user. This is especially true of the codes which are capable of performing a nonlinear analysis.

Although it is impossible to check out a candidate program in every aspect, it helps to try to anticipate some of the pitfalls that one might encounter so that there will not be too many surprises.

RESEARCH AND DEVELOPMENT NEEDS

We list here what we believe to be the most prominent needs for research in the ship structures field. Because the analysis and design of these types of structures are very similar to other structures, it turns out that some of the needs are not peculiar to this field. However, there is one fundamental task associated with naval structures that is more difficult than for any other type of structure. This is in the prediction of loads and thus presents a unique research need.

As in all structural design, a fundamental understanding of the applied loads is mandatory before the geometry and material can be selected to resist these loads. The loading is often dependent on the ship response, and conversely, load prediction is probably the most difficult challenge the structural engineer faces in designing naval structures. Our present-day capability to predict the response for given loads, especially in the linear range, possibly exceeds our capability for predicting the loads for ship structures. Thus we have to be careful and be sure that the sophistication of our analysis does not exceed the accuracy of the predicted loads. Two recent references, [29] and [30], cover the state-of-the-art in load prediction and make recommendations for needed research. This area will continue to receive a large portion of our attention in the future, especially for high-performance craft.

The most common criticism of the use of numerical methods in the analysis of structures is the difficulty in preparing input data. This is especially true of the finite element method. Not only is there usually a great deal of work involved, the type of effort is very prone to mistakes. More effort has to be put into the automatic generation of this data as is given in reference [4]. The most promising approach for doing this is through the use of interactive graphic systems. The primary use of these systems at the present is for checking and not for generation of the input data, and thus they are not being used to their full potential.

The committee report on numerical methods in reference [30] gives the results of a cost-benefit opinion poll taken on interactive graphics systems. The primary object of the poll was to determine what factors are currently responsible for holding back progress in this field and to determine what technical developments are needed. One interesting result of the poll was that it is difficult to relate the use of the graphics systems to monetary benefits. It turns out that there are other benefits, such as better understanding of the physical behavior of the structure being analyzed. This is a result of plotting such output as the deformed shape. It was concluded that methods are needed to assess the value of computer programs in general, other than their monetary benefit.

The evaluation of computer programs is needed to aid analysts in the process of selecting programs for their applications. Not only information on the accuracy is needed but also information on other characteristics such as the running time required for well-defined problems. This information would have to be supported by a detailed description of the machine and its particular configuration.

The evaluation of available programs which perform nonlinear analyses is particularly important. The nonlinear analysis of structures requires much more of the analyst, and he needs help in selecting and setting the parameters that these programs require. Such documentation and evaluation would be valuable. The effective use of these programs now rests in the hands of the limited number of experts in the field and those persons who have had direct contact with these experts.

The computing times for the nonlinear structural programs are excessive, especially in application to dynamic problems. Extensive evaluation and publication of a broad range of problems would help in decreasing running times since such information would provide guidance on the optimum idealization and modeling of the physical problem. Also, in a related matter, improved nonlinear formulations would aid in decreasing these computing times. Reference [31] presents a formulation for decreasing running times for nonlinear static problems along with a brief account of the development of the nonlinear formulation for implementation by numerical methods. Although the nonlinear programs are being used extensively, both the evaluation and improvement of the efficiency of these programs is required to realize their full potential in solving everyday practical problems.

Most analyses are performed on ideal structures which do not recognize the presence of residual stresses and imperfections caused by welding and cold forming. The analysis of the real structure is particularly important in assessing its ultimate load-carrying capacity and stability. There is a need to simulate accurately the fabrication process, probably by numerical methods, and then include this predicted state as the initial state of the structure before the analysis is begun. This capability would allow for the reduction in the reliance on expensive large or full-scale experimental programs now required for structural certification.

Over the past five years, the development of numerical methods for structural applications has been primarily in the refinement of the theory. A major portion of the subsequent research will be directed toward implementation of this theory and developing methods for application to real structures.

ACKNOWLEDGMENT

The author would like to thank those persons who took the time to contribute to this survey. The opinions given by Lieutenant E. A. Chazal of the U.S. Coast Guard, Professor D. Faulkner of the University of Glasgow, and Mr. E. A. Fernald of the Portsmouth Naval Shipyard are appreciated.

REFERENCES

1 Meijers, P., "Review of the ASKA Program," Numerical and Computer Methods in Structural Mechanics, Academic Press, New York, 1973, pp. 123-149.

2 Bushnell, D., "Finite-Difference Energy Models versus Finite-Element Models: Two Variational Approaches in One Computer Program," Numerical and Computer Methods in Structural Mechanics, Academic Press, New York 1973, pp. 292-335.

3 Kamel, H. A., et. al., "Some Developments in the Analysis of Complex Ship Structures," Advances in Computational Methods in Structural Mechanics and Design, University of Alabama Press, Huntsville, 1972, pp. 703-726.

4 Kamel, H. A. and McCabe, M. W., "A Graphics Oriented Interactive Finite Element Time Package," Office of Naval Research Report, Contract Number N00014-67-A-0209-0016, Jul. 1973.

5 Ayres, D. J., "Elastic-Plastic and Creep Analysis via the MARC Finite-Element Computer Program," Numerical and Computer Methods in Structural Mechanics, Academic Press, New York, 1973, pp. 247-263.

6 Tocher, J. L. and Herness, E. D., "A Critical View of NASTRAN," Numerical and Computer Methods in Structural Mechanics, Academic Press, New York, pp. 151-173.

7 Wilson, E. L., "SAP - A General Structural Analysis Program for Linear Systems," Advances in Computational Methods in Structural Mechanics and Design, University of Alabama Press, Huntsville, 1972, pp. 703-726.

8 Almroth, B. O., Brogan, F. A. and Zele, F., "User's Manual for STAGS," AFFDL-TR-71-8, Vol. II, Wright-Patterson Air Force Base, Ohio.

9 Dainora, J., "An Evaluation of the STARDYNE System," Numerical and Computer Methods in Structural Mechanics, Academic Press, New York, 1973, pp. 211-226.

10 Chu, S. L., "Analysis and Design Capabilities of STRUDL Program," Numerical and Computer Methods in Structural Mechanics, Academic Press, New York, 1973, pp. 229-245.

11 Yamamoto, Y., et. al., "A Program System of Structural Analysis of a Bulk Carrier," Journal of the Society of Naval Architecture of Japan, Vol. 131, 1972.

12 Adamchak, J. C., "User's Manual for the Modified Finite Element Program FINEL," Report No. 3609, Nov. 1970, Naval Ship Research and Development Center, Bethesda, Maryland.

13 Nagamoto, R., "Theoretical Calculations on the Motions, Hull Surface Pressures and Transverse Strength of a Ship in Waves," (in Japanese), Journal of the Society of Naval Architects in Japan, Vol. 129.

14 Nagamoto, R., "Theoretical Calculation of Lateral Shear Force, Lateral Bending Moment, and Torsional Moment Acting on the Ship Hull Among Waves," Journal of the Society of Naval Architects of Japan, Vol. 132.

15 Nappi, N. S. and Lev, F. M., "Midship Section Design for Naval Ships," Report No. 3815, Oct. 1972, Naval Ship Research and Development Center, Bethesda, Maryland.

16 Yoshiki, M., "On the Development of the Passage Program," Proceedings of the 1973 Tokyo Seminar on Finite Element Analysis, University of Tokyo Press.

17 Egeland, O. and Araldsen, P. O., "SESAM-69 - A General Purpose Finite Element Method Program," Computers and Structures, Vol. 4, Jan. 1974, pp. 41-68.

18 Chang, P. Y., "Structural Analysis of Longitudinally Framed Ships," Report to the Ship Structure Committee, U. S. Navy Contract No. 00024-70-C-5219, Naval Ship Engineering Center, 1971.

19 Suzuki, T., "Experimental Test on the Transverse Strength of a Large Bulk Carrier," Journal of the Society of Naval Architects of Japan, Vol. 132.

20 Nagamoto, R., "On the Strength of Horizontal Girder on Transverse Watertight Bulkhead," Journal of the Society of Naval Architects of Japan, Vol. 123.

21 Nagamoto, R., "On the Transverse Strength of Oil Tankers," Journal of the Society of Naval Architects of Japan, Vol. 121.

22 Watanabe, M., "On the Transverse Strength of Ore Carriers," Journal of the Society of Naval Architects of Japan, Vol. 120.

23 Yamaguchi, I. and Nitta, A., "Total System of Analysis on the Longitudinal Strength of Ships in NIPPON KAIJI KYOKAI," Preprint of ICCAS, Aug. 1973.

24 Nagamoto, R., "Longitudinal Distribution of Wave Bending Moments and Shearing Forces of a Gigantic Tanker in Regular and Irregular Head Waves," Journal of the Society of Naval Architects of Japan, Vol. 121.

25 Niki, I., et. al., "General Purpose Program of Plane Stress Analysis by Finite Element Method and Its Application," IHI Engineering Review, Vol. 11, No. 5, 1971.

26 Shimizu, S., "On the Structural Analysis of Tankers by the Transfer Matrix Method," IHI Engineering Review, Vol. 11, No. 2, 1971.

27 Neki, I., "Matrix Method of Structural Analysis of Framed Structures and Examples of Its Application," IHI Engineering Review, Vol. 9, No. 3, 1969.

28 Neki, I., "Matrix Method of Vibrational Analysis of Framed Structures and Its Application," IHI Engineering Review, Special Issue No. 4, 1970.

29 Lewis, E. V., et. al., "Load Criteria for Ship Structural Design," Report to the Ship Structure Committee, U. S. Navy Contract No. N00024-71-C-5372, Naval Ship Engineering Center, 1973.

30 Fifth International Ship Structures Congress - Committee Reports, Hamburg, Sep. 1973.

31 Jones, Jr., R. F., "Incremental Analysis of Nonlinear Structural Mechanics Problems with Applications to General Shell Structures," Report No. 4142, Sep. 1973, Naval Ship Research and Development Center, Bethesda, Maryland.

SHOCK WAVE PROPAGATION IN SOLIDS

J. F. Mescall
Army Materials and Mechanics Research Center
Watertown, Massachusetts

ABSTRACT

A review is made of the finite difference codes available for the calculation of shock wave propagation in solid materials. For problems in one spatial dimension Lagrangian and characteristic codes are discussed. For two-dimensional problems the relative merits of Eulerian and Lagrangian codes are assessed in terms of their ability to obtain effective solutions at reasonable cost. Requirements for future developments are discussed.

INTRODUCTION

This paper is a brief survey of present-day capabilities in the prediction of shock wave propagation in solid materials. It is intended to be of assistance to a potential user faced with an application involving strong stress waves and considering which computer program would best serve his purpose. The governing equations for this problem area are highly nonlinear, and accordingly, meaningful solutions for all but the simplest situations can be obtained only by numerical methods. Much attention has been devoted to the development of these methods, and a variety of techniques now exists to handle problems of interest. While the methods are reasonably well established, their intelligent use demands a high degree of technical knowledge and skill. Although large amounts of computer time are often required, the results appear to be accurate enough for engineering purposes provided material behavior is sufficiently well described.

The largest uncertainty in a problem often attaches to the description of material behavior, particularly to the details of dynamic fracture. The earliest codes developed for the description of very strong stress waves dealt with situations in which the hydrostatic component of stress (pressure) is overwhelming--pressures on the order of several million atmospheres. Clearly the most appropriate model was a hydrodynamic one--solids were treated as compressible fluids with no shear strength. As the pressure level descends from the megabar range to that of hundreds or tens of kilobars, strength of material considerations become very important; thus elastic-plastic effects were incorporated. While such topics as strain hardening, strain-rate dependence, pressure, and temperature dependence of the flow stress can be fairly readily treated in such numerical solutions, the choice of a specific model for these effects can turn out to be surprisingly difficult in view of a scarcity of experimental information.

A collection of articles by the developers of various codes is given in the volume Methods in Computational Physics [1]. To describe the approach of these codes as compactly as possible, one might say that they begin with the conservation laws for mass, momentum, and energy, couple to these an equation of state which is realistic for the dynamic high-pressure regimes involved, cast the entire assembly into a finite difference formulation, and integrate step-by-step in time (explicit procedures). The length of the time step between each cycle through the finite difference mesh is chosen on the basis of a stability criterion. In physical terms the time step is required to be less than the time required for any signal to propagate across the smallest zone of the problem, taking any grid distortion into account.

Input to the codes consists of a description of geometry, velocities, special boundary conditions, and the equation of state parameters. Output can be massive depending on the detail desired and consists of a full-field description of physical quantities of interest such as stress, strain, displacement, velocity, density, internal and kinetic energy, yield strength, and material condition (spalled or continuous, solid or porous) throughout the problem as a function of time. These codes thus provide an extremely useful complement to experimental programs by providing information not generally amenable to direct measurement.

The range of problems to which these codes has been applied is impressive. Some of the earliest studies involved high-pressure equation of state measurements on solids and high explosives for atomic energy laboratories. Hypervelocity impact problems for spacecraft protection against meteorites attracted considerable attention. Military applications are voluminous and long-standing. A partial listing would include terminal ballistic problems of kinetic energy and shaped charge penetrators (as well as the inverse problem of improved armor design to resist penetration), fragmentation rounds, blast wave loadings on surface and underwater structures. Geophysicists have rapidly adapted shock wave techniques to their studies of materials under extremely high pressures and to the examination of meteor craters produced by very high-speed impact. Commercial applications include explosive forming, explosive welding, shock synthesis of new materials (e.g., diamonds), mining, and massive earth removal.

Before proceeding to a detailed description of the specific programs available, a brief overview of the general procedures employed is in order. This not only provides the best frame of reference in terms of which we may compare individual codes but turns out to have enormous practical consequences. Hydrodynamic codes may be generally classified as one of four basic types: Lagrangian, Eulerian, characteristic, or mixed. Codes of the Lagrangian type follow the motion of fixed elements of mass, and the finite difference grid is thought of as fixed in the material. Lagrangian codes easily handle boundary conditions at free surfaces and contact surfaces between different materials. They encounter severe difficulties, however, when there is large distortion of the grid (as in turbulent flow). In Eulerian codes the grid is fixed in space and the material passes through it. The material can be represented either by hundreds of discrete point masses, as in the particle-in-cell (PIC) method, or in a continuous manner. Eulerian codes are generally not as accurate for comparable grid sizes, particularly in calculating free surface motion. When the problem of interest contains several different materials, the Eulerian formulation tends to produce diffusion across material interfaces at nonphysical rates. Eulerian codes have no difficulty in treating problems with violent distortion (turbulent flow).

For problems in which the precise details of the shock front are of paramount importance, codes based upon taking finite differences along characteristic lines offer an advantage because they do not employ the so-called artificial viscosity approach. In order to treat shock waves (discontinuities) propagating through a finite difference grid, Von Neumann and Richtmeyer [2] proposed modifying the governing equations by including an additional pressurelike viscosity term "q". This approach ensures numerical stability for the finite difference procedure by smoothing a shock front over several zones of the computational mesh. Suitable choice of the specific form of "q" ensures that distortion of the solution is significant only in regions of high stress gradient (i.e., at the shock front) and is negligible elsewhere. Use of the artificial viscosity thus sets a limit on resolution obtainable near the shock; however, reduction of mesh size reduces the width of a smeared shock to acceptable limits for most engineering applications.

Since no one computational technique will handle all situations, hybrid programs involving two techniques simultaneously have been developed. For example, programs have been developed in which the early part of the motion, which may involve violent distortions, is handled by Eulerian methods. At a suitable point in the computation, results are mapped onto a Lagrangian grid and the remainder of the problem is handled by Lagrangian techniques. Alternatively, Lagrangian grids may be embedded in an Eulerian flow field and both Lagrangian and Eulerian computations are performed side by side. One version of this concept takes the form of a set of Lagrangian tracer particles used to delineate the boundary between different materials. The tracers are massless and do not disturb the basic flow but are carried along with it.

The next section will summarize some of the important characteristics of specific programs available in each of the above categories for one and two spatial dimensions. A critical review of the relative merits of each is included. Finally, an indication is given of what we may realistically expect in the way of future developments.

ONE-DIMENSIONAL CODES

Many problems in shock wave propagation can be legitimately treated as motion in one spatial dimension. This includes plane, cylindrical, and spherical geometry if a sufficient degree of symmetry is present. By far the most widely used technique for this problem class is the Lagrangian formulation. In one spatial dimension its advantages over the Eulerian are overwhelming; in certain special circumstances codes based on the method of characteristics offer an advantage.

Lagrangian Codes

The most prominent one-dimensional Lagrangian codes are (in alphabetical and chronological order) KO, PUFF, and WONDY. Each of these names actually refers to a family of codes. The present generation of the first two families seems to have originated around 1958 with the SHARP code [3], developed by M. Wilkins and his colleagues at Lawrence Livermore Laboratories, Livermore, California. KO is the version presently used there and is also the basis of one of the wave propagation codes marketed under the group name PISCES by Physics International, San Leandro, California. KO has also been implemented on the UNIVAC 1106 at AMMRC. With the addition of later developments at the Air Force Weapons Laboratory, Kirtland Air Force Base, Albuquerque, New Mexico, the generic name PUFF was given to the program in 1963. Later versions include PUFF I through PUFF IV, PUFF V-EP (EP for elastic-plastic), PUFF VTS (variable time step), FOAM PUFF, PUFF 66 (for the CDC6600 version), and P PUFF 66 (plate geometry). Most of the PUFF codes have been described in classified reports, which limits their utility to the general user. The most widely used version is described and documented in [4]. Whereas the KO code uses a simple explicit finite difference procedure, the PUFF codes use a leapfrog integration scheme. Similarly, the WONDY family (I through IV) is very parallel in approach but is based on a slightly different finite difference representation. These codes were developed by W. Herrmann and his colleagues at Sandia Laboratories, Albuquerque, New Mexico, and are described in [5,6]. Another series of codes of this type was developed by L. Seaman and his associates at Stanford Research Institute, Menlo Park, California [7]. The SRI-PUFF series began as modifications of the PUFF 66 and P PUFF 66 codes. An important special feature of SRI PUFF is its numerical integration scheme: the leapfrog method of Von Newmann and Richtmeyer and the Lax-Wendroff method. The resulting procedure, a "double leapfrog" method, is similar to the leapfrog method except that two steps are made. The main advantages are: (1) a direct solution (rather than iterative as in leapfrog) and therefore faster running; (2) better stability than Lax-Wendroff at interfaces and in regions of high density.

In terms of many of the features of interest to a potential user, these codes are surprisingly similar. They are all highly reliable and well tested. They are readily available at nominal, if any, cost from their developer, from experienced users who have made their own modifications or from COSMIC. All are available in FORTRAN IV versions. The program listings are lengthy--four to six thousand statements. Documentation for WONDY series is good [5,6]; for the SRI PUFF code it is outstanding [7]. All fit comfortably on machines with 65K words of core storage and, with this amount of memory, can treat any reasonable problem falling within their domain. This core requirement can be significantly relaxed if a satisfactory compromise with numerical resolution can be made, and in most engineering problems this is the case. Running times are quite comparable. These are on the order of several minutes for most applications but, naturally, depend significantly on the complexity of the physical problem. Specifically, running times scale with the number of mesh points required for adequate resolution and with the time step between cycles through the mesh. Since the time step is proportional to the mesh size, halving the space increment results in a fourfold increase in computational time to reach the same point in physical time. (Twice the volume of computation, twice as often). The advantage then of an improved finite difference procedure such as that of SRI PUFF is twofold. It can reduce the maximum core size required and shorten the running time at the price of a slight increase in computational time per cycle for the added complexity of the numerical integration. Interestingly enough, the use of more sophisticated material property descriptions (e.g., work-hardening vs. perfectly plastic models, or the inclusion of fracture criteria, and so on) adds relatively little computational time.

All the major Lagrangian codes use the same form of artificial viscosity—usually a combined linear plus a quadratic term--whose amplitude can be adjusted at input time to provide a desired degree of smoothness. Their stability criteria are essentially the same; so time steps chosen by the separate codes are very close. The latest versions of these codes (WONDY IV, P PUFF 66, SRI-PUFF, and KO) all contain a rezone feature which is activated automatically during the course of the calculation. This feature improves the efficiency of the calculation significantly by providing greater definition in areas where things are changing most rapidly and less definition in quiet areas. Zones are halved or doubled automatically on the basis of the rate of change of quantities across the zones. The maximum difference in zone size has to be limited to prevent spurious error signals from propagating.

Finally, this group of codes has similar routines available for graphic display of output either on CALCOMP plotters or the more elaborate plotters typified by the SC.4000. Plots of stress vs. time for a fixed coordinate or of stress vs. linear dimension for a specified time are typical of graphical output.

These codes were all developed primarily on Control Data equipment. The latest versions of each are run primarily on CDC6000 machines (on the CDC7600 in the case of KO). However versions of each have been run on the Univac 1108 and on the IBM 360 series. User experience in adapting these codes to new machines (or to a new configuration or operating system on the same machine) indicates that this is sometimes a nontrivial problem and can take, typically, a man-month.

In these codes the flow of the computation is structured so that the constitutive equations are contained in a clearly defined subroutine. Thus changes to material models can be more readily made, without affecting the remainder of the program. For elastic-plastic behavior, e.g., an incremental stress-strain relation is used. Pressure is obtained as in the hydrodynamic case as a function of density and internal energy. Changes in stress deviators are calculated elastically from changes in strain deviators, which are obtained kinematically. If the elastic increment of strain deviators would cause the stress to fall outside the yield surface, then it is relaxed in the direction normal to the yield surface so that it will be on the surface. Usually only isotropic work hardening is considered, but there are provisions in certain versions of all three codes for anisotropic hardening, the Bauschinger effect, and so on. The primary reason for not routinely employing such models is a lack of reliable experimental data.

In trying to formulate guidelines for potential users of these codes, we were motivated primarily by considerations of the probability of obtaining a successful solution to the physical problem addressed. With this in mind, the rather subtle differences between these codes become more significant. These differences reflect to a large extent the experience acquired by developers and users on specific classes of problems. In general the shock waves being computed are initiated by the impact of one material upon another (plate impact), or by the detonation of high explosives adjacent to a material specimen, or by the deposition of radiated energy in the material. The extent to which the codes described above have been applied to each of these problem classes varies, and this is reflected in their relative strengths. Thus, the LRL KO code is particularly strong in the quality of its equations of state for high explosives. The PUFF family has been used most extensively in energy deposition problems. Both P PUFF 66 and SRI PUFF and their derivatives have superior subroutines for the time-dependent absorption of radiant energy, as well as for the complex material response to this energy. This includes highly expanded (even vaporized) material near the surface. A special difficulty in this problem class is the need for very fine zoning near the irradiated surface in order to provide an accurate description of the deposition, expansion, and spallation which occur in that region. After deposition is complete and the stress wave moves into the interior of the specimen, there is less need for such fine zoning. Consequently, a geometric zoning of the problem is advisable. This option is standard in the PUFF codes. Another consideration in this connection is the following. The specific details characterizing the energy distribution being deposited by scattering, flourescence, and the photoelectric effect must be calculated by special codes externally and then the results inserted into a stress wave code. For most deposition codes the material is divided into zones of uniform thickness, a zoning not optimal for the stress calculation. Therefore, interpolation procedures to implement the transfer more efficiently must be incorporated into the stress code. The PUFF codes, particularly SRI PUFF, appear to be strongest in this feature although both WONDY and KO have provisions to effect the transfer of such data.

A special class of material behavior--that of porous materials --generates special difficulties for shock wave propagation codes. The details of the shock loading of such a material through compaction, followed by unloading and reloading, are considerably different from those of a homogeneous material. SRI PUFF and WONDY contain superior subroutines for the description of this problem class.

A final special feature--and perhaps the single most important feature--is that of the description of fracture of spallation. Simple time-dependent spallation criteria such as that proposed by Tuler and Butcher [8] have been implemented on all these codes. The most elaborate and most promising description of spall phenomena is the Nucleation and Growth (NAG) model developed at SRI. This model for ductile and brittle materials has been implemented as special subroutines DFRACT and BFRACT on the SRI-PUFF and on the KO code at AMMRC. Among its special characteristics is the fact that it provides more realistic detail of the gradual relaxation of the stress field surrounding a newly nucleated or a growing crack.

A summary statement regarding guidelines for selecting from among these one-dimensional Lagrangian codes, then, is that the major consideration is probably dictated by the user's application. None of the codes is genuinely simple to implement, despite the fact that input usually consists of between ten and twenty cards. A realistic summary of user opinion is that no new user can expect to run meaningful problems without frequent and repeated discussions with the code developer and/or experienced users. On the other hand, a repeated observation is that the users of these codes form a small, close community who maintain an active dialogue and who share experience (and subroutines) readily.

Characteristic Codes

In the method of characteristics approach the finite difference mesh is varied during the running of a problem to provide detail where needed. The time step is also varied, rather than being determined for the entire mesh by the poorest conditioned zone. Because it does not use an artificial viscosity technique to ensure numerical stability, but instead uses an explicit treatment of discontinuities, it provides greater details of the solution on such regions.

The drawbacks of the characteristics approach include the facts that its logic is relatively cumbersome and one generally needs to have some a priori knowledge of the nature of the solution. For linear problems without discontinuities the computational mesh can be established without knowledge of the solution. For nonlinear problems the mesh must be determined as the solution proceeds, thus at material interfaces or discontinuities. Thus there can be a substantial increase in the level of programming difficulty associated with problems involving multiple materials, multiple shocks, etc., with the result that the probability that a new problem will run successfully at first attempt is not good.

Two of the more prominent codes in this category are the MCDIT series (now up to MCDIT-4) and the SWAP family (now at SWAP-9). MCDIT, which is an acronym for "Method of Characteristics - Drexel Institute of Technology," is a series of codes developed by P. C. Chou and his colleagues at Drexel [9, 10, 11]. It has been applied to problems of shock wave propagation involving one-dimensional strain, including plate impact studies and energy deposition. The material property description has included elastic-perfectly plastic behavior. The code was developed on the IBM 360 series and uses FORTRAN IV. Run times are on the order of minutes, depending on problem complexity. The code is considered fully operational and well tested.

The SWAP codes use a method of numerical integration which is very similar, though not identical, to the method of characteristics. Rather than treat regions of continuous flow divided by lines of discontinuity (as in MCDIT) SWAP represents all waves shapes by a series of discrete shock waves. Since the shock equations and those of continuous flow approach each other in the limit of weak waves, the error introduced in this manner can be controlled. The use of the same equations for shocks and continuous waves results in a simplification of the programming comparable to that obtained by artificial viscosity methods in the Lagrangian codes. The code was developed by L. Barker at Sandia Laboratories, Albuquerque, New Mexico [12]. It has been used extensively by workers there for the analysis of plate-slap experiments, energy deposition problems, and other shock wave problems involving one-dimensional strain and elastic-plastic material response. Written in FORTRAN IV, it was developed on the CDC3600 and has been implemented also on the CDC6600, Univac 1108, and IBM 360-65. It is available through COSMIC or the author.

A general guideline for potential users might be this: Given the availability of both a Lagrangian and a characteristic code, one would use the Lagrangian when the final state of the material was desired but a high resolution of details of shock fronts leading to that state was not critical; one would use the characteristic code where an understanding of the detailed wave motion was paramount. In most engineering applications Lagrangian codes are used for their flexibility.

TWO-DIMENSIONAL CODES

For problems of two spatial dimensions characteristic codes are generally quite cumbersome, to the point where users generally regard them as unsuitable. Lagrangian codes in two dimensions carry over their advantages over Eulerian codes--but only for problems of modest grid distortion. For very high-velocity impact problems, for example, large local deformations of the Lagrangian grid are encountered. Cells can actually fold over upon themselves, resulting in negative masses and general pathological behavior. Since the time step chosen for stability considerations is governed by such zones, computer time can become prohibitive, requiring the calculation to be stopped and a rezoning procedure applied. An illustration of this grid distortion is given in Fig. 1 which contains a blow-up of the impact region in a projectile penetration problem. The initial conditions were that of a blunt steel cylinder impacting a steel target at 2500 ft/sec. Figure 1 shows the grid at 1.75 microseconds after impact. The problem is cylindrically symmetric about the x axis. Eulerian codes, on the other hand, can handle flows with

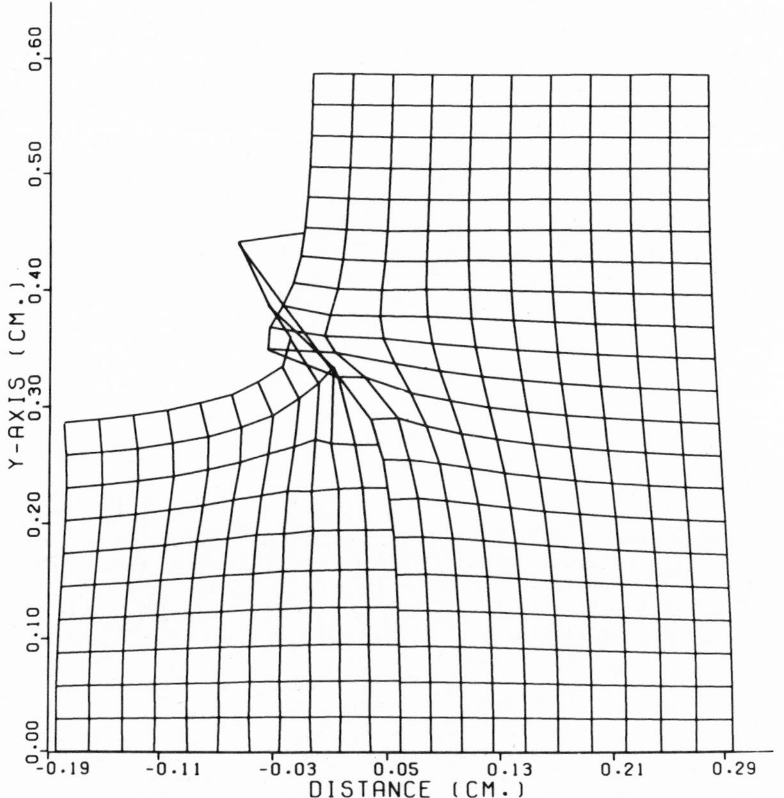

Fig. 1. Blowup of Lagrangian grid distortion at impact face of steel projectile striking steel target

very large material distortion and thus are suitable for very high-velocity impact problems. The region of hypervelocity impact (say 20,000 ft/sec) is an almost exclusively Eulerian domain. Mixed codes, those combining Eulerian and Lagrangian features, are of special interest for many problems.

Some general comments about common operational characteristics are in order before discussing distinctive features of specific codes. The two-dimensional geometry considered may be either planar (plane strain) or may involve three-dimensional problems with rotational symmetry. The selection is made at input time. Input data for any of the prominent codes is very modest--generally on the order of two dozen cards. This is a reflection, of course, of the relative simplicity of the geometries most frequently treated. Output is massive in volume--stacks of paper often too difficult to carry, let alone read. A natural consequence of this is that all these codes have associated auxiliary programs for graphical display of output. Periodically during the computation dumps are taken on tape of the information required for graphical display of the computational grid (or the density of materials in the Eulerian cells), plots of particle velocity, and contours of quantities of physical interest such as pressure or plastic strain. User reaction to these graphical display routines is that they are sufficiently peculiar to the individual computational center's hardware and software packages that a new user is well advised to tailor his needs to his own installation. This is generally not difficult.

As a first step in the flow of the computation, a generator routine develops the finite difference grid and sets up the initial conditions from the input data. All this information--grid coordinates, particle velocites, pressures on grid boundaries, identification of the proper equation of state for each material, etc.--is printed out before the cycle-by-cycle computation begins. Thus, if the user desires, he may make a preliminary run to check the problem set-up at essentially no cost. As soon as it has finished its function, the generator is no longer needed and is overlaid with the routine which controls the iterative cycling through the finite difference grid.

Most codes have provisions for variable zoning, allowing larger zones to be placed in relatively quiet areas. Experience has shown that best results are obtained if the aspect ratio of zones is close to 1:1. Aspect ratios greater than 2:1 are generally not to be encouraged, although certain problems have been run successfully with such zoning. Provisions are made for using polar input for problems involving spherical shapes.

Considerable effort has been put into making the codes flexible and yet reliable. In general these objectives have been met to a highly satisfactory degree. Nevertheless occasionally a new application gives rise to unexpected or anomalous results which can be diagnosed and resolved quickly only by individuals having considerable familiarity with the entire code. It is also true that none of these codes can be utilized effectively without investing a considerable effort - one or two people for several months at least--in becoming familiar with the basic methods and details of the coding and in gaining experience with running the code. Copies of a code may generally be obtained from the developers (whom we shall indicate in discussing the specific codes) or from experienced users. The codes are quite large - from six to twelve thousand FORTRAN statements. They generally require 65K words of core storage; more is sometimes desirable, but techniques exist, as we shall point out, to use disk conveniently for running large problems.

Running times are clearly problem dependent. A convenient rule of thumb is that the number of mesh-point-cycles per hour is fairly constant. Depending on CPU speed, from one million to ten million mesh-point-cycles per hour of CPU time is representative on machines ranging from the UNIVAC-1106 through the CDC-6600 to the CDC-7600. Even modest problems frequently take over an hour of CPU time. An important consideration then is that these codes all have "restart" capability. Periodically during the calculation (typically every fifty cycles) a "dump" is taken on tape; all the information on the current state of the problem is stored together with the identifying cycle number. Should the problem be aborted for any reason, it is then possible to restart the problem from the most convenient cycle. Changes to some of the parameters can also be made when restarting the problem, e.g., frequency of edits as well as some of the physical problem parameters.

Lagrangian Codes

Most two-dimensional Lagrangian calculations are done with one of two codes (or derivatives of these): HEMP [13] developed by M. Wilkins of Lawrence Livermore Laboratories, Livermore, California, or TOODY [14] developed by W. Herrmann and his colleagues at Sandia Laboratories, Albuquerque, New Mexico. Both of these codes employ an unusual finite difference representation which was initially developed by Wilkins. In essential features and capabilities the codes are very similar. In certain respects (which probably have an impact on only new rather than experienced code users) there are important differences. Until quite recently, for example, there was very little documentation to assist a HEMP user, whereas the TOODY code has been fairly well documented [14]. An excellent user's manual for HEMP [15] has recently appeared, however, which removes this difficulty. As it is implemented at LLL on the CDC-7600, the format for input data is rigidly structured; once learned it is easy to use, but it is considered awkward by new users. The input format for the latest version of TOODY on the CDC-6600 is generally considered somewhat easier to use. HEMP has been implemented on the UNIVAC 1106 at AMMRC using the namelist option; this has been found to be very convenient for input. Another version of HEMP is commercially marketed at Physics International, San Leandro, California, where it is the two-dimensional Lagrangian member of the PISCES family of codes. PI has an arrangement with CDC whereby the code can be utilized at local CDC centers. Versions of HEMP which really differ only in those details required to adapt the basic code to the specific machine configuration employed have appeared under several names: e.g., CRAM [16] is the name used by General Electric Space

Sciences Laboratory, Philadelphia, Pa., while SHEP [17] is the name used by Shock Hydrodynamics, Sherman Oaks, California.

An interesting feature of the programming of all these Lagrangian codes is that they make extensive use of intermediate disk storage. The finite difference grid is considered to be divided into "K" lines and "J" lines for organizational purposes, and the computation proceeds by marching along K lines. As the data for zones along a given K line are being updated, information needs to be drawn only from the two adjacent K lines. Hence only the data for three K lines needs to be in core at any time. By storing the data for the remaining K lines on disk or drum and reading it in and out as needed, much more efficient use of core memory is made. In practice space for data for five K lines is provided for in core, the additional two lines corresponding to data which is about to be read in or out. In this way the operations of reading and writing from disk can be overlapped with the computation proceeding in the processor. (To be efficient this presumes that these operations can be performed simultaneously and that separate channels exist for the transport of data.) The procedure allows much finer zoning of problems. There remains a practical limit, however, on the mesh size. At LLL and at AMMRC the maximum number of J points of a K line is 100. The number of K lines of course is limited only by economics. The main portion of the program uses about 22K words of core.

The version of HEMP at LLL is somewhat special in that it implemented on the CDC-7600 which has special features not generally available. The 7600 memory for example is composed of 65K words of fast small-core memory (SCM) and 512K words of slower large-core memory (LCM). Use is made also of special service routines available only at an outstanding facility such as that of LLL.

As in the one-dimensional case, the shock waves whose effects are being calculated are initiated generally by the impact of one body upon another, by the detonation of high explosives adjacent to a material, or by the deposition of radiant energy in the material. Because of the fineness of the zoning which would be required, not many applications of these two-dimensional codes are made in the last category. In the area of high explosive applications HEMP has the most extensive set of well-tested equations of state for H.E. TOODY on the other hand has incorporated the equation of state for porous materials which were developed and tested at Sandia for the WONDY code. For impact problems and for related problems involving, say, pressure or velocity conditions over a portion of a boundary both HEMP and TOODY have considerable generality and flexibility. Problems involving multimaterial regions are handled routinely. HEMP has a total of five different forms of the artificial viscosity "q" which are useful in suppressing different types of undesirable grid distortions. The selection of one of these forms in preference to the standard is a matter of experience with individual problems.

All the common Lagrangian codes employ a technique known as a "slide line," which is actually a decoupling of the finite difference grid along certain specified lines. These are used for both physical and numerical purposes. An example of physical need would be the interface between two different materials which may "slide" relative to one another. Special treatment is afforded the zones on either side (master and slave) of a slide line. Generally only normal forces are considered to be transmitted across a slide line, even after distortion may have removed an initial parallelism. Friction forces may also be accounted for. Another physical application employs a tied slide line, where mesh points along a prescribed slide line are considered fixed (the sliding option is not operative) until a critical value of a specified parameter such as stress or plastic strain is reached. The points along such a tied slide line may be released separately.

A slide line is actually composed of a double set of mesh points. One set of points on each slide line is taken and defined as a master surface; calculation of points on it proceed according to the normal routines, considering the force field from both the slave material and the material within the master surface. The slave surface is advanced along the master surface by treating the (already moved) master surface as fixed.

Slide lines must be positioned along K lines during the problem set-up. An illustration of a nonphysical application of a slide line would be a change of zoning within a given material. Here the only interest would be in decoupling the grid numerically, and a rigidly tied slide line would be used.

Provision is made for allowing a void to open between master and slave surfaces of a slide line. The void is treated on a point-by-point basis so that part of the two materials may be in contact while part is not. The void may be generated as open or allowed to open later. It may be allowed to reopen after closing or may be forced to remain closed.

A number of external service routines exist which process the output of these Lagrangian codes. There are three general applications of this type:

1. Diagnostics: Plots of various quantities such as grid lines, particle velocities or stress contours are made.

2. Links: The output from a run may be processed and supplied to another code.

3. Rezone: Attempts to extend the applicability of the two-dimensional Lagrangian codes have resulted in a variety of rezone features which attempt to straighten out a tangled Lagrangian grid without distorting the physics of the problem. Methods range from simple provisions for slight adjustment of a single point through addition or deletion of entire rows or columns, to complete redesign of the integration grid according to a specified formula. The rezone routine maps the old grid into the new and recalculates physical quantities associated with each zone so that the new data is a reproduction of the old but with nicely formed zones.

A general comment is that rezoning a problem ordinarily buys only small extensions of time before another rezone is called for. The process is external to the basic Lagrangian calculation which must be interrupted and restarted. A reliable and "automatic" rezoning routine does not appear to be available yet.

Eulerian Codes

The earliest versions of an Eulerian approach to a two-dimensional hydrodynamic code involved the use of the particle-in-cell (PIC) method. The fixed Eulerian grid is peppered with hundreds of particles which represent the mass and whatever physical quantities the mass is carrying with it. Accelerations are computed by direct differencing, but convective terms (which appear in the governing equations as soon as Lagrangian coordinates are abandoned) are treated by moving the particles in accordance with local velocities and then transferring a lump of mass, momentum, and energy whenever a particle moves from one cell to another. A significant feature of this technique is that the particles may be used to maintain a record of (a) material identity and (b) material history, two of the troublesome aspects of general Eulerian methods. They also introduce a dissipative term in the difference equations which is equivalent to a viscosity; thus PIC codes do not usually employ the artificial viscosity approach.

Two difficulties arise with PIC applications to real problems; they are expensive and tend to lose accuracy for all but the very highest pressure levels. In order to obtain an accuracy of 1 percent in density, say, one needs approximately 100 particles in each cell. Since each particle is moved individually in each time step, running times can quickly become excessive. Nevertheless, the method was found to be satisfactory in calculating flows with very high pressures (above a megabar) and associated high energy levels; but for late times in such problems or for problems involving lower pressure levels from the outset, the PIC results are subject to large fluctuations caused by the discrete change in the number of particles in a cell when a particle crosses a cell boundary.

These difficulties led to the development of a series of codes of Eulerian flavor but in which the mass is treated in a continuous manner. Johnson developed the OIL code [18] as a continuous version of the PIC transport equations. He was able to follow with good resolution hydrodynamic flows in which the pressures were as low as 10 kilobars. By eliminating the particles, however, one method of keeping track of material identification and history was lost. To recover these features Johnson developed TOIL [19] to treat two materials in a problem, and later DORF [20] which handles up to nine (but at most two in a single cell). The amount of each material present in each cell is calculated, and then the convention is adopted that the mixture to be transported to the next cell is in the same proportion as the mixture in the receiving cell whenever possible.

Boundary conditions are difficult for a purely Eulerian code to treat. Frequently the boundaries are simply placed far enough away from the active region that they have no influence on the solution. In order to keep the boundary comfortably far removed, Eulerian codes usually incorporate a rezone procedure which simply doubles zone widths in each direction. The active region of the grid now occupies a smaller portion of the overall grid, and there is an attendant loss in accuracy.

An important consideration in this context is the following.

By combining a stable first-order difference procedure with a rezone technique, an Eulerian code will run nearly any appropriate problem without difficulty. Unlike Lagrangian codes, which cease to run whenever the error becomes excessive, the Eulerian code need not necessarily signal a deteriorating situation. Thus the user must assume the additional burden of paying greater attention to the details of the calculation to be assured that the results are meaningful.

At the lower stress levels for which the continuous Eulerian codes were developed, strength of material considerations are important. Consequently a rigid-plastic material model was incorporated into OIL and the resulting code was called RPM [21]. This code was still capable of treating only a single material. The HELP code [22], the latest entry in the field, was developed to combine all the desirable features described above. It is an Eulerian code which treats multimaterial problems with a continuous description of the moving mass and incorporates strength of material descriptions. In addition, however, massless tracer particles are introduced which define initial surface positions and then move across the Eulerian grid. Thus, the Eulerian code provides a Lagrangian-type definition of the moving surfaces. Transport from cells containing free surfaces or interfaces between two or more materials is done in a manner which takes account of these surface positions.

The HELP code is written in FORTRAN IV. It was developed by J. M. Walsh and L. Hageman Walsh of Systems, Science and Software of La Jolla, California, on the UNIVAC 1108. It has also been implemented at Ballistics Research Laboratories, Aberdeen, Md., by several users including John Harrison and J. Zukas. With 65K words of core storage about 2000 cells may be used in running a problem. In that version the code is memory contained and does not utilize external storage as outlined above. The writer and his colleagues at AMMRC found the code relatively easy to implement, operate, and modify in minor ways. Documentation [22] is good. Input is somewhat structured (namelist is not employed) but easy to use once learned. For simple problems only about two dozen cards are required. Output is similar to that of Lagrangian codes, and the availability of graphical display equipment is a decided advantage. However the HELP printout has one additional feature of special interest. Summary graphs of the compression, pressure, velocities, and internal energies are printed out along with the tabular values of the quantities of physical interest. These "contour maps" are of considerable assistance in quickly diagnosing a problem's progress.

The equations of state employed in the HELP code are slightly different in form than those employed in HEMP or TOODY. This is presumably because the HELP code authors anticipated going to much higher pressure regimes then the Lagrangian codes. The two forms are completely compatible; or alternatively, one could easily insert the forms used by the Lagrangian codes if it were thought advisable. The other details of the equations of state, such as stress deviator calculations, and details of yield surfaces are completely parallel to those of the HEMP and TOODY codes.

Mixed Codes

There have been several successful attempts to solve the dilemma of physical problems which require the advantages of both Lagrangian and Eulerian codes by linking the output from one code to the input for another. Physics International provides such a code, ELK, in its Pisces family. The Shock Hydrodynamics code SHAPE couples a PIC type code with the HEMP code.

It is difficult to provide an assessment of the long-term significance of this approach. Mechanically the procedure can be made to work, as the existence of the codes mentioned demonstrates. There are difficulties, however, inherent in the Eulerian treatment of material history which we shall discuss momentarily. It is not clear that the process of simply linking Eulerian codes to Lagrangian codes in order to calculate strength effects at late times gets to the heart of the problem. It may be preferable, for example, to extend the HELP concept of embedding a Lagrangian grid within an Eulerian code. In this way one has, for example, a complete record of the material history.

GENERAL OBSERVATIONS

A potential user faced with the problem of selecting the best code with which to attack a given problem will probably be struck by the apparent degree of similarity of the available codes. For example, they all treat the same class of geometries, deal with the same physical quantities, consider nonlinear material response; all require extensive training to use effectively; input data are not the problem they are for finite element codes; reduction of output data is slightly easier for Eulerian codes but is still a significant problem. Thus the selection of a code is generally not based on the above criteria.

Some of the guidelines for selection are more obvious (or at least less apt to generate controversy) than others. For example, the selection of an Eulerian over a Lagrangian code may be made easy by the expected nature of the physical problem. An impact velocity of 20,000 ft/sec or the desire to follow a collapsing cone through enormous deformations to the formation of a jet makes Eulerian treatment almost mandatory, unless, e.g., quasi-automatic rezoning routines become available for Lagrangian codes.

Another major consideration in many problem classes is the description of the stress-deformation history of an element of material. This is crucial, e.g., in dynamic fracture considerations. Keeping a cumulative record of the history of a zone is trivial in Lagrangian systems. It poses a problem for the Eulerian procedure, however, since the material which is presently located in a specific (fixed) cell may not have been found there on previous cycles. Thus, although one can describe the instantaneous state of the material in a cell or the cumulative state of the material in a region (total plastic work in a target, for instance), the more important and incisive question of tracing the historical material response of a particular element is outside the scope of at least the conventional Eulerian technique.

This is an especially significant point when considering the development of suitable fracture criteria. In general the state-of-the-art of fracture criteria in any of the codes discussed is perhaps best described as simplistic. However considerable attention is finally being focused in this direction. As may seem intuitively obvious, suitable fracture criteria for the dynamic situations being discussed are strongly time dependent, and thus one needs to know the material history at potential fracture sites. In this context, then, Lagrangian codes have a strong advantage.

Figures 2 and 3 show some outstanding applications of Lagrangian and Eulerian codes. Figure 2 was taken from [23], a part of an extensive study of armor response to ballistic impact conducted by M. Wilkins of LLL. Figure 2 depicts events which take place shortly after impact of a 30-calibre high-strength steel projectile upon a target composed of a ceramic facing and an aluminum backup plate. Impact velocity was 2500 ft/sec. Of particular interest are the details of the fracture pattern within the ceramic. Calculation revealed that the most damaging set of cracks began at the aluminum interface and progressed forward toward the projectile. Details of the associated energy distribution revealed that ballistic performance could be reasonably expected to improve if the onset of this crack could be prevented or at least delayed by improving material performances.

Figure 3 is taken from [24] and illustrates a problem class clearly within the domain of an Eulerian code (in this case, the HELP code). The successive steps in the collapse of a copper liner under the influence of the high-pressure field of a high explosive are dramatically illustrated. The formation of both a jet and a slug are clearly seen. Details of the mass and velocity distribution are provided by the numerical output.

Some of the other guidelines are a bit "softer." A potential user is well advised to consider the frequency with which a particular code has been applied to problems similar in character to his. The availability of a pool of experienced users is indispensable for a new user; it is helpful even to the most experienced user. Some thought in the selection process should be given to the question of how long-range an effort is being considered. Requirements for development of an in-house expertise are clearly different from those for projects requiring an immediate solution. It is also perhaps advisable to point to the need to consider supporting experimental programs to obtain the maximum amount of reliable information from calculations.

FUTURE CONSIDERATIONS

Eulerian and Lagrangian methods can be extended in a relatively straight-forward manner (conceptually, at least) to three-dimensional problems, and in fact a few such codes have been implemented. Resolution becomes of prime concern in this case, however, and the capacity of today's most powerful computers is so sorely taxed that

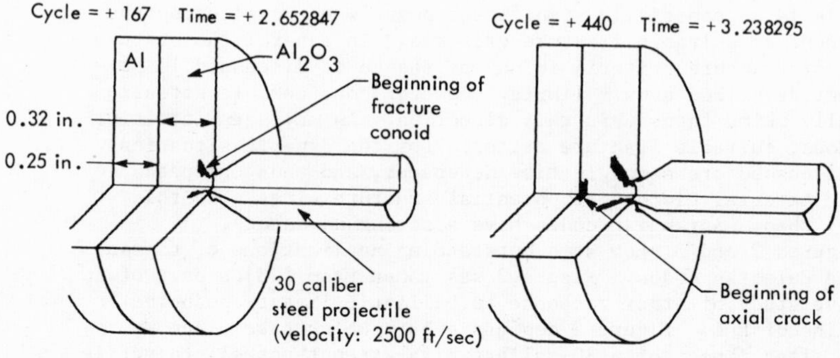

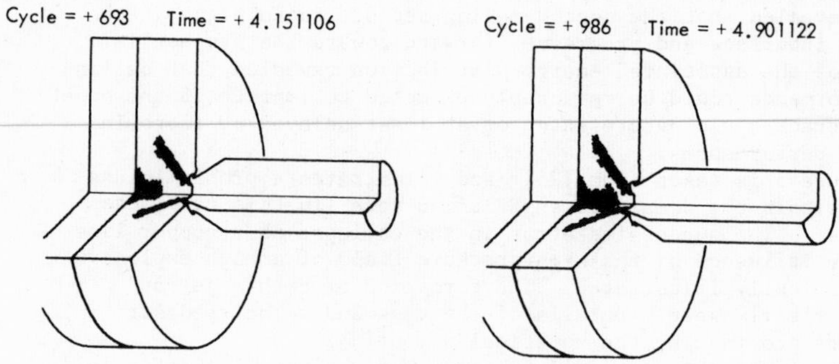

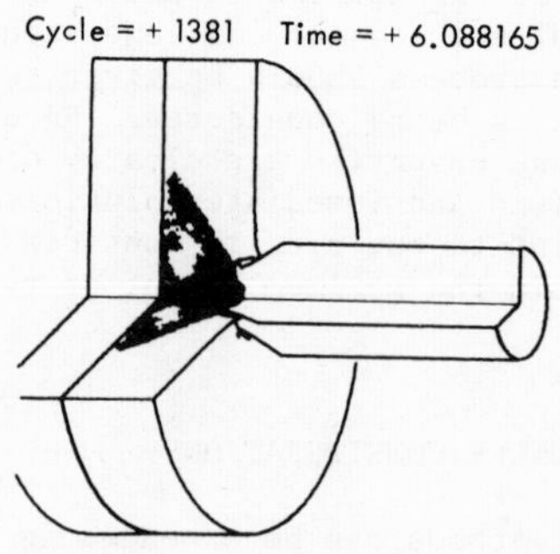

Fig. 2. Calculation of the development of fracture in alumina (Courtesy M. Wilkins [23])

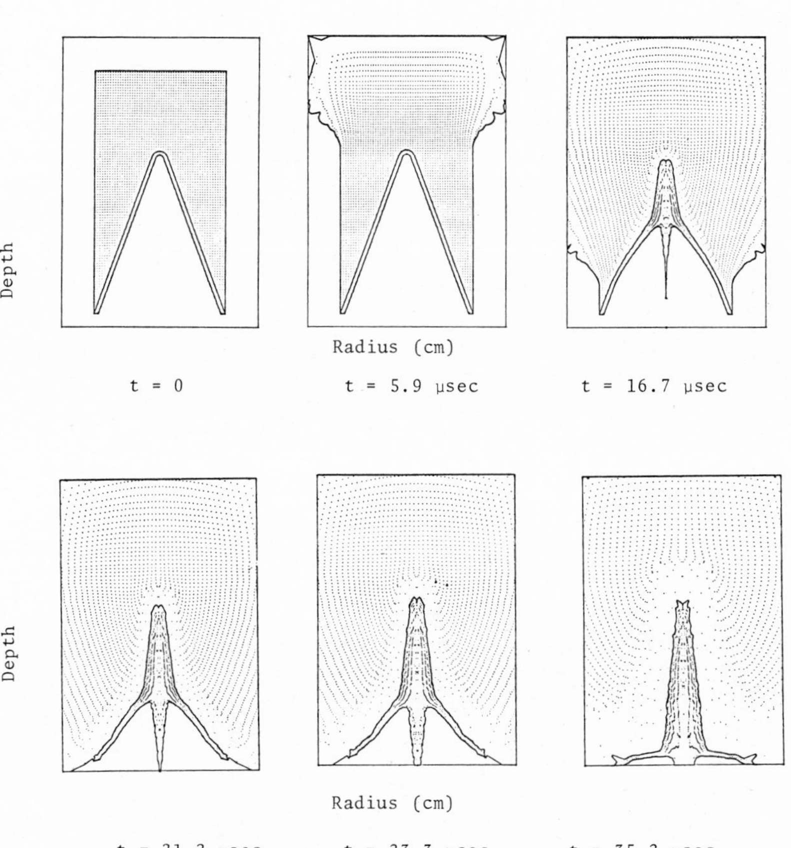

Fig. 3. Numerically predicted high explosive metal liner configurations of a shaped charge at various times after detonation initiation (Courtesy R. Sedgwick [24])

at this stage of development results tend to be somewhat limited in scope. One interesting associated development is the imminent installation of a new breed of computer at LLL. A parallel processor machine, the STAR--developed by CDC--is expected to have a significant impact on large scientific calculations such as those performed by Eulerian and Lagrangian codes. Special programming is required to utilize the machine; HEMP has already been reprogrammed in anticipation of delivery.

Another area in which considerable effort is being expended is that of higher-order accuracy in the finite difference method.[1] For the Eulerian approach this has special significance. In addition to the inherent diffusion difficulty which makes material boundaries hard to define, another problem occurring in Eulerian codes is the difficulty in choosing a grid which will provide resolution where it is needed, and at reasonable cost. Eulerian codes must, of course, place zones in regions where initially there may be no mass, but into which material may later move. If the zoning is not uniform, mass may move from a region in which it is well defined into another region in which it is poorly defined. Uniform zoning over the entire grid may result either in inadequate description of initial conditions, or in prohibitive expense.

In general the greater the accuracy of a finite difference procedure the lower the diffusion or truncation error introduced. Because of higher accuracy, smaller core sizes are needed for adequate resolution. An example of such an effort is a new Eulerian code called SMITE [25] developed by Burstein and his associates at Mathematical Applications Group, Inc. (MAGI), Elmswood, N.Y. This code also embeds Lagrangian tracer particles at material interfaces to enhance resolution. At its present state of development it has been addressed primarily to the problem of the response of a metallic liner material acted upon by a shaped charge of high explosive. The results for this problem are very promising.

Another second-order accurate non-Lagrangian two-dimensional hydrodynamic code called ADAM is being developed at SANDIA Laboratories, Albuquerque. It too is in the early stages of development.

As was discussed earlier, the area of development in two-dimensional code calculations which most needs refinement is probably that of fracture criteria. Several centers are making a concerted effort to improve this situation. It is not an exaggeration to say that even given the capability and computer capacity to handle three dimensional models little will be gained in quantitative engineering unless the modes of deformation and failure of real, inhomogeneous anisotropic materials is far better understood and unambiguously related to the composition, state, and history of the material. Until this (admittedly ambitious) goal is achieved, too many problems will rely for their solution upon expensive full-scale tests or upon an equally undesirable iterative procedure of matching imperfectly understood experiments with theoretical computations based on incomplete models.

---

[1] Finite element procedures for low amplitude stress waves (linear elastodynamics) are just beginning to emerge [26].

REFERENCES

1 Alder, B., Fernaback S., and Rotenberg, M., (eds.), Methods In Computational Physics, Vol. 3, Academic Press, New York and London, 1964.

2 Von Neumann, J., and Richtmeyer, R. D., "A Method for the Numerical Calculation of Hydrodynamic Shocks," Journal of Applied Physics, Vol. 21, 1950, p. 232.

3 Nance, O., Wilkins, M., Blandford, I., Smith, H., "Sharp--A One-Dimensional Hydrodynamics Code for the IBM 704," UCRL-5206-T, University of California, Lawrence Radiation Laboratory, Livermore, California, April 1958.

4 Brodie, R. N., Hormuth, J. E., "The PUFF 66 and P PUFF 66 Computer Programs," Research and Technical Division, Air Force Weapons Laboratory, Kirtland Air Force Base, New Mexico, May 1966.

5 Herrmann, W., Holzhauser, P., Thompson, R. J., "WONDY - A Computer Program for Calculating Problems of Motion in One Dimension," SC-RR-66-601, Sandia Corporation, Albuquerque, New Mexico, Feb 1967.

6 Lawrence, R. J., Mason, D. S., "WONDY IV - A Computer Program for One Dimensional Wave Propagation with Rezoning," SC-RR-710284, Sandia Corporation, Albuquerque, New Mexico, Aug 1971.

7 Seaman, L., "SRI PUFF 3 Computer Code For Stress Wave Propagation," Air Force Weapons Laboratory, AFWL-TR-70-51, Kirtland Air Force Base, New Mexico, Sept 1970.

8 Tuler, F., and Butcher, B. M., "A Criterion for the Time Dependence of Dynamic Fracture," International Journal of Fracture Mechanics, Vol. 4, 1968, p. 431.

9 Chou, P. C., Karpp, K. R., and Huong, S. L., "Numerical Calculation of Blast Waves by the Method of Characteristics," AIAA Journal, Vol. 5, No. 4, April 1967, p. 618.

10 Burns, B. P., "A Numerical Method for One-Dimensional Dissimilar - Material Impact Problems," Ph.D. Thesis, Drexel Institute of Technology, June 1969.

11 Chou, P. C., and Tuckmontel, D., "Method of Characteristics Calculation of Energy Deposition and Impact Problems," the Winter Annual Meeting of the American Society of Mechanical Engineers, Washington D. C., Dec 2, 1971.

12 Barker, L. M., "SWAP-9: An Improved Stress-Wave Analyzing Program," Sandia Corporation, SC-RR-69-233, Albuquerque, New Mexico, Aug 1969.

13 Wilkins, M. L., "Calculations of Elastic-Plastic Flow," Methods In Computational Physics, Vol. 3, Academic Press, New York and London, 1964; See also Lawrence Radiation Laboratory Report UCRL-7322, Rev. I, Livermore, California, 1969.

14 Thorne, B. J., and Herrmann, W., "TOODY - A Computer Program For Calculating Problems of Motion in Two Dimensions," SC-RR-66-602, Sandia Corporation, Albuquerque, New Mexico, July 1967.

15 Giroux, E. C., "HEMP Users Manual," Lawrence Livermore Laboratory, UCRL-51079, Livermore, California, June 1971.

16 Sedgwick, R. T., and Wolfgang, D. C., "CRAM - A Two-Dimensional Lagrangian Code for Elastic-Plastic Hydrodynamic Material Behavior," General Electric, T1S69SD9, Philadelphia, Pennsylvania, Feb 1969.

17 Bjork, R. L., Kreyenhagen, K. N., and Wagner, M., "The Shape Code for Two Dimensional Analysis of Response to Impulsive Loadings," Shock Hydrodynamics Report, Feb 1966.

18 Johnson, W. E., "Oil: A Continuous Two-Dimensional Eulerian Hydrodynamic Code,"General Atomic Report, GAMD-5580, 1965.

19 Johnson, W. E., "Code Correlation Study," AFWL TR 70-144, Kirtland Air Force Base, New Mexico, April 1971.

20 Johnson, W. E., "Development and Application of Computer Programs Related to Hypervelocity Impact," Systems Science and Software Report No. 3SR-749,1971.

21 Dienes, J. K., Evans, M. W., Hageman, L. J., Johnson, W. E. and Walsh, J. M., "An Eulerian Method for Calculating Strength Dependent Deformation," General Atomic Report GAMD-8497, Parts I. II and III, AD678 565, 678 566 and 678 567, Feb 1968.

22 Hageman, L., and Walsh, J. M., "Help, A Multi-Material Eulerian Program for Compressible Fluid and Elastic-Plastic Flows In Two Space Dimensions and Time," BRL Contract Report No. 39, 3SR-350, May 1971.

23 Wilkins, M. L., "Third Progress Report on Light Armor," Lawrence Livermore Laboratories, Report UCRL - 50460, 1968.

24 Sedgwick, R. T., "The Application of Numerical Techniques to Ballistic Problems," Proceedings of the Fourth Army Symposium on Solid Mechanics, Army Materials and Mechanics Research Center, MS73-2, Sept 1973, p. 209.

25 Burstein, S., Scheckter, H. S., and Turkel, E. L., "Eulerian Computations of Second Order Accuracy For Explosively Loaded Compressible Materials," Final Report for Picatinny Arsenal, Contract DAAA 21-71-C-0431.

26 Verner, E. A., and Becker, E. B., "Finite Element Stress Formulation for Wave Propagation," International Journal for Numerical Methods in Engineering, Vol. 7, 1973, p. 441.

WELD PROBLEMS

Pedro V. Marcal
Brown University
Providence, Rhode Island

ABSTRACT

Physical phenomena which occur during the welding and subsequent cooling process are discussed. Considerations are given to the models required for numerical analysis of the weld problem. The complexity of models depends on the focus of interest being in the melted zone. Case studies are provided of the different models. It is concluded that the models provide good results and are consistent with the current ability to define material behavior.

INTRODUCTION

The welding process is of fundamental technological importance because of its universal use as a fabrication technique. By the very nature of the process, welding subjects parts of the structure to an extremely severe thermal history so that weld sections are often the critical determinants of the life and working strength of a fabricated structure. The welding process leaves very large residual stresses in the joint that has been formed so that weld sections are prone to cracking and subsequent failure under cyclic loading. For these reasons, weld fabrication has been the subject of much experimental and analytical investigation both from the metallurgical and the structural points of view.

In terms of a numerical analysis, welding is perhaps the most nonlinear problem encountered in structural mechanics. The physical phenomena involve an initial nonlinear temperature behavior which is started off with a thermal shock and followed by subsequent melting. This is then followed by a solidification process where the structure begins to acquire strength and both the yield stress and the Young's modulus increase in the presence of a very large thermal gradient.

At present, there is much empirical knowledge of the problem although few analytic models have been developed. The work in this area was surveyed in a 1970 paper by Masubuchi [1]. That paper discusses the numerical models of Tall [2] and Masubuchi, Simmons, and Monroe [3], both of which are one-dimensional.

For such a complex problem then, many workers have tried to apply existing tools of analysis and, depending on the point of reference, have introduced different levels of complexity. Masubuchi [4] and co-workers have used an elastic-plastic analysis together with a quasi-steady-state heat transfer analysis by Tall [2]. The analytical results have been correlated with experiments at each stage. Attention was paid primarily to the region away from the melting zone, and agreement was found with the experiments.

Ueda and Yamakawa [5] have also used elastic-plastic analysis to study the stresses in welds. Friedman [6] has also studied the effects of welding with the BESTRAN program. Hibbitt and Marcal [7] have tried to account for most of the important factors in nonlinear weld analysis. Subsequent work by Nickell and Hibbitt [8] has shown that the rate of input is critical in determining the melting zone.

In this survey I shall first consider the various physical phenomena encountered in the welding process, then discuss the type of analysis required to account for these phenomena. In this survey, attention will be restricted to programs and studies that have actually been made of the welding process. It should be remembered that with the current concepts of program interfacing, it should not be too difficult for a determined user to obtain a welding analysis as a composite analysis on different programs. Thus, he may choose to perform thermal analysis on one program, elastic-plastic analysis on another, and a creep analysis on a third.

NOTE ON THE MODELING OF PHYSICAL PHENOMENA ACCOMPANYING WELDING

In this section the physical phenomena and boundary conditions accompanying welding are discussed. Particular attention is given to the computational aspects required to account for the thermal and then the mechanical phenomena.

Phase change and latent heat. The most important phenomenon is that of phase change and latent heat because it represents an initial store for the heat input. Most workers [9 - 13] have assumed that the phase change takes place at some specific temperature. Weiner [14] has pointed out that general alloys change phase over a finite temperature range and has suggested a uniform release of the latent heat over the range of temperature of the phase change. This requires the location of two distinct discontinuity interfaces (solidus and liquidus) and the application of an increased "specific heat" to simulate the latent heat in the interface. Friedman and Boley [15] have introduced an implicit heat absorption feature to simulate the phase change over a temperature range.

Prescribed external heat flow. The external flux is assumed to be completely known as a function of position and time. The flux is the most critical parameter in a welding analysis and in principle may be obtained as a function of the welding parameters. However, there is immediate loss of heat which is accounted for by empirical efficiency factor.

Surface heat loss as a function of surface temperature. This is an important boundary condition which determines much of the intermediate and long-term behavior of the model. Both the linear Newton convective cooling and quartic Stefan-Boltzmann radiative cooling must be employed simultaneously.

Addition of the molten filler material to the base metal. The molten filler has a different temperature from that of the welded structure. This introduces a thermal shock problem at the instant of contact. In finite element analysis it is not economical to refine the mesh near the surface to capture this behavior. Hibbitt and Marcal [7] have introduced an impulse type equation which forces the filler and the structure to take on an intermediate temperature on contact.

Location of the solidus and liquidus surfaces. The phase change introduces a jump in the "specific heat" value. This jump is of the order of 10 for steel. This causes serious problems in developing a good model with the conventional finite elements since discontinuous temperature fields are difficult to represent within an element. Fortunately, mechanical resistance is extremely low in the temperature range close to the melting phase. The primary goal of the temperature analysis is to maintain a proper local energy balance during the fusion part of the transient. It was found that the discontinuities [7] were better modeled by a fine mesh of low-order elements because of the discontinuous temperature gradients.

I now turn my attention to the mechanical part of the analysis. The greatest savings in computing effort come from separating or uncoupling the thermal analysis from the mechanical analysis. It is possible to justify uncoupling for steel and by extension for all metals by considering the differences of heat energy levels in heat conduction and the heat generated by mechanical stresses. It is also useful to note that the nonlinear part of the thermal transient is almost over when the temperature is sufficiently low for the metals to acquire significant yield strength. There is thus a convenient division in time between the significant thermal and mechanical behavior.

Solidification of weld zone. During this period the molten metal gradually gains strength and elasticity. It is necessary to account for yield stress, Young's modulus, and change of Poisson's ratio with temperature.

Yielding due to thermal gradients. The shock thermal gradients emanating from the weld zones cause yielding in the surrounding material. This, together with the cooling of the specimens, results in a high residual stress being locked into the structure.

Creep. The high temperatures involved even after significant cooling result in creep of the structure. This creep also plays an important role on subsequent stress relief. It is, of course, difficult to separate creep from plastic behavior during welding, and in principle the two should be calculated simultaneously. However, from a macro-mechanical point of view, it appears justified to let creep begin only after the initial phase of solidification of the melted zone, since this time is measured in a hundred seconds, as opposed to three or four hours of high-strain creep on cooling.

Fracture of cracks. The frequency with which cracking is exhibited during welding requires that some criteria be included for introducing cracks during cooling of the welds.

Weld dressing. The removal of material during weld dressing relieves some of the residual stresses and introduces other stresses. This process should also be modeled in order to predict subsequent high-cycle fatigue of the structure.

COMPUTER PROGRAMS FOR WELDING ANALYSIS

It can be seen from the above that many physical phenomena must be included in modeling the welding process. An examination of the variation of thermal and mechanical properties with temperature as shown in Fig. 1 will show the nonlinear nature of the problem. In addition to this, special problems of phase change, addition of molten filler, cracking, and dressing should also be considered. The work is sufficiently new to require specialized attention with each analysis. In addition, so few analyses have been made that it is possible to consider all the analyses of which I am aware.

Broadly speaking, in the literature workers have tried to obtain two types of results. In the first type of results, Masubuchi [1] has paid attention to the region away from the melted zone, and stress is placed on correlation with experiment.

In the second type of results, exemplified by the work of Ueda and Tamakawa [5] and by Hibbitt et al. [7, 8], attention is focused on the melted zone as well as the region away from the melted zone. In the first type of analysis, it appears reasonable to neglect all melting effects and just ensure that the total heat transferred is distributed correctly. With this in mind, a steady-state temperature distribution is obtained, and a one-dimensional section of the structure is moved through the temperature field to calculate the corresponding stress transient. Results obtained in the region away from the weld are in good agreement with experiment. In the second type of analysis as many of the above phenomena that were required were included in the analysis. Ueda and Tamakawa [5] did not appear to include the melted zone and the thermal contact problem. Ueda and Tamakawa [5] and Hibbitt et al. [7, 8] did not include either cracking or subsequent dressing of the weld. One of Corrigan's [16] experimental results for welding of grooved discs was studied by Hibbitt and Marcal [7]. In this particular test the results were unlike the usual welding results in that the residual stresses were not near yield. This residual stress pattern appeared to have been caused by cracking.

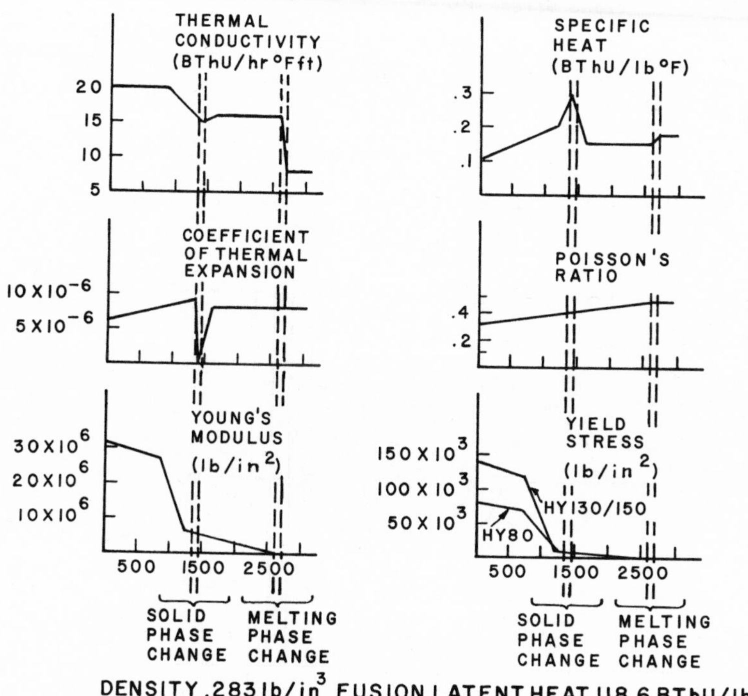

Fig. 1 Assumed material properties, HY130/150, HY80

In Table 1 I summarize the thermal analyses, classifying them according to the phenomena discussed above. Table 2 summarizes the mechanical analysis of the welding phenomena by the various types of mechanical behavior reviewed here. The analysis of Rybicki [17] has been included since the analysis, though originally applied to flame forming, would also apply here. In Table 3 I note the actual specimens analyzed by the various workers and their comments on the results obtained.

CASE HISTORIES

In this section I illustrate my discussions of the welding process with examples taken from the literature. These are intended to give the reader an idea of the various analyses that have been performed.

Plane Stress Plate

The first example illustrates the results of Masubuchi [4]. Figure 2 shows the temperature field obtained around point A. This temperature field is a steady-state field using a closed form analytical solution. When this temperature distribution is obtained, the stresses are subsequently obtained by division into transverse strips assuming plane strain in the plane. And the stress is obtained by moving a strip through the thermal field.

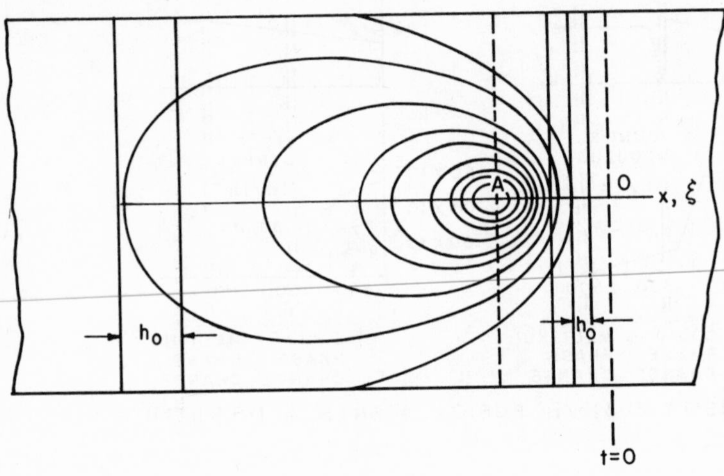

Fig. 2 Dividing the stress field into transverse strips for calculating thermal and residual stresses

Figure 3 gives the experimental and calculated results for a bead on plate weld of 2219-0 aluminum alloy. The strain vs. time plot is given for an assumed efficiency of 0.7. The results in the far field were found to be in good agreement with experiment.

Rigid Restrained Cracking Specimen

Ueda and Yamakawa [5] report on a study of rigid restrained cracking specimen (RRC specimen). The specimen is shown in Fig. 4 and is used to examine the possibility of weld cracking at butt weld connections. The specimen provides a constant intensity of external restraint during the entire course of welding. This critical restraint can be found by altering the length of the specimen and thus its intensity of restraint. The analysis was carried out on a typical RRC specimen. Its length was 200 mm.

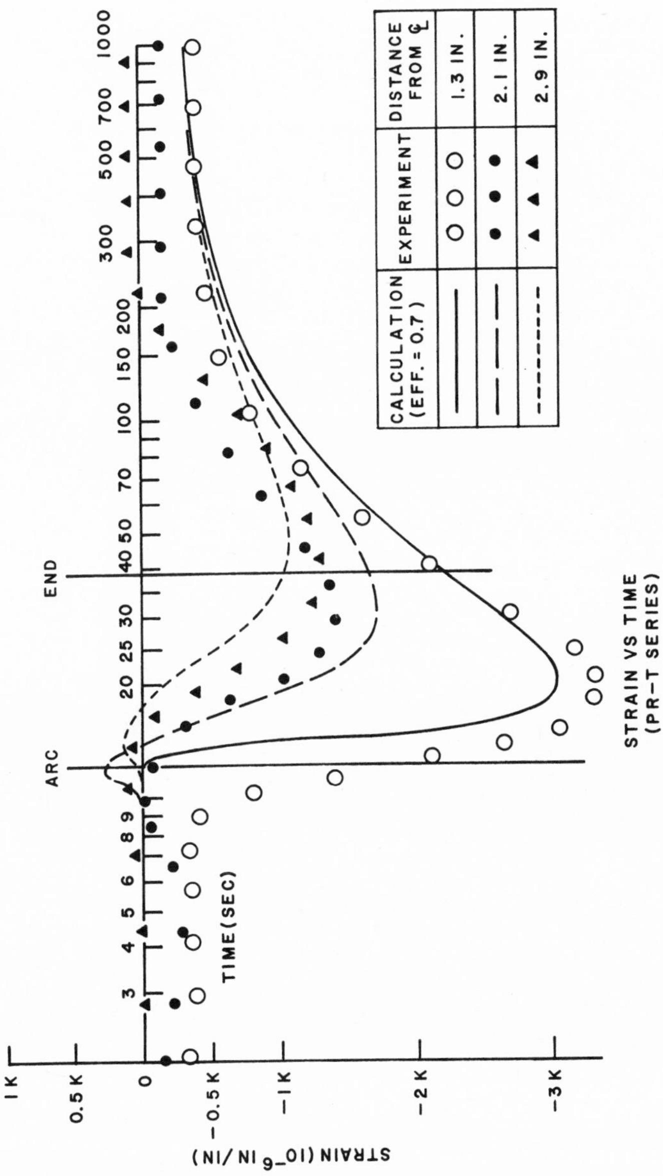

Fig. 3 Measured and computed longitudinal strains for the series 1 experiments

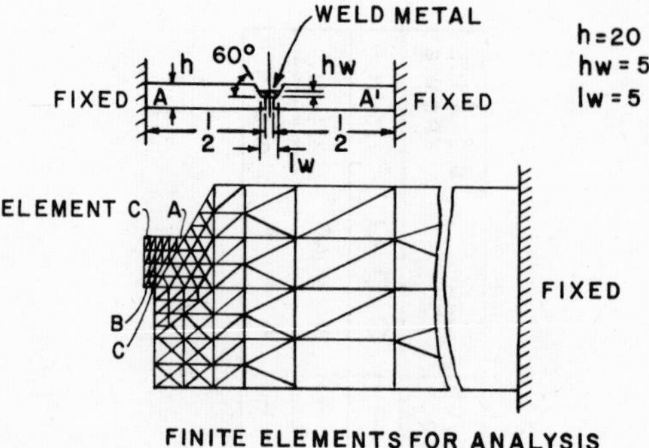

Fig. 4 Rigid restrained weld cracking test specimens and idealization for analysis

This specimen was found to crack during welding. The width of the specimen is 1 mm so that it allowed for a plane stress two-dimensional analysis. A heat input of 4000 joules/sec. was given to the weld metal for 0.4 seconds. The analysis provided the stress and strain results of Fig. 5. It is thought that these results show too small a strain to cause the cracking that was found. This suggests that a finer mesh and the inclusion of a phase change may be required. However, the results are sufficient for an empirical measure of weld cracking much in the same way that K is used in elastic fracture mechanics. By repeating future analysis with the same size mesh, one can hope to establish a criterion for cracking during welding.

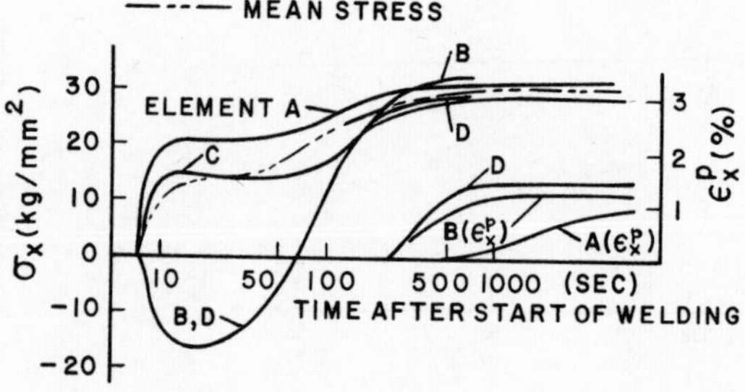

Fig. 5 Stress and strain histories in RRC test specimen of 200 mm length

Welding of HY80, 1-Inch Grooved Disk

In this final example, I illustrate the results of Hibbitt [7] by considering the study that was performed on a V-grooved disc. The specimen was studied experimentally by Corrigan [16] and was found to have a quasi-axisymmetric manner. In this analysis, axisymmetric behavior was assumed. This meant that all of the filler was assumed to be deposited at the same time.

Figure 6 gives a temperature history through the thickness at the weld. In this fairly thick specimen the melted zone did not go through the thickness.

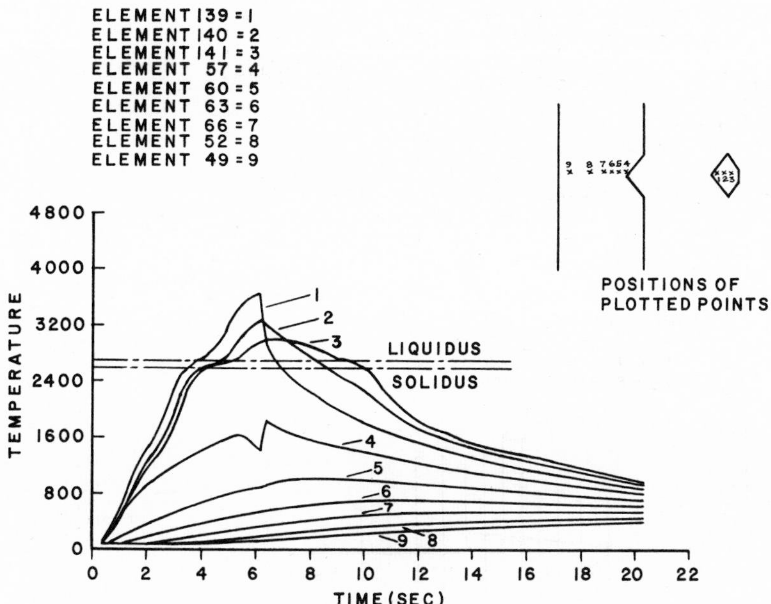

Fig. 6 Temperature history through the thickness

Figure 7 gives the distribution of the residual stress at the top and bottom of the disk together with the residual stress obtained by Corrigan [16] by means of a Sachs boring procedure. To complete the picture, I also show the welded radial stresses obtained for the HY80, 1-inch disk in Fig. 8.

Figure 9 gives a contour plot of the calculated equivalent plastic strains. Here again, as in the previous case study, a finer mesh would have produced larger strains. In the analysis of a thinner specimen, the theoretical and experimental results did not compare as well as those here, and as mentioned earlier, this was attributed to a cracking process. The importance of mentioning it here is that it appears to provide some experimental work on which to build a cracking criterion for welding analysis.

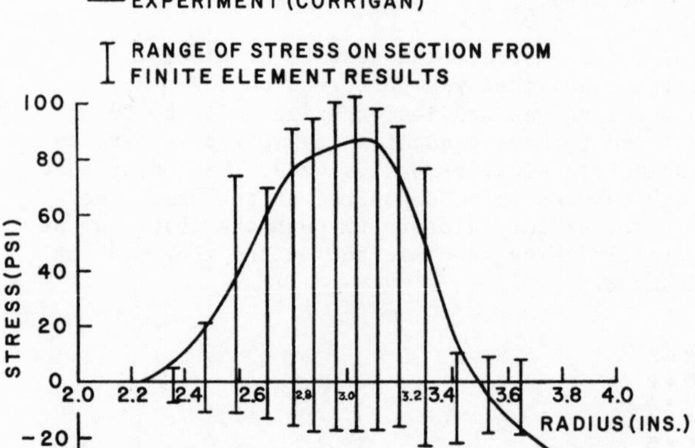

Fig. 7 As welded tangential stress, HY80, 1 in. disc

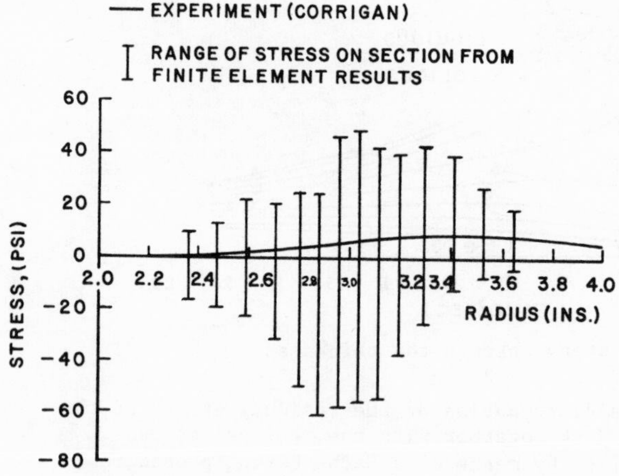

Fig. 8 Welded radial stress, HY80, 1 in. disc

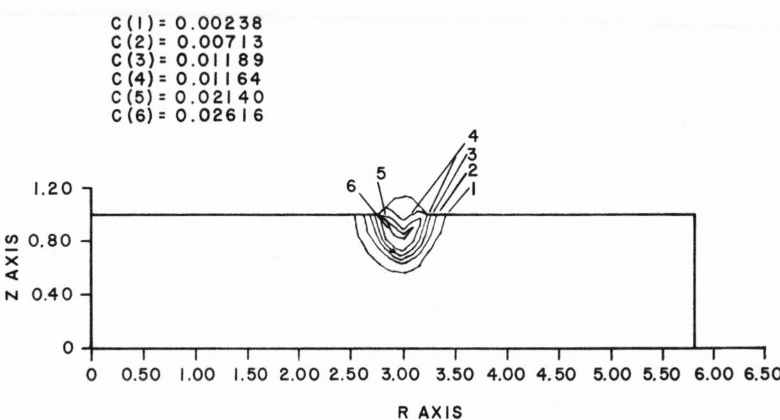

Fig. 9 Equivalent plastic strain, 1 in. disc as welded

DISCUSSION OF RESULTS

The results discussed above are typical of those obtained in the literature. On the whole the agreement with experiment is surprisingly good considering the highly nonlinear nature of the problem. This good agreement is thought to come about because of three factors. First, the uncoupling of the thermal and mechanical behavior means that only a reasonable model of each phenomenon is sufficient to produce a good result. Second, only residual stresses have been compared, and since these appear always to be close to yield, the results cannot stray too far from the experiment. Finally, the really stringent test of comparison of maximum strains and prediction of cracking has not yet been carried out.

CONCLUSIONS AND RECOMMENDATIONS FOR FUTURE WORK

The conclusion is that reasonable agreement exists between analysis and experiment for the specimens studied to date. Many more programs appear to be capable of accounting for welding in addition to those discussed here. The requirement in these cases is that the authors of the programs and their colleagues actually carry out a welding analysis. More work should be carried out on the analysis of test specimens such as the RRC in order to predict the conditions for cracking of welds. Further complications such as cracking and weld dressing should be included in future analysis.

ACKNOWLEDGMENTS

The author is indebted to the Office of Naval Research for support of this research under Contract N00014-67-A-0191-0007.

REFERENCES

1 Masubuchi, K., "Control of Distortion and Shrinkage in Welding", Welding Residual Council Bulletin, No. 149, April 1970.

2 Tall, L., "Residual Stresses in Welded Plates--A Theoretical Study," Welding Journal, Volume 43, No. 1, 1964.

3 Masubuchi, K., Simmons, F. B., Monroe, R. E., "Analysis of Thermal Stresses and Metal Movement During Welding," RSIC-820, Redstone Scientific Information Center, July 1969, Redstone Arsenal, Alabama.

4 Masubuchi, K., "Residual Stress Analysis Techniques for Weldment Studies," A Summary of MIT Studies, Proc. Spring Meeting Society of Experimental Stress Analysis, May 1972.

5 Ueda, Y., Yamakawa, T., "Thermal Stress Analysis of Metals with Temperature Dependent Mechanical Properties," Proc. of 1971 International Conference on Mechanical Behavior of Materials, Society of Materials Science, Japan, 1972, pp. 10-20.

6 Friedman, E., "Private Communication, July 1971.

7 Hibbitt, H. D., Marcal, P. V., "Numerical, Thermo-Mechanical Model for the Welding and Subsequent Loading of a Fabricated Structure," Journal Computers and Structures, Vol. 3, 1973, pp. 1145.

8 Nickell, R. E., Hibbitt, H. D., "Thermal and Mechanical Analysis of Welded Structures," Proc. 2nd Int. Conference on Structural Mechanics in Reactors Technology, Vol. II, F, Berlin, 1973.

9 Carlslaw, H. S., Jaeger, J. C., "Conduction of Heat in Solids," Pergamon Press, Oxford, 1959.

10 Murrae, W. D., Landis, F., "Numerical and Machine Solutions of Transient Heat-Conduction Problems Involving Melting and Freezing-I, Method of Analysis and Sample Solution," Transactions ASME Journal Heat Transfer, May 1959.

11 Lazaridis, A., "A Numerical Solution of the Multidimensional Solidification (or Melting) Problem," International Journal Heat Mass Transfer, Vol. 13, 1970, pp. 1459.

12 Sikarskie, D. L., Boley, B. A., "The Solution of a Class of Two-Dimensional Melting and Solidification Problems," International Journal Solids Structures, Vol. I, 1965, pp. 207.

13 Boley, B. A., "Temperature and Deformation in Rods and Plates Melting Under Internal Heat Generation," Cornell University, Report No. 4, ORN Technical. NR064-401.

14 Weiner, J. H., "Transient Heat Conduction in Multiphase Media," British Journal of Applied Physics, Vol. 6, 1955.

15 Friedman, E., Boley, B. A., "Stresses and Deformations in Melting Plates," Journal of Spacecraft, Vol. 7, 1970, pp. 324-333.

16 Corrigan, D. A. "Thermomechanical Effects in Fusion Welding of High Strength Steels," Ph.D. Thesis, MIT, 1966.

17 Rybicki, E., "A Transient Elastic-Plastic Thermal Stress Analysis of Flame Forming," Transactions ASME, Journal of Engineering for Industry, Vol. 163, February 1973.

APPENDIX

Professor Koichi Masubuchi, Department of Ocean Engineering, Massachusetts Institute of Technology, Cambridge, Massachusetts 02139, U. S. A.

Yukio Ueda, Faculty of Engineering, Osaka University, Suita, Japan.

Taketo Yamakawa, Kawasaki Heavy Industries Co., Kakogawa, Japan.

Dr. H. D. Hibbitt, Senior Research Engineer, Marc Analysis Research Corporation, 105 Medway Street, Providence Rhode Island, U. S. A.

Professor Pedro V. Marcal, Professor of Engineering, Brown University, Providence, Rhode Island, U. S. A.

E. F. Rybicki, Battelle Columbus Laboratory, 505 King Avenue, Columbus, Ohio, U. S. A.

Mr. Edward Friedman, Bettis Atomic Power Laboratory, Westinghouse Electric Corporation, P. O. Box 79, West Mufflin, Pennsylvania, U. S. A.

Table 1 Features of Thermal Analysis

Program	Phase change	External heat source	Surface heat loss	Addition of filler	Heat conduction transient	Melting range
Masubuchi [1]	No	Yes	No	No	No	No
Ueda and Yamakawa [5]	No	Yes	Yes	No	Yes	No
Hibbitt and Marcal [7]	Yes	Yes	Yes	Yes	Yes	Yes
Friedman [6]	Yes	Yes	Yes	Yes	Yes	Yes

Table 2 Features of Mechanical Analysis

Program	Temperature-dependent yield stress	Temperature-dependent elastic properties	Capability for addition and removal of material	Type of geometry	Creep	Interface with thermal analysis
Masubuchi [1]	Yes	Yes	No	1-D and 2-D	No	Yes
Ueda and Yamakawa [5]	Yes	Yes	No	2-D	No	Yes
Hibbitt and Marcal [7]	Yes	Yes	Yes	General purpose	Yes	Yes
Rybicki [17]	Yes	Yes	No	2-D	No	Yes
Friedman [6]	Yes	Yes	Yes	2-D	Yes	Yes

Table 3 Specimens Studied and Comparison with Experiment

Type of specimen	Bead on plate	Butt-weld	Flame heating	Grooved disk	One-dimensional specimen
Masubuchi [1]	Al and steel reasonable agreement with experiment	Al and steel reasonable agreement with experiment	1-D results not in agreement with experiment	HY steel results agree with experiment	Results agree with experiment
Ueda and Yamakawa [5]	Steel reasonable agreement with experiment	Steel reasonable agreement with experiment	No results	No results	Results agree with experiment
Hibbitt and Marcal [7]	Copper reasonable agreement with theory	No Results	No Results	HY steel results agree with experiment	Results agree with experiment
Rybicki [17]	None	No results	Results in reasonable agreement with experiment	No results	Results agree with experiment
Friedman [6]	Results unavailable	Results unavailable	No results	No results	Results agree with experiment

VISCOELASTIC STRUCTURES

K. K. Gupta and E. Heer
Jet Propulsion Laboratory
California Institute of Technology
Pasadena, California

ABSTRACT

The primary purpose of this article is to present a review of the readily available computer software for the analysis of viscoelastic structures. Both general purpose and specific problem-oriented finite element computer programs are included in the review. A brief description of various related analysis procedures is summarized at the beginning to afford an overall picture of such solution techniques. This is followed by a summary of each of the computer programs surveyed, including vital information as capability, usage, and availability. A critical review of the computer programs, outlining their relative merits, is also presented. Based on this information, a user guideline is outlined to aid potential users in the selection of the appropriate program for the analysis of a particular problem. Finally, shortcomings in the present state-of-the-art in viscoelastic structures analysis programs are discussed, followed by a further discussion of expected future developments in this field of analysis.

NOMENCLATURE

$\{e\}$ = Strain vector

$E_{ijk\ell}$ = General anisotropic relaxation moduli

$\{F_n\}$ = Body force vector at step n

$[K_{ij}]$ = Stiffness matrix derived from material matrix computed for the reduced time difference

m,n = Integers used to represent various time steps

$\{P_n\}$ = External load vector at step n

t,τ = Time domain

$\{T_{n,m-1}\}$ = Forces due to temperature changes

$\{u\}$ = Element displacement vector

$\{U_n\}$ = Nodal deformation vector at step n

$\{\alpha\}$ = Coefficients of thermal expansion

ξ = Reduced time scale

σ_{ij} = Stress components

INTRODUCTION

In recent years, much emphasis has been placed on the analysis of structures with materials that exhibit viscoelastic characteristics. Such problems range from solid-propellant rocket motors in aerospace engineering to concrete pressure vessels in nuclear and civil engineering. With the growing complexity in their structural forms, arising out of an increasing demand for more exacting performance, it becomes evident that sophisticated analysis tools are required for safe and economical design of these structures.

In viscoelasticity, the stress-strain equations are dependent not only on the current stress and strain state but also on the entire history of its development. The stress-strain relationships in viscoelasticity are usually expressed in terms of rate operators involving differentiation with respect to time. When such operators are linear, the material is known as linearly viscoelastic, implying validity of the superposition principle. The superposition principle is used [1] to obtain total stress at time t by approximating strain variations by a series of step functions that corresponds to a series of relaxation displacement inputs. For thermoviscoelastic materials with thermorheologically simple behavior, the concept of "reduced time" ξ is introduced by

$$\xi(x_h,t) = \int_0^t \frac{d\tau}{a[T(x_h,\tau)]} \tag{1}$$

where a(T) is the time shift function usually determined experimentally as a function of temperature T only. Such shift function dependence on time t and position x_h within the material region is implicit through T and can often be described by the well-known Williams-Landel-Ferry (WLF) equation. The constitutive equations for the present thermoviscoelastic case may then be derived as [2]

$$\sigma_{ij} = E_{ijk\ell}(\xi)\, e_{k\ell(0)}(x_h) + \int_0^t E_{ijk\ell}(\xi - \xi')\, \frac{\partial}{\partial\tau}\,(e_{k\ell} - \alpha_{k\ell}\theta)\, d\tau \tag{2}$$

in which $e_{k\ell(0)}$ is the initially induced strain at t = 0, $E_{ijk\ell}$ being the general anisotropic tensorial relaxation moduli having 21 independent constants defined as relaxation stress per unit applied strain, and τ is the time variable. The kernel of the hereditary integral may be simply considered as the memory function transforming the influence of pulse strain at time τ to the time instant t. Similar relationship also exists when the material is subjected to creep loading. At this stage it may be pointed out that the integral form of Eq. (2) is equivalent to the original differential operator form. Materials that follow such a linear hereditary law are defined as linear viscoelastic. Nonlinear effects, in which the material property tensor depends on the stress or strain state, can be included, requiring only the step-by-step determination of the relevant material properties.

Numerical analysis of viscoelastic problems may be achieved by either a step-by-step solution procedure or by the integral transform approach. However, for complicated loading and material property relationships, the latter method proves ineffective. As such, most general-purpose computer programs adopt the first method in the solution procedure. Such a step-by-step analysis technique involving finite element formulation in space and finite difference equations in time has been developed earlier [2] for the quasistatic analysis of linear thermoviscoelastic structures, where the material has been assumed to behave in a thermorheologically simple way, when the temperature-time equivalence is expressed in terms of the reduced time $\xi(t)$. Also the concept of synchronized material properties has been adopted in the analysis, when such properties and external loadings are assumed to be a function of only one independent parameter.

The numerical analysis procedure for quasistatic problems in linear thermoviscoelasticity is formulated by a step-by-step incremental process. The basic assumption of thermorheologically simple material behavior enables characteristic functions to be singly defined for the entire temperature range in the time domain. The usual field equations for viscoelastic materials may then be extended for the thermoviscoelastic case. This is achieved by introducing the concept of "reduced time," as defined by Eq. (1), when all characteristic functions fulfill the same time-temperature shift and can be represented as a function of reduced time. The relationship of Eq. (1) signifies that all characteristic functions, such as the relaxation moduli of a thermoviscoelastic material at any arbitrary temperature T corresponding to time t, may then be expressed by their behavior at reference temperature T_0 on the new reduced time scale ξ. Each relaxation modulus signifying relaxation stress variation for unit strain applied initially is then expressed as

$$E^{T}_{ijk\ell}(t) = E^{T_0}_{ijk\ell}(\xi) \qquad (i = j = k = \ell = 1,2,3) \tag{3}$$

$E_{ijk\ell}(t)$ being the general anisotropic relaxation moduli. The constitutive equations are derived from superposition principles in the form of hereditary integrals as defined by Eq. (2). In addition, the following equations are required to define completely the field equations:

Equilibrium equations

$$\sigma_{ij,j} + f_i = 0 \tag{4}$$

Strain-displacement equations

$$e_{ij} = \frac{1}{2}(u_{i,j} + u_{j,i}) \tag{5}$$

in which f_i is the body force component per unit volume. The field equations may next be expressed in incremental forms by subdividing the time domain into arbitrary intervals $\Delta t(m)$. Equations (4), (5) and the stresses defined by Eq. (2) then assume the following form:

$$\Delta\sigma_{ij(m),j} + \Delta f_{i(m)} = 0 \tag{6}$$

$$\Delta e_{ij(m)} = \frac{1}{2}(\Delta u_{i(m),j} + \Delta u_{j(m),i}) \tag{7}$$

$$\sigma_{ij(n)} = E_{ijk\ell}(\xi_{(n)})\, e_{k\ell(0)} + \int_0^{t(n)} E_{ijk\ell}(\xi_{(n)} - \xi')\, \frac{\partial}{\partial\tau}(e_{k\ell} - \alpha_{k\ell}\theta)\, d\tau \tag{8}$$

Then Eq. (8) may finally be approximated and expressed in matrix form as

$$\left\{\sigma_n\right\} = \left[\mathbf{E}(\xi_n)\right]\left\{\mathbf{e}_0\right\} + \sum_{m=1}^{m=n}\left[\mathbf{E}(\xi_n - \xi_{m-1})\right]\left\{\Delta\mathbf{e}_m - \Delta(\alpha\theta)_m\right\} \tag{9}$$

The continuum is divided into small finite elements, when piece-wise continuous displacement fields are prescribed for each such elements in terms of their time-dependent nodal function values. Incremental equilibrium load deflection equations of the entire structure are then obtained by minimizing the total potential energy with respect to such nodal parameters.

The step-by-step matrix incremental equation in global coordinate system is given by

$$[\mathbf{K}_{n,n-1}]\{\Delta\mathbf{U}_n\} = \{\mathbf{P}_n\} - \sum_{m=1}^{m=n-1}[\mathbf{K}_{n,m-1}]\{\Delta\mathbf{U}_m\} - [\mathbf{K}_n]\{\mathbf{U}_0\} + \sum_{m=1}^{n}\{\mathbf{T}_{n,m-1}\} + \{\mathbf{F}_n\} \tag{10}$$

with

$[\mathbf{K}_{i,j}]$ = stiffness matrix derived from material matrix computed for the reduced time difference $\Delta\xi_{ij} = \xi_j - \xi_i$

$\{\mathbf{P}_n\}$ = external load vector at step n

$\{\mathbf{T}_{n,m-1}\}$ = forces due to temperature changes

$\{\mathbf{F}_n\}$ = body forces vector

and where the summation, as usual, signifies the memory of the material. Expressions for stresses and other relevant results are given in [2]. However, it becomes apparent from the nature of Eq. (10) that computation time for Eq. (10) may be excessive after a few time steps. This is due to the fact that at each time step, recomputation of solution results is required for all preceding time steps, which are then added to obtain the final solution. A general purpose computer program VISCEL (VISCoELastic analysis program) [3,4] developed in this connection, however, minimizes such computation efforts by providing an option by which time steps may be so chosen that previous time intervals become a subset of the following time intervals. In such cases, the time intervals tend to remain constant in the logarithmic scale; thus it is then possible to cover a long time domain with relatively small computational effort.

The element stresses may then be obtained from Eq. (9), when element strains are derived from the usual relationship:

$$\left.\begin{aligned} \{\bar{u}^e\} &= \lambda\{U^e\} \\ \{u\} &= a\{\bar{u}^e\} \\ \{e\} &= b\{u\} \end{aligned}\right\} \tag{11}$$

$\{\bar{u}^e\}$, $\{U^e\}$ being element nodal displacements in the local and global coordinate system, respectively.

Another analysis procedure [2,5] assumes that the components of the material property matrix inherent in the formulation of Eq. (2) may be represented by suitable exponential series of the following form:

$$[\mathbf{E}(\xi_n - \xi_{m-1})] = [\mathbf{E}_0] + \sum_{r=1}^{r=s} [\mathbf{E}_r] \exp\{-(\xi_n - \xi_{m-1})\,\tau_r\} \tag{12}$$

in which the coefficients $[E_r]$ and the characteristic relaxation times τ_r are chosen such that experimental data or a particular discrete linear viscoelastic model is represented with sufficient accuracy. In such cases, the evaluation of the associated displacement increments requires information from only the immediately preceding time step [2].

The above formulation denoted by Eq. (2) is based on the assumption that the material behaves in a thermorheologically simple manner; thus, for temperature changes, the characteristic functions, both creep and relaxation, show pure shift when they are plotted against the logarithm of time. However, the shift function may also be a function of the applied stresses and the induced strains and their time derivatives. Such nonlinear material behavior, on the other hand, requires only the step-by-step determination of the shift function after each time step in the numerical computations, while the finite element formulation remains unchanged. Apart from the usual nonlinear viscoelastic cases, the viscoelastoplastic materials also behave in a nonlinear fashion when the superposition principle is no longer valid. This review deals with computer programs relating to the analysis of linear and nonlinear viscoelastic structures. Thus, such a structure, when subjected to stresses/deformations, may exhibit linear or nonlinear creep/relaxation phenomena. Nonlinearities may also be encountered due to viscoelastoplastic behavior and when the material properties are stress dependent.

DESCRIPTION OF COMPUTER PROGRAMS

A large number of computer programs, capable of performing viscoelastic structural analysis, were reviewed in connection with the current study. This study is based on firsthand knowledge of some programs and also several relevant program manuals, as well as the information gathered from standard questionnaires provided by developers and users of various computer programs. Table 1 provides important features of such programs. The bulk of the information was gathered from the various questionnaires as well as from extensive personal contacts established with the developers and users alike of the relevant programs. Our direct experience with such programs has been limited to only a few, which is perhaps typical of most program developers. Also, in selecting the major programs presented herein, a number of small programs with rather limited capabilities and scope have been omitted from the present consideration. In that process it is conceivable that some relevant programs may have been inadvertently omitted.

In this connection, it is to be noted that no single program may be used for effective and economical analysis of all viscoelastic problems. Each program, on the other hand, may prove suitable for some specific range of problems. The computer programs presented herein may be divided into two distinct categories. Programs such as VISCEL [3,4], VISCO-3D [6], and others were specifically designed for the analysis of viscoelastic structures, whereas some such analysis capabilities are often "spin-offs" of some larger general (multi-)purpose computer programs such as MARC [7] or ANSYS [8]. The salient features of such programs are described next.

PROGRAMS SPECIFICALLY DEVELOPED FOR ANALYSIS OF VISCOELASTIC STRUCTURES

VISCEL — A General-Purpose Computer Program for Analysis of Linear Viscoelastic Structures

The VISCEL computer program was developed in connection with the stress analysis of solid-propellant rocket motors [9]. The program is capable of solving equilibrium problems associated with one-, two-, or three dimensional linear thermoviscoelastic structures. Since the program is an extension of the linear equilibrium problem solver ELAS [10], it yields, at the beginning of the initial time step, elastic solutions of structures. A synchronized material property concept utilizing incremental time steps and the finite element matrix displacement approach has been adopted for the current analysis. Resulting recursive equations of the form of Eq. (10) incorporating the memory of material properties are solved at the end of each time step of the general step-by-step procedure in the time domain. A special option enables employment of constant time steps in the logarithmic scale, thereby reducing computational efforts resulting from accumulative material memory effects. A wide variety of structures with elastic or viscoelastic material properties can be analyzed by VISCEL.

VISCEL is written in FORTRAN V language to run on the UNIVAC 1108 computer under the EXEC 8 system. Dynamic storage allocation is automatically effected and a user may request up to 195K core memory in a 260K UNIVAC 1108/EXEC 8 machine. The program, consisting of about 7200 instructions, is divided into four distinct links, namely, input, generation, deflection, and stress links; the compiled program occupies a maximum of about 11,700 decimal words of core storage. VISCEL is stored on a magnetic tape and is available from the Computer Software Management and Information Center (COSMIC), the NASA agency for distribution of computer programs. The present program is an updated and extended version of its earlier form [11] written for the IBM 7094 machine. Essential features of the program are summarized next.

Structural Geometry

Almost any structure with linear viscoelastic material properties can be handled by VISCEL. Such structures may be one-, two-, or three-dimensional in nature. Since the program adopts an in-core solution procedure, the magnitude of the problem that can be solved by VISCEL is dependent on the available core storage of the particular machine being used. Moreover, the program uses various Fastrand (drum) file storage units as additional stores during execution of the program [3]. Among others, the program can handle such structures as frames, plates, shells, and solids and their appropriate combinations.

Material Properties

Material properties may be temperature dependent in the thermorheologically simple sense. Thus for temperature changes the characteristic functions, both creep and relaxation, are assumed to show pure shift when they are plotted against the logarithm of time. Such materials are well suited for a reasonably complete characterization over a rather large range of time and temperature scale since the rheological behavior of the materials can be described for the entire temperature range as a single function of time and temperature. Thus when any characteristic function, such as the relaxation modulus, is plotted against reduced time, all curves will fall on the single curve for initial temperature T_0. As such, it is then necessary to determine relaxation/creep functions for one temperature only.

Associated computational schemes adopt the concept of synchronized material properties for thermorheologically simple materials. Thus material properties and external loadings, both mechanical and/or thermal, are expressed as a function of one parameter ξ, which may represent time, reduced time, or any other suitable variable. Such material and load data are assumed to exist in functional form (Fig. 1), which are to be presented at each time step in the shape of predetermined tabulated values obtained experimentally or derived from analytical considerations; any interaction between them is assumed to be either ignored or included in such values. The material can be isotropic, orthotropic, or general, provided appropriate material definition is available from experimental results in which rheological properties of the material are considered. For this analysis, it is required to have a knowledge of the modulus functions (relaxation-type functions). Also, the material is assumed to be at least slightly compressible.

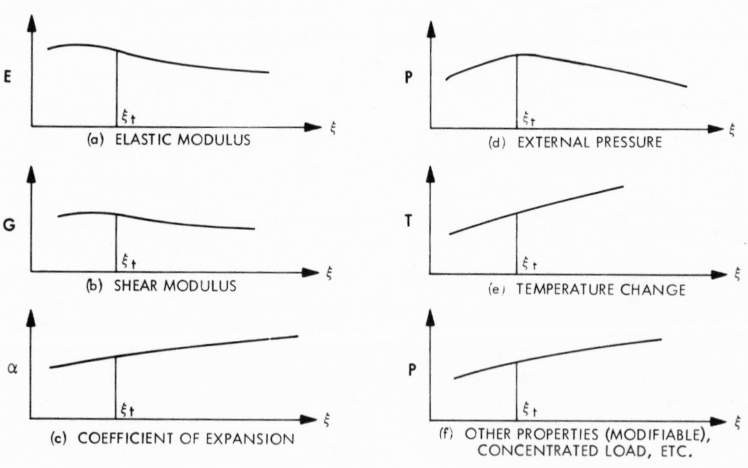

Fig. 1 Schematic representation of the material properties and external disturbances at ξ_T

Linear Viscoelastic Behavior

In this analysis the strains are assumed to be linear functions of stresses, but are strongly dependent on the loading history, which implies that when all loads are doubled the resulting deformations are doubled, too. Thus, the creep/relaxation laws are linear in stress/strain, and as such, the principle of superposition is valid for these cases. Furthermore, geometric nonlinearities, such as large deformations or large strains, are not considered for the present analysis.

Deflection Boundary Conditions

Deflection boundary conditions are assumed to remain unaltered throughout an entire time domain of computation that is defined initially at the beginning of the initial time step. The usual linear elastic analysis of the structure is effected at the beginning of the initial time step.

Special Incremental Procedure

From the nature of the incremental Eqs. (9) and (10), it becomes apparent that computation time may be excessive after a few time steps. This is because, at each time step, recomputation of solution results is required for all preceding time steps, which are then added to achieve the final solution. The program, however, provides a special option to minimize these computational efforts, when the time steps may be so chosen that previous time intervals become a subset of the following time intervals. Thus the parameter ξ is expressed as the summation of incremental value $\Delta\xi$:

$$\xi^i_j = \sum_{i=1}^{M} \sum_{j=1}^{N(i)} \left(N(i)^{i-1} \, \Delta\xi \right)_j \tag{13}$$

in which ξ^i_j is the value of the relevant parameter at the end of $n = i \times j$ steps, M defines the total number of time step groups, and N(i) is the number of time steps in the ith group. The integer values for M and N can be chosen optionally by the user, and this scheme may be conveniently employed for solution of the relevant recursive equations, such as Eqs. (9) and (10), provided material properties and external disturbances are evaluated for each ξ^i_j. It is then possible to cover a long time domain with relatively small computational effort since in such cases the time intervals tend to remain constant in the logarithmic scale. Details of such a typical computation scheme for values of M = 3, N(1) = 2, N(2) = 3, and N(3) = 2 are shown in Fig. 2.

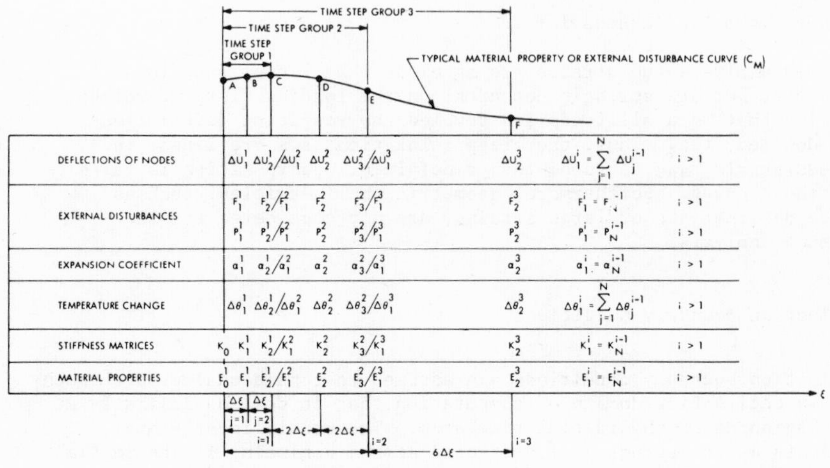

Fig. 2 Typical ξ interval setup

General Discussion

Several important features render the VISCEL program rather attractive. These features include dynamic memory allocation, optional node relabeling scheme, and boundary condition imposition during assembly of the stiffness matrix and its storage within a variable bandwidth. A variable bandwidth Cholesky scheme is employed to solve the set of linear simultaneous equations at the end of each time step. Since the program is divided into four segments, core requirement for the program is rather modest. A plotting program, with a large number of options, has now been added to the present program, which enables relevant plotting of the structural geometry as well as nodal deformations and stresses during any stage of computation. Also, various error messages and diagnostics prove to be most useful in checking and debugging the input data. The average run time for a typical problem involving approximately 1000 degrees of freedom is about 4.5 minutes per time step. Since the program is specifically developed for in-core solution of viscoelastic structures, it proves to be most efficient for analysis of such structures within the bound of available computer core storage. The VISCEL User's Manual [3] provides details of the input/output procedure as well as explanations for various error messages in addition to other general essential program details. The VISCEL Program Manual [4], on the other hand, provides information on the software and its interaction with the machine.

VISCO-3D — A Computer Program for Analysis of Three-Dimensional Viscoelastoplastic Structures

This computer program [6] was developed for viscoelastoplastic analysis of three-dimensional solids using tetrahedral finite elements. Initially, the structural material type is restricted to isotropic ones only. The program, consisting of about 6000 instructions, has been written in FORTRAN IV for the UNIVAC 1108 computer utilizing a 65K core, the compiled program occupying about 13K of the core storage. Both in-core and out-of-core problem solutions may be effected, and as such, there is practically no limitation to the size of the problem that can be analyzed by the program; the program developer quotes maximum problem size to be limited to 10,000 unknowns and 750 bandwidth. External storage consists of a maximum of four tapes and 1.2 million words of disc or drum scratch area.

The structural material may be either linear or nonlinear in nature. Thermoviscoelastic problems can be handled without difficulty. A memory theory of creep has been adopted for the analysis; nonlinear creep analysis under high temperature may be performed by the program. Furthermore, the program caters to such nonlinearities as viscoplasticity and stress dependence of material properties.

An alternating component iterative method has been adopted for numerical solution of the incremental equations. The components of the material property matrix are represented by an appropriate exponential series, which enables computation of displacement increments from information obtained from the previous time step only. The structures may be subjected to both mechanical and thermal loadings. Other attractive features of the program include internal mesh generation and input and output interfacing with plotting programs. Typical computer run time pertaining to a 3000-order problem for a 10-step incremental solution appears to be of the order of 3.5 hours on a UNIVAC 1108 machine. The program is available from its developer on a lease/purchase basis.

CREEP-PLAST — A Computer Program for Viscoelastoplastic Analysis of Two-Dimensional Axisymmetric Structures

The CREEP-PLAST [12] program has been primarily designed for the viscoelastoplastic analysis of two-dimensional structures including axisymmetric shells and solids of revolution. Basic features of this program are similar to that of VISCO-3D [6] described earlier. The program, written in FORTRAN IV for the IBM, CDC, and UNIVAC 1108 machines, consists of about 5000 instructions. Since the program effects out-of-core solutions, if necessary, it can be conveniently utilized for solution of rather large practical problems. Both tape and disc/drum storage are used by the program. External loading may be mechanical and/or thermal in nature.

Both linear and nonlinear materials can be handled by the program, assuming the materials are isotropic. Temperature and stress dependence of material properties as well as the viscoelastoplastic behavior of the material are considered by the program. Numerical solutions of the incremental equilibrium equations are obtained by an alternating component iterative procedure. The program also possesses such important features as internal mesh generation and plotting of various analysis results and structure geometry. Typical problem solution time is similar to the VISCO-3D program. The program is available from the developer as well as from the Oak Ridge National Laboratory, who funded the program development.

VISPAX — Nonlinear Analysis of Two-Dimensional Structures

The computer program VISPAX [13] is capable of solving viscoelastoplastic problems of two-dimensional structures like axisymmetric shells and plane stress problems. A step-by-step incremental solution procedure with equilibrium check has been adopted in conjunction with the finite element discretization procedure. The associated computer program, written in FORTRAN IV for the CDC 6400 machine, consists of about 3500 programmed instructions and is available from the University of California at nominal cost. Since the program effects in-core solution of problems, the magnitude of the problem is limited by the available core size of the computer.

Linear as well as nonlinear problems may be considered by the program. Thus the material may be viscoelastoplastic in nature. Also, the program caters to large deformations as well as temperature and stress dependence of the material. The incremental solution procedure may adopt variable time steps, if required.

ELAS 55 — A General Purpose Computer Program for Viscoelastic Analysis of Structures

This recently developed general purpose computer program [14] is designed for the in-core solution of almost any one-, two-, or three dimensional viscoelastic structures discretized by the finite element method. A step-by-step incremental solution procedure has been adopted for such analysis when it is possible to use the Runge-Kutta type extrapolation by input control. The program is written in FORTRAN IV for the IBM 370/165 computer and is available from Duke University at nominal cost. About 7000 source statements are involved in the program.

The viscoelastic material may be either linear or nonlinear in nature. Such materials include viscoelastoplastic and also stress-dependent materials. Further, the material could be temperature-dependent in the general sense; isotropic, orthotropic, as well as anisotropic materials, may be handled by the program. Other important features, such as automatic mesh generation and plotting capability, render the program useful and attractive. Rather large problems can be solved by the program, the maximum size of such problems being limited only by the core storage of the computer. The structures may be subjected to both thermal and mechanical loading. Relevant program documents are expected to be released by the authors by June 1974. No user comments are available at this stage since the program is rather new.

MULTIPURPOSE PROGRAMS WITH SPECIFIC VISCOELASTIC ANALYSIS CAPABILITIES

ANSYS — A General Purpose Computer Program

ANSYS (Engineering ANalysis SYStem [8] is an efficient general purpose finite element computer program developed for the static and dynamic analysis of almost any practical structure. A special option enables the analysis of creep problems. The material includes plasticity and, furthermore, may be temperature dependent. Both geometric and physical nonlinearities are handled by the program. The external loading may be mechanical and/or thermal, and the program possesses automatic mesh generation and plotting capabilities.

The program consists of about 30,000 FORTRAN IV instructions and can be run on the CDC, IBM, and UNIVAC 1108 machines. Usage of the program is effected on a rental basis. Since the program possesses out-of-core solution capability, rather large practical problems may be solved. Both isotropic and anisotropic materials may be considered by the program.

MARC — Nonlinear Finite Element Analysis Program

The MARC [7] finite element computer program is general purpose in nature and proves to be particularly attractive for solution of nonlinear problems. Thus the program is capable of solving both linear and nonlinear creep problems. Such material properties can be temperature or stress dependent. Also, geometrical nonlinearities, such as large deformations, may be considered along with creep. Furthermore, viscoplastic materials are also handled by the program. The material may be isotropic or anisotropic, whereas external disturbances are mechanical and/or thermal in nature.

The program is written in FORTRAN IV to run on the CDC, IBM, and UNIVAC 1108 machines. For large problems, the program employs an out-of-core solution technique. Other important features of the program include automatic mesh generation and plotting facilities. The program is available from the developer on a rental basis.

Other General Purpose Programs

Several general purpose finite element computer programs, capable of performing creep analysis, have also been reviewed in connection with the present work. The programs ASAS, ASKA [15], PAFEC 70 [16], and SESAM-69 C [17] were found to be the most suitable for primary creep analysis. These programs have been designed for the analysis of almost any practical structure, when the material is isotropic or anisotropic in nature. Another general purpose program, NEPSAP [18], on the other hand, is suitable for the analysis of secondary creep problems. Automatic mesh generation and plotting routines form an integral part of such programs. The programs are written in FORTRAN IV to run on the CDC, IBM, and UNIVAC machines, among others.

Most of these programs are available on a rental/purchase basis. The program developers should be contacted for further information in this regard.

GENERAL DISCUSSION

Table 1 summarizes the essential features of the viscoelastic structural analysis programs reviewed above. The table should prove useful to potential users of viscoelastic programs in choosing the appropriate one that will solve his particular problem in the most economical and efficient manner. The programs are available for public usage on a rental/lease basis, or they may be purchased at a nominal price of around $500. In this connection, it may be observed that no single program may ever prove to be the optimum one for the entire range of problems encountered in practice. The program to be chosen for a particular problem depends on a number of factors, such as size of problem, material, and loading type. Apart from economic considerations from the point of view of run time and actual cost incurred in the solution process, other relevant factors must also be taken into consideration when selecting a program. Thus automatic data handling facilities that include automatic mesh generation and plotting and also the ease with which the user may interact with such facilities are important considerations in the selection process.

The present review suggests the existence of a rather limited number of general purpose computer programs developed solely for the analysis of viscoelastic structures. Thus VISCEL [3,4] appears to be one of the earliest general purpose programs that has also been extensively used for the solution of a large number of practical problems [19]. Other viscoelastic programs, such as VISCO-3D [6], CREEP-PLAST [12], and VISPAX [13], are limited to specific structure geometries as explained earlier. For linear viscoelastic problems, VISCEL will prove to be economic and efficient, particularly for medium-sized structures. When nonlinear problems are involved, VISCO-3D and CREEP-PLAST may be effectively utilized for relevant analysis of three-dimensional solids and two-dimensional axisymmetric problems, respectively. The information provided by the authors of the ELAS 55 program in the standard developer's questionnaire indicates that it may be capable of handling both linear and nonlinear viscoelastic problems. Documentation of this new program is expected to be published by June 1974.

Most large general purpose programs reviewed for this work possess the creep analysis capability as a spin-off in the context of their much broader scope. Thus the programs MARC [7] and ANSYS [8] are capable of solving both linear and nonlinear creep problems when both geometrical and physical nonlinearities can be considered. Similar analysis capabilities exist for the programs ASAS [15] and ASKA [16]. Generally speaking, for the solution of a typical problem, these large programs will tend to be more expensive when compared to the viscoelastic programs developed specifically for analysis of viscoelastic structures.

Future developments will, perhaps, be concerned with the extension in the capabilities of existing general purpose viscoelastic analysis computer programs to include all practical linear/nonlinear effects and also efficient out-of-core solution techniques for very large-order problems.

ACKNOWLEDGMENT

This paper presents the results of one phase of research carried out at the Jet Propulsion Laboratory, California Institute of Technology, under Contract No. NAS 7-100, sponsored by the National Aeronautics and Space Administration.

REFERENCES

1 Lee, E. H., "Viscoelasticity," Handbook of Engineering Mechanics, Chapter 53, McGraw-Hill, New York, 1962, pp. 53-1 to 53-22.

2 Heer, E., and Chen, J. C., "Finite Element Formulation for Linear Thermoviscoelastic Materials," Technical Report 32-1381, Jet Propulsion Laboratory, Pasadena, Calif., June 1, 1969.

3 Gupta, K. K., Akyuz, F. A., and Heer, E., "VISCEL — A General-Purpose Computer Program for Analysis of Linear Viscoelastic Structures: Vol. I, Revision 1, User's Manual," Technical Memorandum 33-466, Jet Propulsion Laboratory, Pasadena, Calif., October 1, 1972.

4 Gupta, K. K., Akyuz, F. A., and Heer, E., "VISCEL — A General-Purpose Computer Program for Analysis of Linear Viscoelastic Structures: Vol. II, Program Manual," Technical Memorandum 33-466, Jet Propulsion Laboratory, Pasadena, Calif., July 15, 1972.

5 White, J. L., "Finite Elements in Linear Viscoelasticity," Proc. 2nd Conference on Matrix Methods in Structural Mechanics, AFFDL-TR-68-150, 1968, pp. 489-516.

6 Rashid, Y. R., VISCO-3D — A Computer Program for Analysis of Three-Dimensional Viscoelastic Structures, User's Manual, General Electric Co., Los Gatos, Calif., November 1972.

7 Marcal, P., MARC-CDC User Information Manual, Control Data Corporation, Minneapolis, Minn., 1973.

8 DeSalvo, G. J., and Swanson, J. A., ANSYS: Engineering Analysis System Users Manual, Swanson Analysis Systems, Inc., Elizabeth, Pa., 1972.

9 Akyuz, F. A., and Heer, E., "Stress Analysis of Solid Propellant Rocket Motors," Technical Report 32-1253, Jet Propulsion Laboratory, Pasadena, Calif., July 1, 1968.

10 Utku, S., and Akyuz, F. A., "ELAS — A General-Purpose Computer Program for the Equilibrium Problems of Linear Structures: Vol. I, User's Manual," Technical Report 32-1240, Jet Propulsion Laboratory, Pasadena, Calif., Feb. 1, 1966.

11 Akyuz, F. A., and Heer, E., "VISCEL — A General-Purpose Computer Program for Analysis of Linear Viscoelastic Structures, Vol. I, User's Manual," Technical Memorandum 33-466, Jet Propulsion Laboratory, Pasadena, Calif., February 15, 1971.

12 Rashid, Y. R., "Theory Report for CREEP-PLAST Computer Program, Part I, Users Manual," AEC Research and Development Report, GEAP-13262-1, March 1973.

13 Nagarajan, S., "Nonlinear Static and Dynamic Analysis of Shells of Revolution Under Axisymmetric Loading," Doctoral Thesis, UC SESM 73-11, University of California, Berkeley, 1973.

14 Utku, S., Tarn, J. Q., and Dvorak, G. J., ELAS-55 Computer Program, Structural Mechanics Series 21, School of Engineering, Duke University, due to be published in June 1974.

15 User's Manual, ASKA I; also Program Developer's Questionnaire, December 1973, prepared by the ASKA Group, Stuttgart, W. Germany, for the "International Symposium on Structures Software," University of Maryland, June 1974.

16 Henshell, R. D., PAFEC 70+, User's Manual, Mechanical Engineering Department, University of Nottingham, England, 1972.

17 Roren, E. M. Q., et al., "The Application of the Superelement Method in Analysis and Design of Ship Structures and Machinery Components," presented at the National Symposium on Computerized Structural Analysis and Design, George Washington University, Washington, D. C., March 27-29, 1972.

18 Sharifi, P., and Yates, D. N., "Nonlinear Thermo-Elastic-Plastic and Creep Analysis by the Finite Element Method," AIAA Paper 73-358, presented at the AIAA 14th Structures Conference, Williamsburg, Va., March 1973. Also NEPSAP User's Manual.

19 Salama, M. A., Rowe, W. M., and Yasui, R. K., "Thermoelastic Analysis of Solar Cell Arrays and Their Material Properties," Technical Memorandum 33-626, Jet Propulsion Laboratory, Pasadena, Calif., September 1, 1973.

Table 1 Summary of Important Features of Viscoelastic Structural Analysis Computer Programs

Item	Program name	Developer	Availability	Cost	Structure geometry		Viscoelastic material					Material symmetry	
							Linear	Nonlinear			Temp. dependence		
					General	Other		Creep/ relax.	Stress depen- dence	Viscoelas- toplastic		Isotropic	Anisotropic
1	VISCEL	K.K. Gupta, E. Heer, & F.A. Akyuz (JPL)	COSMIC, Univ. of Georgia, Athens, Georgia	Nominal	x		x				Thermo- rheologi- cally simple	x	x
2	VISCO-3D	Y.R. Rashid	Y.R. Rashid, GE Co., Los Gatos, Calif.	Lease/ purchase basis		3-D solids	x	x	x	x	x	x	
3	CREEP-PLAST	Y.R. Rashid	Y.R. Rashid; also J. Crowell, Oak Ridge Natl Lab., Tenn.	Nominal		2-D axisymm. structures	x	x	x	x	x	x	
4	VISPAX	S. Nagarajan	Univ. of Calif., Berkeley	Nominal		Axisymm. shells	x		x	x		x	
5	ELAS 55	S. Utku, et al.	Duke Univ., Durham, N.C.	Nominal	x		x	x	x	x	x	x	x
6	ANSYS	J. Swanson	Swanson Analysis Systems, Inc., Elizabeth, Pa.	Rental basis	x		Creep			x	x	x	x
7	MARC	P. Marcal	MARC Analysis Research Corp., R.I.	Rental basis	x		x	x	x	x	x	x	x
8	ASAS	Atkins R&D	Atkins & Partners Epsom, Surrey, Eng.		x		Creep			x	x	x	x
9	ASKA	ASKA Group	E. Schrem, Stuttgart Univ., W. Ger.		x		Creep			x	x	x	x
10	PAFEC 70	R.D. Hen- shell, et al.	Univ. of Nottingham, England		x		Creep			x	x	x	x
11	NEPSAP	P. Sharifi	Lockheed, Loads & Structures Dept., 8-12, Sunnyvale, Calif.	Rental basis	x		Sec- ond- ary creep			x	x	x	
12	SESAM-69 C	Det Norske Veritas	Det Norske Veritas, Olso, Norway		x		x					x	x

Table 1 (cont.)

Item	Structure type			Finite element structural discretization	Loading			Load data input		Material data input		Solution procedure		Solution		Problem size limitations (approx)
	Linear	Nonlinear			Mech.	Thermal	Function of time	Discrete data set	Function form	Discrete data set	Function form	Step-by-Step incremental	Other	In core	Out-of-core	
		Geom.	Matl.													
1	x			x	x	x		x		x		x	Variable time step	x		Machine core dependent
2	x		x	x	x	x		x		x		x		x	x	None
3	x		x	x	x	x		x		x		x		x	x	None
4	x	x	x	x	x			x		x		x	Variable time step	x		Machine core dependent
5	x		x	x	x	x		x		x		x		x		Machine core dependent
6	x	x	x	x	x	x		x		x		x		x	x	None
7	x	x	x	x	x	x	x	x	x	x	x	x		x	x	None
8	x	x	x	x	x	x		x		x		x		x	x	None
9	x		x	x	x	x		x		x		x		x	x	None
10	x			x	x	x		x		x		x		x	x	None
11	x	x	x	x	x	x		x		x		x		x	x	None
12	x			x	x	x		x		x		x		x	x	None

Table 1 (cont.)

Item	Automatic mesh generation	Plotting capability	Computer		Storage		Number of source statements (approx)	Minimum core requirement (approx)	Average run time a/b/c*	Documentation	
			Type	Language	Core	External storage				Available	Latest date
1		x	UNIVAC 1108	FORTRAN V	x	x	7,200	20K	4.5^m/1000/1	x	Oct 1972
2	x	x	UNIVAC 1108	FORTRAN IV	65K	x	6,000	20K	3.5^h/3000/1/10	x	June 1972
3	x	x	CDC, IBM UNIVAC 1108	FORTRAN IV	x	x	5,000	20K		x	1971
4			CDC 6400	FORTRAN IV	x	x	3,500	20K		x	1973
5	x	x	IBM	FORTRAN IV	x	x	7,000	20K			June 1974 (expected date)
6	x	x	CDC, IBM, UNIVAC 1108	FORTRAN IV	x	x	30,000	50K		x	Jan 73
7	x	x	CDC, IBM, UNIVAC 1108	FORTRAN IV	x	x	30,000	30K		x	July 1973
8	x	x	CDC, IBM, UNIVAC 1108, ICL	FORTRAN IV	x	x	50,000	40K		x	1973
9	x	x	CDC, IBM, UNIVAC 1108	FORTRAN IV	x	x	220,000	32K		x	1973
10	x	x	CDC, IBM, GE, UNIVAC 1108, ICL, KDF9	FORTRAN IV	x	x	60,000	10K + data		x	1972
11	x	x	CDC, UNIVAC 1108	FORTRAN IV	x	x	14,000	40K	5^m/1000/1	x	1973
12	x	x	CDC, IBM, GE, UNIVAC 1108	FORTRAN IV	x	x	100,000	64K		x	1972

*a = run time; b = number of degrees of freedom; c = number of time steps.

STRUCTURAL OPTIMIZATION

V. B. Venkayya and N. S. Khot
Air Force Flight Dynamics Laboratory
Wright-Patterson Air Force Base, Ohio

ABSTRACT

This paper is a compendium of structural optimization computer programs which are available from government, industry, and university sources. This information is intended for practicing engineers and researchers who are looking for a readily available computer program in structural optimization. The data are organized in such a way that the engineer can determine the required computer resources for these programs. Much of the information was collected by sending a questionnaire to the authors. A brief description of the program objectives is given along with the names and addresses of the persons in charge of the program.

INTRODUCTION

The purpose of this paper is to compile information on structural optimization programs and to present it in an appropriate form for ready use by a practicing engineer looking for this capability. This is a review article consisting of a detailed summary of the computer programs in structural optimization including the particulars on capacity, availability, language, appropriate machine configurations, underlying technology, reliability, and cost.

The review includes both public and commercial programs that are readily available. We recognize that programs are of interest to the potential user only if they are well documented and the stated capability is well corroborated by adequate test problems and preferably by some user attestation. Unfortunately, there are not too many programs in structural optimization that can satisfy these stringent but important requirements.

The area of optimization of structures subjected to static loads with stress, displacement, and size constraints is well explored. In most cases the objective is to minimize the weight of the structure while satisfying the design requirements. There are a large number of computer programs that can handle these requirements. Many of these are research oriented and not suitable for practical use. However, a few general purpose capabilities exist for design of structures with over a thousand design variables and degrees of freedom under multiple loading conditions. Since many of these programs use finite element codes for analysis, the ability to handle large numbers of variables becomes important.

These programs, though not perfect, can be improvised to handle practical design situations. On the other hand, there are some very practical programs that are designed primarily for handling special purpose structures such as plate girders, columns, transmission towers, and stiffened panel optimization. These programs can do a very good job for the intended purpose. The overall situation, in case of design for static loads, can be judged as being satisfactory.

The developments in optimization of structures subjected to dynamic loads are scant, judging at least from the practical point of view. The reason for this slow progress is twofold. The dynamic analysis of practical structures is at present very expensive and has not developed adequately for use in optimization programs. Secondly, it is much more difficult to establish specific design requirements for the dynamic loads. A number of investigators treated dynamic optimization as a frequency—constrained problem. The objective in this case is to minimize the weight of the structure while not violating a specific frequency constraint. Sometimes it involves raising or lowering the particular frequency with least weight penalty. This narrow view of structural optimization as being a frequency-constrained problem has only limited validity in practice. A true dynamic optimization program should have dynamic response in its formulation. If the response is aggravated by a specific value of a frequency, the program should have the capability to correct this situation. Such an approach to dynamic optimization is much more complex. As a result, methods for optimization of structures subjected to periodic and aperiodic forces are just emerging, and the computer programs based on these methods are research oriented at best.

Another area that is active in research is the optimization of structures with aeroelastic requirements. Most of the optimization programs in this area treat the flutter velocity constraint as being the single design requirement. This approach is akin to frequency constraint approach to dynamic load problems. These simplifying assumptions are necessitated by the complexity of the problem. At present, research efforts are underway to develop optimization programs that include static loads along with aeroelastic requirements. Even though most of the dynamic optimization programs are primarily research oriented, some of them are included here because nothing more practical is available in this area.

EVALUATION OF COMPUTER PROGRAMS

The usefulness of the computer program can be determined to a large extent by the statement of the design objectives, test problems and their results, input-output organization, and the built-in diagnostics in case of a large program. Then the best way to make evaluation is to procure the program and the documentation and use it on some representative problems in addition to the author's test problems. But both monetary resources and the available time preclude such a testing on any more than one or two programs. In the first place, few programs have comprehensive documentation. This is particularly true in the cast of structural optimization programs because most of them are still being developed. The next best approach is to obtain information from as many users as possible. This information is just as hard to get because in many cases the author and his close associates are the only users. This is not to imply that the author's experience is not trustworthy, rather the author is often too familiar with the program and the things that are obvious to him may not be so for the user. Moreover,if the author is aware of the shortcomings he usually corrects them. Since this review is intended for the potential users of a program, the experience of the past users is the best measure of its capability. The reviewers have this kind of information for evaluation on less than 25 percent of the programs listed in this review. For the remaining programs the author's statement of the capability is the main source.

Since the method of optimization is one of the items included in the description of the program capability, a brief statement of the optimization problem and the method is included here. A true structural optimization problem involves finding a minimum weight (or cost) structure to shelter a given space. The sheltering of the space means effective transmission of the attendant forces without endangering the safety of the things that are being sheltered. This is the most lofty objective in structural optimization but it is also the most difficult task to achieve because of the infinite number of possible solutions. A more reasonable definition of structural optimization problem is as follows: given a structural configuration, what combination of the component sizes produce the least weight (or cost) structure without violating the specified design requirements. The loading conditions, material properties, and minimum and maximum sizes are some of the design requirements.

Symbolically, structural optimization may be treated as minimization of an objective function (such as weight or cost) which is given by

$$F(A) = F(A_1, A_2, \ldots A_m) \qquad (1)$$

subject to the constraint conditions as defined by

$$G_i(A) = G_i(A_1, A_2, \ldots A_m) \leq G_i \qquad (2)$$

where A_1 to A_m are the m design variables. The element sizes can be treated as design variables. The limits on stresses, displacements sizes, and so on represent the constraint conditions.

There are two approaches for the solution of this optimization problem. The first approach is referred to as the direct method of optimization. It includes all the numerical search procedures of linear and nonlinear programming. Gradient projection methods and feasible direction methods are both examples of numerical search procedures. All these are iterative methods, and the essential features can be explained by the following recursion formula:

$$A^{\nu + 1} = A^{\nu} + \tau^{\nu} D^{\nu} \quad (3)$$

where $A^{\nu + 1}$ and A^{ν} are the new and old design variable vectors and ν refers to the cycle of iteration. The vector D^{ν} determines the direction of travel, and the scalar τ^{ν} represents the step size in the search. The travel matrix D^{ν} is in general a function of the gradients of the objective function (Eq. 1) and the constraint functions (Eq. 2). It is obvious from the brief description that the direct methods are the most logical approach to the structural optimization problem. They have the generality to consider any objective function or constraint functions. An even more attractive feature is the transferability of optimization routines from one discipline to the other. In spite of these advantages the direct methods have not proved to be very successful in structural optimization. This is especially true in the case of problems with large number of design variables. There are two main reasons for failure. For large problems the gradient information to determine the travel direction becomes expensive. There is no rational way to determine the starting point and the step size in the travel, but both of these effect the outcome of the design.

The second approach is an indirect method; it is also referred to as an optimality criterion approach. The method consists of deriving an optimality criterion and establishing a recursion relation to achieve this criterion. The optimality criteria are usually related to some energy function in the structure. This approach lacks the generality of numerical search methods. Different optimality criteria are necessary for static and dynamic loads. It is not always possible to derive an adequate recursion relation. However, when they are applicable, they are far more efficient than direct methods, particularly in case of problems with large number of variables. In addition, they are not so sensitive to the starting point and the step size in iteration. These conclusions are based on past experience, and it is conceivable that the defects can be remedied in both cases.

From the description of the method the user can make at least a qualitative judgment on the suitability of the program to his particular needs.

COMPUTER PROGRAM DATA

Table 1 contains the data for the structural optimization programs. By no means is this list complete or the information comprehensive. The data are organized into seventeen items; many have subitems. The tabulated data for each program refer to the item numbers given in the key to the table. Finite element programs with a capability to optimize for static loads are listed first, followed by dynamics, aeroelasticity, and component optimization. There are exceptions to this rule in one or two cases. The description of the objectives, and in some cases, comments are given in the next section.

Item No.	Description
I	Name of the Program
II	Author
III	Type of Structures (a) Civil Engineering (b) Aircraft/Aerospace (c) Reactor (d) Other
IV	Loading (a) Static (b) Dynamic (1) free vibration (2) transient response (3) flutter (c) Thermal (d) Other
V	Method of Analysis (a) Finite elements (b) Finite Differences (c) Other
VI	Type of Elements (a) Bars (b) Beams (c) Membrane Triangle (d) Membrane Quadrilateral (e) Shear Panel (f) Plate Bending (g) Thin Shell (h) Axisymmetric (i) Solid (j) Other
VII	Method of Optimization (a) Linear Programming (b) Nonlinear Programming (c) Optimality Criteria (d) Structural Index (e) Other
VIII	Type of Constraints (a) Stresses (b) Displacements (c) Frequencies (d) Response (e) Flutter (f) Minimum and Maximum Sizes

(g) Linking of Variables
(h) Discrete Variables
(i) Other

IX Program Size - Source Statements
(a) Under 2000
(b) 2000-4000
(c) 4000-10,000
(d) Over 10,000

X Minimum Core Requirements

	(1)	(2)	(3)	(4)
(a) IBM (Bytes)	120K	120-250	250-400	over
(b) CDC (Octal)	100K	100-200	200-300	____
(c) Other	____	____	____	____

XI Problem Size

	(1)	(2)	(3)	(4)
(a) No. of DOF	Under 200	200-500	500-1000	1000-3000
(b) No. of Elem.	Under 200	200-500	500-1000	1000-3000
(c) Variables	____	____	____	____
(d) Core Req.	____	____	____	____

XII Approximate Computer Running Time

	(1)	(2)	(3)	(4)
(a) CPU Time	____	____	____	____
(b) I/O Time	____	____	____	____
(c) Other Time	____	____	____	____

XIII Program is Operational On
(a) IBM 360 or 370
(b) CDC 6600
(c) UNIVAC 1108
(d) Other

XIV Program Status
(a) Fully Operational
(b) Being Developed
(c) Other

XV Programming Language
(a) Fortran
(b) ALGOL
(c) Other

XVI Sources of Public Information
(a) Papers
(b) Report
(c) Auxiliary Material of Other Users

XVII Documentation Available

		Available *Free (1)	Available At Cost (2)
(a)	Technical Description	( )	( )
(b)	Source Listing	( )	( )
(c)	Test Pack Listing	( )	( )
(d)	Test Results	( )	( )
(e)	User Manuals	( )	( )
(f)	Deck of Example Problems	( )	( )

* A nominal cost may be involved for reproduction of documentation and decks.

Table 1. Summary of Program Capability

PROG. NO.	I NAME OF THE PROGRAM	II AUTHOR OR PERSON	III TYPE OF STRUCTURE	IV LOADING	V METHOD OF ANALYSIS	VI ELEMENTS	VII OPT METHOD	VIII TYPE OF CONSTRAINTS	IX PROG SIZE
1	ASOP	W. DWYER	a, b	a	a	a, b, c, d, e, f	c	a, b, f	d
2	OPTSTATIC	V. VENKAYYA	a, b	a	a	a, c, d, e	c	a, b, f,	a
3	OPTIM	R. GELLATLY	a, b	a	a	a, c, d, e	c	a, b, f, g	c
4	OPTCOMPOSITE	N. S. KHOT	b	a	a	a, b, c, d, e, i	c	a, b, f, g	b
5	ASDP	R. N. KARNES	a, b	a, c	a	a, b, c, d, e, j	b	a, b, f, g, i	c
6	TRUSSOPT 2	J. S. ARORA	a	a, b-1	a	a	b	a, b, c, g	a
7	FRAMOPT	J. S. ARORA	a	a b-1	a	b	b	a, b, c, g	a

Table 1 (cont.)

PROG. NO.	I NAME OF THE PROGRAM	X CORE REQUIREMENTS	XI PROB. SIZE	XII COMPUTER TIME	XIII OPERATING SYSTEM	XIV PROGRAM STATUS	XV LANGUAGE	XVI SOURCE OF INFORMA-TION	XVII DOCUMENTA-TION
1	ASOP	a-1, b-1	a-4, b-4 c(2000) d(260)	a(50) b(100)	a, b	a	a	a, b, c	ALL-1
2	OPTSTATIC	a-1, b-1 c-1	a-3, b-3 c(800) d(260)	a(60) b(30)	a, b, c, d	a	a	a, b	ALL-1
3	OPTIM	a-2, b-2	a-3, b-3 c(1000) d(200)	a(100) b(200)	a, b	b	a	a, b	ALL-1
4	OPTCOMPOSITE	a-1, b-1	a-2, b-2 c(300) d(260)	a(100) b(30)	a, b	a	a	a, b	ALL-1
5	ASDP	b-2	a-4, b-2 c(140) d(200)	-	b	a	a	a, b	e-1, g
6	TRUSSOPT 2	a-2	a-1, b-1 c(96), d(184)	a(250)	a	a	a	b	a, b, d, e
7	FRAMOPT	a-2	a-1, b-1 c(96), d(184)	a(250)	a	a	a	b	a, b, d, e

Table 1 (cont.)

PROG. NO.	I NAME OF THE PROGRAM	II AUTHOR OR PERSON	III TYPE OF STRUCTURE	IV LOADING	V METHOD OF ANALYSIS	VI ELEMENTS	VII OPT METHOD	VIII TYPE OF CONSTRAINTS	IX PROG SIZE
8	SAFER	R. J. MELOSH	a, b	a	a	a, b, c, e, f	c	a, b, f, h	d
9	PARADES	R. J. MELOSH	a, b	a	a	a, b	a	a	d
10	AUTOTIER	WM. WEAVER	a	a	a	b	e	a	b
11	AUTOTRUSS	WM. WEAVER	a	a	a	a	e	a	b
12	ECI-ICES-STRUDL/DYNAL	E. L. GHENT	a, b c, d	a	a	a, b, c, d, f, g, i, j	e	a, f, i	d
13	ARROW	A. J. DODD	b	a	a	a, e	b, c	a, b, f, g	c
14	TRAN-TOWER	D. LO	a	a	a	a	e	a, f, i	c

Table 1 (cont.)

PROG. NO.	I NAME OF THE PROGRAM	X CORE REQUIREMENTS	XI PROB. SIZE	XII COMPUTER TIME	XIII OPERATING SYSTEM	XIV PROGRAM STATUS	XV LANGUAGE	XVI SOURCE OF INFORMA-TION	XVII DOCUMENTA-TION
8	SAFER	a-2, b-2	a-2	-	c	a	a	a, b	a-2, b-2 d-2, e-2
9	PARADES	a-2, b-2	a-2	-	c	b	a	a, b	a-1, b-2 e-1, f-2
10	AUTOTIER	a-1	a-4, b-4	-	a	a	a	a, b	ALL-2
11	AUTOTRUSS	a-1	a-4, b-4	-	a	a	a	a, b	ALL-2
12	ECI-ICES-STRUDL/DYNAL	a-2	a-4, b-4	-	a	a	a	a, b, c	d-1, e-1 e-2, f-1
13	ARROW	a-2	a-4, b-4 c(2500) d(400)	a(720) b(1800)	a	b	a	b	-
14	TRAN-TOWER	c-1	a-4, b-4 c-6 d(75)	a(3600) b(7200)	c	a	a	a, b	ALL-2 g

Table 1 (cont.)

PROG. NO.	I NAME OF THE PROGRAM	II AUTHOR OR PERSON	III TYPE OF STRUCTURE	IV LOADING	V METHOD OF ANALYSIS	VI ELEMENTS	VII OPT METHOD	VIII TYPE OF CONSTRAINTS	IX PROG SIZE
15	SAD	G. VANDERPLAATS	a, b	a, b-1	a	a, b, c d, e	b, c	a, b, c, f, g	c
16	OPTDYNAMIC	V. B. VENKAYYA	a, b	b-1, b-2	a	a, c, d, e	c	a, d, f	a
17	SOAR	L. GWINN	b	b-3	a	a, b, c, d, e	b	e, f, g	b
18	TSO	L. A. McCULLERS et al	b	a, b-3	b	b, i	b	a, b, c, e, f, g, h, i	b
19	WIDOWAC	R. T. HAFTKA	b	a, b-3	a	a, c, d, e	b	a, e, f,	-
20	PLATE-GIRDER	R. T. DOUTY	a	a	c	b, e, f	b	a, b, f, g, h	d
21	PLATE-GIRDER	S. L. CHU	a	a	c	b	e	a, f	c

Table 1. (cont.)

PROG. NO.	I NAME OF THE PROGRAM	X CORE REQUIREMENTS	XI PROB. SIZE	XII COMPUTER TIME	XIII OPERATING SYSTEM	XIV PROGRAM STATUS	XV LANGUAGE	XVI SOURCE OF INFORMA-TION	XVII DOCUMENTA-TION
15	SAD	a-3, b-2	a-2, b-2 c(200) d(270)	-	a, d	b	a	-	b-1, d-1 f-1
16	OPTDYNAMIC	a-1, b-1	a-2, b-2 c(400) d(260)	a(100) b(30)	a, b	b	a	a, b	ALL-1
17	SOAR	b-2	a-2, b-1 c(100) d(300)	a(2000) b(60)	a, b	a	a	a, b	ALL-1 g
18	TSO	b-3	a-1	a(300) b(150)	a, b	a	a	a, b, c	ALL-1 g
19	WIDOWAC	-	-	-	-	-	-	-	-
20	PLATE-GIRDER	a-3	c(22) d(350)	a(15)	a	a	c	b, c	a-1, d-2
21	PLATE-GIRDER	a-1, c-1	a-1	-	c	a	a	-	g

Table 1 (cont.)

PROG. NO.	I NAME OF THE PROGRAM	II AUTHOR OR PERSON	III TYPE OF STRUCTURE	IV LOADING	V METHOD OF ANALYSIS	VI ELEMENTS	VII OPT METHOD	VIII TYPE OF CONSTRAINTS	IX PROG SIZE
22	CONCRETE	P. SHUNMUGAVEL	a	a	a	b	b	a, f, g	b
23	STFSHL(1)	M. PAPPAS	d	a	c	–	b	a, c, f, h, i	a
24	GMSP	H. H. DIXON	b	a, c	c	b	c	a, f	d
25	CONMIN	G. VANDERPLAATS	d	–	–	–	b	–	a
26	COMAND	G. VANDERPLAATS	b	a	c	i	b	f, g, i	a
27	CYLINDER	W. M. MORROW	b	a	c	–	b	a, f	b

Table 1 (cont.)

PROG. NO.	I NAME OF THE PROGRAM	X CORE REQUIREMENTS	XI PROB. SIZE	XII COMPUTER TIME	XIII OPERATING SYSTEM	XIV PROGRAM STATUS	XV LANGUAGE	XVI SOURCE OF INFORMA-TION	XVII DOCUMENTA-TION
22	CONCRETE	a-3	a-1, b-1 c(100) d(290)	-	a	a	a	-	-
23	STFSHL(1)	a-1	c-8	a(900)	a, d	a	a	a, b	a-1, b-1 d-1, e-1 f-1
24	GMSP	b-2	c(13) d(40)	a(60)	b	a	a	b	a, b, d, e f-2
25	CONMIN	a-1, b-1	c(100)	-	a, d	a	a	a, b	ALL-1
26	COMAND	a-1, b-1	c(18)	a-1, b-2	a, d	a	a	a	ALL-1
27	CYLINDER	c-1	c(7)	-	c	a	a	a, b	ALL-1

PROGRAMS AND THEIR OBJECTIVES

This section gives brief descriptions of the objectives of each program, supplementing the data given in the last section. The names and addresses of the persons in charge of the programs are given in the Appendix I. Additional information can be found in the References listed in the numerical order of the programs in Table 1.

1. ASOP (Automated Structural Optimization Program). This program designs structures of minimum weight under static loading conditions. It is a finite element program and uses a modified stress ratio approach along with a numerical search for resizing the structural elements. Structural optimization problems with 2,000 to 3,000 degrees of freedom and a comparable number of elements can be handled with this program. It is extremely efficient for problems with stress constraints only but is not economical for large problems with displacement constraints. The procedure for obtaining displacement gradient information for numerical search is inefficient and needs improvement. Only when there are displacement constraints does the program enter the numerical search mode before completion of the design. The program has provisions for inputing stability tables for each structural element to guard against local buckling. The output is organized in a convenient form for the stress analyst. It is operational on IBM 360/370 and CDC 6000 and 7000 series machines. Versions with different core requirements are available for different size problems. Incorporation of a more efficient algorithm for displacement constraints and a better form for stability tables are some of the improvements planned for release in the near future.

2. OPTSTATIC. This program is intended for minimum weight design of structures subjected to static loading conditions. Optimization is based on combined optimality criteria and numerical search. The optimality criteria are based on strain energy distribution in the structure. The main program contains bars, membrane triangles, quadrilaterals, and shear panels. A version of this program with bars only and another with rectangular box beams are also available. These programs can handle problems of up to 800 to 900 degrees of freedom and a comparable number of elements on machines with memory resources comparable to CDC 6000 series. The program is extremely efficient for stress constraint problems. It is not very economical for large problems with displacement constraints because it enters into numerical search mode before completion of the design. Replacement of numerical search mode by an algorithm based on optimality criterion (virtual strain energy criterion) is one of the improvements planned for the near future. It is an experimental program suitable for research studies in structural optimization.

3. OPTIM. This program is intended for design of minimum weight structures subjected to static loading conditions. It is a finite

element program with minimization procedure based on optimality criteria. It contains bars, membrane triangles, quadrilaterals, and shear panels. It is extremely efficient for both stress and displacement constraint problems. An earlier version of this program is available for distribution at present. This version can handle only problems of up to 300 degrees of freedom and a comparable number of elements. The new version which is being checked out is expected to be more efficient, better organized for practicing engineers, and able to handle much larger problems.

4. OPTCOMPOSITE. This program is intended for optimization of layered, fiber-reinforced, composite structures. It is a finite element program with minimization procedure based on optimality criteria. The layered composite skin is treated as stacked equivalent anisotropic membrane elements. The stiffness of each element is obtained by summing the stiffnesses of individual layers. The optimality criteria is applied at the element level and for the overall structure. It is designed to handle structures with a mixture of metal and composite elements. For displacement constraint problems the program enters numerical search mode as in OPTSTATIC. It is being modified to replace numerical search by an algorithm based on optimality criterion. It is an experimental program suitable for research studies in optimization of composite structures.

5. ASDP (Automated Structural Design Program). This program is suitable for designing structures of minimum weight under static loading conditions. In addition to the elements listed in the table the program contains ORTHO PLATE and special SPAR element. The program is based on numerical search called method of feasible directions. The input must include an initial design. However, it need not be a feasible design. The program is suitable for built-up structures. It is considered one of the best examples of the application of nonlinear programming methods to structural optimization. The computer costs become quite expensive for large problems unless the variables are linked to reduce their number. Distribution of this program is limited. Release is decided on a case-by-case basis.

6. TRUSSOPT 2 (TRUSS OPTimization program version 2). The program is suitable for minimum weight design of two- and three-dimensional trusses under static loads. It is based on modified steepest-descent method. The documentation contains a number of test problems which are compared with the published results. The program is quite efficient for small problems. It may be possible to extend its capability to larger problems by improving the solution scheme and by proper use of variable linking option.

7. FRAMOPT. This program is meant for minimum weight design of plane frames subjected to static loads. It is similar to TRUSSOPT2 in most other respects. Some of the AISC code requirements are included in the design requirements.

8. SAFER (Structural Analysis Fraility Evaluation and Redesign). The program is intended primarily for minimum weight design of structures subjected to static loads. It has also provisions for fail-safe and vulnerability analysis. It is based on modified fully stressed design method, and it is suitable for problems with stress constraints. The member sizes can be treated as either continuous or discrete variables. It can handle problems up to 500 degrees of freedom and that many variables.

9. PARADES (PArabolic Reflector Analysis and DEsign Subsystem). This program seeks to minimize RMS (root mean square) error of the deformed surface of the reflector from the ideal paraboloid. The weight of the structure is the primary constraint, and stresses in the elements are treated as side constraints. Control of structural displacements due to gravity loading is the means used for minimizing RMS error. The program provides for repetitive and nearly repetitive substructures in a rectangular or a cylindrical coordinate system. It is based on linear programming and can handle structures with up to 500 degrees of freedom.

10. AUTOTIER. This program is for design of rectangular frames subjected to static loads made up of dead and live loads on a building. It is a finite element analysis program with provision for selecting the wide-flange sections to satisfy 1969 AISC specifications. It is a very practical program for the design of multistoried buildings. Plotter capability to draw framing plans and beam and column schedules showing final sizes along with the governing loading conditions are some of the special features of the program. The program can be purchased at a cost of $200.00 from Stanford University.

11. AUTOTRUSS. The program is for the design of truss structures. Its objectives and capabilities are similar to AUTOTIER. The program is also available from the same source.

12. ECI-ICES-STRUDL/DYNAL. This is basically a structural analysis program with a variety of provisions for successful design of large, practical structures. It is not really an optimization program but has the capability to select design from ordered tables based on minimum weight criteria. It is more appropriate to designate it as a structural analysis and design system rather than as a program. It is available from McDonnell Douglas Automation Company in three forms: (a) rent on monthly basis $855 to $1235/month, (b) available on an as-used basis from McDonnell Douglas Automation Company network, (c) for sale at $16,000 - $23,000.

13. ARROW (Automated Reanalysis and Redesign Optimum Weight). This program is being developed at Douglas Aircraft Company (McDonnell Douglas). The program is not fully operational for release, but it is being used in-house. It is intended for automated design of large aerospace structures subjected to static loading. Both nonlinear programming and optimality criteria are being used

for optimization method. It incorporates a large finite element code, FORMAT, for analysis of the structure.

14. TRAN-TOWER (TRANsmission TOWER Design and Analysis). This is essentially a finite element analysis program with number of practical provisions for selecting appropriate sizes for the transmission tower. The program selects the member sizes that satisfy the ASCE tower design code. The program has provisions for generation of data, plotting, investigation, and selection of member sizes for plane and space truss problems. It is a proprietary program of Sargent and Lundy, Consulting Engineers. It is available for sale.

15. SAD (Structural Analysis and Design). This program is intended for minimum weight design of structures subjected to static loads. It is a finite element program with bars, beams, and membrane plate elements. For optimization it uses a hybrid approach. For stress constraint problems it uses a modified stress ratio method for resizing the elements. For displacement constrained problems it starts with modified stress ratio method and completes the design in the numerical search mode. For numerical search it uses the subroutine CONMIN which is based on the method of feasible directions. (See 25 below.) It is extremely efficient for stress constraint problems and not so economical for displacement constraint problems. It is a small in-core program suitable for research studies in structural optimization.

16. OPTDYNAMIC (OPTimization of Structures for DYNAMIC loads). This program is being developed in the Air Force Flight Dynamics Laboratory. Its objective is the design of minimum weight structures with transient response constraints under periodic and aperiodic forces. It is based on optimality criteria for dynamic stiffness. The method is referred to as designing in the dynamic mode, which consists of a single natural mode or a combination of a set of natural modes, depending on the nature of the forcing function. Two preliminary versions of this program are available at present. One contains only bar elements, and the other has simple box beam elements. In both cases only the periodic forces are operational at present. They are primarily research programs and not very suitable for practical structures.

17. SOAR (Structural Optimization for Aeroelastic Requirements). This program is intended for sizing the structural elements for minimum weight with a fixed flutter speed constraint. It uses the method of feasible directions for searching the optimum and finite element analysis for determining the dynamic properties. The flutter speed gradient information is obtained by the method of Van de Vooren. At present the program can handle problems of up to 200 degrees of freedom and 100 design variables. The program is ideally suited for research in structural optimization with aeroelastic requirements. Addition of static strength requirements would improve its practical utility.

18. TSO. The program is a wing aeroelastic synthesis procedure for obtaining optimum skin thickness distributions and ply orientations for composite or metal wing skins that satisfy flutter constraints and strength constraints using aeroelastic loads. The wing box is modeled as an equivalent flat plate, and the stiffness and mass matrices are obtained by Raleigh-Ritz procedure. The optimization is by a nonlinear programming method called Sequential Unconstrained Minimization Technique (SUMT) with a penalty function. The program is primarily tailored for wing design, and it can be used for preliminary design purposes.

19. WIDOWAC (WIng Design Optimization with Aeroelastic Constraints). This program is intended for minimum weight design of wing structures with strength and aeroelastic constraints. It is a finite element program with a symmetric shear web and membrane plate elements. Optimization is by a search technique which uses approximate second derivatives in conjunction with Newton's method. Linking procedures have to be used to reduce number of variables in a finite element idealization. It is a pilot program intended for research studies in combined strength and aeroelastic optimization.

20. PLATE GIRDER. This program is intended for the design of hybrid plate girders in accordance with the 1969 code requirements. The term hybrid is used because it allows different strength steels for flange and web. It also allows wrinkles in the web due to local buckling and the resulting load redistribution is accounted for by tension field theory. The program was developed with private funds and is available on commercial time sharing systems at a cost.

21. PLATE GIRDER. This program is for minimum weight design of plate girders using AISC code requirements as design criteria. As in many component optimization programs, it places emphasis on adequate treatment of practical design considerations. It is a proprietary program of Sargent & Lundy Engineers and can be purchased.

22. CONCRETE. The program is intended for optimization of two-dimensional reinforced concrete building frames. It uses the force method for analysis and method of feasible directions for optimization. The structure is designed to satisfy the 1971 ACI building code with ultimate strength requirements. The strength and sizes of the elements are constraints in optimization. The program was developed by the author as part of doctoral thesis requirement at University of Illinois under N. Khachaturian.

23. STFSHL(1). The author of this program, M. Pappas, has indicated that there are three programs in this series: (1) STFSHL(1), (2) SUBSI(2), and (3) SUBSL4(3). The programs are intended for automated design of frame, stiffened, submersible, circular, and cylindrical shells. The programs use stiffened cylindrical shell buckling formulas for predicting the response. For optimization the "Direct Search Design Algorithm" was used. The stiffener flange and web thicknesses, width of the flange, and skin thickness are the

design variables, and the weight of the shell represents the objective function to be minimized. In general the variables are treated as continuous. However, some of these programs are extended for discrete variables.

24. GMSP (General Missile Sizing Program). This program is intended for preliminary design of stiffened shell structures. At present it can handle shells with cylindrical or conical or elliptic cross sections. The shells are designed for shear and moment loads. Handbook formulas for standard shapes are used for response determination. Designs are essentially strength limited, and they are similar to simultaneous failure mode types. The program has built-in honeycomb, rib-stiffened, and basic monocoque constructions. It considers ten time points for each trajectory.

25. CONMIN. This is a constrained minimization routine and not a program by itself. However, this routine can be used in a structural optimization program if the analysis module can furnish the necessary gradient information. It is primarily based on the method of feasible directions for constrained minimization. It has also some provision for unconstrained minimization. It is suitable for developing a small structural optimization capability for research studies.

26. COMAND. This program is for minimum weight design of layered composite flat plates. The values of the in-plane stress resultants N_x, N_y, and N_{xy} are the inputs to the program. It allows multiple loading conditions. The longitudinal, transverse, and shear strains are the constraints. The program provides for imposing an overall stiffness constraint. The fiber orientations are input to the program. It determines the number of layers of each orientation. The routine CONMIN directs the optimization.

27. CYLINDER. This program is intended for minimum weight design of stiffened cylinders subjected to multiple loading conditions. The cylinders are stiffened in the longitudinal and circumferential directions with rectangular stiffeners. The skin thickness, spacing of the stiffeners, and their dimensions are the design variables in optimization. The program uses stiffened cylindrical shell buckling formulas for determining the failure modes. The design loads are axial force and internal or external pressure. The optimization is by a nonlinear programming method called the sequential unconstrained minimization technique with a penalty function.

REFERENCES

1. ASOP

Dwyer, W. J., Emerton, R. K., and Ojalvo, I. U., "An Automated

Procedure for the Optimization of Practical Aerospace Structures," AFFDL-TR-70-118 (Vol. I and Vol. II) 1970, Air Force Flight Dynamics Laboratory, W-PAFB, Ohio.

Dwyer, W. J., "Finite Element Modeling and Optimization of Aerospace Structures," AFFDL-TR-72-59, 1972, Air Force Flight Dynamics Laboratory, W-PAFB, Ohio.

2. OPTSTATIC

Venkayya, V. B., Khot, N. S., and Reddy, V. S., "Energy Distribution in an Optimum Structural Design," AFFDL-TR-68-156, 1968, Air Force Flight Dynamics Laboratory, W-PAFB, Ohio.

Venkayya, V. B., "Design of Optimum Structures," An International Journal, Computers and Structures, Vol. 1, No. 1/2, Aug. 1971, pp. 265-309.

3. OPTIM

Gellatly, R. A., and Berke, L., "Optimal Structural Design," AFFDL-TR-70-165, 1970, Air Force Flight Dynamics Laboratory, W-PAFB, Ohio.

Berke, L., "An Efficient Approach to the Minimum Weight Design of Deflection Limited Structures," AFFDL-TM-70-4-FDTR, 1970, Air Force Flight Dynamics Laboratory, W-PAFB, Ohio.

4. OPTCOMPOSITE

Khot, N. S., Venkayya, V., Johnson, C. D. and Tischler, V. A., "Application of Optimality Criterion to Fiber-Reinforced Composites," AFFDL-TR-73-6, 1973, Air Force Flight Dynamics Laboratory, W-PAFB, Ohio.

5. ASDP

Karnes, R. N., Tocher, J. L., and Twigg, D. W., "Automated Analysis and Design of General Engineering Structures," Boeing Document D6-24387, 1970, Boeing Computing Center, Seattle, Washington.

Tocher, J. L., and Karnes, R. N., "The Impact of Automated Structural Optimization on Actual Design," ASME/AIAA Structures and Materials Conference, Anaheim, California, April 19-21, 1971.

6. TRUSSOPT2

Arora, J. S., and Rim, K., "Computer Program for Truss Optimization by Generalized Steepest Descent," Dept. of Mechanics and Hydraulics, College of Engineering, Technical Report No. 6, 1973, The University of Iowa, Iowa City, Iowa.

7. FRAMOPT

Arora, J. S., and Rim, K., "Optimal Design of Elastic Structures under Multiple Constraint Conditions," Dept of Mechanics and Hydraulics, College of Engineering, Technical Report No. 4, 1971, The University of Iowa, Iowa City, Iowa.

8. SAFER

Melosh, R. J., "Structural Analysis, Fraility Evaluation and Redesign," AFFDL-TR-70-15, 1970, Air Force Flight Dynamics Laboratory, W-PAFB, Ohio.

9. PARADES

Levy, R., and Melosh, R., "PARADES Structural Design System Capabilities," The Deep Space Network Progress Report, Technical Report 32-1526, Vol. XII, pp. 68-73, Jet Propulsion Laboratory, Pasadena, California, 1972.

10. AUTOTIER

Agaskar, V. L. and Weaver, W., "Automated Design of Tier Building," International Journal, Computer and Structures, Vol. 2 No. 5/6, Dec. 1972, pp. 991-1011.

11. AUTOTRUSS

Weaver, Jr. W. and Patton, F. W., "Automated Design of Space Trusses," AISC Engng J. Vol. 5, 1968, pp. 26-36.

12. ECI-ICES-STRUDL/DYNAL

User's Manual ECI-ICES-STRUDL/DYNAL McDonnell Douglas Automation Co. P. O. Box 516, St. Louis, MO.

13. ARROW

Dodd, A. J., "Specification for a Static Structural Optimization Capacity," Douglas Aircraft Division IRAD Report MDC-J5442, 1972, Douglas Aircraft Company, Long Beach, California.

14. TRAN-TOWER

Lo, D., Morcos, A. and Goel, S. K., "Computer Aided Design of Steel Transmission Towers," Presented at the ASCE National Structural Engineering Meeting in Cincinnati, Ohio, April 22-26, 1974.

15. SAD

Vanderplaats, G., "Structural Analysis and Design Program, Users Manual," NASA-TMX (in preparation), NASA Ames Research Center, Moffett Field, California.

16. OPTDYNAMIC

Venkayya, V. B., Khot, N. S., Tischler, V. A., and Taylor, R. F., "Design of Optimum Structures for Dynamic Loads," Presented at the 3rd Air Force Conference on Matrix Methods in Structural Mechanics, Wright-Patterson AFB, Ohio, 1971.

17. SOAR

Gwin, L. B., and McIntosh, Jr., S. C., "A Method of Minimum-Weight Synthesis for Flutter Requirements," AFFDL-TR-72-22 (Vol. I and II), 1972, Air Force Flight Dynamics Laboratory, W-PAFB, Ohio.

18. TSO

McCullers, L. A., Lynch, R. W., "Dynamic Characteristics of Advanced Filamentary Composite Structures," AFFDL-TR-73-111 (Vol. I and II), 1973, Air Force Flight Dynamics Laboratory, W-PAFB, Ohio.

19. WIDOWAC

Haftka, R. T., "Automated Procedure for Design of Wing Structures to Satisfy Strength and Flutter Requirements," NASA-TN-D-7264, 1973, NASA Langley Research Center, Hampton, Virginia.

20. PLATE-GIRDER

Douty, R. T., "Welded Hybrid Girder Program - 1969, AISC Specifications," Informal User's Manual can be obtained from the author.

21. PLATE-GIRDER

Beck, S., and Chu, S. L., "Minimum Weight Design of a Plate Girder," Presented at ASCE Joint Speciality Conference on Optimization and Nonlinear Problems, Chicago, Illinois, April 18-20, 1968.

Chu, S. L., "An Integrated Computer System for Structural Analysis and Design," Presented at Summer Institute in Structural Design held at Illinois Institute of Technology, Chicago, Illinois, July 10 - Aug 4, 1972.

22. CONCRETE

Shunmugavel, P., "Optimization of Two-Dimensional Reinforced Concrete Building Frames," Thesis submitted to the University of Illinois in partial fulfillment of the degree of doctor of philosophy, January 1974.

23. STFSHL (1)

Pappas, M. and Allentuch, A., "Extended Structural Synthesis Capability for Automated Design of Frame-Stiffened Submersible, Circular, Cylindrical Shells," NCE Report No. NV8, 1973, Newark

College of Engineering, Newark, New Jersey.

24. GMSP

Dixon, H. M., "Theory, Coding and Operation of the General Missile Sizing Program," MDC Report SM-48752, 1965, McDonnell Douglas Astronautics Company, Huntington Beach, California.

25. CONMIN

Vanderplaats, G., "CONMIN, A Fortran Program for Constrained Function Minimization - User's Manual," NASA-TMX-62282, 1973, NASA Ames Research Center, Moffett Field, California.

26. COMMAND

Vanderplaats, G., "COMMAND, A Fortran Program for Simplified Composite Analysis and Design," NASA Ames Research Center, Moffett Field, California.

27. CYLINDER

Morrow, W. M. and Schmit, Jr., L. A., "Structural Synthesis of a Stiffened Cylinder," NASA CR-1217, 1968, NASA Langley Research Center, Hampton, Virginia.

APPENDIX I

1. ASOP

V. B. Venkayya
AFFDL/FBR
WPAFB, Ohio 45433

2. OPTSTATIC

V. B. Venkayya
AFFDL/FBR
WPAFB, Ohio, 45433

3. OPTIM

L. Berke
AFFDL/FBR
WPAFB, Ohio 45433

4. OPTCOMPOSITE

N. S. Khot
AFFDL/FBR
WPAFB, Ohio 45433

5. ASDP

R. N. Karnes
Boeing Computer Center
P. O. Box 24346
Seattle, Washington 98124

6. TRUSSOPT2

J. S. Arora
Dept of Mechanics
The Univ. of Iowa
Iowa City, Iowa 52242

7. FRAMOPT

J. S. Arora
Dept of Mechanics
The Univ. of Iowa
Iowa City, Iowa 52242

8. SAFER

L. Berke
AFFDL/FBR
WPAFB, Ohio 45433

9. PARADES

R. Levy
Telecommunication Division
JPL, California Institute
of Technology
Pasadena, California 91103

10. AUTOTIER

W. Weaver
Prof of Structural Engr
Stanford University
Stanford, Calif. 94305

11. AUTOTRUSS

W. Weaver
Prof of Structural Engr
Stanford University
Stanford, Calif. 94305

12. ECI-ICES-STRUDL/DYNAL

E. L. Ghent
Engineering Product Manager
McDnnell Douglas
Automation Company
P. O. Box 516
St. Louis, Missouri 63166

13. ARROW

A. J. Dodd
Senior Engineer Scientist
C1-250 (35-42)
Douglas Aircraft Company
3855 Lakewood Blvd
Long Beach, Calif. 90846

14. TRAN-TOWER

C. F. Beck
Head, Computer Services
Sargent & Lundy Engineers
140 South Dearborn Street
Chicago, Illinois 60603

15. SAD

G. Vanderplaats
NASA Ames Research Center
MS 227-2
Moffett Field, Calif. 94035

16. OPTDYNAMIC

V. B. Venkayya
AFFDL/FBR
WPAFB, Ohio 45433

17. SOAR

R. A. Taylor
AFFDL/FYS
WPAFB, Ohio 45433

18. TSO

Lt K. E. Griffin
AFFDL/FYS
WPAFB, Ohio 45433

19. WIDOWAC

J. Starnes
NASA LRC
Mail Stop 362
Hampton, Virginia 23665

20. PLATE GIRDER

National CSS Computer Networks
(for details contact:
R. T. Douty
1412 Ridgemount Court
Columbia, Missouri 65201)

21. PLATE GIRDER

C. Beck
Head, Computer Services
Sargent & Lundy Engineers
140 South Dearborn Street
Chicago, Illinois 60603

22. CONCRETE

P. Shunmugavel
Sargent & Lundy Engineers
or Prof N. Khachaturian
University of Illinois
Urbana, Illinois 61801

23. STFSHL(1)

M. Pappas
Dept of Mechanical Engr
Newark College of Engr
323 High Street
Newark, New Jersey 07102

24. GMSP

H. H. Dixon
McDonnell Douglas
Astronautics Company
5301 Bolsa Avenue
Huntington Beach, Calif. 92647

25. CONMIN

G. Vanderplaats
NASA Ames Research Center
MS 227-2
Moffett Field, Calif. 94035

26. COMAND

G. Vanderplaats
NASA Ames Research Center
MS 227-2
Moffett Field, Calif. 94035

27. CYLINDER

W. J. Stroud
NASA LRC
Mail Stop 362
Hampton, Virginia 23665

TRANSIENT ANALYSIS

Ted Belytschko
University of Illinois at Chicago
Chicago, Illinois

ABSTRACT

Computer software and algorithms for the transient analysis of both linear and nonlinear structures and continua are reviewed. Modal superposition and direct explicit and implicit integration methods are described and evaluated, particularly from the aspect of efficiency. Both general purpose and special purpose programs with transient capabilities are listed, described and evaluated from the viewpoints of algorithm effectiveness and user response.

NOMENCLATURE

$\cdot$ = Time derivative
$\{D\}$ = Nodal displacement
$\{F^{int}\}$ = Internal resistance nodal forces
$\{F^{ext}\}$ = External load nodal forces
$[K]$ = Stiffness matrix
$[M]$ = Mass matrix
n = Number of degrees of freedom in mesh
n_b = Semibandwidth
n_d = Number of degrees of freedom per node
n_T = Number of time steps
$\{X\}$ = Eigenvector or mode
$t, \Delta t$ = Time and time increment, respectively
β = Parameter in Newmark β-method
λ = Eigenvalue or square of frequency

INTRODUCTION

Transient analysis software presents a particular challenge to the developer, for the additional dimension of time poses many difficulties in the formulation of input and output and in program efficiency. The latter is particularly important, for the transient analysis of problems of even moderate scale often tax contemporary computers. Furthermore, as will be shown in this paper, the most appropriate algorithm is very much a function of the problem; so unless a variety of methods are at the disposal of the user, the analysis of some problems is hopelessly uneconomical.

However, in spite of these problems, some excellent transient programs have been developed. Before describing these programs, I will review the most widely used algorithms for transient analysis and discuss their effectiveness. The available programs will then be described and discussed from the viewpoints of the algorithms employed and, where available, the user's response. Because of limited resources, this list is far from complete; however, it gives a representative sample of what is available and should give the user some guidance in his search for a suitable program.

METHODOLOGY OF TRANSIENT ANALYSIS

The governing equations for a discrete model can be expressed in the form

$$[M]\{\ddot{D}\} = \{F^{ext}\} - \{F^{int}\} \tag{1}$$

where $\{F^{ext}\}$ are the applied discretized loads, $\{F^{int}\}$ the internal nodal forces of the continuum or structure, $[M]$ the discrete mass, and $\{D\}$ the nodal displacements; superscript dots denote time derivatives.

For small displacement, linear elastic problems

$$\{F^{int}\} = [K]\{D\} \tag{2}$$

where $[K]$ is the stiffness matrix; so we obtain

$$[M]\{\ddot{D}\} + [K]\{D\} = \{F^{ext}\} \tag{3}$$

For the sake of clarity, we will not consider the role of damping in these equations.

Two types of mass matrices are used: (1) consistent, non-diagonal mass matrices and (2) lumped, diagonal mass matrices. Consistent mass matrices tend to yield more accurate frequencies for beams, though this advantage becomes small whenever the wavelength of the mode spans 5 or more elements. For direct integration procedures, the type of mass matrix that should be used depends on the method of temporal integration; this will be discussed later. In addition, it might be mentioned that consistent mass matrices tend to overestimate the frequencies, whereas lumped masses underestimate the frequencies. An interesting aspect of NASTRAN [1] is its use of the average of the lumped and consistent mass for bars. The resulting mass matrix is nondiagonal and converges more rapidly.

Two strategies are used in dealing with the transient problem: (1) modal superposition and (2) direct integration. The choice of method depends on the frequency content of the load and what portion of the frequency response is of interest. If the cutoff for these frequency domains is far above the highest frequencies of the discrete model, modal superposition techniques are of advantage in linear problems. This criterion may also be expressed in terms of wave propagation effects, for these effects involve the upper end of the discrete model's spectrum. Thus, when wave propagation effects are of interest, direct integration methods are appropriate because all of the modes of the discrete model must be treated.

Modal Superposition

In modal superposition methods, the mass and stiffness equations are diagonalized by finding the eigenvalues and eigenvectors of the equation

$$[M]\{X\} = \lambda[K]\{X\} \tag{4}$$

These eigenvalues and eigenvectors correspond to the natural frequencies and modes. Once these are known, the transient equations can be put in uncoupled form and easily solved. Generally only a portion of the eigenvalues at the lower end of the spectrum are used.

In modal superposition procedures, the eigenvalue determination is the most critical and time-consuming aspect of the computation. Three types of eigenvalue methods are currently used in software: (1) transformation methods - (a) Householder's method, (b) Givens's method, (c) Jacobi's method; (2) vector iteration; and (3) determinant search (usually with Sturm sequences).

In the transformation methods, the matrices are transformed to tridiagonal (Householder and Givens) or diagonal form (Jacobi), so that all eigenvalues can be extracted. The methods are listed here in order of effectiveness (see [2]). Most versions of these algorithms found in general purpose programs cannot take advantage of bandedness. Therefore, in order to cope with large problems, eigenvalue extraction by transformation methods is usually preceded by condensation, which replaces the mass and stiffness matrices by approximately equivalent matrices of reduced size.

Two condensation methods are used: (1) Guyan's condensation [3], and (2) zero mass or static condensation. In the second method, the stiffness matrix is reduced by a transformation which eliminates degrees of freedom associated with zero mass. This procedure is identical to static condensation and neglects the inertial forces associated with the specified zero mass degrees of freedom. Usually rotatory degrees of freedom are eliminated. In Guyan's condensation, the reduction transformation is applied to both the stiffness and mass matrices; it is generally the preferred method. However, it can only be used when the program has provisions for a nondiagonal mass matrix, since the Guyan reduction of a diagonal matrix yields a nondiagonal matrix.

Vector iteration methods are the many variations of the classical power method. They may be classified as single vector iteration and multivector iteration. In the first method, the lowest eigenvalue of a system is obtained by iterating on a trial vector until convergence is achieved. By using shift points, the eigenvalue nearest to any shift point can also be obtained. Iteration combined with a sweep, which eliminates the previously found eigenvectors from the vector iterate, permits several eigenvectors to be found in the vicinity of any shift point. By combining shifts and sweeping, it is possible to find all of the eigenvalues of a system. However, the method is uneconomical whenever more than a few modes are needed.

Multivector iteration, or subspace iteration [4], is a variation of the power method in which the iteration is simultaneously performed on a group of trial vectors, yielding many of the eigenvectors after one series of iterations. When a substantial number of modes are needed (i.e. 4 to 20 in a 500 d.o.f. system), the method is more efficient than single vector iteration. Furthermore, both single and multivector iteration methods can take advantage of bandedness. Hence, when iteration methods are used, it is generally not desirable to employ condensation because it destroys bandedness. Some comparisons of subspace iteration and Householder's method have been reported by Clough and Bathe [5]. Because the comparisons were not for identical problems and were not preceded by condensation, firm conclusions are difficult, but it appears that Householder's method preceded by condensation is more efficient when 50 or more eigenvalues of a 1000 d.o.f. system are desired.

Modal superposition is used very little in nonlinear analysis. It appears that it is not very suitable for path-dependent materials, for the stress time history of every point in a structure must be followed accurately for these materials. Hence, it seems that a large number of modes would be required. However, some of the shell programs, such as DYNAPLAS, have used Fourier superposition in the circumferential direction with considerable success; so the method may prove useful in some nonlinear problems.

Direct Integration

In direct integration methods, the coupled equations of motion are integrated directly without any preliminary uncoupling by modes. There are two major classes of direct integration methods: (1) implicit integration in time and (2) explicit integration in time.

Difference formulas and stability. Both the explicit and implicit methods are developed from difference formulas that relate the accelerations, velocities, and displacements. For example, the Newmark β-method [6] employs the following difference formulas

$$\{\dot{D}(t+\Delta t)\} = \{\dot{D}(t)\} + \Delta t[\{\ddot{D}(t)\} + \{\ddot{D}(t+\Delta t)\}] \tag{5a}$$

$$\{D(t+\Delta t)\} = \{D(t)\} + \Delta t\{\dot{D}(t)\} + \Delta t^2[(1/2 - \beta)\{\ddot{D}(t)\} + \beta\{\ddot{D}(t+\Delta t)\}] \tag{5b}$$

When $\beta = 1/4$, this formula may be derived by assuming that the acceleration between t and $t+\Delta t$ is constant and equal to the average of the accelerations at the ends of the interval. For $\beta = 1/6$, it corresponds to a piecewise linear acceleration field, while $\beta = 0$ corresponds to acceleration pulses at intervals of Δt. It can also be shown that the Newmark β-method with $\beta = 0$ is equivalent to the central difference formulas

$$\{\dot{D}(t+\Delta t/2)\} = \{\dot{D}(t-\Delta t/2)\} + \Delta t\{\ddot{D}(t)\} \tag{6a}$$

$$\{D(t+\Delta t)\} = \{D(t)\} + \Delta t\{\dot{D}(t+\Delta t/2)\} \tag{6b}$$

provided that

$$\{\dot{D}(t+\Delta t/2)\} = \{\dot{D}(t)\} + 1/2\, \Delta t\, \{\ddot{D}(t)\} \tag{7}$$

Other difference formulas often used in programs are the Houbolt formulas [7], which are based on a piecewise cubic interpolation of the displacements, and the Wilson θ-formulas [8]. When applied to an implicit method, the following formulas yield unconditionally stable algorithms: the Newmark β-method with $\beta \geq 4$, the Houbolt method, and the Wilson θ-method. Therefore, the time step for these implicit methods is limited only by convergence. In addition, another class of temporal operators, known as stiffly stable methods, has been developed [9]. These methods have limited regions of stability that can be varied to accommodate the problem being solved. Of the programs reviewed here, only STAGS includes this method, for the method has come to the attention of software developers only recently.

The central difference method is only conditionally stable; so it is used only in explicit procedures. The time step must always be below the stability limits, which are

$$\Delta t^2 \leq 4/\lambda \tag{8}$$

where λ is the square of the highest frequency of the mesh. For finite element meshes with linear translational and cubic flexural fields, λ may be estimated by

$$\lambda = 4c^2/\ell^2 \qquad \lambda = 12D/\ell m \tag{9}$$

where ℓ is the element length, c the elastic wave speed, D the flexural rigidity (EI for elastic problems), and m the rotatory lumped mass. The highest value of λ associated with the mesh must be used. Furthermore, because of roundoff errors and other reasons that are poorly understood, Δt is usually chosen to be from 0.5 to 0.8 of the value given by Eq. (8).

If the stability limits are exceeded in linear problems, the results usually rapidly reach the overflow limits on the computer, so that the instability is easily recognized by the user. However, in elastic-plastic problems and in other materials capable of dissipating much energy, it is possible for the instability to be arrested, so that the erroneousness of the results is not obvious to the user. The occurrence of an arrested instability can be detected by an energy balance check, but this feature is unfortunately absent in most finite element programs; so caution is advisable in these problems when conditionally stable methods are used.

The convergence of explicit integration schemes has been the topic of two recent papers of considerable interest, Fujii [10] and Krieg and Key [11]. The former proved the convergence of lumped mass-explicit integration procedures. The latter showed by numerical studies that the spectral errors of lumped mass, explicit integration and consistent mass, implicit integration are compensatory, while that of lumped-implicit and consistent-explicit combinations are cumulative. Thus, from the viewpoint of convergence, the first two schemes are equally good, while the latter two are undesirable.

Implicit integration. To develop the equations used in implicit integration, we consider the equations of motion, Eq. (3), for time t+Δt and eliminate any velocities or accelerations at t+Δt by means of Eqs. (5). For the Newmark β-method this yields

$$[K^{eff}]\{D(t+\Delta t)\} = \{F^{eff}\} \tag{10}$$

where

$$[K^{eff}] = [M] + \beta\,\Delta t^2\,[K]$$

$$[F^{eff}] = \beta\,\Delta t^2\,\{F^{ext}(t+\Delta t)\} + [M]\Big((1/2-\beta)\,\Delta t^2\,\{\ddot{D}(t)\} + \Delta t\{\dot{D}(t)\} + \{D(t)\}\Big) \tag{11}$$

Thus the implicit method directly solves for the displacements at each time step, and the equation solver is an important component of implicit transient analysis. A flow chart of the implicit method is given in Fig. 1. For linear problems $[K^{eff}]$ is triangulated only once; for a system with n degrees of freedom and semibandwidth n_b, this requires nn_b^2 multiplications. Equations (10) are then solved by simple back-substitution, which requires $2nn_b$ multiplications for each time step. Table 1 gives an estimate on the number of computations. Though implicit integration techniques are not as fast per time step as the explicit method, their unconditional stability makes possible the use of large time steps, so they are usually more economical for linear problems when the bandwidth is not too large.

The treatment of nonlinear transient problems for both explicit and implicit methods introduces a bewildering array of alternatives, many of which are topics of current research. Some of these choices concern the description used: Lagrangian, updated Lagrangian, convected coordinates and mixtures of the above have been developed. The user should bear in mind that updated coordinate procedures or procedures which take advantage of small-strain, large rotation properties are generally faster [12,13,14]. We will confine our attention here to the types of stiffness formulations, which may be classified as (1) tangential stiffness methods and (2) pseudoforce methods.

In the tangential stiffness method, the constitutive equation is linearized in each time step so that the internal forces are given by

$$\{F^{int}(\{D+\Delta D\})\} = F^{int}(\{D\})\} + [K_T]\{\Delta D\} \tag{12}$$

where $[K_T]$ is the tangential stiffness. For large deflection problems, the initial stress matrix or geometric stiffness must

Table 1 Estimates for Computations in Direct Integration

	Number of multiplications per time step	Example 1; multiplications per time step	Example 2; multiplications per time step
Explicit for linear problems	$n_N n_C n_D^2$	7×10^4	3×10^3
Implicit for linear problems	$n_D n_N (2n_b + n_D n_C)$ $+ n_b^2 n_D N_N / n_T$	8×10^5	6×10^3
Explicit for nonlinear problems	$(2m_B + m_M) e m_I$	1×10^5	2×10^4
Implicit for nonlinear problems	$n_D n_N (n_b^2 + 2n_b)$ $+ (2m_B + m_M) e m_I$ +(stiffness comp.)	1.5×10^7	3.5×10^4

Example 1: 31 x 62 node two-dimensional plane mesh, constant strain elements; 1000 time steps; $n_c = 9$; $m_I = 1$; $m_B = 24$; $m_M = 12$.

Example 2: 100-node cylindrical, axisymmetric shell; 1000 time steps; $n_c = 3$; $m_I = 10$; $m_B = 12$; $m_M = 8$.

Nomenclature in table:

e = number of elements
m_B = number of multiplications in strain-displacement equations
m_I = number of integration points per element
m_M = number of multiplications in constitutive equation
n_C = connectivity of mesh; g for 2 dimensional plane, 3 for line, 27 for 3 dimensional solid
n_D = degrees of freedom per node
n_N = number of nodes

Explicit Algorithm

1. Initial conditions; t = 0
   Compute [K] and [M]
2. Find $\{D(t+\Delta t)\}$ by Eq. (5b)
3. Find $\{\ddot{D}(t+\Delta t)\}$ by Eq. (14), equations of motion
4. Find velocity $\{\dot{D}(t+\Delta t)\}$ by Eq. (5a)
5. $t = t + \Delta t$, to to 2.

Implicit Algorithm

1. Initial conditions, t = 0
   Compute [K] and [M]
2. Triangulate $[K^{eff}]$
3. Solve Eq. (10 for $\{D(t+\Delta t)\}$
4. From Eqs. (5), find velocities and accelerations, $\{D(t+\Delta t)\}$ and $\{\dot{D}(t+\Delta t)\}$
5. $t = t + \Delta t$, to to 3.

Fig. 1 Flow charts of implicit and explicit integration algorithms for linear problems

also be included in $[K_T]$. When Eq. (12) is inserted in Eq. (1) for $t+\Delta t$, an equation similar to Eq. (10) in $\{\Delta D\}$ can be developed.

The tangential stiffness matrix changes after each time step; so it must be recomputed and triangulated in each time step. The number of computations per time step (given in Table 1) is very large. Furthermore, large time steps often cannot be used because in path-dependent materials the stress history must be followed closely. Hence, when the bandwidth is large, the method is far more expensive than explicit methods.

Equation (13) is only an approximation to $\{F^{int}(\{D+\Delta D\})\}$ based on the current state of the material. In many cases it is quite inaccurate, as, for example, when a plastic material unloads during a time step. To compensate for these errors, some programs compute the error in Eq. (1) at the end of each time step. These error forces, which are often called residual forces, are then added to the external forces in the next time step. Failure to include residual forces can lead to substantial errors.

In pseudoforce methods, the nodal forces are subdivided into linear and nonlinear components in the form

$$\{F^{int}\} = [K]\ \{D\} + \{N\} \tag{13}$$

where $\{N\}$ are the nonlinear portions of the forces, i.e., the difference between the actual resistance forces and the linear estimate, $[K]\{D\}$. If the nonlinear forces are placed on the right hand side of Eq. (3), the resulting system is identical to the elastic equations with an additional external force, $-\{N\}$. This method therefore has the advantage that the stiffness matrix need only be triangulated once. It has been used successfully in shell programs by Stricklin et al. [15], but it is not used in any general purpose programs. This is unfortunate, for the method is very economical when only portions of the structure are nonlinear or when the nonlinearities are mild.

Explicit integration. In the explicit integration of linear problems, the accelerations are found at each time step by

$$\{\ddot{D}\} = [M]^{-1}\ (\{F^{ext}\} - [K]\{D\} \tag{14}$$

The displacements and velocities at that time step are then found by Eqs. (5); note that this is only possible for the Newmark β-method when $\beta = 0$. A flow chart of the method is given in Fig. 1. If $[M]$ is diagonal, the method requires no matrix inversion, only matrix multiplications. As mentioned previously, the use of diagonal mass matrices in explicit methods is also desirable from the viewpoint of accuracy; so they should always be used in explicit schemes. The multiplication of the stiffness and nodal displacement matrices can be performed by sparse matrix techniques so that the number of computations depends linearly on the number of degrees of freedom and is independent of bandwidth; a computation estimate is given in Table 1. The explicit integration methods are therefore advantageous for large bandwidth meshes and short duration loadings. However, for long duration loadings, they become uneconomical because of the need to use small time steps. Only a few finite element programs use explicit integration for linear problems; SLAM and SLADE D are two included in Table 3. None of the general purpose programs in this country currently provide an explicit integration option.

The explicit method is particularly suited for nonlinear problems, for unlike the implicit method, no significant additional costs are introduced by nonlinearities. In the formulation of explicit, nonlinear methods, the stiffness matrix is quite redundant. Instead, the equations of motion, Eqs. (1), can be treated directly in the form

$$\{\ddot{D}\} = [M]^{-1} (\{F^{ext}\} - \{F^{int}\})$$

A flow chart of the procedure is given in Fig. 2 for small deformation nonlinear problems; the equations for large deformations are similarly straightforward and may be found in [16] and [12].

The number of computations depends principally on the number of elements and varies linearly with this number, regardless of the bandwidth. Though a time step consistent with numerical stability limits must be used, in path-dependent materials, such as elastic-plastic materials, the time step required to follow the material's stress history adequately is often of the order of the stability limit. Therefore the explicit method is much more economical for large meshes than implicit methods.

Yet, whereas finite difference methods have usually employed explicit integration, very few finite element programs use it. In fact, none of the general purpose nonlinear programs include this option, thus making large-scale nonlinear analysis prohibitively expensive. It should be stressed here that explicit integration should be an alternative, rather than a replacement, for implicit techniques, because in many structural problems, such as rotationally symmetric shells, implicit integration is the better method.

The importance of choosing the appropriate technique for a problem can best be appreciated by considering the two examples given in Table 1: a large, two-dimensional, plane mesh and a cylindrical shell. For the first example, a nonlinear analysis by the implicit method would require 150 times as many computations per time step as the explicit method. Furthermore, the stiffness matrix for problems of this scale must be stored out-of-core, which reduces the speed of the implicit method even more. So if this example takes one hour when solved by the explicit method, it would not be feasible for an implicit algorithm. For linear problems, as can be seen from Table 1, the disparity is much smaller. On the other hand, in the second example the computational requirements of the explicit and implicit method are of the same order; so because of the unconditional stability of the implicit method, it would be preferable.

The computation figures in Table 1 can give only a rough estimate of computer time, for they include only the dominant parts of the computation and neglect overhead, such as output. However, they are quite indicative of the magnitudes of computer effort for various types of problems and algorithms, and the examples, though hypothetical, are typical in contemporary analyses. Therefore, a variety of methods is obviously mandatory in any general purpose program.

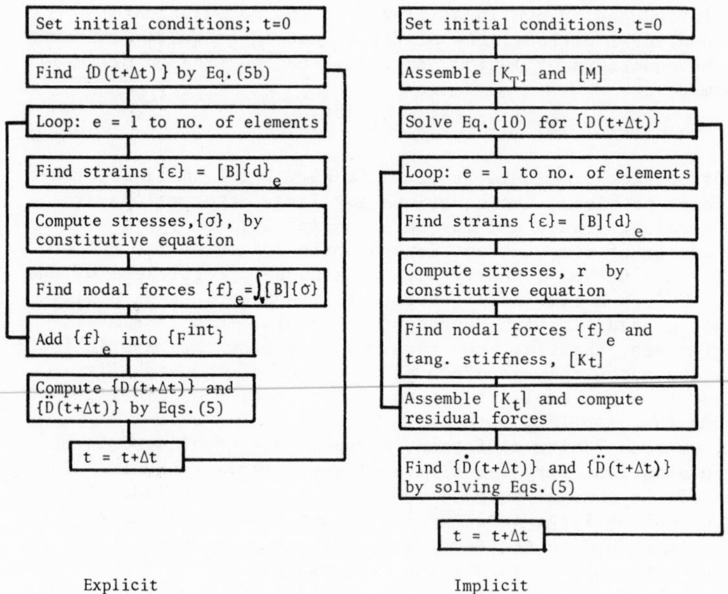

Nomenclature for this figure

[B] = strain-nodal displacement matrix

$\{d\}_e$ = nodal displacements of element e

V = volume of element

$\{\varepsilon\}$= strains

$\{\sigma\}$= stresses

Fig. 2 Flow charts for explicit and implicit algorithms for nonlinear material, small deformation problems

COMPUTER PROGRAMS

The computer programs have been classified in two categories: (1) general purpose programs, which are programs that can handle a wide variety of structures and continua, and (2) special purpose programs, which are programs restricted to one or two classes of problems, such as rotationally symmetric shells or two-dimensional continua. The general purpose programs are listed and described from the aspect of transient analysis in Table 2. This information is taken from survey forms submitted by developers; due to ambiguities in questions and other reasons, these responses were sometimes erroneous. Some errors have been corrected, but undoubtedly others remain. Also included in Table 2 are the user ratings for these programs. These ratings should be judged with care when comparing programs because very few users rated more than one program. The tone of the ratings and written comments were often quite inconsistent. In fact, the written comments provided the most useful information, and they make up a large part of the following evaluations. It should also be borne in mind that the user response for the general purpose programs is for the entire program, not just for the transient algorithms.

General Purpose Programs

NASTRAN is probably the most powerful linear analysis system available from the viewpoint of large-scale capability. It is advertised that it has no limitations with respect to problem size other than cost (or roundoff errors). An extensive evaluation of NASTRAN has been published by Tocher [17]. His comments on the transient capabilities were brief, and he concluded that NASTRAN'S dynamic capabilities exceeded those of most programs available.

The users surveyed had few criticisms about the capabilities of NASTRAN, but comparison with other programs shows several shortcomings. The element library is quite uneven: neither thick shells, curved shells, or higher-order isoparametrics are included. The choice of eigenvalue extraction techniques is also somewhat limited and dated (e.g., it uses the Givens transformation method). Users praised the thorough documentation and good output formats; however, it was generally agreed that the documentation is poorly indexed and organized, which, combined with its great volume, makes it overwhelming for the beginner. The clarity of the output format and excellence of the plotting package were cited by several users, though the absence of stress contour plotting option was noted. The absence of a preprocessor or any mesh generation has been the cause of considerable complaint. Though preprocessors have been written for NASTRAN, they evidently must be run separately, which is quite unwieldy for small and moderately sized problems.

Other problems associated with the size of NASTRAN are: difficulties of installing NASTRAN on in-house systems, relatively large overhead on computer time, and the virtual impossibility of modifying it or linking it with other programs. It is almost impossible to make small modifications where needed. This is being rectified somewhat by the latest version, which includes a dummy element that can programmed by the user.

Table 2 General Purpose Programs

					EIGENVALUE EXTRACTION						DAMPING			TEMPORAL INTE-GRATION				EQUATION SOLVER		
	L-LINEAR, NL-NONLINEAR	DIAGONAL MASS	NONDIAGONAL MASS	CONDENSATION (G-GUYAN, S-STATIC, O-OTHER)	HOUSEHOLDER	GIVENS	JACOBI	MULTIVECTOR ITER	SINGLECEVTOR ITER	STURM SEQUENCE	VISCOUS	STRUCTURAL	LINEAR	MODAL SUPER.	IMPLICIT*	EXPLICIT*	TYPE+	MAX D.O.F. (UNL-UNLIMITED)	MAX BAND	PROGRAM SOURCE OR DISTRIBUTOR
ANSYS	NL,L		X	G			X				X	X	X		U		5	20k+	0.4k	Swanson Analysis, Inc. 870 Pine View Drive Elizabeth, Pa. 15037
ASKA	L	X	X	G	X			X	X	X	X			X			4	40k	40k	ISD Stuttgart Pfaffenwaldring 27 Stuttgart, 80 Germany
ICES-STRUDL (DYNAL)	L	X	X	G	X		X		X	X	X			X			4,5	12k+	0.8k	ICES Group, P.O. Box 8243 Cranston, R.I. 02920
MARC	NL,L	X	X						X		X	X	X	X	N			$\sim$15k		CDC Cybernet
MINIELAS	L	X	X						X	X	X		X	X			3			Alpay & Utku Duke University Durham, N.C. 27706
NASTRAN	L	X	X	G		X			X		X	X	X	X	N		3	UNL	UNL	COSMIC
PAFEC	L,(NL)	X	X		X					X				X	U	U	2,5	8k+		R. D. Henshell University Park Nottingham NG7 2RD U.K.
SAMIS	L	X	X	O						X	X	X	X	X	N		4,5,7	10k	UNL	COSMIC
SAP IV	L	X						X	X	X	X		X	X	W		3	UNL	UNL	NISEE, 729 Davis Hall Univ. of Calif., Berkeley Berkeley, California
STARDYNE	L	X		S	X				X		X		X	X			3	15k	1.2k	CDC Cybernet

*N: Newmark β +1: full matrix; 2:const. band;
W: Wilson θ 3:variable band; 4:partitioning
U: Unknown   5:wave front; 7:iterative

Table 2 (cont.)

	PROPRIETARY	COMPUTER (C:CDC; I:IBM; U:UNIVAC)	PREPROCESSOR	POSTPROCESSORS	ELEMENTS: BEAM (S:STRAIGHT, C:CURVED)	ELEMENTS: MEMBRANE (L:LINEAR C:CURVED; I:ISOPARA*)	ELEMENTS: ARB. PLATE (F:FLAT; S:SHALLOW SHELL)	ELEMENTS: 3D SOLID (L:LINEAR, I:ISOPARA*)	ELEMENTS: AXISYM. SOLID (I:ISOPARA*)	ELEMENTS: THICK PLATE(P) OR SHELL(S)	USER RESPONSE⁺: EASY TO LEARN TO USE	EASY TO USE	DOCUMENTATION: SELF CONTAINED	DOCUMENTATION: PITFALLS	DOCUMENTATION: TEST CASES	DOCUMENTATION: WELL ORGANIZED	INPUT DATA	OUTPUT	GOOD PLOTTING (N - NO PLOTTING)	NUMBER OF RESPONDING USERS
ANSYS	X	I,C,U	X	X	S	L	F,S	L	L	S,P	2.6	1.9	2.1	3.0	3.0	2.5	2.3	2.1	1.2	6
ASKA	X	I,C,U	X	X	S	L,C,I	F,S	L,I	L,I	S,P	3.7	2.1	3.0	3.5	3.2	3.8	2.5	2.1	N	12
ICES-STRUDL (DYNAL)		I,U		X	S	L,I	F,S	I			2.6	2.0	3.2	3.9	4.2	3.0	2.1	2.1	1.6	27
MARC	X	C,I,U	X	X	S,C	L	F,S	I	L	S,P										
MINIELAS		I			S	L,I	F			S,P										
NASTRAN		I,C,U		X	S	L	F	L	L		2.0	2.2	2.0	4.0	3.4	1.7	1.9	2.0	1.9	6
PAFEC	X	I,C,U	X	X	S,C	L,C,I	F,S	L,I	L,I	S,P										
SAMIS		C,U	X		S	L	F,S				4.3	4.0	3.0	4.0	3.3	3.7	3.3	3.3	3.0	3
SAPIV		C,(I,U)			S	L	F,S	L	L	S,P	2.3	1.6	2.8	3.2	3.0	2.4	2.1	2.5	N	21
STARDYNE	X	C	X	X	S,C	L	F	L			1.8	1.6	1.4	2.5	2.1	1.7	1.7	1.5	1.6	10

*I-ISOPARA refers only to higher order isoparametrics

+ -Ratings are from 1.0 (highest rating) to 5.0 (lowest rating)

NASTRAN includes many of the components required for nonlinear, transient analysis: the geometric stiffness and piecewise linear material behavior. However, I do not know of anyone who has solved nonlinear transient problems with NASTRAN, for evidently there are several basic problems in the implementation of these modules to nonlinear analysis: the geometric stiffness cannot be defined in terms of the current stress; the nodal coordinates cannot be updated; the material law assumes path independence, i.e.,it unloads along the same path as the loading path. Hence NASTRAN is a linear transient program. For linear problems, in spite of several shortcomings, it is one of the better packages available.

SAP IV is rather unique in this group in that it is the only genuine large-scale program that is not proprietary and is sufficiently straightforward so that it can be modified by any reasonably competent programmer. It has an exceptionally good library of elements, including isoparametrics. Users had several complaints: (1) poor output formats, (2) incomplete and ambiguous input documentation, (3) difficulties in installing it on IBM and UNIVAC computers, and (4) the availability of only rudimentary preprocessors. The users evidently ignored both the very low cost and low development budget of the program. In my experience, I have found the program to be an excellent framework for developing in-house systems, for the architecture of the program is easily comprehended and changes can be made. The only basic shortcomings are the lack of a consistent mass or of a good variety of eigenvalue methods. Subspace iteration is nice, but many users prefer condensation with a transformation method.

ANSYS is the only program which is used extensively for both linear and nonlinear analysis. The principal feature of ANSYS is a wide variety of pre- and postprocessors which were lauded by many users. There were also many favorable comments on the overall usability of this system. From the viewpoint of methodology the system is somewhat limited. Its only eigenvalue technique is the Jacobi transformation method, which is generally quite slow. For direct integration of nonlinear problems, the user is restricted to implicit integration with a tangential stiffness; yet there are no residual force computations. The element library is also quite small, though it appears to be sufficiently versatile for most engineering problems.

The MARC program is a sophisticated and versatile nonlinear program with an extensive library of elements, including isoparametrics and curved shells, and a large library of nonlinear materials. It can treat the nonlinear transient problem by both direct implicit integration and modal superposition, though the latter option is evidently not fully developed. Users report that the program is not easily learned or used, but the latest version is much improved.

ICES-STRUDL (and the McDonnell proprietary version, DYNAL) is part of the ICES package. Its main distinction is the use of a problem-oriented language which can be used for a variety of data manipulations. Users were quite pleased with this and other aspects of the code. The major objection concerned the inefficiency of the program; although the proprietary version is advertised to be much faster than the free version, S. L. Chu [18] reported that STRUDL analyses are quite expensive, especially for programs of small and moderate size. The system has an excellent, versatile eigenvalue extraction system, but no provisions for direct integration.

STARDYNE is another general purpose program which is widely used. Its capabilities are quite limited: there is no direct integration and very few elements. However, users were very happy with the program. They praised the clarity of the user's manual, ease of use, and its efficiency for both small and large problems. Other useful features are modal damping and provisions for error analysis. The major complaints concerned (1) the absence of isoparametrics, (2) lack of a consistent mass, and (3) difficulties in using the plotter postprocessor.

ASKA has not been used much in this country. It has an extensive element library, a variety of eigenvalue methods, and large-scale capabilities. However, only modal superposition algorithms are included. Implementation of these capabilities requires FORTRAN programming, and most users rated the program as very hard to learn and use.

Special Purpose Programs

The special purpose programs are listed in Table 3. The majority of these programs are for the analysis of axisymmetric shells. Of these programs, the SAMMSOR and SABOR programs received the most user reports and are evidently the most familiar.

SAMMSOR/DYNASOR and SAMMSOR/DYNAPLAS are for large deflection analysis with linear and nonlinear materials, respectively, and both can treat asymmetric loads. These programs have been used quite widely. They are almost trouble-free, well documented, and quite easy to use. They must be run in two parts: SAMMSOR generates the stiffness and mass, whereas DYNASOR or DYNAPLAS perform the temporal integration. Some users have found this partitioning inconvenient and have combined the programs.

A similar group of programs is SABOR/DRASTIC 3A and SABOR/DRASTIC 6, which are linear and nonlinear respectively. The former has been widely used for a number of years with satisfactory results; SABOR/DRASTIC 6, the nonlinear version, has just recently been released.

The other programs listed are less widely known or less versatile. It might be noted that for this class of problems, bandwidths are generally small; so implicit methods are preferable. The above four programs use pseudoforce methods and Fourier superposition to treat arbitrary variations of loads in the circumferential direction. These techniques impair the stability of implicit methods, and Houbolt's method appears to suffer the least. If the user intends to use any of these programs for very large deflection problems, he should check the equations used: most of the programs employ equations that are applicable only to moderate rotations.

For arbitrary shell problems, the available software is more limited. Only five programs were located: STAGS, PETROS 3, WHAM, IMPS and SLADE D. SLADE D is derived from the static program SLADE, which was widely and successfully used, though not easy to learn. The shell surface must be cylindrical, and only explicit integration is available; so it is probably useful only for short duration loads. It contains an automatic time step computation to ensure numerical stability. IMPS is designed for small, nonlinear problems; though it uses explicit integration it employs a stiffness matrix, so that large problems quickly exhaust core. WHAM is for problems of moderately large scale (up to 3000 d.o.f.) and short duration loads; it has not been widely used. The most impressive of these programs is STAGS. It includes a better variety of methods than even the general purpose programs, which should make it suitable for a large range of problems. A feature of STAGS is its combination of finite difference and finite element methods. The finite difference equations are formulated for arbitrary, nonorthogonal grids, although the surface must be defined by a mathematical expression. STAGS is open-ended, and 20,000 d.o.f. solutions have been reported. Only a few user responses to STAGS were obtained. They praised the capabilities of the program, but they indicated that the program is not easily learned and that the documentation is quite involved. PETROS 3 is similar to STAGS but less versatile and more difficult to obtain.

The third class of problems for which special purpose programs have been written are two-dimensional and three-dimensional axisymmetric continuum problems. Programs are available for both nonlinear materials and large deformations. Five of these programs are finite element programs: SLAM, SAMSON, WHAM, NOFEAR, and NONSAP. The latter two use implicit integration and are particularly suited for moderately sized meshes and loads of long duration. NONSAP, which should be released soon, is especially versatile and has several good features: (1) residual force computations, (2) a choice of Lagrangian and updated Langrangian descriptions, (3) efficient procedures for element stiffness generation, which are essential in dynamic nonlinear analysis, and (4) a large library of nonlinear materials.

The remaining three programs all use explicit integration. SLAM is a linear program with very efficient procedures; on a CDC 6600, running time is 1 msec/cycle/element. SAMSON is a similar program that can treat nonlinear materials. It is also well suited for large—scale problems: there are no size limitations and about 4000 d.o.f. systems can be solved in-core. Problems of this size have been solved: running time is 2 msec/cycle/element (CDC 6600). WHAM is restricted to in-core solutions, but it is faster, has more elements, and can treat large deflection problems; running time is 0.2 to 1 msec/cycle/element, depending on the element (IBM 370/195). None of these three programs is very user-oriented, although SAMSON has a posprocessor.

Another alternative for this class of problems are the finite difference programs, TOODY and PISCES. Both programs have an impressive array of capabilities. TOODY is open-ended and quite fast: 3 msec/cycle/element (CDC 6600). It is well tested and quite reliable. Although finite difference methods are widely believed to be quite inflexible in modeling arbitrary geometries, these programs are designed so that they can treat almost any geometry, although complicated geometries are not treated as easily as with finite elements. Inexperienced users have reported many difficulties in using PISCES; the data is rather involved, and there are evidently few internal data checks, so the isolation of data errors is very frustrating.

Table 3 Special Purpose Programs

PROGRAM NAME	TYPE	PROBLEM CLASS OR ELEMENTS	INTEGRATION	OTHER INFORMATION	DEVELOPER (DISTRIBUTOR)
DYNASOR/ SAMMSOR	FE	axi.shell, arb. loads	I(H)	large deflection; linear material	J. A. Stricklin W. E. Haisler Aerospace Eng. Dept. Texas A&M Univ. College Station, Texas 77843
DYNAPLAS/ SAMMSOR	FE	axi.shell, arb. loads	I(H) E	large deflection; elastic-plastic, isotropic and White-Besseling hardening	
SABOR/ DRASTIC 3A	FE	axi.shell,	I(N)	small deflection; linear material	S. Klein Philco-Ford Corp. Ford Road Newport Beach, California 92663
SABOR/ DRASTIC	FE	axi.shell,	I(N)	large deflection; elastic-plastic, isotropic hardening	
SHORE	FD	axi.shell, arb. loads	E	large deflection; elastic-plastic, isotropic hardening	P. Underwood Lockheed Palo Alto 3251 Hanover St. Palo Alto, California 94304
SATANS	FD	axi.shell, arb. loads	I(H)	large deflection; elastic	R. E. Ball Naval Postgrad. School COSMIC
DEPICS		axi.shell,		large deflection; elastic-plastic imperfection sensitivity	L. Mente, J. C. Manzelli Kaman Avidyne Burlington, Massachusetts

axi: axisymmetric, arb.: arbitrary; FD: finite difference; FE: finite element; E: explicit integration; I: implicit integration; H: Houbolt; N: Newmark; MSFI: multi-segment forward integration; MS: modal superposition.

Table 3 (cont.)

PROGRAM NAME	TYPE	PROBLEM CLASS OR ELEMENTS	INTEGRATION	OTHER INFORMATION	DEVELOPER (DISTRIBUTOR)
STARS	MSFI	axi.shell	I	small deflection (axi. large deflections); elastic-plastic, kinematic and isotropic hardening	V. Svalbonas Dept. 461 Grumman Aerospace Bethpage, New York 11714
KSHEL	MSFI	axi.shell	MS	small deflections (axi. large deflections); elastic	A. Kalnins Dept. Mech. Eng. Lehigh University Bethlehem, Pennsylvania 18015
UNIVALVE	FD	2D beams, axi.shell,	E	large deflections; elastic-plastic, isotropic hardening	R. Krieg, T. Duffey Analytical Development Div. Sandia Laboratories Albuquerque, New Mexico 87115
TACOS	FE	thick axi. shell	E	small deflection, elastic; coupled thickness response	S. E. Benzley Sandia (see above)
SLADE D	FE	doubly curved arb. quadrilateral	E	small deflection, elastic; cylindrical reference surfaces only	S. W. Key, Z. E. Beisinger Sandia (see above)
IMAN	FE	arb. flat plate quadrilateral	E	elastic-plastic, large deflection	K. K. Gupta Jet Propulsion Lab. COSMIC
STAGS	FD, FE	arb. FD shell linked to FE truss and membrane	I, E	large deflection; elastic-plastic, White-Besseling hardening; large scale	F. A. Brogan, B. O. Almroth Dept. 52-53, Bldg. 205 2751 Hanover Palo Alto, California

Table 3 (cont.)

PROGRAM NAME	TYPE	PROBLEM CLASS OR ELEMENTS	INTE-GRA-TION	OTHER INFORMATION	DEVELOPER (DISTRIBUTOR)
PETROS 3	FD	arb. shell	E	large deflection; elastic-plastic, kinematic hardening	S. Atluri, E. A. Witmer, et al M.I.T. (government users)
NONLIN 2	FE	rod; 2D curved and straight beam	I(N)	large deflection, elastic-plastic; friction; hinge lines	J. C. Anderson Sargent & Lundy 140 S. Dearborn Chicago, Illinois 60603
WHAM	FE	flat plate-shell, 2D & 3D beam, axi.shell, 3D axi & 2D continua	E	large deflection, elastic-plastic, isotropic hardening, hydrodynamic	T. Belytschko Dept. of Materials Engineering Univ. of Illinois, Chicago Circle 60680
SAMSON	FE	3D axi. & 2D continua, 2D bar	E	small deflection; elastic-plastic; cap-soil model; sliding interface; large scale	T. Belytschko Univ. of Illinois, Chicago Circle R. L. Chiapetta, J. Rouse IIT Research Institute (government users)
NOFEAR	FE	3D axi. & 2D continua	I	small deflection, nonlinear material	I. Farhoomand E. L. Wilson Univ. of California Berkeley
NONSAP	FE	2D & 3D curvilinear, cubic isoparametrics, 3D axi.	I	large deflection, large strain; many material laws	K. J. Bathe and E. L. Wilson NISEE, 729 Davi Hall Univ. of California Berkeley, California Attn: Ken Wong

Table 3 (cont.)

PROGRAM NAME	TYPE	PROBLEM CLASS OR ELEMENTS	INTEGRATION	OTHER INFORMATION	DEVELOPER (DISTRIBUTOR)
TOODY	FD	3D axi. & 2D continua	E	large deflection, large strain; elastic-plastic; hydrodynamic; sliding interface; large scale	C. E. C. Bertholf, S. E. Benzley Sandia gov. users: B.Morris Co, USAMERDC Attn: STSFB-XN Ft. Belvoir, Virginia 22060
PISCES	FD	3D axi. & 2D continua	E	large deflection, large strain; elastic-plastic; hydrodynamic; compaction material; sliding interface; rezoning	Physics International 2750 Merced San Leandro, California (proprietary)
SLAM	FE	3D axi. & 2D continua	E	elastic, small deflection; large scale, high speed	C. Costantino City College of New York
AFTON*	FD	3D axi. & 2D continua	E	large deflection, nonlinear materials incl. soil and concrete	J. G. Trulio Applied Theory Inc. Los Angeles, California (proprietary)
HONDO*	FE	3D axi. continua	E	large deflection, elastic-plastic with isotropic and kin. hard.; viscoelastic	S. W. Key Sandia (see above)

*Received too late to be included in text

CONCLUSIONS

For the analyst primarily interested in linear transient analyst, a good choice of programs is available. To a large extent, the choice is determined by whether or not the user wants a proprietary program. If not, NASTRAN and SAP are the best alternatives; NASTRAN is slightly more powerful, but it must be used as is, whereas SAP can be tailored to meet the specific needs of the user.

For nonlinear analysis, the only general purpose programs are proprietary. Of these, ANSYS is more user-oriented, MARC the more technically sophisticated and versatile. Neither program includes explicit integration; therefore large-scale analyses of nonlinear transient problems are very expensive. MARC, unlike ANSYS, includes residual forces; so it is probably somewhat more reliable.

If the user desires nonproprietary programs or economical large-scale nonlinear analyses, he must resort to the special purpose programs. Unfortunately, as can be seen from the previous section, except for the axisymmetric shell programs, these programs are not easy to use. Documentation, particularly for the more versatile of these programs, is often quite involved; one user employed the term "labyrinthine." These programs are appropriate for specialists who can devote several weeks to familiarization. This is probably only worthwhile when many problems of one class are to be solved.

The user survey showed that an overriding consideration is that the program be user-oriented: easily learned, well documented, self-checking, with good input and output formats and processors. Users tended to overlook major algorithm weakness far more readily than items such as troublesome input. To many software people, this is well known, but the uniform emphasis on convenience was nevertheless striking.

One item related to usability that has not yet been mentioned here is the choice of time step in implicit methods. The appropriate time step is intimately related to the spectra of the discrete model and the load, and particularly for linear problems, a suitable time step could be automatically computed from these spectra. Yet all of the general purpose programs leave the choice of time step to the user: if he chooses too small a time step, the costs are great, whereas when the time step is too large, the peaks are missed. A preprocessor to compute the time step would therefore be very beneficial.

In summary, for the purpose of increased user orientation and economy, the following items should be considered in future developments of general purpose transient software:

1. Preprocessors to (a) compute the time step appropriate for the given load history and material and consistent with stability limits (when applicable) and (b) automatically pick the most economical transient method
2. Inclusion of explicit and implicit integration, perhaps with the option of combining the two in a single analysis
3. Energy balance checks in finite element programs
4. Output filtering to eliminate spurious oscillations and extraneous response time history points which tax buffers and output channels

ACKNOWLEDGMENT

The assistance of Dr. R. L. Chiapetta, who made available the documentation of the IIT Research Institute Computer Library, is gratefully acknowledged.

REFERENCES

1 MacNeal, R. H., The NASTRAN Theoretical Manual, NASA, Washington, 1970.

2 Wilkinson, J. H., "Calculation of Eigensystems of Matrices," Numerical Analysis, An Introduction, ed. by J. Walsh, Thompson Book Co., Washington, D.C., 1967.

3 Guyan, R. J., "Reduction of Stiffness and Mass Matrices," A.I.A.A. Journal, Vol. 3, No. 2, February 1965, pp. 380-381.

4 Bathe, K. J., Wilson, E. L. and Peterson, E., SAP IV: A Structural Analysis Program for Static and Dynamic Response of Linear Systems, Report No. EERC-73-11, University of California, Berkeley, June 1973.

5 Clough, R. W. and Bathe, K. J., "Finite Element Analysis of Dynamic Response," Advances in Computational Methods in Structural, ed. by J. T. Oden, et al., University of Alabama Press, 1972, pp.

6 Newmark, N., "A Method of Computation for Structural Dynamics," J. Eng. Mech. Div., Proc. of A.S.C.E., 1959, pp. 67-94.

7 Houbolt, J. C., "A Recurrence Matrix Solution for the Dynamic Response of Elastic Aircraft," J. of Aero. Sciences, Vol. 17, 1950, pp. 540-550.

8 Wilson, E. L., Farhoomand, L. and Bathe, K. J., "Nonlinear Dynamic Analysis of Complex Structures," Earthquake Eng. and Struct. Dyn., Vol. 1, 1973, pp. 241-252.

9 Gear, C. W., "The Automatic Integration of Stiff Ordinary Differential Equations," Information Processing, Vol. 68, 1967, pp. 187-193.

10 Fujii, H., "Finite Element Schemes: Stability and Convergence," Advances in Computational Methods in Struct. Mech. and Design, ed. by J. T. Oden, et al., University of Alabama Press, Alabama, 1972, pp. 201-218.

11 Krieg, R. D. and Key, S. W., "Transient Shell Response by Numerical Time Integration," Int. J. Num. Methods in Eng., Vol. 17, 1973, pp. 273-286.

12 Belytschko, T. B. and Hsieh, B. J., "Nonlinear Transient Finite Element Analysis with Convected Coordinates," Int. J. Num. Methods in Eng., Vol. 7, 1973, pp. 255-272.

13 Hibbitt, H., Marcal, P. and Rice, J., "A Finite Element Formulation for Problems of Large Strain and Large Displacement," Int. J. of Solids and Structures, Vol. 6, 1970, pp. 1069-1086.

14 Bathe, K. J., Wilson, E. L. and Ramm, E., "Finite Element Formulation for Large Displacement and Large Strain Analysis," Report UCSESM73-14, University of California, Berkeley, September 1973.

15 Stricklin, J. A., Haisler, W. E. and Von Riesemann, W. A., "Computation and Solution Procedure for Nonlinear Analysis by Combined Finite Element-Finite Difference Methods," Computers and Structures, Vol. 2, 1972, pp. 955-974.

16 Oden, J. T., Finite Elements of Nonlinear Continua, McGraw-Hill, New York, 1972.

17 Tocher, J. L. and Herness, E. D., "A Critical View of Nastran," Numerical and Computer Methods in Structural Analysis, ed. by S. J. Fenves, et al, Academic Press, New York, 1973, pp. 151-174.

18 Chu, S. L., "Analysis and Design Capabilities of STRUDL Program," ibid, pp. 229-246.

THIN SHELLS

David Bushnell
Lockheed Palo Alto Research Laboratory
Palo Alto, California

ABSTRACT

The objective of this paper is to give the analyst a physical appreciation of the behavior of thin shells so that he will be able to choose wisely and use effectively computer programs for structural analysis. Examples are selected to illustrate various linear and non-linear stress, buckling, and dynamic phenomena of practical engineering structures of which thin shells are major components. Hints on setting up analytical models for shells are given, including how to distribute nodal points for accurate determination of stress concentrations and buckling or vibration modes, how to handle intersections of shells, how to treat discrete stiffeners in various ways, and how to model concentrated loads.

NOMENCLATURE

- b = Imperfection sensitivity parameter
- C = 6 x 6 symmetric matrix of constitutive equation coefficients for shell wall
- D = Matrix relating analytical displacement variables to nodal point degrees of freedom
- d = Vector of displacement components
- E = Stiffener eccentricity matrix, Eq. (5)
- I = Mass moment of inertia matrix
- K = 4 x 4 symmetric matrix of constitutive equation coefficients for discrete stiffener
- L = Length variable of discrete stiffeners
- M = Moment resultants or number of degrees of freedom, depending on context
- m = Mass/area or mass/length, depending on context
- N = Vector of thermal forces and moments
- N = Stress resultants
- n = Number of circumferential waves
- p = Vector of distributed loads at a point (e.g., pressure and surface traction components)
- P = Matrix for live load (pressure-rotation) effect
- q = Vector of line loads acting on stiffener or nodal point displacement vector, depending on context
- Q = Matrix for live load (line load-rotation) effect
- δq = Infinitesimal buckling modal displacement vector

R = Matrix relating rotation components of shell wall to nodal point degrees of freedom
S = Surface or stability, depending on context
U = Strain energy
X = Matrix for prebuckling shape change due to eigenvalue loads
ε = 6-element strain vector containing ε_1, ε_2, ε_{12}, η_1, η_2, η_{12}
λ = Lagrange multiplier or eigenvalue, depending on context
η = Change in curvature
ω = Rotation vector for discrete stiffener
Γ = Warping constant for discrete stiffener
ψ = Gradient of energy with respect to nodal point variables

Superscripts

$(\ )^T$ = Transpose of a vector or matrix
$(\dot{\ })$ = Differentiation with respect to time

Subscripts

$(\ )_1$ = Surface coordinate first direction
$(\ )_2$ = Surface coordinate, second direction
$(\ )_{12}$ = Shear or twist
$(\ )_r$ = Pertains to discrete stiffener
$(\ )_L$ = Pertains to linear terms
$(\ )_{NL}$ = Pertains to nonlinear terms
$(\ )_i$ = ith point in discrete model
$(\ )_o$ = Initial value or prebuckling value
$(\ )_f$ = Fixed
$(\ )_{cr}$ = Critical value

INTRODUCTION

Two factors enter into the correct choice and effective use of a computer program for structural analysis:

1. The engineer must understand the physics of the problem well enough to be able to define his needs.
2. The engineer must be informed which programs will satisfy these needs.

This paper addresses the first factor. The second factor is addressed in another paper in this volume [1]. There, tabulated descriptions of existing computer programs and evaluations by users are given.

The purpose of this paper is to help the engineer recognize what he needs to know in order to choose and use wisely a computer program for thin shell analysis or to analyze any structure of which thin shells are important components. What qualities should a computer program have in order to allow accurate prediction of the behavior of thin shells? What qualities should an analytical model of a shell structure have in order to best predict this behavior? To answer these questions, the engineer must first understand how shells behave. Thus, the objective here is to convey a "feel" for what a shell is and how is acts. Emphasis is placed on practical shell structures which may be stiffened, segmented, branched, discontinuous, have complex wall constructions, and so on. Descriptions are provided of some of the peculiarities of the behavior of shells: nonlinear behavior due to plasticity and large deflections, stress redistribution effects, stiffener and load path eccentricity effects, bifurcation buckling, local vs. general instability, imperfection sensitivity, low frequency bending vs. high frequency extensional dynamic response, and fluid-shell interaction effects. Many examples are introduced with emphasis on the features a computer program should have in order to permit accurate prediction of the behavior of an actual structure and on the phenomena an engineer should be aware of in order to use a given program effectively. Sections on equations and modeling are included to illustrate what is and what is not important in setting up a discrete model. Such considerations as where to put mesh points; how to model discrete stiffeners; the importance of stiffener eccentricity; load eccentricity; and reference surface discontinuities; how to model concentrated loads; and how to determine the effect of local distortion of built-up wall constructions are discussed, with demonstrations by means of specific examples. From the facts and opinions presented here, and from the charts of [1], the reader will, it is hoped, be guided to acquire those computer programs that best suit his needs and to use them effectively.

The Broad Picture

The discipline of solid mechanics can be divided into subdisciplines according to the geometrical characteristics of deformable bodies, according to their material properties, and according to the nature of environmental influences acting upon them.

Figure 1 shows the field of solid mechanics divided for the reader's convenience into subdisciplines which are specifically identified in the papers of this volume. Some of the papers focus on a given geometrical subclass and cut across the subdisciplines involving environmental influences and material properties; others concentrate on an environmental influence or a quality of material and cut across the subdisciplines involving various classes of geometry. This paper focuses on the particular class of structures known as thin shells. Some attention is given to each of the subdisciplines marked with an X.

The division of the field of solid mechanics according to the geometrical characteristics of bodies can be summarized as follows: if the space occupied by material can be identified in such a way that two dimensions are small compared to the third, then the "one dimensional" body may be called a string, a cable, a spring, a rod, a bar, a beam, a column, or an arch, depending on its shape, properties, and application. If the space occupied by a material can be described such that one dimension is small compared to the other two, then the "two-

GEOMETRY	Linear Statics	Nonlinear Statics	Linear Dynamics	Nonlinear Dynamics	Shock & Vibration	Seismic Interaction	Stability	Optimization	Thermal Stress & Creep	Viscoelasticity	Plasticity	Welding	Fracture	Composites
General Structures														
Ship Structures														
Concrete & Steel Bldgs.														
Bridge & Girder Systems														
Kinematic Systems														
Thick Shells														
Thin Shells	x	x	x	x	x		x		x		x			x
Piping Systems														
Cables														
Shafts														

Influence of Environment/Material Properties

Fig. 1 Solid mechanics and the subdisciplines involving thin shells

dimensional" body may be called a membrane, a plate, a panel, or a shell, depending on its flexural rigidity and curvature. Bodies in which all dimensions are of the same magnitude are simply called "three-dimensional" or solids. Actually, by far the majority of the structures of interest to the engineer consist of assemblages of one-dimensional and two-dimensional bodies. It is often necessary, however, to treat parts of these as three dimensional because of the localized nature of forces or other external effects acting upon them or because of stress concentrations arising from local peculiarities in the allocation of material.

The division of the field of solid mechanics into subdisciplines according to geometrical qualities is logical because one-dimensional, two-dimensional, and three-dimensional bodies behave differently, and because the methods used to analyze them are different. For example columns, plates, and shells, being thin in at least one dimension, can deform a large amount while the strains remain very small. This simple kinematic fact gives rise to the very extensive fields of elastic stability and nonlinear elastic static and dynamic response. In contrast, stability and other nonlinear theories involving the deformation of three-dimensional solids must incorporate large strain effects and nonlinear material properties. Bodies which are thin in a certain dimension or dimensions are analyzed differently from general solids because the displacement fields normal to the "long" dimensions can be specified by means of certain hypotheses such as "normals remain normal," with consequent reduction of the number of independent variables required.

Some History

First use of shell structures. Shell structures were used well before there existed any theory to predict their behavior. The oldest preserved shell is the 140-foot concrete hemispherical dome of the Pantheon in Rome, constructed in 27 B.C. Another interesting example is the light artillery piece designed and built under the direction of Gustavus Adolphus of Sweden in 1626. Weight was saved and mobility thereby increased by wrapping the thin copper bore with leather, which shrunk to guarantee compressive hoop stresses in the metallic bore--a very early example of a composite, prestressed wall construction. Another example, of particular interest to the Navy, is the first submarine, an egg-shaped contrivance, designed and built in 1775 by David Bushnell, a sapper in George Washington's army. The submarine, dubbed Turtle, was used on only one occasion against His Majesty's ships--without success because of strong tidal currents. Other early applications come to mind--in particular cathedral domes and bells.

First analysis of shells. The first known analysis of a shell structure is "De Sono Campanarum" (1766) [2] in which Euler attempted to predict the tones of a bell by modeling it as a stack of thin annuli, each acting as a curved bar. His theory neglected meridional bending stiffness. In 1789 James Bernoulli [3] modeled the bell as a latticework of beams, a sort of primitive Hrennikoff grid in which curvature was neglected. These early applications of the finite element method might be partially responsible for its long neglect, since both failed to predict the correct frequencies! The first successful prediction of the tones of a vibrating plane structure was that of Mlle. Sophie Germain in 1815 [4]. She investigated the flexural modes of a flat plate, predicting the nodal patterns which had been observed some years earlier by Chladni [5].

The first theories for the behavior of thin two-dimensional structures were based upon earlier work on rods and beams by James Bernoulli [6], Euler [7], and Coulomb [8], who established the concept of a beam consisting of a bundle of longitudinal fibers which elongate and contract during flexure. The amount of elongation or contraction varies proportionally with the distance from the neutral axis according to the hypothesis that planes normal to the neutral axis remain unextended and normal to this axis as the beam deflects. This kinematic model, coupled with Hooke's law, discovered in 1660 [9], led to predictions of behavior which agreed with observations. Kirchhoff extended the concept of "normals remain normal" to the analysis of flat plates [10].

Modern shell theory. Exactly 100 years ago H. Aron [11] first approached the problem of thin shells from the point of view of the general equations of elasticity. His main motivating interest, typical of much of the work in those days, was the prediction of natural frequencies.

One of the important contributions of Aron's shell theory was the reduction of the dimensionality of an elasticity problem from three to two. The distribution of displacements in a direction normal to a reference surface follows from a hypothesis that a normal to this surface remains normal and unextended as the surface deforms. Therefore, the displacement field everywhere in such an elastic body is a function of only two independent variables--the distances along two coordinate lines embedded in the reference surface.

Early researchers in shell theory perceived two very different types of shell behavior--one in which the reference surface stretches without much change in curvature and another in which this surface remains unextended but in which its curvature changes. The general theory of Aron was specialized by Mathieu [12] to the case of pure extensional (membrane) vibrations and by Lord Rayleigh [13] to the case of inextensional (bending) vibrations. The more general theory of Love [14], applicable to both extensional and bending phenomena, followed shortly.

Love's approximate strain energy expression forms an acceptable foundation for modern elastic shell analysis, in which large deflections are accounted for. It is based on the following assumptions:

1. The shell is thin, that is $t/R<<1$, where t = thickness and R = smallest principal radius of curvature.

2. The strains are small.

3. The state of stress is plane.

4. The extensional and bending energies are not coupled.

As pointed out by Koiter [15], assumption 3 in effect incorporates what are now called the Love-Kirchoff assumptions: normals to the reference surface remain unextended and normal during deformation of the shell wall. If Love's theory is broadened in a simple way so that assumption 4 is no longer required, it becomes applicable to situations in which the reference surface need not be the middle surface, the material behavior is nonlinear, and the shell is eccentrically stiffened or composed of an arbitrary array of orthotropic layers. The last three of the above situations lead to coupling between extensional and bending energy which in general cannot be eliminated by a shifting of the location of the reference surface relative to the shell wall material.

<u>When is thin shell theory valid?</u> Koiter [15] emphasizes that any refinement regarding the first three of Love's approximations is meaningless in general, unless the effects of transverse shear deformation and normal stresses are also taken into account. These refinements are beyond the scope of this paper and are discussed in the article in this volume on thick shell analysis [16].

The question that naturally arises is; when can thin shell theory be used? For what geometries and loadings can transverse shear strain safely be neglected? The general rule of thumb is that the half-wavelength of deformation should be at least ten times the wall thickness. This rule is useful for bifurcation buckling and modal vibration problems, but not very meaningful for stress analysis, particularly if stress concentrations are present. In a discontinuity stress field near a clamped edge or other structural singularity, for example, there is no clearly defined single wavelength of deformation. Fortunately, neglect of the transverse shear deformation effect is usually conservative in stress analysis problems.

The importance of transverse shear deformations depends also on details of the wall construction of a shell. For example, the behavior of a shell with sandwich wall construction is more strongly influenced by this effect than is a similar monocoque shell because the effective transverse shear modulus of the core is small compared to Young's modulus, and the normal and in-plane shear forces are concentrated near the extreme fibers of the wall. Neglect of the effect of transverse shear deformations can lead to seriously unconservative predictions for buckling loads of sandwich shells.

Recent work on shell theory. Many recent papers have been written on thin shell equations. Koiter [15] tabulates the differences between several of the formulations. However, much of the debate about which shell equations must be used can be safely ignored by the practicing engineer. An exception is the "shallow shell" formulation, set forth for general shells by Green and Zerna [17] and for cylinders by Donnell [18]. The shallow shell equations lead to accurate results only if the wavelength of deformation is small compared to the smallest radius of curvature or if the shell is physically shallow. These equations can be used as a basis for generation of the local stiffness matrices of curved shell finite elements if the elements are shallow.

Computer programs. In the sixties and early seventies many computer programs were written for the analysis of general structures and for shells. Hartung [19] gives a rather complete survey of shell analysis codes in existence as of 1971. The proceedings of several recent conferences are good sources of information on structural analysis programs in general [20, 21, 22]. Program abstracts are published regularly by NASA [23], and Project STORE has released a volume containing abstracts with confidence levels in the codes indicated by the developers [24]. In addition, there exist many software service organizations, each of which maintains several programs. The aim of this paper is to describe certain aspects of the behavior of thin shells and to give the engineer some tips on modeling. With a physical appreciation of phenomena typical of shells, and with the data on existing codes from [1], and from other sources such as [19] through [24], the engineer will be guided toward an appropriate choice of program and effective use thereof.

GOVERNING EQUATIONS

The majority of computerized analyses of thin shells are based on an energy formulation, important exceptions being the programs for shells of revolution based on forward integration and finite difference equilibrium methods. Energy expressions can be used to demonstrate the kinds of terms that should be included in a reasonably comprehensive computer program intended to be widely used for the analysis of stress, buckling, and vibration of practical engineering shell structures.

Strain Energy

If the displacement method is used, the strain energy of the shell is expressed in terms of the strains and changes in curvature of the reference surface, which is not necessarily the middle surface or the neutral surface

$$U_{shell} = \frac{1}{2} \int_S (\varepsilon^T C \varepsilon + 2N\varepsilon)\, dS \tag{1}$$

where dS is the elemental area of the reference surface. The six element vector ε represents the reference surface strains ε_1, ε_2, ε_{12}, and changes in curvature η_1, η_2, η_{12}; and C is a 6 x 6 symmetric matrix of coefficients which depends on the location of the wall material relative to the reference surface, on the details of the wall construction, on the temperature, and, if plasticity is present and the tangent stiffness method is used [25], on the stress-strain curve and flow law of the material. The quantity N is a vector containing thermal expansion effects, creep strains, and plastic strains. If plasticity is present or if the material properties depend on the temperature, the elements of C and N at a point on the reference surface must in general be determined by numerical integration through the thickness. Stricklin et al. [26] and Bushnell [25] point out that Simpson's rule should be used for the integration.

Coupling between Bending and Extensional Behavior

If the middle surface is the reference surface, and if the properties of the wall are symmetric with respect to this surface, then all those elements of C are zero through which stress resultants N_1, N_2, N_{12} cause changes in curvature η_1, η_2, η_{12}, and through which moment resultants M_1, M_2, M_{12} cause normal strains of the reference surface. Generally, however, there exists coupling of bending and extensional behavior which cannot be eliminated by a shifting of the reference surface. Three common examples are shells reinforced in one direction by stiffeners that are eccentrically located with respect to the shell's neutral surface, shells stressed into the plastic region by a combination of stretching and bending, and nonuniformly heated shells constructed of temperature-dependent material. In the first example the neutral surface with regard to bending and stretching in one direction is in a different plane from the neutral surface with regard to bending and stretching in an orthogonal direction. In the second and third examples the location of the neutral surface changes with strain and temperature distribution. In all three of the examples it is not possible to choose a priori a reference surface location in order to eliminate coupling between extensional and bending terms in the energy expression. In addition, it is often advantageous not to have to choose the middle surface as a reference surface, since practical monocoque shell structures often have variable wall thicknesses and doublers which make the middle surface difficult to describe mathematically and which cause its position to change abruptly in space. Coupling between bending and stretching energy is also present in shells with composite walls such as layered orthotropic, fiber-wound, or semisandwich corrugated construction.

One of the first requirements of a computer program for shell analysis, therefore, is to permit arbitrary location of the reference surface with respect to the wall material and to include in the mathematical model the energy coupling between changes in curvature and normal strains of this surface. If the engineer is interested in performing analyses of many different kinds of shell structures, he is advised to obtain a computer program or programs which include coupled membrane and bending behavior. The traditional finite element model of a shell in which membrane and bending behavior are introduced through separate elements is not adequate.

Kinematic Relations and Nonlinear Terms

In Eq. (1) the strain vector ε can be expressed in terms of displacement and rotation components and derivatives of these quantities. Kinematic relations have been given by many authors. The nonlinear strain-displacement equations of Love [14], Novoshilov [27], or Sanders [28] are acceptable as a basis for the displacement method. In general, nonlinearities need not be retained in the change-incurvature-displacement relations as long as the largest reference surface rotations are less than about 20 degrees, which is usually the case. The normal strains ε_1, ε_2 and in-plane shear strain ε_{12} can always be written so that the highest order nonlinearities are quadratic.

There is a good physical explanation for the need to retain nonlinear terms in the strain-displacement but not in the curvature-displacement relations. If a thin shell deflects a large amount, let us say an amplitude many times the thickness, the strains can still be small even though the deflections are rather large. Hence, the linear terms in the strain-displacement relations will tend to cancel each other, and the nonlinear terms will become significant for much smaller displacements than they would have if the linear terms had not tended to cancel. The linear terms in the expressions for the change in curvature, however, do not tend to cancel, and the wall rotations must be large indeed before nonlinear terms have to be included.

Discretization

Finite element method. By far the majority of computer programs for shell analysis are based on the finite element method. Gallagher [29, 30, 31] gives surveys of the use of finite elements for linear and nonlinear analysis of general shells. Brombolich and Gould [32] present such a survey for shells of revolution. A detailed description of the various elements with evaluation will therefore not be presented in this paper. Gallagher encapsulates the state-of-the-art as of 1972: "Three alternative forms of finite element representation of thin curved shells are popular: (1) in 'faceted' form via the use of flat elements, (2) by means of isoparametric solid elements which have been specialized to represent curved thin shells, and (3) via the theories formulated directly for shallow or deep curved shells"[30].

Until very recently finite element experts using the displacement method have been most insistent that the displacements of adjacent elements be fully compatible at the common boundary. Maintenance of slope and displacement compatibility does have the advantage of guaranteeing that convergence of displacements is monotonic from below and that eigenvalues for bifurcation buckling and modal vibrations converge monotonically from above (assuming that in the case of vibrations a consistent mass matrix is used). However, the enforcement of full interelement compatibility results in an overestimation of the stiffness of the structure, which tends to decrease the rate of convergence as the nodal point density is increased. Wilson [33] introduces incompatible displacement functions in order to improve the convergence properties.

Finite difference energy method. A few shell analyses have been performed and computer programs written based on the finite difference energy method, in which the displacement derivatives appearing in ε (Eq. 1) are replaced by finite difference expressions. Johnson [34] was the first to perform such an analysis with use of an arbitrary quadrilateral finite difference mesh. The two most widely used computer programs based on this approach are STAGS, which treats linear and nonlinear stress, buckling, and dynamic response of general shells [35] and BOSOR4, which treats stress, buckling, and vibration of axisymmetric shells [36].

A good test case. Bushnell [37] presents a comparison of the finite element method and the finite difference energy method, showing that in certain cases the finite difference energy method is actually a rapidly convergent kind of finite element method in which the element displacements and rotations are incompatible at interelement boundaries. Figures 2 and 3, taken from [37], show the results of a convergence study involving a free hemisphere pinched by a cos 2θ pressure distribution. This rather ill-conditioned problem is a very good test of various methods of discretization. The problem is ill-conditioned because small forces cause large displacements. Thus, the predicted reference surface strains are very small differences of relatively large numbers. The dotted line in Fig. 3 is obtained with use of a half-station finite difference energy method, which is equivalent to a finite element method based on linear functions for u and v and a quadratic function for w. Detailed descriptions of the finite elements are given in [37] and [38]. Users and developers of computer programs for shell analysis and for general structures are urged to employ this case in order to determine the adequacy of the shell elements in the finite element libraries of their programs.

Discretized kinematic relations. With use of a discretization method, analytical kinematic relations $\varepsilon = L(d)$, where L is a nonlinear differential operator and d is the displacement vector, can be expressed in the algebraic form

$$\varepsilon_i = B_L d_i + d_i^T B_{NL} d_i \tag{2}$$

The vector ε_i represents reference surface strains and changes in curvature at some point i; d_i is the local nodal point displacement vector associated with i; and B_L and B_{NL} are 6 x m matrices dependent on the local reference surface geometry and mesh spacing at i. The

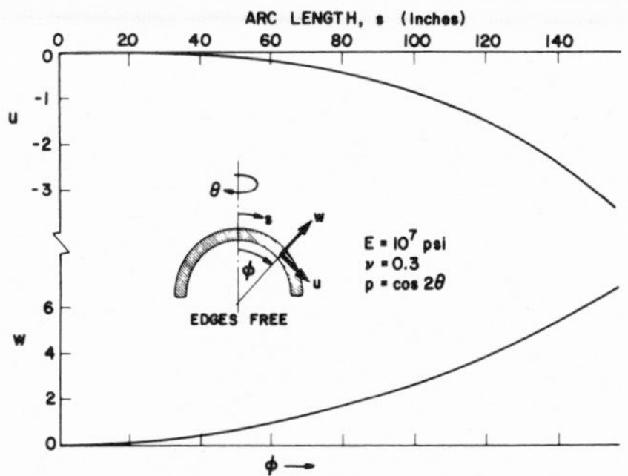

Fig. 2 Meridional and normal displacements at θ = 0 of a hemisphere with a free edge submitted to pressure p = cos 2θ

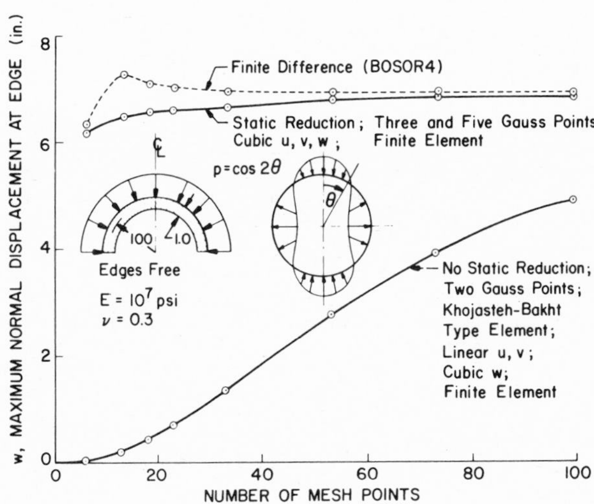

Fig. 3 Comparison of convergence of finite element method with finite difference energy method

number of columns m of B_L and B_{NL} depends on how many nodal degrees of freedom are used in the discrete model. If Eq. (1) is expressed in discrete form, and if the right-hand side of Eq. (2) is substituted into it, the strain energy expression for the thin shell becomes a quartic algebraic form.

Stiffener Strain Energy

Most practical shell structures are reinforced by stiffeners. Depending on the configuration these might be "smeared out" or treated as discrete elastic structures.

"Smeared" stiffeners. If there exists a regular pattern of reasonably closely spaced stiffeners, their contribution to the wall stiffness of the shell or plate might be modeled by an averaging of their extensional and bending rigidities over arc lengths equal to the local spacings between them. Thus, the actual wall is treated as if it were orthotropic. This "smearing" process accounts for the fact that the neutral axes of the stiffeners do not in general lie in the plane of the reference surface of the shell wall. Predictions of buckling loads and vibration frequencies of stiffened cylinders have been found to be very sensitive to this eccentricity effect. A general rule of thumb for deciding whether to smear out the stiffeners or to treat them as discrete is that for smearing there should be about 2 to 3 stiffeners per half-wavelength of the deformation pattern. It may be appropriate to smear out stiffeners in a buckling or vibration analysis but, because of local stress concentrations caused by the stiffeners, not in a stress analysis. The stiffeners can be smeared as an analytical device to suppress local buckling and vibration modes. In order to handle problems involving smeared stiffeners, a computerized analysis must include coupling between bending and extensional energy as described earlier. The paper by Baruch and Singer [39] is a classic in the field of stiffened shell analysis.

Discrete stiffeners. If the stiffeners are so far apart that significant variations of displacement and stress occur between them, then they cannot be averaged over the entire shell surface but must be treated as discrete one-dimensional bodies. The standard approach is to assume that the cross section of the stiffener does not deform but that it translates and rotates in a fashion compatible with the shell to which it is attached. If plane sections of the stiffener remain planar and normal to the reference axis, the strain energy can be written in a form analogous to that for the shell:

$$U_{stiffener} = \frac{1}{2} \int_L (\varepsilon_r^T K \varepsilon_r + 2N_r \varepsilon_r)\, dL \qquad (3)$$

where dL is the incremental length along the reference axis of the stiffener. The four-element vector ε_r represents the reference axis normal strain ε_{r1}, changes in curvature η_{r1}, η_{r2} in two orthogonal planes, and twist η_{r12}. K is in general a full 4 x 4 symmetric matrix of coefficients which depends on the location of the stiffener material relative to the reference axis, on details of the stiffener construction, on the temperature, and if plasticity is present and the tangent stiffness method is used, on the stress-strain curve and the flow law of the stiffener material. The vector N_r is analogous to N in Eq. (1). If plasticity is present or if the material properties of the stiffener are temperature dependent, the elements of K and N_r at a point on the reference axis must in general be determined by numerical integration over the stiffener area. The reference axis strain ε_{r1}, changes in curvature η_{r1}, η_{r2}, and twist η_{r12} can be expressed in terms of the displacement and rotation components and derivatives of these quantities referred to the stiffener reference axis. With appropriate discretization, the stiffener strain vector ε_r can be expressed in algebraic form as

$$\varepsilon_{ri} = B_{Lr} d_{ri} + d_{ri}^T B_{NLr} d_{ri} \tag{4}$$

in which all quantities are analogous to those in Eq. (2). Since the reference axis of the stiffener does not in general lie in the plane of the reference surface of the shell, the local displacement vector d_{ri} must be expressed in terms of the local shell reference surface displacement d_i

$$d_{ri} = E d_i \tag{5}$$

so that Eq. (4) in terms of the dependent variables d_i becomes

$$\varepsilon_{ri} = B_{Lr} E d_i + d_i^T E^T B_{NLr} E d_i \tag{6}$$

If Eq. (3) is expressed in discrete form and if the right-hand side of Eq. (6) is substituted into it, the strain energy expression for the stiffener becomes a quartic algebraic form which is added to the shell strain energy.

Various Discrete Stiffener Models

Various models have been used for predicting the behavior of stiffeners which are treated as discrete.

Shear center vs. centroid. Many analysts neglect warping of the cross section, assume that the axis of shear centers and axis of centroids coincide, and, in buckling or vibration analyses, assume that any prestress in the stiffener is distributed uniformly over the cross section. Figure 4, taken from [40], shows buckling pressures for a ring-stiffened cylinder predicted by means of three models of the same configuration. The results for the most accurate model are represented by the solid line. In this model the two parts of the T-shaped ring are treated as flexible shell segments rather than as a lumped, one-dimensional elastic structure with undeformable cross section. The curve labeled "Old Ring" corresponds to a model in which the shear center and centroid are assumed to coincide, the hoop prestress has the value calculated at the centroid and is uniformly distributed over the cross section, and the warping constant Γ is zero. "New Ring" refers to a more elaborate discrete ring model in which the difference in location between the ring shear center and centroid is accounted for, the variation in ring prestress in the radial direction over the ring cross section is included, and Γ is nonzero. The most important effect in this comparison is that of the offset between shear center and centroid. Both discrete ring models are inadequate in this case because the ring cross section actually deforms as shown in the insert labeled "Buckling Mode."

Stiffener eccentricity effect. Predictions of stress, buckling, and vibration are often extremely sensitive to apparently minor changes in the mathematical model of a discrete stiffener and its location relative to the shell reference surface. Figure 5, taken from [41], shows the axisymmetric prebuckling hoop compression at the bifurcation buckling pressure of a spherical cap loaded by a fixed edge moment. The test, described in [42], most resembled case 1. In case 2, the shell is considered to penetrate the ring and terminate at the ring centroid. In the two cases shown, the external pressure is assumed to be reacted upon by an axial load acting through the ring centroid. Buckling occurs at the pressures p_{cr} indicated, and the predicted 18 wave buckling pattern is concentrated in the area near the edge, where the hoop stress resultants are maximum compressive. In this case, the predicted buckling pressure is most sensitive to the axial component e_2 of the stiffener eccentricity. Because of this component, the meridional resultant N_{10} produces a clockwise moment about the ring centroid, which acts to reduce the destabilizing hoop compression near the edge of the cap.

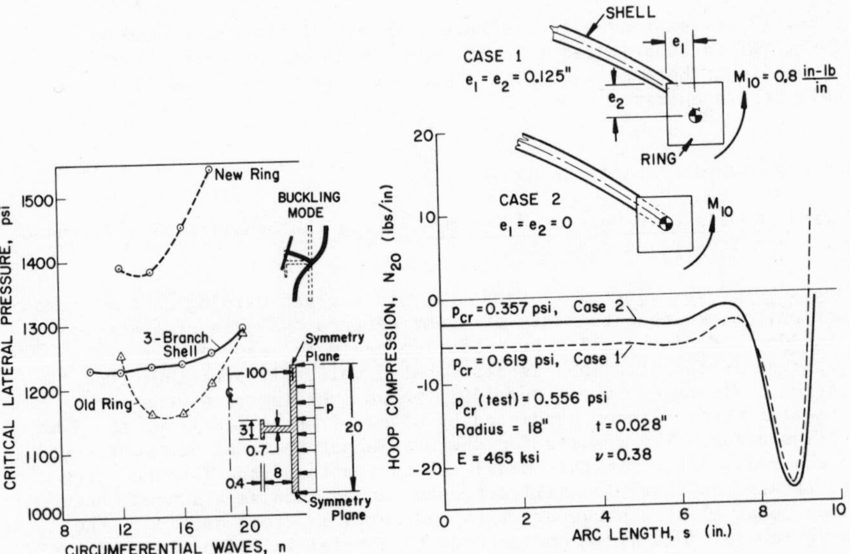

Fig. 4 Comparison of local buckling pressures for various ring models

Fig. 5 Effect of edge ring eccentricity on predicted buckling pressures of cap

Users of computer programs for the analysis of stiffened shells should be aware of the effects just described in order to use these sophisticated analytical tools effectively. More will be said about the modeling of discrete stiffeners in a following section.

Loading

The various types of loading on a structure are listed in the Questionnaire for Program Developers in [1]. Thermal loading refers to the terms $N\varepsilon$ and $N_r\varepsilon_r$ in Eqs. (1) and (3), respectively. Two aspects of loading are of particular interest when thin shells are involved:

1. "live" or following loads vs. "dead" or constant-directional loads, and
2. loading by means of enforced displacement vs. loading by prescribed external forces.

Live loads. A "live" or following load is a load whose direction changes as the shell surface rotates. The expressions for work done by the external forces distributed over the shell surface and along the discrete stiffeners are, respectively

$$W_{shell} = \int_S (pd + d^T Pd)\,dS \tag{7}$$

$$W_{stiffener} = \int_L (qd_r + d_r^T Qd_r)\,dL \tag{8}$$

The second terms in each integrand represent the live load effect. This effect should be included if the deflections or rotations are moderately large or, in bifurcation buckling problems, if the half-wavelength of the buckling mode is the same order of magnitude as the smallest principal radius of curvature. Two examples in which the live load effect is significant are the bifurcation buckling or nonlinear collapse of a very long cylinder under external pressure and that of a ring under external radial compression. Inclusion of the live load effect lowers the predicted failure loads by about 30 percent in these cases.

Displacement vs. force loading. Loading may be applied by means of a controlled displacement distribution (such as uniform end shortening of a cylinder) or by means of a controlled force distribution (such as uniformly applied axial force). A given thin shell structure may behave very differently under these two loading conditions. If a boundary displacement is imposed, a significant amount of stress redistribution can occur, and flexible or "soft" parts of the structure deform considerably with more load subsequently being taken up by the stiff or "hard" parts. Figure 10 shows a system in which this type of behavior occurs. Less stress redistribution can take place if the boundary forces are imposed, leading in general to earlier failure of the structure.

Kinetic Energy

In thin shell analysis it is not necessary to include rotatory inertia of the shell wall. Rotatory inertia of the discrete stiffeners probably should be included however. The kinetic energy of the shell and the stiffeners has the analytical form

$$K.E. = \frac{1}{2} \int_S \dot{d}^T m \dot{d} \, dS + \frac{1}{2} \int_L (\dot{d}_r^T m_r \dot{d}_r + \dot{\omega}_r^T I_r \dot{\omega}_r) dL \tag{9}$$

in which $(\dot{\ })$ indicates differentiation with respect to time, m is the mass/area of the shell reference surface, m_r is the mass/length of discrete stiffener reference axis, ω_r is the rotation vector of the stiffener reference axis, and I_r is a matrix of rotatory inertia components of the stiffener referred to its reference axis. As before, various transformations are used in order to express all quantities in terms of the shell wall displacements. Whether or not the mass matrix is diagonal depends, of course, on the discretization model and the choice of nodal degrees of freedom.

Boundary and Other Constraint Conditions

The energy minimization problem (displacement method) is subject to constraint conditions corresponding to behavior at the boundaries of the shell or other locations within the domain where certain relationships between nodal point displacements are postulated to hold. These conditions may be linear or nonlinear. Two types of nonlinearity may exist: the first may result from continually changing geometry as loads are varied, the second may result from a sudden change in behavior as one part of a structure contacts another. Other types of constraint conditions are listed in the Questionnaire for Program Developers in [1].

The constraint conditions might be introduced into the analytical model by means of Lagrange multipliers, by appropriate elimination of rows and columns of stiffness matrices, or by incorporation into the model of rigid connections. If the Lagrange multiplier method is used, for example, a general nonhomogeneous, nonlinear constraint condition might assume the form

$$U_C = \lambda[d_a - T_L d_b - d_b^T T_{NL} d_b - d_o] \tag{10}$$

in which λ is a vector of Lagrange multipliers; d_a and d_b are displacement vectors at different points, a and b, in the structure; and d_o is an applied displacement.

Constraint Condition Problems to Be Wary Of

There are certain commonly occurring situations in which the program user should take great care with regard to constraint conditions.

These involve rigid body behavior, symmetric vs. antisymmetric behavior at planes of symmetry in the structure, singularity conditions at poles of shells of revolution, discontinuities between various branches and segments of a complex shell structure, and unexpected sensitivity of predicted behavior to changes in boundary conditions.

Rigid body displacement. Rigid body displacement of an analytical model of a structure should not be permitted in static stress and buckling problems. In such problems the shell must be held in such a way that no constraints are introduced which are not actually present in the real structure. The proper application of rigid body constraint conditions requires special care in the case of nonsymmetrically loaded shells of revolution. These conditions apply only if the displacements are axisymmetric or if the displacements vary with one circumferential wave around the circumference and must be released for higher displacement harmonics.

Symmetry planes. Many problems are best analyzed by a modeling of a small portion of the actual structure bounded by symmetry planes. In bifurcation buckling and modal vibration problems, important modes may be antisymmetrical at one or more of the symmetry planes. This occurrence implies that symmetry boundary conditions should be applied in the prestress analysis and antisymmetry conditions at one or more of the symmetry planes in the eigenvalue analysis for bifurcation buckling or modal vibration. Unless the program user is certain about the behavior at a symmetry plane, he must make multiple runs on the computer, testing for both symmetrical and antisymmetrical behavior at each symmetry plane. The total number of runs required might be as great as 2^k, where k is the number of symmetry planes in the structure.

Singularity conditions at a pole. The problem of singularity conditions arises only in the case of shells of revolution or flat circular plates. As with rigid body modes, special conditions must be applied for axisymmetric ($n = 0$) displacements or for displacement modes with one circumferential wave ($n = 1$). If $n \geq 2$ the pole condition acts as a clamped boundary. Many programs for shells of revolution take care of the pole conditions automatically.

Constraint conditions for discontinuous domains. Practical shell structures are very frequently assembled so that the combined reference surfaces of the various branches and segments of the analytical model form a discontinuous domain. The program user should be aware that if the constraint conditions governing the compatibility relations between adjacent surfaces imply that a rigid connection exists across the discontinuity, then the analytical model will be stiffer than the actual structure. Buckling loads and vibration frequencies will be overestimated. It is likely that local discontinuity stresses will also be overestimated. Several examples of the effect of introduction of discontinuities in the reference surfaces of segmented shell structures are given later in this paper.

Unexpected sensitivity of predicted behavior to changes in boundary conditions. Frequently,complicated shell structures are designed and manufactured by more than one company or by more than one organization within a company. Each company or organization is responsible for only one particular segment of the entire structure. Often the properties of the adjoining segments are known only approximately if at all. Therefore some conditions must be assumed at the boundaries of each segment during the design phase of that segment.

The purpose of this section is to warn the designer or program user that predictions of stress, buckling, and vibration of shells may be very sensitive to boundary conditions even though intuition dictates otherwise. Figure 6, taken from [43], provides an example. Buckling loads are given for clamped axially compressed cylinders of various lengths with a wall construction consisting of a longitudinally corrugated sheet welded to an inside smooth sheet. The internal Z-shaped ring stiffeners are heavy enough to cause interring buckling as shown. The asymptote represents the predicted buckling load of a simply supported bay 15 inches long. Intuitively, it is surprising that the clamping condition at the edge significantly affects the critical load for cylinders with many bays. Since buckling occurs between rings, one might think that the critical load would appraoch the asymptote much more rapidly as the number of bays is increased. However, the theoretical results shown in Fig. 6 have been confirmed by tests [43].

Figure 7 helps to explain the slow convergence. The nonsymmetric moment applied at the simply supported edge simulates the effect of clamping there during the transition from an axisymmetric prestressed state to a postbuckling state with 16 circumferential waves. Our intuition of what length of cylinder is required before the buckling load becomes independent of length is based on the more familiar but much shorter axisymmetric boundary layer length, $\sim 3\sqrt{Rt_{eff}}$, also shown in Fig. 7.

In general, engineers interested in designing a particular segment of a larger structure should make every effort to determine as accurately as possible the actual boundary conditions at the ends of "their" segment. Portions of the adjoining segments should be included in the model, possibly with a cruder mesh. If little is known about the adjoining structures, sensitivity studies should be performed in which both upper and lower bounds on the degree of boundary constraint are assumed.

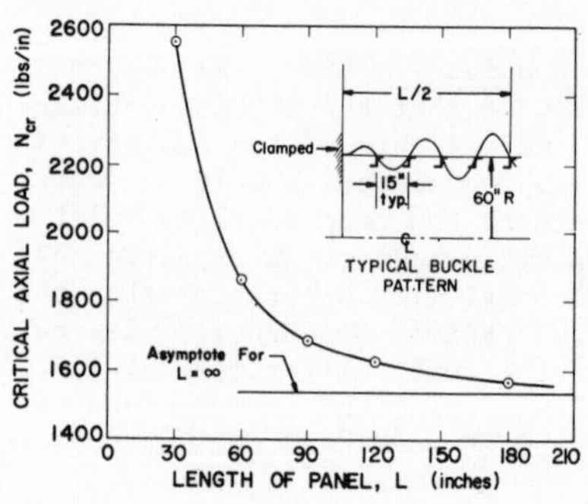

Fig. 6 Critical buckling loads of longitudinally corrugated cylinder

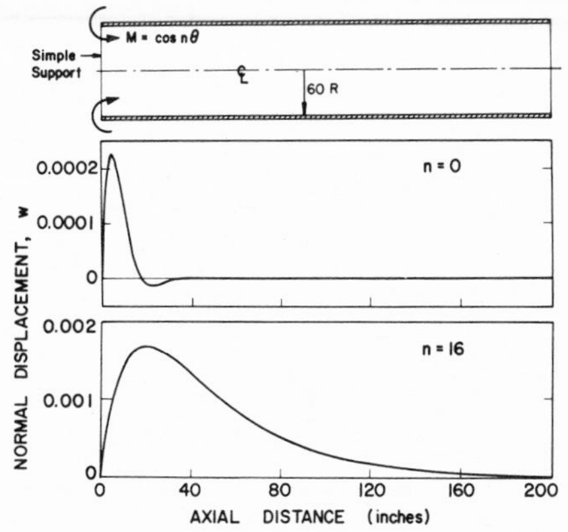

Fig. 7 Decay rates of axisymmetric and nonsymmetric disturbances in longitudinally corrugated cylinder

SOLUTION METHODS

An energy expression can be constructed from the right-hand sides of Eqs. (1) through (10). With appropriate substitutions of discretized displacement components q for continuous variables and numerical integration over the shell reference surface and over the lengths of discrete stiffeners, a nonhomogeneous quartic functional

$$H = \int_{t_1}^{t_2} f(q,\dot{q};t)dt \tag{11}$$

is obtained in which the coefficients as well as the displacements may be time dependent. (Note that damping as well as fluid or soil structure interaction effects have been omitted in the above development.) Numerical solutions for problems involving linear and nonlinear static stress, bifuraction (eigenvalue) buckling, modal vibration with prestress, and linear and nonlinear dynamic response can be based on this functional. Through minimization with respect to the nodal point variables q, a set of simultaneous algebraic equations is generated. The nature of these equations and the numerical method to use for their solution depend on the type of problem that is being solved.

Various methods for matrix decomposition, solution of nonlinear statics problems, eigenvalue extraction, and time integration can be applied. Many of these methods are indicated on the Questionnaire for Program Developers in [1].

Matrix Decomposition

Various techniques of matrix decomposition are described by Schrem [44] and by Irons and Kan [45]. There is nothing in particular about thin shells that leads to a determination as to the method to be used. With problems involving discretization in only one dimension, such as nonlinear axisymmetric stress analysis, linear nonsymmetric stress analysis, and bifurcation buckling or modal vibration analyses of branched shells of revolution, the governing matrices are very narrowly banded except at a relatively few rows corresponding to branch points of the structure. Hence a variable bandwidth or "skyline" method of decomposition is appropriate. The wavefront method or other sparse matrix method may be more efficient for problems involving discretization along two or more coordinate directions.

Solution of Nonlinear Statics Problems

If time dependence is removed from Eq.(11), the static equilibrium condition is

$$\Psi_i = 0 \qquad i = 1, 2, \ldots M \qquad (12)$$

where Ψ_i is the gradient of the quartic form with respect to the ith displacement component q_i, and M is the total number of degrees of freedom in the discrete model. There are many ways of solving these M nonlinear algebraic equations, including the full Newton-Raphson, modified Newton-Raphson, dynamic relaxation, successive substitutions, and the incremental method with and without equilibrium correction. These techniques are described in more detail in other papers in this volume [46, 47].

The purpose here is to emphasize that certain methods are unsuitable for predicting certain kinds of thin shell behavior. Figure 8, taken from [48], shows load-deflection curves from test and theory for a point-loaded spherical cap. The behavior is most nonlinear in a rather small range of dimensionless load between P* = 1 and 2. In this range of load, the extent of the inward dimple expands rapidly with increase in load. The behavior is so extremely nonlinear in this load range that only use of a Newton-Raphson type method will lead to a converged solution at every load step. The method of successive substitutions will not converge even for extremely small load increments. In fact, Archer [49], investigating this problem in 1962, predicted that the shell would fail at about P* = 2 because iterations failed to converge for higher loads.

Figure 9, taken from [50], shows one load-deflection curve predicted with the use of the modified Newton-Raphson method and three curves predicted with the use of the incremental method without equilibrium correction. The points on the modified Newton curve indicate loads for which the stiffness matrix was recomputed and refactored. Note that if the incremental method is used, the collapse load of about 2.3 pounds is approached rather slowly as the load increment is decreased. For finite load increments, the predicted load-carrying capability of the panel is overestimated. In general, the incremental method is unsuitable for problems in which the nonlinear behavior is sudden, that is, for problems in which the effective stiffness of the structure changes significantly within a fairly small percentage of the total load range of interest. The case depicted in Fig. 10 is another

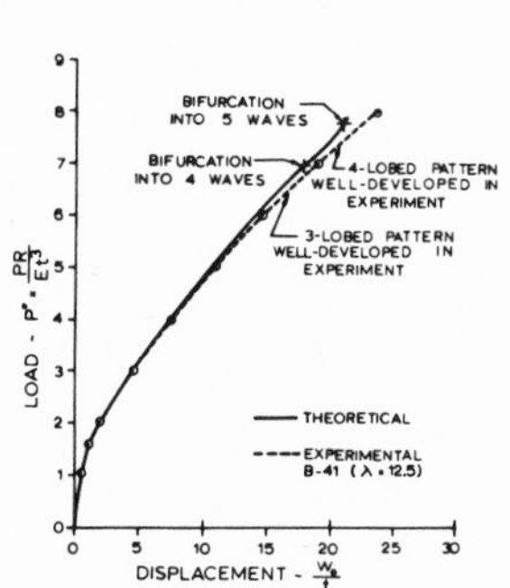

Fig. 8 Load-deflection curve of point-loaded spherical cap

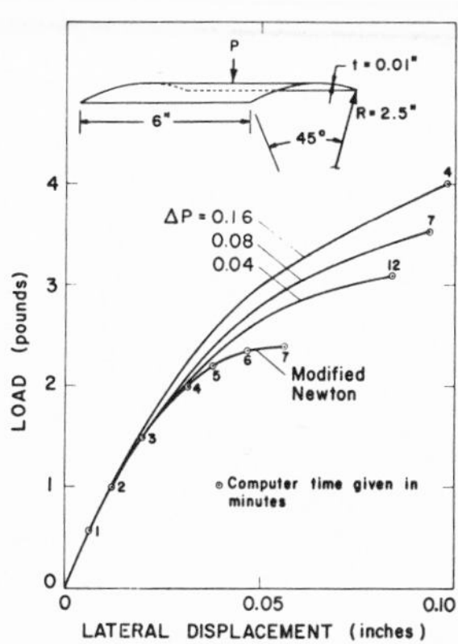

Fig. 9 Load-deflection curves from incremental and modified Newton-Raphson methods

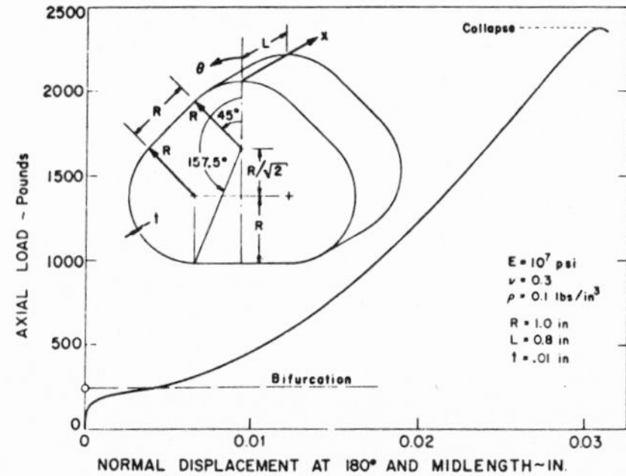

Fig. 10 Noncircular cylinder submitted to uniform end shortening

example for which more sophisticated strategies are required than the simplest incremental method.

Generally one should use more rigorous strategies for the solution of nonlinear problems involving discretization in one dimension (a model of a shell of revolution, for example) than for problems involving discretization in two or more dimensions. The bandwidth of the stiffness matrix for a one-dimensional numerical analysis is very small, and therefore a small percentage of the total computer time is spent decomposing this matrix. For this reason, it is feasible to use the full Newton-Raphson method, which involves decomposition of the stiffness matrix perhaps many times for each load increment. In problems involving more than one discretized dimension, the modified Newton-Raphson method is more suitable. In the modified method the stiffness matrix is recalculated and decomposed only if the method of successive substitutions has failed to converge after a certain predetermined number of iterations. This failure to converge may occur only after several load increments. Thus, the relative balance between matrix decomposition and back-substitution is altered because of the "shape" of the stiffness matrix.

Eigenvalue Problems

Various methods for the calculation of eigenvalues are described in [47] and are indicated in the Questionnaire for Program Developers in [1]. The purpose here is to indicate which terms should be retained in the equations for bifurcation buckling and modal vibration of thin shells.

Bifurcation Buckling

The word "bifurcation" means branching. A bifurcation buckling analysis is a search for points on a primary load-deflection path at which a "branching" of the behavior occurs. For axisymmetrically loaded shells of revolution the primary path is represented by the load-axisymmetric deflection curve, such as in Fig. 8. The smallest bifurcation load is that load for which a nonsymmetric pattern of displacements can first coexist with the primary axisymmetric field. In Fig. 8 such a point is labeled "Bifurcation into 4 Waves." The mathematical model for this behavior is derived from Eq. (12) and is essentially a uniqueness test for equilibrium. If q_o represents an equilibrium state, then $\Psi_i(q_o) = 0$, $i = 1,2,\ldots M$. At the bifurcation load there exists a nontrivial infinitesimal displacement distribution δq, called the buckling mode, such that

$$\Psi_i(q_o + \delta q) = 0 \qquad i = 1,2,\ldots,M \tag{13}$$

The Ψ_i can be expanded in Taylor series about q_o, thus:

$$\Psi_i(q_o + \delta q) = \Psi_i(q_o) + \sum_{j=1}^{M} \left.\frac{\partial \Psi_i}{\partial(\delta q_j)}\right|_{\delta q \to 0} \delta q_j + \ldots = 0,\ i = 1,2,\ldots M \tag{14}$$

Since $\Psi_i(q_o) = 0$, it follows that

$$\sum_{j=1}^{M} \left.\frac{\partial \Psi_i}{\partial(\delta q_j)}\right|_{\delta q \to 0} \delta q_j = 0 \qquad i = 1,2,3,\ldots,M \tag{15}$$

The criterion for the existence of a nontrivial buckling mode δq is that the determinant of the M x M stability matrix $[S] = [\partial\Psi/\partial(\delta q)]$, evaluated in the limit as $\delta q \to 0$, vanish. This matrix can be written in the form

$$\begin{aligned}[S] = {} & [B^TCB + R^TN_{of}R + D^TP_fD] \\ & + \lambda[B^TCX + X^TCB + R^TN_oR + D^TPD] + \lambda^2[X^TCX]\end{aligned} \tag{16}$$

in which N_{of} and P_f represent prestress resultants and live load terms considered to be fixed for this particular eigenvalue problem; B represents the kinematic relations for the shell, including the effect of deformation caused by the fixed loads; C represents the constitutive law at the stress state corresponding to the fixed loads; R represents the shell wall rotation-displacement relations; D is a matrix which transforms the analytical displacements into their discrete forms; λX represents the effect of the change in shape of the shell caused by the loads which are eigenvalue parameters; and λN_o represents the prestress resultants due to the loads which are eigenvalue parameters. The eigenvalues λ can be obtained through calculation of the determinant of [S] for a range of loads or, after the right-hand side of Eq. (16) has been transformed into a standard form as described in [36] and [47], by means of one of the other methods listed in Section 6 of the Questionnaire for Developers [1].

Two features of Eq. (16) should be emphasized:

1. The total loading is considered to be divided into two sets, one fixed and the other with an undetermined magnitude λ.

2. The effect of prebuckling change in shape of the shell is included in the stability equations.

The effects of these features are described in the following subsections.

Two sets of loads. In the analysis of practical shells there often exists a combination of loads which does not vary proportionally during some event that may cause instability. One may need to know the buckling pressure of a nonuniformly heated cylinder, for example. The temperature distribution is known and fixed, but the magnitude λ of the pressure must be determined from a bifurcation buckling analysis. The fixed temperature distribution causes the shell to change shape, affecting the matrices B and R in Eq. (16). The material properties may change, affecting the matrix C. Thermal prestress resultants develop, giving rise to the matrix N_{of}. The applied pressure, an eigenvalue parameter, generates all the terms multiplied by λ and does not affect those not multiplied by λ.

The division of load into two sets is often very helpful in bifurcation buckling problems involving only a single loading parameter. An example is an externally pressurized spherical cap which behaves in a nonlinear fashion in the prebuckling phase, either because of large prebuckling deflections or because of nonlinear material behavior, or because of both. If a lower bound is known for the bifurcation buckling pressure, then this lower bound can be used as a fixed pressure, p_f. The prebuckling meridional rotation χ_{of} which affects the matrix B, the material properties which affect C, and the meridional and hoop stress resultants which constitute N_{of} are known at the pressure p_f. Hence, the stiffness matrix of the deformed shell as loaded by p_f can be calculated. A small increment of pressure p is added, and the matrices X and N_o are calculated. The eigenvalue problem

$$(K_1 + \lambda K_2 + \lambda^2 K_3)q = 0 \tag{17}$$

is solved, and a new estimate of the buckling pressure

$$p_{new} = p_f + \lambda p \tag{18}$$

is available. This estimate, although no longer a lower bound, can be used as a new fixed load. Iterations continue until λp becomes negligibly small compared to p_f. Various strategies for determining buckling eigenvalues in problems which exhibit nonlinear prebuckling behavior are described in more detail for elastic shells in [36] and for elastic-plastic shells in [51].

Effect of prebuckling deformation. The other feature of Eq. (16) to be emphasized is the inclusion of the deformations of the structure due to both fixed and eigenvalue sets of loads. The program user may wonder when these shape changes can be ignored and when they cannot. A general rule of thumb is that the prebuckling deformation effects should be included if the destabilizing prestress resultants are large in regions where the shell wall rotation is moderately large. If an engineer has performed an analysis neglecting prebuckling deformations, he can inspect the buckling mode shape to see if it has a significant amplitude in a region where significant prebuckling rotations and changes in curvature occur at the buckling load. If so, then it is likely that these prebuckling rotations should be included in the stability analysis.

Two examples in which the effect of prebuckling shape change is dramatic are shown in Figs. 11 and 12. If a cylinder is subjected to a bending moment, the cross section flattens. This change in shape of the cross section has three consequences: the maximum compressive stress is higher because the moment of inertia of the cross section required to carry the applied moment has decreased, the stresses are redistributed across the cylinder, and the bifurcation buckling stress

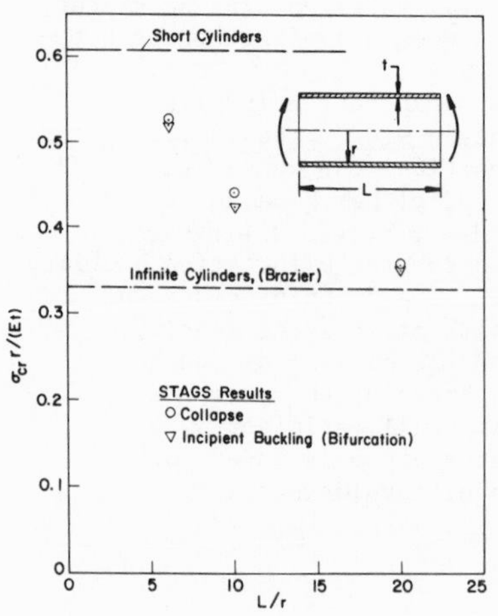

Fig. 11 Critical loads for cylinders under bending

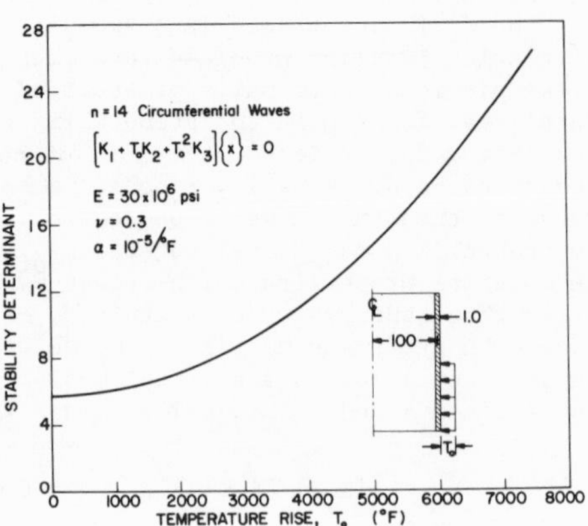

Fig. 12 Determinant plot with shape change effect included

corresponding to a mode with small circumferential waves has decreased because the effective circumferential radius of curvature where buckling begins has increased. Figure 11, taken from [52], shows normalized collapse stress and bifurcation buckling stress as functions of L/r. The upper limit is the classical buckling stress with prebuckling deformations neglected; and the lower limit was calculated by Brazier for infinite cylinders. The intermediate points were calculated with the use of STAGS [35].

The example just described is one in which the effect of the prebuckling shape change is to weaken the structure. Just the opposite effect occurs in the case of an axisymmetrically heated cylinder with a sudden temperature change along its length. If the cylinder is free to expand axially, buckling is caused by a very narrow band of hoop compression in the neighborhood of the abrupt temperature change. If prebuckling rotations are neglected in the stability equations, the cylinder shown in Fig. 12 is predicted to buckle at 4850°F with 14 circumferential waves. However, if prebuckling rotations are included in the stability equations, the prediction is that no buckling will occur. The stability determinant for all circumferential wave numbers increases monotonically with temperature T_o. The destabilizing effect of the local hoop compression is counteracted by the stabilizing effect of the increase in meridional curvature as T_o increases. Thus, the hoop compressive stresses act, not on a cylinder, but on a much stronger doubly curved segment, whose meridional curvature increases with increasing temperature discontinuity. Figure 12 is taken from [53]. Also given in [53] is a comparison between test and theory for buckling of a clamped uniformly heated cylinder. The predicted buckling temperature is within 5 percent of the test value if the prebuckling rotations are included in the stability analysis, but about 30 percent below the test value if they are neglected.

Modal Vibrations

The free vibration problem is governed by the stiffness matrix

$$[K_1] = B^T CB + R^T N_{of} R + D^T P_f D \tag{19}$$

which is the same as K_1 in Eq. (17), and the mass matrix

$$[K_m] = D^T mD \tag{20}$$

The eigenvalue problem

$$\left(K_1 - \Omega^2 Km\right)q = 0 \tag{21}$$

is solved with the use of one of the methods listed in [1] and described in detail in [47]. Note that a reasonably comprehensive computer program for shell analysis should be written such that the effect of prestress on vibration frequencies is included.

In cases involving thin shells it is often necessary to obtain many vibration modes because natural frequencies are rather dense within particular intervals and because frequencies associated primarily with membrane deformations are orders of magnitude higher than those associated primarily with bending deformations. The extreme disparity in frequency between membrane and bending modes and the clustering property are particularly characteristic of very thin shells or of shells which are stiffened at regular intervals in such a way that the stiffeners should not be smeared out in the analytical model. Therefore, the engineer should choose a program with an eigenvalue extraction algorithm that permits the accurate determination of many modes for a given stiffness and mass matrix. The method used in the program should work reliably for cases in which the frequencies are very closely spaced. The techniques of subspace iteration or single vector inverse power iteration with spectral shifts and deflation seem to work well for these types of problems. Determinant plotting may be less reliable, depending on the strategy used, because of the great likelihood of skipping over many roots with a single increment in frequency. The power sweep method should be avoided because its accuracy deterioriates as higher and higher roots are extracted.

Time Integration

Surveys of time integration methods for dyanmic response are presented by Clough and Bathe in [54], by Krieg and Key in [55], and by Belytscho in this volume [56]. Most of the popular methods are identified in the Questionnaire for Program Developers in [1].

What particular properties of thin shells govern the choice of an appropriate method for time integration? The fact that the extensional stiffness is orders of magnitude greater than the bending stiffness has by far the most important consequences for the numerical analyst. This distinguishing property of thin shells causes the frequency spectrum to be divided into two overlapping regimes: the lower regime characterized by bending modes and the upper by membrane modes.

Suppose that a thin cylinder is submitted to a nonuniform pressure blast of short duration. One may wish to calculate the maximum stress in the wall. If the pressure is rather smoothly distributed over the surface, it is likely that considerable membrane stress will build up very rapidly. Bending stresses will rapidly accumulate near local reinforcements, which causes excitation of high frequency, short wavelength bending modes. If the pressure is distributed nonuniformly around the circumference, however, low frequency modes, such as the ovalization mode, will also be excited. The peak response in these slow modes will occur much later than it will in the fast modes, corresponding to membrane or discontinuity bending response. It is not known a priori which phenomenon causes the maximum stress. If modal superposition is used, many modes will have to be calculated to obtain converged results for both early time and late time responses. If forward integration is used, the time step will have to be small enough to permit accurate prediction of the early time response and prevent numerical instabilities from obscuring the behavior in the late time regime. Jensen [57] discusses several "stiffly stable" methods designed especially for the transient analysis of structures that behave in the manner just described. The major objective of these methods is to permit the accurate calculation of late time response while retaining acceptable representation of early time phenomena.

VARIOUS ASPECTS OF THIN SHELL BEHAVIOR

The previous two sections contain descriptions of terms which should be included in the governing equations for shell analysis and the techniques that are most suitable for the solution of these equations. Some examples have been given to illustrate the discussion. The main purpose of these previous sections is to help the analyst to choose a computer program which will best suit his needs in the area of thin shell analysis.

The purpose of this and the following section is to give the analyst a better feel for how thin shells behave, and how they should be modeled so that he will be better equipped to use effectively whatever programs he has acquired or plans to acquire.

In this section various types of linear and nonlinear static and dynamic behavior of thin shells are described. Specific examples are used to illustrate the behavior.

Nonlinear Static Behavior of Shells

Computer program users often wonder when nonlinear effects should be included in an analysis and when they can be neglected. Aside from contact or friction problems, the two basic kinds of nonlinearity are that caused by nonlinear stress-strain relationships and that caused by large or moderately large deformations.

Nonlinear Material Properties

It is relatively easy to determine whether or not nonlinear material behavior should be included in an analysis. If a linear or nonlinear elastic analysis is performed, and it is predicted that the effective stress somewhere in the structure will significantly exceed the proportional limit of the material, and if the tangent modulus varies significantly from the elastic value, then nonlinear material properties must be included.

Several Types of Geometric Nonlinear Behavior

Whether or not geometric nonlinearities need to be included in an analysis depends upon the case. This is not at all easy to decide a priori. However, there are certain different kinds of nonlinear effects, all caused by changes in geometry, which the analyst should understand.

<u>Large deflection nonlinearity.</u> Although all geometric nonlinearities are of this type, in the sense that the nonlinear terms in the strain-displacement relations are included because of moderately large deflection effects, this first and perhaps best appreciated kind of geometric nonlinearity is exemplified by the shallow cap snap-through problem so widely discussed in the literature. The uniformly pressurized shallow cap acts as a spring that continually softens as the external pressure is increased, until at a critical value of pressure it has zero stiffness and snaps through to a nonadjacent inverted equilibrium position. This nonlinearity in behavior is due to the continually

changing shape of the shallow cap. It is like a shallow arch which as the load increases, becomes softer and softer because it becomes shallower. The shallow cap is less and less capable of reacting to further increments in pressure by storing membrane energy, a mode of deformation associated with very small displacement increments, and instead stores more and more strain energy by bending, a mode of deformation associated with rather large displacement increments. Another example of strictly large deflection nonlinearity is that exhibited by a spherical shell with an inwardly directed concentrated load. Experimental and analytically predicted load-deflection curves for such a case are shown in Fig. 8.

Stress redistribution nonlinearity. The redistribution of stress during loading is often caused by the occurrence of large deflections in part of a structure. Figure 10 shows a short pear-shaped cylinder which is submitted to uniform end shortening. The load-deflection curve was obtained with the STAGS computer program [35]. At small values of the axial load the flat sections of the cylinder begin to bow outward, thus forcing the curved sections to take more of the axial load. In the axial load range from about 250 pounds to about 2000 pounds the relatively stiff curved portions are bearing more and more of the flat portions' original share of the load. Failure of the structure occurs at a load for which the curved sections collapse. Note that the lowest bifurcation buckling load is far below the actual failure load. More is said about the adequacy of bifurcation buckling analyses of shells in another section.

Stress redistribution also occurs in shells with cutouts and in shells reinforced by stiffeners. If the region near a cutout is improperly reinforced, it will deform locally more than in regions remote from the cutout. The forces will flow around the highly deformed areas into regions where deformations are small. Similarly, the panels of a stiffened shell may bend considerably, or even buckle as the applied loads are carried by the stiffeners. If the analyst uses a linear analysis, he must assign some effective moduli to the bent or buckled panels or he must design (probably overdesign) the stiffened shell structure so that no or almost no panel bending occurs before failure of the entire structure. If the analyst uses a computer program which includes the nonlinear stress redistribution effect, he will not have to make any a priori assumptions about the effective load carrying capability of the panels, but he must be prepared to use a fair amount of computer time. The purpose of this section is not to advise him on a decision, which depends on the case, but to emphasize the existence of a particular nonlinear phenomenon. Awareness of this phenomenon may affect the analyst's choice of computer program and his ability to use that program effectively.

Nonlinear effects caused by load path eccentricity. Load path eccentricity can be extraordinarily significant even if the deformations are very small. It occurs to some degree in almost all engineering shell structures. That the computer program user should understand and be aware of this type of nonlinear behavior cannot be overemphasized. Perhaps the best way to convey a physical feel for the various phenomena in this category is to give some specific examples analyzed with use of the BOSOR4 computer program [36].

Figure 13 shows part of an internally pressurized elliptical tank which has been thickened locally near the equator for welding. The engineering drawings called for an elliptically shaped inner surface with the thickness varying as shown. The maximum stress occurs at the outer fiber at point C because there is considerable local bending there due to the rather sudden change in direction, or eccentricity, of the load path in the short segment ACB. The nonlinear theory gives lower stresses than the linear theory because the meridional tension causes the tank to change shape in such a way as to decrease the local excursion of the load path, thereby decreasing the effective bending moment acting at point C. If the pressure had been external, the resulting meridional compression would have caused an increase in the local excursion of the load path, thereby increasing the effective bending moment acting at C. Nonlinear analysis would therefore have led to a prediction of higher maximum stress than linear analysis.

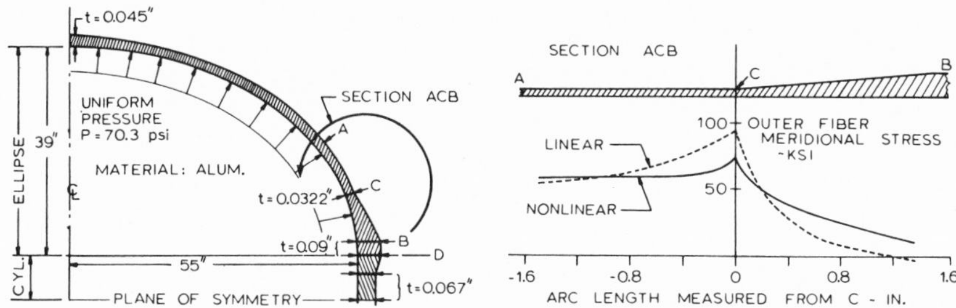

Fig. 13 Linear and nonlinear analysis of internally pressurized elliptical tank

It is generally true that compared to linear treatment, the use of nonlinear analysis in cases involving locally eccentric load paths will lead to predictions of lower maximum stress if the local stress field is tensile and higher stresses if the local stress field is compressive. Tension straightens the load path, leading to a reduction of local bending stresses, whereas compression increases the load path eccentricity leading to an increase in local bending stresses. This nonlinear behavior is very evident for rather small deflections which occur at the design loads of structures fabricated from common engineering materials. In fact, the example shown corresponds to an actual engineering experience: the tank had been built and a linear analysis performed. The user of the tank wanted to know if it would withstand a somewhat higher internal pressure than that for which it had originally been designed. The lower stress predicted with nonlinear theory gave him enough margin of safety to avoid the necessity of redesign.

Rather commonly occurring types of load path eccentricities in shells are caused by strap joints or lap joints. An example is shown in Fig. 14, which is taken from [58]. The cylinder is internally pressurized. The stress factor K is the ratio of the maximum effective (von Mises) stress in the strap to the membrane effective stress in the cylinder far away from the strap joint. Note that as in the previous example, the tensile stress field combined with nonlinear elastic analysis leads to a prediction of reduction of the ratio of bending stresses to membrane stresses as the hoop tension (internal pressure) increases. The design pressure for this cylindrical tank is in the range of 20 to 30 psi. The nonlinear effect is very significant even for much smaller pressures.

A third example of nonlinear behavior caused primarily by load path eccentricity is shown in Fig. 15 taken from [43]. This figure depicts a short section of the generator of a cylinder stiffened by external corrugations. The corrugations are cut away in the neighborhood of a field joint ring to allow for bolting of the two mating flanges of the ring. The cylinder is axially compressed. Far away from the field joint the axial resultant acts through the centroid of

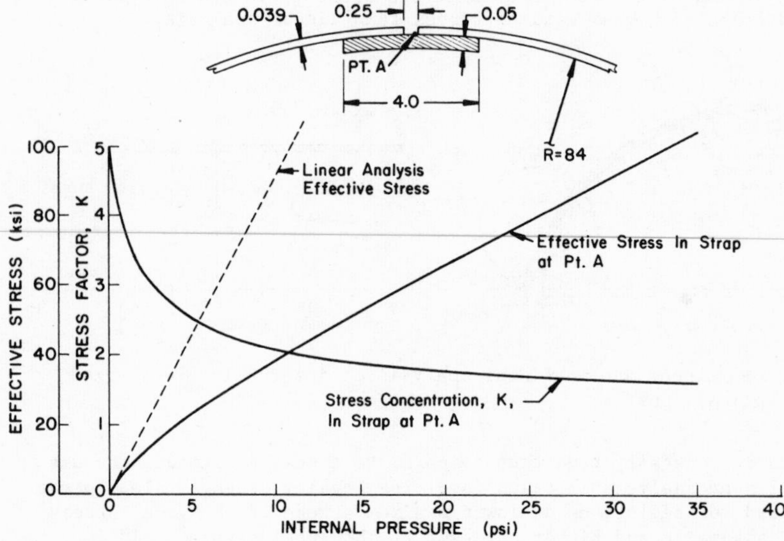

Fig. 14 Nonlinear effect in longitudinal strap joint of internally pressurized cylinder

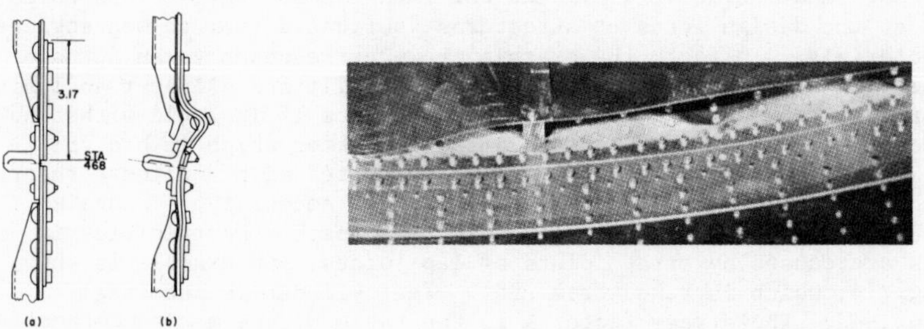

Fig. 15 Field joint geometry and buckle under axial load

Fig. 16 Failure as seen from inside the corrugated cylinder

the corrugation-skin combination. In the neighborhood of the field joint, the load path moves radially inward, effectively causing an axisymmetric dimple. As the axial compression is increased, hoop compressive stresses build up in the regions reinforced by doublers. Slight asymmetry of the assembly causes the ring to roll over axisymmetrically, which generates higher compressive hoop stresses above the ring and eventually leads to buckling there with many small waves around the circumference of the cylinder. Figure 16 shows the actual failure which agrees qualitatively with the prediction by the BOSOR 3 program.

Nonlinear effect of prestress on stress concentrations and on characteristic decay length. The nonlinear effect of prestress need not be associated with large deflections at all. In fact, it can be predicted with the use of linear differential equations. However, it is truly a nonlinear effect in that the superposition principle breaks down because of it. The effect is illustrated in Fig. 17. A cylinder is submitted to various magnitudes of axial tension and compression. For each axial load N_1, an axisymmetric unit meridional bending moment M_1 is applied at the free end, and the distributions of normal displacement and moment are plotted. The results were obtained with the BOSOR 4 computer program [36]. For the particular discrete model of the cylinder used in this analysis the lowest axial buckling load N_{cr} corresponding to an axisymmetric buckling mode is N_{cr} = -302 lb/in, and the second lowest axisymmetric buckling load is -635 lb/in. The buckling modal displacements corresponding to N_{cr} = -302 lb/in are localized near the free edge. Those corresponding to N_{cr} = -635 lb/in extend over the entire shell. Notice that the displacement distribution due to the moment changes sign between the applied loads -295 lb/in and -320 lb/in. For axial tensile loads and compressive loads smaller than -302 lb/in an applied meridional moment causes a meridional rotation in the same sense: One has to force the shell to move by applying an external moment. For loads between -302 lb/in and -635 lb/in an applied meridional moment causes a meridional rotation in the opposite sense: one has to hold the shell to prevent it from moving. For axial loads that are close to the critical value, the small meridional moment causes relatively large displacements. Note that the decay length decreases if the axial load is tensile, and increases if it is compressive. This nonlinear effect has important consequences for the engineer who must decide on how to vary shell wall thicknesses near stress concentrations and where to locate weld lands. He is advised to obtain computer codes in which this nonlinear effect is included.

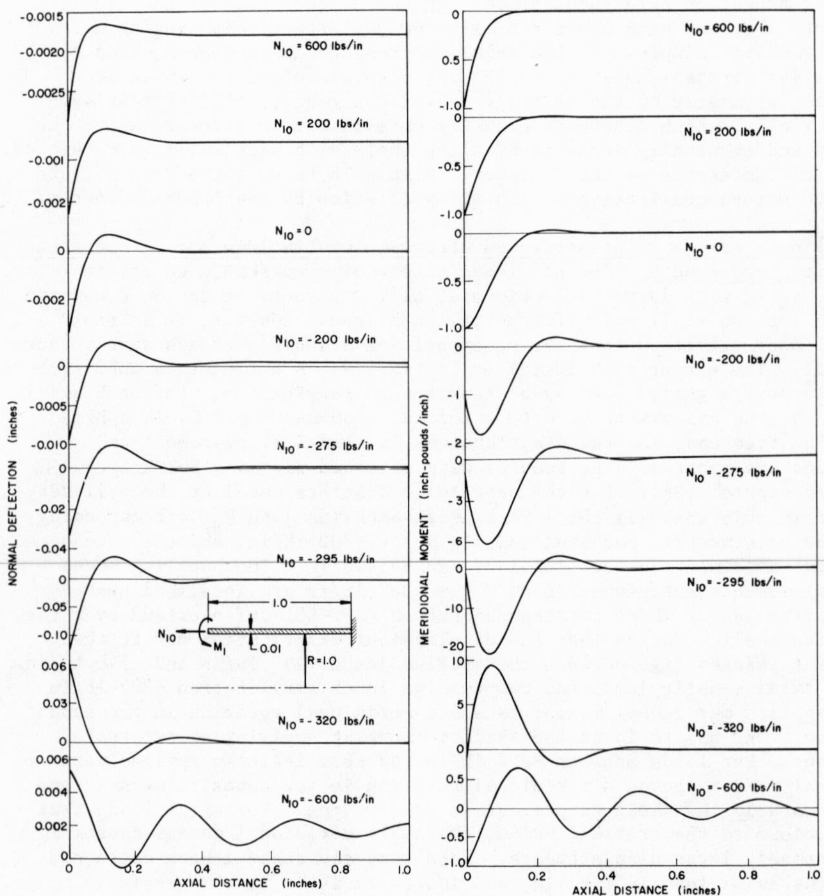

Fig. 17 Decay lengths of edge disturbance in cylinder submitted to axial tension and compression

Buckling of Thin Shells

The subject of stability is covered in a general sense by Almroth and Felippa in this volume [47]. Brush and Almroth have recently written a book on stability [59].

The property of thinness of the shell wall has a consequence that has already been pointed out: the membrane stiffness is in general several orders of magnitude greater than the bending stiffness. A thin shell can absorb a great deal of membrane strain energy without deforming too much. However, it must deform much more in order to absorb an equivalent amount of bending strain energy. If the shell is loaded in such a way that most of its strain energy is in the form of membrane contraction, and if there is a way that this stored-up membrane energy can be converted into bending energy, the shell may fail rather dramatically in a process called "buckling," as it exchanges its membrane energy for bending energy. Very large deflections are required to convert a given amount of membrane energy into bending energy.

The way in which buckling occurs depends on how the shell is loaded and on its geometrical and material properties. The prebuckling process is often nonlinear if there is a reasonably large percentage of bending energy being stored in the shell throughout the loading history. Two types of buckling exist: nonlinear collapse and bifurcation buckling. Nonlinear collapse is predicted by means of a nonlinear stress analysis. The stiffness of the structure, or the slope of the load-deflection curve decreases with increasing load. At the collapse load the load-deflection curve has zero slope, and if the load is maintained as the structure deforms, failure of the structure is usually dramatic and almost instantaneous. The term "bifurcation buckling" refers to a different kind of failure, the onset of which is predicted by means of an eigenvalue analysis. At the buckling load, or bifurcation point on the load-deflection path, the deformations begin to grow in a new pattern which is quite different from the prebuckling pattern. Failure, or unbounded growth of this new flection mode, occurs if the postbifurcation load-deflection curve has a negative slope and the applied load is independent of the deformation amplitude. The difference between nonlinear collapse and bifurcation buckling is clearly explained by Almroth and Felippa in this volume [47].

In the following subsections, the appropriateness of bifurcation buckling analyses of general shells is questioned, and comparisions of results from such approximate treatments with rigorous nonlinear collapse analyses are given. Some examples are shown of bifurcation buckling of shells of revolution. A summary of work done on imperfection sensitivity follows, and the section closes with two examples of prediction of rather complicated postbuckling behavior.

Bifurcation Buckling Analysis vs. Nonlinear Collapse Analysis of General Shells

In the case of axisymmetrically loaded shells of revolution there is obviously considerable merit in the use of a bifurcation buckling model of actual failure. Because of the high degree of symmetry in the structure, it often occurs that the prebuckling strain energy arises mostly from membrane-type deformations. This energy can be transformed into bending energy with the sudden unbounded formation of circumferential waves. In the case of general shells, however, considerable bending deformations usually occur as the load is applied. Thus, a nonlinear analysis is often required to predict the prebuckling behavior accurately. Figures 10 and 18, taken from [60], present two examples of cases for which classical type bifurcation buckling analyses are of little value for predicting failure. The bifurcation load indicated in Fig. 10 corresponds to the load level at which the flat portions of the noncircular cylinder begin to bow outward, causing redistribution of the stress to the stronger curved portions. The collapse load is a factor of ten times the bifurcation load because the curved portions do not fail simultaneously with the flat portions. In Fig. 18, the predicted bifurcation loads are higher than the collapse load because the bifurcation theory does not account for the degrading modifications in geometry caused by nonlinear local deformations. The effect here is similar to that of bending of complete cylinders as discussed by Brazier [61]. The distortion of the cross section reduces the bending stiffness of the shell. Collapse occurs at a load which for the simply supported shell is less than one quarter of the critical load predicted by the classical buckling theory. If the edges are restrained from motion in the axial direction, a tensile stress develops in the axial direction. Hence, for the clamped shell, the load increases monotonically with displacement. Other examples are shown in [60].

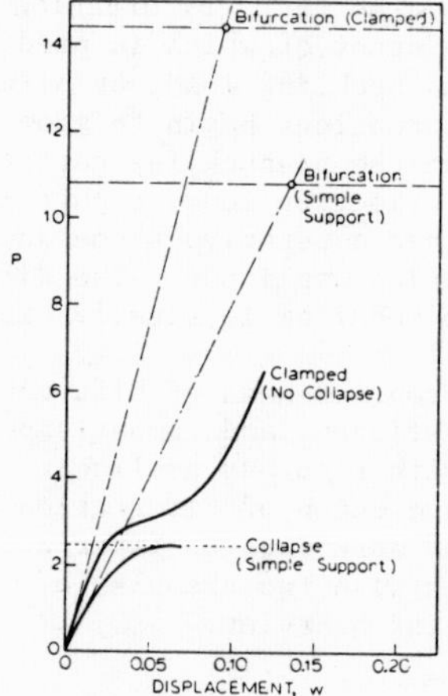

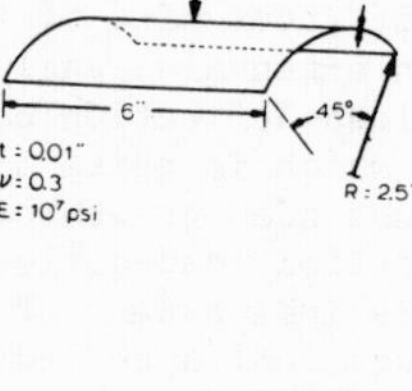

Fig. 18 Load-deflection curves for point-loaded cylindrical panel simply supported and clamped on the curved edge

Buckling of Shells of Revolution

For axisymmetrically loaded shells of revolution, buckling modes are represented by harmonic functions in the circumferential direction. Even if the prebuckling behavior is nonlinear, the entire analysis can be carried out with discretization in only one dimension. The lowest bifurcation buckling load is determined by a search procedure in which the circumferential wave number n is varied. Computer programs based on such analyses have been written by Bushnell [36], Cohen [62], Anderson et al. [63], Kalnins [64], and Svalbonas [65] for the bifurcation buckling of segmented and branched, ring and stringer-stiffened, orthotropic, layered shells of revolution. Descriptions of the properties of these programs are given in [1]. A great deal of engineering detail can be included in the analytical models because the numerical analysis is one dimensional and therefore rapidly executed on the computer. Thus, large systems can be analyzed in one pass through the computer. Discretized models with closely spaced nodal points can be treated, thereby avoiding difficulties associated with convergence. Engineers concerned with the design of thin shells are encouraged to obtain one or more codes for shells of revolution, even if they are ordinarily faced with nonaxisymmetric problems. Many of the problems that occur in the design of practical nonsymmetric shell structures can be solved with axisymmetric models. This major simplification leads to very great savings in computer time. It may be feasible to make parameter studies with a one-dimensional numerical analysis that would be prohibitively expensive with a two-dimensional treatment.

Bifurcation buckling with stable postbuckling behavior. Several examples of bifurcation buckling have been given in previous sections. Fig. 8 shows load-deflection curves from test and theory for a spherical cap with an inward-directed point lead. There are points on the theoretical primary load-deflection curve labeled "Bifurcation into 4 Waves" and "Bifurcation into 5 Waves." On the test curve loads are identified for which nonsymmetry of the deflection pattern was first observed. It is easy to perform a table top experiment to confirm qualitatively the results shown in Fig. 8: puncture a ping-pong ball to allow air to enter and leave freely. Then gently push the tip of a rather blunt pencil perpendicularly against some other point. You will notice that for small loads the dimple is axisymmetric but that as you continue to increase the load the growing dimple gradually assumes a triangular pattern. The bifurcation point corresponds to the load for which a nonsymmetric pattern begins to superpose itself on the axisymmetric dimple. By the time you notice the nonsymmetry, the shell is in its postbuckled state.

Submitted to a concentrated load, the spherical shell is stable in the postbuckling regime, just as the axially compressed pear-shaped cylinder shown in Fig. 10 is stable for loads above the lowest bifurcation load. There is no sudden release of stored-up membrane energy because much of the prebifurcation strain energy has been stored in a bending mode. The gradual growth of the nonsymmetrical pattern in the point-loaded spherical cap is due to the gradual building up of circumferential compressive stresses in a small sector fairly near the load.

Bifurcation buckling due to localized hoop compression. Buckling in the configurations shown in Fig. 5 and Fig. 15 is also caused by the accumulation of circumferential compressive stresses in narrow regions near an edge or near a structural discontinuity. This type of local membrane compression frequently occurs in complex assembled shells. Generally the predicted buckling load is very sensitive to seemingly insignificant changes in the structure or in the analytical model of it. If the analyst perceives that such a buckling phenomenon may occur, he should take great care with the modeling. Local load path eccentricities, meridional discontinuities, prebuckling shape change effects, and prebuckling geometric and material nonlinear behavior should be faithfully modeled and included in the stability analysis. If a stress analysis reveals a local band of circumferential compression, then a bifurcation buckling analysis should be performed. The minimum buckling load will generally correspond to a rather high number of circumferential waves. A reasonably accurate estimate, at least to within an order of magnitude, of the critical circumferential wave number can be calculated from the assumption that the axial and circumferential wavelengths of the buckles will be of approximately the same lengths. If the analytical model of the structure is reasonably good, the predicted buckling load should be fairly close to the test loads. Sensitivity to imperfections is much less important in such cases because the structure has a built-in local imperfection that is generally large compared to any random manufacturing errors. Note that the local stress concentrations implied in this discussion may cause some plastic yielding of the material. Bifurcation buckling loads will be overestimated if this material nonlinearity is neglected.

Bifurcation buckling of ring-stiffened shells. The purpose of this subsection is to emphasize the existence of both local and general modes of instability. Figure 19, taken from [41], shows part of a ring-stiffened cylinder. The shell is submitted to uniform external hydrostatic pressure. Cylinders of this geometry with various sizes of T rings (called "Frames" in Fig. 19) were tested by Blumenberg at the Naval Ship Research and Development Center in 1965 [66]. Mode shapes corresponding to general ($n = 2$ circumferential waves) and local ($n = 4$) instability are plotted. "General instability" denotes buckling in a mode in which both rings and shell deflect. The term "local instability" denotes buckling in a mode in which the rings are at nodes, as shown in Fig. 19 for $n = 4$. If one plotted a curve of critical load vs. circumferential wave number n there might in general be several minima. The general instability load may correspond to a minimum $p_{cr}(n)$ at a low value of n, and minima at higher values of n may occur corresponding to buckling of each bay between adjacent rings. This same phenomenon of multiple minima may occur for shells which buckle locally as described in previous subsection. Users of computer programs such as described in [36] and [62 - 65] should be sure that they cover all ranges of n in which minima may lie.

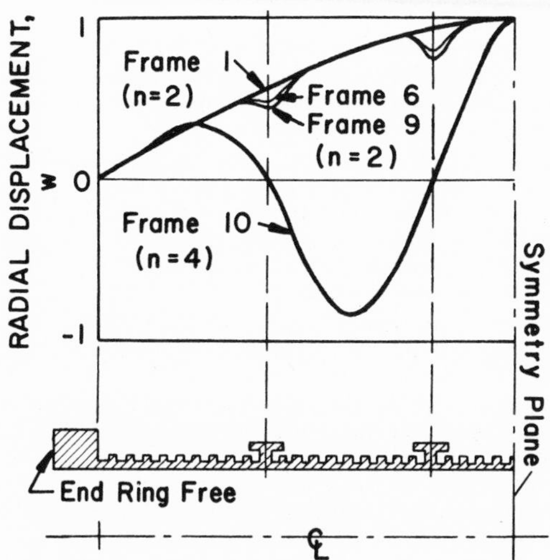

Fig. 19 General and local buckling modes for externally pressurized ring-stiffened cylinder

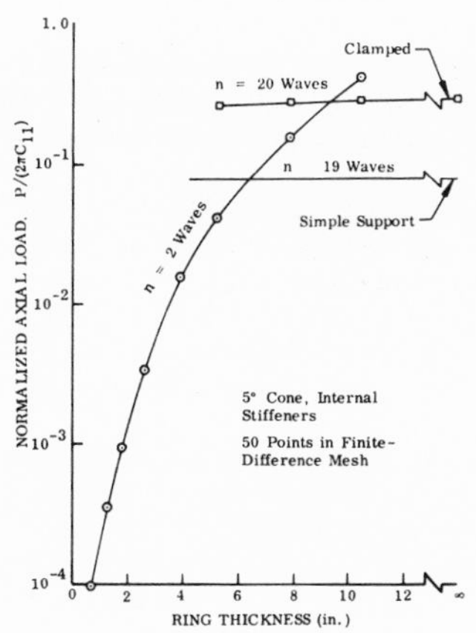

Fig. 20 Critical axial loads for a cone supported at the edges by rings of square cross section

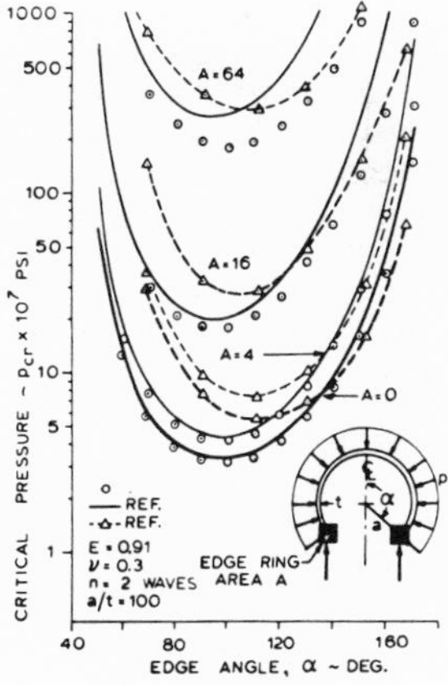

Fig. 21 Critical external pressures for a sphere supported at the edge by rings of square cross section

Inextensional bifurcation buckling. If a shell is to be submitted to destabilizing loads, the designer should avoid a configuration in which inextensional deformations of the wall can easily occur. Buckling loads associated with inextensional mode shapes can be very low indeed, as shown in Figs. 20 and 21. Figure 20, taken from [67], gives buckling loads of an axially compressed 5° cone (almost a cylinder), supported at its edges by rings of square cross section. The buckling mode with n = 2 circumferential waves is inextensional if the edges are free and very close to being inextensional for all ring sizes. Figure 21, taken from [68], shows buckling pressures of incomplete spherical shells with edge rings of various areas. Again, unless the ring is fairly large, buckling loads may be many orders of magnitude smaller than the buckling load for a clamped or simply supported shell of the same geometry.

A physical appreciation of inextensional behavior can be gained by cutting a ping-pong ball in half and squeezing one of the halves between your fingers. Large deflections occur with very small applied force. A coffee cup dispensed from a vending machine is made with a reinforcing ring at the top in order to limit the amplitude of inextensional deformations caused by the squeezing pressure of your fingers required to keep the full cup from dropping to the floor. A conical planetary reentry vehicle, such as the Viking shell, is designed on a similar principle: potentially large inextensional deformations caused by nonsymmetric reentry pressures are prevented by a large edge ring [69].

Various buckling modes. A single structure or class of very similar structures may exhibit all of the types of bifurcation buckling behavior described above. Figure 22 taken from [70] gives buckling modes for an incomplete spherical shell with edge rings of various sizes, such as shown in Fig. 21. If the edge ring is very small, inextensional buckling occurs; if it is very large, edge buckling, similar to that for a clamped shell, occurs; if it is such a size that the prebuckled state is pure membrane compression, classical buckling occurs. The analyst will be able to use a computer program for shell buckling more effectively if he realizes that these various types of buckling can take place, and if he sets up the discrete model accordingly.

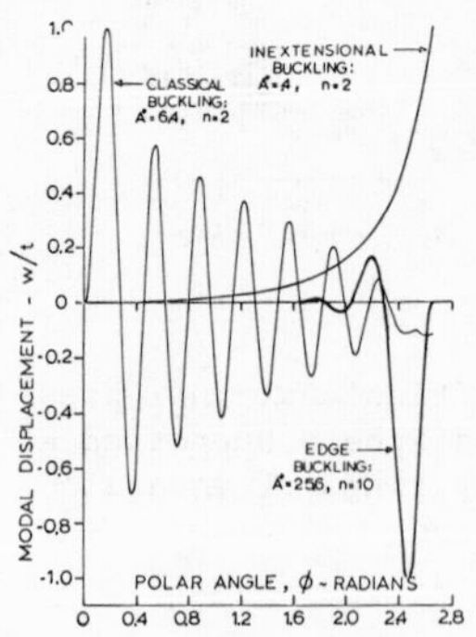

Fig. 22 Various buckling modes for externally pressurized spherical shell with edge rings of various areas

Buckling of nonaxisymmetrically loaded shells of revolution. In most cases a shell of revolution is submitted to nonaxisymmetric loads. Computer programs for the nonlinear analysis of nonsymmetrically loaded shells of revolution and parts of shells of revolution subtending less than 360° of circumference have been written by Stricklin et. al. [71], Klein [72], Ball [73], Underwood [74], Huffington [75], and Hubka [76]. Stricklin's, Klein's, and Ball's programs perform both static and dynamic analyses with expansion of the circumferential variations in trigonometric series. Underwood's, Huffington's, and Hubka's perform dynamic analyses with division of the shell into two-dimensional finite difference grids. All of these programs require the same order of magnitude of computer time as any two-dimensional numerical analysis of a shell of general shape, such as that performed by STAGS [35].

The analyst may wish to embark on a parameter study of buckling of nonsymmetrically loaded shells of revolution but may have a limited budget for computer costs. The following questions arise: When can the nonlinearities be neglected? When can the nonsymmetries be neglected? In this case as in those discussed above, stability phenomena fall into two classes, nonlinear collapse and bifurcation (eigenvalue) buckling. If the structure or loading is such that the shell collapses in a manner similar to that shown in Fig. 18, for example, then one of the above programs or a general shell analyzer such as STAGS [35] must be used for the analysis. If the shell fails by bifurcation buckling, more questions must be asked: Is the behavior prior to bifurcation linear? Does buckling occur locally in some area where the stress field is maximum compressive in some biaxial sense? If the behavior prior to bifurcation is nonlinear, can the nonaxisymmetric nature of the problem be neglected? If the answer to the first two or the last question is affirmative, then a one-dimensional numerical analysis, such as performed by any of the computer programs described in [36] or [62 - 65], can be used. If the prebuckling behavior is linear, the nonaxisymmetrical prestress can be determined by superposition of stresses caused by each Fourier harmonic of the nonaxisymmetric load. The program user can then select the meridian where he thinks buckling will start and, assuming that the stress field along that meridian is axisymmetric, calculate bifurcation loads from the same stability equations used for the treatment of axisymmetrically loaded shells. The prebuckling behavior may be nonlinear, and rotation of the shell wall about a tangent to the meridian may be small. If the analyst feels that this nonlinearity cannot be neglected but that the nonsymmetry can, then the bifurcation buckling analysis can be performed with a one-dimensional numerical analysis as described in the section on Solution of Equations.

Whether or not the prebifurcation behavior is linear depends, of course, on the case. As for bifurcation buckling, it is generally true that if the maximum compressive stresses do not vary much in the circumferential direction within one half of a buckling wave, then the eigenvalue will not be sensitive to the nonsymmetry of the prebuckling stresses. Bushnell and Smith [77] present a limited study on the sensitivity of predicted thermal buckling loads to the circumferential variation of prebuckling compression. The critical loads are surprisingly insensitive to this variation in the cases studied.

Bifurcation buckling and plasticity. Until very recently bifurcation buckling analyses involving plasticity had only been applied to simple structures with uniform prestress. Through the work of Hill and others it is now understood that the nonconservative nature of plastic flow does not prevent the use of bifurcation buckling analysis to predict instability failure of practical structures [78]. The concept of consistent loading of the material in the transition from prebifurcation state to adjacent postbifurcation state, explained clearly by Hutchinson [79], permits the use of instantaneous prebifurcation material properties in the stability equations. An investigation by Onat and Drucker [80] of the effect of very small initial imperfections on the torsion-mode collapse loads of cruciform columns indicates that the reduced shear modulus G obtained from deformation theory should be used in the stability equations even if there is no history of shear along the prebifurcation path. With the high-speed electronic computer it is now feasible to calculate elastic-plastic bifurcation buckling loads of rather complex structures. Bushnell [51] presents a computerized analysis and comparisons with tests of nonsymmetric bifurcation buckling of axisymmetrically loaded branched shells of revolution, including large axisymmetric prebuckling deflections, elastic-plastic effects, and creep.

Imperfection Sensitivity

It is well known that the load-carrying capability of thin shells is in many cases sensitive to initial imperfections of the geometry of the shell wall. The question so often asked by the analyst is: given the idealized structure and loading, and given the means by which to determine the collapse or bifurcation buckling loads, what "knockdown" factor should be applied to assure a reasonable factor of safety for the actual imperfect structure?

We have seen examples (Figs. 8 and 10) in which shells exhibit load-carrying capability considerably greater than that corresponding to the lowest eigenvalue. Postbuckling stability is also exhibited by columns and flat plates. On the other hand, it is well known that the critical loads of axially compressed cylindrical shells and externally pressurized spherical shells are extremely sensitive to imperfections less than one wall thickness in magnitude. Figure 23, taken from [81], gives empirically determined knockdown factors for monocoque axially compressed cylinders as functions of the radius-to-thickness ratio. Similar curves would exist for externally pressurized monocoque spherical shells were there enough test data on which to base them. These highly symmetrical systems are very sensitive to imperfections because many different buckling modes are associated with the same eigenvalue, the structure is uniformly compressed in a membrane state, and the buckling modes have many small waves. Very small local imperfection will tend to trigger premature failure. The buckling loads of most practical shell structures are somewhat sensitive to imperfections, but not this sensitive. How much so is a very important question.

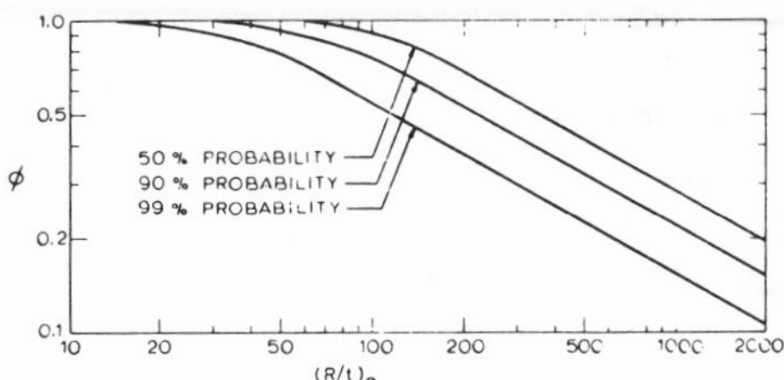

Fig. 23 Empirical knockdown factor ϕ for cylinders subjected to a uniform axial load

Qualitative guidelines for imperfection sensitivity. Hutchinson and Koiter present an excellent survey of postbuckling theory [82]. The engineer concerned with the design of thin shells is encouraged to read that survey and also the very significant papers [81] and [88 - 92]. Some qualitative guidelines are set forth in this subsection.

Buckling loads associated with local failure due to some known peculiarity of the structure which can be modeled a priori are generally less sensitive to imperfections than are loads associated with buckle patterns covering a large percentage of the surface area. Redistribution of the stresses occurs as the load is increased; a serious unknown imperfection is less likely to appear in the local area of the failure, and considerable local prebuckling deformations occur, tending to diminish the significance of the initial unknown imperfections. Failure loads of structures that are submitted to enforced displacements are likely to be less sensitive to initial imperfections than are those for structures submitted to enforced loads. In the former case the growth of an isolated buckle near the worst imperfection tends to cause reduction of the stress in that area, shifting the load to the better parts of the structure. Thicker shells appear to be less sensitive to imperfections than thinner shells simply because it is easier during fabrication to control the quality of the shell. Imperfection amplitude expressed in terms of wall thickness is therefore likely to be smaller the thicker the shell. Cylinders submitted to external pressure are less sensitive to imperfections than are cylinders submitted to axial compression because the axial wavelengths of the buckles are longer in the former case and eigenvalues do not cluster around the critical value. Hence, very small local imperfections do not affect the critical pressure as much as they do the critical axial load.

Koiter's theory. In 1945 Koiter [83] set forth a general theory of stability of elastic systems which includes initial postbuckling behavior. For perfect systems which behave in a symmetric fashion with regard to the sign of the buckling mode and which are sensitive to imperfections, the initial postbuckling load P follows the solid curve shown in Fig. 24a, taken from [84]. The quantity δ is the amplitude of the postbuckling displacement, which is assumed to be proportional to a buckling mode. The quantity t is the shell thickness. The dashed curve pertains to an imperfect shell. The value of b depends on details of the geometry and loading. Buckling loads are sensitive to imperfections if b is negative and insensitive to imperfections if b is positive.

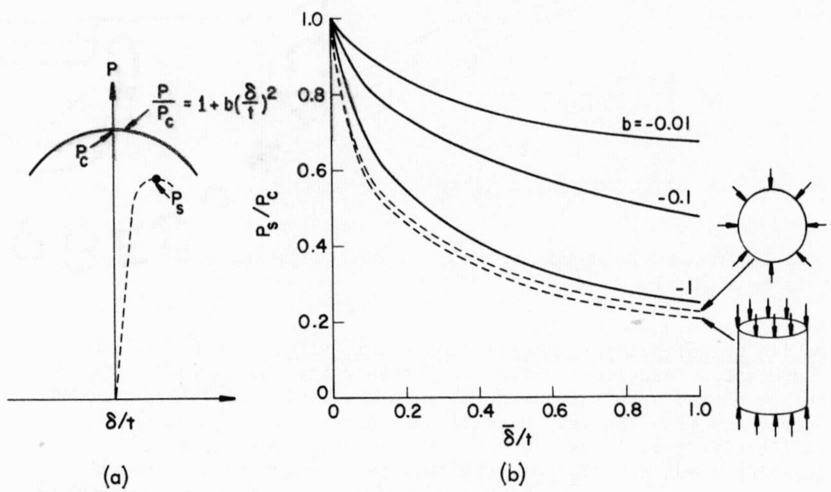

Fig. 24 Imperfection sensitivity as a function of b

If a sufficiently small initial imperfection is assumed to have the shape of one of the buckling modes of the perfect shell, the peak value P_s can be calculated as a function of b from the approximation:

$$(1 - P_s/P_c)^{3/2} = 1.5(3)^{1/2}(-b)^{1/2} \left|\bar{\delta}/t\right| (P_s/P_c) \qquad (22)$$

in which $\bar{\delta}$ is the amplitude of the initial imperfection. Figure 24b shows the dependence of the load-carrying capability P_s of the imperfect shell on the amplitude of the imperfection δ for various values of b. Curves from Koiter's special theory for the postbuckling behavior of cylinders and spheres [85] are also shown for comparison.

Imperfection sensitivity of various shells. The objective of most analyses of imperfection sensitivity, naturally enough, is to obtain b as a function of some geometrical parameter or parameters. Budiansky and Amazigo [86] calculated b as a function of the Batdorf parameter Z for simply supported monocoque and ring-stiffened cylinders (rings treated as discrete) submitted to uniform hydrostatic pressure. The results, shown in Fig. 25, taken from [86], are in qualitative agreement with tests by Dow [87].

Hutchinson and Amazigo [88] calculated b for axially and circumferentially stiffened cylinders under axial compression and hydrostatic pressure. In their analysis the stiffeners are smeared out as described earlier. The sensitivity to imperfections, especially for longitudinally stiffened axially compressed cylinders, is strongly dependent on whether the stiffeners are attached to the outside or inside of the cylinder. This effect is shown in Fig. 26, taken from [88]. Calculations of b for other shells are given in [84].

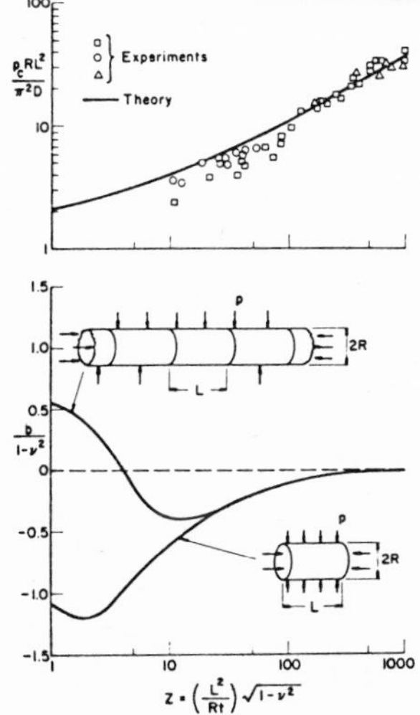

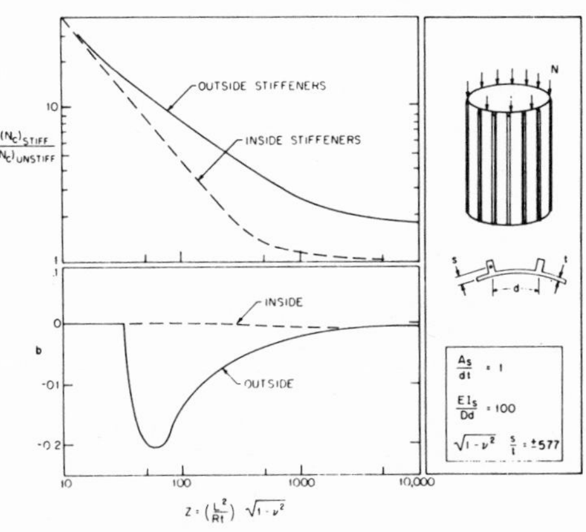

Fig. 25 Comparison between test and classical theory with initial postbuckling predictions for externally pressurized cylinders

Fig. 26 Classical buckling and imperfection sensitivity of simply supported stiffened cylinders under axial compression

Almroth, Burns, and Pittner [81] devised a means of obtaining conservative estimates for buckling of cylinders with many different wall constructions and types of loading. Their estimates, while conservative, are less so than the frequently used wide column buckling load. They introduce, as an alternative to the wide column criterion, one based on Koiter's special theory [85]. The validity of their theory is indicated through comparisons with more than 250 tests of cylinders of different types.

Budiansky and Hutchinson extended the general approach of Koiter to handle the buckling of imperfection sensitive structures under a variety of time-dependent loading conditions [89 - 91]. An example is shown in Fig. 27, taken from [84]. In this case of step loading, the dynamic buckling load is always less than the static buckling load of the imperfect shell.

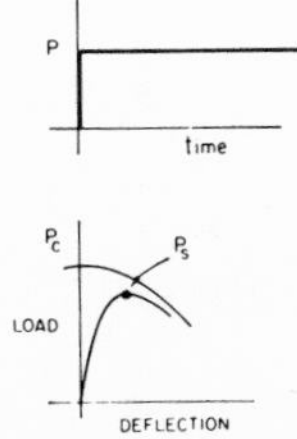

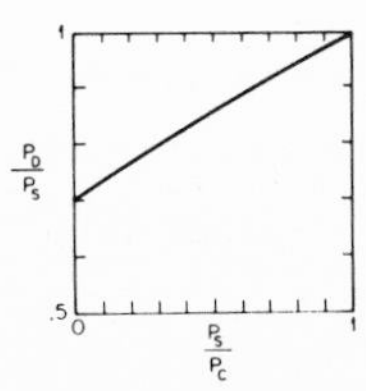

Fig. 27 Dynamic buckling, step loading

Limitations of the Imperfection Sensitivity Theory

The analyst should be aware of the following limitations of the theory of imperfection sensitivity just described:

1. It applies in general only if the imperfections are of small amplitude and only in the immediate neighborhood of the bifurcation point on the load-deflection curve.
2. The assumption is made that the growth of the postbuckling displacement distribution is proportional to a buckling mode of the perfect shell.

The effects of these limitations can perhaps best be illustrated by some examples. In the immediate neighborhood of the bifurcation point the postbuckling behavior of an axially compressed perfect cylinder of elliptic cross section may exhibit the same type of imperfection sensitivity as a circular cylinder. (See Fig. 28, for example.) Hence, the use of the imperfection sensitivity factor b with Eq. (22) would lead to a prediction of failure of an imperfect shell well below the bifurcation point A in Fig. 28. In its postbuckled state, however, this shell can carry more load than the bifurcation load. Therefore, a design based on conventional imperfection sensitivity theory would be overly conservative.

An example of the opposite error also exists: Hutchinson [79] has shown that the initial slope of the postbuckling curve is always positive if at the bifurcation load the material in the shell has yielded. An imperfection sensitivity analysis indicates that the shell will therefore carry at least as much load as the bifurcation load. In the case of a spherical shell submitted to external pressure, however, the peak load is only very slightly higher than the bifurcation load. It is clear that any initial imperfection would grow with increasing pressure, causing additional local plastic flow with consequent further reduction in the local tangent modulus. Obviously, the imperfect shell might easily fail at a pressure considerably lower than the bifurcation pressure for a perfect specimen.

These examples demonstrate the possible disadvantage and error incurred because of the first limitation enumerated above. Another example illustrated the possible harmful effect of the second limitations. Consider an externally pressurized spherical shell supported by a very heavy edge ring. The lowest bifurcation buckling load is associated with edge buckling, which is not particularly sensitive to initial imperfections. The curve labeled A* = 256 in Fig. 22 shows such a mode. The classical buckling pressure may be only about 10 percent higher than this edge buckling pressure. However, the degree of imperfection sensitivity corresponding to the classical buckling mode is considerably greater. Thus, the lowest buckling load and mode shape are irrelevant in this case for determining the true load-carrying capability of the imperfect ring supported spherical shell. Imagine the difficulty of judging the relevance of the lowest buckling loads and modes in a complex practical shell structure! One would have to calculate many buckling modes and associated imperfection sensitivity factors to determine which leads to the minimum peak load P_s of an imperfect shell. Probably more computer time would be involved than in a complete two-dimensional nonlinear analysis.

Behavior in the Far-Postbuckling Regime

An example of postbuckling, or rather, postbifurcation behavior, appears in Fig. 10. The flat portions of the pear-shaped cylinder begin to buckle at a very small load compared to the ultimate collapse load. They deform considerably before the entire structure fails. Other shell structures exhibit similar far-postbuckling behavior. Two examples are given in this section: an axially compressed cylinder with elliptical cross section and a stiffened curved panel submitted to in-plane shearing displacements. The results were obtained with the STAGS computer program [35].

Axially compressed elliptical cylinder. Load vs. end shortening curves for perfect and imperfect elliptical cylinders are shown in Fig. 28, taken from [93]. The cylinder has a length of 1.0 in., a thickness of 0.0144 in., and semiaxes of lengths 1.75 in. and 1.0 in. Young's modulus is 10^7 psi and Poisson's ratio is 0.3. It is submitted to a uniform end shortening with the edges free to rotate but restrained from moving in the radial and circumferential directions. The load-end-shortening curve for the perfect shell is that indicated by OABC. The other curves correspond to imperfect shells with the imperfection shape given by

$$w_{imp}/t = -\zeta \sin(\pi x/L)\cos(6\theta) \tag{23}$$

In a test on this shell, sudden changes in the deflection pattern (buckling) would be noticed at A, B, and C. Notice that the shell may carry more load than the initial peak A indicates. While the primary buckling load A is rather sensitive to imperfections, it appears that the second maximum B is relatively insensitive to imperfections. Hence, it may be suitable as a design limit.

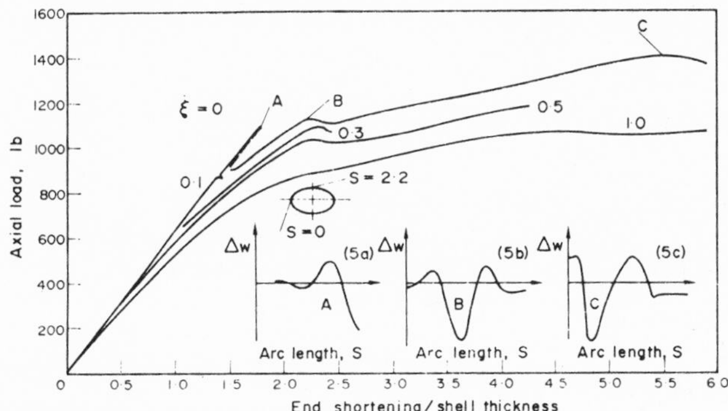

Fig. 28 Load-deflection curves for axially compressed perfect and imperfect cylinders with elliptical cross section

The curves Δw vs. S at the bottom of Fig. 28 are buckling modes calculated by subtraction of displacement vectors obtained in two sequential steps in end shortening and normalization of the result. Such a subtraction yields the shape of the fastest growing displacement component, which might be interpreted as a buckling mode. As one traces one's way along the load-deflection curve OABC, the axial stress in the shell is constantly being redistributed by the local growth of normal displacement. For example, early in the load history the most rapid growth of normal displacement occurs at the point labeled S = 2·2, the area of minimum curvature. This growth relieves the axial stress there and permits loading above the initial peak A. At point B the most rapid growth of normal displacement is about halfway between the ends of the minor and major axes. This growth relieves the axial stress in the corresponding area and thus permits loading to an even higher peak, C, where the rapid growth of normal displacement occurs near the end of the major axis in an area of relatively large curvature.

Stiffened curved panel under shear. The effective stiffness of buckled shear panels has traditionally been estimated semiempirically [94]. With advanced computer programs it is now possible to calculate the postbuckling behavior of such panels rigorously. Figure 29, taken from [95], shows a curved, stiffened panel which was analyzed with the STAGS computer program [35]. The panel was subjected to imposed displacements at the corners A and B. As the imposed displacements are increased, the six subpanels buckle but continue to carry load. Contour plots of normal displacement are shown in Fig. 29, with solid lines indicating outward and dashed lines inward buckles. The modified Newton method was

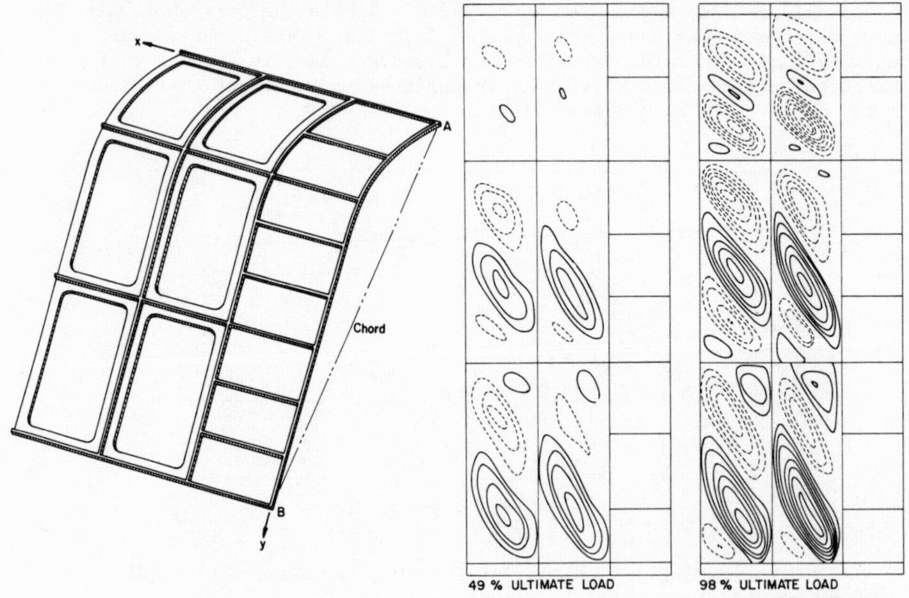

Fig. 29 Complex stiffened shear panel and postbuckling behavior predicted with the STAGS computer program

used, with 198 displacement increments and 35 refactorings of the stiffness matrix being required to reach a displacement slightly in excess of the ultimate value provided as a given quantity. According to this numerical analysis, the effective shear moduli of the buckled subpanels ranges from 36 to 48 percent of that of the unbuckled sheet. The discrete model contained 21 rows and 58 columns, with a total of 4230 degrees of freedom and a stiffness matrix bandwidth of 478. Computer time required on the CDC 6600 was 3.73 hours.

Dynamic Response of Thin Shells

Dynamic behavior of shells is not as well understood as static behavior. Problems are made more complicated by the introduction of an additional independent variable, time. Numerical difficulties occur which are generally absent or more easily circumvented in static numerical analyses. Some of these difficulties, brought about by the fact that the membrane stiffness is much greater than the bending stiffness, have been discussed earlier. The purposes of this section are to give the reader a feeling for the behavior of cylindrical and spherical shells submitted to certain dynamic environments, to indicate the state-of-the-art in computerized dynamic analysis of thin shells, and to illustrate the need for further work in this area.

Dynamic Instability of Cylindrical and Spherical Shells

There is no obvious criterion for dynamic instability, as exists in the case of static bifurcation buckling. The field of dynamic instability is really a subdiscipline of nonlinear dynamic response. Before computer programs existed for the analysis of general shells with large deflections and plasticity included, several studies were made of relatively simple shell dynamics problems. The major purpose of these investigations was to acquire and communicate an understanding of the physics of shell dynamics. The analyst needs to have such an understanding in order to use and to evaluate general purpose programs effectively.

Dynamic instability of cylinders. In [96], Goodier presents a review of the experimental and analytical investigations carried on at the Stanford Research Institute before 1965. A particularly interesting case is the moderately thick cylindrical shell submitted to a uniform radial impulse [97]. Small imperfections in the cylinder material and geometry and circumferential nonuniformity of the initial velocity give rise to nonaxisymmetric perturbations which are driven by periodic axisymmetric motion. The most highly magnified perturbations have very short circumferential wavelengths, giving the plastically deformed cylinder a wrinkled appearance. This driving phenomenon is caused by the interaction of membrane and bending behavior of the shell wall. It is a dynamic nonlinear effect which should be understood by the user of a computer program for the large deflection, elastic-plastic analysis of shells in order that he construct discretized models with enough nodal points, or Fourier harmonics, to capture this unstable short wavelength bending response.

A similar interaction of membrane and bending behavior occurs in the case of a clamped cylinder submitted to an axial impulse [98, 99]. A uniform compressive stress wave travels along the cylinder. As the wave reflects from the clamped end, a region of especially high compressive stress grows in length. For the right combination of this length and the stress amplitude, according to [99], dynamic buckling occurs in a short section near the clamped end.

Dynamic instability of shallow spherical caps. Budiansky and Roth [100] calculated axisymmetric response histories, including large deflection effects, for uniformly pressurized spherical caps with various "shallowness" parameters

$$\lambda = 2[3(1 - \nu^2)]^{1/4} (H/h)^{1/2} \tag{24}$$

where H is the height of the apex above the plane of the boundary and h is the thickness. The pressure is applied as a step load with duration $\bar{\tau}$. Some results of this investigation are shown in Figs. 30 and 31, which are taken from [100]. Figure 30b shows response histories and Fig. 30c shows peak response. The dimensionless parameter τ is ct/R, where c is the speed of sound $\sqrt{E/\rho}$, R is the radius, and t is time. The displacement measure Δ is w_{ave}/Z_{ave}, where w_{ave} is the average normal displacement and Z_{ave} is the average height of the cap. In this case a rather sudden change in behavior occurs for p between 0.50 and 0.55. Thus, $p_{cr} = 0.52$ might be termed the dynamic instability pressure for caps of this geometry with p(t) given in Fig. 30a. Figure 31 shows p_{cr} as a function of the dimensionless load duration $\bar{\tau}$. The limiting value of p_{cr} for long durations is somewhat less than the static collapse load because of the dynamic overshoot effect. The dotted curve labeled α_{cr} shows the variation of critical impulse $p_{cr}\bar{\tau}$ with $\bar{\tau}$. Other more general structures may show similar rather abrupt changes in behavior with small changes in dynamic loading parameters. The analyst should be aware of this possibility as he performs parameter studies with general purpose computer programs which include large deflection effects.

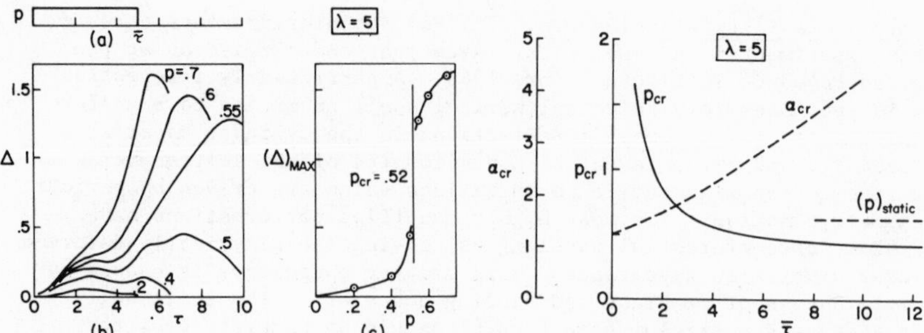

Fig. 30 Dynamic buckling of spherical cap: (a) pressure history, (b) response, (c) variation of maximum response with amplitude of pressure step

Fig. 31 Variation with load duration of critical pressure and critical impulse parameter

Dynamic Response of Cylinders and Cones

During the last five years many computer programs have been written for the large deflection, elastic-plastic dynamic response of shells [35, 101 - 110]. Details of the scope of some of these codes are given in [1]. The development effort on most of them was motivated primarily by interest in the response of reentry bodies to blast and impulsive loadings. Now the concern for the structural integrity of nuclear reactor pressure vessels submitted to dynamic thermal and pressure loads is also a strong motivating factor.

Figure 32, taken from [103], shows the actual [111] and predicted final deflections of an elastic-plastic, six-inch-long cylindrical shell submitted to an impulsive load distributed uniformly along the length and as cos θ over half of the circumference. This comparison is also presented in [112]. In the DYNAPLAS model ten finite elements were used along half of the length, and circumferential variations were represented by use of five trigonometric harmonics. In the SHORE model eleven finite difference stations were used along half of the length and nineteen around 180 degrees of the circumference. The wrinkling of the specimen in the region of the peak load may be due to nonuniformity in the distribution of the explosive used in the experiment or to the nonlinear interaction of membrane and bending effects just described. In any case a finer discretized model would be required in order to predict the short wavelength components of the deflected cross section.

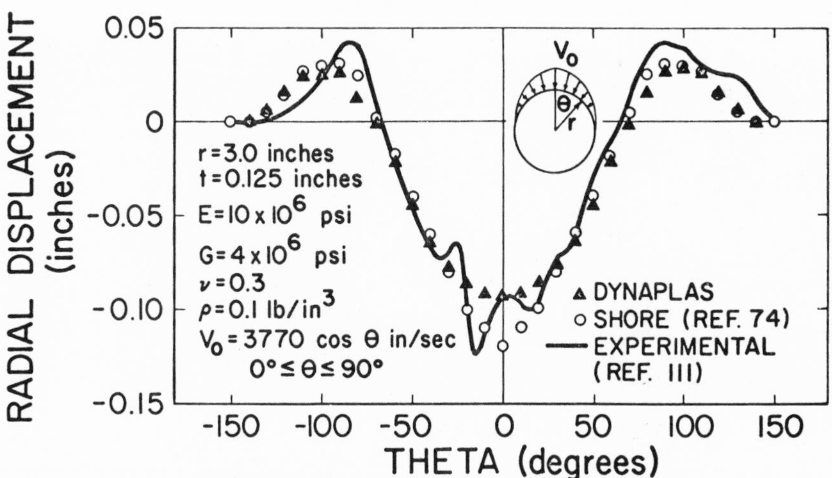

Fig. 32 Dynamic radial displacement around the circumference at the midlength of an elastic-plastic cylinder submitted to an impulse

In spite of the great magnitude of effort expended on the independent and simultaneous development of many computer programs which treat similar phenomena, our understanding of dynamic shell behavior is incomplete. There remain major discrepancies in the predictions from one program to the next, even in the linear range. Figures 33 and 34, taken from [113], illustrate this point. The problem is to predict the linear and nonlinear response of an impulsively loaded elastic conical shell. Two of the computer programs, DYNAPLAS and SATANS, use trigonometric expansion for circumferential variations. SATANS is based on finite difference equations and DYNAPLAS on the finite element method. REPSIL, SHORE, and SMERSH are based on two-dimensional finite difference discretization. In the DYNAPLAS and SATANS models, five trigonometric terms for circumferential variations were used. The meridian was divided into thirty finite elements in the DYNAPLAS model and thiry-one finite difference stations in the SATANS model. The REPSIL, SHORE, and SMERSH models consisted of thirty-one finite difference stations along the length and nineteen around half of the circumference. Results from very recent calculations with an updated version of SHORE are different from those shown in the frame corresponding to displacements predicted with nonlinear analysis. The latest results are in good agreement with those obtained with use of the other codes for times up to about 600 microseconds. While there are indications obtained since publication of the results shown here that the discrepancies are partly due to lack of convergence with spacial discretization, it is felt that this problem remains essentially unsolved. Unfortunately, there are no tests with which to compare these numerical results. Certainly more effort is required in this area, not to produce any more user-oriented computer programs, but to attempt to better understand, primarily from tests at first and then from theory, how thin shells behave when submitted to dynamic loads.

Fluid-Structure Interaction

Generally in problems involving thin shell dynamics the load distribution is given a priori as a function of space and time. The motion of the structure is assumed to have little effect on the applied forces. The title of this subsection implies that the behavior of the structure and the surrounding medium in some cases must be analyzed as an interdependent system. Two aspects of this field are very briefly described here: liquid-shell interaction and flutter. The state-of-the-art in fluid-structure interaction problems as of 1967 is given in [114].

Liquid-shell interaction. In the case of thin plates and shells submerged in a liquid, "the solid mechanics problem of determining the structural response to prescribed forces and the acoustical problem of determining the pressure in the ambient medium in response to a prescribed velocity distribution of the boundary must be solved simultaneously" (Junger [115]). Much of the work on liquid-shell

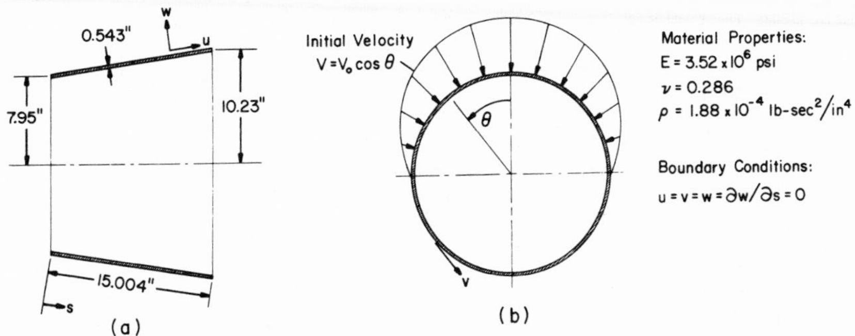

Fig. 33 Geometry and boundary conditions of an impulsively loaded conical shell analyzed with linear and nonlinear theory by many different computer programs

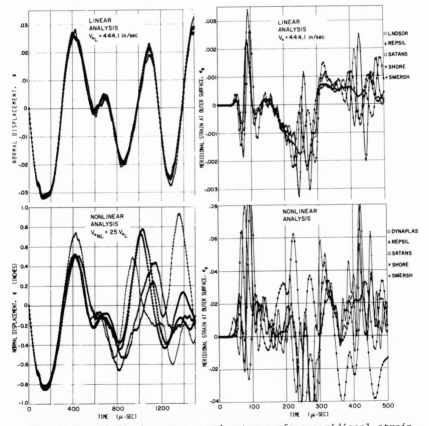

Fig. 34 Normal displacement w and outer surface meridional strain at s = 6.5 in. and θ = 0 deg. from linear and nonlinear analysis of impulsively loaded conical shell shown in Fig. 33

interaction is focused on the problem of response of cylindrical and spherical shells to enveloping shock waves. Geers [116] clearly describes several simplifying approximations that have been used to predict the transient response of submerged shells. Essentially, there are two contrasting kinds of liquid-shell interaction, low frequency or late time phenomena and high frequency or early time phenomena. Late time phenomena are characterized by long wavelength bending motions of the shell and entrained motion of the liquid which can be treated in this case as incompressible. The effect of the liquid is to add virtual mass to the shell, reducing the characteristic frequencies of it. As the shell deforms the liquid circulates around it so that the velocity of the liquid and shell are compatible at the interface. Far away from the shell no motion can be detected in the liquid because the flow in the neighborhood of the shell is self-canceling. Early time phenomena, on the other hand, are characterized by membrane or short wavelength bending motions of the shell and compressible behavior of the fluid through which energy is radiated away from the structure. Further details are given in [116].

Flutter. Flutter is a fluid-structure phenomenon or class of phenomena associated with immersion of a thin plate or shell in a fluid flow. An excellent review is given by Dowell in [117]. He succinctly indicates the objectives of a flutter analysis:

1. To determine under what conditions a disturbance to the system decays or grows
2. If the disturbance grows, to determine what its final state is

These objectives are analogous to those of a bifurcation buckling analysis: Under what load will the equilibrium state of a shell no longer be unique? How does the shell behave in the postbuckling regime? To quote further from Dowell on flutter: "The first question is the classical one of stability and is usually treated within the framework of linear theory. The second necessitates the use of a nonlinear analysis since linear theory predicts that any growing disturbance will continue to grow indefinitely" (Dowell [117]). In his review Dowell emphasized the nonlinear aspects, including effects of prestress and curvature on the flutter characterisitcs of deformed flat plates and cylindrical panels.

The Decay of Edge or Discontinuity Loads in Shells

In previous sections several examples of perhaps unexpected or at least little appreciated edge effects have been given. A slight change in the modeling of an edge ring causes a significant change in the predicted buckling pressure of a spherical cap because alteration of the model significantly affects the maximum predicted prebuckling hoop compression near the edge (Fig. 5). Buckling loads of axially compressed, ring-stiffened, corrugated cylinders are very sensitive to boundary conditions even if the cylinder consists of many bays and if the buckling mode is local rather than general (Figs. 6 and 7). The decay lengths and amplitudes of discontinuity stresses are dependent upon the membrane state of stress in the shell (Fig. 17).

The decay rate of edge disturbances depends upon the nature of these disturbances as well as on the properties of the shell wall. One cannot blindly assume that discontinuity stresses always die away within a small multiple of the axisymmetric "boundary layer" length $\sqrt{Rt}$. Figure 35 shows axial distributions of axial stress resultants N_1 and meridional moments M_1 due to unit applied loads which vary harmonically around the circumference at a free edge of a cylinder. The decay of N_1 is caused by membrane shear lag, which is strongly dependent on the number n of circumferential waves. For small n the decay rate is much more gradual than that characterized by $\sqrt{Rt}$. The decay of M_1, on the other hand, is caused by hoop tension or compression induced by the radial movement of the shell wall--the elastic foundation force which governs the decay rate of axisymmetric bending disturbances. The strength of this force, unlike the shear lag effect, is almost independent of the circumferential wavelength of the disturbance for small numbers of waves. A different kind of shear lag effect, the effect of transverse shear, overshadows the elastic foundation effect only if the rate of change of M_1 in the circumferential direction is sufficiently high. These effects are all clearly described by Flügge [118]. However, because many of us tend to use the computer even for the simplest shell analysis, we are inclined to lose a feeling for the physics of shell behavior.

Figure 36 shows the normal displacement distribution along the meridian $\theta = 0$ for the case in which $M_1 \cos n\theta$ is applied at the edge. Notice that for $1 \leq n < \sim 6$ two modes of displacement are present--a short wavelength edge disturbance and a long wavelength deflection. The stresses are produced almost entirely by the short edge disturbance. The long wavelength mode involves inextensional, low-strain-energy deformation except for n = 1, in which it corresponds to the beam bending, a pure extensional deformation of the entire cylinder.

Edge or discontinuity disturbances in general shell structures are caused by abrupt changes in thickness, terminations of stiffeners, discontinuities in curvature, junctures between shells with dissimilar wall construction, eccentric load application, concentrated loads, and other factors. If the major effect of the disturbance is to cause deflection normal to the shell surface, then it is probable that the decay rate will be governed by the elastic foundation force. The effects illustrated in Figs. 5, 15, and 16 are examples. If the disturbance is reacted primarily by membrane forces in the shell wall, then the decay rate will probably be governed by the shear lag effect. Examples are the distributions of axial force around the circumference of a cylindrical tank resting on pillars or around the circumference of a monocoque section of an axially compressed cylinder with longitudinal stiffeners that terminate at some axial station.

Note that the decay lengths of disturbances depend strongly on the stiffness properties of the shell wall. For example, the decay rate exhibited in Fig. 7 for the longitudinally corrugated shell with a harmonically varying edge moment $M_1 = M_{10} \cos 16\theta$ is much more gradual than that for an axisymmetrically applied moment, whereas the opposite is true in the case of a monocoque isotropic cylinder loaded in a similar manner (Fig. 35.) Clearly, it is not just the elastic foundation rigidity that determines the decay rate, but the relationship of this rigidity to the meridional bending stiffness.

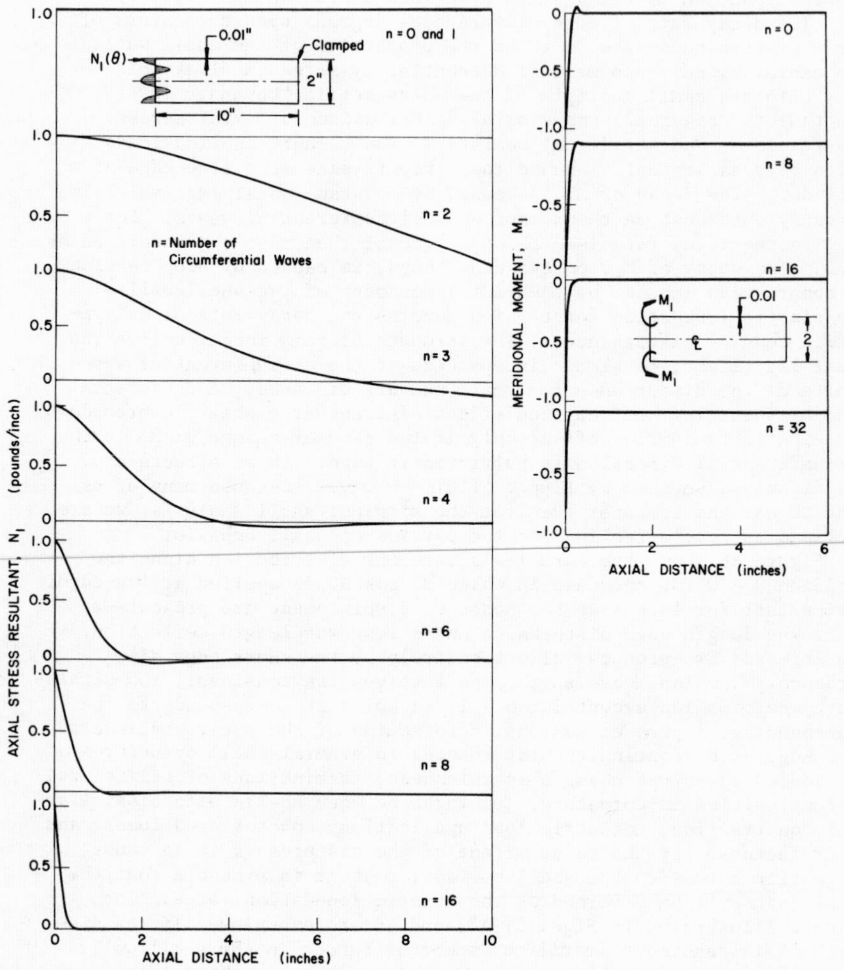

Fig. 35 Decay lengths of edge disturbances applied to free edge of cylindrical shell

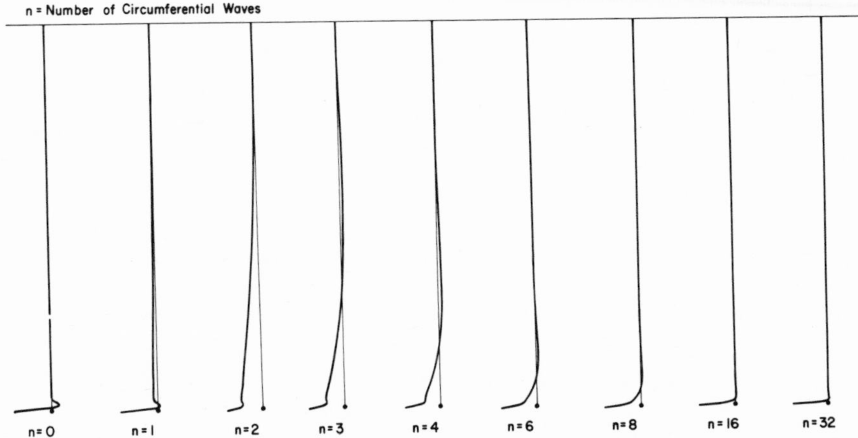

Fig. 36 Edge displacements of cylinder loaded by meridional moment varying harmonically around the circumference at the free edge

SOME ASPECTS OF MODELING SHELLS

Certain aspects of the behavior of thin shells have been summarized in previous sections. The purpose of this section is to give the computer program user some hints about modeling for stress, buckling, and vibration analyses of practical shell structures.

Mesh Point Allocation

The analyst may wish to know what the stresses are in a shell at the bifurcation buckling load. If he sets up a single discretized model for both the stress and the buckling analyses, he must allocate nodal points such that stress concentrations as well as buckling modes can be predicted with reasonable accuracy. It is usually fairly easy to guess where the stress concentrations are, but more difficult to predict where the shell will buckle and the shape of the mode. Peak stresses can generally be predicted with enough accuracy if nodal points are spaced a few wall thicknesses apart. If a higher nodal point density is required for adequate convergence, thin shell theory may not represent a good enough model. Good estimates of buckling loads can usually be obtained with more than four nodal points per half wavelength of the buckling mode. Figure 37 depicts a ring-stiffened cylinder which is submitted to external pressure. The prebuckling normal displacement and meridional moment and the buckling modal displacement distributions are also shown. Notice that mesh points are concentrated near the T-shaped rings and at the boundary where stress concentrations exist. Half the cylinder is modeled with symmetry conditions applied at the symmetry plane.

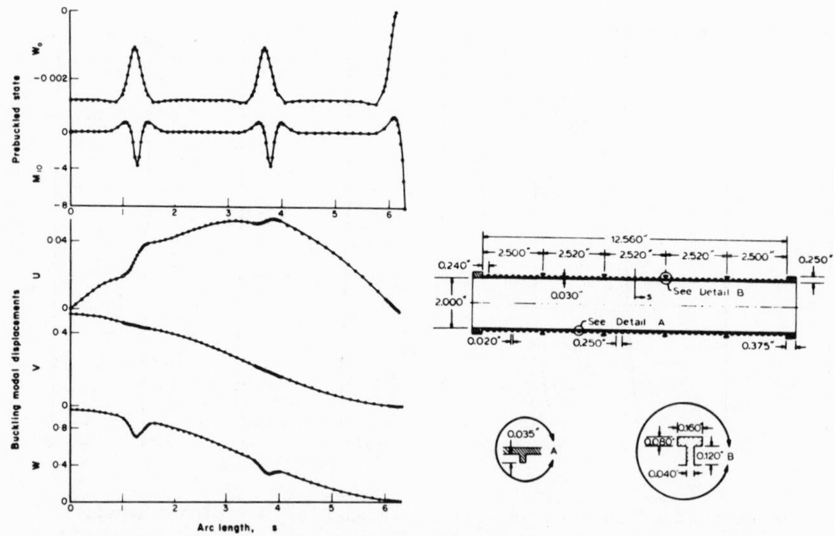

Fig. 37 Prebuckling state and buckling mode of an externally pressurized ring-stiffened cylinder

Modeling a Juncture between Two Shells

Figure 38, which is taken from [51], presents an example in which the predicted bifurcation buckling load is very sensitive to details of the model at a juncture between two segments. The shell is an aluminum torispherical pressure vessel head with a cylindrical nozzle. The loading is uniform external pressure. In model A the two points separated by radial and axial distances d_1 and d_2 are considered to be connected rigidly, free to translate and rotate but not to stretch. In model B the middle surfaces are connected together. The differences in predicted buckling loads are explained by the enlarged views of the two models shown as deformed by a pressure of 140 psi. In model B the short section of cylindrical nozzle immediately adjacent to the spherical head deforms locally a considerable amount within one half the thickness of the head. This physically impossible situation is avoided in model A, for which considerably less prebuckling deformation occurs at a given pressure because a certain length of rather flexible thin walled material has been replaced by a shorter length of relatively stiff thicker walled material. Note, however, that both models contain the same total amount of material at the juncture. In the enlarged view of the deformed models the points of intersection of the middle surfaces have been superposed in order to provide a better comparison of the relative amounts of distortion and rotation of the juncture.

Even with the stiffer model A there is considerable local distortion immediately adjacent to the juncture. A very accurate analysis of the buckling of this vessel requires the inclusion of transverse shear deformations in this area. Thin shell theory is not adequate.

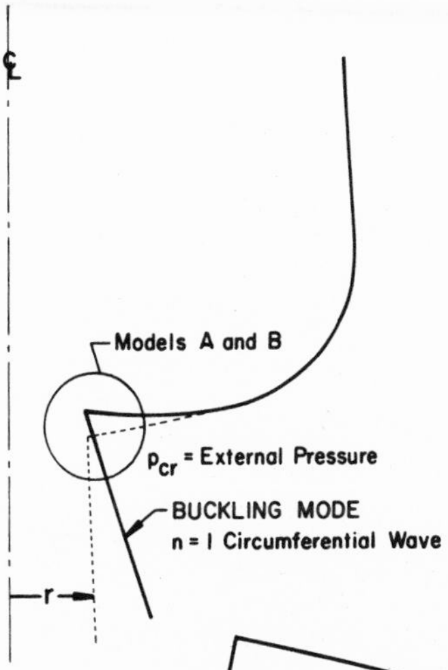

Models A and B

p_{cr} = External Pressure

BUCKLING MODE
n = 1 Circumferential Wave

r

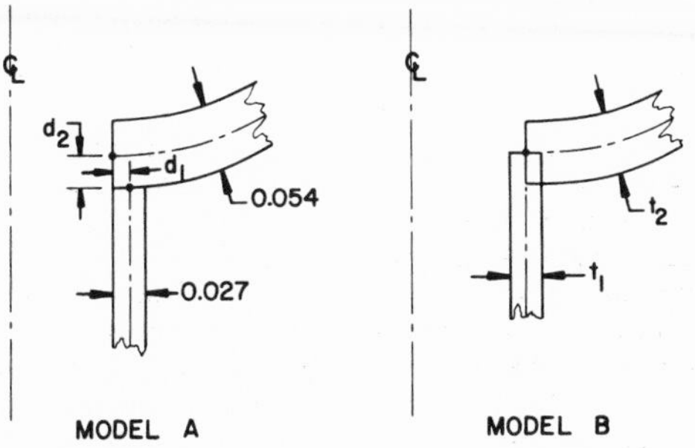

d_2
d_1
0.054
0.027

MODEL A

t_2
t_1

MODEL B

SPECIMEN	A3	A5
r	0.675 in	1.08 in
p_{cr} (Test)	177 psi	168 psi
p_{cr} (Model A)	182 psi	160 psi
p_{cr} (Model B)	164 psi	148 psi

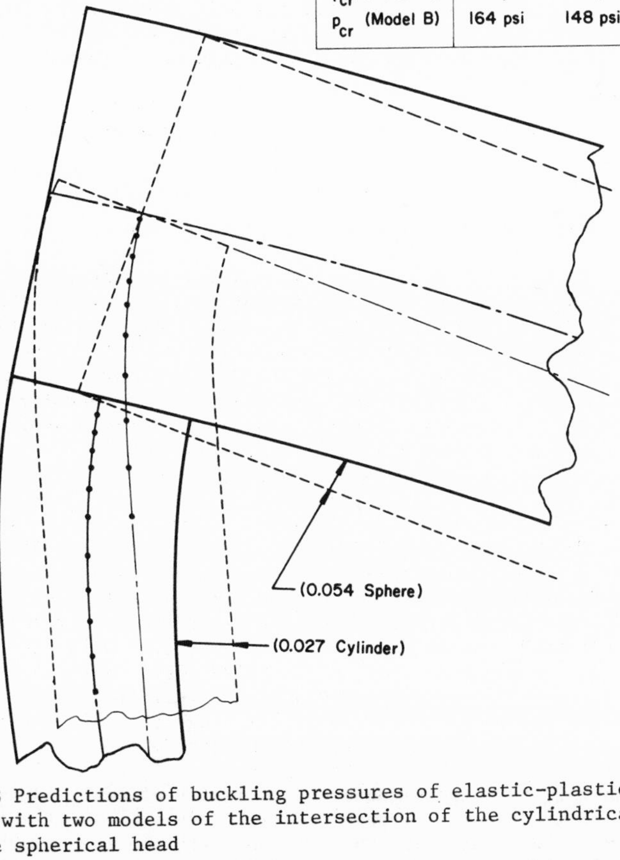

(0.054 Sphere)

(0.027 Cylinder)

Fig. 38 Predictions of buckling pressures of elastic-plastic pressure vessel with two models of the intersection of the cylindrical nozzle and the spherical head

Modeling Structures Consisting of Assemblages of Plate and Shell Segments

Lightweight structures are often composed of curved or flat thin sheets reinforced by fairly deep stiffeners which are welded or riveted to the sheets. Weight limitations dictate that the stiffeners be thin compared to their height. The problem of designing such shell structures is complicated by the existence of many different failure modes and by the fact that the shell wall distorts locally as loads are applied and during buckling and vibration.

Three distinct types of buckling are often investigated, usually in separate analyses. Long wavelength general instability is treated by smearing the stiffeners over the sheet surface. Intermediate wavelength or panel instability is explored with the assumption of certain boundary conditions at stiffener locations or by inclusion of the stiffeners as discrete elastic structures, whose cross sections are not permitted to deform. Short wavelength crippling is usually predicted by analysis of flat or cylindrical panels under axial compression

It appears that there are many cases for which the three types of instability are not distinct. Local distortions of rings and stiffeners may affect to a significant degree general and panel instability predictions.

All of the following numerical results were obtained with use of the BOSOR4 computer program [36]. Many of the examples are taken from [40].

Local Distortions of Axially Compressed Panels

Figure 39 shows predicted normalized buckling loads as functions of the axial length of buckling wave for two types of manufacture, bonded and riveted. The panels are loaded in a direction normal to the plane of the paper. They are treated as branched shell structures, the bonded panel having thirteen segments and the riveted panel having eighteen segments. In classical buckling analyses such panels are treated as equivalent orthotropic sheets with the wall cross section of course not permitted to deform locally. The presence of local deformations makes it very difficult to assign a priori a torsional stiffness per length. The torsional rigidity is particularly important in this case because of the enclosed trapezoidal areas. Local distortions also affect the axial bending stiffness, another significant determinant of the predicted buckling load. The degree of local distortion is largely governed by the way in which the corrugated sheet is fastened to the flat sheet. Bonding along the entire widths of the troughs markedly reduces the amount of distortion with a corresponding increase in the stiffness and buckling load. This difference between bonding and riveting would not be reflected in a classical orthotropic plate analysis, except through the introduction of empirically determined knockdown factors applied to the constitutive law to bring test and theory into agreement.

For the analysis of large structures fabricated of semisandwich corrugated panels, or any other type of complex panel, it is clearly impractical to find buckling loads by division of the entire structure, or even a large section of it, into minute segments such as done for

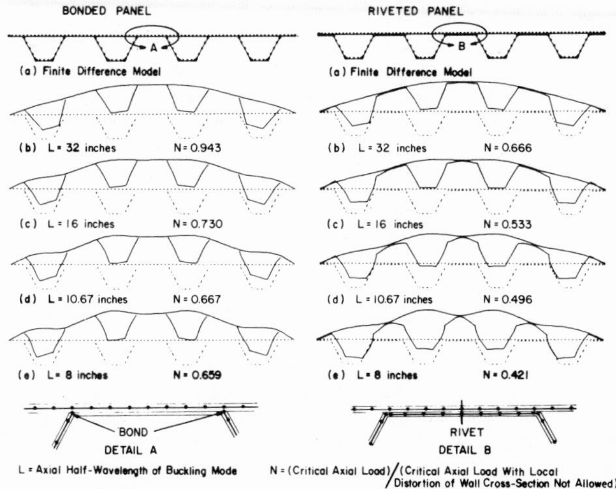

Fig. 39 Buckling modes and loads for axially compressed bonded and riveted corrugated panels

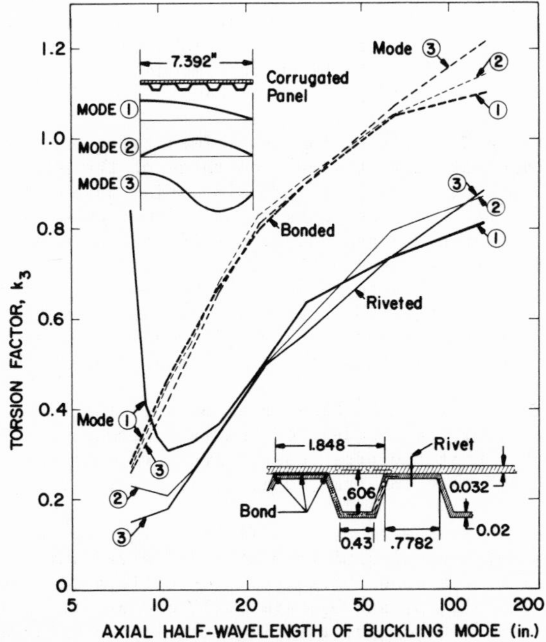

Fig. 40 Analytically determined knockdown factors for the torsional rigidity of riveted and bonded corrugated shells

the models shown in Fig. 39. However, these relatively small, accurate models can be used to calculate appropriate stiffness coefficients as input to a theory in which the corrugations or other stiffeners are smeared out. For example, Fig. 40 shows analytically determined knockdown factors for the calculation of the effective torsional rigidity of the riveted and bonded wall constructions. Notice that these factors depend very strongly on the wave length of the buckling mode in the longitudinal direction. They decrease for shorter wavelengths because as seen from Fig. 39, shorter wavelengths are associated with more local distortion.

Various Models of Ring Stiffeners

Buckling. Figure 41 shows a portion of a cylindrical shell 25 in. long with an internal I ring. Six analytical models (A-F) were constructed and analyzed. Local buckling loads corresponding to uniform external pressure p are given. The boundaries at the ends of the 25-in.-long cylinder are assumed to be symmetry planes, and the buckling mode is assumed to be symmetrical about these planes. Therefore, these models are used to predict local buckling of a cylinder of any length greater than about 50 to 100 in. The encircled numerals represent flexible shell segments in the analysis. Shaded portions represent cross sections treated as discrete rings, permitted to translate and rotate during prebuckling and buckling deflections, but not to deform in cross section. In the riveted model segment 4 is considered to be attached to segment 2 only at the two lines labeled "Rivets." At these two stations the three displacement components and the rotation of segments 2 and 4 are constrained to be equal. This model, labeled "A," is presumed to be the most accurate.

Computer-generated plots of the undeformed reference surface (the middle surfaces of segments 1-6) and three of the buckling modes corresponding to n = 8, 14, and 22 circumferential waves are also shown in Fig. 41. The buckling pressure as a function of n, $p_{cr}(n)$ has two minima, one at n = 8 and one at n = 16. Clearly n = 8 corresponds to failure of the web-flange combination, and n = 14 or 16 corresponds approximately to failure of the cylinder between adjacent rings. Notice that the higher the wavenumber n the less is the excursion of the inner flange from its undeformed position. The out-of-plane bending rigidity of this flange is of course responsible.

Table 1 gives predicted buckling pressures as functions of wave number n for many analytical models of the same structure. The heading "nonlinear" denotes that the prestress analysis is nonlinear. "Old Ring" and "New Ring" have the same meanings as described in connection with Fig. 4.

A few points about the results in Table 1 might be made. The branched shell models reveal a minimum at n = 8 waves which of course the discrete ring models cannot predict. The differences between old and new ring models, D_1 and D_2, are small because for these models the shear center and centroid of the ring actually do coincide, and the differences are thus due entirely to the warping integral and the variation of ring prestress over the cross section of the ring. These two effects are rather small and for internal rings have counteracting effects on the predicted buckling load. The warping integral

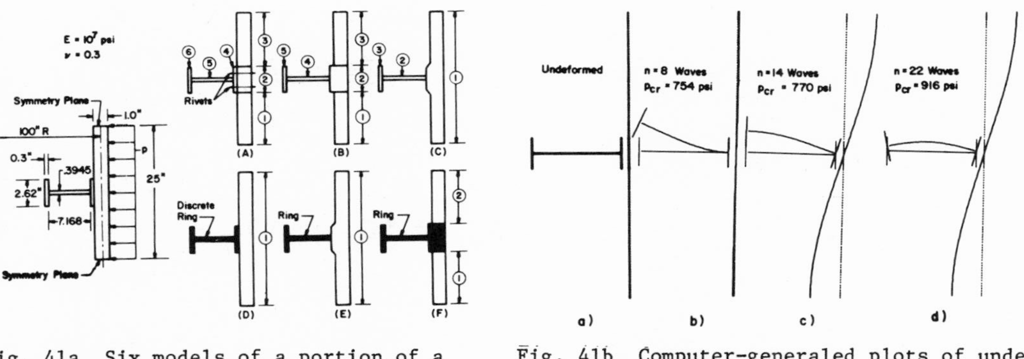

Fig. 41a Six models of a portion of a ring-stiffened cylinder

Fig. 41b Computer-generaled plots of undeformed and buckled cylinder corresponding to model A in Fig. 41a

Table 1 Predictions of Buckling with Various Models of a Ring-Stiffened Cylinder (see Fig. 41)

	Branched Shell Models					Models with discrete rings						
Circumferential waves, n	Model A nonlinear	Model B_1 nonlinear Convergence study 11^a-5^b	Model B_2 nonlinear Convergence study 21^a-11^b	Model B_3 nonlinear Convergence study 51^a-31^b	Model C nonlinear	Model D_1 nonlinear old ring	Model D_2 nonlinear new ring	Model D_3 linear new ring $\chi_0 \neq 0^c$	Model D_4 linear new ring $\chi_0 = 0^c$	Model E_1 nonlinear old ring	Model E_2 nonlinear new ring	Model F nonlinear new ring
4	1043	1077	1082	1082	1083							
6	819				847							
8	754	756	773	775	774	1395	1345	1382	1493			
10	775				793	961	939	958	1023	1055	1059	1065
12	798	806	824	828	828	790	785	796	838	863	884	969
14	770				807	737	743	752	780	794	828	970
16	767	791	797	801	803	744	761	767	788	789	835	1019
18	795				829	786	811	817	832	821	878	1092
20	847	871	875	877	880	852	882	887	898	878	944	1176
22	916				948	932	966	971	978	953	1024	1268
24	999	1021	1026	1027	1030	1025	1061	1065	1070	1042	1117	

[a] Number of mesh points in segments 1 and 3
[b] Number of mesh points in segments 2, 4, and 5
[c] χ_0 is the prebuckling meridional rotation.

has a larger effect at high circumferential wave numbers and the variable prestress effect dominates at lower values of n. In this case the prebuckling behavior is rather linear (compare D_3 with D_2). However, notice that the effect of prebuckling shape change ($\chi_o \neq 0$ vs. $\chi_o = 0$) is quite significant. Models E_1 and E_2 show a bigger difference than do D_1 and D_2 because the shear center and centroid of a T ring do not coincide, the shear center being located approximately at the point of intersection of the flange with the web. Model F is unacceptably nonconservative. It is felt that the results for the different models have converged to three figures, since the mesh point density is the same as that used in model B3 of the convergence study.

Note that the results for model A are between those for model C and model D_1. In model C the faying flange of the ring is considered to be an integral part of the shell wall, an allocation of material which leads to overestimation of the axial bending stiffness of the cylinder. In model D_1 the faying flange is considered to be part of the ring, and the ring is considered to be attached to the shell at a single point. Hence the shell wall can bend under the ring while the ring cross section remains undeformed. The ring stiffness is overestimated because the cross section cannot deform, but the axial bending stiffness of the cylinder is underestimated because the faying flange in this model does not bend with the shell. In this case it appears that the axial bending effect is more significant than the effect of rigidity of the ring cross section. In model A, considered the most accurate, the faying flange bends with the cylinder over part of its length because it is attached at two separate points. However, the total axial bending stiffness of the cylinder wall plus flange is less in this model than it is in model C.

Figure 42 shows buckling pressures vs. circumferential wave number n for an internally stiffened cylinder of the geometry indicated. The solid curve labeled "3-Branch Shell" represents the most accurate model, in which shell, web, and flange are treated as flexible thin shell segments. The minimum at n = 6 corresponds to buckling of the web and flange, a mode similar to (b) in Fig. 41b. The portion of the curve from n = 14 to 18 corresponds to buckling of the ring and shell, a mode similar to (c) in Figure 41b.

The curves labeled "Simple Support" (4, 5, and 6) are generated by replacement of the T ring at buckling by the usual simple-support conditions: normal and circumferential displacements w and v equal zero and no constraint against axial motion or rotation other than that caused by the remainder of the shell. The structure is symmetric about a plane through the ring attachment point. Therefore, simple-support conditions applied at this plane are appropriate, since they force the buckling mode to be antisymmetric. "Exact" prestress in curves 4 and 6 denotes that the prestress distribution is derived from a rigorous theory in which the structure is treated as a three-branch shell. Curve 5 is lower than curve 4 because the rings are spaced closely enough (25 in) that the average prestress predicted from the exact theory is less than the nominal prestress pr. If one replaces the ring at buckling with simple-support conditions at the ring attachment point but retains the destabilizing effect of the ring preload, a compressive force acting approximately at the ring centroid, one obtains curve 6, which considerably underestimates the strength of the structure

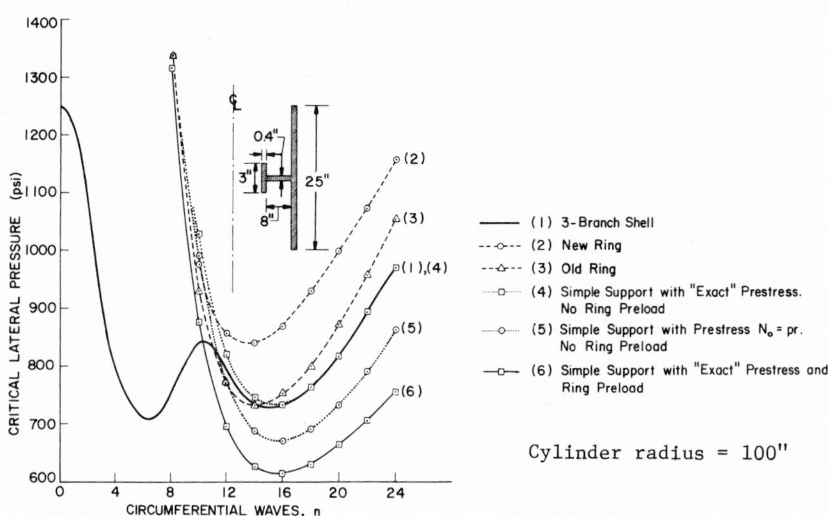

Fig. 42 Local buckling pressures for various models of cylinder with internal T ring

in this antisymmetric mode. A common practice in the analysis of ring-stiffened cylinders is to treat the rings as discrete, to account for the preload of the ring in the stability equations, but to neglect the out-of-plane bending rigidity of the ring. This model is similar to one in which the ring is replaced at buckling by simple support with the ring preload included in the stability analysis (curve 6).

Notice that the old ring model gives results which are closer to the more precise three-branch shell model than the new ring model. Two counteracting errors exist for the old ring model. The stiffness of the ring is overestimated by the restriction that the ring cross section does not deform during buckling but underestimated primarily because the axis of centroids is assumed to coincide with the axis of shear centers.

Vibration. Figure 43 shows the lowest three vibration frequencies and mode shapes for circumferential waves n = 4 - 14 for a ring-cylinder combination with geometry similar to that given in Fig. 42. Symmetry conditions are enforced at the edges of the cylinder. The figure is arranged so that modes of a given class appear in a given row. The modes in the top row resemble fundamental buckling modes in that the low wave number modes correspond to local distortions of the web and the higher wave number modes correspond to movement of the cylinder as if it were simply supported at the ring attachment point. The deflected shapes in the bottom row also resemble fundamental buckling modes. The n = 2 mode (not shown, but virtually identical to n = 4 mode) corresponds to ovalization of the cylinder-T-ring combination, without distortion of the cross section of the structure. The degree

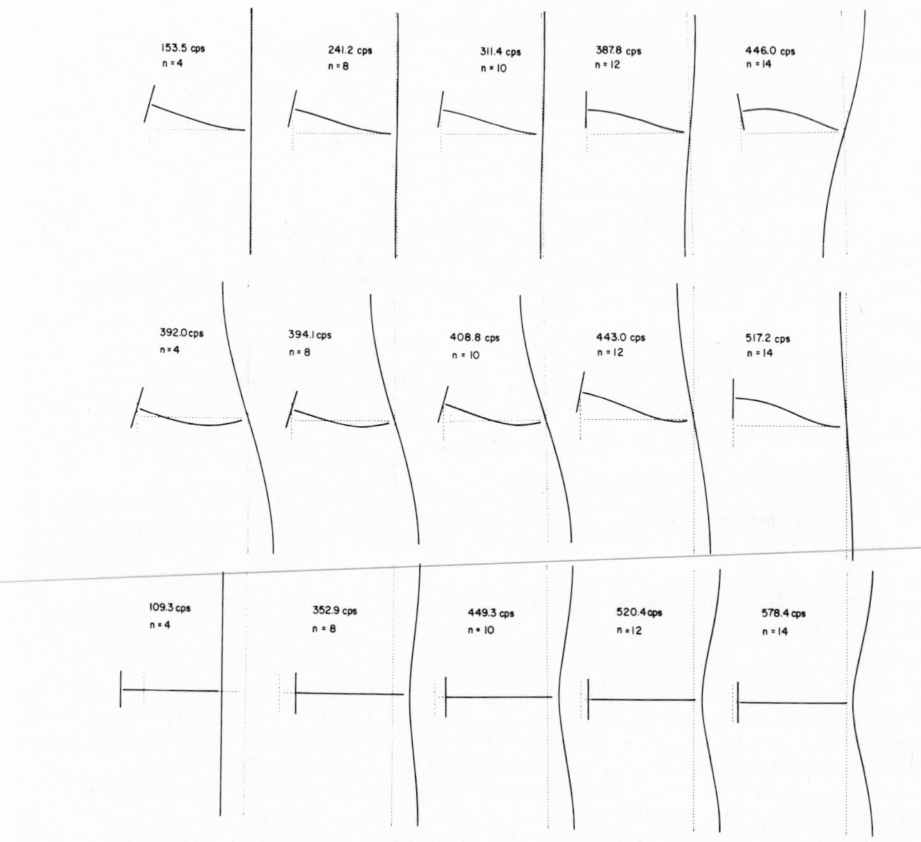

Fig. 43 Local vibration modes of T-ring-stiffened cylinder

of distortion increases as the wave number increases. The middle row does not resemble a fundamental buckling mode for any wave number but does resemble higher buckling modes. For low wave numbers this mode corresponds to cylinder motion with simple support at the attachment point of the ring, and for high wave numbers it corresponds to the second mode of web distortion.

Perhaps the major conclusion to be drawn from Fig. 43 is that the three-branch shell model leads to a prediction of many modes for every one mode predicted by the discrete ring model. Some of the extra modes are obvious; in the three-branch shell model the web can distort. As seen in Fig. 43 the web distortion mode dominates for low circumferential wave numbers n of the fundamental and high wave numbers of the second mode. Another class of extra modes has to do with the phase relationship of the motion of the ring and the cylinder. Notice, for example, the mode shapes for n = 12 waves in the top two rows of

Fig. 43. In the top row the ring is deflecting upward and the top half of the cylinder is moving outward. This in-phase motion is permitted by an analysis in which the ring is treated as discrete. In the second row, however, the ring is deflecting upward but the top half of the cylinder is moving inward--an out-of-phase motion not permitted by the simpler discrete ring model.

Figures 44 and 45 show the geometry and some natural vibration modes of an aluminum ring stiffened cylinder supported by steel end plates. The cylinder was tested by Hayek and Pallett [119] and a previous analysis was performed by Harari and Baron [120]. The experimental results, analytical results from BOSOR4 [36] with eight different models, and analytical results from [120] are given in Table 2.

One of the most important points to be made with regard to Table 2 is that an approximate analysis can by fortuitous coincidence yield very good results because of counteracting errors. Take the bottom row of Table 2, for example. The relatively crude model in which the rings are treated as discrete and the end plates are omitted (modeled as simple supports--v and w restrained, u and rotation free) leads by chance to a very good prediction (2800 cps) of the experimental result (2802 cps). However, the stiffness of each discrete ring is overestimated because its cross section is not permitted to deform. If each ring is treated as a shell segment but no other changes are made in the model, a new frequency of 2663 cps is obtained. This branched model is labeled (1) in Fig. 44b. If an additional refinement is made by the addition of the end plates, frequencies of 2682 or 2724 cps are obtained, depending on the degree of constraint assumed to exist between the end plates and the cylinder. These models are too flexible, however, because axial bending of the cylinder wall is permitted along the 0.375 in. lengths corresponding to the regions of intersection of cylinder and rings. If the cylinder is treated as consisting of six segments with 0.375 in. gaps at the areas where the rings and cylinder intersect, and if the material in each gap is treated as a discrete ring with undeformable cross section, the frequencies of 2750 or 2782 or 2833 cps are calculated, depending on whether the end plates are included and, if they are, on the degree of constraint assumed to exist between them and the cylinder. This segmented cylinder model is labeled (2) in Fig. 44b. The predicted vibration mode shapes with $n = 6$, $m = 3$ for all of the models are given in Fig. 45. The test frequency of 2802 cps is bracketed by the results from the various models. Notice that for other modes the test frequencies are less well predicted by the cruder discrete ring model but that they are still bracketed (with the exception of $n = 5$, $m = 1$) by use of the full range of models as just described. The case $n = 6$, $m = 1$ is an example. In the $n = 1$ case it is important to include the end plates in order to obtain an accurate prediction of the fundamental beam bending mode of the entire free-free cylinder end plate system. This mode is depicted in Fig. 45.

During the study of a particular structure the analyst should set up various models in order to obtain upper and lower bounds on the behavior if possible. Because of imperfections, it is difficult to obtain a lower bound for buckling loads. However, since vibration frequencies and modes are not sensitive to imperfections, vibration test results can usually be regarded as reliable and can therefore be used to determine which models best simulate the actual behavior.

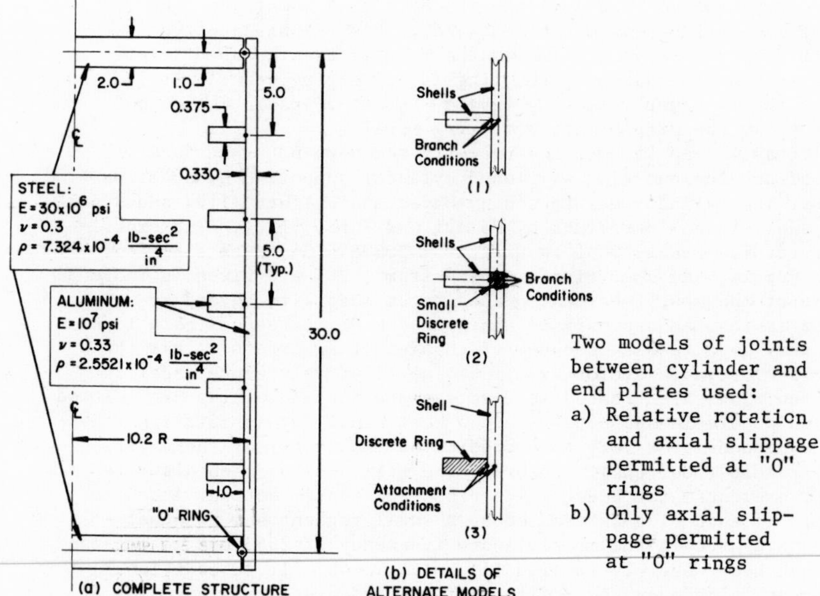

Fig. 44 (a) Geometry of ring stiffened cylinder tested by Hayek and Pallet. (b) Various models of the rings

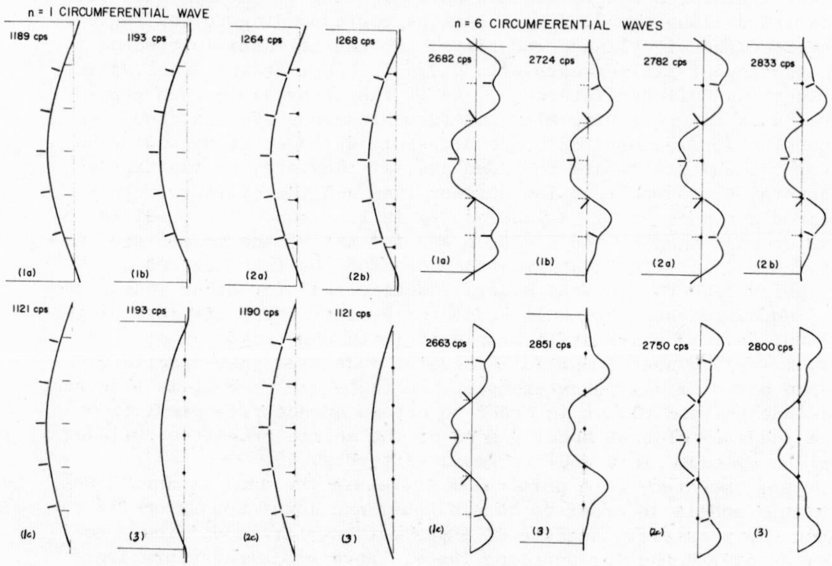

Fig. 45 Vibration modes corresponding to one and six circumferential waves

Table 2 Natural Frequencies for Various Models of Ring-Stiffened Cylinder

Circ. Waves	Axial Waves	Test[a] Results	End Plates Included: Rings Are Shell Branches: Cylinder Is One Segment		End Plates Included: Rings Are Shell Branches: Cylinder Is Six Segments[b]		End Plates Included: Discrete Rings	End Plates Omitted: Rings Treated As Shell Branches, Cylinder Segments: One	End Plates Omitted: Rings Treated As Shell Branches, Cylinder Segments: Six	End Plates Omitted: Discrete Rings	Harari & Baron: Discrete Rings	Harari & Baron: Smeared Rings
n	m		A[c]	B[d]	A	B	B	Cylinder is Simply Supported				
1	r.b.[e]	-	0	0	0	0	0	0	0			
1	r.b.	-	0	0	0	0	0	0	0			
1	p.a.[f]	?	0	67	0	69	67	--	--			
1	p.s.[g]	?	0	70	0	73	70	--	--			
1	1	1232	1189	1193	1264	1268	1193	1121	1190	1121	1124	1133
1	u[h]	?	1714	1714	1819	1819	1716	1738	1848	1740	1743	
1	2	?	2189	2198	2301	2313	2199	2175	2282	2177	2183	
1	3	?	2653	2669	2763	2783	2672	2654	2759	2656	2663	
1	4	2870	2863	2887	2962	2991	2890		2960	2868	2875	2893
2	1	627	607	618	648	660	618	609	647	609	611	640
2	p.a.	?	1040	1047	1040	1047	1047	--	--	--	--	
2	p.s.	?	1041	1048	1041	1048	1048	--	--	--	--	
2	2	?	1378	1395	1469	1489	1396	1385	1468	1386	1389	
2	3	?	1960	1983	2075	2103	1985	1967	2073	1969	1974	
2	4	?	2339	2366	2467	2502	2371					
3	1	787	773	783	803	815	796	773	802	786	786	832
3	2	1190	1137	1160	1203	1230	1166	1136	1197	1143	1145	1194
3	3	1602	1588	1619	1690	1726	1622	1587	1679	1589	1594	1650
4	1	1310	1307	1313	1348	1355	1371	1306	1346	1364	1359	1431
4	2	1503	1453	1475	1509	1535	1525	1450	1503	1501	1497	1575
4	3	1806	1714	1752	1797	1842	1788	1708	1784	1745	1744	1826
5	1	1938	1943	1949	2008	2014	2080	1941	2006	2073	2062	2253
5	2	2059	2020	2041	2088	2113	2163	2015	2080	2137	2126	2331
5	3	2276	2170	2214	2251	2304	2317	2159	2232	2262	2252	2474
6	1	2594	2567	2572	2673	2678	2770	2564	2668	2762	2750	3276
6	2	?	2606	2625	2707	2729	2798	2597	2691	2772	2762	
6	3	2802	2682	2724	2782	2833	2851	2663	2750	2800	2790	3424

[a] Tests performed by Hayek and Pallett.

[b] Gaps between segments of cylinder are "filled" by discrete rings with cross-section dimensions .33 x .375.

[c] Model A: Rotation and axial slippage permitted between end plates and cylinder.

[d] Model B: Axial slippage only permitted between end plates and cylinder.

[e] r.b. = "rigid body mode"

[f] p.a. = "plate antisymmetric" = end plates vibrating in phase.

[g] p.s. = "plate symmetric" = end plates vibrating out-of- "

[h] u = axial motion predominates.

Modeling Concentrated Loads on Shells

The analyst may be interested in several types of concentrated loads which arise in various ways. If the shell structure is to be subjected to concentrated loads in the ordinary course of its service, such as a tank supported on struts or a rocket stage with discrete payload attach points, it is usually provided that the concentrated loads be applied to reinforced areas such as circumferential rings or longitudinal stringers through which these loads are smoothly diffused into the shell. Therefore, deflections are small, and a linear analysis is generally suitable. If, however, the analyst wants to find out what happens if the shell is accidentally poked somewhere, the concentrated load may be applied to an unreinforced area, and the shell may experience large deflections. Prediction of the effect of these accidental loads may therefore require nonlinear analysis. The point-loaded spherical cap, for which a load-deflection curve is shown in Fig. 8, is an example.

The purpose of this section is to present the results of a linear parameter study involving a ring-stiffened cylinder subjected to a concentrated load applied to one of the rings. The structure and applied load are shown in Fig. 46. The results of the parameter study are given in Figs. 46 - 50. Fishlowitz [121] analyzed this system with two programs, NASTRAN and STAGS-A. In both of these analyses the rings were treated as discrete and the shell coordinates were discretized in both axial and circumferential directions.

The results presented here were obtained with STAGS-B [35] and BOSOR4 [36]. STAGS-B is based on a more sophisticated two-dimensional finite difference energy scheme than that used in STAGS-A. BOSOR4 is based on a finite difference energy minimization in which the axial coordinate is discretized and variations in the circumferential direction are obtained by superposition of the responses from each Fourier harmonic of a line-load approximation of the actual concentrated load. As shown in Fig. 46, the point load P is treated in the BOSOR4 model as a triangular pulse subtending an angle θ_L. One of the objectives of this study was to determine the convergence of stress with respect to the load representation parameter θ_L. Also investigated were the rate of convergence with respect to the number of Fourier harmonics used for each value of θ_L and convergence with respect to the density of nodal points in the axial direction. Only one run was made with STAGS-B. The densities of nodal points in the axial and circumferential directions are indicated in Figs. 46 and 47, respectively. The convergence of the maximum stress with decreasing θ_L is indicated in Fig. 48. Directly under the load this convergence is rather slow, but it improves dramatically a very small distance away in either coordinate direction. Figure 49 indicates the rate of convergence with the use of more and more Fourier harmonics for θ_L = 1.6 degrees. Figure 50 shows convergence with both θ_L and the number of Fourier harmonics for three circumferential stations very near the load.

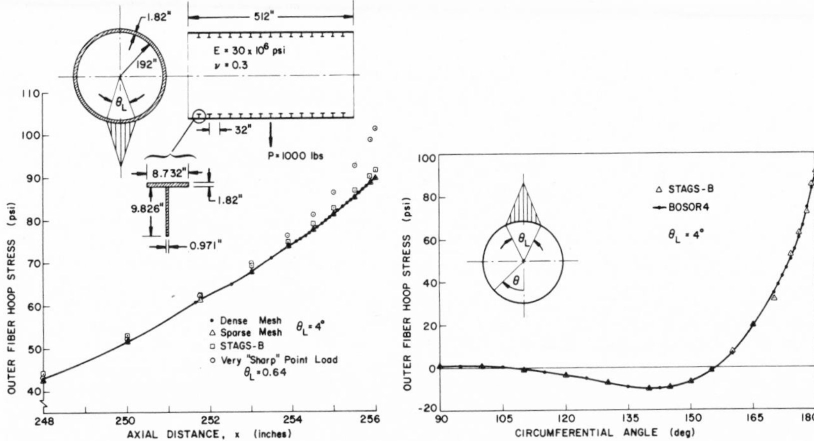

Fig. 46 Maximum stress in ring-stiffened cylinder submitted to a point load

Fig. 47 Stress in point-loaded cylinder at the station where the load is applied

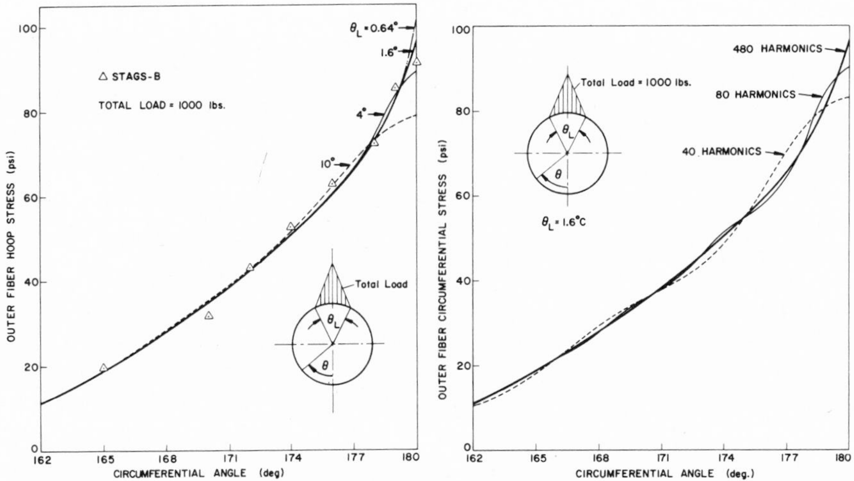

Fig. 48 Maximum stress as a function of angle over which the point load is spread in the analytical model

Fig. 49 Stress distribution as a function of the number of Fourier harmonics used in the expansion of the point load

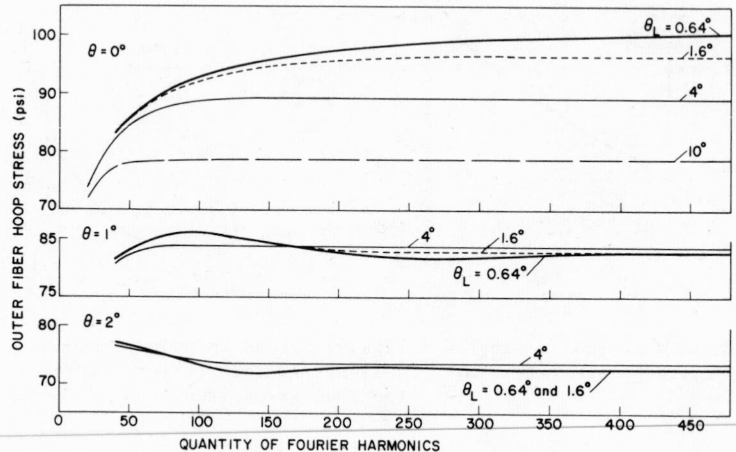

Fig. 50 Convergence with number of Fourier harmonics for three circumferential stations and several models of the point load

This example provides a good illustration of the need for a sub-structuring capability. In the BOSOR4 analysis it was necessary to obtain the solution for every Fourier harmonic for the entire length of cylinder from its simply supported end to the symmetry plane where the load is applied. However, the higher circumferential harmonics decay very rapidly, as seen in Fig. 35. Therefore, operation on the computer would have been much more efficient if the cylinder had been divided into two or more segments with more Fourier harmonics being used in the segments near the applied load. In a model which is discretized in two dimensions it would have been very useful to allocate nodal points in such a way that the grid is sparse in both coordinate directions in areas remote from the load and uniformly dense in areas near the load.

SUMMARY

It is emphasized in the introduction that before the analyst can choose wisely and use effectively a computer program for the analysis of thin shells, he should understand the physics of a problem well enough to be able to define clearly his needs. One of the objectives of this paper, therefore, is to give the analyst a physical feel for the behavior of thin shells. Examples are chosen to illustrate linear and nonlinear stress, buckling, and dynamic phenomena of practical engineering structures of which thin shells are important components.

With regard to the shell wall properties it is pointed out that in the strain energy expression coupling of membrane and bending deformations should be included for the accurate treatment of shells with layered walls, eccentric stiffeners, temperature-dependent or elastic-plastic material, and geometrical discontinuities. Whether to model stiffeners as smeared or as discrete is briefly discussed, and the importance of eccentricity of the stiffeners is demonstrated through an example of buckling of a spherical cap with an edge ring. The significance of discontinuities in the reference surface is illustrated through examples of local buckling and vibration of ring-stiffened cylinders in which the rings are modeled in various ways. The possibility of fortuitously obtaining good predictions with poor models is discussed with some of the counteracting errors identified. Of course, the errors associated with a poor model do not always counteract. It is not recommended that the analyst use a poor model in the hope that they will!

A number of problems are introduced involving linear and non-linear stress analysis. The decay rates of various edge disturbances are shown. These decay rates are determined by two effects, the resistance of the shell against radial expansion--the so-called elastic foundation effect-- and the redistribution of membrane forces due to shear lag. An example is given which demonstrates the sensitivity of behavior to boundary conditions. The analyst is urged to take great care in defining the support conditions for each particular segment of a larger structure for which he has design responsibility. A number of cases are presented which demonstrate various nonlinear effects: nonlinear collapse, stress redistribution, load path eccentricity, effect of prestress on the bending stiffness of a shell wall, and the effect of change in direction of the normal pressure as the reference surface rotates.

A section on buckling is included in which it is shown that a bifurcation buckling analysis may not be meaningful. The actual failure load of a shell may be higher or lower than the bifuraction load, depending on how the shell changes shape as it is loaded and on how the stresses are redistributed. The effect of prebuckling shape change on bifurcation loads is discussed. Also briefly mentioned are the possibility of very low buckling loads associated with inextensional buckling modes, the effect on buckling of very local distortion of built-up walls, and the frequent occurrence of local buckling due to circumferential compression induced by discontinuity forces. The sensitivity of buckling loads to imperfections is described, and results are given from the analysis of various configurations with the use of Koiter's theory for initial postbuckling behavior. The analyst is cautioned about the limitations of this theory as applied to practical engineering structures which may have several instability failure modes, some local and some general, each with its own degree of sensitivity to initial imperfections of shell geometry. Two examples are given of behavior in the far-postbuckling regime predicted with the use of STAGS, a computer program for the nonlinear analysis of general shells.

A shorter discussion of dynamic response is provided. Certain aspects of the dynamic buckling of cylindrical and spherical shells are covered. Comparisons are given of results obtained with several different computer programs applied to the problem of linear and nonlinear dynamic response of a conical shell submitted to a circumferentially nonuniform impulse. It is concluded that more research is needed in this field in order to obtain a better understanding of how dynamically loaded thin shells behave. Fluid-structure interaction with regard to submerged shells and flutter is very briefly summarized.

Another objective of the paper is to give the analyst some hints on modeling shell structures. The section on modeling includes examples of how to distribute nodal points for accurate determination of stress concentrations and buckling or vibration modes, how to handle intersections of shells, how to treat discrete stiffeners in various ways, and how to model concentrated loads. The analyst is encouraged to use many models of a given structure in an attempt to obtain upper and lower bounds for actual behavior.

ACKNOWLEDGMENT

The author is grateful to Dick Hartung, Oscar Hoffman, Jörgen Skogh, Bo Almroth, and Phillip Underwood, who provided many helpful suggestions during the course of this work. Jesse Vosti drew most of the figures. The effort in writing the manuscript was supported by the Lockheed Independent Research Program.

The following figures have been reprinted by permission:

Fig. 2 from D. Bushnell, "Finite-Difference Energy Models versus Finite-Element Models: Two Variational Approaches in One Computer Programs," Numerical and Computer Methods in Structural Mechanics, pp. 291-336, © 1973, Academic Press, New York.

Fig. 3 from D. Bushnell, "Finite-Difference Energy Models versus Finite-Element Models: Two Variational Approaches in One Computer Programs," Numerical and Computer Methods in Structural Mechanics, pp. 291-336, © 1973, Academic Press, New York.

Fig. 4 from D. Bushnell, "Evaluation of Various Analytical Models for Buckling and Vibration of Stiffened Shells," AIAA Journal, Vol. 11, No. 9, 1973, pp. 1283-1291 © 1973, AIAA.

Fig. 5 from D. Bushnell, "Buckling and Vibration of Ring-Stiffened, Segmented Shells of Revolution: Numerical Results," Proceedings of the First Pressure Vessels and Piping International Meeting, pp. 255-268, © 1969, American Society of Mechanical Engineers.

Fig. 6 from D. Bushnell, "Crippling and Buckling of Corrugated Ring-Stiffened Cylinders," AIAA Journal of Spacecraft and Rockets, Vol. 9, No. 5, 1972, pp. 357-363, © 1972, AIAA.

Fig. 7 from D. Bushnell, "Crippling and Buckling of Corrugated Ring-Stiffened Cylinders," AIAA Journal of Spacecraft and Rockets, Vol. 9, No. 5, 1972, pp. 357-363, © 1972, AIAA.

Fig. 8 from D. Bushnell, "Bifurcation Phenomena in Spherical Shells Under Concentrated and Ring Loads," AIAA Journal, Vol. 5, No. 11, 1967, pp. 2034-2040, © 1967, AIAA.

Fig. 9 from F. A. Brogan and B. O. Almroth, "Practical Methods for Elastic Collapse Analysis for Shell Structures," AIAA Journal, Vol. 9, No. 12, 1971, pp. 2321-2325, © 1971, AIAA.

Fig. 10 from B. O. Almroth and F. A. Brogan, "Bifurcation Buckling as an Approximation of the Collapse Load for General Shells," AIAA Journal, Vol. 10, No. 4, 1972, pp. 463-467, © 1972, AIAA.

Fig. 11 from B. O. Almroth, "Nonlinear Behavior of Shells," Numerical Solution of Nonlinear Structural Problems, AMD Vol. 6, American Society of Mechanical Engineers, 1973, pp. 1-15, © 1973, ASME.

Fig. 12 from D. Bushnell, "Nonsymmetric Buckling of Cylinders with Axisymmetric Thermal Discontinuities," AIAA Journal, Vol. 11, No. 9, 1973, pp. 1292-1295, © 1973, AIAA.

Fig. 13 from D. Bushnell, "Nonlinear Analysis for Axisymmetric Elastic Stresses in Ring-Stiffened, Segmented Shells of Revolution," AIAA/ASME 10th Structures, Structural Dynamics and Materials Conference, pp. 104-113, © 1969, ASME.

Fig. 15 from D. Bushnell, "Crippling and Buckling of Corrugated Ring-Stiffened Cylinders," AIAA Journal of Spacecraft and Rockets, Vol. 9, No. 5, 1972, pp. 357-363, © 1972, AIAA.

Fig. 16 from F. A. Brogan and B. O. Almroth, "Practical Methods for Elastic Collapse Analysis for Shell Structures," AIAA Journal, Vol. 9, No. 12, 1971, pp. 2321-2325, © 1971, AIAA.

Fig. 18 from B. O. Almroth and F. A. Brogan, "Bifurcation Buckling as an Approximation of the Collapse Load for General Shells," AIAA Journal, Vol. 10, No. 4, 1972, pp. 463-467, © 1972, AIAA.

Fig. 19 from D. Bushnell, "Buckling and Vibration of Ring-Stiffened, Segmented Shells of Revolution: Numerical Results," Proceedings of the First Pressure Vessels and Piping International Meeting, pp. 255-268, © 1969, American Society of Mechanical Engineers.

Fig. 20 from B. O. Almroth and D. Bushnell, "Computer Analysis of Various Shells of Revolution," AIAA Journal, Vol. 6, No. 10, 1968, pp. 1848-1855, © 1968, AIAA.

Fig. 21 from D. Bushnell, "Inextensional Buckling of Spherical Shells with Edge Rings," AIAA Journal, Vol. 6, No. 2, 1968, pp. 361-364, © 1968, AIAA.

Fig. 22 from D. Bushnell, "Buckling of Spherical Shells Ring-Supported at the Edges," AIAA Journal, Vol. 5, No. 11, 1967, pp. 2041-2046, © 1967, AIAA.

Fig. 23 from B. O. Almroth, A. B. Burns, and E. V. Pittner, "Design Criteria for Axially Loaded Cylindrical Shells," Journal of Spacecraft and Rockets, Vol. 7, No. 6, 1970, pp. 714-720, © 1970, AIAA.

Fig. 24a and 24b from Bernard Budiansky and John W. Hutchinson, "A Survey of Some Buckling Problems," AIAA Journal, Vol. 4, 1966, pp. 1506-1510, © 1966, AIAA.

Fig. 25 from Bernard Budiansky and John C. Amazigo, "Initial Post-buckling Behavior of Cylindrical Shells Under External Pressure," Journal of Mathematics and Physics, Vol. 47, No. 3, 1968, pp. 223-235, © 1968.

Fig. 26 from John W. Hutchinson and John C. Amazigo, "Imperfection-Sensitivity of Eccentrically Stiffened Cylindrical Shells," AIAA Journal, Vol. 5, No. 3, 1967, pp. 392-401, © 1967, AIAA.

Fig. 27 from Bernard Budiansky and John W. Hutchinson, "A Survey of Some Buckling Problems," AIAA Journal, Vol. 4, 1966, pp. 1506-1510, © 1966, AIAA.

Fig. 28 from D. Bushnell, B. O. Almroth, and F. Brogan, "Finite-Difference Energy Method for Nonlinear Shell Analysis," Computers & Structures, Vol. 1, 1971, pp. 361-387, © 1971.

Fig. 29 from J. Skogh and P. Stern, "Postbuckling Behavior of a Section Representative of the B-1 Aft Intermediate Fuselage," AFFDL-TR-73-63, May 1973, Air Force Flight Dynamics Laboratory, Wright-Patterson Air Force Base, Ohio.

Fig. 30 from B. Budiansky, R. S. Roth, "Axisymmetric Dynamic Buckling of Clamped Shallow Spherical Shells," Collected Papers on Instability of Shell Structures - 1962, NASA TN D-1510, National Aeronautics and Space Administration, Washington, D. C., 1962, pp. 597-604.

Fig. 31 from B. Budiansky, R. S. Roth, "Axisymmetric Dynamic Buckling of Clamped Shallow Spherical Shells," Collected Papers on Instability of Shell Structures - 1962, NASA TN D-1510, National Aeronautics and Space Administration, Washington, D. C., 1962, pp. 597-604.

Fig. 32 from W. E. Haisler, J. A. Stricklin and W. A. Von Riesemann, "DYNAPLAS-A Finite Element Program for the Dynamic Elastic-Plastic Analysis of Stiffened Shells of Revolution," Texas A & M University, College Station, Texas, TEES-RPT 72-27, and Sandia Laboratories, Albuquerque, New Mexico, SLA 73-0127, December 1972.

Fig. 33 from R. E. Ball, W. F. Hubka, N. J. Huffington, Jr., P. Underwood and W. A. Von Riesemann, "A Comparison of Computer Results for the Dynamic Response of the LMSC Truncated Cone," Computers & Structures, 1974

Fig. 34 from R. E. Ball, W. F. Hubka, N. J. Huffington, Jr., P. Underwood and W. A. Von Riesemann, "A Comparison of Computer Results for the Dynamic Response of the LMSC Truncated Cone," Computers & Structures, 1974

Figs. 39 through 43 from D. Bushnell, "Evaluation of Various Analytical Models for Buckling and Vibration of Stiffened Shells," AIAA Journal, Vol. 11, No. 9, 1973, pp. 1283-1291, © 1973, AIAA.

REFERENCES

1 Bushnell, D. W., "A Computerized Information Retrieval System," in this volume.

2 Euler, "De Sono Campanarum," Nov. Comm. Acad. Petropolitanae, t. 10, 1766.

3 Bernoulli, J., "Essai Théorique sur les Vibrations des Plaques Élastiques," Nov. Acta Petropolitanae, t. 5, 1789.

4 Germain, Sophie, "Recherches sur la Theorie des Surfaces Élastiques," Paris, France, 1821.

5 Chladni, E. F. F., Die Akustik, Leipzig, 1802.

6 Bernoulli, J., "Veritable Hypothèse de la Résistance des Solides, avec la Demonstration de la Courbure des Corps qui Font Ressort," (found in Bernoulli's collected works, t. 2, Geneva, 1744).

7 Euler, see the Additamentum "De Curvis Elasticis" in the Methodus Inveniendi Lineas Curvas Maximi Minimive Proprietate Gaudentes, Lausanne, 1744.

8 Coulomb, "Essai sur une Application Des Règles de Maximis et Minimis à quelques Problèmes de Statique, Relatifs à l'Architecture," Mém....par Divers Savants, 1776.

9 Hooke, Robert, De Potentia Restitutive, London, 1678.

10 Kirchhoff, J. F. Math. (Crelle), Bd. 40 (1850).

11 Aron, H., J. F. Math. (Crelle), Bd. 78, 1874.

12 Mathieu, E., J. de l'École Polytechnique, t. 51, 1883.

13 Lord Rayleigh, London Math. Soc. Proc., Vol. 13, 1882.

14 Love, A. E. H., Phil. Trans., Roy. Soc. (Ser. A), Vol. 179, 1888.

15 Koiter, W. T., "A Consistent First Approximation in the General Theory of Thin Elastic Shells," IUTAM Proceedings of the Symposium on the Theory of Thin Elastic Shells, W. T. Koiter, ed., North-Holland Publishing Company, Amsterdam, 1960, pp. 12-33.

16 Bathe, Klaus-Jurgen and Wilson, E. L.,"Thick Shell Structures," (in this volume).

17 Green, A. E., and Zerna, W., Theoretical Elasticity, Oxford University Press, Oxford, 1954.

18 Donnell, L. H., "A New Theory for the Buckling of Thin Cylinders Under Axial Compression and Bending," Trans. Am. Soc. Mech. Engrs., Vol. 56, 1934, p 795.

19 Hartung, Richard F., "An Assessment of current Capability for Computer Analysis of Shell Structures," AFFDL-TR-71-54, April 1971, Air Force Flight Dynamics Laboratory, Wright-Patterson Air Force Base, Ohio. See also Computers & Structures, Vol. 1, 1971, pp 3-32.

20 Fenves, Steven J., Perrone, Nicholas, Robinson, Arthur R. and Schnobrich, William C., editors, Numerical and Computer Methods in Structural Mechanics, Academic Press, Inc., New York and London, 1973.

21 Oden, J. T., Clough, R. W., and Yamamoto, Y., editors, Advances in Computational Methods in Structural Mechanics and Design (Proceedings of 2nd U.S. - Japan Seminar on Matrix Methods of Structural Analysis and Design), University of Alabama Press, Huntsville, Alabama, 1972.

22 Rowan, William H., Jr. and Hackett, Robert M., editors, Proceedings of the Symposium on Application of Finite Element Methods in Civil Engineering, Vanderbilt University and the American Society of Civil Engineers, 1969.

23 National Aeronautics and Space Administration, Computer Program Abstracts, issued periodically.

24 McCormick, J. M. and Baron, M. L., and Perrone, N., "The STORE Project (The Structures Oriented Exchange)," 1973, pp 439 - 458 of [20].

25 Bushnell, David, "Large Deflection Elastic-Plastic Creep Analysis of Axisymmetric Shells,"Numerical Solution of Nonlinear Structural Problems, American Society of Mechanical Engineers, AMD - Vol. 6, 1973, pp 103-137.

26 Stricklin, J. A., Haisler, W. E., and von Riesemann, W. A., "Formulation, Computation, and Solution Procedures for Material and/or Geometric Nonlinear Structural Analysis by the Finite Element Method," SC-CR-72 3102, July 1972, Sandia Laboratories, Albuquerque, New Mexico.

27 Novoshilov, V. V., Foundations of the Nonlinear Theory of Elasticity, Graylock Press, 1953, pp. 186-198

28 Sanders, J. Lyell, Jr., "Nonlinear Theories for Thin Shells," Quart. Appl. Math., Vol. 21, No. 1, 1963, pp 21-36.

29 Gallagher, Richard H., "Analysis of Plate and Shell Structures," 1969, pp. 155-203 of [22].

30 Gallagher, R. H., "Applications of Finite Element Analysis," 1972, pp 641-678 of [21].

31 Gallagher, R. H., "Geometrically Nonlinear Finite Element Analysis," Specialty Conference on the Finite Element Method in Civil Engineering, McGill University, 1972.

32 Brombolich, Lawrence J. and Gould, Phillip L., "Finite Element Analysis of Shells of Revolution by Minimization of the Potential Energy Functional," 1969, pp. 279-307 of [22].

33 Wilson, E. L., Taylor, R. L., Doherty, W. P., and Ghaboussi, J., "Incompatible Displacement Models," 1973, pp 43-57 of [20].

34 Johnson, D. E., "A Difference-Based Variational Method for Shells," International Journal of Solids & Structures, Vol. 6, 1970, p 699-724.

35 User's Manual for STAGS: Almroth, B. O. and Brogan, F. A., Vol. 1; "Scope of the Code," LMSC-D358197, June 1973; Almroth, B. O., Brogan, F. A., Meller, E., Petersen, H. T., Vol. 3, "User's Instructions," LMSC-D358197, January 1974, Lockheed Missiles & Space Company, Palo Alto, California.

36 Bushnell, David, "Stress, Stability, and Vibration of Complex Branched Shells of Revolution: Analysis and User's Manual for BOSOR4," NASA CR-2116, October 1972.

37 Bushnell, David, "Finite-Difference Energy Models versus Finite-Element Models: Two Variational Approaches in One Computer Program," 1973, pp 291-336 of [20].

38 Khojasteh-Bakht, M., "Analysis of Elastic-Plastic Shells of Revolution Under Axisymmetric Loading by the Finite Element Method," Ph.D. Dissertation, Dept. of Civil Engineering, University of California, Berkeley, California, 1967 (also published as SESM Report 67-8).

39 Baruch, M and Singer, J., "Effect of Eccentricity of Stiffeners on the General Instability of Stiffened Cylindrical Shells Under Hydrostatic Pressure," Journal of Mechanical Engineering Sciences, Vol. 5, 1963, pp 23-27.

40 Bushnell, David, "Evaluation of Various Analytical Models for Buckling and Vibration of Stiffened Shells," AIAA Journal, Vol. 11, No. 9, 1973, pp 1283-1291.

41 Bushnell, David, "Buckling and Vibration of Ring-Stiffened, Segmented Shells of Revolution: Numerical Results," Proceedings of the First Pressure Vessels and Piping International Meeting, 1969, American Society of Mechanical Engineers, pp 255-268.

42 Wang, L. R. L., Rodriguez-Agrait, L., and Little, W. A., "Effect of Boundary Conditions on Shell Buckling," J. Eng. Mech. Div., ASCE, Vol. 92, EM6, 1966, pp 101-116.

43 Bushnell, David, "Crippling and Buckling of Corrugated Ring-Stiffened Cylinders," Journal of Spacecraft and Rockets, Vol. 9, No. 5, 1972, pp 357-363.

44 Schrem, Ernst, "Computer Implementation of the Finite-Element Procedure," 1973, pp 79-121 of [20].

45 Irons, Bruce M and Kan, D. K. Y., "Equation-Solving Algorithms, for the Finite-Element Method," 1973, pp 497-511 of [20].

46 Stricklin, J. A., Von Riesemann, W. A., and Haisler, W. E., "Nonlinear Continuum Software," paper in this ONR/NSF volume on Nonlinear Analysis.

47 Almroth, B. O. and Felippa, C. A., "Structural Stability," (in this volume).

48 Bushnell, David, "Bifurcation Phenomena in Spherical Shells Under Concentrated and Ring Loads," AIAA Journal, Vol. 5, No. 11, 1967, pp 2034-2040.

49 Archer, R. R., "On the Numerical Solution of the Nonlinear Equations for Shells of Revolution," J. Math. Phys., Vol. 14, 1962, pp 165-178.

50 Brogan, F. A., and Almroth, B. O., "Practical Methods for Elastic Collapse Analysis for Shell Structures," AIAA Journal, Vol. 9, No. 12, 1971, pp 2321-2325.

51 Bushnell, David, "Bifurcation Buckling of Shells of Revolution Including Large Deflections, Plasticity and Creep" and "Comparisons of Test and Theory for Nonsymmetric Elastic-Plastic Buckling of Shells of Revolution," (co-authored with G. D. Galletly), to appear Int. J. of Solids and Structures, 1974.

52 Almroth, B. O., "Nonlinear Behavior of Shells," Numerical Solution of Nonlinear Structural Problems, AMD Vol. 6, American Society of Mechanical Engineers, 1973, pp 1-15.

53 Bushnell, David, "Nonsymmetric Buckling of Cylinders with Axisymmetric Thermal Discontinuities," AIAA Journal, Vol. 11, No. 9, 1973, pp 1292-1295.

54 Clough, R. W. and Bathe, K. J., "Finite Element Analysis of Dynamic Response," 1972, pp 153-179 of [21].

55 Krieg, R. D. and Key, S. W., "Transient Shell Response by Numerical Time Integration," 1972, pp 237-258 of [21].

56 Belytscho, T. "Linear and Nonlinear - Transient Analysis," (in this volume).

57 Jensen, Paul S., "Transient Analysis of Structures by Stiffly Stable Methods," Computers & Structures, 1974.

58 Skogh, J., "Nonlinear Stress Analysis of Longitudinal Strap Joints in Weld Bond LH_2 Droptank," LMSC Engineering Memorandum L2-12-01-M1-15, 17 June 1971, Lockheed Missiles & Space Co., Space Shuttle Project, Sunnyvale, California.

59 Brush, D. O. and Almroth, B. O., Buckling of Bars, Plates, and Shells, McGraw-Hill, New York, 1974.

60 Almroth, B. O. and Brogan, F. A., "Bifurcation Buckling as an Approximation of the Collapse Load for General Shells," AIAA Journal, Vol. 10, No. 4, 1972, pp 463-467.

61 Brazier, L. G., "On the Flexure of Thin Cylindrical Shells and Other 'Thin' Sections," Proceedings of the Royal Society, Series A, Vol. CXVI, 1926, pp 104-114.

62 Cohen, G. A., "User Document for Computer Programs for Ring-Stiffened Shells of Revolution," NASA CR-2086, 1973; "Computer Analysis of Ring-Stiffened Shells of Revolution," NASA CR-2085, 1973; "Computer Program for Analysis of Imperfection Sensitivity of Ring-Stiffened Shells of Revolution," NASA CR-1801, 1971, National Aeronautics and Space Administration, Washington, D. C.

63 Anderson, M. S., Fulton, R. E., Heard, W. L., Jr., and Walz, J. E., "Stress, Buckling and Vibration Analysis of Shells of Revolution," Computers & Structures, Vol. 1, 1971, pp 157-192.

64 Kalnins, A., "User's Manual for KSHEL Computer Programs," 1970, available from Dr. Kalnins, Lehigh University, Pa. Also see "Free Vibration, Static and Stability Analysis of Thin Elastic Shells of Revolution," AFFDL-TR-68-144, March 1969, Wright-Patterson Air Force Base, Ohio.

65 Svalbonas, V., "Numerical Analysis of Stiffened Shells of Revolution," NASA CR-2273, 1973, National Aeronautics and Space Administration, Washington, D. C.

66 Blumenberg, W. F., "The Effect of Intermediate Heavy Frames on the Elastic General-Instability Strength of Ring-Stiffened Cylinders Under External Hydrostatic Pressure," David Taylor Model Basin Report 1844, February 1965.

67 Almroth, B. O. and Bushnell, D., "Computer Analysis of Various Shells of Revolution," AIAA Journal, Vol. 6, No. 10, 1968, pp 1848-1855.

68 Bushnell, David, "Inextensional Buckling of Spherical Shells with Edge Rings," AIAA Journal, Vol. 6, No. 2, 1968, pp 361-364.

69 Heard, W. L., Jr., and Anderson, M. S., "Design, Analysis, and Tests of a Structural Prototype Viking Aeroshell," Part I, Aeroshell Design and Analysis;" Part II by Anderson, J. K. and Card, M. F., "Test Procedure and Test Results," AIAA Paper No. 72-370, AIAA/ASME/SAE 13th Structures, Structural Dynamics, and Materials Conference, San Antonio, Texas, April 1972.

70 Bushnell, David, "Buckling of Spherical Shells Ring-Supported at the Edges," AIAA Journal, Vol. 5, No. 11, 1967, pp 2041-2046.

71 Stricklin, James A., Haisler, Walter E., Von Reisemann, Walter A., Leick, Roger D., Hunsaker, Barry, and Saczalski, Kenneth J., "Large Deflection Elastic-Plastic Dynamic Response of Stiffened Shells of Revolution," TEES-RPT-72-25 and SLA-73-0128, December 1972, Texas A&M University, Texas.

72 Klein, S., see ref. [108].

73 Ball, R. E., "A Program for the Nonlinear Static and Dynamic Analysis of Arbitrarily Loaded Shells of Revolution," Computers and Structures, Vol. 2, 1972, pp 141-162.

74 Underwood, P.G., see ref. [102].

75 Huffington, N. J., Jr., "Large Deflection Elastoplastic Response of Shell Structures," U.S. Army Ballistic Research Laboratories Report No. 1515, 1970, Aberdeen Proving Ground, Maryland.

76 Hubka, W. F., Windholz, W. M., and Karlsson, T., "A Calculation Method for the Finite Deflection, Anelastic Dynamic Response of Shells of Revolution," Kaman Sciences Corporation, Report KN-69-660(R), January 1970, Kaman Sciences Corporation, Colorado Springs, Colorado.

77 Bushnell, David and Smith, Strether, "Stress and Buckling of Nonuniformly Heated Cylindrical and Conical Shells," AIAA Journal, Vol. 9, No. 12, 1971, pp 2314-2321.

78 Hill, R., "A General Theory of Uniqueness and Stability in Elastic-Plastic Solids," J. Mech. Phys. Solids, Vol. 6, 1958, pp 236-249.

79 Hutchinson, J. W., "Plastic Buckling," Advances in Applied Mechanics, Vol. 14, edited by C. S. Yih, Academic Press, 1974.

80 Onat, E. T. and Drucker, D. C., "Inelastic Instability and Incremental Theories of Plasticity," J. Aero. Sci., Vol. 20, 1953, pp 181-

81 Almroth, B. O., Burns, A. B., and Pittner, E. V., "Design Criteria for Axially Loaded Cylindrical Shells," Journal of Spacecraft and Rockets, Vol. 7, No. 6, 1970, pp 714-720.

82 Hutchinson, J. W. and Koiter, W. T., "Postbuckling Theory," Applied Mechanics Reviews, Vol. 23, 1970, pp 1353-1366.

83 Koiter, W. T., "Over de Stabiliteit van het Elastisch Evenwicht," (On the Stability of Elastic Equilibrium), Thesis, Delft, H. J. Paris, Amsterdam, 1945. English translation issued as NASA TT F-10, 833, 1967.

84 Budiansky, Bernard and Hutchinson, John W., "A Survey of Some Buckling Problems," AIAA Journal, Vol. 4, 1966, pp 1506-1510.

85 Koiter, W. T., "The Effect of Axisymmetric Imperfections on the Buckling of Cylindrical Shells Under Axial Compression," Koninkl. Nederl. Akademie van Wetenshappen-Amsterdam, Series B, Vol. 66, No. 5, 1963, pp 265-279.

86 Budiansky, Bernard and Amazigo, John C., "Initial Postbuckling Behavior of Cylindrical Shells Under External Pressure," Journal of Mathematics and Physics, Vol. 47, No. 3, 1968, pp 223-235.

87 Dow, D. A., "Buckling and Postbuckling Tests of Ring-Stiffened Cylinders Loaded by Uniform External Pressure," NASA TN-D-3111, November 1965, National Aeronautics and Space Administration, Washington, D. C.

88 Hutchinson, John W., and Amazigo, John C., "Imperfection-Sensitivity of Eccentrically Stiffened Cylindrical Shells," AIAA Journal, Vol. 5, No. 3, 1967, pp 392-401.

89 Budiansky, B. and Hutchinson, J. W., "Dynamic Buckling of Imperfection-Sensitive Structures," Proceedings of the XI International Congress on Applied Mechanics, Julius Springer-Verlag, Berlin, 1966, pp 636-651.

90 Hutchinson, J. W. and Budiansky, B., "Dynamic Buckling Estimates," AIAA Journal, Vol. 4, No. 3, 1966, pp 525-530.

91 Budiansky, Bernard, "Dynamic Buckling of Elastic Structures: Criteria and Estimates," Proceedings of an International Conference on Dynamic Stability of Structures, edited by George Herrmann, Pergamon Press, London, 1967, pp 83-106.

92 Hutchinson, J. W., "Imperfection Sensitivity of Externally Pressurized Spherical Shells," Journal of Applied Mechanics, Vol. 34, 1967, pp 49-55.

93 Bushnell, David, Almroth, B. O., and Brogan, Frank,"Finite-Difference Energy Method for Nonlinear Shell Analysis," Computers & Structures, Vol. 1, 1971, pp 361-387.

94 Kuhn, P., Stresses in Aircraft and Shell Structures, McGraw-Hill, New York, 1956.

95 Skogh, J. and Stern, P., "Postbuckling Behavior of a Section Representative of the B-1 Aft Intermediate Fuselage," AFFDL-TR-73-63, May 1973, Air Force Flight Dynamics Laboratory, Wright-Patterson Air Force Base, Ohio.

96 Goodier, J. N., "Dynamic Plastic Buckling," Proceedings of International Symposium on Dynamic Stability of Structures, George Herrmann, editor, Pergamon Press, 1965, pp 189-211.

97 Abrahamson, G. R. and Goodier, J. N., "Dynamic Plastic Flow Buckling of a Cylindrical Shell from Uniform Radial Impulse," Proc. Fourth U. S. National Congress of Applied Mechanics, Vol 2, 1962, pp 939-950.

98 Lindberg, H. E. and Herbert, R. E., "Dynamic Buckling of a Thin Cylindrical Shell Under Axial Impact," J. Appl. Mechanics, Vol. 33, 1966, pp 105-112.

99 Humphreys, John S. and Sve, Charles, "Dynamic Buckling of Cylinders Under Axial Shock Tube Loading," AIAA Paper No. 66-82, AIAA 3rd Aerospace Sciences Meeting, January 1966.

100 Budiansky, Bernard and Roth, Robert S., "Axisymmetric Dynamic Buckling of Clamped Shallow Spherical Shells," Collected Papers on Instability of Shell Structures - 1962, NASA TN D-1510, National Aeronautics and Space Administration, Washington, D. C., 1962, pp 597-604.

101 Morino, L., Leech, J. W., and Witmer, E. A., "An Improved Numerical Calculation Technique for Large Elastic-Plastic Transient Deformations of Thin Shells," Part I, "Background and Theoretical Formulation," Part 2, "Evaluation and Applications," J. Appl. Mech., 1971, pp 423-436. Also see: "PETROS 3: A Finite Difference Method and Program for the Calculation of Large Elastic-Plastic Dynamically-Induced Deformations of Multilayer Variable-Thickness Shells," Atluri, Satyanadham, Witmer, Emmett A., Leech, John W., and Morino, Luigi, BRL CR 60, ASRL TR 152-2, November 1971, Aeroelastic and Struc-

tures Research Laboratory, Massachusetts Institute of Technology, Cambridge, Massachusetts.

102 Underwood, P. G., "User's Guide to the SHORE Code," LMSC-D244589, November 1971, Lockheed Missiles & Space Company, Palo Alto, California.

103 Haisler, W. E., Stricklin, J. A., and Von Riesemann, W. A., "DYNAPLAS-A Finite Element Program for the Dynamic Elastic-Plastic Analysis of Stiffened Shells of Revolution," Texas A&M University, College Station, Texas, TEES-RPT 72-27, and Sandia Laboratories, Albuquerque, New Mexico, SLA 73-0127, December 1972.

104 Santiago, J. M., Wisniewski, H., and Huffington, N. J., Jr., "A User's Manual for the REPSIL Code," U.S. Army Ballistic Research Laboratories Report (in preparation).

105 Hubka, W. F., Keefe, R. E., and Eamon, J. C., "Final Report, Kaman Science Corporation IR&D Project 538-14," December 1972, Kaman Sciences Corporation, Colorado Springs, Colorado.

106 Zudans, Z., Reddi, M.M., Fishman, H.M., and Tsai, H. C., "Elastic-Plastic Creep Analysis of High Temperature Nuclear Reactor Components," Proceedings of 2nd International Conference on Structural Mechanics and Reactor Technology, Berlin, 1973.

107 Ball, R. E., "A Computer Program for the Geometrically Nonlinear Static and Dynamic Analaysis of Arbitrarily Loaded Shells of Revolution, Theory and User's Manual," NASA CR-1987, April 1972, National Aeronautics and Space Administration, Washington, D. C.

108 Klein, S., "SABOR/DRASTIC 6 - User's Manual," The SABOR Manual, Vol. III, Poulter Laboratory Technical Report (in preparation), 1974, Stanford Research Institute, Menlo Park, California.

109 Emery, A. F. and Cupps, F. J., "RIBSTEAK: A Computer Program for Calculating the Dynamic Motion of Cylindrical and Conical Shells," SCL-DR-720019, April 1973, Sandia Laboratories, Livermore, California.

110 Underwood, P. G., "TROCS (Transient Response of Coupled Shells) User's Manual," LMSC-D266238, 1973, Lockheed Missiles & Space Company, Inc., Sunnyvale, California. Also see Underwood, P. G., Bonner, C. J., Lindow, D. W., and Rankin, C. R., "Transient Response of Soft-Bonded Multi-Layered Shells," presented at AIAA/ASME/SAE 15th Structures, Structural Dynamics and Materials Conference, Las Vegas, Nevada, April 17-19, 1974.

111 Lindberg, H. E. and Sliter , G. E., "Response of Reentry-Vehicle-Type Shells to Transient Surface Pressures," AFWL-TR-68-56, June 1969, Stanford Research Institute, Menlo Park, California.

112 Underwood, Philip, "Transient Response of Inelastic Shells of Revolution," Computer & Structures, Vol. 2, 1972, pp 975-989.

113 Ball, R. E., Hubka, W. F., Huffington, N. J., Jr., Underwood, P., and Von Riesemann, W. A., "A Comparison of Computer Results for the Dynamic Response of the LMSC Truncated Cone," Computers & Structures, 1974.

114 Greenspon, Joshua E., editor, Fluid-Solid Interaction, The American Society of Mechanical Engineers, New York, 1967.

115 Junger, Miguel C., "Normal Modes of Submerged Plates and Shells," 1967, pp 79-119 of [114].

116 Geers, Thomas L., "Residual Potential and Approximate Methods for 3-Dimensional Fluid-Structure Interaction Problems," The Journal

of the Acoustical Society of America, Vol. 49, No. 5 (Part 2), 1971, pp 1505-1510.

117 Dowell, E. H., "Nonlinear Analysis of the Flutter of Plates and Shells," 1967, pp 160-187 of [114].

118 Flugge, Wilhelm, Stresses in Shells, Springer-Verlag, Berlin, 1960.

119 Hayek, S. and Pallett, D. S., "Theoretical and Experimental Studies of the Vibration of Fluid Loaded Cylindrical Shells," Symposium on Application of Experimental and Theoretical Structural Dynamics, Southampton University, England, April 1972.

120 Harari, A and Baron, M. L., "Analysis for the Dynamic Response of Stiffened Shells," ASME Paper No. 73-APM-FFF, to appear J. Appl. Mech.

121 Fishlowitz, Ely, G., "Analysis of Stiffened Cylinders Under Concentrated Loads Using the STAGS-A Computer Program," NSRDC Report Report, January 1973, Naval Ship Research and Development Center, Bethesda, Maryland.

BRIDGE AND GIRDER SYSTEMS

W. McKeel and J. Korf
Virginia Highway Research Council
Charlottesville, Virginia

D. Vannoy and W. Pilkey
University of Virginia
Charlottesville, Virginia

ABSTRACT

This paper presents a review of the software systems available for bridge design. Also, the individual girder design programs are summarized. An actual bridge which was designed by conventional means is used to compare the bridge design systems.

INTRODUCTION

The primary objective of a computer bridge design system is to perform the complete structural design of a standard highway crossing in one operation with one set of data input. In order to establish criteria for reviewing the available bridge systems, we will consider what might constitute a good bridge system.

The design engineer should have the option of inputing information into the system on any of the following categories to suit his particular problem:

1. Designation of AASHO or other loading
2. Limiting design stresses of materials
3. Superstructure framing
4. Substructure component dimensions
5. Foundation conditions

A complete bridge system should then design the bridge, including determination of

1. General plan of bridge
2. Framing plan
3. Transverse section
4. Beam and girder schedules
5. Abutment details
6. Pier details
7. Slab bar plan for straight bridges

The subprograms of a complete system would encompass the major phases of highway bridge design as illustrated in Fig. 1.

In addition to the many various design programs, a bridge system should be controlled by a master control program. This program would allow at the discretion of the designer the execution of the various programs within the system individually or the preparation of a complete bridge design. The control program should allow information to be passed from a given program to any other program in the system.

Since only a limited number of bridge systems are available, we have chosen also to summarize the individual programs available for the design of girders. Basically, girder design programs can perform the first five functions listed in Fig. 1, excluding the geometry programs.

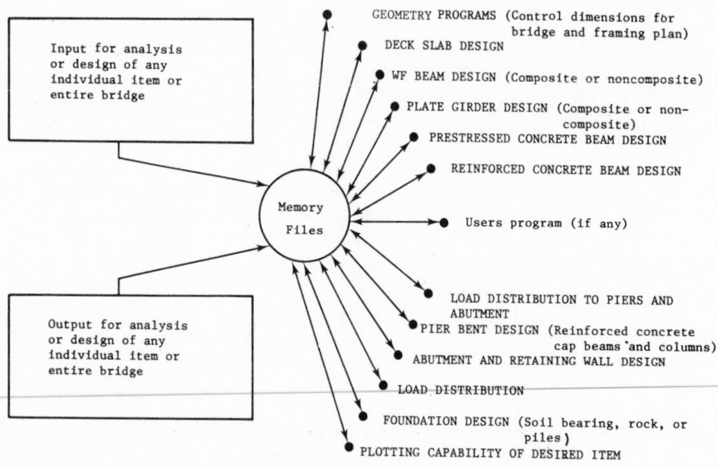

Fig. 1 Bridge design subsystems

BRIDGE SYSTEMS

From the literature it would appear that four bridge systems are available.

1. BEST, Bridge Engineering Subsystem of TIES
2. STRC1 and STRC2, Omnidata Services, Inc.
3. BRIDGE, Subsystem of the ICES (Integrated Civil Engineering System)
4. BRASS, the Wyoming Bridge Rating and Analysis Structural System

We will study each of these systems in some detail to see how they measure up against the criteria set forth in the introduction.

The development of these systems began in the mid 1960s. At that time most bridge design by computer imitated the way a bridge designer used to do it manually. This same method is still used today; the bridge design is put together component by component, starting with geometry and terminating with quantities and costs. Each component is designed separately on the computer after the engineer supplies the necessary input data. The interaction between the component designs is performed by hand; that is, the designer transfers from one computer program to another data coupling the design of two components.

PRELIMINARY QUESTIONNAIRE

As an initial step in our review of bridge systems, questionnaires were sent to all of this country's state highway departments asking which, if any, of these systems they used. They were also asked if they had developed or planned to develop their own systems. Responses from 49 departments were received. New York and Vermont responded that they were using all or part of the BEST system. No states said they used the OMNIDATA programs. California, Hawaii, and West Virginia responded that they are using parts of BRIDGE. Arkansas, Kentucky, North Carolina, North Dakota, West Virginia, and Wyoming stated that they are using or investigating the use of BRASS.

As expected, the responses to the questionnaires also indicated that almost all states are involved in the development of individual programs. This means, in effect, that if a state is not satisfied with a program developed by another state, e.g., because of differences in practices, they either modify someone else's program or write their own program. As a result, there are almost as many composite rolled beam design programs, for example, as there are states. Many of the programs are good, but the waste in development cost is incredible considering how much in common the different states have in terms of standards and practices. The same situation is true of programs developed by consultant firms and other private companies.

Oklahoma and Michigan indicated in their replies that they were developing their own integrated systems of analysis or design. The Oklahoma system (BRDESIGN), "will allow the engineer to combine geometric and structural programs with connecting and plotting programs, enabling him to produce a set of bridge construction plans from preliminary sketches to final drafting documents--in one continuous process. The first phase of the system covers simple prestressed concrete I-Beam bridges."

Michigan's system, Bridge Design, applies to the superstructure of the bridge. It will design cantilevered simple spans. The subprograms of this system will perform beam layout, rolled beam or girder design, determination of all elevations required to form and pour decks, bearing design, and pier design. All of these functions can be executed with a single page of input. An abutment design and plotting routines for superstructure and structural steel details are being added now. In order to minimize input, the Bridge Design program is highly specialized to meet Michigan's standards. The output they provided for an example design indicated that the system lives up to the claims made by the developers.

Many states indicated that they used ICES STRUDL for various analysis and design tasks. These were Alabama, Alaska, Arkansas, Oregon, Tennessee, Washington, and West Virginia. STRUDL is basically an analysis program, and it may be applied to a wide range of structural problems, including two-dimensional trusses, frames, plates, and grids, as well as three-dimensional trusses and frames. STRUDL also allows the proportioning and checking of reinforced concrete beams and columns of various shapes and of flat plate floor slabs. The user may control the design of elements by specifying some of the cross section design parameters or by placing upper and lower bounds on these parameters. The proportioning procedure takes into account biaxial bending for columns and flexure, shear, bond, and deflection criteria for beams and slabs and uses the ultimate strength theory. The output consists of cross section dimensions and required reinforcement at the critical design sections. The adequacy of a given member may also be checked against any or all of the criteria.

Georgia, Iowa, Montana, Ohio, Oklahoma, and South Dakota indicated they used the Control Data Corporation's Bridge Analysis and Rating System (BARS). With this program, bridge ratings are produced at required stress levels for the normal AASHO lane and truck live loads, state legal loads, and special permit loads. BARS is available to users through Control Data's CYBERNET Services network and on a license basis.

Another system used is the Portland Cement Package Program, PCSPAN. PCSPAN performs the analysis and design of simple span, precast, prestressed highway or railway bridges. The designer selects and inputs the main structural parameters, and the computer makes the routine calculations. The program will accommodate the composite and noncomposite sections included in Fig. 2 and will compute and print out the following: section properties, dead load and live load reactions, shears and moments, stresses for various loading conditions, ultimate moments required and provided, spacing of shear reinforcement, horizontal shear stress between the composite slab and precast member, midspan elastic deflections for various loading conditions, and the number and center of gravity of prestressing strands required. The program can be operated in any one of the following modes: (1) analysis and design of standard sections with a composite deck slab (sections 1 through 5 inclusive in Fig. 2), (2) analysis and design of noncomposite standard sections (sections 6, 7, and 8 in Fig. 2), (3) analysis and design of noncomposite single- and double-celled box beams (sections 9 and 10 in Fig. 2), and (4) analysis and design of all sections illustrated in Fig. 2 when the number and location of prestressing strands is provided as input data.

A byproduct of the questionnaire was information on the computer hardware used by the states for most of their bridge design work. Three states use UNIVAC equipment, three states use a Burrough machine, one state uses a CDC computer, and the remainder use IBM equipment, mostly IBM 370 configurations.

EVALUATION OF THE MAJOR SYSTEMS

Of the four bridge systems BEST, STRC, BRIDGE, and BRASS, only the first two approach being truly comprehensive design systems. All four will be discussed in this section. A comparison of BEST and STRC1 is tabulated in Table 1.

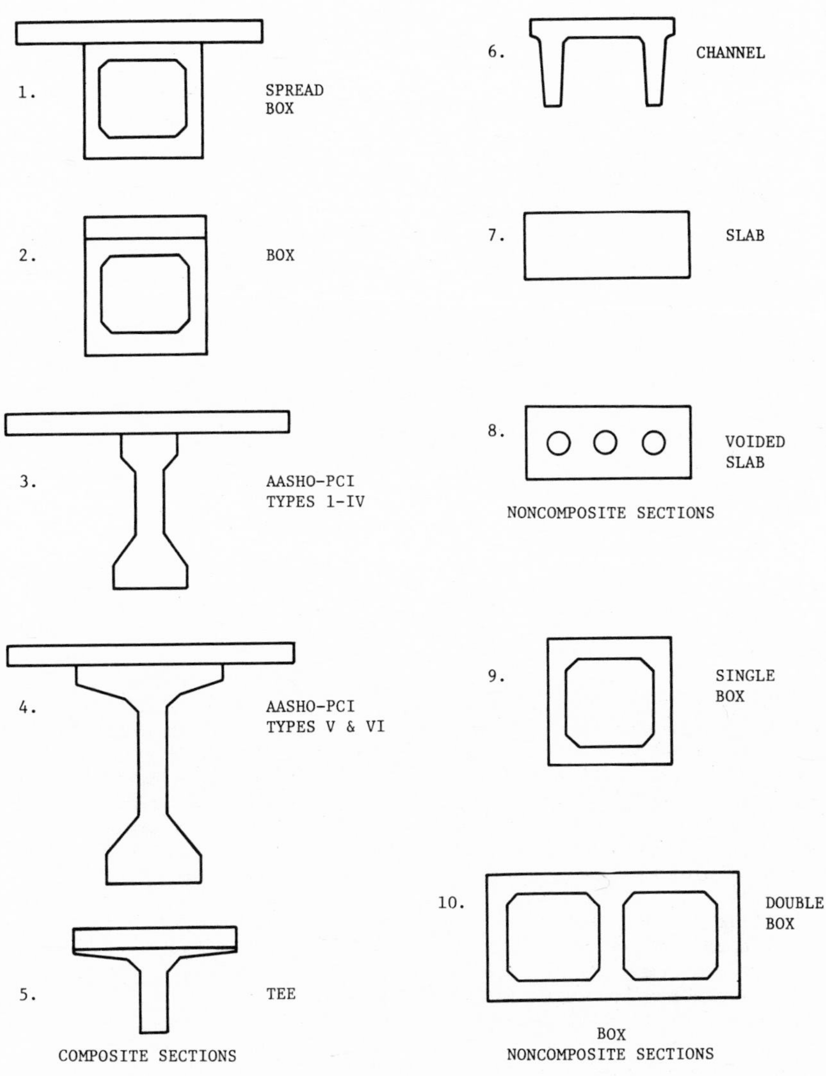

Fig. 2 Sections accommodated by PCSPAN

Table 1 Comparison of Capabilities of BEST and STRC1

Subprogram of system	BEST	STRC1
Bridge geometry	X	No
Deck slab design	X	X
WF beam design (composite or noncomposite)	X	X
Plate girder design (composite or noncomposite)	X	X
Prestressed conc. beam design	No	Pre- or posttensioned
Reinforced conc. beam design	X	No
Load distribution to piers and abutments	X	X
Bridge bearing design	X	No
Pier bent design	X	X
Abutment and retaining wall design	X	X
Load distribution to foundations	X	X
Foundation design	X	X
Bridge quantities	X	X
Plotting		
General plan of bridges	X	INGO[1]
Framing plan	X	INGO
Transver section	X	INGO
Beam and girder schedules	X	INGO
Abutment details	X	INGO
Pier details	X	INGO
Slab bar plan for str. bridges	X	INGO

[1]INGO, Integrated Geometry System, is a separate program for solving problems in geometrics and used for plotting.

Best. BEST is an acronym for Bridge Engineering Subsystem of TIES, and TIES is an acronym for Total Integrated Engineering System. The TIES concept was proposed by the Federal Highway Administration as part of its National Program for Research and Development in Highway Transportation. The primary objective of the New York Department of Transportation in developing BEST was to put together an integrated bridge design subsystem that included computer programs to solve computational problems and generate necessary parameters for plotting [1].

The limitations of BEST are:

1. The bridge can span up to two roadways.
2. The maximum number of spans is 4.
3. The maximum number of beams in cross section is 10.
4. The bridge must be of constant width with tangent or circular curve alignment.
5. Only simple spans can be designed with composite rolled beams or welded plate girders.

Subprograms of BEST

Control geometry. This subprogram produces control dimensions for a bridge. This program is designed in such a way that an infinite

BEST is developed for a Burroughs 5500 and is closely tied to this machine. This is considered to be a major shortcoming of the system. It appears that BEST cannot be readily converted to equipment that is in more common use by bridge engineers.

A comparison of the application of BEST and STRC1 to a common bridge is given in the next section. This application provides an impressive demonstration of the case of using BEST. Particularly important is the minimal input required for a bridge design and the capability provided the designer of interacting with the design process to almost any extent desired.

Another example of bridge design, including drawings, is given in [2] . This design, which deals with a bridge that was built to carry a local road over SH5274, near Syracuse, New York, will give the user information on the complete output of BEST. The structure was a two-span bridge crossing two proposed roadways. The abutments were placed 30 feet from the edge of the pavement, the minimum clearance under the highway safety program. The structure had solid abutments on spread footings and one two-column pier on a combined spread footing. The bridge carried a straight roadway and was placed at the crest of a vertical curve. The analysis and design took 35 minutes of B5500 time. Plots were processed off line on a Calcomp plotter using magnetic tape. Plotter time for the drawings shown in [2] was approximately three hours. The cost of this complete design was about $120, based on estimated computer and plotter usage. Comparison of the BEST system design with the design as built indicated one major difference. The design using the BEST system would have resulted in spans approximately 11 feet longer than those used in the actual structure, with a resultant reduction in abutment height. A net cost reduction of $25,000 would have resulted by using the BEST design.

BEST is now used for the design of approximately 20 percent of the bridges being built in New York State. The New York Department of Transportation intends to supplement the present BEST system to a point where it will be applicable to about 80 percent of the bridges being designed.

<u>STRC</u>. Omnidata Services, Inc. offers the STRC1 and STRC2 systems for the design of simple span and continuous span bridges, respectively.

STRC1 is an integrated program consisting of many subprograms which can be used in succession to analyze and design any deck type simple span bridges for the superstructures and the substructures. The superstructures may consist of reinforced concrete slabs with steel beams of rolled wide flange or built-up sections or prestressed reinforced concrete beams. The beams may have composite or noncomposite action. The substructures may include reinforced concrete piers and abutments on spread footings or piles.

STRC1 observes provisions in AASHO specifications in loading and stress computations as well as in pertinent design details. The subprograms may be used together to design a complete simple span bridge or individually to design parts of a bridge. When used to design a complete bridge, intermediate answers and solutions relevant to each phase of design are stored in data files and carried automatically from one subprogram to another. However, during the design process automatic cycling from subprogram to subprogram is not done. This system can be executed on the IBM 1130 with 32k core memory and 3 disk drives.

Subprograms of STRC1

The developer provides the following definitions of the subprograms in STRC1.

BSLAB. This designs reinforced concrete bridge slabs with main reinforcement perpendicular to traffic within the span limits given by the AASHO specifications.

WBEAM. This designs composite or noncomposite WF beam stringers in a simple span bridge by choosing automatically the minimum steel weight of a WF beam according to AASHO specifications.

SGIRD. This designs composite and noncomposite steel girder stringers in a simple span bridge by automatically choosing the minimum steel weight of the steel girder according to AASHO specifications.

PBDES. This analyzes and designs prestressed concrete beams, both pretensioned and posttensioned, in a simple span bridge according to AASHO specifications.

DKLDS. This solves all bridge deck loads from the superstructure to the substructure of pier bents or abutments according to the requirements imposed by the input data as well as the AASHO specifications.

REWAD. This analyzes and designs bridge abutments and retaining walls resting on soil or rock according to AASHO specifications.

COLMN. This designs reinforced concrete column of rectangular or circular sections subjected to combined axial load and two-directional bending moments, according to the AASHO specifications.

PRFDN. This analyzes and designs pier footings resting on soil or rock. The footings may be the individual type carrying a single column or the continuous type carrying multiple columns (maximum 6) from the pier bent.

PBENT. This analyzes and designs cap steel for a pier bent for all different combinations of loads according to AASHO specifications.

The Omnidata-provided documentation illustrates the use of STRC1 to design a 3-span bridge, with input of 75 lines, in less than one minute of CPU time on an IBM 360/67 or a Univac 1108. It takes 15 minutes to do the same job on a slower, smaller IBM 1130 with 32k core and 3 disks.

number of layouts are possible. The layout finally selected will be based on the input of control variables such as span-to-depth ratio, lateral clearances, minimum end span length, skew to be equalized, etc., making it possible to select alternate designs.

Framing plan. This program seeks the most desirable framing plan for any given conditions, i.e., a plan which results in economy of fabrication by causing as much duplication or similarity of details as possible.

Reinforced concrete slab design. This designs reinforced concrete bridge slabs with main reinforcement parallel to the supports or normal to the stringers satisfying AASHO specifications. A designer's option to establish whether the design should have equal tension and compression steel, compression steel equal to half the torsion steel, or a balanced design is present.

Beam and girder design. This designs composite and noncomposite beams or girders in a simple span bridge, with or without cover plates, that will satisfy stress and deflection requirements. It processes many designs to select the most economical section within the specified depth range according to AASHO specifications.

Bridge bearing design. This designs fixed and expansion type bearings for bridge stringers based on the stringer's length and reaction at the bearing according to AASHO specifications. The program also acts as a collector and provides the necessary input information for the substructure design programs.

Abutment and retaining wall design. This analyzes and designs stub or pedestal type abutments, solid or high type of abutments or retaining walls on spread footings or on piles in accordance with AASHO specifications. It processes a series of designs to select the most economical section based on concrete volume and number of piles, if applicable.

Pier design. This analyzes and designs each pier component, including rectangular and circular concrete columns; combined footings on soil, rock, or piles; individual footings on soil, rock, or piles; and the pier cap beam reinforcing steel in accordance with AASHO specifications. Subprograms detail the reinforcing steel in the beams, columns, and footings.

Bridge quantities. This accumulates and combines quantities determined in all other design and plot programs.

Another feature of BEST are the subprograms for plotting as shown in Table 1.

In addition to the various design and plot programs described BEST also will allow the execution of the various programs within the system in a preestablished order. Information can be passed from a given program to any other program in the system. By cycling from subprogram to subprogram, BEST attempts to find the minimum cost design.

The system STRC2 designs deck-type continuous span bridges of up to ten spans with any random distribution of moment of inertia along the length of the spans. The end spans may be cantilever spans; the bridge may be symmetrical or unsymmetrical in arrangement; and the deck section construction may be of any of the following types: 1) noncomposite, 2) prestress-composite, 3) steel-composite. STRC2 does not consider curved girders. It was written for the IBM 1130 with 32k core memory and 3 disk drives.

BRIDGE. The BRIDGE system is part of the ICES system developed at MIT. Because of the publicity given to the ICES system, the impression is frequently given that BRIDGE is a complete, versatile, and available system. This is not the case at the moment. However, if substantial development of BRIDGE is made, it could be converted into the highly useful system it was intended to be.

Presently, BRIDGE is limited to:

1. A preliminary bridge planning phase called the Geometry Program Block for determining the intersection geometry and the possible alternative configurations of spans, piers, and abutments of a nonsuperelevated, nonhorizontally curved bridge crossing over a highway, with span arrangements conforming to the standards of the Massachusetts Department of Public Works.

2. The design, or analysis, of a noncomposite concrete bridge deck based on design standards as per AASHO or the Massachusetts DPW specifications called the Concrete Deck Supported by Steel Stringers Program Block.

3. The Preliminary Girder Design Program Block, which will determine maximum moments, shears, reaction design values, and required section moduli for prismatic and nonprismatic simple or continuous girders. An iterative process, based on relative stiffness, which in turn is based on the required section moduli, is also available.

The present BRIDGE subprogram does not select girder dimensions (design) or analyze a given girder (analysis of a design), in the composite, noncomposite, or intermediate composite state. At the moment, BRIDGE has limited application since it is primarily an analysis system with few design capabilities.

BRASS. The Wyoming system has just been completed. It is referred to as Bridge Rating and Analysis Structural System (BRASS). It has been developed so that a user may analyze, review, or load rate structures. BRASS cannot be used for the complete design of a bridge as done by BEST or STRC 1. In fact, BRASS is esentially an analysis system. Similar to BEST and STRC1, it does possess a common data base for the subprograms. That is, data from one subprogram can be passed automatically to another.

The system will handle steel girders, concrete girders, concrete slabs, timber beams, and composite concrete-steel girders. When one of these components, e.g., a wide flange girder or built-up steel girder, is to be designed, the designer must first enter dimensions of the flange, the web, and the fillets as required. Thus, the "design" program will analyze a prescribed configuration. It will not automatically choose the minimum steel weight of a wide flange beam or plate girder. The user would have to first make preliminary layouts of his structural elements to be designed. For the complete design of a bridge from deck to footings, the system BRASS is quite inferior as compared to BEST and STRC1. As its name implies, BRASS is basically a bridge rating and analysis system.

The components of the system listed by the developers are bridge design, structural inventory, deck design and review, structural analysis, structural loading, and girder section design and review. The system consists of a set of 45 computer programs. BRASS will be used by state highway departments mainly to determine the safe load-carrying capacity for a highway bridge and the structural rating for the bridge.

EXAMPLE OF BRIDGE DESIGN

The structure redesigned in this paper was a fairly typical highway bridge carrying an access ramp over a street in the city of Richmond, Virginia. Plans for the bridge were completed in August 1965, and it was subsequently constructed. As shown in Fig. 3, the structure was a heavily skewed, three-span bridge, lying on a tangent and a vertical curve. The span lengths are nominally 60, 105, and 82 feet, and the supporting members are composite steel beams, including a 55-inch welded plate girder for the longest span and 36-inch rolled beams with cover plates for the shorter spans. The superstructure is supported on four-column piers and abutments on piles. Further details are provided in Figs. 4-6, which show the framing plan and transverse section of the bridge, beam and girder details, and pier details. The structural details will be discussed more fully when they are contrasted with those developed by the design systems.

The original structure was designed by conventional methods, perhaps with the use of individual computer programs.

The redesign of the Richmond bridge in this paper was performed in New York using BEST, through the courtesy of the New York State Department of Transportation, and by STRC1, through the courtest of Omnidata Services, Inc. A review of the system designs and a comparison with the original design should serve to indicate the capabilities of BEST and STRC1, as well as the degree of flexibility built into the systems.

Input Data for BEST

The BEST system requires minimal input data, essentially road profile and curvature data and design ground rules, such as the vertical clearance and type of steel to be used. Specifically, the input information consisted of those items shown in Table 2 and Fig. 7.

This information totaled only 15 lines for input into the system, as shown in Table 3. The first program to operate with the input data is the Bridge Control Geometry program, which establishes the basic layout including the stations of the various substructure elements, the

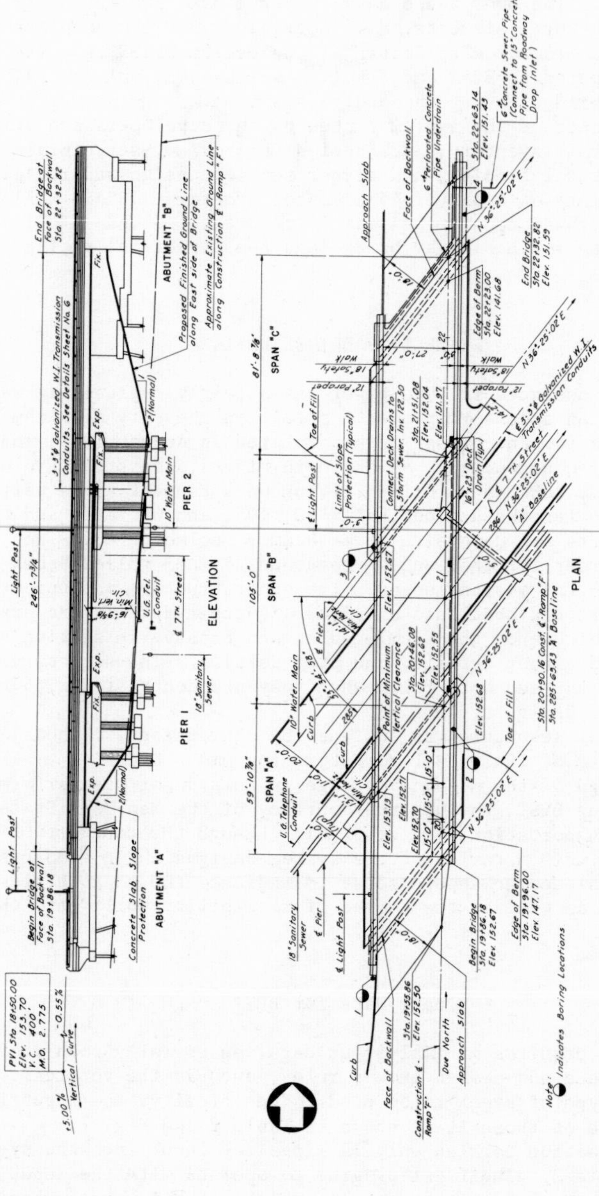

Fig. 3 Plan view of Richmond bridge

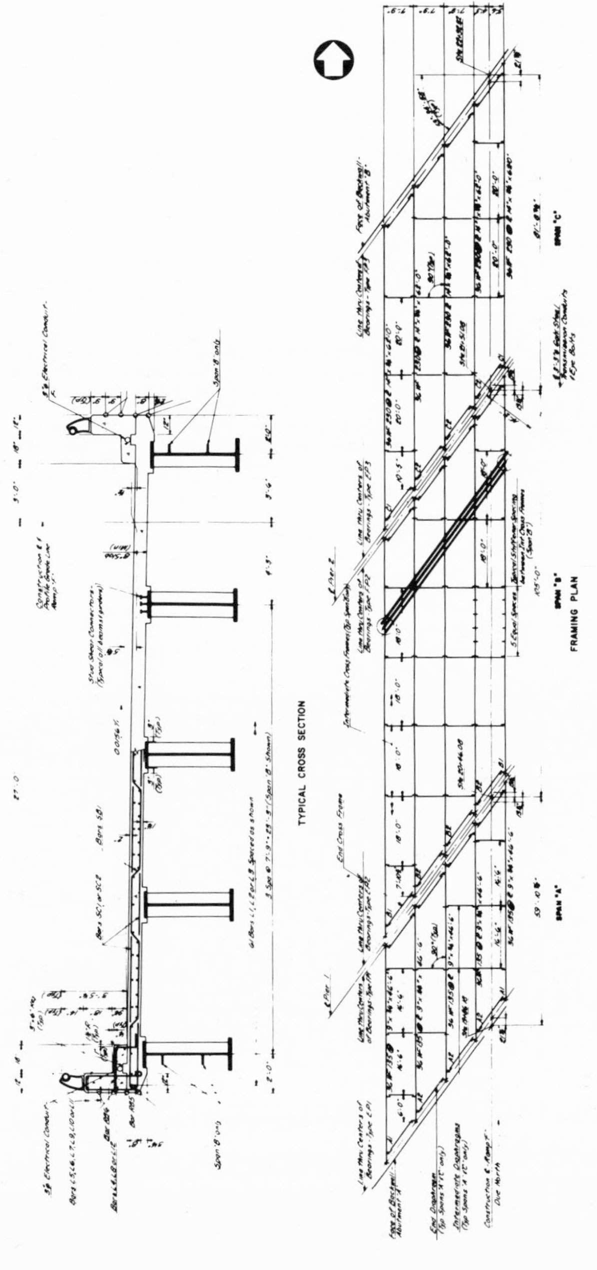

Fig. 4 Framing plan and transverse section of Richmond bridge

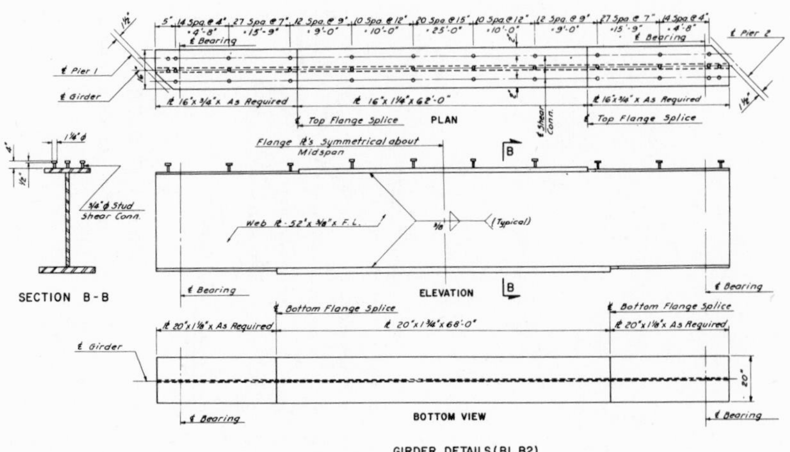

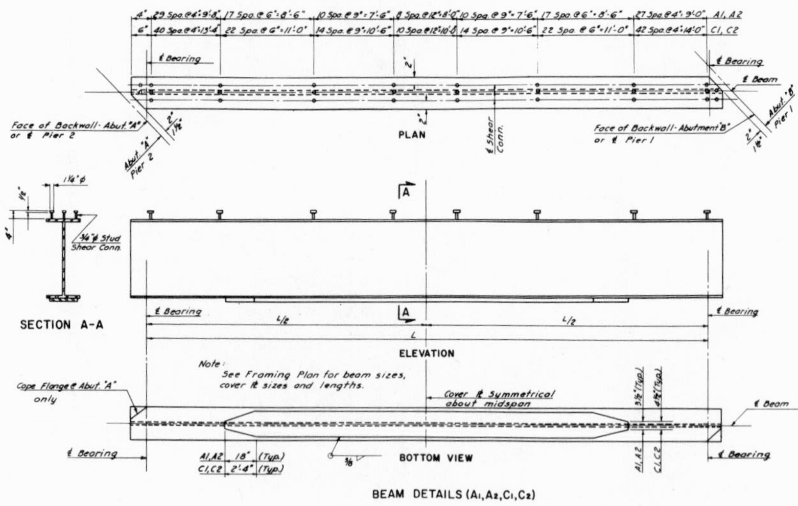

Fig. 5 Beam and girder details of Richmond bridge

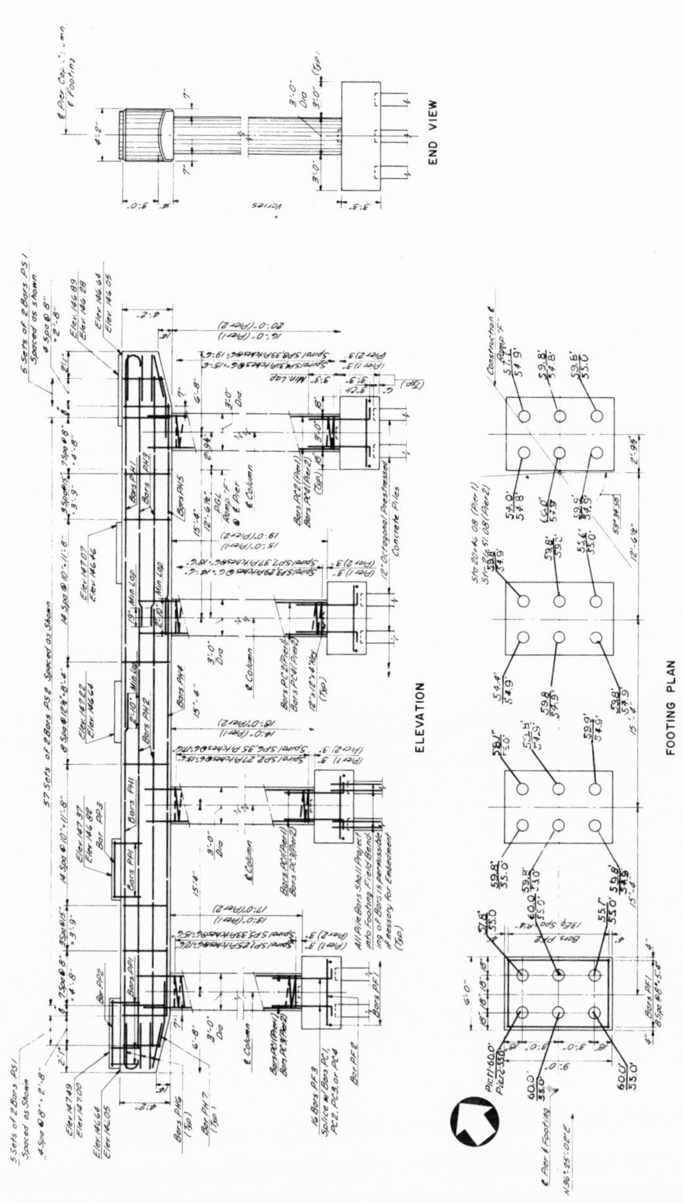

Fig. 6 Pier details of Richmond bridge

Table 2 BEST System, Input Data for Ramp F over Seventh Street, City of Richmond

	Data	Description
Upper roadway data		
	00	Radius of upper roadway
	2099.16	Intersection station with lower roadway
	0°	Azimuth at intersection station
	1850.0	Station P.V.I.
	153.7	Elevation at P.V.I.
	400.0	Length of vertical curve
	5.0	Grade #1 Dfn (lower station values)
	-.55	Grade #2 Dfn (higher station values)
	-29.5	Horizontal offset from construction center-line to left fascia
	5.5	Horizontal offset from construction center-line to right fascia
	-30.67	Horizontal offset from construction center-line to left wall
	6.67	Horizontal offset from construction center-line to right wall
	-27.5	Range #3 - See Fig. 7
	-1.56	Slope #3 - See Fig. 7
	3.5	Range #4 - See Fig. 7
	-1.56	Slope #4 - See Fig. 7
Lower roadway data		
	00	Radius of lower roadway
	28563.43	Intersection station with upper roadway
	36°25'2"	Azimuth at intersection station
	28590.	Station P.V.I.
	124.42	Elevation at P.V.I.
	0.0	Length of vertical curve
	-6.57	Grade #1 (approaching)
	-6.35	Grade #2 (leaving)
Lower road offsets		
	-20.	Left pavement edge - Horizontal
	-.25	Left pavement edge - Vertical
	11.	Right pavement edge - Horizontal
	-.25	Right pavement edge - Vertical
	-4.5	Left shoulder - Horizontal
	0.	Left shoulder - Vertical
	11.	Right shoulder - Horizontal
	-.25	Right shoulder - Vertical
	-38.5	Left ditch - Horizontal
	.75	Left ditch - Vertical
	28.25	Right ditch - Horizontal
	.75	Right ditch - Vertical
Bridge geometry and design input data		
	1.083	Distance from center line bearings to bridge end
	2.5	Sidewalk width (ft.) both left and right
	-.016	Left shoulder slope in ft/ft
	-.014	Right shoulder slope in ft/ft
	2.0	Overhand min. and max. in feet
	.25	Parapet load in (K/ft) left and right
	2.0	Number of lanes on bridge
	16.5	Minimum vertical clearance in feet
	0.0	Distance from tangent grade line to station line
	1.0	Amount slab cut in feet
	0.0	Maximum skew for normalizing ends of slab
	14.08	Minimum lateral clearance (left) lower road
	13.08	Minimum lateral clearance (right) lower road
	20.0	Skew to be equalized
	25.0	Ratio of span to steel depth
		Maximum Allowable Bottom of Footing Elevation
	143.17	Beginning abutment
	123.22	Pier 1
	118.63	Pier 2
	137.12	Ending abutment
	A36	Steel type
	9.0	Height of curb in inches (left and right)
	4.17	Cap beam width in feet
	72.0	Foundation loads allowed in K/ft^2 for both abutments and piers

Note: Concrete footing support to be used throughout

Data	Description
1.0	Distance from fascia to face of railing in feet (left and right)
127.	Elevation at top of ground - pier 1
123.	Elevation at top of ground - pier 2
45	Design speed of traffic in miles per hour
1.	Fraction of full wind

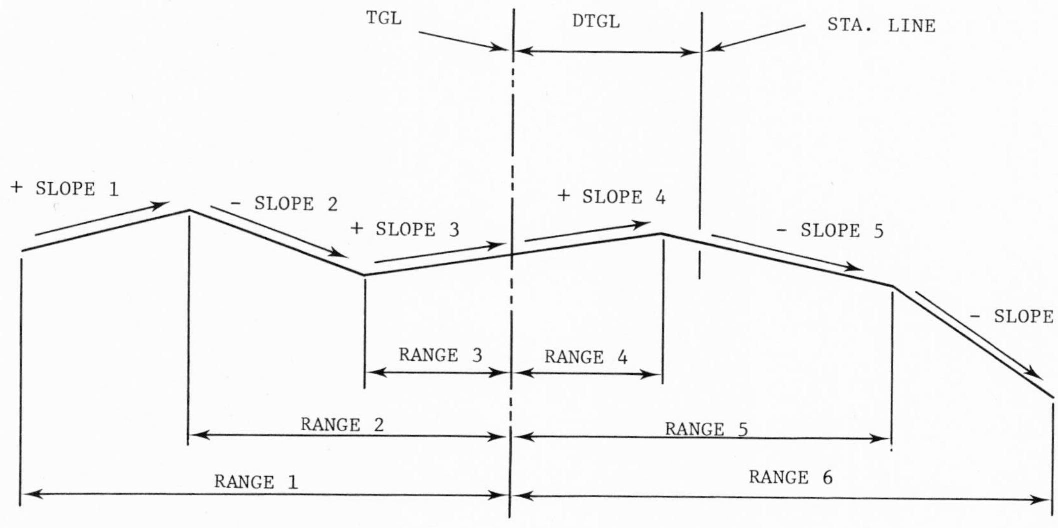

NOTE: RANGES 1, 2, AND 3 WILL ALWAYS BE NEGATIVE.
IF THE TGL IS TO THE RIGHT OF THE G.L. BRIDGE THE DTGL is (-).

Fig. 7 Cross section geometry input required by BEST

point of minimum vertical clearance, and the depth available for the steel stringers, given the minimum vertical clearance. The locations of the abutments are established from the vertical clearance and profile information, and the piers are situated on the basis of horizontal clearance and maximum span length.

Table 3 Final Input Data for BEST

```
RAMP F OVER 7TH STREET - CITY OF RICHMOND
           T     2090.16      0 0 0.0
           T    28563.43     3625 2.
  1850.          153.7        400.            5.          -.55
 28590.00        124.42         0.          -6.57         -6.35
 -29.5         5.5      -30.67      6.67
 -27.5        -1.56       3.5      -1.56
  -20.       -.25     11.       -.25     -4.5   0.        -20.       -.25
  11.          -.25     -38.25       .75      28.25        .75
   1.083       2.5        2.5   -.016-.0142    2.          .25        .25
   2.          0.         2.        16.5       0.          1.         0.
  14.08       13.08                            0.         20.        25.
 143.17  123.22           118.63  137.12A36               9  9 4.17
 72.     72.     72.     72.     72.  CONCCONCCONCCONCCONCCONC1.   1.
0.  0.  0.   127.               123.  45.  1.
```

Result of BEST Design

Figure 8 shows the basic layout of the bridge developed by the BEST system. A comparison with Fig. 3 shows the system-designed bridge to have small differences (5 ft. maximum) in the stations at its ends and practically no difference in the pier locations. Little significant difference is apparent between the two designs. The depth available for the steel beams was computed to be 5.83 feet. This figure is of importance, because the New York design policy is to maximize spacing between beams, using the available maximum depth, to require the least number of lines of beams.

The framing plan, Fig. 9, developed by BEST uses only four lines of stringers, as opposed to the five lines used on the original design (Fig. 4). As shown by a comparison of the two transverse sections, Figs. 4 and 10, the beam spacing has been increased from 7' 9" to 10' 4". The framing plan program, like others in the system, allows the engineer the flexibility to override any of the chosen variables to obtain a desired design.

Increased beam spacing requires deeper members, normally thought to result in a longer bridge, but the integrated system, which considers all variables together instead of one set at a time, can apparently optimize the layout. Thus, although the system-designed girders are 69 1/8" in maximum depth as opposed to 55" in the original design

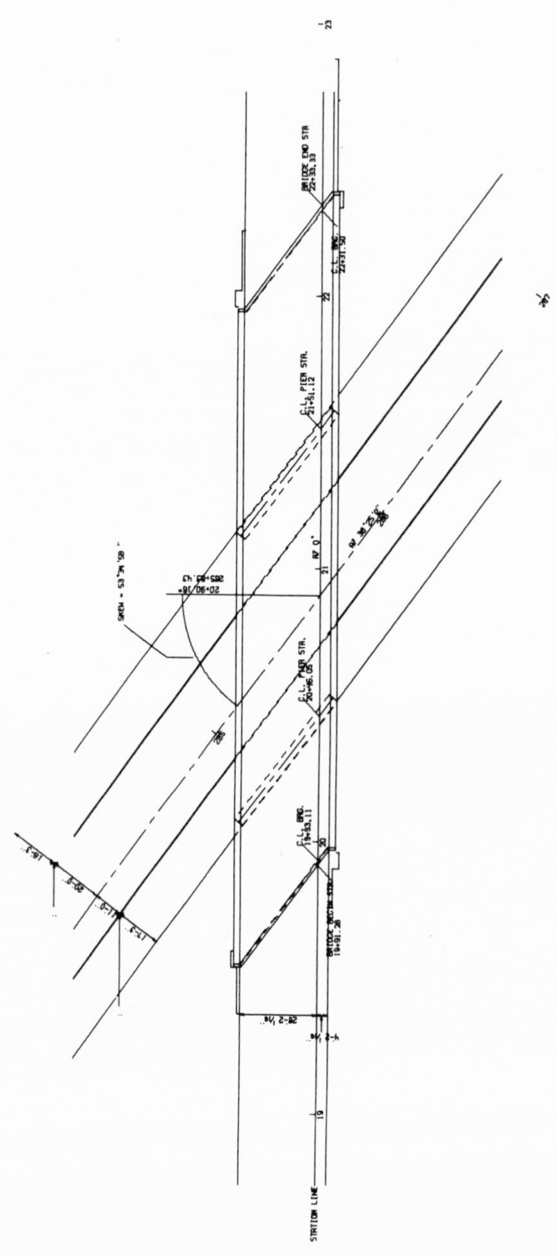

Fig. 8 Plan view of bridge plotted by BEST system

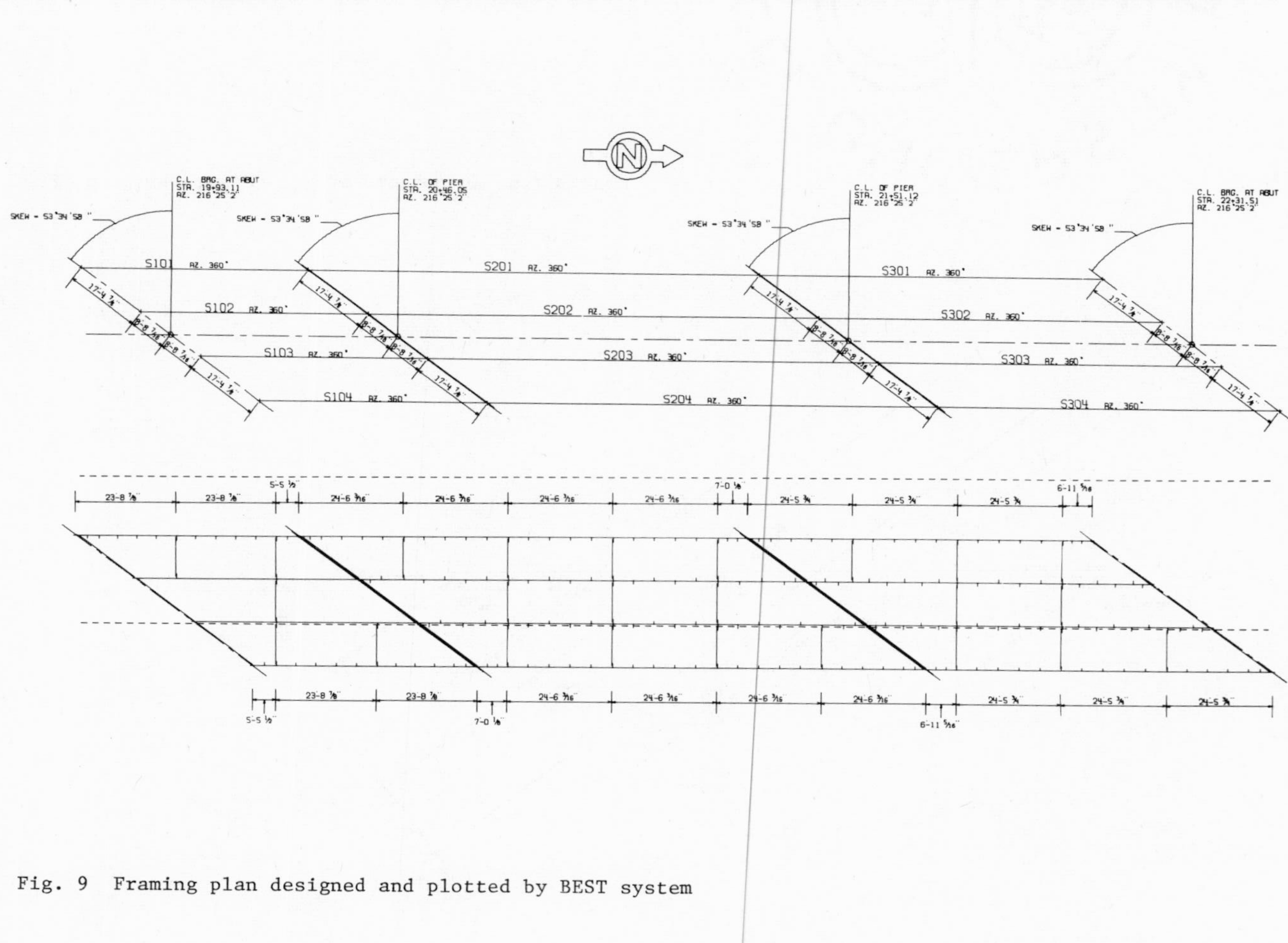

Fig. 9 Framing plan designed and plotted by BEST system

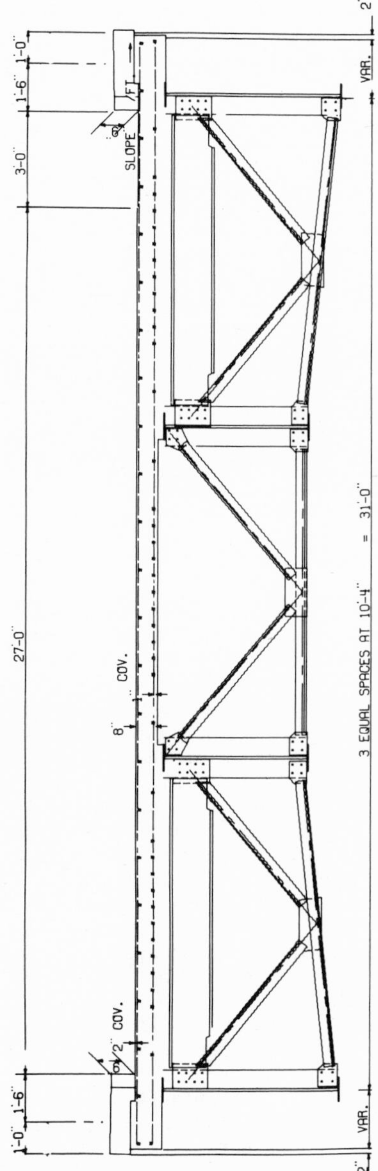

Fig. 10 Transverse section designed and plotted by BEST system

(Figs. 5 and 11), the overall bridge length is about 3 feet less for the system design.

The beneficial effect of the use of a larger beam spacing, with fewer lines of deeper beams, is most important. Steel quantities, based on the weight of the beams only and neglecting the cross bracing, show the total weight of the system's girders to be approximately 175 kips, while the original girders, without cross bracing, weigh nearly 270 kips. This difference would probably offset the greater amount of fabrication required by the BEST design, which uses plate girders throughout all of the spans.

The use of welded plate girders, even on the short approach span, results from the New York design policy, which requires that the fascia beams in all spans have a common depth. It follows that the use of a very deep fascia girder on a short span requires a fairly deep interior member for compatability of deflections. The interior members in the short spans of the BEST-designed bridge, thus, had to be deeper than the 36 inches available in rolled sections. Conversely, it is Virginia's practice to use rolled sections on step haunches if span length permits. The approach spans on the original bridge are supported on heavy wide flange rolled sections with cover plates. This practice could be incorporated into BEST, if desired. However, the original beams are heavier, in total weight, than the more widely spaced girders in the redesigned bridge.

Some slight disadvantages were noted in the BEST system. First, in the original output data, the bottom center plate in the lower flange of beams 101 and 104 extended for a length of only 2 feet. This did not seem to reflect good design practice. However, the BEST designers agreed that tests for these marginal conditions should be incorporated in the programs, and the condition was subsequently remedied. Also, the original plot of the girder details showed a disparity between the overall length of the beam and the sum of the lengths of the plates; the sum of the plate lengths being 6 inches short of the overall length in each case. The overall length shown on the plot was also in error in the case of the end spans. It is apparent that the program for the plot of girder details is not performing properly, and difficulty was also encountered in running the abutment plot program; however, the difficulties were apparent only in some of the plots, which were later corrected, not in the printed data. Unfortunately, the user's faith in the computer system is often shaken by minor errors such as these.

The BEST system also computes shear connector spacing. Both 4-inch and 6-inch studs were called for on different beams in the redesign, and in contrast to Virginia's practice, a uniform spacing was employed over the length of a beam. Stiffener plates were sized and located, and cross-bracing was located along the beam. The cross bracing itself is a standard configuration sized by l/r ratio for the given beam spacing; it is not designed by the integrated system.

A deck thickness of 8 inches was called for in both designs. Transverse reinforcement, number 5 bars in both designs, is spaced at 6-inch centers on the original bridge and at 4-inch centers on the BEST redesign. The depths of cover, top, and bottom are shown in Figs. 4 and 10. It is likely that the BEST system slab design, which spans a greater distance between beams, may be less conservative than the practice of the Virginia Department of Highways. The increased quantity of deck reinforcement required in the BEST design to span the larger beam spacing would slightly diminish the material savings realized in the girder design.

The final superstructure program is the bearing design segment, which designs the required fixed and expansion bearings and sets the bridge seat elevations at the piers and abutments. The program can design either low, sliding bearings or high, rocker-type bearings. The latter are generally provided for plate girders, as they were for the BEST redesign of the Ramp F bridge. Thus the bearings used in the redesign are larger, heavier, and more difficult to fabricate than those originally used. Both designs vary the bearing dimensions from span to span. The bearing design subprogram, because it accumulates from the preceding programs the information pertinent to the substructure design, is an important part of the BEST system. It is possible that the module could be replaced by another bearing design program of the user's choice, but few such programs apparently exist. Thus, a user might be forced to accept, at least initially, the New York bearings.

Significant differences were also found in the sizes of the pier elements, compared in Figs. 6 and 12. It can be seen that the BEST redesign had a smaller beam (3' 6" wide by 3' 9" deep versus 4' 2" by 4' 2" for the original bridge), larger columns (3' 6" by 3' 0" rectangular columns versus 3' 0" diameter round columns), and significantly larger footings with 19 more piles required. The BEST designed footings, all 3' 6" in depth, were 9' 0" by 15' 0" under each of the three columns of Pier 1 and 9' 0" by 12' 0" under the columns of Pier 2, while the footings under each of the four columns of the original piers were 9' 0" by 6' 0" by 3' 3". The actual analysis procedure was not completely shown in the printed data, and it is difficult to say why the larger footings occurred at the juncture of the relatively short Span A and Span B rather than between Spans B and C. It was apparent, however, that different philosophies were used in the original pier design and the BEST system design. Under BEST, the pier was regarded as a rigid frame, and the thrusts and moments of the frame were carried to the footings. Note that the long dimension of the BEST footings is transverse to the direction of traffic flow, while the long dimension of the original footings lies in the direction of traffic flow. Certainly, use of the BEST pier design subprogram resulted in a more expensive pier.

This somewhat disconcerting pier design was discussed with the developers of BEST. They stated that, under some conditions, the pier design program produced an overly conservative structure, and they have discontinued its use in their office. They are nearly finished developing new pier programs, which they feel give more consistent and reasonable results. They redesigned the footing for one of the piers. The design was a continuous footing 56' 10" by 8' 0" by 3' 0" on 16 piles (8 piles/pier less than original). These results seem much more reasonable. These new programs have not yet been integrated into the BEST system, but they will be in the near future.

Unfortunately, plots of the system designed abutments are not available. Both designs called for stub or shelf abutments on piles.

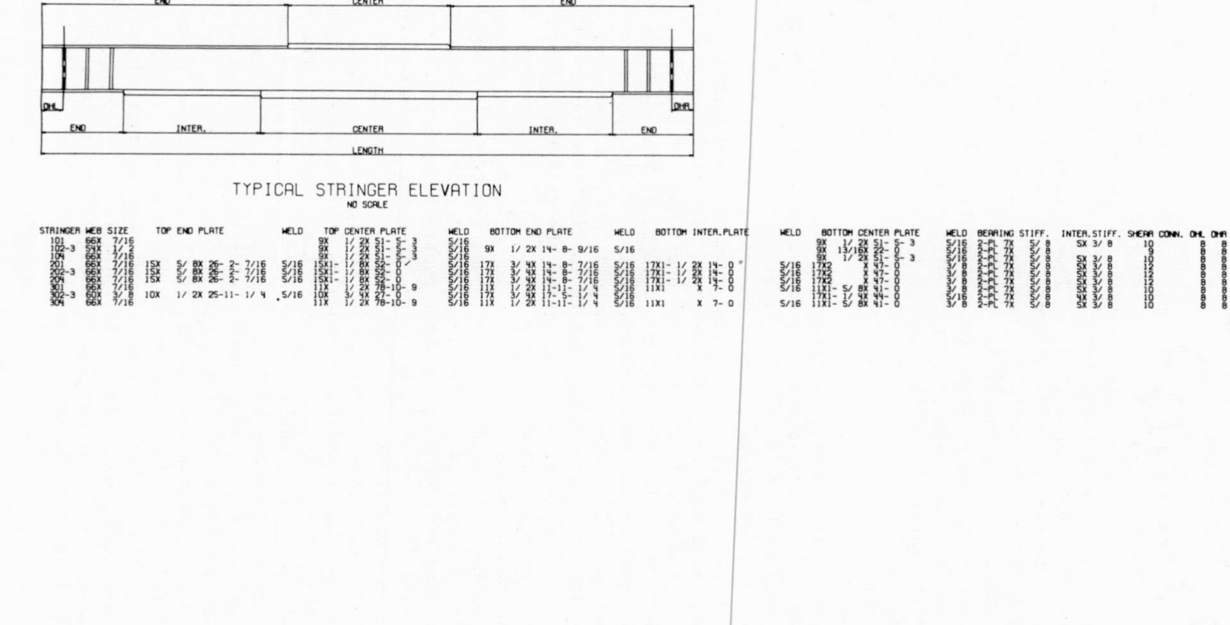

STRINGER	WEB SIZE	TOP END PLATE	WELD	TOP CENTER PLATE	WELD	BOTTOM END PLATE	WELD	BOTTOM INTER.PLATE	WELD	BOTTOM CENTER PLATE	WELD	BEARING STIFF.	INTER.STIFF.	SHEAR CONN.	DHL	DHR	LENGTH
101	66X 7/16			9X 1/ 2X 51- 5- 3	5/16					9X 1/ 2X 51- 5- 3	5/16	2-PL 7X 5/ 8	5X 3/ 8	10	8	8	51-11 3/16
102-3	54X 1/ 2			9X 1/ 2X 51- 5- 3	5/16	9X 1/ 2X 14- 8- 9/16	5/16			9X 13/16X 22- 0	5/16	2-PL 7X 5/ 8		9	8	8	51-11 3/16
104	66X 7/16			9X 1/ 2X 51- 5- 3	5/16					9X 1/ 2X 51- 5- 3	5/16	2-PL 7X 5/ 8	5X 3/ 8	10	8	8	51-11 3/16
201	66X 7/16	15X 5/ 8X 26- 2- 7/16	5/16	15X1- 1/ 8X 52- 0	5/16	17X 3/ 4X 14- 8- 7/16	5/16	17X1- 1/ 2X 14- 0	5/16	17X2 X 47- 0	3/ 8	2-PL 7X 5/ 8	5X 3/ 8	12	8	8	104-10 7/ 8
202-3	66X 7/16	15X 5/ 8X 26- 2- 7/16	5/16	15X1- 1/ 8X 52- 0	5/16	17X 3/ 4X 14- 8- 7/16	5/16	17X1- 1/ 2X 14- 0	5/16	17X2 X 47- 0	3/ 8	2-PL 7X 5/ 8	5X 3/ 8	12	8	8	104-10 7/ 8
204	66X 7/16	15X 5/ 8X 26- 2- 7/16	5/16	15X1- 1/ 8X 52- 0	5/16	17X 3/ 4X 14- 8- 7/16	5/16	17X1- 1/ 2X 14- 0	5/16	17X2 X 47- 0	3/ 8	2-PL 7X 5/ 8	5X 3/ 8	12	8	8	104-10 7/ 8
301	66X 7/16			11X 1/ 2X 78-10- 9	5/16	11X 1/ 2X 11-11- 1/ 4	5/16	11X1 X 7- 0	5/16	11X1- 5/ 8X 41- 0	3/ 8	2-PL 7X 5/ 8	5X 3/ 8	10	8	8	79- 4 9/16
302-3	60X 3/ 8	10X 1/ 2X 25-11- 1/ 4	5/16	10X 3/ 4X 27- 0	5/16	17X 3/ 4X 17- 5- 1/ 4	5/16			17X1- 1/ 4X 44- 0	5/16	2-PL 7X 5/ 8	4X 3/ 8	10	8	8	79- 4 9/16
304	66X 7/16			11X 1/ 2X 78-10- 9	5/16	11X 1/ 2X 11-11- 1/ 4	5/16	11X1 X 7- 0	5/16	11X1- 5/ 8X 41- 0	3/ 8	2-PL 7X 5/ 8	5X 3/ 8	10	8	8	79- 4 9/16

Fig. 11 Beam and girder details designed and plotted by BEST system

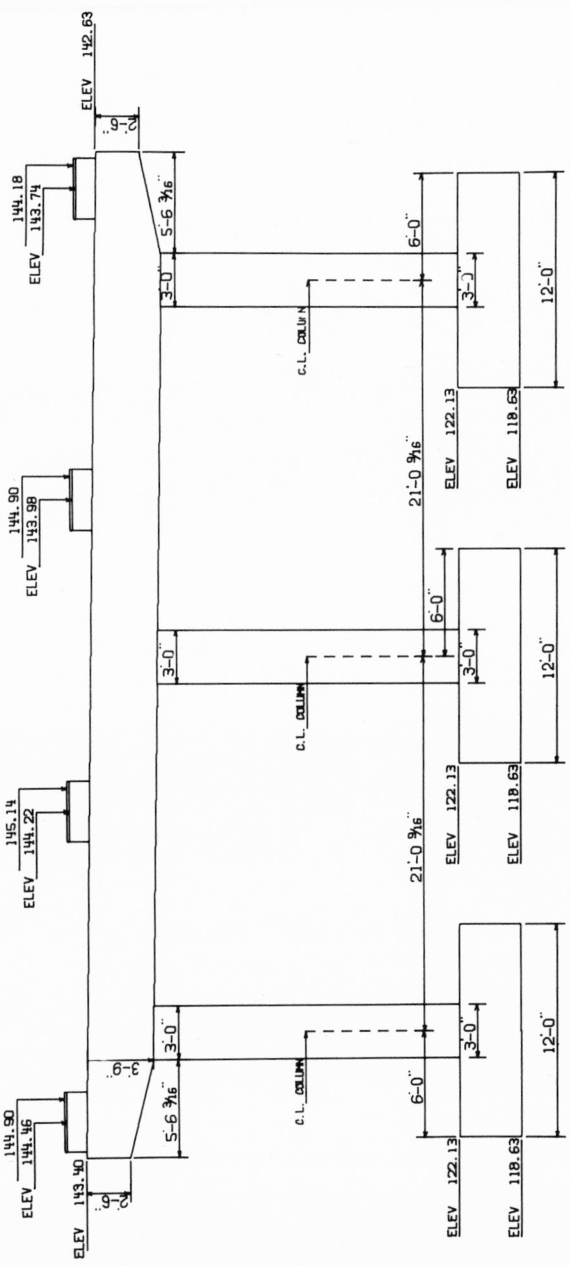

Fig. 12 Pier details designed and plotted by BEST system

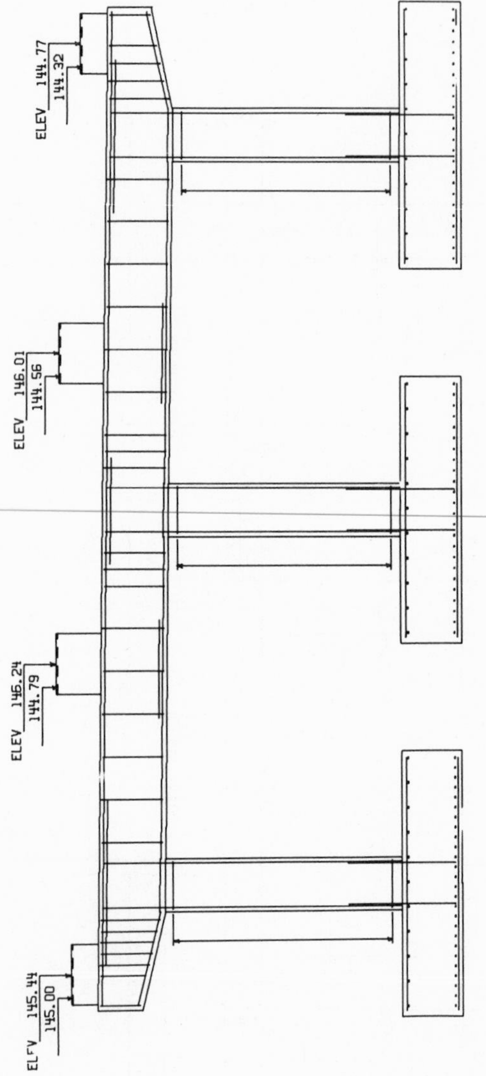

Fig. 13 Pier reinforcing details designed and plotted by BEST system

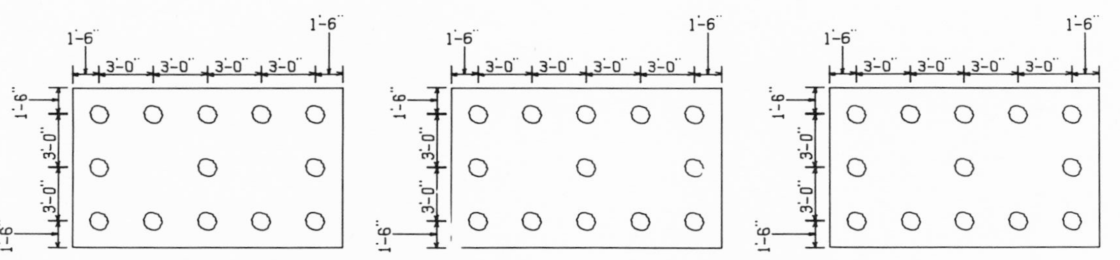

Fig. 14 Footings and pile details designed and plotted by BEST system for one of the piers

A detailed comparison was not performed, but great differences are not apparent.

Input Data for STRC1

The input data required for STRC1 are quite different from those for BEST. Whereas the first program to operate in BEST is the Control Geometry program, STRC1 does not contain a geometry program. Instead, Omnidata has a system INGO (integrated geometry system) which must be run before STRC1 to develop information which is then passed into STRC1. The input information provided Omnidata consisted of those items shown in Table 2. This information was used by Omnidata personnel to obtain the 75 lines of input for the system shown in Table 4.

We see from the input that the designer enters the STRC1 system with a firm idea of the bridge layout and geometry, including the beam spacing and lengths. The system will then carry the design through to the foundations. Beam lengths of 58' 11", 103' 0", and 80' 9" were input. The original beam spacing of 7' 9" was input. This resulted in five lines of stringers.

Results of STRC1 Design

The system-designed girders are 57" in maximum depth as opposed to 55" in the original design. The system-designed flange plates are 14" and 16" as opposed to 16" and 20".

The STRC1 system optimizes the weight of the members using either welded girders or rolled sections, as specified by the designer; welded plate girders were used on the Ramp F bridge, even on the short approach spans. For the 58' 11" span the system designed a 32 3/8" maximum depth welded plate girder as opposed to a 36" rolled beam with cover plates. Also, for the 80' 9" span the system designed a 44 4/5" maximum depth welded plate girder as opposed to the 36" rolled beam with cover plates. Steel quantities, based on the weight of the beams only and neglecting the cross bracing, show the total weight of the system's girders to be approximately 190 kips, as compared to 175 kips for BEST and 270 kips as built. As with the BEST design, the plate girders selected by STRC1 would involve more fabrication cost than if all rolled beams were chosen.

Significant differences are not found in the sizes of the pier elements. This is due mainly to the fact that data were input into the pier program to design the pier using 4 columns 3' 0" in diameter. Also, the pier cap beam was input to be 4' 2" by 3' 0". This is similar to the original design as shown in Fig. 6. However, significant differences did occur in the footings. The STRC1 designed footings all 3' 0" in depth, and all vary in plan dimensions. The four footings of Pier 1 are 11' 6" by 11' 6", 7' 3" by 7' 3", 7' 0" by 7' 0", and 10' 6" by 10' 6", while the original four footings are all 9' 0" by 6' 0" by 3' 3" as shown in Fig. 6.

Table 4 Final Input Data for STRC1

```
 OMNIDATA PAGE 0001

// XEQ STRC1
SAMPLE BRIDGE DESIGN - RAMP F OVER 7TH STREET, CITY OF RICHMOND
SGIRD
STEEL GIRDER DESIGN
  20      16        3         12      10
  -1      -1                                          875     2       100     1000        10
 165        375    24         5                       8
INTERIOR GIRDER - SPAN 58 FT 11 IN
                    5         75        1             5       775                58917     30
  12        25                                       12       25
   050      250                                                                  2    2     1
EXTERIOR GIRDER   SPAN 58 FT 11 IN
  12        2       5         75        1             5       775       2        58917     30
  12        25                                       12       25
   050      250                                                                  2    2     1
INTERIOR GIRDER - SPAN 103 FT
                              75        1             5       775       2       103        54
  14        35      25         125                   16       35        25        125
   050      250                                                                  4    4     1
EXTERIOR GIRDER - SPAN 103 FT
  12        2                 75        1             5       775       2       103        54
  14        35      25         125                   16       35        25        125
   050      250                                                                  4    4     1
INTERIOR GIRDER - SPAN 80 FT 9 IN
                              75        1             5       775       2       18075      42
  12        3       15                               14       3         15
   050      250                                                                  3    3     1
EXTERIOR GIRDER - SPAN 80 FT 9 IN
  12        2                 75        1             5       775       2       18075      42
  12        3       15                               14       3         15
   050      250                                                                  3    3     1    1
DKLDS
RAMP F DECK LOADS
  20      16
PIER 1 DECK LOADS
  58917  103     -5334580               3         6
  1                 4    1    1    1    5         1    1
 -15               30
 -155      775
 -27       4       153333  153333  153333    4
  -302     -301
  -304     -303
PIER 2 DECK LOADS
 103      8075   -5334580               3         6
  1                 4    1    1    1    5         1    1                                    1
 -15               30
 -155      775
 -27       4       153333  153333  153333    4
  -304     -303
  -306     -305
PBENT
PIER BENT DESIGN RAMP F
   3       12      20
PIER 1 RAMP F
  -501     1                      11    8    4    2   10    9    5    2
   4167    3        2        2         -1         3         19
PIER 2 RAMP F
  -502     1                      11    8    4    2   10    9    5    2                      1
   4167    3        2        2         -1         3         19
COLMN
COLUMN DESIGN RAMP F
   3       20       16      40                    2         1         2         1    4
PIER 1 COLUMN DESIGN
  -701
PIER 2 COLUMN DESIGN
  -702
PRFDN
PIER FOUNDATION DESIGN RAMP F
   3       12      20                                                 72        4         1
PIER 1 FOOTING
  -701     1                 3         2    1    8    8    5    5    1
PIER 2 FOOTING
  -701     1                 3         2    1    8    8    5    5    1                       1
```

Unfortunately, plots of the system designed components were not produced for this example problem. The total fee for this design, three superstructure spans and one pier, would be $310 using STRC1.

Summary of BEST and STRC1

A major difference between BEST and STRC1 is the geometry block that is included in BEST and not in STRC1. This subprogram makes it possible to select optimal control dimensions for a bridge from among an infinite number of layouts. The layout finally selected will satisfy constraints as established by the input of such variables as span to depth ratio, lateral clearances, and minimum end span length.

Both systems select girders or rolled beams by automatically choosing the minimum steel weight. BEST selects a rolled beam design if the span is less than 80 feet or available depth is less than 42 inches. If the span is over 80 feet or if the depth available is over 42 inches, a girder design is automatically selected. For the STRC1 system, the designer must make an initial choice as to whether a rolled beam or plate girder is to be used in the design. This difference between BEST and STRC1 is demonstrated in the example bridge design that was designed using the two systems. The total weight of the system girders designed were: Best - 175 kips; STRC1 - 190 kips; as built - 270 kips. By BEST choosing the optimum layout and framing plan itself, the BEST design resulted in a savings of 15 kips over the STRC1 system and 95 kips over the as built design in structural steel weight. This savings by BEST is probably due to the recycling capability of BEST. BEST system attempts to find the minimum cost design by cycling from subprogram to subprogram as opposed to STRC1 which optimizes only within a subprogram.

Another difference noted is that BEST will design bridge bearings and STRC1 does not. By comparing Table 3 and Table 4, we see that BEST requires less input than STRC1.

It can be concluded that either BEST or STRC1 produced better designs than was actually used for the Virginia bridge.

GIRDER ANALYSIS AND DESIGN PROGRAMS

Brief summaries of some of the analysis and design programs for girders are presented in Appendix II. Some of this information was collected using a questionnaire. Other data are taken from the computer systems index [4].

The data is organized under the following six categories:

1. Simple span wide flange beam girder programs
2. Continuous span wide flange beam girder programs
3. Simple span plate girder programs
4. Continuous span plate girder programs
5. Simple span prestressed concrete beam programs
6. Continuous span prestressed concrete beam programs

Table 5 Summary of Simple Span Wide Flange Beam Girder Design Program

Name	Complete analysis	Composite design	Noncomposite design	Designs coverplates	AASHO specs	Capability for adding constraints	Selects W from table in computer	Minimum cost design	Hybrid designs	Quantities of steel given	Displays loading configuration	Displays design stresses	Displays deflections	Moments displayed	Reactions displayed	Designs shear connector spacing
Structure analysis (Wyoming)		X	X		X				X			X	X			
Composite wide flange beam design (Pennsylvania)	X	X	X	X	X		X	X		X		X	X		X	X
Bridge deck (N.C. State Univ.)	X	X	X								X	X	X			
SIMPBM01 (Georgia)	X	X	X	X	X		X					X	X	X	X	X
(Beam design) (Vermont)	X	X	X	X	X								X	X	X	
WBEAM (Omnidata)	X	X	X	X	X	X	X	X		X	X	X	X	X	X	X
Rolled beam (New York)	X	X	X	X	X	X	X	X		X		X	X	X	X	X

Table 6 Summary of Continuous Span Wide Flange Beam Girder Design Programs

Name	Maximum no. of spans	Complete analysis	Ends may be cantilevered	Composite design	Noncomposite design	Designs coverplates	AASHO specs	Capability for adding constraints	Selects W from table in computer	Hybrid designs	Displays loading configuration	Displays design stresses	Displays deflections	Moments displayed	Reactions displayed	Designs shear connector spacing
Analysis of cont. beam for highway bridge IV (Georgia)	8	X		X	X	X	X				X	X	X	X	X	X
Structure analysis (Wyoming)	19			X	X		X	X		X		X	X			
BRDG1 (Van Doren, Hazard)	5	X	X	X	X	X	X					X	X	X	X	X
Cont. beam design (DeLeuw Cather)	5	X		X	X	X	X		X			X		X	X	X
STRC2 (Omnidata)	10	X	X	X	X		X		X		X	X	X	X	X	X

Table 7 Summary of Simple Span Girder Design Programs

Name	Complete analysis	Composite design	Noncomposite design	AASHO specs	Capability for adding constraints	Minimum cost design	Flange plates may be different sizes	Max. no. of flange thickness within span	Hybrid designs	Quantitives of steel given	Straight haunch	Parabolic haunch	Displays loading configuration	Displays design stresses	Displays deflections	Moments displayed	Reactions displayed	Designs shear connector spacing
Structure analysis (Wyoming)		X	X	X			X	10	X		X	X		X	X			
Composite welded girder design (Pennsylvania)	X	X	X	X		X	X	7		X				X	X		X	X
Composite girder design (California)		X		X			X							X	X	X	X	
Simple span beam design (Georgia)	X	X	X	X	X	X	X							X	X	X	X	
SGIRD (Omnidata)	X	X	X	X	X	X	X			X	X		X	X	X	X	X	X
Plate girder (New York)	X	X	X	X	X	X	X			X	X			X	X	X	X	X

Table 8 Summary of Continuous Span Plate Girder Design Programs

Name	Maximum no. of spans	Complete analysis	End spans may be cantilevered	Composite design	Noncomposite design	AASHO specs	Capability for adding constraints	Flange plates may be different sizes	Max. no. of flange thickness within span	Hybrid designs	Straight haunch	Parabolic haunch	Displays loading configuration	Displays design stresses	Displays deflection	Moments displayed	Reactions displayed	Designs shear connector spacting
Analysis of cont. beams for highway bridge IV (Georgia)	8	X		X	X	X		X			X	X	X	X	X	X	X	X
Structure analysis (Wyoming)	19			X	X	X	X	X	10	X	X	X		X	X			
BRDG1 (Van Doren, Hazard)	5	X	X	X	X	X					X	X		X	X	X	X	
DC-S93 (DeLeuw Cather)	5	X		X	X	X		X	19		X	X		X		X	X	X
STRC2 (Omnidata)	10	X	X	X	X	X		X			X			X	X	X	X	X

Table 9 Summary of Simple Span Prestressed Concrete Beam Design Programs

Name	Pretensioned	Posttensioned	Complete analysis	AASHO or ACI specs	Analysis using transformed section	I, T, or box section	Composite	Noncomposite	Max. no. of steel strand profile rows	Calculates opt. no. of steel strands	Calculates complete set of stress calculations	Gives prestressing steel strand profile	Gives moments, shears and stirrup spacing	Gives section properties	Gives stresses for top of slab	Gives stresses for top of beam	Gives stresses for bottom of beam	Gives deflections for beam cambers	Stresses given for both service conditions and at transfer	Quantities of conc. and steel listed
Prestressed beam design (Tennessee)	X		X	X	X	I and box	X	X	12	X	X	X	X	X		X	X	X	X	
Prestressed girder (Louisiana)	X		X	X	X	T	X			X	X	X	X	X	X	X	X	X	X	
PSGDES (Washington)	X		X	X	X	X				X	X	X	X	X		X	X		X	
FINLBM (New Mexico)	X		X	X	X	X							X	X		X	X	X	X	
HYBDØ19 (Kansas)	X			X		I and T	X	X	3	X	X	X	X	X	X	X	X	X	X	
Prestressed concrete girder design (Pennsylvania)	X		X	X		X	X	X	20	X	X		X	X	X	X	X	X	X	
Frame system (California)	X	X	X	X		X	X	X			X		X	X		X	X	X	X	
PCSPAN (UCS)	X		X	X	X	X	X	X		X	X	X	X	X	X	X	X	X	X	
SSCPBD (Oregon)	X	X	X	X	X	I and T	X	X		X	X	X	X	X	X	X	X	X	X	
Prestressed concrete girder design (Texas)	X		X	X		I	X	X		X	X	X	X	X	X	X	X	X	X	
PRESTRO1 (FHWA)	X			X		X				X	X	X	X	X		X	X	X	X	
PRESTRO2 (Georgia)	X	X	X	X		X				X	X	X	X	X		X	X	X	X	
PBDES Omnidata)	X	X	X	X	X	I and box	X	X	20	X	X	X	X	X	X	X	X	X	X	X

Table 10 Summary of Continuous Span Prestressed Concrete Beam Design Programs

Name	Max. no. of spans	Pretensionsed	Posttensioned	Complete analysis	AASHO or ACI specs	Analysis using trans-formed section	I, T, or Box section	Composite	Noncomposite	Max. no. of steel strand profiles	Calculates opt. no. of steel strands	Calculates complete set of stress calculations	Gives prestressing steel strand profile	Gives moments, shears, and stirrup spacing	Gives section properties	Gives stresses for top of slab	Gives stresses for top of beam	Gives stresses for bottom of beam	Gives deflections for beam cambers	Stresses given for both service conditions and at transfer
Prestressed beam design (Tennessee)		X			X	X	I and box	X	X	12	X	X	X	X	X		X	X	X	X
Prestressed girder design (Minnesota)	any	X	X	X	X				X	7		X		X	X	X	X	X	X	
PSGDES (Washington)		X		X	X	X	X				X	X	X	X	X		X	X		X
FINLBM (New Mexico)	10	X		X	X	X	X							X	X		X	X	X	X
Frame system (California)	15		X	X	X		X		X	X		X		X	X		X	X	X	X

The programs listed under each of the six categories are described in the following manner:

Title

Agency: No. in [4] : Date available:

Computer: Language:

Abstract

The number provided for some of the programs refers to a number assigned to the program in [4]. The address of each agency is given in Appendix III.

Details of the capabilities of some of the girder design programs are provided in Tables 5-10. These are separated into the six categories mentioned above.

REFERENCES

1 The Bridge Design Segment of TIES, New York State Department of Transportation, Albany, New York, 1974.

2 Hourigan, E. V., "The Bridge Design Segment of TIES," Proceedings Committee on Computer Technology, American Association of State Highway Officials, May 1970, pp. 187-282.

3 STRUC I Users Manual, Omnidata Services, Inc., New York, New York, 1974.

4 "Computer Systems Index," American Association of State Highway Officials, Administrative Sub-Committee on Computer Technology, September 1973, pp. 127-167.

APPENDIX I

MANAGERS OF BRIDGE SYSTEMS

1. BEST

   C. K. Bartholomew
   Asst. Deputy Chief Engineer
   1220 Washington Avenue
   State Campus
   Albany, New York 12226

2. STRC1 and STRC2

   Eugene L. Sheninger
   Manager of Engineering Services
   Omnidata Services, Inc.
   2 Park Avenue
   New York, New York 10016

3. BRIDGE

   Leroy Empkin
   Department of Civil Engineering
   Georgia Institute of Technology
   Atlanta, Georgia 30304

4. BRDESIGN - Oklahoma

   Y. S. Yang
   Senior Structural Engineer
   Bridge Division
   Oklahoma Highway Department
   Jim Thorpe Building
   Oklahoma City, Oklahoma 73105

5. BRIDGE DESIGN - Michigan

   Maurice Van Auken
   Computer Coordinating Engineer
   Department St. Hwys. & Transportation
   Lansing, Michigan 48904

6. BRASS - Wyoming

   Ralph R. Johnston
   Wyoming Highway Department
   P. O. Box 1708
   Cheyenne, Wyoming 82001

   or

   Web H. Collins
   HDV-21
   Office of Development
   Federal Highway Admin.
   Washington, D.C. 20590

7. BARS

Charles W. Beilfuss, Manager
Systems & Computer Applications
Control Data Corporation
Professional Services Division - Meiscon
2021 Spring Road
Oak Brook, Illinois 60521

8. McDonnell-ECI ICES STRUDL

Dr. E. L. Ghent
Engineering Product Manager
McDonnell Douglas Automation Co.
P. O. Box 516
St. Louis, Missouri 63166

9. PCSPAN

United Computing Systems, Inc.
Technical Publications
3130 Broadway
Kansas City, Missouri 64111

APPENDIX II

GIRDER ANALYSIS AND DESIGN PROGRAMS

Simple Span Wide Flange Beam Girders

Minimum moment of inertia for maximum live load
Agency: Arkansas; No. in [4] : 209; Date available: 1973;

This program computes and lists minimum moment of inertia for simple span beams with maximum live load deflections of S/800 and S/1000 using H, HS, and lane loadings for various span lengths and girder spacings.

Simple span beam design
Agency: Georgia; No. in [4] : 12; Date available: 1973;
Computer: IBM 360/50; Language: FORTRAN IV

This program designs composite or noncomposite rolled beams or plate girders, or reinforced concrete T-beams. Input span lengths, loads, and design specs. Output consists of moments, shears, reactions, dead and live load deflections, beam size, plate sizes, or reinforcing steel requirements. Fatigue is considered. Can handle railroad live load. Can analyze a given beam for overstress.

Composite rolled beam analysis
Agency: Minnesota; No. in [4] : 13; Date available: 1973;
Computer: IBM 370/155; Language: FORTRAN

This program analyzes I-beams with plates and complete sections. It computes moment of inertia, section modulus, and centroid data for composite beams with cover plates.

Simple span moments and shears
Agency: Nebraska; No. in [4] : 264; Date available: 1973;
Computer: IBM 370/145; Language: FORTRAN

This program computes maximum moments and shears at each one-tenth point of a simple span girder where live loads are variable.

Structural design system
Agency: New Jersey; No. in [4] : 42; Date available: 1973;
Computer: IBM 360/50; Language: FORTRAN

This system consists of a battery of standalone structural design systems including overhead sign design, ground-mounted sign design, pile bent analysis, pile group analysis, stress, composite rolled beam design, abutment and retaining wall design, prestressed concrete beam design, rolled beam design, and continuous beam analysis.

Moments and shears in simple spans
Agency: Pennsylvania; No. in [4] : 14; Date available: 1973;
Computer: B6700/196K; Language: ALGOL

This program computes moments and shears due to dead load and live load plus impact. For remote job entry only. It is written in Burroughs extended Algol.

Composite wide flange beam analysis and design
Agency: Pennsylvania; No. in [4] : 18; Date available: 1973;
Computer: B6700/196K; Language: FORTRAN

This program will analyze or design a composite wide flange section, with or without a cover plate.

Continuous beam design
Agency: Vermont; No. in [4] : 211; Date available: 1972;
Computer: IBM 1130/M2D; Language: FORTRAN

This system designs or analyzes composite or noncomposite rolled beams or plate girders, continuous or simple span structures. Also, beam profiles may be obtained for simple or continuous spans for any type of alignment. Other calculations are end shear and influence lines, moments, dead load deflections, live load plus impact deflections, and flange and cover plate dimensions.

Steel beam design
Agency: Technical Programs, Inc.; Date available: 1973;
Computer: Wang 700; Language:

This program is available from Technical Programs, Inc., for a fee of $150.

Continuous Span Wide Flange Beam Girders

Analysis of continuous beams for highway bridges
Agency: Arkansas; No. in [4] : 205; Date available: 1973;
Computer: IBM 360/40; Language: FORTRAN IV

This program does a complete analysis of a continuous beam for a highway bridge and reports the moments, shears, reactions, deflections, and shear connector spacings produced by the dead loads and the standard highway live loads.

Continuous beam analysis
Agency: Georgia; No. in [4] : 17; Date available: 1973;
Computer: IBM 360/50; Language: FORTRAN

This program analyzes continuous rolled beams, plate girders, or reinforced concrete beams of constant or variable depth. Steel beams may be composite or noncomposite. There may be two to eight spans. Various live and dead loads are considered and moments, shears, reactions, deflections, and stresses are output. Provides for allowable fatigue stresses and shear connector spacing.

Influence lines for a 2- to 8-span continuous beam
Agency: Illinois; No. in [4] : 203; Date available: 1968;
Computer: IBM 370/155; Language: FORTRAN IV

This program computes influence coefficients for moment, shear, and deflection for nonprismatic beams. Coefficients are furnished at tenth points of spans. It handles cantilever moment at extreme right support.

Continuous beam analysis
Agency: Minnesota; No. in [4] : 11; Date available: 1973;
Computer: IBM 370/155; Language: FORTRAN

This program furnishes moments, shears, and deflections for dead load and live load section moduli for composite and noncomposite sections and influence lines. The program plots a continuous beam rating sheet.

Continuous wide flange beam analysis
Agency: Missouri; No. in [4] : 1; Date available: 1971;
Computer: IBM 370/145; Language: FORTRAN IV

This program analyzes wide flange beam bridges. It finds fatigue, plate cutoff lengths, and load factors.

Continuous structures
Agency: Nebraska; No. in [4] : 260; Date available: 1973;
Computer: IBM 370/145; Language: FORTRAN

This program computes maximum moments and shears at each one-tenth point of a continuous girder. The number of spans may very from one to six.

Continuous bridge analysis
Agency: Oregon; No. in [4] : 3; Date available: 1973;
Computer: IBM 370/155; Language: FORTRAN

This program will analyze a continuous structure with up to five spans with or without cantilevers. Output includes influence lines, dead load moments, shears, deflections, maximum and minimum live load, plus impact moments and shears. Maximum column reactions and moments are given for dead and live load.

Utah beam
Agency: Utah; No. in [4] : 600; Date available: 1973;
Computer: IBM 370/155; Language: FORTRAN

This program performs complete analysis of continuous beams used in highway bridges. It computes and prints out moments, stresses, shears, stiffener spacing, shear connector spacing, reactions, and deflections due to both live and dead loads. Meets AASHO 1969 specifications.

Continuous beam
Agency: Wyoming; No. in [4] : 158; Date available: 1971;
Computer: IBM 360/40; Language: FORTRAN IV

This program calculates the beam properties and beam characteristics of each span of a 2- to 5-span continuous bridge. The moments, shears, and reactions are calculated for dead and live loads. The deflections are calculated for dead loading.

Continuous beam analysis
Agency: Technical programs, Inc.; Date available: 1973;
Computer: Wang 720 Programmable Calculator

This program is available from Technical Programs, Inc., for $325.

BRIDGE/1
Agency: Genesys Centre; Date available: 1973;

This subsystem allows an engineer to analyze bridges that can be considered as continuous beams by ignoring the effect of transverse load distribution. A bridge may be of constant or varying cross section with an inclined or horizontal surface. Reinforced, prestressed, or plain concrete, steel, timber, and composite construction can all be used to suit the engineer's particular specification.

A finite element method of analysis for composite beams
Agency: Sci-Tek Incorporated; Date available: 1973;
Computer: UNIVAC 1108

This program performs a complete analysis of a composite beam with any degree of hoirzontal shear. The beam being analyzed may be subjected to any longitudinal distribution of transverse or axial loads and may be supported in any reasonable manner.

Simple Span Plate Girders

Bridge-horizontal curved girder
Agency: Arizona; No. in [4] : 212; Date available: 1973;
Computer: IBM 370/155; Language: FORTRAN

This program analyzes and designs horizontally curved bridges such as calculating bending and torsional moments, shear forces at ten points in each span, and torsional reactions at the supports.

Composite girder design
Agency: California; No. in [4] : 606; Date available: 1973;
Computer: IBM 370/165; Language: FORTRAN

This program designs a composite girder including flange areas.

Composite girder analysis
Agency: California; No. in [4] : 607; Date available: 1973;
Computer: IBM 370/165; Language: FORTRAN

This program does the analysis of a composite girder.

Simple span plate girder analysis
Agency: Province of Manitoba; No. in [4] : 420; Date available: 1973;
Computer: IBM 370/155; Language: FORTRAN IV

Given dead and live loads and estimated web and flange dimensions for a specific number of cross sections, the program determines allowable plate lengths, shear bending moments, and stresses at the point of maximum stress and at plate size transition points. Also evaluated are required web stiffener size and spacing, slab shear connector spacing,and principal and construction stresses.

Composite welded girder analysis and design
Agency: Pennsylvania; No. in [4] : 19; Date available: 1973;
Computer: B6700/196K; Language: FORTRAN

This program analyzes or designs a composite welded plate girder for 1969 AASHO specifications.

Girder analysis and design
Agency: West Virginia; No. in [4] : 265; Date available: 1973;
Computer: IBM 360/65; Language: FORTRAN IV

This is a series of programs to analyze and design a variety of typical beams. Included are reinforced concrete, prestressed concrete, and steel. Reinforced concrete and steel beams may be simple or continuous.

Guider computations
Agency: County of Santa Clara; No. in [4] : 559; Date available: 1973; Computer: IBM 130/16K; Language: FORTRAN IV

Theoretical flange areas and stresses and deflections for simple span girders are calculated. Cross section computations for simple or continuous girders are made.

Steel beam, welded plate girder
Agency: Systems Professional; Date available: 1974;
Computer: IBM 360; Language: FORTRAN IV

This designs simple spans using steel beams, composite beam design, and welded plate girder designs using 1969 AISC specifications.

Continuous Span Plate Girders

Bridge design programs
Agency: Alaska; No. in [4] : 510; Date available: 1973;
Computer: IBM 360/40; Language: FORTRAN

This contains stress, continuous bridge design, analysis and design of continuous beams, column analysis and design, and prestressed girder analysis programs.

Continuous girder analysis
Agency: Province of Alberta; No. in [4] : 10; Date available: 1973;
Computer: IBM 360/65; Language: FORTRAN

This combines user-defined (prisms) and (sections) from which support-moment influence lines are calculated. These are loaded with user or AASHO live loadings and dead loadings to determine reactions, moments, shears (including impact), and dead load deflections, all at tenth panel points. Supplementary influence lines for shears and moment at any point are available.

Continuous steel plate girder analysis
Agency: Province of Manitoba; No. in [4] : 410; Date available: 1970; Computer: IBM 370/155; Language: FORTRAN IV

Dead and live load shears, bending moments and stresses, and dead load deflections at cross sections ranging in number from 100 to 200 over total girder length are determined with this program. Spacing of the sections is optional within limits. The permissible number of continuous spans may vary from 2 to 9.

Analysis of continuous bridge girder
Agency: Illinois; No. in [4] : 213; Date available: 1973;
Computer: IBM 370/155; Language: FORTRAN IV

The program analyzes welded continuous plate girder, box girders, or wide flange beam for 2 to 4 spans by either working stress design or load factor design. Computes envelopes for moment, shear, reactions, deflections, flexural stress, and shear stress. Computes stress in slab reinforcing and shear connector spacing. Compares the results with the allowable standards to determine sufficiency of the design per 1971 AASHO specs.

Bridge analysis - continuous girder
Agency: Kansas; No. in [4] : 10; Date available: 1973;
Computer: IBM 370/155; Language: FORTRAN IV

This is a structural analysis of continuous nonprismatic bridge members with columns, one hinge per span, maximum of 16 spans, and maximum of 25 sections per span. The output for spans is: moments, shears, reactions, and deflections.

Bridge analysis - continuous, haunched, skewed slab
Agency: Kansas; No. in [4] : 11; Date available: 1973;
Computer: IBM 370/155; Language: FORTRAN IV

This is a structural analysis of continuous, parabolically haunched, skewed concrete slab bridges with edge curbs.

BRDG1
Agency: Van Doren, Hazard, Stallings; Date available: 1974;
Computer: IBM 1130/BK; Language: FORTRAN

The program designs of up to five continuous spans for wide flange girders or welded plate girders, using composite, noncomposite, and designs coverplates if needed.

Continuous beam design
Agency: De Leuw Cather; Date available: 1974;
Computer: IBM 360; Language: FORTRAN

The program designs of up to five continuous spans for wide flange beam girders or welded plate girders, using composite, noncomposite, and designs coverplates if needed.

Simple Span Prestressed Concrete Beams

Pretensioned girder design
Agency: Arkansas; No. in [4] : 218; Date available: 1973;
Computer: IBM 360/40; Language: FORTRAN IV

This program designs pretensioned girders, AASHO type I through IV, for either draped or straight strands.

Prestress girder analysis
Agency: California; No. in [4] : 604; Date available: 1973;
Computer: IBM 370/165; Language: FORTRAN

This is an analysis of a prestressed girder by elastic theory.

Simple span prestressed beam design
Agency: Georgia; No. in [4] : 14; Date available; 1973;
Computer: IBM 360/50; Language: FORTRAN

This program designs and analyzes prestressed beams. Cross section, span length, and other design specifications are input. Output is number of strands, strand arrangement, and hold-down point. Also moments, stresses, shears, ultimate moments, and maximum and minimum eccentricity envelope plot are output. Program can handle highway, sidewalk, military, and railroad live loads.

Prestressed beam analysis
Agency: Idaho; No. in [4] : 303; Date available: 1973;
Computer: IBM 370/155; Language: FORTRAN

Using input values of girder properties, girder spacing, span and slab thickness, this program computes and prints out composite qualities and stresses at various stages of fabrication and under service load. Service load stresses are computed only for simple span conditions.

Prestressed girder design
Agency: Illinois; No. in [4] : 212; Date available: 1973;
Computer: IBM 370/155; Language: FORTRAN IV

This program computes the strand layout for a simple span, I-shaped prestressed girder. It handles standard or nonstandard beams. It uses AASHO or railway (cooper) loading and 1971 AASHO Specifications.

Prestressed girder design
Agency: Louisiana; No. in [4] : 5; Date available: 1973;
Computer: IBM 370/145; Language: FORTRAN

This program performs a systematic standard design or analysis procedure on prestressed concrete girders. Procedures are those used by the Louisiana Department of Highways and AASHO.

Prestressed girder design
Agency: Minnesota; No. in [4] : 15; Date available: 1973;
Computer: IBM 370/155; Language: FORTRAN

This program performs the design and analysis of prestressed concrete girders by calculating stresses and comparing them to allowable limits.

Composite prestressed concrete channels
Agency: Missouri; No. in [4] : 5; Date available: 1970;
Computer: IBM 360/50; Language: FORTRAN IV

This program computes the dead load moments, deflections, horizontal shear, critical stresses, and live load stresses.

Prestressed girder design
Agency: Nebraska; No. in [4] : 261; Date available: 1973;
Computer: IBM 370/145; Language: FORTRAN

This program performs the analysis of a simple span prestressed girder which may be any of four AASHO standard sections or an AASHO modified section. The size of strands may be varied.

FINLIBM
Agency: New Mexico; Date available: 1973;
Computer: IBM 370; Language: FORTRAN IV

This program performs the analysis and design of pretensioned simple and continuous beams. It is set to use a trial and error procedure for finding the strand pattern.

Prestressed concrete beams continuous for live load
Agency: New York; No. in [4] : 6; Date available: 1973;
Computer: B5500/32K; Language: FORTRAN

This program designs prestressed concrete beams to be erected as simple spans and made continuous for live load through the use of mild steel reinforcement placed in a composite cast in place concrete deck slab.

Prestressed I-beam design
Agency: Oklahoma; No. in [4] : 101; Date available: 1972;
Computer: IBM 370/155; Language: FORTRAN

This program performs the design and analysis of a simple span prestressed I-beam girder of AASHO type I to IV.

Prestressed girder
Agency: Pennsylvania; No. in [4] : 17; Date available: 1973;
Computer: B6700/196K; Language: FORTRAN

This program can analyze or design a prestressed concrete beam, 140 feet maximum span length, and plot family of span curves for design standards giving initial prestressing forces.

Prestressed concrete design package
Agency: Technical Programs, Inc.; Date available: 1973;
Computer: Wang 700 Series Calculator

This program is available from Technical Programs for $625.

Prestressed concrete beam design
Agency: Systems Professional; Date available: 1974;
Computer: IBM 360; Language: FORTRAN IV

This program performs the design of a simple span prestressed concrete beam uniformly loaded of either rectangular, single T, double T, or box shape. Shored or unshored simple span programs.

Continuous Span Prestressed Concrete Beams

Continuous prestressed girder analysis
Agency: California; No. in [4] : 602; Date available: 1973;
Computer: IBM 370/165; Language: FORTRAN

The program will analyze a continuous prestressed girder (post-tensioned).

Bridge and structure analysis
Agency: Colorado; No. in [4] : 20; Date available: 1973;
Computer: CDC 6400; Language: FORTRAN IV

These programs analyze continuous beams, prestressed concrete girder analysis, multiple column piers for bridges, and retaining wall design.

Continuous prestressed concrete girder
Agency: Nebraska; No. in [4] : 262; Date available: 1973;
Computer: IBM 370/145; Language: FORTRAN

This does an analysis of a continuous prestressed girder which may be AASHO standard sections, type 2, 3, or 4, or these sections modified. The size of the strands may be varied.

Prestressed concrete beam design
Agency: New York; No. in [4] : 3; Date available: 1972;
Computer: B5500/32K; Language: FORTRAN

This program designs pretensioned prestressed concrete beams which are AASHO types.

Multispan continuous posttensioned series
Agency: Oregon; No. in [4] : 6; Date available: 1973;
Computer: IBM 370/155; Language: FORTRAN

This program gives the dead and live load, prestress and maximum combined stress envelopes, final prestressing force, and moment coefficients at tenth points for two thru five spans in a continuous posttensioned series. Variable I-members are allowed.

Continuous prestressed girder program
Agency: South Dakota; No. in [4] : 107; Date available: 1974;
Computer: IBM 370/145; Language: FORTRAN IV

This program designs or analyzes continuous prestressed girder structures utilizing continuous composite action for the live loading.

Prestressed beam design
Agency: Tennessee; No. in [4] : 612; Date available: 1973;
Computer: IBM 370/155; Language: FORTRAN

This program designs or checks the design of simple and continuous prestressed box and I-beams, and also can rate them.

APPENDIX III

AVAILABILITY OF GIRDER ANALYSIS AND DESIGN PROGRAMS

Arkansas

Frank Harrison
Bridge Design Division
Arkansas Highway Department
P. O. Box 2261
Little Rock, Arkansas 72209

California

Robert C. Cassano
Chief Engineer
Structures Planning Branch
P.O. Box 1499
Sacramento, California 95807

Colorado

Lloyd W. Farnsworth
Data Processing Engineer
Staff Bridge Engineer
Division of Highways
4201 E. Arkansas Avenue
Denver, Colorado 80222

Georgia

Jose M. Nieves
Systems Development Administrator
Georgia Department of Transportation
2 Capitol Square
Atlanta, Georgia 30334

Idaho

R. B. Jarvis
Bridge Engineer
Idaho Department of Highways
P. O. Box 7129
Boise, Idaho 83707

Illinois

Walter Gussman
Bureau of Computer Science & Information
Department of Transportation
Rm. 019
2300 S. 31st Street
Springfield, Illinois 62764

Kansas

Kenneth Hurst
Bridge Design
916 N.
State Office Bldg.
Topeka, Kansas 66612

Louisiana

James C. Porter
Assistant Bridge Design Engr.
P. O. Box 44245
Capitol Station
Baton Rouge, Louisiana 70804

Minnesota

A. E. Holmboe
Bridge Plans Engineer
Room 610F
Minnesota Highway Bldg.
St. Paul, Minnesota 55101

Missouri

Warren D. Vanderslice
Computer Engineer
Missouri State Highway Dept.
119 West Capitol Avenue
Jefferson City, Missouri 65101

Nebraska

Charles D. Smith
Bridge Engineer
Nebraska Department of Roads
P. O. Box 44759
Lincoln, Nebraska 68509

New Jersey

Simeon Chanley
Bureau of Structural Design
New Jersey Dept. of Transportation
1035 Parkway Avenue
Trenton, New Jersey 08625

New Mexico

Herman Tachua
Bridge Engineer
New Mexico State Highway Dept.
1120 Cerrillos Road
P. O. Box 1149
Santa Fe, New Mexico 87501

New York

John G. Ruby
N.Y.S. Dept. of Transportation
Albany, New York 12226

Oklahoma

Y. S. Yang
Senior Structural Engineer
Oklahoma Highway Department
Jim Thorpe Bldg.
Oklahoma City, Oklahoma 73105

Oregon

Thomas Bricher
Sr. Structural Designer
State Highway Bldg.
Rm. 320
Salem, Oregon 97310

Pennsylvania

K. R. Patel
Room 1008
Transportation & Safety Bldg.
Pennsylvania Dept. of Trans.
Harrisburg, Pennsylvania 17120

South Dakota

Clyde H. Jundt
Bridge Section
Division of Highways
Pierre, South Dakota 57501

Tennessee

A. S. Mallory
Manager of Engr. Systems
960 Highway Annex
Nashville, Tennessee 37219

Utah

David L. Christensen
Department of Highways
State Office Bldg.
Salt Lake City, Utah 84114

Vermont

Robert L. Oatley
Bridge Division
Vermont Highway Department
133 State Street
Montpelier, Vermont 05602

West Virginia

Rusty R. Allen
West Virginia Dept. of Highways
1900 Washington Street East
Charleston, West Virginia 25305

Wyoming

Charles H. Wilson
State Bridge Engineer
Wyoming Highway Department
P. O. Box 1708
Cheyenne, Wyoming 82001

Technical Programs, Inc.

604 Park Drive
University Park
Boca Raton, Florida 33432

The GENESYS Center

University of Technology
Loughborough, Leicestershire
LE11 3TU, UK

SCI-TKE Incorporated

1707 Gilpin Avenue
Wilmington, Delaware 19806

Systems Professional

3055 Overland Avenue
Los Angeles, California 90034

Van Doren, Hazard, Stallings

Box 719
Topeka, Kansas 66601

DeLeuw Cather

La Salle & Wacker
Chicago, Illinois

PCSPAN

United Computing Systems, Inc.
Technical Publications
3130 Broadway
Kansas City, Missouri 64111

COMPOSITE STRUCTURES

D. M. Purdy
Douglas Aircraft Company
McDonnell Douglas Corporation
Long Beach, California

ABSTRACT

Special purpose computer programs for use in composite structure analysis which are available for public distribution to qualified organizations are surveyed and assessed. Programs which are currently in existence fall into four basic categories: strength and stiffness analysis, stability analysis, laminate optimization, and bonded joint analysis. It is observed that the status of available programs generally parallels the state of composite structure usage. That is, most composite structure activities are R&D oriented with a few production applications starting to appear. Most of the programs were developed as part of R&D programs and require varying levels of modifications to make them useful on a production basis. In most cases improved manuals and more refined input and output would be required in addition to complete checkout and correlation of the programs' predictive capability.

INTRODUCTION

In 1963 the U.S. Air Force, as a result of its Project Forecast, introduced the advanced composites era. In the 11 years that have followed Project Forecast there has been steady progress in the development of engineering methods, structural concepts, and improved materials. When consistent-quality boron filaments became available in 1966 a number of programs [1] were initiated to develop aircraft structures using boron/epoxy. Four such structures which were developed and flown are shown in Fig. 1.

A 16.0-pound A-4 flap was fabricated, static and fatigue tested, and flown demonstrating 21 percent weight saving. A 42-pound F-4 rudder showed 35 percent weight saving, a 419-pound F-111 horizontal stabilizer showed 27 percent saving, and a 190-pound C-5A slat, a 21 percent saving. Programs were so successful that contracts were let to produce F-4 rudders, C-5A slats, and F-111 stabilizers for production and flight-life demonstration.

As a result of these and other programs, boron/epoxy aircraft structures were developed to the point that boron/epoxy was selected for use on several components in both the F-14 and F-15.

Graphite/epoxy aircraft structures have not reached the level of usage that exists with boron/epoxy structures. However, programs such as the DC-10 Composite Upper Aft Rudder Program (Fig. 2) should provide the experience necessary for widespread usage of graphite/epoxy aircraft structure.

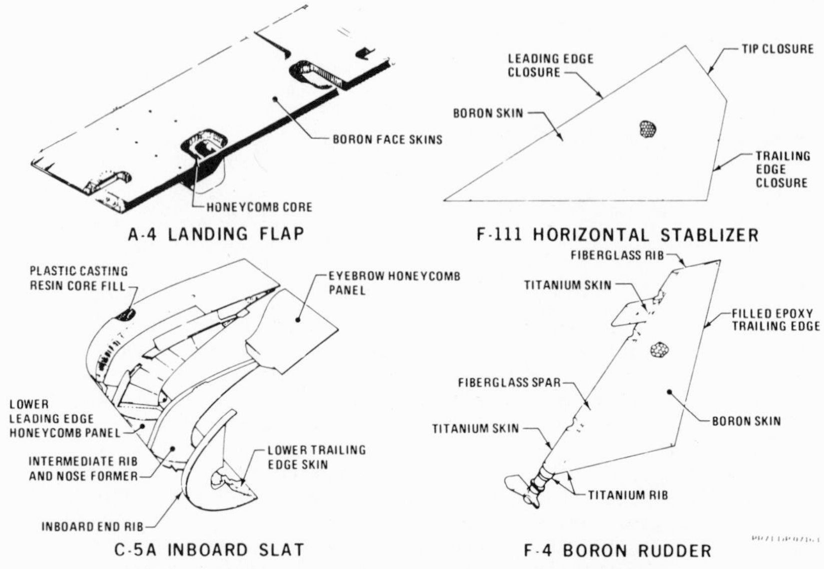

Fig. 1 Boron/epoxy aircraft structures

Fig. 2 DC-10 composite upper aft rudder

The development of these structures has required the development of engineering methods which account for the differences [2] between composites and conventional metals (Table 1). The methods that have been developed to date have enabled the development of composite structures on a fairly small scale. A considerable amount of engineering methods and computer software development remains before widespread production applications of composite structures take place. In general most of the special purpose programs that are available for composite structures are not suitable for production engineering since they have not been completely checked out and are not well documented.

Table 1 Design Complications Associated with Composites

BRITTLENESS
- o LOAD TRANSFER AT ATTACHMENT
- o STRESS CONCENTRATION
- o STRENGTH VARIABILITY
- o FAILURE THEORIES

HETEROGENEITY
- o LOAD TRANSFER AMONG CONSTITUENTS
- o STRAIN COMPATIBILITY
- o CONSTITUTIVE PROPORTION

NONISOTROPY
- o DIRECTION DEPENDENT PROPERTIES
- o NUMEROUS ELASTIC CONSTANTS
- o RELATIVE ORIENTATION TO LOADS
- o FAILURE THEORIES

AVAILABILITY OF DATA
- o QUANTITY
- o TEST TECHNIQUES
- o ANALYTICAL THEORY

Programs which are used for the analysis of nonisotropic structure fall into two general categories: first the large general purpose programs such as NASTRAN and, second, the smaller special purpose programs developed specifically for composite structure analysis. There are a number of special purpose composite analysis programs which have been developed and are available to the general public. These programs fall into four basic categories:

1. Strength stiffness analysis
2. Composite structure stability analysis
3. Laminate optimization
4. Bonded joint analysis

In this presentation a number of special purpose programs which are available to the public and are finding various degrees of use in the advanced composite structures community are discussed. Programs which have been developed under NASA sponsorship can be obtained from COSMIC (Table 2). There are two basic Air Force Sources of programs. First there are programs documented in the Design Guide (Table 3), and second there are programs which are on file with the Air Force Flight Dynamics Laboratory Composites Branch (Table 4). There are also a number of programs of interest which were developed under government sponsorship which are apparently available only from the original contractor (Table 5).

Table 2 Composite Programs from Cosmic

PROGRAM NUMBER	PROGRAM TITLE
M71-10112	ANALYSIS OF MULTILAYERED FIBER COMPOSITES
	BUCKLAS – STIFFENED PLATE STABILITY

Table 3 Air Force Advanced Composite Design Guide Computer Program

PROGRAM NAME	PROBLEM TYPE
AC-3	POINT STRESS ANALYSIS
AC-5	HONEYCOMB SANDWICH PANEL STABILITY UNDER INPLANE BIAXIAL LOADING
AC-7	ANALYSIS OF SIMPLY SUPPORTED ORTHOTROPIC HONEYCOMB SANDWICH PANELS UNDER PRESSURE WITH OR WITHOUT INPLANE LOADS
AC-10	ANISOTROPIC PLATE ANALYSIS
AC-11	HONEYCOMB SANDWICH PANEL STABILITY FOR INPLANE SHEAR LOADING
AC-20	BONDED SYMMETRIC STEPPED LAP JOINT ANALYSIS
AC-31	LAMINATE DESIGN FOR STRENGTH AND STIFFNESS – MULTIPLE LOAD CONDITION, UPPER BOUND
AC-32	LAMINATE DESIGN FOR STRENGTH AND STIFFNESS – MULTIPLE LOAD CONDITIONS, OPTIMAL DESIGN
AC-33	LAMINATE DESIGN FOR ELASTIC STABILITY – MULTIPLE LOAD CONDITION, BIAXIAL INPLANE LOADING

Table 4 AFFDL Composite Computer Programs

PROGRAM NAME	PROBLEM TYPE	SOURCE
BONJO	ELASTIC-PLASTIC ANALYSIS OF SINGLE- AND DOUBLE-LAP BONDED JOINTS	LOCKHEED
BONJOI	ELASTIC ANALYSIS OF SINGLE- AND DOUBLE-LAP BONDED JOINTS	LOCKHEED
STPS4	STEP-LAP BONDED JOINT DESIGN AND/OR ANALYSIS	GRUMMAN
A4EA	SINGLE-LAP BONDED JOINTS	MCDONNELL DOUGLAS
A4EB	ELASTIC-PLASTIC DOUBLE-LAP BONDED JOINTS	MCDONNELL DOUGLAS
A4EC	ELASTIC, BALANCED, SCARF JOINTS	MCDONNELL DOUGLAS
A4ED	ELASTIC-PLASTIC, UNBALANCED, SCARF JOINTS WITH DESIGN CAPABILITY	MCDONNELL DOUGLAS
A4EE	ELASTIC-PLASTIC, UNBALANCED, SCARF JOINTS	MCDONNELL DOUGLAS
A4EF	ELASTIC, STEP-LAP BONDED JOINTS	MCDONNELL DOUGLAS
A4EG	ELASTIC-PLASTIC, STEP-LAP BONDED JOINTS	MCDONNELL DOUGLAS
TM1	ADVANCED COMPOSITE PANEL OPTIMIZATION	GENERAL DYNAMIC
RD5	ADVANCED COMPOSITE LAMINATE ULTIMATE STRENGTH ANALYSIS	GENERAL DYNAMIC
SQ5	LAMINATE POINT-STRESS ANALYSIS	GENERAL DYNAMIC

Table 5 Public Programs with No Central Source

PROGRAM NAME	PROBLEM TYPE	SOURCE
PANBUCK	PANEL BUCKLING	GRUMMAN
OPLAM	LAMINATE OPTIMIZATION	GRUMMAN
STOP3	LAMINATE OPTIMIZATION	NORTHRUP
SPADE	SANDWICH PANEL DESIGN	NORTHRUP
SX8	PANEL BUCKLING	GENERAL DYNAMICS
S00	TRAPEZOIDAL PLATE ANALYSIS	GENERAL DYNAMICS

GENERAL ANALYSIS

In order to evaluate the difference between the analyses of composite structures and that of isotropic structures, it is appropriate to examine the form of the basic equations governing the behavior of the material. The fundamental difference in the problem lies in the form of the constitutive equations. Hook's law for the nth layer of a composite laminate is given by

$$\begin{bmatrix} \sigma_x^{(n)} \\ \sigma_y^{(n)} \\ \sigma_{xy}^{(n)} \end{bmatrix} = \begin{bmatrix} C_{11}^{(n)} & C_{12}^{(n)} & C_{16}^{(n)} \\ C_{21}^{(n)} & C_{22}^{(n)} & C_{26}^{(n)} \\ C_{61}^{(n)} & C_{62}^{(n)} & C_{66}^{(n)} \end{bmatrix} \begin{bmatrix} \varepsilon_x \\ \varepsilon_y \\ \varepsilon_{xy} \end{bmatrix}$$

Using the classical assumptions of thin plate theory the constitutive equations become

$$\begin{bmatrix} N \\ M \end{bmatrix} = \begin{bmatrix} A & B \\ B & D \end{bmatrix} \begin{bmatrix} \varepsilon^\circ \\ \kappa \end{bmatrix}$$

where N and M are the force and moment result, ε° and κ are the midplane strains and curvatures, and

$$(A_{ij}, B_{ij}, D_{ij}) = \int_{-\frac{h}{2}}^{+\frac{h}{2}} C_{ij}^{(m)} (1, z, z^2)\, dz$$

The specific form of the constitutive equations varies widely depending upon the particular laminate in question. Most laminates encountered can be classified as pseudo-orthotropic laminates, and (Fig. 3). Pseudo-orthotropic laminates have midplane symmetry, for every ply with an orientation of θ there exists a ply at $-\theta$. This type of laminate is widely used and has a coupling between bending and twisting. For the most part this coupling (D_{16}, D_{26}) becomes small for 16 or more plies and can be neglected. For these cases the laminate may be treated as a classical orthotropic laminate (Fig. 4).

$$\begin{bmatrix} N_x \\ N_y \\ N_{xy} \end{bmatrix} = \begin{bmatrix} A_{11} & A_{12} & 0 \\ A_{21} & A_{22} & 0 \\ 0 & 0 & A_{66} \end{bmatrix} \begin{bmatrix} \epsilon_x \\ \epsilon_y \\ \epsilon_{xy} \end{bmatrix}$$

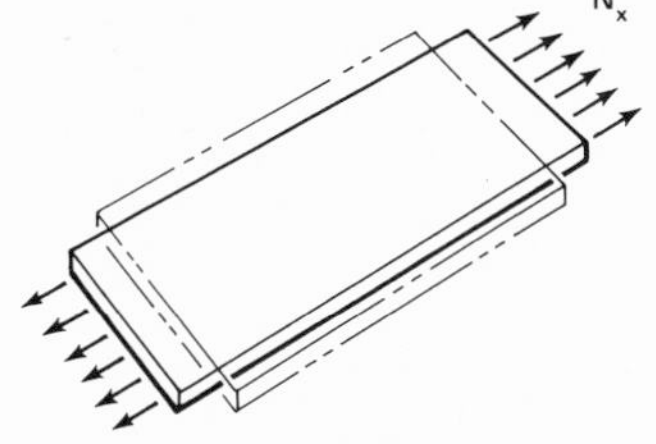

$$\begin{bmatrix} M_x \\ M_y \\ M_{xy} \end{bmatrix} = \begin{bmatrix} D_{11} & D_{12} & D_{16} \\ D_{21} & D_{22} & D_{26} \\ D_{61} & D_{62} & D_{66} \end{bmatrix} \begin{bmatrix} \kappa_x \\ \kappa_y \\ \kappa_{xy} \end{bmatrix}$$

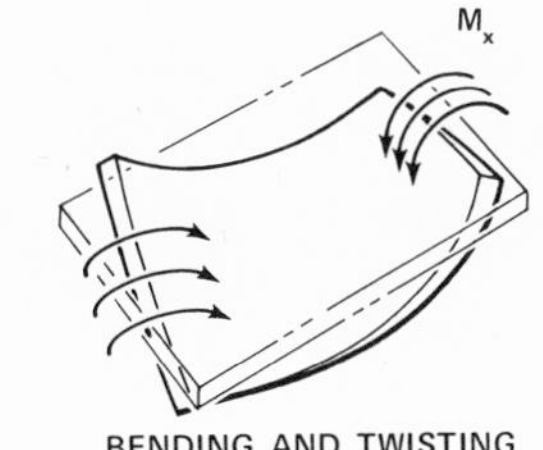

Fig. 3 Pseudo orthotropic laminates

$$\begin{bmatrix} N_x \\ N_y \\ N_{xy} \end{bmatrix} = \begin{bmatrix} A_{11} & A_{12} & 0 \\ A_{21} & A_{22} & 0 \\ 0 & 0 & A_{66} \end{bmatrix} \begin{bmatrix} \epsilon_x \\ \epsilon_y \\ \epsilon_{xy} \end{bmatrix}$$

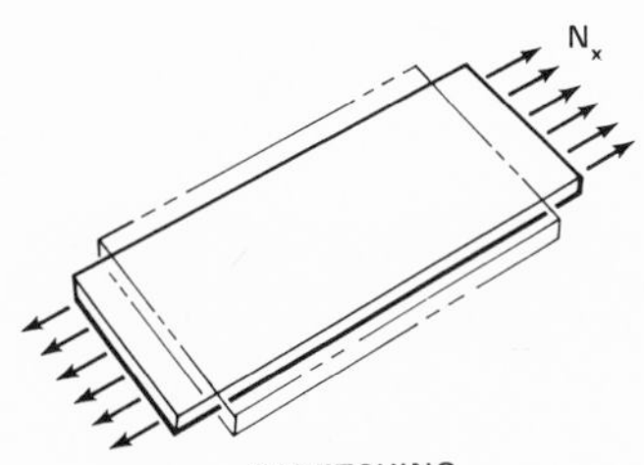

$$\begin{bmatrix} M_x \\ M_y \\ M_{xy} \end{bmatrix} = \begin{bmatrix} D_{11} & D_{12} & 0 \\ D_{21} & D_{22} & 0 \\ 0 & 0 & D_{66} \end{bmatrix} \begin{bmatrix} \kappa_x \\ \kappa_y \\ \kappa_{xy} \end{bmatrix}$$

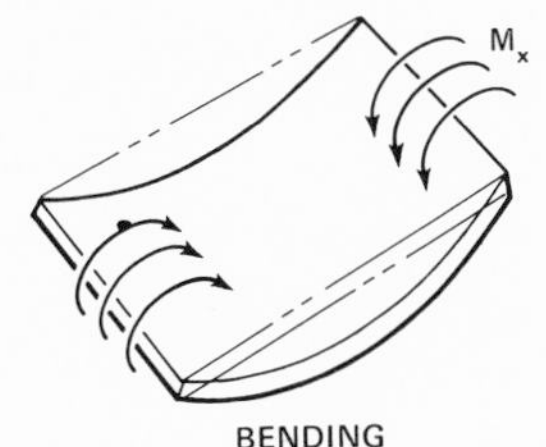

Fig. 4 Classical orthotropic laminates

SPECIAL PURPOSE COMPUTER PROGRAMS

Strength and Stiffness Analysis

Laminate strength and stiffness analysis is normally accomplished by assuming a state of the plane stress

$$\sigma_{xz} = \sigma_{yz} = \sigma_{zz} = 0$$

The following notation for stress and strain is employed:

$$\sigma^i_{LT} = \begin{Bmatrix} \sigma_L \\ \sigma_T \\ \tau_{LT} \end{Bmatrix}_{\text{lamina } i} \qquad \varepsilon^i_{LT} = \begin{Bmatrix} \varepsilon_L \\ \varepsilon_T \\ 1/2\ \gamma_{LT} \end{Bmatrix}_{\text{lamina } i}$$

$$\sigma^i_{xy} = \begin{Bmatrix} \sigma_x \\ \sigma_y \\ \tau_{xy} \end{Bmatrix}_{\text{lamina } i} \qquad \varepsilon^i_{xy} = \begin{Bmatrix} \varepsilon_x \\ \varepsilon_y \\ 1/2\ \gamma_{xy} \end{Bmatrix}_{\text{lamina } i}$$

$$\sigma_{xy} = \begin{Bmatrix} \sigma_x \\ \sigma_y \\ \tau_{xy} \end{Bmatrix}_{\text{laminate}} \qquad \varepsilon_{xy} = \begin{Bmatrix} \varepsilon_x \\ \varepsilon_y \\ 1/2\ \gamma_{xy} \end{Bmatrix}_{\text{laminate}}$$

Here, the L,T axes are aligned respectively with the longitudinal and transverse directions of the monolayer lamina. The x,y axes are aligned with the laminate's directions of orthotropy, if present, and, if not present, with any orthogonal reference directions. The strength and stiffnesses of the laminate are determined relative to the x,y axes.

The stress and strain states of the lamina obey the following laws of transformation:

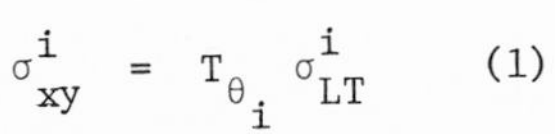

$$\sigma^i_{xy} = T_{\theta_i} \, \sigma^i_{LT} \qquad (1)$$

$$\varepsilon^i_{xy} = T_{\theta_i} \, \varepsilon^i_{LT} \qquad (2)$$

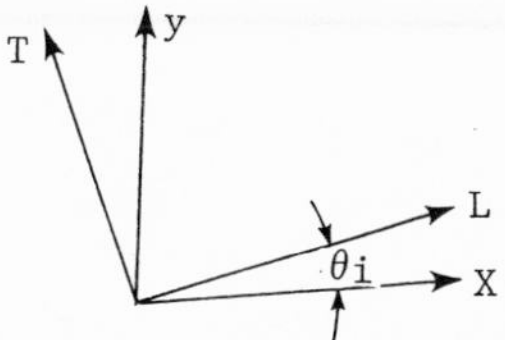

where T_θ is a transformation matrix with elements being functions of θ.

The inverse transformation matrix is denoted by

$$T^{-1}_{\theta}$$

$$T_{\theta_i} = \begin{bmatrix} \cos^2\theta_i & \sin^2\theta_i & \sin^2\theta_i \\ \sin^2\theta_i & \cos^2\theta_i & -\sin^2\theta_i \\ -1/2\ \sin^2\theta_i & 1/2\ \sin^2\theta_i & \cos^2\theta_i \end{bmatrix}$$

$$T^{-1}_{\theta_i} = \begin{bmatrix} \cos^2\theta_i & \sin^2\theta_i & -\sin^2\theta_i \\ \sin^2\theta_i & \cos^2\theta_i & \sin^2\theta_i \\ 1/2\ \sin^2\theta_i & -1/2\ \sin^2\theta_i & \cos^2\theta_i \end{bmatrix}$$

The monolayer constitutive matrix relates stress to strain and, for the lamina, is denoted by C_{LT}. The significance of the superscript is to permit lamina of different materials. The lamina constitutive relations are expressed as follows:

$$\sigma^i_{LT} = C^i_{LT} \, E^i_{LT} \qquad (3)$$

where

$$C^i_{LT} = \begin{bmatrix} E_L & M_{LT}E_T & 0 \\ M_{TL}E_L & E_T & 0 \\ 0 & 0 & 2\lambda_{LT}G_{LT} \end{bmatrix} \frac{1}{\lambda_{LT}}$$

E_L = monolayer longitudinal modulus

E_T = monolayer transverse modulus

G_{LT} = monolayer shear modulus

μ_{LT} = monolayer major Poisson's ratio

$$\lambda_{LT} = 1 - \mu_{LT}\mu_{TL} = 1 - \frac{E_T}{E_L}\mu^2_{LT}$$

A lamina's constitutive matrix can alternatively be defined relative to the x,y axes of the laminate. The lamina's constitutive relation employing this matrix can be derived from Eqs. (1), (2), and (3) as follows:

$$\sigma^i_{LT} = T^{-1}_{\theta_i}\,\sigma^i_{xy}$$

$$\sigma^i_{LT} = C^i_{LT}\,\varepsilon^i_{LT} = C^i_{LT}\,T^{-1}_{\theta_i}\,\varepsilon^i_{xy}$$

$$T^{-1}_{\theta_i}\,\sigma^i_{xy} = C^i_{LT}\,T^{-1}_{\theta_i}\,\varepsilon^i_{xy}$$

$$\sigma^i_{xy} = T_{\theta_i}\,C^i_{LT}\,T^{-1}_{\theta_i}\,\varepsilon^i_{xy}$$

The lamina's constitutive matrix relative to the x,y axes can now be identified from the last equation as

$$C^i_{xy} = T_{\theta_i} C^i_{LT} T^{-1}_{\theta_i} \tag{4}$$

For the whole laminate, Kirchoff's hypothesis is used to relate the strain of individual lamina to the laminate strain and curvature:

$$\varepsilon^i_{xy}(\zeta) = \varepsilon_{xy} + (z_i + \zeta)\,\kappa_{xy} \tag{5}$$

The parameter z_i denotes the distance from the laminate mid-line to the lamina mid-line, and ζ denotes the distance from the mid-line to a point on the lamina. The laminate strain and curvature are defined to be

$$\varepsilon_{xy} = \begin{Bmatrix} \varepsilon_x \\ \varepsilon_y \\ 1/2\ \gamma_{xy} \end{Bmatrix} \qquad \kappa_{xy} = \begin{Bmatrix} -\omega_{xx} \\ -\omega_{yy} \\ -\omega_{xy} \end{Bmatrix}$$

The parameter represents the lateral displacements of the laminate mid-line, and the subscripts denote differentiation. The laminate membrane and bending loads are defined to be, respectively,

$$N_{xy} = \begin{Bmatrix} n_x \\ n_y \\ n_{xy} \end{Bmatrix} \qquad M_{xy} = \begin{Bmatrix} m_x \\ m_y \\ m_{xy} \end{Bmatrix}$$

The laminate loads may be expressed by summoning the loads on the individual lamina

$$N_{xy} = \sum_i \int_{-1/2\,t_i}^{+1/2\,t_i} \sigma^i_{xy}\, d\zeta = \sum_i \int_{-1/2\,t_i}^{1/2\,t_i}$$

$$C^i_{xy} \left[\varepsilon_{xy} + (z_i + \zeta)\,\kappa_{xy} \right] d\zeta$$

$$= \varepsilon_{xy} \sum_i t_i C^i_{xy} + \kappa_{xy} \sum_i z_i t_i C^i_{xy}$$

$$M_{xy} = \sum_i \int_{-1/2\, t_i}^{+1/2\, t_i} (z_i + \zeta)\, \sigma^i_{xy}\, d\zeta = \sum_i \int_{-1/2\, t_i}^{1/2\, t_i} (z_i + \zeta)$$

$$C^i_{xy} \left[\varepsilon_{xy} + (z_i + \zeta)\, \kappa_{xy} \right] d\zeta$$

$$= \varepsilon_{xy} \sum_i z_i t_i\, C^i_{xy} + \kappa_{xy} \sum_i (z_i^2 t_i + t_i^3/12)\, C^i_{xy}$$

or,

$$\begin{Bmatrix} N_{xy} \\ M_{xy} \end{Bmatrix} = \begin{bmatrix} A & B \\ B & D \end{bmatrix} \begin{Bmatrix} \varepsilon_{xy} \\ \kappa_{xy} \end{Bmatrix} \tag{6}$$

where

$$A = \sum_i t_i\, C^i_{xy}$$

$$B = \sum_i z_i t_i\, C^i_{xy}$$

$$D = \sum_i (z_i^2 t_i + t_i^3/12)\, C^i_{xy}$$

For balanced laminates, B = 0. Eq. (6) is the laminate stiffness equation relating laminate loads to laminate strain and curvature.

For a balanced, orthotropic laminate, the laminate membrane stiffness matrix, A, may be defined in terms of laminate moduli E_x, E_y, G_{xy} and the major Poisson's ratio μ_{xy}:

$$\sigma_{xy} = C_{xy}\, \varepsilon_{xy}$$

where

$$C_{xy} = \begin{bmatrix} E_x & \mu_{xy}E_y & 0 \\ \mu_{yx}E_x & E_y & 0 \\ 0 & 0 & \partial\lambda_{x_y} \; G_{xy} \end{bmatrix} \frac{1}{\lambda_{xy}}$$

$$\lambda_{xy} = 1 - \mu_{xy}\ \mu_{yx} = 1 - \frac{E_y}{E_x}\mu_{xy}^2$$

and since

$$\sigma_{xy} = \frac{1}{t} N_{xy} = \frac{1}{t} A\ \varepsilon_{xy}$$

$$A = t\ C_{xy} = t \begin{bmatrix} E_x & \mu_{xy}E_y & 0 \\ \mu_{yx}E_x & E_y & 0 \\ 0 & 0 & \partial\lambda_{xy} \; G_{xy} \end{bmatrix} \frac{1}{\lambda_{xy}}$$

The laminate moduli and major Poisson's ratio may be expressed in terms of the elements in matrix A.

$$E_x = \frac{1}{t}\left\{A_{11}\ A_{22} - A_{12}^2/A_{22}\right\}$$

$$E_y = \frac{1}{t}\left\{A_{11}\ A_{22} - A_{12}^2/A_{11}\right\}$$

$$G_{xy} = \frac{1}{2t}\ A_{33}$$

$$\mu_{xy} = A_{12}/A_{22} \qquad (7)$$

where

F_L, F_T, F_{LT} are the basic lamina stress allowables

σ_L, σ_T, σ_{LT} are the lamina stresses

$\bar{\sigma}_L$, $\bar{\sigma}_T$, $\bar{\sigma}_{LT}$ are the lamina stresses corresponding to $E_T = 0$

ϕ is the specified fraction of ultimate load below which transverse tensile fracture must be precluded

Hill's failure criterion with the first modification is defined by the inequality $F_1 \geq 1$. The adopted failure criterion is governed by the inequality $F_2 \geq 1$.

It may be observed that the equation for F_2, corresponding to positive transverse tensile stress, is simply a comparison of three terms. The first term accounts for Hill's criterion with the first modification of single-stress cutoffs, as illustrated by the solid lines of Fig. 5. The second term is an allowance for transverse tensile strain beyond fracture and is illustrated by the dotted lines of Fig. 5. The third term checks for increased stresses accompanying possible transverse fractures and is indicated in Fig.5 by the dashed cutoff lines. It should be realized that the dashed line cutoffs are dependent on the particular laminate design as well as the laminate loading. But, of course, the loadings are being restricted to the three single-stress allowables.

The stress state of each lamina is a function of the single-stress allowables (F_x, F_y, F_{xy}) being sought for the laminate as a whole. These single-stress allowables are determined once any corresponding lamina stress state exceeds the failure criteria of that lamina. Once the single-stress allowables are determined for a laminate, a laminate interaction formula can be used to assess the factor of safety associated with that laminate for any loading case. Loading cases may consist of any combination of panel loads, N_x, N_y, and N_{xy}.

Most computer programs which have been developed for predicting the strength of laminates use these basic equations combined with a definition of failure. The definition of failure or criterion is the basic building block of a strength analysis procedure. The following six basic types of procedures have been considered.

1. Failure occurs when any lamina stress (strain) exceeds the uniaxial lamina allowable stress (strain) in the direction of the filaments. This criterion assumes that filament fracture dominates all in-plane failure modes of the laminate.

2. Failure occurs when any of the three in-plane lamina stress (strain) components exceeds the corresponding allowable stress (strain).

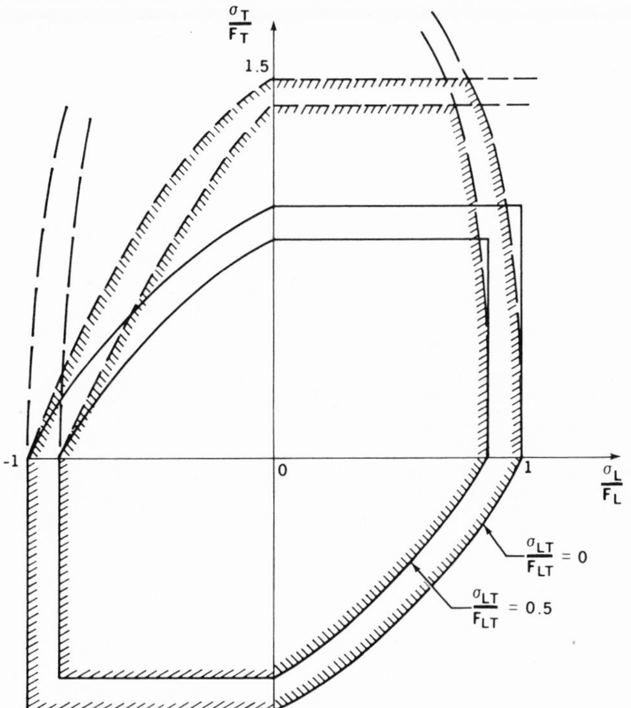

Fig. 5 Lamina failure criteria

3. Failure occurs when the stress (strain) state of any lamina violates a selected interaction inequality.

4. Failure occurs when the laminate stress state violates an interaction inequality which is based on single-stress allowable established from criterion [1]. For this criterion, the three in-plane laminate allowable stress components are, first, independently established using criterion [1]. A selected laminate interaction relation is then utilized to predict failure of the laminate.

5. Failure occurs when the laminate stress state violates a selected interaction inequality which is based on single-stress allowables established from criterion [2].

6. Failure occurs when the laminate stress state violates a selected interaction inequality which is based on single-stress allowables established from criterion [3].

It should be noted that procedures 3 and 6 require the use of an interaction formula on the lamina, while procedures 4, 5, and 6 require the use of an interaction formula on the laminate. There are several interaction formulas or criteria which have found use in laminate strength analysis. Examples of interaction formulas which have been proposed follow.

1. Tsai Criterion [4]

The strength criterion for plane stress is given for any layer with respect to the L-T coordinate system as:

$$(\sigma_L/F\ )^2 - (\sigma_L\sigma_T)/F_L{}^2 + (\sigma_T/F_T)^2 + (\sigma_{LT}/F_{LT})^2 = 1$$

It is assumed that this criterion remains applicable when the material properties are different in tension and compression. Therefore, it is only necessary to modify the allowable stresses F_L and F_T, depending on the sign of σ_L and σ_T. The four possibilities are presented in Table 3.

This strength criterion gives no indication as to the type of failure (e.g., transverse, longitudinal). In most cases, it would appear that a transverse failure would occur first and that the layer would still be capable of carrying load in the longitudinal direction. There is no positive way of ensuring that this is the case. However, the ration of σ_L/F_L should give an indication as to the mode of failure. Thus, if the ratio of σ_L/F_L is less than some prescribed value, it could be assumed that the failure was transverse. For initial design purposes this ratio is set at approximately 0.95, although this value is tentative and may have to be modified as test data are generated.

2. Hoffman Criterion [5]

An alternative strength criterion was proposed by Hoffman. This criterion is basically the same as Tsai's, except that terms are included that are odd functions of σ_L, σ_T, and σ_{LT}. These terms were chosen as linear terms and yield the following equations for the case of plane stress:

$$(\sigma_L{}^2 - \sigma_L\sigma_T)/(F_cF_t)_L + \alpha_T{}^2/(F_cF_t)_T + (F_c - F_t)/F_cF_{t_L}\ \sigma_L$$

$$+ (F_c - F_t)/F_cF_{t_T}\ \sigma_T + (\sigma_{LT}/F_{LT})^2 = 1$$

When $F_{c_L} = F_{t_L}$ and $F_{c_T} = F_{t_T}$, this equation is identical to Tsai's.

3. Modified Tsai Criterion [6]

The three allowable single-stress components are determined using a Hill-type quadratic interaction formula for each of the individual lamina. This criterion, however, has been modified in several ways. The first modification is a limitation to ensure that neither of the basic normal stress allowables is exceeded.

Another modification has been incorporated to counter premature failure predictions of the laminate due to low transverse tensile allowables, characteristic of most composite lamina. The ultimate strength of a laminate is not necessarily recorded at the first sign of a lamina transverse tensile fracture. Such a mode of failure is triggered by a breakdown of the epoxy rather than the filaments and the lamina. It is, therefore, felt that as long as this mode of failure is precluded until a specified fraction of the ultimate load is reached, these failure modes are permissible in a limiting sense for the ultimate laminate strength predictions. This specified fraction of ultimate load is generally taken as 2/3, corresponding to limit load for aircraft structures. Of course, the other lamina stresses can be expected to increase as transverse tensile fractures occur. A final modification of Hill's interaction formula accounts for these increases. The finally adopted failure criterion for the individual lamina may be defined mathematically by the use of two functions, F_1 and F_2.

$$F_1 (\sigma_L, \sigma_T, \sigma_{LT}) = \max\left\{ \left|\frac{\sigma_L}{F_L}\right| , \left|\frac{\sigma_T}{F_T}\right| , \text{SQRT}\left[\left(\frac{\sigma_L}{F_L}\right)^2 + \left(\frac{\sigma_T}{F_T}\right)^2 - \left(\frac{\sigma_L}{F_L}\right)\left(\frac{\sigma_T}{F_T}\right) + \left(\frac{\sigma_{LT}}{F_{LT}}\right)^2\right]\right\}$$

$$F_2(\sigma_L, \sigma_T, \sigma_{LT}, \bar{\sigma}_L, \bar{\sigma}_T, \bar{\sigma}_{LT}) = \min\left\{ F_1 (\sigma_L, \sigma_T, \sigma_{LT}), \max\left[F_1 (\sigma_L, \phi\sigma_T, \sigma_{LT}), F_1 (\bar{\sigma}_L, \bar{\sigma}_T, \bar{\sigma}_{LT})\right]\right\} \quad \ldots \text{ when } \sigma_T > 0$$

$$= F_1 (\bar{\sigma}_L, \bar{\sigma}_T, \bar{\sigma}_{LT}) \quad \ldots \text{ when } \sigma_T \leq 0$$

Although there would seem to be an almost endless confirmation of possibilities, apparently there are only three programs which are available for distribution. These three programs are discussed in the following paragraphs together with a strength approach which has been developed by Grumman. This approach is mentioned even though no program is available for distribution; it is possible to develop a short simple program from the equations in [7].

Grumman Approach

The Grumman approach to strength analysis uses procedure 4 where single stress allowables are established considering only failure modes involving filament fracture. The basic tension and compression allowables are established by relating the laminate allowables to the unaxial lamina allowables. From the equations at the start of this section:

$$\sigma_{xy} = C^{i}_{xy} \varepsilon_{xy} = C_{xy} \varepsilon_{xy} = C_{xy} T_{\theta_i}^{-1} \varepsilon_{LT}^{i}$$

$$= C_{xy} T_{\theta_i}^{-1} {C_{LT}^{i}}^{-1} \sigma_{LT}^{i}$$

In the Grumman approach an attempt is made to account for the fact that for certian laminates the matrix may not be completely effective and a matrix effectiveness factor is used. The laminate shear allowables are established from

$$F_{xy} = n F_{xy}^{\pm 45} + (\ell + m) F_{xy}^{1}$$

where

- ℓ = Fraction of 0-degree plies
- m = Fraction of 90-degree plies
- n = Fraction of 45-degree plies
- $F_{xy}^{\pm 45}$ = Test-established allowable for 45-degree laminate
- F_{xy}^{1} = Shear stress in the 0-degree and 90-degree plies at the failing shear strain of the 45-degree plies

AC-3 Point Stress Analysis

This program is listed in the Air Force Design Guide and has an input and output description. Laminate flexural and extensional stiffnesses and margins of safety are calculated for in-plane and out-of-plane loads. Two theories of failure are available in the program. The first is a strain criterion where lamina strains are found and compared with the allowable lamina strains to determine a margin of safety. The second is a strength criterion, and margins of safety are found from:

$$R_L \leq 1$$

$$R_T^2 + R_{LT}^{2} \leq 1$$

where

$$R_L = \frac{\sigma_L}{F_L}$$

and

$$R_T = \frac{\sigma_T}{F_T}$$

$$R_{LT} = \frac{\sigma_{LT}}{F_{LT}}$$

RD5 Advanced Composite Laminate Ultimate Strength Analysis

This program [8] incrementally produces the stress-strain curve to ultimate for laminate consisting of orthotropic lamina which have nonlinear stress-strain behavicr. The program is applicable only to plane anisotropic laminates with mid-plane symmetry subjected to membrane loads and uses a maximum strain criterion. The lamina stress-strain characteristics are data base information within the program. This program was developed by General Dynamics and is maintained in the file of the Air Force Flight Dynamics Laboratory Composites Branch. The nonlinear feature of this program is particularly useful for materials which exhibit highly nonlinear transverse behavior such as boron/aluminum. A linear version of this program, named SQ5, has also been developed by General Dynamics.

M71-10112 Analysis of Multilayered Fiber Composites

This program [9] is available from COSMIC. A semiempirical failure theory is used to conduct a laminate failure test based upon a given set of loads. The program incorporates theory-experiment correlation factors to account for variables which cannot be directly accounted for.

Composite Structure Stability Analysis

Stability analysis of composite structures is a straightforward extension of methods used in isotropic structures. The formulation of the problem is generally more complex but does not represent a new area of analysis as is the case with laminate strength and stiffness analysis. For most practical laminates the existence of the D_{16}, D_{26} terms in the constitutive equations introduces added complexities not found in classical orthotropic analysis. These terms can be significant for thin laminates and should be considered.

As is the case with isotropic structures, stability problems fall into three general classes:

1. Panel stability
2. Section stability
3. General stability

Although several excellent programs exist for panel stability, the areas of section and general stability are only recently drawing attention.

PANBUC Panel Buckling

The PANBUC program [10] was developed by Grumman under an Air Force contract and uses classical solutions for stability analysis of classical orthotropic laminates. Although there were problems with the earlier versions of the progam, the current versions are well checked out, have straightforward input and well-labeled output, and require only nominal run times. The program predicts the critical elastic buckling loads and modes for rectangular laminated composite plates with midplane symmetry and for honeycomb sandwich panels subjected to inplane biaxial compression and shear loads applied parallel to the panel's axes of orthotropy. Edge conditions may be fully clamped, simply supported, or a combination of these. There is apparently no central source for this program, and it would have to be obtained from Grumman after obtaining approval from the Advanced Composites Division of the Air Force Materials Laboratory.

SX8 Panel Buckling

The SX8 program [11] was developed by General Dynamics and uses the Ritz approach to minimize the potential energy. The program is considerably more powerful than the PANBUC program in that normal loads and varying edge loads may be included. Provisions for elastic boundary conditions and panel stiffening are also included. The program accounts for the D_{16} and D_{26} terms and uses the reduced stiffness approach to analyze unbalanced laminates. The user may desire to modify the input and output routines as both as fairly awkward in the present form. This program would have to be obtained from General Dynamics after obtaining approval from the Advanced Composites Division of the Air Force Materials Laboratory. A version of the program which can be used for trapezoidal panels is named SOO. Another version of this program is available in the Air Force Composites Design Guide [3] as program AC-10 Anistropic Plate Analysis.

AC-5 Honeycomb Sandwich Panel Stability under Inplane Biaxial Loading

The AC-5 program is available in the Air Force Advanced Composites Design Guide [3]. Buckling strengths of honeycomb sandwich panels with classical orthotropic face sheets subjected to biaxial loads are calculated. The program is set up for parametric studies of various load ratios and panel aspect ratios.

AC-11 Honeycomb Sandwich Panel Stability for Inplane Shear Loading

The AC-11 program is available in the Air Force Advanced Composites Design Guide [3]. Buckling strengths of honeycomb sandwich panels with classical orthotropic face sheets subjected to inplane shear loads are calculated.

BUCKLASP

The BUCKLASP program [12] is available through COSMIC. This program calculates buckling loads for structures of uniform cross section that may be idealized as an assemblage of laminated flat and curved plate-strips and beams. Arbitrary boundary conditions may be specified on any of the external longitudinal sides of the plate-strips. The program provides a powerful capability for the analysis of stiffened compression panels. One drawback exists, however, in that there are currently no provisions for inplane shear loads.

OPTIMIZATION OF COMPOSITE LAMINATES

Although the discipline of optimization in structural design has become ever important in the past 10 years for conventional design, the advent of advanced composite materials in design has made this discipline even more necessary. In conventional design, the sizes and locations of fabricated parts are important in least-weight designs. With composite materials, not only are these factors still important, but a new dimension is added. The orientation in which the built-up plies are patterned is also vitally important to the weight of the structure. Hence, some sort of optimization is always incorporated in designs using the advanced composites.

The most frequently encountered optimization problem in designing composite structures is the following. Given a set of loading conditions, each consisting of combined membrane panel loads, and a set of minimum stiffnesses, what is the optimum pattern of ply orientations? In practice, it has been found that a reasonably good design can be determined if only 0-, ±45-, and 90-degree orientations are treated. In this case, it is only necessary to determine X = (L, M, N) from a 3-dimensional design space, where L, M, N denote the number of 0-, 90-, and ± 45-degree plies, respectively. Several of the best-known optimization programs are described in some detail in the following paragraphs.

RC7

This is a computer program which was coded by General Dynamics [13]. This program determines for up to 20 plies, the optimum pattern of orientation at 5-degree increments. Combined membrane loads for a set of loading conditions are treated, but no stiffness constraints are treated. Either of two strength criteria may be used, the Hill Tsai stress criterion or the maximum strain criterion.

The search for the optimum layup for a given set of load conditions is quite exhaustive and takes place essentially as follows. First, the best location for a single lamina is determined by considering every orientation from 0 to 180 degrees (in 5-degree increments). If one lamina is insufficient to sustain the applied loads, a second lamina is added (with the first ply locked in position) and swept through 180 degrees (in 5-degree increments) to determine the best location for the second ply. A key feature of the program now comes into play. Realizing that the location of the second ply as just determined was found with the constraint imposed that the first ply was locked in position, this constraint is now relaxed. That is, a lamina readjustment or reorientation is performed. The program performs two coarse reorientations followed by two fine reorientations. A coarse reorientation, for a 2-ply laminate consisting of plies initially at θ_1 and θ_2, consists of trying the 25 layups defined by all possible combinations of $\theta_1 + \Delta\theta_1$, $\theta_2 + \Delta\theta_2$ where $-30° < \Delta\theta_1 \leq 30°$ and $-30° \leq \Delta\theta_2 \leq 30°$ and 15° increments are used.

After the two coarse reorientations are performed and two angles α_1 and α_2 are determined as best for the two plies, two fine reorientations are performed. For the 2-ply laminate, a fine reorientation consists of trying the 25 layups defined by all possible combinations of $\alpha_1 + \Delta\alpha_1$, $\alpha_2 + \Delta\alpha_2$ where $-10° \leq \Delta\alpha_1 \leq 10°$ and $-10° \leq \Delta\alpha_2 \leq 10°$ and 5° increments are used. The layup finally arrived at after performing the two coarse and two fine reorientations is then tested to determine if positive margins of safety exist for the loadings applied. If not, another ply is added and the process is repeated.

To see the necessity of having the reorientation feature in the program, consider the case of equal biaxial loading. The best location for a single ply is found to be at 45 degrees (i.e., the lamina orientation bisects the loading directions). When a second ply is added, it is initially placed at 135 degrees. The series of two coarse and two fine reorientations finally results in the plies being oriented in the load directions as expected intuitively. If the reorientation feature were not present, there would be no way of placing the fibers in the load directions for the case of equal biaxial loading. (In the discussion the loads are assumed to be tensile; so stability is not a consideration.)

The procedure described above for the case of two plies is repeated for a laminate consisting of up to six plies. Note that for a 3-ply laminate, each coarse reorientation consists of 125 trial orientations so that the total reorientation routine, containing two coarse and two fine reorientations, consists of 500 trial orientations. For a 4-ply laminate on the other hand, the total number of reorientations considered is 2500. It has been found that program runs which employ four lamina readjustment (as described above) can be executed in about a minute on the IBM 360/65 computer.

In addition to the lamina readjustment feature just described, the RC7 program possesses an additional capability which expedites the search for the optimum configuration. Whenever an additional ply is added to a laminate and placed in the best position (with the constraint imposed that all other plies remain locked in position), the margins of safety for the laminate are computed and stored. The current number of plies in the laminate is then broken down into prime numbers, and the optimum orientations for these prime numbers are checked to see if they comprise a more efficient layup. For example, suppose one is at the point where a sixth ply has just been added to a laminate. With the five previously considered plies locked in position, the best location (i.e., the orientation giving rise to the highest safety margins) is determined for the sixth ply. These margins of safety are then compared to those which would obtain if one utilized the optimum 2-ply orientation (e.g., +45-, +135-degree) as a triply repeated element in a 6-ply layup (resulting in a +45-, +135-, +45-, +135-, +45-, +135-degree layup), and also to those one would obtain from the optimum 3-ply orientation (e.g., +60-, +90-, +130-degree) repeated twice in a 6-ply layup (resulting in a 60-, 90-, 130-, 60-, 90-, 130-degree configuration). The best of these choices is used when subsequent plies are added, if found to be necessary.

The RC7 program is not available through any of the central sources and would have to be obtained by General Dynamics with the approval of the Advanced Development Division, Air Force Materials Laboratory. An updated version of the program named TMI is available from the Advanced Composites Branch of the Air Force Flight Dynamics Laboratory.

OPLAM

OPLAM [14] is another optimization program which is cast as a constrained optimization problem, except the search is conducted in the feasible design space. This program was coded by Grumman Aircraft, and like the Douglas program, is restricted to a search for the optimum number of 0-, ±45-, and 90-degree orientations. The initial, feasible design point is determined by assuming that failure occurs when the longitudinal rupture strain is exceeded and by using the maximum of the applied loads. Layers are then removed such that the excess strength potential is reduced in a maximum degree. After the initial feasible design point has been found, the Hill-Tsai strength criteria are used. Sets of combined membrane loads for up to five different temperatures are satisfied by the resulting laminate. Although the program was initially coded for boron-epoxy laminates, it can easily be extended to cover other composites.

AC-31

This program is available in the Air Force Advanced Composites Design Guide [3]. It is a classic example of a constrained optimization problem. The method employed here is the method of centered circles, and is much more general a method than those which have been described heretofore. This method is a search in the feasible design space, where the centers of hyperspheres positioned tangent to the constraint planes force the sequence of design points to funnel down the middle of the acceptable region. This program allows for sets of combined membrane loads as well as for stiffened constraints.

STOP3 AND SPADE

These programs were coded by Northrup [15] and are essentially an extension of RC7. In addition to membrane loading requirements and minimum stiffness requirements, these programs determine laminates which must also satisfy a stability requirement associated with an additional specified loading condition. In the case of STOP3, a laminate plate is determined, while in the case of SPADE, a sandwich panel with composite face sheets and one of several core designs is established as the optimum design. Associated with the stability option, one of four different boundary conditions may be specified for the rectangular panel, and then all combinations of compression and shear may be entertained.

BONDED JOINT ANALYSIS

Hart-Smith [16] identified 12 types of bonded joints of interest (Fig. 6). The large number of joint types and the complex analysis associated with each has resulted in the development of a number of computer programs. Each of the programs which are discussed here is on file with the Advanced Composites Branch of the Air Force Flight Dynamics Laboratory. The programs A4EA-G were developed under contract to NASA [17].

STEPS

This first computer program developed by Grumman to analyze stepped-lap bonded joints started as a purely elastic analysis of a composite-to-metal joint. It has now been modified by the inclusion of an empirical modification of the adhesive elastic stress concentration for each step to incorporate an allowance for adhesive nonlinearity in terms of the secant modulus at failure. The method does account for adherend stiffness imbalance but not for the thermal mismatch and necessarily associated with composite-to-metal joints.

BONJO 1

The Lockheed-developed program applies to single- and double-lap bonded joints. It accounts for either or both adherend transverse tension and transverse shear deformation as well as the usual in-plane adherend deformations and adhesive shear and peel stresses. The significant contribution is the softening of the shear stress peaks by accounting for finite transverse moduli in the adherends. The solution is basically elastic, and the originators of the program have applied the empirical plastic-zone approach to account for adherend nonlinear behavior.

A4EA - Single-Lap Joints

This program computes both the joint strength and efficiency (with respect to the basic adherend strength) for each of three possible failure modes. These are adherend bending due to the eccentric load path, adhesive shear failure, and adhesive peel (or composite interlaminar tension) failure mode. Each prevails over a different range of thicknesses with the order given above being associated with progressive increases in thickness. The analysis is based on elastic-plastic adhesive behavior in shear and elastic behavior in peel. Full account is taken of the eccentricity in the load path, but provision for dissimilar adherends is not included.

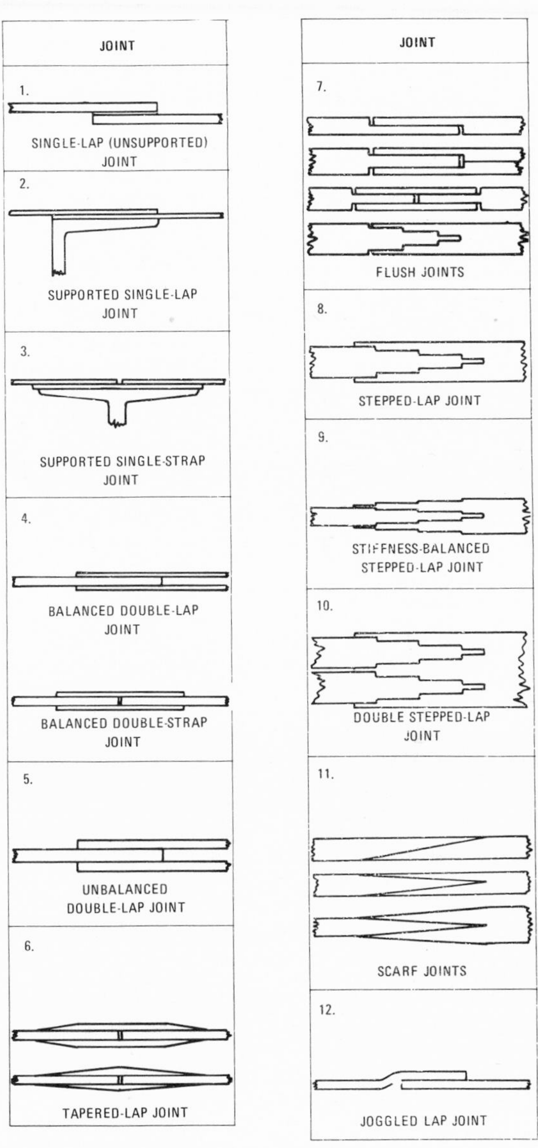

Fig. 6 Bonded joint concepts

A4EB - Double-Lap Joints

The nondimensionalized adhesive shear joint strength and average bond shear stress are computed as a function of three nondimensionalized joint parameters. These three characterize the adhesive plastic-to-elastic strain behavior, any adherend stiffness imbalance, and thermal mismatch between adherends of dissimilar materials. The shear strength and average shear stress are computed as functions of the nondimensionalized overlap. A peel stress cut-off, governed by an explicit formula, must be applied for thick sections, the failure of which occurs at loads less than those predicted by the shear analysis.

A4EC - Elastic Scarf Joints

This power-series solution for the adhesive shear stresses between nonidentical adherends estimates the joint strength but not internal stress distribution. The nondimensionalized joint strength and ratio of average-to-peak adhesive shear stress are computed in terms of the nondimensionalized overlap. The joint parameters account for adherend thermal mismatch and stiffness imbalances. Peel stresses are ignored on the basis of the small scarf angles used in practice.

A4ED - Elastic-Plastic Scarf Joints (Lower-Bound Solution)

A pair of explicit algebraic solutions are solved by iteration to provide a very close lower bound estimate of the nondimensionalized joint strength and average adhesive shear stress for bonded scarf joints between dissimilar adherends. The independent variable is the nondimensionalized overlap and the joint parameters accounted for are adhesive plasticity, adherend stiffness imbalance, and adherend thermal mismatch. This simple solution is adequate for most practical configurations.

A4EE - Elastic-Plastic Scarf Joints (Precise Solution)

A power series solution is obtained for the nondimensionalized joint strength and average bond shear stress for scarf joints. The factors accounted for are adhesive plasticity, adherend thermal mismatch, adherend stiffness imbalance, and joint overlap. With the high strengths predicted by the inclusion of adhesive plasticity, it is necessary also to check on the adherend stress at the tip of the stiffer adherend.

A4EF - Elastic Stepped-Lap Joints

This program solves, in dimensional form, for the internal adhesive and adherend stresses and load capacity of arbitrary stepped-lap joints. The step increments are all independent. Factors accounted for include different gross adherend stiffness and thermal properties, adhesive properties, and change in strength with load direction.

Peel stresses are excluded because the end steps are typically too thin to be prone to peel failures.

A4EG - Elastic-Plastic Stepped-Lap Joints

The inclusion of adhesive plasticity in the analysis of stepped-lap joints having arbitrary independent step sizes provides a means for successfully optimizing the joint detail proportions. Dissimilar adherend stiffnesses and thermal characteristics are provided for. Internal load distributions for both the adhesive and adherend can be computed at both ultimate and partial load levels.

CONCLUSIONS

Special purpose computer programs in the four basic composite categories of strength and stiffness analysis, stability analysis, laminate optimization, and bonded joint analysis have been reviewed. The programs discussed are public programs and can be obtained by qualified organizations. Most of the programs have been developed as part of government-sponsored research and development programs and as such may require additional development to enable use on a production engineering basis.

ACKNOWLEDGMENT

Significant contributions were made to the paper by Mr. C. G. Dietz and Dr. L. J. Hart-Smith of the Douglas Aircraft Company in compiling information on laminate optimization and bonded-joint analysis computer programs, respectively. Verbal or written contributions were also made by Mr. P. E. Parmley, Air Force Flight Dynamics Laboratory, Dr. D. L. Reed, General Dynamics Corporation, and Dr. D. Y. Konishi, Rockwell International Corporation.

REFERENCES

1 Schjelderup, H. C. and Purdy, D. M., "Advanced Composites - The Aircraft Material of the Future," AIAA Third Aircraft Design and Operations Meeting, July 1971.

2 Purdy, D. M. and Schaeffer, H. G., "Stress Analysis of Composite Structures," Design Engineering Conference American Society of Mechanical Engineering, April 1971.

3 Advanced Composites Design Guide, Advanced Development Division, Air Force Materials Laboratory, Wright-Patterson Air Force Base, Ohio, Third Edition January 1973.

4 Tsai, S. W., "Strength Characteristics of Composite Materials," NASA Report CR-224, 1965.

5 Hoffman, O., "The Brittle Strength of Orthotropic Materials," Journal of Composite Materials, 1, n2, p. 200, 1967.

6 Purdy, D. M., Dietz, C. G., and McGrew, J. A., "Optimization of Laminates for Strength and Flutter," Air Force Conference on Fibrous Composites in Flight Vehicle Design, Dayton, Ohio, September 1972.

7 Hadcock, R., et al., "Advanced Composite Wing Structures Boron-Epoxy Design Data, Volume II, Analytical Data," Advanced Composite Division, Air Force Materials Laboratory, Wright-Patterson Air Force Base, Ohio, November 1969.

8 Petit, P. H., "Ultimate Strength of Laminated Composites," AF33615-5257, December 1967.

9 Chavis, C. C. and Delivuk, T., "Multilayered Filamentary Composite Analysis Computer Code-User's Manual," Rep. 26, Case Western Reserve University, 1968.

10 Cairo, R. P. and Hadcock, R. N., "Optimum Design and Strength Analysis of Boron-Epoxy Laminates," Advanced Composites Division, Air Force Materials Laboratory, Wright-Patterson Air Force Base, Ohio, December 1968.

11 Ashton, J. E., "Anisotropic Plate Analysis," FZM-4899, Contract AF33(615)-5257, General Dynamics/Fort Worth Division, October 1967.

12 Tripp, L. L., Tamekuni, M., and Viswanatham, A. V., "A Computer Program for Instability Analysis of Biaxially Loaded Composite Stiffened Panels and Other Structures," NASA CR-112, 226, March 1973.

13 Weddoups, M. E., McCullers, L. A., Olsen, F. O., and Ashton, J. E., "Structural Synthesis of Anisotropic Plates," AIAA/ASME 11th Structures, Structural Dynamics, and Materials Conference, Denver, Colorado, April 1970.

14 Hadcock, R. N., et al., "Preliminary Analysis and Optimization Methods," Contract F33615-68-C-1301, Advanced Composite Division, Air Force Materials Laboratory, Wright-Patterson Air Force Base, Ohio, August 1968.

15 Verette, R. M., "Stiffness, Strength, and Stability Optimization of Laminated Composites," Northrop Report No. NOR70-138, August 1970.

16 Hart-Smith, L. J., "Advances in the Analysis and Design of Adhesive-Bonded Joints in Composite Aerospace Structures," The Society for the Advancement of Material and Process Engineering's 19th National Symposium and Exhibition, Anaheim, California, April 1974.

17 Hart-Smith, L. J., "Analysis and Design of Advanced Composite-Bonded Joints," NASA CR2218, January 1973.

AEROELASTICITY

John Kenneth Haviland
University of Virginia
Charlottesville, Virginia

Dale E. Cooley
Air Force Flight Dynamics Laboratory
Wright-Patterson Air Force Base, Ohio

ABSTRACT

This paper presents a review of computer programs available to the general user for aeroelastic calculations. Included is a brief description of the important static and dynamic aeroelastic phenomena, with a short mathematical treatment of these problems, in which the aeroelastic equations are formulated. There is also a short review of the aerodynamic flow equations in use, described in terms of the free-field Green's function. In discussing both the aeroelastic equations and the flow equations, the work of many authors has been condensed into a simple brief notation for the sake of clarity and conciseness. Following this, some of the successful methods of analysis are discussed, including the versatile and now widely used numerical procedures for three-dimensional lifting surface aerodynamic representations. Some comments on applications to structural optimization with aeroelastic constraints are also given. Finally, some of the available computer programs are described. These include a variety of steady and unsteady procedures for subsonic, transonic, and supersonic flow, some with aerodynamic interference effects. The various programs cover the prediction of flutter, divergence, reversal, and aeroelastic deflections and will, in some cases, optimize the weight to satisfy flutter speed constraints. Information is provided regarding methods used, program capabilities, published references, and points of contact for the computer programs. The latter information was based on a survey, the results of which indicated that much of the software used in the aerospace industry is considered to be proprietary and is not therefore generally available, so that most of the programs described in this paper were developed under government-sponsored efforts and some are available only to qualified users.

NOMENCLATURE

Roman

B = Matrix defined by Eq. (6)
c = Speed of sound
$[C]$ = Flutter air force matrix
C'_{RR}, etc. = Submatrices of $[C]$
F = Cutoff function defined under Eq. (10)
g = Structural damping ratio
$G, G(\bar{r},\bar{\rho})$ = Free-field Green's function

$[K]$ = Stiffness matrix
K_{SS} = Submatrix of $[K]$
$K(\bar{r},\bar{\rho})$ = Kernel function
M = Mach number V/c
$[M]$ = Mass matrix
M_{RR}, etc. = Submatrices of $[M]$
n = Local normal coordinate to a one-sided surface in $\bar{r}$ space
$\{q\}$ = Matrix vector of generalized coordinates
q_R, etc. = Subvectors of $\{q\}$
$\{Q\}$ = Matrix vector of generalized forces
Q_A, etc. = Subvectors of Q
R = Acoustical distance defined under Eq. (10)
$\bar{r}$ = Coordinate vector
s = One-sided surface area
s_A, s_W = Airfoil and wake surfaces, respectively
V = Free-stream velocity
$\bar{V}$ = Free-stream velocity vector
w = Local flow velocity normal to surface
X = Eigenvalue of flutter equation
X,Y,Z = Components of vector $\bar{\rho}-\bar{r}$, with X parallel to $\bar{V}$

Greek

α = Time constant of motion
β = $\sqrt{1 - M^2}$
δ = Local deflection normal to surface
μ = Local normal coordinate to boundary surface in $\bar{\rho}$ space
ν = Local normal coordinate to one-sided surface in ρ space
ρ = Air density
$\bar{\rho}$ = Dummy coordinate for $\bar{r}$
σ = Boundary surface area in $\bar{\rho}$ space
τ_R, τ_A = Retarded and advanced times of transit of acoustical waves, respectively
ϕ = Velocity potential
ψ = Acceleration potential
ω = Angular frequency

Mathematical Symbols

$L\{\ \}$ = Wave operator
$\delta(\ )$ = Dirac delta function
$\Delta\phi$ = Difference of ϕ across one-sided surface (positive when quantity decreases in positive ν direction)
$U(\ )$ = Step function
∇ = Grad operator
$\{\ \}$ = Matrix vector
$\{\ \}^T$ = Transposed matrix vector
$[\ ]$ = Matrix
$\bar{V}$ = Denotes vector V
C' = Denotes real part of C
C'' = Denotes imaginary part of C
i = $\sqrt{-1}$

Subscripts

A = Aerodynamic, airfoil, or advanced
E = External
R = Rigid body, or retarded
S = Elastic (referring to modes)
W = Wake

INTRODUCTION

The body of knowledge comprising aeroelasticity has accumulated over the past sixty years, starting with the historic paper by Lancaster [1] in 1916. Many individuals have made significant contributions, but it is impossible to describe them in a brief review of this nature. Aeroelasticity mainly concerns aircraft but has obvious application to launch vehicles passing through the atmosphere, as well as to underwater hydrofoils, galloping power lines, and even a celebrated bridge failure.

For background reading, there are texts by, to name a few, Scanlan and Rosenbaum [2], Fung [3]; Bisplinghoff, Ashley and Halfman [4]; Bisplinghoff and Ashley [5]; and the five-volume NATO manual [6]. A selection of significant papers has been edited by Garrick [7]. Also, there have been many review articles [8-24].

Starting with the work of Frazer, Duncan, and Collar [25], matrix notation has been used by many workers in the field. Of special signficance was the paper by Loring [26], who made extensive use of matrix algebra to derive the flutter equation, and who was first to use the truly generalized mode approach, in contrast to such terminology as "torsion-bending flutter." Development of expressions for airforces on two-dimensional oscillating wings in the 1940s gave way to methods of calculating airforces on oscillating three-dimensional wings in the 1950s, which have since been refined and applied to interacting, out-of-plane surfaces.

Development of finite element methods in structural analysis has come much more recently and has been geared to the thinking induced by large high-speed computers. Presently, there appears to be a need for parallel developments in unsteady aerodynamics, so that aerodynamic forces can be calculated in a manner consistent with structural component breakdowns.

In the present paper, the important aeroelastic phenomena are briefly described, and reasons for the present emphasis on treating the aerodynamics of harmonically oscillating airframes by linear perturbation of uniform flows are given. Next, formulations of aeroelastic equations are described, followed by a short review of the derivation of the aerodynamic flow equations. In both cases, a simple brief notation is used, with emphasis on the role of the free-field Green's function, to describe several decades of work by many researchers, who used many forms of notation. Following this, some of the successful methods of analysis are discussed, and finally, some of the available computer programs are described.

In order to obtain the necessary background information for the discussion of the computer programs, a somewhat incomplete survey was conducted. It was evident from the response that, although there are many excellent aeroelastic programs in use within the aerospace industry, they are not generally available. Therefore, most of the programs described in this paper have been developed under the sponsorship of a government agency, and may be available to qualified users only.

A list and description of computer programs is included in the paper based on the brief survey conducted. These computer programs include a variety of steady and unsteady aerodynamics methods such as subsonic doublet-lattice, subsonic collocation method (kernel function), finite difference transonic theory, modified strip theory, supersonic Mach box, and piston theory. Some of the procedures include provisions for control surface terms and aerodynamic interference effects between lifting surfaces in close proximity. Programs are listed which perform analyses to predict flutter instabilities, reversal, divergence, and/or aeroelastic distortions. Optimization of certain structural elements is automatically performed by some programs. Indications of the computer program capabilities, size, type of computer, points of contact for availability, and documented references are given in the survey.

AEROELASTIC PHENOMENA

Since the early days of flight, certain aircraft have exhibited unwanted behavioral characteristics traceable to the interaction of aerodynamic forces and elastic structural deflections. Such phenomena have been termed "aeroelastic phenomena." Other interactions of a more favorable nature, such as the aerodynamic damping on a wing passing through a gust or on an aerospace vehicle subject to shock or vibration, may still be classified as aeroelastic.

The aeroelastic phenomena may be loosely classified as static, quasi-static, or dynamic. Of the last, the best example is that of flutter, the unstable coupling of two or more structural vibration modes, under the influence of aerodynamic forces, which can lead to self-destruction of an aircraft. Whereas flutter is a homogeneous problem, two related nonhomogeneous problems are of great interest. They are, first, the response to oscillatory forces, such as from a control surface, which is important in the study of feedbacks in automatic control systems, and second, the response to a gust spectrum. A related problem is that of the response to arbitrary gust or shock loads, in which the forces are not oscillatory.

Of the static phenomena, the most significant are the modification of the aerodynamic loading and of the overall dimensionless force coefficients of an airplane, due to the deflection of the structure under load and the accompanying redistribution of aerodynamic loads. Since neither the deflection nor the aerodynamic loading is known a priori, the phenomenon must be described by an integral equation or, in the case of finite elements or generalized coordinates, by simultaneous equations. Because stability derivatives are important in defining the motions of aircraft at frequencies well below the lowest structural frequencies, the notion of quasistatic phenomena is introduced to determine the aeroelastic effects on these derivatives. Two critical phenomena may be observed under static conditions. One is divergence, at the onset of which a state of equilibrium can exist between the elastic forces in a deflected structure and the aerodynamic forces induced by the deflections. From another point of view, this is a special case of flutter in which a zero frequency root is found. The other is control reversal, in which the motion of an aircraft due to the deflection of a control surface is the opposite of what is normally intended and expected.

Oscillatory Aerodynamic Forces

It can be readily shown that, given the ability to determine the aerodynamic forces on a harmonically oscillating structure, many of the important aeroelastic phenomena can be investigated analytically. For example, conditions of incipient flutter can be treated, and distinctions can be made between stable and unstable situations, even though the diverging oscillations of a wing toward destruction cannot be treated properly. Again, response to oscillatory forces and gust spectra can be calculated, while the response to an arbitrary gust can, in principle, be obtained by first finding the transform of the motion along the imaginary axis and then computing the inverse transform. Where questions of structural fatigue are involved, the response to the gust spectrum is the more useful, because it can be used in conjunction with cumulative damage theories to predict structural lifetime. Again, by treating static aerodynamic forces as special cases of oscillatory forces, all static phenomena are encompassed, while quasi-static phenomena are readily handled by considering oscillations of very low frequency.

Linearized Aerodynamic Forces

Because structural deflections within the elastic limit are generally small, and because interest generally centers on the incipient condition rather than on the ultimate behavior before destruction, the aerodynamics of most aeroelastic phenomena can be handled satisfactorily by linear perturbations of the fluid dynamical equations.

Uniform Flows

The solutions for linear perturbations of nonuniform flows, such as over thick wings or bodies, present formidable problems which have not been overcome satisfactorily. Fortunately, except in the transonic case, adequate results are often obtained from perturbations of uniform flows; i.e., the aerodynamic surfaces are treated as thin, with a uniform flow velocity equal to the velocity at infinity. Thus the main thrust of the effort has been toward the development of methods which treat uniform perturbations. As might be expected, methods specializing in the static case are generally more advanced than those treating the oscillating case. In fact, some of the latter have been adaptations of methods originally developed for the static case.

FORMULATION OF THE AEROELASTIC EQUATIONS

It is readily shown that the aerodynamic forces acting on a dynamically deflected airframe do not constitute a conservative system of forces. Because the internal forces are inertial and elastic, both of which are conservative, together with small structural damping forces, the Lagrangian equations for harmonic oscillatory motion can be written in terms of the complex vector {q} of generalized coordinate amplitudes[1] in the matrix form:

$$-\omega^2 [M]\{q\} + (1+ig)[K]\{q\} = \{Q_A\} + \{Q_E\} \tag{1}$$

where ω is the angular frequency, [M] is the inertia, or mass matrix, g is the structural damping ratio, [K] is the stiffness matrix, and {Q} is the vector of generalized forces due to aerodynamic forces (subscript A) and external forces (subscript E).

The linear aerodynamic forces can be expressed in terms of a flutter airforce matrix [C], in the form:

$$\{Q_A\} = -\tfrac{1}{2}\rho V^2 [C(\omega/V,M)]\{q\} \tag{2}$$

where ρ is the air density and V is the uniform flow velocity. The airforce matrix [C] is a function of the reduced frequency ω/V and of the Mach number M, which is equal to V/c, where c is the speed of sound.

The Flutter Equation

The flutter equation is obtained by combining Eqs. (1) and (2) in homogeneous or eigenvalue form

$$[-[M] + \tfrac{1}{2}\rho (V/\omega)^2[C(\omega/V,M)] + X[K]]\{q\} = 0 \tag{3}$$

where X is the eigenvalue $(1 + ig)/\omega^2$. Alternatively, if diverging motions are admitted, X can be interpreted as $1/(\alpha + i\omega)^2$, where α is the time constant of the motion. However, this is inconsistent with the assumption of pure harmonic motion used in deriving the airforce matrix.

Generally, solutions for the eigenvalues X are obtained as functions of reduced frequency ω/V and Mach number M and are reduced to values of V, ω, g, and M. By cross-plotting, or other means, the flutter speed is found at the point at which g is equal to the small value attributed to available structural damping. At this speed, a condition of incipient flutter exists. When g is greater, the airframe in unstable. If a root is found for which ω is zero, i.e., if the determinant

---

[1] Symbols for dependent variables, such as {q}, ϕ, and ψ, are generally to be understood as referring to complex amplitudes of harmonic motion at an angular frequency ω, unless defined to the contrary.

$$\left|\tfrac{1}{2}\rho V^2[C(0,M)] + [K]\right| = 0$$

then this can be interpreted as a case of incipient divergence.

The Nonhomogeneous Equation

The nonhomogeneous equation is generally written without the structural damping term in the form

$$[-\omega^2[M] + \tfrac{1}{2}\rho V^2[C(\omega/V,M)] + [K]]\{q\} = \{Q_E\} \tag{4}$$

which can be readily solved as a simultaneous equation for each value of the frequency ω, at any given density ρ, airspeed V, and Mach number M.

The Static and Quasi-Static Cases

The vector of generalized coordinates $\{q\}$ can be subdivided into two vectors, the "rigid body" coordinates $\{q_R\}$ and the elastic coordinates $\{q_S\}$. Under the "quasi-static" assumption, the inertial and aerodynamic damping terms due to the elastic coordinates are ignored. Thus part of the mass matrix $[M]$ becomes void and is replaced by the matrix:

$$\left[\begin{array}{c|c} [M_{RR}] & 0 \\ \hline [M_{SR}] & 0 \end{array}\right]$$

Also, the airforce matrix $[C]$ is replaced by

$$\left[\begin{array}{c|c} [C_{RR}'] + i\omega/V[C_{RR}''] & [C_{RS}'] \\ \hline [C_{SR}'] + i\omega/V[C_{SR}''] & [C_{SS}'] \end{array}\right]$$

where the values of the real matrices $[C']$ and $[C'']$ are readily found by deriving the airforce matrix for a small reduced frequency ω/V.

Then, since the terms in the stiffness matrix $[K]$ corresponding to the rigid body coordinates are zero, the quasi-static form of Eq. (4) becomes:

$$\left[\begin{array}{c} [M_{RR}] \\ \hline [M_{SR}] \end{array}\right]\{\ddot{q}_R\} + \tfrac{1}{2}\rho V\left[\begin{array}{c} [C_{RR}''] \\ \hline [C_{SR}''] \end{array}\right]\{\dot{q}_R\} +$$

$$\tfrac{1}{2}\rho V^2\left[\begin{array}{c|c} [C_{RR}'] & [C_{RS}'] \\ \hline [C_{SR}'] & [C_{SS}'] + [K_{SS}]/\tfrac{1}{2}\rho V^2 \end{array}\right]\left\{\begin{array}{c} q_R \\ \hline q_S \end{array}\right\} = \left\{\begin{array}{c} Q_{ER} \\ \hline Q_{ES} \end{array}\right\} \tag{5}$$

The elastic coordinates $\{q_S\}$ are readily eliminated from this equation, leading to the following equation in the rigid body coordinates,

$$[[M_{RR}] - [B][M_{SR}]]\{\ddot{q}_R\} + \tfrac{1}{2}\rho V[[C_{RR}''] - [B][C_{SR}'']]\{\dot{q}_R\}$$

$$+ \tfrac{1}{2}\rho V^2[[C_{RR}'] - [B][C_{SR}']]\{q_R\} = \{Q_{ER}\} - [B]\{Q_{ES}\} \qquad (6)$$

where

$$[B] = [C_{RS}'][[C_{SS}'] + [K_{SS}]/\tfrac{1}{2}\rho V^2]^{-1}$$

By nondimensionalizing the forces according to the convention normally used for stability derivatives, the coefficients in the above equation can be interpreted as force coefficients and stability derivatives, properly corrected for aeroelastic effects, and therefore functions of $\frac{1}{2}\rho V^2$ and M. A surprising feature of this equation is the apparent aeroelastic correction of the mass and moments of inertia.

Very often, the terms in the matrices C_{RR}' and C_{RR}'' are already known from wind tunnel data, or from other sources, to greater accuracy than can be predicted analytically. In such cases, the better values should be used. Also, if the aerodynamic loadings are to be found, they can be expressed in terms of the generalized forces, which can be then calculated from Eq. (5) once the rigid body coordinates have been determined.

AERODYNAMIC FORCES

It has been the objective of the preceding section to demonstrate that, given the capacity to calculate the aerodynamic forces for pure harmonic motion by linear perturbation of a uniform flow, the majority of aeroelastic problems can be handled analytically. Because the perturbations are linear, the contributions of the different modal deflections are decoupled from each other, so that the matrix formulation of Eq. (2) is possible. Also, within reasonable limits, the steady state lift forces can be considered to act independently of the harmonic forces due to airfoil oscillations, and do not appear in the formulation of the flutter equation. The assumption of uniform form is equivalent to treating every aerodynamic body as infinitely thin and parallel to a uniform flow field. These thin bodies, together with their wakes, which are surfaces of discontinuity, form the bounding surfaces of the fluid dynamics problem, on which certain known boundary conditions apply.

An integral formulation for the general harmonic perturbation problem can be derived in the form of the Green's equation. It is very similar to the formulation given for the Helmholtz equation by Morse and Feshbach [27], Eq. (7.2.7), which is frequently referred to elsewhere as the Helmboltz Integral Equation.

The convected wave operator for a uniform flow velocity $\bar{V}$ can be written as

$$L = \nabla^2 - [i\omega/c - (\overline{V}/c)\cdot\nabla]^2 \qquad (7)$$

then the wave equation is

$$L\{\phi\} = 0 \qquad (8)$$

where $\phi(r)$ is the velocity potential at the point $\bar{r}$. The Green's equation for the convected wave problem is

$$\phi(\bar{r}) = \int_\sigma \{\phi(\bar{\rho}) \frac{\partial G(\bar{r},\bar{\rho})}{\partial\mu} - G(\bar{r},\bar{\rho}) \frac{\partial\phi(\bar{\rho})}{\partial\mu}\} \, d\sigma \qquad (9)$$

where $\bar{\rho}$ is a point on the bounding surface σ, with local normal coordinate μ, which is positive when directed into the boundary. $G(\bar{r},\rho)$ is the free-field Green's function, otherwise known as the unit or source solution. Physically, it can be thought of as the velocity potential at $\bar{r}$ due to a unit source at $\bar{\rho}$. It satisfies the equation:

$$L\{G(\bar{r},\bar{\rho})\} = \delta(\bar{r} - \bar{\rho}) \qquad (10)$$

and has the following forms

<u>Subsonic</u>: $G(\bar{r},\bar{\rho}) = G_R(\bar{r},\bar{\rho})$

<u>Supersonic</u>: $G(\bar{r},\bar{\rho}) = U(F) \{G_R(\bar{r},\bar{\rho}) + G_A(\bar{r},\bar{\rho})$

where

$$G_R(\bar{r},\bar{\rho}) = \frac{-e^{-i\omega\tau_R}}{4\pi R} \; ; \qquad G_A(\bar{r},\bar{\rho}) = \frac{-e^{-i\omega\tau_A}}{4\pi R}$$

in which subscripts R, A are for 'retarded' and 'advanced' wave, respectively. If X, Y, and Z are the components of $\bar{\rho} - \bar{r}$, with X parallel to $\bar{V}$, then the cutoff function U(F), following Watkins and Berman [28], is the step function, with

$$F = -X - \sqrt{-\beta^2 Y - \beta^2 Z}$$

The times of transit τ_R and τ_A are given by

$$\tau_R = (MX + R)/\beta^2 c; \qquad \tau_A = (MX - R)/\beta^2 c$$

where

$$R = \sqrt{X^2 + \beta^2 Y^2 + \beta^2 Z^2}$$

and

$$\beta^2 = 1 - M^2$$

The corresponding transonic terms can be derived as limits of the subsonic expressions. However, the uniform flow assumption applied to the transonic case does not yield results which are in good agreement with experimental values. The solutions for the non-uniform case present many difficulties and none have been based on the Green's equation.

The pressure p and the acceleration potential ψ are found as functions of ϕ from the equation

$$p = -\rho\psi = -\rho(\bar{V} \cdot \nabla\phi + i\omega\phi) \tag{11}$$

while the inverse equation is obtained by integrating with respect to X along a streamline parallel to V from upstream infinite ($X = -\infty$) to zero

$$\phi(\bar{r}) = \frac{1}{V}\int_{-\infty}^{0} \psi(\bar{\rho})\, e^{i\omega X/V}\, dX \tag{12}$$

Because G is the solution for a unit source and its derivative $\partial G/\partial\mu$ is the solution for a unit doublet, reference is often made in the literature to source or doublet distributions when describing formulations of the integral expressions.

In problems relating to the perturbations of flows over airfoils, the airfoil surfaces and their wakes become interior boundaries, while the only condition on the exterior boundary at infinity is that it does not radiate. If the motion of the surface in the direction of its local normal coordinate μ is expressed by the displacement amplitude δ, then the fluid boundary condition, which is set by the requirement that the flow be tangential, becomes

$$w = \bar{V} \cdot \nabla\delta + i\omega\delta \tag{13}$$

where w is the local normal flow velocity in the direction of the local normal coordinate μ.

In order to obtain the airforce matrix, δ, ω, ϕ, and ψ are expressed in terms of the generalized coordinates, i.e.,

$$\phi = \{\phi\}^T \{q\}$$

Then the generalized forces can be expressed in terms of the generalized coordinates and of the airforce matrix [C] as

$$\{Q\} = -\tfrac{1}{2}\rho V^2 [C] = -\rho\int_{\sigma}\{\psi(\bar{\rho})\}\{\delta(\bar{\rho})\}^T\, d\sigma \tag{14}$$

If deflections at structural nodes are used as generalized coordinates, then the generalized forces are the forces at these nodes. It should be noted, however, that the airforce matrix is full and that it is a complex function of reduced frequency ω/V and Mach number M. Therefore, use of a large number of nodal deflections as generalized coordinates imposes severe storage requirements on the computer.

Integrated Potential Form

Each of the thin aerodynamic bodies treated in the last section can be considered as a pair of back-to-back surfaces. In the integrated potential method, otherwise referred to as the velocity potential formulation by Landahl and Stark [21], the paired surfaces are replaced by the one-sided surfaces s_A for airfoils and s_W for their wakes, with local normals ν. Then, in the absence of "breathing" of the airfoils, the two paired surfaces move together, resulting in cancellation of one term in Eq. (9). The result is then differentiated with respect to the local normal coordinate n at the point $\bar{r}$, to give the known normal flow component w on the left-hand side of the equation. The result is an integral equation in the velocity potential differential $\Delta\phi$, which is positive if ϕ decreases in the positive ν direction

$$w(\bar{r}) = \int_{s_A+s_W} \Delta\phi(\bar{\rho})\ \partial^2 G(\bar{r},\bar{\rho})/\partial n\ \partial\nu\ ds \tag{15}$$

On the airfoil surface s_A, w is known, while $\Delta\phi$ is unknown. At any point in the wake, which cannot support a pressure differential, $\Delta\phi$ is given by Eq. (12) in terms of its value on the trailing edge point immediately ahead.

The use of the two-sided surfaces makes some redefinition of the expression for the airforce matrix [C] in Eq. (14) possible. It now is

$$[C] = \frac{2}{V^2} \int_{s_A} \{\Delta\psi(\bar{\rho})\}\{\delta(\bar{\rho})\}^T\ ds \tag{16}$$

while from Eq. (11)

$$\Delta\psi = \bar{V} \cdot \nabla\ \Delta\phi + iw\ \Delta\phi \tag{17}$$

It is possible in many instances to avoid the use of Eq. (17) by combining it with Eq. (16) and then integrating by parts.

Integrated Pressure Form

If Eq. (12) is redefined as

$$\Delta\phi(\bar{r}) = \frac{1}{V} \int_{\infty}^{0} \Delta\psi e^{iwX/V}\ dX \tag{18}$$

and is then combined with Eq. (15), the result is the integrated pressure form

$$w(\bar{r}) = \int_{s_A} \Delta\psi(\bar{\rho})\ K\ (\bar{r},\bar{\rho})\ ds \tag{19}$$

This was first used by Watkins et al. [29], for planar wings, when it was known as the "kernel" method, K being the kernel of the integral equation. Because K contains an integral to infinity, the method presents some difficulties in evaluation. However, this form has been used almost exclusively in the subsonic case and has also been adapted to the supersonic case [28,30]. One advantage of the method is that integration is restricted to the airfoil s_A.

Direct Potential Method

For supersonic wings, the concept of flexible, non-load-carrying membranes, first introduced by Evvard [31], has proved valuable. By this means, the different surfaces of the airfoils can be isolated aerodynamically, so that the continuous region of air stretching out to infinity and reached by all of the unit solutions is replaced by separate regions, each with its own solution.

In the most general case, it is not possible to simplify Eq. (9) further, so that values for ϕ must be obtained separately for each side of the airfoil. From these, using Eq. (11), values for ψ are obtained, or they may be obtained directly by first combining Eq. (9) and (11). Initially, boundary conditions on the airfoil are given in the form of the normal flow components $\partial\phi/\partial\mu$. However, although this quantity is unknown on the membranes, it is known that the membranes cannot support a pressure differential so that $\Delta\psi$ is zero.

A method for solving this problem was described by Ashley [32], for folded wings, and this was extended to arbitrary configurations of out-of-plane airfoils by Moore and Andrew [33] and by Andrew [34].

Direct Potential Method for Planar Wings

In this case, considerable simplification of the equation is possible, leading to the one-sided integral

$$\Delta\phi(\bar{r}) = 2\int_{s_A+s_W} w(\bar{\rho}) \quad G(\bar{r},\bar{\rho})\nu \, ds \tag{20}$$

Boundary conditions on the membranes covering the area s_W are as before.

NUMERICAL METHODS

The most comprehensive evaluation of numerical methods published to date in the NATO comparison of in-plane oscillating airfoil methods edited by Woodcock [35], which covered thirty different methods applied to a number of inplane wings in subsonic, transonic, and supersonic flow. An extension of this study to cover out-of-plane cases is underway. A similar, though much less extensive investigation of methods applicable to the steady-state case was undertaken by Thomas and Wang [36], which covered fifteen different methods applied to two planforms in subsonic flow. The latter did not include the subsonic method developed by Woodward [37,38].

Classification of Lifting Surface Methods

Essentially, all of the three-dimensional lifting surface methods used either must be variations of the integrated potential or pressure methods of Eqs. (15) or (19) or, in supersonic flow, may be variations of the direct potential method of Eq. (20) or a mixed method derived from Eq. (9). Except in the case of the purely supersonic wing solved by the direct method, there is an integral equation to solve. In the case of the integrated pressure method the unknown pressure is restricted to the airfoil surfaces alone, while in the direct supersonic methods, the normal flow component (often referred to as downwash) is unknown over the diaphragm regions.

It is impractical to seek analytical solutions; therefore, it is necessary to assume that the unknown distributions can be represented by a finite number of prescribed functions. Then their amplitudes become the unknowns of the problem, which are to be solved in such a manner that the prescribed boundary conditions of known normal velocity on the wing and of zero pressure distributions across wakes or diaphragms are met in the most accurate manner.

Although the classifications into integral and direct methods are often referred to, other classifications become more descriptive when numerical methods are to be considered. These methods are described next.

Collocation Methods. These include applications of the integrated pressure method in which overall functions are prescribed for the pressure distributions and their unknown amplitudes are solved by simultaneous equation from the condition that known values for the normal velocity must be obtained at selected collocation points. These methods originated with the subsonic kernel function method of Watkins et al. [29] and the corresponding supersonic formulation of Watkins and Berman [28]. A considerable effort has been expended in overcoming some of the inherent difficulties of the methods, which include uncertainties about the optimum location of collocation points and sensitivity of the solution to the forms of prescribed functions, especially over control surfaces. Some of the numerical methods used have resulted in poorly conditioned sets of simultaneous equations and have therefore been limited in the numbers of unknown functions which could be permitted. The kernel function for application of the collocation method to nonplanar configurations has been given by Ashley, Widnall, and Landahl [19] and has been condensed considerably by Yates [39] and Landahl [40]. Nine subsonic collocation methods, one sonic method, and two supersonic methods are included in the comparison by Woodcock [35].

Lattice Methods. There have been many applications of the horseshoe vortex concept to represent steady-state or quasi-steady lift lines, two of the more recent being the use by Hedman [41] of nested vortices to represent a lifting surface and the use of Rubbert and Saaris [42] of vortex loops enclosing surfaces at uniform potential to represent aerodynamic bodies of all kinds. The equivalent use of lift lines formed from doublets in the doublet-lattice method of Albano and Rodden [43] is a further application of the integrated pressure method to the oscillating case. In all of these applications, the strength of the horseshoe vortex, vortex loop, or lift line is an unknown, to be matched by the known normal velocity at a collocation point. One of the disadvantages of these methods is that if relative locations are not carefully controlled, unreliable results are obtained. This is because a vortex or doublet line is a line of singularity for the normal flow component. The methods are readily applied to out-of-plane cases.

Panel Methods. These have been developed by Woodward [37,38] for the steady-state subsonic cases; they are further examples of the integrated pressure method, in which unknown pressures (or potentials) are distributed over panels, with a collocation point near the center of each panel. In a sense, the vortex lattice and doublet loop methods are particular variations of the panel method.

Supersonic Box Methods. The integrations of velocity potential required in the direct methods are generally obtained by using standard expressions for the contributions of rectangles, diamonds, or triangles, each with locally constant values for the normal velocity boundary condition. They may be further classified as "Mach Box" methods if the diagonals of the rectangles or edges of the diamonds are Mach lines. The original application of box methods was by Pines et al. [44], later applications to out-of-plane cases having been made by Moore and Andrew [33] and by Andrew [34]. Four supersonic box methods are included in the comparison by Woodcock [35].

Transonic Methods. These are all essentially based on a limiting expression for the integrated potential case. The general theory was given by Landahl [45]. An application of the kernel method was developed by Runyan and Woolston [46], while a transonic box method (actually a panel method according to the notation in this report) was developed by Rodemich and Andrew [47] and later extended to wings with control surfaces by Stenton and Andrew [48].

Methods Not Based on Lifting Surface Theory

Two-Dimensional Aerodynamics. The theory of the lift on an oscillating two-dimensional wing in incompressible flow was independently arrived at by Theodorsen [49], Cicala [50], and Kussner [51], while the most complete account of its application to aeroelasticity has been given by Smilg and Wasserman [52]. Numerical results for calculations with compressibility effects have been given by many authors but have the disadvantage, when applied to computer solutions, that they require very large tables. The supersonic case was given by Garrick and Rubinow [53]. A comprehensive method for performing aeroelastic calculations using two-dimensional derivatives, taking full advantage of matrix algebra, was given by Loring [26].

One-Dimensional Aerodynamics. A method of computing supersonic derivatives on supersonic wings of finite thickness was developed by Ashley and Zartarian [54]. It is in quite good agreement with lifting surface theories at high Mach numbers and has the added advantage that it accounts for thickness effects. The method, which is based on the analogy between the normal flow component on the surface of a wing and conditions adjacent to a moving piston, is referred to as "piston theory." Because of the very great simplification achieved in the aerodynamic calculations, it has had early applications in structural optimization techniques, such as that by Stroud, Dexter, and Stein [55].

Nonuniform Flow Solutions. Indications are that the uniform flow methods may not give acceptably accurate results in the transonic case, where flows can vary locally from subsonic to supersonic conditions. Although some attempts have been made to develop methods which will properly account for nonuniform flow conditions, these are still in the highly speculative stage and are not covered in this review of methods. One such computer program is included in Table 1; this is the SEIDEL program, item number 15.

APPLICATIONS TO STRUCTURAL OPTIMIZATION

The nature of the airforce matrix, which is a function of both reduced frequency and Mach number, greatly complicates the flutter analysis problem and necessitates interpolation or cross-plotting to find the flutter speed. Suppose now that a given margin on flutter speed is a constraint in a structural optimization program, then repeated checks on the flutter speed must be made during the process of iteration toward an optimum structure. Formally, this requires the following sequence of calculations; mass matrix, stiffness matrix, natural modes of vibration, modal components of normal velocity boundary conditions at collocation points by interpolation, transformation of stored air force matrices to conform to natural modes, calculations of flutter eigenvalues, and final interpolation to obtain flutter speeds. Such a procedure has been incorporated into a program known as WIDOWAC by Haftka [55]. An investigation into alternative sequences of calculation was undertaken by Haftka and Yates [57], who compared the basic sequence to one in which fixed modes were used, using a subsonic collocation method derived from the kernel function method of Watkins et al.[29].

SURVEY OF AEROELASTIC COMPUTER PROGRAMS

A survey of available aeroelastic computer programs was made; a brief description is given in subsequent paragraphs, and a list of pertinent features is given in Table 1. From the response it appears that many excellent computer programs are used in the aerospace industry which are not generally available. Most of the programs described have been developed under studies sponsored by government agencies and some are available only to qualified users. The programs cover several steady and unsteady aerodynamic procedures for the subsonic, transonic, and supersonic speed regimes. Some include aerodynamic interference effects between lifting surfaces in close proximity. Other programs will predict aeroelastic behavior including flutter, reversal, divergence, and aeroelastic distortions. Structural optimization programs for flutter constraints and/or strength constraints are also included. The following program descriptions together with Table 1 cover computer program capabilities, size, type of computer, points of contact for availability, and documented references.

1. C81 - Rigid Body Dynamics Analysis Digital Computer Program. This program was developed by the Bell Helicopter Company for the U.S. Army Aviation Material Research and Development Laboratory, Ft. Eustis, Virginia. The program simulates helicopter flight and can be used to analyze single-rotor, compound, tandem, or side-by-side helicopter configurations in hover, transition, cruise, or high-speed flight. The analysis, with a uniform level of complexity for its different phases, can calculate performance, stability and control, or rotor loads. Its inputs are organized to make the program easy to use, and the output format facilitates comparison of computed results with flight and tunnel test data. Three major parts of the analysis are a mathematical model of an elastic rotor, rotor aerodynamics, and basic rigid-vehicle flight mechanics. The program is used in support of four phases of rotor system design and evaluation: rotor blade frequency placement, wind-tunnel simulation, steady-state flight simulation, and transient or maneuvering flight simulation. The stability and control section calculates trim positions (including control positions), gradients, and margins in level, climbing, diving, turning, or accelerated flight. It uses linear analysis to compute response characteristics and locations of stability roots for coupled flight modes. It can also use fully coupled nonlinear equations to calculate and plot variables against time. Disturbances in the form of gusts, sinusoidal control motions, and weapon recoil can be inserted in this section of the program. The rotor blade loads portion of the program includes a fully coupled, time-variant aeroelastic analysis. This portion uses equations to calculate beam, chord, and torsional loads during either steady or maneuver flight, and prints out the results. This program is described and applied in available documentation by Bennet [58] and in AFFDL/FGC TM 72-10 [59], and is available through the U.S. Army.

2. DEAL 17 & 18 - Folding Proprotor VTOL Aircraft Dynamics. This computer program, developed by Bell Helicopter Co. under a program sponsored by the Air Force Flight Dynamics Laboratory, predicts the dynamic stability and response during the feathering and folding of the blades of a folding-proprotor VTOL. Correlation of the theory with measured dynamic stability and response characteristics during feathering and folding was good and indicates that the analytical methods can be used with confidence. The inputs consist of the geometry, mass, and stiffness distributions of the wing, pylon, and blades. The following section properties can be varied with blade fold angle: strip width, chord, reference semispan and chord, cosine of blade fold angle, main aerodynamic chord, distance between dumbbells, blade sweep, distance from elastic axis to dumbbells, mass unbalance, and inertia. The aerodynamic interference between proprotor and aeroelastic wing is properly represented. The AFFDL TR-71-7, Vol. I [60] contains the development of the theory, correlation of theory with experimental data, the dynamic response and parametric study, while Vol. II [61] covers the computer program.

3. AIC-INT - Supersonic Unsteady Consistent Aerodynamics for Interfering Parallel Wings. This computer program was developed by Bell Aerospace Company for NASA Langley Research Center. The analytical development of unsteady supersonic aerodynamic influence coefficients for isolated and nearly parallel interfering coplanar and noncoplanar wings is described in NASA CR-2168 [62]. Numerical

formulations based on triangular discretizations of wing and diaphragms are handled in a kinematically consistent manner. The basic elements used to represent the planform can be of different sizes and orientations to "best fix" (independent of Mach number diaphragms). Computer programs for the interfering case are described in a user's manual, NASA CR-112184, and presented in a programmer's manual, NASA CR-112185 [63].

4. AEDERIV - Longitudinal Stability Derivatives for Elastic Airplanes. This program, developed at the NASA Langley Research Center, calculates the longitudinal stability derivatives of elastic airplanes [64]. This development is subject to the assumptions of small perturbations from a steady reference flight condition, structural deflection constrained to a direction normal to a structural reference plane, and the quasi-steady aeroelastic assumption of structural deflections proportional to applied loads. The development avoids any constraint on the forward-speed degree of freedom, and the resulting stability derivatives exhibit two important departures from past practice: (1) the aeroelastic contributions of dynamic pressure perturbations are included, and (2) the aeroelastic contributions of normal acceleration appear primarily in the derivatives with respect to pitching velocity and angle-of-attack rate and, for an unaccelerated reference flight condition, do not influence the derivatives with respect to angle of attack. The aerodynamic and structural influence coefficient matrices and the airplane jig shape are not calculated in the program but must be supplied as inputs.

5. RHØIII - Prediction of Unsteady Aerodynamic Loadings Caused by Trailing Edge Control Surface Motions in Subsonic Compressible Flow. The theoretical analysis and computer program [65,66] were developed by the Boeing Company for NASA Langley Research Center and cover prediction of unsteady lifting surface loadings caused by motions of trailing edge control surfaces having sealed gaps in subsonic flow. The final form of the downwash integral equation was formulated by isolating the singularities from the nonsingular terms and establishing a preferred solution process to remove and evaluate the downwash discontinuities in a systematic manner. The method is capable of numerically predicting the unsteady loading caused by control surface motions and is relatively insensitive to locations of collocations stations in the surface. Spanwise symmetry or antisymmetry of motion and up to four control surfaces on each half span can be accommodated.

6. SØAR - Method of Minimum-Weight Synthesis for Flutter Requirements. SØAR has been developed by Stanford University under an Air Force Flight Dynamics Laboratory program to size structural members of finite element generated lifting surface design automatically to achieve minimum weight without violation of a fixed flutter speed constraint. Zoutendyk's method of feasible directions is used to determine resizing steps. Since this method requires gradient information, a method due to Van de Vooren was implemented. After the variables for the starting design have been specified, the program calculates the weight and flutter speed of the starting design. The flutter constraint and minimum gauge constraints are evaluated to determine whether the starting design point is a feasible or an infeasible design point and, if feasible, whether any of the constraints are critical. If the starting design point is infeasible, the procedure

moves in the direction of increasing weight to a point where no constraints are positive. If the starting design point is feasible and the flutter constraint is not critical, the procedure takes a step in the direction of decreasing weight. The method is described in AFFDL TR 72-22, Part I [67] and Part II [68], AFFDL TR-73-91, Part I [69] and Part II [70], and AFFDL TM-73-19 [71].

7. SUBSØNC - Subsonic Flutter Analysis Program. Three tasks have been required to perform a flutter analysis when free vibration mode shapes were available: (a) a polynomial or spline fit of deflection shapes, (b) calculation of generalized aerodynamic forces, and (c) solution of the complex eigenvalue problem. Programs which perform these tasks have been combined by the Air Force Flight Dynamics Laboratory into a single program, SUBSØNC, for efficient operation on the CDC 6600 computer. The generalized aerodynamic forces are calculated by the Albano-Rodden [43] doublet-lattice method. Calculation of the optional two-dimensional spline coefficients is based on the small deflection theory of an infinite plate. The flutter analysis is performed by the V-g method. The program is described in AFFDL TM-74-27-FYS [72].

8. FACES - Flutter of Aircraft Carrying External Stores. The FACES system is a collection of modular computer programs for the rapid performance of wing/store flutter calculations, for the efficient storage of the results of these calculations, for the rapid retrieval of these results, for the estimation of flutter speeds of new stores, and for the utilization of diagnositics for increasing comprehension of results. The program is available in IBM 360 FORTRAN IV and CDC 6600 FORTRAN extended version 3.0. The program was developed by McDonnell Aircraft Company for the Air Force Flight Dynamics Laboratory and is described in AFFDL TR 73-74, Vol. I [73], Vol. II [74], and Vol. III [75]. Flutter computations can be based on either of two processes. The first process features a finite section approach with vibration calculated inside the FACES program based on programmed equations. Configuration applicability includes:

a. Clean wings, or lifting surfaces, with serially kinked elastic axes or reference axes

b. Wings with up to five pylons per side

c. Any combination of single store and multiple (TER, MER) loadings

d. Flexible pylons and racks

e. Cantilever, symmetric, and antisymmetric restraints.

The second process requires vibration to be calculated outside of the FACES program and may include more complex idealizations such as a finite element approach, or it may include measured vibration data The unsteady aerodynamic routines allow for:

a. The inclusion of aerodynamic forces on the wings using the combination of modified strip theory and piston theory to cover essentially all Mach numbers

b. The inclusion of subsonic lifting surface aerodynamics on the wing, and three-dimensional aerodynamics on the pylon/rack/store using the doublet-lattice method

9. PERTURB - Perturbation Technique for Rapid Flutter Clearance of Aircraft Carrying External Stores. The perturbation program calculates vibration and flutter characteristics for aircraft carrying external stores using matrix eigenvalue and eigenvector perturbations

up to third order. The program was developed by Northrop Corporation for the Air Force Flight Dynamics Laboratory and is described in AFFDL TR-72-114, Part I [76] and Part II [77]. It is in modular form and the only significant departure from ANS (American National Standard) FORTRAN is reflected in a limited number of extended FORTRAN instructions employed for retrieving baseline case data from the files defined in a basic flutter solution. The program is presently operational on the IBM 370/165 and is being converted to operation on the CDC 6600. Basic operation of the program is as follows:

a. Read perturbed mass on flexibility controls and data card input

b. Retrieve baseline vibration data from disc files

c. Compute perturbed vibrations data, including generalized modal inertia and stiffness matrices

d. Retrieve baseline aerodynamics and flutter data

e. Compute perturbed flutter data

f. Print out and/or plot perturbed flutter data summaries

10. TWOS - Subsonic Oscillatory Aerodynamics for Wing/Horizontal-Tail Configurations. This computer program is based on the kernel function method of Laschka [78] and provides the capability to analyze the subsonic, oscillatory lift distributions on wing/horizontal-tail combinations. It was developed by Northrop Corp. for the Air Force Flight Dynamics Laboratory and is described in AFFDL-TR-70-59, Part I [79] and Part II [80]. It can account for longitudinal and vertical separation between wing and tail surfaces. However, configurations with small vertical separation between wing and tail cannot be analyzed with this program. The surfaces to be analyzed must be composed of trapezoidal sections. Each semiwing is a segment of a plane; the wing and/or tail may be folded at the middle chord. The tail must not intersect the wing wake, but it may lie entirely in it. In the coplanar case, the tail semispan must be less than or equal to that of the wing. The computer program operates within core on the IBM 7094 or CDC 6600 computers.

11. H7WC - Nonplanar Doublet-Lattice Method for General Configurations. The doublet-lattice method is a simple, versatile, and accurate lifting surface theory capable of analyzing lifting surfaces with arbitrary planform and dihedral. Control surfaces, either full or partial span, may be included. Problems of intersecting and/or interfering nonplanar configurations, such as wing-pylon combination, a T- or V-tail, a wing-tail combination, etc., may be analyzed. The present method developed by Douglas Aircraft Co. for the Air Force Flight Dynamics Laboratory is also capable of solving problems involving lifting surfaces and bodies where the bodies may be in motion. It is presented in AFFDL TR 71-5, Part I, Vol. I [81], and Part I, Vol. II [82]. The following options are included in the program:

a. Aerodynamic data including lifting pressures, spanwise lift and moment distributions, aerodynamic center locations, total lift and side force coefficients, and total pitching, yawing and rolling moments

b. Generalized forces for polynomial modes of motion specified by the user

c. Aerodynamic influence coefficients

d. Gust loads from a harmonic gust field

e. Symmetry and ground effects

12. N5KA - Application of the Doublet-Lattice Method and the Method of Images to Lifting Surface/Body Interference. The N5KA program is based on the doublet-lattice method and the method images and is useful for predicting steady and oscillatory loads on very general configurations. It was developed by Douglas Aircraft Co. for the Air Force Flight Dynamics Laboratory and is described in AFFDL TR-71-5, Part II, Vol. I [83], and Part II, Vol. II [84]. Configurations may include a combination of any or all of the following components: (1) lifting surfaces such as wings, pylon, stabilizer, fin, etc. with arbitrary dihedral, (2) partial or full span control surfaces, and (3) bodies such as fuselages, nacelles, stores, etc. with elliptic cross-sectional shapes. The operating conditions are also very general: (1) all frequencies of practical interest and all subsonic Mach numbers, (2) symmetry and ground effect, (3) mutual interference of lifting surfaces, and (4) multiple modes of oscillation (described by polynomials). The following list gives the program limits:

a. The maximum number of unknowns, i.e., the total number of all the lifting surface elements plus the interference body elements, is 500

b. The maximum number of modes is either 50 or 10,000 divided by the total number of unknowns for the case, whichever is smaller

c. The maximum number of panels is 99, while the maximum number of bodies is 10

d. The maximum number of spanwise strips per panel is 50, while the maximum number of chordwise boxes per strip is 50

e. The maximum number of body interference elements (for all bodies) is 100, while the maximum number of slender body elements is 200

f. The maximum number of modal coefficients for panels is 150. The maximum number of modal coefficients for bodies is 150 for z-motions and 150 for body y-motions

g. The maximum number of reduced frequencies is six

The maxima outlined above are tailored to allow the computer program to fit into a core (360/65) of 260 K bytes. If more core is available, the user may wish to increase the dimensions in order to accommodate larger cases in the program.

13. MBØX - A Computer Program for Unsteady Supersonic Aerodynamic Coefficients. A digital computer program, MBØX, has been developed and written for surfaces with trailing edge control surfaces. The program was developed by Rockwell Corp. for the Air Force Flight Dynamics Laboratory and is reported in AFFDL TR-78-30 [85]. It is a very versatile and reliable method for obtaining steady and unsteady supersonic aerodynamic coefficients that may be used in response and flutter analyses of rather general lifting surfaces. Wing configurations can include those with folded tips, cranked leading and trailing edges, and supersonic or subsonic leading and trailing edges. At the option of the user, MBØX will also calculate steady or unsteady lifting pressure distributions, and if the generalized mass and stiffness matrices are provided, it will obtain solutions of flutter equations in one computer run.

14. MBOX - Computer Program for Supersonic Unsteady Aerodynamic Analysis. This computer program was developed by the Boeing Company for the Air Force Flight Dynamics Laboratory and is reported in AFFDL TR-71-108, Part I [85] and Part II [86]. It is a three-dimensional

extension of the Mach Box technique for the unsteady aerodynamic analysis of nonplanar wings and wing-tail configurations in supersonic flow. Various refinement procedures have been included to improve the accuracy of the results. The program is capable of treating wing-tail combinations with or without vertical separation, longitudinal separation, and dihedral on either surface. A nonintersecting wing-vertical tail combination may be examined. If a wing alone is treated, perturbation velocity components in the flow field may be found.

15. SEIDEL - Transonic Thickness Problem. SEIDEL was developed in the Air Force Flight Dynamics Laboratory and solves the transonic small disturbance equations for nonlifting flow by the "mixed difference" method of Murman and Cole. The user provides the Mach number ($M<1$), ratio of specific heats, thickness ratio, body shape, maximum extent of the flow region being considered, convergence tolerance, and a relaxation parameter. The program then automatically subdivides the flow region into a 17 X 10 gridwork of the finite difference points and iterates the nonlinear equations by point relaxation to solution. If the user desires, the program will continue to subdivide the region to 33 X 19, 65 X 37, and 129 X 73 gridworks, using the results from the prior subdivision as starting values. The program computes subcritical solution in about 3 to 5 seconds; it computes supercritical solutions with embedded shocks in 5 to 15 seconds. The method is described in AFFDL TM-FYS-73-85 [88].

16. TSO - Wing Aeroelastic Synthesis Procedure. The TSO computer program was developed by General Dynamics, Fort Worth, for the Air Force Flight Dynamics Laboratory. It uses a Fletcher-Power one-dimensional minimization algorithm to reduce to a minimum a Fiacco-McCormick penalty function formulation of 30 structural polynomial coefficients and ply orientations of a composite, metal, or combination wing. Active parameters used in the design process include flutter speed, divergence speed, aircraft angle of attack (including wing flexibility effects and required tail trim) for a specified load factor, strains and relative strain margins for that load factor, flexible lift, flexible to rigid lift ratio, fundamental frequency, and structural weight. These parameters can be included either as values of merit to be adjusted to an optimum, as constraints, or both. Ten design variables representing mass balances for flutter prevention are also available. The structural representation uses a variable-thickness Rayleigh-Ritz plate formulation, the steady aerodynamics uses the NASA-Ames subsonic/supersonic box method, and the unsteady aerodynamics uses the doublet-lattice box method. The procedure is described in AFFDL TR 73-111, Vol. II [89].

17. Aerodynamic Analysis of Wing-Body-Tail Configurations in Subsonic and Supersonic Flow. This method was developed by F. A. Woodward, Analytical Methods, Inc. for Aerodynamic Research Corp. under a contract for NASA Langley Research Center. The method calculates the pressure distribution and aerodynamic characteristics of wing-body-tail combinations in subsonic and supersonic potential flow. A computer program has been developed to perform the numerical calculations and is described in NASA CR-2228, Part I [38] and Part II [90]. The configuration surface is subdivided into a large number of panels, each of which contains an aerodynamic singularity distribution. A constant source distribution is used on the body panels, and a vortex distribution having a linear variation in the streamwise

direction is used on the wing and tail panels. The normal components of velocity induced at specified control points by each singularity distribution are calculated and make up the coefficients of a system of linear equations relating the strengths of the singularities to the magnitudes of the normal velocities. The singularity strengths which satisfy the boundary condition of tangential flow at the control points for a given Mach number and angle of attack are determined by solving this system of equations using an iterative procedure. Once the singularity strengths are known, the presssure coefficients are calculated, and the forces and moments acting on the configuration are determined by numerical integration.

18. Modified Strip Analysis (Program LAR-10199). A modified strip analysis has been developed by NASA Langley Research Center for rapidly predicting flutter of finite span swept or unswept wings at subsonic to hypersonic speeds [91, 92]. The method employs distributions of aerodynamic parameters which may be evaluated from any suitable linear or nonlinear steady-flow theory or from measured steady-flow load distributions for the undeformed wing. The method has been shown to give good flutter results for a broad range of wings at Mach number from 0 to as high as 15.3. The principles of the modified strip analysis may be summarized as follows: Variable section lift-curve slope and aerodynamic center are substituted, respectively, for the two-dimensional incompressible-flow values of 2π and quarter chord which were employed by Barmby, Cunningham, and Garrick [93]. Spanwise distributions of these steady-flow section aerodynamic parameters, which are pertinent to the desired planform and Mach number, are used. Appropriate values of Mach number-dependent circulation functions are obtained from two-dimensional unsteady compressible flow theory. Use of the modified strip analysis avoids the necessity of reevaluating a number of loading parameters for each value of reduced frequency, since only the modified circulation functions, and of course the reduced frequency itself, vary with frequency. It is therefore practical to include in the digital computing program a very brief logical subroutine, which automatically selects reduced-frequency values that converge on a flutter solution. The problem of guessing suitable reduced-frequency values is thus eliminated, so that a large number of flutter points can be completely determined in a single brief run on the computing machine. If necessary, it is also practical to perform the calculations manually. Flutter characteristics have been calculated by the modified strip analysis and compared with results of other calculations and with experiments for Mach numbers up to 15.3 and for wings with sweep angles from 0° to 52.5°, aspect ratios from 2.0 to 7.4, taper ratios from 0.2 to 1.0, and center-of-gravity positions between 34 percent chord and 59 percent chord. These ranges probably cover the great majority of wings that are of practical interest with the exception of very low-aspect-ratio surfaces such as delta wings and missile fins. The program is available through COSMIC.

SUMMARY

The trend continues in flight vehicle design toward higher speeds; lighter-weight, more flexible structures; more complex aerodynamic configurations; and control systems which are more closely coupled with the elastic response. This clearly requires that the prevention of aeroelastic problems be an increasingly important and challenging consideration in new vehicle designs. These aeroelastic problems can include phenomena such as flutter instabilities, static divergence, control reversal, extreme gust response, and interactions with control systems. The development of finite element methods in structural analyses geared to the large, high-speed computers has provided significant improvements in recent years to the analysis of stiffness and vibration characteristics of large and complex structural configurations. Aerodynamic representations have likewise improved based on several methods using large, high-speed computers. While the more simplified, faster, two-dimensional strip theories or strip theories modified with aspect ratio and compressibility corrections are still used for some preliminary design or for some applicable configurations, the three-dimensional lifting surface numerical procedures such as the subsonic doublet-lattice method, collocation methods (kernel function), and supersonic Mach box are now commonly used in flutter safety investigations of new or modified flight vehicles. It appears that further significant improvements in design procedures will be possible in automated opimization procedures for rapid redesign for minimum weight structures with constraints which include flutter, strength, and possibly other requirements. Such rapid design procedures also indicate a need for more efficient, less lengthy aerodynamic prediction methods which can be calculated in a manner consistent with structural component breakdowns.

REFERENCES

1 Lanchester, F. W., Torsional Vibrations of the Tail of an Aeroplane, British A.R.C. Reports and Memoranda No. 276, Volume 11 for the Year 1916-17, 1920, pp. 458-460.

2 Scanlan, R. H., and Rosenbaum, R., Introduction to the Study of Aircraft Vibration and Flutter, The Macmillan Company, New York, 1951. (Also Dover Publications, New York, N. Y.)

3 Fung, Y. C., An Introduction to the Theory of Aeroelasticity, John Wiley and Sons, New York, 1955.

4 Bisplinghoff, R. L., Ashley, H., and Halfman, R. L., Aeroelasticity, Addison-Wesley Publishing Company, Cambridge, Mass., 1955.

5 Bisplinghoff, R. L., and Ashley, H., Principles of Aeroelasticity, John Wiley and Sons, Inc., New York, 1962.

6 Jones, W. P., General Editor, Manual on Aeroelasticity, Vols. 1-5, NATO Advisory Group for Aeronautical Research and Development, (available in USA from Report Distribution and Storage Unit, NASA Langley Research Center) 1959.

7 Garrick, I. E., editor, Aerodynamic Flutter, Vol. V, AIAA Selected Reprint Series, 1969, American Institute of Aeronautics and Astronautics, New York, N. Y.

8 Collar, A. R., "The Expanding Domain of Aeroelasticity," Journal of the Royal Aeronautical Society, Vol. L, 1946, pp. 613-636.

9 Williams, J., "Aircraft Flutter," British A.R.C. Reports and Memoranda No. 2492, 1951.

10 Templeton, H., "A Review of the Present Position on Flutter," NATO Advisory Group for Aeronautical Research and Development, Report 57 (see reference 6 for availability), 1956.

11 Bisplinghoff, R. L., "Some Structural and Aeroelastic Considerations of High Speed Flight," The Nineteenth Wright Brothers Lecture, Journal of Aeronautical Sciences, Vol. 23, No. 4, 1956, pp. 289-321.

12 Garrick, I. E., "Some Concepts and Problem Areas in Aircraft Flutter," the 1957 Minta Martin Aeronautical Lecture, Sherman Fairchild Paper No. FF-15, Institute of the Aeronautical Sciences, 1957.

13 Goland, M., "An Appraisal of Aeroelasticity in Design, with Special Reference to Dynamic Aeroelastic Stability," presented at the Sixth Anglo-American Aeronautical Conference, London, September, 1957.

14 Collar, A. R., "Aeroelasticity-Retrospect and Prospect," The Second Lanchester Memorial Lecture, Journal of the Royal Aeronautical Society, Vol. 63, No. 577, 1959, pp. 1-15.

15 Laidlaw, W. R., "The Aeroelastic Design of Lifting Surfaces," Notes for the M.I.T. Summer Course on Aeroelasticity, June-July 1958. (Available from the North American Rockwell Co., Downey, Calif.)

16 Bisplinghoff, R. L., "Aeroelasticity," Applied Mechanics Reviews, Vol. 11, No. 3, 1958, pp. 99-103.

17 Jones, W. P., "Research on Unsteady Flow," The 1961 Minta Martin Lecture, Massachusetts Institute of Technology with the Institute of Aeronautical Sciences, 1961.

18 Rodden, W. P., and Revell, J. D., "The Status of Unsteady Aerodynamic Influence Coefficients," S.M.F. Paper No. FF-23, Institute of Aeronautical Sciences, 30th. Annual Meeting, New York, 1962.

19 Ashley, H., Widnall, S., and Landahl, M. T., "New Directions in Lifting Surface Theory," AIAA Journal, Vol. 3, No. 1, 1965, pp. 3-16.

20 Greidanus, J. H., and Yff, J., "A Review of Aeroelasticity," Applied Mechanics Surveys, ed. H. N. Abrahamson, Spartan Books, Washington, D. C., 1966.

21 Landahl, M. T., and Stark, V. J. E., "Numerical Lifting Surface Theory-Problems and Progress," AIAA Journal, Vol. 6, No. 11, 1968, pp. 2049-2060.

22 Ashley, H., "Aeroelasticity," Applied Mechanics Reviews, Vol. 23, No. 2, 1970, pp. 119-129.

23 Garrick, I. E., "Perspectives in Aeroelasticity," (5th. Theodore von Karman Memorial Lecture), Israel Journal of Technology, Vol. 10, No. 1-2, 1972, pp. 1-22.

24 Widnall, S., "Subsonic Aerodynamics," Aeronautics and Astronautics, Vol. 11, No. 4, 1973, p. 14.

25 Frazer, R. A., Duncan, W. J., and Collar, A. R., Elementary Matrices, Cambridge University Press, Cambridge, England, 1957.

26 Loring, S. J., "Outline of a General Approach to the Flutter Problem," SAE Transactions, August 1941, pp. 345-355.

27 Morse, P. M., and Feshback, H., Methods of Theoretical Physics, McGraw-Hill, New York, 1953, p. 806.

28 Watkins, C. E., and Berman, J. H., "On the Kernel Function of the Integral Equation Relating Lift and Downwash Distributions of Oscillating Wings in Supersonic Flow," NASA Report No. 1267, 1956.

29 Watkins, C. E., Runyan, H. L., and Woolston, D. S., "On the Kernel Function of the Integral Equation Relating the Lift and Downwash Distributions of Oscillating Finite Wings in Subsonic Flow," NASA Report No. 1234, 1955.

30 Cunningham, H. J., "Improved Numerical Procedure for Harmonically Deforming Lifting Surfaces from the Supersonic Kernel Function Method," AIAA Journal, Vol. 4, No. 11, 1966, pp. 1961-1968.

31 Evvard, J. C., "Distribution of Wave Drag and Lift in the Vicinity of Wing Tips at Supersonic Speeds," NASA TN 1382, 1947.

32 Ashley, H., "Supersonic Airloads on Interfering Lifting Surfaces by Aerodynamic Influence Coefficient Theory," Boeing Report D2-200-67, 1962, Seattle, Washington.

33 Moore, M. T., and Andrew, L. V., "Unsteady Aerodynamics for Advanced Configurations, Part IV-Application of the Supersonic Mach Box Method to Intersecting Planar Lifting Surfaces," Air Force Flight Dynamics Laboratory FDL-TDR-64-152, Part IV, 1965, Wright-Patterson Air Force Base, Ohio.

34 Andrew, L. V., "Unsteady Aerodynamics for Advanced Configurations, Part VI-Application of the Supersonic Mach Box Method to T-Tails, V-Tails, and Top-Mounted Vertical Tails," Air Force Flight Dynamics Laboratory FDL-TDR-64-152, Part VI, 1965, Wright-Patterson Air Force Base, Ohio.

35 Woodcock, D. L., editor, "A Comparison of Methods Used in Lifting Surface Theory," supplement to Manual on Aeroelasticity, Part VI (see ref. 6), NATO Advisory Group for Aeronautical Research and Development, Report No. 583, 1971.

36 Thomas, J. L., and Wang, H. T., "Evaluation of Lifting Surface Programs for Computing the Pressure Distribution on Planar Foils in Steady Motion," Report No. 4021, 1973, Naval Ship Research and Development Center, Bethesda, Maryland.

37 Woodward, F. A., "Analysis and Design of Wing-Body Combinations at Subsonic and Supersonic Speeds," Journal of Aircraft, Vol. 5, No. 6, 1968, pp. 528-534.

38 Woodward, F. A., "An Improved Method for the Aerodynamic Analysis of Wing-Body-Tail Configurations in Subsonic and Supersonic Flow, Part I-Theory and Application," NASA Contractor Report NASA CR-2228, 1973, Aerophysics Research Corporation for NASA-Langley Research Center, Hampton, Va.

39 Yates, E. C., "A Kernel Function Formulation for Non-Planar Lifting Surfaces Oscillating in Subsonic Flow, "AIAA Journal, Vol. 4, No. 8, 1966, pp. 1486-1488.

40 Landahl, M. T., "On the Kernel Function for Non-Planar Oscillating Surfaces in Subsonic Flow," AIAA Journal, Vol. 5, No. 5, 1967, pp. 1045-1046.

41 Hedman, S., "Vortex Lattice Method for Calculation of Quasi Steady State Loadings on Thin Elastic Wings in Subsonic Flow," Report No. 105, 1966, Flygtekniska Foroksanstalten (FAA), Stockholm, Sweden.

42 Rubbert, P. E., and Saaris, G. R., "Review and Analysis of a Three-Dimensional Lifting Potential Flow Analysis Method for Arbitrary Configurations," AIAA Paper No. 72-188, presented at the AIAA 10th Aerospace Sciences Meeting, San Diego, California, 1972.

43 Albano, E., and Rodden, W. P., "A Doublet-Lattice Method for Calculating Lift Distributions on Oscillating Surfaces in Subsonic Flows," AIAA Journal, Vol. 7, No. 2, 1969, pp. 279-285.

44 Pines, S., Dugunji, J., and Nueringer, J., "Aerodynamic Flutter Derivatives for a Flexible Wing with Supersonic and Subsonic Edges," Journal of Aeronautical Sciences, Vol. 22, No. 10, 1955, pp. 693-700.

45 Landahl, M. T., Unsteady Transonic Flow, Pergamon Press, New York, 1961, Chaps. 4, 6, and 7.

46 Runyan, H. L., and Woolston, D. S., "Method for Calculating the Aerodynamic Loading on an Oscillating Finite Wing in Subsonic and Sonic Flow," NACA TN 3694, 1956.

47 Rodemich, E. R., and Andrew, L. V., "Unsteady Aerodynamics for Advanced Configurations, Part II-A Transonic Box Method for Planar Lifting Surfaces," FDL-TDR-64-152, Part II, 1965, Air Force Flight Dynamics Laboratory, Wright-Patterson Air Force Base, Ohio.

48 Stenton, T. E., and Andrew, L. V., "Transonic Unsteady Aerodynamics for Planar Wings with Trailing Edge Control Surfaces," AFFDL-TR-67-180, 1968, Air Force Flight Dynamics Laboratory, Wright-Patterson Air Force Base, Ohio.

49 Theodorsen, T., "General Theory of Aerodynamic Instability and the Mechanism of Flutter," NACA Report No. 496, 1935.

50 Cicala, P., "Aerodynamic Forces on an Oscillating Profile in a Uniform Stream," L'Aerotecnica, Vol. 16, 1936, p. 635.

51 Kussner, H. G., "Zusammenfassender Bericht uber den instationaren Auftrieb von Tragfugeln," Luftfahrtforschung, Vol. 13, 1936, pp. 410-424.

52 Smilg, B., and Wasserman, L. S., "Application of Three-Dimensional Flutter Theory to Aircraft Structures," Technical Report No. 4798, 1942, Air Force Flight Dynamics Laboratory, Wright-Patterson Air Force Base, Ohio.

53 Garrick, I. E., and Rubinow, S. I., "Flutter and Oscillating Air Force Calculations for an Airfoil in a Two-Dimensional Supersonic Flow," NACA Report No. 846, 1946.

54 Ashley, H., and Zartarian, G., "Piston Theory-A New Aerodynamic Tool for the Aeroelastician," Journal of Aeronautical Sciences, Vol. 23, No. 12, 1956, pp. 1109-1118.

55 Stroud, W. J., Dexter, C. B., and Stein, M., "Automated Preliminary Design of Simplified Wing Structures to Satisfy Strength and Flutter Requirements," NASA TN D-6534, 1971.

56 Haftka, R. T., "Automated Procedure for the Design of Wing Structures to Satisfy Strength and Flutter Requirements," NASA TN D-7264, 1973.

57 Haftka, R. T., and Yates, E. C., "On Repetitive Flutter Calculations in Structural Design," AIAA Paper No. 74-141, presented at the AIAA 12th Aerospace Sciences Meeting, Washington, D.C., Jan. 30-Feb. 1, 1974.

58 Bennett, R. L., "Rotor System Design and Evaluation Using a General Purpose Helicopter Flight Simulation Program," AGARD Conference Proceedings No. 122, Specialist Meeting on Helicopter Rotor Loads Prediction Methods, Milan, Italy, 30 March 1973.

59 Stoddart, S. A., 1/Lt, USAF, "Supplementary User's Guide For The Bell Helicopter Rigid Body Dynamics Analysis Digital Computer Program (C-81, ASAJ02)," AFFDL/FGC-TM-72-10, June 1973, Air Force Flight Dynamics Laboratory, Wright-Patterson Air Force Base, Ohio.

60 Yen, J. G., Weber, G. E., and Goffey, T. M., "A Study of Folding Proprotor VTOL Aircraft Dynamics, Volume 1, Analytical Methods," AFFDL-TR-71-7, Vol. I, September 1971, Bell Helicopter Company for Air Force Flights Dynamics Laboratory, Wright Patterson Air Force Base, Ohio.

61 Losey, H. E. and Hsieh, P. Y., "A Study of Folding Proprotor VTOL Aircraft Dynamics, Volume II, Computer Programs ," AFFDL-TR-71-7, Vol. II, Sept. 71, Bell Helicopter Company tor Air Force Flight Dynamics Laboratory, Wright-Patterson Air Force Base, Ohio.

62 Appa, K. and Smith, G. C. C., "Development and Applications of Supersonic Unsteady Consistent Aerodynamics for Interfering Parallel Wings," NASA CR-2168, March 1973, Bell Aerospace Company for NASA Langley Research Center, Hampton, Va.

63 Paine, A. A., "Development and Applications of Supersonic Unsteady Consistent Aerodynamics for Interfering Parallel Wings," User's Manual, Bell Aerospace Rept. No. 2471-956003. Programmer's Manual, Bell Aerospace Rept. No. 2741-956004 (Available as NASA-CR-112184 User's Manual and NASA CR-112185 Programmer's Mannual).

64 Kemp, W. B., "Definition and Application of Longitudinal Stability Derivatives for Elastic Airplanes," NASA TN D-6629, March 1972, NASA Langley Research Center, Hampton, Va.

65 Rowe, W. S., Winther, B. A., and Redman, M. C., "Prediction of Unsteady Aerodynamic Loadings Caused by Trailing Edge Control Surface Motions in Subsonic Compressible Flow - Analyses and Results," NASA CR-2003, June 1972, The Boeing Company, Renton, Wash., for NASA-Langley Research Center, Hampton, Va.

66 Redman, M. C., Rowe, W. S., and Winther, B. A., "Prediction of Unsteady Aerodynamic Loadings Caused by Trailing Edge Control Surface Motions in Subsonic Compressible Flow - Computer Program Description," NASA CR 112015, June 1972, The Boeing Company, Renton, Wash., for NASA-Langley Research Center, Hampton, Va.

67 Gwin, L. B. and McIntosh, S. C., Jr., "A Method of Minimum-Weight Synthesis for Flutter Requirements, Part I - Analytical Investigations," AFFDL-TR-72-22, Part I, June 1972, Air Force Flight Dynamics Laboratory, Wright-Patterson Air Force Base, Ohio.

68 Gwin, L. B and McIntosh, S. C., Jr., "A Method of Minimum-Weight Synthesis for Flutter Requirements, Part II - Program Documentation," AFFDL-TR-72-22, Part II, June 1972, Air Force Flight Dynamics Laboratory, Wright-Patterson Air Force Base, Ohio.

69 Gwin, L. B. and McIntosh, S. C., Jr., "Large Scale Flutter Optimization of Lifting Surfaces, Part I - Analytical Investigations," AFFDL-TR-73-91, Part I, January 1974, Air Force Flight Dynamics Laboratory, Wright-Patterson Air Force Base, Ohio.

70 Gwin, L. B. and McIntosh, S. C., Jr., "Large Scale Flutter Optimization of Lifting Surfaces, Part II - Program Documentation," AFFDL-TR-73-91, Part II, January 1974, Air Force Flight Dynamics Laboratory, Wright-Patterson Air Force Base, Ohio.

71 Andries, R. A., Batill, S. M., and Taylor, R. F., "Documentation and Application of a Minimum-Weight Synthesis Procedure for Flutter Requirements," AFFDL-TM-73-19, February 1973, Air Force Flight Dynamics Laboratory, Wright-Patterson Air Force Base, Ohio.

72 Andries, R. A., "A User's Guide to a Minimum-Weight Synthesis Procedure for Flutter Requirements, Part I - Program Input Instructions," AFFDL TM-74-27-FYS, March 1974, Air Force Flight Dynamics Laboratory, Wright-Patterson Air Force Base, Ohio.

73 Ferman, M. A., "A Rapid Method for Flutter Clearance of Aircraft with External Stores - Volume I, Theory and Application," AFFDL-TR-73-74, Vol. I, September 1973, McDonnell Aircraft Company for Air Force Flight Dynamics Laboratory, Wright-Patterson Air Force Base, Ohio.

74 Unger, W. H., "A Rapid Method for Flutter Clearance of Aircraft with External Stores, Volume II - User's Manual for FACES Computer Program," AFFDL-TR-73-74, Vol. II, September 1973, McDonnell Aircraft Company for Air Force Flight Dynamics Laboratory, Wright-Patterson Air Force Base, Ohio.

75 Jennings, M. E., and Wells, J. R., "A Rapid Method for Flutter Clearance of Aircraft with External Stores, Volume III - Programmer's Manual for FACES Computer Program," AFFDL-TR-73-74, Vol. III, September 1973. McDonnell Automation Company for Air Force Flight Dynamics Laboratory, Wright-Patterson Air Force Base, Ohio.

76 Cross, A. K. and Albano, E. A., "Computer Techniques for the Rapid Flutter Clearance of Aircraft Carrying External Stores - Part I Perturbation Theory and Application," AFFDL-TR-72-114, Part I, February 1973, Northrop Corporation, Aircraft Division for Air Force Flight Dynamics Laboratory, Wright-Patterson Air Force Base, Ohio.

77 Cross, A. K. and Albano, E. A., "Computer Techniques for the Rapid Flutter Clearance of Aircraft Carrying External Stores - Part II Documentation of Data Retrieval System and Perturbation Program," AFFDL-TR-72-114, Part II, February 1973, Northrop Corporation, Aircraft Division, for Air Force Flight Dynamics Laboratory, Wright-Patterson Air Force Base, Ohio.

78 Laschka, B., "Zur Theorie der harmonisch schwingenden tragenden Fläche bei Unterschallanströmung," Zeitschrift für Flugwissenschaften, Vol. 11, No. 7, 1963, pp. 265-292.

79 Albano, E., Perkinson, F., and Rodden, W. P., "Subsonic Lifting-Surface Theory Aerodynamic and Flutter Analyses of Interfering Wing/Horizontal-Tail Configurations, Part I - Subsonic Oscillatory Aerodynamics for Wing/Horizontal-Tail Configurations," AFFDL-TR-70-59, Part I, September 1970, Northrop Corporation, Aircraft Division, for Air Force Flight Dynamics Laboratory, Wright-Patterson Air Force Base, Ohio.

80 Albano, E., Perkinson, F., and Rodden, W. P., "Subsonic Lifting-Surface Theory Aerodynamics and Flutter Analysis of Interfering Wing/Horizontal-Tail Configurations, Part II Wing/Tail Flutter Correlation Study," AFFDL-TR-70-59, Part II, September 1970, Northrop Corporation, Aircraft Division, for Air Force Flight Dynamics Laboratory, Wright-Patterson Air Force Base, Ohio.

81 Giesing, J. P., Kalman, T. P., and Rodden, W. P., "Subsonic Unsteady Aerodynamics for General Configurations, Part I, Vol. I - Direct Application of the Nonplanar Doublet-Lattice Method," AFFDL-TR-71-5, Part I, Vol. I, November 1971, McDonnell-Douglas Corp. for Air Force Flight Dynamics Laboratory, Wright-Patterson Air Force Base, Ohio.

82 Giesing, J. P., Kalman, T. P., and Rodden, W. P., "Subsonic Unsteady Aerodynamics for General Configurations - Part I, Vol. II - Computer Program H7WC," AFFDL-TR-71-5, Part I, Vol. II, November 1971, McDonnell-Douglas Corp. for Air Force Flight Dynamics Laboratory, Wright-Patterson Air Force Base, Ohio.

83 Giesing, J. P., Kalman, T. P. and Rodden, W. P., "Subsonic Unsteady Aerodynamics for General Configurations, Part II, Volume I - Application of the Doublet-Lattice Method and the Method of Images to Lifting-Surface/Body Interference," AFFDL-TR-71-5, Part II, Vol. I, April 1972, McDonnell-Douglas Corp. for Air Force Flight Dynamics Laboratory, Wright-Patterson Air Force Base, Ohio.

84 Giesing, J. P., Kalman, T. P. and Rodden, W. P., "Subsonic Unsteady Aerodynamics for General Configurations - Part II, Vol. II - Computer Program N5KA," AFFDL-TR-71-5, Part II, Vol. II, April 1972, McDonnell-Douglas Corp. for Air Force Flight Dynamics Laboratory, Wright-Patterson Air Force Base, Ohio.

85 Donato, V. W. and Huhn, C. R., Jr., "Supersonic Unsteady Aerodynamics for Wings with Trailing Edge Control Surfaces and Folded Tips," AFFDL-TR-68-30, August 1968, Rockwell International, for Air Force Flight Dynamics Laboratory, Wright-Patterson Air Force Base, Ohio.

86 Morito, Jack II, Borland, C. J. and Hogland, J. R., "Prediction of Unsteady Aerodynamic Loadings of Non-Planar Wings and Wing-Tail Configurations in Supersonic Flow - Part I, Theoretical Development, Program Usage, and Application," AFFDL-TR-71-108, Part I, March 1972, The Boeing Company, Commercial Airplane Group for Air Force Flight Dynamics Laboratory, Wright-Patterson Air Force Base, Ohio.

87 Krammer, G. D., and Keylon, G. E., "Prediction of Unsteady Aerodynamic Loadings on Non-Planar Wings and Wing-Tail Configurations in Supersonic Flow - Part II Computer Program Description," AFFDL-TR-71-108, Part II, March 1972, The Boeing Company, Commercial Airplane Group for Air Force Flight Dynamics Laboratory, Wright-Patterson Air Force Base, Ohio.

88 Olsen, J. J., and Batill, S. M., "Application and Improvement of Finite-Difference Methods in Transonic Nonlifting Flow," AFFDL-TM-73-85-FYS, July 1973, Air Force Flight Dynamics Laboratory, Wright-Patterson Air Force Base, Ohio.

89 McCullers, L. A., and Lynch, R. W., "Dynamic Characteristics of Advanced Filamentary Composite Structures, Volume II Aeroelastic Synthesis Procedure Development," AFFDL-TR-73-111, Volume II, March 1973, Convair Aerospace Division of General Dynamics, Fort Worth Operation for Air Force Flight Dynamics Laboratory, Wright-Patterson Air Force Base, Ohio.

90 Woodward, F. A., "An Improved Method for the Aerodynamic Analysis of Wing-Body-Tail Configurations in Subsonic and Supersonic Flow, Part II - Computer Program Description," NASA CR-2228, Part II, 1973, Aerophysics Research Corporation for NASA-Langley Research Center, Hampton, Va.

91 Yates, E., "Modified-Strip-Analysis Method for Predicting Wing Flutter at Subsonic to Hypersonic Speeds," Journal of Aircraft, Vol. 3, No. 1, 1966, pp. 25-29.

92 Yates, E. C., "Calculations of Flutter Characteristics for Finite-Span Swept or Unswept Wings at Subsonic and Supersonic Speeds by a Modified Strip Analysis," NASA RM L57L10, 1958, NASA-Langley Research Center, Hampton, Va.

93 Barnby, J. O., Cunningham, H. T., and Garrick, I. E., "Study of the Effects of Sweep on the Flutter of Cantilever Wings," NASA Rept. 1014, 1951.

Table 1 Summary of Important Features of Aeroelastic Analyses Computer Programs

Item	Program Name	Developer	Availability[1]
1	C81	R.L. Bennett, Bell Helicopter Co. P.O. Box 482, Ft. Worth, Texas 76101 (817) 280-3956	U.S. Army Aviation Material R&D Laboratory Ft. Eutis, Virginia (E. Austin)
2	DFAL 17 & 18 (Proprotor Aeroelastic Analysis. Bell DYN-5)	J.G. Yen, T.Gaffey, Bell Helicopter Co., P.O. Box 482, Ft. Worth, Texas 76101 (817) 280-3956	Air Force Flight Dynamics Laboratory AFFDL/FYS, A. Basso, WPAFB 45433 (513) 255-5236
3	AIC-INT	Kari Appa and G.C.C. Smith, Bell Aerospace Co., P.O. Box 1, Buffalo, N.Y. 14240 (716) 297-1000 Ext 7867 or 7447	Developer
4	AEDERIV	W. Kemp, Jr. NASA-Langley, MS 403, Hampton, Virginia 23365, (804) 827-2961	Developer
5	RHØIII Unsteady Aero Loadings caused by trailing-edge control surface motions.	W. Rowe, B. Winther, M. Redman, The Boeing Co., Renton, Washington	NASA Langley Research Center Hampton, Virginia 23365, Technical Monitor - H. Cunningham (804) 827-2661
6	SØAR (Structural Optimization for Aeroelastic Reqmts)	L.B. Gwin and S.C. McIntosh, Jr., Stanford University	Air Force Flight Dynamics Laboratory (FYS), R.F. Taylor, Capt S.M. Batill, R.A. Andries, WPAFB, Ohio 45433 (513) 255-3297
7	SUBSØNC (Subsonic Flutter Analyses)	R. Andries, AF Flight Dynamics Laboratory, (AFFDL/FYS)	Developer
8	FACES (Flutter of Aircraft Carrying External Stores)	M. Ferman, W. Unger, McDonnell Aircraft Co., St. Louis, Missouri	AFFDL/FYS, S. Pollock, WPAFB, Ohio 45433 (513) 255-3297
9	PERTURB (Perturbation Method for Store Flutter)	E. Albano, A. Cross, Northrop Corp. Aircraft Div., Hawthorne, Calif.	AFFDL/FYS, S. Pollock, WPAFB, Ohio 45433 (513) 255-3297
10	TWOS (Wing-Horizontal Tail Flutter)	E. Albano, W. Rodden, Northrop Corp. Aircraft Div., Hawthorne, Calif.	AFFDL/FYS, S. Pollock, WPAFB, Ohio 45433 (513) 255-3297
11	H7WC (Nonplanar Doublet-Lattice)	J. Giesing, W. Rodden, T. Kalman Douglas Aircraft Company, Long Beach, California	AFFDL/FYS, S. Pollock, WPAFB, Ohio 45433 (513) 255-3297
12	N5KA - Doublet-Lattice & Images	J. Giesing, W. Rodden, T. Kalman Douglas Aircraft Company, Long Beach, California	AFFDL/FYS, S. Pollock, WPAFB, Ohio 45433 (513) 255-3297
13	MBØX (AFFDL TR 68-30)	V. Donato and C. Huhn Rockwell Corporation	AFFDL/FYS, L. Huttsell, WPAFB, Ohio 45433 (513) 255-3297
14	MBØX (AFFDL TR 71-108)	W. Rowe, J. Ii, C. Borland, J. Hogley, G. Kramer, G. Keylon, The Boeing Company, Seattle, Washington	AFFDL/FYS, L. Huttsell, WPAFB, Ohio 45433 (513) 255-3297
15	SEIDEL	J. Olsen, AFFDL/FYS, WPAFB, Ohio 45433 (513) 255-3434	Developer
16	TSO (Wing aeroelastic synthesis procedure)	L. McCullers, R. Lynch, M. Waddoups General Dynamics, Ft. Worth, Texas	AFFDL/FYS, Lt K. Griffin, WPAFB, Ohio 45433 (513) 255-3297
17	Aerodynamic Analysis of Wing-Body-Tail Configurations in Subsonic and Supersonic Flow	F. A. Woodward, Analytical Methods, Inc., Bellevue, Wash 98004 and Aerophysics Research Corp, Box 187, Bellevue, Wash 98009	NASA Langley Research Center, Hampton, Virginia 23365
18	Modified Strip Analysis (Program No. LAR-10199)	E.C. Yates, NASA Langley Research Center, Hampton, Virginia 23365	Developer or COSMIC

[1]Potential users of computer programs should check with the listed organization regarding qualifications, costs, etc.

Table 1 (cont.)

Item	Aerodynamic Representation				
	Mach range	Type	Planar surfaces	Control surfaces	Interfering surfaces
1	Subsonic, Transonic, Supersonic	Uniform, steady flow and harmonic oscillation, also from data table			
2	Subsonic	Non-linear, 2-D, quasi-steady	X	X	X (wing upwash effects on blade included)
3	Supersonic	Uniform, 3-D, harmonic oscillation, vel. potential from integral of downwash over arbitrary triangular elements	X	X	X
4	Depends only on AIC Matrix Supplied	Uniform, steady flow, requires AIC matrix generated externally, forces generated internally; externally supplied force distribution due to rigid shape and rigid angle of attack may be used optimally.	X	X	X
5	Subsonic (Compressible)	Perturbation of non-uniform flow, harmonic oscillations, 3-D, integral equation in acceleration potential (Kernel function)	X	X	
6	Subsonic, Supersonic	3-D Doublet-lattice, subsonic; and supersonic Mach box	X		
7	Subsonic	3-D Doublet-lattice, subsonic (polynomial or spline mode fit)	X		
8	Subsonic, High Supersonic	Modified strip theory, doublet-lattice, and piston theory	X		
9	Subsonic	Doublet-lattice	X		
10	Subsonic	Collocation method (Kernel function)	X		X (Wing and horizontal tail with dihedral and separation)
11	Subsonic	Doublet-lattice	X	X	X (Intersecting surfaces such as T-tails, also lifting surface & body combinations, and wing/tails)
12	Subsonic	Doublet-lattice and method of Images	X	X	X (Same as 11 except also uses method of images for body aero)
13	Supersonic	Mach Box	X	X	
14	Supersonic	Mach Box	X		X Nonplanar wing-tail
15	Subsonic, Transonic	Steady, perturbation of non-uniform flow, 2-D, finite-difference	X	X	
16	Subsonic, Supersonic	Subsonic steady & unsteady (doublet-lattice) Supersonic steady (Woodward's method)	X	X (Steady only)	X (Steady only)
17	Subsonic, Supersonic	Lifting surface theory using constant source distribution on body panels and vortex distribution on wing and tail panels.	X		X
18	Subsonic to Hypersonic	Aero parameter distributions may be calculated from any suitable theory or measured data and employed in the method	X		

Table 1 (cont.)

Item	Computer		Plotting	Number of Source Statements (approx)	Core Storage-Decimal Words (approx)	Documentation
	Type	Language	Capability			
1	IBM 360/65 CDC 6600		X	30,000	206K fast core 400K slow core	Paper by R.L. Bennet in AGARD Conference Preceedings #122, [58], & AFFDL/FGC TM 72-10, [59].
2	IBM 360/65 CDC 6600	Fortran IV	X	4,000	50K	AFFDL TR 71-7, Vol I (Dev. of Method) [60], Vol II (Computer Programs) [61].
3	IBM 360	Fortran IV	X		160K	NASA CR-2168 (Dev. & Application of Method) [62], NASA CR-112185 (Programmers Manual) [63] NASA CR-112184 (Users Manual) [63].
4	CDC 6000 series	Fortran IV		520	50K	NASA TN D-6629 [64].
5	CDC 6400 CDC 6600			10,500	28K	NASA CR-2003 (Analyses & Results) [65], NASA CR-112015 (Computer Programs) [66].
6	IBM 360/67 CDC 6600	Fortran IV		1,968	76K	AFFDL TR-72-22, Part I [67], Part II [68], AFFDL TR-73-91, Part I [69], Part II [70].
7	CDC 6600	Fortran IV		2,689	35K	AFFDL TM-73-19 [71] and AFFDL TM-74-27-FYS, Part I (Program Instruction) [72].
8	IBM 360 CDC 6600	Fortran IV	X	12,000	41K	AFFDL TR-73-74, Vol I (Theory & Appl.) [73], Vol II (User's Manual) [74], Vol III (Programmer's Manual) [75].
9	IBM 370/165	Fortran IV	X	2,500	39K	AFFDL TR-72-114, Part I (Theory & Appl.) [76], Part II (Documentation of Program) [77].
10	CDC 6600	Fortran IV		2,000	32K	AFFDL TR-70-59, Part I (Aerodynamic Method) [79], Part II (Correlation Study) [80].
11	IBM 360/65 CDC 6600 GE 635	Fortran IV		5,000	32K	AFFDL TR-71-5, Part I, Vol I (Application) [81], Vol II Computer Program) [82].
12	IBM 360/65 CDC 6600	Fortran IV		6,000	57K	AFFDL TR-71-5, Part II, Vol I (Application) [83], Vol II (Computer Program) [84].
13	IBM 7094 CDC 6600	Fortran IV		4,000	32K 35K	AFFDL TR-68-30 [85].
14	CDC 6600	Fortran IV		10,000	50K	AFFDL TR-71-108, Part I (Theoretical Dev.) [86], Part II (Computer Program) [87].
15	CDC 6600	Fortran IV		500	20K	AFFDL TM-FYS-73-85 [88].
16	CDC 6600	Fortran IV		10,700	220K	AFFDL TR-73-111, Vol II [89].
17	CDC 6000 series	Fortran IV		6,594	29K	NASA CR-2228, Part I (Method) [38], Part II (Computer Program) [90].
18	CDC 6000 IBM 360	Fortran IV	X	850	29K	Journal of Aircraft, Vol 3, No. 1, Jan-Feb 1966, Page 25-29 [91], NACA RML57L10, [92].

Table 1 (cont.)

Item	System Analysis Capability						
	Airloads	Flutter	Divergence	Reversal	Control System Feedback	Optimization	Other
1	X	X	X		X		Aeroelastic effects, dynamic loading
2	X	X	X				Dynamic stability, vibration, and loads - for folding - prop-rotor during starting and stopping and folding
3	X						
4	X		X	X	X		Generates quasi-steady elastic, longitudinal stability derivitives and equilibrium plus perturbed load and shape distributions. Derivatives are based on 3-degrees of freedom (forward speed, pitch & plunge)
5	X						Computes generalized aero forces for subsequent use in flutter and aeroelastic analyses
6	X	X				X	
7	X	X					
8	X	X					Includes up to five external store stations per semi-span
9	X	X					Provides flutter solutions for store parameter variations based on initial flutter solution
10	X	X					
11	X						
12	X						
13	X	X					
14	X						
15	X						
16	X	X	X			X	Optimizes weight and/or aeroelastic parameters of either composite or metal wings
17	X						
18	X	X	X				Dynamic Loading

ROTOR-BEARING SYSTEMS

Neville F. Rieger
Rochester Institute of Technology
Rochester, New York

ABSTRACT

A review of the nature and functioning of computer programs for rotor-bearing dynamic analysis is presented. The programs discussed are available for general use by lease, time sharing, or purchase. The contribution of each type of program to rotor system analysis is described, and then current program approaches to critical speed, unbalance response, stability, torsional analysis, and balancing are reviewed. Detailed tabular comparisons of programs in each of these aspects are given. Summary comments on the state-of-the-art for each program category are listed.

INTRODUCTION

This paper is a review of certain commercially available computer programs for the dynamic analysis of rotor-bearing systems. The programs discussed are intended for use in the design analysis of rotating machinery, rather than for basic research. The purpose of the paper is to discuss the functioning and the capabilities of the available programs. To do this, information has been obtained from program authors, vendors, and the users of these programs. All programs are available from the sources indicated, either on request, for sale, or on a long-term lease arrangement.

This comparison is presented in the format which would be encountered by a designer during the analysis of a rotor system. Questions which would naturally arise are:

What aspects is it important to analyze?
Which programs are available to provide this information?
What are the relative merits of these programs?
How do these programs differ?
What do other users say about these programs?

A full comparison study is not possible at this time because identical and comprehensive data are not available for each program. For example, no rotor-bearing system has been analyzed through each category using each of the programs discussed, and then been built and tested for correlation. Secondly, the number and nature of those sample calculations from which program evaluation can be made is also quite small. The best documented program verification data are those which have been developed by the program vendors. Most users seem to be satisfied to accept this and check their program functioning against data provided in the program manual.

Certain rotor-bearing system concepts and methodology will first be given to develop a frame of reference. Program discussion and comparison then follows.

THE ROTOR SYSTEM

A typical rotor-bearing system is shown in Fig. 1. It consists of a rotor, bearing supports, gas seals, and foundation. The properties attributed to each aspect of this system are quite general and correspond to the most it is likely to possess in customary industrial practice. In most instances the programs discussed possess similar generality. This generality can, of course, be readily adapted to suit specific cases; e.g., a flexible foundation becomes a rigid foundation when its stiffness is made sufficiently great. Any exception to this generality is indicated where it occurs.

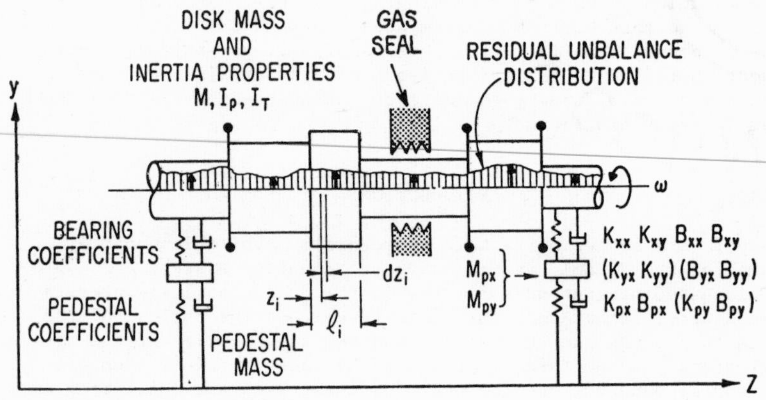

Fig. 1 Rotor-bearing system with flexible, damped pedestals and seals

Basic procedure for modeling of system components is:

1. Rotor. Most rotors may be adequately represented by a series of connected cylindrical shaft sections. The more advanced programs permit distributed mass and elastic effects and axial taper. Older programs require the rotor to be represented by equivalent discrete masses interconnected by massless springs. The effects of any impellers or disks are usually included as discrete masses and inertias (polar and transverse) at the ends of the shaft sections.

2. Bearings. Bearings are customarily modeled as an arrangement of springs and dashpots for which the properties are known or may be computed. Alternatively, bearing forces may be computed point-by-point in time to determine the whirl orbit, within the response program (predicator-corrector approach). The bearings are modeled to

occur at the ends of shaft sections, as for the disks.

3. Seals. Seal dynamic forces are modeled in a similar manner to bearing forces. A special program is required to calculate the seal force coefficients.

4. Foundation. The bearing support shown in Fig. 1 has concentrated mass, elastic, and damping properties and is attached to a rigid substructure by a further level of damped flexible elements. Such a foundation is called a one-level program, with level 1 being the rotor. Where the distributed bearing support itself is attached to other bearing supports by a continuous member (e.g., casing), a two-level program is involved. Other one-level programs exist which do not allow discrete pedestal mass, stiffness, and damping effects to be included. The dynamic properties of the foundation may be constants or may be frequency-dependent inputs.

IMPORTANT ASPECTS OF ROTOR SYSTEM DYNAMICS

Before making any system dynamics computations it is necessary to have the following input information available:

1. Rotor. Simulated distribution of mass and elastic properties to suit the rotor model, such as that shown in Fig. 1.

2. Bearing Static Properties.[1] Details of size, fluid, load, and speed are needed to determine the steady-state operating properties, e.g., eccentricity vs. Sommerfeld number.

3. Bearing Dynamic Properties.[1] These properties may be given as input or computed by the dynamics program for each operating condition.

4. Seal Forces. These forces are computed where needed, in a similar manner to bearing forces, for the given geometry with input as dynamic coefficients.

5. Foundation Properties. Foundation details covering related mass, stiffness, and damping properties are input as required by problem and program level.

To understand the dynamic behavior of any high-speed machine it is necessary to have information on the following aspects of its behavior, throughout the prescribed range of speed and operating conditions:

1. Critical Speed Analysis. Includes effect of bearing stiffness on location of each critical speed in operating speed range.

2. Unbalance Response Spectrum. Amplitude response of rotor to specified distribution of residual unbalance throughout operating speed range.

3. Stability Analysis. Location of threshold speed for onset of any unstable whirl motions of rotor in bearings.

4. Torsional Critical Frequency Analysis. Location and distribution of torsional critical frequencies of rotor drive train in operating range.

---

[1]Nonlinear programs often combine these steps in performing time step orbital calculations of system response.

After the above aspects of rotor system dynamics have been satisfied, the design may be completed and manufactured. Finally the rotor system will require balancing before it will run smoothly. This may be accomplished by computer program, as follows:

5. Rotor Balancing Program. A format procedure for determining required rotor balancing weights for low-vibration operation of rigid and flexible rotors.

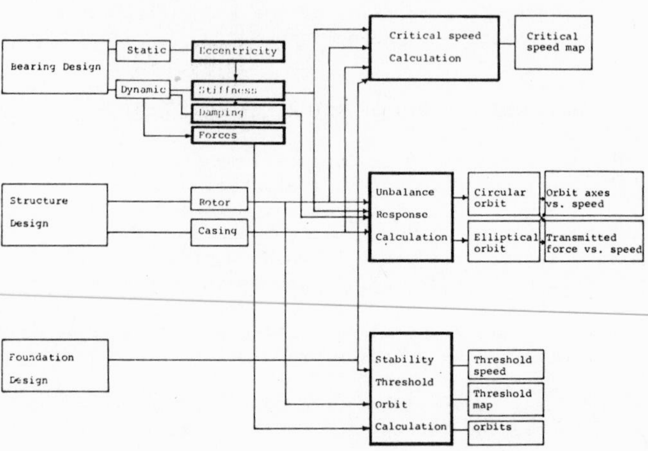

Fig. 2 Interrelationship between various aspects of rotor-bearing analysis

Figure 2 shows the various aspects of the above process by which a typical rotor system would be analyzed. The lines between the various computational blocks demonstrate the interrelationship and sequencing of the computations involved in this process.

FUNCTIONING OF COMPUTER PROGRAMS

Component Properties Data Needed for Calculations

Rotor

The primary need is to define the rotor as a mass-elastic system in the specified program format. Details of internal damping are required by only a few unbalance response and stability programs. Details of residual unbalance magnitude and distribution are required for response calculations.

Bearing Steady-State Analysis

Steady-state bearing data is needed at operating conditions throughout the speed range as a prerequisite for determining the corresponding bearing dynamic properties. Bearing type (plain cylindrical, tilting pad, partial arc, etc.) is specified in advance. Bearing geometry is usually represented as a series of partial arcs over which Reynold's equation and the energy equation (temperature effects on viscosity) are solved simultaneously using a prescribed finite difference mesh over the bearing surface, with suitable film rupture criteria. Important program outputs are eccentricity ratio vs. Sommerfeld number, temperature profiles, and minimum film thickness. A number of programs for performing such calculations are presently available from the major program vendors listed in the tables of this paper.

Bearing Dynamic Properties

Two procedures are currently in use for obtaining bearing dynamic properties. The first is to linearize the bearing force equations with respect to dynamic displacements and velocities about the steady-state position. Displacement and velocity coefficients are then computed which are constant for any specified operating condition but which vary with Sommerfeld number, bearing geometry, turbulence, and so on. The second procedure is to solve Reynold's equation and the energy equation simultaneously at each dynamic displacement location and to use a numerical method (predictor-corrector or Runge-Kutta) to identify the next equilibrium position. This determines the rotor orbit. Both methods have advantages. The first method is simple and appears to work well, but it may preclude the prediction of certain subharmonic rotor motions within the bearing.

Seal Dynamic Properties Data

There is apparently no commercially available program for calculating the forces which occur in either stage seals or tip shroud seals. The open data which exist for such forces are simple formulas or empirical coefficients derived from test.

Foundation Properties

In most instances machine foundation effects may be included directly into the main program or by a separate finite element calculation where necessary. On rare occasions it may be necessary to input foundation characteristics as frequency- or displacement-dependent stiffness and damping coefficients to stimulate observed foundation conditions, e.g., turbo-generation on marshy foundation.

Rotor-Bearing System Properties

Critical Speed

Available critical speed programs calculated undamped natural frequencies of rotors in isotropic flexible supports, the so-called circular orbit programs. All damping and unbalance forces are absent from the system model. The program output is a mode shape, usually normalized on the largest amplitude, and a value for the corresponding natural frequency. Such calculations are useful for rotor systems possessing little damping, e.g., rolling element bearings rather than fluid film bearings, or those which have a very flexible rotor in relatively stiff, damped bearings. Where substantial damping is distributed throughout the system, large resonant peaks cannot occur, and the rotor response does not follow the resonant buildup indicated by the critical speed calculation.

Unbalance Response

Three kinds of programs exist: (1) synchronous circular orbit, (2) synchronous elliptical orbit, and (3) transient orbital response. In each case the response of a rotor in bearings to unbalance forcing is sought, though the analytical procedure of the first two cases differs from that of the third case. In the first two cases rotor amplitude at specified speeds is calculated throughout a range of speeds and loads. In the third case the whirl orbit itself is calculated in detail at selected speeds, using a time-increment procedure. Several successive orbits are usually involved. Orbit diameter at each main station along the rotor is given by the first type of program, along with phase angle and transmitted force details. Details of maximum and minimum orbit diameters and phase angles at mass stations are given by programs (2), together with the force transmitted to the foundation by the bearings with force phase angle. Orbital options also exist for cases (1) and (2) with several programs. Both one-level and two-level program versions exist for case (1) and (2) calculations.

The unbalance response programs include bearing damping which limits the resonant or critical speed amplitudes. Damping is included as velocity-bearing coefficients in cases (1) and (2), or it may be computed as a nonlinear instantaneous velocity-dependent force, case (3). Seal force effects are usually omitted from response calculations unless there is a special need for including them.

Stability

Two types of stability analysis are in present use. The first is a linearized stability threshold speed calculation procedure, in which the positive real part of the lowest complex eigenvalue of the system characteristic equation is sought. The second method is a transient orbit analysis in which the growth or decay of the rotor whirl is studied. Programs for both types of analysis exist which utilize a flexible rotor model. Seal forces may be included in the more general programs along with some degree of foundation dynamics (usually as discrete mass-stiffness-damping units). The principal differences between such programs lie in the bearing formulation, i.e., linear vs. nonlinear, and the nature of the results obtained. The linear analysis finds a threshold speed and a whirl frequency, whereas the nonlinear analysis presents the orbital motion for a number of successive cycles, based on some suitable starting condition. The orbits show whether the whirl is growing or decaying, and so indicate stable or unstable operation.

Torsional Critical Frequency Analysis

Many high-speed rotors are driven (or drive) through a gearbox, e.g., multistage compressor drive train. A torsional vibration analysis of such systems is usually required to avoid torsional resonances. Recently large turbine generators have been subjected to detailed torsional analyses to avoid possible last-stage blade excitation. The torsional programs which perform such calculations are usually based on the Holzer method. In the past a discrete mass, weightless shaft formulation was used, but more recently distributed mass elastic programs have been introduced. The overall simplicity of such programs has led to the writing of a wide variety, especially those used to calculate single shaft and simple geared system torsional frequencies. Both Holzer and matrix methods have been utilized. More recently, forced damped torsional response programs have become available.

Rotor Balancing

Programs for balancing rigid rotors in two planes have been available for a number of years. More recently programs for multiplane balancing of flexible rotors in flexible supports have been developed and are available. Both methods use the so-called influence coefficient method, in which the magnitude and phase of discrete unbalance forces corresponding to selected balancing planes are found from measurements of rotor response, using trial weights. Installation of suitable eccentric weights causes equal and opposite balancing forces, which annul the discrete unbalances. The rotor will then run smoothly. The function of the computer program is to interpret the trial weight response data and to prescribe the required balance weights.

CRITICAL SPEED PROGRAMS

Details of programs reviewed are given in Table 1. These programs may be classified into those which seek undamped eigenvalues of the

Table 1 Comparison of critical programs

Feature	ROTDYN	LINK III	CADENSE 26
Vendor	Franklin Institute	Com/Code Corp.	Mechanical Technology, Inc.
Address	Benj. Franklin Pkwy Phila., Penn.	2550 Huntington Ave Alexandria, VA 22302	968 Albany-Shaker Rd. Latham, NY 12110
Contact	W. Shapiro	Inf. & Publ. Dept.	A. J. Smalley
Available/price	Yes/$10,000 (package)	Yes/ negotiable	Yes/$2100
Rent/lease/sale	Yes/yes/no	Lease/time sharing	Yes/yes/no
Cards/tape	Yes/yes	Yes/yes	Yes/yes
Machines	CDC 6600, UNIVAC 1108, etc.	UNIVAC 1108, GE 605 CDC 6600, IBM 360	CDC 6600, UNIVAC 1100 series IBM 360, 370 . Period rent.
System		UCS, UCC, Al-Com, surcharge	Cybernet, GE/TS
Language	FORTRAN IV	FORTRAN	FORTRAN IV
Storage, words	65,000	36,000	25,000
Manual/check case	Yes/yes	Yes/yes	Yes/yes
Run time	200-300 syst. secs (response)	3 modes 2 CPU min (own compiler)	2 to 3 critical speeds per CPU sec on CDC 6600
Comment	Critical speeds obtained as part of response output option of ROTDYN package	User oriented time-sharing program	
Author	M. Reddi	P. Y. Chang	J. Lund
Publication	Machine Design, 1972	-	ASME 1967 (shaft theory)
GENERAL			
Lumped mass rotor	Yes	Yes & distributed	No, distributed
Circular orbit	No, elliptical	Not necessarily	Yes
Bending effects	Yes	Yes	Yes
Bending-torsion effects	No	No	No
Number critical speeds possible	Unlimited	Unlimited	Unlimited
BEARINGS			
Maximum number bearings	15	Unlimited	30
Bearing type	Optional	Optional	Optional
Linear/non-linear	Linear	Linear	Linear
Stiffness coefficients	Kxx, Kxy, Kyx, Kyy	Kxx, Kyy	K-effective
Damping coefficients	Dxx, Dxy, Dyx, Dyy (damped C.sp)	None	None
Source bearing data	Other than this program	Other than this program	Other than this program
ENVIRONMENT			
External direct forces	Yes	Yes	No
Gravity/field	Yes	Yes	No
Geared to other rotors	Yes (part of response option)	No	No

Table 1 (cont.)

Feature	ROTDYN	LINK III	CADENSE 26
ROTOR			
Maximum number stations	400 deg./freedom	Unlimited	200
Formulation	Finite element	Distributed mass	Distributed
Discrete masses	Yes	Yes	Yes, at section end stations
Discrete inertias	Yes	Yes	Yes, same
Beam shear effects	Yes	Yes	Yes
Gyroscopics-shaft	No	Yes	Yes, discrete
-disks	Yes	Yes	Yes
Translatory inertia-shaft	No	Yes	Yes, discrete
-disks	Yes	Yes	Yes
Axial thrust	No	Yes	No
Internal damping	Yes (damped C sp)	No	No
PEDESTAL			
Stiffness radial/angular	Yes/no	Yes/no	Yes/no
Damping radial/angular	Yes/no	No/no	No/no
Mass/inertia	Yes/no	Yes/no	Yes/no
Discrete/distributed	Yes/no	Yes/no	Yes/no
Input data source	Other than this program	Other than this program	Other than this program
FOUNDATION	One level only	One level only	(Multilevel version) (CADENSE 26)
Stiffness radial/angular	↑ Corresponds to pedestal data ↓	Yes/no	Yes/yes
Damping radial/angular		No/no	No/no
Mass/inertia		Yes/no	Yes/yes
Discrete/distributed		No/yes	Yes/yes
Input data source		Other than this program	Other than this program
OUTPUT			
Printout, critical speed	Yes	Yes	Yes
mode shape	Yes	Yes	Yes
orbit	Yes	No	No
static bearing load	Yes	Yes	Yes (two-bearing rotor)
GRAPHICS CAPABILITY			
with program	Option	Yes	Yes
mode shape	Yes	Yes	Yes
orbit	Yes	No	No
critical speed map	No	No	Yes
other chart	Yes (see Fig. 3)	No	Printer plot mode shape

rotor system using a root search technique (e.g. Newton-Raphson) such as CADENSE 21, MTI-WPAFB, and LINK III and other programs such as ROTDYN which find damped resonant speeds as part of the unbalance response calculation. This latter procedure of finding critical speeds from unbalance response results appears to be gaining favor, particularly with more heavily damped systems where it provides meaningful data for a minor cost penalty. Differences which exist between the eigenvalue programs are mostly related to the degree of system complexity the program can handle, e.g., one-level, two-level, shear deformations, etc., and to the manner in which the rotor and bearings are formulated.

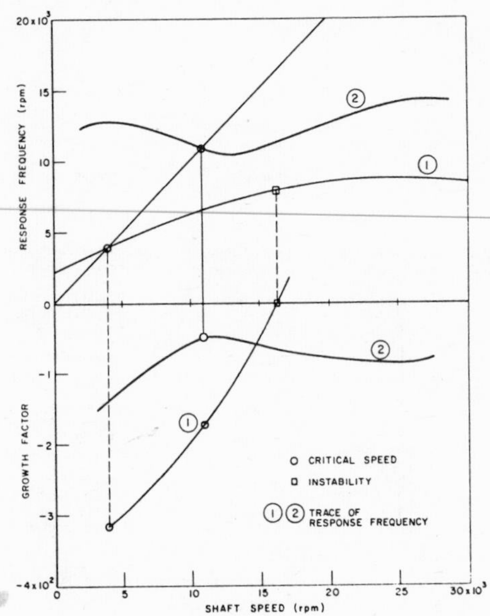

Fig. 3 Critical speed and stability map: ROTDYN program

Machine computation of critical speeds was initiated by Prohl [1], who applied the Holzer method to shaft vibrations. Many subsequent programs have been written to perform this calculation, most of which are proprietary. The three programs listed in Table 1 are proven commercially available versions which differ as follows:

ROTDYN will find damped critical speeds as part of an unbalance response calculation. Results appear to be valid based on both owner and user experience. Typical response results are shown in Fig. 3 from the ROTDYN user's manual. Output is easy to read and interpret.

LINK III is a time-sharing program available on teletype which finds undamped eigenvalues. Ease of input and data change is an advantage. The program seems likely to be suitable for shaftlike rotors.

CADENSE 26 is a commercially available eigenvalue program which accommodates sophisticated shafts in linear bearings. One- and two-level versions are available. The program has been proven in practice. The basic theory of the program has been published, and earlier versions have been documented with full input instructions, test case, and output details. Operating costs are claimed to be small as the root-search routine is sophisticated. There is no automatic data generation (as exists in LINK III), and input is on cards, which adds significantly to overall cost of program operation.

UNBALANCE RESPONSE PROGRAMS

General

Details of the programs reviewed herein are given in Table 2. Two types of program have been developed, e.g., synchronous programs which calculate rotor response to unbalance centrifugal forces only and nonsynchronous programs which calculate overall response to both unbalance forces and other forces such as gravity, bearing and seal forces, applied harmonic forces, etc. Nonsynchronous programs thus give the overall response of the rotor to its environment, and so may also be used to calculate system stability threshold speed. The synchronous programs assume that the rotor remains stable throughout its operating range and no nonsynchronous effects such as backward and subharmonic whirls appear in the rotor motions. Another difference between the programs detailed in Table 2 is in the program level. This refers to the complexity of the rotor support structure. Thus a one-level program, Fig. 4, deals with a rotor which is supported in damped, flexible bearings, which are supported in pedestals possessing mass, stiffness, and damping. The pedestals are discrete and attach to a rigid (motionless) foundation. No interconnections exist between pedestals. A two-level program applies to a rotor in damped, flexible bearings which are attached to a continuous structure possessing mass and flexibility. This structure is mounted to a rigid foundation through damped flexible supports, Fig. 5. A three-level program would have a second continuous structure similar to the first, and so on.

Documentation

In most instances program documentation is confined to user's manuals and some brochure pages. User's manuals describe the theory on which

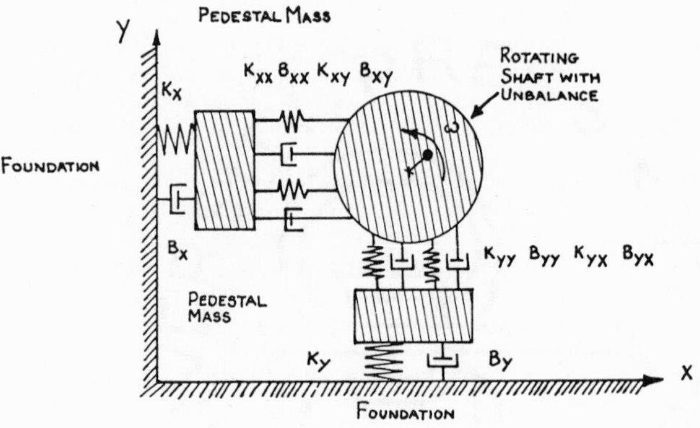

Fig. 4 Representation of one-level rotor system using displacement and velocity coefficients

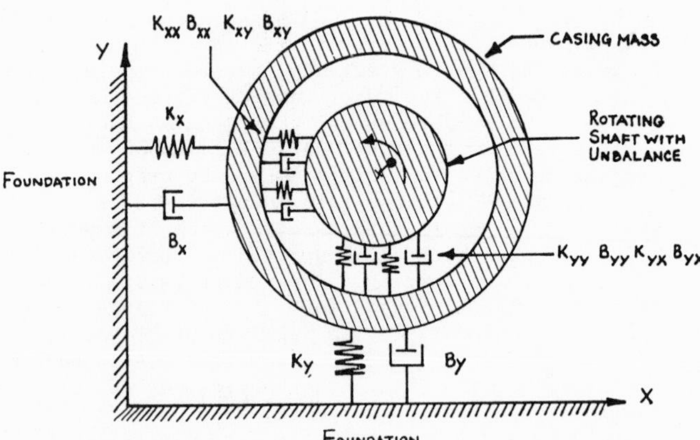

Fig. 5 Representation of two-level rotor system using displacement and velocity coefficients

Table 2 Details of unbalance response programs

Feature	ROTDYN	CADENSE 21,27	MTI-WPAFB	GIBERSON	LINK III
Vendor	Franklin Inst. Res. Labs	MTI-CAD 21, 27	MTI-WPAFB	Turboresearch	Com/code Corp.
Address	Benj. Franklin Pkwy. Phila., Penna.	968 Albany-Shaker Road, Latham, N.Y.	968 Albany-Shaker Road, Latham, N.Y.	1440 Phoenix Ave. W. Chester, Penna.	2550 Huntington Ave. Alexandria, Va.
Contact	W. Shapiro	A. J. Smalley	A. J. Smalley	M. Giberson	Inf. & Publ. Dept.
Available/Price	Yes/$10,000(package)	Yes/$4200	Yes/WPAFB report	Yes/N.F.S.	Yes/negotiable
Rent/lease/sale	Yes/yes/no	Yes/yes/no	Arrangement	Yes/no/no	T S/yes/yes
Cards/tape	Yes/yes	Yes/yes	Yes/no	-	-
Machine	UNIVAC 1108, CDC 6600, etc.	CDC 6600, IBM 360,370 UNIVAC 1100 series	-	-	IBM 360, CDC 6600 UNIVAC 1107
Language	Fortran IV	Fortran IV	Fortran IV	Fortran	Fortran IV
System	No	Cybernet, G.E/TS	-	-	UCS, UCC, Al-Com
Storage, words	65,000(package)	63,000	2782 Fortran statements	20K Fortran statements	
Manual/check case	Yes/yes	Yes/yes	Yes	Yes/yes	Yes/yes
Typical run time	200-300 syst. secs.	2 to 4 response speeds per CPU sec on CDC 6600	1-2 response speeds per CPU sec on CDC 6600		
Author	M. Reddi	J. Lund	J. Lund	M. Giberson	P. Y. Chang
Publication	Machine Design 1972	ASME 1967	WPAFB 65-TR-45	-	
Comment	One option of general rotordynamics program	Specific program or part of package	Specific programs	Advanced comprehensive program package	One option of general package
GENERAL					
Synchronous response only	No	Yes (non-synch. special option)	Yes	No	Option
Lumped mass rotor?	Yes	No, distributed	Yes	Yes	Yes and distributed
Circular orbit?	No, computed	No, elliptical	No, elliptical	No, computed	No
Orbital response caculation?	Yes	No, optional elliptical display.	No, axes	Yes	No
BEARINGS					
Maximum number bearings	15	120	25	10	20
Bearing type	Optional	Optional	Optional	Optional	Optional
Linear/non-linear	Linear	Linear	Linear	Non-linear	Linear
Eight-coefficient representation	Matrix of coeffics. related to all brg. d.o.f.	Yes. Options for other brg. d.o.f.	Yes. 8 tilt coeffc. also	No	Yes
Source of bearing coefficient data	Other program (or package)	Other program (or package)	Other than program	Routine in program	Other program
ROTOR					
Maximum number rotor stations	100 400 d.o.f.	60	80	100	unlimited
Rotor formulation	Lumped mass, finite element	Distributed mass-elasticity	Lumped mass	Discrete mass	Lumped and Distributed mass elasticity
Shear effects	Included	Yes	No	No	Yes
Concentrated disks	Yes	Yes	Yes	Yes	Yes
How input?	Mass, inertia	Mass, inertia	Mass, inertia	Mass, inertia	Mass, inertia
Shaft taper included	Yes	No	No		Yes
Gyroscopic effects -shaft	No	Yes, discrete	Yes, discrete	Yes, discrete	Yes
--disks	Yes	Yes	Yes	Yes	Yes
Translatory inertia -shaft	No	Yes, discrete	Yes, discrete	Yes, discrete	Yes
-disks	Yes	Yes	Yes	Yes	Yes
Axial thrust	No	No	No		Yes
Internal damping	No	No	No	Yes	No
PEDESTAL					
Stiffness	Yes	Yes	Yes	Yes	Yes
Damping	Yes	Yes	Yes	Yes	Yes
Mass	Yes	Yes	Yes	Yes	Yes
					Discrete

Table 2 (cont.)

Feature	ROTDYN	CADENSE 21,27	MTI-WPAFB	GIBERSON	LINK III
Discrete/distributed	Discrete	Discrete/distributed	Yes/no	Discrete	Discrete
Input data source	Other than program	Other than program	Other than program	Other than program	Other than program
FOUNDATION	ONE LEVEL PROGRAM	CAD 21 One level / CAD 27 Two level	ONE LEVEL PROGRAM	ONE LEVEL PROGRAM	ONE LEVEL PROGRAM
stiffness	↑	- / Yes	↑	Special options: Misalignment, Settlement, Tables of foundation properties, etc.	↑
damping		- / Yes			
mass	Foundation corresponds to pedestal	- / Yes	Foundation corresponds to pedestal		Foundation corresponds to pedestal
Discrete/distributed		- / Distrib.			
Input data source	↓	- / Other than program	↓		↓
ENVIRONMENT					
Gas forces					
-static	Yes	No	No	Yes	No
-dynamic	Yes	Yes, coefficients	Yes, coefficients	Yes	Yes, coefficients
Gravity	Yes	No	No	Yes	No
Field force	Yes, linear	No	No	Yes	No
Eccentric gear load	Yes	No	No	Yes	No
Arbitrary rotating load/moment	As function of time	Special option	No	Yes, linear or non-linear	No
OUTPUT					
Printout					
-orbit details	Yes	Yes	Yes	Yes	Yes
-response ampl.	Yes	Yes	Yes	Yes	Yes
-transmitted force	Yes	Yes	Yes	Yes	Yes
-stress	Yes	Option	No	No	No
Graphics capability	Yes	Yes	No	Yes	No
-with program	Option	Option	No	Yes	No
-selected orbit	Yes	No	No	Yes	No
-response amplitude vs. speed	No	Yes	No	No	No
-transmitted force	No	Yes	No	No	No

the program is based and give input instructions, one or two sample cases with input and output, and some comments on program functioning. This format is followed by all programs except NASA, for which a descriptive outline and source listing is given in [2], and the GIBERSON program,[2] for which no manual is available to the user.

Operation

Reports from users and vendors indicate satisfaction with the functioning of ROTDYN and CADENSE [21, 27]. The original version of MTI-WPAFB [15] , has some statement errors for which a correction sheet is available.[3] The program is somewhat cumbersome due to size and runs less efficiently than the previous two programs. NASA is a simple program which runs cheaply, from which useful guidance can be obtained. The single-disk rotor and bearing/pedestal formulation with four linear coefficients is inadequate for complex machine calculations. The LINK program does not appear to be widely used, but with teletype operation should offer maximum user convenience when accustomed to the format. Cost of operation is said to be high, as the program has its own compiler. The GIBERSON program appears to be the most comprehensive rotor system program in existence, but no user's manual is available. User reports suggest that it is more costly to run than the other programs mentioned.

Program Validation

Generally speaking, it appears that most vendors have validated their programs initially by checking textbook cases and then by checking against test data from industrial rotor systems. With rare exceptions, the purchaser will find no true validation in the user's manuals, and it seems doubtful if many programs have received a rigorous checkout. The sample calculations given usually describe a rotor system with input-output data, against which program functioning may be checked by a new user.

An exception to the above is CADENSE 21 for which published details of correlation with test results are available; see Fig. 6 from Ref. [3] . In a careful series of experiments, accurate response correlation was obtained below the bending critical speed (which was accurately identified). There was less accurate amplitude correlation above the critical. Experiments such as this require care and are costly, and many potential sources of inaccuracy exist, such as inaccurate bearing data, foundation resonances, and instrumentation problems. Reports from several independent users indicate that results from both ROTDYN and CADENSE 21 correlate well with critical

---

[2] Program is available for supervised in-house operation on 100 percent surcharge basis. Not available for purchase or lease.

[3] From MTI Latham, New York or Wright Patterson AFB APL, Dayton Ohio.

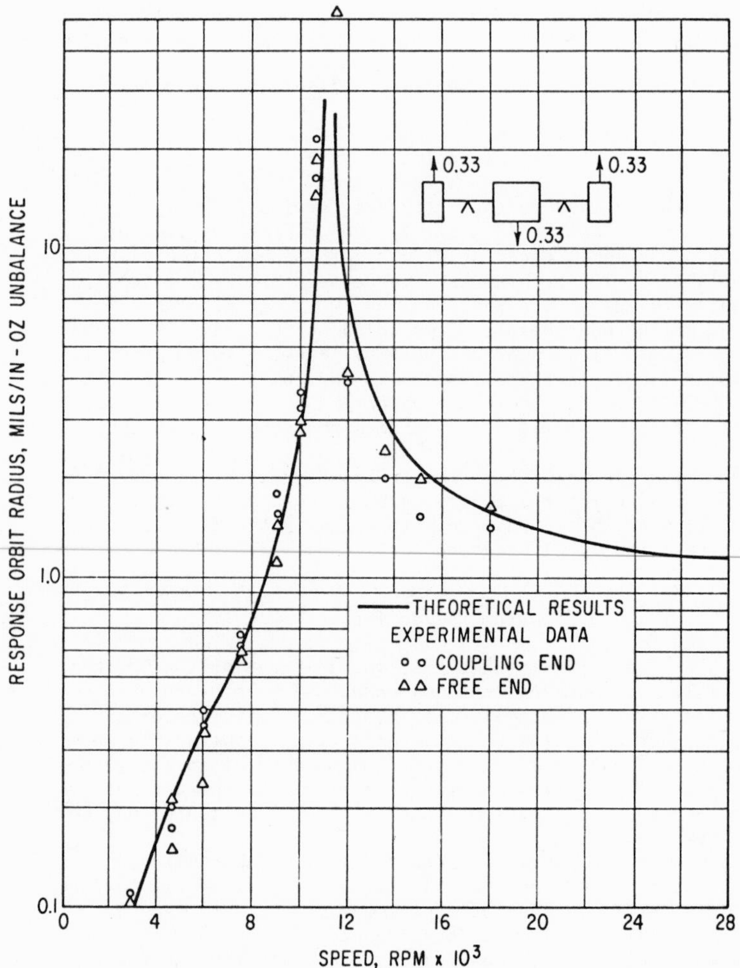

Fig. 6 Correlation of theoretical results from unbalance response program with experimental results near critical speed (Ref. [3])

speeds measured during tests, but no independent amplitudes correlation has yet been obtained. The amplitude response of ROTDYN is reported to be in good agreement with that of CADENSE 21.

No correlation of response results appears to be available for the other programs in Table 2, either in the open literature or from the program users who were contacted in this survey. It appears that vendors could enhance confidence in their programs by including details of some experimental correlations in their user's manuals.

STABILITY PROGRAMS

General

Several techniques are in use for finding the instability threshold speed of rotors operating in linear bearings:

1. Modal growth factor analysis, based on sign of real part of complex eigenvalue. Sign change, negative to positive on any mode signals onset of instability.
2. Orbit plots to determine conditions at which the rotor whirl will begin to grow without bonds. Figure 7 shows a typical result.
3. Routh-Hurwitz criterion applied to matrix characteristic equation coefficients.

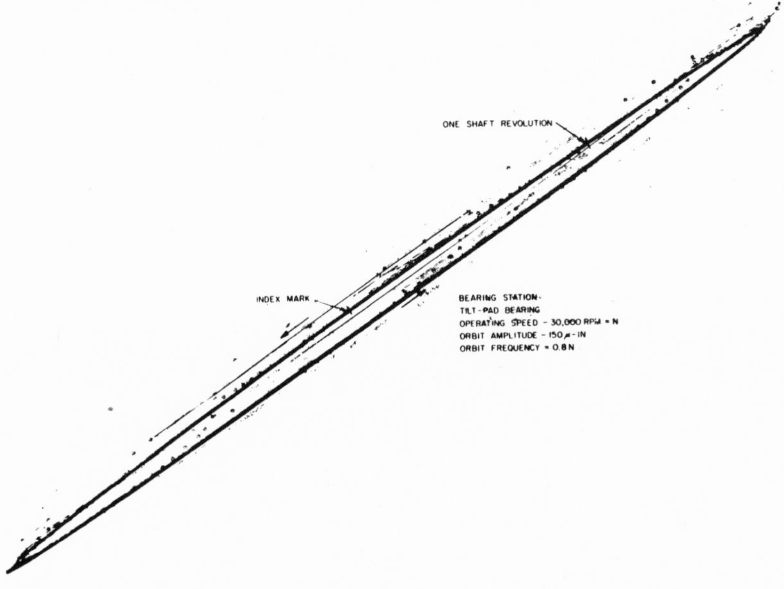

Fig. 7 Orbit response at bearing station: ROTDYN program

ROTDYN and CADENSE 25 utilize modal growth factors. MTI-WPAFB uses the complex real part technique. Plot options are also available with the first two programs to display the whirl orbits of the rotor at selected stations along its length. The NASA program and LINK III use the Routh-Hurwitz criterion. An analog procedure is also described in the NASA program report for obtaining whirl orbits.

Stability of rotors in nonlinear bearings may be examined with the GIBERSON program. This program also uses orbital plots to check whirl growth or decay under given conditions. The full nonlinear bearing forces are used to calculate the orbital motion of the rotor in its supports. This program provides a stability analysis which may include an extremely broad range of system parameters.

All the above calculations consider flexible rotors which operate in damped, flexible bearings. Differences exist between programs in the degree of complexity of the rotor and support structure in the type of bearing formulation and in the instability check technique, as mentioned. The NASA rotor is a simple single disk rotor whereas the ROTDYN rotor may have up to 100 masses, and the CADENSE 25 rotor may have up to 60 mass stations.

Documentation

ROTDYN Stability is another part of the ROTDYN package, which is obtained by utilizing a program option. The rotor-bearing formulation is linear and is similar to that used in the unbalance response portion except that unbalance and other applied external forces are omitted in this option. Figure 3 shows how the growth factors from each mode vary and indicates instability at a certain speed.

CADENSE 25 is a separate program. The user's manual gives an explanation of the program and adequate input instructions with a test case to validate operation on a machine system.

MTI-WPAFB is an older program (available free) which predicts threshold speeds, whirl frequencies, and mode shapes. The input and output data are described in detail, along with program functioning and several test cases in the user's manual, which contains a program listing; see Ref. [15] .

NASA stability program is described in Ref. [2]. Input information is given in the report with sample output and test cases to cover the program stability options. The report reviews the published work on stability and contains studies of several types of stability problems, mainly concentrating on rotor internal hysteresis instability. A second NASA program, Ref. [4] , gives a program for instability of rigid rotors in fluid-film bearings.

No documentation is available on the GIBERSON program, although there is an in-house manual. Output data center around whirl orbits at specified rotor stations, but tabular printout of rotor and pedestal motions is also generated.

Validation

Early analyses were developed for rigid rotor gas bearing machines (e.g. gyros), and the predicted threshold speeds were verified experimentally. Flexible rotor stability programs are customarily tested on the same examples, but a rigorous and general flexible rotor system stability program validation against adequate experimental data similar to that for unbalance response (Ref. [3]) does not seem to have been performed for rotor stability. Some stability experiments were conducted by Pinkus [5], Tondl [6, 7], and others, and

Duckworth and Rieger [8] have demonstrated that linear bearing theory is able to predict threshold speeds for these cases with good accuracy. However, for the rotor stability programs listed in Table 3 there appears to be no open literature data describing their validation. This does not mean these programs have not been verified, but independent validation of any program must presently rely on the user's possessing his own high-quality test results. Most users and vendors contacted in this survey believed the programs were functioning reliably, based on industrial comparisons.

TORSIONAL PROGRAMS

A large number of torsional programs are available, ranging from single discrete mass single-shaft Holzer programs for critical frequencies to transient torsional dynamics programs for amplitude vs. time response of complex geared systems. Those programs discussed in Table 4 cover this range of options. Almost every industry requiring a torsional program now has at least a simple version available. Most torsional programs calculate critical frequencies, and more advanced versions apply to geared systems. Here, special purpose programs of great complexity have been developed on a special need basis (planetary gear programs, torsional kinematics programs, etc.) Most

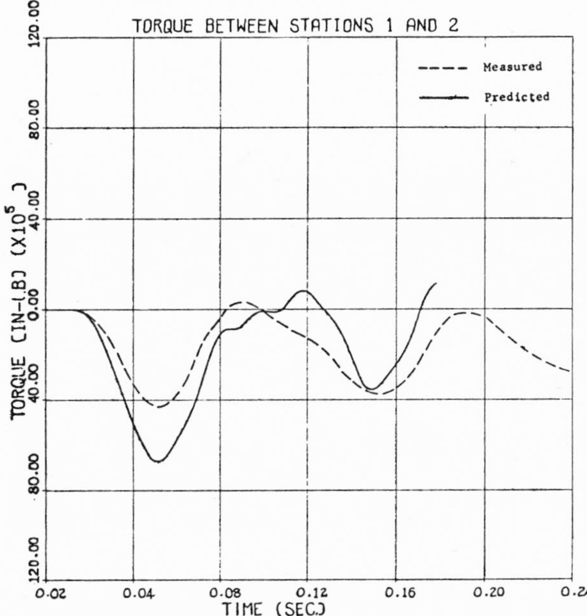

Fig. 8 Correlation of predicted transient torques with measured transient torques in rolling mill drive (Ref. [10])

Table 3 Comparison of stability programs

Feature	ROTDYN	CADENSE 25	MTI-WPAFB	NASA	GIBERSON	LINK III
Vendor	Franklin Inst. Res. Labs.	Mechanical Technology Inc.	Mechanical Technology Inc.	NASA-Lewis	Turboresearch	Com/code Corp.
Address	Benj. Franklin Pkwy. Phila., Penna	968 Albany-Shaker Road, Latham, N.Y.	968 Albany-Shaker Road, Latham, N.Y.	M.E. Dept., Univ. of Va. Charlottesville, Va.	1440 Phoenix Ave., W. Chester, Pa.	2550 Huntington Ave. Alexandria, Va. 22302
Contact	W. Shapiro	A. J. Smalley	A. J. Smalley	E. J. Gunter	M. Giberson	Inf. & Publ. Dept.
Available/price	Yes/$10,000(package)	Yes/$8,000, lease	Yes/WPAFB report	Yes/NASA	Yes/N.F.S.	Yes/negotiable
Rent/lease/sale	Yes/yes/no	Yes/yes/no	Arrangement	Free	Yes/no/no	Sale/T.S.
Cards/tape	Yes/yes	Yes/yes	Yes/no	-	-	Yes/yes
Manual/check case	Yes/yes	Yes/yes	Yes/yes	NASA report	-	Yes/yes
Machines	UNIVAC 1108 CDC 6600, etc	CDC 6600, IBM 360,370 UNIVAC 1100 series	IBM 360 H-GE 635	Most		IBM 360 UNIVAC 1107, CDC 6600
System	No	Cybernet, G.E/T.S.	No	No	No	UCS, UCC, Al-Com
Storage, words	65,000(package)	60,000	521-Fortran statements	150 Fortran statements	20,000 Fortran statements	
Run time	200-300 system secs.	2 CPU sec per damped natural frequency	-	-	-	
Author	M. Reddi	J. W. Lund	J. W. Lund	E. J. Gunter	M. Giberson	P. Y. Chang
Publication	Machine Design 1972	ASME 1973	WPAFB 65 TR 45	NASA SP113 1966	-	-
GENERAL						
Nonsynchronous response	Yes	No	No	No	Yes	No
Threshold eigenvalue	Indirectly	Indirectly	Yes	No	No	No
Other solution procedure	-	-	-	Routh-Hurwitz	-	Routh-Hurwitz
Orbit growth factors	Yes	Yes	No	No	No	No
Circular Orbit	No	No	No	No	No	No
Orbital response calculation	Yes	No	No	No	Yes	No
BEARINGS						
Maximum number bearings	15	20	10	2	10	20
Bearing type	Optional	Optional	Optional	Optional	Optional	Optional
Linear/non-linear	Linear	Linear	Linear	Linear	Non-linear	Linear
Eight coefficient type	Matrix of coeffics. related to all brg. d.o.f.	Yes, Individ. tilt coeffic. also	Yes, 8 tilt coeffics. also	No, 4	No	Yes
Source of bearing data	Other program (or package)	Other program (or package)	Other than this program	Other than this program	Routine in program	Other than this program
ROTOR						
Maximum number rotor stations	100 400 d.o.f.	60	30	1	100	4
Formulation	Finite element discrete mass	Discrete mass and elasticity	Discrete mass	Discrete mass massless shaft	Discrete mass massless shafts	Discrete mass
Discrete mass	Yes	Yes	Yes	Yes	Yes	Yes
Discrete inertia	Yes	Yes	Yes	No	Yes	No
Beam shear	Yes	Yes	No	No	No	Yes
Gyroscopics, shaft	No	Yes	Yes, discrete	No	Yes, discrete	No
disks	Yes	Yes	Yes	No	Yes	Yes
Trans. inertia, shaft	No	Yes	Yes, discrete	No	Yes, discrete	No
disks	Yes	Yes	Yes	No	Yes	Yes
Axial thrust forces	No	No	No	No	No	No

Table 3 (cont.)

Feature	ROTDYN	CADENSE 25	MTI-WPAFB	NASA	GIBERSON	LINK III
Internal damping	Yes	Yes	No	Yes	Yes, material and fluid coefficients	No
PEDESTAL						
Stiffness, radial/angular	Yes/no	Yes/yes	Yes/yes	Yes/no	Yes/no	Yes/no
Damping, radial/angular	Yes/no	Yes/yes	Yes/yes	Yes/no	Yes/no	Yes/no
Mass/inertia	Yes/no	Yes/yes	Yes/yes	Yes/no	Yes/no	Yes/no
Discrete/distributed	Yes/no	Yes/no	Yes/no	Yes/no	Yes/no	Yes/no
Input data source	Other than this program	Other than this program	Other than this program	Other than this program	Other than this program	Other than this program
FOUNDATION	One level only	One level only	One level only	One level only	One level	One level
Stiffness radial/ angular	↑	↑	↑	↑	↑	↑
Damping radial/ angular						
Mass/inertia	Corresponds to pedestal	Corresponds to pedestal	Corresponds to pedestal	Corresponds to pedestal	Corresponds to pedestal	Corresponds to pedestal
Discrete/distributed						
Input data source	↓	↓	↓	↓	↓	↓
ENVIRONMENT						
External forces --unbalance	Yes	No	No	Yes	Yes (All as linear or non-linear forces)	No
-gas force static	Yes	No	No	No	Yes	No
-gas force, dynamic	Yes, coeffics.	Yes, coeffics.	Yes, coeffics.	No	Yes, coeffics.	Yes, coeffics.
-gear load	Yes	No	No	No	Yes	No
-magnetic load	Yes, indirectly linear	No	No	No	Yes	No
-gravity	Yes	No	No	Yes	Yes	No
OUTPUT						
Printout						
-Selected orbits	Yes	No	No	No	Yes	No
-Threshold speed	Yes, indirectly	Yes, indirectly	Yes	Yes, indirectly	Indirectly	Indirectly
-Mode shape	Yes	Yes	Yes	No	No	No
Graphics capability	Yes	No	No	No	Yes	No
-with program	Option	No	No	No	Yes	No
-selected orbits	Yes	No	No	No	Yes	No
-Mode shape	Yes	Printer plot	No	No	No	No
ADDITIONAL FEATURES		• Dimensionless bearing input data, including heat balance.			• Any nonlinear forces • Misaligned bearings • Nonsymetrical fluid force effects option	

Table 4 Comparison of torsional programs

Feature	TABU	CADENSE-22	TASS	CADENSE-23	CADENSE 24
Vendor	Southwest Research Institute	Mechanical Technology, Inc.	Univ. Cincinatti M.E. Dept.	Mechanical Technology, Inc.	Mechanical Technology, Inc.
Address	San Antonio, Texas	968 Albany-Shaker Road, Latham, N.Y.	Cincinatti Ohio	968 Albany-Shaker Road, Latham, N.Y.	968 Albany-Shaker Road, Latham, N.Y.
Contact	J.C. Wachel	A. J. Smalley	I. E. Morse	A. J. Smalley	A. J. Smalley
Available/ Price	No/-	Yes/$2100	Yes/Arrangement	Yes/$4200	Yes/$4000
Rent/lease/ sale	No/Yes/No	Yes/Yes/No	No/No/Yes	Yes/Yes/No	Yes/Yes/No
Cards/tape	Yes/Yes	Yes/Yes	Yes/Yes	Yes/Yes	Yes/Yes
Machine	CDC 6600	CDC 6600 UNIVAC 1100 Series	CDC 6600	CDC 6600 UNIVAC 1100 Series	CDC 6600 UNIVAC 1100 Series
Language	Fortran IV	Fortran IV	Fortran IV	Fortran IV	Fortran IV
System	--	Cybernet	--	Cybernet	Cybernet
Storage words	12000 C.Freq. 30000 Resp.	55,000	32,000	55,000	65,000
Manual check/case	No/Yes	Yes/Yes	Yes/Yes	Yes/Yes	Yes/Yes
Typical run time	4 cpu Sec. for 10 mass	2 CPU sec/crit spd. 60 sta. rotor	--	1.8 CPU Sec/resp. 40 station rotor	800 Station-timesteps per CPU sec on CDC 6600
Author	Wachel/Szeasi/ Saathoff	J. Lund	S. M. Wang	J. Lund	A. J. Smalley
Publication	--		ASME 1972		ASME 1972
Comment	Nested system. Fourier diagnosis	Planetary gear sets.	Various gearbox stiffness effects	Planetary gear sets.	
GEARS					
Inertialess	No	No	No	No	No
Stiffness of mesh	Yes	Yes	Yes	Yes	Yes
Damping of mesh	No	No	No	No	No
Gear error excitation	Complex harmonics input	No	No	Yes	Not directly
Backlash	No	No	Not directly	No	Yes
INPUT					
Inertia-stiffness model	Yes	Yes	Yes	Yes	Yes
Harmonic excitation	Yes	No	Yes	Yes	No
Velocity-time	Yes	No	No	No	Yes
Torque-time	Yes	No	No	No	Yes
Torque-speed	No	No	No	No	Yes

Table 4 (cont.)

Feature	TABU	CADENSE-22	TASS	CADENSE-23	CADENSE 24
OUTPUT					
Printout: -critical freq:	Yes	Yes	Yes	No	No
-mode shape	Yes	Yes	Yes	No	No
-ampl. vs. freq.	No	No	Yes	Yes	No
-displ. vs. time	Fourier ampl.	No	No	No	Yes
-shaft torque	Yes	Yes	Yes	Yes	Yes
-shaft stress	Yes	Yes	Yes	No	No
-Tooth Forces	No	No	No	Yes	Yes
General					
Purpose of Program	Crit. freq. Mode shapes Response	Crit. freq. Mode shapes	Crit. freq. forced undamped response.	Forced response	Transient response
Number of branches	10	Unlimited	Unlimited	Unlimited	Unlimited
INERTIA					
Number of Conc. inertias	50	200	120	200	20
Distributed inertias	No	Yes	Yes	Yes	No
SHAFTS					
Inertialess flexible shafts	Yes	No	No	No	Yes
Distributed stiffness	No	Yes	Yes	Yes	No
Distributed damping	Modal damping matrix	No	No	No	No
Conc. damp. betw. inertias		No	No	Yes	Yes
Torsional-to -ground stiffness	Yes	Yes	Yes	Yes	Yes
Damping to ground	No	No	No	Yes	Yes
GRAPHICS					
-mode shape	No	No	No	No	No
-angle vs. time	No	No	No	No	No
-veloc. vs. time	No	No	No	No	No
Torque vs. time	No	No	No	No	Yes
Torque vs. speed	No	No	No	No	No
Options		• Modal response option			

programs appear to have been checked out on textbook examples and on industrial applications, and there does not appear to have been any classical test devised against which a new complex program could be proven. The closest to this seems to be the transient dynamics validation carried out with CADENSE 24 (see Fig. 8).

TABU calculates critical frequencies and mode shapes of discrete mass branched systems. The stiffness matrix method (presumably the stiffness matrix for the entire structure) is used, with a matrix iteration procedure for eigenvalues. A forcing function and modal damping may be included to determine stresses at selected frequencies.

CADENSE 22 also calculates critical frequencies. The system may include branches, gears, epicyclic gears, and an elastic torsional connection to ground. Normalized mode shapes and torque distributions are calculated at each critical frequency. This program uses a distributed inertia-elastic representation of shaft sections with concentrated inertias (gears) at their ends, as needed. Ability to handle planetary gears is an option.

TASS calculates torsional critical frequencies and forces dynamic response of undamped systems, also using a uniform distributed inertia-elastic approach with end inertias and springs to ground. Any number of gear pairs may be used. Torque and rotation at each junction are calculated in the response to specified forcing. Shear stresses may also be output.

CADENSE 23 calculates damped torsional response to specified input torque excitations, using the same type of system as in CADENSE 22. Drive train damping may be input in parallel to any spring. Gears may be rigidly or elastically mounted. A feature of this program is its ability to accept excitation in the form of gear manufacturing errors.

CADENSE 24 calculates the damped transient response of a geared drive train. A fourth-order Runge-Kutta method is used to obtain angular displacements at specified time intervals. Inertia, stiffness, and damping in shafts and gear meshes may be included. Initial conditions of the motion may be torque history, velocity history, or initial displacement and velocity at specified locations in the drive. The program has a plotting routine so that time-varying torque and displacement results may be obtained. This is an efficient, advanced program which has been verified and runs effectively on those cases studied [9, 10] .

BALANCING PROGRAMS

Few general purpose flexible rotor-balancing programs exist because at present this is a specialized topic for which users develop their own programs on a need basis. Rigid rotor programs suitable for a programable calculator have been available for a long time, based on the influence coefficient method [11] . A2Z-SWRI is a commercially available rigid rotor balancing program. Shaft amplitude and phase angle data are required as program input, based on results obtained from runs with trial weights in the balancing planes. The program calculates the amount and orientation of correction weights to balance the rigid rotor, using a least-squares procedure. Vendor is SouthWest

Research Inst., San Antonio, Texas.

CADENSE 10 is a flexible rotor-balancing program which uses the influence coefficient method. This balances the rotor as though the input data related to a circular whirl orbit. The program is easy to use and has been thoroughly tested under laboratory conditions for up to four balancing planes and through one strong bending critical speed. Several applications of the method are usually necessary to eliminate any flexible rotor, and there are definite relationships between the number of balancing planes available and the number of critical speed peaks to be eliminated. Verification details are given in Refs. [12, 13, 14].

CONCLUSIONS

1. Of the many critical speed programs available, the best documented and most efficient appears to be CADENSE 26. Critical speed eigenvalue programs are most useful for analysis of lightly damped systems, such as rotors in rolling element bearings with small, localized discrete damping. Present trends appear to favor calculation of damped critical speeds using an unbalance response program.

2. A variety of unbalance response programs exist. Several of these have excellent generality with regard to the number of rotor sections permitted, number of bearings, and numbers of levels. These programs have been validated and widely user tested. Orbital printout options and convenient input-output operations make them highly functional codes.

3. A variety of stability programs exist with similar excellent generality and user-convenience options. Validation of these programs is presently restricted to user experience.

4. Although a wide range of torsional vibration programs exist, these programs are restricted mainly to critical frequencies of geared systems. Independent versions of the excellent advanced torsional response programs and torsional transient programs now existing should be developed.

5. The available range of flexible rotor-balancing programs is quite limited, and additional needs for programs which will balance rotors in nonisotropic bearings and rotors with nonisotropic shaft stiffness still remain.

6. The recent development of user program packages which include both response and stability options is an important convenience for users as it avoids having separate program decks/tapes, often with different input/output formats.

7. Accuracy of all programs operating with linear bearings depends on having available accurate bearing dynamic coefficient data. A definitive study of the data available in this regard should be undertaken. Similar studies are needed to compare programs for seal coefficients and for foundation data.

8. In general it appears that the development of rotor dynamics programs has lacked the support afforded to other areas of structural mechanics. Very few excellent programs exist; validation has been confined to a few thorough tests; only recently have sophisticated graphics been adopted; and the number of highly talented programmer-analysts working in the field appears to be remarkably few.

ACKNOWLEDGEMENTS

The author thanks those program authors, vendors, and users who provided him with information on their experiences with the programs discussed in this paper.

REFERENCES

1 Prohl, M. A., "A General Method for Calculating Critical Speeds of Flexible Rotors," Journal of Applied Mechanics, Vol. 12, 1945, pp. 142.

2 Gunter, E. J., "Dynamics Stability of Rotor-bearing Systems," NASA-SP-113 Contract NAS 3-6473, 1966.

3 Lund, J. W., Orcutt, F. W., "Calculations and Experiments on the Unbalance Response of a Flexible Rotor in Fluid Film Bearings," ASME paper 67 Vibr. 27, Vibrations Conference, Boston, Mass., 1967.

4 Gunter, E. J., Kirk, R. G., "Transient Journal Bearing Analysis," NASA Cr-1549, June 1970, NGR-47-005-050.

5 Pinkus, O., "Experimental Investigation of Resonant Whip," Trans. ASME, Vol. 78, 1956, pp. 975.

6 Tondl, A., "Notes on the Problem of Self-excited Vibrations and Nonlinear Resonance of Rotors Supported in Several Journal Bearings," Wear, Elsevier Publishing Company, Amsterdam, Vol. 8, 1965, pp. 349-357.

7 Tondl, A., Some Problems of Rotor-dynamics, Publishing House Czechoslovakian Academy of Sciences, Prague, Czechoslovakia, 1966.

8 Duckworth, S., Rieger, N. F., "Oil Whip Threshold Speed Analysis of a Flexible Rotor in Bearings," Independent Study Report, Union College, Schenectady, New York, 1970.

9 Rieger, N. F., Smalley, A. J., "Dynamic Loads on Gear Teeth from Transient Shock", paper presented at AGMA semi-annual meeting, St. Louis, Missouri, October 23-25, 1970.

10 Kashay, A. M., Voelker, F. C., Smalley, A. J. "Dynamic Shock Phenomena in Rolling Mills," Journal of Engineering for Industry, Vol. 94, May 1972, pp. 47.

11 Myklestad, N. O., Fundamentals of Vibration Analysis, McGraw-Hill Book Co., Inc., New York, 1956.

12 Tessarzik, J., Badgley, R. A., Anderson, W. J., "Flexible Rotor Balancing by the Exact Point-speed Influence Coefficient Method," Journal of Engineering for Industry, Vol. 94, No. 1, 1972, pp. 233-242.

13 Rieger, N. F., Badgley, R. H., "Flexible Rotor Balancing of a High-Speed Gas Turbine Engine," SAE paper presented at National Powerplant meeting, Milwaukee, Wisconsin, September 11, 1972.

14 Badgley, R. H., Rieger, N. F., "Effects of Multi-plane Balancing on Flexible Rotor Whirl Amplitudes," SAE paper Automotive Congress and Exposition, January 8-12, 1973, Detroit, Michigan.

15 Lund, J. W., "Rotor-bearing Dynamics Design Technology. Part V. Computer Program and Manual for Rotor Response and Stability," USAF Technical Report, AFAPL-TR-65-45, Research and Technology Division, AFSC WPAFB, Dayton, Ohio, May 1965.

STRUCTURAL STABILITY

B. O. Almroth and C. A. Felippa
Lockheed Palo Alto Research Laboratory
Palo Alto, California

ABSTRACT

The purpose of this paper is to present information that may aid a user in the selection of a computer program for structural stability analysis. A discussion of the general approach to the stability problem is included. Most emphasis is placed on the evaluation of procedures as this tends to have more permanent value than an evaluation of specific computer programs.

The paper begins with an introductory discussion of the general concept of structural stability and of the type of structures for which stability must be considered. The approach to the stability problem is then discussed in detail for specific structural systems.

Appendix A reviews procedures for discretization of the continuum problem, algorithms for numerical solution of algebraic systems arising in nonlinear stability analysis, and numerical methods for extraction of eigenvalues of buckling eigensystems. As deemed practical, these methods are evaluated with regard to their usefulness in stability analysis.

Appendix B contains a brief description of those computer programs with capabilities for structural stability analysis that enjoy a relatively wide distribution. An evaluation table summarizing users' reactions is provided.

NOMENCLATURE

A = Area of column
B = Arcsin (b/L)
E = Young's modulus
I = Moment of inertia
L = Length of column
L' = Length of deformed column
N = Axial load in column
N^* = N/λ
P = Applied load (Fig. 1)
P^* = P/λ
c = Normalized spring constant (Eq. 6)
k = Spring constant
λ = Euler load $EI/(\pi/L)^2$
ρ = $(I/A)\ (\pi/L)^2$

GENERAL CONCEPT OF STABILITY

It seems reasonable to define as stability failure the situation in which a small increase in applied load leads to a disproportionately large increase in the deformation of the structure. It should be noted that the critical load according to the classical method of stability analysis does not always correspond to stability failure according to the definition above. Therefore a brief discussion is included here of the general concept of structural stability. More comprehensive discussions are available in the literature [1, 2, 3].

The discussion of the basic concepts involved in structural stability considerations is made less abstract by reference to a very simple example. Figure 1 shows a structure consisting of two columns and a linear spring. The load carried by the structure is

$$P = k\delta + 2N \sin\beta \tag{1}$$

where k is the spring constant and N is the axial load in each of the two columns. The quantities δ and β are shown in the figure. Denoting the length of the deformed column by L' we have

$$\frac{N}{EA} = \frac{1}{L}(L - L') = 1 - \cos B/\cos\beta \tag{2}$$

where $B = \arcsin(b/L)$. We set

$$\lambda = EI(\pi/L)^2 \tag{3}$$

and introduce the nondimensional parameters

$$P^* = P/\lambda;\ N^* = N/\lambda \tag{4}$$

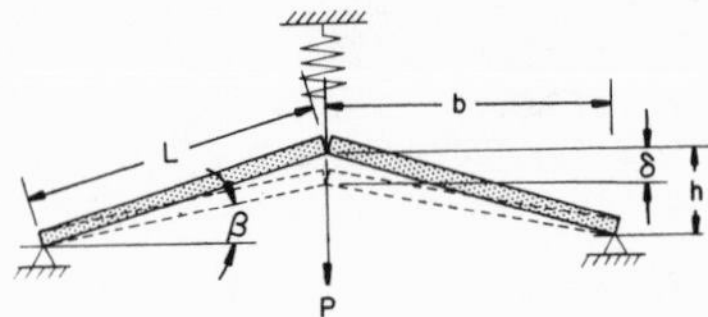

Fig. 1 Two column structure with spring

The relation between deformation δ/L and load P^* is then readily obtained in the parametric form

$$\delta/L = \sin B - \sin\beta$$
$$P^* = \frac{2}{\rho}\sin\beta(1 - \cos B/\cos\beta) + c\,\frac{\delta}{L} \tag{5}$$

where

$$\rho = (I/A)(\pi/L)^2, \quad c = kL/\lambda \tag{6}$$

In the derivation of Eqs. (5) it was assumed that $|L - L'|/L \ll 1$. This assumption is valid for shallow structures, i.e., if B is small.

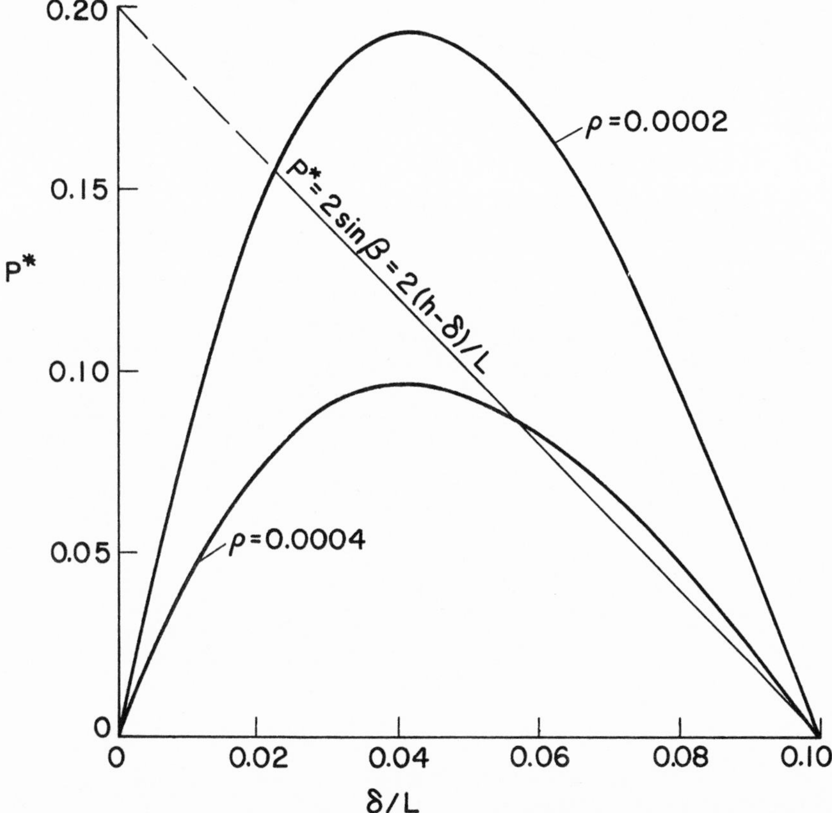

Fig. 2 Load displacement diagrams for a structure with h/L = 0.1

Substituting a series of values of δ in the Eqs. (5), with $h/L = 0.1$ and $c = 0$, we obtain the two curves shown in Fig. 2 for $\rho = 2 \times 10^{-4}$ and $\rho = 4 \times 10^{-4}$ respectively. With increasing load the stiffness $\partial P^*/\partial(\delta/L)$ decreases, and at a deformation corresponding to $\delta/L = 0.04$ ($\delta/h = 0.4$) a maximum occurs in each of the two load displacement curves. If the load P^* is increased beyond this maximum, there is no equilibrium configuration available in the immediate neighborhood, but the structure must snap over into such a position that the two columns are subjected to tension. At that point the structure undergoes a large change in deformation with a small change in load.

We notice that if $N^* > 1$, i.e., $N > EI(\pi/L)^2$, column buckling occurs. For relatively shallow structures ($h \ll L$) the shortening of the column is moderate. For all practical purposes it can be assumed that the column shortens under a constant axial load. A secondary equilibrium form with slightly bent bars is then represented by

$$P^* = 2 \sin\beta + c\delta/L \tag{7}$$

This equilibrium form corresponds to $N^* = 1$ and exists only for values of δ larger than that for which column buckling occurs.

The straight line corresponding to Eq. (7) is also shown in Fig. 2. The slender columns start to buckle at $P^* = 0.155$. The load cannot be increased beyond that point. The structure snaps through, and the columns are bent in the process. For the structure with $\rho = 4 \times 10^{-4}$ the point of intersection occurs beyond the maximum. In this case stability is lost at the point of maximum loads, while for the structure with slender columns the loss of stability occurs at the intersection with a secondary branch in the load displacement diagram.

In analyses of precritical configurations it is often assumed that effects of structural deformation can be omitted, i.e., that the prebuckling behavior is linear. As the columns buckle at $N^* = 1$, the critical load in that case (with $c = 0$) would be

$$P^* = 2 \sin B = 2h/L \tag{8}$$

For shallow structures this is a poor estimate of the critical load but for larger values of B it gradually becomes better.

Figure 3 shows the behavior of two different structures, both with $h/L = 0.1$ and $\rho = 2 \times 10^{-4}$. In one case the spring constant equals zero, and in the other $c = 2.5$. With a spring, buckling still occurs at the same value of δ/L. However, if the spring constant is sufficiently large, the slope of the curve for the secondary solution becomes positive. The increase in the load in the spring is more than sufficient to compensate for the decrease in the load carried by the column. The structure does not snap through as the intersection with the secondary path is reached.

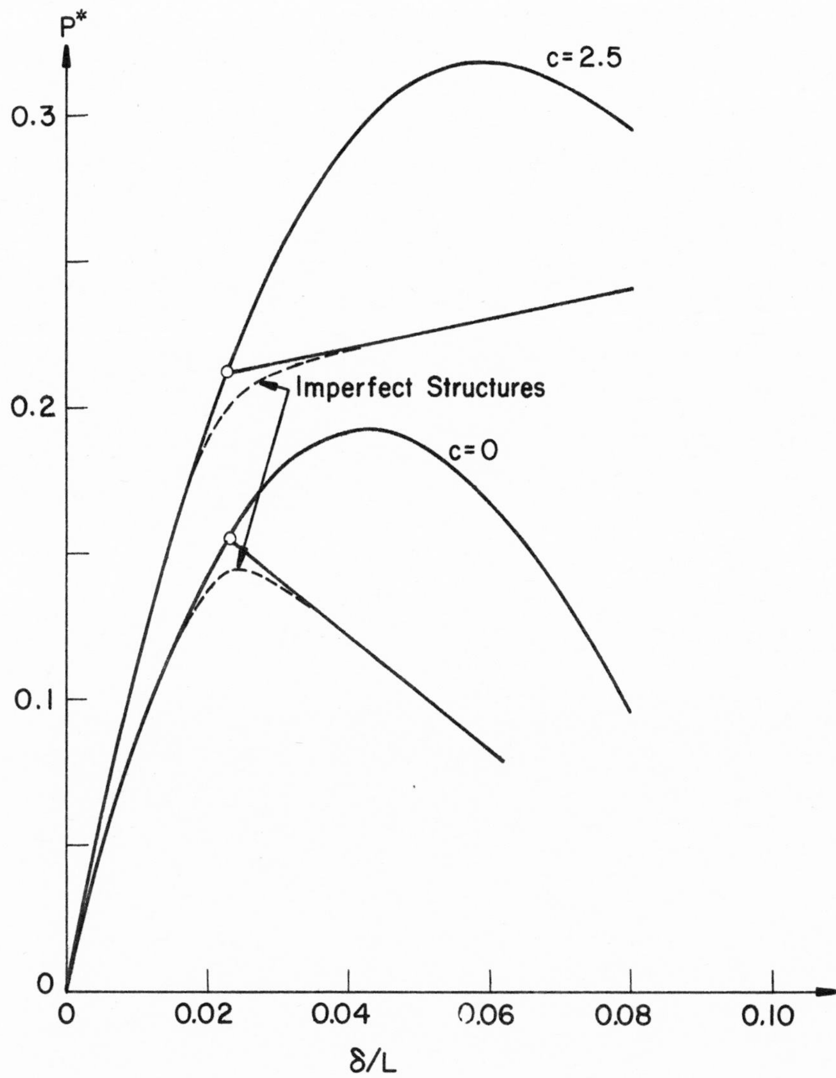

Fig. 3 Load displacement diagram for structures with ρ = 0.0002

The two-column structure discussed here serves well to illustrate the behavior of more complex structures. A load displacement curve, such as the P* vs. δ/L relation shown in the figures, is generally referred to as an equilibrium path. As we can see, such a path may, but need not, pass through the origin. It can be shown that only one of the equilibrium paths passes through the origin. This path is referred to as the primary path. Any other equilibrium path is called a secondary path.

The primary path corresponds to linear behavior if there exists a solution to the governing equations that is rotation free. Examples of structures with a linear primary path are axially loaded columns, flat plates with inplane loading, rings, and complete spherical shells under uniform pressure. However, the governing equations do contain nonlinear terms, and secondary solutions may exist. Secondary equilibrium paths may intersect the primary path.

Intersections of the primary path and a secondary path are called bifurcation points. It can be shown that the equilibrium on the primary path becomes unstable at the first bifurcation point.

If the primary path is nonlinear it may exhibit a maximum, as was the case with the two-column structure. The structure loses its stability at the maximum in the load displacement curve. Such a state is called a limit or snap through point. With a nonlinear primary path our stability analysis must allow for the possibility of a maximum in the load displacement curve as well as the possibility of bifurcation.

For simplicity, we restrict the discussion to dead weight loading. In that case the structure loses stability at a maximum in the primary path. Under a load exceeding this maximum there exists no equilibrium configuration in the immediate neighborhood. The structure is set in motion, and the process of buckling is violent. On the other hand, the existence of a bifurcation point indicates only that the equilibrium on the primary path loses its stability. The consequences of this loss of stability are not immediately clear. With loss of stability of equilibrium on the primary path at the bifurcation point it follows that the structural behavior is governed by the conditions on the secondary path. Thus a bifurcation point signifies a load level at which a new deformation pattern begins to develop. The equilibrium on the secondary path may be unstable. This is the case in the example with a two-column structure ($h/L = 0.1$; $\rho = 2\text{x}10^{-4}$) without a spring. In this case the loss of stability on the primary path results in the loss of stability of the structure. Buckling is violent, and in addition the critical load is more or less sensitive to imperfections. With some initial crookedness the columns begin to bend in the prebuckling regime. There is no real bifurcation, but the primary path gradually approaches the secondary path for a perfect structure. The behavior of imperfect structures is indicated by the broken lines in Fig. 3.

On the other hand, if equilibrium on the secondary path is stable, as in the case with $c = 2.5$ in Fig. 3, the structure can take an additional load beyond the bifurcation point. However, a new deformation pattern begins to develop and the stiffness of the structure may be considerably reduced.

APPROACH TO THE STABILITY PROBLEM

As the discussion in the preceding section shows, stability analysis is not simple and straightforward. The approach to the stability problem differs from case to case. It is generally not sufficient to acquire and apply a computer program for stability analysis. In most cases the user needs a reasonably good understanding of the rather complex structural behavior that leads to stability failure. This section contains a discussion that may be useful with regard to the choice of analysis method for a given type of structure.

We will exclude cases with large strains (e.g. plastic necking). Then loss of stability is caused by the nonlinear terms in the kinematic relations. These nonlinear terms result from rotations of structural elements. If a compact body is constrained against rigid body motions, large rotations can occur only in connection with large strains. The compact body therefore, in the regime of small strains, behaves linearly, and stability is not a problem. Possible loss of stability need be considered only if the structure or a portion of it is thin. In that case stability failure is possible if the thin part is subjected to compressive stresses.

The simplest structure for which a stability failure must be considered is a truss composed of bars that are linked at the node points and whose centers of gravity lie on the straight line connecting two nodes. If the bars are flexible, a local mode of stability failure is possible in addition to general collapse. Such local loss of stability is a bifurcation buckling phenomenon, and obviously it will occur when the Euler load is first reached in one of the members. If the load in the member is statically determinate, the local instability will result in collapse of the structure. If the load in the buckling member is statically indeterminate, the consequences of local instability are not immediately clear. This simple example shows that in some situations loading can be continued and in others local instability results in total collapse even if the system is indeterminate. However the latter can occur only if the structure is susceptible to general stability failure, that is, a collapse mode in which the nodes are displaced.

A general mode of stability failure is possible if all the node points are located close to some reference surface; i.e., if the truss is thin. In that case it may happen that the geometry of the truss changes in an unfavorable way with increasing load. The behavior is nonlinear, and collapse at a maximum in the load displacement curve is possible. General instability can also occur as a result of bifurcation into a displacement mode including displacements of node points.

For a truss in which the members are curved or eccentrically loaded, or for a truss in which moments can be transferred from one member to another (rigidly joined), beam-column effects will cause nonlinearity at the onset of loading. With moderate eccentricities this effect can usually be neglected, and local as well as general bifurcation buckling loads may serve as approximate estimates of the critical loads.

If flat plates are subjected to inplane loads only (without eccentricity), the primary path will be free from lateral displacements. All secondary paths (within the regime of small strains) must correspond to deformation modes that include bending of the plate. Any loss of structural stability must therefore occur at a bifurcation point. However, it has long been known that for plates with supported edges, the equilibrium on the secondary path is stable and they do generally have considerable postbuckling strength. This postbuckling strength is accounted for in the effective width theory for plates under compression or in the so-called diagonal tension field theories for plates under shear. The information from such theories is not only approximate but also limited,and in many practical cases it may be necessary to perform a complete nonlinear analysis to determine accurately stresses and deformations on the secondary path. An example of such an analysis is given in [4] and discussed in another paper in this volume [5].

If the plate is subjected to transverse or eccentric as well as inplane loading,the primary path will be nonlinear. Stability failure is most likely to occur at a point of zero slope on the primary path.

For a shell of revolution with axisymmetric loading, the independence of the precritical displacement of the circumferential coordinate results in considerable advantages from a computational point of view. The generally nonlinear primary path corresponds to axisymmetric behavior, and this path can be determined from a one-dimensional analysis. Furthermore, all the secondary branches correspond to displacement patterns that are either axially symmetric or at the bifurcation point harmonic with respect to the circumferential coordinate. A bifurcation point can therefore also be found from a one-dimensional analysis. This analysis generally has to be repeated for a few different values of the wave number in the circumferential direction, so that the bifurcation point corresponding to the lowest load can be found.

Most shell of revolution computer programs offer at least one analysis branch in which nonlinear terms are neglected in the prebuckling analysis (the primary branch). Use of this approach does not take possible axisymmetric collapse into account. In addition, the value of the load corresponding to the bifurcation point is only an approximation. For axially loaded cylinders with the edges constrained from radial deformation, inclusion of the nonlinear terms in the prebuckling analysis results in a reduction of the critical load of 10 to 20 percent depending on the boundary conditions (in comparison to the results from an analysis with a membrane solution for the prebuckling state). For clamped spherical caps under external pressure this effect is somewhat more significant.

The bifurcation buckling analysis gives somewhat limited information about the structural behavior. It gives no information about the shape of the secondary branch. In the case of a spherical shell subjected to a point force, for example, the equilibrium on the secondary path (corresponding to two or three circumferential waves) is stable, and bifurcation results only a slight change in the deformation pattern. In other cases the critical load is imperfection sensitive, and the critical load according to the bifurcation analysis cannot be carried by the structure.

Even for the axisymmetrically loaded shell of revolution, analysis of conditions on the secondary branch requires a two-dimensional analysis. The deformation pattern on the secondary branch is harmonic only at the intersection with the primary branch (at the limit as the amplitude of the incremental deformation pattern approaches zero). Because a two-dimensional analysis is so much more cumbersome, the bifurcation approach is most frequently used. Often the analyst has some idea about the imperfection sensitivity of his particular configuration from experience with similar shells. Sometimes it may be worthwhile to undertake a two-dimensional nonlinear analysis with imperfections included in order to avoid unnecessary reductions of the allowable loads. An example of such an analysis is given in [6]. Clearly the inclusion of imperfection is a valuable addition to a computer code for nonlinear shell analysis.

The bifurcation approach for the axisymmetrically loaded shell of revolution is a procedure that, in comparison with a complete nonlinear analysis, saves computer time while giving accurate but somewhat limited information about the structural behavior. If either the geometry of the shell or the loading is not axisymmetric, a similar time-saving procedure is not available. In tracing the nonlinear primary path, the nonlinear partial (two-dimensional) differential equations must be solved for each step. The existence of a bifurcation point along this path is indicated by a change of sign in the coefficient determinant. This determinant is readily available if the regular Newton-Raphson method or one of the incremental methods is used. The bifurcation point may be found from a plot of the value of the determinant versus the applied load. If the modified Newton-Raphson method is used, the factored matrix is updated only occasionally and the value of the determinant usually is available only for a few load steps.

Notice that bifurcation occurs only if there exists some mode of deformation that is orthogonal to the prebuckling displacement pattern. In the general case bifurcation is not likely. However, if such a mode exists and is recognized, inclusion of a small imperfection that resembles the shape of the buckling mode will expedite the analysis. Since points on the primary path must be computed from a nonlinear two-dimensional analysis, even for the perfect structure, the incorporation of an initial imperfection does not change the computation time. An advantage of this approach is that the problem of possible bifurcation does not arise. In addition some information is obtained about the secondary path.

If the structure contains a plane of symmetry, with respect to geometry as well as loading, it is easy to recognize possible buckling patterns as those that are antisymmetric with respect to that plane. In the analysis of an elliptic cylinder in [7], an antisymmetric imperfection was used in order to induce antisymmetric displacements on the primary path. Another possible procedure in such a case is to take advantage of the symmetry plane in the computation of points on the primary path and to consider possible bifurcation into antisymmetric modes by use of the linearized analysis. Notice that in this case, a determinant has to be evaluated at a number of load steps in the neighborhood of a bifurcation load. This determinant is not readily available because it corresponds to the adjacent equilibrium equations with boundary conditions that are different from those corresponding to the primary branch. It is not clear in this case that much can be gained in computer time by use of the linearized analysis.

In order to save on computer time for the analysis of shells of general shape it is necessary to accept an approximation. It may be assumed that the primary path can be computed by use of a linear analysis. In that case loss of stability is always due to bifurcation. This approach must be used with great caution. Clearly if the shell geometry changes considerably in the prebuckling range, as is the case with long cylinders under bending [8], the critical load may be drastically overestimated. On the other hand, if the structure allows redistribution of stress, as is the case when cylinders with cutouts are subjected to an axial load [9], the collapse load may be much above the load indicated by the linearized theory. Frequently the experienced analyst can judge a priori whether the linearized analysis will lead to satisfactory results. For example, if a relatively short cylinder is subjected to bending, it seems rather obvious that neither a substantial change in cross section of the cylinder nor a significant redistribution of stresses will occur in the prebuckling range. Some examples in which the bifurcation buckling results for general shells are compared to results from nonlinear analysis are given in [10].

Special problems arise if the structure is so rapidly loaded that the effects of inertia must be included in the analysis. If the duration of loading is sufficiently short, the response of the shell depends exclusively on the size of the impulse. The other extreme is the static case in which the loading is so slow that the response depends exclusively on the size of the load. For intermediate cases the response can be determined accurately only if the loading history is known.

In [11], Budiansky and Roth present an analysis of spherical caps under external dynamic pressure loading. Aided by the results they find a way to determine a critical value of the impulse. Subjected to the normal displacement corresponding to static snap through (the maximum in the load displacement curve), the shell loses stiffness and the maximum response as a function of impulse increases rapidly. As the shell stiffens again in a configuration with reversed curvature, the curve levels off. Thus the maximum response as a function of impulse shows a threshold at some value of the impulse. This value is defined as the critical impulse.

The behavior exhibited by the spherical cap under impulsive loading may be typical for a number of structures. However, for others a critical impulse cannot be so readily defined because the response as a function of the impulse increases in a more gradual fashion. Frequently the loading, even if short in duration, cannot be considered as impulsive. The load may be rapidly applied and then maintained at a constant level. For such cases a critical load or a critical value of the duration of a fixed load may be defined in a similar way.

The problem of stability under rapid loading is in most cases handled adequately only by a transient response analysis in which nonlinear terms are included. Procedures for integration of the equations of motion are beyond the scope of this paper. These are discussed by Belytschko in [12]. The load as a function of time is defined as accurately as possible, and the analysis yields stresses and displacement components as functions of time. If these response quantities are within the limits that can be tolerated, then the structure is adequate. It may be of interest to determine in addition whether a critical load or impulse in the sense discussed above exists at which the response quantities grow rapidly. A critical value so defined does not necessarily represent the limit of structural capability. The merit of the approach lies in the fact that it lends itself to parameter studies that may be particularly useful in a preliminary stage of design.

Inelastic deformation introduces another complication,in the structural stability analysis. Material behavior in the inelastic range, such as plasticity and creep, is very complicated and no mathematical model has been devised that adequately represents the real material in all situations. Plasticity is discussed in this volume by Armen [13] and creep by Nickell [14]. Therefore we will discuss here only the special problems associated with bifurcation buckling in the inelastic range.

The first controversy in the field of inelastic stability concerned possible unloading of the fibers on one side of an axially loaded column. If the column suddenly, without change in the axial load, passes into a slightly bent configuration, the axial load will decrease on one side. Thus, the resistance against bending would be underestimated if the critical load is computed by use of the tangent modulus instead of Young's modulus. However, experiments indicated that the tangent modulus load gives a good estimate of the critical load. Shanley [15] first explained this paradox by assuming that the transition from a straight to a slightly bent equilibrium configuration was associated with an increase in axial load large enough to preclude any unloading. A more rigorous and general treatment of this problem is presented by Hutchinson in [16].

Another problem is connected with the dependence of the plastic strain on the loading path. Since the classical stability equations are based on the assumption that the system is conservative, the bifurcation approach becomes questionable. However, Hill in [17 and 18] and Hutchinson [16] have presented convincing arguments in favor of the bifurcation approach.

It was early noticed that use of the deformation theory of plasticity leads to results that generally are more in line with experimental buckling loads than the results from the presumably more accurate flow theory are. The reasons for this circumstance are also discussed in [16].

It may be concluded that inelastic bifurcation buckling analysis represents an adequate approach. In a bifurcation buckling analysis, the stiffness matrix corresponding to the incremental deformations must be based on the assumption that any point in the structure that is loading (the effective stress is increasing) just before the bifurcation point is reached will be loading also in the transition to the buckled configuration. The deformation theory usually leads to reasonably good estimates of the critical load. The flow theory can also be successfully applied if some special assumptions are made. These problems are discussed in detail in [18, 19, and 20].

Bifurcation buckling analysis is useful in the one-dimensional case (shells of revolution). This is because the nonlinear prebuckling analysis is one dimensional. In the general case, it is questionable if it can offer any advantages over nonlinear analysis. Of course, in the inelastic case we cannot save computations by linearization of the prebuckling analysis.

Procedures and algorithms that are useful for implementation on the computer of stability analyses are discussed and, as far as we feel reasonable, evaluated in Appendix A. A few computer programs and some of the users' reactions are discussed in Appendix B.

NOMENCLATURE FOR APPENDIXES

A = Area; general matrix (Eq. A17)
B = Structural boundary
C = Stress-strain matrix
D = Structural domain
D_e = Finite element domain (Fig. A2)
D_m^e = Integration subdomain in energy finite difference method (Fig. A1)
F = Vector of governing nonlinear equations; force residual vector in displacement formulation
F_L = Linear portion of F
F_N = Nonlinear portion of F
F_q = Matrix of partials of F with respect to q (Jacobian of F)
K = Tangent stiffness matrix
K_0 = Linear conventional stiffness matrix
K_1 = Geometric stiffness matrix
P = External load potential
Q = Generalized force vector
U = Strain energy
V = Total potential energy
e = Finite element identification index
k = Iteration cycle index
m = Node or grid point identification index (Fig. A1); stiffness matrix reevaluation period in modified Newton's method (Eq. A9)
n = Number of degrees of freedom of discrete model
p = Distributed load
q = State vector; generalized displacement vector in displacement formulation
$\tilde{q}$ = Buckling mode in eigenstability problem
$r(k)$ = See Eq. (A9)
s = Arc length coordinate in one-dimensional problems
t = Timelike parameter
u = Solution field
u_n = N-freedom approximation to u
x = Vector of space coordinates
y = See Eq. (A17)
α,β = Parameters in self-correcting initial value procedures, (Eq. A13 - A14)
ε = Vector of strain field components
λ = State parameter (prescribed load or displacement amplitude)
ω = Step-control coefficient in damped Newton's method
$(\ )^T$ = Matrix or vector transpose
$(\ )^{-1}$ = Matrix inverse
$(\dot{\ })$ = Path derivative (total λ-derivative)
$(\ )_\lambda$ = Partial derivative with respect to λ
$\Delta(\ )$ = Increment of

APPENDIX A - PROCEDURES AND ALGORITHMS

DISCRETIZATION TECHNIQUES

General

Few practical stability problems can be successfully attacked by purely analytical methods. Stability analysis programs with some claim to generality of application must resort to numerical solution techniques. This approach is based on approximating the behavior of the actual structure by that of a simpler model selected by the analyst.

The process by which the mathematical model representing a continuous structure is replaced by a discrete model with a finite number of degrees of freedom is termed discretization. We shall be primarily concerned with spatial discretization techniques, since discretization in the time domain (e.g., for dynamic stability problems) is discussed in [12].

The choice of discretization technique represents one of the fundamental decisions to be made prior to the development of a structural analysis program. It also has a significant impact on the modeling (data preparation) and interpretation of results. This Appendix reviews those discretization procedures that have been used, or are likely to be used in the immediate future, in the computer analysis of stability problems. Although the current trend in general purpose structural analysis software is toward variational (energy) formulations, other approaches have been utilized--with varying degrees of success--in special purpose programs and therefore deserve mention.

Discretization Technique Classification

General discretization techniques may be classified from two different viewpoints:

1. The level at which the mathematical model is discretized: differential equations or variational principle
2. The discretization mechanism used: finite difference expressions or trial function expansion

Combination of these approaches gives rise to four general categories, as shown in Table 1. These four classes are discussed in the following sections, and their strengths and weaknesses noted. Sometimes a combination of two discretization techniques (or of discrete and analytic methods) may be used to advantage on special problems; some of these methods are noted in a later section of this appendix.

Table 1 Classification of Discretization Procedures

Discretization Mechanism	Discretization Level	Method Designation(s)
Finite Difference Expressions	Differential Equation	Classical finite difference methods ("equilibrium" finite differences)
	Variational Principle	Variational finite difference methods ("energy" finite differences)
Trial Function Assumptions	Differential Equation	Weighted residual methods Subclasses: Galerkin, least squares, collocation, adjoint function, etc.
	Variational Principle	Rayleigh-Ritz methods Subclasses: classical Ritz, finite element methods (conforming, hybrid, nonconforming, etc.).

Finite Difference Methods

The term finite difference (FD) methods embodies those discretization techniques in which the derivatives in the governing equations are directly replaced by difference expressions. Methods in which the discretization is carried out on the differential (equilibrium) equations are called classical FD, or equilibrium FD. If the discretization is effected on the governing variational principle, the approach is known as variational FD, or energy FD.

Finite Difference Equilibrium Methods

This method was the dominant discretization approach in the precomputer era, especially after the development of relaxation methods for solving sparse difference equation systems in the 1930s [21,22]. It is still the most widely used method for boundary value problems not amenable to variational formulation and for solving initial value problems. The mechanics of this discretization approach will not be discussed, since it is well covered in the literature [22 - 26]. Furthermore, this method is primarily of historical importance in structural problems, having been displaced by energy FD and finite element procedures. For completeness, the merits and disadvantages of this approach are noted as follows.

Advantages:

1. Generality of application
2. Fast generation of discrete equations
3. Produces sparse discrete equations
4. Well-developed discretization error theory
5. Very simple to program for regular grids

Disadvantages:

1. Awkward implementation of accurate difference schemes in problems with irregular boundaries and/or physical interfaces
2. Often requires use of phantom grid points (outside the problem domain) for modeling boundary conditions (BC)
3. Relatively poor characterization of overall or mean quantities such as eigenvalues in comparison to discrete variational methods
4. Requires that all BC be accounted for
5. Generates unsymmetric equations even for self-adjoint problems if natural BC are present

It should be observed that the last three disadvantages are valid with respect to variational-based formulations and therefore inoperative if the problem does not admit an energy principle.

Finite Difference Energy Methods

This variant of the FD approach is relatively recent. It was apparently first used in 1953 [27] and is treated in some detail in only one textbook [24]. Because of its expanding application to nonlinear structural mechanics, a short description of the discretization mechanism is presented here.

Consider the two-dimensional domain D with boundary B of Fig. A1, over which a linear elasticity problem is posed in the usual total potential energy (V) formulation:

$$V(u) = U(u) - P(u) = \text{minimum}$$
$$U = \int_D \varepsilon^T C \varepsilon \, dA, \quad P = \int_D p \, u \, dA \tag{A1}$$

where U is the strain energy, P the potential of the external load density p, u the unknown displacement field satisfying essential BC on B, $\varepsilon = \varepsilon(u)$ the associated strain field stored as a column vector, and C the stress-strain matrix. The essential steps of the discretization process are:

1. A node or grid point set is placed upon D by the analyst (Fig. A1(a)).

2. FD expressions for u and $\varepsilon(u)$ are constructed at each grid point m in terms of grid values q_i of u (and possibly u-derivatives) at m and neighboring nodes. These expressions are inserted in the integrands of Eq. (A1).

3. D is subdivided into disjoint integration subdomains D_m, as illustrated by Fig. A1(b). Integrals over D are thus replaced by the sum of integrals over the D_m.

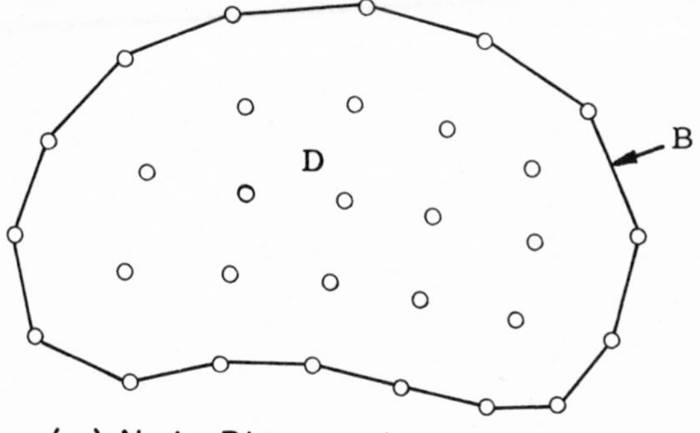

(a) Node Placement

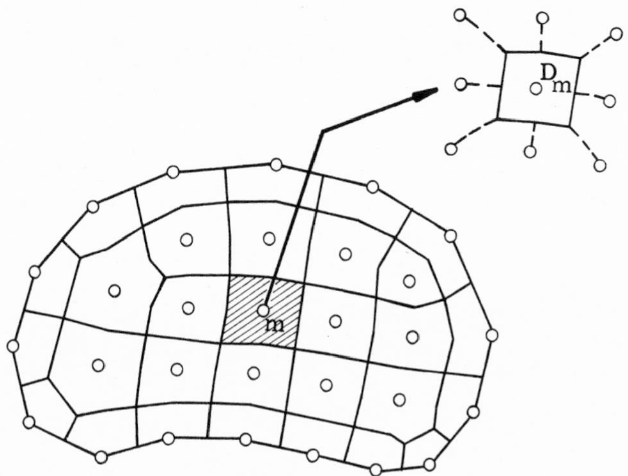

(b) Element Integration Area Subdivision

Fig. A1 Finite difference energy discretization of a two-dimensional problem

4. The D_m integrals are evaluated by numerical quadrature, most implementations using one-point formulas. The integration points (often called energy-evaluation or energy-sampling points) do not coincide, in general, with the grid points, and appropriate interpolation schemes have to be devised.
Extremizing the discrete functional $V(q_i)$ with respect to the unknown q_i yields the discrete linear system

$$K\, q = Q \tag{A2}$$

where vector q collects all q_i, Q contains associated generalized forces ($Q_i = \partial P/\partial q_i$), and the stiffness matrix K with entries $K_{ij} = \partial^2 U/\partial q_i \partial q_j$ is symmetric.

This procedure eliminates disadvantages 3, 4, and 5 of the equilibrium FD approach and lessens the second one since only essential BC must be imposed.[1] The most serious computational disadvantage associated with this method is the problem of characterizing the D_m subdomains given an arbitrary grid point layout. In this context, it should be noted that in contrast to finite element methods (Fig. A2) the user supplies only node definition data to the FD code; the input preparation effort is therefore considerably reduced.

The application of this method to structural mechanics problems has been largely restricted to linear and nonlinear analysis (stress, stability, and vibration) of thin shells [28 - 31]. Application to general shells has been limited to shapes in which grid layout can be analytically mapped onto a rectangular mesh. More development work is needed for problems with irregular geometries [32]. The relationship of energy FD and finite element models is discussed in [31, 33].

Trial Function Methods

These methods are based on assuming the form of the unknown solution u(x) (x = vector of spatial coordinates):

$$u \sim u_n(x) = \sum_{i=1}^{n} q_i\, \phi_i(x) \tag{A3}$$

over the entire structure, or part thereof. In Eq. (A3), $\phi_i(x)$ are the trial functions (also called base, coordinate, or shape functions), and q_i are the generalized coordinates of the n-freedom discrete model. As in the case of FD methods, the generation of discrete equations in the q_i may proceed at two levels.

Weighted Residual Methods

Equation (A3) is inserted into the differential equilibrium

---

[1] The use of phantom grid points (nodes outside D) can in fact be completely eliminated by selection of appropriate u-derivatives as q_i unknowns at boundary nodes.

equations, and an overall error measure $E(q_i)$ is constructed by multiplying the local equilibrium error (the residual) by a weighting function and integrating over the structure volume. Discrete equations result upon equating E to zero [34 - 36]. Different choices of weighting functions precipitate a number of well-known methods listed in Table 1. Of these, the Galerkin method (obtained by selecting (A3) as the error-weighting function) is commonly regarded as the most convenient in the applications [36]. Since for structural stability formulations derivable from a variational principle the Galerkin and Rayleigh-Ritz methods coalesce, further treatment of the Galerkin method is unnecessary.

Rayleigh-Ritz Method

This time-honored procedure consists of inserting Eq. (A3) into the governing functional, e.g., (A1). Extremizing $V(u_n)$ with respect to q_i produces symmetric discrete equations formally similar to (A2) in the linear case. In the classical Ritz method [23, 24 - 26] the trial functions $\phi_i(x)$ are nonzero over the entire structure and satisfy all essential boundary conditions exactly. This method possesses all of the typical attributes of variational techniques, namely

1. Accurate characterization of overall quantities, for instance stability eigenvalues
2. Relaxed boundary condition satisfaction requirements (Eq. (A3) need not verify natural boundary conditions)
3. Symmetric discrete equations

In addition, the classical Ritz method provides rigorous bounds on energy and eigenvalues of the mathematical model.

The classical Ritz method is a hand-computation oriented procedure largely confined to relatively simple problems. The main difficulty in extending the method to more complicated cases lies in the construction of simple yet admissible trial functions. The difficulty was overcome with the advent and rapid development of the finite element method.

Finite Element Method

The finite element method (FEM) is a computationally convenient way of generating admissible trial functions for regions of arbitrary geometry. The method is based on the concept of local (piecewise) approximation. The problem domain D exemplified in Fig. A2(a) is partitioned into disjoint subdomains D_e called finite elements. The geometry of each element is defined by user-specified node points located on the element boundary (see Fig. A2(b)). The degree of refinement of the element geometry may be controlled by adding more node points per element, as depicted in Fig. A2(c). Such "refined" elements permit better representation of curved boundaries.

The generalized coordinates q_i of Eq. (A3) are selected as values taken by u (and possible u-derivatives) at nodal points in such a way that minimal continuity requirements are maintained across

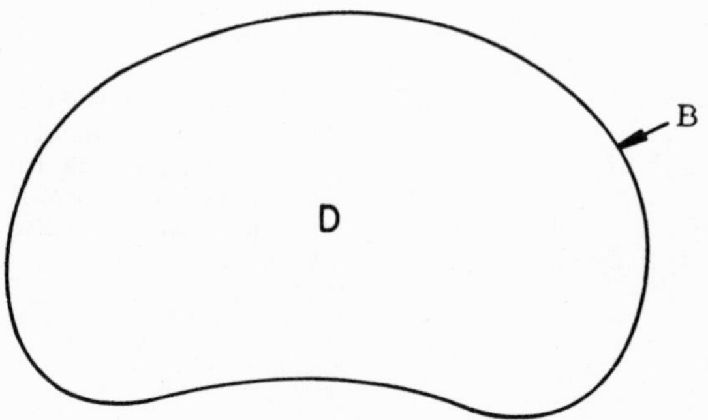

(a) Original Domain

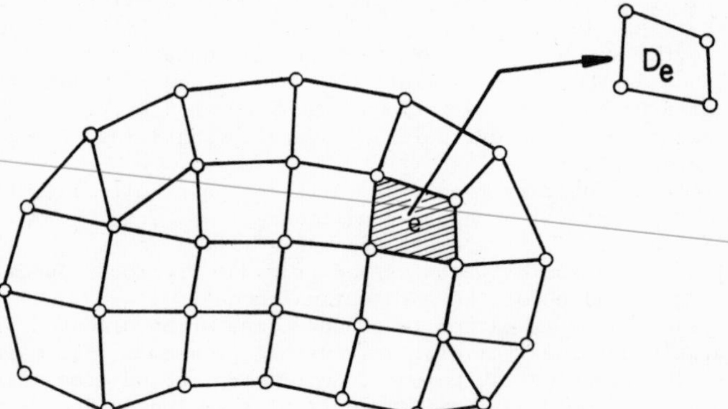

(b) Discretization By Simple Elements

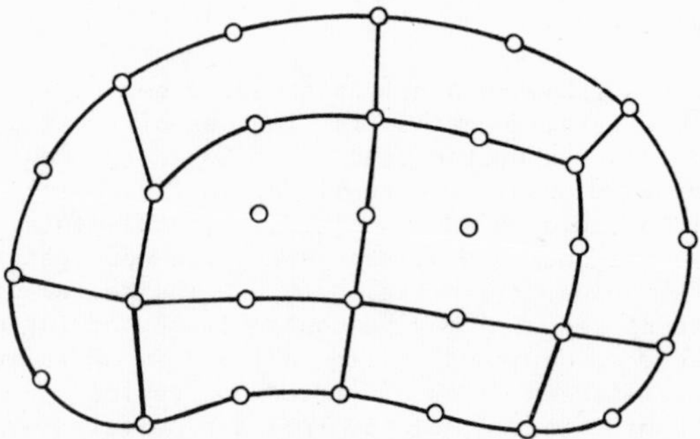

(c) Discretization By Refined Elements

Fig. A2 Finite element discretization concept

element interfaces.[2] A trial function ϕ_i associated with a q_i prescribed at node m vanishes on all elements not connected to m. A consequence of the local support of ϕ_i is that the coefficient matrix of the discrete equations is sparse, as is the case with difference methods. The automatic generation of the finite element equations parallels in many respects the process outlined in Section 3 for FD energy models; the main difference is that the integration subdomains are now the element domains D_e, and these are defined a priori by the user. The current trend is to evaluate all D_e integrals by Gaussian quadrature rules; this process is computationally effective but has the disadvantage of losing the eigenvalue bounding attribute of classical Ritz methods.

The success of the FEM is to a great extent due to the fact that it combines the best attributes of variational methods (listed above) and of FD schemes (generality of application, sparse discrete operators). From the user's viewpoint, the most outstanding feature is the ease with which arbitrary geometries, boundary conditions, physical interfaces, and coupling of structural components can be handled. The FEM is now well described in several introductory and medium-level textbooks [37 - 41], most of which contain adequate bibliographies on applications to structural stability analysis. Those interested in the mathematical foundations (still under intensive development) may consult Refs. [42 - 45].

Special and Combined Methods

In many practical situations a combination of two or more discretization strategies (or of numerical and analytical techniques) may be used to effect considerable computational savings. Some examples relevant to structural stability analysis follow.

Separation of Variables

Problems exhibiting certain geometric-loading BC symmetries are often amenable to a separation of space variables. For instance, in trial function methods

$$\phi_i(x) = \phi_i(x_1, x_2) = \phi_{1i}(x_1)\phi_{2i}(x_2) \tag{A4}$$

may uncouple the discrete equations. The solution of the complete problem is then reduced to solving a set of lower dimensionality problems, each of which may be attacked by numerical or analytical procedures.

---

[2]Such elements are said to be conforming or compatible. If certain continuity requirements are relaxed, nonconforming or incompatible elements result. A detailed discussion of the convergence behavior of these models is beyond the scope of this review.

An important example is the stability analysis of arbitrarily loaded axisymmetric shells [46]. A Fourier series displacement expansion in the circumferential direction permits a one-dimensional eigenstability analysis to be carried out for each circumferential wave number. Thus a very detailed modeling of the shell in the meridional direction is possible.

Numerical Integration

This technique is applicable to one-dimensional boundary value problems (BVP)[3] characterized by a single space coordinate s. The BVP is transformed into an initial value problem (IVP) in s by some imbedding procedure [47]. The system of ordinary differential equations can then be solved by standard numerical integration methods.

The chief advantages of this method with respect to standard BVP methods are effective discretization error control (if a step-adjusting integration package is used) and economy in computer storage. However, the method is prone to cancellation and roundoff error propagation due to exponential growth of parasitic solutions introduced in the conversion to IVP; this often necessitates problem segmentation. The method had its heyday on the small computers of the late 1950s, when it was known (for linear structural analysis) as the transfer matrix method [48]. Modernized versions have proved to be very effective in the linear analysis of shells of revolution; in Ref. [49] the relative merits of numerical integration and FD schemes for such applications are reviewed.

Boundary Integral Technique

This is a dimensionality reduction method not based on a separation of variables but on a source potential formulation. The structural problem is reduced to an integro-differential equation over its boundary, which is then discretized by boundary elements. This approach is due to Massonnet and co-workers [50] and predates the computer. It has been revived recently [51, 52] and appears to be competitive with finite element techniques for elasticity problems in infinite domains or those exhibiting singularities (e.g., cracks).

Combination of Finite Element and Finite Difference Energy Models

Numerical experiments have indicated that for problems involving linear and nonlinear analysis of thin shells, FD energy formulations run significantly faster and produce more accurate results than FE idealizations with similar number of degrees of freedom [21, 31, 53]. On the other hand, FE modeling has a substantial edge on versatility

---

[3]Either a bona fide one-dimensional problem or the product of a separation of variables technique (as in the case of linear axisymmetric shells.

of application. It follows that a hybridization process in which both discretization mechanisms are used concurrently to model different positions of a complex shell-type structure seems to offer many computational advantages. FD energy modeling could then be reserved for regular portions of the shell where their superior convergence characteristics are most effective, whereas finite elements in the form of boundary layers would take care of boundary conditions and interface constraints. (It should be noted that at the assembled discrete equation level there is no difference between the two modeling approaches, so that the same sparse matrix solution packages can be employed.) This combination process will undoubtedly be implemented in future general purpose codes.

NONLINEAR STABILITY ANALYSIS

Background

The investigation of static stability limits of a discrete structural model by means of a full nonlinear analysis requires that large systems of nonlinear algebraic equations be repeatedly solved. Numerical procedures for dealing with such systems are reviewed in the next sections, with particular emphasis on their suitability to nonlinear analysis codes based on a displacement formulation.

Since 1960 a systematic development of general computer-oriented techniques for solving systems of nonlinear equations has taken place in the field of numerical analysis. Results of this intense activity are reflected in the publication of several monographs [54 - 57] and conference proceedings [58 - 61] devoted entirely to the subject. Concerning applications to nonlinear structural analysis, excellent surveys by Stricklin and co-workers [62 - 64] and Oden [65 - 66] are available. However, it should be pointed out that some of the newest methods have not been extensively tried on large-scale problems.

The proliferation of algorithms strongly suggests that a universal solution procedure is not yet available (as is the case with linear systems). The success of a nonlinear solution algorithm depends primarily on two factors:

1. The type of problem to which applied
2. Details of the computer implementation (the set of procedural rules embodied in what is often referred to as solution strategy)

In nonlinear structural stability analysis, the type of problem is largely determined by the source of nonlinearity: geometry changes configuration-dependent loads, varying kinematic boundary conditions, or nonlinear material law. This is because the physical source of nonlinearity determines the mathematical structure of the governing equations.

The second factor (implementation details) has a significant influence on the degree of efficiency and reliability of a particular method. However, a discussion of this topic is beyond the scope of this review.

Formulation

This review of nonlinear static analysis techniques will center around the parametric solution of a nonlinear algebraic system symbolized by the vector equation

$$F(q,\lambda) = 0 \tag{A5}$$

where λ is a user-varied state parameter (e.g., a load parameter)[4] and the vector $q(\lambda)$ collects the degrees of freedom $q_j(\lambda)$ of the discrete model. It is assumed that Eq. (A5) incorporates all essential constraints.

The critical states of the nonlinear structure modeled by Eq. (A5) are defined by the condition that the Jacobian matrix $K = F_q$, whose entries are

$$K_{ij} = \frac{\partial F_i}{\partial q_j} \tag{A6}$$

be singular for $\lambda = \lambda cr$, i.e.,

$$\det K(q,\lambda cr) = 0 \tag{A7}$$

Most nonlinear analysis codes are based on a displacement formulation. In such cases, the governing equations (A5) are discrete equilibrium equations, in which

q = generalized displacement vector
F = residual or unbalanced force vector
K = tangent stiffness matrix

The singularity condition (A7) indicates that multiple equilibrium solutions may exist for the same value of λ, if λ is a prescribed load parameter. Condition (A7) does not correspond to a maximum in the load-displacement curve (usual definition of limit point) if λ is a prescribed displacement parameter.

Further investigation into the mathematical structure of Eq. (A5) is required to ascertain whether the state (A7) represents a bifurcation or limit point. However, in view of the discussion in the main body of the paper, such a distinction is largely academic.

Solution Procedure Classification

In Table 2 numerical techniques for solving nonlinear algebraic

[4]The case of multiple parameters is discussed in detail in Ref. [67].

systems are arranged into four major groups identified in accordance to current usage in the numerical analysis literature. The best known algorithms within each group are then tabulated. Some of them are known by different names in the engineering literature (where they are periodically rediscovered).

The treatment of the four general groups in the next four sections is necessarily superficial and incomplete, since only those techniques currently used in nonlinear structural analysis software are described. A more detailed description of a broader class of algorithms is presented in Ref. [67].

Table 2 Algorithms for numerical solution of nonlinear algebraic equations

Major Groups	Algorithms	Identifiers often used in engineering literature
Newton-like	Standard Newton Modified Newton Damped Newton Quasi-Newton Steffensen	Newton-Raphson
Successive substitution	Picard iteration Perturbation Contraction	Initial stress Initial strain Pseudoload
Initial value (continuation, imbedding, parameter differentiation)	Uncorrected integration, Corrected inte-integration, Infinite interval	Incremental Step-by-step Self-correcting Dynamic relaxation
Minimization	Random search Sequential search Steepest descent Conjugate gradient Variable metric Gauss-Newton	Energy search

Newton-like Methods

This is a class of one-step iterative methods based on Taylor series linearization of Eq. (A5) in the neighborhood of a solution (q, λ). A sufficiently close initial estimate q^0 is required to initiate the iteration process.

Standard Newton's Method

The well-known Newton's method [55 - 58] is based on the iteration

$$q^{k+1} = q^k - \Delta q^k$$
$$K(q^k,\lambda)\Delta q^k = F(q^k,\lambda) \qquad \text{(A8)}$$
$$k = 0, 1, \ldots$$

where k is an iteration cycle index and λ is fixed. Convergence of q^k to q can be ascertained by monitoring the magnitude of F (force equilibrium residual), Δq (configuration change), or $F^T \Delta q$(energy variation.

The standard Newton's method is very accurate and quadratically convergent in the vicinity of the solution. In addition, the singularity condition Eq. (A7) may be tested at each iteration cycle, since the determinant of K is obtained as a byproduct when the linear system $(A8)_2$ for Δq^k is solved. The primary disadvantages of this method are:

1. Lack of convergence, or convergence to an unwanted solution, if the initial estimate q^o is outside the domain of attraction of q.
2. Recalculation of K and factorization to solve for Δq^k is required at each iteration cycle.
3. Convergence difficulties may be encountered in certain nonlinear problems for which a unique Taylor series expansion at equilibrium solutions does not exist [68]. An example is provided by incremental plasticity analysis with elastic unloading.

The standard way of circumventing the first difficulty is to increase λ in a stepwise fashion so that the last converged solution is used as initial estimate for the next step. A slightly more refined approach consists of combining Newton's method with a predictor technique, such as extrapolation of previous solutions or an initial value method which can supply reasonably good starting values.

The second disadvantage is lessened by reevaluating K occasionally, as discussed below.

The third disadvantage is more serious, as it pertains to the basic assumption on which the algorithm Eq. (A8) is based. One remedy is a nested iteration procedure in which nonsmooth nonlinear effects affecting the validity of Eq. (A8) are removed from the inner iteration loop [69]. For two- and three-dimensional problems with large K-bandwidths, such a procedure is probably too costly as K must be factored several times in the various iteration levels. The underrelaxed or damped Newton's method (described below) may be used to alleviate oscillatory behavior. Since experience with such cases is limited, it is not possible to be more specific at this time.

Variants

In the modified Newton's method [55 - 56], K is reevaluated only every r(k) steps ($r(k) \leq k$). Usually

$$r(k) = k/m \qquad \text{(A9)}$$

which corresponds to periodic reevaluation every m steps (m = 1 yields the standard Newton's method). The optimal choice of m is primarily controlled by the problem dimensionality (affecting the relative cost of evaluation of K^{-1} and F) and the degree of nonlinearity of the structural response to changes in λ. This technique has been proved extremely effective in the nonlinear analysis of general (two dimensional) shells [70]. It appears that for moderate to large K bandwidths the best choice of m ranges from 4 to 10, whereas for small bandwidths (e.g., shells of revolution), m should be 1 or 2.

The damped Newton's method incorporates a step-control coefficient ω^k in the displacement correction Δq^k. This factor is adjusted during the early iteration stages to ensure that the magnitude of the residual force vector F decreases. Once the process begins to converge rapidly, ω is set equal to one. This trick helps to prevent early solution blowup due to an unfortunate choice of q^o, especially near a critical point.

Other Newton-like methods listed in Table 2, although promising, have not been extensively tested on structural problems.

Successive Substitution Methods

The successive substitution methods embody iterative procedures in which all nonlinear terms are transferred to the right-hand side as pseudoloads. A typical recursive process is

$$
\begin{aligned}
q^{k+1} &= q^k + \Delta q^k \\
K_o \Delta q^k &= \lambda F_L + F_N(q^{k+1}, \lambda) \qquad \text{(A10)} \\
k &= 0, 1, \ldots
\end{aligned}
$$

in which K_o is a constant stiffness matrix (usually the tangent stiffness at the undeformed state) and the nonlinear terms are collected and the pseudoforce vector F_N. These methods have been very popular in the early stages of development of nonlinear finite element codes, particularly in plasticity analysis. Variants known as initial stress, initial strain, etc., correspond to different ways of predicting F_N from previous solutions.

Substitution methods have the merit of using a constant K_o throughout the recursion process and of being very easy to implement in existing linear codes. Such advantages are outweighed by the following serious weaknesses:

1. The process (A10) converges only under certain conditions [56, Ch. 12] associated with the "degree of nonlinearity" of the problem. If the convergence requirements are not met, (A10) diverges no matter how close the initial estimate q^o is to the solution.

2. Even if the process converges, the asymptotic rate of convergence is only linear (compared to the superlinear convergence of Newton-like methods).

3. The stability determinant (A7) is not available.

These disadvantages are sufficient to disqualify these methods for use in general purpose nonlinear stability programs. However, useful variants may be obtained in special situations if (a) only selected nonlinear terms are moved to F_N and (b) the result of such an operation widens the domain of convergence of the method up to (or close to) the stability limit. One such case is the geometrically nonlinear analysis of asymmetrically loaded shells of revolution in which nonlinear terms coupling circumferential harmonics are treated as pseudoloads [71].

Initial Value Methods

Initial value techniques reduce the problem of solving the governing nonlinear equation (A5) for a fixed λ to an initial value problem in λ, for which standard forward integration methods are available. These procedures are also known as continuation, imbedding, Davidenko, or parameter differentiation methods in the mathematical literature [56, 61, 66, 72]. In engineering publications, terms such as incremental, step-by-step, marching, and dynamic relaxation are often used to identify members of this large family [62 - 65].

Three implementation variants are of particular relevance to structural applications.

Uncorrected Integration

Eq. (A5) is converted into an initial value problem by differentiation with respect to λ (dot denotes λ-path derivative):

$$\dot{F} = F_q \dot{q} + F_\lambda = 0 \qquad \text{(A11)}$$

or, explicitly (with $K = F_q$):

$$\frac{dq(\lambda)}{d\lambda} = -K(q, \lambda)^{-1} F_\lambda(q, \lambda) \tag{A12}$$

with the initial condition $q(0) = 0$ if $\lambda = 0$ corresponds to the undeformed configuration. The system (A12) may be attacked by any standard forward integration scheme such as Euler, Runge-Kutta, or predictor-corrector linear multistep methods. The main source of concern in the applications of uncorrected integration is the tendency of the computed solution $q(\lambda)$ to deviate from the equilibrium path (A5) because of propagated integration error. This is particularly true of the lowest order schemes such as Euler and second-order Runge-Kutta. Reduction of this error to engineering accuracy often forces the use of very small λ-increments, and since the nonlinear stiffness matrix K must be continuously updated and refactored, excessive computer times are likely to result.

Corrected Integration

The simplest way to compensate for drifting from the equilibrium path is to apply a Newton-like method using the integrated solution as initial estimate. The modified Newton's method is particularly attractive in this respect, since recomputation of K^{-1} is avoided. The Newton correction may be applied after each or several integration steps, depending on the type and degree of nonlinearity. In Ref. [70] it is shown that application of a modified Newton correction to a one-step incremental (Euler) method does not appreciably change the computer time, but the improved accuracy permits larger load steps to be taken.

Another equilibrium-correcting approach, advocated by Stricklin and co-workers[62 - 64], consists of forming a linear combination of Eqs. (A5) and (A11),

$$\dot{F} + \alpha F = 0 \tag{A13}$$

where α is a scalar adjusted according to the integration stepsize. The method (A13) is called a first order self-correcting method. The second order self-correcting method is based on the second order system.

$$\ddot{F} + \alpha\dot{F} + \beta F = 0 \tag{A14}$$

Infinite Interval Integration

The infinite interval integration methods result from the variable transformation $\lambda = \lambda^*(1 - e^{-t})$ for each target load level λ^*. The solution $q(\lambda^*)$ may be interpreted as the steady-state component ($t \to \infty$) of a pseudodynamical system. This technique eliminates the accuracy (equilibrium drifting) problem and provides an automatic error control scheme. However, the problem of numerical stability for large t-steps becomes of paramount importance. It should be noted that Euler's scheme applied to this formulation is equivalent to the standard Newton's iteration (A8), so that no new method results in this case. The advantages of this formulation are not fully realized unless an unconditionally stable integration scheme is used [73].

Evaluation

Among the advantages of initial value methods are:

1. Implementation of low order schemes in existing linear codes is straightforward.
2. Much of the numerical integration software can be applied to nonlinear dynamic analysis.
3. The stability determinant is readily available.

The chief disadvantage of uncorrected schemes is the problem of assessing the accumulated integration error. This difficulty can be overcome by using an equilibrium iteration or a self-correcting formulation. The fact that a correcting feature can be obtained with very little computational effort strongly suggests that the use of uncorrected schemes is not justifiable.

All standard forward integration schemes, whether corrected or not, exhibit inherent numerical instability near a critical point, a fact that limits the usefulness of those methods in postbuckling analysis. This problem can be overcome only by including dynamic terms (real or fictitious) in the equilibrium equations or, alternatively, utilizing an infinite interval formulation in conjunction with an A-stable or stiffly stable [74] integration scheme. A systematic evaluation of the relative merits of postbuckling analysis techniques is still unavailable.

Direct Minimization Methods

These methods convert the problem of solving Eq. (A5) into the minimization of a related function, e.g., the energy functional (A1). Although a rapid development of sophisticated minimization techniques has been recorded in the past five years [56, 57, 59 - 61], applications to structural analysis problems have been largely disappointing (cf. bibliography in [64]). One possible explanation of the poor computational performance is the difficulty of implementing and tuning up procedures that have been originally developed for nonsparse, small-order systems arising in optimization problems. Nonetheless, direct minimization methods may eventually prove to be suitable for dealing with certain types of nonlinearity such as crack propagation, penetration, and contact problems, which are not handled effectively by any other method [68].

STABILITY EIGENVALUE ANALYSIS

Linearized Stability Problem

As noted in the discussion of approaches to the stability problems, there is a large class of structures for which all nonlinear effects prior to loss of stability can be neglected [3, 75]. In such cases Eq. (A7) can be linearized to the form

$$\det (K_0 + \lambda K_1) = 0 \tag{A15}$$

where $K_o = K(0, 0)$ is the conventional linear stiffness matrix and $K_1 = K(\overset{o}{0}, 0)$ is called the initial stress, geometric stiffness, or stability matrix. K_1 can be assembled from the prestress distribution obtained from a static linear analysis. Eq. (A15) can be presented as an algebraic eigenvalue problem

$$(K_0 + \lambda K_1)\tilde{q} = 0 \tag{A16}$$

where the eigenvectors $\tilde{q}$ are buckling modes. The following characteristics of the eigen problem (A16) should be noted:

1. Both K_o and K_1 are sparse symmetric matrices with similar connectivity structure (unlike the vibration eigenvalue problem, K_1 may not be diagonal). All eigenvalues are real.
2. In most cases, only the eigenvalue λ_{min} closest to zero is of interest; λ_{min} may be positive or negative.

The following numerical procedures are commonly utilized to solve (A16).

Direct Reduction

Eq. (A16) is reduced to the standard symmetric eigenproblem

$$Ay = \lambda y \tag{A17}$$

following a Cholesky factorization of K_o (assumed to be positive definite) as described in [76]. The problem (A17) is then submitted to a standard symmetric eigensolver. This procedure is reliable and requires little specialized software. However, it is limited to system orders not exceeding 300 on most computers, since matrix A is full and the solution time for (A17) grows as the cube of its order.

Determinant Tracking

Eq. (A15) is solved directly by searching for its smallest root along the real axis.[5] Once λ_{min} is separated, a faster root extraction procedure [54] may be invoked. This technique is safe and easy to

[5]In order to avoid bypassing an even number of closely spaced roots, it is necessary to examine not only the sign of the stability determinant, but also the number of negative diagonal elements in the factorization of $K_0 + \lambda K_1$.

program, but requires a factorization of $(K_o + \lambda K_1)$ for each sample λ. Furthermore, the determination of the buckling mode $\tilde{q}$ involves additional calculations.

Inverse Power (Wielandt) Iteration

The inverse power iteration method is described in Refs. [76, 77, Ch. 9, and 78, Ch. 7]. It is by far the most efficient method for solving (A16) in the case of large systems (over, say, 500 freedoms), since full advantage is taken of the sparseness of K_o and K_1 and the fact that only one root is of interest. It also has the important advantage of providing the buckling eigenvalue λ_{min} and associated buckling mode simultaneously. Periodic spectral shifting is an essential adjunct of this technique in the case of close eigenvalues. Stability eigensystems of over 20,000 freedoms have been handled with the use of this method [6].

Quadratic Stability Problem

If the effect of prebuckling rotations from the reference state $(q = 0)$ is not neglected, a quadratic term appears [5] in the λ-polynomial approximation to (A7):

$$\det(K_0 + \lambda K_1 + \lambda^2 K_2) = 0 \tag{A18}$$

where K_o, K_1 and K_2 are symmetric matrices (K_1 is not identical to K_1 in Eq. (A16)). Numerical procedures for handling (A18) are similar to those quoted for the linear eigenproblem. Inverse iteration is again recommended for large systems [31, 36, 49]. However, the convergence of the inverse iteration process to a real root of (A18) may be vitiated by the presence of spurious complex roots near the real root.

APPENDIX B - AVAILABLE COMPUTER PROGRAMS

It is hoped that the discussion of nonlinear structural behavior and of the stability problem will help a structural analyst to establish what his needs are in terms of computer programs. In this appendix we intend to give some information about the more widely used computer programs for stability analysis. Together with the evaluative discussion of pertinent solution procedures, in Appendix A, this information may help the analyst to find a computer program that is suitable for his purpose.

In the development of a computer program, conflicts frequently occur between the demands for generality, efficiency, and ease of use. It is possible to write one computer program that is very easy and can handle all stability problems. However, such a program would in most cases be inefficient. The user must be given the option to choose the discretization procedures and solution algorithms that are most suitable for his particular problem. If the user is given the option to select among various discretization procedures and solution algorithms, the program would become more efficient but more difficult to use (in the sense that it would require a deeper knowledge of the subject on the part of the user). Future computer programs will probably have more strategy decisions embedded in the code. In many cases this can be done with little loss in efficiency. More efficient multipurpose codes will certainly be available in the future,but any user who expects to face a variety of stability problems is advised to acquire and become acquainted with more than one code.

It is, of course, wasteful to use a code based on a two dimensional discretization to solve one-dimensional problems. For bifurcation buckling of shells of revolution with axisymmetric loading, one should use a code that has been specialized to this purpose, or one that contains a special branch for the buckling of axisymmetrically loaded shells of revolution. Likewise, a computer code restricted to shell structures is likely to solve shell problems more efficiently than a general purpose code does.

Important points to consider are the degree and cost of assistance offered by the developer and whether the program is periodically updated. Some codes are available free of charge. Generally, in such a case the user cannot expect to obtain free assistance or information about updates. Computer programs partially or totally developed under government funding are often available at a nominal fee. This fee is intended to cover updates of the program and a moderate amount of assistance. Some program developers are recovering development expenses by sale of the program. In these cases the user pays a sizable amount and should expect considerable assistance. Finally, some programs are not distributed, but they can be used at a surcharge at the computer center.

Below are listed some programs that perform either a nonlinear static analysis or a bifurcation buckling analysis. Programs have been included for shells of revolution, shells of general shape, and general structures. This information has been extracted from questionnaires distributed by Bushnell [79]. Special purpose programs have not been included, and the discussion of each entry is brief. More information about such codes can be obtained from Bushnell [79].

User reaction to some of the codes is recorded in Table 3. Only those codes have been included for which at least six users returned the questionnaires. In the questionnaire section "Evaluation of Program" the user is given five options for each answer. These were coded as follows:

1 = strongly agree
2 = agree
3 = neutral
4 = disagree
5 = strongly disagree

The answers to some of the questions are summarized in Table 3. For each of the questions included, Table 3 gives the number of answers on the particular question, the range of the answers (in the same code as used in the questionnaire), and their average.

Programs for Shell Revolution

SRA. Developed by G. A. Cohen, Structures Research Associates, 456 Forest Avenue, Laguna Beach, California. Branched or segmented shells. Discrete ring stiffeners. Layered, anisotropic shell wall. Axisymmetric collapse, bifurcation with linear or nonlinear prestress. Multisegment forward integration method. Eigenvalue by inverse power iteration. Nonlinear prebuckling behavior by standard Newton. CDC or UNIVAC systems. Price $750, available through developer, COSMIC.

KSHEL. Developed by A. Kalnins, Dept. Mech. Eng. Lehigh U., Bethlehem, Pa. Branched and segmented shells. Discrete ring stiffeners. Layered anisotropic shell wall. Bifurcation with linear prestress. Multisegment forward integration method. Eigenvalue by inverse power iteration method. IBM, CDC, UNIVAC, and GE systems. Price to be negotiated. Available through developer.

BOSOR. Developed by D. Bushnell, Dept. 52-33, Bldg. 205, Lockheed Missiles & Space Co., Inc., 3251 Hanover Street, Palo Alto, Calif. Branched or segmented shells. Layered anisotropic shell wall. Simple input for special shell walls. Discrete or smeared stiffeners, axisymmetric collapse, bifurcation with linear or nonlinear prestress. Finite difference energy method. Eigenvalues by inverse power iteration method. Nonlinear prebuckling behavior by regular Newton. IBM, CDC, and UNIVAC systems. Price $300, available through developer, COSMIC.

SATANS. Developed by R. E. Ball, Naval Postgraduate School, Code 57 Bp, Monterey, Calif. Monocoque, isotropic, shell wall transient or static loading. Nonlinear (nonsymmetric) collapse. Finite difference equilibrium method. Fourier analysis in circumferential direction. Nonlinear equation solution by successive substitutions. IBM or CDC systems. Price not quoted. Available from COSMIC. It is also available from the Aerospace Research Applications Center, Indiana University Foundation, Indiana Memorial Union, Bloomington, Indiana 47407.

STAGS. Developed by B. O. Almroth and F. A. Brogan, Dept. 52-33, Bldg. 205, Lockheed Missiles & Space Co., Inc., 3251 Hanover Street, Palo Alto, Calif. Branched or segmented shells. Cutouts, framework shell combination. Layered anisotropic shell wall. Simple input for special shell walls. Discrete or smeared stiffeners. Plasticity. Nonlinear collapse, bifurcation with linear prestress. Finite difference energy method for shell structure. Modified Newton for solution of nonlinear equations. Inverse power iteration for eigenvalues. Restart capability. Max. no. of freedoms and bandwidth limited only by available auxiliary storage. CDC and UNIVAC systems. Price $1,000. Available through developer.

Programs for General Structures

All programs in this class are based on the finite element method.

ASEF. Developed by Lab. Techniques Aerospatiates, U. of Liege, 75 Rue du Val Benoit, Liege, Belgium. Anisotropic material, fluid-structure interaction, bifurcation with linear prestress. Inverse power iteration for eigenvalues. Elements: rod, straight beam, conical shell, flat membrane, flat plate, shallow shell, deep shell, axisymmetric solid, thick plate, thick shell, solid 3-D (hybrid elements). Max. no of freedoms 12,000, max. bandwidth 450 unless further limited by available auxiliary storage. IBM and CDC systems. Price $40,000. Available through developer.

MINI ELAS. Developed by I. B. Alpay & S. Utku, School of Engineering, Duke U., Durham, N.C. Anisotropic material, bifurcation with linear prestress. Inverse power iteration for eigenvalues. Elements: rod, straight beam, flat membrane, curved membrane, flat plate, shallow shell, deep shell, thick plate, thick shell. Maximum no. of freedoms and maximum bandwidth limited only by available auxiliary storage. IBM and Burroughs systems. Available from developer at handling cost.

MSC-NASTRAN. Developed by C. W. McCormick, MacNeal-Schwendler Corp., 7742 No. Figueroa St., Los Angeles, Calif. Anisotropic material. Fluid-structure interaction. Bifurcation with linear prestress. Inverse power iteration for eigenvalues. Elements: rod, straight beam, conical shell, axisymmetric shell, flat membrane, flat plate, axisymmetric solid, solid 3-D. Max. no. of freedoms and bandwidth limited only by available auxiliary storage. IBM, CDC, and UNIVAC systems. Available on lease, MacNeal-Schwendler Corp.

ASKA. Developed by ASKA-group, ISD Stuttgart, Pfaffenwaldring 27, St Stuttgart 80, Germany. Anisotropic material. Bifurcation with linear prestress. Inverse power iteration for eigenvalues. Elements: rod, straight beam, flat membrane, curved membrane, flat plate, shallow shell, deep shell, axisymmetric solid, thick plate, thick shell, solid 3-D. Maximum number of deg. of freedom and bandwidth 40,000 unless further limited by available auxiliary storage. IBM, CDC, and UNIVAC systems. Price not quoted. Available through developer.

ANSYS. Developed by Swanson Analysis Systems Inc., 870 Pine View Drive, Elizabeth, Pa. Anisotropic material. Soil-structure interaction. Nonlinear collapse analysis. Nonlinear equations solved by use of incremental method without equilibrium check. Elements: rod straight beam, conical shell, axisymmetric shell, flat membrane, flat plate, shallow shell, axisymmetric solid, thick plate, solid 3-D. Maximum no. of freedoms 20,000, maximum bandwidth 420 unless further limited by available auxiliary storage. IBM, CDC, UNIVAC, & Honeywell systems. Available on lease or against royalty through developer.

MARC. Developed by H. D. Hibbitt and P. V. Marcal, March Analysis Res. Corp., 105 Medway St., Providence, R.I. Anisotropic material. Bifurcation with linear or nonlinear prestress. Nonlinear collapse analysis. Inverse power iteration for eigenvalues. Nonlinear equations solved by incremental method with equilibrium check. Elements: rod, straight beam, curved beam, conical shell, axisymmetric shell, flat

Table 3 User's Reaction to Some Computer Programs

QUESTION \ PROGRAM	BOSOR			STAGS			ASKA			ANSYS			MSC-NASTRAN		
	No. of Answ.	Range	Ave.	No. of Answ.	Range	Ave.	No. of Answ.	Range	Ave.	No. of Answ.	Range	Ave.	No. of Answ.	Range	Ave.
User has confidence in this program	18	1-3	1.6				12		1.8						
User's manual is self contained	19	1-4	2.1												
User's manual is well organized	18	1-4	2.6												
Input is easy to prepare	19	1-5	2.5												
Output is easy to understand	19	1-5	2.2												
Program runs efficiently on our equipment	16	1-3	2.1												
There is good communication with program developer	19	1-3	1.8												
User has encountered few bugs in the program	19	1-3	2.0												
The program has good plotting capability	13	1-5	2.5												

membrane, curved membrane, flat plate, deep shell, axisymmetric solid, thick plate, thick shell, solid 3-D, crack tip. Max. no. of freedoms and bandwidth limited only by available auxiliary storage. IBM, CDC, and UNIVAC systems. Price not quoted. Available through developer, against royalty through CDC Cybernet.

Other programs for stability analysis for which information is available in [79] are CABLE3, NOSTRA, STARS, FARSS, SAMIS, ISTRANS, NEPSAP, SADADS, and NONLIN. These have not been listed here because at the present time they have not been extensively used outside of the developer's organization; they are new programs or proprietary programs.

The summary of the questionnaires from the users is generally based on too few responses. Therefore no distinction has been made between BOSOR3 and BOSOR4, between STAGSA and STAGSB, etc. This may tend to distort the results somewhat; for example, if the two questionnaires that apply to ASKADYNAM were excluded, the average of the answers would be 1.4 rather than 1.8 for the question, User has confidence in this program?

ACKNOWLEDGMENT

The preparation of this paper was supported by the Lockheed Missiles & Space Company Independent Research program.

REFERENCES

1 Koiter, W. T., "On the Stability of Elastic Equilibrium" (in Dutch), Thesis, Delft, H. J. Paris, Amsterdam, 1945. English translation, Air Force Flight Dynamics Laboratory, Technical Report AFFDL-TR-70-25, Wright-Patterson Air Force Base, Ohio 1970.

2 Hutchinson, J. W. and Koiter, W. T., "Postbuckling Theory", Applied Mech. Rev., Vol. 23, 1970, pp. 1353-1366.

3 Brush, D. O. and Almroth, B. O., Buckling of Bars, Plates, & Shells, McGraw-Hill New York, to be published.

4 Skogh, J. and Stern, P., "Postbuckling Behavior of a Section of the B-1 Aft Intermediate Fuselage," AFFDL Report, to be published.

5 Bushnell, D., "Shells," paper included in this volume.

6 Skogh, J., Stern, P., and Brogan, F. A., "Instability Analysis of Skylab Structure," presented at the National Symposium on Computerized Structural Analysis and Design, George Washington U., Washington, D. C., 1972.

7 Almroth, B. O., Brogan, F. A., and Marlowe, M. B., "Collapse Analysis for Elliptic Cones," AIAA Journal, Vol. 9, 1971, pp. 32 - 37.

8 Stephens, W., Starnes, J. H., and Almroth, B. O., "Bending of Pressurized Cylindrical Shells," to be presented at the 15th AIAA-ASME Structures, Structural Dynamics and Materials Conf., Las Vegas, Nev., 1974.

9 Almroth, B. O. and Holmes, A. M. C., "Buckling of Shells with Cutouts, Experiment and Analysis," Int. Jour. of Solids and Structures, Vol. 8, 1972, pp. 1057-1071.

10 Almroth, B. O. and Brogan, F. A., "Bifurcation Buckling as an Approximation of the Collapse Load for General Shells," AIAA Jour., Vol. 10, 1972.

11 Budiansky, B., and Roth, R. S., "Axisymmetric Dynamic Buckling of Clamped Shallow Spherical Shells," Collected Papers on Instability of Shell Structures, - 1962, NASA TN D-1510, National Aeronautics and Space Administration, Washington, D. C., pp. 597-604, 1962.

12 Belytschko, T., "Transient Analysis," paper included in this volume.

13 Armen, H., "Capabilities of Available Software for Plastic Analysis," included in this volume.

14 Nickell, R. E., "Thermal Stress and Creep" paper included in this volume.

15 Shanley, F. R., "Inelastic Column Theory," J. Aero. Sci., Vol. 14.

16 Hutchinson, J. W., "Plastic Buckling," Advances in Applied Mechanics, Vol. 14, Edited by C. S. Yih, Academic Press, 1974.

17 Hill, R., "A General Theory of Uniqueness and Stability in Elastic/Plastic Solids," J. Mech. Phys. Solids, Vol. 6, 1958, pp. 236-249.

18 Hill R., "Bifurcation and Uniqueness in Nonlinear Mechanics of Continua," Muskhelishvili Volume, Soc. Indust. Appl. Math., Philadelphia, 1961, pp. 155-164.

19 Bushnell, D., "Bifurcation Buckling of Shells of Revolution Including Large Deflections, Plasticity, and Creep," to be published in Int. Jour. of Solids and Structures.

20 Bushnell, D., and Galletly, G. D., "Comparisons of Test and Theory for Nonsymmetric Elastic-plastic Buckling of Shells of Revolution," to be published in Int. Jour. of Solids and Structures.

21 Southwell, R. V., Relaxation Methods in Engineering Sciences, Oxford Univ. Press, Oxford, 1940.

22 Shaw, F. S., Relaxation Methods, Dover Publications, New York, 1960.

23 Kantorovich, L. V. and Krylov, V. I., Approximate Methods of Higher Analysis, Wiley (Interscience), New York, 1958.

24 Forsythe, G. E. and Wasow, W. R., Finite Difference Methods for Partial Differential Equations, John Wiley and Sons, New York, 1960.

25 Collatz, L., The Numerical Treatment of Differential Equations, Springer-Verlag, Berlin, 1957.

26 Mikhlin, S. G. and Smolistkiy, K. L., Approximate Methods for Solution of Differential and Integral Equations, American Elsevier, New York, 1967.

27 MacNeal, R. H., "An Asymmetric Finite Difference Network," Quart. Appl. Math., Vol. 11, 1953, pp. 295-310.

28 Johnson, D. E., "A Dfiference-Based Variational Method for Shells," Int. J. Solids Structures, Vol. 6, 1970, pp 699-724.

29 Forsberg, K., An Evaluation of Finite Difference and Finite Element Techniques for the Analysis of General Shells, in B. Fraeijs deVeubeke (ed.) Proc. IUTAM Symposium on High-Speed Computing of Elastic Structures (Liege, Aug. 1970), Univ. de Liege, Liege, Belgium 1971.

30 Bushnell, D., Almroth, B. O. and Brogan, F. A., "Finite Difference Energy Methods for Nonlinear Shell Analysis," J. Comp. Struct., Vol. 1, 1972, pp. 361-387.

31 Bushnell, D., "Finite Difference Energy Models vs. Finite Element Models: Two Variational Approaches in One Computer Program," Proc. ONR Symposium on Numerical Methods in Engineering, Univ. of Illinois, Urbana, 1971.

32 Jensen, P. S., "A Finite Difference Technique for Arbitrary Grids," J. Comp. Struct., Vol. 2, 1972, pp. 17-29.

33 Felippa, C. A., "Finite Element and Finite Difference Energy Techniques for the Numerical Solution of Partial Differential Equations," Proc. 1973 Summer Comp. Simulation Conf., Montreal, Canada, 1973, pp. 1-14.

34 Crandall, S. H., Engineering Analysis, McGraw-Hill, New York, 1956.

35 Becker, M., The Principles and Applications of Variational Principles, MIT Press, Cambridge, Massachusetts, 1964.

36 Finlayson, B. A., The Method of Weighted Residuals and Variational Principles, Academic Press, New York, 1972.

37 Zienkiewicz, O. C. and Cheung, Y. K., The Finite Element Method in Structural and Continuum Mechanics, McGraw-Hill, New York, 1967 (expanded 2nd edition 1971).

38 Przeminiecki, J. S., Theory of Matrix Structural Analysis, McGraw-Hill, New York, 1968.

39 Meek, J. L., Matrix Structural Analysis, McGraw-Hill, New York, 1972.

40 Desai, C. S. and Abel, J. F., An Introduction to the Finite Element Method, Van Nostrand, New York, 1972.

41 Martin, H. C., and Carey, G. F., Introduction to Finite Element Analysis: Theory and Applications, McGraw-Hill, New York, 1973.

42 Aubin, J. P., Approximation of Elliptic Boundary Value Problems, Wiley (Interscience), New York, 1971.

43 Hubbard, B., (ed.), Numerical Solution of Partial Differential Equations - II, Academic Press, New York, 1971.

44 Aziz, A. K., (ed.), The Mathematical Foundation of the Finite Element Method with Application to Partial Differential Equations, Academic Press, New York, 1973.

45 Strang, G., and Fix, G., An Analysis of the Finite Element Method, Prentice-Hall, Englewood Cliffs, New Jersey, 1973.

46 Bushnell, D., "Stress, Stability, and Vibration of Complex Branched Shell of Revolution," Proc. AIAA/ASME/SAE 14th Structures, Structural Dynamics and Materials Conference, Williamsburg, Va., April 1973, to appear also in J. Comp. Structures.

47 Meyer, G. H., Initial Value Methods for Boundary Value Problems, Academic Press, New York, 1973.

48 Pestel, E. C., and Leckie, F. A., Matrix Methods in Elastomechanics, McGraw-Hill, New York, 1963.

49 Anderson, M. S., Fulton, R. E., Heard, W. L., and Walz, J. E., "Stress, Buckling and Vibration Analysis of Shells of Revolution," AFFDL-TR-71-79, AFIT Wright-Patterson AFB, Ohio, 1971, pp 1173-1249.

50 Massonet, C. E., "Numerical Use of Integral Procedures," in Stress Analysis, ed. by O. C. Zienkiewicz and G. S. Holister, John Wiley and Sons, London, 1964, pp. 198-235.

51 Rizzo, F. J., "An Integral Equation Approach to Boundary Value Problems of Classical Elastostatics," Quart. Appl. Math., Vol. 25, 1967, pp. 83-95.

52 Cruse, T. A., "Application of the Boundary Integral Equation Method to Three-Dimensional Stress Analysis," Int. Journal Comp. and Structures, Vol. 3, No. 3, 1973, pp. 509-527.

53 Hartung, R. F., and Ball, R. E., "A Comparison of Several Computer Solutions to Three Structural Shell Analysis Problems," AFFDL-TR-73-15, AFIT, Wright-Patterson AFB, Ohio, 1973.

54 Traub, J., Iterative Methods for the Solution of Equations, Prentice-Hall, Englewood Cliffs, New Jersey, 1964.

55 Rall, L. B., Computational Solution of Nonlinear Operator Equations, John Wiley and Sons, New York, 1969.

56 Ortega, J. M., and Rheinboldt, W. C., Iterative Solutions of Nonlinear Equations in Several Variables, Academic Press, New York, 1970.

57 Daniel, J. W., The Approximate Minimization of Functionals, Prentice-Hall, Englewood Cliffs, New Jersey, 1971.

58 Rabinowitz, P. (ed.), Numerical Methods for Nonlinear Algebraic Equations, Gordon and Breach, London, 1971.

59 Murray, W. (ed.), Numerical Methods for Unconstrained Optimization, Academic Press, London, 1972.

60 Lootsma, F. A. (ed.) Numerical Methods for Nonlinear Optimization, Academic Press, London, 1972.

61 Byrne, G. D., and Hall, C. A., (eds.), Numerical Solution of Systems of Nonlinear Algebraic Equations, Academic Press, New York, 1973.

62 Stricklin, J. A., Haisler, W. E., and von Riesemann, W. A., "Evaluation of Solution Procedures for Material and/or Geometrically Nonlinear Structural Analysis by the Direct Stiffness Method," AIAA/ASME 13th Structures, Structural Dynamics and Materials Conference, San Antonio, Texas, April 1972.

63 Stricklin, J. A., von Riesemann, W. A., Tillerson, J. R., and Haisler, W. E., "Static Geometric and Material Nonlinear Analysis," Proc. 2nd U. S.-Japan Seminar on Advances in Computational Methods in Structural Analysis and Design, ed. by J. T. Oden, R. W. Clough, and Y. Yamamoto, UAH Press, Univ. of Alabama, Huntsville, Alabama, 1972, pp. 301-324.

64 Tillerson, J. R., Stricklin, J. A., and Haisler, W. E., "Numerical Methods for the Solution of Nonlinear Problems in Structural Mechanics," in Numerical Solution of Nonlinear Structural Problems, ed. by R. F. Hartung, 1973 ASME Winter Annual Meeting, AMD Vol. 6, ASME, New York, pp. 67-102.

65 Oden, J. T., "Finite Element Applications in Nonlinear Structural Analysis," Proc. ASCE Symposium on Application of Finite Element Methods in Civil Engineering, Vanderbilt Univ., Nashville, Tenn., pp. 419-456.

66 Oden, J. T., Finite Elements of Nonlinear Continua, McGraw-Hill, New York, 1972.

67 Felippa, C. A., "Numerical Techniques for Geometrically Nonlinear Structural Analysis," to be presented at the International Symposium on Discrete Methods in Engineering, Milan, Sept. 1974.

68 Newman, J. B., "Analysis of Problems Involving Nonlinear Boundary Conditions and Nonlinear Material Properties," in ASME Proc. quoted in Ref. 64, pp. 51-66.

69 Bushnell, D., "Large Deflection Elastic-Plastic Creep Analysis of Axisymmetric Shells," in ASME Proc. quoted in Ref. 64, pp. 103-138.

70 Brogan, F. A., and Almroth, B. O., "Practical Methods for Elastic Collapse Analysis of Shell Structure," AIAA J. Vol. 9, No. 12, 1971, pp. 2321-2325.

71 Ball, R. E., "A Geometrically Nonlinear Analysis of Arbitrarily Loaded Shells of Revolution," NASA CR-909, Jan. 1968.

72 Wasserstrom, E., "Numerical Solutions by the Continuation Method," SIAM Review, Vol. 15, No. 1, 1973, pp. 89-119.

73 Boggs, P. T., "The Solution of Nonlinear Systems of Equations by A-stable Integration Techniques," SIAM J. Numer. Anal., Vol. 8, 1971, pp. 767-785.

74 Gear, C. W., Numerical Initial Value Problems in Ordinary Differential Equations, Prentice-Hall, Englewood Cliffs, New Jersey 1971.

75 Britvec, S. J., The Stability of Elastic Systems, Pergamon Press, New York, 1973.

76 Wilkinson, J. H., and Reinsch, C., Handbook for Automatic Computation, Vol. II - Linear Algebra, Springer-Verlag, Berlin, 1971.

77 Wilkinson, J. H., The Algebraic Eigenvalue Problem, Oxford Univ. Press (Clarendon), London, 1965.

78 Householder, A., The Theory of Matrices in Numerical Analysis, Ginn (Blaisdell), Boston, Massachuestts, 1964.

79 Bushnell, D., COMSTAIRS:, Computerized Structural Analysis Information Retrieval System", paper included in this volume.

STRUCTURAL MEMBERS AND MECHANICAL ELEMENTS

W. D. Pilkey and A. Jay
University of Virginia
Charlottesville, Virginia

ABSTRACT

Computer programs used for the analysis of structural members and mechanical elements are surveyed. Included are programs dealing with extension members, springs, torsional systems, beams, rectangular plates, grillages, circular plates, thin-walled beams, disks, and cross section properties and stresses.

INTRODUCTION

In recent years the development of general purpose structural analysis programs has accelerated rapidly. It is frequently painfully discovered that large-scale programs are very expensive to use for the analysis of a single structural member. Consequently, programs that efficiently perform specific tasks are of great value to the engineer. It is this type of program that is considered here.

Programs for structural member analyses abound. Those treated here are sufficiently well documented to be of value to the general public. Thus, the so-called "one shot" programs developed by an engineer for a very specialized problem are avoided.

The following subsections treat the areas of emphasis:

1. Beams
2. Cross section properties and stresses
3. Plates and grillages
4. Torsional systems
5. Extension members
6. Disks
7. Springs

BEAMS

Program Summaries

This section contains abstracts describing the capabilities and availability of beam analysis programs. Some of the summaries below are taken from developer supplied information.

MULTISPAN. This program performs a static, Euler-Bernoulli analysis of multiple span beams. There can be up to 10 spans having constant or piecewise variable cross sections. Interior supports are pinned; end supports can be fixed, pinned, or free. A numerical integration scheme is used to solve the equations of motion. Data can be supplied interactively or from a data file. Because the output is entirely interactive, the user can select output for the entire beam or for individual spans. The program can be recycled to specify new geometry or to apply new load cases.

Developer: TRW Systems Group, STRU-PAK
1 Space Park
Redondo Beach, Calif. 90278
Marketed: CDC Cybernet Service
Machine: CDC 6000 Series
Mode: Time sharing
Contact: R. W. Farnsworth (TRW)
(213) 535-1250
Mode available: Use basis

BEAMRESPONSE. This is a general beam analysis program. Static, stability, and dynamic analyses can be performed for beams of uniform or variable cross section. The beams may be ordinary Euler-Bernoulli beams, or the effects of axial forces, shear deformation, rotary inertia, and gyroscopic moments can be included. The beam can lie on Winkler elastic or higher-order foundations. Any number of in-span supports are acceptable, including extension springs, rotary springs, rigid supports, guides, shear releases, and moment releases.

The program calculates the deflection, slope, bending moment, and shear force for static and steady state conditions, the critical load and mode shape for stability, and the natural frequencies and mode shapes for transverse vibrations. If desired the results can be plotted.

This is a transfer matrix program taken from [1]. This reference contains the theory and many example problems.

Developers: P.Y. Chang
George C. Sharp, Inc.
New York, New York 10007

W.D. Pilkey
Dept. of Engr.Sci.& Systems
University of Virginia
Thornton Hall
Charlottesville, Virginia 22901

Marketed: Developer or COSMIC
Machine: CDC 6000 Series, UNIVAC 1108, IBM 370
Mode: Batch or time sharing
Contact: W. D. Pilkey
(804) 924-3291
Mode available: The batch program can be taken from [1] or obtained from the developer at nominal cost. Interactive pre- and postprocessors that can be coupled to the program are available through COSMIC. These processors permit on-line modifications to be made.

SPIN. This program calculates the critical speeds of rotating shafts and the natural frequencies in bending of multispan beams of arbitrary cross sections. In addition it will calculate the response due to sinusoidally applied forces. The deflections, bending moments, shear forces, and stresses created by static forces can also be found by forcing the shaft at zero speed.

SPIN uses a distributed mass method for dynamic analysis, but additional mass and rotary inertia can be lumped at points for proper modeling of gears, disks, etc. External springs to ground, both linear and rotary, can be included in the analysis to represent supports or bearings. Two segments of the beam or shaft can also be joined with linear and rotary springs to represent a flexible coupling, gear mesh, or hinge. A forced response can be found due to concentrated forces and moments, distributed loading, rotating unbalance loads, weight loading, or any combination of the above. Both forced static deflections and dynamic frequency response curves may be generated. Transfer matrices are used for the analysis.

Developer: Structural Dynamics Research Corp. (SDRC)
5729 Dragon Way
Cincinnati, Ohio 45227
Marketed: Same
Machine: Honeywell 400 & 6000, XDS Sigma 9, and CDC 6000 Series
Mode: Time sharing
Contact: Edward Carl (SDRC)
(513) 272-1100
Mode available: Use basis

<u>LINKI</u>. This is a general beam program. It performs static, stability, and free dynamic analyses of Euler-Bernoulli, Rayleigh, or Timoshenko beams. It accepts in-span supports, variable cross sections, and foundations.

LINKI has a highly sophisticated user-oriented pre- and postprocessor. Using the LINKI language a beam of any complexity can be simply input to the program. Modifications to data can be made with ease.

The program is based on the transfer matrix method.

Developer: COM/CODE Corp.
Alexandria, Virginia
Marketed: UCS, AL/COM, UCC Computer Services
Machine: CDC, UNIVAC, GE
Mode: Time sharing, remote batch, or batch
Contact: COM/CODE Corp.
Mode available: Sale or use basis

<u>FAMSUB</u>. This program determines the transverse natural frequencies and mode shapes of uniform beams subjected to any of the following boundary conditions: simply supported, cantilever, simply supported-free, simply supported-clamped, free-free, and clamped-clamped. Frequencies and mode shapes of these beams are obtained by finding the roots of the frequency function for any of the above boundary conditions, putting the values of these frequencies into the corresponding modal equation, and calculating the relative displacements

of evenly displaced points along the axis of the beam. The frequency functions and modal equations are derived from the Timoshenko theory of the transverse vibrations of uniform beams. The Timoshenko theory accounts for the effects due to bending, rotary inertia, and shear flexibility. Provisions are made for deleting certain terms from the frequency and modal equations so that other cases accounting for bending only, bending and rotary inertia, or bending and shear flexibility can also be considered.

Developer: NASA, Goddard
Goddard Space Flight Center
Greenbelt, Maryland 20771
Marketed: COSMIC, Prog. No. GSC-10429
Machine: IBM 7094
Mode: Batch
Contact: COSMIC
Mode available: For sale, $275.00

Critical Speed and Natural Frequencies. The purpose of this program is to compute the critical frequency of a rotating shaft with rotary inertia and shear effect and the lateral frequency of a beam vibrating in plane motion. The analysis is identical in both cases. The shaft or beam is divided into an arbitrary number of sections. Each section is selected so that there is a linear relation between the parameters (deflection, slope, moment, shear) at the two ends of each section. The parameters of two adjacent sections are also connected by a linear relation. This leads to a reduced transfer matrix from which the critical speed or lateral frequencies are obtained.

Developer: Aerojet-General Corp.
P.O. Box 15847
Sacramento, California
Marketed: COSMIC, Prog. No. NUC-10090
Machine: IBM 7094
Mode: Batch
Contact: COSMIC
Mode available: For sale, $275.00

Critical Speeds. The purpose of this program is to determine the critical speeds or the lateral vibrations of uniform or nonuniform continuous shafts with any number of supports. The beam is represented by a system of lumped parameters in matrix form using the transfer matrix method. The conditions at one end of the beam are related to those at the other end. The mass of each shaft is assumed to be concentrated at the middle with or without a spring support. The deflection, slope, moment, and shear of the left end are expressed in terms of those at the right end in matrix form. Based on static and dynamic equilibrium conditions similar matrices can be put together according to the structural system and are reduced one by one to a single four-by-four matrix. Equating the determinant to zero yields the desired lateral frequencies or critical speed of the system.

Developer: Aerojet-General Corp.
P.O. Box 15847
Sacramento, California
Marketed: COSMIC, Prog. No. NUC-10091
Machine: IBM 7094
Mode: Batch
Contact: COSMIC
Mode available: For sale, $275.00

STANBEAM. This program performs the static bending analysis of single-span beams. The beam can have a piecewise variable cross section, and any stable combination of support conditions can be handled. Internal forces and displacements are found using an integration procedure. Maximum shear and bending stresses are calculated by the usual VQ/IB and MC/I formulas. Data can be entered interactively or from a data file. Because the output is interactive, the user can select only those items of interest for printout. The program can be readily recycled to specify new geometry or new load cases.

Developer: TRW Systems Group, STRU-PAK
1 Space Park
Redondo Beach, California 90278
Marketed: CDC Cybernet Service
Machine: CDC 6000 Series
Mode: Time sharing
Contact: Rex W. Farnsworth (TRW)
(213) 535-1250
Mode available: Use basis

Natural Frequencies. Given information about beam loading, moment of inertia, and modulus of elasticity, the program calculates the simple beam deflection for each individual load and determines the natural beam frequency from these deflections. Beams of any type material may be analyzed by the input of proper material properties. This program will calculate the natural beam frequency for simple beams only. Input loads are limited to dead loads attached to the beam.

Developer: Dow Engineering Co.
3636 Richmond Avenue
Houston, Texas 77027
Marketed: Same
Machine IBM 360
Mode: Batch
Contact: Bill Frazure
(713) 623-3011
Mode available: For sale, $350.00

General Analysis. This is a versatile program able to statically analyze continuous beams, beams on elastic foundations, and simple frames. It uses a general two-dimensional beam analysis procedure which requires relatively simple input. A plot option will draw the shear diagram, the moment diagram, and the deflection diagram. The solution method is by finite differences. The fourth difference, deflection, is estimated, compared to the input conditions, and adjusted in recursive fashion until assumptions match input.

Developer: Dow Engineering Co.
3636 Richmond Avenue
Houston, Texas 77027
Marketed: Same
Machine: IBM 360
Mode: Batch
Contact: Bill Frazure
(713) 623-3011
Mode available: For sale, $750.00

DANAXX0. The program calculates the frequencies and eigenvectors of a beam with lumped masses. Rotary inertia may be included. The program may also be used to calculate the response due to static loads. The beam is subdivided into segments, within each segment, the bending stiffness, EI, is constant. The program allows the user to include effects of both linear and rotational external elastic springs at one or more joints. The program uses a stiffness matrix method of analysis.

Developer: Southwest Research Institute
8500 Culebra Road
P.O. Drawer 28510
San Antonio, Texas 78284
Marketed: Same
Machine: Unknown
Mode: Batch
Contact: T. R. Jackson
Manager, Computer Laboratory (SRI)
Mode available: For sale, nominal charge for handling

DANAXX4. This program calculates the time history of the response of a beam to applied force pulses and applied torque pulses. The beam is represented by a lumped parameter system which is essentially equivalent to a finite difference approximation of the governing equations. In addition to solving the general case of coupled bending and torsion, the program can be used for uncoupled bending and torsion, for torsion alone, or for bending alone. The program allows any combination of hinged, clamped, free, or guided flexural boundary conditions. The applied forces and torques are functions of time. The program provides for inelastic behavior by assuming that both the moment curvature and the torque angle of twist relations are of the bilinear type with hysteretic recovery. Shear and rotary inertia are neglected,and no damping is included. The response is determined by a step-by-step integration of the equations of motion using the linear acceleration method.

Developer: Southwest Research Institute
8500 Culebra Road
P.O. Drawer 28510
San Antonio, Texas 78284
Marketed: Same
Machine: Unknown
Mode: Batch
Contact: T. R. Jackson
Manager, Computer Laboratory
Mode available: For sale, nominal charge to cover handling

BMPLAT. This is a computer program which utilizes a finite element method to determine the transverse linear deflections of a vibrating beam or plate. The solution for the beam and plate are separate formulations. Both solutions permit arbitrary variations in bending stiffness, mass density, and dynamic loading. The static equations have been included in the development so that the initial deflections can be conveniently established. In the beam the difference equations are solved by a recursive procedure, while for the plate the same procedure is combined with an alternating direction technique to obtain an iterative solution.

Developer: SCI-TEK Incorporated
1707 Gilpin Avenue
Wilmington, Delaware 19806
Marketed: Same
Machine: UNIVAC 1108
Mode: Time sharing
Contact: S. F. Sarsfield
(302) 658-2431
Mode available: Use basis

LEBMCL. This program can be used to analyze static linearly elastic beam and beam column problems. It employs a finite mechanical analog to allow very general loading and elastic restraint conditions to be considered. A finite element solution is used.

Developer: SCI-TEK Incorporated
1707 Gilpin Avenue
Wilmington, Delaware 19806
Marketed: Same
Machine: UNIVAC 1108
Mode: Time sharing
Contact: S. F. Sarsfield
(302) 265-2431
Mode available: Use basis

BMCOL. This is a computer program that analyzes a beam column subjected to movable static loads. A beam column can be analyzed for any pattern of transverse loads that move across the member. These load patterns may be any diverse system of loading such as a highway truck, a series of trucks, or possibly a train on a railroad structure. In addition, the effect of fixed loads can be included.

Developer: SCI-TEK Incorporated
1707 Gilpin Avenue
Wilmington, Delaware 19806
Marketed: Same
Machine: UNIVAC 1108
Mode: Time sharing
Contact: S. F. Sarsfield
(302) 658-2431
Mode available: Use basis

NUBWAM (Non-Uniform Beam With Attached Masses). This program computes upper and lower bounds to bending frequencies and estimates mode shapes of nonuniform beams with elastically attached masses. The Rayleigh-Ritz procedure is used to obtain the upper bounds and the mode shapes; the lower bounds are obtained using the method of intermediate problems. Upper and lower bounds that bracket the true frequencies are found, whereas other methods give estimates of the frequencies but do not give error bounds.

Developer: APPLIED PHYSICS LAB
Johns Hopkins University
8621 Georgia Avenue
Silver Spring, Maryland 20910
Marketed: Same
Machine: IBM 360/91
Mode: Batch
Contacts: N. Rubinstein or J. T. Stadter
Mode available: For sale, only expenses incurred in reproduction of program are charged.

Rotating, Twisted Beam. This program determines the natural frequencies and normal modes of a lumped parameter model of a rotating, twisted beam, with nonuniform mass and elastic properties. The end of the beam near the center of rotation may have one of four types of boundary conditions which are common to helicopter rotor systems; the outboard end has zero forces and moments, i.e., free boundary conditions. Six types of motion coupling may be modeled: fully coupled torsional-flatwise-edgewise motion; partially coupled torsional-flatwise motion or flatwise-edgewise motion; and uncoupled torsional motion, flatwise motion, or edgewise motion. Three frequency search methods have been implemented including an automated search technique which allows the program to find up to fifteen lowest natural frequencies without the necessity for input estimates of these frequencies by the user.

Developer: Rochester Applied Science Assoc., Inc.
Rochester, New York
Marketed: COSMIC, Prog. No. LAR-11461
Machine: CDC 6000 Series
Mode: Batch
Contact: COSMIC
Mode available: For sale, $350.00

BEAMCOL. This program performs the bending analysis of single-span beam columns. The beam column can have a piecewise variable cross section and any stable combination of support conditions. The solution is obtained using integration and finite difference techniques. The analysis yields internal forces, displacements, and stresses. Input data can be supplied interactively or from a data file. Because the output is interactive, the user can select only those items of interest for printout. The program can be readily recycled to enter new geometry or new load cases.

Developer: TRW Systems Group, STRU-PAK
1 Space Park
Redondo Beach, California 90278
Marketed: CDC Cybernet Service
Machine: CDC 6000 Series
Mode: Time sharing
Contact: Rex W. Farnsworth (TRW)
(213) 535-1250
Mode available: Use basis

ELASTCOL. This program determines the smallest buckling load for elastic end-loaded columns. The column can have a piecewise variable moment of inertia, and any stable combination of support conditions can be handled. A finite difference technique is used to determine the buckling load and the associated mode shape. Data can be supplied interactively or from a data file.

Developer: TRW Systems Group, STRU-PAK
1 Space Park
Redondo Beach, California 90278
Marketed: CDC Cybernet Service
Machine: CDC 6000 Series
Mode: Time sharing
Contact: R. W. Farnsworth (TRW)
(213) 535-1250
Mode available: Use basis

Simple Column Analysis. This is a program for stability analysis of structural columns. The centroidal axis of the column is assumed to be a straight line before loading. Variable cross sections are allowed. Output data consist of the magnitude and position of the column deflections, column moments, and maximum stress on each side of the column. In addition, the input data and the midpoint deflection at the next-to-final iteration are also printed out.

Developer: Rocketdyne
6633 Canoga Avenue
Canoga Park, California 91304
Marketed: COSMIC, Prog. No. MFS-2230
Machine: IBM 360
Mode: Batch
Contact: COSMIC
Mode available: For sale, $25.00

Column Analysis. The program analyzes a column of variable cross section subjected to externally applied end movements and tranverse end loads that lie in two perpendicular planes. The method provides for the computation of the deflection normal to the column axis, the internal moments along the column, and the resulting bending and axial stresses. This program determines deflections, moments, stresses, and interactions for a specific column or performs a stability analysis of columns.

Developer: Rocketdyne
6633 Canoga Avenue
Canoga Park, California 91304
Marketed: COSMIC, Prog. No. MFS-1633
Machine: IBM 7094
Mode: Batch
Contact: COSMIC
Mode available: For sale, $25.00

Tabulation of Program Capabilities

The capabilities of several of the more general beam programs are tabulated in Table 1. The four programs included correspond to the first four programs abstracted in the previous subsection. The conditions of availability are in these summaries. A comparison of capabilities of four beam codes leads to some interesting conclusions. The MULTISPAN code marketed by TRW offers no dynamic analysis, and LINK, SPIN, and BEAMRESPONSE all offer this capability. BEAMRESPONSE and LINK are very similar and offer the most varied analysis capabilities. All four codes analyze only straight beams. Only MULTISPAN offers a stress analysis postprocessor that is directly coupled to the program.

Usability

To demonstrate the relative ease of the use of the beam programs listed in Table 1, the input for each of the programs has been set up for the problem given in Fig. 1. These input samples are given in Figs. 2 through 5. The example problem was chosen to show the relative ease of use of the programs. This problem does not indicate the difficulty of input for a complex problem, but it may aid a potential user in deciding on a specific code. The input for MULTISPAN, LINK, and BEAMRESPONSE may be fully interactive. All codes operate in the time-sharing mode, but only SPIN requires the user to enter his data from a data file. If a user has a large amount of data to enter for a complex problem it may be cheaper to input data in this mode, but for a very simple problem the savings may be questionable. The LINK, MULTISPAN, and BEAMRESPONSE codes allow the user several different input options.

With MULTISPAN, BEAMRESPONSE, and SPIN the user must define his problem in terms of questions asked of him. The most

Table 1 Capabilities of General Beam Programs

BEAM PROPERTY \ COMPUTER CODE	MULTISPAN	BEAMRESPONSE	SPIN	LINKI
ANALYSIS TYPE				
Static	x	x	x	x
Stability		x		x
Free vibration		x	x	x
Steady state dynamic		x	x	x
Transient				
Wave propagation				
Random vibration				
SUPPORT CONDITIONS				
Pinned-pinned	x	x	x	x
Pinned-fixed	x	x	x	x
Fixed-pinned	x	x	x	x
Fixed-fixed	x	x	x	x
Fixed-free	x	x	x	x
Free-fixed	x	x	x	x
Free-free		x		x
INSPAN SUPPORTS				
None				
Rigid (continuous beam)	x	x		x
Elastic extension springs		x	x	x
Elastic rotary springs		x	x	x
Plastic springs				
Dashpots				
Guides		x		x
Shear releases		x		x
Moment releases		x		x
Branches				
FOUNDATIONS				
Winkler (elastic)		x		x
Shear (rotary)		x		x
Viscoelastic				
Plastic				
MATERIAL				
Linear elastic	x	x	x	x
Nonlinear elastic				
Rigid		x		x
Plastic				
Viscoelastic				
Viscous damping				
Creep				
Isotropic		x		x
Anisotropic				
GEOMETRY				
Straight	x	x	x	x
Curved				
Semi-infinite length				
Infinite length				
THEORY FOR EQUATIONS OF MOTION				
Engineering theory		x		x
Euler-Bernoulli beam	x	x		x
Shear beam		x		x

Table 1 (cont.)

BEAM PROPERTY \ COMPUTER CODE	MULTISPAN	BEAMRESPONSE	SPIN	LINKI
Rayleigh (rotary inertia) beam		x		x
Timoshenko (shear & rotary inertia) beam		x	x	x
Elasticity		x		x
Plasticity				
DESIGN				
Design capability				
STRESS ANALYSIS				
Post processors for stress-analysis	x			
Automated yield criteria				
ANALYSIS METHOD				
Transfer matrices		x	x	x
Numerical integration	x			
COMPUTATIONAL STABILITY				
Unstable for long members		x		x
Unstable for higher frequences		x		x
Algorithm for overcoming numerical problems				
LIMITATIONS				
Max. number of degrees of freedom		NONE	NONE	NONE
Max. number of loadings	25	NONE	NONE	NONE
CROSS SECTION				
Any shape	x	x	x	x
Restricted to particular geometry				
Thin-walled		x		x
Shear center and c.g. coincide		x		x
SPATIAL VARIATION OF PROPERTIES				
No variation				
Cross section can be piecewise constant	x	x	x	x
No restrictions				
Continuously variable cross section				
Foundation constants can vary		x		x
Restrictions on number of variations	25	NONE	NONE	NONE
LOADING				
Concentrated forces	x	x	x	x
Concentrated moments	x	x	x	x
Uniform forces	x	x	x	x
Force gradients	x	x	x	x
Uniform moments		x		x
Moment gradients		x		x
Thermal loading		x		x
MODELING FOR DYNAMIC RESPONSE				
Lumped mass		x	x	x
Continuously distributed mass		x	x	x
Lumped and/or continuous mass		x	x	x
Concentrated rotary inertia		x		x
Distributed rotary inertia		x		x
Gyroscopic effects for rotating shaft		x	x	x
NATURAL FREQUENCIES METHOD				
Iterative search		x	x	x
STABILITY				
Ordinary buckling		x		x
Lateral instability				

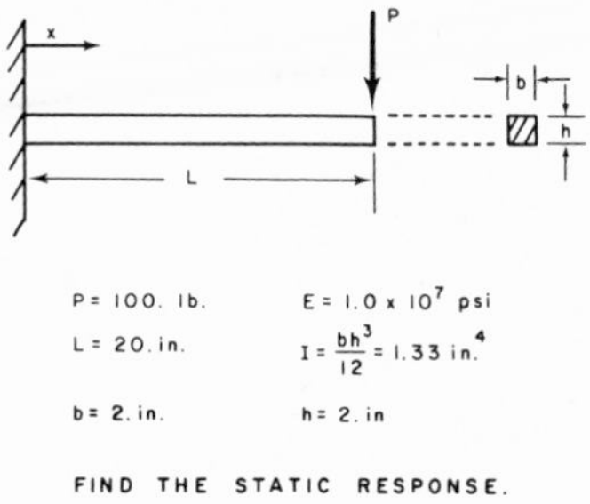

x

P

b

h

L

P = 100. lb. E = 1.0×10^7 psi

L = 20. in. $I = \frac{bh^3}{12} = 1.33\ in.^4$

b = 2. in. h = 2. in

FIND THE STATIC RESPONSE.

Fig. 1 Example problem for comparing input of beam programs

```
**** MULTISPAN ****

PERFØRMS THE BENDING ANALYSIS ØF MULTI-SPAN BEAMS

- I N P U T -
* * * * * * * * * * * * * * * * *

DØ YØU WANT TØ ENTER DATA FRØM A DATA FILE (YES ØR NØ)  ?NØ

   - SPAN IDENTIFICATIØN -

ENTER THE NUMBER ØF SPANS =   ?1

ENTER THE X-STATIØNS (IN) ØF THE 2 SUPPORTS (BEGIN AT 0.0) -

   =   ?0.0,20.

SUPPØRT CØNDITIØN CØDE NUMBERS:

   1- PINNED
   2- FIXED
   3- FREE

ENTER THE SUPPØRT CØNDITIØN CØDE NUMBERS FØR
THE LEFT AND RIGHT ENDS ØF THE MULTI-SPAN BEAM

   (JLEFT,JRIGHT) =   ?2,3

   - GEØMETRY -

ENTER THE NUMBER ØF GEØMETRIC SEGMENTS =   ?1

   ENTER A CRØSS SECTIØN DATA SET AT THE END
   ØF EACH GEØMETRIC SEGMENT.

   1 (XEND,E,I,Z,B,Q) =   ?20.,1.E7,1.33,2.66,.5,.15

   - APPLIED LØADS -

ENTER THE NUMBER ØF CØNCENTRATED LØADS =   ?1

   ENTER THE X-STATIØN (IN) AND SIGNED MAGNITUDE (LBS)
   FØR EACH CØNCENTRATED LØAD.

   1 (XLØAD,P) =   ?20,-100

ENTER THE NUMBER ØF DISTRIBUTED LØADS =   ?0

ENTER THE NUMBER ØF APPLIED MØMENTS =   ?0
```

Fig. 2 Simulated input of MULTISPAN for problem of Fig. 1

Line No.	IDENTIFICATION
100	CANTILEVER BEAM — STATIC RESPONSE

	No. of Spans	Type of Analysis	Whirl Constant	Density	E	G	Type of Cross-section
110	1	0	0	.1	1.E7	3.75E6	1.2

	FORE END BOUNDARY CONDITIONS	
	Type	Value
120	1	0
130	2	0

BOUNDARY CONDITION TYPES

1 = Displacement
2 = Twist (Slope)
3 = Moment
4 = Shear

TYPE OF ANALYSIS

-1 = Forced Response with weight loading.
0 = Forced Response without weight loading.
1 = Natural Frequency Search.

	AFT END BOUNDARY CONDITIONS	
	Type	Value
140	3	0
150	4	100

WHIRL CONSTANT - H

H = 1 For Whirl in same direction as beam rotation.
H = -1 For Bending only.
If Greater than 10 H = Rotational Speed.

CROSS-SECTION TYPES

0 = Circular

For General (non-circular) cross-sections, enter value of shear stress ratio.

	FREQUENCY RANGE (RPM)				
	Initial Frequency	Frequency Interval	Final Frequency	Tolerance	No. of Iterations
160	0	0	0		

	OUTPUT SELECTION			
	Type	Station Nos. for Selective Output		
170	1			

TYPES OF OUTPUT

1 = Displacement, Forces, & Stresses
2 = Displacement & Forces
3 = Displacement Only

SPAN PROPERTIES				
Line No.	Length	O.D. / Area	I.D. / Inertia	Equivalent Density
180	20	4	1.33	.1

190	0	NO. OF STATIONS WITH LUMPED WEIGHT AND/OR INERTIA
200	0	NO. OF STATIONS ELASTICALLY RESTRAINED TO GROUND
210	0	NO. OF STATIONS WITH ELASTICALLY CONNECTED SPANS

APPLIED LOADS— NOT REQUIRED FOR NATURAL FREQUENCY SEARCH

220	1	NO. OF STATIONS WITH AN APPLIED LOAD(S)

	Station	Force	Moment
230	2	100	0

240	0	NO. OF SPANS WITH A DISTRIBUTED LOAD
250	0	NO. OF STATIONS WITH A ROTATING UNBALANCE FORCE

Fig. 3 Simulated input of SPIN for problem of Fig. 1

```
DØ YØU WANT A PRØMPTING (ENTER 1) ØR A NØN-PRØMPTING (ENTER 2) INPUT,
ØR DØ YØU WISH TØ ENTER DATA FRØM YØUR DATA FILE (ENTER 3)  ?1

INDICATE THE TYPE ØF ANALYSIS DESIRED BY ENTERING
        1 FØR STATIC RESPØNSE
        2 FØR STEADY STATE
        3 FØR STABILITY
ØR      4 FØR FREE DYNAMICS   ?1

DØ YØU HAVE A SIMPLE BEAM (NØ CHANGES IN CRØSS SECTION ØR MØDULUS ØF
ELASTICITY, NØ INSPAN SUPPØRTS, NØ ELASTIC FØUNDATIØN, FØRCE AND MØMENT
LØADING ØNLY; YES ØR NØ)    ? YES

LENGTH ØF BEAM =    ? 20

        *********** END AND INSPAN SUPPØRTS **********
USING 1=FIXED, 2=PINNED, 3=FREE, 4=GUIDED, SPECIFY THE END SUPPØRTS:
[NL,NR], WHERE NL=TYPE ØF LEFT END SUPPØRT
               NR=TYPE ØF RIGHT END SUPPØRT
[NL,NR]   ? 1,3

ENTER THE MØDULUS ØF ELASTICITY AND MØMENT ØF INERTIA FØR THE BEAM

 ? 1.E7,1.33

ENTER THE DATA SET [NCF,NCMØ], WHERE
        NCF =NØ. ØF CØNCENTRATED FØRCES
        NCMØ=NØ. ØF CØNCENTRATED MØMENTS

[NCF,NCMØ]   ? 1,0

ENTER, CØNSECUTIVELY, THE DATA SET [PØSITIØN,MAGNITUDE] FØR ALL
CØNC. FØRCES (PØSITIVE DØWN):

  ? 20,100

ENTER THE DATA SET [NDF,NRAF], WHERE
        NDF =NØ. ØF UNIFORM FØRCES
        NRAF=NØ. ØF RAMPED FØRCES

[NDF,NRAF]  ? 0,0

NØ. ØF AXIAL LØADS (INCLUDING REACTIØN AT LEFT END; 0 IF NØNE)  ? 0

DØ YØU WANT A DEBUG PRINTØUT (YES ØR NØ)  ? NØ

NØ. ØF PØINTS BETWEEN ØCCURRENCES FØR PRINTING RESULTS (MAX.200)  ? 3

DØ YØU WISH TØ HAVE THE RESULTS PLØTTED (YES ØR NØ)  ? YES

DØ YØU WANT A SUMMARY ØF YØUR INPUT DATA (YES ØR NØ)  ? YES
```

Fig. 4 Simulated input of interactive BEAMRESPONSE for problem of Fig. 1. This input corresponds to that generated using the preprocessor (COSMIC) for BEAMRESPONSE. Versions with no preprocessor are also available.

```
STATICS,BEAM
LENGTH=20,LEFT END=FIXED,RIGHT END=FREE
E@0=1.0E+07,I@0=1.33
FØRCE@20=100
ØUTPUT=ALL,@INCREMENTS=1
GØ
```

Fig. 5 Simulated input of LINKI for problem of Fig. 1

important feature of the LINK code is its simplicity of user input. The user need only describe his problem in straightforward engineering-oriented language. If LINK is used in the time-sharing mode, the data input for any problem, large or small, may be easily accomplished in just a few lines. The only drawback for the LINK code appears to be its cost for usage. For a very small problem, such as this cantilever beam, the input and run costs may be relatively high. But for a very complex problem the ease of data input could be valuable.

One problem that could occur with the input for SPIN is the inability to model easily an intermediate support condition. The user's manual for SPIN indicates that one must use a spring with a sufficiently high enough (k) value to yield a zero displacement at the support. Unfortunately, the user is not told how stiff is stiff enough. Severe numerical problems can occur if this value is not properly chosen.

CROSS SECTION PROPERTIES AND STRESS ANALYSIS

Program Summaries

This section contains abstracts describing programs that can compute such cross-sectional properties as torsional constants. Cross-sectional stress analysis programs are also discussed. Some of the summaries are taken from developer supplied information.

SASA. This is a dual-purpose program that calculates both cross-sectional properties and stresses. Such properties as moments of inertia, torsional constant, and warping constant can be found for a cross section of any shape. For given internal torque, shear forces, and bending moments, SASA can calculate the normal and shear stress distributions over the cross section.

SASA is based on the finite element method as proposed by Herrmann in [2,3]. In fact, SASA is an improved version of a program originally written by Herrmann [3].

Developer: Structural Dynamics Research Corp. (SDRC)
5729 Dragon Way
Cincinnati, Ohio 45227
Marketed: Same
Machines: Honeywell 400 & 6000, XDS Sigma 9, CDC 6000 Series
Mode: Time Sharing
Contact: Edward Carl (SDRC)
(513) 272-1100
Mode available: Use basis

BEAMSTRESS. This program determines the section properties and stresses in an arbitrary cross section of a bar. For section properties it is necessary to input the geometry of the cross section perimeter. Properties computed include area, centroid, moments of inertia, product of inertia, shear center, shear deformation coefficients, torsional constant, warping constant, radii of gyration, and principal moments of inertia. Modulus-weighted properties are calculated for composite sections.

For given internal shear forces, bending moments, axial force, and axial torque, BEAMSTRESS will compute the normal and shear stress distributions. The effect of warping on the stresses can be taken into account. Composite sections are acceptable.

The finite element method of [2,3] is employed. The original version of the program was written by Herrmann [3]. Modifications were made by A. Jay and C. Thasanatorn of the University of Virginia.

Developers: P.Y. Chang, George C. Sharp, Inc., New York, New York 10007
W.D. Pilkey, Dept. of Engr. Sci. & Systems, University of Virginia, Charlottesville, Virginia 22901

Marketed: Same
Machine: CDC 6000 Series
Modes: Batch or time sharing
Contact: W. D. Pilkey (804) 924-3291
Mode available: The batch version can be taken from [1] or obtained from the developer at nominal cost. Interactive pre- and postprocessors that can be coupled to the program are available from COSMIC. These processors permit on-line modifications to be made.

GENSECT1. This program uses an integration technique to compute the section properties of arbitrary planar cross sections. The user must input the x,y coordinates for points which define the outline of the section to be analyzed. Voids in a section are handled similarly. At user option, the program computes section properties with respect to a new coordinate system. Input data can be supplied interactively or from a data file.

Developer: TRW Systems Group, STRU-PAK
1 Space Park
Redondo Beach, California 90278
Marketed: CDC Cybernet Service
Machine: CDC 6000 Series
Mode: Time sharing
Contact: R. W. Farnsworth (TRW)
(213) 535-1250
Mode available: Use basis

GENSECT2. This program computes the section properties of arbitrary planar cross sections that can readily be subdivided into rectangles, right triangles, and circles. Voids can be composed of these same geometric figures. The user must divide the cross section into rectangles, right triangles, and circles and specify their dimensions and centroids. At user option, the program computes section properties with respect to a new coordinate systems. Input data can be supplied interactively or from a data file.

Developer: TRW Systems Group, STRU-PAK
1 Space Park
Redondo Beach, California 90278
Marketed: CDC Cybernet Service

Machine: CDC 6000 Series
Mode: Time sharing
Contact: R. W. Farnsworth (TRW)
(213) 535-1250
Mode available: Use basis

STANSECT. The section properties of 18 types of standard cross sections (i.e. I's, Z's, T's, and others) can be calculated. At user option, the program computes section properties at intermediate stations along a linearly varying member having one of the 18 different cross sections. The torsional constant is computed as part of the output.

Developer: TRW Systems Group, STRU-PAK
1 Space Park
Redondo Beach, California 90278
Marketed: CDC Cybernet Service
Machine: CDC 6000 Series
Mode: Time sharing
Contact: R. W. Farnsworth (TRW)
(213) 535-1250
Mode available: Use basis

PREBEST. This program calculates the beam section properties such as areas, centroid locations, principal axes, and moments of inertia. There are two completely different methods of defining the cross section to the computer. The user may divide the cross section into a number of standard geometric shapes, or one may approximate the cross section's peripheral contour by a series of connected straight line segments.

Developer: Structural Dynamics Research Corp. (SDRC)
5729 Dragon Way
Cincinnati, Ohio 45227
Marketed: Same
Machines: Honeywell 400 & 6000, XDS Sigma 9, CDC 6000 Series
Mode: Time sharing
Contact: Edward Carl (SDRC)
(513) 272-1100
Mode available: Use basis

AREA$$. This program enables the user to calculate cross-sectional properties for any area bounded by straight lines and circular arc segments. The program uses a problem oriented language designed specifically for geometric property calculations.

Developer: General Electric
Info. Services Bus. Division
7735 Old Georgetown Road
Bethesda, Maryland 20014
Marketed: Same
Machines: GE-235-MARKI, GE-635-MARKII

Mode: Time sharing
Contact: Bowman Irani (GE)
(301) 654-9360
Mode available: Use basis

SECTON1. This program calculates from basic cross section data, the center of gravity, moment of inertia, total area, extreme fiber distance, section modulus, and radius of gyration of any structural section. The only restriction is that it must be possible to resolve the structural section into rectangles.

Developer: United Computing Systems (UCS)
Manger, Lib. Applications & Support
3130 Broadway
Kansas City, Missouri 64111
Marketed: Same
Machine: CDC 6000 Series, CDC CYBER
Mode: Time sharing
Contact: Ron Kogan (UCS) (816) 753-4500
Mode available: Use basis

SHERLOC. This is a program for determining the location of the shear center of 14 different types of standard geometric cross sections. The shear center location is computed with respect to the centroid.

Developer: Structural Dynamics Research Corp. (SDRC)
5729 Dragon Way
Cincinnati, Ohio 45227
Marketed: Same
Machine: Honeywell 400 & 6000, XDS Sigma 9, CDC 6000 Series
Mode: Time sharing
Contact: Edward Carl (SDRC)
(513) 272-1100
Mode available: Use basis

SPOTS. There are two SPOTS programs. They calculate the basic section properties of any thin-walled open cross section or any single-celled closed cross section whose centerline can be described as a single continous straight or curved line. If requested, the SPOTS programs will also plot the cross section as defined by the user. SPOTS1 will plot the cross sections on an incremental plotter, and SPOTS2 will plot the cross sections on the terminal. This is the only difference between SPOTS1 and SPOTS2. In addition, the location of the centroid and shear center will be noted on the plot and SPOTS1 will also plot the principal axes.

Developer: Structural Dynamics Research Corp. (SDRC)
5729 Dragon Way
Cincinnati, Ohio 45227
Marketed: Same

Machine: Honeywell 400 & 6000, XDS Sigma 9, CDC 6000 Series
Contact: Edward Carl (SDRC)
(513) 272-1100
Mode available: Use basis

ETC. This program evaluates the torsional constant of 19 different types of standard geometric cross sections.

Developer: Structural Dynamics Research Corp. (SDRC)
5729 Dragon Way
Cincinnati, Ohio 45227
Marketed: Same
Machine: Honeywell 400 & 6000, XDS Sigma 9, CDC 6000 Series
Mode: Time sharing
Contact: Edward Carl (SDRC)
(513) 272-1100
Mode available: Use basis

Torsion Analysis of Open Sections. This program is for the torsional analysis of thin-walled open sections for both unrestrained and restrained torsion. Torsional shear stress, angle of twist, and warping deformations are determined for unrestrained torsion. Torsional shear stress, warping shear stress, warping normal stress, angle of twist, and the first, second, and third derivatives of angle of twist are determined for restrained torsion.

Developer: Teledyne-Brown Engineering
Research Park
Huntsville, Alabama 35807
Marketed: COSMIC, Prog. No. MFS-20648
Machine: IBM 7094
Mode: Batch
Contact: COSMIC
Mode available: For sale

TORCELL. This is a time shared computer program which computes the shear flows and stress distributions for multicellular cross sections loaded in torsion. The angle of twist and the torsional constant for the cross section are also computed. All input data for TORCELL are read from a data file. Input consists of the applied torque, the modulus of rigidity, the geometry of each cell, and the wall thicknesses.

Developer: TRW Systems Group, STRU-PAK
1 Space Park
Redondo Beach, California 90278
Marketed: CDC Cybernet Service
Machine: CDC 6000 Series
Mode: Time sharing
Contact: R. W. Farnsworth (213) 535-1250
Mode available: Use basis

Stress Analysis of Torsional Members. This program computes the angle of twist per unit length, the components of shear stress at each point in a solid, elastic, prismatic torsional member of arbitrary cross-sectional shape. The program uses a numerical solution along with finite difference techniques, since it is impractical to attempt a general analytical solution for an arbitrary, simply connected region. The maximum grid possible using the program is 33 by 65 with a running time of approximately 3 minutes.

Developer: Rocketdyne
6633 Canoga Avenue
Canoga Park, California 91304
Marketed: COSMIC, Prog. No. MFS-1487
Machine: IBM 7094
Mode: Batch
Contact: COSMIC
Mode available: For sale

Section Properties. This program determines the area, centroid location,principal axis orientation, external moment of inertia, and section moduli of any area defined by peripheral coordinates. Areas are restricted to those bounded by convex curves. A complete set of data including a description card must be submitted for each case. All points must be referred to in the same arbitrary coordinate axis system. Multiple cases may be run. To ensure a high degree of accuracy, double precision is used.

Developer: North American Aviation
Canoga Park,California 91304
Marketed: COSMIC, Prog. No. MFS-2224
Machine: IBM 360
Mode: Batch
Contact: COSMIC
Mode available: For sale, $275.00

Geometric Properties. Computation of area properties (area centroid, moments of inertia, and principal axes) of any section is provided utilizing input parameters that specify location, orientation, and dimensions of each standard geometrical shape comprising the section. Three types of various geometrical shapes are considered, using separate subroutines for different shapes to facilitate ease in future program expansion. Input data consist of shape, identity, size, location of point in common references axis system,and direction numbers. An indicator is also given for one- or two-card formulae,specifying subgroup and negative or positive area properties. Output consists of area, centroid, and inertia delta values about axes parallel to a common reference axis. Also, composites for specified subgroups; tabulation of total area, centroid area, moment and product of inertia location of principal axes system; and maximum principal axis with value of moment of inertia about this axis are included.

Developer: Rocketdyne
6633 Canoga Avenue
Canoga Park, California 91304
Marketed: COSMIC, Prog. No. MFS-2229
Machine: IBM 360
Mode: Batch
Contact: COSMIC
Mode available: For sale, $275.00

CROS. This program calculates the cross-sectional characteristics: area of the cross section, position of the center of gravity, position of the principal axis of inertia, and shear coefficients of the section.

Developer: Research Center of the Belgian
Metal Working Industry
Structural Analysis Group
CAMPUS ARENBERG
Celestigneniaan 3006
B-3030 Hevelee
BELGIUM
Marketed: Same
Machine: Unknown
Mode: Time Sharing
Contact: P. Van Loon, Engineer
Mode available: Use basis

Section Properties. A computer program to calculate cross-sectional properties for sections of arbitrary, simple connected shapes. It calculates section area, coordinates of centroid, moments of inertia, angle to principal axis, principal moments of inertia, coordinates of shear center, and the torsional constant by arbitrarily specifying coordinates of a number of points on a section boundary.

Developer: G. R. Cowper
National Aeronautical Establishment
National Research Council of Canada
Ottawa, Canada KIA OR6
Marketed: Same
Machines: IBM 360/67, 360/50
Mode: Batch
Contact: G. R. Cowper
Mode available: For sale, $100.00

Roark's Formulas for Stress and Strain. All of the formulas from Professor Roark's Formulas for Stress and Strain, McGraw Hill, 1965, reside in this program. To solve a problem, the engineer finds the applicable formula in the book and enters data interactively.

Developer: United Computing Systems (UCS)
Manager, Lib. Applications & Support
3130 Broadway
Kansas City, Missouri 64111

Marketed: Same
Machine: CDC 6000 Series
Mode: Time sharing
Contact: Ron G. Kogan (UCS)(816) 753-4500
Mode available: Use basis

Tabulation of Program Capabilities

Table 2 compares the capabilities offered by eight of the most general section properties and stress analysis computer codes. The first two, SASA and BEAMSTRESS, provide section properties and stress analyses, while the next six programs calculate the section properties only.

Usability

To illustrate the relative ease of the use of the beam programs listed in Table 2, the input for each program has been set up for the problem given in Fig. 6. These input samples are given in Figs. 7 through 14. Each of the section analysis computer codes GENSECT1, GENSECT2, STANSECT, PREBEST, AREA$$, and SECTON1 determines virtually the same section properties with almost the same type of input format.

SASA and BEAMSTRESS appear to be very similar. The user must model the cross sections of interest with arbitrary quadrilateral finite elements. This can be time consuming and cumbersome since neither program possesses mesh generation capabilities. Both programs would be easier to use if a mesh generator or a library of standard shapes were made available. The drawback to SASA is that the data input must come from a data file. This mode of data input is cumbersome. The time-shared version of BEAMSTRESS with prompted input is easier to use but may be more costly for simple problems or standard sections.

PLATES AND GRILLAGES

This section contains abstracts describing the capabilities and availabilities of plate and grillage analysis programs. Some of the summaries below are taken from developer-supplied information.

Circular Plates

CIRCPLAT. This program performs a static elastic analysis of symmetrically loaded circular plates. Allowable loadings include combinations of radially varying pressure loads and concentrated line loads. Capabilities exist to handle linear variation in thickness and arbitrary boundary conditions at the inner and outer

Table 2 Some Programs for Cross Section Properties and Stress Analyses

PROPERTIES \ PROGRAMS	SASA	BEAMSTRESS	GENSECT1	GENSECT2	STANSECT	PREBEST	AREA$$	SECTON1
SECTION DEFINITION								
Cross sections of standard shapes					x	x		
Built-up from standard shapes					x	x		
Built-up from square elements	x	x		x	x	x	x	x
Built-up from quadrilateral elements	x	x		x	x	x	x	
Built-up from circular or semi-circular elements				x	x	x	x	
Built-up from arbitrary elements of user's choice			x		x	x		
SECTION PROPERTIES: NON-SHEAR RELATED								
Section area	x	x	x	x	x	x	x	x
Coordinates of centroid	x	x	x	x	x	x	x	x
Moments of inertia	x	x	x	x	x	x	x	x
Angle to principal axis	x	x	x	x	x	x	x	
Radii of gyration	x	x	x	x	x		x	x
Polar moment of inertia		x				x		
Principal moments of inertia	x	x	x	x				
Moments of inertia W/R/T arbitrary axes			x	x				
Modulus-weighted section properties		x					x	
SECTION PROPERTIES: SHEAR RELATED								
Coordinates of shear center	x	x						
Torsional constant	x	x			x			
Warping constant	x	x			x			
Shear deformation coefficients		x						
STRESS ANALYSIS								
Biaxial bending stresses								
Simple bending stress	x	x						
Composite sections								
Curved bar								
Straight bar	x	x						
Constant cross section along bar								
Variable cross section along bar								
Shear stresses due to torsional loads	x	x						
Shear stresses due to shear forces in 1 direction	x	x						
Shear stresses due to shear forces in 2 directions								
Contribution of constrained warping to shear stresses		x						
Contribution of constrained warping to normal stresses		x						
Principal stresses	x	x						
Von Mises combined stress	x							
Tresca combined stress								
ANALYSIS METHOD								
Engineers' theory of bending stresses	x	x						
Strength of materials theory for shear stresses								
Finite element theory for shear stresses	x	x						
Finite difference theory for shear stresses								
Elasticity theory for shear stresses								
SPECIAL PROGRAM FEATURES								
Mesh generation								
Bandwidth minimization								
Ability to determing section properties W/R/T a new coordinate axes			x					
Shear stress plots								
Bending stress plots								
Principal stress plot								

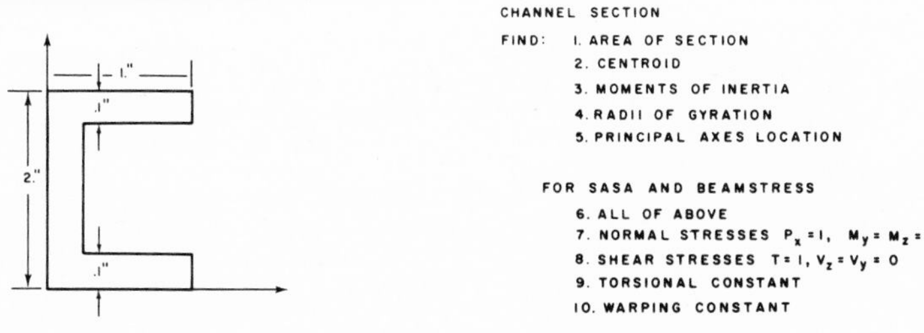

1."
.1"
2."
.1"
CHANNEL SECTION
FIND: 1. AREA OF SECTION
2. CENTROID
3. MOMENTS OF INERTIA
4. RADII OF GYRATION
5. PRINCIPAL AXES LOCATION
FOR SASA AND BEAMSTRESS
6. ALL OF ABOVE
7. NORMAL STRESSES $P_x = 1$, $M_y = M_z = 0$
8. SHEAR STRESSES $T = 1$, $V_z = V_y = 0$
9. TORSIONAL CONSTANT
10. WARPING CONSTANT

Fig. 6 Example problem for input study of programs in Table 2

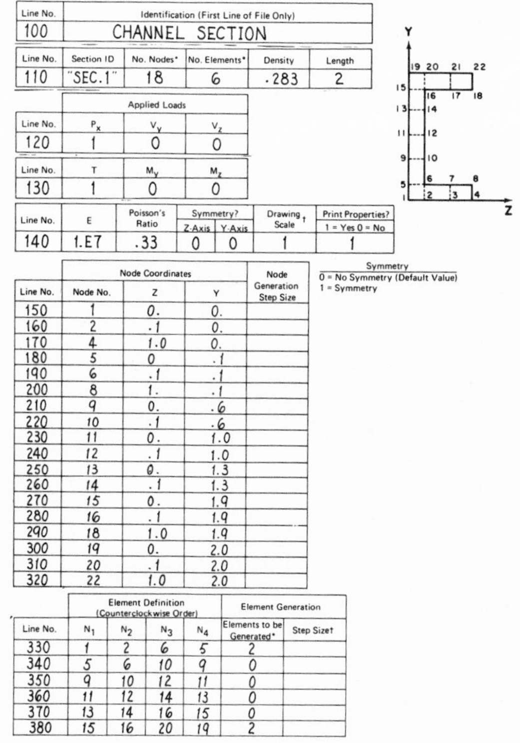

Line No.	Identification (First Line of File Only)
100	CHANNEL SECTION

Line No.	Section ID	No. Nodes*	No. Elements*	Density	Length
110	"SEC.1"	18	6	.283	2

	Applied Loads		
Line No.	P_x	V_y	V_z
120	1	0	0
Line No.	T	M_y	M_z
130	1	0	0

Line No.	E	Poisson's Ratio	Symmetry? Z-Axis	Symmetry? Y-Axis	Drawing Scale†	Print Properties? 1 = Yes 0 = No
140	1.E7	.33	0	0	1	1

Symmetry
0 = No Symmetry (Default Value)
1 = Symmetry

		Node Coordinates		
Line No.	Node No.	Z	Y	Node Generation Step Size
150	1	0.	0.	
160	2	.1	0.	
170	4	1.0	0.	
180	5	0	.1	
190	6	.1	.1	
200	8	1.	.1	
210	9	0.	.6	
220	10	.1	.6	
230	11	0.	1.0	
240	12	.1	1.0	
250	13	0.	1.3	
260	14	.1	1.3	
270	15	0.	1.9	
280	16	.1	1.9	
290	18	1.0	1.9	
300	19	0.	2.0	
310	20	.1	2.0	
320	22	1.0	2.0	

	Element Definition (Counterclockwise Order)				Element Generation	
Line No.	N_1	N_2	N_3	N_4	Elements to be Generated*	Step Size†
330	1	2	6	5	2	
340	5	6	10	9	0	
350	9	10	12	11	0	
360	11	12	14	13	0	
370	13	14	16	15	0	
380	15	16	20	19	2	

Fig. 7 Simulated input of SASA for problem of Fig. 6

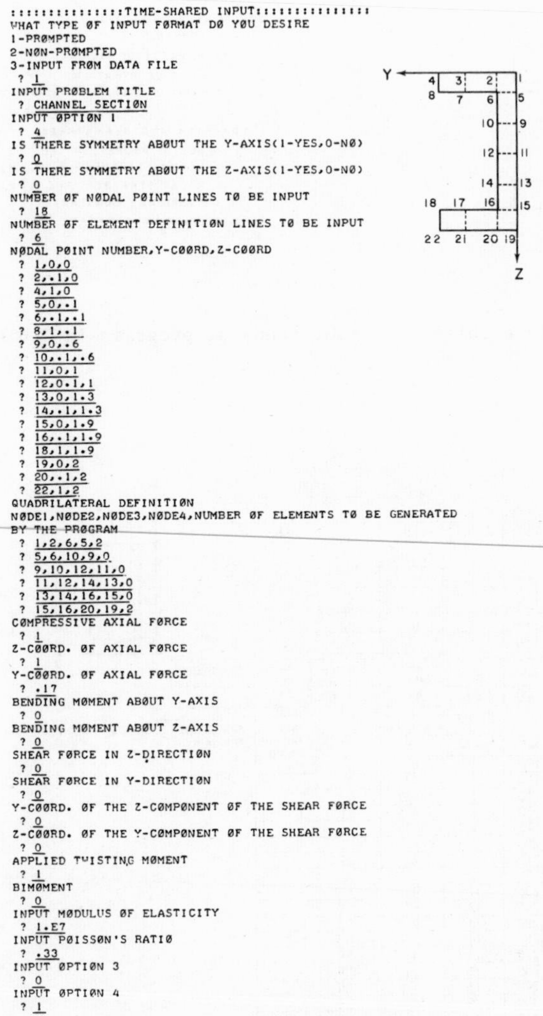

```
::::::::::::::TIME-SHARED INPUT::::::::::::::
WHAT TYPE ØF INPUT FØRMAT DØ YØU DESIRE
1-PRØMPTED
2-NØN-PRØMPTED
3-INPUT FRØM DATA FILE
 ? 1
INPUT PRØBLEM TITLE
 ? CHANNEL SECTIØN
INPUT ØPTIØN 1
 ? 4
IS THERE SYMMETRY ABØUT THE Y-AXIS(1-YES,0-NØ)
 ? 0
IS THERE SYMMETRY ABØUT THE Z-AXIS(1-YES,0-NØ)
 ? 0
NUMBER ØF NØDAL PØINT LINES TØ BE INPUT
 ? 18
NUMBER ØF ELEMENT DEFINITIØN LINES TØ BE INPUT
 ? 6
NØDAL PØINT NUMBER,Y-CØØRD,Z-CØØRD
 ? 1,0,0
 ? 2,.1,0
 ? 4,1,0
 ? 5,0,.1
 ? 6,.1,.1
 ? 8,1,.1
 ? 9,0,.6
 ? 10,.1,.6
 ? 11,0,1
 ? 12,0.1,1
 ? 13,0,1.3
 ? 14,.1,1.3
 ? 15,0,1.9
 ? 16,.1,1.9
 ? 18,1,1.9
 ? 19,0,2
 ? 20,.1,2
 ? 22,1,2
QUADRILATERAL DEFINITIØN
NØDE1,NØDE2,NØDE3,NØDE4,NUMBER ØF ELEMENTS TØ BE GENERATED
BY THE PRØGRAM
 ? 1,2,6,5,2
 ? 5,6,10,9,0
 ? 9,10,12,11,0
 ? 11,12,14,13,0
 ? 13,14,16,15,0
 ? 15,16,20,19,2
CØMPRESSIVE AXIAL FØRCE
 ? 1
Z-CØØRD. ØF AXIAL FØRCE
 ? 1
Y-CØØRD. ØF AXIAL FØRCE
 ? .17
BENDING MØMENT ABØUT Y-AXIS
 ? 0
BENDING MØMENT ABØUT Z-AXIS
 ? 0
SHEAR FØRCE IN Z-DIRECTIØN
 ? 0
SHEAR FØRCE IN Y-DIRECTIØN
 ? 0
Y-CØØRD. ØF THE Z-CØMPØNENT ØF THE SHEAR FØRCE
 ? 0
Z-CØØRD. ØF THE Y-CØMPØNENT ØF THE SHEAR FØRCE
 ? 0
APPLIED TWISTING MØMENT
 ? 1
BIMØMENT
 ? 0
INPUT MØDULUS ØF ELASTICITY
 ? 1.E7
INPUT PØISSØN'S RATIØ
 ? .33
INPUT ØPTIØN 3
 ? 0
INPUT ØPTIØN 4
 ? 1
```


Fig. 8 Simulated input of BEAMSTRESS for problem of Fig. 6. This input corresponds to that generated using the preprocessor (COSMIC) for BEAMSTRESS. Versions with no preprocessor are also available.

```
          ***** GENSECT1 *****
PLANAR SECTIØN PRØPERTIES ØF ARBITRARY CRØSS-SECTIØNS
           USING AN INTEGRATIØN TECHNIQUE
```

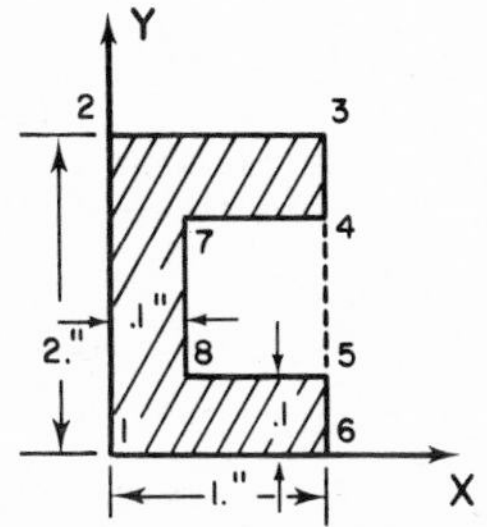


```
DØ YØU WANT TØ USE A PREVIØUSLY GENERATED
DATA FILE, (YES ØR NØ)  ?NØ


          ***** SECTIØN 1 *****

NUMBER ØF VØID CØMPØNENTS IN SECTIØN 1 =  ?1

NUMBER ØF VERTICES IN THE ENVELØPE ØF SECTIØN 1 =   ?6

       -ENVELØPE ØF SECTIØN 1-
CØØRDINATES, VERTEX 1 (ENTER X,Y) =  ?0,0
CØØRDINATES, VERTEX 2 (ENTER X,Y) =  ?0,2
CØØRDINATES, VERTEX 3 (ENTER X,Y) =  ?1,2
CØØRDINATES, VERTEX 4 (ENTER X,Y) =  ?1,1.9
CØØRDINATES, VERTEX 5 (ENTER X,Y) =  ?1,.1
CØØRDINATES, VERTEX 6 (ENTER X,Y) =  ?1,0

ANY CØRRECTIØNS TØ THE VERTICES ØF THE ENVELØPE, (YES ØR NØ)  ?NØ

NUMBER ØF VERTICES IN VØID 1 ØF SECTIØN 1 =  ?4

       -VØID 1 ØF SECTIØN 1-
CØØRDINATES, VERTEX 1 (ENTER X,Y) =  ?.1,.1
CØØRDINATES, VERTEX 2 (ENTER X,Y) =  ?.1,1.9
CØØRDINATES, VERTEX 3 (ENTER X,Y) =  ?1,1.9
CØØRDINATES, VERTEX 4 (ENTER X,Y) =  ?1,.1

ANY CØRRECTIØNS TØ THE VERTICES ØF VØID 1, (YES ØR NØ)   ?NØ

DØ YØU WANT TØ MULTIPLY THE CØØRDINATES BY
SØME NUMBER, (YES ØR NØ)  ?NØ

DØ YØU WANT THE SECTIØN PRØPERTIES CØMPUTED WITH
RESPECT TØ A NEW CØØRDINATE SYSTEM, (YES ØR NØ)  ?NØ

DØ YØU WANT THE INPUT DATA FØR SECTIØN 1
PRINTED ØUT, (YES ØR NØ)  ?YES
```

Fig. 9 Simulated input of GENSECT1 for problem of Fig. 6

```
          ***** GENSECT2 *****
   PLANAR SECTIØN PRØPERTIES ØF ARBITRARY SECTIØNS


DØ YØU WANT TØ USE A PREVIØUSLY GENERATED
DATA FILE, (YES ØR NØ)  ?NØ


          ***** SECTIØN  1 *****

NUMBER ØF ELEMENTS IN SECTIØN 1 =   ?3

     -ELEMENT DATA FØR SECTIØN 1-
ELEMENT 1 (ENTER C,X,Y,B,H) =   ?1,.5,.05,1,.1

ELEMENT 2 (ENTER C,X,Y,B,H) =   ?1,.05,1,.1,1.8

ELEMENT 3 (ENTER C,X,Y,B,H) =   ?1,.5,1.95,1,.1

DØ YØU WANT SECTIØN PRØPERTIES CØMPUTED WITH
RESPECT TØ A NEW CØØRDINATE SYSTEM, (YES ØR NØ)  ?NØ

DØ YØU WANT THE INPUT DATA FØR SECTIØN 1
PRINTED ØUT, (YES ØR NØ)  ?YES
```

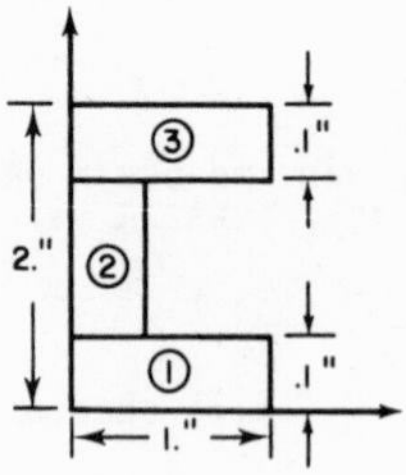

Geometric Configuration	Code Number, C	X, Y, B, H Referred to 1st Quadrant of Reference Coordinate System
Rectangle	1 Rectangular Element -1 Void Rectangular Element	
Triangle	2 Triangular Element -2 Void Triangular Element	
Circle (Note 3)	3 Circular Element -3 Void Circular Element	

Note 1. B and H must always be positive. X and Y can be positive, negative, or zero depending upon the location of the element with respect to the X-Y axes.

Note 2. X, Y, B, and H must be assigned the same dimensional units.

Note 3. Both B and H must be input for a circle.

Fig. 10 Simulated input of GENSECT2 for problem of Fig. 6

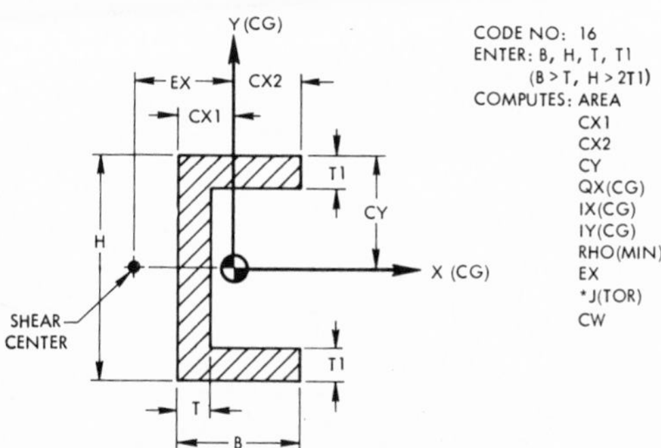


```
          ***** STANSECT *****
   PLANAR SECTIØN PRØPERTIES FØR STANDARD SECTIØNS


-PRØGRAM ØPTIØNS-

   ØPTIØN 1 .... FØR SØLVING STANDARD CRØSS SECTIØNS.
   ØPTIØN 2 .... FØR SØLVING LINEARLY VARYING MEMBERS,
                 HAVING A STANDARD CRØSS SECTIØN.

ENTER PRØGRAM ØPTIØN CØNTRØL NUMBER, (1 ØR 2) =  ?1

ENTER CRØSS SECTIØN CØDE NUMBER =  ?16

            *** SECTIØN 1 ***
         -CHANNEL CRØSS SECTIØN-

SECTIØN PARAMETERS, (ENTER B,H,T,T1) = ?1,2,.1,.1

DØ YØU WANT SECTIØN PRØPERTIES CØMPUTED WITH
RESPECT TØ A NEW CØØRDINATE SYSTEM, (YES ØR NØ)  ?NØ
```

Fig. 11 Simulated input of STANSECT for problem of Fig. 6

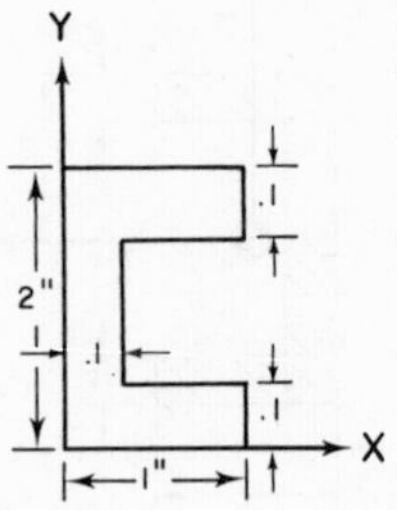


```
  ENTER DATA FILE NAME
?NØNE

DØ YØU WANT THE CRØSS - SECTIØNAL PRØPERTIES FØR ØNE DIRECTIØN ØNLY?
   (YES ØR NØ)

?YES

INPUT: JØB IDENTIFICATIØN

?CØNVERSATIØNAL INPUT EXAMPLE

INPUT: "SECTIØN ID",NØ. ØF FIGURES, DENSITY, LENGTH & REPEAT CØDE
?"SEC B1",1,.283,2,0

INPUT: FIGURE TYPE, Z-CØØRD, Y-CØØRD,AND 4 PARAMETERS
                  ØR
Z-CØØRD & Y-CØØRD (TYPE B INPUT)
?8,.27,.04,0,.38

INPUT: "SECTIØN ID",NØ. ØF FIGURES, DENSITY,LENGTH & REPEAT CØDE
     (ENTER 0 IF  FINISHED)

?0
```

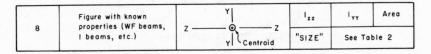

8	Figure with known properties (WF beams, I beams, etc.)	Z —— Y —— Z, Y, Centroid	I_{zz}	I_{YY}	Area
			"SIZE"	See Table 2	

Fig. 12 Simulated input of PREBEST for problem of Fig. 6

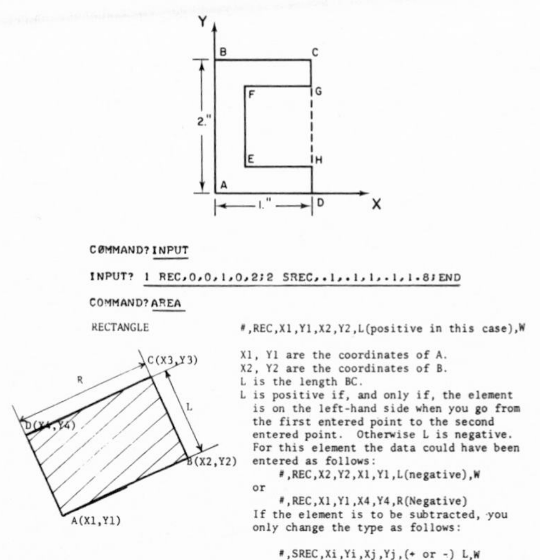

Fig. 13 Simulated input of AREA$$ for problem of Fig. 6

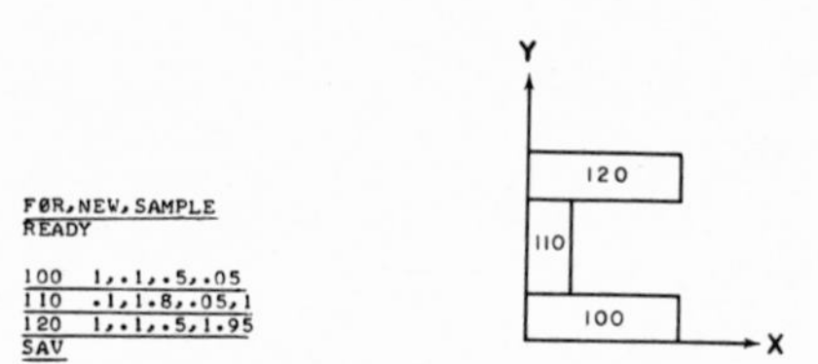

To simplify preparing the input data file, the following steps should be followed:

1. Draw a coordinate axis below and to the left of the section.
2. Divide the section into rectangles and assign a line number to each sub-section.
2. This number will identify the area.
3. Determine the following:
   A. The base dimension of each sub-area.
   B. The height dimension of each sub-area.
   C. The X and Y dimension to the center of gravity of the sub-area as shown.

For further explanation, see the sample problem, above.

Note: A maximum of 50 areas may be used, with area numbers designated as 9999 or less.

Fig. 14 Simulated input of SECTON1 for problem of Fig. 6

edges. Data can be entered interactively or from a data file. Output options allow the user to select only those items of interest for printout. A finite difference method is used for the analysis.

Developer: TRW Systems Group, STRU-PAK
1 Space Park
Redondo Beach, California 90278
Marketed: CDC Cybernet Service
Machine: CDC 6000 Series
Mode: Time sharing
Contact: R. W. Farnsworth (TRW)
(213) 535-1250
Mode available: Use basis

CIRCULARPLATE. This is a general circular plate analysis program. Static, steady state, stability, and free dynamic responses can be calculated. The loading can be symmetric or unsymmetric. Plates of variable thickness are acceptable. The plate can be supported by discrete extension or rotary springs, rigid supports, or continuous elastic foundations. It can rotate about a central axis. The plate can be continuous at the center or can have a central hole or rigid insert. Thermal loads can be applied.

The deflection, slope, radial moment, shear force, and twisting moment are calculated for static and steady state loading. The critical load and modeshape are found for stability. Natural frequencies and modeshapes are computed for symmetric and unsymmetric freed dynamic modes.

This is a transfer matrix program taken from [1]. This reference contains the theory and many example problems.

Developers: P. Y. Chang
George C. Sharp, Inc.
New York, New York
10007

W. D. Pilkey
Dept. of Engr. Sci. & Systems
University of Virginia
Charlottesville, Virginia
22901

Marketed: Developer or COSMIC
Machine: CDC 6000 Series, UNIVAC 1108, IBM 370
Mode: Batch or time sharing
Contact: W. D. Pilkey
(804) 924-3291
Mode available: The batch program can be taken from [1] or obtained from the developer at nominal cost. Interactive pre- and postprocessor that can be coupled to the program are available through COSMIC.

DEPROSS-3. This program calculates the dynamic responses of circular plates and spherical shells that are subjected to an initial axisymmetric impulsive loading. Assumed are uniform plate thickness and elastic material. Strain hardening and strain-rate effects may be included. Finite differences are used for the analysis.

Developer: Hans A. Balmer
Dept. of Aeronautics and Astronautics
M.I.T.
Cambridge, Massachusetts 02139
Marketed: Southwest Research Inst.
8500 Culebra Road
P.O. Drawer 28510
San Antonio, Texas 78284
Machine: IBM 7094
Mode: Batch
Contact: T. R. Jackson
Manager, Computer Laboratory (SRI)
Mode available: For sale, nominal charge for handling

Rectangular Plates

Stress Analysis of Rectangular Plates. This program computes stresses, displacement, and stress resultants in solid rectangular plates which are subjected to arbitrary nominal loads, thermal gradients, and edge loads. The program also provides for arbitrarily varying elastic foundation and edge supports, as well as thickness. A finite difference method of solution is utilized and is limited to those loads, stress, and displacement components intrinsic to plate bending theory. In-plane tractions or displacements cannot be calculated.

Developer: Rocketdyne
6633 Canoga Avenue
Canoga Park, California 91304
Marketed: COSMIC, Prog. No. MFS-18688
Machine: IBM 360
Mode: Batch
Contact: COSMIC
Mode available: For sale

RECTANGULARPLATE. This is a general analysis of a rectangular plate with two sides simply supported. Static, steady state, stability, and free dynamic response can be calculated. The boundary conditions at the edges other than those that must be simply supported can be simply supported, free, guided (no rotation), or clamped. The thickness, modulus of elastic foundation, Young's modulus, Poisson's ratio, thermal expansion coefficient, density, and in-plane loads can be varied in different panels. A quite arbitrary system of concentrated forces and moments and distributed forces and moments can be placed on the plate. The plate can be supported by discrete extension or rotary springs, rigid supports, or continuous elastic foundations. Thermal loads can be applied.

The plate can be orthotropic. Stiffened plates with large stiffeners in both the x and y directions can be analyzed by RECTANGULARPLATE if the spacing of the stiffeners is small and the torsional rigidity of the stiffeners is important. It is better to use a grillage program if the spacing of the stiffeners is large and their effects of torsional rigidity are small. In general, the effect of torsional rigidity is negligible for open section members.

The deflection, slope, moments, and shear forces are calculated for static and steady state loading. The critical load and modeshapes are found for stability. Natural frequencies and modeshapes are computed for free dynamic response.

This is a transfer matrix program taken from [1]. This reference contains the theory and many example problems.

Developers: P.Y. Chang, George C. Sharp, Inc., New York, New York 10007

W. D. Pilkey, Dept. of Engr.Sci.& Systems, University of Virginia, Charlottesville, Virginia 22901

Marketed: Developer or COSMIC
Machines: CDC 6000 Series, UNIVAC 1108, IBM 370
Modes: Batch or time sharing
Contact: W. D. Pilkey
(804) 924-3291
Mode available: The batch program can be taken from [1] or obtained from the developer at nominal cost. Interactive pre- and postprocessors that can be coupled to the program are available through COSMIC.

Finite Difference Plate Program. This program computes the dynamic response of the plate based on the Reuss-Mises plastic flow rule and the isotropic or kinematic strain-hardening assumption.

Developer: School of Engineering and Applied Science
The George Washington University
Washington, D. C. 20006
Marketed: Same
Machine: IBM 360/40
Mode: Batch
Contact: Ted Toridis
(202) 676-6530
Mode available: For sale

Nonlinear Static Analysis. This program is for static nonlinear analysis by an iterative application of the stiffness method according to the finite element approach for plates, idealized in a grid of up to 260 rectangular finite elements subjected to 10 loading conditions. Consideration of different thicknesses and different material properties of finite elements as well as consideration of elastic supports and of equal node deformation is possible. The program uses a fixed input format.

Developer: Electronic Calculus, Inc. (ECI)
468 Park Avenue
New York, New York 10019

Marketed: Same, Prog. No. 803
Machine: UNIVAC 1108
Mode: Batch
Contact: G. Kostro
(212) 532-1545
Mode available; Use basis or lease

Linear Static Analysis. This program is for static linear analysis by the stiffness method according to the finite element approach for plates, idealized in a grid of up to 600 rectangular finite elements subjected to up to 15 loading conditions and 15 loading combinations. Consideration of different thicknesses and different material properties of finite elements and of equal node deformations is possible. A fixed input format is employed.

Developer: Electronic Calculus, Inc. (ECI)
468 Park Avenue
New York, New York 10019
Marketed: Same, Prog. No. 801
Machine: UNIVAC 1108
Mode: Batch
Contact: G. Kostro
(212) 532-1545
Mode available: For sale or lease

Large Plates. This program is for the static linear analysis by the stiffness method according to the finite element approach of plates idealized into a grid of up to 1,600 rectangular finite elements, subjected to up to 50 loading conditions and 30 loading combinations. Consideration of different thicknesses and different material properties of finite elements, as well as consideration of elastic supports, equal node deformations, and an automatic elimination of the solution round-off errors, is possible. A fixed input format is used.

Devèloper: Electronic Calculus, Inc. (ECI)
468 Park Avenue South
New York, New York 10019
Marketed: Same, Prog. No. 831
Machine: UNIVAC 1108
Mode: Batch
Contact: G. Kostro
(212) 532-1545
Mode available: For sale or lease

Buckling of Anisotropic Plates. This program computes the buckling loads of simply supported rectangular anisotropic and layered composite material plates. The plates can be subjected to combinations of in-plane loads (normal and shear). The solution method is based on the Galerkin method for the plate displacement solution and the power method of obtaining the smallest buckling load. The user need only input the loading conditions, bending (flexural), rigidities (stiffness), and dimensions of the plates. Computed are the buckling load and printed out are the following: composite system input data, load condition, buckling load, terms in the series expansion, relative convergence error, and a normalized array of the buckling shape. The program will analyze several types of composites and several load conditions for each plate, both compressive or tensile membrane loads, and bending-stretching coupling via the concept of reduced bending rigidities.

Developer: NASA, Lewis Research Center
Cleveland, Ohio
Marketed: COSMIC, Prog. No. LEW-11961
Machine: IBM 7091
Mode: Batch
Contact: COSMIC
Mode available: For sale $300.00

Series Solution of Irregular Plates. This program is designed to calculate the stresses and deflections of laterally loaded thin flat plates of irregular geometry, for which analytical solutions are not available. The program determines the best coefficients in the least-squares sense) for a truncated series solution of the deflection of a flat plate under a lateral load. At each boundary point two boundary conditions are selected. The allowed boundary conditions are specified deflection, specified slope, specified moment, specified shear force, and a free straight edge. From these conditions the usual cases of fixed edges, simply supported edges, free straight edges, and lines of symmetry can be obtained. After the best coefficients for the series solution are obtained, the deflections shapes and stresses can be calculated for any point on the plate.

Developer: Westinghouse Electric Corp.
Pittsburgh, Pennsylvania 15221
Marketed: COSMIC, Prog. No. NUC-10170
Machine: CDC 6600
Mode: Batch
Contact: COSMIC
Mode available: For sale

SASP. This program performs the stress analysis of reinforced rectangular plates under a lateral loading. The plate differential equations are solved by Fourier series. The program is limited to simply supported rectangular plates with one-directional stiffeners parallel to one side.

Developer: M. Higuchi
Ship Research Section
Technical Research Center
Nippon Kokan K.K.
Keihin Works
Kawasaki, Japan

Marketed: Same
Machine: IBM 370/155
Mode: Batch
Mode available: Contact developer for information

VASP. This program performs a vibration analysis of a reinforced rectangular plate subjected to in-plane loads. Direct energy minimization by nonlinear programming is used to obtain the solution in terms of Fourier or power series. Only plates with one-directional stiffeners can be analyzed. Various loadings and boundary conditions may be considered.

Developer: M. Higuchi
Ship Research Section
Technical Research Center
Nippon Kokan K.K.
Keihin Works
Kawasaki, Japan
Marketed: Same
Machine: IBM 370/155
Mode: Batch
Contact: M. Higuchi
Mode available: Contact developer for information

LAMPS. This is a computer program which presents a finite element solution of laminated orthotropic plates and shallow shells under static loads. The program is limited to linearly elastic structures. A general procedure for the finite element displacement method for a homogenous shell element is extended to a laminated shallow shell element, where each lamina may have different thickness and material properties.

Developer: Dept. of the Navy
NSRDC
Carderock Labs.
Bethesda, Maryland 20034
Marketed: Same
Machines: CDC 6000 Series, IBM
Mode: Batch
Contact: Ben Whang
Mode available: Available to qualified individuals. There is no warranty, expressed or implied, as to the correctness of this program or its documentation. The government accepts no liability for the results of the program. While a reasonable effort will be made to answer questions about the program, and a defective copy will be replaced on request, the government does not undertake to provide maintenance in the future.

Grillages

DANAXX1. This program analyzes a plane gridwork composed of beams which are rigidly connected at joints. It calculates the deflections due to static loads applied at the joints and normal to the plane of the gridwork and the frequencies and eigenvector of the gridwork with lumped inertias at the joints. The stiffness and flexibility matrices are generated by the program. Each beam member is assumed to have a uniform bending stiffness, EI, and a uniform torsional stiffness, GJ.

Developer: Southwest Research Institute
8500 Culebra Road, P.O. Drawer 28510
San Antonio, Texas 78284
Marketed: Same
Machine: Unknown
Mode: Batch
Contact: T. R. Jackson
Mode available: For sale, nominal charge to cover handling

GRIDWORK. This program is for the study of a set of mutually perpendicular members, rigidly connected at the intersection. The members in one direction are called girders, and in the other direction stiffeners. Either set of beams can be designated as stiffeners. The girders must be identical in size, end conditions, and spacing. The stiffeners can vary in size. The stiffeners are simply supported while the girders may have fixed, simply supported, or free ends. The loading may be uniform, hydrostatic, or concentrated forces placed anywhere on the girdwork. The girders may vary in moment of inertia along their lengths. They may rest on rigid or flexible supports or foundations.

The program calculates the deflection, slope, bending moment, and shear force for static conditons. The documentation provides information sufficient to calculate buckling loads and natural frequencies for transverse vibrations.

This program is based on a technique that reduces the grillage to a sequence of beams. The program is taken from [1], where the theory is developed and many example problems are given.

Developers: P. Y. Chang
George C. Sharp, Inc.
New York, New York
10007

W. D. Pilkey
Dept.of Engr.Sci.& Systems
University of Virginia
Charlottesville, Virginia
22901

Marketed: Developer or COSMIC
Machine: CDC 6000 Series, UNIVAC 1108, IBM 370
Mode: Batch or time sharing
Contact: W. D. Pilkey
(804) 924-3291
Mode available: The batch program can be taken from [1] or obtained from the developer at nominal cost. Interactive pre- and postprocessors that can be coupled to the program are available through COSMIC.

GRID. This program computes the deflections, slopes, bending moments, and shear forces of grillages. A simplified finite-element method of solution is used. The capability is similar to the previous program except the girders can be of variable spacing and size.

Developer: COM/CODE Corp.
Alexandria, Virginia
Marketed: AL/COM Service of Applied Logic Corp.
900 State Road
Princeton, New Jersey 08540
Machine: IBM
Modes: Time sharing or batch
Contact: W. E. Dillmeier (AL/COM)
(609) 924-7800
Mode available: Use basis

Static Analysis. This program is for the static linear analysis by the stiffness method of rigid plane grids with straight members of constant or variable cross section, having up to 400 joints or 600 members, subjected to up to 24 loading conditions, and 24 loading combinations. Member bending and shear deformability are taken into account. Consideration of elastic hinges, elastic supports, equal joint deformations, and members of different materials is possible. Fixed and free input formats are accepted.

Developer: Electronic Calculus, Inc. (ECI)
468 Park Avenue
New York, New York 10019
Marketed: Same, Prog. No. 401
Machine: UNIVAC 1108
Modes: Batch or time sharing
Contact: G. Kostro
(212) 532-1545
Mode available: Use basis or lease

DANAXX6. This program calculates the time history of the response of an orthotropic rectangular plate due to applied triangular force pulses. The analysis is based on the finite difference representation of the classical theory of flexure of thin plates. The plate is divided into a network of rectangular grids of uniform spacing. To account for the boundary conditions, the network is extended one additional mesh point beyond each of the actual plate boundaries. The mass is lumped at the nodal point and the force pulses are applied there also. Any combination of simply supported, clamped, free, and guided boundary conditions is allowed. Inelastic behavior is approximated by assuming that each of the moment-curvature relations are of the bilinear type with hysteretic recovery. The response is determined by a stepwise integration of the equations of motion of the lumped masses at the nodal points using the linear acceleration method.

Developer: Southwest Research Institute
8500 Culebra Road
P.O. Drawer 28510
San Antonio, Texas 78284
Marketed: Same
Machine: Unknown
Mode: Batch
Contact: T. R. Jackson,
Manager, Computer Laboratory
Mode available: For sale, nominal cost to cover handling

Curved Grillages. This program is for static linear analysis by the stiffness method of rigid plane grids with straight members of constant and variable cross section and curved members with constant cross sections having up to 400 joints or 600 members, subjected to up to 24 loading conditions and 24 loading combinations. Member bending and shear deformability is taken into account. Consideration of elastic hinges, elastic supports, equal joint deformations,and members of different materials is possible. Fixed and free input formats are accepted.

Developer: Electronic Calculus, Inc. (ECI)
468 Park Avenue
New York, New York 10019
Marketed: Same, Prog. No. 401C
Machine: UNIVAC 1108
Modes: Time sharing or batch
Contact: G. Kostro
(212) 532-1544
Mode available: Use basis or lease

Large Curved Grillages. This program is for the static linear analysis by the stiffness method of rigid plane grids with straight members of constant or variable cross sections and curved members with constant cross sections having up to 4,000 members, subjected to up to 50 loading conditions and 50 loading combinations. Member bending and shear deformability is taken into account. Consideration of elastic hinges, finite size joint dimensions, elastic supports, equal joint deformations, members of different materials and an automatic elimination of the solution round-off errors are possible. The input is in a fixed format.

Developer: Electronic Calculus, Inc. (ECI)
468 Park Avenue
New York, New York 10019
Marketed: Same, Prog. No. 431C
Machine: UNIVAC 1108
Mode: Batch
Contact: G. Kostro
(212) 532-1545
Mode available: Use basis or lease

TORSIONAL SYSTEMS

Abstracts are presented describing capabilities for finding the internal torque and twisting moment in bars and systems undergoing torsion.

Natural Frequencies. This program calculates the torsional vibration, natural frequencies, and corresponding mode shapes of a physical system under free vibration that can be idealized to N lumped mass polar moments of inertia connected by weightless shafts possessing torsional stiffness. Both free-free and free-fixed shafts can be accommodated. The Holzer method is used to determine the desired frequencies.

Developer: Rocketdyne
6633 Canoga Avenue
Canoga Park, California 91304
Marketed: COSMIC, Prog. No. MFS-2485
Machine: IBM 7094
Mode: Batch
Contact: COSMIC
Mode available: For sale

TWIST. This program is for the twisting motion of torsional systems. It calculates the angle of twist and twisting moment in a bar for static and steady state loading. Natural frequencies and modeshapes are computed for free dynamic response. The mass can be modeled as lumped or continuous, or as a combination of lumped and continuous. Geared branches are acceptable.

This is a transfer matrix program taken from [1]. This reference contains the theory and many example problems.

Developers: P. Y. Chang, George C. Sharp, Inc., New York, New York 10007
W. D. Pilkey, Department of Engr. Sci. & Sys., University of Virginia, Charlottesville, Virginia 22901
Marketed: Developer or COSMIC
Machines: CDC 6000 Series, UNIVAC 1108, IBM 370
Modes: Batch or time sharing
Contact: W. D. Pilkey (804) 924-3291
Mode available: The batch program may be obtained from the developer at nominal cost. Interactive pre- and postprocessors that can be coupled to the program are available through COSMIC.

TOFA (TOrsional Frequency Analysis). This program is designed for studying torsional vibrations of a mechanical system consisting of a set of connecting shafts with disks concentrated along the lengths of these shafts.

TOFA is designed to handle systems of not more than 50 stages, where a stage consists of a disk and a shaft.

Developer: Atomic Energy of Canada Limited
Chalk River Nuclear Laboratories
Chalk River, Ontario
Marketed: Same
Machine: CDC 6600
Mode: Batch
Contact: P. Y. Wong
Mode available: Unspecified

CADENSE-22. This program calculates the critical frequencies of a torsional system. It may include branches, gears, epicyclic gears, and elastic torsional connection to ground. At each critical frequency the normalized mode shape and corresponding torque distribution are computed. The program employs the Holzer method extended to account for continuously distributed shaft sections.

Developer: Mechanical Technology, Inc. (MTI)
968 Albany-Shaker Road
Latham, New York 12110
Marketed: Same
Machines: CDC 6000 Series, IBM, etc.
Mode: Batch
Contact: Applied Technology Dept. (MTI)
Mode available: Use basis, for sale or lease

TABU. This program is used for the torsional analysis of branched systems. Stiffness matrix method is utilized to determine the natural frequencies and mode shapes of the system. A forcing function and model damping may be included to determine stresses. Shaft masses and stiffnesses must be input.

Developer: Southwest Research Institute (SRI)
8500 Culebra Road
P.O. Drawer 28510
San Antonio, Texas 78284
Marketed: Same
Machine: CDC 6600
Mode: Batch
Contact: T. R. Jackson
Manager, Computer Laboratory (SRI)
Mode available: For sale, minimal charge to cover handling

CADENSE-23. This program computes the damped torsional response of a system to exitation in the form either of torques or angular displacements (gear errors). The system may include branches, single reduction gears, and constraints to ground with both stiffness and damping. The gears may be rigidly or elastically mounted. The system amplitudes, torque distribution, and gear tooth meshing forces are calculated. For elastically mounted gears the transmitted forces and gear tooth displacements are evaluated. A Holzer method of solution is used.

Developer: Mechanical Technology, Inc. (MTI)
968 Albany-Shaker Road
Latham, New York 12110
Marketed: Same
Machines: CDC 6000 Series, IBM, etc.
Mode: Batch
Contact: Applied Technology Dept. (MTI)
Mode available: Use basis, for sale or lease

CADENSE-24. This program computes the transient response of a torsional system to specified input at one or more locations of the system. The system may include branches, subbranches, and gears. Backlash may be specified at any point. The program takes into account both stiffness and damping in each elastic member and also allows stiffness and damping between any station and ground. The torque at each station as a function of time may be plotted. The system of dynamic equations is solved using a fourth-order Runge-Kutta method to give angular displacements.

Developer: Mechanical Technology, Inc. (MTI)
968 Albany-Shaker Road
Latham, New York 12110
Marketed: Same
Machines: CDC 6000 Series, IBM, etc.
Mode: Batch
Contact: Applied Technology Dept. (MTI)
Mode available: Use basis, for sale or lease

TASS. This program calculates the torsional critical frequencies and the forced dynamic response in torsion of undamped shaft systems. The program uses a distributed mass approach. It can analyze any single-branched gear train system with any number of gear trains. External forces, lumped inertias, and torsional springs to ground may be included. Data are entered from a data file.

Developer: Structural Dynamics Research Corp. (SDRC)
5729 Dragon Way
Cincinnati, Ohio 45227
Marketed: Same
Machines: Honeywell 400 & 6000, XDS Sigma 9, and CDC 6000 Series
Mode: Time sharing
Contact: Edward Carl (SDRC)
(513) 272-1100
Mode available: Use basis

EXTENSION MEMBERS

EXTEND. This program is for the calculation of extension problems of bars and longitudinal motion of spring mass systems. The axial displacement and force are found for static and steady state conditions. Natural frequencies and mode shapes are computed for free dynamics. The mass can be modeled as lumped or continuous.

This program is superimposed on the TWIST program described previously.

Developers: P. Y. Chang, George C. Sharp, Inc., New York, New York 10007
W. D. Pilkey, Dept. of Engr. Sci. & Sys., University of Virginia, Charlottesville,Virginia 22901

Marketed: Developer or COSMIC
Machines: CDC 6000 Series, UNIVAC 1108, IBM 370
Modes: Batch or time sharing
Contact: W. D. Pilkey (804) 924-3291
MOde available: The batch program can be taken from [1] or obtained from the developer at nominal cost. Interactive pre- and postprocessors that can be coupled to the program are available through COSMIC.

DISKS

DISK. This program performs a two-dimensional stress analysis on variable geometry rotors of symmetrical cross sections subjected to symmetric boundary forces. The program is also capable of analyzing disks with integral shafts. The method of solution is based on replacing the variable geometry rotor by a series of finite concentric rings of constant axial width. The compatability conditions of the interfaces of the adjacent rings provide a set of linear simultaneous equations in terms of unknown interface pressures. The tangential and radial stresses can be computed using these pressure values.

Developer: Structural Dynamics Research Corp. (SDRC), 5729 Dragon Way, Cincinnati, Ohio 45227
Marketed: Same
Machines: Honeywell 400 & 6000, XDS Sigma 9, CDC 6000 Series
Mode: Time sharing
Contact: Edward Carl (513) 272-1100
Mode available: Use basis

DISK. This program is for the thick elastic solids problems of disks, cylinders, and spheres. For the disk it calculates the radial displacement and the radial and tangential force per unit length of r face and θ face respectively for static and steady state conditions and the natural frequencies and mode shapes of radial vibration. The disk is based on a plane stress assumption. For the cylinder, DISK calculates the radial displacement, radial stress, tangential stress, and axial stress for static and steady state conditions and the natural frequencies and mode shapes of radial vibration. The cylinder is based on a plane strain assumption. In the case of the sphere, the radial displacement, radial stress, and tangential stress are found for static and steady state conditions as well as the natural frequencies and mode shapes of radial vibration. For the disk and cylinder all loadings and responses are axially symmetric. For the sphere, loadings and responses are spherically symmetric.

Developers: P. Y. Chang
George C. Sharp, Inc.
New York, New York 10007

W. D. Pilkey
Dept. of Engr. Sci. & Sys.
University of Virginia
Charlottesville, Virginia 22901

Marketed: Developer or COSMIC
Machines: CDC 6000 Series, UNIVAC 1108, IBM 370
Modes: Batch or time sharing
Contact: W. D. Pilkey (804) 924-3291
Made available: The batch program can be obtained from [1] or from the developer at a nominal cost. Interactive pre- and post-processors that can be coupled to the program are available through COSMIC.

SPRINGS

SPRINGS. This can be used to design, analyze, and document the following type of springs:

1. Helical compression springs with round or rectangular wire
2. Helical extension springs with round wire
3. Helical torsion springs with round or rectangular wire

Data input for the program is interactive.

Developer: General Electric Infonet, Business Div.
7735 Old Georgetown Road
Bethesda, Maryland 20014
Marketed: Same
Machines: GE-235-MARKI, GE-635 - MARKII
Mode: Time sharing
Contact: Bowman Irani (301) 654-9360
Mode available: Use basis

Stress Analysis of Belleville Springs. This program computes deflections, membrane forces, bending moments, stresses, and the load deflection history for shallow, truncated, thin, axisymmetric, conical shells of uniform thickness. The program has the following limitations (1) the shell must be shallow, (2) the shell must be thin, (3) the program is limited to the case of symmetrical axial loads applied and reacted at the edges of the shell, and (4) the program is limited to the specific boundary conditions that no axial, radial, or rotational constraints are enforced at either boundary.

Developer: Rocketdyne
6633 Canoga Avenue
Canoga Park, California 91304
Marketed: COSMIC, Prog. No. MFS-2405
Machine: IBM 360
Mode: Batch
Contact: COSMIC
Mode available: For sale, $25.00

COMSPRING. This program performs the design and analysis of a straight, open-coil, round-wire, helical spring subjected to axial compressive loads. Certain basic factors predetermined by the desired spring performance, such as load, movement, or space limitations, are input to the program. The user may specify end-turn conditions or end-fixity constraints when spring instability is investigated. Also, output are nominal spring weight and coil helix angle.

Developer: TRW Systems Group, STRU-PAK
1 Space Park
Redondo Beach, California 90278
Marketed: CDC Cybernet Service
Machine: CDC 6000 Series
Mode: Time sharing
Contact: R. W. Farnsworth (TRW)
(213) 535-1250
Mode available: Use basis

BELSPRING. This program provides a simplified design and analysis solution for conical-disk springs or, as they are sometimes called, Belleville springs. The program will perform an analysis of a given spring design, or it will determine the spring design characteristics when load-deflection and geometric constraints are imposed.

Developer: TRW Systems Group, STRU-PAK
1 Space Park
Redondo Beach, California 90278
Marketed: CDC Cybernet Service
Machine: CDC 6000 Series
Mode: Time sharing
Contact: R. W. Farnsworth (TRW)
(213) 535-1250
Mode available: Use basis

Contact Spring System. This program computes the forces and stresses acting upon a contact spring system as a function of the design parameters. An elimination scheme using the largest pivotal divisor is used to solve the linear simultaneous equations.

Developer: John R. Wolberg
Technion - Israel Inst. of Technology
Haifa, Israel
Marketed: National CSS, Inc., time sharing services
Machine: IBM 360/40
Mode: Time sharing
Contact: John R. Wolberg or Reference [16]
Mode available: Use basis

REFERENCES

1 Pilkey, W. D., and Chang, P. Y., Formulas and Methods for Simple and Complex Structural Member Analysis-Including Computer Programs for Bars, Plates, and Shells, McGraw-Hill, New York, to appear.

2. Herrmann, L. R., "Elastic Shear Analysis of General Prismatic Beams," Journal of the Engineering Mechanics Division: ASCE, Vol. 94, No. EM4, Aug. 1968, pp. 965-983.

3 Herrmann, L. R., "Elastic Torsional Analysis of Irregular Shapes," Journal of the Engineering Mechanics Division: ASCE, Vol. 91, No. EM6, Dec. 1965, pp. 11-19.

4 Butler, M. K., Hegan, Marianne, Ranzini, L., "Argonne Code Center: Compilation of Program Abstracts," ANL-7411, Act. 1968, Argonne National Labs, Argonne, Ill.

5 Berman, I., ed., Engineering Computer Software: Verification, Qualification, Certification, A.S.M.E., New York, 1971.

6. Kraus, H., ed., The Software User: Education and Qualification, A.S.M.E., New York, 1972.

7 Kraus, H., ed, Use of the Computer in Pressure Vessel Analysis, A.S.M.E., New York, 1969.

8 Schiffman, R. L., "Report on the Special Workshop on Engineering Software Co-ordination," Computing Center Report 72-2, March 1972, University of Colorado, Boulder, Col.

9 Schiffman, R. L., "Papers Prepared for the Special Workshop on Engineering Software Co-ordination," Computing Center Report 72-4, April 1972, University of Colorado, Boulder, Col.

10 Medearis, K., "An Investigation of the Feasibility of Establishing a National Civil Engineering Software Center," Report to N.S.F., January 1974.

11 Kraus, H., "Atitude Toward Computer Software and its Exchange in the Pressure Vessel Industry," RPI Report, Oct. 1973, Rennselear Polytechnic Inst., Hartford, Conn.

12 Hrennikoff, "A Solution of Problems in Elasticity by the Framework Method," Journal of Applied Mechanics, Vol. 8, No. 4, Dec. 1941 pp. 169-175.

13 Yettram, A. L., Husain, H. M., "Grid-Framework Method for Plates in Flexure," Journal of the Engineering Mechanics Division: ASCE, Vol. 91, No. EM3, June 1965, pp. 369-371

14 Timoshenko, S., Woinowski-Krieger, S., Theory of Plates and Shells, 2nd ed., McGraw Hill, New York, 1959, pp. 369-371.

15 Pestel, E. C., Leckie, F. A., Matrix Methods in Elastomechanics, McGraw Hill, New York, 1963, pp. 51-148.

16 Wolburg, J. R., Application of Computers to Engineering Analysis, McGraw Hill, New York, 1971, pp. 216-235.

II. SOFTWARE DEVELOPMENT AND DISSEMINATION

CURRENT SOFTWARE DISSEMINATION PRACTICES AND ORGANIZATIONS

Robert L. Schiffman
University of Colorado
Boulder, Colorado

ABSTRACT

This paper is directed at the assessment of current activities with regard to software distribution and dissemination of software information as a mechanism for enhancing computer-based technology transfer. Emphasis is placed on defining organizations which attempt to provide active participation between the user and the library. The paper critically reviews several of the better-known program libraries and information centers.

It is concluded that software distribution via off-the-shelf libraries is less than optimal as a means of technology transfer. It is further concluded that present trends in the development of computerized information systems to locate software are in a state of chaos. It is finally concluded that technology transfer efforts should be directed toward the implementation of the software center concept.

INTRODUCTION

Computer-based technology transfer and utilization is the process by which applications software developed in a single environment is made available and is used on an operational basis in a multiplicity of environments.

A variety of public, quasi-public, and private groups have an interest in viable computer-based technology transfer and utilization. Government has an interest on two levels. Computer-based techniques developed under sponsorship of the research support arms of public agencies should be transferred to practice or other research areas. The operating arms of government are likewise concerned with the use of computer-based techniques. Universities have a concern in their mission of providing students and faculty with the most up-to-date technology. The concern of private industry is one of providing services to its clients. Properly, a private firm would desire to apply the latest technology in engineering practice.

Computer program development is currently unrestricted with regard to language, dialect, hardware and software environment, documentation, and maintenance. There are no standards or requirements which must be followed to permit the programs and systems developed in the course of a research and development effort to be easily passed on to others. In addition, there are legal problems associated with user responsibility, copyrights, patents, and trademarks which are not clearly defined. The effects of unbundling and the development of a market for proprietary software packages place constraints on the user which impact professional usage of computer-based technology. Furthermore, the swift and recent growth of interactive (time-sharing) computer systems and communication networks have compounded the problems associated with computer-based technology transfer and utilization. These systems have brought the power of the computer to a point of easy access to all segments of the user community. In addition to providing access, interactive systems have removed many of the psychological barriers in computer use which are inherent in batch process computing.

As a result of the above factors, programs are repeatedly remanufactured with ensuing large wastes in manpower and money. Furthermore, programs are often used without adequate validation by competent professionals with ensuing large wastes in inefficient usage and losses of confidence in computer-based techniques.

Problems of technology transfer and utilization are of concern to all engineering disciplines. There are particularly acute problems in the civil engineering and building construction community since applications cover the full range of machine types from large number crunchers to minicomputers. The program developers are a diverse group, including individuals in professional practice, government, and university-affiliated groups. The user community ranges from totally unsophisticated (in computer usage) to sophisticated systems designers. In addition, the program requirements are diverse, including algorithmically based programs and design programs which manipulate large data bases. The legal and professional problems are also particularly acute due to the diversity of civil engineering projects, ranging from large publicly financed projects involving interactions with several government agencies, to small projects involving essentially interpersonal relationships.

The extensive, uncontrolled, and unmonitored growth of civil engineering applications software has been such that the situation with regard to the usability and availability of this software is in a thoroughly chaotic state. This state is exhibited by the efforts which are under way in various parts of the world. The GENESYS project in the United Kingdom is a concentrated effort to apply computer-based standards and procedures to civil engineering. APEC, CEPA, COSMIC, and STORE are program distribution efforts based in this country. ICES and NASTRAN are examples of large programming and computer systems designed to widen the base of computer usage among professional groups. In addition, a technology transfer effort has been proposed in Canada.

It has been suggested that the establishment of a national software center for the civil engineering and building construction community might be an appropriate focal group for the solution of the diverse technical, legal, professional, and commercial problems associated with computer-based technology transfer and utilization [1]. The establishment of a national software center could provide a focus for the evaluation, verification, and validation of the documentation of algorithms, data structures, and computer programs. It could also maintain and distribute programs. In addition, by reason of its expertise, it could serve as a central organism for the resolution of legal and professional problems associated with the development and use of civil engineering software. Conceptually, such a center could, in its more mature stages, develop and document programs and systems based upon professional need [2].

It has been suggested that the needs for computer-based technology transfer can be served by an expansion of currently active user groups and/or distribution agencies. In order to assess this suggestion, this paper reviews some of the more active efforts which are currently providing a mechanism for software dissemination.

SOFTWARE DISSEMINATION

It is almost axiomatic with users of engineering software that experience in program exchange, except in specialized cases, has been bad. It is quite common for engineering users to develop their own systems in lieu of using other "available" systems [3]. While the generality of this expression is by itself a measure of proof of a serious problem, the hard facts supporting these conclusions are not readily available. All too often the factual documentation is distorted by emotionalism.

Software distribution activities of both an organized and an ad hoc nature can be classified in five ways. These are:

1. Vendor-maintained libraries and user groups
2. University activities
3. Disciplinary user groups
4. Government activities
5. Professional societies and industry

These libraries, groups, and activities were brought into existence because of need and demand. They have served their community of users with varying degrees of success.

The following exposition attempts to describe various current efforts directed toward the distribution of software and information concerning software. The efforts described below are directed toward software for civil engineering and, in particular, structural mechanics. For reasons of space and applicability such efforts as the Argonne Code Center, the International Mathematical and Statistical Library (IMSL), the Quantum Chemistry Program Exchange (QCPE), and other essentially nonengineering-oriented activities will not be discussed [4].

Vendor-Maintained Libraries and User Groups

There are three broad categories of software distribution via hardware-vendor—maintained libraries and user groups. First, the hardware vendors maintain software libraries as an adjunct to their machine marketing operations. In addition, some of the hardware firms maintain data centers where application software processing services are sold. The third category is the hardware-oriented user groups

Engineering software via vendor-maintained libraries is rare, and probably nonexistent. This was not always the case. The emergence, however, of unbundling, in which hardware and software are sold separately, has led to the demise of vendor-sponsored application libraries.

Some of the hardware firms maintain processing centers, which in turn contain software libraries. The software maintained at these service bureaus is primarily for the purpose of bureau customers and is usually not distributed.

Just about all hardware manufacturers foster hardware-oriented user groups. These user groups maintain libraries for the sharing of software. The user groups of the major main frame manufacturers are:

1. COMMON - International Business Machines Corporation. This group is concerned with the IBM 360/20, IBM 1130, IBM System/3, and IBM System/7 hardware applications.
2. CUBE - Burroughs Corporation.
3. FOCUS - Control Data Corporation. This group is concerned with CDC 3000 series machines.
4. Honeywell Users Group - This is the user group for Honeywell Information Systems.
5. SHARE - International Business Machines Corporation. This group started as a user group for the IBM 704/709/7090/7094 machines. It currently serves the IBM 360/System and 370/System machines.
6. UNIVAC Scientific Exchange - This is the user group for UNIVAC hardware.
7. VIM - Control Data Corporation. This group is concerned with CDC 6000 series machines and the CDC 7600

Currently, applications software exchange via vendor-sponsored user group libraries is only moderately successful within a series of fairly rigid constraints. First, the software is machine dependent. Second, the state of documentation is highly variable. Since submission to these libraries is purely voluntary, without standards, formal recognition, or prestige value, the program documentation is generally inadequate for proper transfer. The third restraint on vendor library programs is their complete lack of maintenance. Future projections of computer-based technology transfer must, by necessity, exclude consideration of this mechanism of program exchange.

University Activities

Software exchange between universities and software distribution by universities is a common occurrence and generally operates in a relatively informal manner. This section describes three efforts which have been formalized.

University of California, Berkeley

The National Information Service - Earthquake Engineering (NISEE) has, as part of its research and information service, a computer application arm, NISEE/Computer Applications. This is a service through which computer programs related to earthquake engineering are distributed at cost of reproduction. The service is funded by the National Science Foundation and user fees. Approximately 50 percent of the funding originates from NSF.

The Service relies on voluntary contributions of programs, both from within the University of California, Berkeley, and from outside sources, usually other universities.

The operation of the Service is relatively passive. Contributed programs are informally checked for the adequacy of the documentation and for portability. In the final analysis, however, there is no control exercised over contributors.

The Service tends to be bound by portability considerations. If a program is operational on a CDC 6400 it will be distributed from the University of California, Berkeley. If not, the Service, through its program abstracts, directs the user to the author. Currently the Service is cooperating with the University of Southern California to provide both CDC 6400 and IBM 370/System versions of SAPIV.

The largest single problem faced by the Service is its inability to receive voluntary contributions of programs. It appears that potential contributors are reluctant to place their programs in this library. The most often stated reason for this reluctance is the loss of distribution control on the part of the program author.

University of Colorado

The University of Colorado Computing Center has, as an informal part of its efforts in computer-based technology transfer, established the Engineering Program Library (EPL). This library contains a selective set of programs which were either developed by the Computing Center staff or were provided to the Center. The EPL serves two main functions. First, it provides a centralized engineering software source for the University of Colorado community. In addition, it provides a means of distribution for software developed outside the University. In the latter case, the EPL is distributing a series of programs developed by the Commonwealth Scientific and Industrial Research Organization of Australia. This particular distribution effort was established under the aegis of the Committee on Earth Masses and Layered Systems of the Highway Research Board.

It should be noted that software deposited in the EPL is not maintained on a uniform basis. Maintenance is a function of the author. While no program is placed in the EPL without documentation, the completeness of the documentation varies widely. The minimal required documentation is a user's manual. This is the only documentation available in most cases, although some of the programs are extensively documented.

All EPL programs operate on the Center's CDC 6400 system. Portability, however, is a prime consideration. All EPL programs developed by Center staff are written with portability in mind. Programs received from outside the Center are converted to run on Center hardware (CDC 6400) and are thus made more portable.

The measure of success of the operation of the EPL is currently passive, in the sense that a quiet user is assumed to be satisfied.

Where problems have occurred, attempts have been made to respond quickly and positively to user problems. The major problem areas are the portability of programs to smaller machines, and the scaling of control constants to machines with different word lengths. It should be noted, however, that almost all of the EPL programs are algorithmically based and do not rely on file and string manipulation. Thus, programs of this type, written with strict adherence to ANSI FORTRAN standards [5], are more than normally portable. Where EPL programs rely heavily on data manipulation, the implementation on stranger hardware is discouraged.

The lack of negative response to this semipassive library effort is attributed to two primary factors.

1. The user documentation is complete. Further, the documentation is as specific as possible as to the portability requirements for a given program.
2. Each distributed program is accompanied by a set of test data. Furthermore, the distributed card deck or tape is compiled and executed, and the recipient of the program is provided with this output.

The first major source of concern with this effort is in the breadth of the EPL holdings. Being a controlled informal effort, the library holdings are generally confined to localized areas of interest. In addition, the documentation and portability constraints generally restrict the library emphasis to algorithmically based programs.

University of Illinois

A computer-based technology transfer effort, sponsored by the now defunct Office of State Technical Services of the Department of Commerce, involved the development of a conversational user-oriented system for engineering design and construction management [6]. Over twenty-five engineering and construction firms participated directly in the project by participating in the definition of the scope and capability of the system and by verifying and running the programs from their respective offices.

The major objective was to develop a large, flexible data base and algorithms which would immediately enable the participants to transfer their routine processing tasks onto the computer, but which would also support higher-level management and design techniques as the participants' confidence and sophistication increased.

Although portability of the programs was not a prime design criterion, portions of the system have, in fact, been transferred to other machines and different operating environments. There are no figures available to quantify the cost and effort involved, but it appears that the transfers could be performed relatively easily because of three factors.

1. The data structure was designed separately from the file structure required to map the data structure into secondary storage.
2. The individual algorithms were generally small and independent.

They referenced the conceptual data structure rather than the actual file structure.

3. The conversational input and output commands referenced the data structure only, and not the algorithms.

In addition to this effort, the University of Illinois is actively engaged in developing civil engineering software for its Burroughs B6700 computer. As such, it is a major contributor of this software to the CUBE library.

Canadian Universities

The Committee of Chairmen and Heads of Civil Engineering Departments in Canada has established a Subcommittee on Computation. This subcommittee currently maintains a program abstract bulletin which provides information on a variety of programs available from university sources in Canada. The subcommittee is considering the following as a series of further actions to promote computer use and information exchange.

1. The establishment of specialist subcommittees to undertake specific tasks, such as soliciting and evaluating programs in specific areas, and the establishment of guidelines and standards.
2. The establishment of cooperation with trade organizations and government agencies.
3. The consideration of technical assistance to program contributors and users.

Disciplinary User Groups

Disciplinary user groups have developed to exchange pertinent software and to serve as a common meeting ground. Although these groups are providing a form of technology transfer, their impact is limited by budgetary restrictions and by the lack of interdisciplinary contacts. Within the civil engineering and construction community, three prominent groups are currently operating.

Automated Procedures for Engineering Consultants

APEC (Automated Procedures for Engineering Consultants) is a nonprofit association of groups mutually concerned with the use of computer technology in the design of buildings. APEC members cooperatively define the scope and environment of needed programs. The development and maintenance of these programs is then commissioned to software firms. These programs are then distributed to the membership at costs substantially below the normal "market price" for such systems. In addition to program development, APEC operates in the manner of a professional society in providing a forum and workshops on the use of its programs.

APEC's current membership is 190 firms and individuals. These firms have staffs from zero (principals only) to 70,000 employees. Approximately half of these firms have reported that they maintain their own in-house computers. Some firms maintain more than one computer.

The APEC library currently contains six major programs, all of which are related to mechanical, electrical, or environmental considerations for buildings.

Originally APEC programs were developed for in-house batch processing operations usually centered around the IBM 1130. This has expanded to include other hardware as demand increased. More recently there has been a tendency to move from in-house operation to terminal operation. APEC is currently contracting with Control Data's CYBERNET system for the use of its software.

Civil Engineering Program Applications

CEPA (Civil Engineering Program Applications) is a nonprofit body whose principal objective is to further the effective application of computers in civil engineering and related fields. To this end, CEPA provides a means for the exchange of information and for the exchange and cooperative development of computer programs and systems pertaining to civil engineering and related fields.

Membership is open to organizations and/or individuals utilizing computers in the practice of civil engineering or in related fields and subscribing and contributing to the cooperative purposes and efforts of CEPA. Most member companies are from North America, but there are participants from such places as Britain, Italy, Sweden, Israel, Saudi Arabia, Ethiopia, Australia, Japan, Hong Kong, Switzerland, and Panama.

From its beginning in March 1965 as an IBM 1130 user's group, CEPA has grown to a maximum size of 250. The present active membership is just over 230. Of these, 82 percent are consulting engineering companies. Out of the top 500 design firms in the U.S., from Engineering News Records' annual survey, 52 are members of CEPA, including 4 out of the top 5.

One of the important facilities offered by CEPA to its membership is the operation of a program library. The library consists of programs contributed by members, presently numbering more than 450, and classified generally under Geometry, Bridges, Buildings, Hydraulics and Sanitary, Traffic, Management, Environmental, and Soils. Programs are obtained by members based on a point system similar to a bank account. Members are credited with points when their programs are requested, debited points when they request a program.

In operating the program library, CEPA acts as a broker by providing information on availability and in keeping an account of the point system. The costs associated with program acquisition are matters of concern between the owner of the program and the requester. CEPA cares only about points.

Neither CEPA nor, generally, the original authors accept responsibility for use of programs. This is a user responsibility. Authors, however, are responsible for disseminating information on "fixes" should any "bugs" be discovered.

Programs in the CEPA library range from systems containing thousands of source cards to small subroutines containing a few source statements and cover most aspects of civil engineering. There is a considerable amount of duplication in the library, although not all of it can be termed unnecessary.

CEPA encourages cooperative development of programs by members. Two systems have been completed in this way: bridge pier design and curved bridge girder design.

A major function of CEPA is to serve as a forum for its members. A three-day conference is held twice a year.

Recently CEPA undertook a National Science Foundation Project to define a national effort for civil engineering and building construction software coordination.

ICES User's Group

ICES (Integrated Civil Engineering System), originally developed at the Massachusetts Institute of Technology, is a problem-oriented language system designed to provide relatively large, open-ended application packages. The goal of the system was the provision of facile user-oriented languages for the performance of a variety of application efforts and for the handling of large data bases.

The system, together with a number of example subsystems (or application packages), was released to the public in 1967. Several of the application packages (subsystems) were accepted reasonably well by the profession and are extensively used in practice.

As the use of ICES grew, a user's group was formed for the purpose of exchanging information and providing a focal point for the interface and training of new users. When IBM, on whose computer the initial implementation of ICES was made, decided to abandon the distribution of programs which had been provided to its Program Information Department, the ICES User's Group arranged to provide for the public domain distribution of donated ICES programs. This arrangement took the form of designating a single source for the programs and of ensuring that this source received all new ICES programs which the group was given. The source was selected through a competition based on the lowest distribution costs to users. The present distribution source is Pacific International Computing, a subsidiary of the Bechtel Corporation.

The ICES User's Group has a primary mission of providing information on and encouraging the use of ICES. As such, it has funded several grants to educational institutions who are desirous of developing civil engineering programs using the ICES concept.

While the Group does not attempt to develop new subsystems or to provide maintenance for the system or new or existing subsystems, its Newsletter distributes fixes. Additionally, it provides the membership with data on groups maintaining specific subsystems.

ICES was initially implemented on IBM 360/System hardware and is highly machine dependent. RCA implemented the system on its virtual memory computers. The system has also been implemented on the UNIVAC 1108.

Government Activities

Almost every U.S. Government agency concerned with construction has had some experience relating to the portability and adaptability of engineering software. A study of program duplication in this area has been conducted by the Building Research Advisory Board [7]. The information related below represents a cross section of the activities of several agencies and groups.

Corps of Engineers

The Hydrologic Engineering Center of the U.S. Army Corps of Engineers is located in Davis, California, and has as a primary mission the development, dissemination, and maintenance of programs in hydrologic engineering. The Center, established in 1964, has three major missions: training, research, and project assistance. Its 26-person staff, consisting of 17 professionals and 9 subprofessionals, serves 50 Corps of Engineer offices. The operating budget is approximately $1,000,000 per year. Of the 17 professionals on the staff, 6 to 8 are engineer/programmers who spend more than one-half of their time in program development.

Hardware facilities currently available to HEC are a UNIVAC 1108 connected via an M&M terminal. HEC uses the Information System Design Corporation for this service. A CDC 6600 and a CDC 7600 at the Berkeley Radiation Laboratory are also available via a batch terminal located at HEC.

HEC is supporting approximately 39 programs varying in size from 100 to 7,500 source statements. Over 1,000 programs have been distributed to other Corps of Engineer Offices. There were 300 programs distributed in 1973 alone. In addition, over 600 programs have been distributed outside the Corps. Of the 39 active programs, the distribution varies from 1 to over 200 requests.

HEC is currently training engineers in private practice on the use of its software. This effort is part of a larger federally supported program to encourage private industry to engage actively in flood plain planning.

A more general Corps effort is being coordinated through the Office of the Chief of Engineers. This effort is establishing programming and documentation standards and regulations to assist in the portability and adaptability of applications software. In addition, a set of FORTRAN callable system library routines for engineering functions is being developed which will facilitate the development of programs with a minimum of redundancy.

A library of programs available to all Corps offices is located at the Waterways Experiment Station in Vicksburg, Mississippi. These programs are resident on a GE 600 series time-sharing machine located at WES. The programs have been reviewed by the Office of the Chief of Engineers and have been certified for technical accuracy.

COSMIC

An effort to disseminate computer programs publicly has been established by the National Aeronautics and Space Administration under its Technology Utilization Program. Through COSMIC (CØmputer Software Management and Information Center), located at the University of Georgia in Athens, Georgia, programs which are developed for and by the National Aeronautics and Space Administration, the Department of Defense, and the Atomic Energy Commission are filed in a central depository. These programs can be secured by the general public for a nominal price. Information concerning the availability of these programs is disseminated through a monthly publication entitled COMPUTER PROGRAM ABSTRACTS. This program has been critically reviewed, primarily because of the documentation on the user and systems levels. It is, however, among the best software distribution activities currently available to the profession. COSMIC is defined to act exclusively as a depository and distribution center. No responsibility is assigned to evaluate, document, or maintain programs. Thus, COSMIC is dependent on the authors for these latter services. Although COSMIC has no responsibility for the workability of a program, it has, by reason of its sponsorship and prominence, assumed to be responsible for program quality. As a means of upgrading its image COSMIC is expanding its service to perform some program evaluation and has published recommended standards for program submittal [8]. IN FY 1973 COSMIC accepted approximately 50 percent of the offered programs. This percentage is increasing yearly. The checkout procedure for candidate programs consists of compilation and link-editing (loading). In order to cover the resulting costs, the program-charging structure has undergone an upward adjustment. Initially, documentation was distributed without charge, and the cost of programs was nominal. Typical current charges range from $50 to $1,700 per program and $1.50 to several hundred dollars for documentation. Experienced COSMIC users generally agree that program quality tends to improve with increasing documentation cost.

As of July 1971, the COSMIC library consisted of a total of 1,030 programs. This has increased to 1,300 as of July 1973. There are between 90 and 100 structural mechanics programs in the library.

The COSMIC budget is approximately $300,000 per year. NASA underwrites this budget in the amount of $100,000. The remainder is derived from user fees. In FY 1973 direct NASA support amounted to $50,000.

The COSMIC staff contains 12 full-time equivalent staff members. There are actually 14 persons on the staff including some part-time people.

COSMIC distributed 2,500 packages in FY 1970. A package includes programs, documents, or both of these items. In FY 1973, a total of 6,238 packages were distributed and invoiced. Of this number, 335 programs and 5,903 documents were distributed. The largest single distribution item is NASTRAN. In FY 1973, COSMIC distributed 60 NASTRAN programs and 4,300 items of NASTRAN documentation.

Department of the Navy

Two known efforts undertaken by the Department of the Navy are directly related to problems associated with computer-based technology transfer and utilization. The first is project STORE (STructures ORiented Exchange). This system was developed for the Office of Naval Research as a user-oriented system based upon interactive capability [9]. STORE is an interconnection of applications programs and an engineering and systems oriented information retrieval mechanism. The information retrieval system contains the documentation for each program in the system. Thus, the source documentation is decentralized and computerized. The use of STORE requires a two-page manual which provides all of the instructions on how one enters the system and secures the information desired to use a given program. This use of information retrieval techniques eliminates the necessity of maintaining large libraries of printed material. It places the documentation responsibility at the source program level. A feature of STORE permits any user to enter comments in a program file. These comments provide a necessary link between the user and the author or maintainer of a given program. They also provide a medium for useful information exchange between users. STORE has demonstrated its effectiveness and viability as a mechanism of technology transfer among a group of reasonably sophisticated users. It is presently in operation on a limited basis within the Navy. A commercial spin-off of Project STORE is the STRU-PAK system, a proprietary structural mechanics package being marketed by TRW, Inc.

The Naval Civil Engineering Laboratory, Naval Construction Battalion Center, has recently undertaken a project to develop a computer-aided design system for the Naval Facilities Engineering Command. The overall objectives of the system are to improve the cost effectiveness of Navy shore facilities during their useful life and to reduce design costs and time required to design these facilities.

The initial effort is to develop information on computer-aided design technology, a set of goals for computer-aided design, and a plan of action to achieve these goals.

Air Force Flight Dynamics Laboratory

The Air Force Flight Dynamics Laobratory of Wright-Patterson Air Force Base has recently established the Aerospace Structures Information and Analysis Center (ASIAC). This Center is operated by Anamet Laboratories, Inc., in San Carlos, California [10].

The Center is designed to serve as a central agency for the collection, analysis, and dissemination of information for aerospace structures engineering. The Center operates primarily as a hard-copy library. The library is using existing data bases obtained from a number of agencies and several information retrieval systems.

In addition to its library functions, ASIAC serves as a facility to provide consulting services in structures. ASIAC also collects and disseminates structural computer programs. There are 17 programs currently available. Several of these programs are also available from other sources.

National Aeronautics and Space Administration

The National Aeronautics and Space Administration is one of the world's largest developers, managers, and users of engineering software. In the area of structural mechanics, it is concerned with programs and systems whose costs range from a few hundred to several million dollars. Probably the largest single civil engineering related development is the NASTRAN system.

NASTRAN was developed by the Computer Science Corporation for NASA in an effort spanning approximately five years. The first general public release of NASTRAN was in September 1970. The development cost was approximately $3,000,000. There have been two subsequent releases. The net cost of all versions was approximately $5,000,000.

The first release of NASTRAN consisted of 130,000 cards and 1,000,000 instructions. It is comparable in size to a large operating system and consists of 750 specialized programs, 300 major programs, 14 small operating systems, and an executive. The current version is approximately 200,000 source card images and contains 1,246 subroutines. NASTRAN has been implemented on the CDC 6000 series machines, CDC 7600, IBM 360 and 370/Systems, and the UNIVAC 1106, 1108, and 1110.

In 1971 NASTRAN was installed at 70 facilities. Today there are 240 facilities with a NASTRAN capability. During the first year of its release there were 1,000 to 1,200 users. Today there are at least 2,400 users. In all probability the current number of NASTRAN users is of the order of 5,000.

NASA maintains a systems management office with 6 professionals for the exclusive purpose of managing NASTRAN. Maintenance and improvement costs are minimally of the order of $500,000 per year. The improvements to NASTRAN have been in three general areas.

1. Implementation of new elements
2. Increasing efficiency
3. Adaptations to user needs

Federal Highway Administration

Probably the largest computer user group in civil engineering is the Federal Highway Administration (FHWA) and the 50 state highway departments. These groups have vast experience in the development, operation, and distribution of computer programs in all civil engineering fields related to highways.

Expect as noted later, program exchange between highway departments and FHWA is a relatively informal operation for most of the smaller programs. It is reasonably successful because the distribution activity is generally between computer professionals. FHWA acts as a central distribution agency for the larger systems which have been developed with a substantial input of federal funds.

Department of Commerce

There are two major dissemination activities within the Department of Commerce. The first of these is the National Technical Information Service (NTIS). NTIS publishes Weekly Government Abstracts which abstracts publications submitted to its library and which are generally available. The program card decks, tapes, and documentation will be distributed by NTIS. The availability of this information is publicized in the Abstracts. Needless to say, NTIS is solely a distribution agent. No attempt is made to evaluate or maintain its program holdings.

Under Executive Order 11717 [11], the responsibility for federal automated information-processing standards was placed under the purview of the Department of Commerce. The National Bureau of Standards (NBS), Institute for Computer Sciences and Technology, has developed as part of its standards activity a brief software abstract form. It is planned that all federal agencies will be required to complete the abstract form. The General Services Administration will maintain a general indexed catalogue of abstracts. The publication mechanism is currently in the discussion and planning stage.

The next phase of the standards activity is directed toward the establishment of mandatory federal standards for program documentation. Current efforts are directed toward consideration of the form and content of user documentation. NBS is examining currently available standards used by several federal agencies in an attempt to arrive at a uniform standard.

Another related NBS project is supported by the National Science Foundation. This project is designed to improve national awareness of software and computer services [12]. This is coupled with a survey project on available interactive data bases [13].

Professional Societies and Industry

In addition to the groups and activities previously described, professional societies, trade associations, and industrial groups have concerned themselves with software distribution and dissemination of information on software. Some of the more active groups are discussed below.

American Society of Civil Engineers

Activities of the American Society of Civil Engineers (ASCE) date to the early 1950s with the organization of the Electronic Computation Committee within the Structural Division. Since that time every technical division has, in some form, been engaged in computer based activities.

Recently, the Task Committee on Problem-Oriented Languages in Hydrology of the Hydraulics Division issued a report recommending that ASCE establish, jointly with government groups and private engineering firms, an international study group on program libraries [14]. This group would engage in the following activities.

1. Make plans for the establishment of one or more central repositories of programs
2. Recommend procedures for review and assurance of quality of programs incorporated in the library and certified for it
3. Propose government, legal, and professional actions necessary to facilitate the use of certified programs (in terms of proprietary rights, acceptance of programs, users' responsibility, etc.)
4. Initiate an intense educational effort to generate the fullest possible awareness of the role and implications of a central repository
5. Encourage and support developments leading to the establishment of a fully operational problem-oriented language for hydraulic engineering or, more generally, a family of problem-oriented languages for pertinent areas of civil engineering

The Committee on Computer Applications of the Geotechnical Engineering Division is engaged in several activities contributory to better mechanisms of technology transfer and utilization. First, they have developed a set of program documentation standards and procedures as a means of providing a mechanism for the publication of program information in the journal of the Division [15]. A second effort concerns dissemination of programs. Members of the committee are experimenting with various techniques of disseminating programs.

The specialty conference entitled "Analysis and Design in Geotechnical Engineering," sponsored by the Geotechnical Engineering Division, was held at the University of Texas, Austin, on June 9-12, 1974. A major feature of this conference was the holding of workshops describing the use of geotechnical software. This software is being made available to the profession at cost of reproduction.

The Electronics Computation Committee of the Structural Division is the oldest (mid 1950s) ASCE committee on computers. This Committee has held five Conferences on Electronic Computation. The first conference was held in 1958. The Sixth Conference is being held in August 1974. In addition, the Committee has been active in developing program standards and documentation, program abstracts, bibliographies, etc. In recent years the Committee has broadened its activities to include a concern with a variety of professional problems associated with the use of computers. This is reflected by the papers presented at the more recent Electronic Computation Conferences.

The need for an effective collective effort to meet the full impact of the computer in civil engineering practice, research, and education has been recognized by the Society by the establishment of the Research Council on Computer Practices and the Technical Council on Computer Practices (TCCP).

Among the activities of the TCCP are the coordination of the intersociety computer activities (i.e., Hydraulics Division, Structural Division, Geotechnical Engineering Division, etc.), and cooperation with civil engineering user groups such as CEPA, HEEP, APEC, and others, as well as with other societies and trade groups such as ASME, IEEE, ACM, ACI, and PCA. One of the major activities currently in progress is the evaluation of a proposal to establish an index and abstracting service for computer programs. The TCCP is a cosponsor of the Sixth Conference on Electronic Computation.

The Research Council on Computer Practices is responsible for stimulating research in computing activities. It recently monitored the National Science Foundation funded study on the professional feasibility of a national software center [16]. The Council is currently studying means of enhancing computer-based technology transfer in civil engineering.

American Society of Mechanical Engineers

The American Society of Mechanical Engineers (ASME) maintains five committees which are directly related to computer activities.

1. Research Committee on Computer Software - Policy Board Research
2. Computing in Applied Mechanics - Applied Mechanics Division
3. Committee on Computer Technology - Pressure Vessel and Piping Division
4. Computer Technology Policy Committee - Policy Board Communications
5. Design Technology Transfer Committee - Design Engineering Division

To date these committees have sponsored meetings and special publications on the use and verification of computer programs [17, 18, 19, 20, 21]. The Research Committee on Computer Software is currently working on three projects.

1. An information retrieval system for program information.
2. A market analysis is being conducted of the computer-related needs of the various technical groups within ASME. This study is being conducted for planning purposes.
3. The committee is considering writing a set of guidelines for software development.

The Design Technology Committee held a panel discussion on the present and future role of interactive computer systems in technology transfer from research to design at the 1974 Design Engineering Conference. A major theme of the National ASME meeting in October 1974 will be the role of computers in technology transfer.

American Association of State Highway and Transportation Officials

For several years preceding 1969, the American Association of State Highway Officials (AASHO) maintained, through its subcommittee on Computer Technology, an index of programs available from various state highway departments.[1] The 1969 issue contained 3,262 entries, many

---

[1] In late 1973, the American Association of State Highway Officials (AASHO) was renamed the American Association of State Highway and Transportation Officials (AASHTO). Also the Subcommittee on Computer Technology changed its name to the Subcommittee on Data Processing.

of these being duplications and programs written for obsolete machines. The index was officially discontinued in 1969, presumably for reasons of inadequate funding.

In 1971 the Subcommittee on Computer Technology circulated a questionnaire among its member groups concerning the advisability of reactivating a program exchange mechanism. The favorable response motivated the updating of the 1969 index [22].

Highway Engineering Exchange Program

The Highway Engineering Exchange Program (HEEP) is a professional user's group whose purpose is the exchange of programs and information concerning applications in highway design. HEEP recently circulated a questionnaire querying the membership's desires concerning an organized mechanism of program exchange. The response of this questionnaire represented ten State highway departments, ten Federal agencies, four County agencies, two cities, and one province in Canada. Ninety-seven percent of the respondents expressed their willingness to contribute programs and documentation to a central library.

Highway Research Board

The Highway Research Board has maintained the Highway Research Information Service (HRIS) since 1964. This is a technical information service initially implemented for highways. More recently this service has been expanded to other modes of transportation.

HRIS is cooperating with the International Road Research Documentation Scheme housed within the Organization for Economic Cooperation and Development (OECD). This Scheme involves 17 centers in 15 countries. The computer subcommittee of the Scheme is concerned with the portability of information considering hardware and software differences between different centers. To this end, a set of specifications, or communications format tape (CFT), has been established. The specification is based on the ISO draft standard, and includes the following.

1. Physical characteristics
2. Character set
3. Tape labels
4. Record labels
5. Record directories, field tags, addresses, lengths
6. Subfield tags, data, separators
7. Data formats (coding rules)

The exchange system was successfully tested in 1971-72 and is now operational. Tapes are being exchanged on a monthly basis and are compatible to a wide variety of information retrieval systems and machines.

Recently HRIS established categories for publishing program abstracts. At the present time there is an exploration of expanding this effort to an international exchange of program abstracts.

Draft specifications and working rules for abstracts are being developed.

Illinois Institute of Technology Research Institute

For the past seven years the Structural Analysis section of the Illinois Institute of Technology Research Institute (IITRI) has maintained the Structural Mechanics Computer Programs Library. The Library is maintained for a user group in the Chicago area. The function of the Library is to obtain, validate, and distribute existing computer programs to its member organizations.

This Library functions in the following manner:

1. A search is made of a variety of technical publications, trade journals, dissertation abstracts, etc. to ascertain the availability of programs.
2. A list of programs is compiled and distributed to the Library clientele.
3. Based upon inputs from the clientele, the Library will attempt to secure the desired programs.
4. The Library validates the program for use by the client. The validation consists of implementing the program on the client's machine and running the test problems. The Library will document the program, if necessary, and if additional user fees are available.
5. An operational program is turned over to the client.
6. Continuous feedback is maintained between the Library and the client. If bugs are found in a program the Library will contact the author for a fix.

The Library maintains approximately 250 programs, mostly in shells. Its initial membership was 45 firms. There are currently 10 firms who contribute $4,000 apiece for membership in the group. The Library clientele are primarily industrial firms in the manufacturing business. It is estimated that after approximately two years of membership the potential resources of the Library are exhausted to the client.

The operation of the Library is reasonably successful for several reasons. First, there is continuous communication between Library personnel and the client. In addition, it is made clear by the Library that client firms must have an in-house programming support staff.

The clientele of the Library is directly related to the reliability of the programs involved. Were the candidate programs in such a form with regard to the code and documentation that in-house support would be minimal, the Library could increase its membership threefold.

Other Groups

There are a variety of other groups actively distributing programs and program information and who are professionally involved with computers. The American Concrete Institute (ACI) sponsors Committee 118, the Use of Computers. This committee is currently studying a proposed "ACI Certification" of existing concrete design applications systems.

Various trade associations distribute programs pertinent to their areas of expertise. Some of these are the Portland Cement Association, the American Institute for Steel Construction, the American Iron and Steel Institute, and the Prestressed Concrete Institute.

In addition to the public and semipublic groups, there are several private firms which distribute software or provide a technology transfer service via their proprietary software. These firms have varying degrees of success directly related to their breadth of application and the service they provide their clients.

DISCUSSION

An overview of program distribution efforts clearly points to a pattern of behavior. Initially the development of computer programs was treated as an extension of the slide rule. That is, they were calculation tools that were used on a unique basis. Each program was developed for a particular project, and little or no consideration was given to the generality and possibilities of technology transfer. Unlike slide rule calculations, however, computer program development consumes substantial resources in time and money. Furthermore, their inherent capability of breadth of application became a dominating force which could not be ignored. This force developed in two streams. First, management within an organization, being cost conscious, looked toward multiuse of programs as a means of saving resources and reducing the engineering costs of a given project. This was coupled with the realization by the engineer/programmer that programs developed for one application could often be adapted to another application with a small amount of effort. A third cost aspect was in the use of computers to solve new and challenging problems. These newer problems usually required large and complex computer programs. Time and money constraints forced the adaptation and cannibalization of existing software.

The second stream of activity involved the hardware vendor. The availability of software to fill potential customers' needs became a prime marketing point in the sale of hardware. With software made available to a customer, the buyer of computers could potentially reduce the cost of a computerized operation and, in addition, could shorten the lead time between hardware acquisition and production.

These factors inevitably led to the establishment of the vendor-supplied library and then the development of the user groups. As noted earlier the vendor-supplied libraries grew in a haphazard manner with little or no thought to documentation or maintenance. The result was a high level of frustration for the computer user. Often the user spent more time and effort in searching for a nonexistent, nonadaptable, or nonportable program than would have been consumed in developing the program from scratch. As a means of self-protection the user group was established. These groups were initially centered on a common hardware base. Thus, the user of a particular hardware configuration could reasonably expect that shared software was portable.

In addition to hardware-based user groups, it was realized that user interaction was necessary on a disciplinary basis. Thus, user groups were organized to serve a user industry as well as the users of a particular brand and model of hardware.

The various combinations and permutations of this growth, with professional society, private industry, and government involvement, has led to an almost unending list of software libraries and distribution agencies, many of them serving redundant functions. It has been seriously suggested that an effort be undertaken to coordinate program libraries. The coordination would be based on the use or development of information retrieval systems which would provide a user with program library information as well as program capability information. Because of the diversity of data bases, and the diversity of information retrieval systems, this type of coordination implies that the user should have access to a large number of information systems or that a master information system be developed to access the particular information system desired by the user.

It is ironic that efforts to centralize software information have, in fact, led to its decentralization. There is at present, or under consideration, a hierarchy of information which one must traverse before hard technical facts concerning a desired program can be obtained.

The distribution efforts previously described have a series of common denominators. First, there are no universal standards for programs or documentation. While most libraries make an attempt to maintain a standard of excellence, they are fundamentally at the mercy of the originator. Some programmers and engineers take pride in offering well-structured programs and good documentation. Others could care less. The library, by and large, has little control, but gets all the blame.

Because program libraries historically have been developed using hard-copy libraries as a prototype, the maintenance function is usually divorced from the library function. Almost all libraries disclaim any responsibility for their holdings. Maintenance is the responsibility of the originator. Since software maintenance is a time-consuming, unrewarding task, it is more often than not ignored.

With a few notable exceptions none of the existing software distribution efforts pays heed to problems associated with the portability and adaptability of its distributed software. Usually the most that can be expected is a statement of the originating hardware. It is always the user's responsibility to implement the program and to adapt it to a particular need.

In spite of the fact that passive software libraries perform less than optimal service, and despite the widespread and uniform consensus as to their lack of service, these libraries continue to increase in number. The patterns of birth of new software libraries are remarkably consistent. Each new library is launched as a unique entity which will cure the ills of the existing structure. The new library, however, usually functions almost identically to its maligned predecessors, since the concept of a passive library is by its very nature limited.

Thus, the situation today is one in which there are a multitude of passive libraries holding a multitude of nonportable and non-adaptable software. As mentioned earlier, the newest twist in passive library operation is the development, or use, of information retrieval systems which will index the passive libraries. This effort is bound to fail for a variety of reasons. First, current information is distributed over a variety of systems. Thus, either the user must interrogate a multitude of systems or a new system must be developed. This would be an information system which would retrieve information systems. This, however, is fraught with difficulties since information retrieval systems are notoriously nonportable.

The cure for the chaos of software libraries is not another library. In fact, it is suggested that the time has come to stop inaugurating off-the-shelf software distribution. Coordinating existing software distribution methods will not add to the solution; it will only add to the problem and the cost. The cure is the establishment of an active coordination and distribution effort which will actively participate in the maintenance and use of software. In short, the solution is a software center [2].

ACKNOWLEDGMENT

The preparation of this paper was made possible by a National Science Foundation collaborative research grant made to Carnegie-Mellon University, the University of Colorado Computing Center, and Weidlinger Associates.

REFERENCES

1. Schiffman, R. L., (1972), "Report on the Special Workshop on Engineering Software Coordination," Computing Center, University of Colorado, Report No. 72-2.

2. Baron, M. L., Schiffman, R. L., and Fenves, S. J., (1974), "A National Software Center" International Symposium on Structural Mechanics Software.

3. Allen, W., (1973), "Shared Software in the Production Environment," Computers and Structures, Vol. 3, pp. 59-76.

4. Fosdick, L. D., (1972), "Proceedings of the Conference on the Validation and Distrubtion of Computer Software," Department of Computer Science, University of Colorado, Report No. CU-CS-004-72.

5. American National Standards Institute, (1966), "USA Standard FØRTRAN," United States of America Standards Institute, X3.9-1966.

6. Brooks, A. C., and Fenves, S. J., (1970), "A Computer-Aid d Construction Management System," Construction Review, U.S. Department of Commerce, Vol. 16, No. 7, pp. 4-7.

7. National Academy of Sciences, (1970), "Directory of Computer Programs of Federal Construction Agencies," Building Research Advisory Board, Survey of Practice Report No. 9.

8. University of Georgia, (1973), "CØSMIC-Software Submittal Guidelines," Information Services Division, Office of Computing Activities.

9. McCormick, J. M., Baron, M. L., and Perrone, N., (1973), "The STØRE Project (The Structures Oriented Exchange), "Numerical and Computer Methods in Structural Mechanics (edited by S. J. Fenves, N. Perrone, A. R. Robinson, W. C. Schnobrich), Academic Press, New York, N.Y., pp. 439-458.

10. Citerley, R. L., Flora, E. B., and Bader, R. N., (1974), "ASIAC - The Center with Structures Information," International Symposium on Structural Mechanics Software.

11. Federal Register, (1973), Tuesday, August 28, 1973, Washington, D.C., Vol. 38, No. 166, p. 2299.

12. Marron, B., Fong, E., Fife, D. W., and Rankin, K., (1973), "A Study of Six University-Based Information Systems," National Bureau of Standards Technical Note 781.

13. Fife, D. W., Rankin, K., Fong, E., Walker, J. C., and Marron, B., (1974), "A Technical Index of Interactive Information Systems," National Bureau of Standards Technical Note 819.

14. Bugliarello, G., Chmn., (1972), "Computer Languages and Program Libraries, by the Task Committee on Problem Orientation Languages in Hydrology of the Committee on Surface-Water Hydrology of the Hydraulics Division," Journal of the Hydraulics Division, ASCE, Vol. 98, No. HY7, Proceedings Paper 9072, pp. 1243-1253.

15. Reti, G. A., Chmn, (1973), "Engineering Computer Program Documentation Standards, by the Subcommittee on Program Documentation of the Committee on Computer Applications of the Soil Mechanics and Foundations Division," Journal of the Soil Mechanics and Foundations Division, ASCE, Vol. 99, No. SM3, Proceedings Paper 9614, pp. 249-266.

16. Medearis, K., (1973), "Report on an Investigation of the Feasibility of Establishing a National Civil Engineering Software Center," American Society of Civil Engineers.

17. American Society of Mechanical Engineers, (1969a), Use of the Computer in Pressure Vessel Analysis.

18. American Society of Mechanical Engineers, (1969b), Computational Approaches in Applied Mechanics.

19. American Society of Mechanical Engineers, (1970a), Computer Software in Structural Analysis.

20. American Society of Mechanical Engineers, (1970b), On General Purpose Finite Element Computer Programs.

21. American Society of Mechanical Engineers, (1971), Engineering Computer Software-Verification, Qualification, Certification.

22. American Association of State Highway Officials, (1973), "Computer Systems Index," Administrative Subcommittee on Computer Technology.

23. National Academy of Sciences, (1974), "A Program for Facilitating the Use of Computers in Federal Construction Agencies," Building Research Advisory Board, Federal Construction Council, Technical Report No. 64.

APPENDIX A

The following is a selected list of names and addresses of some of the groups and organizations described in this paper.

1. AASHTO - Mr. K. F. Kohler
   Federal Highway Administration
   400 7th Street S.W.
   Washington, D.C. 20590

2. APEC - Ms. Doris J. Wallace
   Grant-Deneau Tower
   Suite M-15
   Fourth and Ludlow Streets
   Dayton, Ohio 45402

3. CEPA - Mr. J. C. Rodgers
   Rodgers and Associates
   Rockville, Maryland 20850

4. COMMON - Mr. William L. Hogan
   Suite 1717
   Tribune Tower
   435 North Michigan Avenue
   Chicago, Illinois 60611

5. CUBE - Mr. G. R. Kennedy
   Brock University
   St. Catharines
   Ontario, Canada

6. FOCUS - Mr. C. F. White
   Singer-Librascope
   808 Western Avenue
   Glendale, California 91201

7. HEEP - Mr. J. C. Bridwell
   Kentucky Department of Transportation
   State Office Building
   Frankfort, Kentucky 40601

8. Honeywell User's Group - Mr. Wallace Juntunen
   Honeywell Admin. Service
   2701 4th Avenue South
   Minneapolis, Minnesota 55408

9. HRIS - Dr. Paul E. Irick
   Highway Research Board
   National Academy of Sciences
   2101 Constitution Avenue, N.W.
   Washington, D.C. 20418

10. ICES User's Group - Mr. F. E. Hajjar
    PO Box 231
    Wooster, Massachusetts 01613

11. IITRI - Mr. Anatole Longinow
    IIT Research Institute
    Chicago, Illinois 60616

12. SHARE - Mrs. J. A. Johnson
Suite 750
25 Broadway
New York, New York 10004

13. UNIVAC - Mr. Jean Frable
PO Box 500
Bluebell, Pennsylvania 19422

14. VIM - Mr. A. L. Siegel
Computer Center
Battelle Memorial Institute
505 King Avenue
Columbus, Ohio 43201

APPENDIX B

Subsequent to the preparation of this paper it was learned that the Federal Construction Council of the Building Advisory Board is implementing a technology transfer effort within the federal construction agencies [23]. This program will pool the computer program resources of the various federal construction agencies and will coordinate program development.

The project will make available a wider variety of computer programs through a time-sharing network called FACTS (Federal Agencies Computer Time-sharing Systems). The programs will be maintained by the Atlanta Data Processing Center of the General Services Administration. The development, adaptation, maintenance, and validation of computer programs in the FACTS library will be the responsibility of the contributing agency. Contributions will be voluntary. The development of the executive library system and associated documentation will be the responsibility of the Office, Chief of Engineers, U.S. Army Corps of Engineers.

The programs to be included in the FACTS library will be selected by the Federal Construction Council Standing Committee on Computer Technology in concert with specialists in the various engineering disciplines. It is anticipated that approximately 100 currently available programs will be placed in the library during the first year of operation. It is expected that the library will contain approximately 250 programs at the end of five years of operation.

In addition to the library time-sharing functions, the project plans to call for the holding of training sessions and workshops on the use of the FACTS system and the programs resident in the system.

A NATIONAL SOFTWARE CENTER FOR ENGINEERING

M. L. Baron
Weidlinger Associates
New York, New York

S. J. Fenves
Carnegie-Mellon University
Pittsburgh, Pennsylvania

R. L. Schiffman
University of Colorado
Boulder, Colorado

ABSTRACT

This paper discusses several efforts currently in progress which may lead to recommendations for the establishment of a national software center for engineering. The paper outlines past and current studies which are directed toward determining the feasibility of a center for engineering software coordination. In addition, the paper describes the possible functions of a center and comments on possible implementation criteria for successful center operation.

INTRODUCTION

The concept of a national engineering software center is discussed in this paper, together with a description of several investigative efforts currently being made, under National Science Foundation sponsorship, on the methodology and feasibility of organizing such a center in civil engineering. It is perhaps in order at this point to ask why should such a center be organized. The answer to such a question is necessarily complex and cannot fully be answered here. However, two primary arguments can be made for the establishment of such a center. First, it would speed the process of the transfer of public domain research in engineering and the related sciences to the practicing profession. This research is largely in the public domain by virtue of its sponsorship by various public agencies. In addition, the national engineering software center would, if properly organized, support and further the use of computer-based techniques and associated software by engineers.

In the first instance, the concern is primarily with technology transfer. It is no secret that there is a delay, often as much as five years or more, between the development of new results and methodology in engineering research and its implementation on a practical basis by the engineering profession. One reason for this lies in the fact that much of the meaningful research is often conducted by graduate students seeking advanced degrees in universities. This research is usually concerned only with solving physical problems, and the software, however potentially useful, is viewed as a tool to produce specific numerical results. Such work is often published, and the results are noted. The practical use of the research via the software, however, is just as often not feasible, since the programs involved are very often not usable by anyone but their authors [1]. By assuring in a variety of ways, as discussed later in this paper, that appropriate software associated with publicly sponsored research is portable, adaptable, maintained, and well documented, a national software center can perform one of its most important and useful tasks.

These comments on the need for portability, adaptability, and good documentation apply equally well to the direct development of public domain software under various government auspices.

The second task that has been mentioned is basically educational in nature. Although the use of computers in engineering has progressed tremendously in the last ten years, their acceptance as a tool is by no means universal in the engineering industries. In fact, there are still many engineers who look upon the computer as a potential rival, and it is only when they must react to an emergency and get answers which could not be obtained by any other means that they begin to look on the computer as a useful tool. Many of the larger and more progressive engineering firms practicing today are, of course, computer oriented, and in fact many have their own computer departments for developing and using software. In the past, it has been relatively difficult for small firms to do this since they could afford neither a computer nor the personnel to run it. Today, however, the use of remote batch terminals and time sharing has brought the computer within the means of all engineering firms and even individual practicing engineers. Simple but efficient portable terminals enable the engineer to run programs equally well from office or from home. The tremendous progress in such procedures is of course well known to many engineers, but there is still a large and important group of practicing professionals to whom such information would come as a surprise. It is contended that since public agencies sponsor a large amount of theoretical and applied research in engineering, they have a public responsibility to foster the means for making this information readily and efficiently available to the entire profession, i.e., both the sophisticated and the nonsophisticated computer user, in as short a time as possible. Also, a mechanism is necessary to provide incentive and the means by which both small and large engineering firms can readily use the developed software. If a national software center is to be a viable tool for aiding and promoting the use of computer technology by engineering firms in a wide variety of general disciplines, it will necessarily have to respond to the needs of a group of users with widely varying degrees of understanding of computers and computer-based technology. This does not imply that every engineer will run programs on home terminals. In fact, an important outlet for the programs of the software center will undoubtedly be the private software companies who will make the programs available to the smaller engineering firms.

In summary, it is the authors' contention that it is not sufficient for the government merely to sponsor research, but rather, it must also pay heed to the technology transfer aspects of getting the research to the profession in an efficient and usable form. One important tool for effecting this computer-based technology transfer could be a national engineering software center.

In a workshop on engineering software coordination held at the University of Colorado in 1971, under the sponsorship of the National Science Foundation [2], a resolution was adopted which called for "a national effort to optimize common use of engineering software." The workshop also recommended that further studies be carried out on: (1) the desire and need by the profession for such a center, (2) the methodology for establishing efficient procedures for technology transfer for use by the center, and (3) the establishment of a pilot program involving a study of the methods and problems inherent in establishing such a center.

Two questionnaire studies on the professional feasibility of a national software center for the engineering profession have been completed under the sponsorship of the National Science Foundation. The first study was conducted by Medearis [3], under a project monitored by the Research Office of the American Society of Civil Engineers. While the responses to the questionnaire were somewhat inconclusive and the respondees varied widely in their interpretations of the function of a center, there appeared to be considerable sentiment for establishing some type of national effort. A second study by Kraus [4] directed at the pressure vessel industry indicated considerable favorable sentiment.

An overview of these and other efforts has recently been articulated by Babendreier [5]. This exposition reviews the current status of National Science Foundation initiatives and reports the findings of a special meeting held by a working group of the National Academy of Engineering.

The technical feasibility (methodology) of a center is currently being studied under a collaborative research grant from the National Science Foundation. This collaborative effort by the University of Colorado Computing Center, Carnegie-Mellon University, and Weidlinger Associates has reported various aspects of methodology [6].

The definition of a national effort in software coordination for the civil engineering and building construction industry is currently being studied by Civil Engineering Program Applications (CEPA). This study is specifically directed toward that part of the profession that is in the private sector. It is funded by the National Science Foundation and is expected to provide specific recommendations on the need, organization, and implementation of a national software center.

A discussion of several aspects of a center is presented here, as well as a description of some of the work in progress. This exposition should not be considered complete. Rather, it is intended to stimulate discussion and an exchange of ideas between the members of the engineering profession who would be served by a national software center.

SOME BASIC REQUIREMENTS FOR A NATIONAL SOFTWARE CENTER

Starting with the a priori premise that the use of computers will enhance engineering productivity, it appears that the overall objective of an engineering software center is twofold: (1) to increase the base of computer applications and improve the user efficiency of that portion of the profession who already use computers, and (2) to develop an incentive for computer use by the segments of the profession who do not presently use computers.

Conceptually, the national software center, or perhaps more accurately the center for the coordination of engineering software, would serve as an information exchange, dissemination, and validation agency for computer-based engineering technology. Some of its duties would include:

1. The definition of those elements of computer-based engineering technology (software) which have professional utility. The center, with the aid and advice of professional consultants, would determine which programs and program areas should be supported.
2. The collection and storage of useful and available software.
3. The enhancement, verification, maintenance, and updating of those programs which are most useful and in most demand by the profession.
4. The dissemination and distribution of those elements of computer-based engineering technology which fall under the center's purview. Programs distributed by the center would be tailored to operate on the clients' target machines.
5. The dissemination of information, including the distribution of supported and unsupported programs, and educational activities with regard to computer usage.

It should be noted that the five areas of activity listed above, when examined in the large, envision a center which operates in a substantially different manner than existing program distribution libraries [7]. At the present time, most libraries and program distribution centers are essentially passive operations which serve almost exclusively as distribution agents between an author and a user. The concepts of technology transfer and utilization as applied to a national software center, on the other hand, envision the center as an active participant in the development, upgrading, distribution, and use of software packages.

A question arises regarding the practicability and economic feasibility of a center as defined above to coordinate all of the available computer-based engineering technology. This question is addressed later. However, for purposes of establishing requirements it is assumed that the center would serve as an institution for effecting technology transfer in a somewhat restricted range, such as civil or mechanical engineering. It may even be further narrowed to one or more disciplines within a professional group.

For each discipline or set of disciplines, the center would then serve as a focal point in the definition of those elements of computer-based technology transfer which will have the greatest potential for professional use and for advancing the state-of-the-art of engineering. The selection of programs and their continued maintenance in the reference program file of the center must be a dynamic process. The initial selection for programs in a particular discipline would be achieved by means of a review board in each discipline as well as a board of consultants from engineering societies with an interest and capability in the discipline in question. The information gained from the continued use and experience of the center's users with each program must be entered into the general consideration of whether the program is indeed good and whether it should be kept in the center inventory file. Periodic reviews of the center programs with respect to the experience of the users is considered essential. Such a procedure was quite successfully used in Project STORE [8], the STructures ORiented Exchange, which contains many of the technical elements which would be useful to a prototype software center.

The basic central library of reference programs of the center would necessarily be a continuously changing entity, which would respond to the indicated needs of the profession. The center would in all probability maintain both a supported and an unsupported library of programs. The latter would consist of a fairly complete listing and a brief description of all known public domain software in a particular discipline, including programs not in the library but available from other known sources. Many current research projects require that project software be made generally available. Such software could be deposited directly into the unsupported file, or a notation of its existence and place of availability could be given.

The main focus of the center's activities would be the supported activities. These would consist of the collection, verification, enhancement, maintenance, and periodic updating of the software most in demand and most used professionally. There are a variety of operational considerations connected with this activity. First, a mechanism for defining those items of software which would qualify as a part of the center's supported library would be required. Sources of identification of software needs would include: boards of advisors with disciplinary expertise; the professional staff of the center; users of the center, such as individual firms or groups; national (and international) engineering societies and their various computer-related subgroups; public agencies; and private, commercial software houses. These groups would act in identifying useful programs and program areas, as well as in providing information on the availability and source of already existing or "in-development-phase" programs in the identified areas. One can visualize all requests for new development in a particular area passing through an advisory board in that area and finally being given a priority by a central planning board of the center. At this point it is necessary to develop the source of the required need. First, it should be ascertained if a required program exists in a form adaptable to the center's needs. If not, the center would be responsible for developing the program or encouraging other competent professionals to develop the desired program.

It is anticipated that many of the programs will probably originate from publicly funded grants and research contracts or from private firms under some type of purchase arrangement or royalty payment. A separate developmental capability for the center itself is desirable, however, if the center is to serve as and remain a viable useful entity.

A major function of the center would be the verification of engineering software. This is a multiphase operation consisting of the following components:

1. The engineering competency of the algorithms used in the software must be verified. This activity would be a joint effort between the center, its user groups, and the professional groups advisory to the center. This activity requires a close working relationship between center personnel and technical committees and consultants advisory to the center. The ultimate decisions concerning the engineering competence of a program must reside with the engineering community through its technical groups.

2. The competency of the code must be verified. The center staff, using software tools, would determine if the previously verified algorithms are properly programmed.

3. Center personnel would test each supported program with respect to the manner of performance and the calculated results. This activity would initially rely on the use of originator-supplied test problems. With advancing technology in the computer sciences, it is conceivable that software will be available which will automatically generate test problems.

Once a program has been verified the center would be responsible for ensuring that the program is properly documented. The documentation must be prepared so as to be useful to a wide range of potential users from the most sophisticated to the novice.

The center would be responsible for providing portable and adaptable software to its user community. The portability and adaptability effort would be concerned with providing a user with a given program from the center's library, which would run on a specified target machine, using a specified operating system and a specified input/output mode (time sharing, batch process, and so on).

A significant part of the activity of the center would be the enhancement of supported programs. In this effort the professional programming staff of the center would apply its skills along with the latest developments in the computer sciences to improve, rework, and polish the code of a given program, so as to produce versions of the program which operate in a cost-effective manner on a specified target machine in a specified operating environment.

Those programs residing in the center's supported library would be maintained by center staff. Maintenance would be performed in the customary manner within a feedback loop to the active user community.

The center would have an updating responsibility with regard to its supported programs. This would include the implementation and development of program and algorithm modifications. Program modifications are largely concerned with the removal of bugs and improvements in operating efficiency. This function would be largely the responsibility of the center staff. Algorithm modification would be mostly generated by the user groups. The center staff would have an implementation and testing role.

The growth of the center would be evolutionary. In its initial stages, it would confine itself exclusively to managing programs supplied by the user community. As maturity develops, the center could serve as an agency for in-house program development or could contract for these developments by outside groups.

The activities described above view the software coordination center as playing a role similar to that of standards associations in the physical testing world. The support of a program by the center would provide a basic professional acceptance of the software, without in any way hindering developments of new software. To the contrary, it would free time currently being spent unproductively on remanufacturing software for the development of new programs.

The center could make programs available for use by several different methods:

1. The dissemination of card decks and/or program tapes together with full documentation, especially users manuals.
2. Through remote terminals in both time-sharing or batch modes, using the programs which would be available on a large central computer or on a series of large central computers operated or leased by the center at several local centers; some of these centers could be those operated by private software companies.
3. Through network systems, which interconnect groups of large computers, each of which might have several of the programs entered. This could include private, semiprivate, and public network systems.

Because of the wide variety of computers on which the programs must be run, the question of portability of the programs becomes important. The active program library of the center will consist of programs which have been made portable (or programs with a series of separate versions) for use on several of the available machines. It is anticipated that portability guidelines can be set up by the center for use by program developers, but nevertheless, it is probably unreasonable to expect the original program developers to write completely portable programs. The actual programming for portability will probably be done by the center staff, or perhaps by outside firms under contract. At best, it may be possible to have the original program developers adhere to certain general portability guidelines which will be aimed at helping the center staff who may later work on specific portability requirements for the active program library items.

IMPLEMENTATION

A national software center should be a nonprofit entity. Within this constraint the center can be organized by government, industry, or a university or consortium of universities. If the center is established by government it must be housed within an agency which can maintain working relationships with other agencies. If the center is housed within an industrial firm, it must be a nonprofit effort, and the firm should be one which has the respect of the profession. A university setting for the center would be feasible if the center staff is full time. In all cases the image of the parent group must be one which inspires confidence in the user community.

There are three possibilities for funding the center. First, the center can be wholly funded by government as a public service. Since the major source of the center software would have been developed with public support, it follows that there is a government responsibility to implement the technology transfer aspects of the software. The second possible mechanism of support is by means of government-guaranteed funding. Under this plan the center would attempt to be self-sustaining. Government, however, would underwrite any deficits. The third funding alternative would be one in which the center would be funded entirely by its user community. Certainly, the various agencies of government would be subscribers to the center.

There is a serious question as to the immediate feasibility of a national software center which encompasses all the engineering disciplines, or even all of the many disciplines in a single branch of engineering. This is a problem of scale. If a software center is to cover all engineering disciplines or even, for example, all of the civil engineering disciplines and have as its clientele the full complement of these professions, it is questionable whether the center would be economically viable under current technology. This is not to say that such a center could not improve on present mechanisms of computer-based technology transfer; however, the improvement would in today's technology probably be of questionable cost/benefit. Considerably more should be known about the technological and professional problems associated with software center operation before an efficient center could be established on the very large scale mentioned above.

On a smaller scale, it is currently feasible to establish a software center. An appropriate set of disciplines could be chosen, and the center would serve the entire professional community of users.

Some of the requirements for a successful operation are:

1. The subdisciplines should be relatively cohesive so that a confidence level can be established between the user and the center.

2. The software objectives of the subdisciplines must be readily definable.

3. The subdisciplines should be of such a nature that professional support is likely.

4. The scale should be of such a nature as to make the operation economically viable.

5. The subdisciplines should be relatively free of proprietary software interests. Furthermore, they should be ones in which there is a relatively free flow of information between people.

There is some precedent for the above statements. A survey of current software distribution efforts [7] indicates that those activities which best serve their users are ones confined to a recognized discipline and/or serving a limited geographical area.

WORK IN PROGRESS

One of the studies currently in progress is a National Science Foundation grant for collaborative research on computer-based technology transfer in civil engineering. This project is a collaborative effort involving Carnegie-Mellon University, the

University of Colorado Computer Center and Weidlinger Associates.

The purpose of the research program is to study, develop, and test general techniques for making computer applications programs and systems portable and adaptable. The overall objectives of this research program are to develop techniques which will enable new programs to be written possessing the properties of portability and adaptability and to permit existing programs to be made more portable and adaptable.

The anticipated products of this research program are threefold:

1. A set of standards will be established to enhance portability and adaptability of application programs and systems written in higher-order languages. These standards will apply at the level of the initial development of a program or system.

2. A set of production procedures will be established to enable the programs and systems written under the constraints of item 1 to be restructured for a variety of engineering options, hardware configurations, and input/output environments. The procedures will apply at the software distribution center level and will facilitate the reconstruction of programs with a minimum of effort.

3. A second set of production procedures will be established to enable existing programs written for a fixed set of engineering options and for a single operating environment to be dissected into a structure satisfying the constraints of item 1. The procedures will apply at the software center level with, it is hoped, a minimal interaction with the application-oriented initial developer.

The techniques being developed to accomplish the three goals listed above will make possible the dissection of a collection of problem-solving capabilities embodied in a program consisting of algorithms, input/output philosophy, and data structure. The selected capabilities of the original program can then be implemented on a variety of machines and input/output configurations in a standardized production environment. The techniques will enable programs and systems to be designed with minimal hardware and operating system constraints, so that the assumptions, limitations, and capabilities of these programs may be redefined for different environments.

Next to the problem of portability and adaptability there is the equally important problem of program interaction. It is expected that the procedures developed by this study will influence future designs of operating systems to facilitate portability, adaptability, and program interaction.

The primary activities associated with this project have been directed toward the development of software tools which a software coordination center could use to process programs and algorithms submitted to the center in order to produce custom programs requested by center clientele [9, 10]. These include: (1) the design and development of a generalizer/specializer system (GSS) [11, 12, 13], (2) the manual decomposition of programs in order to gain empirical experience in program decomposition and recomposition [14, 15, 16], and (3) the development of software tools to aid in the evaluation and machine decomposition of application programs [17, 18, 19, 20].

GENERALIZER/SPECIALIZER System (or alternatively referred to as the MASTER FILE System) is designed as a software system for building, maintaining, and customizing programs and modules within a software center. In particular, it provides a mechanism whereby programs and algorithms submitted to the center can be added to or removed from the center's holdings in an orderly manner. Equally importantly, it provides a means by which a user of the center can order a custom program tailored to suit particular requirements with respect to both engineering options and machine environment. This system, conceptually shown in Fig. 1, consists of two major software entities, the GENERALIZER and the SPECIALIZER, and a data base, the MASTER FILE.

The MASTER FILE is a nonexecutable system consisting of program modules and other information in a form that can be assembled by a SPECIALIZER to an executable custom program. There is a corresponding MASTER FILE DIRECTORY of program areas and program modules which are available for specialization to a custom program.

The GENERALIZER is a body of software which will decompose a debugged prototype program into modules defining the input and output environments, the algorithms embodied in the program, the machine-dependent code, and the data structure.

The GENERALIZER is sketched in Fig. 2. This figure shows the software entities and the flow of information through the GENERALIZER. The software entities are:

1. ANSI CHECKER: This unit of software checks source program statements for conformance with ANSI standards. In its present implementation the ANSI CHECKER will flag non-ANSI code. In future implementations this unit of software will alter the input program to ANSI text.

2. The RECOGNIZER [21] is a program which takes an input of an ANSI FORTRAN program and produces, as output, an equivalent expanded FORTRAN program and associated data structures. The output of the RECOGNIZER is:

   a. Expanded source text including the FORTRAN source statements, a coding of the type of FORTRAN statements, and line number information

   b. Identification, in tabular form, of the blocks comprising the source program

   c. Tabular information about the program as a whole

3. The GRAPH BUILDER is a program which uses the block table to generate a data structure representing the flow of control of the input program, producing a logic table.

4. The GRAPH MODIFIER/EXTRACTOR operates on the expanded FORTRAN text and the logic table. It modifies particular elements of the graph to extract and construct new modules. The program produces the data structure for the program modules to be stored in the MASTER FILE. The output of this program is the module logic table and the module text.

5. The PROGRAM AREA BUILDER collects the logic tables of all modules comprising a given program area. It produces a global directory for the specification of user options.

The SPECIALIZER is the software which produces the customized executable program requested by the software center client. A flow

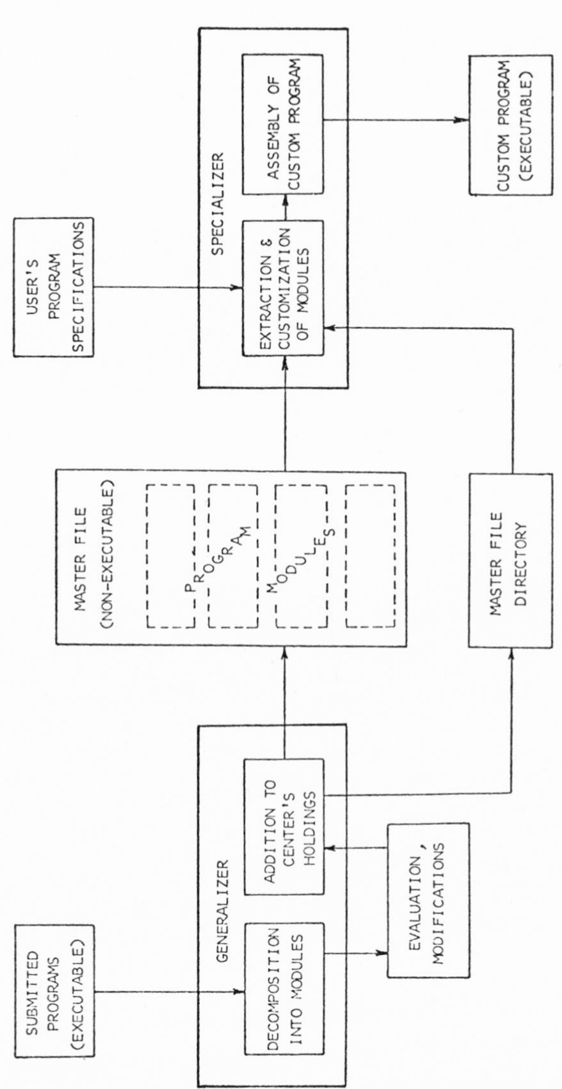

Fig. 1 Generalizer/Specializer System

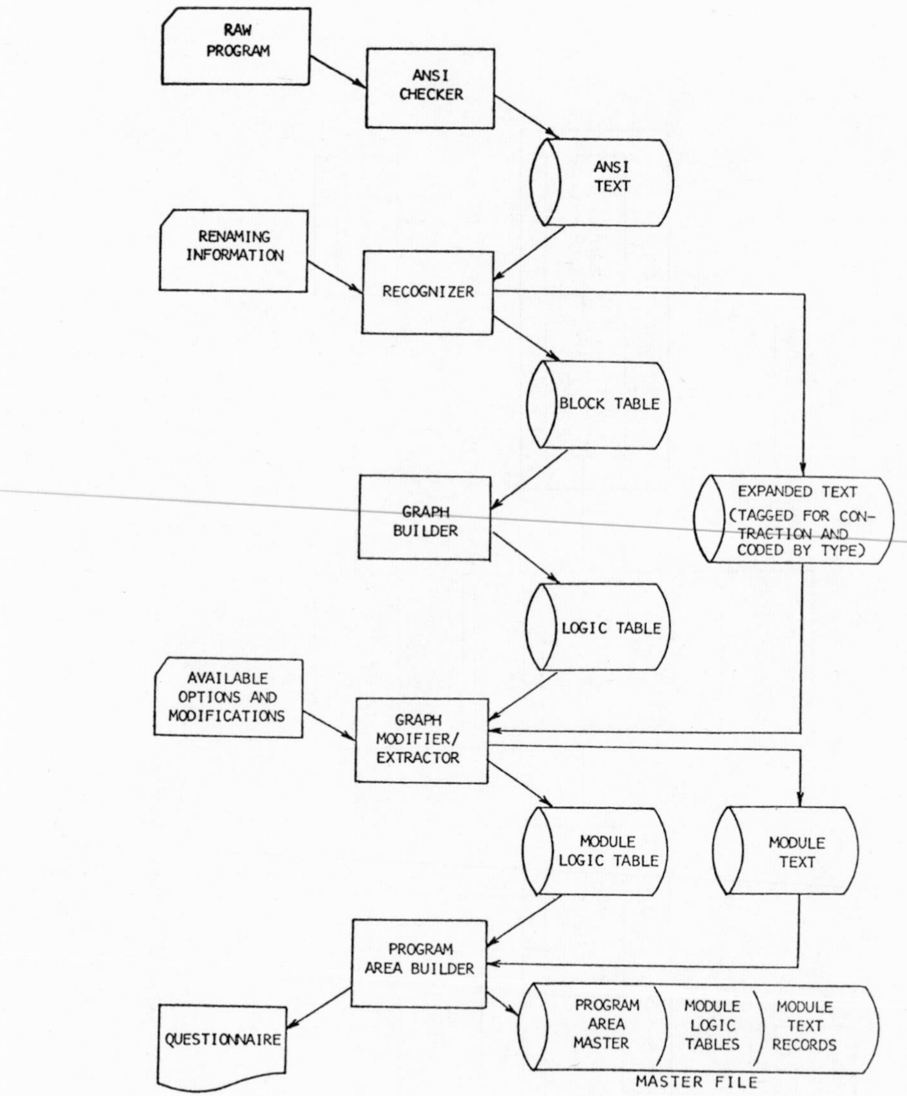

Fig. 2 Generalizer

of the SPECIALIZER is shown in Fig. 3. The SPECIALIZER consists of the following software entities:

1. The DIRECTORY SEARCH program processes user-supplied descriptions to locate the program area.

2. The SELECTION GENERATOR converts user-option specifications into specifications for extracting a target program.

3. The LOGIC EXTRACTOR processes the selection specifications against the module logic tables to determine which segments of which modules are to be included in the target program and provides customized statements for module segments to be modified.

4. The TEXT EXTRACTOR retrieves and stacks the required program text.

5. The POST-PROCESSOR prepares the text for final output.

Another study currently being funded by the National Science Foundation is seeking a definition of a national effort to promote effective application of computer software in the practice of civil engineering and building construction. This project is in its initial phases and is being conducted by Civil Engineering Program Applications (CEPA). The study is divided into five phases:

1. The detailed definition of the problem and the preparation of a work flow diagram and task schedule.

2. The collection of data and the establishment of the criteria which will provide the basis for the selection of activities in later phases. Specifically, this phase is planned to:

a. Assemble and review existing pertinent information. Among these data are previous studies, reports, and current surveys.

b. Determine, from available sources, the approximate number of practitioners who are computer users and their predominant application and to forecast additional users and applications for the foreseeable future.

c. Ascertain the types of sources of civil engineering application software and collect them into groups by varying classifications (i.e., source, availability, support).

d. Investigate past and present cooperative efforts which may contribute experience to the determination or limitation of center activities.

e. Review academic efforts, as they relate to the professional, to develop and to transfer it to industry use and to determine the problems involved with technology diffusion and methodology and how a national effort can best fit this need.

f. Review past and present efforts of government agencies to develop software, implement it internally, and distribute it externally and to determine how a national effort such as the one proposed can best assist in this task.

g. Examine hardware and software vendors and service bureau activities past and present as related to the problems of software development, distribution, hardware compatibility, documentation, liability, and advertising and to determine how a national center could help or hurt them.

h. Examine the methods of professional information systems (library automation and information retrieval, cataloging and abstracting service) for possible implementation by a software center.

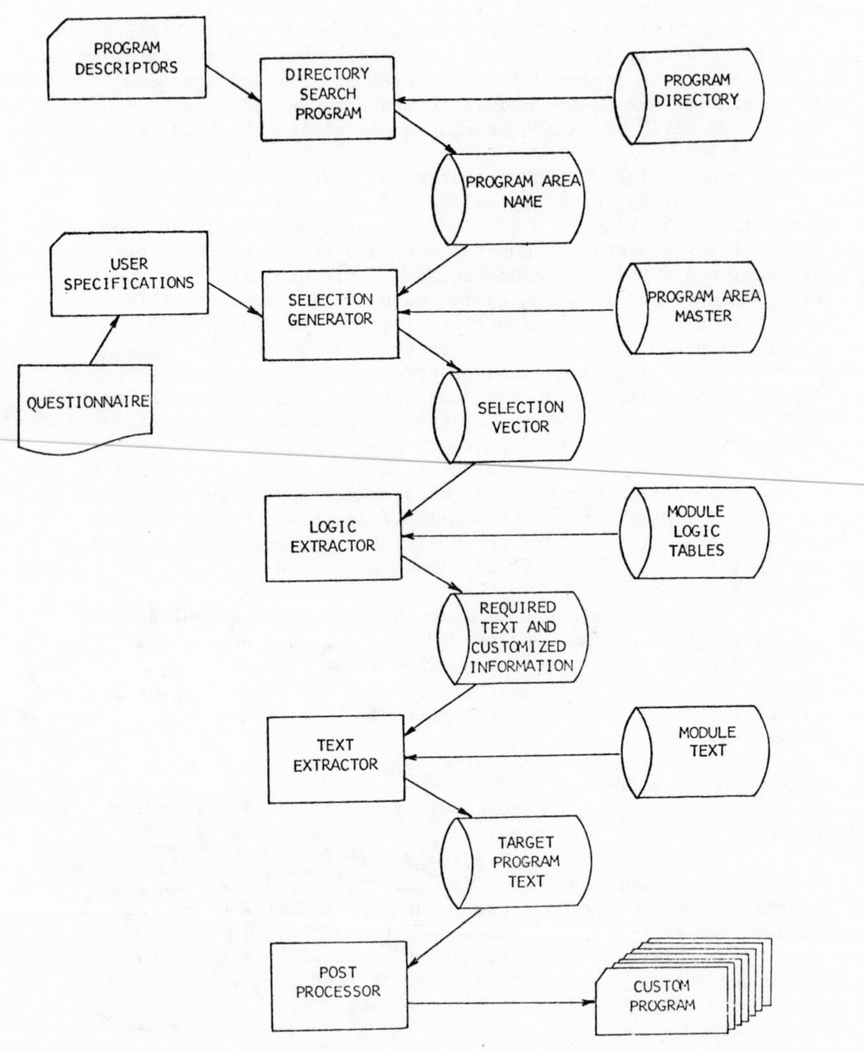

Fig. 3 Specializer

i. Add CEPA's experience (public and private engineering practice) to the findings of the examination of academic, governmental, and commercial experience to arrive at their interrelationship.

3. This phase will consolidate and analyze the information assembled above. In order to formulate a solution, feasible activities will be weighed against the criteria established above, rejecting some of the functions examined and selecting others and combining these into an effective whole.

4. The establishment of a set of definitive conclusions and recommendations.

5. The formulation of implementation details based on the conclusions and recommendations developed under item 4. These will include, but will not necessarily be limited to the following areas:

a. Management-Administration: Recommend the number, structure and type of the center's governing body. Recommend the number, structure, and classification of the staff necessary to administer and operate the center. Investigate and recommend procedures for continuously evaluating center effectiveness.

b. Technical: Recommend from information gathered in earlier phases the activities and services to be undertaken with details of how they may be accomplished. Consider, for instance, software development, software distribution, education, and maintenance. Investigate and make recommendations as to the need, type, and size of hardware necessary to provide the above activities and services.

c. Financial: Investigate and make recommendations concerning the continuing support of such a center including possible support by the National Science Foundation, other government funding, user fees, hardware and software vendors, service bureaus, or any combination of these.

d. Legal: Investigate and report the findings concerning the legal aspects of the following as they affect the center or its operation: liability for software, proprietary rights, tax structure, information data banks, antitrust laws, and advertising.

e. Budget: Prepare an annual operating budget. Break it down in parallel to the center structure recommended in the study and incorporate the above categories where applicable.

CONCLUDING COMMENTS

In summary, this paper has attempted to review the past and current activities related to the possible establishment of a national center for the coordination of engineering software. It further has attempted to bring into focus the possible scope of activity of a national software center and some of the problems concerning the implementation of such a center.

It should be noted that the entire question of software coordination is large and complex. Further, there is not universal agreement as to the most appropriate mechanism for enhanced software coordination. It is agreed, however, that engineering productivity can be increased by better means of computer-based technology transfer.

To this end, the authors hope that this paper will encourage meaningful discussions which will ultimately lead to an optimum solution of this pressing problem.

ACKNOWLEDGMENT

The preparation of this paper was made possible by a National Science Foundation collaborative research grant made to Carnegie-Mellon University, the University of Colorado Computing Center, and Weidlinger Associates.

REFERENCES

1 McCormick, J. M., and Baron, M. L., "On the Implementation of Computer Software," Guidelines for Acquisition of Computer Software for Stress Analysis, ASCE National Structural Engineering Meeting, April 9-13, 1973, San Francisco, California, Meeting Preprint 2022, 1973.

2 Schiffman, R. L., "Report on the Special Workshop on Engineering Software Coordination," Computing Center University of Colorado, Report No. 72-2, 1972.

3 Medearis, K., "Report on an Investigation of the Feasibility of Establishing a National Civil Engineering Software Center," American Society of Civil Engineers, 1973.

4 Kraus, H., "Attitudes Toward Computer Software and its Exchange in the Pressure Vessel Industry," Hartford Graduate Center, Rensselaer Polytechnic Institute, Report, 1973.

5 Babendreier, C. A., "Technology Transfer, Requirements in Engineering Sciences," ASCE Annual and National Environmental Engineering Meeting, October 29 - November 1, 1973, New York, New York, Meeting Preprint 2096, 1973.

6 Fenves, S. J., Schiffman, R. L., and Baron, M. L., "A Position Paper on Computer Based Technology Transfer in Civil Engineering," Colloque International sur Les Systèmes Intégres en Génie Civil, CEPOC, Université de Liège, Belgique, 1972.

7 Schiffman, R. L., "A Review of Current Software Dissemination Practices and Organizations," International Symposium on Structural Mechanics Software, 1974.

8 McCormick, J. M., Baron, M. L., and Perrone, N., "The STORE Project (The Structures Oriented Exchange)," Numerical and Computer Methods in Structural Mechanics, Edited by S. J. Fenves, N. Perrone, A. R. Robinson, W. C. Schnobrich, Academic Press, New York, 1973.

9 Ewald, R. H., Goodspeed, C. H., Jones, R. A., Smith, R. R., Voreades, J. and Wong, Y. C., "Computer Based Technology Transfer," Paper to be submitted to ASCE Sixth Conference on Electronic Computation, 1974.

10 Fenves, S. J., "Methods Appropriate to an Engineering Software Center," Paper presented at ASCE Annual and National Environmental Meeting, October 29 - November 1, 1973, reprinted in APEC Journal, 8, No. 5, November 1973, pp. 7-12.

11 Ewald, R. H., Schiffman, R. L., and Smith, R. R., "Composition/Decomposition System," Computing Center University of Colorado, Report No. 72-29, 1972.

12 Fenves, S. J., Goodspeed, C. H., Wong, Y. C., "Master File Structure for Software Decomposition/Composition System," Department of Civil Engineering, Carnegie-Mellon University, Report R73-1, 1972.

13 Ewald, R. H., Fenves, S. J., Goodspeed, C. H., Jones, R. A., Poole, P. C., Schiffman, R. L., Smith, R. R., and Wong, Y. C., "Generalizer/Specializer," Joint Report, Carnegie-Mellon University and University of Colorado Computing Center, 1974.

14 McCormick, J. M., "Modularization of AISC Column Design Programs," Weidlinger Associates Report, 1974.

15 McCormick, J. M., and Ranlet, D., "FRAME3D An Experiment in the Composition of a Portable and Adaptable Structural Analysis Program," Weidlinger Associates Report, 1972.

16 McCormick, J. M., and Ranlet, D., "FRAME3D Input Files and Sample Problems," Weidlinger Associates Report, 1973.

17 Ewald, R. H., "T06400," Computing Center, University of Colorado, Report No. 72-30, 1972.

18 Jones, R. A., and Schiffman, R. L., "T01130, A Program to Transfer an Algorithmically Based Program Written for a Large Machine to a Small Machine," Computing Center, University of Colorado, Report No. 72-32, 1972.

19 Jones, R. A., and Taylor, D. P., "FR1108, A Program to Transfer Certain Machine Dependent UNIVAC 1108 Statements to ANSI FORTRAN," Computing Center, University of Colorado, Report No. 73-25, 1973.

20 McCormick, J. M., "Guidelines for Designing Portable and Efficient Large Scale FORTRAN Programs," Paper presented at ASCE Annual and National Environmental Meeting, October 29 - November 1, 1973.

21 Poole, P. C., Smith, R. R., and Winograd, S. F., "The RECOGNIZER, Version 2.0," Computing Center, University of Colorado, Report No. 74-2, 1974.

SOFTWARE EXCHANGE WITHIN THE EUROPEAN COMMUNITY

C. Mongini-Tamagnini and G. Gaggero
Commission of the European Communities
Joint Research Centre - Ispra Establishment
CETIS - COPIC
Ispra, Italy

ABSTRACT

An outline of the software exchange situation in Europe is presented. Various partial solutions (program libraries, indexes, service bureaus, large integrated program systems) applied to the problem of software sharing are discussed. The role of the Computer Program Information Centre (COPIC), set up in 1971 at CETIS, a Division of the Joint Research Centre of the Commission of the European Communities, is explained in detail. The aim of COPIC is to provide computer users with complete and objective information on programs and, by the most suitable means, to facilitate access to programs. COPIC endeavors to provide various forms of support (maintenance of a computerized program-information data base, automatic information retrieval, program testing, study, advice, maintenance, distribution, and so on); it operates within the European Community and covers, in principle, the main computer application fields. Finally, CETIS and some of its achievements and projects related to COPIC activity are also presented (e.g., the development of a fully automatic information retrieval system and the participation of CETIS in the European Informatics Network).

INTRODUCTION

Software Exchange Situation in Europe

In Europe as in the rest of the world, the use of computers has grown up rapidly during the past decade, both in the commercial and in the scientific and technical fields. Consequently, computer applications have advanced quantitatively, many programs being written by software companies, computer firms, universities, industries, and governmental bodies. Nevertheless, the development of new programs involves increasing costs, and any action aimed at promoting software sharing is therefore considered with favor.

For an efficient software exchange, the first need is for complete and objective information on the software available on the market; second, suitable ways to access the chosen programs must be available to avoid unsuccessful attempts and time delays; finally, as programs are frequently being corrected, improved, modified, or extended in scope, it is highly desirable that some maintenance should be provided.

To solve, at least partially, these and other problems related to software sharing various attempts have been made in Europe. They may be classified, for simplicity, into four major types:

1. Creation of program libraries which collect and distribute programs, usually within a particular field of computer application and/or in certain geographical areas

2. Organization of programs indexes, which are often compiled and maintained by national institutions, professional associations, computer manufacturers, or specialized publishing companies

3. Establishment of centers offering programs for use on a computer service bureau basis, possibly with teleprocessing facilities

4. Development of large integrated systems supporting comprehensive collections of problem-solving programs related to specific but wide areas of computer application.

To the first category may be assigned the following libraries operating in Europe: the Nuclear Energy Agency Computer Program Library (NEA-CPL) in the field of nuclear reactors, OECD area; the Queen's University of Belfast Library with programs in physics; and the University of Edinburgh Library for Scientific and Social Sciences.

Program indexes are maintained by the National Computing Centre (NCC), Manchester, U.K., for the U.K. area, mainly ADP and management field; the Institution of Electrical Engineers (IEE), London, U.K.; the Centre pour l'Aide au Calcul Technique (CACT), Paris, France, in the construction engineering field, France area; the University of East Anglia, U.K., with programs in geography and environmental sciences; Software International (Guide Européen du Software), Paris, France, mainly system software and ADP; and IRI (Istituto per la Ricostruzione Industriale), whose index is still in the planning phase, Italian area.

In addition to the above reported indexes, various program lists or compilations are published in scientific and technical journals.

In the third category, apart from service bureaus belonging to computer firms and operating in various places in Europe, one should note in the field of engineering software the computer service center of CACT in France, and in the U.K., the CADC (Computer Aided Design Centre) of Cambridge University and the GENESYS Centre at the University of Loughborough. The last two rely on several interlinked service bureaus in the U.K.

To the fourth category may be assigned such developments as ICES, the MIT Integrated Civil Engineering System; the GENESYS System, developed, maintained, and marketed by the GENESYS Centre; the ASKA System, developed and maintained by ISD, University of Stuttgart.

These and other systems differ in several aspects, but all are based on the single assumption that the development, maintenance, and sharing of a few large general purpose integrated systems result in a substantial reduction of efforts and costs, compared with the creation of many heterogeneous programs.

In spite of the number and variety of solutions applied to the problem of software sharing in Europe, the situation is far from satisfactory. The principal reasons for this seem to be the following:

1. Each of the above described approaches has intrinsic merits but represents only a partial contribution.

2. Many activities have been developed on a national basis and are confined to specific aspects of computer applications

3. Little or no form of coordination exists at a European level.

In conclusion, it is evident that none of the described organizations provides a service which

1. is exhaustive as a form of support (points 1, 2, 3, and 4),

2. operates on a full European basis, and

3. covers in principle all computer application fields.

For these reasons, an activity with the long-term scope described was initiated in 1971 at the European Information Processing Centre (CETIS) of the Joint Research Centre of the Commission of the European Communities. The CETIS program library group, entrusted since 1962 with the maintenance of computer programs , was the nucleus of the Computer Program Information Centre (COPIC) taking charge of this activity.

The Computer Program Information Centre (COPIC)

The Computer Program Information Centre (COPIC) was set up at the European Information Processing Centre (CETIS), a Division of the Joint Research Centre (JRC), Ispra Establishment, of the Commission of the European Communities. The aim of COPIC is to permit computer users easier access to the large and ever increasing quantity of existing computer programs and to diminish the enormous waste of energies and means spent in the analysis and development of new programs.

COPIC is financed by the public funds of member countries of the European Community. The type of information that it supplies is aimed at a wide range of users (laboratories, universities, industries, and any scientific, technological, or commercial organization belonging to the European Cmmmunity member countries) and takes their experience into account.

The services offered are free of charge for the time being. However, after a certain period it is hoped that a form of revenue can be arranged which will partially finance the Centre.

The aims of COPIC are:

1. To provide all users with complete and objective information on computer programs and later on, if possible, also on other matters relating to the use of computers

2. To collect programs and, by the most suitable means, to facilitate access to the programs registered in the Centre, for all users.

For the purposes of achieving the aims defined above, the work of the Centre consist of:

1.1 Collecting, recording, and disseminating information on computer programs available inside and outside the Centre

1.2 Developing software tools for the processing of information collections

2.1 Facilitating all contacts between the authors and owners of programs and the computing centers entitled to use them, on the one hand, and potential users on the other hand;

2.2 Offering a support as complete as possible (test, study, advising, computer time, distribution) for the rapidly expanding collection of programs in scientific/technological fields

2.3 Developing procedures and systems to facilitate the use of programs

The description of the day-by-day work of COPIC is recorded on disk memory, and a special program can perform statistical analyses mainly concerning the various COPIC correspondents and the nature and frequency of their interactions with COPIC.

The COPIC Environment: JRC and CETIS

The Joint Research Center (JRC)

The Joint Research Centre (JRC) was set up in execution of Article 8 of the Treaty of Rome, establishing, in 1957, the European Atomic Energy Community (EURATOM).[1] The aim is to ensure the implementation of the direct action of the European Commission in the field of research and teaching and to act as general scientific support to the Commission.

The JRC, with a staff of about 1800, consists of four establishments located in four different Community countries: the ISPRA establishment, Northern Italy, for general purposes; the PETTEN establishment, The Netherlands, for the study of nuclear materials; the European Transuranium Institute, Germany, for the study of plutonium and transplutonic elements; the Central Bureau for Nuclear Measurements, Belgium.

The Ispra Establishment contains the following scientific divisions: Biology, European Scientific Information Processing Centre (CETIS), Chemistry, Nuclear Studies, Materials, Physics, Research Reactors, Technology.

The CETIS

CETIS was set up in 1960 with a staff of about 90. It is concerned with the methods and techniques of automatic information processing (informatics) together with their applications to fields of activity of the various Commission services.

This implies two complementary activities:

1. Placing its skills and computing equipment at user's disposal for the solution of scientific and administrative problems. This involves management of the computing installations, maintenance

---

[1]As is known, there exist two other European Communities, the Steel and Coal Community and the Common Market. Since 1965, only one Executive board (the European Commission) and only one Council of Ministers have been established for these three Communities.

of the available application software, and analysis and programming work on request

2. Research into informatics in order to extend the range of methods and techniques for improving the man-computer interaction.

A certain amount of CETIS potential can also be made accessible, on request, to universities, organizations, and enterprises of the Community by means of service-rendering contracts.

The current medium term activity covers the following fields:

Numerical mathematics: development and application of numerical methods and algorithms concerning linear algebra, multidimensional integration, ordinary differential equations, partial differential equations, and function approximation. Work in the field of applied mathematics and statistics, strictly connected with the Institution's research program, essentially covers the fields of fluid dynamics, magnetohydro-dynamics, structural analysis, experimental data analysis, and process simulations.

Information Science: the present activity in this field aims to employ the existing know-how for the creation of operational systems. Three such systems are being implemented: a generalized software package and programming language for natural-language processing (SLC- II Simulated Linguistic Computer), a fully automatic information storage and retrieval system (FAIRS), a Russian-English machine translation system.

Administrative Data Processing: development of integrated systems for the administration and management of the Ispra Establishment.

Programming languages and systems: the definition and development of specialized languages and systems to be used, especially in an interactive context.

CETIS Participation in the European Informatics Network (EIN)

History

An informatics network involving a number of data-processing centers in various European countries is to be constructed. These centers will be linked by a communications subnetwork of computers and data links. The project is known as:

COST project 11-A European Informatics Network (EIN)

The COST group (Co-operation Européenne dans le Domaine de la Recherche Scientifique et Technique) is a committee formed of senior officials from 10 European nations. This committee coordinates work on proposals for European cooperation in a variety of advanced scientific and technological projects. Proposals for a number of projects in several areas were originally put forward in 1969 by the EEC Working Group for research in science and technology. The Working Group set up a number of expert study groups charged with examining the various proposals and with making detailed plans for implementing these cooperative projects.

The study group for Project 11 recommended the establishment of an experimental informatics network to explore the problems of communications between computers. A treaty was signed on 23 November 1971 by the following signatories: Euratom, France, Italy, Norway, Portugal, Sweden, Switzerland, United Kingdom, and Yugoslavia.

The Purpose of the Network

The European Informatics Network will link data-processing centers situated in various European countries and allow the rapid interchange of information between these centers. This will facilitate investigations into problems associated with linking up computers to form large multiple-computer data-processing systems. The network will be used for experiments in resource sharing between centers and will promote the development of techniques and standards for communication between programs running on two or more geographically separated computers.

Initially the experimental network will link together CETIS and centers in France, Italy, Switzerland, and the United Kingdom, but it can later be extended to include centers in other countries.

COPIC ACTIVITIES

The Handling of the Information Data Base

At earlier stages of COPIC existence, program information was handled manually. Program descriptions were kept in drawers, and answers to the queries were sought by specialized persons. But by 1969, the quantity of programs dealt with has increased to several hundred, and we thought of the possiblity of relying on computer-handled data collection.

In particular, the aim of COPIC was to have available an automatic storing retrieval system for handling information, operating both in batch and in teleprocessing mode, to permit direct access to the data base to remote users. From the very beginning COPIC considered the possibility of an automatic indexing-retrieval scheme (i.e., a scheme in which both texts describing programs and queries on programs could be examined and indexed automatically). However, at that time no such automatic indexing-retrieval scheme was operating, and not even the related studies had reached a sufficiently advanced state.[2]

COPIC then set up an automatic interactive system of storing and retrieval with a manual indexing scheme. The system called SIMAS was developed during 1970 and 1971 and has been operating for two years.

[2]At present, COPIC is setting up an experiment for utilizing the FAIRS program, developed at CETIS for the automatic construction of a thesaurus and automatic indexing of texts. FAIRS program will be described in a later paragraph.

As for the data collection, COPIC, having contacts with various program libraries, editors of program catalogues, computer manufacturers, and computer centers, has collected a data base of about 20,000 program abstracts. However, only 1500 of them are presently computerized, i.e., stored in the SIMAS files.

The Automatic Information Storing/Retrieval System, SIMAS

After a study of program descriptions reported in various catalogues, COPIC has arrived at a standard program definition, comprised of the following items:

Source	author date of development . . . .
Purpose	problem solving limitations . . . .
Programming/ Operating Information	operative computer programming language . . . .
Availability	acquisition conditions conditions of use . . . .

For each item a list of keywords has been prepared, which constitutes an open-ended thesaurus. Inside an item, keywords may be related to each other by relations of hierarchy or synonymity.

For the description of the programs according to the program definition, forms have been prepared on which each item is to be completed with extended information (in English). This description is further indexed by using keywords extracted by the above mentioned thesaurus.

SIMAS performs the following main automatic activities:

1. Storing-updating of the item definitions and of the thesaurus of keywords subdivided in accordance with the items
2. Storing-updating of information on programs
3. Retrieval of programs by name or by an appropriate Boolean expression of keywords, with checks on possible restrictions concerning the release of particular information
4. Editing of selected bulletins

The SIMAS system is written mainly in FORTRAN and operates on an IBM 370/165 computer. The retrieval part of SIMAS can operate both in batch and in interactive mode by means of CRT or teletype-like terminals (IBM 2260, IBM 2741). The CRT version of SIMAS is provided with a flexible, tutorial dialogue which permits users to consult the thesaurus to select the keywords for the best formulation of the query. At any stage of the query formulation, the user is supplied with the number of programs retrieved at that time. Use of this version does not require more than one hour of training, even for persons without programming experience.

The Program Testing, Study, Advising

A subset of scientific and technical programs, judged to be of interest to a large class of users, are centralized at COPIC. At the beginning of its existence, the JRC was oriented almost entirely toward nuclear energy studies, and hence the programs at COPIC were concerned mainly with nuclear reactor design. Since then, however, the JRC interest has shifted gradually toward other fields, although these are often related in some way to nuclear energy. These fields include the design of conventional and nuclear power generating plants, the static and dynamic analysis on continuous and framed structures, the statistical evaluation of geological exploration data, the simulation of reprocessing plants, and the long-term optimization of energy economy in given countries by means of a planned start-up of conventional or nuclear power plants.

At present the sectors considered to be of major interest for the Joint Research Centre are the following: nuclear reactor design, engineering, earth sciences, chemistry, management science/operational research, biomedical sciences, mathematics, general utility.

There are about 800 programs centralized at COPIC of which 300 are implemented on the CETIS computer installation (IBM 370/165), and 100 have been thoroughly studied.[3] COPIC personnel give technical as well as programming advice on this last category of programs.

The advice consists of:

1. Preparation of user's manuals
2. Counsel on the choice of the programs likely to solve the user's problem
3. Advice on input preparation/output interpretation
4. Implementing of program modifications as requested

The programs centralized at COPIC come from a variety of sources; some have been:

1. Developed by JRC researchers
2. Freely released by the originating establishments (by means of exchange, or gratis)
3. Libraries of programs (by means of exchanges, or by payment of special fees, or gratis)
4. Software-producer houses, software licenses (by purchase or renting)

The libraries from which programs are obtained are the following: Biomedical Computer Program Library, UCLA, USA; Computer Physics Library, Queen's University Belfast, U.K.; QCPE, Quantum Chemistry Program Exchange, Chemistry Dept., Indiana University; IMSL, International Mathematical and Statistical Library, Houston, USA; SHARE P.L.A., Triangle Universities, USA; ICES Distribution Agency, USA; Computer Program Library of the Nuclear Energy Agency (NEA) of the OECD.

The last-mentioned library, supported by OECD countries, and entrusted with nuclear energy programs, is located at Ispra, in the

---

[3]It should be noted that the setting up of the European computer network described in the following paragraphs, will greatly increase the number of programs accessible to COPIC users.

CETIS building. It works in close collaboration with its host Center CETIS, and shares the CETIS computing center as an outside user.

The programs rented or bought include: SIMULA, LIBRARIAN, AUTOFLOW, BERSAFE, FLHE, SIMPL/1, MPSX, PMS-4, CSMP 3, IMS, and DYNAMO.

Dispatching of Programs

Programs collected at COPIC may all be dispatched to JRC requestors. As for dispatching outside the JRC, the following conditions have to be fulfilled:

1. The programs must be free from license restrictions.
2. The programs must not be officially disseminated by existing program libraries.

However, COPIC is in contact with several existing libraries in the U.S. with the aim of becoming their European partner for program distribution. This is the case with the QCPE library. COPIC will also probably be made responsible for distributing programs inside the recently founded European ICES User's Group.

The program library management (i.e., the storing, testing, retrieval, and reproduction of programs) is achieved through a set of fully automatic procedures written to speed up and improve the services. Source programs to be tested are stored on on-line disk memory and the test is performed via the FILE EDITOR system, developed at CETIS, which permits by means of terminals (teletypewriter IBM 2741) the correction of source programs and a direct request for compilation, link-edit,and execution.

Tested programs are then stored in their source form on disk memory. Depending on the frequency of demands for reproduction,the programs are stored on an on-line or off-line disk memory, the allocation being dynamic. The system utilized for the management of source programs (storing, retrieval, reproduction) is LIBRARIAN developed by Applied Data Research. The set of untested programs centralized at COPIC are directly stored on off-line disk memory, and managed by the LIBRARIAN system. These programs are also dispatched on request, provided the requestor is aware that the programs are untested.

In-House Use of COPIC Programs

All programs centralized at COPIC may be used on the CETIS computer. In fact, for the programs rented or bought, the type of contract signed permits the running of programs on the CETIS computer to all organizations belonging to the European Community member countries. Users have to sign a contract with CETIS. The CPU equivalent time is then invoiced to the client.

As far as program use is concerned, COPIC has been setting up a number of facilities for clients, which are described in the following.

Tested programs in link-edited form are stored on disk memory ready for use.

At present two systems have been set up at COPIC, CTE, and CARONTE, for the benefit of users. The first one permits storage, retrieval, and batch execution of programs. Statistical analyses are performed concerning the relative use of the programs. The second one is a far more sophisticated modular calculation system obtained after several man-years of work.

CARONTE controls the automatic execution of a sequence of interdependent programs. The programs have to be chosen by the user from a given group of logically related programs coming from heterogeneous sources. The programs of the group do not have to be standardized in any way in their I/O interfaces since the main feature provided by the system is the automatic control of software interface adapters (transfer programs) between any two programs which have to communicate. The system can be utilized for groups of programs (libraries) of any type. So far CARONTE has been applied to libraries dealing with (1) nuclear reactor design, (2) editing of texts in various forms, and (3) theoretical reconstruction of ESR (electron spin resonance) spectra. The CARONTE system operates both in batch mode and by the use of a CRT terminal (IBM 2250-graphical terminal). The CRT interactive version permits the user to examine the intermediate results at any moment of the sequence; to modify accordingly the input data, the program sequence, and even the intermediate results; and to recommence the execution.

COPIC is now setting up another facility for the benefit of its clients, i.e., the possibility of utilizing the programs centralized at COPIC (isolated or in chain) by using remote terminals (IBM 2741 or IBM MC 72/T teletypewriter).

The user will then be able to select the programs to be executed, to specify input data, to command the program execution in batch or TS depending on the program size, to examine selected sets of results, to correct input data, to iterate the process as many times as required, and to instruct COPIC to dispatch complete results, on a terminal.

This project (COREA, COPIC Remote Access) would require the development of a conversational control system and the adaptation of the COPIC batch programs to terminal I/O operations (modulization and restructuring of input and output data). A preliminary version will be operative before the end of 1974.

COPIC Support to the Use of Integrated Systems with Problem-Oriented Languages

A research and development activity is carried out at COPIC in the sector of application-oriented programming systems with problem-oriented languages. Among the most advanced and widely used engineering systems, GENESYS and ICES have been chosen as preliminary reference points. The major efforts are, at present, devoted to ICES, since it is already implemented on the CETIS computer. Within the European Section of the ICES User's Group, COPIC is involved in the plans for effective collaboration on the maintenance of existing subsystems.

In addition, the development of new ICES subsystems, based

on existing nonproprietary programs in use at COPIC and of value in fields other than civil engineering, is underway.

This activity is aimed at promoting a wider use of integrated modular systems and providing COPIC clients with up-to-date software, allowing the use of problem-oriented languages.

Similar work is planned for the GENESYS system, as soon as a license contract, at present under study, is signed and the system is operative on the CETIS computer.

Recording of COPIC Activity

A special feature of the SIMAS system permits storage of a description of COPIC correspondents and the daily record of interactions that COPIC has with them. The system can perform as a sort of automatic secretary for the addressing of any material to selected classes of correspondents and can also perform statistical analyses of the various classes of correspondents, the type of requests, and cross connections between classes of users and requests.

More particularly, correspondents are identified in extended form by: name of institute; address; and in codified form by: country; fields of interest (chemistry, physics, etc.); type of institution (bank, hospital, industry, software house, university, etc.); type of parent organization (public, private); type of relationship with COPIC (supplier of information, supplier of programs, requestor of programs, cooperation); right to address queries concerning the different classes of programs.

As for the interactions between the correspondents and COPIC, they are described in codified form by: name of the correspondent; date of interaction; nature of interaction (dispatching of material to COPIC, request for material, visits, cooperation etc.); in the case of dispatching/requesting programs or program information: the name of program; in the case of inquiry on programs matching given characteristics: keywords utilized.

The SIMAS program can provide the following: extracting addresses (in the form of listing with autoadhesive labels) related to a subset of correspondents; extracting lists, histograms, concerning a subset of correspondents; extracting lists, histograms, concerning a subset of interactions; and combination of the second and third items.

The definitions of the subsets are given by the use of any Boolean expression of keywords.

CETIS ACTIVITY

CETIS Support Activity

The CETIS guarantees the support activity through its Computing Centre, its Computer Program Centre, and its Russian-English Automatic Translation Service.

The services offered by the Computing Centre are based on a team which runs computing installations and on a team of experts capable of analyzing problems in various fields of applied mathematics and of automatic data processing.

Computing Facilities

In December 1973, CETIS had at its disposal an IBM 370/165 system (1024 K-bytes), the access of which is available to users in: local batch; remote batch, by means of remote job entry type terminals; and time sharing, by means of conversational terminals.

System Configuration

The following is a list of the principal features available at the computing installation: CPU 370/165 (1024 K bytes of main memory); 8 disk storage drivers IBM 3330 (800 M bytes); 16 disk storage drivers IBM 2319 (480 M bytes) and 2 spare drivers; 7 magnetic tape drivers (9 tracks); 1 magnetic tape driver (7 tracks).

The teleprocessing equipment is subject to rapid evolution, so that the following list is merely indicative: 2 lines control units (IBM 2703 and 2701); 16 teletype terminals (IBM 2741); 4 CRT Terminals (IBM 2260); 1 graphic terminal (IBM 2250); 1 remote batch terminal (IBM 2770); and 1 process communication system (IBM 1070) connected to process data acquisition devices.

In addition, the Computing Centre is connected to outside remote users; 3 in Italy, and 1 in Belgium via remote batch terminals of their own (IBM 1130, IBM 360/30, IBM 2770, IBM 2780).

Standard System Software

Programs run under the control of OS-MFT II (release 21) and HASP (Houston Automatic Spooling Program). The system software library is enriched by an open-ended list of IBM program products and a teleprocessing facility developed at CETIS.

Languages available are: Assembler E, F; FORTRAN E, G, H; Watfor; Cobol E, F, ASA; RPG; Sort/merge; PL1; Algol.

Special Systems or Services

CETIS is the only institution in Europe and one of three all over the world to provide a Russian-English machine translation service. The system in operation at Ispra, originally had been developed by the Institute of Languages and Linguistics of the Georgetown University, Washington, D.C., and it has been steadily improved and enlarged by CETIS.

A teleprocessing facility was implemented in 1969, and performs the following basic functions:

1. Terminals input/output
2. Program dispatching under time-slicing discipline
3. Program swapping in a dedicated partition

The system is in effect a small time-sharing processor which allows an easy terminal interface.

A conversation file-editor (FILEDIT) which runs under the control of the teleprocessing facility has been developed. FILEDIT performs, by the use of a command language, the following basic functions:

1. File creation, deletion and updating
2. Submission of jobs for batch execution

CETIS Research Activity

Some CETIS achievements and projects could be of interest in this context. They are carried on by the Information Science group, which is subdivided into three distinct but complementary branches:

1. A software branch, charged with the implementation of a generalized system for natural-language data processing (SLC-Simulated Linguistic Computer)
2. A documentary branch, which deals with automatic indexing, automatic thesaurus compilation, formulation of questions, and information retrieval (FAIRS System)
3. A linguistic branch, which investigates the automatic analysis of natural language

In the following paragraphs SLC-II and FAIRS will be described in some detail.

Simulated Linguistic Computer (SLC-II)

SLC-II is a generalized software package and programming language for natural language processing which could form a common basis for such different applications as IR, scientific data base management, and machine translation.

The basic principle of SLC-II is that each transformation of information can be conceived as performed by an abstract machine which controls a set of recurrent basic functions. SLC-II is able to perform the following basic functions:

1. Source text input
2. Morphological analysis
3. Syntactic and semantic analysis and homography recognition
4. Transfer to target language features for source language features where a metalinguistic description of the transformation is not available
5. Target text generation starting from a metalinguistic description of the target text

The subsequent functions are specific to an IR environment:

6. Creation and updating of the document base and appropriate search devices
7. Retrieval algorithms

Modules 1 and 2 are fully operational in a generalized environment. Modules 3, 6, and 7 are operational and are used on an experimental basis in FAIRS.

Fully Automatic Information Retrieval System (FAIRS)

The characteristics of FAIRS are automatic thesaurus construction; indexing; query formulation and retrieval.

a. Thesaurus construction. The thesaurus construction procedure needs a grammar of the characterstring representing the source text and a morphological source language dictionary, that is, a statistically representative text collection.

The process of automatic thesaurus construction is subdivided into the following cycles: detection of words not in dictionary; homographies detected during dictionary search, determination of relevant concepts by statistical decision; construction of compound expressions by statistical evaluation methods; automatic classification using a modified multipass-type algorithm.

The lexicographer may influence the flow by changing several parameters interactively. The dicitonary thus created is used in a fully automatic IR system.

b. Indexing. The automatic indexing procedure is strictly connected to the thesaurus construction algorithm and is primarily based on a statistical approach.

c. Query formulation and retrieval. A query is considered to consist of a formal and a textual part. For the formal part some conventions have to be established. The textual part is processed in a manner similar to that used for indexing.

Feedback during retrieval allows modification of the natural language query using text parts of relevant documents or their internal search file representation. In either case query reformulation is done completely automatically.

The retrieval results are presented in ranked order. The rank is based on the similarity of document and query. At present a batch version of FAIRS can be used.

THE EUROPEAN INFORMATICS NETWORK

The Initial Centres

The National Nodal Centres so far nominated for the initial network are listed below, with the computing facilities currently available and the functions they will perform. Participating nations may nominate further Nodal Centres at a later date.

Centre	Computer equipment	Function
Zurich Federal Institute of Technology (Switzerland)	CDC 6400 CDC 6500	Scientific Computation, education Administration, Library automation
Paris IRIA (France)	CII 9080 CII 10070 CII IRIS 50 CII IRIS 80	Interface with possible French network and data banks
ISPRA-EURATOM J.R.C.-CETIS(Italy)	IBM 370/165	Scientific Computation, Data Banks, Information Services, Network Research
Milan, Politechnic school of Milan (Italy)	2 X UNIVAC 1180	Scientific Computation Network Research
Teddington National Phys. Laboratory (U.K.)	2 X ICL KDF9 3 X Honeywell DDP 516	Scientific Computation, Network Research, Terminal Processor Design

The Characterizing Features

As opposed to existing networks, which operate on the scheme of packet switching (see for instance ARPA network, in USA), the main features of the planned European network are the following:

1. Many countries are involved, so there will be lines belonging to different PTT.
2. The choice of the contractee who will undertake the planning of the NSC (Network Switching Centres), and the NSC themselves, should attempt to favor the foundation of European consortia.
3. The velocity will be 9,600 bit/sec (while ARPA is for instance 48,000 bit/sec).
4. The software related to the NCS will be high-level language with great emphasis on portability and flexibility (for ARPA the language is Honeywell Assembler).
5. The possibility of internetworking with existing national networks such as Cyclades (France), and NPL (England) based on packet switching should be kept in mind.
6. The possibility of connection between the subscriber (SC) and two NSC will be inserted.

The scheme below gives an idea of the project.

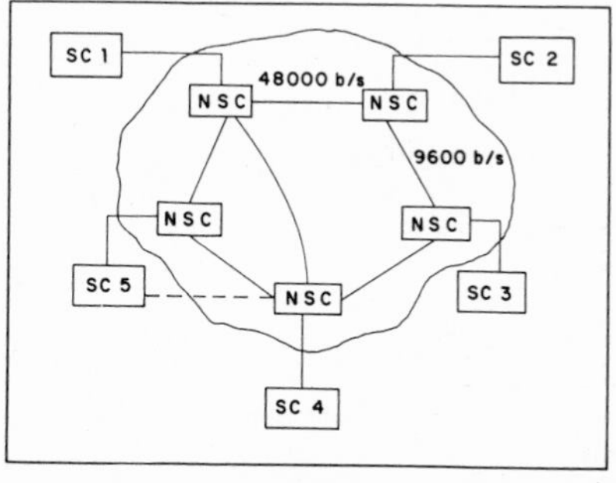

The Schedule

A first prototype should be in operation by May 1975 having at least 3 NSC. The end of the project is foreseen before the end of 1977.

CETIS AND COPIC EVOLUTION TRENDS

CETIS New User Profile

The CETIS Computing Centre's conventional user is somewhat obsolete. He writes his programs in FORTRAN, submits card decks to batch processors, miserably scans large core dumps, and eventually calls for programming support. But the advent of interaction is going to attract a growing number of "second-generation users." Features generally considered to be progressive are: standard software for terminal I/O management; time sharing capabilities; extended use of high-level languages for system and application software implementation; accurate human engineering interface to abstract machines; symbolic level debugging aids; and data base management systems.

It is quite obvious that a great deal of expectation exists in relation to the prospective connection with the European Information Network (EIN). One is tempted to predict the new user's behavior: He will access the system from a time-sharing terminal. His JCL will be an accurately designed command language. He will write his programs using PL/1. He will test and debug his program on line. He will store his files on an access-controlled file system. He will try to share EIN resources.

Hardware and System Software

It is felt that the evolution of the computing facility should, in the short-medium range, pass through the following steps:

1. Implementation of a generalized control program for terminal I/O
2. Installation of a time sharing subsystem, possibly on a virtual memory system
3. Installation of a data base management system
4. Adoption of a programmable front-end processor for local network and EIN support

All products have to be standard or at least supported by the manufacturer.

Lines of COPIC Evolution

We can foresee in the coming years the following lines of evolution:

1. The abstracts of programs will be examined by a program capable of automatically extracting a thesaurus of keywords and of indexing the abstracts accordingly.
2. The queries meant to extract the list of the programs capable of solving classes of problems will be formulated to the computer in natural language. The same algorithm used for the indexing of abstracts will be used for indexing queries.

3. The users wishing to consult the COPIC data base will still utilize letters, telexes, and telephone lines, but will increasingly make use of remote terminals, probably of teletypewriter type.

4. Once the data base has been consulted the terminal users will be able to instruct COPIC for receiving chosen programs or to specify the input for a selected program and to send commands for its execution.

The program will be executed in time-sharing (if very fast and small) or in batch, and the terminal user will be able to examine partial results, to correct input data, and to instruct COPIC for receiving complete results.

5. The number and type of programs and data bases available to the COPIC user will be greatly augmented by the presence of the EIN network.

REFERENCES

1 Gaggero, G., Mongini-Tamagnini,C. and Lunghi, G., "SIMAS - An Interactive Information System for a Library of Software Packages", Proceedings of the International Computing Symposium, Venice, 12-14 April, 1973, pp. 172-180.

2 Buccari, G. Fattori, G., Mongini-Tamagnini,C., "CARONTE - The EURATOM System for Automatic Control of Linked Calculations", Proceedings of Conference of the Effective Use of Computers in the Nuclear Industry, Knoxville, Tex., April 21-23, 1969, pp. 297-303.

3 "COPIC, Computer Program index, May 1972".

4 "COPIC, Computer Program Index, April 1973 .

5 "Joint Research Centre, Ispra Establishment Annual Report 1971", EUR-4842 e.

6 "Joint Research Centre, Ispra Establishment, Annual Report 1972", EUR-5060 e.

PROGRAMMING FOR EFFECTIVE INTERCHANGE

J. M. McCormick
Weidlinger Associates
New York, New York

ABSTRACT

The paper presents suggestions and some guidelines for the development of large FORTRAN programs which can easily be interchanged among the readily available large computing systems. Some previously published sources of such guidelines are examined; many of these guidelines are found to be too restrictive for the developers of powerful software. The importance of some techniques, such as the use of modular coding and dynamic dimensioning, is stressed. A new technique for making large in-core programs operate efficiently (on certain large multiprocessors) is presented. Some dangerous but common coding techniques are discussed. The coding in question gives different results when executed on different computing systems. Finally, some suggestions are made for developing portable and adaptable code.

INTRODUCTION

It is often desirable to try to transfer a large-scale FORTRAN program from one computer system to another or to adapt a large program to solve a slightly different or larger problem than the one envisioned by the original developers of that program. This transfer and/or adaptation of large FORTRAN programs is usually extremely time consuming unless the originators of the program consciously tried to make their program portable and adaptable. This paper will examine and comment on some published guidelines for the construction of such programs, suggest additional guidelines derived from experience, and make some suggestions for making large portable FORTRAN programs efficient (in the sense of an overall efficiency which considers memory charges, input/output charges, and turnaround time, as well as central processor charges). Examples of some coding to be avoided will be given, and some prototype programs will be discussed. These programs illustrate some new techniques for handling adjustable length arrays by using a data manager and FORTRAN callable subroutines to change the memory requirements of the program during execution. These techniques can greatly reduce average turnaround time and the cost of processing large FORTRAN programs on certain large computing systems, e.g., CDC 6600, which allow multiprocessing and use a charging algorithmn based on amount of core used at each job step during execution.

SOME THOUGHTS ON GUIDELINES

It is a thankless task to set up guidelines for designing portable and efficient large-scale FORTRAN programs because a guideline that may be helpful for a novice programmer may be counterproductive for a very experienced program developer and a technique which may be efficient on one computing system may be wasteful on another. The guidelines that are sought are those that will be most appropriate for large (say 1000 or more FORTRAN statements) programs to be executed on large "number crunchers" (e.g.,CDC 6000 series or 7600, IBM 360, model 75 or higher, UNIVAC 1108, and processors of similar capability), though some of them will be useful for smaller programs and valid on smaller processors. If any readers can recognize that a particular guideline is inappropriate for what they are trying to do either because of some special knowledge or because of some characteristic of their computing system, so much the better.

Many of the guidelines that have been published and that will be reviewed here are of the "motherhood" type, i.e., they seem to be obvious and beyond serious criticism, and indeed many of them are. But, it is contended by the author that some of these apparently "motherhood" guidelines are inappropriate for the special objective considered here, the designing of large, portable, efficient FORTRAN programs to be executed solely on large processors.

PUBLISHED GUIDELINES

Several compilations [3, 6, 7, 8, 10, 12, 13] of suggested guidelines for FORTRAN programming have been published, and numerous authors have made contributions in the form of a chapter here or a section there. Many of the published suggestions are redundant and most are inappropriate for the developers of large FORTRAN programs to be run on large processors. Nevertheless, a review of some of these guidelines is clearly the best starting point in the search for useful suggestions.

The book by Kreitzberg and Shneiderman [10] contains many suggestions and techniques for effective programming. In fact, it is such a rich store of information that part of their book will be discussed later in a separate section.

Fleiss et al. [6] presents useful guidelines, compares the results from specific coding when executed on several large processors, and has an excellent annotated bibliography.

Morrison [17] presents general guidelines for virtual storage programming. Many of his guidelines and suggestions are quite useful in the development of efficient and portable code.

Schrem [19] discusses many of the design objectives in the development of large-scale programs. He pays particular attention to the modular structure of both the program and the internal data.

Ghan [7] discusses techniques for developing large-scale FORTRAN programs from a somewhat different point of view than the position presented here, which is basically to develop highly portable large-scale programs. He defines a large program as having (a) 100 or more subroutines, (b) 300 or more COMMON variables, (c) 10,000 or more FORTRAN statements, and then concentrates on higher-level software, interprogram communication techniques, and basic programming practices. Much of what Ghan suggests is quite useful to the developer of any large FORTRAN program, portable or not.

Perry and Sommerfeld [18] and McCracken and Weinberg [14] put their emphasis on techniques for making the program readable and self-documenting. There are many interesting suggestions which are general enough to be of use to any programmers.

Larson [11], on the other hand, writes from the point of view of efficient coding, the efficiency usually being a trade-off of the programmer's time for increased execution speeds by such techniques as "unrolling" loops, equivalencing multidimensional arrays to one-dimensional arrays, avoiding implied loops, avoidance of data type conversions, and the like.

Cohn [4] also writes from the point of view of efficiency and gives techniques to decrease the running time and memory space requirements for FORTRAN programs, with emphasis on replacement statements, optimization of polynomials, use of local variables in subprograms, and so forth.

There is some advice in these papers that it is perhaps better not to heed. For example, there is much conflicting advice on the use of subprograms; this topic will be discussed separately next. It is useful to be aware of many of the techniques for reducing execution time; some of these should be routinely used since they cost little or nothing beyond the need to be aware of them. Examples of this kind of advice are (a) use meaningful mnemonic names for all variable names and subroutine names since names often provide the memory stimulus that makes it unnecessary to refer to external documentation for the definition and (b) use implicit FORTRAN naming conventions for integer and real variable names. But other techniques do cost something, either in length or clarity of coding and, usually, in programming time. These techniques should not be routinely used, especially when developing large portable FORTRAN programs, since many of them are compiler dependent.

On Modularity

One of the few guidelines that everyone seems to agree on is that programs should be constructed in a modular manner. Modularity is one of the four design goals proposed by Kreitzberg and Shneiderman [10]. The advantages of modular coding are many and often obvious. It has long been common practice to make subroutines out of segments of code that perform a well defined sequence of operations that may be repeated several times in a program. Knuth [9], page 225, mentions that subroutines existed as early as 1944 and speculates that they may even have been envisioned by Babbage.

Liskov [12] and three recent connected articles in Datamation [15, 5, 16] give an excellent account of the structured programming which results from the intelligent use of modular coding.

An important question is how long a subprogram should be and when a subroutine should be broken into two or more smaller subroutines. On these points there is no commonly accepted conventional wisdom.

Ghan [7] suggests that the bulk of all subroutines be in the 100-300 card size range, while Cohn [4] points out that if a subprogram is called from only one place in the main program, it is most efficient to eliminate its separate identity as a subprogram and incorporate it directly into the main program. Some program developers favor a somewhat stricter guideline, which holds that no subroutine should be longer than a two-page printer listing (about 120 lines of coding), the size of listing that fits easily on most desks. This rather arbitrary rule has often provided the impetus that led to effective subroutining of large programs, because careful examination of blocks of coding which are longer than 120 lines will usually uncover at least one quite independent section that should be subroutinized. Such subroutining does make the logic of the program more evident.

On Kreitzberg and Shneiderman

Reference [10] contains many guidelines for programmers who wish to develop programs which are machine independent and hence portable or transferable. It might be instructive to comment upon them here because this paper is written from a quite different viewpoint than that of [10], where the authors are clearly trying to give guidelines which are transferable among most computing systems, including both small- and large-scale processors. In contrast, the guidelines proposed here are meant only to enable large programs to be portable among very large systems. Consequently, many of the guidelines of [10] will be found to be unnecessarily restrictive from this point of view. In what follows, the guidelines from [10] will be stated, then followed by a comment.

Guideline: Adhere to the ANSI ([1] and [2]) definition of Basic FORTRAN IV.

Comment: Much too restrictive from the point of view of large processors. It gives away too much of the power of the bigger compilers.

Guideline: The maximum length of variables should be <u>five</u> alphanumeric characters.

Comment: Compilers on all large processors allow variable names to be at least six alphanumeric characters; so this guideline is unnecessary and is possibly counterproductive from the point of view of creating a self-documenting program. The programmer should give much thought to his choice of variable names; when the variable names suggest the variable they represent, it is much easier to follow the logic of the program, and it is often possible to give a more representative name with six characters than with five. However, the use of seven (and longer) character variable names, allowed by some compilers, should be avoided since many compilers allow only six characters. It it true that one can often choose a better variable name with seven characters available, but the portability problems introduced outweigh this advantage.

Guideline: Do not use DATA statements.

Comment: Unnecessarily restrictive. Compilers for all large processors allow DATA statements. The DATA statement is preferred as a method of initializing variables over a sequence of arithmetic statements. There are, however, two guidelines that should be suggested for DATA statements, one is that the use of literals (e.g., 'A' or *A*) should be avoided since the delimiters are different on different compilers. Second, run-on Hollerith specifications in DATA statements should be avoided since their implementation depends on the word length (in characters) of the central processor being used.

Guideline: Do not use labeled COMMON.

Comment: Too restrictive. Labeled COMMON is often useful in avoiding lengthly argument lists in special purpose subroutines.

Guideline: Use only the arithmetic IF statement; avoid the logical IF statement.

Comment: On larger processors, the use of the logical IF is generally preferred because it leads to better articulation of the logic of the program.

Guideline: Do not use the PRINT or PUNCH statements; stick to the READ (unit,format) and WRITE (unit, format) statements.

Comment: A useful guideline. Many compilers, even for larger processors, do not compile PRINT or PUNCH statements. Further, the use of these statements prevents the use of variables for I/O units (see following proposed guideline).

Guideline: Specify variables for I/O units so that they may be easily changed. For example,

```
IN  =  1
JOUT  =  3
READ(IN,1001)  SALES
WRITE(JOUT,1002)  PROFIT
```

Comment: An extremely useful guideline. It not only enables one to change quickly between systems where the mandatory input and output devices are on different units but also allows one to cut down selectively on output by changing the value of JOUT from the output unit to a scratch file unit and back again at will during the course of the computation.

Guideline: Use only INTEGER, REAL, and DOUBLE PRECISION type declarations. Avoid COMPLEX, LOGICAL, INTEGER*N, REAL*N, CHARACTER, and IMPLICIT declarations.

Comment: A mixed bag; too restrictive in part. COMPLEX and LOGICAL are too powerful not be used with large processors. INTEGER *N is used only to save storage (for N=2), and it is usually possible to do without it. On the other hand, it is fairly simple conversion to change INTEGER*2 to INTEGER when converting from one processor to another. So whether to use INTEGER*2 or not is a trade-off involving the amount of space saved against the effort of changing all the INTEGER*2 statements. The REAL*N, CHARACTER (and ASCII), and IMPLICIT declarations are best avoided since the compilers for some larger processors do not allow them.

Guideline: Avoid multiple-entry subroutines.

Comment: A close decision. Multiple-entry points allow important savings in coding of similar subroutines, but they have different forms on different processors. Some processors do not allow the ENTRY statement to have the argument list as part of an ENTRY statement. Its argument list is implicitly the same as that of the host routine. It is possible to write subroutines with ENTRY statements which still are portable by clearly including with each ENTRY statement comments which show the alternate form that must be used on other processors. Whether the savings are worth all this trouble makes this a close decision.

Guideline: Keep one statement per card even if one's present compiler permits more than one statement with a delimiter character.

Comment: A good guideline, even though software is available with some processors to unpack the statements. One typical compiler allows $ as a delimiter; the program which removes those $'s is called TAXES.

Guideline: Avoid octal or hexadecimal operations or printout.

Comment: Efficient use of large programs on large processors requires that this guideline be rejected. The technique suggested in this paper (see Dynamic Dimensioning) for changing the memory requirements of a program three or more times during the course of the execution of the program depends on being able to instruct the central processor controller as to the memory required at each step. This process often requires octal or hexadecimal operations, and it is useful to output the memory requirements at each step in octal or hexadecimal, as well as decimal, locations, so that the programmer can follow the memory changes.

Guideline: Use A1 formats for character input/output even if the computer hardware permits A4, A6, or A10 formats. Following this rule wastes a great deal of storage, so it should be carefully considered.

Comment: So it should. All large processors allow at least A4, so that is the basis for a more appropriate guideline for large processors. But even this wastes a lot of storage on hardware allowing A10, so if there is much character input/output in a program intended for such hardware, A10 formats can be used, but they should be marked clearly so they may be changed when the program is transferred to hardware allowing less than A10.

Guideline: Consider output line width. Most high-speed printers have 136 characters per line, but some are limited to 120 characters per line. Teletype or display terminals may have 80, 72, or fewer characters per line.

Comment: It is most convenient to sit at a display terminal shortly after a run is completed and be able to sample selectively the output from a stored file. For this reason, it is well to format an output file based on a maximum of 72 characters per line. On the other hand, such formats are crowded and awkward looking when printed on a line printer on which at least 120 characters per line are permitted. The suggested guideline is to allow the user to save either or both of two output files, one formated 72 characters per line, the other formated 120 characters per line.

Guideline: Use Hollerith fields in FORMAT statements. Many compilers permit literals to be enclosed within a pair of apostrophes or asterisks. These special features are extremely convenient since they eliminate the need to count the length, but they are not universally accepted.

Comment: A good guideline. The one major modification that I have had to make in converting many large-scale programs from one large processor to another was changing all those apostrophes or asterisks, depending on the system the original code was written for. The guideline should be expanded to include suggesting the use of Hollertith fields in DATA statements where many compilers also permit literals to be enclosed within a pair of apostrophes or asterisks.

Guideline: Avoid arrays with more than two dimensions.

Comment:Probably too restrictive. It is often convenient to use arrays with three dimensions, and occasionally useful to have four dimensional arrays available. However, arrays of dimensions larger than one should always be replaced by one-dimensional arrays when it is convenient, e.g., when initializing large multidimensional arrays.

Guideline: Subscripts should be in one of the following seven formats: v, c_1, c_1*v, $v + c_1$, $v - c_1$, $c_1*v + c_2$, $c_1*v - c_2$, where v is a nonsubscripted integer variable and c_1 and c_2 are integer constants.

Comment: The purpose of this guideline in the past has been to allow compilers to do some optimization on subscripting. Since so much time is spent, in the typical program, in calculations involving subscripted variables, this guideline is still useful. In particular, the use of subscripted subscripts should be avoided, even though it is permitted by some compilers. Many other compilers do not provide for the use of an array element as a subscript and this technique has historically caused problems in the transfer of the code.

Guideline: Do not use free-format output statements or NAMELIST; they are not standard.

Comment: A good guideline. It is surprising how the details of free-format output and NAMELIST vary, even among very large processors.

DYNAMIC DIMENSIONING

Probably the most important thing to keep in mind, when developing a large FORTRAN program for a large computing system, is that all adjustable dimensions should be in the executive routines. An adjustable dimension is one whose length is not independent of all possible data sets. Burying such dimension statements in subroutines is a common cause of difficulties when a program is being adapted to a new data set or to a slightly different purpose. All such subroutines must have variable dimensions. For example, part of the relevant coding for a typical finite element input routine might be:

```
      SUBROUTINE INP(NJ,  NLC,  IC,  X,  P,  NP)
      REAL  X(3,  1),  P(6, NP, 1)
```

where X is the array of the three-dimensional coordinates, P is the array of the generalized nodal forces for any number, NLC, of loading conditions and any number, NJ, of modes, IC is the maximum length of the stiffness matrix (computed from other input quantities not shown here), and NP is the value of the second index as it appears in the original dimension statement or as it is defined during allocation of space within another array. It is assumed here that the program, or each overlay of the program, consists of a set of subroutines, called in sequence by an executive routine. An executive routine, program ONE, for a typical program with an input phase (reading of data, initializing variables, etc.), a preprocessing phase (setting up equations, etc.), an algorithmic phase (construction of the solution, solving equations, etc.), and a postprocessing phase (calculation of additional results, output, saving of files, etc.) is shown below.

```
C...PROGRAM ONE
      REAL  X(3, 100), P(6, 100, 4),  C(33000)
      NP  = 100
      CALL  INP(NJ, NLC, IC, X, P, NP)
      IF(IC.GT.33000)  CALL TOOBIG(IC)
      CALL  PRE(NJ, NLC, IC, X, P, NP, C)
      CALL ALGO(NJ, NLC, IC, P, NP, C)
      CALL POST(NJ, NLC, IC, X, P, NP)
      CALL EXIT
      END
```

The input routine, subroutine INP, will generate the elements of the arrays X and P (as well as many others) and calculate the length, IC, of the array C, into which the stiffness matrix is to be mapped. If IC, as calculated in INP, is larger than the dimension of C, an error routine, TOOBIG, is called, which writes a message to the effect that IC is too large and ends the processing. If $C \leq 33000$, the preprocessor, subroutine PRE, calculates the array C using many arrays not shown here. Next, the algorithmic subroutine ALGO solves for the displacements and returns them in the array P. The postprocessor subroutine POST then computes forces and stresses from the displacements. A typical large-scale program differs from this small example in no important way. There would be two dozen or more arrays rather than the three exhibited here, and INP, PRE, ALGO, and POST would usually be broken into additional subroutines (each of which calls many other subroutines), but the essentials remain the same - a sequence of subroutines transferring arrays.

Program ONE satisfies the requirements that all adjustable dimensions should be in the executive routine. Someone who wishes to solve a larger or smaller data set need only change the dimension statements in the executive routine. It is easy to overlook a dimension statement buried in a lower level subroutine. But in the case of program ONE the user must make sure that the number of nodes to be generated is ≤ 100 and that the length of the stiffness matrix is ≤ 33,000, and change the dimension statements appropriately if they are not.

```
C...PROGRAM TWO
      REAL A(35700)
      CALL INP1(NJ, NLC)
      J2 = 1 + 3*NJ
      J3 = J2 + 6*NJ*NLC
C...A(1)   IS FIRST ELEMENT OF COORDINATES
C...A(J2)  IS FIRST ELEMENT OF NODAL FORCES
C...A(J3)  IS FIRST ELEMENT OF STIFFNESS MATRIX
      CALL INP2(NJ, NLC, IC, A(1), A(J2))
      IF ((J3+IC-1).GT.35700) CALL TOOBIG(IC)
      CALL PRE(NJ, NLC, IC, A(1), A(J2), A(J3))
      CALL ALGO(NJ, NLC, IC, A(J2), A(J3))
      CALL POST(NJ, NLC, IC, A(1), A(J2))
      CALL EXIT
      END
```

A better technique is illustrated by the program TWO. There is now only one array, A, which has the same size as the total of the three arrays from program ONE. All three arrays will be stored within array A; their locations within A are generally not known at the beginning of execution. Notice that the input subroutine has been broken into two subroutines, INP1 and INP2. This is necessary because the second array cannot be read into A until the length of the first array is determined. So INP1 just reads, or computes, the parameters (in this case, these parameters are NJ, the number of joints or nodes, and NLC, the number of loading conditions) which are necessary for the calculation of the lengths of the individual arrays to be stored within A. Note that NP can now be made equal to NJ, and hence NP is dropped from the argument list of INP2 and the other routines of program TWO. In a large-scale structural program, the routine corresponding to INP1 would determine the number of nodes, of each kind of element, of restraints, and so on, and calculate the length of each array required. There are many more arrays than the two used in this example. The executive routine would than allocate storage within the array A, In this example the coordinates are stored first in A, beginning at location A(1). The length of this array is 3*NJ; so the first location of the second array should be J2 where J2 = 1 + 3*NJ. Thus the address of the first element of the second array is A(J2). This array is 6*NJ*NLC elements long; so the third array should begin at location J3 where J3 = J2 + 6*NJ*NLC. There are only three arrays in this example; so that is the end of the space allocation within array A. However, in larger programs, this process is continued until all arrays whose lengths are known at this stage have been assigned space within A. Now that the space has been assigned, INP2 is called to read, or generate, the first array into A beginning at A(1) and the second array beginning at A(J2). INP2 can now compute the length required for the stiffness matrix, and this value is returned to the executive routine in IC. With the length of all three arrays now known, a test is made to determine whether A is long enough. If not, TOOBIG is called, an explanatory message is written, and the processing is terminated. If A is long enough, PRE is called to set up the stiffness matrix in an array beginning at location A(J3). ALGO then solves for the displacements, and POST computes the forces, as was the case for program ONE.

The advantages of program TWO over program ONE are obvious. The subroutines they call are identical, except that INP has been divided into INP1 and INP2 and NP has been dropped from the argument list. But program TWO is easier to adapt to problems of different sizes, and there are no dimension statements to change except the master array. No longer will a program be unusable because it can allow only 75 joints and the problem required 85 (even though the stiffness matrix required only 5,000 locations and the offending program allowed the use of up to 12,000 locations) or 80 eigenvalues were needed and the program only allowed 50 (even though only 200 field points were used to describe the structure and the offending program allowed up to 450 field points). In each of these cases there is excess room in most of the arrays used by the program, but not enough locations in one or more key arrays. This is always the difficulty when each array is dimensioned separately. In most such large programs, it is very difficult for an adapter of the program to change a given limitation because it is difficult to trace how many different arrays in how many different subprograms are tied to that particular limitation. Any modern program can avoid these difficulties by using the techniques illustrated by program TWO, where the only adjustable length array is located in the main program and its length can be trivially changed. No modern program should place separate limits on individual variables (e.g., number of joints, elements, load cases, eigenvalues, field points, and so on) but only a single limit (which may be a function of the system's available memory) on the combination of all such variables. Even this single limit can be varied automatically on some large systems, as program THREE illustrates.

```
C...PROGRAM THREE
      COMMON A(1)
      CALL INP1(NJ, NLC)
      J2 = 1 + 3*NJ
      J3 = J2 + 6*NJ*NLC
C...A(1)   IS FIRST ELEMENT OF COORDINATES
C...A(J2)  IS FIRST ELEMENT OF NODAL FORCES
C...A(J3)  IS FIRST ELEMENT OF STIFFNESS MATRIX
      CALL FINDM(JOLD)
      CALL CHANGE(JOLD+J3-1)
      CALL INP2(NJ, NLC, IC, A(1), A(J2))
      CALL CHANGE(JOLD+J3+IC-1)
      CALL PRE(NJ, NLC, IC, A(1), A(J2), A(J3))
      CALL ALGO(NJ, NLC, IC, A(J2), A(J3))
      CALL CHANGE(JOLD+J3-1)
      CALL POST(NJ, NLC, IC, A(1), A(J2))
      CALL EXIT
      END
```

Program THREE uses much of the same technique as program TWO with one very important exception: the all-inclusive array A is in COMMON and occupies only one location. Its length will be adjusted automatically and internally by the program as it determines its storage requirements at each step. Further, the total core memory that the program requires at each step will be automatically changed as required (either more or less memory) by FORTRAN callable subroutines. Such routines are available on some of the large processors, and program THREE, as shown, has been compiled and successfully executed many times to test various techniques. This technique is very cost effective on large processors which allow variable memory requirements (e.g., on the CDC 6600 and similar processors) for several reasons:

1. Typical large processors often have a complicated charging algorithm in which the user is charged not only for central processor time but also for peripheral processor time (or data transfers or some other measure of read/write activity) multiplied by the fraction of core used. Thus, the smaller the core requested at any state of processing, the less the user is charged.

2. Turnaround time is usually inversely proportional to the memory size requested. Thus, a job which requires nearly all of memory will often not be processed for many hours. If it turns out that the job aborts at a stage before all that memory was required, then much human time has been wasted and usually much money. Thus, a program which uses the minimum memory possible at each step can be very cost effective.

Consider the ways in which program THREE makes efficient use of a large processor which allows the user to request memory as needed and charges the user based on the amount of memory used. The numbers given to fix ideas somewhat will be from experience with executing a space frame program on several 6600s each with a maximum available memory of over 300,000 octal words. Program THREE is modeled on the essentials of this space frame program. The typical execution of any program like program THREE includes these steps:

1. Compilation (if the source code has been modified): The memory required for compilation is usually small (less than 50,000 octal words), and the user will request the required memory at the compile step in the control stream. The job will generally get to this stage quite quickly after submission (though at many installations it does depend on the priority chosen), and if there is a compiler abort, the user will quickly find out. This would not be the case if the user could not change his memory requirements during execution and was forced to request immediately the greatest memory required at any step during his job. In that case he might wait a long time for his large memory request job to begin execution, only to have it fail before it ever got to the point where it needed all that memory. This does not happen with programs like program THREE.

2. Link-editing and loading: For a program like program THREE this step usually takes even less core than compilation because there is at this stage only the one array which requires only one word of memory. This array will be expanded as necessary to accommodate all the arrays required for the solution of the problem, but none of this space is required at this point.

3. Reading input parameters: This step is represented by the call to INP1 in program THREE. The program reads parameters which

enable it to compute the lengths of certain arrays. In program THREE, these parameters are NJ and NLC, from which the lengths of X and P can be determined. Notice that X and P themselves cannot yet be read in because space cannot be allocated until their lengths have been determined. In large structural problems, the parameters read in INP1 will be the number of elements, and so on. After the lengths of all arrays have been determined, their starting locations within array A are calculated. This is accomplished in program THREE by the calculation of J2 and J3. Next FINDM is called to determine how many words of memory the program is using at this instant, which value is returned in JOLD. FINDM calls different macros on different systems, but they function similarly. CHANGE is now called to change the memory assigned to this job during execution. Again, there are macros available that perform this function on several common large processors. The memory required now must be whatever was necessary for the coding and all the scalar quantities (and fixed size arrays, if wished) plus what will be required for all the arrays whose lengths have been determined to this point, i.e., JOLD + 3*NJ + 6*NJ*NLC in this case, or, equivalently, JOLD + J3 - 1, with which argument CHANGE is now called.

4. Reading input arrays: Now that CHANGE has increased the memory assigned to this job (which might take some time if the additional memory called for is large), INP2 can be called to read in or generate certain arrays and initialize others. In large programs the joint loads, finite element properties, and so on are read in or calculated (from mesh generators and the like) at this step.

One thing requires an explanation at this point. How is all this done with array A still dimensioned as occupying one location? There are several possible approaches. The simplest one, which works on many processors, is that illustrated by program THREE. Put the variable in COMMON, which is assigned to higher core, and let it overflow as needed. As long as the core is expanded at each step, by CHANGE, and the overflow goes into this expanded region, the effect is the same as if array A were properly dimensioned. This approach does work on several large systems (but I do not suggest that this is a very desirable technique from the point of view of portability, and the trade-off between efficiency and independence must always be considered carefully). The other techniques involve ways which actually change the dimension of the array A. This can be done on some systems by using editing systems during execution.

5. Setting up calculated arrays: There are often arrays in a program whose length cannot be determined until other input arrays have been processed. A typical example is the stiffness matrix in a finite element program. Its length depends on the bandwidth of the equations, and that is not known until the connectivity data set is processed. Such an array is represented in program THREE by the array beginning at A(J3) whose length, IC, is determined by INP2. The routine CHANGE is then called again to change the memory allocation, this time increasing it by IC. On large problems, this step may require much memory, and the job may wait for a long while for the memory request to be granted. But at least it is not necessary to wait until the job actually requires all that memory. If the job aborted at an earlier stage, it ended without ever making a large memory request. By requesting only exactly as much memory as the data set for the particular job requires and by requesting it

only as it is needed during execution, the turnaround time is decreased and the computer costs are reduced. Small data sets on programs that are designed to accommodate large data sets run very quickly using this procedure, a phenomenon not observed when the program uses fixed dimensions.

6. Central algorithm: This is the part of the program which usually requires the largest amount of core, and CHANGE has provided just the required amount.

7. Postprocessing: At some point in most programs, the large (stiffness) matrix is no longer required, and CHANGE is called to reduce the memory to that required. This can save on memory-related costs since most of the output (displacements, forces, stresses, and so on) takes place after the stiffness matrix is no longer needed.

Program THREE is an elementary illustration of the use of a "data manager" in which space is allocated explicitly within the master array in the main program. A more general approach allows this allocation to be performed by a set of subprograms which can, among other things:

1. Maintain a directory of all array names and their locations within the master file
2. Assign, during execution, an array of any length to an entry in the directory and a location within the master array
3. Associate the array name with the proper location at any later time
4. Remove any array name from the directory and close up the space vacated within the master file

This ability to generate space for arrays only when and if they are needed during execution and to recover this space when they are no longer needed greatly increases the power of dynamic dimensioning. This technique is illustrated by program FOUR:

```
C...PROGRAM FOUR
      COMMON A(1)
      CALL INP1(NJ, NLC)
      J1 = ISPACE(1HX, 3*NJ)
      J2 = ISPACE(1HP, 6*NJ*NLC)
      CALL INP2(NJ, NLC, IC, A(J1), A(J2))
      J3 = ISPACE(1HC, IC)
      CALL PRE(NJ, NLC, IC, A(J1), A(J2), A(J3))
      CALL ALGO(NJ, NLC, IC, A(J2), A(J3))
      CALL REMOVE(1HC)
      CALL POST(NJ, NLC, IC, A(J1), A(J2))
      CALL EXIT
      END
```

Program FOUR is the same as program THREE except that the space allocation is done by a function subprogram, ISPACE, which includes its first argument in a dictionary of Hollerith names, increases the length of the master array by the amount indicated by the second argument, and returns the first location of the named array. The routines FINDM and CHANGE are used by ISPACE. The subroutine REMOVE deletes its argument from the array name directory and decreases the length of the master array and the memory being used by the length of the array that was removed.

All of the data managing can be relegated to the main subroutines and the main program can be simplified to program FIVE:

```
C...PROGRAM FIVE
      COMMON A(1)
      CALL INP5
      CALL PRE5
      CALL ALGO5
      CALL POST5
      CALL EXIT
      END
```

VIRTUAL MEMORY SYSTEMS

Program TWO is a simple example of dynamic storage allocation in which the user is constrained to problems where data storage is less than 35,700 decimal words unless he recompiles. Programs THREE, FOUR, and FIVE are examples of dynamic storage allocation in which the user does not need to recompile the program for various size problems. However, in order to run problems larger than the maximum core memory available, the code developer is constrained to read and write on external devices (tape, disk, drum). Systems based on virtual memory concepts avoid even this constraint. The user simply dimensions his array as large as he wishes, and the system takes care of the external data management implicitly. FORTRAN programming guidelines for effective use of virtual memory systems have been appearing for several years. Many of these guidelines are relevant to the dynamic dimensioning techniques mentioned above (see [17]).

DANGEROUS TECHNIQUES

Many guidelines can be put in the form of a suggestion that the program developer avoid certain practices. In particular, there are many examples of code which compile and execute on different processors without any diagnostic messages whatever, yet give completely different answers. Several examples of these types of coding were discussed in [13]. In particular:

1. Avoid calling subroutines with constant values, if there is any chance that the associated parameter in the subroutine can have its value changed. This is a not uncommon practice that can be very difficult to detect when it is buried in a large program. See [13] for three example programs with results.
2. Avoid nonstandard initialization, i.e., relying on the processor or some coding tricks to initialize variables to certain values. See [13] for an example program.
3. Avoid testing of alphameric data. These tests often fail because of variable type differences, different numbers of characters in a word, or different fill procedures.
4. Avoid mixed-mode arithmetic since different compilers treat mixed-mode calculations in different ways. See [13] for an example program and results.
5. Avoid hidden input, i.e., input in the form of arithmetic statements or hidden in data statements. Make sure all input is in

input lists or clearly marked within the program.

6. Avoid hidden dimensions, i.e., dimensions of adjustable length arrays in any level lower than the executive layer. This point was treated in some detail in the section on dynamic dimensioning.

7. Avoid ENTRY statements without arguments. I have examples of such code which give completely different results on two large processors.

8. Avoid ambiguous computations, e.g., A = 2**3**4 gives different answers on different processors.

9. Avoid making calculations in functions which are part of logical OR expressions. Some compilers evaluate the simpler expression first, and the function may never be evaluated.

10. Watch out for missing SUBROUTINE cards when adapting a program. Some compilers compile such routines with a START entry and transfer control immediately to that subroutine upon beginning execution. Other compilers compile such incomplete subprograms but do not transfer control to them, and some compilers give a diagnostic message and abort execution.

SUGGESTIONS

Developers of large FORTRAN programs who would like to make life pleasant for those who will someday adapt their programs to other purposes and/or other large computing systems should:

1. Use modular programming. Each section of coding which performs a single function should be coded as a subroutine even if that subroutine is called only once. This enables the user to replace that subroutine for any reason (different system, different function, and so on) and often helps to clarify the overall logic of the program.

2. Carefully choose appropriate names for these subroutines so that the name gives the reader some idea what the subroutine does. This is often quite difficult,but it is well worth the time spent thinking of a good name.

3. Carefully document each subroutine with meaningful comments; not "I IS REPLACED BY I + 1" but useful information that is not readily apparent.

4. Choose meaningful mnemonic variable names whenever possible. Again, it is worth the effort. The author is aware that some program developers believe that this technique (heavy use of mnemonics) is dangerous because it encourages programmers to rely on their memories; rather they propose including all variables in a common array and identifying the elements of this array by a program variable dictionary. The proponents of this technique argue that the programmer cannot then depend on his memory and must therefore consult the dictionary, thus eliminating the problem of misidentifying one or two variables with similar names.

5. Place all system-dependent coding in separate subroutines and clearly explain what they do so that an adaptor or implementor can simply replace each routine with an equivalent one for his system.

6. Place all input in separate subroutines. It is difficult and not nearly so useful to try and segregate all output in a similar manner.

7. Place all dimensioned variables in the main program and use a single master array for all the adjustable length arrays, as was exhibited in the section on dynamic dimensioning.

8. Avoid the types of coding that were identified as treacherous in the section on dangerous techniques.

9. All of the foregoing are more important than the complete external documentation that many people claim is so valuable. A carefully constructed program with well-thought-out subroutine and variable names, with a sufficient number of pertinent comments, and with much exhibition of good technique can often be read, understood, and modified by an experienced person without ever consulting external documentation. This is not to say that documentation is unnecessary. It is essential for the average user of any program who may not be able to code well in FORTRAN and it is often of great use to the sophisticated user. But program documentation often lags behind program development and is sometimes inadequate. For the experienced programmer, the code itself, if well done, is probably the most reliable documentation.

ACKNOWLEDGMENT

The preparation of this paper was made possible by a National Science Foundation collaborative research grant made to Carnegie-Mellon University, the University of Colorado Computing Center, and Weidlinger Associates.

REFERENCES

1 American National Standard FORTRAN, ANSI X3.9 - American National Standards Institute, New York, 1966.

2 American National Standard Basic FORTRAN, ANSI X3.10 - American National Standards Institute, New York, 1966.

3 Cartwell, W. J., "Helpful Programming Guidelines," Data Processing Magazine, Spring 1972, pp. 29-30.

4 Cohn, C. E., "Efficient Programming in FORTRAN," Software Age, June 1968, pp. 22-31.

5 Donaldson, J. R., "Structured Programming," Datamation, December 1973, pp. 52-54.

6 Fleiss, J. E., Phillips, G. W., Edwards, A., and Rieder, L., "Programming For Transferability," International Computer Systems, Inc., RADC-TR-72-234, September 1972. Available from NITS, U.S. Department of Commerce.

7 Ghan, L. "Better Techniques for Developing Large Scale FORTRAN Programs," Proceedings ACM 1971 Annual Conference ACM, New York, pp. 520-537.

8 Hofmeyr, J. F., "Preparation of User-Oriented Computer Programs with a Modular Structure Based on FORTRAN," SCIR guide K-25, NTIS, Pretoria, July 1972.

9 Knuth, D. D., The Art of Computer Programming, Addison - Wesley Publishing Company, Reading, Mass., 1968.

10 Kreitzberg, C. B., and Shneiderman, B., The Elements of FORTRAN Style, Techniques for Effective Programming, Harcourt Brace Jovanovich, New York, 1972.

11 Larson, D., "The Efficient Use of FORTRAN," Datamation, August 1971, pp. 24-31.

12 Liskov, B. H., "Guidelines for the Design and Implementation of Reliable Software Systems," The MITRE Corporation, ESD-TR-72-164, February 1973. Available from NITS, U.S. Department of Commerce.

13 McCormick, J. M., and Baron, M. L., "On the Implementation of Computer Software," Guidelines for Acquisition of Computer Software for Stress Analysis, ASCE National Structural Engineering Meeting, April 9-13, 1973, San Francisco, California, Meeting Preprint 2022.

14 McCracken, D. D., and Weinberg, G. M., "How to Write a Readable FORTRAN Program," Datamation, October 1972, pp. 73-77.

15 McCracken, D. D., "Revolution in Programming: An Overview," Datamation, December 1973, pp. 50-51.

16 Miller, E. F., Jr., and Lindamood, G. E., "Structured Programming: Top-down Approach," Datamation, December 1973, pp. 55-57.

17 Morrison, J. E., "User Program Performance in Virtual Storage Systems," IBM Systems Journal, Volume Twelve, Number Three, 1973, pp. 216-237.

18 Perry, G. L., and Sommerfeld, J. T., "FORTRAN Programming Aids," Software Age, October/November 1970.

19 Schrem, E., "Computer Implementation of the Finite Element Procedure," in Numerical and Computer Methods in Structural Mechanics, Edited by Fenves, S. J., Perrone, N., Robinson, A. R., and Schnobrich, W. C., Academic Press, Inc., New York, 1973, pp. 79-122.

DEVELOPMENT AND MAINTENANCE OF LARGE FINITE ELEMENT SYSTEMS

E. Schrem
Institut für Statik und Dynamik
University of Stuttgart
Stuttgart, Germany

ABSTRACT

The present state-of-the-art of sharing large-scale software systems for finite element analysis is discussed with respect to machine independence and program maintenance. Manpower required for maintaining a large system during its operating life is compared to the investment for its development. For a one-year maintenance period of a major software system for structural analysis, the effort required for postimplementation debugging is contrasted to the effort spent for software enhancements. Training and consultation of users, continuous software maintenance, and charges for software usage are discussed as the principal factors of software sharing. The danger of mushrooming variants and undisciplined extensions, on the one hand, and the tendency to lose the ability of responding to changing requirements, on the other hand, are considered as the major problem of large software systems. After a brief look at integrated systems, the programming systems approach to improving software sharing is presented. Basic principles of a clean modular design are mentioned, and the higher levels of programming systems are reviewed, with particular consideration of a recent concept of a programming system for finite element analysis. In order to provide a firm basis for further improving the user-system interface, a virtual computer model is proposed.

INTRODUCTION

During the past few years there has been an ever-growing interest in software sharing as a means of efficient transfer of technology in the engineering profession [1, 2, 3, 10]. Historically, structural analysis has been the pioneering discipline, both with regard to the intensity of computer utilization in the domain of engineering as well as in the development of comprehensive software packages. There is now a broad variety of widely distributed software systems, based upon the finite element method, and a rich stock of experience has been accumulated about the problems specific to software sharing. The time has come to reflect critically on our achievements in order to find an improved methodology for software dissemination that could lead to more flexibility and higher quality of shared software and at the same time would help to avoid unnecessary duplication of effort due to an abundancy of uncoordinated software projects. In this paper main emphasis will be given to large scale software systems for finite element analysis; however, it is hoped that some of the aspects will be of general interest for users and developers of shared application software in engineering.

STATE-OF-THE-ART OF SHARING LARGE SOFTWARE SYSTEMS

Machine Independence: A Prerequisite for Wide Distribution

In creating large software systems the most characteristic phenomenon is that development costs are larger than manufacturing costs by several orders of magnitude. The actual manufacturing of a software package simply consists of copying one or several data sets. On the other hand, the development of a large-scale software system for finite element analysis requires considerable effort, in terms of both financial resources and manpower (there are examples exceeding $1 million and 100 man-years [4]). In order to justify the high development costs, sharing the system in a very large user community becomes a vital necessity. As a consequence, it is necessary in the system development phase to devote much attention to the requirement that the system should be available on a large variety of computer systems.

This leads to the well-known problem of machine independence, a property which is claimed for many existing systems but has been achieved to a reasonable extent by only a few (typical examples are GENESYS [5], ASKA [6], and NASTRAN [7]). The claim of machine independence is used to describe the fact that the system can be implemented, with little additional effort, on a broad spectrum of computer systems, comprising a large variety of both hardware and software configurations. Clearly, machine independence is a relative quality that is affected by each configuration variable. Therefore, it has become customary to present an articulation in several categories, such as representation independence, transferability, portability, and adaptability [8, 9, 10, 11]. Programs have to be planned for a high degree of machine independence. In general, the programs can be safely transferred to such computer systems only if they are very closely related to the configurations considered in the basic design. Coding in ANSI-Fortran alone is not a sufficient condition for attaining a high degree of machine independence; one could even doubt whether it is the most important one. A considerable number of additional features and constraints must be considered in the basic design because they vary from one computer system to the other. The most important ones are:

1. Data organization in various levels of storage hierarchies
2. Program organization in executable memory (binders, linkage editors, loaders; planned overlays versus virtual memory)
3. Interfaces with the operating systems, both internally (system macros) and externally (job control language)
4. Considerable ranges of configuration parameters (optimal record sizes, available central memory and long-term storage, utilization of special storage
5. Utility packages for organizing and updating large libraries, both for program and data

These constraints could only be reduced if one decided to use specifically designed, highly standardized utility programs for updating and binding the programs (which would be feasible), as well as a standardized operating system interface (which in general is not possible).

On the other hand, attempting to be machine independent in the wrong place has led quite frequently to design decisions that affected the macro-efficiency of the system to an intolerable extent. Well-known examples are:

1. Second-generation computers were often restricted to sequential access for backing storage, whereas direct access mass storage is a standard feature of all third-generation machines. If one adheres to the requirement that sequential access backing storage has to suffice for operating the system, either considerable advantages of third-generation computers are inevitably lost (when only sequential access data organization is chosen) or the system becomes extremely inefficient on second-generation machines (when direct access is emulated on sequential access devices).

2. There are computers (e.g., IBM 360/370) with a relatively small word length for representing floating point numbers, so that double precision arithmetic becomes mandatory. For this reason a major system for structural analysis was required to perform all critical calculations in double precision, with the result that efficiency on machines with a decent size of the single precision mantissa (e.g., CDC 6000 series) became intolerably bad.

Continuous System Support and Maintenance

It is a common experience that large software systems are not really ready to be used in the field just after coding, documentation, and acceptance tests have been completed. There should be at least another full year of actual experience with their continuous support and maintenance before a sufficient level of confidence can be credited to them. This also gives a chance to experience a soft implementation phase [8] in which the system is being run in a trial mode (see Fig. 1). During that period the users are receiving results and can try out interfaces and investigate the available analysis features. At the same time the implementors have an opportunity for turning the system for better responsiveness and efficiency. Only after having been subject to such a soft implementation phase will the system have matured enough for wide distribution and shearing in a larger user community.

In addition, intensive and widespread software sharing requires considerable effort for adequate maintenance and support. Supporting and maintaining a large-scale software system during its operating life could very well be considerably more expensive than the initial investment for developing and coding the program. There is a most important consequence, which unfortunately has been overlooked in many cases: large program systems must be designed for good maintainability!

In Fig. 1 the curves of total manpower spent for software development and maintenance are plotted over the lifetime of the linear static analysis part of a major software system for structural analysis [6]. The additional effort for developing the theoretical background (e.g., finite elements) is not included in the diagram. The curves demonstrate in an impressive way that the total manpower spent for maintenance will, even in the near future, exceed the initial manpower investment for developing the system.

Adequate software support and maintenance comprises a broad spectrum of activities which could be classified according to the following categories:

1. Tracking down of program errors
   Defensive: Continuous effort to find flaws and deficiencies by analyzing program behavior in the field
   Responsive: Reaction to complaints from the user community
2. Removal of program errors
   Conservative: Correct the source code
   Amputative: Restrict program application
3. Program modification
   Additive: Extend capabilities by pure additions of software modules
   Reformative: Reorganize existing code (danger of regression!)
4. Program adaptation to a new environment
   Conservative: Adaptation of an existing implementation to modest changes of hardware configuration or operating system
   Hazardous: Changeover to an entirely new computer system
5. Update documentation

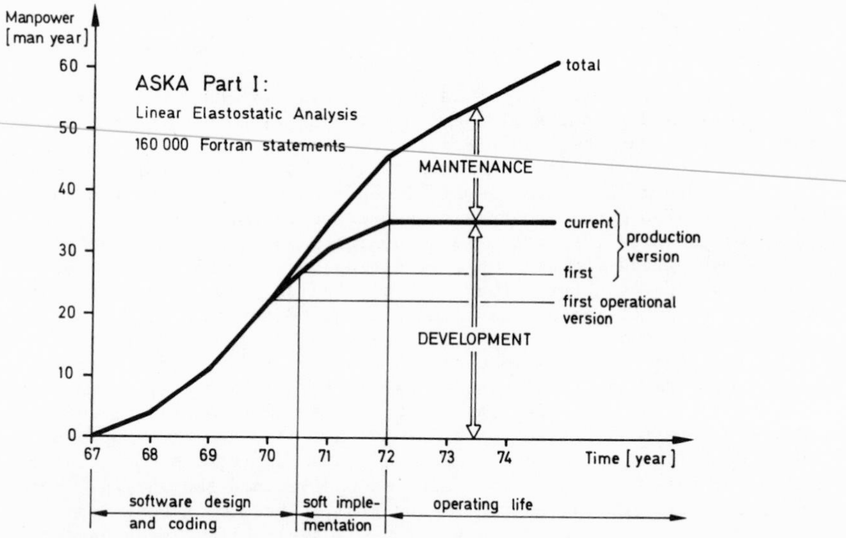

Fig. 1 Total manpower spent for development and maintenance of a large-scale structural analysis software system

Activities 1 and 2 represent postimplementation debugging. If the program system turns out to be faulty in relation to its original specifications, either the errors have to be removed or the specifications must be restricted. Unfortunately it is not feasible to prove the correctness of complex program systems. Therefore, postimplementation debugging must be provided throughout the operating life of the system. Software packages for which this minimum support is not guaranteed are virtually useless for sharing.

The situation is different for activities 3 and 4, which yield extensions and alterations in order to enhance the usefulness of the software or to react to changing requirements. This kind of software enhancement is complicated not only by every computer system on which the software is installed but also by every specialized domain within its scope. As a consequence the effort required for a certain degree of software enhancement grows exponentially with the extent to which the software system is being shared. Also, the inertia of a large user community toward accepting changes to the software specifications should not be underestimated. When the number of users exceeds a certain critical value, any modifications that cannot be treated as mere optional additions become virtually prohibitive because they would require changes in the user-system interface.

Figure 2 shows maintenance statistics derived from the activities depicted in Fig. 1 for the period from September 1972 to September 1973 as compiled by Knapp [12]. During this period the code was already consolidated: all features had been fully implemented for at least nine months. The number of source statements involved in post-implementation debugging is by a full order of magnitude smaller than the number of statements affected by software enhancement. Only 3 severe malfunctions and 8 minor deficiencies were brought to the attention of the maintenance centre, resulting in corrections to about 60 program statements. This corresponds to an annual fraction of about 4×10^{-4} faulty statements. (This remarkably good error rate is the rare exception rather than the general rule in contemporary application software.) After a malfunction had been reported in sufficient detail, tracing down the incorrect program statements was normally a matter of a few days; however, completion of the corrections in all installations was sometimes delayed by several months.

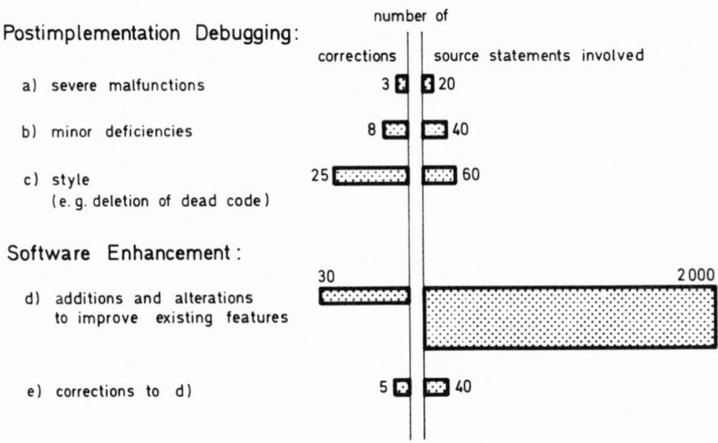

Fig. 2 Maintenance statistics for ASKA 1 for the period September 1972 to September 1973

Due to the relatively wide distribution of the system (more than 30 installations, most of them with highly differing software/hardware configurations), we restricted software enhancements to a minimum that was compatible with the original software specifications and released them for general distribution at intervals of about one year. Nevertheless, additions and alterations to improve existing features represented the dominate part of the total maintenance activities. Also, postimplementation debugging of software enhancements required another 5 corrections to a total of 40 statements. The relatively large error fraction of 2 percent is another confirmation of the general experience that modifications to a living program system are far more susceptible to errors than might result during the initial coding of the system.

Main Factors in Software Sharing

Conventional sharing of large software systems for structural analysis is determined by three main factors:

1. User training and consultation. Structural analysis software based upon the finite element method is a most flexible tool and therefore constitutes a permanent temptation to use to tackle more and more complex problems. The success in more exacting applications depends critically on the availability of adequate support at the user's site. Therefore, at least two experts should be available at the user's site for consultation about software engineering aspects as well as the theoretical background of the program system. This kind of direct consultation can hardly be substituted by training courses or written documentation.

2. Continuous maintenance. In every installation there should be a competent authority for properly maintaining the system. Even quite complex and large systems can be maintained adequately without depending too much on the program authors, provided that

   a. The modular organization of the system is sufficiently clear.

   b. The source code together with adequate internal documentation is accessible.

   c. Highly qualified maintenance experts are available.

For instance, in a few installations in the United States and Japan the ASKA system has been kept running successfully over several years without intensive contacts with the authors of the program. Of course there are a large number of other installations where the above conditions for autonomous maintenance are not provided. In these cases the existence of a powerful maintenance network is a most vital requirement.

3. Charges for using the system. A broad spectrum of agreements is customary, ranging from rights to use to outright purchase. In the former, only an absolute or relocatable version of the program is normally available, whereas the latter also comprises the source code with all internal documentation. The appropriate choice depends primarily upon the extent to which the system is to be used. If the system is used quite frequently or for solving relatively complex problems, access to the updated source code becomes mandatory.

There is a natural hierarchy of various modes of using a large program system for structural analysis:

1. Having problems solved by a commercial software center that already has sufficient expertise in using the system
2. Using the system out of house on a surcharge basis and paying for additional consultation services
3. In-house usage under own responsibility on a royalty or leasing agreement
4. Acquiring the source code and executing autonomous control of all in-house applications

An Example of a Software Distribution Network

Due to its wide distribution, the ASKA system can also serve as an illustration of present practice in the distribution of large-scale application software. This system is installed on more than 30 computer systems in the United States, Japan, and Europe, including the CDC 6000 series under SCOPE and KRONOS operating systems, the UNIVAC 1100 series under EXEC 8, and the IBM 360/370 under OS. Every implementation is derived from a nonexecutable standard source code (approx. 220,000 card images ANSI Fortran IV) by adding machine-dependent and installation-dependent code (normally a few hundred card images in assembly language and/or Fortran dialect) (see Fig. 3).

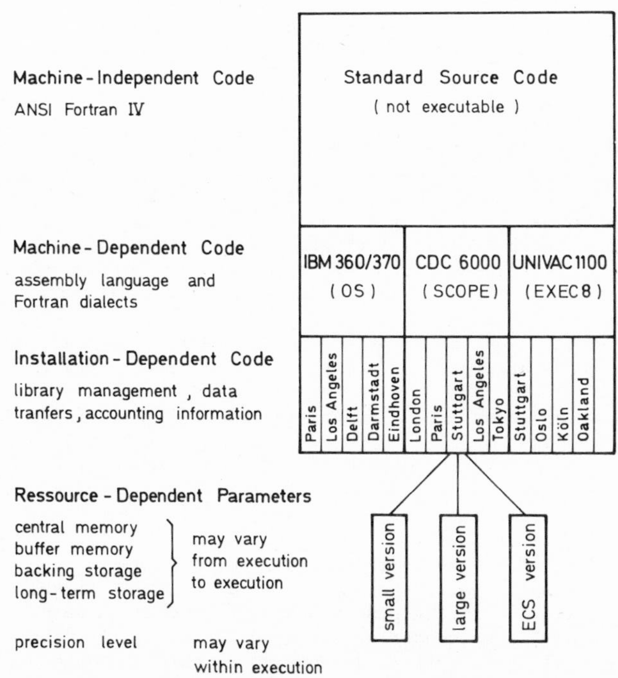

Fig. 3 Standard system without local modifications

A maintenance center in Stuttgart is responsible for a standard version of the system on CDC 6600, UNIVAC 1108, and IBM 370/155. The maintained system, as well as any documentation, is directly distributed to 11 primary installations. Most of the primary installations are responsible for further distributing the program system to a varying number of secondary installations. Roughly once in a year a maintenance cycle is completed, yielding a new version of the system in relocatable format. In the intermediate periods urgent corrections to the source code are communicated to the primary installations in the form of newsletters.

As a consequence of this maintenance procedure there are three different kinds of users:

1. Users at the maintenance centre. These users constitute a source of valuable information about field behavior and system responsiveness. Most of the program errors that came to the attention of the maintenance center had been discovered in applications at the maintenance center itself. Naturally, users at the maintenance center enjoy good support, which is also facilitated by the opportunity of informal and immediate exchange of information.

2. Users at primary installations. Due to the relatively large user community at a primary installation, good access to experts is usually provided. As a rule, there are also maintenance professionals who stay in regular contact with the maintenance centre. Any difficulty that appears can therefore be resolved within a short time.

3. Users at secondary installations. Being separated from the maintenance center by relatively long communication paths, these users' situations are critically determined by the quality of their own local support and by the intensity of communication with their primary installation center.

Mushrooming Variants, Incompatibility, and Regression

A program system could be compared to a living organism which continuously must adjust to a changing environment. Therefore there is a strong temptation to modify existing software system due to:

a. The large variety of problems to which the system could be applied in principle, which - unfortunately enough - were not considered during the system's design

b. The need for better adaptation of the system to specific problems (the whole preprocessor/postprocessor complex falls into this category)

c. Further progress in the underlying theory and in information processing technology

Thus it is little wonder that as soon as the source code of a comprehensive software package becomes publicly available, numerous attempts at further improvements will be started, eventually resulting in mushrooming variants of highly differing quality, all offered to the perplexed end-user under the sufficiently prominent name of the original system. To be fair, there are several examples of such after-the-fact polishing of existing systems which result in quite substantial upgradings, some rendering the original system all the more useful. Unfortunately the price that has to be paid for enterprises of this kind is not low. Often enough the modifier has severe difficulties in obtaining sufficiently comprehensive information in order to understand the internal structure of the software system in all its parts. Thus his alterations to the program are quite likely to violate certain hidden internal conventions and constraint, which might become evident only after a longer period of actual system utilization. Such internal inconsistencies are likely to spoil the overall reliability of the "improved" system to an extent where it is less useful than the original one. This is the well-known phenomenon of system regression - a nightmare to every software developer.

There are good reasons to enforce a standard system which is frozen in its basic concepts and has a centralized maintenance which forces all further modifications upon the user community in its entirety. As a consequence, conventional widely distributed systems tend to become very rigid, losing the ability to respond in a flexible and rapid manner to further progress in technology. Eventually, after an operating life of five to ten years, they will lose the struggle for existence because more recent, better-adjusted systems will have entered the field.

STRUCTURED SOFTWARE SHARING

Program Systems, Programming Systems, Integrated Systems

In the preceding section we encountered a rather unsatisfactory situation of large-scale program systems: on the one hand they require a wide distribution in order to justify the considerable investments into their development, and on the other hand they tend to become too rigid just as a consequence of being successful in the field. This situation is one aspect of the software crisis in general and has been commonly recognized for some five years.

There is little evidence that integrated systems, following the lines of the pioneering ICES project [13, 14], could ever offer a cure to this problem. Nevertheless, in building upon fundamental ICES concepts considerable effort has been invested in the development of two integrated systems, GENESYS in the UK [5, 15, 6] and more recently ISB/IST in Germany [17, 18, 19]. Another integrated system, SYSFAP [20], is based on a somewhat different approach, particularly in the user-software interface which is oriented toward a question-answering system. Integrated systems seem to be able to provide some relief to another serious problem of large software packages which has its origin in the fact that frequently the input data for describing the problem are quite voluminous and thus are handled best inside the computer as the result of some preprocessing by other software packages. Similarly, the bulky output data normally have to be subject to comprehensive postprocessing until they really represent those answers to the problem that the user is interested in seeing. Eventually, integrated systems will offer a sufficiently rich menu of subsystems such that all project data could be kept inside the integrated system's data base and processed by a series of subsystem executions. To observe how integrated systems will perform in the accomplishment of this important objective will be most instructive, especially for improving our understanding of the field behavior of very large software systems.

Another promising attempt to tackle the present software crisis is the development of so-called programming systems. The basic objective is not to develop one large general purpose software system from the beginning but rather to devise a comprehensive set of tools and software modules that can be combined in a very flexible way to various software packages with a high degree of differentiation. This approach has become feasible due to further progress in software engineering. In particular, concepts like "modularization by information hiding" [21, 22] and "structured programming" lead to a better understanding of how complex systems should be structured into modules with a minimum of interface specifications.

An example of a programming system for structural analysis is the Norwegian NORSAM project [23]. A more recent project has been started by our group, based upon the experience gathered during design, implementation, and maintenance of the ASKA system. At present, the project is far from completion; however, the results obtained during the first two years are quite encouraging. The time required for the implementation of software modules for matrix algebra with large, sparsely populated operands was reduced by more than one order of magnitude, as compared to the corresponding modules in the ASKA system. At the same time, overall software reliability seems to be improved by several orders of magnitude, reducing post-implementation debugging to almost nil.

In developing programming systems, the ultimate goal is to be able to select cheap and highly reliable modules from an order list and to integrate them with special purpose software in as much as today minicomputers and peripherals are built into hand-crafted specialized data processing equipment. A drastic reduction of the number of interface specifications will also promote sharing of software development. In order to enforce or complete existing capabilities, "plug-compatible" software modules could be developed in various places with relative ease. In that way, sharing of integral, ready-to-use software packages will be complemented by structured software sharing, where specifically selected modules are combined from various sources with in-house software to produce highly differentiated operational software. In the following, a few basic concepts of programming systems will be presented.

Modular Structure

A modular structure of a large program is not at all peculiar. Due to its size alone it can only be described, understood, and programmed after a suitable decomposition into smaller parts. We then have a program system consisting of modules connected to each other through interfaces. According to Parnas [21], interface information consists of the assumptions which the modules make about each other. The precise meanings of "module" and "interface information" depend on the particular description of the system. To conclude, the question is not whether, but rather, how a program should be decomposed into modules. The answers to this question have to be based upon two important but independent principles [22]:

1. Minimization of interface information. In order to simplify intermodule connections, important design decisions should be hidden as far as possible inside modules. The interface information should reveal only the absolutely necessary minimum about the inner workings of a module.

2. Hierarchical structure. A hierarchical modular structure is formed when there is a relation between modules, like "module a uses module b"; that is, a partial ordering [20]. The different levels in the hierarchy can be considered as increasing levels of abstraction. In higher levels, there is an increased volume of potential data processing (as could be measured in time spent in activities, number of machine operations, amount of storage required, size of data structures, and so on).

A well-designed hierarchical structure is an essential requirement for a programming system because it is the basis for assembling a particular program system from various subsystems in the most straightforward manner. A subsystem is an assembly of modules that can be used as an independent software package. As a consequence, modules have to be combined into subsystems by progressing from the lowest level in the hierarchy upward to the higher levels (see Fig. 4). In that way the number of visible levels can also be drastically reduced.

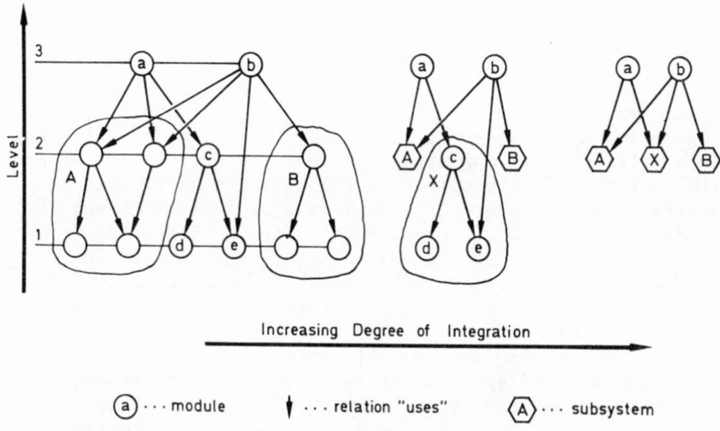

Fig. 4 Hierarchical structure of a programming system

Between subsequent hierarchical levels there are three aspects of modularity that are inseparably linked to each other [4]:

1. Modularity of the data: the objects that are created, referenced, or destroyed on the lower level upon request of the higher level and their internal structure that is visible to the higher level

2. Modularity of the operations: the operations in the lower level that are activated from the upper level and their functional specifications that are visible from the higher level

3. Modularity of the problem: the way in which operations and data have to be combined in order to solve a particular problem in terms of subproblems

These three aspects correspond to the general view of a data-processing system in terms of data, operations, and control.

The Processor/Book Level

This brief review is restricted to the higher levels of a programming system that are most interesting to the user. On the lowest level where the modules could be directly used for solving a practical problem in engineering, e.g., finite element analysis, there is a set of operations, each representing a distinct solution step, such as generating a table of nodal point coordinates, calculating a structural stiffness matrix from elemental stiffness matrices and connectivity information, calculating a displacement matrix from a stiffness and a load matrix, and so on. The operands represent logical entities from the viewpoint of the underlying finite element theory, such as nodal point coordinates of a mesh, kinematic connection matrices, stiffness, load and displacement matrices of a structure, and so on. Due to the importance of this level for understanding the system from the user's point of view, we introduced into the ASKA system special terms for the operations and operands. For lack of better words we called the former "processors," and the latter "books" [4, 6]. In order to simplify the presentation, we shall use the same terms here for analogous objects in programming systems.

In programming systems, books play a key role comparable to that of files in a file-oriented operating system. The quality of the book concept will therefore determine the usefulness of a particular programming system. The structure of a book must be general enough to be suitable for operands of all processors under consideration in a straightforward manner, and at the same time, it must be as simple as possible in order to avoid unnatural complexity in the processor programs. These are two conflicting requirements which imply very difficult design decisions.

In the LPR2 programming system [24], which has been designed as a general tool for a broad spectrum of engineering applications, there is a very clean and consistent concept of organizing all books in tabular form. Such clarity is somewhat lacking in the NORSAM programming system [23], which is intended for supporting finite element analysis. Here, information on book level is primarily organized in hypermatrix form, which, however, has to be complemented by so-called control data. A similar concept was used about a decade ago in ASKA 102, a predecessor of the ASKA system, and it soon became obvious that a clean modularization cannot be achieved in this way because there are severe violations of the principle of hiding information inside modules.

In our present project, the book structure visible to the processor programs is represented by two-dimensional orthogonal lists, organized in two levels [25]. This data structure is equally well suited for organizing large tables, as well as large structural matrices with arbitrary population pattern. Our books are fully self-descriptive, i.e., all parameters governing their internal structure are contained inside the book in a small descriptor block which is accessible both for the data management subsystem and the processor programs. A book is identified by an alphanumeric name, which is the only information about the book that has to be submitted as a parameter when calling a processor program.

Below the processor/book level there is a data management subsystem which itself is structured in several levels. On its lowest levels, the data management system takes care of all aspects of physical storage management, including both central memory and peripheral storage. In that way, we achieved a high degree of flexibility in adapting the data management to different computer systems without affecting the processor programs at all. For the processor programs, the physical organization of data in the various levels of the storage hierarchy is entirely hidden in the data management subsystem. This leads to a considerable simplification of the processor programs, as compared to the ASKA system, where addresses of submatrices have to be handled in two levels inside the processor programs. Now it has become feasible that engineers who are not highly qualified data-processing experts can write processor programs of remarkably good quality. This is due to the drastic reduction of interface information to exactly the minimum that is inherent to the problem itself at the processor/book level.

Higher Levels: The Virtual Computer Model

For simpler applications of finite element analysis the processor/book level of the programming system is already adequate. In this case the various processors are called as subroutines from a FORTRAN program, written directly by the user, and the names of the operand books are specified as subroutine parameters. Here the problem-oriented language is just FORTRAN, the user having at his disposal all FORTRAN features for combining modules of the programming system with his own programs. One major disadvantage of this method is the fact that in writing such FORTRAN programs, the user has to perform all book administration by himself, i.e., inventing names for newly created books and inserting the correct names into the processor calls. This is analogous to writing assembly language programs.

This approach is inadequate in more complex applications. Additional system levels above the processor/book level are certainly needed in order to provide better support to the user. The data management subsystem offers the possibility of handling books as the elements of a data base which resided on long-term storage. Books together with their names thus are able to survive particular program executions. In order to exploit this feature fully, either the user is required to remember the operand books and their names from execution to execution, or an additional software mechanism has to be set up for supporting the user in this task. Furthermore, there exist numerous relations between the books that are specific to the particular analysis project. As an example, in substructure analysis, every book representing structural information has a well-defined place in the hierarchy of substructures, and at the same time, the sequence of processor calls has to reflect this hierarchy. If there are a large number of substructures or more than two levels of substructuring, the necessary coordination of operations and operands is not a trivial task. Another example is the analysis of modified structures, where a particular book appears in modified form in several instances during the modification history.

Unfortunately, there is no uniform solution that fits all interesting applications equally well because there is a large variety of structures in the problems themselves. One approach that is general enough to provide a powerful support for the implementation of different solutions of the user-system interface is presently being investigated by our group. The basic idea evolved from the history that there is a certain analogy to machine languages of general purpose computers:

processors	...	machine instructions
books	...	data items
book names	...	addresses of data items

In order to complete the analogy, we just have to add control structure, including a notion of storage locations for instructions, and we obtain a virtual computer on the processor/book/control structure level. In exploiting the analogy, we can benefit from the rich stock of experience accumulated in the development of high-level computer languages. In that way, the implementation of problem-oriented languages via macroprocessors, compilers, or interpreters is established on a firm basis.

Operations on Books:

[label] operation operand$_1$ operand$_2$ ··· operand$_n$

Control Instructions:

[label] SET $Ri expression

[label] GOTO label

[label] IF expression label

Macro - Calls

[label] CALL macro-name [fp$_1$ = ap$_1$, ···, fp$_n$ = ap$_n$]

[label] CALL macro-name ap$_1$, ap$_2$, ap$_3$, ···, ap$_n$

fp ···formal parameter   ap ···actual parameter

Explanations:

[·····] ··· string in brackets is optional

operation ··· processor name

operand ··· book name or register denotation

expressions are formed of:

constants <u>bool</u> 0 (false) | 1 (true)

<u>int</u> 2 14 256

<u>real</u> 3.14 12.3E-2

<u>name</u> ABC1 XYZ3

register denotations $R1, $R2, $R3, ···

monadic operators + | − | ¬

dyadic operators + | − | ∗ | /

= | ≠ | < | ≤ | ≥ | >

∧ | ∨

Fig. 5 Symbolic instructions of the virtual computer

As this approach seems to be quite attractive, an incomplete and informal outline of our virtual computer model will be appropriate. Figure 5 shows the format of the instruction set in symbolic machine language. The instructions are partitioned into two classes: control instructions and operations on books. Apart from the redundant unconditional jump, there are just two control instructions: a conditional jump and an assignment to a register. There is an arbitrary but fixed number of registers, denoted as R1, ... , Rn. The expressions appearing in control instructions can be built from register denotations and denotations of constants (of type Boolean, integer, real, name), connected by Boolean, arithmetical, and comparison operators. At present, we are also investigating the impression of a linear storage structure to the registers which would allow indexing in the register space. The interpreter has two functional sections, one for evaluating the expressions in control statements and one for decoding the operations on books and activating the appropriate processor calls.

This concept of a virtual computer allows the programming system to be extended in two dimensions (see Fig. 6): in the horizontal direction by adding processors and thus enlarging the instruction set and in the vertical direction by adding further levels of software. Our first development in the vertical direction is a macroprocessor that will use both predefined and user-defined macros. Such macro definitions allow simplification of the user specification of the solution path to a considerable extent. In particular, it will be possible to build into the macros fixed definitions of book names that can be easily overruled by the user in the rare circumstances when this might be necessary.

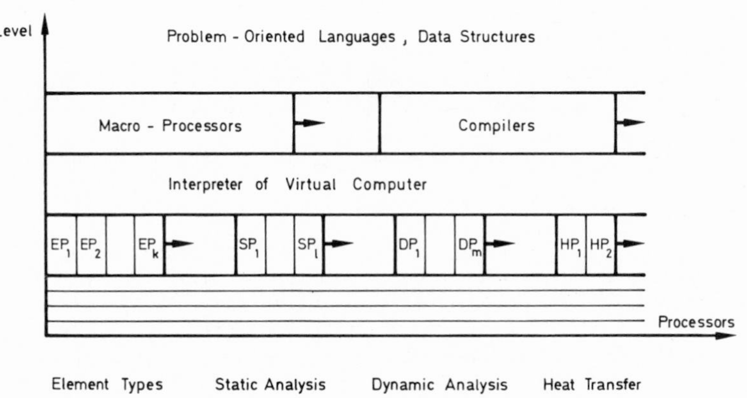

Fig. 6 Two-dimensional growth of the programming system

CONCLUSIONS

In the case of large-scale systems for finite element analysis, successful software sharing is primarily a question of adequate user support and program maintenance. In large, widely distributed software systems, the costs for maintenance and support can easily exceed the original expenses of system development. Program systems, therefore, must be designed for good maintainability. Another serious problem with current software systems is the danger of either getting out of control due to undisciplined growth and alterations or becoming too rigid, losing the ability to adjust themselves to a changing environment, be it caused by progress in hardware or due to new requirements from the user community.

Programming systems, provided that their design corresponds to the present state-of-the-art in software engineering, could offer some cure to this problem. Their advent will also alter the current practice of software sharing. Conventional sharing of integral, ready-to-use software packages will lose its importance in favor of structured software sharing, where cheap and highly reliable modules of a programming system are elected from an order list and combined with in-house software to highly differentiated operational software. In that way, software sharing will be turned from a hazardous enterprise into a creative performance.

REFERENCES

1 Schiffman, R. L., "Special Workshop on Engineering Software Co-ordination," Computing Center Reports 72-2, 72-4, and 72-17, University of Colorado, Boulder, 1972.

2 Medearis, K., Civil Engineering Software Center, ASCE, New York, 1973.

3 Kraus, H., Attitudes Toward Computer Software and its Exchange in the Pressure Vessel Industry, R.P.I. of Connecticut, Hartford, 1973, (NSF Grant GK-36084).

4 Schrem, E., "Computer Implementation of the Finite-Element-Procedure," in: Fenves, S. J., Perrone, N., Robinson, A. R., and Schnobrich, W. C. (Eds.), Numerical and Computer Methods in Structural Mechanics, Academic Press, New York, 1973, pp 79-121.

5 Alwood, R. J., and Maxwell, T. O'N, "GENESYS-a Machine-Independent System", International Colloquium on Integrated System in Civil Engineering, CEPOC, Université de Liège, Belgium, August 1972.

6 Schrem, E., and Roy, J. R., "An Automatic System for Kinematic Analysis, ASKA Part I", in: B. Fraeijs de Veubeke (ed.), High Speed Computing of Elastic Structures, Congrès at Colloques de l'Université de Liège, Vol. 69, 1971, pp. 477-507.

7 Butler, T. G., "Considerations for the Design of a General-Purpose Structural Analysis Program," in: Neubert, V. H., Raney, J. P. (eds), Synthesis of Vibrating Systems, ASME, New York, 1971.

8 Armstrong, R. M., Modular Programming in COBOL, Wiley-Interscience, New York, 1973.

9 Mealy, G. H., "Another Look at Data," Proc. AFIPS 1967 FJCC, Vol. 31, AFIPS Press, Montvale, pp. 525-534.

10 Fenves, S. J., Schiffman, R. L., and Baron, M. L.,"A Position Paper on Computer Based Technology Transfer in Civil Engineering," International Colloquium on Integrated Systems in Civil Engineering, CEPOC, Université de Liège, Belgium, August 1972.

11 Ward, J. A., Bemer, R. W., Gosden, J. A., Hopper, G. M., Morenoff, E., Sable, J. D., "Software Transferability," SJCC 1969, AFIPS Conf. Proc., Vol. 34, 1969, pp. 605-612.

12 Knapp, H., "Maintenance of Large, Widely Used Software Systems," Workshop on Computational Aspects of the Finite Element Method, Stuttgart, September 17-18, 1973.

13 Roos, D., "ICES - an Integrated Computer Based System for Engineering Problem Solving," International Colloquium on Integrated Systems in Civil Engineering, CEPOC, Université de Liège, Belgium, August 1972.

14 Roos, D., ICES System Design, M.I.T. Press, Boston, 1966.

15 The GENESYS Reference Manual, GENESYS Centre, University of Technology, Loughborough, England, 1971.

16 Alcock, D. G., and Shearing, B. H., "GENESYS - an Attempt to Rationalize the Use of Computers in Structural Engineering," The Structural Engineer, Vol. 48, 1970, pp. 143-152.

17 Pahl, P. J., Informationssystem für das Bauwesen, Zentralstelle für Atomenergie-Dokumentation, Leopoldshafen, Germany, 1972.

18 Pahl, P. J., "Systemkerne für CAD-Informationssysteme," CAD-Mitteilungen 1/1973, Gesellschaft für Kernforschung, Karlsruhe, Germany, 1973, pp. 322-338.

19 Informationssystem Technik: Programmierhandbuch, Siemens-System 4004, D 14/40313, 1973.

20 Deprez, G. , "SYSFAP: Description Generale de Systemes," International Colloquium on Integrated Systems in Civil Engineering, CEPOC, Liège,1972.

21 Parnas, D. L., "Information Distribution Aspects of Design Methology," IFIP Congress 71, TA-3, pp. 26-30.

22 Parnas, D. L., "One the Criteria to be Used in Decomposing Systems in Modules," Comm. ACM, Vol. 15, 1972, pp. 1053-1058.

23 Bell, K., Hatlestad, B., Hansteen, O. E., and Araldsen, P. O., NORSAM - A Programming System for the Finite Element Method, User's Manual, Trondheim, Norway, 1973.

24 Becque, A. J., "LPR2 - List Processing Version 2," International Colloquium on Integrated Systems in Civil Engineering, CEPOC, Université de Liège, Belgium, August 1972.

25 Schrem, E., "Experience with Two-Level Indexed Orthogonal List Structures for Internal Data Presentation in an Integrated System for Structural Analysis," International Colloquium on Integrated Systems in Civil Engineering, CEPOC, Université de Liège, Belgium, August 1972.

THE DEVELOPMENT OF GENERAL PURPOSE SOFTWARE OR WHAT IS A SOFTWARE SUPPLIER?

John A. Swanson
Swanson Analysis Systems, Inc.
Elizabeth, Pennsylvania

ABSTRACT

This paper explores possible sources for structural mechanics computer software. The emphasis is placed on the development of large-scale general purpose computer programs. The paper separates programs into classifications, ranging from desk calculator programs to large-scale analysis systems. The flow of technology from basic research in numerical analysis to production programs is traced, and the scale of present development programs is illustrated. The difference between project program development and evolutionary development is presented, with the emphasis on evolutionary development. The concept of a technical software supplier is introduced, and the contrast between government, industry, university, and technical software supplier produced programs is discussed. The paper concludes that the technical software supplier is a desirable alternative to the government in the production and marketing of large-scale general purpose structural analysis programs.

INTRODUCTION

Much of the engineering community seems to be working under the delusion that someone is going to supply it with the computer programs that it needs, that these programs will be modified to suit the needs of each organization, converted to run on the machine which the company has in the basement, that these programs will be perfect, verified, certified, and blessed, and that all this will be done at no cost to the program user. This paper explores the development of structural analysis software and is intended to shed some light on the development process and attack the myth of free computer software.

STRUCTURAL MECHANICS COMPUTER PROGRAM CLASSIFICATIONS

The first six figures of this paper divide the spectrum of structural analysis computer programs into six general classes, ranging from the smallest size, which runs on a programmable desk calculator, up to the largest analysis system, which requires the full capacity of a third-generation computer system.

Desk Calculator Programs

The smallest programs used in structural analysis are those written by an engineer to use on his programmable desk calculator (Fig. 1). These programs are usually used for single repetitive calculation, such as the evaluation of a Roark formula. The development cost is trivial, and there is little incentive or need to share such programs. No documentation or maintenance is required, since these are often reprogrammed for each use.

SIZE	- 1 - 10 STATEMENTS
DEVELOPMENT GROUP	- ONE PERSON
DEVELOPMENT COST	- < $100.00
USER DOCUMENTATION	- NONE TO ONE PAGE
USER COMMUNITY	- DEVELOPER AND ASSOCIATES
DIAGNOSTICS	- NONE
MACHINE REQUIREMENTS	- DESK CALCULATOR (10 - 15 WORDS)
COST PER RUN	- ≈ $0.0
LANGUAGE	- MACHINE
MAINTENANCE	- NONE
CONVERSION COST	- REPROGRAM FOR EACH NEW MACHINE
EXAMPLE	- ROARK FORMULA EVALUATION

Fig. 1 Desk calculator program characteristics

Time Share or Small Batch Programs

A large number of programs fall into the class which I call Time Share or Small Batch Programs (Fig. 2). This class of programs is usually written by a single person. These programs are written in either FORTRAN or BASIC and are usually simple enough that a few pages of user documentation are sufficient. Development costs are usually small, and there are a large number of these programs which do the same functions. Sharing of this type of program is common, but it is often cheaper and easier to develop your own program than to try to use a program developed by another person. The computer storage requirements and cost per run are small with these programs. Some of these types of programs are made available to time share customers and provide a rapid evaluation of small problems. I would include beam bending with redundant supports and column design programs in this class.

SIZE	- 10 - 500 STATEMENTS
DEVELOPMENT GROUP	- ONE - TWO PERSONS
DEVELOPMENT COST	- \$100.00 - \$1,000.00
USER DOCUMENTATION	- 1 - 10 PAGES
USER COMMUNITY	- DEVELOPER AND DIRECT CONTACTS TIME SHARE CUSTOMERS
DIAGNOSTICS	- MINIMAL
MACHINE REQUIREMENTS	- 500 - 5,000 WORDS
COST PER RUN	- \$0.10 - \$1.00
LANGUAGE	- BASIC OR FORTRAN
MAINTENANCE	- NONE TO MINIMUM
CONVERSION COST	- SMALL
EXAMPLES	- BEAM BENDING WITH REDUNDANT SUPPORTS COLUMN DESIGN PROGRAMS SIMULTANEOUS EQUATIONS

Fig. 2 Time share or small batch program characteristics

Single-Element Type Finite Element Programs

In increasing program size, the next class is the Single-Element Type Finite Element Program (Fig. 3). These are numerous examples of this type of program, ranging from simple two-dimensional plane and axisymmetric stress analysis programs to three-dimensional shell or solid analysis programs. The development of these programs is a significant undertaking, usually with an effort of one to four persons. These programs are most used by the developing organization, but many are available for the cost of reproduction through software distribution centers. Many of these programs have become quite complete, with preprocessors and postprocessors. Graphics are often included. The publicly available version is usually considered obsolete by the developing organization, although it may still be very useful for the person obtaining the program. The developing organization has little interest in the public version once it has been released. There are many from structure programs and piping programs in this class.

Multiple-Element Type Static Finite Element Programs

The next step after a single-element type finite element program is the Multiple-Element Type Static Finite Element Program (Fig. 4). These programs are usually developed by a group of people, because the use of a library of finite elements makes their construction

SIZE	- 500 - 2000 STATEMENTS
DEVELOPMENT GROUP	- ONE TO FOUR PERSONS
DEVELOPMENT COST	- \$5,000 - \$50,000
USER DOCUMENTATION	- 20 - 100 PAGES
USER COMMUNITY	- DEVELOPING ORGANIZATION PUBLIC DISTRIBUTIONS (USUALLY FOR REPRODUCTIVE COST)
DIAGNOSTICS	- MINIMAL TO ADEQUATE
MACHINE REQUIREMENTS	- 10,000 - 30,000 WORDS
COST PER RUN	- \$1.00 - \$100.00
LANGUAGE	- FORTRAN
MAINTENANCE	- CONVERSION TO USER MACHINE
CONVERSION COST	- \$100 - \$500
EXAMPLES	- FRAME STRUCTURES PIPING PROGRAMS 2-D FINITE ELEMENT PROGRAMS

Fig. 3 Single-element type finite element program characteristics

SIZE	- 2,000 to 10,000 STATEMENTS
DEVELOPMENT GROUP	- TWO TO TEN PERSONS
DEVELOPMENT COST	- \$25,000 - \$200,000
USER DOCUMENTATION	- 50 - 500 PAGES
USER COMMUNITY	- DEVELOPING ORGANIZATION SERVICE BUREAUS
DIAGNOSTICS	- MINIMAL TO ADEQUATE
MACHINE REQUIREMENTS	- 20,000 - 50,000 WORDS
COST PER RUN	- \$1.00 - \$500.00
LANGUAGE	- FORTRAN
MAINTENANCE	- ADDITION OF NEW DEVELOPMENTS CORRECTION OF BUGS
CONVERSION COST	- \$1,000.00 - \$5,000.00
EXAMPLES	- SHELL PROGRAMS 3-D SOLID PROGRAMS

Fig. 4 Multiple-element type static finite element program characteristics

very modular. The development cost is considerably higher, because of the higher manpower and debugging costs. Documentation becomes more extensive, because of the larger number of elements and options. The investment in a program of this size precludes a public distribution of these programs; so their primary usage is within the developing organization, where they are spoken of in glowing terms but are jealously guarded from any outside release. A few of these programs may be made available to the public through computer service bureaus, but often competitive advantage dictates that the public version be less comprehensive than that available in the developing organization. The diagnostics with the program may be minimal to adequate, depending on the number of users and the availability of the program developers. The programs require a reasonably large computer and can run very large static structures. The language here is still FORTRAN, with minimal machine language or system-dependent routines. A significant amount of maintenance is required for these programs, for adding new elements and options and in the correction of the ever-present bugs. This class of programs is usually used for large structures where good modeling requires a variety of element types.

Multiple Element, Multiple-Analysis Type Program

There are now several programs available which I choose to call Multiple Element Multiple-Analysis Type Programs (Fig. 5).

SIZE	- 10,000 - 50,000 STATEMENTS
DEVELOPMENT GROUP	- FOUR TO THIRTY PERSONS
DEVELOPMENT COST	- $150,000 - $2,000,000
USER DOCUMENTATION	- 400 - 2,000 PAGES
USER COMMUNITY	- DEVELOPING ORGANIZATIONS DEVELOPING INDUSTRY SERVICE BUREAUS
DIAGNOSTICS	- EXTENSIVE - CHECK RUNS, GRAPHICS
MACHINE REQUIREMENTS	- 30,000 - 120,000 WORDS
COST PER RUN	- $1.00 - $5,000.00
LANGUAGE	- FORTRAN
MAINTENANCE	- ADDITION OF NEW DEVELOPMENTS COMPENSATION FOR SYSTEM CHANGES CORRECTION OF BUGS
CONVERSION COST	- $2,000.00 - $20,000.00
EXAMPLES	- ANSYS, STARDYNE, MARC

Fig. 5 Multiple-element, multiple-analysis type program characteristics

These programs contain a library of finite elements for two and three-dimensional structures. They can do a variety of static and dynamic analyses with the same model. Some also do steady-state and transient heat transfer analyses. These are large computer programs, with development staffs ranging from four to twenty persons. These programs represent a large investment in man and machine time; this investment is going to be protected, and attempts will be made to maximize the return on this investment. This user community is usually a large developing organization, a coordinated industry, or a service bureau with a significant customer base.

This class of program requires extensive documentation because of the size of the user group and the inaccessibility of the program developer. It must have preprocessors and postprocessors to facilitate data input and output. Graphic display of the input and results is mandatory. The diagnostics must be complete so that the program and documentation help the user locate his errors.

Maintenance of a program of this size is expensive. New developments must be added--new elements, new analysis types, new options, and new pre- and postprocessing routines. In addition, these programs must be modified for new computer systems; they represent too large an investment to die due to machine changes.

A single program of this type can often support the entire structural analysis requirement of a corporation.

These programs are coded in FORTRAN, with a minimum number of machine-dependent routines to improve performance. No modifications to the computer operation system are required to run these programs so they do not disrupt the normal flow of work through the computer center.

I would include in this class such programs as ANSYS, MARC, and STARDYNE. I suspect that it will be a long time before someone provides software of this caliber to the engineering community at no charge.

Large-Scale Analysis Systems

The largest class of programs are those which I call Large-Scale Analysis Systems (Fig. 6). These are the giants of the structural software business, and as is typical of giants, they are often clumsy and difficult to manage. The investment in these programs is enormous, and a bureaucracy forms to administer the investment. Strange things can happen in circumstances of this sort, such as having contracts issued which require the use of these programs, inflating the contract costs, spending additional money to do studies of the amount of usage of the software, and holding conferences to discuss the usage of this software. It seems to be a combination of a bureaucracy and a cult.

These programs represent the state-of-the-art at the time of their conception but may be obsolete at the time of their release because of the rapidly developing technology. Additional infusions of money can keep them from becoming more obsolete than they were when they were released.

These programs are demanding of machine resources, often taking over the entire machine and/or the operating system as well. They cause serious disruptions to the normal computer work load and are thus left to the late night hours.

These programs may be developed with public funds, and thus are available to anyone. However, this availability and subsequent modification by many user sites leads to new questions of reliability and verification.

SIZE	- 100,000 - 300,000 STATEMENTS
DEVELOPMENT GROUP	- TWENTY TO ONE HUNDRED PERSONS
DEVELOPMENT COST	- $2,000,000.00 - $10,000,000.00
USER DOCUMENTATION	- 2,000 - 5,000 PAGES
USER COMMUNITY	- DEVELOPING INDUSTRY SERVICE BUREAUS
DIAGNOSTICS	- EXTENSIVE - CHECK RUNS, GRAPHICS
MACHINE REQUIREMENTS	- 50,000 - 150,000 WORDS
COST PER RUN	- $10.00 - $10,000.00
LANGUAGE	- FORTRAN AND MACHINE LANGUAGE
MAINTENANCE	- ADDITION OF NEW DEVELOPMENTS COMPENSATION FOR SYSTEM CHANGES CORRECTION OF BUGS WRITING OF PROCEDURES
CONVERSION COST	- $20,000.00 - $200,000.00
EXAMPLES	- NASTRAN, ASKA

Fig. 6 Large-scale analysis system characteristics

TYPES OF STRUCTURAL MECHANICS PROGRAM DEVELOPERS

Figure 7 is a list of possible developers of structural mechanics programs.

The design engineer will usually write desk calculator or small time share or batch programs to solve a specific problem.

The analysis engineer will write small batch programs, and sometimes will develop single-element finite element programs.

The university student will usually write enough of a batch or single-element program to solve the problem class which he has chosen for his project. The program rarely has general usefulness.

A consultant will write a batch or finite element program in response to a specification issued by a customer. The program will tend to be documented and have some general usefulness.

An analysis group may pool its resources and develop a multiple-element program to solve a variety of problems which it faces. This requires some foresight and coordination by the group manager. The resulting program is usually used by that group and by other design engineers which that group supports. The larger the user group, the better the documentation.

An industry research laboratory usually has on its staff people capable of developing general purpose computer programs. Many laboratories have or are developing general purpose structural analysis programs. Most such efforts are privately funded, and the results are proprietary to the developing firm.

A technical software supplier develops computer software for the use of as large a user community as possible, and so tries to write software that is as general in purpose as possible. This software is supplied to the user at a price.

Some structural analysis software has been developed by government funding of industrial corporations. Such software tends to be very expensive to develop but can be obtained by the user at a minimum cost. The user must adapt the program to his computer and needs. Some very sophisticated software can be obtained by this means.

DESIGN ENGINEER

ANALYSIS ENGINEER

UNIVERSITY STUDENT

CONSULTANT

ANALYSIS GROUP

UNIVERSITY STAFF

INDUSTRY RESEARCH LABORATORY

TECHNICAL SOFTWARE SUPPLIER

INDUSTRY-GOVERNMENT COOPERATIVE

Fig. 7 Types of structural mechanics program developers

FLOW OF TECHNOLOGY

Figure 8 shows the normal flow of technology from basic research through a final product. The intermediate steps included an applied research step and a pilot project or proof-of-principle demonstration.

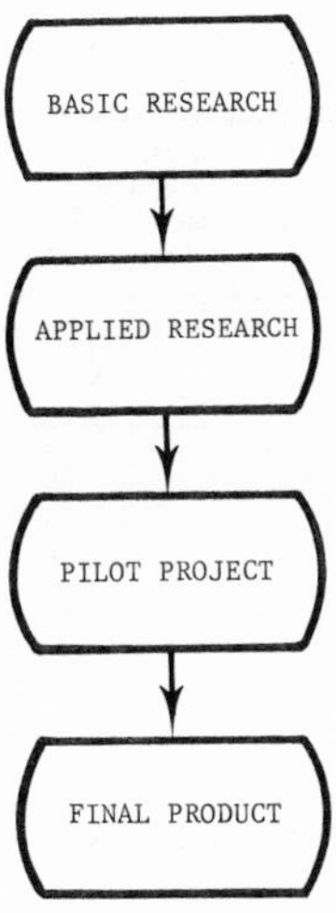

Fig. 8 Flow of engineering technology

Figure 9 is an adaptation of Fig. 8 for the special case of analysis technology. The basic research in the field is in the area of numerical methods and the mathematical characteristics of material behavior. The applied research includes experimental work to correlate with analysis research and the development of new finite elements. The pilot projects are the special purpose single-element type programs written to demonstrate the validity of the approach or to solve the particular industrial problem. The final step in the flow of analysis technology is the incorporation of the new technology (along with that of a wide variety of other projects) in a computer program which is marketed to a wide audience of users. The new technology is of little use until it is in a form which can be accessed by anyone needing to use it.

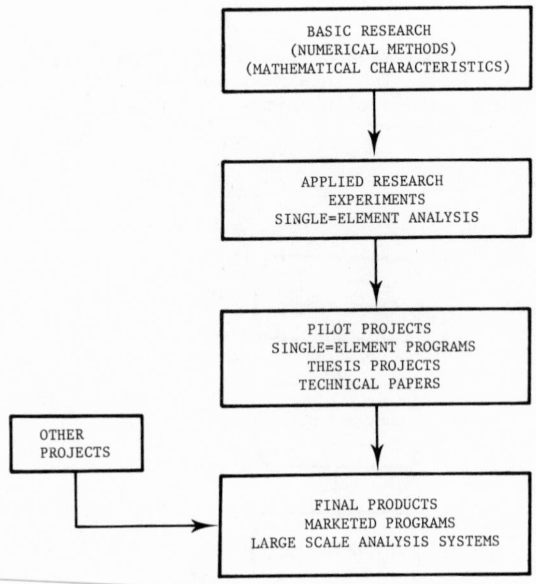

Fig. 9 Flow of analysis technology

SCALE OF DEVELOPMENT TASK

The reason for the appearance of program development specialists is illustrated by Fig. 10. The ability of computer programs and computer hardware to solve simultaneous linear equations has been increasing by about a factor of ten every four years. It would be

SOLUTION OF SIMULTANEOUS LINEAR EQUATIONS

METHOD	YEAR	CAPACITY
IN-CORE SOLUTIONS	1960	200 Degrees of Factor
BANDED MATRIX SOLUTIONS	1964	2,000 Degrees of Factor
WAVE FRONT EQUATION SOLUTIONS	1968	20,000 Degrees of Factor
SUBSTRUCTURE SOLUTIONS	1972	200,000 Degrees of Factor
MACROSTRUCTURE SOLUTIONS	1976	?
	1980	?

Fig. 10 Scale of development task

unrealistic to ask a university student or a consultant to develop a computer program with the capacity to solve 200,000 simultaneous equations; yet this is the size of the analysis which is being solved by current software.

FORMS OF PROGRAM CREATION

There are two distinct forms of program creation, as shown in Fig. 11. There are many examples of programs developed in each form.

PROJECT

- PRESPECIFIED GOALS, REQUIREMENTS, CAPABILITIES
- FIXED COMPLETION SCHEDULE
- FIXED PRICE
- LITTLE INCENTIVE TO EXCEED SPECIFICATIONS
- INCORPORATES STATE-OF-ART AT BEGINNING OF PROJECT

EVOLUTION

- BASE CAPABILITY DEVELOPED OR OBTAINED (BIRTH)
- PROGRAM APPLIED TO PROJECTS, WITH GENERALIZED ENHANCEMENTS (GROWTH)
- SPECIALIZED vs. GENERALIZED GROWTH (RANGE OF EDUCATION)
- PERIODIC REWRITES (REPRODUCTION)
- OBSOLESCENCE OR REPLACEMENT (DEATH)

Fig. 11 Forms of program creation

Project

Programs which are written for use by an organization outside the developing organization tend to be done on a project basis. These programs are developed to a set of prespecified goals, requirements, and capabilities. They are usually done to a fixed completion schedule at a fixed price. There is usually little or no incentive to exceed the original specifications; so the resulting program incorporates the state-of-the-art at the time the project was conceived. If the project is for longer than two years, there is a high probability that the resulting product will be obsolete before it is completed.

Evolution

Other programs have a life cycle which is quite similar to that of a living organism. Birth is when the capability is developed or obtained and the first real problem is solved. The growth phase occurs as the program is applied to a variety of projects and problems, with generalized enhancements added as necessary. The range of education of the program is dependent on the size of problem to which it is applied. Restriction to a single company or industry leads to specialized growth, while multiindustry usage leads to generalized growth. Specialized growth may be more useful to the specific industry but may prove cumbersome for other industries.

Any program which is growing must suffer the periodic agony of a rewrite or major revision. No amount of foresight seems to be sufficient to avoid a periodic reassessment and consolidation of program capabilities. This periodic rewrite is a form of reproduction. Other more specialized programs may be developed from a common ancestor, as in the case of two-dimensional finite element programs in the United States, again a clear case of reproduction.

Computer programs, like all living organisms, must face the prospect of their own death. The death of a computer program is quite like the death of a god--when the last true follower ceases to believe, the program is dead. Death may be due to replacement by a newer program (often a descendant), loss of the developer without sufficient documentation, changing of computer hardware and a slow death by emulation, or ultimate solution of the problem for which the program was written.

A program which evolves usually has a much larger life than one which is developed on a project basis. A project program may become an evolving program, if there is continuing support for its maintenance and advancement.

WHAT IS A TECHNICAL SOFTWARE SUPPLIER?

An individual or corporation which writes, markets, maintains, and expands one or more technical computer programs is referred to in this paper as a technical software supplier.

The supplier observes a need for a certain computer software capability, develops a program which he believes will meet that need, and then markets the resulting program to the user community. The supplier strives to service as large a user community as possible to maximize the return on this investment in the program.

The programs are usually marketed on a royalty (computer time based surcharge), lease (fixed price per month), or restricted sale basis.

The supplier usually supplies associated services, such as training, consulting, and specialized development projects.

The supplier should not do analysis work for profit using his own software in competition with other analysis firms, because this leads to the withholding of new developments for competitive advantage, and the software supplier becomes another analysis firm with its own proprietary software.

The supplier competes with other suppliers and other program sources for the opportunity to provide analysis tools to the engineering community.

SOURCES FOR LARGE-SCALE ANALYSIS SYSTEMS

There are four major sources for structural analysis computer programs. These are governments, private industries, universities, and technical software suppliers (Fig. 12). We will examine each in turn to determine which is likely to supply structural analysis software to the engineering community and on what basis.

GOVERNMENT

- ENGINEERING ANALYSIS A WORLDWIDE PROBLEM
- HAS FUNDING RESOURCES
- WORKS ON PROJECT BASIS
- SHARING BETWEEN GOVERNMENTS UNLIKELY

PRIVATE INDUSTRY

- ONLY A LARGE INDUSTRY HAS THE RESOURCES
- TENDS TO HOLD INTERNAL DEVELOPMENTS FOR COMPETITIVE ADVANTAGE
- LACK OF COORDINATION PRODUCES MUCH DUPLICATION OF EFFORT
- PROGRAMS EVOLVE RATHER THAN BEING PROJECTS

UNIVERSITY

- FUNDING WOULD BE REQUIRED
- CONTINUITY OF PERSONNEL MAY BE DIFFICULT
- TENDENCY TO BE REMOTE FROM THE ACTUAL ANALYSIS PROBLEMS AND PRESSURES

TECHNICAL SOFTWARE SUPPLIERS

- NEW PHENOMENON
- PRODUCTS MADE AVAILABLE FOR A CHARGE
- COMPETITION ALIVE AND FUNCTIONING
- DEVELOPMENT IN DIRECT RESPONSE TO NEEDS
- MUST MAXIMIZE PRODUCTIVITY OF HIS PRODUCTS
- PROVIDES SYSTEM MAINTENANCE AND MACHINE INDEPENDENCE
- MUST ANTICIPATE NEW REQUIREMENTS
- USER HAS READY ACCESS TO PROGRAM DEVELOPER
- HAS INCENTIVE TO ASSURE SIMPLICITY, RELIABILITY, AND USABILITY
- PROGRAM AVAILABILITY ASSURED
- DEVELOPMENT COST SPREAD ACROSS MANY USERS
- CAN ATTRACT CREATIVE PERSONNEL
- EFFORT CONCENTRATED ON A SINGLE PRODUCT (OR FEW)
- NOT TIED TO A SINGLE PROJECT OR CONTRACT
- ABLE TO GENERALIZE DEVELOPMENT REQUIREMENTS
- TRANSFER OF TECHNOLOGY FROM DISCIPLINE TO DISCIPLINE
- ORIENTED TO KEEP PACE WITH RAPIDLY EXPANDING TECHNOLOGY

Fig. 12 Organizations which might develop large-scale analysis and pressures

Government

The government has the funding resources required to develop large scale structural analysis computer programs, and many people seem to feel that this should be the source of software. However, there are several major drawbacks to government production of computer software.

Engineering analysis is a worldwide problem, and the software should be available on a worldwide basis.

Government development programs are usually obtained on a fixed price project basis, and the longer the project, the longer the development time. There is little opportunity for creativity above that requested in the original proposal.

Governments tend to impose the usage of their programs on projects which are government funded, whether or not the program is applicable.

Programs developed by the government are paid for by tax revenues, so that everyone is forced to support the development of the software, whether they desire it or not.

Private Industry

Many corporations have the capability within their organizations to develop large-scale computer programs. Programs developed by private industry tend to evolve to meet the changing needs of that corporation, and these are more flexible than those developed by government contract. However, private industry tends to hold internal developments for competitive advantage; so there is little likelihood of any program sharing, and the lack of coordiantion and sharing produces a tremendous amount of duplication of effort.

University

A possible, but unlikely, third possibility for large-scale structural software is university staff. This source has no advantage and several major disadvantages.

A university should be a source of basic research, whereas large-scale software is highly marketing oriented and involves little basic research.

A large amount of funding is required to develop large analysis systems, and the source of this funding would put restrictions on the program equivalent to those which would apply if the funding source developed the program.

Continuity of personnel is difficult in a university. Much of the work is done by graduate students, and they leave in a short time. Continuity of personnel is mandatory for a large-scale analysis system.

Universities tend to be remote from the actual analysis problems and pressures. The main thrust of the university effort is (and should be) in the development of new concepts and demonstration of their correctness and applicability. They should not get involved in repetitive analyses.

Technical Software Suppliers

The technical software supplier is becoming the chief source of production quality structural analysis software, and this trend will continue because there are many reasons for it.

Technical software suppliers are a new phenomenon, but they have appeared because no other software source was meeting the needs of the industry.

The computer software supplied by these suppliers will be made available for a charge. It is completely unrealistic to expect to obtain access to large-scale state-of-the-art software without carrying some of the cost of the development of such software. Organizations and societies which develop lists of available software and omit publicly available proprietary software are taking an unrealistic view, are completely out of touch with the state-of-the-art in structural analysis software, and may even be doing a disservice to the readers of the program compilation. Even so-called free programs such as NASTRAN are made available to the public with a surcharge to reflect the cost of maintaining and supporting them.

With the appearance of a group of strong and aggressive structural analysis software suppliers, the software user now finds himself in a buyer's market. Let it be known that you want to do a large piping analysis, and at least five suppliers will compete for your attention and business. They will be happy to adapt their software to your requirements (usually at no charge).

Software development is very responsive to the needs of the users, since meeting these needs results in increased usage of the software.

The software supplier has an incentive to work with the user to maximize the engineering productivity, since the more an engineer produces the more analysis he does, and the more usage there will be of the software. It is not profitable to have an engineer spend a month to prepare the input, ten minutes to run the problem, and another month to interpret the output. The supplier will work to cut the input and interpretation time to a minimum so that more analysis can be done.

The software supplier maintains his software and provides machine independence. He will adapt his software to all machines possible to maximize the usage of his product. It will be converted to each new generation of machine, usually before the machine becomes generally available.

The software supplier tries to anticipate new analysis needs so that the development is complete when the need arises. The supplier incorporates the technology developed by university and research projects and makes it available to the general user community.

The user has ready access to the developer of the proprietary software, because the supplier wants to maximize the usage of his product.

The supplier tries to make his product simple to use, reliable, and usable. A product which is complex, unreliable, or difficult to use will not generate the revenue needed to support it.

The user has assurance that the software will be available, because no revenue is developed if the analysis cannot be run.

The royalty or lease financing of the development costs means that the costs are paid only by those who wish to use the software. Also, the software user has available, at no development cost, a wide range of sophisticated structural analysis software.

By concentrating on a very specific field of endeavor, the software supplier can attract and support a staff of capable and creative personnel. The total effort is concentrated on a single product (or only a few).

The development is not tied to a single product or contract, and the technical personnel are not shifted from project to project but continue on the development of the software. They are relatively immune to project cutbacks in a specific corporation or industry.

The software supplier is in contact with many industries and is able to generalize development requests into broad general capabilities, rather than restricting his attention to a specific request.

The supplier is also able to transfer technology rapidly from discipline to discipline. An integration scheme which works well in dynamics is simultaneously available for heat transfer analysis. Elements which are used for heat transfer can be easily modified to fluid wave analyses.

The supplier must keep pace with the rapidly expanding structural analysis technology, or he will not survive.

SUMMARY

The cost of developing general purpose programs is expanding. The cost of maintaining programs must be spread over a wide market.

Government, private industry, and university programs have limitations due to their sources.

Technical software suppliers' goals are consistent with the goals of the analysis organization; that is, to maximize engineering productivity.

Only the user of software developed by a supplier pays for the development cost. The general and/or technical community is not taxed for software which is does not want or use.

Software supplied by technical software suppliers will be modified to take advantage of the efficiency of the next generation(s) of computers.

If large-scale analysis systems are needed, the technical software supplier will create them, provided they will be used.

THE DEVELOPMENT OF GENERAL PURPOSE PROGRAMS AND SYSTEMS

G. Ruoff and E. Stein
Technical University of Hanover
Germany

ABSTRACT

The primary purpose of this paper is to discuss guidelines for the development of general purpose programs and systems. Detailed discussions of particular programs are not included. The first part of the paper deals with the characteristics used for the classification of the various program types. The general purpose programs are compared to large general purpose systems.

INTRODUCTION

In recent years digital computers have been used in all branches of techniques and have proved very useful, especially in computed-aided design and computation of structures. A large number of computer programs--still increasing enormously--have been developed, which vary considerably with respect to the range of applicability, the permissible problem size, which can be worked out with the hardware requirements, and the basic program philosophy.

In the framework of this paper we have to define certain terminology as a common base for further discussions. Therefore some terms are defined in order to classify the different types of programs.

a. A "small" program is defined by the fact that the possible problem size (e.g., the number of unknowns) is primarily limited by the core size of the computer available.

b. A "large" program is not restricted in this way. The limits of the problem size are usually determined by the capacity of the background storage and, perhaps, by the relation between the computing costs for a certain problem and the value of the results.

In addition the distinction between "large" and "small" programs comprises implicitly a scale for the size of the source decks of the programs, although there does not exist an accurately defined limit for the number of source statements which indicates exactly when the program begins to become "large."

The terms "special purpose" (S.P.) and "general purpose" (G.P.) state the distinctive features of the programs. They cannot be defined in as simply a way as "small" and "large."

c. An S.P. program stands out because of the fact that it is conceived for the computation of a special type of structures (e.g., frames, shells of revolution, plates, without the facilities for the combination of different types of structures), with exactly defined loading types (static, dynamic, or thermal) or types of material properties (linear elastic, rigid-plastic, nonlinear elastic, any kind of strain hardening, etc).

d. In addition, a G.P. program provides the facilities to combine several types of geometry, loading cases, and some of the following topics:

1. Combinations of loading types like
   - static loadings
   - free dynamic modes
   - forced motions with or without damping
   - transients
   - random excitations
   - thermal loadings (stationary or unstationary)
2. Nonlinear material properties like
   - nonlinear elastic material
   - elastic-plastic materials with or without any kind of strain hardening
   - rheological material behavior
   - temperature-dependent material properties
3. Buckling problems like
   - linear buckling via eigenvalue calculation
   - nonlinear buckling
   - postbuckling behavior
4. General geometrical nonlinear behavior like
   - large displacements
   - large strains (in conjunction with, e.g., 2 and 3)
5. Nonlinear dynamic behavior like
   - dynamic behavior together with geometric nonlinearities
   - dynamic behavior together with nonlinear material properties

In addition to these mechanical distinctions we have to compare the differences in the S.P. and G.P. programming philosophies. These differences are discussed in the following section.

The third very important distinction which has to be drawn here is the distinction between a program (P) and an integrated program system (IPS). Usually a program consists of several modules, whose execution is arranged and supervised solely by the operational system of the computer. An IPS generally comprises the following parts:

1. An integrated system, which replaces parts of the usual operating system during the preparation and the execution of subsystems

2. A problem-oriented language (IPOL), combined with a precompiler which translates the IPOL-coded programs into a code which is familiar to the computer (e.g., FORTRAN IV)

3. A succession of subsystems for various problems, which usually contain the formulations for solving the problems, while the system operates the data management and the control of the subsystem execution

4. A set of modules, usually occurring in the range of problems covered by the IPS, which is part of the system's object library and may be compared with macros in an assembler language

The programs may be based on various numerical methods. We do not consider them as criteria. In our opinion it is mostly insignificant whether the solution of a problem is approximated by finite element methods (FEM), finite difference methods, point-matching methods, methods of transfer matrices, direct integration techniques, or any other method of numerical analysis. The fact that the majority of the existing programs are based on the FEM, may be explained (1) by the lack of programs based on other methods (there are only few of different conception), (2) by the high standard of the FEM programs from the point of view of the programs' technical development and their proven value in nearly all fields, (3) by the fact that it is a method which enables the programmer to connect different types of geometry and loading cases in the easiest possible way, (4) because this method permits programming on a low stage of inner logics and meets the requests for inner and outer program frequency perfectly.

Taking into account these criteria we can now group the programs into five different types:

1. Small S.P. programs
2. Large S.P. programs
3. Small G.P. programs
4. Large G.P. programs
5. Integrated G.P. program systems

The following sections will mainly deal with a comparative study of the programs of types 4 and 5. We will give a more detailed presentation of the criteria for the classification. Moreover we will try to discover rules ensuring an optimal use of the different types of programs. Finally some tendencies concerning the future development of programs and program types will be compared with the prospective progress in the development of general computer software and operational systems.

Some Remarks about the References:

The information used in this paper originates mainly from two different sources: (a) books, e.g., [1] to [8] and papers published in the most important journals (which are too numerous to be listed here completely) and (b) program manuals and results of inquiries of the authors and members of the University of Virginia. Appendix I also contains a list showing the addresses of developers/software centers, who might give more detailed information on the listed programs.

DESCRIPTION AND DETAILING OF THE FIVE PROGRAM CLASSES

In this section we will give a detailed definition of those criteria which permit the classification of a program precisely according to one of the five types.

Type 1: Small Special Purpose Programs

In this group can be classed all programs which

1. Deal with only one special type of structure
2. Limit the possible problem size by means of the core storage available
3. Permit only a few different loading types
4. Deal only with static, dynamic, or thermal problems

With regard to these qualities all programs for testing new algorithms, e.g., in the field of FEM, in the case of plasticity problems, etc., must be classified in this category too. Programs of this type are mainly used in firms and universities for parameter studies of special kinds of structures, e.g., shells of revolution and frameworks.

The most important advantages provided by these programs are:

1. A short time for the development (order of magnitude: 2 man-years)
2. Simple logics in programming philosophy
3. A high degree of independence of the computer type
4. A very high efficiency

The main disadvantages of this program are:

1. A special program must be provided for every type of problem (or parts of the program, e.g., new subroutines for other finite elements).
2. Except for a few practical cases, the high dependence of the problem types makes it impossible to obtain programs of this type in a suitable form from a software center of other institutions.

The institutions, mostly firms which are interested in these programs, have to develop them on their own, a process whose efficiency generally cannot be judged. The efficiency depends on various factors like availability of other types of programs which cover the range of problems required, efficiency of these other programs, frequency of applications, cost comparisons concerning the application (including the cost for development, research), and computing capacity--small programs frequently can be handled on the developer's computer, whereas the use of general purpose programs often necessitates the use of a general software center.

Some typical program examples are nos. 1, 19, 30, 36, 43 in Table 1.

Type 2: Large Special Purpose Programs

The programs of this type have the same characteristics as those of type 1 except that the possible problem size is not limited by the core size. They can involve program parts which work interactively with the background storage (e.g., equation solvers or eigenvalue subroutines with intermediate storage of parts of the total matrices). Furthermore, programs of this type may prove more satisfactory if pre- or postprocessors like data check facilities and plot routines for geometry or results are used. Typical examples for programs of this type can be found in the region of FEM programs. There exist

Table 1 Survey of the Evaluated Programs

DESCRIPTION OF THE MATH.-MECHANICAL MODEL																														COMPUTER REQUIREMENTS					SOFTWARE INFORMATION									
	SYMBOLS: KEY: + = YES - = NO U = UNDE-FINED	DISCRETIZATION METHOD								TYPES OF GEOMETRY				LOADING CASES						TYPES OF MATERIALS LINEAR ELASTIC																								
REFERENCE		FE-MODEL DISPLACEMENT	FE-MODEL HYBRID	FE-MODEL MIXED	FE-FOURIER SERIES	FINITE DIFFERENCE	POINT MATCHING	FORWARD INTEGRATION	OTHER	1 DIMENSIONAL	2 DIMENSIONAL	2 DIM. CURVED	3 DIMENSIONAL	STATIC	FREE MODES	FORCED MOTION	TRANSIENT (GENERAL)	STATIONARY TEMP. - FIELDS	INSTAT. TEMP. - FIELDS	ISOTROPIC	ANISOTROPIC	LAYERED	SANDWICH	TEMP. - DEPENDENT	NONLIN. ELASTIC	RIGID-PLASTIC	ELAST. - STRAIN HARD.	OTHER TYPES	LARGE DEFLECTIONS	IN-CORE PROGRAM	MIN.CORE SIZE (K WORDS)	NO. OF DISK-FILES	NO. OF TAPES	RANDOM ACCESS	PROGRAM LANGUAGE	PREPROCESSOR	POSTPROCESSOR	PLOT ROUTINES	INTERACTIVE MODES	RESTART CAPABILITY	ERROR DIAGNOSTIC	SUBSTRUCTURING	DISTR. BY DEVELOPER	DISTR. BY SOFTWARE CENTER
		1	2	3	4	5	6	7	8	9	10	11	12	13	14	15	16	17	18	19	20	21	22	23	24	25	26	27	28	29	30	31	32	33	34	35	36	37	38	39	40	41	42	43
1	AC50-A	0	0	0	0	0	0	0	0	-	+	-	-	+	-	-	-	+	-	+	+	+	+	-	-	-	-	-	-	+	20	-	-	-	F4	-	-	-	-	-	-	-	+	-
2	ADEPT	-	-	-	-	-	+	-	-	+	-	-	-	+	-	-	-	-	-	+		-	-	-	-	+	-	-	-	-	76	-	-	-	AT	-	-	-	-	+	+	+	U	U
3	AMSA 20	+	-	-	-	-	-	-	-	+	+	-	+	+	+	-	-	-	-	+	-	-	-	-	-	-	-	-	-	-	35	0	0	0	F5	-	-	-	-	-	-	+	U	-
4	ANSYS	+	-	-	+	-	-	+	-	+	+	+	+	+	+	+	+	+	0	+	+	+	+	+	-	-	+	+	+	-	50	1	-	+	F	+	+	+	+	+	+	+	+	-
5	ASAS	+	-	-	-	-	-	-	-	+	+	-	+	+	-	-	-	-	-	+	+	-	-	-	+	-	+	+	+	-	32	U	U	U	F	-	-	-	-	+	+	-	+	-
6	ASKA	+	-	-	+	-	-	-	-	+	+	+	+	+	+	+	+	+	-	+	+	-	+	-	+	+	+	+	+	-	32	1	U	+	F4	+	+	+	-	+	+	+	+	+
7	BASY	+	+	-	-	-	-	-	-	+	+	+	+	+	+	-	-	-	-	+	-	-	-	-	-	-	-	-	-	-	16	U	U	U	F	-	-	-	-	+	+	-	+	-
8	BERSAFE	+	-	-	-	-	-	-	-	+	+	+	+	+	-	-	-	-	-	+	-	-	-	+	-	-	+	+	+	-	-90	9	+	+	F4	+	+	+	-	-	+	-	+	-
9	BOSOR 4	-	-	-	-	+	-	-	-	-	-	+	+	+	+	-	+	-	-	+	+	+	+	+	-	-	-	-	-	-	64	1	-	+	F4	-	-	+	-	-	+	-	+	+
10	COSA	+	-	-	-	-	-	-	-	+	+	-	+	+	+	+	-	-	-	+	+	-	+	-	-	-	-	-	+	-	64	1	-	+	F	-	-	-	-	+	+	+	+	-
11	DYNAS	U	U	U	U	U	U	U	+	+	+	-	-	+	+	+	+	-	-	+	-	-	-	-	-	-	-	-	U	-	80	8	-	+	F	-	-	+	-	+	+	+	+	-
12	EASE 2	+	-	-	-	-	-	-	-	+	+	+	+	+	-	-	-	+	-	+	-	-	-	-	-	-	-	-	-	-	30	1	-	+	F/A	+	-	+	-	+	+	+	-	+
13	ELAS 75	+	-	-	-	-	-	-	-	+	+	+	+	+	-	-	-	+	-	+	+	+	+	-	-	-	-	-	-	+	20	-	-		F4	-	+	+	-	-	+	-	-	+
14	FARSS	+	-	-	-	+	-	-	-	-	-	+	+	+	+	+	+	-	-	+	+	-	-	+	-	-	-	-	-	-	20	1	-	+	F4	+	+	+	+	+	+	-	+	-
15	FESAP	+	-	-	-	-	-	-	-	+	+	+	+	+	+	+	+	-	-	+	+	-	-	+	-	-	-	-	-	-	20	8	-	1	F4	+	+	+	-	-	+	-	-	-
16	FLHE	+	-	-	-	-	-	-	-	+	+	-	+	-	-	-	-	+	+	+	-	-	-	-	-	-	-	-	-	-	60	U	U	U	F	-	-	-	-	-	+	U	+	-
17	ISOPAR SHL	+	-	-	-	-	-	-	-	-	+	+	+	+	-	-	-	-	-	+	+	+	+	-	-	-	-	-	-	-	65	9	-	+	F	+	-	-	-	-	+	-	+	-
18	ISTRAN/S	+	-	-	-	-	-	-	-	+	+	+	-	+	+	-	-	-	-	+	-	-	-	-	+	-	-	-	+	-	64	20	-	+	F/A	+	-	+	-	+	+	-	+	-
19	KSHEL	-	-	-	-	-	-	+	-	-	-	+	-	+	+	+	+	-	-	+	+	+	+	+	+	-	-	-	+	+	33	-	+	-	F	-	-	-	-	-	+	-	+	-
20	MARC	+	-	-	-	-	-	-	-	+	+	+	+	+	-	-	+	-	-	+	-	+	-	-	-	-	+	-	+	-	60	U	U	U	F	+	-	-	-	+	U	-	-	+
21	MINIELAS	+	-	-	-	-	-	-	-	+	+	+	-	+	+	+	-	+	-	+	+	+	+	-	-	-	-	-	+	+	32	-	-	-	F4	-	-	-	-	-	-	-	+	-
22	NASTRAN	+	-	-	-	-	-	-	-	+	+	+	+	+	+	+	+	+	-	+	+	-	+	+	-	-	+	-	+	-	50	U	-	+	F	+	+	+	-	+	+	+	-	+
23	NEPSAP	+	-	-	-	-	-	-	-	+	+	+	+	+	-	-	-	-	-	+	+	+	+	+	-	-	+	-	+	-	45	12	+	+	F	+	+	+	-	+	+	-	+	+
24	NONLIN 2	+	-	-	-	-	-	-	-	+	-	-	-	+	-	+	-	-	-	+	-	-	-	-	-	-	+	-	+	-	57	1	-	+	U	-	-	-	-	-	-	-	+	-
25	NONSAP	+	-	-	-	-	-	-	-	+	+	+	-	+	+	+	-	-	-	+	+	-	+	+	+	+	+	-	+	+	128	-	-	-	F	-	-	-	-	+	-	-	+	-
26	NOSTRA	+	-	-	-	-	-	-	-	+	+	+	+	+	-	-	-	-	-	+	+	-	-	+	-	-	-	-	+	-	28	6	-	+	F	+	+	+	+	+	+	-	U	U
27	PAFEC 70	+	+	-	-	-	-	-	-	+	+	+	+	+	+	-	-	+	-	+	+	+	-	-	-	-	-	-	-	+	10	1	-	+	F	+	+	+	-	-	-	-	+	-
28	FRAKSI	+	+	-	-	-	-	-	-	+	+	-	-	+	-	-	-	-	-	+	+	-	-	-	-	-	-	-	+	-	33	4	-	+	F4	-	-	+	-	-	+	+	-	-
29	REXBAT	+	-	-	-	-	-	-	-	+	+	-	+	+	-	-	-	+	-	+	+	-	-	+	-	-	-	-	-	-	65	10	-	+	F	+	+	+	-	+	+	+	+	-
30	SABOR/DRASTIC 6	+	-	-	+	-	-	-	-	-	-	+	-	+	-	+	+	-	-	+	+	-	-	-	+	+	+	-	+	+	45	13	1	+	F	-	+	+	-	+	+	-	+	-
31	SAMBA	+	-	+	-	-	-	-	-	+	+	-	-	+	-	-	-	+	-	+	+	+	-	-	-	-	-	-	-	-	24	1	-	+	AL	-	-	+	-	+	+	+	+	-
32	SAP IV	+	-	-	-	-	-	-	-	+	+	+	+	+	+	+	+	+	-	+	+	+	+	+	-	-	-	-	-	-	48	10	-	+	F4	-	-	-	-	+	+	-	-	+
33	SATANS	-	-	-	+	-	-	-	-	-	-	+	-	+	-	+	+	+	-	+	-	-	+	-	-	-	-	-	+	+	36	-	-	-	F4	-	-	-	-	+	-	-	-	+
34	SESAM 69	+	-	-	-	-	-	-	-	+	+	+	+	+	+	-	-	-	-	+	+	+	-	-	-	-	-	-	+	-	40	+	+	+	F4	-	-	-	-	-	-	+	+	-
35	SHORE	-	-	-	-	+	-	-	-	-	-	+	-	+	-	+	-	-	-	+	+	-	-	+	-	-	+	-	-	-	62	1	+	+	F5	-	-	+	+	+	+	-	+	-
36	STARDYNE	+	-	-	-	-	-	-	-	+	+	-	+	+	+	+	+	+	-	+	+	+	+	-	-	-	-	-	-	-	32	4	+	+	F	+	+	+	-	+	+	+	-	+
37	STARS	-	-	-	-	+	-	-	-	+	-	+	-	+	+	+	+	+	-	+	+	+	+	+	+	-	+	-	+	U	U	U	U	U	F	+	-	+	-	+	+	-	U	U
38	STRIP	-	+	-	-	-	-	-	-	+	+	+	-	+	-	-	-	-	-	+	-	+	-	-	-	-	+	-	+	-	64	+	U	+	F4	+	+	-	-	+	+	+	+	-
39	TEXGAP	+	+	-	+	-	-	-	-	-	+	+	+	+	-	-	-	+	-	+	+	+	-	-	-	-	-	-	-	-	25	9	-	+	F4	+	-	+	-	+	+	-	+	-
40	TIRE	-	+	-	-	-	-	-	-	-	-	+	-	+	-	-	-	+	-	+	+	+	-	-	-	-	-	-	+	-	105	U	U	+	F/A	-	-	-	-	-	-	-	-	+
41	TITUS	+	-	-	-	-	-	-	-	+	+	+	+	+	-	-	-	+	+	+	+	-	-	+	-	-	-	-	-	-	27	U	U	+	F	-	+	+	-	+	+	-	+	-
42	VISCEL	+	-	-	-	-	-	-	-	+	+	+	+	+	-	-	-	+	-	+	+	-	-	+	-	-	-	-	-	-	20	15	-	+	F	+	-	+	-	-	+	-	-	+
43	ZP 26	+	-	-	-	-	-	-	-	-	+	-	+	+	-	-	-	-	-	+	+	-	+	-	-	-	+	-	+	+	98	-	-	-	F4	+	+	+	-	-	+	-	+	-

Key:

+ ≙ Yes

- ≙ No

u ≙ Undefined

except in columns 30 and 34.

Column 30 contains the number of K words necessary for the implementation of the program (for byte-structured machines the scale 4 bytes to 1 word is used).

Column 34 contains the following abbreviations for the program language:

F	FORTRAN
F4	FORTRAN IV
F5	FORTRAN V
AL	ALGOL
A	ASSEMBLER
AT	ATLAS, to be translated to AL of F.

Combinations are given by combinations of the symbols, e.g., F/A for F programs combined with assembler parts.

A list of names and addresses of those people who can give further information on the programs follows.

programs, e.g., the EASE type, which do not provide full G.P. facilities but use only a few different elements for the discretization of the whole structure instead of many different elements or even element families found in today's large G.P. programs. Although the range of applicability for this type may cover a large class of structures, one cannot consider them to be equivalent to the general purpose programs.

The time for development is of the order of magnitude of 10 man-years. The machine dependency is naturally much higher than in programs of type 1 and the efficiency is lower.

Another very interesting class of programs of this type is of steadily increasing importance. In the field of linear finite element methods the computation gets segmented into three main parts:

Part 1: prepares and checks the input and generates data files which contain element matrices and the associated topological data. This part is a typical example for a program of type 1.

Part 2: uses the data generated by part 1, assembles the stiffness or mass matrix for the total structure and solves the system of linear equations or the eigenvalue problem. With a suitable equation or eigenvalue solver this part is a typical example of a large S.P. program. It can be used for arbitrary types of finite elements, if the conditions of the standardized intersections on data base are observed in the different parts 1.

Part 3: executes the computation of all dependent variables which are not part of the solution of part 2, e.g., strains and/or stresses (forces) in the displacement method; sorts the results; and selects the desired output data for the specified output device such as line printer, tape, or plotter. This part too has to be classed as type 1.

This type of program provides such advantages as:

1. Large parts of the program reach a very high efficiency. Data generated in part 1 or 2 can be stored in very easy way for other computations of slightly modified structures without running the whole problem again (some kind of substructuring).

2. The difficult linkage of the parts to a large segmented program can be avoided.

The main disadvantage lies in the fact that the described intersection does not allow for easy iterative execution of nonlinear problems.

A further, relatively rare field of application may be seen in the programs for computation of very large structures which consists of only a few different types of elements (e.g., space trusses, cable nets: the structure built for the Olympic games in Munich shows an example for the necessity of programs of this type. Several programs have been developed for the special nonlinear static and dynamic analysis). A large G.P. program is usually too unwieldy for such investigations.

Type 3: Small General Purpose Programs

This classification includes all programs which handle combinations of at least two different types of geometry, handle several distinct types of loadings, as listed in the first part of this paper (linear and/or nonlinear), and limit the possible problem size by the available core size.

Only a few programs of this type are available for general use. Programs belonging in this class are usually preliminary studies within the development phase of large G.P. programs; see, for example, NONSAP [8], the general nonlinear version of SAP IV or SOLID SAP.

Programs of this type usually are not used commercially. They prove most advantageous in testing new algorithms in certain parts of the whole program flow and in basic scientific investigations and research. The time of development lies in the order of magnitude of 4 man-years.

Type 4: Large General Purpose Programs

To this type of programs belong all those with type 3 characteristics but without the limitation of the problem size according to the core size. The only limitation of the problem size is the ultimate storage capacity of the computer and perhaps considerations of the computing cost in relation to the cost of the structure or safety requirements (e.g., for airplanes and nuclear plants). The large versatility of applications requires

1. The program to involve a high degree of computer systems logic, which, in reverse, reduces the efficiency of the program
2. An elaborate data management and data retrieval system, comparable to the virtual storage techniques
3. Massive manpower for development in the order of magnitude of 50 to 100 man-years
4. An elaborate training program for the possible users which will enable the user after a reasonable initial period to use all facilities of the program efficiently

Only few organizations like software centers, institutions associated to universities, or other (public) institutions have the possibilities of investigating the necessary money for developing and maintaining programs of this type.

Programs of this type depend to a high degree on the operational systems of the computer. Every change in this system can lead to many other changes in the program.

From the user's point of view, it is on the one hand useful to have programs of this kind for covering the peak of problems to be computed. On the other hand, only very high and different computing requirements justify the implementation of such a program in a firm's computer center, if the hardware is large enough to run it efficiently. The necessary size is usually a great deal larger than the minimal requirements for installation.

In the most practical cases it is sufficient to have any access to a program of this type, for example in a commercial computer center. Only few users are able to cover the costs for the hardware required and the rent for the software package, which lies in the range of $500 (e.g. NASTRAN) to $1,300 (e.g., ASKA) per month. These prices usually include maintenance of the object code.

In the following we give some further remarks on the development tendencies of this type of program:

1. A great deal of the development time has to be used for the design and implementation of efficient data management utilities and the suitable data structure. The general software development tendencies in the field of operational systems are strongly oriented toward virtual storage (vs) techniques, which mask for the programmer the boundary between primary core storage and second-level storage (like ECS, LSC, disk, drum); in the programmer's view, the core seems to be infinite. In principle type 3 programs can thus have the same capacity as the type 4 programs, although the question of the efficiency cannot be answered in general form; the up-to-date vs techniques do not reach the efficiency of specially coded data management utilities. On the other hand the vs techniques manage the segmentation of the code, too, and the programmer does not have to worry about doing this.

The loss of efficiency has to be compared with the advantages of the resulting shorter time of development and the saving of manpower, the costs of which increase steadily in contrast to the specific computing costs, which show decreasing tendencies.

2. The lifespan of a large general purpose program lies in the order of magnitude of 15 years. In comparison to the developing time, this lifespan is very short and justifiable in only a few regions of applications, especially in the fields of structures with very high safety requirements.

3. The use of this type of program over a number of years makes it necessary to adapt the programs to refinements of the operational system of the computer. In many cases it takes a very great effort to apply the improvements in a way that allows their full use.

4. None of the existing large G.P. programs enable the user to work in a real interactive mode with, e.g., computer-aided design. In this field of application much effort still has to be exerted in the range of G.P. programs. Some of the large G.P. programs include restart capabilities, which permit the user to interrupt a run. The restart usually continues the run without any change.

Considerations of this kind make it urgently necessary to clarify carefully the basic questions of commercial and scientific efficiency of programs of this type before the development is started.

The following section shows a way to achieve similar computing capacities with less programming effort by the engineer using digital computers for the design and layout of structures.

Type 5: Integrated General Purpose Program Systems (IPS)

This section deals with a completely different tool for using computer facilities for the design and layout of structures. The basic philosophy is characterized by a certain kind of division of labor: a team of system programmers develops a system core (nucleus), which administrates the data management and--together with the operational system--the general program flow, e.g., the execution of the subsystems.

The IPS usually contain a private problem-oriented language (IPOL) for the coding of subsystems, together with a suitable pre-compiler, which translates the IPOL code into a code familiar to the computer (e.g. FORTRAN, ALGOL).

The formulation of the algorithm for solving the problems handled with the IPS has to be given in a set of subsystems, which replace the programs discussed in the previous sections.

Typical examples are:

ICES: Integrated Civil Engineering System, developed at Massachusetts Institute of Technology, Civil Engineering Systems Laboratory in cooperation with many other public institutions

IST: Integrierter Systemkern für die Technik, developed by the IST project-team at the Technical University of Berlin (leader: Prof. P. J. Pahl). The system may be better known under its former title ISB (B stood for "Bauwesen," i.e. civil engineering)

GENESYS: GENeral Engineering SYStems, developed at the Genesys Centre, University of Technology in Loughborough/U.K.

We restrict our discussion to these three examples. A wider discussion may be found in [1].

The development of such systems was started in about 1964 mainly for civil engineering with the ICES system. In the meantime many other branches have been using systems of this kind for nearly all design problems. The ICES system, one of the first available systems, has found a very broad distribution all over the world.

The practical use of all these systems is exclusively defined by the available subsystems, which define the range of applicability of a system.

These systems are summarized below.

	ICES	IST	GENESYS
Coding language	Assembler	Assembler	FORTRAN + assembler
Name of the IPOL	ICETRAN	ISTRAN	GENTRAN
Precompiler translates to	FORTRAN	FORTRAN	FORTRAN
available from	ICES-USER'S GROUP INC. P.O. Box 8243 Cranston, RI 02920	Prof. P. J. Pahl IST-Projektgr. T.U. Berlin D-1 Berlin 30 Europa-Center	Genesys Centre Univ. of Technology Loughborough/ U.K.

A program coded in IPOL language runs according to the following scheme:

Step 1. The precompiler translates the IPOL code into a FORTRAN code.

Step 2. The usual FORTRAN compiler translates the generated FORTRAN code into an executable code.

Step 3. Either the usual loader of the operational system or a special loader of the IPS links the modules to an executable program.

Step 4. The execution of the program is supervised by parts of both systems (IPS and operational system).

The following is a list of available subsystems for ICES and GENESYS. Subsystems are not yet available for IST.

ICES:

STRUDL I, II, III: STRUctural Design Languages
COGO I: COordinate GeOmetry
TABLE I, II: ICES File Storage subsystem
SEPOL: SEttlement Problem-Oriented Language
ROADS: ROadway Analysis and Design System
TRANSNET: TRANSportation NETwork Analysis
BRIDGE: BRIDGE Design System
PROJECT: PROJect Engineering ConTrol
OPTECH: OPTimization TECHniques System
LEASE: Limit Equilibrium Analysis of Slopes and Embankments

GENESYS:

RC-Building: Reinforced concrete buildings
BRIDGE: Straight bridges (cont. beams)
SLAB-BRIDGE: Skewed and curved bridges
FRAME-ANALYSIS: Plane or space frames, grillage
SLIP-CIRKLE: Safety or slopes along circles

The following subsystems will be available soon. UBM (Ultimate Bending Moment/axial load for concrete cross sections), HIGHWAYS, STORM-WATER, WATER-DISTRIBUTION COMPOSITE CONSTRUCTION (reinf. concrete/slabs/steel beams), NONCIRCULAR-SLIP, DAMWAND, DRAINAGE, FLAT-SLAB, PORTAL-FRAMES, PILE-2D, and BACKWATER.

As mentioned above, a code in this language is translated into FORTRAN by the precompiler. All IPOLs are defined so that FORTRAN is a subset of the IPOL. The extension of FORTRAN to IPOL consists in the definition of macro-statements which are replaced by usual call statements for subroutines or groups of modules contained in the IPS library. Thus the result of the precompilation is a pure FORTRAN-code enriched with many call statements. The modules or subroutines cover a wide span of the engineer's particular problems in order to simplify the description of the complete problem by a sequence of macro-statements and as few as possible intermediate classical FORTRAN statements.

From this discussion it is evident that at least several parts of an IPS depend on the computer type and the operational system as well: parts like the precompiler, at least some parts of the system core, of the data management, and others. The degree of machine dependency of a system may be a possible characteristic distinction. ICES- and IST-type systems are coded with full dependency on a certain computer type, whereas the GENESYS system is conceived in a manner which provides the greatest possible independence from the computer type. This independence is achieved by the fact that as much as possible is coded in standard FORTRAN, while the rest is coded in the assembler language of a specific computer. The efficiency of a system certainly depends to a high degree on its basic philosophy. The "machine-independent" concept certainly slows down the execution speed of a subsystem. One cannot give general characteristics which depend on a lot of various factors. On the other hand, such a concept makes possible easier transfer of the system between different computer types: the machine-dependent part of the GENESYS system is said to have the volume of about 200 FORTRAN statements--not including the necessary precompiler.

The IST system is a further development of the ICES system. It is based on the same principles that are implemented in ICES and have proved widely useful, e.g.,principles like segmentation in the system core, separate subsystems for the problem description, development of a command definition language, extension of FORTRAN to the ISTRAN language, dynamic data fields with tree structure stored separately from the procedures, and dynamic linkage between procedures of different level.

In addition to these facilities, a certain kind of vs techniques for the dynamic data field is implemented in the form of infinite common regions. Furthermore, the changes with respect to ICES involve, among other things, the extension facilities of the data regions (e.g., separation of the pointers from the data pools, transfer of variable record sizes by one transfer call) and the administration of the subsystem (e.g., complete control of the program region, reservation of the static subsystem region) by means of the programmer (user).

The first release of IST for a Siemens 4004 computer was in June 1973. This first version does not utilize the whole concept of data management. It was released only to aid in developing subsystems in organizations outside the IST center. The completion of the system for the Siemens 4004 is anticipated by the end of 1974. The inner structure of this computer is very similar to IBM computers; therefore, it should be possible to transfer the system to IBM machines without too much effort. Information on the question of transferring to other computer types will have to wait until the system is tested in connection with some subsystems.

Systems of this kind are by definition more suitable for interactive working modes than the large G.P. programs of type 4, although none yet contains real interactive modes. The IST concept contains implicitly a feature of interpreting successively every single statement of the SBW (Sprache des Bauwesens, "language of civil engineering"), a sort of command definition language. This feature can be used within a time-sharing operational system for interactive communication (e.g., via display).

The decision on implementing and using an IPS depends mainly on:

1. The costs for rent and installation
2. The reliability of the maintenance service
3. The availability of suitable subsystems
4. The possibilities of getting at reasonable costs subsystems which are developed by other users of the system

A very good example of maintenance and distribution of such a system is given by the ICES User's Group. The "rent" of the system is covered by the fees for the membership in the User's Group and for computer time and tape reels used for copying the required parts of the system and/or subsystems. Nearly all subsystems can be drawn from the same group, except some special developments of few users.

There is a certain difficulty in the dependence of the IPS on the operational system. The conditions of distribution should include service possibilities for adaption of the IPS to changes in the operational system.

CONCLUSIONS

General purpose software is mostly directed toward providing generally applicable data management utilities and flexibility of the programs (systems) with respect to the addition of new modules (subsystems).

The same aims have governed the development of general computer software within the past few years. This development tends in the directions of vs techniques, which still have to be really refined, and loader/linkage utilities within the usual operational systems, which for several important computer types today give very good segmentation facilities for the generation of large, multipartitioned program codes.

No final judgment is possible now about the future development of G.P. programs of IPSs. We tend to feel that the chances for the development of IPS-type communication systems are better than those of large G.P. programs. Further development of the concepts existing today cannot be avoided for the near future.

All these considerations depend strongly on the further development of computer hardware and software. An IPS is certainly more suitable for a time sharing system with terminals for many users. In such a system the implemented data pools can be used in the most efficient way.

ACKNOWLEDGMENTS

The authors want to express their thanks to all the firms and especially to the University of Virginia for the information they provided.

REFERENCES

1 Proceedings of the international CEPOC-conference "Colloque International sur les Systèmes Intégrés en Génie Civil," Liege, August 1972. To be published.

2. Perrone, N. (ed.),"Compendium of Structural Mechanics Computer Programs," International Journal of Computers and Structures, Vol. 2, No. 3, April 1972.

3 Buck, K. E., Scharpf, D. W., Schrem, E., Stein, E.,"Einige allgemeine Programmsysteme für finite Elemente" In: Buck, Scharpf, Stein, Wunderlich (ed.), Finite Elemente in der Statik, W. Ernst & Sohn, Berlin-München-Düsseldorf, 1973.

4 Jaeger, T. A., (ed.), Preprints of the 2nd International Conference on Structural Mechanics in Reactor Technology, Vol. V, Berlin, Germany, September 1973, Publication Management: Commission of the European Communities, DG XIII, CID-Publications, Luxembourg.

5 Przemieniecki, J. S., et al. "Matrix Methods in Structural Mechanics,"AFFDL-TR-66-80, Nov. 1966.

6 "Proceedings of the 2nd Conference in Matrix Methods in Structural Mechanics," AFFDL-TR-69-150, Dec. 1968.

7 Brebbia, C., Tottenham, H., (ed.), Variational Methods in Engineering, Vol I a. II, Southampton University Press, 1973.

8 Bathe, K. J., Ramm, E., Wilson, E. L.,"Finite Element Formulations for Large Displacements and Large Strains,"Rep. No. UCSESM 73-14, Structural Engineering Laboratory, University of California, Berkeley, Sept. 1973.

APPENDIX I

Developers of the Programs Evaluated in Table 1

No.	Name	Developer's name and affiliation
1	AC50A	L. Tantalo/D. Y. Konishi c.o. Rockwell International, LAAD Division Int. Airport, Los Angeles, Calif. 9009
2	ADEPT	J. L. Morris c.o. Simon Engineering Laboratories, University of Manchester, Manchester/England Owner: National Research Development Corp.
3	AMSA20	Fiat-Divizione Aviazione-S.C.V. C. so March, 41 10100 Turin, Italy
4	ANSYS	Swanson Analysis Systems, Inc. 870 Pine View Drive, Elizabeth, Pa. 15037
5	ASAS	ATKINS Research and Development Woodcote Grove Ashley Road, Epsom, Surrey, England
6	ASKA	Aska-Group, ISD Stuttgart Pfaffenwaldring 27, 7000 Stuttgart 80, Germany Or: IKO - Software - Service Vaihinger StraBe 49, 7000 Stuttgart 80, Germany
7	BASY	T-Programm GmbH. Technischer Software und Datendienst Gustav-Werner-StaBe 3, 741 Reutlingen, Germany
8	BERSAFE	T. K. Hellen C.E.G.B. Berkeley Nuclear Laboratories Berkeley, Glos., U.K.
9	BOSOR4	David Bushnell, Dept. 5233 Lockheed Missiles and Space 3251 Hanover Street, Palo Alto, Calif.
10	COSA	Dornier A.G. Hauptabteilung Strukturberechnung und theoretische Akustik Postfach 317, 799 Friedrichshafen, Germany

No.	Name	Developer's name and affiliation
11	DYNAS	Charles F. Beck Computer Service Sargent & Lundy Engineers 140 South Dearborn Chicago, Ill. 60603
12	EASE	Engineering Analysis Systems 1611 S. Pacific Coast Highway Redondo Beach, Calif.
13	ELAS 75	S. Utku Duke University School of Engineering Durham, N. C. 27706
14	FARSS	M. B. Marlowe See No. 9
15	FESAP	Don B. Van Fossen Babcock & Wilcox R & D P.O. Box 835 Alliance, Ohio 44601
16	FLHE	K. Fullard Central Electricity Generaling Board Berkeley Nuclear Laboratories Berkeley, Gloucestershire Gl. 13 9PB England
17	ISOPARSHL	Dr. A. K. Gupta Cf. no. 11
18	ISTRAN/S	Ishikawajima-Harima Research Institute 1 - 15 Toyosu 3 - Chome, Koto-Ku Tokyo, 135 - 91, Japan
19	KSHEL	Arturs Kalnins Lehigh University Dept. Mech. Eng., Bethlehem, Pa. 18015
20	MARC	J. Ries Engineering Sciences Control Data Corp. Minneapolis, Minn.
21	MINIELAS	I. B. Alpany, S. Utku Cf. no. 13

No.	Name	Developer's Name and affiliation
22	NASTRAN	C. W. McCormic Mac Neal-Schwendler Corp. 7442 No. Figueroa St. Los Angeles, Calif. 90041
23	NEPSAP	Parviz Sharifi Lockheed Dept. 81 - 12, Bldg. 154 P.O. Box 504 Sunnyvale, Calif.
24	NONLIN2	J. C. Anderson Cf. no. 11
25	NONSAP	E. L. Wilson, K. J. Bathe Structural Engineering Laboratory University of California Berkeley, Calif.
26	NOSTRA	C. A. Fellippa Cf. no. 9
27	PAFEC70	R. D. Henshell Mech. Ing., Notts. Univ. University Park Nottingham NG7 2RD, U.K.
28	PRAKSI	Rechen - und Entwicklungsinstitut für EDV im Bauwesen - RIB e.V. 7000 Stuttgart 80 Schulze-Delitzsch-Str. 28
29	REXBAT	W. A. Loden Cf. no. 9
30	SABOR-DRASTIC6	Stanley Klein Philco Ford Corp. Ford Road Newport Beach, Calif. 92663
31	SAMBA	H. Jareland FKVH-27, SAAB-Scania AB S 58188 Linkoping, Schweden
32	SAPIV	K. J. Bathe University of Califoria, Berkeley 410 Davis Hall, Univ. of Cal. Berkeley, Calif.

No.	Name	Developer's name and affiliation
33	SATANS	R. E. Ball Naval Postgraduate School Code 57 Bp Monterey, Calif. 93940
34	SESAM69	A/S Computas Okerveien 145 Oslo 5/Norway
35	SHORE	P. Underwood Cf. no. 9
36	STARDYNE	R. Rosen Mechanics Research Inc. 9841 Airport Blvd. Los Angeles, Calif. 90045
37	STARS	V. Svalbonas Grumman Aerospace Dept. 461 Bethpage New York, N. Y. 11714
38	STRIP	Digital AG Seilergraben 53 CH-8001 Zürich
39	TEXGAP	R. S. Dunham Texas Institute for Computational Mechanics University of Texas Austin, Texas 78712
40	TIRE	A. L. Deak Mathematical Sciences Northwest, Inc. 3030 NE 45th Street Seattle, Wash. 98105
41	TITUS	Vouillon-Citra 13 Av. Morane Squlnier 78 Velizy/France
42	VISCEL	K. K. Gupta Jet Propulsion Laboratory 4800 Oak Grove Pasadena, Calif. 91103
43	ZP26	L. Nash Gifford, Jr. Naval Ship R & D Center Code 1725, Bethesda, Md. 20034

DISSEMINATION OF GRAPHICS-ORIENTED PROGRAMS

L. J. Feeser and R. H. Ewald
University of Colorado
Boulder, Colorado

ABSTRACT

Problems associated with the dissemination of graphics-oriented computer programs are outlined with respect to devices, languages, and installation dependencies. Some approaches to the solution of these problems are suggested with evaluations of their probability of success.

INTRODUCTION

The general usefulness of computer programs is limited by their availability to a wide user audience. The development of high-level languages such as FORTRAN has allowed a broader dissemination of computer applications programs. The standardization of languages such as FORTRAN [1] has also helped to create a basis for dissemination of programs in scientific/engineering application areas. However, standardization in languages does not, in itself, solve all the problems associated with software dissemination [2, 3, 4].

Problems of word size and peculiarities of compiler implementations have led to some differences of results from identical programs run at different installations, although the programs were in ANSI Standard FORTRAN [5]. Additional problems are found in programs in the data declaration, alphanumeric character representation, and file-handling areas. Severe problems may be encountered in attempting to run a program written for a virtual memory machine on a non-virtual memory computer. One might expect to experience compounded problems, therefore, when trying to define dissemination mechanisms for programs which contain graphical input and output.

The proliferation of both graphics devices and graphics languages has considerable effect upon the ability to disseminate software containing graphics capability. This paper will describe the problems which arise in the areas of devices, languages, and installation dependencies and will propose some techniques for dissemination which have some possibility of success in overcoming these problems.

GRAPHICS DEVICES

Due to the number of different types of graphics devices and the functional differences between these types, the problem of disseminating graphics-oriented programs becomes more complex than for non-graphic programs. Functionally, graphic devices may be categorized in three broad classifications:

1. Printer plotters - alphanumeric output devices which may be made to plot by the positioning of characters on a page.
2. Plotters - flatbed, drum, and film plotters. A recent study indicated over thirty hardware manufacturers producing over seventy-five plotters of these types [6].
3. CRT - storage tubes, randomly written refresh, and raster scan devices. Currently there are more than thirty-five hardware and system suppliers of these types of devices offering over sixty terminals [7].

Graphical input devices may be connected directly to graphical output hardware, or they may be stand-alone units. Included in the list of graphical input devices are digitizers, tablets, light pens, joysticks, and several other pieces of hardware. A recent survey listed nineteen manufacturers of digitizers and tablets alone, producing over twenty-five devices [8].

Printer Plotters

Perhaps the simplest graphical output device is a standard alphanumeric printer or terminal. By the positioning of print characters on a page, a plot may be produced to give a rough representation of a curve or surface.

Recently, electrostatic printer/plotters have been introduced. These devices usually print between sixty and one hundred points per inch, and so they may be used to produce graphical output in a manner similar to half-tone picture printing. These printer/plotters may also be used to produce standard alphanumeric listings or a mixture of an alphanumeric listing with a plot.

Moving a program written for one printer plotter to another may be a relatively simple task since the page size is usually defined and the plotting is performed by positioning characters.

Plotters

The plotter classification used in this paper includes drum, flatbed, film, and microfilm plotters. Plotters of these types generally receive digital information indicating pen or beam positioning and convert the digital information into signals which cause the pen or beam to move to the correct location. These plotters may be on-line or off-line, and some of the on-line plotters provide for interactive operation.

Drum plotters move paper, film, or some other recording medium under a pen, light beam, or electron beam. The recording medium moves in one direction, and the drawing instrument usually moves in a perpendicular direction. Thus the pen or beam is able to move to any point on the recording medium by appropriate movements of both pen and paper.

Flatbed plotters hold the recording medium in place while the drawing instrument moves. The pen or beam is capable of moving in orthogonal directions and is able to reach any point on the recording medium. This type of plotter is generally capable of greater accuracy than the drum plotter since only the pen moves. Figure 1 shows a small flatbed plotter used in time-sharing applications.

Fig. 1 Time-sharing flatbed plotter

Microfilm plotters generally draw an image on a small CRT which is then photographed or draw the image directly on a piece of film. A hard copy of the microfilm may be made on a standard copying device. Microfilm may also be used as a medium for creating an animation of some process. For example, if the deflected shape of a structure under a dynamic loading is known for many time steps, a few frames of microfilm of the deflected shape for each time step may be recorded. When the film is projected via a movie projector, the structure appears to move dynamically. Figure 2 shows a copy of a microfilm plot.

Implementing a program written for one plotter on another plotter of that type may generally be accomplished within a particular operating environment. However, moving a program written for one type of plotter to another type (i.e., drum to microfilm) may be much more complex due to the differences in hardware.

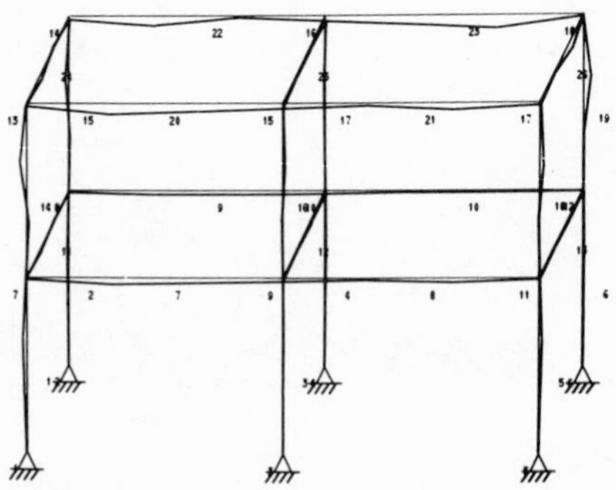

Fig. 2 Microfilm plot

Interactive Cathode Ray Tubes

Interactive cathode ray tubes (CRTs) are of three types: storage tubes, randomly written refresh devices, and raster scan devices. All provide for interactive graphic operation with varying degrees of sophistication, usually indicated by cost. The terminals range in price from $4,000-$5,000 for low-cost storage tubes to $150,000-$200,000 for randomly written refresh or raster scan devices with associated processors and communications equipment. Some of these devices are intelligent and contain a minicomputer capable of performing some local computation. Other devices operate much like an alphanumeric terminal, but with line-drawing capabilities. Figure 3 presents a general schematic of the three types of devices showing the components of the display system.

In a storage tube the display memory and the CRT are combined in a single unit, the display CRT. When a drawing is made on the screen of a storage tube, the image is retained on the tube with no refreshing (redrawing) necessary. These devices are comparatively inexpensive and are generally connected to a large computer's time sharing network. Some devices of this type have a small scratch pad area which may be erased without erasing the rest of the screen. The graphical input devices attached to storage tubes usually are tablets, joysticks, or other built-in devices.

Storage tubes are capable of maintaining very complex images since no refreshing of the screen is necessary. Since the display is not refreshed, the image will never flicker. However, if some portion of the display is to be changed, the entire display must be

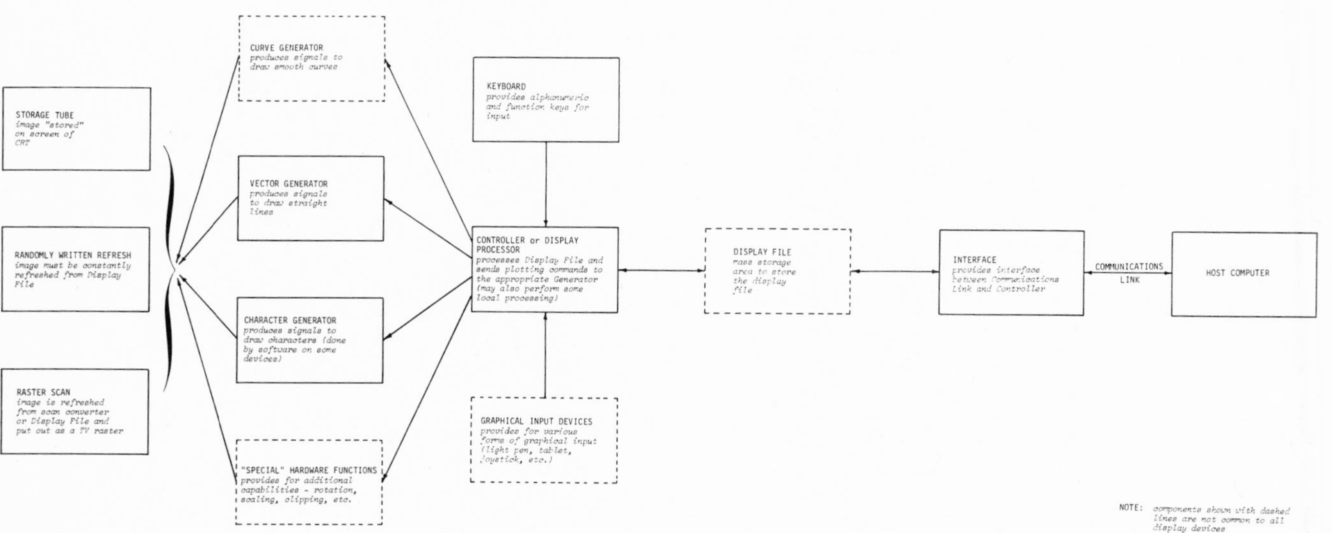

Fig. 3 Typical CRT display system schematic

erased and redrawn. The redrawing is time consuming for complex images when the storage tube is connected to a computer over a slow communications link.

Many storage tubes have an accessory called a hard-copy device which records the image on the screen onto paper, so that the terminal may provide interactive graphics with a hard-copy capability as well. Figure 4 shows a storage tube terminal with a tablet for graphical input and a hard-copy unit.

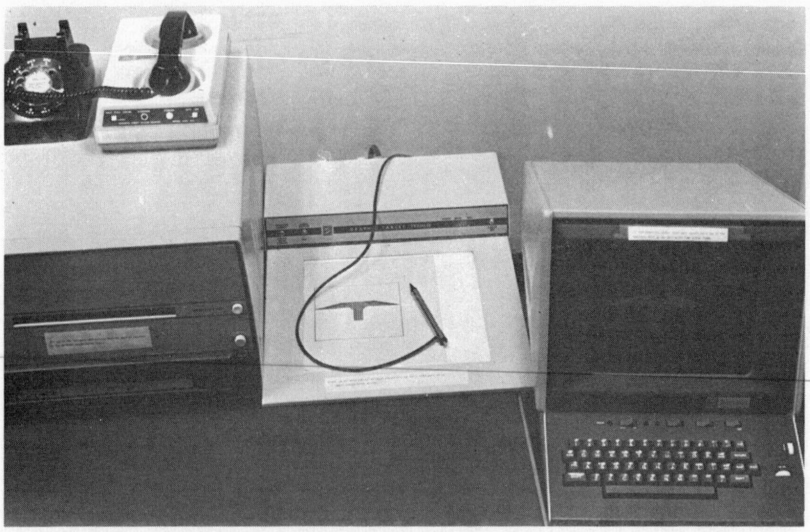

Fig. 4 Storage tube, tablet, and hard-copy unit

In a randomly written refresh device, the beam traces an image based on the data in a display file. When the CRT phosphor is hit by the beam, it is illuminated for a very short period of time. If the image is to be retained on the screen, it must be painted periodically (refreshed). Refresh rates are typically in the range of ten to one hundred times per second. Because the display is refreshed so often, the display file becomes an important part of the refresh system. The display processor must be able to access the display file (often called a buffer) repetitively to generate the image on the tube; so in many systems the display buffer is the storage system of a minicomputer located in the display system. On other systems the display file is stored on a mass storage device which the display processor can access, while still other devices require that the host computer provide the display buffer.

Since the display processor draws an image based on the contents of the display file, if the display file is changed, the image also changes. This allows randomly written refresh devices to erase selectively part of the display or to provide for dynamic graphics. Objects may be moved, rotated, zoomed, and transformed on these devices. On certain devices, the dynamic capabilities are implemented in hardware, while in others they are performed by software.

Display systems which provide for some local processing in addition to display processing are termed intelligent terminals. This capability is usually implemented by including a minicomputer in the display system. These devices allow some processing to be done at the terminal, freeing the host computer from these processes.

Historically, graphical input devices for randomly written refresh units have been light pen oriented. Recently some of these devices have allowed tablets and other means of graphical input to be used. Hard—copy units generally are not attached directly to these devices.

Because the randomly written refresh system does refresh the screen, it provides a dynamic capability. However, the refresh technology does have some drawbacks. The complexity of a display is limited to the size of the display buffer and, more practically, by the number of lines which may be displayed before the image begins to flicker. Figure 5 shows a randomly written refresh device.

Fig. 5 Randomly written refresh graphics device

Raster scan devices are similar to randomly written devices because they both utilize refresh methods, but the raster scan devices form an image by making a number of horizontal sweeps across the tube in the same manner as a television set. Raster scan units generally create a layout of the display area in the display file, which is then output by the display processor to the device's video monitor.

Raster scan systems are able to perform dynamic graphics in much the same manner as randomly written refresh devices. As the layout of the display area in the display file is changed, the image displayed on the screen is also changed.

In addition to providing some dynamic graphics capability, raster scan units may output color displays or black-and-white displays with a large number of gray levels. Since the display is output in a raster scan manner, the resolution of some of these devices is not as good as storage tubes or some randomly written refresh units. However, flicker usually is not as big a problem as in the other refresh units. Some ragged edges are occasionally noted on lines or surfaces with edges approaching the horizontal because of the discontinuity introduced in the raster scan method.

Implementing a program written for one type of interactive CRT on another CRT may be very difficult, regardless of the types of the two CRTs.

At the hardware level, the difficulties encountered in disseminating graphics programs are due to a lack of standardization. The various graphics devices use different codes and protocols for display generation, and the manufacturers provide different capabilities for the various devices. Many of these problems could be overcome with the appropriate software techniques.

GRAPHICS SOFTWARE

The software which is generally used in a graphic computer system is composed of three parts:

1. The applications program
2. The graphic drivers
3. The computer's operating system

These three sets of software interact to produce the desired graphical output and user interaction.

The applications program is the program which performs some specified algorithm and calls the graphic drivers to produce the graphical output. The graphic drivers are a set of routines which determine the action to be taken as a result of a call to a particular routine. The driver routine converts the call into the appropriate data to be sent to the graphics device to cause the desired action. The operating system sends the code to the device and monitors its status to allow for graphical input and user interaction.

Typically, the applications program is written in some high-level language, which for many scientific and engineering problems is FORTRAN. Imbedded in the applications program are calls to the graphic drivers, which in many cases are written in the same high-level language. For most graphic devices the manufacturer provides the drivers as a set of primitive operators which may be expanded by the implementer.

Thus, the problem of dissemination of graphics—oriented software may be focused in the graphic drivers. Since each manufacturer provides different software, a program written using one such driver will not run using another set of graphic driver routines. Since the capabilities of the various graphic devices are different, the drivers also provide for varying capabilities. There exists another difference in the driving routines based on whether or not the terminal is intelligent and whether certain operations are to be performed by hardware or software.

It is apparent from the above discussion that the graphic driving software may be more diverse than the graphics hardware. Currently there is no standard for the development of this software, so each implementer may design his own scheme. If some standard were developed, it is conceivable that graphics drivers for the various devices could be written so that at some level an applications program could be run on a variety of devices. This scheme has been implemented at a few installations on a somewhat limited scale [9, 10]. With the appropriate software, a graphics device may be made to emulate another device and hardware options on one terminal may be implemented via software on another.

Some programs have combined the application program and the graphic drivers into one package. Graphic drivers for several different plotters or CRTs may be included in the system so that the user may specify the graphical output device. However, the major disadvantage of this approach is that it requires the application program to contain all of the graphic drivers when only one device is actually used for each run. By using overlay techniques only the required code is loaded into the computer at execution time, but the execution overhead may be increased.

There have been several attempts to create graphic languages [11]; however, most have been implemented for specific application areas. A universal graphics language has been under discussion for some time, but one has not been implemented. Any standard graphic language will require the support of the computer manufacturers, graphic device manufacturers, and the graphic users. This unified support is not yet in evidence.

The computer's operating system and the graphic drivers interact to send and receive code from the graphics device. As a result, the dissemination of graphics-oriented programs is also dependent upon the particular computer installation.

INSTALLATION DEPENDENCIES

The installation dependencies for graphics-oriented programs are a superset of those for nongraphic programs. The standard program portability problems of word size, file handling, language dialect, and so on are all encountered in disseminating graphics-oriented programs. In addition there are some other problems.

The host computer must communicate with the graphics device using some standard communications code. Different computer and graphic devices use different codes; so at the hardware level the computer and graphics device must be compatible with regard to the communication code. When a program is moved from one computer to another, or from one graphics device to another, any communication code dependencies in the program or the graphic drivers must be accounted for.

On some computers and graphics devices the rate at which the communication code is transmitted is variable. The graphics program and drivers may have to incorporate the communications rate in their procedures to provide for appropriate delays during screen erasure, hard-copy generation, and so on. Thus, when a program is implemented on a computer which has different communication rates than the original implementation, the effects of the rate change on the software must be considered.

The communication protocol required by the different graphic devices varies widely. Some devices require a special code to indicate that a display is to be created and another special code to terminate the display, while other devices require no special codes. Some computer systems automatically send the carriage return and/or line feed character at the end of a buffer of data. The hardware of some graphics automatically disregards these characters, while on other devices they may be used for x and y coordinate data. Alternately, these characters are not automatically sent by other computers, so that this problem does not exist.

All the problems listed in this section must be solved by the combined applications program, graphics drivers, and operating system. Because of the diversity of computers and graphics devices, the solution to these problems for one computer configuration may not be optimal for another configuration, and in fact it may never work.

POSSIBLE SOLUTIONS

Standardization

Many of the problems identified above could be solved with reasoned standardization of both hardware and software. However, since the development of devices is taking place very rapidly, standardization of hardware seems remote at this time. The hope for communications protocol standardization is high since this interfacing is a necessity in other areas.

One of the trends which we see developing is the tendency toward de facto standardization brought about by certain hardware capturing a large share of the market. In particular, in passive graphics, the CALCOMP plotter has become the device to be simulated within applications programs which are desired to be portable to many other installations. In a similar fashion, the Tektronix direct view storage tube graphics terminal is becoming very widespread, and, therefore, programs which are written for these devices have considerable portability to identical devices at other installations.

One can argue the merits and demerits of de facto standardization; however, just as the terminology denotes, there is little that can be done about it unless there is considerable cooperation among the various manufacturers to arrive at standards which are acceptable to all.

This ability of certain manufacturers to capture significant portions of markets with devices also leads to de facto standardization towards their devices' software. In this area there are some possibilities for providing automatic translation of one piece of software for a particular device to equivalent operations for another device without destroying the high-level language-calling codes. Indeed there are a number of examples of partially successful attempts in this direction [9, 10]. If agreement could be reached on basic names for FORTRAN callable graphics subroutines which do similar operations, considerable progress would be made toward easing the problem of dissemination of software which contains graphics capability.

There are some attempts under way to develop standards for graphics—oriented software. Some of the efforts of the special interest group of ACM, SIGGRAPH, are described in [12].

Networks

The possible availability of large-scale computer networks has been proposed as a solution to the dissemination problem of graphics related software. The network would be available either on a dial-up basis or on a leased line basis, allowing many users access to the same software operating on a single piece of computer hardware at some remote site. If each of the user groups has a graphics terminal which meets certain defined specifications, theoretically pieces of maintained software are available to them, and they do not have to worry about the problems of moving software from one central processor to another. Indeed there are commercially available time-sharing networks which work successfully for devices such as the CALCOMP plotter and the Tektronix DVST.

The question then arises, what does the individual user who needs more capability than that offered by the above two devices do for his problem? It appears that at this time he must still go through the painful exercise of handcrafting useful applications software to run on his own device and/or computer.

Distribution Organizations

Organizations such as COSMIC [13] and other proposed software centers [2, 3, 4] provide hope for the dissemination of graphics-oriented software.

In the case of COSMIC, they do disseminate software containing graphics capability. At this time, they carefully specify the hardware (both computer and graphics device) and operating system as well as graphics software packages used so that if the potential user has access to a similar configuration, the software he obtains may be usable. He may have to do some minor modifications, since it is very seldom that one finds two identical operating systems.

In the case of some of the proposed software centers, the possibility of machine-translated and machine-modified software, tailored toward particular uses, hardware, and operating environments, appears feasible [14, 15]. Although most of these proposals at this stage are only for nongraphic applications programs, it appears that the extension to graphics capability has promise.

CONCLUSIONS

There are many problems associated with attempts to disseminate graphics-oriented software. The solutions to these problems fall into three basic approaches: standardization, networking, and distribution organizations. Each of these approaches offers something toward solving certain of the dissemination problems, but no one appears to solve all the problems all of the time. For the foreseeable future, it appears that work must continue in all three areas in order to maximize the availability of graphics capability to the user community.

ACKNOWLEDGMENT

The work described in this paper was partially supported by the National Science Foundation and the Control Data Corporation through grants to the University of Colorado Computing Center.

REFERENCES

1 American National Standard FORTRAN, ANSI X3.9-1966, American National Standards Institute, New York, 1966.

2 Schiffman, R.L., "Report on the Special Workshop on Engineering Software Coordination," Report No. 72-2, University of Colorado Computing Center, March 1972, Boulder, Colorado.

3 Schiffman, R. L., "Papers Prepared for the Special Workshop on Engineering Software Coordination," Report No. 72-4, University of Colorado Computing Center, April 1972, Boulder, Colorado.

4 Schiffman, R. L., "Transcript of the Special Workshop on Engineering Software Coordination," Report 72-17, University of Colorado Computing Center, June 1972, Boulder, Colorado.

5 McCormick, J. M. and Baron, M. L., "On the Implementation of Computer Software," Guidelines for Acquisition of Computer Software for Stress Analysis, Meeting Preprint 2022, ASCE National Structural Engineering Meeting, April 1973, San Francisco, California.

6 Stiefel, M. L., "Plotter Terminals and Systems," Modern Data, Vol. 4, No. 9, September 1971, pp. 48-57.

7 Machover, C., "Interactive CRT Terminal Selection," Society of Information Display Journal, Vol. 1, No. 4, 1972, pp. 10-22.

8 Stiefel, M. L. and Murphy, J. A., "Graphics Digitizers," Modern Data, Vol. 4, No. 8, August 1971, pp. 38-44.

9 Dohrmann, R. C., Ewald, R. H., and Kopolow, R. N., "CUG/T4000, Colorado University Graphics," Report No. 73-7, University of Colorado Computing Center, May 1973, Boulder, Colorado.

10 "Primer on Computer Graphics Programming," Report No. 73-2, United States Military Academy, July 1973, West Point, New York.

11 Nake, F., and Rosenfeld, A., Graphic Languages, North-Holland Publishing Co., Amsterdam, The Netherlands, 1972.

12 Al-Banna, S., and Feeser, L. J., "Report on the Special Workshop on Computer Graphics as Related to Engineering Design," Report No. 74-3, University of Colorado Computing Center, March 1974, Boulder, Colorado.

13 "COSMIC, Software Submittal Guidelines," Information Services Division, Office of Computing Activities, The University of Georgia, July 1973, Athens, Georgia.

14 Ewald, R. H., Schiffman, R. L., and Smith, R. R., "Composition/Decomposition System - First Progress Report," Report 72-29, University of Colorado Computing Center, December 1972, Boulder, Colorado.

15 Ewald, R. H., "FLAGBR, Version 1.0, A Program to Aid in the Modularization of a FORTRAN Program or Procedure," Report No. 73-8, University of Colorado Computing Center, May 1973, Boulder, Colorado.

A COMPUTERIZED INFORMATION RETRIEVAL SYSTEM

David Bushnell
Lockheed Palo Alto Research Laboratory
Palo Alto, California

ABSTRACT

A computerized structural analysis information retrieval system, COMSTAIRS, is described, including a survey of several widely used programs. COMSTAIRS is an interactive system designed to help those who have structural analysis problems and need the best large-scale computer programs to solve them. The system contains detailed descriptions of the capabilities of many programs with regard to geometry, material properties, type of loading, phenomena, discretization, methods of solution, and program distribution, documentation, operation, and maintenance. Also given are evaluations by users of several programs based on responses to questionnaires sent out in October 1973 - March 1974. The user of COMSTAIRS can find out which programs will solve problems identified by various combinations of key words, compare the capabilities of two or more programs, obtain detailed evaluations by users including user comments, and obtain detailed capability summaries of specified programs. The COMSTAIRS user can also provide input at the terminal which is then stored in an accumulating data bank. Recommendations for further work are given.

INTRODUCTION

A number of surveys of computer programs for structural analysis have been conducted recently [1 - 6]. In these surveys the developers of computer programs were asked to supply information on their programs, and the results were published in notebook or bound form. Evaluations of the programs were supplied by the developers or close associates of the developers. No attempt was made, as far as the author knows, to include user evaluations of the programs. Some user evaluations of major programs have been obtained through such forums as the NASTRAN users' colloquia [7], ASME symposia sessions on program validation [8], and at a session at the last ONR conference held at the University of Illinois in 1971 [9]. Other kinds of surveys concerning computer software for structural analysis include that of Kraus [10] on attitudes toward software exchange in the pressure vessel industry and that of Medearis [11] on the establishment of a national civil engineering software center. Other papers on software information dissemination and exchange appear in this volume.

The two main purposes of this paper are:

1. To present an interactive information retrieval system for

computerized structural analysis

2. To summarize the results of the survey upon which the data bank of the retrieval system is based

ACCUMULATION OF THE DATA BANK

Late in 1973 separate questionnaires were drafted for developers of structural analysis computer programs and users of such programs. These questionnaires were distributed to about 3000 individuals under the following cover letter:

QUESTIONNAIRES FOR

INFORMATION RETRIEVAL SYSTEM FOR

COMPUTERIZED STRUCTURAL ANALYSIS

Dear Colleague:

> As the manager of a structural analysis group, have you ever wondered which computer programs have the proven capability to solve your structures problems?
>
> As an engineer with a specific structures problem to solve, have you ever wondered which computer program would best do the job?
>
> As the developer of such a program, have you ever wondered what users think of your code?

With these questions in mind, I have decided to create an information retrieval system based on responses to two questionnaires: One questionnaire is designed for developers of computer programs and the other is designed for users of computer programs. In the questionnaire for program developers, the author of a program indicates what kinds of problems his program will solve, what methods are used, which computers the program runs on, how the program is maintained, whether or not it has been superseded by another program, etc. In the questionnaire for program users, the user indicates in which fields he has applied the program, the intensity of use at his facility, his evaluation of the program, which branches of the program he has experience with, etc.

The data obtained from this survey will be useful to both program developers and program users. Developers will be able to improve and extend their programs with the practical problems of users in mind. Users will be able to identify new programs which can handle their increasingly sophisticated problems--problems involving such things as composite materials, large deflections, plasticity, and creep.

Here is an example of how the information retrieval system would work: An engineer might encounter a problem which he cannot solve with the computer programs currently available to him. He would provide as input data a sequence of numbers indicating problem characteristics such as "shell of revolution, branched, layered orthotropic, nonlinear axisymmetric stress analysis, bifurcation buckling, etc." The information retrieval system would list the names of all those computer programs which include the capability of solving such a problem, such as BALORS, BOSOR4, SRA, KSHEL, STARS, etc. The engineer could then focus the system on each of these programs individually to obtain evaluations by users, information on what computer the program runs on, price of the program, whether or not it can be executed through a computer software service bureau, references possibly with similar applications; and a list of names, addresses, and telephone numbers of experienced users and of the developer whom he can contact for further information.

I would prefer not to receive replies regarding very special purpose programs applicable, for example, only to rectangular flat plates or simply supported cylinders. A respondent might ask himself before replying: "Is this computer program of sufficient scope as to be widely used, and is it thoroughly enough documented to be used without the developer's frequent assistance?"

Respondents are asked to fill out one copy of the developer's questionnaire for each program developed by themselves and one copy of the user's questionnaire for each program used by themselves. For each computer program recorded in the information retrieval system, there must be evaluations by many users. Therefore, please have several copies of the user's questionnaire made for every copy of the developer's questionnaire. Keep as many as you need for your own use and forward copies of this cover letter and user's questionnaires to several of your colleagues, asking them to do the same. Please make a special effort to see that structural analysts in all departments and branches of your institution receive copies of this material. The greater the number of responses, the more complete the data base will be for the information retrieval system.

I hope that you are willing to fill out copies of the questionnaires and return them to me. PRELIMINARY TESTS OF THE QUESTIONNAIRES INDICATE THAT THEY ARE VERY EASY TO COMPLETE. Your effort will facilitate the flow of new technology from the developer to the user, provide useful feedback to the developer, permit the engineer with a structural problem to locate the best available technology to solve it, and minimize duplication of effort in the field of computerized structural analysis. Respondents will receive the results of the survey. Thank you in advance for your cooperation.

Sincerely yours,

David Bushnell
Lockheed Missiles & Space Co., Inc.
Dept. 52-33/Bldg. 205
3251 Hanover Street
Palo Alto, California 94304
U. S. A.

Questionnaire for Computer Program Developers

The questionnaire for developers of structural analysis computer programs is given on the following pages. It consists of seven sections through which the characteristics of a particular computer program are specifically pinpointed and three "free form" sections for comments, references, and a list of users.

While the questionnaire for developers is fairly complete, it suffers from one major drawback: It inadequately reveals limitations of computer programs. For example, the developer may indicate that his program accounts for elastic-plastic effects and performs bifurcation buckling analysis. However, the elastic-plastic effects may be included only in stress problems but not in bifurcation buckling problems. The developer is not given the opportunity specifically to reveal this limitation. If he is the modest sort, he may do so in the "Comments" section, but most developers understandably are loath to provide information with a negative flavor.

Another justifiable criticism of the questionnaire is that some of the items in the first seven sections are apparently unclear. In particular, many developers had difficulty with item 1.06 "prismatic structure," and items 1.30 "transition from 3-dim. to 2-dim. region" and 1.31 "transition from 2-dim. to 1-dim. region." By prismatic structure is meant a body such as a corrugated sheet, the geometry of which is independent of an axial coordinate. Many developers think that frameworks or smooth shells modeled as assemblages of flat plates are prismatic structures. By items 1.30 and 1.31 are meant changes in dimensionality modeled by means of constraint conditions such as that specified in 1.28. Since many general purpose programs can handle beams or rods attached at single points to surface elements or solids, these items were often circled even though 1.28 was not. The items 2.33 and 2.34 were sometimes not understood, presumably because the developer had one of his associates fill out the questionnaire. Other items, such as 1.07 "segmented (in series)" and 1.08 "branched" are meaningful only for special purpose programs such as those restricted to analyses of shells of revolution. Item 2.32 "arbitrary variation" was frequently misunderstood to mean arbitrary variation within a single finite element, whereas no such meaning was intended.

On the whole, however, the responses received so far indicate that the questionnaire for program developers serves its intended purpose reasonably well. The author is indeed grateful to those program developers who responded to the rather lengthy questionnaire.

Questionnaire for Program Users

The questionnaire for program users as originally composed was much too long. An abbreviated version is given following the developer's questionnaire. The data from program users included in the data bank of COMSTAIRS corresponds to the items shown. Any further information from users will be obtained with a form similar to this one or directly from the user at an interactive terminal. Out of approximately 3000 user's questionnaires mailed, 276 responses had been received at the time of this writing. Of these, 28 correspond to programs for which no developer's questionnaire has been filled out. The author is very grateful to those users who responded to the lengthy questionnaire. It is hoped that more will respond to the abbreviated form.

Table 1 shows the names of programs for which there was information in the COMSTAIRS data bank as of March 15, 1974. The numbers of respondents to the user's questionnaire as of March 15, 1974,are also given. The differences in the numbers of user respondents from program to program do not necessarily indicate the relative popularity of the programs. For example, there are only about 100 users of BOSOR3 and BOSOR4, out of which 31 responded. On the other hand, there are many thousands of NASTRAN users, out of which only 6 responded. Programs for which the users and the developer are at the same institution or for which the users are at the institution which funded the development of the program are indicated by an asterisk so that the reader will be warned of possible bias in the response. The reader is also warned that the developers were sent copies of the questionnaire for users and asked to urge users of their programs to fill them out. It is possible, of course, that some developers "hand picked" users whom they knew would give favorable evaluations.

The Original Long User's Questionnaire

The original questionnaire for program users consisted of four pages with a total of 7 sections, including a section for function, motivation, and application, a section for operation of the program at the user's facility, an evaluation section, a section on prior experience and experience with the particular program in question, and three "open-ended" sections for comments, references, and lists of other users at the same institution as the respondent. Many of the items pertinent to operation of the program at the user's facility were bewildering to some users, who unfortunately were not told to skip over whatever mystified them. The section on prior experience and current experience with the program was far too long and in addition was heavily slanted toward thin shell analysis. One respondent wrote, justifiably so, "You must think we're all a bunch of aerospace engineers." Another commented, "Wow! This is hard to keep up with!" The author received some feedback to the effect that "Management won't let us fill out these questionnaires because it takes too much time from work." The author apologizes for unduly burdening respondents and potential respondents.

The original thought behind the long form was to try to establish some kind of weight to the user's response by determining the fit between the prior experience of the user and the use to which he put the program that he was evaluating. The response rather strongly indicates that users perceive themselves as being adequately prepared to use the programs as intended, even for programs which treat phenomena only very recently investigated by researchers.

Table 2 shows the average responses by users to five statements that appeared in the long questionnaire for program users. Only those programs are included in Table 2 for which at least ten indi-

DEVELOPERS

Page 1 of 5

QUESTIONNAIRE FOR DEVELOPERS OF STRUCTURAL ANALYSIS COMPUTER PROGRAMS

Directions: Fill out one questionnaire for each computer program that you have developed. Circle all appropriate items in each Section.

COMPUTER PROGRAM NAME____________________ DATE__________

DEVELOPER'S NAME_______________ AFFILIATION__________________

ADDRESS____
(Street, City, State, Zip) (Country)

TELEPHONE____________________(Area Code, Number, Extension)

Section 1
Geometry/Boundary Conditions

Geometry:
01 general structure
02 solid of revolution
03 general shell
04 part of shell of revolution ($<360^\circ$)
05 shell of revolution (360°)
06 prismatic structure
07 segmented (in series)
08 branched
09 shell with cutouts
10 thick shell
11 framework-shell combinations
12 flat plate of general shape
13 other geometry__________

Imperfections:
14 axisymmetric
15 general

Support Conditions:
16 axisymmetric
17 general
18 at boundaries only
19 at internal points

Deformation-dependent Support:
20 elastic foundation
21 stiffeners
22 contact
23 friction
24 sliding without friction
25 other__________

Other Constraint Conditions:
26 juncture compatibility
27 singularity conditions
28 general linear: $\overline{u}_i = [T]\,\overline{u}_n$ $(n \neq i)$
29 general nonlinear
30 transition from 3-dimensional region to 2-dimensional region
31 transition from 2-dimensional region to 1-dimensional region
32 other__________
33 Lagrange multiplier method used
34 other comments pertinent to this section (include in Section 8)

Section 2
Wall Construction/Material Properties

Wall Construction:
01 monocoque
02 layered
03 sandwich
04 composite material (explain, Section 8)

Stiffeners Treated as Discrete:
05 meridional
06 circumferential
07 general direction
08 stiffeners treated as smeared

Stiffener Theory:
09 general open-section theory
10 cg and shear center coincide
11 other wall construction (Section 8)

Material Properties:
12 isotropic
13 anisotropic
14 linear elastic
15 nonlinear elastic
16 temperature-dependent material
17 rigid-perfectly plastic
18 elastic-perfectly plastic
19 elastic-linear strain hardening
20 many straight line segments
21 Ramberg-Osgood stress-strain law
22 isotropic strain hardening
23 kinematic strain hardening
24 White-Besseling hardening law
25 other post-yield law__________
26 primary creep
27 secondary creep
28 viscoelastic (describe in Section 8)
29 strain-rate effects included

Spatial Variation of Properties:
30 no variation
31 axisymmetric variation only
32 arbitrary variation

Integrated Constitutive Law:
33 coupling between reference surface strains and curvature changes retained
34 coupling between normal and shear strains retained
35 other comments pertinent to this section (comment in Section 8)

DEVELOPERS

QUESTIONNAIRE FOR DEVELOPERS, page 2 of 5

Section 3: Loading

Spatial Variation of Loads:
01 uniform
02 axisymmetric variation only
03 general variation
Time-Variation of Loads:
04 static or quasi-static
05 general transient
06 periodic (dynamic)
07 impulsive
08 random
09 other________________
Combined Loads Varying During the Case:
10 proportionally varying
11 some varying, some constant
12 with different time histories
13 cyclic (quasi-static)
14 other________________
Deformation-dependent Loads:
15 live (following) loads
16 gyroscopic loads
17 inertial loads
18 electromagnetic loads
19 fluid-structure interaction
20 acoustic loading
21 soil-structure interaction
22 rigid solid-structure interaction
23 other________________
Types of Loads:
24 point loads, moments
25 line loads, moments
26 normal pressure
27 surface tractions
28 body forces
29 dead-weight loading
30 thermal loading
31 initial stress or strain
32 centrifugal loading
33 non-zero displacements imposed
34 other________________
35 other comments pertinent to this section (here and/or in Section 8)

Section 4: Phenomena

Stress Analysis:
01 axisymmetric small deflections
02 general small deflections
03 axisymmetric large deflections
04 general large deflections
05 axisymmetric plasticity
06 general plasticity
07 large strains
08 transverse shear deformations
09 thermal effects
10 radiation effects
11 automated yield criterion
12 automated fracture criterion
13 automated fatigue criterion
14 automated buckling criterion
15 other built-in criteria (Section 8)
Stability Analysis:
16 nonlinear collapse
17 post-buckling phenomena
Bifurcation (eigenvalue) Buckling:
18 linear axisymmetric prestress
19 linear general prestress
20 nonlinear axisymmetric prestress
21 nonlinear general prestress
22 including prebuckling rotations
23 including transverse shear deformations
24 other________________
Vibrations and Dynamic Response:
25 dynamic buckling
Modal Vibrations:
26 no prestress
27 linear axisymmetric prestress
28 linear general prestress
29 nonlinear axisymmetric prestress
30 nonlinear general prestress
31 including transverse shear deformations
32 other________________
33 nonlinear vibrations
Dynamic Response:
34 linear axisymmetric
35 linear general
36 nonlinear axisymmetric
37 nonlinear general
38 wave propagation
39 including transverse shear deformations
40 other________________
Damping:
41 viscous
42 structural
43 linear
44 nonlinear
45 other________________
46 flutter
47 other dynamic phenomena____________
48 other phenomena (comment in Section 8)

DEVELOPERS

QUESTIONNAIRE FOR DEVELOPERS, page 3 of 5

Section 5: Discretization

01 multisegment forward integration
02 Galerkin method
03 Rayleigh-Ritz method
04 boundary point matching
05 finite difference method based on equilibrium equation formulation
06 finite difference energy method
Finite-difference Discretization:
07 one-dimensional discretization
08 orthogonal 2-dimensional grid
09 non-orthogonal 2-dimensional grid
10 general quadrilateral 2-d grid
11 other discretization scheme (comment)
12 Fourier series in circumferential direction
13 maximum number of Fourier harmonics___
Finite Element Library: Give (corners/nodes/unknowns) per element

	Type Structure	Formulation: displacement	Formulation: hybrid
14	rod	(/ /)	(/ /)
15	straight beam	(/ /)	(/ /)
16	curved beam	(/ /)	(/ /)
17	conical shell	(/ /)	(/ /)
18	axisym. shell	(/ /)	(/ /)
19	flat membrane	(/ /)	(/ /)
20	flat membrane	(/ /)	(/ /)
21	curved memb.	(/ /)	(/ /)
22	curved memb.	(/ /)	(/ /)
23	flat plate	(/ /)	(/ /)
24	flat plate	(/ /)	(/ /)
25	shallow shell	(/ /)	(/ /)
26	shallow shell	(/ /)	(/ /)
27	deep shell	(/ /)	(/ /)
28	axisym. solid	(/ /)	(/ /)
29	axisym. solid	(/ /)	(/ /)
30	thick plate	(/ /)	(/ /)
31	thick shell	(/ /)	(/ /)
32	solid (3-D)	(/ /)	(/ /)
33	solid (3-D)	(/ /)	(/ /)
34	solid (3-D)	(/ /)	(/ /)
35	crack-tip	(/ /)	(/ /)
36	other___		

37 force method
38 incompatible displacement functions used
39 substructuring
40 repeated use of identical substructures
41 maximum number of nodal points___
42 maximum number of deg. of freedom___
43 maximum bandwidth of matrices___
44 nodes can be introduced in arbitrary order
45 automatic renumbering of nodes for optimum efficiency of solution
46 non-diagonal mass matrix
47 diagonal mass matrix
48 other comments pertinent to this section (comment in Section 8)

Section 6: Solution Methods

Linear Equation Solution:
01 full matrix
02 constant bandwidth
03 skyline method
04 partitioning
05 wave front method
06 other sparse matrix method (comment)
07 iterative method
08 conjugate gradient
09 forward integration
10 other
11 condensation (identify method)___

Eigenvalue Calculation Methods:
12 LR, QR methods___
13 Lanczos method
14 Sturm sequence method
15 determinant "plot"
Inverse Power Iteration Method:
16 single vector with spectral shift and deflation
17 multivector (subspace) iteration
18 other___

Nonlinear Equation Solving Strategies:
19 full Newton-Raphson method
20 modified Newton (comment)
Incremental Methods:
21 no equilibrium correction
22 with equilibrium correction
23 Runge-Kutta type
24 other___
25 direct energy search (identify method)___
26 dynamic relaxation
27 successive substitutions in load vector (nonlinear terms appear on right-hand side [load vector] only)
28 other nonlinear strategy (comment)
29 Lagrangian formulation
30 updated Lagrangian formulation
31 Eulerian formulation

Time Integration for Dynamic Response:
32 modal superposition
33 explicit method
34 implicit method
35 Euler (constant acceleration)
36 Newmark (Beta = ___)
37 Runge-Kutta of order___
38 Houbolt
39 Wilson
40 central difference
41 stiffly-stable (Gear, Jensen)
42 predictor-corrector of order___
43 multi-step
44 other___
45 other comments pertinent to this section (comment in Section 8)

DEVELOPERS

QUESTIONNAIRE FOR DEVELOPERS, page 4 of 5

Section 7: Computer Program Distribution/Documentation/Organization/Maintenance

Program Distribution:
01 source program tape available
02 source program cards available
03 object program or absolute element available but source program not
04 program completely proprietary
05 program price in dollars________

Program is available through:
06 developer
07 COSMIC
08 other software center (give name) ________

09 program can be executed through service bureaus (list in Section 10)
10 number of institutions using this program is about ______ (ATTACH LIST W/ USERS' NAMES & ADDRESSES, Sec. 10)
11 this program has been superseded by another (give name)________
12 an improved, extended version of this program is now being prepared and will be completed by ________

Documentation:
13 program is not completed
User's Manual(s):
14 are self-contained
15 have list of pitfalls
16 have test cases
17 have flow charts
18 give theory
19 other________

Program Organization:
20 program language is ________

Program Runs On: (Give Type)
21 IBM ( )
22 CDC ( )
23 UNIVAC ( )
24 GE ( )
25 other________
26 number of source statements________
27 number of subroutines________
28 number of primary overlays________
29 man-years required to develop________
30 core storage (decimal words) required for execution________
31 if dynamic allocation is used, give the minimum core space required________
32 program runs entirely in core

Program Requires Auxiliary Storage:
33 direct access disk or drum. Number of units required________
34 tape(s) required
35 low speed core or virtual memory required (how much?)________

General Program Capabilities:
36 restart capability
37 options for various output
38 direct access data base
39 free-field input
40 runs in interactive mode
41 automatic mesh generator(s)
42 automatic loading generator
43 section properties calculated

Interaction With Auxiliary Programs:
44 Pre-processor (describe)
45 Post-processor (describe)
46 eigenvalue solver
47 modal superposition routine
48 other time-integration routine
49 other capability (describe - Section 8)

Plot Routines For:
50 undeformed/deformed geometry
51 various views and sections
52 contour plots
53 ordinary $y = f(x)$ plots
54 stress-on-structure plots
55 other

Plotting Capability For:
56 CALCOMP
57 Stromberg-Carlson 4020/4060/FR80
58 other________

Debugging and Checkout Aids:
59 optional detailed print output
60 equilibrium checks
61 alternative branches for same problem
62 matrix conditioning estimation
63 automatic error control

Input Data Checking:
64 graphical preprocessor
65 applied loading checks
66 boundary condition checks
67 section properties checks
68 geometry checks
69 other checking aids (comment, Sec. 8)

Program Maintenance: For a reasonable fee the developer or other will help user:
70 get program running at user's facility
71 modeling of specific cases
72 find suspected bugs
73 circulate notices of bugs found to all users for whom he has addresses
74 set up and run cases
75 conduct workshops
76 make special program changes
77 keep users informed of improvements
78 other________
79 other comments pertinent to this section (comment in Section 8)

DEVELOPERS

QUESTIONNAIRE FOR DEVELOPERS, page 5 of 5

SECTION 8: ADDITIONAL COMMENTS (Continue on reverse side if necessary)

SECTION 9: REFERENCES (User's manual(s) and articles based on method and results. Continue on reverse side if necessary.)

SECTION 10: LIST OF USERS OF THIS COMPUTER PROGRAM (This section is especially important. Users will be sent questionnaires and their evaluations of your program will be sent to you. Please give complete mailing addresses. Your cooperation is appreciated. Continue on reverse side if necessary.)

Please return filled out form to:

DAVID BUSHNELL
LOCKHEED PALO ALTO RESEARCH
LABORATORY (O/52-33 - B/205)
3251 HANOVER STREET
PALO ALTO, CALIFORNIA 94304

USERS

QUESTIONNAIRE FOR USERS OF STRUCTURAL ANALYSIS COMPUTER PROGRAMS

COMPUTER PROGRAM NAME (use name given by developer)____________

USER'S NAME AND AFFILIATION (only if you wish to give it)____________

Section 1. Amount of Use/Hardware
(check space or fill in as appropriate)

01	Haven't used code enough to comment
02	No. of people using code in this dept.
03	No. of departments using this code
04	No. of years I have used this code
05	Code used intensively all the time
06	Code receives intermittently intensive use
07	Code is used daily
08	Code is used about once a week
09	Code is used about once a month
10	Code is rarely used by us
11	We use software service bureau for execution of program (identify)
12	Code runs here on IBM____
13	Code runs here on CDC____
14	Code runs here on UNIVAC____
15	Code runs here on GE____
16	Code runs here on other____
17	Plot output with CALCOMP____
18	Plot output with SC____

Section 2. Evaluation

Indicate your opinion of the following statements by entering the appropriate numeral in the space provided:
0 = not applicable or insufficient data
1 = strongly agree 4 = disagree
2 = agree 5 = strongly disagree
3 = neutral

19	User has confidence in this program
20	Agreement with test results is good
21	Easy to learn how to use this program
22	Easy to use this program once learned
23	User's manual(s) are self-contained
24	Manual(s) have good list of pitfalls
25	Manual(s) have enough test cases
26	Manual(s) are well organized
27	Input data are easy to prepare
28	Output is easy to understand
29	Program has good plotting capability
30	Program was easy to get running here
31	Developer keeps us up to date
32	Developer helps find bugs in program
33	Good communication between user and developer
34	User has encountered very few bugs

Section 3. What This Code Has Been Used For

Indicate the extent of your experience with this code by writing the appropriate integer in the space provided:
0 = none
1 = limited
2 = moderate
3 = extensive
4 = not applicable

35	Preliminary design
36	Final design evaluation
37	General structures
38	Solid of revolution
39	General shell
40	Part of axisymmetric shell (less than 360 degrees)
41	Axisymmetric shell (360 degrees)
42	Branched shells
43	Shells with cutouts
44	Thick shells
45	Framework-shell combinations
46	Flat plate of general shape
47	Layered wall construction
48	Composite material
49	Shells with stiffeners
50	Elastic-plastic material
51	Creep
52	Static or quasi-static loads
53	General transient loads
54	Periodic dynamic loads
55	Linear stress analysis
56	Nonlinear stress analysis
57	Nonlinear collapse
58	Bifurcation (eigenvalue) buckling
59	Modal (eigenvalue) vibrations
60	Linear dynamic response
61	Nonlinear dynamic response

Section 4. Comments (continue on reverse)

Section 5. References (continue on reverse)

Table 1 Computer Programs Represented in COMSTAIRS Data Bank as of March 15, 1974

More than 10 Users Responded		Between 5 and 10 Users Responded		Between 1 and 5 Users Responded		No Users Responded
Program	Users	Program	Users	Program	Users	
ANSYS	11	NASTRAN	6	ADEPT	1*	AC-3
ASKA	13	SAMIS	6	ARCHDAM2	2*	AC-50A
BOSOR4	31	TEXGAP	8	ASAS	1*	ARCHDAM1
ELAS75	12			ASOP	1	ASEF
SAPIV	32			BERSAFE	1	BOLS
STAGS	13			DYNAL	3	BOND4
STARDYNE	19			DYNAPLASII	4	BOPSII
STRUDLII	31			DYNASORII	1	CABLE3
				DYNAS	1*	COBA
				EASE	3	CYDYN
				FARSS	4*	DANUTA
				FESAP	4*	EASE2
				ISOPAR-SHL	1*	ELAS55
				KSHEL	1	ELAS65
				MARC	3	FAMSOR
				MATCOMP	1*	HONDO
				NEPSAP	1*	ISTRAN/S
				PCALS	2*	MINIELAS
				REXBAT	4*	NONLIN
				SABOR/ DRASTIC3A	2	NONLIN2
				SALORS	2	NOSTRA
				SATANS	1	PAFEC70
				SATURN	1*	PCAP
				SHORE	4*	PDIL
				SNAP	3*	PETROS3
				SNASORII	1	PILEFEM
				SOLIDSAP	4	REGULA
				SRA	1*	REPSIL
				STARS	2*	SABOR/ DRASTIC6
				STMFR-60	1*	SADAOS
				STRESS	3	SAMBA
				SWBFR	1*	SAMMSORIII
				VISCEL	1	SESAM69C
						SINGUL
						SLADE
						SLADE-D
						SPADAS
						STRESSIII
						TABLES
						TANKER
						TIRE
						TRANSTOWER
						TROCS
						ZP26

*Users responding so far are <u>all</u> at same institution as developer or at an institution which funded the program development.

Table 2 How Adequately Do Users Feel That They Are Prepared to Use These Programs?

Users were asked to indicate their agreement or disagreement with the statements below.
1=strongly agree 4=disagree
2=agree 5=strongly disagree
3=neutral

	STRUDLII	ELAS75	ANSYS	SAPIV	STARDYNE	ASKA	BOSOR4	STAGS
Approximate Number of Responses to Statement	28	10	10	24	15	13	24	11
S T A T E M E N T S								
1. This program treats phenomena that are adequately understood by engineers/designers.	2.0	2.3	2.5	2.5	2.6	2.7	2.8	3.6
2. This program treats phenomena that are adequately understood by researchers.	2.0	2.0	1.9	1.9	1.9	2.1	2.1	2.2
3. This program treats phenomena that are adequately taught in universities.	1.9	2.1	2.4	2.1	2.1	2.1	2.8	3.4
4. User's ability to use this program effectively has been enhanced by prior formal education or experience on the job concerning phenomena treated by this program.	1.9	1.7	1.7	1.5	1.5	2.6	1.9	2.2
5. User's understanding of the phenomena treated by this program results primarily from experience with this program and not from prior education or job experience.	3.3	3.4	3.8	4.0	3.7	3.2	3.7	3.1

viduals responded. Those who provided the data in Table 2 represent a variety of institutions and occupations, as shown in Table 3.

Table 3 Program User Institutions and Occupations

No. of Users	Type of Institution	No. of Users	Occupation
112	Private Firm	104	Engineering/ Design
17	University	37	Research
13	Government Agency	20	Supervision/ Management
10	Other	8	Teaching

In Table 2 the eight programs are arranged from left to right according to the users' perceptions of the adequacy of understanding by engineers and designers of phenomena treated by these programs. For example, users of STRUDLII, which is applied most frequently to the linear analysis of frameworks, agree on the average that the phenomena treated by STRUDLII are adequately understood by engineers and designers. Users of STAGS, which is applied mostly to the nonlinear static analysis of thin shells, disagree that phenomena treated by STAGS are well understood by engineers and designers.

Evaluations of programs by users are likely to be related to the degree of understanding of the phenomena treated by these programs. Hence, a person trying to find out which programs to acquire through COMSTAIRS should not decide against a program just because users found it difficult to provide proper input or difficult to understand the output. The important thing is that the program cover the critical aspects of the problem which must be solved. Programs for linear analysis generally received more favorable evaluations than those for nonlinear analysis. It would be absurd, of course, to avoid nonlinear analyzers on this ground alone.

Similarly, some programs for the analysis of nonlinear problems are easier to use than others because the method of solution, perhaps grossly inadequate for certain kinds of problems, is not susceptible to "numerical failure." The use of incremental analysis for the solution of problems involving the large-deflection collapse of thin shells is an example. An incremental analysis will always give <u>some</u> answer, which is comforting to a user faced with an imminent deadline. If the user does not perform a convergence study of resolving the problem with smaller and smaller load steps, however, he will never know how good or bad the answers are, and he will be left with the impression that a program which generated the possibly erroneous data is very easy to apply. On the other hand, a program in which an iterative method is used for the solution of nonlinear equations may fail to yield a solution at some load because of failure of the iterations to converge. This failure to converge may be due to a bug in the program, a bug in the input data, roundoff error, use of an initial load which cannot be carried by the structure, use of a load increment which is too large given the degree of nonlinearity exhibited by the structure, or increase of load above the maximum value for which a solution exists. In any of these cases, the failure to converge is naturally somewhat disturbing to the user. He is perhaps left with the impression that the program is hard to use, the output difficult to understand. In spite of these difficulties, however, the program may be the best one for solution of the particular problem being treated.

On the user's long questionnaire the respondent was asked to identify himself. While he could have remained anonymous had he wanted to, he was not explicitly advised of this possibility and in fact all response from users received by the author as of March 15, 1974, contained the name of the user or at least of the using institution. The fact that most of the evaluative response was of a generally positive nature might be related to this.

Additional Results of the Survey

Tables 4-6 give a summary of the response from users of the eight programs for which more than ten questionnaires had been returned by March 15, 1974. The three tables correspond to the first three sections of the new questionnaire for program users reproduced in this paper.

Table 4 indicates the number of respondents, the intensity of use of the programs, and what kinds of hardware are employed. Table 5 gives the "average" evaluations of these programs. Table 6 indicates what types of problems have been solved by the users of these programs.

As a whole, the evaluative response from users was positive, as seen in Table 5. Presumably, if a person is continuing to use a program, he likes it reasonably well, since there are many competing programs in each subdiscipline of structural mechanics. If in the past the user tried a program and, not having confidence in it, dropped it in favor of another, he would be unlikely to fill out a questionnaire on the first program. Thus, the survey is undoubtedly biased in favor of the programs since it represents current more or less satisfied users rather than past disgruntled ones. For more information of a negative nature use the "EVALUATE" option of the COMSTAIRS system and look for user comments. (See Appendix, for example.)

Some doubtful areas. Taking averages, as was done to generate the data in Table 5, is somewhat questionable. The programs are used more intensively at some institutions than at others, but each response was given equal weight in the average. Some of the items on which the user bases his evaluation may be of questionable value. For example, the second item, "good agreement with tests," may reflect the quality of the tests and the ability of the user to set up a

Table 4 Amount of Use of Program and Type of Hardware

ITEM	ANSYS	ASKA	BOSOR4	ELAS75	SAPIV	STAGS	STARDYNE	STRUDLII
total number of responses	11	13	31	12	32	13	19	31
no. different institutions	9	7	25	11	26	6	18	23
average number of users in this department	5.2	8.0	2.5	3.1	7.0	6.8	3.9	3.2
average number of departments using program	2.6	1.5	1.5	2.2	2.4	1.6	1.9	2.3
average number of years of use	2.7	2.7	1.4	1.8	1.6	1.6	2.7	2.8
	The quantities below give the number of respondents indicating the items at the left.							
intensive use of program	0	4	1	0	6	2	2	2
intermittently intensive use	7	6	15	5	6	8	10	16
daily use	1	2	0	2	10	4	2	1
used once/ week	2	2	4	4	3	2	5	2
used once/ month	0	0	3	1	1	1	2	4
used rarely	2	0	4	2	2	0	1	5
execution through software center	3	0	2	0	2	1	5	15
runs on IBM	1	9	9	5	7	0	0	15
runs on CDC	4	6	10	3	13	7	18	1
runs on UNIVAC	3	1	7	1	5	8	0	0
runs on OTHER	0	0	0	1	6	0	0	0
CALCOMP plotter	9	7	3	3	11	2	15	12
Stromberg-Carlson	0	1	5	0	1	3	9	0

Table 5 Evaluation of Computer Programs with Most Response from Users

Users were asked to indicate their agreement or disagreement with the statements below... 1=strongly agree 2=agree 3=neutral 4=disagree 5=strongly disagree	ANSYS	ASKA	BOSOR4	ELAS75	SAPIV	STAGS	STARDYNE	STRUDLII
	Average response to statement on left							
user has confidence in program	1.6	1.8	1.7	2.0	1.5	2.1	1.4	1.6
good agreement with tests	1.7	1.9	1.7	2.4	1.7	1.9	1.6	2.1
easy to learn how to use program	2.7	3.6	3.0	2.3	2.3	2.6	1.8	2.1
easy to use program once it is learned	1.8	2.1	2.2	1.7	1.6	2.1	1.6	1.7
user's manual is self-contained	1.8	3.1	2.1	2.9	2.6	2.8	1.7	2.6
good list of pitfalls	3.2	3.3	2.4	3.3	3.4	3.2	2.6	3.1
enough test cases	2.9	3.1	2.0	2.7	3.1	2.7	2.9	3.5
manual well organized	2.9	3.5	2.8	2.6	2.3	2.3	1.8	2.5
input data easy to prepare	2.5	2.4	2.7	2.3	2.3	1.7	1.9	1.8
output easy to understand	2.3	2.1	2.3	2.9	2.4	2.2	1.6	1.8
good plotting capability	1.5	4.0	2.8	2.0	3.8	4.2	2.2	2.2
easy to get program running here	2.4	2.8	2.4	2.4	2.4	3.3	1.8	2.4
developer keeps us up-to-date	1.9	2.1	1.8	1.7	2.4	2.2	1.7	2.3
developer helps find bugs in code	1.7	1.8	1.9	2.0	3.2	1.9	1.5	2.4
good communication with developer	1.4	2.1	2.0	2.1	2.7	1.3	1.7	2.1
user has encountered very few bugs in code	2.6	2.0	2.0	2.3	1.9	3.8	1.8	2.7

Table 6 Purpose of Use of Computer Codes with Most Response from Users

What Program Was Used For User was asked to indicate the extent of his experience with the program by means of an appropriate integer... 0 = none 1 = limited 2 = moderate 3 = extensive for each of the items below.	ANSYS	ASKA	BOSOR4	ELAS75	SAPIV	STAGS	STARDYNE	STRUDLII
	PROGRAM NAME — Percentage of Users Indicating Moderate to Extensive Use of Code							
preliminary design	64	46	40	45	46	69	67	59
final design evaluation	64	77	44	45	50	31	72	76
general structures	55	85	12	55	69	38	83	79
solid of revolution	55	54	12	18	35	8	11	3
general shell	55	46	28	45	38	69	50	14
part of shell of rev. (<360°)	36	31	8	9	4	77	6	7
shell of revolution	55	46	68	18	19	23	22	0
branched shells	18	46	48	0	8	31	22	7
shells with cutouts	36	62	0	0	19	31	28	3
thick shells	45	46	0	18	15	0	17	0
framework-shell combinations	18	46	8	0	27	23	72	10
flat plate of general shape	27	77	0	36	46	23	56	21
layered wall construction	0	31	24	0	8	31	17	0
composite material	0	15	16	36	4	0	11	3
shells with stiffeners	9	62	44	9	27	85	50	17
elastic-plastic material	27	15	4	0	0	15	0	0
creep	18	0	0	0	0	0	0	0
static or quasi-static loads	64	92	60	55	69	85	72	76
general transient loads	45	15	4	0	38	0	50	17
periodic dynamic loads	45	31	0	0	31	0	56	14
linear stress analysis	73	92	68	64	69	85	83	72
nonlinear stress analysis	9	15	56	0	4	69	0	10
nonlinear collapse	18	0	24	0	0	61	6	0
bifurcation(eigenv.)buckling	0	0	72	0	0	77	6	0
modal(eigenvalue)vibrations	45	38	28	0	65	0	61	3
linear dynamic response	45	31	4	0	38	8	50	7
nonlinear dynamic response	36	0	0	0	0	8	0	0

proper model, as well as the adequacy of the computer program. It is well known, for instance, that predictions of buckling loads for externally pressurized spherical shells and axially compressed cylindrical shells are often high by a factor of four or five because of small imperfections in the test specimens. This discrepancy should not be cause for throwing out a program based on classical buckling analysis. The item "good plotting capability" is questionable because respondents often wrote "5," meaning that the plot routines are not operational at their facility, not that the wrong data are plotted or that the plots are otherwise deficient in some way. The item "easy to get program running here" is questionable because the person responsible for the conversion of a code to a particular facility is generally not the respondent. "User has encountered very few bugs" should be interpreted with care. What is really significant is the current rate of "bug encounters," not the total number of bugs found in the past. Some users obtain a program before it is really checked out, while it is still in its trial stage. Others obtain it after it has been widely used for several years. The reader needs more information than this brief summary of the survey gives.

Table 6 shows what the eight programs, for which the most response from users was received, were used for. Often users indicated that they had used a program in a way that, according to the developer, is beyond the scope of that program. For example, as the author of BOSOR4, a special purpose program for the elastic analysis of thin shells of revolution, I find it interesting that 12 percent of the users treated to a moderate or extensive degree general structures and solids of revolution with it. One respondent indicated that he had used BOSOR4 extensively for the analysis of solids of revolution. In the evaluation section he responded, understandably, that he did not like the program.

THE COMSTAIRS SYSTEM

The data bank accumulated by means of the questionnaire for program developers and the questionnaire for program users is the foundation of COMSTAIRS. Sample input and output for this system are given in the Appendix. The user-supplied input is indicated by boldface italic print.

The system is limited as of March 1974 because there are many fairly widely used programs for which no developer's questionnaire has been filled out and the response from users is rather small, except for the eight programs previously noted.

The COMSTAIRS system is interactive. The user's manual is built into it, so that all the user has to know is how to "log on," what the file name is, what command is required to begin execution of COMSTAIRS, and how to terminate the run. The user is asked to supply some feedback on the COMSTAIRS system and on programs that he has used or written. Although he can use COMSTAIRS without having to supply any such information, he is urged to do so. It appears that some dynamic information exchange system is needed by the community of structural analysts. The more feedback obtained from this community, the better designed will be the system to meet its needs.

Survey of Program Capabilities

Many of the pages of the Appendix contain lists of the capabilities of various programs represented in the data bank. Three sets of programs are listed separately, corresponding to (1) general purpose programs, (2) programs for the analysis of solids of revolution, and (3) programs for the analysis of shells. These data were obtained from the filled-out questionnaires for program developers. Some program developers were more modest than others. For example, the developer of SNAP checked off only "general structures" under geometry, realizing that the various classes of structures listed therein fall into this class. Most other developers of general purpose programs circled all the items under geometry and even felt it necessary to write "any" or "you choose" under 1.13 "other." All other things being equal, I would tend to choose the program corresponding to the more modest response. Such a developer is less likely to make extravagant claims as to his program's capability.

In order to judge the adequacy of a code for a given purpose, the user of COMSTAIRS must look at all of the sections of the response. For example, many of the programs will presumably perform nonlinear collapse analysis of thin shells. However, several of these programs are based on the incremental method, with and without equilibrium correction. Incremental methods are often not suitable for nonlinear thin shell analysis, as pointed out in another paper in this volume [12].

Six Basic Commands

There are six basic commands that the COMSTAIRS user can provide. These commands with definitions are listed on the first page of the Appendix. Although any sequence is permitted, they are given in the order that I expect most engineers would naturally use COMSTAIRS. An engineer would first identify a problem class by means of certain key words. COMSTAIRS will tell him which programs will supposedly solve this class of problems. Then he will ask COMSTAIRS to compare the programs as to their capabilities. After perusing this comparison, the engineer will ask for user evaluations of certain of the programs. Narrowing down his search still further, he will ask for the scope of a few of the codes. Certain items, such as the finite element library and developer comments, are given with use of the "SCOPE" option but not with use of the "COMPARE" option.

User-provided Feedback

After the engineer has finished his search, he can do it again with another problem, or he can punch "FINISHED," whereupon COMSTAIRS will ask him if he wants to evaluate 11 simple statements about itself. He will then be asked to add any comments he wishes. The user's input will be stored on a permanent file, to be edited and perhaps (if pertinent) added to the main COMSTAIRS data bank at a later date. I urge all users to respond to this request for information. The comments will not be associated with your name, so you can be frank. You are especially encouraged to supply information on programs you have used or developed or problems for which you need solutions but cannot find adequate programs.

CONCLUSIONS AND RECOMMENDATIONS

COMSTAIRS should be regarded as a pilot effort only. The author worked on it essentially alone (see acknowledgments, however) for a couple of months while the questionnaires were being designed and again for a couple of months while the information retrieval system was being set up.

The ultimate goal is to facilitate information exchange with regard to computer programs covering a broad spectrum of disciplines. A continued, perhaps expanded, effort is a good investment because it will lessen duplication of work, identify areas in which more research is needed, and help those who need computerized analysis to choose the proper programs. The following tasks are suggested:

1. Identify which disciplines are involved. Some examples are:
    a. Structures and materials
    b. Heat transfer and thermodynamics
    c. Fluids and hydraulics
    d. Control systems and mechanical engineering system
    e. Interdisciplinary fields
2. Identify experts in each major discipline and in each subdiscipline. Ask them to participate in this effort, explaining its purpose.
3. Set up a format for questionnaires through which the data for the information retrieval systems will be accumulated. Have this proposed format checked by experts identified in Task 2 and by those who will be involved with the actual programming of the information retrieval systems. Hopefully, it will be possible to follow the same format for all disciplines. Make sure to draft separate questionnaires for developers of programs and users of programs. It would be beneficial if the user's questionnaire was also appropriate for those who need programs but do not currently have them. Developers should be asked to answer very specific questions requiring a high degree of specific knowledge. Users should not be asked to answer such questions.
4. Have each expert design the specific questions to be included which cover his field of expertise. Allow him to choose "subexperts" for designing questions in pertinent "subareas" within a questionnaire. The designs of the questionnaires are crucial. They should be drawn up so as not to become obsolete, since the same questionnaires should be used from year to year as the information retrieval systems are kept up-to-date. Psychologists and statisticans should be consulted to check for inherent bias, possibilities of statistical correlation, etc. Computer-oriented committees of professional societies should be consulted for review of the questionnaires. Preliminary versions of the questionnaires should be sent out to a limited number of respondents to check for bugs, inconsistencies, ambiguities, and suggestions from others who did not participate in the

drafting. Preliminary questionnaires should also be checked carefully by those who are going to set up the information retrieval systems. It is important that the form and content be such as to ensure easy transfer of data from the filled-out questionnaires to the data banks of the information retrieval systems.

5. Distribute the final questionnaires:

a. Identify those to whom questionnaires should be sent

b. Look into the possibility of having professional societies or others do the distribution

c. Publicize the questionnaires at professional meetings and in professional publications

d. Get software service bureaus to help with distribution

e. Send follow-up notes urging respondents to reply

6. Set up the information retrieval systems:

a. Design the information retrieval systems. What information do you want?

b. Decide which system or systems the IRS will be set up on

c. Perform the necessary programming

d. Allow developers of programs to respond to comments by users of those programs. Edit in an appropriate manner all comments. Decide how much data to include on which programs

e. Feed the resulting data into the data banks for the information retrieval systems

f. Test and allow others to test the systems by trying to retrieve different types of information

g. Write reports on the designs of the systems giving examples of how they work and user's manuals. Distribute these reports to respondents

7. Keep the information retrieval systems up-to-date:

a. Send out the questionnaires described above periodically or otherwise provide for continuous or periodic updating by respondents. Perhaps respondents can put new data on programs into "waiting" banks. These data would then periodically be reviewed. Developers would be permitted to review and rebut new comments by users, and in fact would be permitted to review the entire portions of the data banks relevant to the programs they developed. New data would be edited in an appropriate way for final incorporation into the up-to-date data banks. Obsolete data would be dropped

b. Issue periodic reports and distribute them to respondents, publish them at technical meetings or in progessional journals

ACKNOWLEDGMENTS

Many individuals were of indispensable help during this effort. First and foremost I would like to express my appreciation to Professor Walter D. Pilkey at the University of Virginia for distributing the questionnaires and for continually during the past year feeding me as well as others information on the use and development of computer programs for structural analysis.

I would also like to thank my colleagues at Lockheed, in particular Jorgen Skogh, Lee Coven, Don Whetstone, and Parvis Sharifi, for willingly acting as guinea pigs during the development of the questionnaires. They offered many good suggestions which were incorporated into the final versions. As far as making COMSTAIRS operational on an interactive terminal, Bill Loden, Rose Cervo, and Danielle Colvin at Lockheed were invaluable. I learned more about control cards in a few short weeks from them than I have ever known. Their patience with a "systems-lubber" is commendable. Thanks, also, to Harry Schaeffer, who put COMSTAIRS "up" on the UNIVAC 1108 at the University of Maryland.

This work was performed with the financial assistance of the Lockheed Independent Research Program.

REFERENCES

1 Scharpf, B., Schrem, E., and Stein, "Einige Allgemeine Programmsysteme fur Finite Elemente," Finite Elemente in der Statik, Wilhelm Ernst and Sohne, 1973.

2 McCormick, J.M., Baron, M.L., and Perrone, N., "The STORE Project (The Structures Oriented Exchange)," Numerical and Computer Methods in Structural Mechanics," edited by Fenves, Perrone, Robinson, and Schnobrich, Academic Press, Inc., New York and London, 1973, pp 439-458.

3 Marcal, P.V., "Survey of General Purpose Programs for Finite Element Analysis,"Advances in Computational Methods in Structural Mechanics and Design, edited by Oden, J.T., Clough, R.W. and Yamamoto, Y., published by University of Alabama, Huntsville, Ala., 1972, pp 517-528.

4 Anonymous, CASDAC--Abstracts of Computer Programs, NAVSHIPS 0900-009-7012, Naval Ship Engineering Center, Department of the Navy, Washington, D.C., 1972.

5 Potts, J., Poyner, V., Roth, P., Siegrist, F. and Meyer, P., "Compendium of Computer Programs Used by the Structures Department," Naval Ship Research and Development Center, Bethesda, Md., April 1973.

6 Hartung, R.F., "An Assessment of Current Capability for Computer Analysis of Shell Structures," AFFDL-TR-71-54, April 1971, Air Force Flight Dynamics Laboratory, Wright-Patterson AFB, Ohio. See also Computers & Structures, Vol. 1, 1971, pp 3-32.

7 NASTRAN Users' Colloquium, Langley Research Center, Hampton, Va., September, 1971.

8 Program Verification and Qualification Subcommittee, a subcommittee of the Committee on Computer Technology, Pressure Vessel and Piping Division of the ASME, Subcommittee charter dated Jan. 1972. PROGRAM VERIFICATION AND QUALIFICATION PROBLEM LIBRARY, "Example Problem on the Dynamic Model of a Triangular Wing," Test problem posed by James C. Robinson of NASA Langley Research Center at the Structural Mechanics Seminar held at the Westinghouse Tele-Computer Center, Nov. 18-19, 1972.

9 Fenves, Steven J., Perrone, Nicholas, Robinson, Arthur R., and Schnobrich, William C., editors, Numerical and Computer Methods in Structural Mechanics, Academic Press, New York and London, 1973, pp 123-263.

10 Kraus, Harry, "Attitudes toward Computer Software and Its Exchange in the Pressure Vessel Industry," unnumbered RPI report out-

lining work supported by the National Science Foundation Grant GK-36084, Oct. 1973.

11 Medearis, Kenneth, "An Investigation of the Feasibility of Establishing a National Civil Engineering Software Center," American Society of Civil Engineers, New York, April 1973.

12 Bushnell, David, "Thin Shells," paper in this volume.

APPENDIX

Output from the COMSTAIRS Systems Including Comparisons of Programs for General Structures, Solids of Revolution, and Thin Shells and User Evaluations of STRUDLII, STARDYNE, SAPIV, and BOSOR4

```
@XQT DB*COMSTAIRS.GO

                         WELCOME TO

                         'COMSTAIRS'

COMPUTERIZED STRUCTURAL ANALYSIS INFORMATION RETRIEVAL SYSTEM
---          --         -        -           -         -


                            BY

                       DAVID BUSHNELL

   LOCKHEED PALO ALTO RESEARCH LABORATORIES, CALIFORNIA
   DEPT. 5233, BLDG. 205, 3251 HANOVER ST.(415)-493-4411


THERE ARE 6 BASIC COMMANDS THAT YOU CAN GIVE..

'PROBLEM'
'COMPARE'
'EVALUATE'
'SCOPE'
'ALL'
'FINISHED'

DO YOU NEED DEFINITIONS OF THE COMMANDS
'PROBLEM', 'COMPARE', 'EVALUATE', 'SCOPE', 'ALL', 'FINISHED'?

  PUNCH  'YES'   IF YOU DO.
  PUNCH  'NO '   IF YOU DON'T.
YES

NEVER PUNCH THE QUOTATION MARKS.

IF YOU PUNCH 'PROBLEM'  THE NAMES OF PROGRAMS WILL BE LISTED
                        WHICH SOLVE THE CLASS OF PROBLEMS
                        THAT YOU IDENTIFY BY MEANS OF
                        CERTAIN KEY WORDS AND PHRASES

IF YOU PUNCH 'COMPARE'  THE CAPABILITIES WILL BE GIVEN OF
                        WHATEVER PROGRAMS YOU SUBSEQUENTLY
                        IDENTIFY BY NAME.

IF YOU PUNCH 'EVALUATE' THE EVALUATIONS AND COMMENTS OF USERS
                        OF A PARTICULAR PROGRAM WILL BE GIVEN
                        FOR THE PROGRAM SUBSEQUENTLY SPECI-
                        FIED BY YOU.

IF YOU PUNCH 'SCOPE'    THE CAPABILITIES WILL BE GIVEN OF A
                        PARTICULAR PROGRAM TO BE IDENTIFIED
                        BY YOU.

IF YOU PUNCH 'ALL'      THE NAMES OF ALL PROGRAMS IN THE DATA
                        BANK WILL BE LISTED. (DON'T PUNCH THE
                        QUOTES HERE OR ANYWHERE).

IF YOU PUNCH 'FINISHED' YOU WILL BE ASKED TO PROVIDE  RESPONSES
                        TO A FEW SIMPLE QUESTIONS AND BE
                        GIVEN A CHANCE TO ADD YOUR OWN
                        COMMENTS INTO A DATA BANK.

--------------------------------------------------------
```

NOW THE COMPLETE DATA BANK WILL BE READ IN. THIS PROCESS TAKES SOME TIME (ABOUT 5 SECONDS). DON'T WORRY, THE MACHINE WILL TELL YOU WHEN IT NEEDS INPUT FROM YOU. IN FACT, NEVER WORRY ABOUT SUPPLYING INPUT DATA---YOU WILL ALWAYS BE GIVEN APPROPRIATE INSTRUCTIONS JUST BEFORE YOU ARE EXPECTED TO PROVIDE INPUT. IGNORE ANY MYSTERIOUS OUTPUT THAT YOU DO NOT UNDERSTAND. THIS IS LIKELY TO BE ONE OR MORE OF THE MANY CONTROL CARDS RE- QUIRED TO KEEP THINGS MOVING. JUST RELAX AND HAVE FUN.....

NOW PLEASE WAIT 5 SECONDS FOR THE DATA BANK TO BE LOADED. THE NEXT INSTRUCTION YOU WILL RESPOND TO BEGINS WITH

'NOW PUNCH EITHER.....'

PLEASE DON'T TOUCH THE KEYBOARD UNTIL YOU ARE ASKED SPECIFICALLY TO GIVE ONE OF THE 6 BASIC COMMANDS

```
NOW PUNCH EITHER   'ALL' ,   'SCOPE' ,   'COMPARE' ,
                   'EVALUATE' ,  'PROBLEM' ,  OR   'FINISHED'
```

PROBLEM

```
NOW THE NAMES OF PROGRAMS WILL BE LISTED WHICH
SOLVE A CLASS OF PROBLEMS IDENTIFIED BY CERTAIN
KEY WORDS AND PHRASES TO BE PROVIDED BY YOU.
EACH KEY WORD OR PHRASE PROVIDED BY YOU MUST
MATCH EXACTLY ONE OF THE KEY WORDS OR PHRASES
PROGRAMMED INTO THE COMSTAIRS SYSTEM. AS OF THIS
DATE THERE ARE 59 KEY WORDS AND PHRASES IN THE
COMSTAIRS SYSTEM.          ABOUT A PAGE OF OUTPUT IS
REQUIRED TO LIST THEM.

IF YOU WISH TO HAVE THE PROPER KEY WORDS AND
PHRASES LISTED NOW, PUNCH  'LIST WORDS' .

IF YOU DO NOT WANT THEM LISTED, PUNCH 'NO LIST' .
```

LIST WORDS

```
   GEOMETRY AND          WALL CONSTRUCTION AND
BOUNDARY CONDITIONS       MATERIAL PROPERTIES
-------------------     ----------------------

GENERAL STRUCTURE          MONOCOQUE
SOLID OF REVOLUTION        LAYERED
GENERAL SHELL              COMPOSITE MATERIAL
SHELL OF REVOLUTION        DISCRETE RINGS
PART OF SHELL OF REV.      STIFFENED
THICK SHELL                DISCRETE STRINGERS
FRAMEWORK-SHELL            ISOTROPIC
FLAT PLATE                 ORTHOTROPIC
IMPERFECTIONS              ANISOTROPIC
ELASTIC FOUNDATION         STRAIN RATE EFFECTS
CONTACT                    NONLINEAR ELASTIC
FRICTION                   TEMPERATURE-DEPENDENT MATERIAL
SLIDING WITHOUT FRICTION   ELASTIC-PLASTIC
                           CREEP
-------------------------------------------------


     LOADING               PHENOMENA
     -------               ---------

STATIC LOADS               SMALL DEFLECTIONS
QUASI-STATIC LOADS         LARGE DEFLECTIONS
GENERAL TRANSIENT LOADS    LARGE STRAINS
PERIODIC LOADS             TRANSVERSE SHEAR DEFORMATION
IMPULSIVE LOADS            NONLINEAR COLLAPSE
RANDOM LOADS               BIFURCATION BUCKLING
COMBINED LOADS             MODAL VIBRATIONS
CYCLIC STATIC LOADS        LINEAR DYNAMIC RESPONSE
FOLLOWING LOADS            NONLINEAR DYNAMIC RESPONSE
FLUID-STRUCTURE            DAMPING
SOIL-STRUCTURE             FLUTTER
SOLID-STRUCTURE            FRACTURE
BODY FORCES                FATIGUE
THERMAL STRESS             INITIAL YIELD

INITIAL STRESS
INITIAL STRAIN
CENTRIFUGAL LOADING
PRESCRIBED DISPLACEMENT
-------------------------------------------------


NOW PUNCH A SERIES OF KEY WORDS OR PHRASES.
THE SERIES MUST TERMINATE WITH THE WORD 'END'.

AN EXAMPLE IS THE SERIES OF KEY CONCEPTS--

              GENERAL SHELL
              ELASTIC-PLASTIC
              LARGE DEFLECTIONS
              GENERAL TRANSIENT LOADS
              END

A LIST OF PROGRAM NAMES AND DEVELOPERS WILL BE
PROVIDED. THE PROGRAMS IN THIS LIST ARE SUPPOSEDLY
CAPABLE OF SOLVING ALL PROBLEMS WHICH FALL INTO
THE CLASS PRESCRIBED BY THE KEY CONCEPTS PROVIDED
IN THE ABOVE EXAMPLE.  FOR MORE DETAILS ON THESE
PROGRAMS THE USER CAN NOW USE ONE OR MORE OF THE
OTHER OPTIONS OF THE COMSTAIRS SYSTEM.
```

```
NOW PUNCH IN KEY CONCEPTS, ONE CONCEPT TO A LINE

        GENERAL STRUCTURE
        END

THE FOLLOWING PROGRAMS WILL PERFORM THE ABOVE-SPECIFIED ANALYSIS

ANSYS, SWANSON ANALYSIS SYSTEMS, 870 PINEVIEW DR.,ELIZABETH, PA
ASAS,ATKINS RES. AND DEV., ASHLEY ROAD, EPSOM, SURREY, ENGLAND
ASEF,     LAB. TECH.AEROSPATIALES, LIEGE, BELGIUM
ASKA,     ASKA-GROUP,           ISD STUTTGART, GERMANY
ASOP, DWYER,W.J. ET AL, GRUMMAN AEROSPACE.DEPT.461,BETHPAGE,NY
BERSAFE, HELLEN,T.K., CEGB, BERKELEY NUCLEAR LABS, BERKELEY,GLOS
DANUTA,CHACOUR,S., ALLIS CHALMERS, BOX 712, YORK, PA 17405
DYNAL, DAVIS, R.E.,MDAC, PO BOX 516,ST.LOUIS, MO 63166
EASE,    ENGINEERING ANALYSIS CORP., REDONDO BEACH, CA 90277
EASE2,   ENGINEERING ANALYSIS CORP., REDONDO BEACH, CA 90277
ELAS55, UTKU, TARN, DVORAK, DUKE UNIV. SHC OF ENGR., DURHAM, NC
ELAS65, UTKU, RAO, DVORAK, DUKE UNIV. SCH OF ENGR., DURHAM,NC, 9
ELAS75, UTKU, S., DUKE UNIV., SCH. OF ENGR, DURHAM, N.C. 27706,
FESAP, VANFOSSEN,D.B., BABCOCK$WILCOX R$D.PO BOX 835,ALLIANCE,OH
MARC,HIBBITT,MARCAL, MARC ANAL. RES. CORP.,105 MEDWAY,PROV. RI
MSC/NASTRAN, MCCORMICK, C.W.      MACNEAL-SCHWENDLER CORP., LA.
NEPSAP, SHARIFI, P., DPT 81-12/154, LMSC, SUNNYVALE, CALIF
NOSTRA, FELIPPA,C.A., LMSC DEPT 5233, PALO ALTO, CALIF 94304
PAFEC70,HENSHELL,ET AL, NOTTS UNIV,UNIV. PARK,NOTTINGHAM,NG7 2RD
REGULA, DEAK,A.L., MATH.SCI.INC., 3030 NE 45TH, SEATTLE,WASH
REXBAT, LODEN,W.A., LMSC DEPT 5233, PALO ALTO, CALIF 94304
SAMIS,   MELOSH, R.J.,            VIRGINIA TECH.,BLACKSBURG,VA
SAPIV, BATHE, K.J., 410 DAVIS HALL, UNIV. OF CALIF,BERKELEY
SATURN,KATNIK,R.B.,CHEVROLET ENG.,30001 VAN DYKE, WARREN,MICH
SESAM69C, DET NORSKE VERITAS,   NO ADDRESS GIVEN,   NO PHONE GI
SINGUL, DEAK,A.L., MATH. SCI., 3030 NE 45TH, SEATTL(, WASH 98105
SNAP, WHETSTONE, W. D., LOCKHEED MSC.62-62,SUNNYVALE, CA  408-74
SOLIDSAP,WILSON, E.L.,720 DAVIS HALL,U OF CALIF, BERKELEY
SPADAS, PETYT,M., I.S.V.R.,UNIV. SOUTHAMPTION, SO95NH, ENGLAND
STARDYNE, ROSEN,R, MRI. INC., 9841 AIRPORT BLVD.,LOS ANGELES, CA
STRESS,FENVES,LOGCHER,REINSCHMIDT,MIT-CARNEGIE-MELLON UNIV.,PA.
STRUDLII,LOGCHER, R.D.          MIT,CAMBRIDGE, MASS
TRANSTOWER, SHIEH,W.Y., HARZA ENG.,150S.WACKER DR.,CHICAGO,ILL
VISCEL, GUPTA,AKYUZ,HEER, JPL, 4800 OAK GROVE, PASADENA, CA
```

```
DO YOU WISH TO DO THE KEY WORD BIT AGAIN?

IF YOU DO, PUNCH  'YES',  IF NOT, PUNCH  'NO'.
```

YES

```
NOW PUNCH IN KEY CONCEPTS, ONE CONCEPT TO A LINE

        GENERAL SHELL
        END

THE FOLLOWING PROGRAMS WILL PERFORM THE ABOVE-SPECIFIED ANALYSIS

ANSYS, SWANSON ANALYSIS SYSTEMS, 870 PINEVIEW DR.,ELIZABETH, PA
ASAS,ATKINS RES. AND DEV., ASHLEY ROAD, EPSOM, SURREY, ENGLAND
ASEF,     LAB. TECH.AEROSPATIALES, LIEGE, BELGIUM
ASKA,     ASKA-GROUP,           ISD STUTTGART, GERMANY
ASOP, DWYER,W.J. ET AL, GRUMMAN AEROSPACE.DEPT.461,BETHPAGE,NY
BERSAFE, HELLEN,T.K., CEGB, BERKELEY NUCLEAR LABS, BERKELEY,GLOS
DANUTA,CHACOUR,S., ALLIS CHALMERS, BOX 712, YORK, PA 17405
DYNAL, DAVIS, R.E.,MDAC, PO BOX 516,ST.LOUIS, MO 63166
EASE,    ENGINEERING ANALYSIS CORP., REDONDO BEACH, CA 90277
EASE2,   ENGINEERING ANALYSIS CORP., REDONDO BEACH, CA 90277
ELAS55, UTKU, TARN, DVORAK, DUKE UNIV. SHC OF ENGR., DURHAM, NC
ELAS65, UTKU, RAO, DVORAK, DUKE UNIV. SCH OF ENGR., DURHAM,NC, 9
ELAS75, UTKU, S., DUKE UNIV., SCH. OF ENGR, DURHAM, N.C. 27706,
FESAP, VANFOSSEN,D.B., BABCOCK$WILCOX R$D.PO BOX 835,ALLIANCE,OH
ISOPAR-SHL, GUPTA, SARGENT-LUNDY, 140 S.DEARBORN, CHICAGO,ILL
ISTRAN/S,ISHIKAWAJIMA-HARIMA RES. INST., TOKYO, JAPAN     03 53
MARC,HIBBITT,MARCAL, MARC ANAL. RES. CORP.,105 MEDWAY,PROV. RI
MINIELAS,ALPAY,IB, UTKU, S, DUKE UNIV. SCH OF ENGR., DURHAM,NC
MSC/NASTRAN, MCCORMICK, C.W.      MACNEAL-SCHWENDLER CORP., LA.
NEPSAP, SHARIFI, P., DPT 81-12/154, LMSC, SUNNYVALE, CALIF
NOSTRA, FELIPPA,C.A., LMSC DEPT 5233, PALO ALTO, CALIF 94304
PAFEC70,HENSHELL,ET AL, NOTTS UNIV,UNIV. PARK,NOTTINGHAM,NG7 2RD
PETROS3,HUFFINGTON, US ARMY BRL, ABERDEEN PROVING GRD,MD 21005
REGULA, DEAK,A.L., MATH.SCI.INC., 3030 NE 45TH, SEATTLE,WASH
REPSIL,SANTIAGO, US ARMY BRL, ABERDEEN PROVING GRD, MD 21005
REXBAT, LODEN,W.A., LMSC DEPT 5233, PALO ALTO, CALIF 94304
SAMBA, JARELAND, SAAB-SCANIA, FKVH-27, AB,S58188 LINKOPING, SWED
SAMIS,   MELOSH, R.J.,            VIRGINIA TECH.,BLACKSBURG,VA
SAPIV, BATHE, K.J., 410 DAVIS HALL, UNIV. OF CALIF,BERKELEY
SATURN,KATNIK,R.B.,CHEVROLET ENG.,30001 VAN DYKE, WARREN,MICH
SESAM69C, DET NORSKE VERITAS,   NO ADDRESS GIVEN,   NO PHONE GI
SINGUL, DEAK,A.L., MATH. SCI., 3030 NE 45TH, SEATTL(, WASH 98105
SNAP, WHETSTONE, W. D., LOCKHEED MSC.62-62,SUNNYVALE, CA  408-74
SOLIDSAP,WILSON, E.L.,720 DAVIS HALL,U OF CALIF, BERKELEY
SPADAS, PETYT,M., I.S.V.R.,UNIV. SOUTHAMPTION, SO95NH, ENGLAND
STAGSAI, BROGAN,FA AND ALMROTH,BO LMSC, PALO ALTO, CALIF.
STAGSB,  BROGAN,FA AND ALMROTH,BO LMSC, PALO ALTO, CALIF
STARDYNE, ROSEN,R, MRI. INC., 9841 AIRPORT BLVD.,LOS ANGELES, CA
STRESS,FENVES,LOGCHER,REINSCHMIDT,MIT-CARNEGIE-MELLON UNIV.,PA.
STRUDLII,LOGCHER, R.D.          MIT,CAMBRIDGE, MASS
TIRE, DEAK, A.L., MATH.SCI.,INC.,3030 NE 45TH,SEATTLE, WASH 9810
TRANSTOWER, SHIEH,W.Y., HARZA ENG.,150S.WACKER DR.,CHICAGO,ILL
VISCEL, GUPTA,AKYUZ,HEER, JPL, 4800 OAK GROVE, PASADENA, CA
```

```
DO YOU WISH TO DO THE KEY WORD BIT AGAIN?

IF YOU DO, PUNCH  'YES',  IF NOT, PUNCH  'NO'.
YES

NOW PUNCH IN KEY CONCEPTS, ONE CONCEPT TO A LINE

        LARGE DEFLECTIONS
        ELASTIC-PLASTIC
        GENERAL SHELL
        COMPOSITE MATERIAL
        GENERAL TRANSIENT LOAD

GENERAL TRANSIENT LOAD   IS NOT A KEY WORD. PLEASE RETYPE ENTRY

        GENERAL TRANSIENT LOADS
        END

THE FOLLOWING PROGRAMS WILL PERFORM THE ABOVE-SPECIFIED ANALYSIS

ANSYS, SWANSON ANALYSIS SYSTEMS, 870 PINEVIEW DR.,ELIZABETH, PA
MARC,HIBBITT,MARCAL, MARC ANAL. RES. CORP.,105 MEDWAY,PROV. RI
STAGSA1, BROGAN,FA AND ALMROTH,BO LMSC, PALO ALTO, CALIF.
STAGSB,  BROGAN,FA AND ALMROTH,BO LMSC, PALO ALTO, CALIF

DO YOU WISH TO DO THE KEY WORD BIT AGAIN?

IF YOU DO, PUNCH  'YES',  IF NOT, PUNCH  'NO'.

NO

NOW PUNCH EITHER   'ALL' ,   'SCOPE' ,    'COMPARE' ,
                   'EVALUATE' ,   'PROBLEM' ,  OR   'FINISHED'
COMPARE

NOW PUNCH THE NAMES OF SEVERAL PROGRAMS,ONE NAME
AT A TIME.

EXAMPLE--

   BOSOR4
   KSHEL
   SABOR/DRASTIC6
   SRA
   STARS
   DYNAPLASII
   END

ALWAYS TERMINATE THE SERIES OF NAMES WITH 'END'

ABOUT 8 PAGES OF OUTPUT WILL FOLLOW. THE CAPABILI-
TIES  OF THE PROGRAMS WHOSE NAMES YOU SUPPLY WILL
BE COMPARED.
```

```
        SECTION 1.  GEOMETRY / BOUNDARY CONDITIONS

                                          READ PROGRAM NAMES VERTICALLY

                                                    AAAABDDEEMMNNPRSSSSSSSS
                                                    NSSSEAYALASEOAEAAFNOPTT
                                                    SAEKRNNSARCPSFXMPSALAAR
                                                    YSFASUAESC/STEBIIAPIDRU
                                                    S...ATL27.NARCASVM.DADD
                                                    ....FA..5.APA7T..6.SSYL
                                                    ....E.....S..0...9.A.NI
                                                    ..........T......C.P.EI
                                                    ..........R............
                                                    ..........A............
                                                    ..........N............
                                                    .......................
01 GENERAL STRUCTURE .............................. XXXXXXXXXXXXXXXXXXXXXXX
02 SOLID OF REVOLUTION............................. XXXXXX..XXXX.X..XX.X...
03 GENERAL SHELL................................... XXXXXX.XXXXX.XXXXX.X..X
04 PART OF SHELL OF REVOLUTION ( L.T. 360 DEG.      X.X.XX..XXXX.X.XXX.X...
05 SHELL OF REVOLUTION (360 DEG.) ................. XXX.XX..XXXX.XXXXX.X...
06 PRISMATIC STRUCTURE............................. XXXXXX..XXXX.X..XX.X..X
07 SEGMENTED (IN SERIES)     ...................... XXXX....XXXX.XX.XX.X..X
08 BRANCHED ....................................... XXXX....XXXX.XXXXX.X..X
09 SHELL WITH CUTOUTS ............................. XXXX.X..XXXX.XXXXX.X..X
10 THICK SHELL .................................... XXXXXX..XX.X.X.XXX.X...
11 FRAMEWORK-SHELL COMBINATIONS    ................ XXXX...XXXXX.XXXXX.X..X
12 FLAT PLATE OF GENERAL SHAPE     ................ XXXXXX.XXXXX.XXXXX.X..X
13 OTHER GEOMETRY--SEE SECT.8, 1.13        ........ XX...X.XXXX..X..XX.....
14 AXISYMMETRIC IMPERFECTIONS      ................ X.X......X.X.X..X..X...
15 GENERAL IMPERFECTIONS     ...................... X.X.X.X..X.X.X..X.XX...
16 AXISYMMETRIC SUPPORT CONDITIONS................. XXXXXX..XXXX.X.XXX.X...
17 GENERAL SUPPORT CONDITIONS      ................ XXXXXXXXXXXXXXXXXXXXXXX
18 SUPPORTS AT BOUNDARIES ONLY     ................ X.XXX...XXX..X.XXX.X...
19 SUPPORTS AT INTERNAL POINTS     ................ XXXXX...XXXX.X.XXX.X..X
20 ELASTIC FOUNDATION ............................. XXXXXXXXXXXX.X..XXXXXXX
21 STIFFENERS   ................................... XXX.X.X.XXXX.XXXXX.XX..
22 CONTACT......................................... XX.X.X...XX..X.......X.
23 FRICTION     ................................... XX.......XX............
24 SLIDING WITHOUT FRICTION ....................... XXXXXX...XXX.XXXXX.X...
25 OTHER DEFORMATION DEPENDENT SUPPORT--SEE SE      .........X.......X.....
26 JUNCTURE COMPATIBILITY    ...................... XXXX.X...XX.XXX.XX.X...
27 SINGULARITY CONDITIONS    ...................... .X....X..X...X...X...X.
28 GENERAL LINEAR CONSTRAINT COND. U(I)= TU(N)      XXX.X..XXXXX.X.XXX.XX..
29 GENERAL NONLINEAR CONSTRAINT CONDITIONS          X........X.............
30 TRANSITION FROM 3-D TO 2-D REGION       ........ XXXX...XXXX.....XX.X...
31 TRANSITION FROM 2-D TO 1-D REGION       ........ XXXX..XXXXX.....XX.X...
32 OTHER CONSTRAINT CONDITIONS--SEE SECT.8, 1.      ....XX..X........X.....
33 LAGRANGE MULTIPLIER METHOD USED................. ..X.......X............
34 OTHER COMMENTS PERTINENT TO SECT. 1--SEE SE      .......................
```

SECTION 2, WALL CONSTRUCTION/MATERIAL PROPERTIES

READ PROGRAM NAMES VERTICALLY

	AAAABDDEEMMNNPRSSSSSSSS
	NSSSEAYALASEOAEAAENOPTT
	SAEKRNNSARCPSFXMPSALAAR
	YSFASUAESC/STEBIIAPIDRU
	S...ATL27.NARCASVM.DADD
	....FA..5.APA7T..6.SSYL
	....E.....S..O...9.A.NI
	..........T......C.P.EI
	..........R............
	..........A............
	..........N............
	....
01 MONOCOQUE WALL CONSTRUCTION	XXXX.X..XXXXXXXXXX.XXX.
02 LAYERED WALL CONSTRUCTION	X.X..X..XX.X.X.XXX.X.X.
03 SANDWICH WALL CONSTRUCTION	XXX.....XXXX...XX..X.X.
04 COMPOSITE MATERIAL	X.X.....XX.X....X..X...
05 MERIDIONAL STIFFENERS TREATED AS DISCRETE	X.XX.....XXX.X.XXX.XX..
06 CIRCUM. STIFFENERS TREATED AS DISCRETE	X.XX.....XXX.X.XXX.X...
07 GENERAL STIFFENERS TREATED AS DISCRETE	XXXXXXXXXXXXXXXXXXXX.X.
08 STIFFENERS TREATED AS SMEARED	XXX......X...X...X.....
09 GENERAL OPEN-SECTION THEORY FOR STIFFENERS	X..X..X..X........X.X..
10 CG, SHEAR CENTER OF STIFFENER MUST COINCIDE	XXXX.X.XXXXX.XX..X.....
11 OTHER WALL CONSTRUCTION--SEE SECT.8, 2.11	.........X.X...........
12 ISOTROPIC MATERIAL PROPERTIES	XXXXXXXXXXXXXXXXXX.XXXX
13 ANISOTROPIC MATERIAL PROPERTIES	XXXX....XXXXXXXXXXXX.XX
14 LINEAR ELASTIC MATERIAL PROPERTIES	XXXXXXXXXXXX.XXXXXXXXXX
15 NONLINEAR ELASTIC MATERIAL PROPERTIES	.X.X.....X............X
16 TEMPERATURE-DEPENDENT MATERIAL PROPERTIES	XX..X..X.XXXX.XXXX.X...
17 RIGID-PERFECTLY PLASTIC MATERIAL	.........X.......X.....
18 ELASTIC-PERFECTLY PLASTIC MATERIAL	XX.XXX...X.......X.....
19 ELASTIC-LINEAR STRAIN HARDENING MATERIAL	XX.XXX...X.......X.....
20 STRESS-STRAIN CURVE OF MANY LINE SEGMENTS	XX..XX...XXX.....X.....
21 RAMBERG-OSGOOD STRESS-STRAIN LAW	.X.......X.............
22 ISOTROPIC STRAIN HARDENING	XX.X.X...X.X.....X.....
23 KINEMATIC STRAIN HARDENING	X..X.....X.X...........
24 WHITE-BESSELING HARDENING LAW	....
25 OTHER POST-YIELD LAW--SEE SECT.8, 2.25	X..X.X...X.......X.....
26 PRIMARY CREEP	XX.X.....X.......X.....
27 SECONDARY CREEP	X........X.X.....X.....
28 VISCOELASTIC MATERIAL	.........X.......X.....
29 STRAIN-RATE EFFECTS INCLUDED	....
30 NO SPATIAL VARIATION OF PROPERTIES	...X..........X.X..X...
31 AXISYMMETRIC VARIATION OF PROPERTIES ONLY	................XX.X...
32 ARBITRARY VARIATION OF PROPERTIES	XXXX.X.XXX.XXX.XXXXXX.X
33 COUPLING OF STRAINS AND CURVATURE CHANGES	..X......X.X.X.........
34 COUPLING OF NORMAL AND IN-PLANE SHEAR	..X......X.X...X.......
35 OTHER COMMENTS PERTINENT TO SECTION 2--SEE	.............X.........

```
         SECTION 3, LOADING

                                   READ PROGRAM NAMES VERTICALLY

                                                  AAAABDDEEMMNNPRSSSSSSSS
                                                  NSSSEAYALASEOAEAAENOPTT
                                                  SAEKRNNSARCPSFXMPSALAAR
                                                  YSFASUAESC/STEBIIAPIDRU
                                                  S...ATL27.NARCASVM.DADD
                                                  ....FA..5.APA7T..6.SSYL
                                                  ....E.....S..O...9.A.NI
                                                  ..........T......C.P.EI
                                                  ..........R............
                                                  ..........A............
                                                  ..........N............

                                                  .......................
 01 UNIFORM DISTRIBUTION OF LOADS  ............   XXX..X.......X.XX..X..X
 02 AXISYMMETRIC VARIATION OF LOADS ONLY ......   XXX..X..X....X..X..X...
 03 GENERAL VARIATION OF LOADS     ............   XXXXXX.XXX.XXXX.XXXX..X
 04 STATIC OR QUASI-STATIC LOAD VARIATION......   XXXX.X.XXXXXX.XXXXXX.XX
 05 GENERAL TRANSIENT VARIATION OF LOADS ......   XXXX.XX..XX....XXX...XX
 06 PERIODIC (DYNAMIC) VARIATION OF LOADS......   XXXX.XX..XX....XXX...XX
 07 IMPULSIVE LOADS    ........................   X.X..XX..XX....XX....X.
 08 LOADS VARYING RANDOMLY IN TIME ............   XX.......XX.....X....X.
 09 OTHER TEMPORAL VARIATION OF LOADS--SEE SECT   .........X...........X.
 10 LOADS VARYING PROPORTIONALLY DURING CASE      XXXX.X...XX.....X....X.
 11 SOME LOADS VARY, SOME CONSTANT DURING CASE    XX.X.X...X.XX..XX....X.
 12 DIFFERENT LOADS HAVE DIFFERENT TIME HISTORI   XX.X.X...X.....XX....X.
 13 QUASI-STATIC LOADS VARYING CYCLICLY DURING    XX.X.X...X......X....X.
 14 OTHER MANNER OF LOAD VARIATION DURING CASE-   .........X.............
 15 LIVE (FOLLOWING) LOADS   ..................   X........XX............
 16 GYROSCOPIC LOADS   ........................   .X.......X.............
 17 INERTIAL LOADS     ........................   XX.X..X..XX..XXXX....XX
 18 ELECTROMAGNETIC LOADS   ...................   .......................
 19 FLUID-STRUCTURE INTERACTION    ............   ..X.......X......X.....
 20 ACOUSTIC LOADING   ........................   .....................X.
 21 SOIL-STRUCTURE INTERACTION     ............   XX.......X...X..X..X.X.
 22 RIGID SOLID-STRUCTURE INTERACTION    ......   .X.......X...X..XX.X...
 23 OTHER DEFORMATION-DEPENDENT TYPE OF LOADING   .......................
 24 POINT LOADS, MOMENTS     ..................   XXXXXXXXXXXXXXXXXXXX.XX
 25 LINE LOADS, MOMENTS........................   XXXXXX.XXXXX.X.XXXXX.XX
 26 NORMAL PRESSURE    ........................   XXXXXX.XXXXXXXXXXX.X.XX
 27 SURFACE TRACTIONS  ........................   XXXXXX.X.XXXXX.XXX.X.XX
 28 BODY FORCES  ..............................   XXXXXX.XXXX..X.XXX.X.X.
 29 DEAD-WEIGHT LOADING........................   XXXX.X.XXXX..X.XXX.X.XX
 30 THERMAL LOADING    ........................   XXXXXX.XXXXX.XXXXXXX.XX
 31 INITIAL STRESS OR STRAIN ..................   XX.XX....X...X..XX.X..X
 32 CENTRIFUGAL LOADING........................   XX..XX..XXXX.X...X.....
 33 NON-ZERO DISPLACEMENTS IMPOSED ............   XXXXXX.XXXX.XXXXXX.X.X.
 34 OTHER TYPE OF LOADING--SEE SECT.8, 3.34       .................X.....
 35 OTHER COMMENTS PERTINENT TO THIS SECTION--S   .................X.....
```

SECTION 4, PHENOMENA

READ PROGRAM NAMES VERTICALLY

```
                                                     AAAABDDEEMMNNPRSSSSSSSS
                                                     NSSSEAYALASEOAEAAENOPTT
                                                     SAEKRNNSARCPSFXMPSALAAR
                                                     YSFASUAESC/STEBIIAPIDRU
                                                     S...ATL27.NARCASVM.DADD
                                                     ....FA..5.APA7T..6.SSYL
                                                     ....E.....S..O...9.A.NI
                                                     ..........T......C.P.EI
                                                     ..........R............
                                                     ..........A............
                                                     ..........N............
                                                     .......................
01 AXISYMMETRIC SMALL DEFLECTIONS ................. XXXXXX..XXXX.XXXXX.X...
02 GENERAL SMALL DEFLECTIONS...................... XXXXXXXXXXXXXXXXXX.X.XX
03 AXISYMMETRIC LARGE DEFLECTIONS ................. X........X.X...X.......
04 GENERAL LARGE DEFLECTIONS...................... X........X.XX..X......X
05 AXISYMMETRIC PLASTICITY  ...................... X...XX...X.X.....X.....
06 GENERAL PLASTICITY ............................ XX.XXX...X.X.....X.....
07 LARGE STRAINS.................................. .X.........X...........
08 TRANSVERSE SHEAR DEFORMATIONS  ................ XXX...X..XXX.X..XX.X.X.
09 THERMAL EFFECTS    ............................ XXXXXX..XXXX.XXXXXXX.XX
10 RADIATION EFFECTS ............................. X........X.............
11 AUTOMATED YIELD CRITERION...................... XX.XXX...X.X..X..X.....
12 AUTOMATED FRACTURE CRITERION   ................ ....X..................
13 AUTOMATED FATIGUE CRITERION    ................ .......................
14 AUTOMATED BUCKLING CRITERION   ................ .........X.X.X.........
15 OTHER BUILT-IN CRITERIA--SEE SECT.8, 4.15       .......................
16 NONLINEAR COLLAPSE ANALYSIS       ............. X........X.XX..........
17 POST-BUCKLING PHENOMENA ....................... .........X.............
18 BIFURC. BUCKLING, LINEAR AXISYM. PRESTRESS      ..X......XX....X.......
19 BIFURC. BUCKLING, LINEAR GENERAL PRESTRESS      ..XX.....XX....X.......
20 BIFURC. BUCKLING, NONLIN. AXISYM. PRESTRESS     .........X.............
21 BIFURC. BUCKLING, NONLIN. GENERAL PRESTRESS     .........X.............
22 BIFURC. BUCKLING, PREBUCKLING ROTATIONS         .........X.............
23 BIFURC, BUCKLING, TRANSVERSE SHEAR DEFORMAT     .........XX............
24 BIFURCATION BUCKLING WITH OTHER EFFECTS--SE     .......................
25 DYNAMIC BUCKLING  ............................. .........X.............
26 MODAL VIBRATIONS WITH NO PRESTRESS   .......... XXXX.XX..X....XXXXX.XXX
27 MODAL VIBRATIONS, LINEAR AXISYM. PRESTRESS      .........XX....X.......
28 MODAL VIBRATIONS, LINEAR GENERAL PRESTRESS      .........XX..X.X.......
29 MODAL VIBRATIONS, NONLIN. AXISYM. PRESTRESS     .........X.............
30 MODAL VIBRATIONS, NONLIN. GENERAL PRESTRESS     .........X.............
31 MODAL VIBRATIONS, TRANSVERSE SHEAR DEFORMAT     ..X..X....X....XXX...X.
32 MODAL VIBRATIONS WITH OTHER EFFECTS--SEE SE     .......................
33 NONLINEAR VIBRATIONS   ........................ .........X.............
34 LINEAR AXISYMMETRIC DYNAMIC RESPONSE .......... XXX..X...XX....XXX.....
35 LINEAR GENERAL DYNAMIC RESPONSE................ XXXX.XX..XX....XXX...XX
36 NONLINEAR AXISYMMETRIC DYNAMIC RESPONSE         X........X.............
37 NONLINEAR GENERAL DYNAMIC RESPONSE      ....... X........X.............
38 WAVE PROPAGATION ............................. .........X.....X.......
39 DYNAMIC RESPONSE, TRANSVERSE SHEAR DEFORMAT     ..X..XX...X....XXX.....
40 DYNAMIC RESPONSE WITH OTHER EFFECTS INCLUDE     .......................
41 VISCOUS DAMPING    ............................ XXXX..X..XX....XXX...X.
42 STRUCTURAL DAMPING ............................ XXX......XX....X.......
43 LINEAR DAMPING     ............................ XXX......XX....XXX...X.
44 NONLINEAR DAMPING  ............................ .........X.............
45 OTHER KIND OF DAMPING--SEE SECT.8, 4.45         .......................
46 FLUTTER........................................ .......................
47 OTHER DYNAMIC PHENOMENA--SEE SECT.8, 4.47       .......................
48 OTHER PHENOMENA OF ANY KIND--SEE SECT.8, 4.     .......................
```

SECTION 5, DISCRETIZATION

```
                                                     READ PROGRAM NAMES VERTICALLY

                                                          AAAABDDEEMMNNPRSSSSSSSS
                                                          NSSSEAYALASEOAEAAENOPTT
                                                          SAEKRNNSARCPSFXMPSALAAR
                                                          YSFASUAESC/STEBIIAPIDRU
                                                          S...ATL27.NARCASVM.DADD
                                                          ....FA..5.APA7T..6.SSYL
                                                          ....E.....S..O...9.A.NI
                                                          ..........T......C.P.EI
                                                          ..........R............
                                                          ..........A............
                                                          ..........N............
                                                          .......................
01 MULTI-SEGMENT FORWARD INTEGRATION               ...... X......................
02 GALERKIN METHOD          ......................... .......................
03 RAYLEIGH-RITZ METHOD         ..................... .........XX.XX..X......
04 BOUNDARY POINT MATCHING      ..................... .......................
05 FINITE DIFF. METHOD BASED ON EQUILIBRIUM EQ      .......................
06 FINITE DIFFERENCE ENERGY METHOD............     .......................
07 1-DIMENSIONAL FINITE-DIFF. DISCRETIZATION         .......................
08 ORTHOGONAL 2-DIM. FINITE-DIFFERENCE GRID          .......................
09 NON-ORTHOGONAL 2-DIM. FINITE-DIFF. GRID           .......................
10 GENERAL QUADRILATERIAL 2-D FINITE-DIFF.GRID       .......................
11 OTHER FINITE-DIFF. DISCRETIZATION SCHEME--        .......................
12 FOURIER SERIES IN CIRCUMFERENTIAL DIRECTION       X......................
14 ROD ELEMENT  ..................................... XXXX.XXXXXXXXXXXXX.X.XX
15 STRAIGHT BEAM ELEMENT        ..................... XXXXX.XXXXXXXXXXXXXXXXX
16 CURVED BEAM ELEMENT................................ .X....X..X...XX..X..XX.
17 CONICAL SHELL ELEMENT        ..................... X.X.....XXX..XX........
18 AXISYM. SHELL ELEMENT        ..................... XX.......XX..X..X..X...
19 FLAT MEMBRANE ELEMENT        ..................... .XXXXXXXXXX..XXXXXXXXXX
20 FLAT MEMBRANE ELEMENT        ..................... XXXXXX.X.XXX.XX..XX.XX.
21 CURVED MEMBRANE ELEMENT      ..................... .X.X.X...X..XX...X.....
22 CURVED MEMBRANE ELEMENT      ..................... .X.X.X...X...X.........
23 FLAT PLATE ELEMENT ............................... XXXXX.XXXXXX.XXXXXXXXXX
24 FLAT PLATE ELEMENT ............................... .XXX...X..X..XX..XX.XX.
25 SHALLOW SHELL ELEMENT        ..................... XXXX...XX..X.X.XXX.XX..
26 SHALLOW SHELL ELEMENT        ..................... .XXX...X...X.X..XX.XX..
27 DEEP SHELL ELEMENT ............................... ..XX.....X...X..X..X...
28 AXISYM. SOLID ELEMENT        ..................... XXXXXX..XXXX.X..XX.X...
29 AXISYM. SOLID ELEMENT        ..................... .X.XXX...XX..X...X.....
30 THICK PLATE ELEMENT................................ XXXX.....X...X.XX..X...
31 THICK SHELL ELEMENT................................ X.XX.....X.X.X..XX.X...
32 SOLID (3-D) ELEMENT................................ XXXXXX.XXXXXXXX.XX.X.XX
33 SOLID (3-D) ELEMENT................................ XXXXXX.X.XX..XX..X...X.
34 SOLID (3-D) ELEMENT................................ .X.XX......X..X...X....
35 CRACK-TIP ELEMENT    ............................. .X..X....X...X.........
36 OTHER TYPE OF ELEMENT        ..................... ..X.X......X..XX.XX....
37 FORCE METHOD OR OTHER METHOD      ............... .....................X.
38 INCOMPATIBLE DISPLACEMENT FUNCTIONS USED          XX.....X...X.XX.X..X.XX
39 SUBSTRUCTURING           ......................... XXXX.X....X....XX.X..XX.
40 REPEATED USE OF IDENTICAL SUBSTRUCTURES           XXXX.....XX...X..X..XX.
44 NODES CAN BE INTRODUCED IN ARBITRARY ORDER        XXXXX.XXX.XXXXXX..X.XXX
45 AUTOMATIC RENUMBERING OF NODES ............       .....X.XXXX.XXX..XX..XX
46 NON-DIAGONAL MASS MATRIX ...................      XXXX..X..XX..XXX.XX.X.X
47 DIAGONAL MASS MATRIX         ..................... .X.X.XX..XX..XXXXXX..XX
48 OTHER COMMENTS PERTINENT TO THIS SECTION--S       .X............X..X.....
```

SECTION 6, SOLUTION METHODS

```
                                                  READ PROGRAM NAMES VERTICALLY

                                                    AAAABDDEEMMNNPRSSSSSSSS
                                                    NSSSEAYALASEOAEAAFNOPTT
                                                    SAEKRNNSARCPSFXMPSALAAR
                                                    YSFASUAESC/STEBIIAPIDRU
                                                    S...ATL27.NARCASVM.DADD
                                                    ....FA..5.APA7T..6.SSYL
                                                    ....E.....S..0...9.A.NI
                                                    ..........T......C.P.EI
                                                    ..........R............
                                                    ..........A............
                                                    ..........N............
                                                    .......................
01 FULL-MATRIX EQUATION SOLVER    .............     .X....X...X..X.X.......
02 CONSTANT BANDWIDTH EQUATION SOLVER    ......     .......X..X..X.......X.
03 SKYLINE METHOD OF SOLUTION USED.............     .........X.XX.X.X..X.X.
04 MATRIX PARTITIONING USED ...................     .X.X......X...XX.X...XX
05 WAVE-FRONT METHOD USED FOR SOLUTION   ......     XXX.X.X..XX..X.X.....XX
06 OTHER SPARSE MATRIX METHOD USED FOR SOL'N--       ..........X....X..X..X.
07 ITERATIVE METHOD FOR SOLUTION OF LINEAR EQU      ...............X.......
08 CONJUGATE GRADIENT METHOD FOR SOL'N OF LIN.      ..X..............X.....
09 FORWARD INTEGRATION USED FOR SOL'N OF LIN.       .X.....................
10 OTHER SOLUTION METHOD FOR LINEAR EQUATIONS       .....X..X.X......X.....
11 CONDENSATION METHOD--SEE SECT.8, 6.11......      XXXX..G...X....X.X...X.
12 LR,QR METHOD FOR EIGENVALUE EXTRACTION           XX.X.XQ......X.......XX
13 LANCZOS METHOD FOR EIGENVALUE EXTRACTION         .......................
14 STURM SEQUENCE METHOD FOR EIGENVALUE EXTRAC      .X....X......X.X.X..X.X
15 DETERMINANT SEARCH FOR EIGENVALUE EXTRACTIO      ..........X.....X......
16 SINGLE VECTOR INVERSE PWR ITER. WITH SHIFTS      .........XX...X......XX
17 MULTI-VECTOR (SUBSPACE) ITERATION FOR EIGEN      ..XX.....X......X......
18 OTHER EIGENVALUE EXTRACTION METHOD--SEE SEC      ...X.............XX....
19 FULL NEWTON-RAPHSON METHOD FOR NONLIN. SOLN      ......................X
20 MODIFIED NEWTON METHOD FOR NONLINEAR SOL'N       ............X..........
21 INCREMENTAL METHOD WITH NO EQUILIB. CHECK        X..X...................
22 INCREMENTAL METHOD WITH EQUILIBRIUM CHECK        .X.XXX...X.X...........
23 RUNGE-KUTTA TYPE OF INCREMENTAL METHOD           .......................
24 OTHER TYPE OF INCREMENTAL METHOD--SEE SECT.      .......................
25 DIRECT ENERGY SEARCH FOR NONLINEAR SOLUTION      .......................
26 DYNAMIC RELAXATION FOR NONLINEAR SOLUTION        .......................
27 NONLIN. EQ. SOLVED BY SUCCESSIVE SUBSTITUTI      .X.X.X...X...X...X.....
28 OTHER NONLINEAR STRATEGY--       .............   .................X.....
29 LAGRANGIAN FORMULATION    ..................     .....X...X.XX..........
30 UPDATED LAGRANGIAN FORMULATION .............     .X..........X..........
31 EULERIAN FORMULATION      ..................     .......................
32 TIME INTEGRATION BY MODAL SUPERPOSITION          .XXX.XX..XX..X.XXX....X
33 TIME INTEGRATION BY EXPLICIT METHOD  ......      .X.......X...X.........
34 TIME INTEGRATION BY IMPLICIT METHOD  ......      X........XX..X.XX......
35 TIME INTEGRATION BY EULER (CONST. ACCEL.)        .X.....................
36 TIME INTEGRATION BY NEWMARK BETA METHOD. SE      .X.......XX....X.......
37 TIME INTEGRATION BY RUNGE-KUTTA OF ORDER--S      .......................
38 TIME INTEGRATION BY HOUBOLT METHOD               ..X......X.............
39 TIME INTEGRATION BY WILSON METHOD    ......      ................X......
40 TIME INTEGRATION BY CENTRAL DIFFERENCE METH      .........X.............
41 TIME INTEGRATION BY STIFFLY-STABLE METHOD        .......................
42 TIME INTEGRATION BY PREDICTOR-CORRECTOR OF       .......................
43 TIME INTEGRATION BY MULTI-STEP METHOD......      .......................
44 TIME INTEGRATION BY SOME OTHER METHOD--SEE       .......................
45 OTHER COMMENTS PERTINENT TO THIS SECTION--S      .......................
```

```
SECTION 7. COMPUTER PROGRAM DISTRIBUTION
 DOCUMENTATION/ORGANIZATION/MAINTENANCE

                                        READ PROGRAM NAMES VERTICALLY

                                                  AAAABDDEEMMNNPRSSSSSSSS
                                                  NSSSEAYALASEOAEAAFNOPTT
                                                  SAEKRNNSARCPSFXMPSALAAR
                                                  YSFASUAESC/STERIIAPIDRU
                                                  S...ATL27.NARCASVM.DADD
                                                  ....FA..5.APA7T..6.SSYL
                                                  ....E.....S..0...9.A.NI
                                                  ..........T......C.P.EI
                                                  ..........R............
                                                  ..........A............
                                                  ..........N............
                                                  .......................
01 SOURCE PROGRAM TAPE AVAILABLE ............. .XXXX...X..X.X.XXX....X
02 SOURCE PROGRAM CARDS AVAILABLE ............ .XX.X...X..X.X.X...XX..
03 PROGRAM ABSOLUTE ELEMENT  ONLY AVAILABLE    .....XX.X..............
04 PROGRAM COMPLETELY PROPRIETARY ............ XX..XXXX.XX.X.X..X...X.
05 PROGRAM PRICE IN DOLLARS--SEE SECT.8, 7.05  XX....X.X.X..XX.XX.X...
06 PROGRAM IS AVAILABLE THROUGH DEVELOPER      XXXXXXX..X.X.XX..X.....
07 PROGRAM IS AVAILABLE THROUGH COSMIC ...... ................X..X....
08 PROGRAM AVAILABLE THRU OTHER SOFTWARE CENTE ......X.XXXX....X..XX.X
09 PROGRAM CAN BE EXECUTED THRU SERVICE BUREAU XXXXXXXX.XX.......X...XX
11 THIS PROGRAM HAS BEEN SUPERSEDED BY ANOTHER .........X........X....
12 IMPROVED VERSION BEING PREPARED--SEE SECT.8 ....X.X..XXX..X..X...XX
13 PROGRAM IS NOT COMPLETED .................. .............X.........
14 USERS MANUALS ARE SELF-CONTAINED    ...... XXXXXXXXXXXX.XX.XX.XXXX
15 USERS MANUALS HAVE LIST OF PITFALLS  ...... ........XXX...X...X...X.
16 USERS MANUALS HAVE TEST CASES  ............ XXXXXXXXXX...X..XX..XXX
17 USERS MANUALS HAVE FLOW CHARTS ............ XXX...X.XX...X.XXX...X.
18 USERS MANUALS GIVE THEORY.................. X.XXX.XXXXX..X.X.X..XX.
19 OTHER QUALITIES OF USERS MANUALS--  ...... .X..X........X...X...X.
20 PROGRAM LANGUAGE IS--SEE SECT.8,7.20 ...... ...X..XX.....X........X
21 PROGRAM RUNS ON IBM........................ XXXXXXX.XXX..X..XX.X..X
22 PROGRAM RUNS ON CDC........................ XXXX...XXXXX.X.XXX.X.X.
23 PROGRAM RUNS ON UNIVAC  ................... XX.X.X..XXXXXXXXXX.X..X
24 PROGRAM RUNS ON GE ........................ X.......X....X.XXX.....
25 PROGRAM RUNS ON COMPUTER IDENTIFIED IN SECT .X.......X....X..XX..X..
28 NO. OF PRIMARY OVERLAYS *.................. 7.LL3L.L4L.72.5!L6...3.
   MAN-YEARS REQUIRED TO DEVELOP PROGRAM *     LLLL3485!LL33LL9.1..55L
31 MINIMUM CORE SPACE REQUIRED FOR EXECUTION   5.43.6X32355X1.216.1.33
32 PROGRAM RUNS ENTIRELY IN CORE ............. .........X....X......X..
33 NO. UNITS OF DIRECT ACCESS MASS STORAGE GIV 1XL1981115VLX2X89X.8.L1
36 RESTART CAPABILITY ........................ XXXX.XX..XXXX.XXXX...XX
37 OPTIONS FOR VARIOUS OUTPUT     ............ XXXXXXXXXXXXX.X.XX...XX
38 DIRECT ACCESS DATA BASE ................... ...X..X.....XXX..X...XX
39 FREE-FIELD INPUT .......................... ......X.....X.........X
40 RUNS IN INTERACTIVE MODE .................. X...........X.........X
41 AUTOMATIC MESH GENERATOR(S)    ............ XX..X....X.XXXX.XX.X.XX
42 AUTOMATIC LOADING GENERATOR    ............ X...X....X.XXXXXXX.X.X.
43 SECTION PROPERTIES CALCULATED ............. .........X.......X...X.
44 PRE-PROCESSOR.............................. XXXXXX.X.X.XXXXX.X..XX.
45 POST-PROCESSOR     ........................ XXXXXX..XX.XXXX..X...X.
46 EIGENVALUE SOLVER ......................... XXX...........X..X...X.
47 MODAL SUPERPOSITION ROUTINE    ............ XXX...........X..X...X.
48 OTHER TIME-INTEGRATION ROUTINE ............ XXX...........X........
49 OTHER CAPABILITY--SEE SECT.8, 7.49  ...... .......................
50 PLOT ROUTINES FOR UNDEFORMED/DEFORMED GEOME XXX.XXX.XXXX.XXX.X...X.
51 VARIOUS VIEWS AND SECTIONS PLOTTED  ...... XXX.XXXXXXXXXXX..X...X.
52 CONTOUR PLOTS.............................. X...X...XXX.XXX..X.....
53 ORDINARY Y = F(X) PLOTS  .................. X.....X.XXXX.X...X...X.
54 STRESS-ON-STRUCTURE PLOTS.................. X...X...XX.XXXX..X.....
55 OTHER KINDS OF PLOTS--SEE SECT.8, 7.55      X.......X.........X.....
56 PLOTTING CAPABILITY FOR CALCOMP............ X...XXXXXXX..X.X.X...X.
57 PLOTTING CAPABILITY FOR STROMBERG-CARLSON   X.....X.XXXX..X.......X.
58 PLOTTING CAPABILITY FOR HARDWARE IDENTIFIED .XX.X.X.X.X.X....X.....
59 OPTIONAL DETAILED PRINT OUTPUT ............ XXXXXXXXXXXXX.X..X...XX
60 EQUILIBRIUM CHECKS ........................ XXXX.X..XX..X.X..X...XX
61 ALTERNATIVE BRANCHES FOR SAME PROBLEM...... .X.X.....X..X.X.......X.
62 MATRIX CONDITIONING ESTIMATION ............ XXXX.....X..X....X.....
63 AUTOMATIC ERROR CONTROL  .................. XX.X.....X.X.....X...X.
64 GRAPHICAL PREPROCESSOR   .................. XX.XXXXX.XXXXXXX.X...X.
65 APPLIED LOADING CHECKS   .................. ....X..X.X....X..X...X.
66 BOUNDARY CONDITION CHECKS.................. .XXXX..X.X..XX....X...X.
67 SECTION PROPERTIES CHECKS.................. ...X...X.X............X.
68 GEOMETRY CHECKS    ....................... .XXXXX.X.X.X.XXX.X...X.
69 OTHER CHECKING AIDS--SEE SECT.8, 7.69...... ......X.......X........
70 DEVELOPER WILL HELP USER GET PROGRAM RUNNIN XXXXXXX.XXX..XX..X..XXX
71 DEVELOPER WILL HELP USER MODELING CASES     XXXX.XXXXXXX.XXX.X..XXX
72 DEVELOPER WILL HELP USER FIND SUSPECTED BUG XXXXXXXXXXXX.XXX.X..XX.
73 DEVELOPER WILL CIRCULATE NOTICES OF BUGS FO XXXXXXXXXXXX.XX..X..XXX
74 DEVELOPER WILL HELP USER SET UP AND RUN CAS XX.X.XXXXXXX.XX..X..XX.
75 DEVELOPER WILL CONDUCT WORKSHOPS    ...... XX.XXXXXXXXX.XX..X..XXX
76 DEVELOPER WILL MAKE SPECIAL PROGRAM CHANGES XXX.XXX.XXXX.XX..X..XXX
77 DEVELOPER WILL KEEP USERS INFORMED OF IMPRO XXXXXXXXXXXX.XX..X..XXX
78 OTHER ASSISTANCE DEVELOPER WILL GIVE USERS- .......XX.X......X...X.
79 OTHER COMMENTS PERTINENT TO THIS SECTION--S .......................

*L = LARGE, X= SEE SECT. 8, (WITH 'SCOPE' OPTION)
```

```
NOW PUNCH EITHER   'ALL' ,   'SCOPE' ,   'COMPARE' ,
                   'EVALUATE' ,   'PROBLEM' ,  OR   'FINISHED'
```

COMPARE

```
NOW PUNCH THE NAMES OF SEVERAL PROGRAMS,ONE NAME
AT A TIME.

EXAMPLE--

     BOSOR4
     KSHEL
     SABOR/DRASTIC6
     SRA
     STARS
     DYNAPLASII
     END

ALWAYS TERMINATE THE SERIES OF NAMES WITH 'END'


ABOUT 8 PAGES OF OUTPUT WILL FOLLOW. THE CAPABILI-
TIES  OF THE PROGRAMS WHOSE NAMES YOU SUPPLY WILL
BE COMPARED.


            SAASII.............. PROGRAM NOT IN DATA BANK


            SEALSHELL........... PROGRAM NOT IN DATA BANK
```

SECTION 1, GEOMETRY / BOUNDARY CONDITIONS

READ PROGRAM NAMES VERTICALLY

	FARSS	FESAP	PILEFEM	SAASII	SEALSHELL	SOLIDSAP	TEXGAP	VISCEL	ZP26
01 GENERAL STRUCTURE	.	X	.	.	.	X	.	X	.
02 SOLID OF REVOLUTION	X	.	X	.	.	X	X	X	X
03 GENERAL SHELL	.	.	.	.	.	X	.	X	.
04 PART OF SHELL OF REVOLUTION (L.T. 360 DEG.	.	.	.	.	.	X	.	X	.
05 SHELL OF REVOLUTION (360 DEG.)	X	.	.	.	.	X	X	X	X
06 PRISMATIC STRUCTURE	.	.	.	.	.	X	.	X	.
07 SEGMENTED (IN SERIES)	.	.	.	.	.	X	.	X	.
08 BRANCHED	.	.	.	.	.	X	X	X	X
09 SHELL WITH CUTOUTS	.	.	.	.	.	X	.	X	.
10 THICK SHELL	.	.	.	.	.	X	X	.	X
11 FRAMEWORK-SHELL COMBINATIONS	.	.	.	.	.	X	.	X	.
12 FLAT PLATE OF GENERAL SHAPE	.	.	.	.	.	X	.	X	X
13 OTHER GEOMETRY--SEE SECT.8, 1.13	.	.	.	.	.	.	.	.	.
14 AXISYMMETRIC IMPERFECTIONS	.	.	.	.	.	X	X	.	X
15 GENERAL IMPERFECTIONS	.	X	.	.	.	X	.	.	.
16 AXISYMMETRIC SUPPORT CONDITIONS	.	.	.	.	.	X	X	.	X
17 GENERAL SUPPORT CONDITIONS	X	X	X	.	.	X	X	X	.
18 SUPPORTS AT BOUNDARIES ONLY	.	.	.	.	.	X	.	.	.
19 SUPPORTS AT INTERNAL POINTS	.	.	.	.	.	X	X	.	X
20 ELASTIC FOUNDATION	X	X	.	.	.	X	.	X	.
21 STIFFENERS	.	X	.	.	.	X	.	.	.
22 CONTACT	X	.	.	.	.	.	.	.	X
23 FRICTION	.	.	.	.	.	.	.	.	.
24 SLIDING WITHOUT FRICTION	X	X	.	.	.	X	X	.	X
25 OTHER DEFORMATION DEPENDENT SUPPORT--SEE SE	.	.	.	.	.	.	.	.	X
26 JUNCTURE COMPATIBILITY	.	.	.	.	.	X	.	.	X
27 SINGULARITY CONDITIONS	.	.	.	.	.	.	.	.	X
28 GENERAL LINEAR CONSTRAINT COND. U(I)= TU(N)	X	.	.	.	.	X	.	X	X
29 GENERAL NONLINEAR CONSTRAINT CONDITIONS	.	.	.	.	.	.	.	.	.
30 TRANSITION FROM 3-D TO 2-D REGION	.	.	.	.	.	X	.	.	.
31 TRANSITION FROM 2-D TO 1-D REGION	.	.	.	.	.	X	.	.	.
32 OTHER CONSTRAINT CONDITIONS--SEE SECT.8, 1.	.	.	.	.	.	.	.	.	.
33 LAGRANGE MULTIPLIER METHOD USED	X	.	.	.	.	.	.	.	.
34 OTHER COMMENTS PERTINENT TO SECT. 1--SEE SE	.	X	.	.	.	.	.	.	X

SECTION 2, WALL CONSTRUCTION/MATERIAL PROPERTIES

READ PROGRAM NAMES VERTICALLY

	FARSS	FESAP	PILEFEM	SAASII	SEALSHELL	SOLIDSAP	TEXGAP	VISCEL	ZP26
01 MONOCOQUE WALL CONSTRUCTION	.	.	.	.	.	X	.	.	X
02 LAYERED WALL CONSTRUCTION	.	.	.	.	.	X	X	.	.
03 SANDWICH WALL CONSTRUCTION	.	.	.	.	.	X	.	.	X
04 COMPOSITE MATERIAL	.	.	.	.	.	X	.	.	.
05 MERIDIONAL STIFFENERS TREATED AS DISCRETE	.	.	.	.	.	X	.	.	.
06 CIRCUM. STIFFENERS TREATED AS DISCRETE	.	.	.	.	.	X	X	.	X
07 GENERAL STIFFENERS TREATED AS DISCRETE	.	X	.	.	.	X	.	.	.
08 STIFFENERS TREATED AS SMEARED	.	X	.	.	.	.	.	.	.
09 GENERAL OPEN-SECTION THEORY FOR STIFFENERS	.	.	.	.	.	.	.	.	.
10 CG, SHEAR CENTER OF STIFFENER MUST COINCIDE	.	X	.	.	.	.	.	.	.
11 OTHER WALL CONSTRUCTION--SEE SECT.8, 2.11	.	.	.	.	.	.	.	.	.
12 ISOTROPIC MATERIAL PROPERTIES	X	X	.	.	.	X	X	X	X
13 ANISOTROPIC MATERIAL PROPERTIES	X	X	.	.	.	X	X	X	X
14 LINEAR ELASTIC MATERIAL PROPERTIES	X	X	.	.	.	X	X	X	X
15 NONLINEAR ELASTIC MATERIAL PROPERTIES	.	.	X	.	.	.	.	.	.
16 TEMPERATURE-DEPENDENT MATERIAL PROPERTIES	X	X	.	.	.	X	.	X	.
17 RIGID-PERFECTLY PLASTIC MATERIAL	.	.	.	.	.	.	.	.	.
18 ELASTIC-PERFECTLY PLASTIC MATERIAL	.	.	.	.	.	.	.	.	X
19 ELASTIC-LINEAR STRAIN HARDENING MATERIAL	.	.	.	.	.	.	.	.	X
20 STRESS-STRAIN CURVE OF MANY LINE SEGMENTS	.	.	.	.	.	.	.	.	X
21 RAMBERG-OSGOOD STRESS-STRAIN LAW	.	.	.	.	.	.	.	.	X
22 ISOTROPIC STRAIN HARDENING	.	.	.	.	.	.	.	.	X
23 KINEMATIC STRAIN HARDENING	.	.	.	.	.	.	.	.	.
24 WHITE-BESSELING HARDENING LAW	.	.	.	.	.	.	.	.	.
25 OTHER POST-YIELD LAW--SEE SECT.8, 2.25	.	.	.	.	.	.	.	.	.
26 PRIMARY CREEP	.	.	.	.	.	.	.	.	.
27 SECONDARY CREEP	.	.	.	.	.	.	.	.	.
28 VISCOELASTIC MATERIAL	.	.	.	.	.	.	.	X	.
29 STRAIN-RATE EFFECTS INCLUDED	.	.	.	.	.	.	.	.	.
30 NO SPATIAL VARIATION OF PROPERTIES	.	X	.	.	.	X	.	.	.
31 AXISYMMETRIC VARIATION OF PROPERTIES ONLY	X	.	.	.	.	X	X	.	X
32 ARBITRARY VARIATION OF PROPERTIES	.	.	.	.	.	X	.	.	.
33 COUPLING OF STRAINS AND CURVATURE CHANGES	.	.	.	.	.	.	.	.	.
34 COUPLING OF NORMAL AND IN-PLANE SHEAR	.	.	.	.	.	.	.	.	.
35 OTHER COMMENTS PERTINENT TO SECTION 2--SEE	X	.	.	.	.	.	.	.	.

SECTION 3. LOADING

READ PROGRAM NAMES VERTICALLY

	FARSS	FESAP	PILEFEM	SAASI	SEALSHELL	SOLIDSAP	TEXGAP	VISCEL	ZP26
01 UNIFORM DISTRIBUTION OF LOADS	.	.	X	.	.	X	.	.	X
02 AXISYMMETRIC VARIATION OF LOADS ONLY	.	.	.	.	.	X	.	.	X
03 GENERAL VARIATION OF LOADS	X	X	.	.	.	X	X	X	.
04 STATIC OR QUASI-STATIC LOAD VARIATION	X	X	X	.	.	X	.	X	X
05 GENERAL TRANSIENT VARIATION OF LOADS	X	X	.	.	.	.	.	.	.
06 PERIODIC (DYNAMIC) VARIATION OF LOADS	X	.	.	.	.	.	.	.	.
07 IMPULSIVE LOADS	X	.	.	.	.	.	.	.	.
08 LOADS VARYING RANDOMLY IN TIME	.	.	.	.	.	.	.	.	.
09 OTHER TEMPORAL VARIATION OF LOADS--SEE SECT	.	.	.	.	.	.	.	.	.
10 LOADS VARYING PROPORTIONALLY DURING CASE	X	.	X	.	.	.	.	.	X
11 SOME LOADS VARY, SOME CONSTANT DURING CASE	X	X	.	.	.	.	.	.	.
12 DIFFERENT LOADS HAVE DIFFERENT TIME HISTORI	.	.	.	.	.	.	.	X	.
13 QUASI-STATIC LOADS VARYING CYCLICLY DURING	X	.	.	.	.	.	.	.	X
14 OTHER MANNER OF LOAD VARIATION DURING CASE-	.	.	.	.	.	.	.	.	.
15 LIVE (FOLLOWING) LOADS	.	.	.	.	.	.	.	.	X
16 GYROSCOPIC LOADS	.	.	.	.	.	.	.	.	.
17 INERTIAL LOADS	X	.	.	.	.	.	.	.	X
18 ELECTROMAGNETIC LOADS	.	.	.	.	.	.	.	.	.
19 FLUID-STRUCTURE INTERACTION	.	.	.	.	.	.	.	.	.
20 ACOUSTIC LOADING	.	.	.	.	.	.	.	.	.
21 SOIL-STRUCTURE INTERACTION	.	.	X	.	.	X	.	.	.
22 RIGID SOLID-STRUCTURE INTERACTION	.	.	.	.	.	X	.	.	X
23 OTHER DEFORMATION-DEPENDENT TYPE OF LOADING	.	.	.	.	.	.	.	.	.
24 POINT LOADS, MOMENTS	X	X	X	.	.	X	X	X	X
25 LINE LOADS, MOMENTS	X	.	.	.	.	X	X	X	X
26 NORMAL PRESSURE	X	X	X	.	.	X	X	X	X
27 SURFACE TRACTIONS	.	.	X	.	.	X	X	.	X
28 BODY FORCES	X	X	X	.	.	X	X	X	.
29 DEAD-WEIGHT LOADING	.	X	X	.	.	X	X	X	.
30 THERMAL LOADING	X	X	.	.	.	X	X	X	.
31 INITIAL STRESS OR STRAIN	.	.	X	.	.	X	.	.	.
32 CENTRIFUGAL LOADING	X	.	.	.	.	.	X	.	.
33 NON-ZERO DISPLACEMENTS IMPOSED	X	X	X	.	.	X	X	.	X
34 OTHER TYPE OF LOADING--SEE SECT.8, 3.34	.	.	X	.	.	.	.	.	.
35 OTHER COMMENTS PERTINENT TO THIS SECTION--S	.	.	.	.	.	.	.	.	X

SECTION 4, PHENOMENA

READ PROGRAM NAMES VERTICALLY

		FARSS	FESAP	PILEFEM	SAASII	SEALSHELL	SOLIDSAP	TEXGAP	VISCEL	ZP26
01	AXISYMMETRIC SMALL DEFLECTIONS	.	.	X	.	.	X	X	.	X
02	GENERAL SMALL DEFLECTIONS	X	X	.	.	.	X	.	X	.
03	AXISYMMETRIC LARGE DEFLECTIONS	.	.	.	.	.	.	.	.	X
04	GENERAL LARGE DEFLECTIONS	.	.	.	.	.	.	.	.	.
05	AXISYMMETRIC PLASTICITY	.	.	.	.	.	.	.	.	X
06	GENERAL PLASTICITY	.	.	.	.	.	.	.	.	.
07	LARGE STRAINS	.	.	.	.	.	.	.	.	.
08	TRANSVERSE SHEAR DEFORMATIONS	.	.	.	.	.	X	.	.	.
09	THERMAL EFFECTS	.	X	.	.	.	X	X	.	.
10	RADIATION EFFECTS	.	.	.	.	.	.	.	.	.
11	AUTOMATED YIELD CRITERION	.	X	.	.	.	.	.	.	.
12	AUTOMATED FRACTURE CRITERION	.	.	.	.	.	.	.	.	.
13	AUTOMATED FATIGUE CRITERION	.	.	.	.	.	.	.	.	.
14	AUTOMATED BUCKLING CRITERION	.	.	.	.	.	.	.	.	.
15	OTHER BUILT-IN CRITERIA--SEE SECT.8, 4.15	.	.	.	.	.	.	.	.	.
16	NONLINEAR COLLAPSE ANALYSIS	.	.	.	.	.	.	.	.	X
17	POST-BUCKLING PHENOMENA	.	.	.	.	.	.	.	.	X
18	BIFURC. BUCKLING, LINEAR AXISYM. PRESTRESS	X	.	.	.	.	.	.	.	.
19	BIFURC. BUCKLING, LINEAR GENERAL PRESTRESS	.	.	.	.	.	.	.	.	.
20	BIFURC. BUCKLING, NONLIN. AXISYM. PRESTRESS	.	.	.	.	.	.	.	.	.
21	BIFURC. BUCKLING, NONLIN. GENERAL PRESTRESS	.	.	.	.	.	.	.	.	.
22	BIFURC. BUCKLING, PREBUCKLING ROTATIONS	.	.	.	.	.	.	.	.	.
23	BIFURC, BUCKLING, TRANSVERSE SHEAR DEFORMAT	.	.	.	.	.	.	.	.	.
24	BIFURCATION BUCKLING WITH OTHER EFFECTS--SE	.	.	.	.	.	.	.	.	.
25	DYNAMIC BUCKLING	.	.	.	.	.	.	.	.	.
26	MODAL VIBRATIONS WITH NO PRESTRESS	X	X	.	.	.	.	.	.	.
27	MODAL VIBRATIONS, LINEAR AXISYM. PRESTRESS	.	.	.	.	.	.	.	.	.
28	MODAL VIBRATIONS, LINEAR GENERAL PRESTRESS	.	.	.	.	.	.	.	.	.
29	MODAL VIBRATIONS, NONLIN. AXISYM. PRESTRESS	.	.	.	.	.	.	.	.	.
30	MODAL VIBRATIONS, NONLIN. GENERAL PRESTRESS	.	.	.	.	.	.	.	.	.
31	MODAL VIBRATIONS, TRANSVERSE SHEAR DEFORMAT	.	.	.	.	.	.	.	.	.
32	MODAL VIBRATIONS WITH OTHER EFFECTS--SEE SE	.	.	.	.	.	.	.	.	.
33	NONLINEAR VIBRATIONS	.	.	.	.	.	.	.	.	.
34	LINEAR AXISYMMETRIC DYNAMIC RESPONSE	.	.	.	.	.	.	.	.	.
35	LINEAR GENERAL DYNAMIC RESPONSE	X	X	.	.	.	.	.	.	.
36	NONLINEAR AXISYMMETRIC DYNAMIC RESPONSE	.	.	.	.	.	.	.	.	.
37	NONLINEAR GENERAL DYNAMIC RESPONSE	.	.	.	.	.	.	.	.	.
38	WAVE PROPAGATION	.	.	.	.	.	.	.	.	.
39	DYNAMIC RESPONSE, TRANSVERSE SHEAR DEFORMAT	.	.	.	.	.	.	.	.	.
40	DYNAMIC RESPONSE WITH OTHER EFFECTS INCLUDE	.	.	.	.	.	.	.	.	.
41	VISCOUS DAMPING	X	.	.	.	.	.	.	.	.
42	STRUCTURAL DAMPING	X	X	.	.	.	.	.	.	.
43	LINEAR DAMPING	X	.	.	.	.	.	.	.	.
44	NONLINEAR DAMPING	.	.	.	.	.	.	.	.	.
45	OTHER KIND OF DAMPING--SEE SECT.8, 4.45	.	.	.	.	.	.	.	.	.
46	FLUTTER	.	.	.	.	.	.	.	.	.
47	OTHER DYNAMIC PHENOMENA--SEE SECT.8, 4.47	.	.	.	.	.	.	.	.	.
48	OTHER PHENOMENA OF ANY KIND--SEE SECT.8, 4.	.	.	.	.	.	.	.	.	.

SECTION 5, DISCRETIZATION

READ PROGRAM NAMES VERTICALLY

	FARSS	FESAP	PILEFEM	SAASII	SEALSHELL	SOLIDSAP	TEXGAP	VISCEL	ZP26
01 MULTI-SEGMENT FORWARD INTEGRATION	.	.	.	.	.	.	.	.	.
02 GALERKIN METHOD	.	.	X	.	.	.	.	X	X
03 RAYLEIGH-RITZ METHOD	.	X	X	.	.	.	.	.	X
04 BOUNDARY POINT MATCHING	.	.	.	.	.	.	.	.	.
05 FINITE DIFF. METHOD BASED ON EQUILIBRIUM EQ	.	.	.	.	.	.	.	.	.
06 FINITE DIFFERENCE ENERGY METHOD	.	.	.	.	.	.	.	.	.
07 1-DIMENSIONAL FINITE-DIFF. DISCRETIZATION	.	.	.	.	.	.	.	.	.
08 ORTHOGONAL 2-DIM. FINITE-DIFFERENCE GRID	.	.	.	.	.	.	.	.	.
09 NON-ORTHOGONAL 2-DIM. FINITE-DIFF. GRID	.	.	.	.	.	.	.	.	.
10 GENERAL QUADRILATERIAL 2-D FINITE-DIFF.GRID	.	.	.	.	.	.	.	.	.
11 OTHER FINITE-DIFF. DISCRETIZATION SCHEME--	.	.	.	.	.	.	.	.	.
12 FOURIER SERIES IN CIRCUMFERENTIAL DIRECTION	X	.	.	.	.	.	X	.	.
14 ROD ELEMENT	.	X	.	.	.	X	.	X	.
15 STRAIGHT BEAM ELEMENT	.	X	.	.	.	X	.	X	.
16 CURVED BEAM ELEMENT	.	.	.	.	.	.	.	X	.
17 CONICAL SHELL ELEMENT	.	.	.	.	.	.	.	X	.
18 AXISYM. SHELL ELEMENT	.	.	.	.	.	X	.	X	.
19 FLAT MEMBRANE ELEMENT	.	X	.	.	.	X	.	X	X
20 FLAT MEMBRANE ELEMENT	.	.	.	.	.	.	.	X	.
21 CURVED MEMBRANE ELEMENT	.	.	.	.	.	.	.	X	.
22 CURVED MEMBRANE ELEMENT	.	.	.	.	.	.	.	X	.
23 FLAT PLATE ELEMENT	.	X	.	.	.	X	.	X	.
24 FLAT PLATE ELEMENT	.	.	.	.	.	.	.	X	.
25 SHALLOW SHELL ELEMENT	.	.	.	.	.	X	.	X	.
26 SHALLOW SHELL ELEMENT	.	.	.	.	.	X	.	X	.
27 DEEP SHELL ELEMENT	.	.	.	.	.	X	.	.	.
28 AXISYM. SOLID ELEMENT	X	X	X	.	.	X	.	X	X
29 AXISYM. SOLID ELEMENT	.	.	.	.	.	.	.	.	.
30 THICK PLATE ELEMENT	.	.	.	.	.	X	.	X	.
31 THICK SHELL ELEMENT	.	.	.	.	.	X	.	X	.
32 SOLID (3-D) ELEMENT	.	X	.	.	.	X	.	X	.
33 SOLID (3-D) ELEMENT	.	X	.	.	.	.	.	X	.
34 SOLID (3-D) ELEMENT	.	.	.	.	.	.	.	X	.
35 CRACK-TIP ELEMENT	.	.	.	.	.	.	.	.	X
36 OTHER TYPE OF ELEMENT	.	.	X	.	.	.	.	.	.
37 FORCE METHOD OR OTHER METHOD	.	.	.	.	.	.	.	.	.
38 INCOMPATIBLE DISPLACEMENT FUNCTIONS USED	X	X	.	.	.	X	.	.	.
39 SUBSTRUCTURING	.	.	.	.	.	.	.	.	.
40 REPEATED USE OF IDENTICAL SUBSTRUCTURES	.	.	.	.	.	.	.	.	.
44 NODES CAN BE INTRODUCED IN ARBITRARY ORDER	X	X	.	.	.	.	X	X	.
45 AUTOMATIC RENUMBERING OF NODES	.	X	.	.	.	.	.	X	.
46 NON-DIAGONAL MASS MATRIX	.	.	.	.	.	.	.	X	.
47 DIAGONAL MASS MATRIX	X	X	.	.	.	.	.	X	.
48 OTHER COMMENTS PERTINENT TO THIS SECTION--S	.	.	.	.	.	.	.	.	X

SECTION 6, SOLUTION METHODS

READ PROGRAM NAMES VERTICALLY

	FARSS	FESAP	PILEFEM	SAASII	SEALSHELL	SOLIDSAP	TEXGAP	VISCEL	ZP26
01 FULL-MATRIX EQUATION SOLVER	.	X	.	.	.	.	.	.	.
02 CONSTANT BANDWIDTH EQUATION SOLVER	.	.	X	.	.	.	.	X	X
03 SKYLINE METHOD OF SOLUTION USED	X	.	.	.	.	X	.	.	.
04 MATRIX PARTITIONING USED	.	.	X	.	.	.	.	.	.
05 WAVE-FRONT METHOD USED FOR SOLUTION	.	.	.	.	.	.	X	.	.
06 OTHER SPARSE MATRIX METHOD USED FOR SOL'N--	.	X	.	.	.	.	.	.	.
07 ITERATIVE METHOD FOR SOLUTION OF LINEAR EQU	.	.	.	.	.	.	.	.	.
08 CONJUGATE GRADIENT METHOD FOR SOL'N OF LIN.	.	.	.	.	.	.	.	.	.
09 FORWARD INTEGRATION USED FOR SOL'N OF LIN.	.	.	.	.	.	.	.	.	.
10 OTHER SOLUTION METHOD FOR LINEAR EQUATIONS	.	.	.	.	.	.	.	.	.
11 CONDENSATION METHOD--SEE SECT.8, 6.11	.	.	.	.	.	.	.	.	.
12 LR,QR METHOD FOR EIGENVALUE EXTRACTION	.	.	.	.	.	.	.	.	.
13 LANCZOS METHOD FOR EIGENVALUE EXTRACTION	.	.	.	.	.	.	.	.	.
14 STURM SEQUENCE METHOD FOR EIGENVALUE EXTRAC	.	.	.	.	.	.	.	.	.
15 DETERMINANT SEARCH FOR EIGENVALUE EXTRACTIO	.	X	.	.	.	.	.	.	.
16 SINGLE VECTOR INVERSE PWR ITER. WITH SHIFTS	.	.	.	.	.	.	.	.	.
17 MULTI-VECTOR (SUBSPACE) ITERATION FOR EIGEN	X	.	.	.	.	.	.	.	.
18 OTHER EIGENVALUE EXTRACTION METHOD--SEE SEC	.	.	.	.	.	.	.	.	.
19 FULL NEWTON-RAPHSON METHOD FOR NONLIN. SOLN	.	.	.	.	.	.	.	.	.
20 MODIFIED NEWTON METHOD FOR NONLINEAR SOL'N	.	.	.	.	.	.	.	.	.
21 INCREMENTAL METHOD WITH NO EQUILIB. CHECK	.	.	.	.	.	.	.	.	X
22 INCREMENTAL METHOD WITH EQUILIBRIUM CHECK	.	.	.	.	.	.	.	.	.
23 RUNGE-KUTTA TYPE OF INCREMENTAL METHOD	.	.	.	.	.	.	.	.	.
24 OTHER TYPE OF INCREMENTAL METHOD--SEE SECT.	.	.	X	.	.	.	.	.	.
25 DIRECT ENERGY SEARCH FOR NONLINEAR SOLUTION	.	.	.	.	.	.	.	.	.
26 DYNAMIC RELAXATION FOR NONLINEAR SOLUTION	.	.	.	.	.	.	.	.	.
27 NONLIN. EQ. SOLVED BY SUCCESSIVE SUBSTITUTI	.	.	.	.	.	.	.	.	.
28 OTHER NONLINEAR STRATEGY--	.	.	.	.	.	.	.	.	.
29 LAGRANGIAN FORMULATION	.	.	.	.	.	.	.	.	.
30 UPDATED LAGRANGIAN FORMULATION	.	.	.	.	.	.	.	.	.
31 EULERIAN FORMULATION	.	.	.	.	.	.	.	.	.
32 TIME INTEGRATION BY MODAL SUPERPOSITION	X	X	.	.	.	.	.	.	.
33 TIME INTEGRATION BY EXPLICIT METHOD	X	.	.	.	.	.	.	.	.
34 TIME INTEGRATION BY IMPLICIT METHOD	.	.	.	.	.	.	.	.	.
35 TIME INTEGRATION BY EULER (CONST. ACCEL.)	.	.	.	.	.	.	.	.	.
36 TIME INTEGRATION BY NEWMARK BETA METHOD, SE	X	.	.	.	.	.	.	.	.
37 TIME INTEGRATION BY RUNGE-KUTTA OF ORDER--S	.	.	.	.	.	.	.	.	.
38 TIME INTEGRATION BY HOUBOLT METHOD	.	.	.	.	.	.	.	.	.
39 TIME INTEGRATION BY WILSON METHOD	.	X	.	.	.	.	.	.	.
40 TIME INTEGRATION BY CENTRAL DIFFERENCE METH	.	.	.	.	.	.	.	.	.
41 TIME INTEGRATION BY STIFFLY-STABLE METHOD	.	.	.	.	.	.	.	.	.
42 TIME INTEGRATION BY PREDICTOR-CORRECTOR OF	.	.	.	.	.	.	.	.	.
43 TIME INTEGRATION BY MULTI-STEP METHOD	.	.	.	.	.	.	.	.	.
44 TIME INTEGRATION BY SOME OTHER METHOD--SEE	.	.	.	.	.	.	.	.	.
45 OTHER COMMENTS PERTINENT TO THIS SECTION--S	.	.	.	.	.	.	.	.	.

SECTION 7, COMPUTER PROGRAM DISTRIBUTION
DOCUMENTATION/ORGANIZATION/MAINTENANCE

READ PROGRAM NAMES VERTICALLY

	FARSS	FESAP	PILEFEM	SAASII	SEALSHELL	SOLIDSAP	TEXGAP	VISCEL	ZP26
01 SOURCE PROGRAM TAPE AVAILABLE	.	.	.	.	.	.	X	X	X
02 SOURCE PROGRAM CARDS AVAILABLE	.	.	.	.	.	X	X	.	X
03 PROGRAM ABSOLUTE ELEMENT ONLY AVAILABLE	.	.	.	.	.	.	.	.	.
04 PROGRAM COMPLETELY PROPRIETARY	.	X	.	.	.	.	.	.	.
05 PROGRAM PRICE IN DOLLARS--SEE SECT.8, 7.05	.	.	X	.	.	X	X	.	.
06 PROGRAM IS AVAILABLE THROUGH DEVELOPER	X	.	.	.	.	.	X	.	.
07 PROGRAM IS AVAILABLE THROUGH COSMIC	.	.	.	.	.	.	.	X	.
08 PROGRAM AVAILABLE THRU OTHER SOFTWARE CENTE	.	.	.	.	.	X	.	.	.
09 PROGRAM CAN BE EXECUTED THRU SERVICE BUREAU	.	.	.	.	.	.	.	.	.
11 THIS PROGRAM HAS BEEN SUPERSEDED BY ANOTHER	.	.	.	.	.	.	.	.	X
12 IMPROVED VERSION BEING PREPARED--SEE SECT.8	.	X	.	.	.	.	X	.	.
13 PROGRAM IS NOT COMPLETED	X	.	X	.	.	.	X	.	.
14 USERS MANUALS ARE SELF-CONTAINED	X	X	X	.	.	X	.	X	X
15 USERS MANUALS HAVE LIST OF PITFALLS	.	X	.	.	.	.	.	.	X
16 USERS MANUALS HAVE TEST CASES	.	X	X	.	.	.	X	X	X
17 USERS MANUALS HAVE FLOW CHARTS	.	.	.	.	.	.	.	X	X
18 USERS MANUALS GIVE THEORY	.	X	X	.	.	.	.	X	X
19 OTHER QUALITIES OF USERS MANUALS--	.	.	.	.	.	.	X	.	.
20 PROGRAM LANGUAGE IS--SEE SECT.8,7.20	.	.	.	.	.	.	.	.	.
21 PROGRAM RUNS ON IBM	.	X	.	.	.	X	.	.	X
22 PROGRAM RUNS ON CDC	X	X	.	.	.	X	X	.	X
23 PROGRAM RUNS ON UNIVAC	X	.	.	.	.	X	.	X	.
24 PROGRAM RUNS ON GE	.	.	X	.	.	.	.	.	.
25 PROGRAM RUNS ON COMPUTER IDENTIFIED IN SECT	.	.	.	.	.	.	.	.	.
28 NO. OF PRIMARY OVERLAYS *	L	8	.	.	.	.	6	4	X
MAN-YEARS REQUIRED TO DEVELOP PROGRAM *	2	2	1	.	.	.	6	2	X
31 MINIMUM CORE SPACE REQUIRED FOR EXECUTION	X	2	.	.	.	1	.	X	.
32 PROGRAM RUNS ENTIRELY IN CORE	.	.	.	.	.	.	.	.	X
33 NO. UNITS OF DIRECT ACCESS MASS STORAGE GIV	1	8	.	.	.	8	9	L	.
36 RESTART CAPABILITY	X	.	.	.	.	.	X	.	.
37 OPTIONS FOR VARIOUS OUTPUT	X	X	.	.	.	.	X	X	X
38 DIRECT ACCESS DATA BASE	.	.	.	.	.	.	.	.	.
39 FREE-FIELD INPUT	.	.	.	.	.	.	X	.	.
40 RUNS IN INTERACTIVE MODE	X	.	.	.	.	.	.	.	.
41 AUTOMATIC MESH GENERATOR(S)	X	X	X	.	.	X	X	.	X
42 AUTOMATIC LOADING GENERATOR	.	.	.	.	.	X	X	.	.
43 SECTION PROPERTIES CALCULATED	.	.	.	.	.	.	X	.	.
44 PRE-PROCESSOR	.	X	.	.	.	.	.	X	X
45 POST-PROCESSOR	X	X	.	.	.	.	.	.	X
46 EIGENVALUE SOLVER	X	.	.	.	.	.	.	.	.
47 MODAL SUPERPOSITION ROUTINE	X	.	.	.	.	.	.	.	.
48 OTHER TIME-INTEGRATION ROUTINE	X	.	.	.	.	.	.	.	.
49 OTHER CAPABILITY--SEE SECT.8, 7.49	.	.	.	.	.	.	.	.	.
50 PLOT ROUTINES FOR UNDEFORMED/DEFORMED GEOME	X	X	.	.	.	.	X	.	X
51 VARIOUS VIEWS AND SECTIONS PLOTTED	X	X	.	.	.	.	.	.	X
52 CONTOUR PLOTS	X	X	.	.	.	.	X	.	X
53 ORDINARY Y = F(X) PLOTS	.	.	.	.	.	.	.	.	.
54 STRESS-ON-STRUCTURE PLOTS	.	X	.	.	.	.	.	.	X
55 OTHER KINDS OF PLOTS--SEE SECT.8, 7.55	.	X	.	.	.	.	.	.	.
56 PLOTTING CAPABILITY FOR CALCOMP	.	X	X	.	.	.	X	X	.
57 PLOTTING CAPABILITY FOR STROMBERG-CARLSON	X	.	.	.	.	.	.	.	X
58 PLOTTING CAPABILITY FOR HARDWARE IDENTIFIED	.	.	.	.	.	.	.	.	.
59 OPTIONAL DETAILED PRINT OUTPUT	X	X	.	.	.	.	X	X	.
60 EQUILIBRIUM CHECKS	.	.	.	.	.	.	.	X	.
61 ALTERNATIVE BRANCHES FOR SAME PROBLEM	.	.	.	.	.	.	.	.	.
62 MATRIX CONDITIONING ESTIMATION	.	X	.	.	.	.	.	.	X
63 AUTOMATIC ERROR CONTROL	.	.	X	.	.	.	.	.	.
64 GRAPHICAL PREPROCESSOR	X	X	.	.	.	.	X	.	X
65 APPLIED LOADING CHECKS	.	.	.	.	.	.	.	X	X
66 BOUNDARY CONDITION CHECKS	.	X	.	.	.	.	.	X	X
67 SECTION PROPERTIES CHECKS	.	.	.	.	.	.	.	.	.
68 GEOMETRY CHECKS	X	X	.	.	.	.	X	X	X
69 OTHER CHECKING AIDS--SEE SECT.8, 7.69	.	.	.	.	.	.	.	.	.
70 DEVELOPER WILL HELP USER GET PROGRAM RUNNIN	X	X	.	.	.	.	X	X	.
71 DEVELOPER WILL HELP USER MODELING CASES	X	X	.	.	.	.	X	X	.
72 DEVELOPER WILL HELP USER FIND SUSPECTED BUG	X	X	.	.	.	.	X	X	.
73 DEVELOPER WILL CIRCULATE NOTICES OF BUGS FO	X	X	X	.	.	.	X	X	.
74 DEVELOPER WILL HELP USER SET UP AND RUN CAS	X	X	X	.	.	.	X	X	.
75 DEVELOPER WILL CONDUCT WORKSHOPS	X	X	.	.	.	.	X	X	.
76 DEVELOPER WILL MAKE SPECIAL PROGRAM CHANGES	X	X	.	.	.	.	X	X	.
77 DEVELOPER WILL KEEP USERS INFORMED OF IMPRO	X	X	X	.	.	.	X	X	.
78 OTHER ASSISTANCE DEVELOPER WILL GIVE USERS-	.	.	.	.	.	.	.	X	.
79 OTHER COMMENTS PERTINENT TO THIS SECTION--S	X	.	.	.	.	.	.	.	.

*L = LARGE, X= SEE SECT. 8, (WITH 'SCOPE' OPTION)

```
SECTION 1,  GEOMETRY / BOUNDARY CONDITIONS

                                        READ PROGRAM NAMES VERTICALLY

                                                 BDDIIKMNPRSSSSSSSSSSTT
                                                 OYYSSSIOEEAAAAHLNRTTIR
                                                 SNNOTHNNTPBDLTOAAAAARO
                                                 OAAPREILRSOAOARDS.GREC
                                                 RPSAALEIOIRORNEEO.SS.S
                                                 4LORN.LNSL/SSS.-R.B...
                                                 .AR-/.A.3.D....DI.....
                                                 .SISS.S...R.....I.....
                                                 .IIH......A...........
                                                 .I.L......S...........
                                                 ..........T...........
                                                 ..........I...........
                                                 ..........C...........
                                                 ..........6...........
                                                 ......................
01 GENERAL STRUCTURE ........................... ......................
02 SOLID OF REVOLUTION.......................... ......................
03 GENERAL SHELL................................ ...XX.X.XX........X.X.
04 PART OF SHELL OF REVOLUTION ( L.T. 360 DEG.  .X.X....XX....XX..XX.X
05 SHELL OF REVOLUTION (360 DEG.) .............. XXX..X.XXXXXXXXXXXXX.X
06 PRISMATIC STRUCTURE.......................... X.....X...........XX..
07 SEGMENTED (IN SERIES)    .................... X....XX...X.X....XXX..
08 BRANCHED     ................................ X..X.XXX..X.....X.XXX..
09 SHELL WITH CUTOUTS .......................... ...X..X........X..X...
10 THICK SHELL  ................................ ...X..X...............
11 FRAMEWORK-SHELL COMBINATIONS  ............... ......X...........X...
12 FLAT PLATE OF GENERAL SHAPE   ............... ...XX.X.XX........X...
13 OTHER GEOMETRY--SEE SECT.8, 1.13       ...... .....................X
14 AXISYMMETRIC IMPERFECTIONS    ............... X....X.X..X.......XXX..
15 GENERAL IMPERFECTIONS   ..................... ...XX.....X.......XX...
16 AXISYMMETRIC SUPPORT CONDITIONS.............. XXX..X.X....XX..XXXX..
17 GENERAL SUPPORT CONDITIONS    ............... .XXXX.XX..XX.X.XX.X.X.
18 SUPPORTS AT BOUNDARIES ONLY   ............... .....XX.XX.XXXXX...X.X
19 SUPPORTS AT INTERNAL POINTS   ............... XXXX..X...X...X.XXXX..
20 ELASTIC FOUNDATION .......................... .....X....X.X.X..X.X.X
21 STIFFENERS   ................................ X..X..X...X.X....XXX..
22 CONTACT...................................... ....................X.
23 FRICTION     ................................ ......................
24 SLIDING WITHOUT FRICTION .................... X.....X.........X..X...
25 OTHER DEFORMATION DEPENDENT SUPPORT--SEE SE .....X.........X.......
26 JUNCTURE COMPATIBILITY   .................... X..X.X.X....X.....XXX..
27 SINGULARITY CONDITIONS   .................... XXX..........X....X..X..
28 GENERAL LINEAR CONSTRAINT COND. U(I)= TU(N) ......X............X....
29 GENERAL NONLINEAR CONSTRAINT CONDITIONS       ............X.........
30 TRANSITION FROM 3-D TO 2-D REGION      ...... ...X..X...............
31 TRANSITION FROM 2-D TO 1-D REGION      ...... ......X...............
32 OTHER CONSTRAINT CONDITIONS--SEE SECT.8, 1. ......X........X.......X
33 LAGRANGE MULTIPLIER METHOD USED.............. X..................X...
34 OTHER COMMENTS PERTINENT TO SECT. 1--SEE SE ......................
```

```
SECTION 2, WALL CONSTRUCTION/MATERIAL PROPERTIES

                                                  READ PROGRAM NAMES VERTICALLY

                                                         BDDIIKMNPRSSSSSSSSSSTT
                                                         OYYSSSIOEEAAAAHLNRTTIR
                                                         SNNOTHNNTPBDLTOAAAAARO
                                                         OAAPREILRSOAOARDS.GREC
                                                         RPSAALEIOIRORNFEO.SS.S
                                                         4LORN.LNSL/SSS.-R.B...
                                                         .AR-/.A.3.D....DI.....
                                                         .SISS.S...R.....I.....
                                                         .IIH......A...........
                                                         .I.L......S...........
                                                         ..........T...........
                                                         ..........I...........
                                                         ..........C...........
                                                         ..........6...........

                                                         ......................
01 MONOCOQUE WALL CONSTRUCTION          ............  XXXXXXXX..XX.XX.X.XX..
02 LAYERED WALL CONSTRUCTION.......................  X..X.XX.X..XX.XX.XXXXX
03 SANDWICH WALL CONSTRUCTION           ............  X..X.XX......X.....X..
04 COMPOSITE MATERIAL ...............................  X.....X...........XXX.
05 MERIDIONAL STIFFENERS TREATED AS DISCRETE          ..................X...
06 CIRCUM. STIFFENERS TREATED AS DISCRETE             XX...X.X..X.X....XXX..
07 GENERAL STIFFENERS TREATED AS DISCRETE             ...XX.X...........X...
08 STIFFENERS TREATED AS SMEARED  ............        X.................X...
09 GENERAL OPEN-SECTION THEORY FOR STIFFENERS         .X................XX..
10 CG, SHEAR CENTER OF STIFFENER MUST COINCIDE        X...X.X.....X....XX...
11 OTHER WALL CONSTRUCTION--SEE SECT.8, 2.11          X....X..X.........XX..
12 ISOTROPIC MATERIAL PROPERTIES  ............        XXXXXXXXXXXX.XXXX.XX.X
13 ANISOTROPIC MATERIAL PROPERTIES...........         X..X.XX...X.X.XX.XXXXX
14 LINEAR ELASTIC MATERIAL PROPERTIES   ......        X.XXXXX.XXX.XXXXXXXX.X
15 NONLINEAR ELASTIC MATERIAL PROPERTIES......        ....XX.X..X........X..
16 TEMPERATURE-DEPENDENT MATERIAL PROPERTIES          X....X..X..X..XX..XX..
17 RIGID-PERFECTLY PLASTIC MATERIAL     ......        ..........X.......X...
18 ELASTIC-PERFECTLY PLASTIC MATERIAL   ......        ........XXX...X...XX..
19 ELASTIC-LINEAR STRAIN HARDENING MATERIAL           ..........X...X....X.X
20 STRESS-STRAIN CURVE OF MANY LINE SEGMENTS          .X......XXX.......X...
21 RAMBERG-OSGOOD STRESS-STRAIN LAW     ......        ...................X..
22 ISOTROPIC STRAIN HARDENING     ............        .X........X...X....X.X
23 KINEMATIC STRAIN HARDENING     ............        ........XX.........X..
24 WHITE-BESSELING HARDENING LAW  ............        .X................X...
25 OTHER POST-YIELD LAW--SEE SECT.8, 2.25             ......................
26 PRIMARY CREEP.............................         ......................
27 SECONDARY CREEP  ..........................        ..................X...
28 VISCOELASTIC MATERIAL   ...................        ......................
29 STRAIN-RATE EFFECTS INCLUDED   ............        .X......XX............
30 NO SPATIAL VARIATION OF PROPERTIES   ......        ........XX.X.........X
31 AXISYMMETRIC VARIATION OF PROPERTIES ONLY          XXX..X.X....XX..XX.X..
32 ARBITRARY VARIATION OF PROPERTIES    ......        ...XX.X...X....X..X.X.
33 COUPLING OF STRAINS AND CURVATURE CHANGES          X..X.X..XX..X.XX.XXXXX
34 COUPLING OF NORMAL AND IN-PLANE SHEAR......        ..............XX..X..X
35 OTHER COMMENTS PERTINENT TO SECTION 2--SEE         ........XX...........X
```

SECTION 3, LOADING

```
                                                READ PROGRAM NAMES VERTICALLY

                                                  BDDIIKMNPRSSSSSSSSSSTT
                                                  OYYSSSIOEEAAAAHLNRTTIR
                                                  SNNOTHNNTPBDLTOAAAAARO
                                                  OAAPREILRSOAOARDS.GREC
                                                  RPSAALEIOIRORNEEO.SS.S
                                                  4LORN.LNSL/SSS.-R.B...
                                                  .AR-/.A.3.D....DI.....
                                                  .SISS.S...R.....I.....
                                                  .IIH......A...........
                                                  .I.L......S...........
                                                  ..........T...........
                                                  ..........I...........
                                                  ..........C...........
                                                  ..........6...........

                                                  ......................
01 UNIFORM DISTRIBUTION OF LOADS  ............... X..X......X.......X...
02 AXISYMMETRIC VARIATION OF LOADS ONLY ......... ......................
03 GENERAL VARIATION OF LOADS     ............... XXXXXXXXXXX.XXXXXXXXXX
04 STATIC OR QUASI-STATIC LOAD VARIATION......... X..XXXXX..X..X.XXXX.X.
05 GENERAL TRANSIENT VARIATION OF LOADS ......... .XX..X..XXXX.XX...XX.X
06 PERIODIC (DYNAMIC) VARIATION OF LOADS......... .....XX...........XX..
07 IMPULSIVE LOADS     .......................... .XX.....XXX..XXX..X..X
08 LOADS VARYING RANDOMLY IN TIME ............... ......X...............
09 OTHER TEMPORAL VARIATION OF LOADS--SEE SECT .. ......................
10 LOADS VARYING PROPORTIONALLY DURING CASE       XXX.XXX......X.X.XX...
11 SOME LOADS VARY, SOME CONSTANT DURING CASE     X.........X......XXX..
12 DIFFERENT LOADS HAVE DIFFERENT TIME HISTORI .. ..........X...X....X.X
13 QUASI-STATIC LOADS VARYING CYCLICLY DURING     ......................
14 OTHER MANNER OF LOAD VARIATION DURING CASE- .. ......................
15 LIVE (FOLLOWING) LOADS   ..................... X...X..X....X....X.X..
16 GYROSCOPIC LOADS   ........................... ......................
17 INERTIAL LOADS   ............................. ......XX.....X...X..X.
18 ELECTROMAGNETIC LOADS   ...................... ......................
19 FLUID-STRUCTURE INTERACTION    ............... ......................
20 ACOUSTIC LOADING ............................. ......................
21 SOIL-STRUCTURE INTERACTION     ............... ......................
22 RIGID SOLID-STRUCTURE INTERACTION    ......... ......................
23 OTHER DEFORMATION-DEPENDENT TYPE OF LOADING .. .....................X
24 POINT LOADS, MOMENTS   ....................... XXXXXXX...X...X.X.XXX.
25 LINE LOADS, MOMENTS........................... XXXXXX.X..X.X.XXXXXXX.
26 NORMAL PRESSURE   ............................ XXXXXXXXXXXXX.XXXXXXXX
27 SURFACE TRACTIONS   .......................... XXXX.X.X..XXXXXXXXXX.X
28 BODY FORCES   ................................ ...XX.XX..X.......XXX.
29 DEAD-WEIGHT LOADING........................... XXXXXXXX.........X.X..
30 THERMAL LOADING   ............................ X.XX.XXXX...XXXXXXXX..
31 INITIAL STRESS OR STRAIN ..................... ....XXX.......X..X.XX.
32 CENTRIFUGAL LOADING........................... ....XX.X............X..
33 NON-ZERO DISPLACEMENTS IMPOSED ............... ...XX.XX..X.......XX...
34 OTHER TYPE OF LOADING--SEE SECT.8, 3.34        ......................
35 OTHER COMMENTS PERTINENT TO THIS SECTION--S .. ........XX............
```

```
         SECTION 4, PHENOMENA

                                         READ PROGRAM NAMES VERTICALLY

                                                BDDIIKMNPRSSSSSSSSSSTT
                                                OYYSSSIOEEAAAAHLNRTTIR
                                                SNNOTHNNTPBDLTOAAAAARO
                                                OAAPREILRSOAOARDS.GREC
                                                RPSAALEIOIRORNEEO.SS.S
                                                4LORN.LNSL/SSS..R.B...
                                                .AR-/.A.3.D....DI.....
                                                .SISS.S...R.....I.....
                                                .IIH......A...........
                                                .I.L......S...........
                                                ..........T...........
                                                ..........I...........
                                                ..........C...........
                                                ..........6...........

                                                ......................
01 AXISYMMETRIC SMALL DEFLECTIONS ............. X....X....X.......XX..
02 GENERAL SMALL DEFLECTIONS................... X..XXXXX..X.X..X.XXXX.
03 AXISYMMETRIC LARGE DEFLECTIONS ............. X....X.X..XX.....XXX..
04 GENERAL LARGE DEFLECTIONS................... .XX.X.X.XXX..X..X.X.X.
05 AXISYMMETRIC PLASTICITY  ................... ..........X.......XX..
06 GENERAL PLASTICITY ......................... ........XXX.......X...
07 LARGE STRAINS............................... ........XX............
08 TRANSVERSE SHEAR DEFORMATIONS  ............. ...X..X....X........X.
09 THERMAL EFFECTS      ....................... X.XX.XXXX..XXX..XXXX..
10 RADIATION EFFECTS    ....................... ......................
11 AUTOMATED YIELD CRITERION................... .X......XX....XX...X..
12 AUTOMATED FRACTURE CRITERION    ............ ..............XX......
13 AUTOMATED FATIGUE CRITERION     ............ ......................
14 AUTOMATED BUCKLING CRITERION    ............ ...........X.....X.X..
15 OTHER BUILT-IN CRITERIA--SEE SECT.8, 4.15    ......................
16 NONLINEAR COLLAPSE ANALYSIS     ............ X......X...X.X..XXXX..
17 POST-BUCKLING PHENOMENA  ................... .................XX...
18 BIFURC. BUCKLING, LINEAR AXISYM. PRESTRESS   X....X............XX..
19 BIFURC. BUCKLING, LINEAR GENERAL PRESTRESS   X...X.X..........XX...
20 BIFURC. BUCKLING, NONLIN. AXISYM. PRESTRESS  X................X....
21 BIFURC. BUCKLING, NONLIN. GENERAL PRESTRESS  ................X.....
22 BIFURC. BUCKLING, PREBUCKLING ROTATIONS      X...............XX.X..
23 BIFURC. BUCKLING, TRANSVERSE SHEAR DEFORMAT  ......................
24 BIFURCATION BUCKLING WITH OTHER EFFECTS--SE  ......................
25 DYNAMIC BUCKLING     ....................... .XX..X..XXX..XX.......
26 MODAL VIBRATIONS WITH NO PRESTRESS   ....... ....X.X...............
27 MODAL VIBRATIONS, LINEAR AXISYM. PRESTRESS   .....X.............X..
28 MODAL VIBRATIONS, LINEAR GENERAL PRESTRESS   ......X...............
29 MODAL VIBRATIONS, NONLIN. AXISYM. PRESTRESS  X................X....
30 MODAL VIBRATIONS, NONLIN. GENERAL PRESTRESS  ......................
31 MODAL VIBRATIONS, TRANSVERSE SHEAR DEFORMAT  ......................
32 MODAL VIBRATIONS WITH OTHER EFFECTS--SEE SE  ......................
33 NONLINEAR VIBRATIONS      .................. ......................
34 LINEAR AXISYMMETRIC DYNAMIC RESPONSE ....... .....X....X...X...X..X
35 LINEAR GENERAL DYNAMIC RESPONSE............. .....XX...X...XX..XX.X
36 NONLINEAR AXISYMMETRIC DYNAMIC RESPONSE      ..........X...X...X..X
37 NONLINEAR GENERAL DYNAMIC RESPONSE   ....... .XX.....XXX..XX...X..X
38 WAVE PROPAGATION      ...................... ...................X..
39 DYNAMIC RESPONSE, TRANSVERSE SHEAR DEFORMAT  ......................
40 DYNAMIC RESPONSE WITH OTHER EFFECTS INCLUDE  ......X..............X
41 VISCOUS DAMPING      ....................... .....XX.......X......X
42 STRUCTURAL DAMPING ......................... ..........X.......X...
43 LINEAR DAMPING       ....................... ......X.......X...X..X
44 NONLINEAR DAMPING    ....................... ......................
45 OTHER KIND OF DAMPING--SEE SECT.8, 4.45      .....................X
46 FLUTTER..................................... ......................
47 OTHER DYNAMIC PHENOMENA--SEE SECT.8, 4.47    ......................
48 OTHER PHENOMENA OF ANY KIND--SEE SECT.8, 4.  ......................
```

```
        SECTION 5. DISCRETIZATION

                                                   READ PROGRAM NAMES VERTICALLY

                                                           BDDIIKMNPRSSSSSSSSSSTT
                                                           OYYSSSIOEFAAAAHLNRTTIR
                                                           SNNOTHNNTPBDLTOAAAAARO
                                                           OAAPREILRSOAOARDS.GREC
                                                           RPSAALEIOIRORNEEO.SS.S
                                                           4LORN.LNSL/SSS.-R.B...
                                                           .AR-/.A.3.D....DI.....
                                                           .SISS.S...R.....I.....
                                                           .IIH......A...........
                                                           .I.L......S...........
                                                           ..........T...........
                                                           ..........I...........
                                                           ..........C...........
                                                           ..........6...........
                                                           ......................
 01 MULTI-SEGMENT FORWARD INTEGRATION               ...... .....X.X.........X.X..
 02 GALERKIN METHOD  ..................................... ...X..................
 03 RAYLEIGH-RITZ METHOD     ......................... ...X..X...............
 04 BOUNDARY POINT MATCHING  ......................... ......................
 05 FINITE DIFF. METHOD BASED ON EQUILIBRIUM EQ .......XX..XXX......X
 06 FINITE DIFFERENCE ENERGY METHOD................ X.................X...
 07 1-DIMENSIONAL FINITE-DIFF. DISCRETIZATION       X..........X.X.....X..
 08 ORTHOGONAL 2-DIM. FINITE-DIFFERENCE GRID        ..............X...X..X
 09 NON-ORTHOGONAL 2-DIM. FINITE-DIFF. GRID         ........XX........X...
 10 GENERAL QUADRILATERIAL 2-D FINITE-DIFF.GRID     ..................X...
 11 OTHER FINITE-DIFF. DISCRETIZATION SCHEME--      ......................
 12 FOURIER SERIES IN CIRCUMFERENTIAL DIRECTION X......X..X.XX...X.X..
 14 ROD ELEMENT ..................................... ......X...........X...
 15 STRAIGHT BEAM ELEMENT    ......................... ....X.X...............
 16 CURVED BEAM ELEMENT.............................. .X....................
 17 CONICAL SHELL ELEMENT    ......................... .XX.......X.....X.....
 18 AXISYM. SHELL ELEMENT    ......................... .XX.......X.....X.....
 19 FLAT MEMBRANE ELEMENT    ......................... ....X.X...............
 20 FLAT MEMBRANE ELEMENT    ......................... ......X...........X...
 21 CURVED MEMBRANE ELEMENT  ......................... ......X...............
 22 CURVED MEMBRANE ELEMENT  ......................... ......................
 23 FLAT PLATE ELEMENT .............................. .XX.X.X.........X...X.
 24 FLAT PLATE ELEMENT .............................. ....X.X...............
 25 SHALLOW SHELL ELEMENT    ......................... ...X..X...............
 26 SHALLOW SHELL ELEMENT    ......................... .XX.............X.....
 27 DEEP SHELL ELEMENT .............................. .XX...X........XX.....
 28 AXISYM. SOLID ELEMENT    ......................... ......................
 29 AXISYM. SOLID ELEMENT    ......................... ......................
 30 THICK PLATE ELEMENT.............................. ...X..X...............
 31 THICK SHELL ELEMENT.............................. ...X..X...............
 32 SOLID (3-D) ELEMENT.............................. ...X..................
 33 SOLID (3-D) ELEMENT.............................. ...X..................
 34 SOLID (3-D) ELEMENT.............................. ...X..................
 35 CRACK-TIP ELEMENT  ............................... ......................
 36 OTHER TYPE OF ELEMENT    ......................... ...X..................
 37 FORCE METHOD OR OTHER METHOD     ................. ....X.................
 38 INCOMPATIBLE DISPLACEMENT FUNCTIONS USED         ......................
 39 SUBSTRUCTURING       ............................. ......................
 40 REPEATED USE OF IDENTICAL SUBSTRUCTURES          ......................
 44 NODES CAN BE INTRODUCED IN ARBITRARY ORDER       ...X..X.............X.
 45 AUTOMATIC RENUMBERING OF NODES ................. ......X...............
 46 NON-DIAGONAL MASS MATRIX ......................... X.X...X...X.....X.....
 47 DIAGONAL MASS MATRIX     ......................... ....X.X........X..X...
 48 OTHER COMMENTS PERTINENT TO THIS SECTION--S X.................X...
```

```
SECTION 6, SOLUTION METHODS

                                          READ PROGRAM NAMES VERTICALLY

                                                   BDDIIKMNPRSSSSSSSSSSTT
                                                   OYYSSSIOEEAAAAHLNRTTIR
                                                   SNNOTHNNTPBDLTOAAAAARO
                                                   OAAPREILRSOAOARDS.GREC
                                                   RPSAALEIOIRORNEEO.SS.S
                                                   4LORN.LNSL/SSS._R.B...
                                                   .AR-/.A.3.D....DI.....
                                                   .SISS.S...R.....I.....
                                                   .IIH......A...........
                                                   .I.L......S...........
                                                   ..........T...........
                                                   ..........I...........
                                                   ..........C...........
                                                   ..........6...........
                                                   ......................
01 FULL-MATRIX EQUATION SOLVER     ............    ..........X......X.X..
02 CONSTANT BANDWIDTH EQUATION SOLVER    ......    .XX..X.X..X.XX..X..XX.
03 SKYLINE METHOD OF SOLUTION USED.............    X.................X...
04 MATRIX PARTITIONING USED ...................    ..........X...........
05 WAVE-FRONT METHOD USED FOR SOLUTION   ......    ...X......X........X..
06 OTHER SPARSE MATRIX METHOD USED FOR SOL'N--    ....X..............X..
07 ITERATIVE METHOD FOR SOLUTION OF LINEAR EQU    ...........X.X........
08 CONJUGATE GRADIENT METHOD FOR SOL'N OF LIN.    ......................
09 FORWARD INTEGRATION USED FOR SOL'N OF LIN.     ......................
10 OTHER SOLUTION METHOD FOR LINEAR EQUATIONS     ......X...............
11 CONDENSATION METHOD--SEE SECT.8, 6.11......    ......................
12 LR,QR METHOD FOR EIGENVALUE EXTRACTION         ...................X..
13 LANCZOS METHOD FOR EIGENVALUE EXTRACTION       ......................
14 STURM SEQUENCE METHOD FOR EIGENVALUE EXTRAC    ...................X..
15 DETERMINANT SEARCH FOR EIGENVALUE EXTRACTIO    .....X.............X..
16 SINGLE VECTOR INVERSE PWR ITER. WITH SHIFTS    X...XX...........XX...
17 MULTI-VECTOR (SUBSPACE) ITERATION FOR EIGEN    ......................
18 OTHER EIGENVALUE EXTRACTION METHOD--SEE SEC    ......X...............
19 FULL NEWTON-RAPHSON METHOD FOR NONLIN. SOLN    X..........X....XX....
20 MODIFIED NEWTON METHOD FOR NONLINEAR SOL'N     .....X.X........X.XX..
21 INCREMENTAL METHOD WITH NO EQUILIB. CHECK      ..................X.X.
22 INCREMENTAL METHOD WITH EQUILIBRIUM CHECK      ....X...........XX.X..
23 RUNGE-KUTTA TYPE OF INCREMENTAL METHOD         ......................
24 OTHER TYPE OF INCREMENTAL METHOD--SEE SECT.    ......................
25 DIRECT ENERGY SEARCH FOR NONLINEAR SOLUTION    ......................
26 DYNAMIC RELAXATION FOR NONLINEAR SOLUTION      ..................X...
27 NONLIN. EQ. SOLVED BY SUCCESSIVE SUBSTITUTI    ..........X..X.....X..
28 OTHER NONLINEAR STRATEGY--      ............   ...................X..
29 LAGRANGIAN FORMULATION    ..................   X.......XXX.....XXX...
30 UPDATED LAGRANGIAN FORMULATION ............    ....X.................
31 EULERIAN FORMULATION      ..................   ......................
32 TIME INTEGRATION BY MODAL SUPERPOSITION        .....XX...............
33 TIME INTEGRATION BY EXPLICIT METHOD  ......    .X......XX....XX..X..X
34 TIME INTEGRATION BY IMPLICIT METHOD  ......    .XX.......X..X....X...
35 TIME INTEGRATION BY EULER (CONST. ACCEL.)      ...................X..
36 TIME INTEGRATION BY NEWMARK BETA METHOD, SE    ..........X...........
37 TIME INTEGRATION BY RUNGE-KUTTA OF ORDER--S    ......................
38 TIME INTEGRATION BY HOUBOLT METHOD   ......    .XX..........X........
39 TIME INTEGRATION BY WILSON METHOD    ......    ......................
40 TIME INTEGRATION BY CENTRAL DIFFERENCE METH    .X......XX....XX.....X
41 TIME INTEGRATION BY STIFFLY-STABLE METHOD      ..................X...
42 TIME INTEGRATION BY PREDICTOR-CORRECTOR OF     ..................X...
43 TIME INTEGRATION BY MULTI-STEP METHOD......    ..................X...
44 TIME INTEGRATION BY SOME OTHER METHOD--SEE     ......................
45 OTHER COMMENTS PERTINENT TO THIS SECTION--S    ......................
```

```
SECTION 7, COMPUTER PROGRAM DISTRIBUTION
 DOCUMENTATION/ORGANIZATION/MAINTENANCE

                                     READ PROGRAM NAMES VERTICALLY

                                                     BDDIIKMNPPSSSSSSSSSSTT
                                                     OYYSSSIOEFAAAAHLNPTTIR
                                                     SNNOTHNNTPRDLTOAAAAARO
                                                     OAAPREILRSOAOARDS.GREC
                                                     RPSAALEIOIRORNEEO.SS.S
                                                     4LORN.LNSL/SSS.-R.B...
                                                     .AR-/.4.3.D....DI.....
                                                     .SISS.S...R.....I.....
                                                     .IIH......A...........
                                                     .I.L......S...........
                                                     ..........T...........
                                                     ..........I...........
                                                     ..........C...........
                                                     ..........6...........
                                                     ......................
01 SOURCE PROGRAM TAPE AVAILABLE  ............ XXX...X...XX..XXXXXX.X
02 SOURCE PROGRAM CARDS AVAILABLE ............ .....XX...XX..X......X
03 PROGRAM ABSOLUTE ELEMENT  ONLY AVAILABLE    ....X.................
04 PROGRAM COMPLETELY PROPRIETARY ............ ...X..................
05 PROGRAM PRICE IN DOLLARS--SEE SECT.8, 7.05  XXX.XXX.........XX....
06 PROGRAM IS AVAILABLE THROUGH DEVELOPER      XXXXXXX.XXXX..X.XXX..X
07 PROGRAM IS AVAILABLE THROUGH COSMIC  ...... X.X.........XX..XX....
08 PROGRAM AVAILABLE THRU OTHER SOFTWARE CENTE .............X.X..X.X.
09 PROGRAM CAN BE EXECUTED THRU SERVICE BUREAU X....X................
11 THIS PROGRAM HAS BEEN SUPERSEDED BY ANOTHER ..X....X..............
12 IMPROVED VERSION BEING PREPARED--SEE SECT.8 ..X.X............XX.X.
13 PROGRAM IS NOT COMPLETED .................. ...X..................
14 USERS MANUALS ARE SELF-CONTAINED     ...... XXX.XX.XX..XXXXXXXXXXX
15 USERS MANUALS HAVE LIST OF PITFALLS  ...... XXX.............X.XX.X
16 USERS MANUALS HAVE TEST CASES  ............ XXX..X.XX..XXXXXXXXX.X
17 USERS MANUALS HAVE FLOW CHARTS ............ XXX........XXXXXXXXXXX
18 USERS MANUALS GIVE THEORY.................. XXX..X.XXX.XXXXXXXXXXX
19 OTHER QUALITIES OF USERS MANUALS--   ...... .......X..X..........X
20 PROGRAM LANGUAGE IS--SEE SECT.8,7.20 ...... ....X......X..X......X
21 PROGRAM RUNS ON IBM........................ XXX..XX.X.X..X..X..X..
22 PROGRAM RUNS ON CDC........................ XXX..X.X..XXXXXXXXX.X.
23 PROGRAM RUNS ON UNIVAC  ................... X..XXX........X..XXX.X
24 PROGRAM RUNS ON GE ........................ .....X....X...........
25 PROGRAM RUNS ON COMPUTER IDENTIFIED IN SECT .....XX.XX............
28 NO. OF PRIMARY OVERLAYS *.................. 6..4611X..4...2..X2.92
   MAN-YEARS REQUIRED TO DEVELOP PROGRAM *     5211L.12..31.2331LL.11
31 MINIMUM CORE SPACE REQUIRED FOR EXECUTION   .5...3.........1..3.L.
32 PROGRAM RUNS ENTIRELY IN CORE ............. ..X..XX.XXXXXX..X.....
33 NO. UNITS OF DIRECT ACCESS MASS STORAGE GIV 1..9L.....L...11.X1.X2
36 RESTART CAPABILITY ........................ .XX.X..XXXX..XXX.XXX..
37 OPTIONS FOR VARIOUS OUTPUT     ............ XXXXXX..XXX.X.XXXXXX.X
38 DIRECT ACCESS DATA BASE  .................. ....X.................
39 FREE-FIELD INPUT  ......................... ............X.......X.
40 RUNS IN INTERACTIVE MODE .................. ...............X......
41 AUTOMATIC MESH GENERATOR(S)    ............ .XXXX.....X.....X.X...
42 AUTOMATIC LOADING GENERATOR    ............ .XXXX.....X...X.X.X...
43 SECTION PROPERTIES CALCULATED  ............ X.........X......X.X..
44 PRE-PROCESSOR.............................. ..................X...
45 POST-PROCESSOR      ....................... ..........X....X..X...
46 EIGENVALUE SOLVER ......................... ......................
47 MODAL SUPERPOSITION ROUTINE    ............ ......................
48 OTHER TIME-INTEGRATION ROUTINE ............ ......................
49 OTHER CAPABILITY--SEE SECT.8, 7.49  ...... ..................X...
50 PLOT ROUTINES FOR UNDEFORMED/DEFORMED GEOME X...X..XXXX...X....X..
51 VARIOUS VIEWS AND SECTIONS PLOTTED   ...... ....X...XX....X.......
52 CONTOUR PLOTS.............................. .........XX.......X...
53 ORDINARY Y = F(X) PLOTS  .................. X.........X...X..XXX..
54 STRESS-ON-STRUCTURE PLOTS.................. ....X..X......X.......
55 OTHER KINDS OF PLOTS--SEE SECT.8, 7.55      .........XX...X......X
56 PLOTTING CAPABILITY FOR CALCOMP............ ....X..XXXX......XXX..
57 PLOTTING CAPABILITY FOR STROMBERG-CARLSON   X.........X...X...XX.X
58 PLOTTING CAPABILITY FOR HARDWARE IDENTIFIED ....X...XX............
59 OPTIONAL DETAILED PRINT OUTPUT ............ XXXXX.X.XXX.....X.XX..
60 EQUILIBRIUM CHECKS ........................ ....X.......X...X..X..
61 ALTERNATIVE BRANCHES FOR SAME PROBLEM...... X.................X...
62 MATRIX CONDITIONING ESTIMATION ............ ....X.....X...........
63 AUTOMATIC ERROR CONTROL  .................. .X..X...........XX....
64 GRAPHICAL PREPROCESSOR   .................. ...XX.............XX..
65 APPLIED LOADING CHECKS   .................. .XX.X.....X.X.........
66 BOUNDARY CONDITION CHECKS.................. ....X.....XX....X.....
67 SECTION PROPERTIES CHECKS.................. ..........X...........
68 GEOMETRY CHECKS   ......................... .XX.X.....X.X...X.....
69 OTHER CHECKING AIDS--SEE SECT.8, 7.69...... ..............X....X..
70 DEVELOPER WILL HELP USER GET PROGRAM RUNNIN XXX..X.X..X.....XXXX..
71 DEVELOPER WILL HELP USER MODELING CASES     X....X.X..X...X..XXX.X
72 DEVELOPER WILL HELP USER FIND SUSPECTED BUG XXX..X.X..X...X.XXXX.X
73 DEVELOPER WILL CIRCULATE NOTICES OF BUGS FO XXX..X........XXXX.X.X
74 DEVELOPER WILL HELP USER SET UP AND RUN CAS X....X.X..X...X..XXX.X
75 DEVELOPER WILL CONDUCT WORKSHOPS     ...... X....X....X...X..XXX.X
76 DEVELOPER WILL MAKE SPECIAL PROGRAM CHANGES X....X.X..X...X..XXX.X
77 DEVELOPER WILL KEEP USERS INFORMED OF IMPRO XXX..X....X.....XXXX.X
78 OTHER ASSISTANCE DEVELOPER WILL GIVE USERS- ......X...............
79 OTHER COMMENTS PERTINENT TO THIS SECTION--S ...X....XX........X...

*L = LARGE, X= SEE SECT. 8, (WITH 'SCOPE' OPTION)
```

```
NOW PUNCH EITHER   'ALL' ,   'SCOPE' ,    'COMPARE' ,
                   'EVALUATE' ,   'PROBLEM' ,  OR    'FINISHED'
```

EVALUATE

```
NOW PUNCH THE NAME OF THE PROGRAM FOR WHICH YOU
WANT EVALUATIONS BY MANY USERS.
```

```
EXAMPLE---'SAPIV'  (DON'T PUNCH THE QUOTES)

ABOUT 2 TO 4 PAGES OF OUTPUT WILL FOLLOW
```

```
STRUDLII.LOGCHER, R.D.            MIT.CAMBRIDGE, MASS

 USER'S STORY..AMOUNT OF USE, EVALUATION, WHAT IT WAS USED FOR

                              READ PROGRAM NAMES AND USER NO. VERTICALLY
                                 (NAMES OF USERS HAVE BEEN WITHHELD)

                                             SSSSSSSSSSSSSSSSSSSSSSSSSSSS
                                             TTTTTTTTTTTTTTTTTTTTTTTTTTTT
                                             RRRRRRRRRRRRRRRRRRRRRRRRRRRR
                                             UUUUUUUUUUUUUUUUUUUUUUUUUUUU
                                             DDDDDDDDDDDDDDDDDDDDDDDDDDDD
                                             LLLLLLLLLLLLLLLLLLLLLLLLLLLL
                                             ....................IIIIII.I..
                                             UUUUUUUUUUUUUUUUUUUUIIIIIIUIUU
                                             SSSSSSSSSSSSSSSSSSSS......S.SS
                                             EEEEEEEEEEEEEEEEEEEEUUUUUUEUEE
                                             RRRRRRRRRRRRRRRRRRRRSSSSSSRSRR
                                                                 EEEEEE E
                                             12345678911111111111RRRRRR2R22
                                             .........0123456789      6 89
   AMOUNT OF USE/HARDWARE                    ....................222222.2..
                                             ....................012345.7..
                                             ..............................

WE HAVEN'T USED CODE ENOUGH TO COMMENT       ..............................
NUMBER OF PEOPLE USING CODE IN THIS DEPT*    124334IL423354241.32115282.13
NUMBER OF DEPARTMENTS USING CODE=  ......    34321.2..41111.11.1244222.L11
NUMBER OF YEARS I HAVE USED THIS CODE=       041454353111122134.41444451412
CODE USED INTENSIVELY ALL THE TIME ......    .......X...................X..
CODE RECEIVES INTERMITTANTLY INTENSIVE US    ...XXXX..XXX.XX.....XXXXXX..X
CODE IS USED DAILY        ................    ...........................X..
CODE IS USED ABOUT ONCE A WEEK      ......    .........X.......X............
CODE IS USED ABOUT ONCE A MONTH     ......    .X................XX..X......
CODE IS RARELY USED BY US     ............    X...........X....X.......X.X.
WE USE SOFTWARE SERVICE BUREAU FOR EXECUT    XXX..X.X....XXXX.XX.XX....X.X
PROGRAM RUNS HERE ON IBM      ............    ..XXXXXX......X.X...XXXXX..XX
PROGRAM RUNS HERE ON CDC      ............    .....X........................
PROGRAM RUNS HERE ON UNIVAC   ............    ..............................
PROGRAM RUNS HERE ON GE...................    ..............................
PROGRAM RUNS HERE ON-- ...................    ..............................
PLOT OUTPUT WITH CALCOMP EQUIPMENT ......    XXXXXX.X.............XX..X..XX
PLOT OUTPUT WITH STROMBERG-CARLSON EQUIPM    ..............................

         EVALUATION SECTION           0=NOT ENOUGH DATA
1=STRONGLY AGREE, 2=AGREE, 3=NEUTRAL, 4=DISAGREE, 5=STRONGLY DISAGREE

USER HAS CONFIDENCE IN THIS PROGRAM......    4311122211222111121C112222111
AGREEMENT WITH TEST RESULTS IS GOOD......    430210.2003321.1200.000002123
EASY TO LEARN HOW TO USE THIS PROGRAM        33212222324411212210223322131
EASY TO USE THIS PROGRAM ONCE LEARNED        2321222221221121221C110222121
USER'S MANUAL(S) ARE SELF-CONTAINED......    22421322224441205330223424322
MANUAL(S) HAVE GOOD LIST OF PITFALLS         4,3344332444325243402222243442
MANUAL(S) HAVE ENOUGH TEST CASES   ......    2.444433435543534342245434I2
MANUAL(S) ARE WELL ORGANIZED ............    2.2222222444215132IC22342442I
INPUT DATA ARE EASY TO PREPARE     ......    2.2222123222121121C222122131
OUTPUT IS EASY TO UNDERSTAND ............    2.2212121433111121C22222222
PROGRAM HAS GOOD PLOTTING CAPABILITY         2.2022.232222143123.332220411
PROGRAM WAS EASY TO GET RUNNING HERE         0.20404232220031300.22232220.
DEVELOPER KEEPS US UP-TO-DATE............    4,154254132211342210222222121
DEVELOPER HELPS FIND BUGS IN PROGRAM         4515524422222141121.222222201
GOOD COMMUNICATION BETW USER $ DEVELOPER     3.155244222221311121C11223221I
USER HAS ENCOUNTERED VERY FEW BUGS ......    452052442222423350I.332223412

       WHAT THIS PROGRAM HAS BEEN USED FOR
0=NO USE, 1=LIMITED USE, 2=MODERATE USE, 3=EXTENSIVE USE, 4=NOT APPLIC.

PRELIMINARY DESIGN      ..................    12422023331122123403223311201
FINAL DESIGN EVALUATION..................    02323323330023132113223321233
GENERAL STRUCTURES      ..................    12323023332123133113223332233
SOLID OF REVOLUTION     ..................    4,400002.00040001 4.0000040000
GENERAL SHELL  ...........................    4,000022.004400014.2000020001
PART OF AXISYMMETRIC SHELL(LT 360 DEG.)      4.0000.1.004400014.2000020000
AXISYMMETRIC SHELL (360 DEGREES)    ......    440000.1.004400014.1000010000
BRANCHED SHELLS  .........................    1,0000.0.000400014.0220001000
SHELLS WITH CUTOUTS     ..................    4,0000.1.000400004.1000021000
THICK SHELLS     .........................    4,0000.0.000400004.0000010000
FRAMEWORK-SHELL COMBINATIONS ............    1.0000.0.000400014.2000020003
FLAT PLATE OF GENERAL SHAPE   ...........    1,3020.0.2004000041233000000I
LAYERED WALL CONSTRUCTION     ...........    0.0000.0.000400004.0000000000
COMPOSITE MATERIAL      ..................    0.100010.000420000.0000000000
SHELLS WITH STIFFENERS ...................    1.001012.100400004.2330000030
ELASTIC-PLASTIC MATERIAL      ...........    0.0000.0.001400004.0000010000
CREEP.....................................    0.0000.0.100401004.0000010000
STATIC OR QUASI-STATIC LOADS ............    133233.3.32122233113333322033
GENERAL TRANSIENT LOADS..................    123000.2.300010004.0000100003
PERIODIC DYNAMIC LOADS ...................    113000.2.00001000410000300003
LINEAR STRESS ANALYSIS ...................    133233.3.221222234.3332231133
NONLINEAR STRESS ANALYSIS     ...........    420000.0.021001104.0000210000
NONLINEAR COLLAPSE      ..................    0.0000.0.000000004.0000010000
BIFURCATION (EIGENVALUE) BUCKLING  ......    0.000010.000000004.0000000000
MODAL (EIGENVALUE) VIBRATIONS............    110000I0.000011004.0000000003
LINEAR DYNAMIC RESPONSE..................    110000.0.000010004.0010000033
NONLINEAR DYNAMIC RESPONSE    ...........    0.0000.0.000000004.0.00000000
ADDITIONAL COMMENTS PROVIDED BY USER         YYYNNYYNYYYNNNYNYYNN.YNNNNNNN
SOME REFERENCES PROVIDED BY USER   ......    00000000000000000000.10000000

*L = LARGE  (GREATER THAN 9)
```

USER COMMENTS ON THIS PROGRAM

STRUDL	WE USE PROGRAM PRIMARILY FOR STIFFNESS ANALYSIS OF PLANE FRAMES, PLANE GRIDS & SPACE FRAMES.
STRUDL	DESIRABLE TO HAVE MANUAL INDICES AS WELL AS TABLE OF CONTENTS . OUTPUT DATA FOR SHELLS WAS FOUND DIFFICULT TO UNDERSTAND AND NOT WELL EXPLAINED IN MANUAL.
STRUDL	USER MANUAL LACKING IN UNDERSTANDABILITY ON SUCH SUBJECTS AS: JOINT RELEASES, MEMBER RELEASES, & LOCAL COORDINATES.
STRUDL	USER MANUAL COULD BE VASTLY IMPROVED BY INCLUDING COMPREHENS-IVE EXAMPLES FOR THE COMMANDS ILLUSTRATING THE VARIOUS OPTIONS. APPENDIX B LISTING COMMANDS SHOULD BE REFERENCED TO THE MANUAL WHERE THE DETAILED EXPLANATION APPEARS. AN INDEX OF COMMANDS SHOULD BE INCLUDED. A TABLE SHOULD BE INCLUDED SHOWING MINIMUM COMMAND STRUCTURE FOR VARIOUS TYPES OF PROB-LEMS. MORE TREATMENT SHOULD BE GIVEN TO EFFICIENT JOINT MEM-BERING. WARNINGS SHOULD BE INCLUDED TO INDICATE RATIO OF MEM-BER STIFFNESS WHICH MAY CAUSE INACCURACIES IN RESULTS. THIS MANUAL IS NOT EASY TO USE & ITS IMPROVEMENT AND EXPANSION WOULD FACILITATE USE OF THE PROGRAM.
STRUDL	WE HAVE HAD EXCELLENT RESULTS WITH STRUDL FOR STRAIGHT ANALYSIS & DESIGN OF SPACE FRAME STRUCTURES. WE FEEL THE ADDITIONAL COST IS JUSTIFIED OVER LESS COSTLY PROGRAMS.
STRUDL	NEED PLOTTING CAPABILITY OF STRESSES AT CENTROIDS OF FINITE ELEMENTS.
STRUDLII	ANALYSIS SECTIONS APPEAR WELL-DEBUGGED BUT DESIGN SECTIONS HAVE NOT BEEN AS SUCCESSFUL IN OUR EXPERIENCE.
STRUDLII	PROGRAM IS VERY EFFECTIVE IN FRAME ANALYSIS BUT HAVE LITTLE CONFIDENCE IN ITS F.E. CAPABILITIES.
STRIDLII	PROGRAM EASY TO USE + HANDLES MAJORITY OF OUR PROBLEMS. WOULD PREFER A MANUAL WHICH IS UPDATABLE (RING-TYPE RATHER THAN BOUND).
STRUDLII	ON OUR SYSTEM, PROBLEM-SOLVING COSTS ARE 5-15 TIMES HIGHER WITH THIS PROGRAM THAN WITH OTHER PROGRAMS (ELAS, SOLID SAP + HOMEMADE). THE NONLINEAR AND BUCKLING PARTS OF THE PROGRAM ARE UNRELIABLE.
STRUDLII	TRIED TO USE PROGRAM ON AXISYMMETRIC SHELL (VIA SPRING FOUNDATION METHOD) + FOUND THE PROGRAM TOO CUMBERSOME COMPARED TO ANSYS.
	USED STRUDL ON FOLDED PLATE ROOF PRELIMINARY DESIGN. HOWEVER, THE PROGRAM DID NOT INCLUDE THE PROPER SHEAR LAG ON PLATES AND FINALLY USED ANSYS TO SOLVE THE PROBLEM.

```
STARDYNE, ROSEN,R, MRI, INC., 9841 AIRPORT BLVD.,LOS ANGELES, CA

 USER'S STORY..AMOUNT OF USE, EVALUATION, WHAT IT WAS USED FOR
                                       READ PROGRAM NAMES AND USER NO. VERTICALLY
                                            (NAMES OF USERS HAVE BEEN WITHHELD)

                                               SSSSSSSSSSSSSSSSSS
                                               TTTTTTTTTTTTTTTTTT
                                               AAAAAAAAAAAAAAAAAA
                                               RRRRRRRRRRRRRRRRRR
                                               DDDDDDDDDDDDDDDDDD
                                               YYYYYYYYYYYYYYYYYY
                                               NNNNNNNNNNNNNNNNNN
                                               EEEEEEEEEEEEEEEEEE
                                               ................T..
                                               UUUUUUUUUUUUUUUUTUU
                                               SSSSSSSSSSSSSSSSUSS
                                               EEEEEEEEEEEEEEEESEE
                                               RRRRRRRRRRRRRRRRERR
                                                               R
     AMOUNT OF USE/HARDWARE                    1234567891111111 11
                                               .........0123456189
                                               ................7..

WE HAVEN'T USED CODE ENOUGH TO COMMENT         ..................
NUMBER OF PEOPLE USING CODE IN THIS DEPT*      25212LL251443154341
NUMBER OF DEPARTMENTS USING CODE=  ......      ..23.3.11.4.112.121
NUMBER OF YEARS I HAVE USED THIS CODE=         1333157101033444431
CODE USED INTENSIVELY ALL THE TIME ......      .....X..X..........
CODE RECEIVES INTERMITTANTLY INTENSIVE US      X.XX......XXXX..XXX
CODE IS USED DAILY      ..................     ......X........X...
CODE IS USED ABOUT ONCE A WEEK      ......     .X..X..XX.....X....
CODE IS USED ABOUT ONCE A MONTH     ......     ..............X....X
CODE IS RARELY USED BY US    ............      .........X.........
WE USE SOFTWARE SERVICE BUREAU FOR EXECUT      X..X...X.X........X
PROGRAM RUNS HERE ON IBM    .............      ...................
PROGRAM RUNS HERE ON CDC    .............      XXXXXXX.XXXXXXXXXXX
PROGRAM RUNS HERE ON UNIVAC .............      ...................
PROGRAM RUNS HERE ON GE..................      ...................
PROGRAM RUNS HERE ON-- ..................      ...................
PLOT OUTPUT WITH CALCOMP EQUIPMENT ......      X..XXXXX.XXXXXX.XXX
PLOT OUTPUT WITH STROMBERG-CARLSON EQUIPM      ...X..XXX.X.XXX..X.

          EVALUATION SECTION              0=NOT ENOUGH DATA
1=STRONGLY AGREE, 2=AGREE, 3=NEUTRAL, 4=DISAGREE, 5=STRONGLY DISAGREE

USER HAS CONFIDENCE IN THIS PROGRAM......      2222211112111211211
AGREEMENT WITH TEST RESULTS IS GOOD......      03.0211222111211310
EASY TO LEARN HOW TO USE THIS PROGRAM          1124412122112311221
EASY TO USE THIS PROGRAM ONCE LEARNED          1123312122112211121
USER'S MANUAL(S) ARE SELF-CONTAINED......      2.22212122112311221
MANUAL(S) HAVE GOOD LIST OF PITFALLS           3432212243212314433
MANUAL(S) HAVE ENOUGH TEST CASES   ......      3..4323243413311444
MANUAL(S) ARE WELL ORGANIZED ............      22.2212222311213212
INPUT DATA ARE EASY TO PREPARE     ......      22.3212322212311221
OUTPUT IS EASY TO UNDERSTAND ............      22.3112222111211221
PROGRAM HAS GOOD PLOTTING CAPABILITY           02.2214122312204322
PROGRAM WAS EASY TO GET RUNNING HERE           21.0401222.12201221
DEVELOPER KEEPS US UP-TO-DATE............      22.2210122111311331
DEVELOPER HELPS FIND BUGS IN PROGRAM           0..1210122211211300
GOOD COMMUNICATION BETW USER $ DEVELOPER       32.1012122212211300
USER HAS ENCOUNTERED VERY FEW BUGS ......      12.2221222112313321

          WHAT THIS PROGRAM HAS BEEN USED FOR
0=NO USE, 1=LIMITED USE, 2=MODERATE USE, 3=EXTENSIVE USE, 4=NOT APPLIC,

PRELIMINARY DESIGN      .................. 0..11331222.2333332
FINAL DESIGN EVALUATION................... 0..11332223.2333333
GENERAL STRUCTURES      .................. 0..2133322332333333
SOLID OF REVOLUTION     .................. 0..00300200.14004.0
GENERAL SHELL    ......................... 0..00300220.3423232
PART OF AXISYMMETRIC SHELL(LT 360 DEG.)    0..00310100.04014.0
AXISYMMETRIC SHELL (360 DEGREES)    ...... 0..10310200.14214.3
BRANCHED SHELLS  ......................... 0..00130200.11301.3
SHELLS WITH CUTOUTS     .................. 0..10020200.24212.1
THICK SHELLS     ......................... 0..00010100.24022.0
FRAMEWORK-SHELL COMBINATIONS ............. 3..10323220.3323332
FLAT PLATE OF GENERAL SHAPE  ............. 0..11222203.33233.0
LAYERED WALL CONSTRUCTION    ............. 0..00000000.03131.3
COMPOSITE MATERIAL      .................. 0..00000000.03111.2
SHELLS WITH STIFFENERS ................... 3..10023120.1323230
ELASTIC-PLASTIC MATERIAL     ............. 0..01004000.0404410
CREEP..................................... 0..00004000.04044.0
STATIC OR QUASI-STATIC LOADS ............. 0..21333223.1333333
GENERAL TRANSIENT LOADS................... 12.1033100033323210
PERIODIC DYNAMIC LOADS ................... 12.0032300133223 2.0
LINEAR STRESS ANALYSIS ................... 0..22333223.3333333
NONLINEAR STRESS ANALYSIS    ............. 0..00004000.0101410
NONLINEAR COLLAPSE      .................. 0..00004000.04012.0
BIFURCATION (EIGENVALUE) BUCKLING   ...... 0..00014000.04012.0
MODAL (EIGENVALUE) VIBRATIONS............. 32.0033300133332 2.1
LINEAR DYNAMIC RESPONSE................... 22.1103300133332 1.0
NONLINEAR DYNAMIC RESPONSE   ............. 0..00104000.01001.0
ADDITIONAL COMMENTS PROVIDED BY USER       YNNYNNYYYNNNYNNYYYY
SOME REFERENCES PROVIDED BY USER    ...... 0001030000000000000

*L = LARGE  (GREATER THAN 9)
```

USER COMMENTS ON THIS PROGRAM

STARDYNE GOOD GENERAL PURPOSE CODE WITH EXCELLENT DOCUMENTATION AND IS EASY TO USE. HAS MANY GOOD FEATURES FOR SIMPLIFYING INPUT. ALSO HAS SOME USEFUL FEATURES THAT ARE NOT OFTEN AVAILABLE IN OTHER CODES.

STARDYNE PROGRAM TREATS PHENOMENA THAT IS BEST UNDERSTOOD BY ENGINEERS & RESEARCHERS WHO HAVE PRACTICAL EXPERIENCE IN PROBLEMS DEALING WITH STATIC STRESS ANALYSIS. DYNAMIC TRANSIENT & DYNAMIC STEADY STATE RESPONSE, AND SEISMIC RESPONSE ANALYSIS.

STARDYNE PROGRAM NEEDS: (1)ABILITY TO TREAT OFFSET BEAMS TO IMPROVE THE ABILITY TO ANALYZE STIFFENED PLATES. (2)ORTHOTROPIC PLATES. (3)IMPROVED RESTART ABILITY TO ACQUIRE & USE RESULTS OF PRIOR ANALYSES STORED ON PRIOR DATA TAPES (NOTE: PLURAL). (4)IMPROVED PLOT OUTPUT TO HANDLE (A)STRESS CONTOURS, ETC. (B)MORE EFFICIENT GEOMETRY PLOT LOGIC. EXISTING LOGIC IS VERY INEFFICIENT (CRUDE) ON ANYTHING OTHER THAN THE SC-4020/4060.

STARDYNE DESIRE IMPROVEMENTS OR EXTENSIONS IN FOLLOWING AREAS: (1)NEED A GENERAL THIN-WALL BEAM ELEMENT (INCLUDING WARPING EFFECTS) THAT IS SUITABLE FOR BOTH STATIC + DYNAMIC ANALYSIS. ONE ELEMENT WILL DO IF THE ABILITY TO HAVE A NON-DIAGONAL MASS MATRIX IS PROVIDED. (2)NEED NON-DIAGONAL MASS MATRIX. (3)NEED ABILITY TO IMPOSE MULTI-POINT CONSTRAINTS. (4)NEED STRESS CONTOUR PLOTS.

STARDYNE PROGRAM COULD BE MADE MORE USEFUL IF A CONTOUR PLOTTING POST-PROCESSOR WERE ADDED.

STARDYNE PROGRAM WELL SUITED TO USE BY ENGINEERS ENGAGED IN DESIGN OF STRUCTURES. FINDS IT QUITE EXPENSIVE TO USE BUT VERY COMPETITIVE IN TERMS OF OVERALL PROGRAM COSTS DUE TO THE EASE + QUICKNESS BY WHICH COMPLEX STRUCTURES CAN BE CODED. THE USER MANUAL IS VERY GOOD + CAN BE READILY APPLIED BY ENGINEERS WITH LITTLE BACKGROUND IN COMPUTER PROGRAMMING. HOWEVER, TO GAIN ACCESS TO THE PROGRAM ONE DOES HAVE TO HAVE EXPERTISE TO FIND THEIR WAY THROUGH THE MAZE OF CONTROL CARDS WHICH ARE LOADED WITH COSTLY BOOBY-TRAPS. WE HAVE ESTIMATED THAT 90% OF THE PROBLEMS ENCOUNTERED DURING THE USE OF THIS PROGRAM HAVE BEEN WITH CONTROL CARDS RATHER THAN THE PROGRAM ITSELF.

STARDYNE FINDS STARDYNE THE BEST PROGRAM AVAILABLE TODAY TO SOLVE LINEAR STRUCTURAL DYNAMIC PROBLEMS AND IS MUCH SUPERIOR TO NASTRAN (FEWER ERRORS, EASIER TO USE, ACCURACY OF ELEMENTS). DUE TO PROGRAM EFFICIENCY, FINDS THE PROGRAM COST EFFECTIVE FOR SMALL TO LARGE PROBLEMS.

STARDYNE THE PLOT3D ROUTINE SHOULD BE MORE GENERAL + EASIER TO USE. IT'S NICE BUT A NUISANCE TO USE.

STARDYNE FINDS STARDYNE ONE OF THE BETTER F.E. PROGRAMS FOR ELASTIC RANGE ANALYSIS.

STARDYNE HIGHER ORDER ELEMENTS SHOULD BE INCLUDED IN F.E. LIBRARY. ESPECIALLY DESIRABLE ARE MEMBRANE TRIANGULAR + QUADRILATERAL ELEMENTS WITH QUADRATIC DISPLACEMENT FUNCTIONS WITH THE POSSIBILITY TO MODEL THESE MEMBRANE ELEMENTS WITH AT LEAST A 10:1 RATIO BETWEEN THE LONGEST + SHORTEST SIDES WITHOUT LOSS OF ACCURRACY.

```
SAPIV, BATHE, K.J., 410 DAVIS HALL, UNIV. OF CALIF,BERKELEY

 USER'S STORY..AMOUNT OF USE, EVALUATION, WHAT IT WAS USED FOR

                              READ PROGRAM AND USER NAMES VERTICALLY

                                          SSSSSSSSSSSSSSSSSSSSSSSSSSS
                                          AAAAAAAAAAAAAAAAAAAAAAAAAAA
                                          PPPPPPPPPPPPPPPPPPPPPPPPPPP
                                          IIIIIIIIIIIIII,,,,,,I,,I,12,
                                          VVVVVVVVVVVVVVVCMASGP.FBVIV,M
                                          ,,,,,,,,,,,,,,,.AEAERHO$,R,MI
                                          ROHBJGMFGRLLBGSERRRNEARWFWWAC
                                          UVEE.EALEULLEEO.YOG.GRD.IIECH
                                          HEBCH.NURHOOLNUALJEENZ.CSNLQ.
                                          R.RH.NMOMRYYL.T.AENLOAMOH.TUS
                                          .AETHUURA-DDAEHANTTEF.OMEJOAT
                                          URWEICN.NUGGELEUD.$CFETPRONRA
                                          NU.LGLCCYNEERERS.NLT,NOACH.IT
                                          IPUPGEHO.IRROCNTUUURMGRNONBEE
    AMOUNT OF USE/HARDWARE                V.NIIAER.VMMSNSRNCNIA..YNSE..
                                          ..ICNRNP..AAPEEAILDCT.C.TTCUU
                                          ..VCS.....NNADRLVEY.H.O.R.K..

WE HAVEN'T USED CODE ENOUGH TO COMMENT    .........................X.XXX
NUMBER OF PEOPLE USING CODE IN THIS DEPT* 515.2.4325LLLLLLLL51423L2....
NUMBER OF DEPARTMENTS USING CODE=  ...... ...22.1L1.33.41.1131.13.1....
NUMBER OF YEARS I HAVE USED THIS CODE=    101.01210132130312202313O1...
CODE USED INTENSIVELY ALL THE TIME ...... .....X....XX.X..XX...........
CODE RECEIVES INTERMITTANTLY INTENSIVE US ....X..X....X.........XX.X...
CODE IS USED DAILY    .................. X.X......XXXX..XX..X..X.......
CODE IS USED ABOUT ONCE A WEEK     ...... ...X..X...................X....
CODE IS USED ABOUT ONCE A MONTH    ...... ..................X........
CODE IS RARELY USED BY US   ............ .X................X.........
WE USE SOFTWARE SERVICE BUREAU FOR EXECUT .X..................X.........
PROGRAM RUNS HERE ON IBM    ............ ..........XXX.X.XX.......X....
PROGRAM RUNS HERE ON CDC    ............ .XXXX.XXX....X....X.XX.X.X...
PROGRAM RUNS HERE ON UNIVAC ............ ................X..X.XX...X...
PROGRAM RUNS HERE ON GE.................. .....X........X...............
PROGRAM RUNS HERE ON-- .................. T.........T.........X..X......
PLOT OUTPUT WITH CALCOMP EQUIPMENT ...... ..X..X.X..X.XXXX.....XXX.....
PLOT OUTPUT WITH STROMBERG-CARLSON EQUIPM .......X......................

         EVALUATION SECTION           0=NOT ENOUGH DATA
1=STRONGLY AGREE, 2=AGREE, 3=NEUTRAL, 4=DISAGREE, 5=STRONGLY DISAGREE

USER HAS CONFIDENCE IN THIS PROGRAM...... 2222221112112122211011102...
AGREEMENT WITH TEST RESULTS IS GOOD...... 22220222121131 2.0000111122...
EASY TO LEARN HOW TO USE THIS PROGRAM     232.31222211222344351221 24...
EASY TO USE THIS PROGRAM ONCE LEARNED     22222212111122122201 11103...
USER'S MANUAL(S) ARE SELF-CONTAINED...... 1234412221213133552513232 5...
MANUAL(S) HAVE GOOD LIST OF PITFALLS      2255333242323243553523440 5...
MANUAL(S) HAVE ENOUGH TEST CASES   ...... 3234333223234243552512520 5...
MANUAL(S) ARE WELL ORGANIZED ............ 3242331212112123552312213 4...
INPUT DATA ARE EASY TO PREPARE     ...... 4234222213113132123421231 4...
OUTPUT IS EASY TO UNDERSTAND ............ 3322422313222132532411422 2...
PROGRAM HAS GOOD PLOTTING CAPABILITY      505004.40533434305003203 05...
PROGRAM WAS EASY TO GET RUNNING HERE      432232.14222443154321112 21...
DEVELOPER KEEPS US UP-TO-DATE............ 100201.13122013124350344 00...
DEVELOPER HELPS FIND BUGS IN PROGRAM      404401.35.000122.4050304 00...
GOOD COMMUNICATION BETW USER $ DEVELOPER  304401.22112.13254050324 03...
USER HAS ENCOUNTERED VERY FEW BUGS ...... 222222.11222.12252321112 02...

       WHAT THIS PROGRAM HAS BEEN USED FOR
0=NO USE, 1=LIMITED USE, 2=MODERATE USE, 3=EXTENSIVE USE, 4=NOT APPLIC.

PRELIMINARY DESIGN      ................. 0.210223.012.222012..31312...
FINAL DESIGN EVALUATION.................. 1.011223.133.322032..31303...
GENERAL STRUCTURES      ................. 210313.3.233.222232.333213...
SOLID OF REVOLUTION     ................. 0.120322.011.22400...22210...
GENERAL SHELL  .......................... 0.0301.2.033.121002.330312...
PART OF AXISYMMETRIC SHELL(LT 360 DEG.)   0.0401.4.044.12400...40.10...
AXISYMMETRIC SHELL (360 DEGREES)   ...... 0.1403.2.044.32400...21.10...
BRANCHED SHELLS ......................... 0.0000.3.000.2.400...40.10...
SHELLS WITH CUTOUTS  .................... 0.0000.2.022.014002..20.10...
THICK SHELLS    ......................... 0.0200.2.000.014011.331.10...
FRAMEWORK-SHELL COMBINATIONS ............ 0.0300.4.033.001201..20202...
FLAT PLATE OF GENERAL SHAPE  ............ 0.0303.3.033.112222.330.12...
LAYERED WALL CONSTRUCTION    ............ 0.0000.1.020.00420...40.0....
COMPOSITE MATERIAL      ................. 0.0000.4.021.01400...40.0....
SHELLS WITH STIFFENERS .................. 0.0200.2.033.00120...2024....
ELASTIC-PLASTIC MATERIAL     ............ 0.4004.4.044.00101...40.00...
CREEP.................................... 0.4004.4.044.00100...40.0....
STATIC OR QUASI-STATIC LOADS ............ 3.321223.333.322232..3331....
GENERAL TRANSIENT LOADS.................. 010202.3.120.30203..33021....
PERIODIC DYNAMIC LOADS .................. 0.0202.3.130.10203...30200...
LINEAR STRESS ANALYSIS .................. 31331323.333.322233..33.13...
NONLINEAR STRESS ANALYSIS    ............ 0.4404.2.044.001.0...40.00...
NONLINEAR COLLAPSE      ................. 0.4404.4.044.001.0...40.0....
BIFURCATION (EIGENVALUE) BUCKLING  ...... 0.0404.4.044.001.0...40.4....
MODAL (EIGENVALUE) VIBRATIONS............ 21032323223323122 01.330310...
LINEAR DYNAMIC RESPONSE.................. 010313.330332314.01.330100...
NONLINEAR DYNAMIC RESPONSE   ............ 0.0404.4.044.001.0...40.00...
ADDITIONAL COMMENTS PROVIDED BY USER      YYYYYNNYNYNNNNYYYYYYNYNYNYY..
SOME REFERENCES PROVIDED BY USER   ...... 0000000000240000010000001000..

*L = LARGE  (GREATER THAN 9)
```

USER COMMENTS ON THIS PROGRAM

SAP4	STRUCTURE OF THE PROGRAM IS GOOD AND THE STANDARD OF PROGRAMMING IS EXCELLENT. PROGRAM CAPABILITY, SPEED, USEFULLNESS IS EXCELLENT. WE THINK THIS PROGRAM HAS GREAT POTENTIAL FOR DEVELOPMENT. HOWEVER, THE PROGRAM IS POOR IN THE FOLLOWING AREAS: (1)MANUAL AND CORRESPONDENCE BETWEEN MANUAL & PROGRAM (2)MUCH IMPROVED INPUT WITH FAR BETTER DATA GENERATION (3)ABILITY TO PREPROCESS INPUT DATA, PRINT FULL GENERATED DATA LISTINGS PRIOR TO EXECUTION IN ORDER TO PREVENT ABORTS DUE TO VERY COMPLEX LOAD COMBINATIONS. (4)SLIGHT IMPROVEMENT IN OUTPUT FORMATS. (5)FAR MORE COMPREHENSIVE MEMBER LOAD CAPABILITY. WE HAVE IMPLEMENTED (2)(3)(4) AND ARE ALSO WRITING A POST-PROCESSOR TO STORE OUTPUT ON TAPE & THEN PERFORM COMBINATIONS AS REQUIRED BY THE USER.
SAPIV	NEED PLOT ROUTINE FOR DEFORMED/UNDEFORMED STRUCTURES. WOULD LIKE MORE FREEDOM IN THE TREATMENT OF RESPONSE SPECTRUM ANALYSIS INPUT.
SAPIV	OBTAINED PROGRAM AS A POSSIBLE TOOL IN STRUCTURAL ANALYSIS & GENERATION OF STIFFNESS MATRICES FOR DYNAMIC ANALYSIS. HAVE NOT ATTEMPTED ANY EVALUATION OF THIS PROGRAM WITH OTHER PROGRAMS SUCH AS NASTRAN OR ANSYS. HOWEVER, THE OVERALL APPROACH TO SOLUTION APPEARS EFFICIENT AND WELL ORGANIZED AND LOOKS TO BE POSSIBLY SUPERIOR FOR OUR NEEDS TO NASTRAN. FIND THE DOCUMENTATION SKETCHY ON MAKING INPUT & INTERPRETING OUTPUT AND THE TERMINOLOGY SO NON-STANDARD THAT ONLY A FEW SIMPLE CASES HAVE BEEN RUN SO FAR & THESE HAVE NOT BEEN FULLY CORRELATED YET. THE EFFORT WAS DISCONTINUED AFTER WHAT APPEARED TO BE EXCESSIVE TIME.
SAPIV	USER MANUAL IS GREATLY IMPROVED OVER THOSE OF EARLIER PROGRAM VERSIONS. A SEPARATE PRE-PROCESSOR PROGRAM WHICH PLOTS F.E. MESH/SECTION WAS DEVELOPED AND IS IN USE IN COMPANY.
SAPIV	HAVE BEEN USING SAP PROGRAMS (1 & 3) EXTENSIVELY AND HAVE BEEN MODIFYING THEM TO SUIT COMPANY'S NEEDS. RECENTLY OBTAINED SAP4.
SAPIV	USER MANUAL IS POORLY WRITTEN. THE CODING MUST BE FOLLOWED TO DETERMINE HOW TO USE SOME INPUT VARIABLES. THE PROGRAM IS LACKING IN GENERAL PURPOSE CAPABILITIES: NO PLOTTING, NO MATRIX INPUT, DOES NOT CALCULATE REACTION OR CHECK FOR STATIC EQUILIBRIUM. HAVE MODIFIED THE IN-HOUSE CODE BY ADDING A 1-POINT RESTART CAPABILITY BY STORING THE DECOMPOSED STIFFNESS MATRIX + ELEMENT STRESS MATRICES SO ADDITIONAL LOAD CASES CAN BE RUN. FINDS THE PROGRAM VERY EFFICIENT.
SAPIV	GOOD PROGRAM WITH BROAD CAPABILITIES. EASY TO ADAPT WITH MINOR PROGRAM CHANGES. OUTPUT MORE USEFUL FOR RESEARCH THAN FOR DESIGN. LACKS GENERALITY IN LOAD COMBINATIONS FOR DESIGN. LACKS COMPUTATION OF TRANSVERSE SHEAR STRESS IN SHELL ELEMENT
SAPIV	POOR PROGRAM DOCUMENTATION. WOULD BE USEFUL TO IMPLEMENT (1)AXISYMMETRIC SHELL ELEMENT. (2)4-POINT, 6-POINT SOLID ELEMENTS. (3)PREPROCESSOR GEOMETRY PLOT SUBROUTINES. HAS POOR CAPABILITY ON LOAD COMBINATION (MAX=4). TO USE AS A PRODUCTION PROGRAM I/O SHOULD BE IMPROVED + DYNAMIC CORE ALLOCATION SHOULD BE INCORPORATED.
SAPIV	ACQUIRED SAP4 TO OBTAIN CAPABILITY OF DYNAMIC ANALYSIS (PARTICULARLY RESPONSE HISTORY ANALYSES) & STATIC ANALYSIS OF LARGE COMPLEX SHELLS. CURRENTLY USE MANY PROGRAMS INCLUDING STRESS & ASKA SYSTEM. ONE OF OUR SPECIFIC CRITICISMS OF MOST DYNAMIC PROGRAMS CONCERNS DAMPING PARTICULARLY IN SOIL-STRUCTURE INTERACTION PROBLEMS. MODAL SUPPOSITION IS INVALID AND, UNLESS IT IS POSSIBLE TO DAMP DIFFERENT MODES, DYNAMIC ANALYSIS IS A WASTE OF MONEY.
SAPIV	FIND A GREAT DEAL OF WORK IS PUT INTO SPECIAL PURPOSE PROGRAMS AS COMPARED TO GENERAL PURPOSE PROGRAMS. CURRENTLY DEVELOPING MODELS, USING SAP4 FOR TESTS & COMPARISONS, IN NON-LINEAR MATERIALS & GEOMETRY.
SAPIV	NO OPTIONS PROVIDED TO ALLOW PUNCHING OF EIGENVECTORS ONTO CARDS FOR OTHER APPLICATIONS. FINDS ALL EIGENVECTORS NORMALIZED WITH RESPECT TO THE MASS MATRIX. THIS MAKES PLOTTING + IDENTIFICATION OF DOMINANT DIRECTION DIFFICULT. SINCE THE MASS MATRIX FORMED-USING NODAL GEOMETRY + MEMBER PROPERTIES-IS NOT PRINTED AS OUTPUT, USER CANNOT VERIFY MODEL
SAPIV	FINDS THE PROGRAM'S USEFULNESS + PROGRAM MODIFICATION HAMPERED BY POOR DOCUMENTATION. MANUAL HAS INSUFFICIENT DESCRIPTIONS OF INPUT DATA, ESPECIALLY THE CONSTRAINTS, RIGID LINKS, LOADS + AUTOMATIC GENERATION FEATURES. DESCRIPTION OF OUTPUT DATA IS TOTALLY LACKING INCLUDING COORDINATE SYSTEM FOR OUTPUT. SAMPLE PROBLEMS WOULD BE WORTH 1000 WORDS. FINDS IT DIFFICULT TO TRACE ERRORS DUE TO COMPLEXITY OF SUB-ROUTINES. NO OVERLAY OF STRUCTURES & INADEQUATE USE OF COMMENT CARDS TO IDENTIFY PURPOSE OF SUBROUTINES. ATTRIBUTES ARE THE AVAILABILITY ON DEMAND WITH A PREPROCESSOR ; HOWEVER, PREFERS NASTRAN BECAUSE I/O IS EASIER AND BETTER UNDERSTOOD.
SAPIV	THE FOLLOWING INFORMATION WOULD BE HELPFUL: (1)THEORETICAL MANUAL DESCRIBING EACH OF ELEMENTS IN DETAIL. (2)SAMPLE PROBLEMS WHICH USE EACH OF THE ELEMENTS AVAILABLE IN PROGRAM. PROGRAM COULD BE MORE FLEXIBLE IN DESCRIBING LOAD CONDITIONS. A PROGRAM COURSE WOULD BE DESIRABLE.
SAPIV	PRE- + POSTPROCESSORS FOR ONLINE PLOTS SHOULD BE DEVELOPED.
SAP	VALUABLE TOOL FOR CONSULTING ENGRS IN AUSTRALIA + N.ZEALAND. USED FOR HIGH-RISE BUILDINGS EXTENSIVELY.
SAP	FOUND IN-HOUSE ADDITION OF A NON-ZERO STIFFNESS MATRIX PACKING ROUTINE RESULTED IN SAVINGS IN RUN TIME.

```
BOSOR4, BUSHNELL, D., DPT 5233/205, LMSC, PALO ALTO, CALIF 94304

 USER'S STORY..AMOUNT OF USE, EVALUATION, WHAT IT WAS USED FOR

                            READ PROGRAM AND USER NAMES VERTICALLY

                                              BBBBBBBBBBBBBBBBBBBBBBBBBBBBB
                                              OOOOOOOOOOOOOOOOOOOOOOOOOOOOO
                                              SSSSSSSSSSSSSSSSSSSSSSSSSSSSS
                                              OOOOOOOOOOOOOOOOOOOOOOOOOOOOO
                                              RRRRRRRRRRRRRRRRRRRRRRRRRRRRR
                                              44444444444444444433333334444
                                              ,,,**,**,*,,,,,,,,,,,,,,,,,,,
                                              SCCNNWLLJNHTCTPNUTFVSSLLCSPAM
                                              WOHSAEMMPAEHORHATHIIHAIIHMRID
                                              RNIRSISSLSRINWIVCIRTERVVAIIRA
                                              ITCDADCC.ACOV.LA.OLRLGEFLTNEC
                                              ...C-L--.-UKASCL.K.OLSRRMHCS.
                                              ....LI65.ILOIYO..OP..LPPEAEEC
                                              .CB.RN22.RELRSFU.LALDUOORSTAA
    AMOUNT OF USE/HARDWARE                    .A$.CG--.CSA..OS...AENOOSTORL
                                              .NI..E63...LA.RC...BVDLL.RNCI
                                              .....R13...AE.D....S.Y...O.HF

WE HAVEN'T USED CODE ENOUGH TO COMMENT        .........................XXXX
NUMBER OF PEOPLE USING CODE IN THIS DEPT*     2318225331215312322112222....
NUMBER OF DEPARTMENTS USING CODE=  ......     1221213411212111110.1.1111....
NUMBER OF YEARS I HAVE USED THIS CODE=        0324012432111212112012222....
CODE USED INTENSIVELY ALL THE TIME ......     .X...........................
CODE RECEIVES INTERMITTANTLY INTENSIVE US     X.XX..XXXX..XXXX....X.XXX....
CODE IS USED DAILY     ..................     .............................
CODE IS USED ABOUT ONCE A WEEK      ......    .....X.X...X..X..............
CODE IS USED ABOUT ONCE A MONTH     ......    ..........X.....XX...........
CODE IS RARELY USED BY US      ............   ....X.............XX.X.......
WE USE SOFTWARE SERVICE BUREAU FOR EXECUT     ..............X.X............
PROGRAM RUNS HERE ON IBM     .............    .XX.......XX.....X.X..XXX....
PROGRAM RUNS HERE ON CDC     .............    X..XXX...X..XXXXX............
PROGRAM RUNS HERE ON UNIVAC  .............    ......XXX......X..X.XX.......
PROGRAM RUNS HERE ON GE...................    .............................
PROGRAM RUNS HERE ON-- ...................    .............................
PLOT OUTPUT WITH CALCOMP EQUIPMENT ......     ..X.......X...........X.......
PLOT OUTPUT WITH STROMBERG-CARLSON EQUIPM     X..X..XX....X................

           EVALUATION SECTION              0=NOT ENOUGH DATA
 1=STRONGLY AGREE, 2=AGREE, 3=NEUTRAL, 4=DISAGREE, 5=STRONGLY DISAGREE

USER HAS CONFIDENCE IN THIS PROGRAM......     2121112110321222021222222....
AGREEMENT WITH TEST RESULTS IS GOOD......     2101110110233003021022212....
EASY TO LEARN HOW TO USE THIS PROGRAM         51323323235 44.422434241 33....
EASY TO USE THIS PROGRAM ONCE LEARNED         412121221 2342.3223232312 3....
USER'S MANUAL(S) ARE SELF-CONTAINED......     2111222123432222232223232....
MANUAL(S) HAVE GOOD LIST OF PITFALLS          3221213323432222233223233....
MANUAL(S) HAVE ENOUGH TEST CASES   ......     2112212112342432132312221....
MANUAL(S) ARE WELL ORGANIZED ............     41212324244344422232.3233....
INPUT DATA ARE EASY TO PREPARE     ......     5121312312442342242235234....
OUTPUT IS EASY TO UNDERSTAND ............     5122222123422322222223234....
PROGRAM HAS GOOD PLOTTING CAPABILITY          52110031252310 0.5430.2005....
PROGRAM WAS EASY TO GET RUNNING HERE          5132010214351012122223235....
DEVELOPER KEEPS US UP-TO-DATE............     2111212111222212122322234....
DEVELOPER HELPS FIND BUGS IN PROGRAM          2111212111222212123322235....
GOOD COMMUNICATION BETW USER $ DEVELOPER      3111112111222213222323225....
USER HAS ENCOUNTERED VERY FEW BUGS ......     2122212221232213223312232....

          WHAT THIS PROGRAM HAS BEEN USED FOR
0=NO USE, 1=LIMITED USE, 2=MODERATE USE, 3=EXTENSIVE USE, 4=NOT APPLIC.

PRELIMINARY DESIGN      ..................    33.2101302120.3212.0C4.40....
FINAL DESIGN EVALUATION...................    33.3302331021.3200.1C1.40....
GENERAL STRUCTURES      ..................    2..34004.01.0.0210.0.4.40....
SOLID OF REVOLUTION     ..................    3..04204.0100.0011.0.4.43....
GENERAL SHELL    .........................    2..34004.1130.0212.0.2.20....
PART OF AXISYMMETRIC SHELL(LT 360 DEG.)       0..34004.0000.0200.0.4.40....
AXISYMMETRIC SHELL (360 DEGREES)    ......    33.3222331031332121 1C1323....
BRANCHED SHELLS   ........................    21103223.1231.3232.1.2.40....
SHELLS WITH CUTOUTS     ..................    0..04004.0000.0000.0.4.40....
THICK SHELLS      ........................    0..04004.0100.0010.0.4.40....
FRAMEWORK-SHELL COMBINATIONS .............    2..04004.0020.0000.0.4.40....
FLAT PLATE OF GENERAL SHAPE   ............    0..11004.0010.0010.0.4.40....
LAYERED WALL CONSTRUCTION     ............    0..1201213311.2210.0.1.00....
COMPOSITE MATERIAL      ..................    0..14000.33.112200.1.4.00....
SHELLS WITH STIFFENERS ...................    22331123.3121.0211.0C1.02....
ELASTIC-PLASTIC MATERIAL      ............    0..04000.01.0.0000.0.4.42....
CREEP.....................................    0..04000.00.0.0000.0.4.40....
STATIC OR QUASI-STATIC LOADS .............    23.33023322.1.2211.1.2333....
GENERAL TRANSIENT LOADS...................    0..00000.00.0.0002.0.4.00....
PERIODIC DYNAMIC LOADS ...................    0..04000.00.0.0000.0.4.00....
LINEAR STRESS ANALYSIS ...................    2..3302333231.3212.1C2323....
NONLINEAR STRESS ANALYSIS     ............    2323300332321.1110.1C0223....
NONLINEAR COLLAPSE     ...................    2.103002.20.0.1210.0.0120....
BIFURCATION (EIGENVALUE) BUCKLING   ......    3123302333231322 12.1C0233....
MODAL (EIGENVALUE) VIBRATIONS.............    0.130203201.022110.0.0210....
LINEAR DYNAMIC RESPONSE...................    0..04100100.0.0000.0.4200....
NONLINEAR DYNAMIC RESPONSE    ............    0..04000000.0.0000.0.4000....
ADDITIONAL COMMENTS PROVIDED BY USER          YNYYNYNYYNYNYYYNYNNYNNNNY..Y.
SOME REFERENCES PROVIDED BY USER    ......    00000002100000000000000401..0.

*L = LARGE  (GREATER THAN 9)
```

USER COMMENTS ON THIS PROGRAM

BOSOR3 THE PLOT SUBROUTINES DO NOT WORK ON THE IBM360/65 COMPUTER IN OUR COMPANY BECAUSE THEY ARE WRITTEN IN A DIFFERENT CODE. IMPROVEMENTS IN THIS DIRECTION IS STRONGLY RECOMMENDED.
THE TEST EXAMPLE NO. 7 (CRYOGENIC TANK) IS INCORRECTLY COM-PUTED ON BOTH THE IBM360/65 & THE IBM360/75.

BOSOR4 THE PROGRAM HAS BEEN USEFUL AND NOT DIFFICULT TO USE ONCE YOU GET THE HANG OF IT. WOULD USE IT MORE IF WE HAD MORE SHELL BUCKLING PROBLEMS. THE MANUAL APPEARS WELL ORGANIZED ON THE SURFACE BUT THE LIST OF DO'S AND DONT'S ARE SCATTERED. A GOOD INDEX WOULD HELP LOCATE PROBLEM AREAS. BECAUSE THE PROGRAM DOES A WIDE VARIETY OF PROBLEMS, ADDITIONAL SAMPLE PROBLEMS WOULD BE HELPFUL. THE PROGRAM DEVELOPER, DR. BUSHNELL, HAS BEEN VERY HELPFUL ON SEVERAL PROBLEMS WHICH HAVE LOOMED UP. HAVE NOT USED THE PLOT ROUTINES BECAUSE WE DO NOT HAVE THE REQUISITE SOFTWARE ON OUR SYSTEM. MOST NEW PROGRAMS IN OUR COMPANY NOW USE NAMELIST INPUT WHICH SIMPLIFIES & ELIMINATES INPUT ERRORS.

BOSOR4 RECOMMEND PROGRAM IMPROVEMENTS IN FOLLOWING AREAS; (1)ELIMINATION OF REDUNDANT INPUT. (2)IMPROVED ORGANIZATION OF MANUAL. I AM CONSTANTLY FLIPPING BACK & FORTH THROUGH THE MANUAL. (3)IMPROVED OUTPUT. START NEW PROBLEMS & TABLES WITH-IN PROBLEMS, AT TOP OF NEW PAGE. ELIMINATE ALL MESSAGES REPORTING ANALYSIS PROGRESS UNLESS REQUESTED. PROVIDE FOR OPTIONAL PRINTOUT OF MODAL STRESSES. (4)ADD SHEAR DEFLECTION CAPABILITY & TRANSVERSE SHEAR PRINTOUT.

BOSOR4 PLOTTER NOT OPERATIONAL ON CDC6600.

BOSOR4 ENGINEERING MANHOURS NEEDED TO PREPARE INPUT DATA TOO COSTLY COMPARED TO OTHER IN-HOUSE PROGRAMS.

BOSOR4 WOULD LIKE TO SEE THE INPUT PREPARATION SIMPLIFIED AND A MACHINE-INDEPENDENT PLOT PACKAGE INCORPORATED TO FACILITATE EASE OF ADAPTING SC-4020 PLOT ROUTINE TO OTHER SYSTEMS--PERHAPS ON THE ORDER OF THE GRAPHIC PACKAGE DISSPLA.

BOSOR4 PROGRAM COULD USE KEY-WORD INPUT FORMAT. FINDS THE LARGE DECKS OF PURELY NUMERICAL, SEQUENTIALLY DEPENDENT INPUT DIFFICULT TO WORK WITH.
INPUT OPTIONS FOR CHANGING MATERIAL PROPERTIES + FIBER ANGLES + THICKNESS WOULD BE USEFUL. FEELS MORE OPTIONS ARE NEEDED TO CONTROL OUTPUT INFORMATION SUCH AS ORIGINAL, NEW + DISPLACED COORDINATES. PERHAPS ADAPTING THE PROGRAM TO USER DEFINED PRE- + POST-PROCESSORS CAN ACCOMPLISH THIS.

BOSOR4 VIKING ORBITER '75.. PROPULSION SUBSYSTEM HAS TWO LARGE PROPE LLANT TANKS. THEIR DESIGN AND STRUCTURAL ANALYSIS HAVE EMPLOYED BOSOR4 EXTENSIVELY. WE USE BOSOR4 TO GENERATE THE STRESS COEFFICIENT MATRIX,THAT IS THE STRESS AT VARIOUS POINTS DUE TO A UNIT LOAD AT A POINT. THIS MATRIX WAS SUB-SEQUENTLY USED IN A FINITE ELEMENT ANALYSIS OF THE TOTAL SPACE CRAFT TO CALCULATE THE STRESSES DUE TO THE FLIGHT LOADS EXTENSIVE TEST DATA HAVE BEEN COLLECTED FOR THE PROPELLANT TANKS WHICH IN GENERAL ARE IN GOOD AGREEMENT WITH THE BOSOR4 RESULTS. WE HAVE VERY EXTENSIVE EXPERIENCES WITH LARGE GENERAL PURPOSE COMPUTER COEES. WE MUST SAY THAT BOSOR4 IS THE ONE THAT GIVES US PLEASURE INSTEAD OF HEADACHES. AS FAR AS DR. BUSHNELL IS CONCERNED. THE USER-DEVELOPER RELATIONSHIP COULD NOT BE BETTER.

BOSOR4 FINDS IT HELPFUL TO LIVE TWO-DOORS DOWN HALLWAY FROM PROGRAM DEVELOPER SINCE THIS CONSIDERABLY ALLEVIATES THE EFFECTS OF THE BYZANTINE USER'S MANUAL.

BOSOR4 PROGRAM HAS VAST POTENTIAL BUT, SO FAR, HAVE ONLY USED THE VIBRATION ANALYSIS CAPABILITY ON RATHER SIMPLE STRUCTURES. COMPUTER RESULTS AGREE WITH TEST RESULTS + RESULTS OBTAINED FROM OTHER CODES WHEN COMPARISONS WERE POSSIBLE.
HAVE MODIFIED THE CODE WITHOUT MUCH DIFFICULTY. FEEL THAT THE INCORPORATION OF *COMDECK & *CALL INTO THE CDC-6600 WOULD HAVE MADE THE MODIFICATIONS EASIER.

BOSOR4 FEELS USER MANUAL IS GOOD BUT WOULD LIKE MORE SAMPLE PROBLEMS CONTACT WITH DR. BUSHNELL HAS BEEN EXCELLENT. LINES OF COMMUNICATION ARE ALWAYS OPEN. DEVELOPER HAS BEEN MOST HELPFUL DURNING ALL STAGES OF USE.

BOSOR4 WOULD LIKE CAPABILITY OF PERFORMING TIME-HISTORY DYNAMIC ANALYSIS PROBLEMS AND ALSO TO BE ABLE TO HANDLE MORE THAN 20 HARMONICS. WOULD ALSO LIKE TO BE ABLE TO INVESTIGATE CIRCUMFERENTIAL IMPERFECTIONS.

BOSOR4 PROGRAM DIFFICULT TO WORK WITH DUE TO POOR I/O. FINDS THE MESH GENERATOR CUMBERSOME AS CARE MUST BE EXERCISED WHEN USING THIS FEATURE IN BRANCHED SHELLS SINCE FICTITIOUS POINTS ARE ADDED WHICH GENERALLY CAUSE THE SHELL INTERSECTION TO BE DISPLACED.
OUTPUT DATA IS MUCH TOO VOLUMINOUS. NORMAL OUTPUT OPTION PRO-DUCES CONSIDERABLY USELESS INFORMATION + THE MINIMUM OUTPUT OPTION OMITS ESSENTIALS SUCH AS APPLIED LOADS.
CORRELATION OF DATA HAMPERED BY CORRELATING THE DATA TO AN ACCUMULATIVE SHELL LENGTH RATHER THAN TO PHYSICAL COORDINATES . ONLY USE THIS PROGRAM WHEN OTHER PROGRAMS ARE INCAPABLE OF TREATING THE PROBLEM; USED FOR BUCKLING PRIMARILY.

```
NOW PUNCH EITHER   'ALL' ,   'SCOPE' ,    'COMPARE' ,
                   'EVALUATE' ,   'PROBLEM' , OR    'FINISHED'
```

SCOPE

```
NOW PUNCH THE NAME OF THE PROGRAM FOR WHICH YOU
WANT A LIST OF THE CAPABILITIES,PRICE,ETC.

EXAMPLE---'ANSYS'  (DON'T PUNCH THE QUOTES)
THE PROGRAM NAME HAS TO BE PUNCHED EXACTLY AS IT
APPEARS IN THE DATA BANK.

FROM 3 TO 6 PAGES OF OUTPUT WILL FOLLOW.

BOSOR4, BUSHNELL, D., DPT 5233/205, LMSC, PALO ALTO, CALIF 94304

SECTION 1,   GEOMETRY / BOUNDARY CONDITIONS   BOSOR4

 05 SHELL OF REVOLUTION (360 DEG.)
 06 PRISMATIC STRUCTURE
 07 SEGMENTED (IN SERIES)
 08 BRANCHED
 14 AXISYMMETRIC IMPERFECTIONS
 16 AXISYMMETRIC SUPPORT CONDITIONS
 19 SUPPORTS AT INTERNAL POINTS
 21 STIFFENERS
 24 SLIDING WITHOUT FRICTION
 26 JUNCTURE COMPATIBILITY
 27 SINGULARITY CONDITIONS
 33 LAGRANGE MULTIPLIER METHOD USED

SECTION 2, WALL CONSTRUCTION/MATERIAL PROPERTIES  BOSOR4

 01 MONOCOQUE WALL CONSTRUCTION
 02 LAYERED WALL CONSTRUCTION
 03 SANDWICH WALL CONSTRUCTION
 04 COMPOSITE MATERIAL
 06 CIRCUM. STIFFENERS TREATED AS DISCRETE
 08 STIFFENERS TREATED AS SMEARED
 10 CG, SHEAR CENTER OF STIFFENER MUST COINCIDE
 11 OTHER WALL CONSTRUCTION--SEE SECT.8, 2.11
 12 ISOTROPIC MATERIAL PROPERTIES
 13 ANISOTROPIC MATERIAL PROPERTIES
 14 LINEAR ELASTIC MATERIAL PROPERTIES
 16 TEMPERATURE-DEPENDENT MATERIAL PROPERTIES
 31 AXISYMMETRIC VARIATION OF PROPERTIES ONLY
 33 COUPLING OF STRAINS AND CURVATURE CHANGES

SECTION 3, LOADING                              BOSOR4

 01 UNIFORM DISTRIBUTION OF LOADS
 03 GENERAL VARIATION OF LOADS
 04 STATIC OR QUASI-STATIC LOAD VARIATION
 10 LOADS VARYING PROPORTIONALLY DURING CASE
 11 SOME LOADS VARY, SOME CONSTANT DURING CASE
 15 LIVE (FOLLOWING) LOADS
 24 POINT LOADS, MOMENTS
 25 LINE LOADS, MOMENTS
 26 NORMAL PRESSURE
 27 SURFACE TRACTIONS
 29 DEAD-WEIGHT LOADING
 30 THERMAL LOADING

SECTION 4, PHENOMENA                            BOSOR4

 01 AXISYMMETRIC SMALL DEFLECTIONS
 02 GENERAL SMALL DEFLECTIONS
 03 AXISYMMETRIC LARGE DEFLECTIONS
 09 THERMAL EFFECTS
 16 NONLINEAR COLLAPSE ANALYSIS
 18 BIFURC. BUCKLING, LINEAR AXISYM. PRESTRESS
 19 BIFURC. BUCKLING, LINEAR GENERAL PRESTRESS
 20 BIFURC. BUCKLING, NONLIN. AXISYM. PRESTRESS
 22 BIFURC. BUCKLING, PREBUCKLING ROTATIONS
 29 MODAL VIBRATIONS, NONLIN. AXISYM. PRESTRESS

SECTION 5, DISCRETIZATION                       BOSOR4

 06 FINITE DIFFERENCE ENERGY METHOD
 07 1-DIMENSIONAL FINITE-DIFF. DISCRETIZATION
 12 FOURIER SERIES IN CIRCUMFERENTIAL DIRECTION
 13 MAXIMUM NO. OF FOURIER HARMONICS GIVEN IN SECT.8, 5.13
 46 NON-DIAGONAL MASS MATRIX
 48 OTHER COMMENTS PERTINENT TO THIS SECTION--SEE SECT.8, 4,48

    MAXIMUM NUMBER OF NODAL POINTS =   450
    MAXIMUM NUMBER OF DEGREES OF FREEDOM =  1500
    MAXIMUM BANDWIDTH OF MATRICES = NO LIMIT
```

SECTION 6, SOLUTION METHODS BOSOR4

03 SKYLINE METHOD OF SOLUTION USED
16 SINGLE VECTOR INVERSE PWR ITER. WITH SHIFTS AND DEFLATION
19 FULL NEWTON-RAPHSON METHOD FOR NONLIN. SOLN
29 LAGRANGIAN FORMULATION

SECTION 7, COMPUTER PROGRAM DISTRIBUTION
DOCUMENTATION/ORGANIZATION/MAINTENANCE BOSOR4

01 SOURCE PROGRAM TAPE AVAILABLE
05 PROGRAM PRICE IN DOLLARS--SEE SECT.8, 7.05
06 PROGRAM IS AVAILABLE THROUGH DEVELOPER
07 PROGRAM IS AVAILABLE THROUGH COSMIC
09 PROGRAM CAN BE EXECUTED THRU SERVICE BUREAUS--SEE SECT.10, 7.09
10 NUMBER OF INSTITUTIONS USING THIS PROGRAM ---- SEE SECT.10, 7.10
14 USERS MANUALS ARE SELF-CONTAINED
15 USERS MANUALS HAVE LIST OF PITFALLS
16 USERS MANUALS HAVE TEST CASES
17 USERS MANUALS HAVE FLOW CHARTS
18 USERS MANUALS GIVE THEORY
21 PROGRAM RUNS ON IBM
22 PROGRAM RUNS ON CDC
23 PROGRAM RUNS ON UNIVAC
26 NO. OF SOURCE STATEMENTS (10,000'S)* SEE SECT.8
27 NO. OF SUBROUTINES (100'S)* SEE SECT.8
28 NO. OF PRIMARY OVERLAYS * SEE SECT.8
MAN-YEARS REQUIRED TO DEVELOP PROGRAM * SEE SECT.8
30 CORE STORAGE REQUIRED FOR EXECUTION(000'S) SEE SECT.8
33 NO. UNITS OF DIRECT ACCESS MASS STORAGE GIVEN IN SECT.8, 7.31
37 OPTIONS FOR VARIOUS OUTPUT
43 SECTION PROPERTIES CALCULATED
50 PLOT ROUTINES FOR UNDEFORMED/DEFORMED GEOMETRY
53 ORDINARY Y = F(X) PLOTS
57 PLOTTING CAPABILITY FOR STROMBERG-CARLSON
59 OPTIONAL DETAILED PRINT OUTPUT
61 ALTERNATIVE BRANCHES FOR SAME PROBLEM
70 DEVELOPER WILL HELP USER GET PROGRAM RUNNING
71 DEVELOPER WILL HELP USER MODELING CASES
72 DEVELOPER WILL HELP USER FIND SUSPECTED BUGS
73 DEVELOPER WILL CIRCULATE NOTICES OF BUGS FOUND
74 DEVELOPER WILL HELP USER SET UP AND RUN CASES
75 DEVELOPER WILL CONDUCT WORKSHOPS
76 DEVELOPER WILL MAKE SPECIAL PROGRAM CHANGES
77 DEVELOPER WILL KEEP USERS INFORMED OF IMPROVEMENTS

SECTION 8, ADDITIONAL COMMENTS

2.04,2.11& LAYERED ORTHOTROPIC, FIBERWOUND, CORRUGATED SEMI-SANDWICH; 5.48& U AND V NODES ARE MIDWAY BETWEEN W NODES
5.13 MAXIMUM NO. OF FOURIER HARMONICS IS 20
7.05 PROGRAM PRICE IS 300.00 DOLLARS
7.26 12000 SOURCE STATEMENTS, 150 SUBROUT., 6 PRIMARY OVERLAYS
7.29 5 MAN-YEARS TO DEVELOP, 64,000 DEC.WORDS CORE REQUIRED
7.33 ONE UNIT DIRECT ACCESS MASS STORAGE REQUIRED

SECTION 9, REFERENCES

BOSOR4 USER'S MANUAL, LMSC-D243605, MARCH 1972 AND NASA CR-2116 OCT. 1972, AIAA PAPER NO. 73-360, PRESENTED AT AIAA/ASME/SAE CONF.,APRIL 1973, TO APPEAR COMPUTERS $ STRUCTURES, SEVEN ADDITIONAL JOURNAL ARTICLES ALSO LISTED UNDER AUTHORSHIP OF DAVID BUSHNELL

SECTION 10, LIST OF USERS OF THIS PROGRAM

SEVERAL SERVICE BUREAUS LISTED...
CONTROL DATA, ROCKVILLE, MD (AL THOMPSON (301)881-5800)
MACDONELL DOUGLAS AUT, BERKELEY, MO(B.FLACHSBART(314)232-5569)
WESTINGHOUSE TELECOMPUTER,PITTSBURGH,PA(DUANE SCHMIEDEL(412) 256-7979
BOEING COMPUTER SERV.,SEATTLE,WASH.(LEONARD TRIPP(206)237-4744
LIST OF SOME 50 ADDITIONAL USERS APPENDED

```
NOW PUNCH EITHER   'ALL' ,   'SCOPE' ,    'COMPARE' ,
                   'EVALUATE' ,   'PROBLEM' , OR    'FINISHED'
```

SCOPE

```
NOW PUNCH THE NAME OF THE PROGRAM FOR WHICH YOU
WANT A LIST OF THE CAPABILITIES,PRICE,ETC.

EXAMPLE---'ANSYS'  (DON'T PUNCH THE QUOTES)
THE PROGRAM NAME HAS TO BE PUNCHED EXACTLY AS IT
APPEARS IN THE DATA BANK.

FROM 3 TO 6 PAGES OF OUTPUT WILL FOLLOW.

SESAM69C, DET NORSKE VERITAS,   NO ADDRESS GIVEN,    NO PHONE GI

SECTION 1,   GEOMETRY / BOUNDARY CONDITIONS   SESAM69C

 01 GENERAL STRUCTURE
 02 SOLID OF REVOLUTION
 03 GENERAL SHELL
 04 PART OF SHELL OF REVOLUTION ( L.T. 360 DEG.)
 05 SHELL OF REVOLUTION (360 DEG.)
 06 PRISMATIC STRUCTURE
 07 SEGMENTED (IN SERIES)
 08 BRANCHED
 09 SHELL WITH CUTOUTS
 10 THICK SHELL
 11 FRAMEWORK-SHELL COMBINATIONS
 12 FLAT PLATE OF GENERAL SHAPE
 13 OTHER GEOMETRY--SEE SECT.8, 1.13
 16 AXISYMMETRIC SUPPORT CONDITIONS
 17 GENERAL SUPPORT CONDITIONS
 18 SUPPORTS AT BOUNDARIES ONLY
 19 SUPPORTS AT INTERNAL POINTS
 20 ELASTIC FOUNDATION
 21 STIFFENERS
 24 SLIDING WITHOUT FRICTION
 25 OTHER DEFORMATION DEPENDENT SUPPORT--SEE SECT.8, 1.25
 26 JUNCTURE COMPATIBILITY
 27 SINGULARITY CONDITIONS
 28 GENERAL LINEAR CONSTRAINT COND. U(I)= TU(N)
 30 TRANSITION FROM 3-D TO 2-D REGION
 31 TRANSITION FROM 2-D TO 1-D REGION
 32 OTHER CONSTRAINT CONDITIONS--SEE SECT.8, 1.32

SECTION 2, WALL CONSTRUCTION/MATERIAL PROPERTIES  SESAM69C

 01 MONOCOQUE WALL CONSTRUCTION
 02 LAYERED WALL CONSTRUCTION
 05 MERIDIONAL STIFFENERS TREATED AS DISCRETE
 06 CIRCUM. STIFFENERS TREATED AS DISCRETE
 07 GENERAL STIFFENERS TREATED AS DISCRETE
 08 STIFFENERS TREATED AS SMEARED
 10 CG, SHEAR CENTER OF STIFFENER MUST COINCIDE
 12 ISOTROPIC MATERIAL PROPERTIES
 13 ANISOTROPIC MATERIAL PROPERTIES
 14 LINEAR ELASTIC MATERIAL PROPERTIES
 16 TEMPERATURE-DEPENDENT MATERIAL PROPERTIES
 17 RIGID-PERFECTLY PLASTIC MATERIAL
 18 ELASTIC-PERFECTLY PLASTIC MATERIAL
 19 ELASTIC-LINEAR STRAIN HARDENING MATERIAL
 20 STRESS-STRAIN CURVE OF MANY LINE SEGMENTS
 22 ISOTROPIC STRAIN HARDENING
 25 OTHER POST-YIELD LAW--SEE SECT.8, 2.25
 26 PRIMARY CREEP
 27 SECONDARY CREEP
 28 VISCOELASTIC MATERIAL
 31 AXISYMMETRIC VARIATION OF PROPERTIES ONLY
 32 ARBITRARY VARIATION OF PROPERTIES

SECTION 3, LOADING                        SESAM69C

 03 GENERAL VARIATION OF LOADS
 04 STATIC OR QUASI-STATIC LOAD VARIATION
 05 GENERAL TRANSIENT VARIATION OF LOADS
 06 PERIODIC (DYNAMIC) VARIATION OF LOADS
 19 FLUID-STRUCTURE INTERACTION
 22 RIGID SOLID-STRUCTURE INTERACTION
 24 POINT LOADS, MOMENTS
 25 LINE LOADS, MOMENTS
 26 NORMAL PRESSURE
 27 SURFACE TRACTIONS
 28 BODY FORCES
 29 DEAD-WEIGHT LOADING
 30 THERMAL LOADING
 31 INITIAL STRESS OR STRAIN
 32 CENTRIFUGAL LOADING
 33 NON-ZERO DISPLACEMENTS IMPOSED
 34 OTHER TYPE OF LOADING--SEE SECT.8, 3.34
 35 OTHER COMMENTS PERTINENT TO THIS SECTION--SEE SECT.8, 3.35
```

```
SECTION 4, PHENOMENA                                    SESAM69C

 01 AXISYMMETRIC SMALL DEFLECTIONS
 02 GENERAL SMALL DEFLECTIONS
 05 AXISYMMETRIC PLASTICITY
 06 GENERAL PLASTICITY
 08 TRANSVERSE SHEAR DEFORMATIONS
 09 THERMAL EFFECTS
 11 AUTOMATED YIELD CRITERION
 26 MODAL VIBRATIONS WITH NO PRESTRESS
 31 MODAL VIBRATIONS, TRANSVERSE SHEAR DEFORMATIONS
 34 LINEAR AXISYMMETRIC DYNAMIC RESPONSE
 35 LINEAR GENERAL DYNAMIC RESPONSE
 39 DYNAMIC RESPONSE, TRANSVERSE SHEAR DEFORMATIONS
 41 VISCOUS DAMPING
 43 LINEAR DAMPING

SECTION 5, DISCRETIZATION                               SESAM69C

 39 SUBSTRUCTURING
 40 REPEATED USE OF IDENTICAL SUBSTRUCTURES
 45 AUTOMATIC RENUMBERING OF NODES
 46 NON-DIAGONAL MASS MATRIX
 47 DIAGONAL MASS MATRIX
 48 OTHER COMMENTS PERTINENT TO THIS SECTION--SEE SECT.8, 4.48

    MAXIMUM NUMBER OF NODAL POINTS =  NO LIMIT
    MAXIMUM NUMBER OF DEGREES OF FREEDOM =  NO LIMIT
    MAXIMUM BANDWIDTH OF MATRICES = NO LIMIT


FINITE ELEMENT LIBRARY
                  DISPLACEMENT        HYBRID
                  VERTICES/NODES/D.O.F.

  ROD                 / 2/ 2
  STRAIGHT BEAM       / 2/12
  CURVED BEAM         / 3/18
  FLAT MEMBRANE     3/ 3/ 6
  FLAT MEMBRANE     4/ 4/ 8
  CURVED MEMBRANE   4/ 8/16
  FLAT PLATE        3/ 3/ 9
  FLAT PLATE        4/ 4/12
  SHALLOW SHELL     4/ 8/40
  SHALLOW SHELL     3/ 6/30
  AXISYM. SOLID     3/ 3/ 6
  AXISYM. SOLID     4/ 4/ 8
  THICK SHELL       4/ 8/40
  SOLID (3-D)       8/20/60
  SOLID (3-D)       4/10/30
  SOLID (3-D)       6/15/45
  Q-19 BENDING AND LINEAR STRAIN MEMBERS

SECTION 6, SOLUTION METHODS                             SESAM69C

 04 MATRIX PARTITIONING USED
 08 CONJUGATE GRADIENT METHOD FOR SOL'N OF LIN. EQ.
 10 OTHER SOLUTION METHOD FOR LINEAR EQUATIONS
 11 CONDENSATION METHOD--SEE SECT.8, 6.11
 14 STURM SEQUENCE METHOD FOR EIGENVALUE EXTRACTION
 18 OTHER EIGENVALUE EXTRACTION METHOD--SEE SECT.8, 6.18
 27 NONLIN. EQ. SOLVED BY SUCCESSIVE SUBSTITUTIONS IN LOAD VECTOR
 28 OTHER NONLINEAR STRATEGY--
 32 TIME INTEGRATION BY MODAL SUPERPOSITION
```

SECTION 7, COMPUTER PROGRAM DISTRIBUTION
DOCUMENTATION/ORGANIZATION/MAINTENANCE SESAM69C

01 SOURCE PROGRAM TAPE AVAILABLE
04 PROGRAM COMPLETELY PROPRIETARY
05 PROGRAM PRICE IN DOLLARS--SEE SECT.8, 7.05
06 PROGRAM IS AVAILABLE THROUGH DEVELOPER
09 PROGRAM CAN BE EXECUTED THRU SERVICE BUREAUS--SEE SECT.10, 7.09
10 NUMBER OF INSTITUTIONS USING THIS PROGRAM ---- SEE SECT.10, 7.10
12 IMPROVED VERSION BEING PREPARED--SEE SECT.8, 7.12
14 USERS MANUALS ARE SELF-CONTAINED
15 USERS MANUALS HAVE LIST OF PITFALLS
16 USERS MANUALS HAVE TEST CASES
17 USERS MANUALS HAVE FLOW CHARTS
18 USERS MANUALS GIVE THEORY
19 OTHER QUALITIES OF USERS MANUALS--
21 PROGRAM RUNS ON IBM
22 PROGRAM RUNS ON CDC
23 PROGRAM RUNS ON UNIVAC
24 PROGRAM RUNS ON GE
25 PROGRAM RUNS ON COMPUTER IDENTIFIED IN SECT.8, 7.25
26 NO. OF SOURCE STATEMENTS (10,000'S)* SEE SECT.8
27 NO. OF SUBROUTINES (100'S)* SEE SECT.8
28 NO. OF PRIMARY OVERLAYS * SEE SECT.8
MAN-YEARS REQUIRED TO DEVELOP PROGRAM * SEE SECT.8
30 CORE STORAGE REQUIRED FOR EXECUTION(0000'S) SEE SECT.8
31 MINIMUM CORE SPACE REQUIRED FOR EXECUTION SEE SECT.8
33 NO. UNITS OF DIRECT ACCESS MASS STORAGE GIVEN IN SECT.8, 7.31
34 NUMBER OF TAPES REQUIRED IS GIVEN IN SEC.8, 7.34
35 LOW SPEED CORE OR VIRTUAL MEMORY REQUIRED--
36 RESTART CAPABILITY
37 OPTIONS FOR VARIOUS OUTPUT
38 DIRECT ACCESS DATA BASE
41 AUTOMATIC MESH GENERATOR(S)
42 AUTOMATIC LOADING GENERATOR
43 SECTION PROPERTIES CALCULATED
44 PRE-PROCESSOR
45 POST-PROCESSOR
46 EIGENVALUE SOLVER
47 MODAL SUPERPOSITION ROUTINE
50 PLOT ROUTINES FOR UNDEFORMED/DEFORMED GEOMETRY
51 VARIOUS VIEWS AND SECTIONS PLOTTED
52 CONTOUR PLOTS
53 ORDINARY Y = F(X) PLOTS
54 STRESS-ON-STRUCTURE PLOTS
55 OTHER KINDS OF PLOTS--SEE SECT.8, 7.55
56 PLOTTING CAPABILITY FOR CALCOMP
58 PLOTTING CAPABILITY FOR HARDWARE IDENTIFIED IN SECT.8, 7.58
59 OPTIONAL DETAILED PRINT OUTPUT
60 EQUILIBRIUM CHECKS
62 MATRIX CONDITIONING ESTIMATION
63 AUTOMATIC ERROR CONTROL
64 GRAPHICAL PREPROCESSOR
65 APPLIED LOADING CHECKS
66 BOUNDARY CONDITION CHECKS
68 GEOMETRY CHECKS
70 DEVELOPER WILL HELP USER GET PROGRAM RUNNING
71 DEVELOPER WILL HELP USER MODELING CASES
72 DEVELOPER WILL HELP USER FIND SUSPECTED BUGS
73 DEVELOPER WILL CIRCULATE NOTICES OF BUGS FOUND
74 DEVELOPER WILL HELP USER SET UP AND RUN CASES
75 DEVELOPER WILL CONDUCT WORKSHOPS
76 DEVELOPER WILL MAKE SPECIAL PROGRAM CHANGES
77 DEVELOPER WILL KEEP USERS INFORMED OF IMPROVEMENTS
78 OTHER ASSISTANCE DEVELOPER WILL GIVE USERS--SEESECT.8, 7.78

SECTION 8, ADDITIONAL COMMENTS

1.13 ANY GEOMETRY
1.25 LINEAR DEPENDENCE BETWEEN DEGREES OF FREEDOM
1.32 3-DIMENSIONAL FRACTURE CALCULATION
2.13 ORTHOTROPIC MATERIAL PROPERTIES
2.16 TEMP-DEPENDENT MATERIAL IN 2-D FIELD PRO.
2.25 ANY POST-YIELD LAW
2.26, 2.27 CREEP IN ROTATIONALLY SYMMETRIC PROBLEMS
3.34 OTHER TYPES OF LOADS--LINEAR DEPENDENCY DISP.
3.35 WAVE LOADS IN AN IRREGULAR OR REGULAR SEA
4.06 GENERAL PLASTICITY IN MEMBRANES
4.11 AUTOMATED YIELD CRITERION IS TRESCA OR VON MISES
5.39 MULTI LEVEL SUBSTRUCTURING
6.00 NORSAM IS A LIBRARY OF SUBROUTINES AVAILABLE IF NEEDED.
6.04 PARTITIONING-- CHOLESKY DECOMPOSITION IN NORSAM
6.08 CONJUGATE GRADIENT METHOD IN NORSAM ONLY
6.10 GAUSSIAN ELIMINATION AND BACK SUBSTITUTUON
6.11 CONDENSATION METHOD-- CONSERVATION OF STRAIN ENERGY (UNIT DISPLACEMENT METHOD)
6.18 GIVENS, HOUSEHOLDERS METHOD FOR EIGENVALUE EXTRACTION,
6.18 JACOBI METHOD IN NORSAM
6.28 OTHER NONLINEAR STRATEGY-- TANGENT STIFFNESS METHOD
7.05 PROGRAM PRICE DEPENDS ON RIGHT OF USE
7.10 ABOUT 10 INSTITUTIONS USING SESAM69C
7.12 AN IMPROVED EXTENDED VERSION WILL BE COMPLETED BY JAN 1975
7.25 VERSION A WILL RUN ON ICL BY DEC 1974
7.26 ABOUT 100,000 SOURCE STATEMENTS, 1000 SUBROUT., 6 OVERLAY
7.29 40 MAN-YEARS TO DEVELOP, 64K DEC. WORDS REQ'D TO EXECUTE
7.31 64K DEC. WORDS MINIMUM, 0.4 MILLION WORDS DIRECT ACCESSREQ
7.34 2-5 TAPES REQUIRED, PROBLEM-DEPENDENT
7.44 PRE-PROCESSOR-- DATA GENERATION
7.44 INPUT DATA GENERATING FACILITATOR, GENERATE NODE NUMB., ELEMENTS, LOADS, BOUNDARY CONDITIONS, SUBSTRUCTURES.
7.45 POST-PROCESSOR-- PLOT DISP., STRESS, MODES, LINEAR COMBINATION OF RESULTS FROM DIFFERENT LOADING CONDITIONS.
7.55 MODAL PLOTS
7.58 PLOT ROUTINES SIMPLY CONVERTABLE (SEPARATE PROGRAM)

SECTION 9, REFERENCES

DET NORSKE VERITAS, 'APPLICATION OF COMPUTERIZED METHODS IN ANALYSIS AND DESIGN OF SHIP STRUCTURES, MARINE STRUCTURES AND MACHINERY,' SEMINAR IN OSLO NORWAY NOV 8-9, 1972
'THE APPLICATION OF THE SUPERELEMENT METHOD IN ANALYSIS AND DESIGN OF SHIP STRUCTURES AND MACHINERY COMPONENTS' PRESENTED AT THE NATIONAL SYMPOSIUM ON COMPUTERIZED STRUCTURAL ANALYSIS AND DESIGN, 27-29 MARCH 1972, GEORGE WASHINGTON UNIV., WASH, DC

SECTION 10, LIST OF USERS OF THIS PROGRAM

USERS-- LIST OF 14 USERS PROVIDED--IRCN IN FRANCE, NCRE IN SCOTLAND, NEL IN SCOTLAND,ASEA IN SWEDEN, SIA IN ENGLAND, ITAL CANTIERI IN ITALY, IFS IN GERMANY, GEOCOM IN TEXAS, THE AKER GROUP IN NORWAY, GOTAVERKEN IN SWEDEN, BURMEISTER AND WAIN IN DENMARK, UDEVALLAVARVED IN SWEDEN, ROYAL NORWEGIAN INST. TEC IN NORWAY, TECH. UNIV. IN FINLAND

NO USERS HAVE YET RESPONDED TO THE QUESTIONNAIRE

```
NOW PUNCH EITHER   'ALL' ,   'SCOPE' ,    'COMPARE' ,
                   'EVALUATE' ,   'PROBLEM' ,  OR    'FINISHED'
```

ALL

```
LIST OF ALL PROGRAMS IN DATA BANK FOLLOWS.

AC-3,              PCAP,
AC-50A,            PDILB,
ADEPT,             PETROS3.
ANSYS,             PILEFEM,
ARCHDAM1,          REGULA,
ARCHDAM2,          REPSIL,
ASAS,              REXBAT,
ASEF,              SABOR/DRASTIC3A,
ASKA,              SABOR/DRASTIC6,
ASOP,              SADAOS,
BERSAFE,           SALORS,
BOLS,              SAMBA,
BOND4,             SAMIS,
BOPSII,            SAMMSORIII,
BOSOR4,            SAPIV,
CABLE3,            SATANS,
COBA,              SATURN,
CYDYN,             SESAM69C,
DANUTA,            SHORE,
DYNAL,             SINGUL,
DYNAPLASII,        SLADE,
DYNASORII,         SLADE-D,
DYNAS,             SNAP,
EASE,              SNASORII,
EASE2,             SOLIDSAP,
ELAS55,            SPADAS,
ELAS65,            SRA,
ELAS75,            STAGSA1,
FAMSOR,            STAGSB,
FARSS,             STARDYNE,
FESAP,             STARS,
HONDO,             STMFR-60,
ISOPAR-SHL,        STRESS,
ISTRAN/S,          STRESSIII,
KSHEL,             STRUDLII,
MARC,              SWBFR,
MATCOMP,           TABLES,
MINIELAS,          TANKER,
MSC/NASTRAN,       TEXGAP,
NEPSAP,            TIRE,
NONLIN,            TRANSTOWER,
NONLIN2,           TROCS,
NOSTRA,            VISCEL,
PAFEC70,           ZP26,
PCALS,

NOW PUNCH EITHER   'ALL' ,   'SCOPE' ,    'COMPARE' ,
                   'EVALUATE' ,   'PROBLEM' ,  OR    'FINISHED'
```

FINISHED

```
PLEASE WAIT FOR THE STATEMENT BEGINNING

     'COMSTAIRS IS AN EXPERIMENTAL.....'

DON'T TOUCH THE KEYBOARD UNTIL ASKED TO. THANKS
```

```
COMSTAIRS IS AN EXPERIMENTAL INFORMATION RETRIEVAL
SYSTEM. FEEDBACK FROM  YOU IS NEEDED IN ORDER TO
IMPROVE IT.

DO YOU WANT TO  INDICATE YOUR AGREEMENT OR
DISAGREEMENT WITH ELEVEN STATEMENTS CONCERNING
COMSTAIRS?

  PUNCH  'YES'  IF YOU DO.
  PUNCH  'NO'   IF YOU DON'T.

YES

PLEASE INDICATE YOUR AGREEMENT OR DISAGREEMENT
WITH THE FOLLOWING STATEMENTS BY PUNCHING....

  '0'  IF YOU DON'T UNDERSTAND THE STATEMENT OR HAVE NO OPINION.
  '1'  IF YOU STRONGLY AGREE WITH THE STATEMENT
  '2'  IF YOU AGREE
  '3'  IF YOU ARE NEUTRAL
  '4'  IF YOU DISAGREE
  '5'  IF YOU STRONGLY DISAGREE


THE ELEVEN (11) STATEMENTS WILL NOW BE PRINTED,
ONE AT A TIME. AFTER EACH STATEMENT THE MACHINE WILL
PAUSE, AWAITING YOUR ANSWER (0,1,2,3,4, OR 5)


1. COMSTAIRS IS EASY TO USE.
2. COMSTAIRS GIVES THE RIGHT KIND OF INFORMATION.
3. I WOULD USE SUCH A SYSTEM AS COMSTAIRS IN MY WORK
4. I HAD A DEFINITE PURPOSE IN USING COMSTAIRS
5. I FOUND OUT WHAT I WANTED TO BY USING COMSTAIRS.
6. THE DATA BANK HAS ENOUGH PROGRAMS.
7. THERE IS ENOUGH RESPONSE FROM USERS.
8. THERE ARE ENOUGH KEY WORDS.
9. AS A PROGRAM USER I WOULD LIKE TO SUPPLY INPUT DATA
   TO THE DATA BANK.
10 AS A PROGRAM DEVELOPER I WOULD LIKE TO SUPPLY INPUT
   DATA TO THE DATA BANK.
11 I HAVE A PROBLEM TO SOLVE AND NEED A PROGRAM TO
   SOLVE IT.


PLEASE DON'T TERMINATE THE RUN.  WAIT FOR
THE STATEMENT BEGINNING

             'NOW YOU CAN ADD....'

DON'T TOUCH THE KEYBOARD UNTIL ASKED TO.  THANKS
```

```
NOW YOU CAN ADD SOME DATA TO THE COMSTAIRS DATA
BANK. FEEL FREE TO TYPE IN ANY COMMENTS YOU WISH ON..

  (1) THIS INFORMATION RETRIEVAL SYSTEM.
  (2) A COMPUTER PROGRAM THAT YOU HAVE USED.
  (3) A COMPUTER PROGRAM THAT YOU HAVE DEVELOPED.

PUNCH  'YES'  IF YOU WOULD LIKE TO TYPE IN COMMENTS
PUNCH  'NO'   IF YOU WOULD NOT LIKE TO COMMENT.
YES

YOUR COMMENTS WILL BE STORED ONE LINE AT A TIME.
JUST KEEP TYPING, PUNCHING THE RETURN KEY WHEN YOU
GET TO THE END OF A LINE.  WHEN YOU ARE THROUGH,
PUNCH  'END'   ON A LINE BY ITSELF.  FOR EXAMPLE...

'I AM A DEVELOPER AND MY PROGRAM IS NOT INCLUDED IN
THE COMSTAIRS DATA BANK. I WOULD LIKE TO RECEIVE A
QUESTIONNAIRE FOR PROGRAM DEVELOPERS. MY NAME AND
ADDRESS ARE...'
END

NOW PUNCH YOUR COMMENTS, PLEASE.
I WOULD LIKE TO FILL OUT SOME QUESTIONAIRES FOR PROGRAM
USERS. PLEASE SEND QUESTIONAIRES TO JOHN DOE, 20 CENTRAL
AVE, DISNEYLAND, CAL 92103.
END
PLEASE DON'T TERMINATE THR RUN.. WAIT FOR THE
MESSAGE SAYING COME AGAIN.  YOU WILL BE INSTRUCTED
WHAT TO DO THEN.  THANKS..
```

```
@ELT,IDL GOODBY,,GOODBY
ELT PROCESSOR LEVEL     4
000001      000     HOPE YOU ENJOYED YOURSELF.  COME AGAIN...
000002      000     YOU CAN NOW TERMINATE THE RUN BY TYPING
                    THE FIN CARD AS INSTRUCTED.
```

```
@FIN
```

III. PRE- AND POSTPROCESSORS AND COMPUTER GRAPHICS

PREPROCESSORS FOR GENERAL PURPOSE FINITE ELEMENT PROGRAMS

L. G. Napolitano, R. Monti, and P. Murino
Istituto di Aerodinamica, Facoltà di Ingegneria
Università di Napoli
Italia

ABSTRACT

This survey on preprocessors for finite element computer programs examines the preprocessors' desirable features, comments on their present-day capabilities, and discusses future trends. A description and evaluation of currently available preprocessor programs is also given.

INTRODUCTION

This work is aimed at answering the following questions: what should a preprocessor do, what the available processors actually do and how well, and what it is reasonable to expect from future generation preprocessors. A section is devoted to each topic, and the description and evaluation of presently available preprocessors is summarized in a number of tables and detailed in the Appendix.

A few preliminary comments are appropriate to explain the philosophy with which the above questions were approached.

By preprocessor, we mean a separately running computer program which can be employed routinely to ease and/or improve the use of finite element (FE) processor programs at an overall cost saving.

The user appreciates the possibility of interacting with the machine and making use of experience and knowledge to overcome difficulties associated with improper problem specification. Substantial reductions in the overall cost of the computation is achieved if he can preprocess his problems at a low cost and on more readily available computer facilities. Consequently, preprocessors should be able to run separately on small computers and not necessarily on the same large computer systems used for processors.

A number of preprocessing capabilities are often included in FE general purpose programs. A survey of 70 such programs has revealed the following distribution of preprocessing capabilities: automatic mesh generation (40%), geometry check (40%), graphical input data check (30%), automatic loading generation (30%), boundary conditions check (20%), relabeling (20%). Preprocessor facilities available in large-capacity FE programs will not be reviewed here because normally they are intimately linked with the processor.

Information on available preprocessors is not abundant in the open literature. Data presented here come from replies to a questionnaire sent to 1,500 universities, public and private research centers, industries, consultant companies, and individuals. Some 200 answers were received; most of them pertained to proprietary programs and will not be reported here. Others did not contain any substantive information. This lack of content was reflected by the repliers' reluctance to describe preprocessors which either dealt with problems of very specific nature or performed simple operations.

This same reason may well help explain the small number of replies received.

Nevertheless, it is felt that the survey is very significant because the answers are mostly related to very recent, relatively unknown programs.

PREPROCESSOR FEATURES

The tasks that a preprocessor should perform can be grouped, in order of increasing sophistication, in three categories: (1) input data generation and checking; (2) improvement of computational efficiency; (3) improvement of structures idealization.

Adequate FE idealization of real structures[1] usually requires a large number of nodes and consequently a large quantity of input data must be specified: nodal points (coordinates), elements (topology, material type, temperature, and elastic and thermal properties), loads, and constraints. In this connection, preprocessors may perform convenient tasks in two ways: by generating data automatically and by checking input and generated data.

The most significant data generation is mesh generation, i.e., computation of nodal point coordinates and of element topology. With it considerable reduction of input data cards can be obtained, much labor saved, and errors avoided.

Data checking can be performed by printing output and/or by plots. Printed output may include the input data cards' image and tables of geometry, topology, loads, constraints, and materials. Correctness of the geometry and the topology data can be checked by a plot of the mesh, which is one of the most reliable and efficient ways of detecting errors in the structure idealization.

Most of FE general processors involve handling of symmetric matrices which are usually very sparse. Computational algorithms of many subroutines try to capitalize on symmetry and sparseness in order to achieve high computational efficiency both in operation count and in storage.

Two types of storage are normally used for such matrices:

1. Diagonal band storage: all the elements of either super- or subdiagonals are stored, up to the corresponding diagonal beyond which all elements are zero, in a two-dimensional array.

2. Wavefront storage: for each row, elements from the first nonzero element to the main diagonal are stored in a one-dimensional array.

---

[1]As is well-known, FE techniques are used also in other fields of engineering and applied physics. However, only structural mechanics terminology is used in this work.

Mixed code methods that can operate on a combination of bandwidth and wavefront storage are also used.

A minimum bandwidth or wavefront can be obtained with a suitable choice of the nodal point labels. When the structure entails many degrees of freedom and automatic data generation is used, it is very difficult and time consuming for the user to provide a numbering of the nodes yielding a small bandwidth or wavefront. Therefore a very useful preprocessor feature is the capability to resequence grid point numbers automatically to obtain reduced bandwidths or wavefronts.

There are more things one may want to ask from a preprocessor as an aid toward improvement of structures idealization. For instance, preprocessors may perform a number of suitably defined optimizations on mesh characteristics--such as grid points location and element shapes, tending to minimize the overall cost for given required solution accuracy.

These are sophisticated tasks, well beyond the capabilities of present day preprocessors. The further elaboration of this point is therefore deferred until the third section dealing with future trends.

PRESENT STATUS

Needs for preprocessors arose only recently as a consequence of the growth of the use of FE programs. The FE analyst, facing a number of practical problems connected with data preparation and with computer memory (and time) requirements, developed preprocessors to overcome these urgent difficulties.

Understandably then preprocessors of the first generation perform, in general only the simplest tasks. General comments on the present state-of-the-art will be given in connection with the following three points: (1) automatic mesh generation, (2) data checking and debugging, and (3) nodal point resequencing.

1. The user usually specifies a few boundary points of the structure, and the preprocessors generate topology and geometry data. Full meshes can be generated automatically for 1-D and 2-D geometries (plane and axisymmetric). 3-D meshes can be generated only for simply shaped bodies; there are few preprocessors which can generate meshes for more complicated but particular geometries. The assignment of nodes to elements is automated only when the generated elements are of the same type and are located on well defined patterns. More sophisticated rules, either to generate isoparametric elements or to follow assigned boundaries (by analytical laws or by table data), can in some case be assigned.

2. Data checking and debugging is often done very satisfactorily. Available preprocessor codes can provide:

a. Line printer and plotter for 1-D and 2-D structures.

b. Perspective or axonometric plots of 3-D meshes for the complete structure or any part of it from a point of view chosen by the user.

c. Stereo plots. By using two close points of view, two perspective plots may be obtained for a stereoscopic pair, which appear through a stereoscopic viewer as a 3-D plot.

3. A number of algorithms are available to perform nodal point resequencing. A comparison among a few of them is reported in [1, 2] and shows that: (1) the Cuthill-McKee resequencing strategy (used for instance by BANDIT preprocessor [3]) and its reverse perform generally best for small bandwidth; (2) the Levy strategy (used for instance by WAVEFRONT preprocessor) performs generally best for small wavefronts; and (3) the Levy and reverse Cuthill-McKee strategies perform generally best for small profiles (sum of wavefronts). Some comparisons between the two preprocessors BANDIT and WAVEFRONT with respect to NASTRAN processing are reported in [4].

The computation time for bandwidth reduction seems to be less than the computational time for wavefront reduction. Relabeling methods employed in the available preprocessors only yield improvements in bandwidth or wavefront. Existing methods ensuring an absolute optimum relabeling would be too expensive (in computer time) to be used currently.

Whereas there is in general little doubt about whether to use automatic mesh generation and checking capabilities, this is not so for resequencing capabilities. Indeed their worthiness depends, more than for the other tasks, on the structure studies, the processor used, and the computer system available.

Of the several preprocessors surveyed only those readily available, fully operational, and nonproprietary will be considered here. The description and evaluation of these preprocessors is presented as follows.

Table 1 summarizes general information features for 22 programs. More detailed information for each of them is given in the Appendix. The evaluation and documentation availability is summarized in Table 2.

Table 1 lists the 22 programs alphabetically and gives the following information, whenever supplied: primary computers, related programs (since these preprocessors were developed primarily for use on prescribed computers and for specific processors), typical core memory, main preprocessor features (relabeling mesh generation, and plotting capabilities). In mesh generation and plotting columns the dimensions of the structure are shown.

Replies were not always exhaustive; this is reflected in the empty boxes of the columns. Some programs have capabilities not falling under the above headings. Reference is made to the Appendix for a more detailed description. In this Appendix the following information is given for each program: program name, author name and affiliation, program language, primary computers for which the processor is designed, machine requirements, main features of the preprocessor, related structural codes, where available, and preprocessor price (to buy).

In the evaluation of a preprocessor performance items thought significant were: confidence in the code, ease of learning how to use it, ease of use once learned, code performance as advertised, developer keeps up to-date, input easy to prepare, easy to get working on the computer system, runs efficiently, and developer helps to find bugs. For each item, four grades, ranging from Poor to Excellent, have been used.

Since almost no documentation (user's manuals, listings, cards, etc.) has been received, in very few cases were the authors able to gain direct experience on the codes. Hence the evaluations reported in Table 2 are based exclusively on users' or developers' answers.

The following documentation is considered helpful for users: technical description, source listing, test pack listing, test results, user's manuals, and a deck of sample problems in their availability for each preprocessor is also shown in Table 2. All the programs reported are available; for some of them availability may be restricted to U.S. Government contractors. Availability of the programs distributed by COSMIC is normally limited to the United States.

FUTURE TRENDS

The foregoing discussion indicates that not all the desirable features of a preprocessor have been properly developed, nor have the existing capabilities of present hardware generation been exploited to their fullest sophistication. For instance, sustained efforts should be made toward an efficient interactive use of available computer graphics systems, even in modest size peripheral facilities. Many other feasible tasks could and should be foreseen for the next generations of preprocessors.

A fundamental continuing challenge in the computer era is to transfer into computer programs an always increasing part of the know-how of specialists in given fields. What results at one time from the "intuition" or the ability of the few must, at subsequent times, be made available to and usable by the many. This transfer process is in a certain sense the trademark of progress and is already inherent to the FE technique. Finite element computer programs have brought the solution of a number of very complex problems within the reach of great many more people with much less conceptual efforts needed. The trend must and will continue, and it is in this perspective that many useful areas for promising developments of new preprocessor tasks can be identified.

Mesh optimization may be one of them. It entails three intimately connected aspects: establishment of the optimization criteria and subsequent formulation of the optimization problem (i.e., with mathematical programming terminology, and definition of the objective functions and of the sets of constraints), development of the appropriate computational algorithm, and implementation into a preprocessor unit.

The first aspect involves the above-mentioned transfer process. Criteria for mesh optimization cannot, in general, be established without connection between desirable mesh properties and behavior of field variables. In many cases the optimal choice of the mesh is left to the specialist who has the due feeling of the problem and, in addition, relies on his experience and/or intuition. No doubt this will necessarily continue to be so in the future; specialists always have to be called upon to solve the new and more complex problems. However the question of interest here is how frequently and how much of this optimal choosing can be translated into a computer program.

Long term answers to this question have deep implications for the future development of all computer programs intended for diffuse and routine usage. The ultimate goal is to compile a taxonomy of the problems more commonly encountered in the different fields of engineering and applied physics. The classification should be made in terms of a priori identifiable problem features exhibiting a certain common set of characteristics relevant from the computer programs viewpoint. Attention could then be focused on each such set, appropriate strategies studied, optimization criteria found, and computational algorithms developed. The end result would be a series of preprocessor units, each corresponding to one class of the compiled taxonomy.

A fraction of the transfer process would then have been accomplished. The user need only see whether his specific problem falls within a class of the taxonomy and use the corresponding preprocessor unit (together with all other recommended programs). The knowledge, experience, and/or intuition which would have been required from him in order to identify the optimal approach for the solution of his problem are no longer needed. They have been transferred from him the very moment in which the compilers of the taxonomy have placed that problem in a given category.

In the near future preprocessors will hopefully have mesh optimization capabilities for subdomains where the unknown field variables and their derivatives, up to the appropriate order, behave smoothly enough, i.e., for the first and simplest class of the taxonomy. Even in this simpler case a preprocessor level procedure can only be based on purely geometric optimization criteria.

Other criteria will indeed be too tightly connected with processor usage. For instance, Pedersen and Giordani, Meola, and Napolitano [5, 6] approach the problem of the most convenient shape and positioning of the elements of a given type by computing the total potential energy and then looking for the subdivision into elements which yields the minimum value of the energy. This is equivalent to finding the most accurate answers for a given processor computation time. These procedures require calculations which are typically performed by the processor program; very often, in fact, one needs a solution for a reference starting mesh which is progressively improved to yield a lower potential energy under given loading conditions.

Uniformity and regularity of the mesh appear to be reasonable optimization criteria based on geometry alone. The accuracy of the finite element method relies, among other things, on the accuracy of interpolating functions which link the field variables, within each element, to their values at the nodes. Everything else being the same, one may say that the more regular the element shape, the better the representation of the field variables inside the element.

Recent research along these lines [9] considers domains to be subdivided into finite elements of assigned type and assumes as constraints the total number of points, the total number of elements, and the position of subdomain boundary nodal points. The objective functions are taken to be the mean square deviations of the distribution over the entire domain of parameters related to the size and shape of elements.

A corresponding algorithm which may conveniently be applied with small computers is the so-called method of point relaxation [6] which consists of displacing the nodes one at a time, by some suitable amount, in order to improve the adjoining elements (the improvement may refer to their shapes and/or dimensions). The displacement may be executed with a variety of prescribed procedures which may account for a number of additional requirements.

Other optimization problems can be formulated by weakening some of the constraints. New strategies may thus be originated; such as a restructuring process which may delete or insert nodes or elements with the aim of improving a local undesirable situation [7]. In all these problems element screening and decision making logics must be provided in the computer program in order to avoid completely wrong answers.

Although none of these programs is as yet operational, the outlook for the near future is positive and promising. In the next couple of years routine programs will be available embodying useful mesh optimization capabilities.

ACKNOWLEDGMENTS

The authors wish to acknowledge the helpful assistance in the preparation of this paper given by W. D. Pilkey and thank all the developers and users of preprocessor programs who sent the requested information.

This work falls within research programs partly sponsored by C.N.R. (Italian Council of Researchers).

REFERENCES

1 Cuthill, E. H., McKee, J. M., "Reducing the Bandwidth of Sparse Symmetric Matrices," Proceedings of the 24th National Conference ACM, 1969.

2 Cuthill, E. H., "Several Strategies for Reducing the Bandwidth of Matrices" in Sparse Matrices and their Applications, Rose, D. J., Wilburghby, R. A., editors Plenum Press, 1972.

3 Everstine, G. C., "The BANDIT Computer Program for the Reduction of Matrix Bandwidth for NASTRAN" Naval Ship Res. & Development Center, Computation and Mathematic Dept. Report, Bethesda, Md.

4 Levy, K., "Structural Stiffness Matrix Wavefront Resequencing Program (WAVEFRONT)" J.P.L. Tech. Rept. 32-1526, vol. XIV.

5 Pedersen, P., "Some properties of linear strain triangles and optimal finite element models" The Danish Center for Applied Math. and Mech. The Technical University of Denmark, Rept. No. 540, January 1973.

6 Giordani, G., Meola, C, Napolitano, L. G., "Optimization models for finite elements" Aerotecnica 1973, to be published.

7 Jones, R. E., "A self organizing mesh generation program," Sandia Labs. Applied Math. Div. Alburquerque, New Mexico, Paper submitted to Miami ASME Conference June 1974.

8 McNally, W. D., "Fortran Program for generating a two dimensional orthogonal mesh between two arbitrary boundaries" NASA TN-D-6766, April 1972.

9 Monti, R., Meloa, C., "Geometric optimization for finite element meshes" to be published.

Table 1

NAME	PRIMARY COMPUTERS	TYPICAL CORE MEMORY (words)	RELATED PROGRAMS	PROGRAM FEATURES Relabell.	Mesh Generat.	Plotting
ASP	I 360/65	35,000	ELAS,SSAP,STRUDL		2-D	2-D
AIDPN	I 360		NASTRAN	B	2-D	
BANDIT	C 6400,6500 C 6600,6700 I 360, 370 U 1108	$50,000_8$	NASTRAN	B		
CUTUP	C 6000	48,000	NASTRAN		2-D	
ELINPT	C 6400,7600	12,000	Any FE program		2-D	2-D
ELSA 31	I 370/158		NASTRAN		2-D	2-D
ES/PREP	I 360,370	10,000	ES/STAT ES/MODE ES/FOUR	B	2-D 3-D	2-D 3-D
FEDGE	I 7094-7044	32,000	SAMIS,ELAS		1-D 2-D 3-D	
FEILG	I 360		ELAS,ASKA,PREP, STARDYNE			
FELAP/ PLOT	I 360 U 1108	32,000	FELAP,PIPDYN		3-D	
FELAP/ RENUM	I 360 U 1108	27,000	FELAP,PIPDYN	B		
GENDA	C 6000 series	$100,000_8$	Any FE program	B	2-D	2-D
HIDE	C 6000 series	$100,000_8$	Any FE program			3-D
IMAGE	I 370/165	30,000	NASTRAN,SATURN, SPEED		1-D 2-D	1-D 2-D
INGA	C 1700	32,000	ASKA,TURBAN,LTS			2-D 3-D
MESH	I 360 U 1108	32,000	Any 2-D FE program		2-D	2-D
MESHG	U 1106	30,000	ELAS,DYNAX,QUAD4, SAP, and others			2-D 3-D
NEPSAP/ DRIVER	C 6600 U 1108		NEPSAP		2-D 3-D	
PING	C 6400	12,000	NASTRAN		2-D	
REX/ CHECK	U 1108	65,000	REXBAT			
SAP/ PREP	U 1108		SAP III			
WAVEFRONT	U 1108		NASTRAN,SAMIS	W		

Key to the symbols:

B = Bandwidth reduction C = CDC D = Dimensions I = IBM
W = Wavefront reduction U = UNIVAC

Table 2 Evaluation of Preprocessors

NAME	EVALUATION									DOCUMENTATION AVAILABILITY					
	1	2	3	4	5	6	7	8	9	a	b	c	d	e	f
BANDIT	E	E	E	E	E	E	E	E	Y	+	+			+	
CUTUP	G	E	E	E	E					x				x	
ELINPT	E	G	G	E	G	G	G	E	Y				x		
ELSA 31	G	E	E	P	G	E	G	G	Y						
ES/PREP	G	G	E	E	G	G	G	G	Y	x	x			x	x
FEDGE										x	x	x	x	x	x
FELAP/PLOT	E	E	E	E	E	E	E	E	Y	+	x	x	+	x	x
FELAP/RENUM	E	E	E	E	E	E	E	E	Y	+	x	x	x	x	x
GENDA	E	F	E	E	G	G	G	G	Y		x				
HIDE	E	E	E	G	F	E	G	F	Y		x				
IMAGE	E	E	E	E	E	E	G	G	Y						
INGA	G	G	G	E	G	E	G	G	Y					x	
MESH	G	E	E	G	F	G	E	G	Y		+	+	+	+	
MESHG	E	G	E	E	F	G	G	F	Y						
NEPSAP/DRIVER		E	G	E	E	G	G	E	Y	x	x	x	x	x	x
PING	E	E	E	E	E	E	E	G		x	x	x	x	x	x
REX/CHECK	E	E	E	E	F	E	E	E	Y						
SAP/PREP	E	E	E	E	F	E	E	E	Y	x	x			x	
WAVEFRONT	G	E	E	G	F	E	E	G		x			x		

Key to the symbols:

1) Confidence in this code
2) Easy to learn how to use
3) Easy to use once learned
4) Code performs as advertised
5) Developer keeps up-to-date
6) Input easy to prepare
7) Easy to get working on the computer system
8) Runs efficiently
9) Developer helps find bugs
a) Technical description
b) Source listing
c) Test pack listing
d) Test results
e) User's manuals
f) Deck of example problems
E = Excellent
G = Good
F = Fair
P = Poor
Y = Yes
+) Available before the purchase of the FE preprocessor
x) Available

APPENDIX

SUMMARY OF THE AVAILABLE PREPROCESSOR COMPUTER PROGRAMS

For all the programs the following information will be reported:

A = Program name
B = Author name and affiliation
C = Program language
D = Primary computers for which the processor is designed
E = Machine requirements
F = Main features of the preprocessor
G = Related structural codes
H = Where available
I = Preprocessor purchase price

Quite often programs are available through COSMIC, whose address is: COSMIC, Barrow Hall, University of Georgia, Athens, Georgia 30601

A ASP
B C. W. Martin, Asst. Prof. University of Nebraska
W. L. Booker, Caterpillar Tractor Company
R. Veys, Graduate Student, University of Nebraska
C Fortran IV
D IBM 360/65, Calcomp Plotter
E 140K or more of core depending on problem size
F Generates and plots mesh coordinates and element numbering. Automatically subdivides triangular element into 4 similar triangular elements and creates input suitable for analysis program. Subdivision can be repeated as many times as desired.
G ELAS, SSAP, STRUDL
H Author
I No effort has been made to sell the program.

A AIDPN - Automated Input Data Preparation for Nastran
B NASA Goddard Space Flight Center, Greenbelt, Md.
C Fortran IV (61%), PL-1 (39%)
D IBM 360
F Five structural analysis programs are provided; the first three programs (AXIS, SHELBY, and COONS) are coded in FORTRAN IV for the IBM 7094 or the IBM 360, the last two programs (BANDAID and MOVE) are coded in PL-1 for the IBM 360. The AXIS program generates data for shells described by the rotation of a plane curve about an axis. SHELBY generates data for shells described by the translation of a plane curve along an arbitrary axis in space. COONS

generates data for free-form shell structures based on the description of four bounding curves. BANDAID automatically re-sequences the grid points of a structural problem to achieve a reduced bandwidth in the stiffness matrix, and MOVE generates data for structures having a number of identical segments.
G NASTRAN
H COSMIC
I $350

A BANDIT
B Dr. Gordon C. Everstine, Naval Ship Research & Development Center, Bethesda, Maryland 20037
C Fortran
D CDC 6400, 6500, 6600, 6700
IBM 360, 370
UNIVAC 1108
E On CDC : 50,000 words of memory
On IBM : 20,000[8] bytes of memory
F Grid point relabeling for matrix bandwidth reduction for NASTRAN. Reads NASTRAN input. Generates SEQGP cards for NASTRAN. Uses Cuthill-McKee algorithm.
G NASTRAN
H COSMIC
I $400

A CUTUP
B James M. McKee, Evangeline T. Marcus, Naval Ship Research and Development Center, Bethesda, Md. 20034
C Fortran
D CDC 6000
E 48,000 words of core memory
SC4020 software + hardware for plotting generated structure
F Modular data generation program which permits the combination of several simple generation programs to model a complete structure. Modular approach-user may add his own simple generators.
Automatic propagation of geometry and mesh parameters between modules.
Generates complete structural idealization from node points and boundary conditions. May use several coordinate systems.
Integral plotting capability.
G NASTRAN
H COSMIC

A ELINPT
B Ruoff, G., Berg A., Ahmad R., Meyer R., Technische Universität Hannover, Lehrstuhl fur Baumechanik, 3000 Hannover, Welfengarten 1, Westdeutschland

C Fortran IV
D CDC 6400/7600
E Core-size 12 ÷ 24K words; disk or drum with random access facilities.
F Minimization of the amount of input data for 2-D problems (plates, disks, shells). Preparation of data groups for the generation of element matrices (displacement or mixed models).
G Finite element programs (triangle elements) with variable shape functions (polynomials invariant against rotation of the reference system). The order of the polynomials is defined by the input. ELINPT is used in connection with: plane stress - plane strain problems, plate-bending analysis, and shell-bending analysis for small and large deflections with or without shear deformations.
H Author

A ELSA 31
B Aeritalia, Torino, Italy
C Fortran IV
D IBM 370/158
F Reduces external distributed loads on frame-type elements to concentrated nodal forces
G NASTRAN
H Author

A ES/PREP
B Espri, Genova, Italy
C Fortran IV
D IBM series 360 and 370
E From 100K up.
F Automatic mesh generation for plane and 3D bodies of regular shape. Bandwidth minimization with automatic relabeling. Interactive computer graphics. Plane and axonometric views can be displayed, rotated, reduced, and plotted.
G ES/STAT static analysis
ES/MODE modal dynamic analysis
ES/FOUR Fourier analysis
H Author

A FEDGE - Finite Element Data Generation
B F. A. Akyuz, Engineering Mechanics Div., Jet Propulsion Lab., Pasadena, Calif.
C Fortran II (97%) FAP (3%)
D IBM 7094-7044
E 32K
F Provides for an automatic means of discretization for continuous domains of any form by using the concept of natural coordinate system.
G ELAS, SAMIS
H COSMIC

I $310.00
A modified version of FEDGE is used at Mathematical Analysis and Computation Group of Mathematics Section, Alliance Research Center of Babcock and Wilcox Company.

A FELAP/PLOT
B Dr. M. M. Reddi, H. M. Fishman, The Franklin Institute Research Laboratories, 20th and Race Streets, Philadelphia, Pa. 19103
C Fortran + basic CALCOMP software package
D UNIVAC 1108, IBM 360
E 32,000 words
F Displays plan or isometrics of plats, shell, and beam finite element model, with full annotation of nodes and elements. Selective plot of portions of model for clarity. Automatic as chosen scale. Any orthonometric view. Hidden elements dashed. Plots deformed model alone or superimposed on original model.
G FELAP, PIPDYN
H Author
I $500

A FELAP/RENUM
B H. M. Fishman, Franklin Institute Research Laboratories, 20th and Race Streets, Philadelphia, Pa. 19103
C Fortran
D UNIVAC 1108, Ibm 360
E 27,000 decimal 36 bit words
F Accepts randomly numbered nodes on finite element model and renumbers in attempt to reduce bandwidth. Punches new cards and performs additional input edit.
G FELAP, PIPDYN
H Author
I $300

A FEILG - Finite Element Inertial Load Generator
B North American Rockwell Corp., Canoga Park, Calif.
C Fortran IV
D IBM 360
F Replaces density specifications per elements on a structure by an equivalent external loading of concentrated forces at the nodal points.
G PREP, STARDYNE, ELAS, ASKA
H COSMIC
I $300

A GENDA

B Robert D. Rockwell, Naval Ship Research & Development Center, Carderock, Md. 20034

C Fortran IV

D CDC 6000 series

E 100,000 octal storage

F After mesh is generated, bandwidth is reduced using Cuthill-McKee method and plots which have node numbered. Also plots with the elements numbered are presented. Using simple input data, the program generates and plots nodes and elements for triangular and/or rectangular elements for 2-D and 3-D structures.

G Any program using triangular and/or rectangular plate or shell elements.

H Mrs. Jackie Potts, Code 1245, Naval Ship Research & Development Center, Carderock, Md. 20034

I $10.00

A HIDE

B Robert D. Rockwell, Naval Ship Research & Development Center, Carderock, Md. 20034

C Fortran IV

D CDC 6000 series

E 100,000 octal storage

F Program generates plots of structures with elements or portions of elements not seen from the user's viewpoint removed.

G Any FE program using triangular and rectangular elements.

H Mrs. Jackie Potts, Code 1745, Naval Ship Research & Development Center, Carderock, Md. 20034

I $10.00

A IMAGE

B Kenneth S. Sohocki and David G. Lebot, Chevrolet Engineering Center, General Motors Corp., C-2-232, 30001, Van Dyke, Warren, Mich. 48084

C Fortran IV

D IBM 370/165 with TSO

E TEKTRONIX 4002A, 4012 or 4014
BENDIX DATAGRID DIGITIZER
TSO 300K REGION
NOMINAL DISK STORAGE

F Interactive finite element modeling, and graphic evaluation of program results. User creates a model by digitizing points on a layout drawing. Beams and plates are then created using a joystick. Element properties, reactions, and loads are all created on-line graphically. Extensive checking is done by the program. The model is then submitted for a batch run. Output from the analysis program is saved on disc. The postprocessor then can be used to display deflections and stresses on line.

G NASTRAN, SPEED, SATURN

H Authors

A INGA
B Ingolf Grieger, Institut für Statik und Dynamik der Luft-und Raumfahrtkonstruktionen, 7 Stuttgart 80, Pffaffenwaldring 27, Westdeutschland
C Fortran, Assembler
D CDC 1700 with 274 Digigraphic
E 32K core memory (16 bit/word)
F Display and debugging of general finite element idealization using 1-D, 2-D, and 3-D elements. Display of results (deformations, stresses, eigenvectors). The pictures can also be plotted. The program allows all linear graphical transformations (rotation, projection, perspective, etc.). The data (nodal point coordinates, elements, etc.) can be displayed on the screen of the CRT and then corrected by the user.
G ASKA, TURBAN, LTS
H Author

A MESH
B L. G. Napolitano, R. Monti, P. Murino, Istituto di Aerodinamica, Facolta di Ingeneria, Piazzale Tecchio 80, 80125 Napoli, Italy
C Fortran IV
D IBM 360, UNIVAC 1108
E 32,000 words
F 2-D mesh generation. Line-printer plot.
G Any 2-D FE program
H Author
I $25
Another version of this program, named MASFLAY, is available from COSMIC

A MESHG
B G. D. Savitzky, Structural Analytical Division & Computer Services Division, Sargent & Lundy Engineers, 140 Dearborn St., Chicago, Ill. 60603
C Fortran IV
D Univac 1106
E 30K core memory
Houston instrument incremental plotter for off-line plotting
F Plots 2-D elements in 2-D space; different views for 3-D representation; rotation of plotting axes; numbers elements and nodes if required; given the mesh finds the boundary and plots it or only number nodes on the boundary.
G HEATRN, ASAL, TORS, ELAS, DYNAX, PLFEM, SAP, LAKFEM, TRAN, SOLIDSAP, QUAD4
H Author

A NEPSAP/DRIVER

B Parviz Sharifi, Lockheed Missiles & Space Co., Dept. 81-12, Bldg. 154, P. O. Box 504, Sunnyvale, Calif. 94088
C Fortran
D Univac 1108, CDC 6600
E Auxiliary storage: disk, drum, or tape
F Generates input data for nodal points (coordinates, freedoms, temperatures, loads) and elements. Program consists of a collection of subroutines each designed to communicate with the user through their argument lists. The user writes his own MAIN-DRIVER program to generate the finite element mesh (nodes, elements, loads, etc.) utilizing the utility subroutines in the program.
G NEPSAP (Nonlinear Elastic Plastic Structural Analysis Program)
H Author

A PING - Planform INput Generator
A NASTRAN preprocessor for lifting surfaces
B Dr. Pao C. Huang and John P. Matra, Jr., U. S. Naval Ordnance Laboratory/White Oak, Silver Spring, Md. 20910
C Fortran
D CDC 6400
E 120K
F Automatically develops finite element models for lifting surfaces and punches NASTRAN bulk data cards. Program can automatically develop the finite element models for many different wing planforms, included are sweptback, delta, diamond, cylindrical, and built-up wings. Eighty to ninety percent of the required NASTRAN bulk data cards are also punched out.
G NASTRAN
H Author

A REX-CHECK
B L. E. Stearns, Lockheed Missiles & Space Co., Inc., D-19-44 B-102, P. O. Box 504, Sunnyvale, Calif. 94086
C Fortran IV
D Univac 1108
E Program is segmented several times with a maximum core size of 65K. Dynamic assignment drum/disk files may be needed depending on the nature of the input.
F Reviews data processed by REXBAT for coding and configuration errors. Permits extensive debugging without using the high-impact REXBAT program until ready. Can translate and randomly renumber all nodal data. Checks that all nodes used to define the various elements, plot data, constraints, and loads were defined in the model coordinate data. Checks section and material properties against expected norms for the industry. Checks for zero length bars and warped panels. Flags use of duplicate elements to guard against accidental usage.
G The preprocessor is designed for use with REXBAT, a structural finite element program (author W. A. Loden of LMSC).
H Author

A SAP/PREP
B A. Schaeffer, Dept. of Aero. Engr., Univ. of Md., College Park, Md.
C Fortran V
D Univac 1108
E Exec 8 operating system
F Create input file for SAP. Free-form input, output formated for SAP.
G SAP III
H Author

A WAVEFRONT
B Roy Levy, Jet Propulsion Laboratory, 4800 Oak Grove Drive, Pasadena, Calif. 91103
C Fortran V
D Univac 1108
F Resequences connectivity matrix to reduce wavefront (active columns) prior to stiffness matrix decomposition. Produces SEQGP cards for insertion in NASTRAN bulk data.
G NASTRAN, SAMIS
H COSMIC or Author

A SURVEY OF PRE- AND POSTPROCESSORS FOR NASTRAN

Gordon C. Everstine and James M. McKee
Naval Ship Research and Development Center
Bethesda, Maryland

ABSTRACT

This paper surveys various preprocessors and postprocessors developed for use with the NASA Structural Analysis Computer Program, NASTRAN. The following topics are included in the survey: automatic data generation, data checking and updating, conversion of data to NASTRAN format, grid point resequencing, partitioning vectors, radiation view factors, contour plotting, NASTRAN's General Purpose Plotter (NASTPLT), data transfer utilities, transient response to input accelerations, rigid links, processors for vehicle dynamics, antenna radiation, and structural modification reanalysis.

INTRODUCTION

The development over the last few years of structural analysis computer programs has enabled the engineer to solve complicated structural engineering problems, in contrast with the previous situation in which he was limited to the study of only very simple approximations to the real structure. Early programs generally addressed either particular types of analyses or limited classes of structures, or both, and hence would now probably be classified as special purpose.

More recently, this fragmented approach to computer structural analysis has given way to the development of so-called general purpose computer programs. Of the programs which address primarily linear problems, the most widely used is probably the NASA Structural Analysis Program, NASTRAN [1-6]. NASTRAN's success is probably due to a combination of its wide-ranging capability, its convenience of use, and its availability.

In the opinion of T. G. Butler [7], the original NASTRAN project manager at NASA, one consequence of the existence of computer programs such as NASTRAN will be a shift in emphasis away from mechanical testing in favor of analysis as a proof of design. This in turn will result in an increased need and justification for the development of preprocessors and postprocessors to aid the user in both the preparation of data and the interpretation of results. With respect to NASTRAN, this prediction has certainly been realized (as evidenced in part by the number of references listed at the end of this paper).

In view of the length of time that NASTRAN has been publicly available (over four years) and the proliferation of NASTRAN-related computer software development, there seemed to be a need for a survey of the work performed to date. It is thus the object of this paper to survey the pre- and postprocessors that have been developed for use with NASTRAN and briefly describe the capabilities of each.

In what follows, it is assumed that the reader has at least a rudimentary knowledge of NASTRAN's capabilities and how to use it. With the exception of the following section, in which the survey's scope is defined, a separate section is devoted to each of 12 areas in which NASTRAN-related processors have been written.

SCOPE OF SURVEY

For the purposes of this survey, we consider a preprocessor or postprocessor to be any computer program peripheral to NASTRAN which interfaces with NASTRAN to perform some useful job not already performed by NASTRAN. In general, a processor must satisfy the following three requirements in order to be included: (1) its interface with NASTRAN must be operational, (2) it must be documented, and (3) it must be available to the general public. In some cases, availability involves interaction with a commercial computer service.

Specifically excluded from this survey are the many so-called one-shot processors which were never intended to be used by anyone other than the developer. For some jobs, such processors are often invaluable. The first requirement above also excludes processors developed for programs other than NASTRAN, even those processors which could be easily adapted to NASTRAN format. The more general scope is addressed by other papers in this volume.

While we have attempted to be as comprehensive as possible in this survey, we were unable to obtain copies of some relevant documentation, and hence had to omit several processors. Even for those programs included, many are receiving continued development, so that our descriptions may soon be out-of-date. Hopefully this survey will have served its purpose if it does nothing more than collect a list of references in one place and possibly identify those hotbeds of NASTRAN-related development.

Finally, since the authors have not personally executed all the computer programs described, the paper is, of necessity, an uncritical survey.

DATA GENERATING PREPROCESSORS

As finite element users have long recognized, the preparation of input data usually involves considerable drudgery. Data generation for finite element programs covers a wide range of applications. The range of data generation computer programs surveyed for this paper is correspondingly wide, both in terms of program capability and in the way that the programs are used.

Background

Data generation programs are labor-saving devices which are used to eliminate the tedium and manual effort usually associated with the preparation of point-by-point structural data for finite element programs like NASTRAN. The effort required to write a program which will generate most of the geometric and connecting data for a structure is seldom more than that required for computing and transcribing the data by hand. Moreover, the resulting idealization will have fewer errors. A useful program may be written which generates only one particular structural idealization, but if certain attributes of the model (e.g., dimensions) are made parameters rather than constants, the program will be capable of generating various families of structural idealizations. The degree to which the parameterization is carried out depends, to some extent, on whether the program developer expects to encounter similar problems in the future.

Some organizations which make heavy use of finite element programs have concluded that significant savings could be realized by developing one general purpose data generation program and making it available to all their engineers, rather than having each engineer put together his own generators.

Although most generation programs have been developed from heuristic considerations, Kamel and Eisenstein [8] have investigated techniques for forming an acceptable finite element mesh from a more theoretical viewpoint. These techniques are used to model arbitrary surfaces with triangular elements. The determination of the best mesh for a particular structure is still largely an art, and the generation programs considered here either constrain the user to a predefined mesh pattern or require that he determine the appropriate mesh. Although in some situations one might like to have a data generation program and an analysis program combined into a foolproof "black box", the fact that engineering judgment is required to create a suitable mesh is sufficient to rule out existing programs for such a combination. Eppink [9] has studied some of the Navy's data generation requirements and has offered guidelines for the development of systems to generate complete models of ships and other large complex structures.

Techniques for generating data vary widely among programs. Most generation programs, however, use a combination of three techniques: (1) a parametric description of the model, (2) a compiler or other facility for user-defined generating algorithms, and (3) a digitization of points on the model by electromechanical means.

A generator which employs the first technique, a parametric description of the model, will accommodate structures with one basic form. Each particular individual structure can be generated by assigning specific values to a set of parameters. For example, a program for generating rectangular plates might have length, width, thickness, and the number of elements on each side as parameters. By themselves, these generators are most applicable when the user has many similar structures to analyze. However, a combination of several of these generators, each generating a basic geometric shape, can be used effectively for general problems.

Facilities for user-defined generating algorithms, the second technique, function as compilers or translators which assemble the required computer instructions from statements given by the user. Although these programs can be similar to FORTRAN or ALGOL compilers, they have additional features which automatically perform sequences of operations which occur frequently in data generation.

A typical compiler could readily process commands such as the following:

> Generate ten grid points beginning with number 105 and increment each grid number by ten. Let the coordinates of the first grid point be (1.0, 1.0, 0.0) and increment the first component by .5 for each successive grid point.

This type of generation program is usually very flexible and can be applied to any structural problem, although it does require the user to invent the generation procedure. A good deal of ingenuity may be required to use this type of program effectively for complex structures.

The third technique used in generation programs employs electromechanical digitizing equipment. Here the user is permitted to scan a scale drawing of the structure with a digitizing device and automatically record the positions of grid points for the model. Stand-alone digitizers can be used for this purpose, but they tend to be cumbersome whenever supplementary information is required. If the digitizing device is connected to a computer and a graphic display device, so as to permit the user and the generation program to interact, generators can be developed which are quite flexible and can be used for a broad range of applications. Interactive computer graphics systems are sometimes used for this type of generator, utilizing a cathode—ray tube (CRT) as both the digitizer and the display device. The high initial cost of digitizing and display equipment usually limits their use to those organizations which have a high volume of applications. Stand-alone digitizers are now used in repetitive applications like ship certification by the U.S. Coast Guard, while interactive graphics equipment tends to be used for combined design and analysis applications.

Regardless of the technique employed by the generation program, some programs will be easy to use and will come close to minimizing the effort required to prepare finite element data, and others will not. For programs which require only small amounts of user data, the input data format and data organization will not be a significant factor. The user will probably find the general purpose programs easier to use if the data are organized so that values are specified in about the same sequence as would occur in manual preparation, or if the order can be chosen by the user. Programs with a variety of data cards invite less confusion if they allow key-word or free-field specification of the data, or if they adopt one general format which is used for all data cards.

Except for some of the large general purpose programs, data generators are usually easy to modify. If a user finds a program that comes close to generating the required structure or if he finds a program that is acceptable but doesn't generate the data in NASTRAN format, he is usually only a few FORTRAN statements away from having the program he needs.

Survey of Available Data Generation Programs

All the programs summarized in this section do generate NASTRAN data, but in the absence of sufficient common characteristics to group the programs, summaries are listed alphabetically by author.[1] All programs will generate grid point and connection cards and, unless noted otherwise, are written in FORTRAN. Little information is given in the references on the cost of running these programs; however, judging from programs used by the authors, the computer costs for generation of the model should be small when compared to either the cost of manual data preparation or to the computer cost of the NASTRAN analysis.

(1) GRIDXY, J. M. Brophy, Frankford Arsenal [10]

This program was developed to idealize small arms cartridge cases using NASTRAN's triangular and trapezoidal ring elements. It may be used, however, for any thick-walled axisymmetric problem. The user is required to divide the structure into subregions which are bounded by four straight lines or polynomial curves. Parameters are then specified to define the idealization for each subregion. Several types of mesh variation are possible within each subregion. Varying pressure loads are converted into the required FORCE cards. The program operates on CDC 6000 computers.

(2) FEM, R. C. Burk and F. H. Held, McDonnell Douglas Astronautics Company - East [11]

In its normal mode of operation the Finite Element Modeling (FEM) program accepts nodal geometry data which have been digitized from scale drawings using an interactive graphics system, although nodal data may also be entered from prepunched cards or other generation programs. Once the nodal geometry has been defined, the user may define the finite element mesh by pointing out the nodal points associated with each element. The user may also specify pressure loads and symmetry constraints in a similar manner. When generation is complete, the model may be transferred to an output file in either NASTRAN format or that of several other finite element programs. The FEM program has been implemented using an inexpensive Computek interactive graphics terminal and digitizing tablet and runs on an XDS SIGMA 7 computer.

(3) AXIS, SHELBY, COONS, and MOVE, W. L. Cook, Goddard Space Flight Center [12]

Three stand-alone data generation programs, AXIS, SHELBY, and COONS,

---

[1] Some of these programs are part of larger pre- and post-processor packages, in which case they will also be reviewed in other appropriate sections of this paper.

generate shells of revolution, shells described by the translation of a plane curve along an arbitrary axis in space, and Coons' surfaces, respectively. The programs generate grid point cards, quadrilateral connection cards, and, if desired, the cards to define a varying pressure load. The fourth program, MOVE, will generate a complete structural model from the data cards for one segment of the model which can then be replicated, translated, and rotated as required to form the complete structure. These programs were written for the IBM 360 computers in FORTRAN IV, except for MOVE which is in PL/1.

(4) PING, P. C. Huang and J. P. Matra, Jr., Naval Ordnance Laboratory [13]

The Planform Input Generator (PING) program is a preprocessor which develops NASTRAN finite element models for missile lifting surfaces. The types of wing planforms which PING handles include sweptback, delta, diamond, cylindrical, and built-up. The program can also generate the transition region between two meshes of different density. One useful application of this capability is the modeling of a cutout patch in which the cutout boundary has more grid points than are on the patch boundaries.

PING was written in FORTRAN for the CDC 6000 series of computers. The second phase of the same NOL project which produced PING will result in a program called BING (body input generator) for the automatic modeling of the axisymmetric shell bodies to which the wings are attached.

(5) SAIL II, M. W. Ice, Boeing Computer Services [14, 15]

The SAIL II language is a FORTRAN-like data generation language which is translated into FORTRAN and compiled into an executable data generation program.

The user normally supplies the program with a complete NASTRAN deck (Executive, Case Control, and Bulk Data) which has SAIL II statements instead of the usual Bulk Data. Any FORTRAN statements included in the deck will become part of the SAIL II generation program. Any NASTRAN Bulk Data contained in the deck will be included with the SAIL II generated data. All the usual compiler-type capabilities are present in the SAIL II language including looping and subroutine definition.

SAIL II operates on BCS's IBM 360, 370 computers and is set up so that the generation phase and the NASTRAN analysis phase may be executed in the same run.

(6) -----, M. S. Katow and B. M. Cooper, Jet Propulsion Laboratory [16]

From a set of NASTRAN GRID cards which describe the surface of a structure, this program permits the user to specify graphically the finite element connectivity of the model. This program has been written using the UNIVAC 1108-EXEC 8 Graphics Programming Library subroutines and operates on the UNIVAC 1557/1558 graphics system on the 1108. An illustration in the referenced paper shows the CBAR connections being defined for a large antenna, but it appears that connections for most of the available NASTRAN elements could be defined similarly. The program may also be used for visual data checking of any NASTRAN model. (See next section.)

(7) IGFES, W. Lorensen, Watervliet Arsenal [17]

The preprocessor portion of this program permits the user to define a two-dimensional finite element mesh using either an interactive graphics terminal or punched data cards. The user divides his structure conceptually into four-sided subregions and defines the bounding curves for each subregion. The program will determine appropriate interpolation functions which are then used to locate grid points in the interior of each subregion. Several rectangular and triangular mesh patterns may be selected for each subregion. The program will ensure that grid points are not duplicated along boundaries which are common to several subregions. To be used interactively, the program requires a Tektronix 4002 storage tube display connected to an IBM 360/44. When the program is run using punched card specifications, the generated mesh can be plotted using CALCOMP plotters.

(8) NARFEM, M. A. Martens, et al., Space Division, North American Rockwell [18,19]

The NARFEM program, while not a compiler, does provide certain incrementing and data manipulation facilities which are under user control. There is also a capability for parametric description of surfaces of revolution. The complete finite element model of a structure can be generated during one application of the program, or the structure can be segmented and the model generated by invoking the program repeatedly. Grid point data supplied for an earlier segment do not need to be redefined for subsequent segments. The program requires data in fixed-field format and can generate data in the format required by several finite element programs including NASTRAN. The program will also accept certain data prepared for one finite element program and translate it to the format required by another program.

The program operates on IBM 360, 370 computers and can optionally produce CRT plots of the generated structure.

(9) DATGEN, P. M. Meyer, Naval Ship Research and Development Center [20]

This program is used primarily to generate deck and hull models for ship structures from a parametric description. The user may locate stiffeners, holes, and rectangular cutouts within a deck as well as specify various loadings and boundary conditions on the structure. The present version of the program will generate both plane surfaces and simply warped surfaces.

The program operates on CDC 6000 series computers and requires data in fixed-field format.

(10) CUTUP, J. McKee and E. Marcus, Naval Ship Research and Development Center [21,22]

The CUTUP program provides a convenient mechanism for linking several simple data generation modules in order to generate NASTRAN Bulk Data for a complete structure. The program automatically manages such details as unique grid point numbering on module boundaries, communication of geometric data and properties between modules, and automatic propagation of the finite element mesh density between modules.

The user normally supplies a complete NASTRAN deck (Executive, Case Control, and Bulk Data) which has generation options in the Executive or Case Control decks and data generation specifications (in NASTRAN Bulk Data format) in the Bulk Data deck. Each of these specifications may be used as a "super" element, thereby generating a variety of data, including plates, spheres, cones, Coons's surfaces, stiffeners, and boundary constraints. Any NASTRAN Bulk Data cards included in the deck will become a part of the CUTUP-generated data. SC-4020 plots of the generated structure are produced upon request.

CUTUP operates on CDC 6000 series computers with the NASTRAN linkage editor and is set up so that the generation phase and the NASTRAN analysis phase may be executed in the same run.

(11) GENDA, R. D. Rockwell, Naval Ship Research and Development Center [23-25]

To use this program, the user defines, in integer coordinates, the boundaries of a finite element mesh which is topologically equivalent to a segment of the model to be idealized. A triangular mesh is implied for this pseudostructure with grid points located at each coordinate point. Then, given the boundary curves on the actual structure, the program maps the integer mesh into the structural space. The process is then repeated until the complete structure has been generated. The program as described is primarily for generating intersecting cylinders but could be easily modified to accommodate other structures.

The program requires fixed field data specifications and generates connection cards and coordinate cards according to user-prescribed formats. The program operates on CDC 6000 series computers and can produce SC-4020 CRT plots.

(12) NASTRAN.LINKØ-2, SCI-TEK, Inc. [26]

The data generation portion of this preprocessor will accept simple user-defined generation procedures which are limited to incrementation of grid point coordinates and identification numbers. The program also has the facility to generate geodesic dome structures from parametric specifications.

Finally, several generators were reviewed which were tied very closely to particular structures. These programs may be used to generate models of structures such as small-water-plane-area twin hull ships [27], threaded connections found at the breech ring in artillery [28], and the fuselage and wings of aerospace vehicles [29].

DATA CHECKING AND UPDATING

Regardless of how NASTRAN input data are generated, the data must be checked for errors prior to a complete NASTRAN execution. The most comprehensive checks on NASTRAN data are performed, of course, by NASTRAN itself. However, at many installations, the central memory requirements to run NASTRAN and the passive nature of the NASTRAN plot package (as extensive as it is) are not conducive to efficient routine data checking and undeformed model plotting. In this section we describe several stand-alone capabilities which aid in the checking, display, and modification of NASTRAN data.

We have identified ten such NASTRAN preprocessors and listed them in Table 1. Since all involve computer graphics and hence are machine-dependent, the computer hardware requirements are listed in the table. However, since the ideas used are applicable to other machines, some of the programs could probably be adapted to different hardware without major effort. For interactive programs, which are usually strongly machine-dependent, the conversion effort required could be considerable.

Four of the programs, the first three and Cronk's, have extracted NASTRAN's input file processor, geometry processor, and structural plotter in order to duplicate both the data checking performed in the NASTRAN preface and the undeformed structural plotting. They thus have all the plotting versatility that NASTRAN has. The principal advantage that this type of program has over NASTRAN is that less central memory is required, which results in faster turn-around time.

However, reduced memory benefits are negated by passive graphics. Thus the most useful preprocessors of this type use interactive graphics. All those listed will perform substructure plotting (with magnification) and allow for view rotation.

One preprocessor listed in Table 1, HIDE, is unique in that it contains a hidden-line capability, i.e., only portions of structural elements that can be seen from the user's viewpoint are plotted. As a NASTRAN preprocessor, HIDE is slightly less convenient to use in that the user must also input the FORTRAN formats indicating how the data are to be read.

Finally, NARFEM, FEM, and the Katow-Cooper preprocessors, while capable of being used alone for data checking and updating, contain additional capability such as data generation and are also covered in that section of this survey.

CONVERSION OF DATA TO NASTRAN FORMAT

Since no single computer program satisfies all the needs of structural analysts, most users have occasion to convert input data from the format used for one program to that of another. The automation of such a conversion by means of a short FORTRAN program is usually a simple job. As a result, most conversion programs are written as the need arises and never documented. Two examples of documented programs can be cited. First, Anderson and Buell [37] have written a preprocessor called NASTIE to convert from SAMIS

Table 1 Summary of Data Checking and Updating Preprocessors

Program Name	Developer	Computer	Graphics Hardware	NASTRAN Preface-Type Checking	Active or Passive Graphics	Modify Data at Console	Hidden Line Capability
Smith I [30]	NASA-LRC	CDC 6000	DD 80 B	Yes	Passive	N.A.	No
Smith II [31]	NASA-LRC	CDC 6000	CDC 250 CRT	Yes	Active	Yes	No
SAGE [32]	Israel Aircraft	XDS Sigma 7	Tektronix 4002A or 4010	No	Active	Yes	No
IDEAL [33,34]	NSRDC	CDC 6000	CDC 1700 CDC 274 CRT	No	Active	Yes	No
FASTDRAW [35]	McDonnell Douglas Automation	IBM 360/ 195	XDS Sigma 7 Computek or Tektronix	No	Active	Yes	No
Cronk [36]	Convair	CDC 6000	CDC 1700 CDC 274 CRT	Yes	Active	Yes	No
NARFEM [18,19]	North Amer. Rockwell	IBM 370	IMLAC PDS-1	No	Active	Yes	No
Katow-Cooper [16]	JPL	UNIVAC 1108	UNIVAC 1557/1558	No	Active	Yes	No
HIDE [24,25]	NSRDC	CDC 6000	SC 4020	No	Passive	N.A.	Yes
FEM [11]	McDonnell Douglas Astronautics	IBM 360 UNIVAC 418	IBM 2250 DEC 340	No	Active	Yes	No

(N.A. = Not Applicable)

format to NASTRAN format. Second, Giles and Dutton [38] wrote a routine to convert input data from McDonnell Douglas's Automated Structural Design (ASD) program [39] to NASTRAN format.

GRID POINT RESEQUENCING

The structural matrices formed during a NASTRAN analysis are normally both symmetric and sparse. For a given structure, the locations of nonzero terms in the matrices are determined solely by the choice of numbers (labels) assigned to the grid points. NASTRAN, like all finite element programs, has a solution algorithm whose speed depends on the grid point sequence. In NASTRAN's case, a combination band/active-column algorithm is used, i.e., the solver operates fastest for those matrix topologies exhibiting small bandwidth and few active columns. For example, in bandwidth-dependent routines, the number of calculations required (and hence the computer running time) is of order NB^2 for large N and B, where N and B are the matrix order and bandwidth, respectively.

Although proper grid point sequencing is essential to the user, NASTRAN burdens the user with supplying his own sequence. Since this is often an excessive burden (particularly when automatic data generators are used), several algorithms have been devised to automatically resequence grid point numbers to reduce both computer running time and core storage.

Three of these algorithms have been coded into NASTRAN preprocessors. One of the most widely used resequencers is the BANDIT program [40,41], which uses the Cuthill-McKee bandwidth reduction strategy [42]. Levy [43,44] has developed an iterative algorithm to reduce matrix wavefront. It has been implemented into a program called WAVEFRONT [45] and applied successfully to NASTRAN data. Another bandwidth reduction algorithm was developed by Cook and called BANDAID [12]. Unlike BANDIT and WAVEFRONT, which are in FORTRAN, BANDAID was written in PL/1. BANDIT will run on all NASTRAN computers; WAVEFRONT requires some conversion to run on machines other than the UNIVAC 1108.

Input to these preprocessors is the NASTRAN data deck. Output includes a set of SEQGP bulk data cards for insertion into the NASTRAN deck.

The algorithms used in these programs, as well as several other approaches, are reviewed in a survey article by Cuthill [46].

Since the resequencing algorithm must be tailored to the approach used for equation solving, these preprocessors will probably have to be modified when NASTRAN's Level 16 is released. That version of NASTRAN will contain a new equation solver [47].

PARTITIONING VECTORS

Substructure analysis using NASTRAN's standard release Level 15 requires that the user generate manually the partitioning vectors used to merge the structural matrices. For realistic nontrivial problems, this approach is both time consuming and error prone.

Although future releases of NASTRAN will provide some automation in the generation of partitioning vectors, at least one such preprocessor has been developed for the current version of NASTRAN. Called PVEC [48], the program uses the Phase I Checkpoint tape and a few additional input cards to generate the partitioning and other matrices used by NASTRAN in Phase II.

RADIATION VIEW FACTORS

The NASTRAN Thermal Analyzer (standard release version 15.5) includes the capability to perform complete thermal analyses on structures. In order to simulate the radiative heat transfer between surfaces, NASTRAN requires the user to input the view factors (also called shape factors, form factors, configuration factors) between those surfaces. A NASTRAN preprocessor called VIEW [49,50] has been developed which automates the computation of these factors. Output from VIEW includes RADLST and RADMTX bulk data cards for inclusion in the NASTRAN deck.

VIEW was adapted from an earlier view factor program called RAVFAC [51], which is not a NASTRAN preprocessor. VIEW was developed for use on the IBM 360 and contains some machine language code.

CONTOUR PLOTTING

To assist in the interpretation of the large volumes of output frequently produced by structural analysis programs, it is useful for the analyst to be able to plot contours of certain variables. For NASTRAN, several contour-plotting postprocessors have been developed.

Giles and Blackburn [29] report one such program which reads NASTRAN Bulk Data and punched output to generate contours for stresses, displacements, or eigenvectors (vibration or buckling).

Another such NASTRAN postprocessor called CONPLT [24,25] can produce two kinds of plots: (1) contour plots for each of several surfaces, including substructure enlargement, and (2) line graphs for user-specified groups of nodal or elemental ID numbers. The latter amounts to X-Y plots where the list of grid or element IDs appears along the X-axis. CONPLT was written for the CDC 6000 computers and the SC-4020 plotter.

The IGFES system [17] discussed earlier with respect to data generation also includes an output package with several alternatives for display of NASTRAN results. The graphics capability consists of contour plots, perspective plots, and X-Y plots of some dependent variable versus one of the principal coordinates. IGFES is implemented on a Tektronix 4002 storage display connected to an IBM 360/44 and interfaces with NASTRAN via an OUTPUT2 file.

Finally, contour-plotting capabilities have also been incorporated directly into the NASTRAN Plot Module by Kelly [52], although this is not a NASTRAN postprocessor. As a NASTRAN enhancement, the program will run on all the NASTRAN computers and plotters.

NASTPLT

No discussion of NASTRAN graphics postprocessors would be complete without some mention of the NASTRAN General Purpose Plotter package, NASTPLT [2,3]. This package is normally used at installations with plotting hardware not recognized by NASTRAN (e.g., plotters attached on-line to a computer). To use NASTPLT, a separate program (a postprocessor) must be written to interpret NASTRAN's plot tape and create the appropriate plotter commands. The plot tape consists of a sequence of elementary plot operations, each of which must, in turn, be translated into the appropriate commands to drive the plotter.

Although the writing of such a translator postprocessor is relatively straightforward, the authors are not aware of any documented NASTPLT applications.

DATA TRANSFER UTILITIES

Since NASTRAN is currently operational on the computers of three different manufacturers (IBM, CDC, UNIVAC), a compatibility problem arises in attempting to transfer data between dissimilar computers. Although BCD punched cards or their images can be used, the numerical precision obtained is inadequate for some applications. For example, the NASTRAN DMI bulk data card can pass a maximum of only ten significant digits. To overcome this problem, Rogers [53,54] has developed a pair of utilities which interface with the NASTRAN user tapes in such a way that no precision is lost. Typically, the RDUSER utility reads the OUTPUT2 binary tape and generates a BCD tape. The latter is then transferred to a dissimilar computer where WRTUSER converts it to an INPUTT2 binary tape readable by NASTRAN. These utilities find application, for example, in substructuring problems being run on two or more different computers.

A second type of utility, available from Boeing Computer Services [55], reads data from the NASTRAN Checkpoint tape (NPTP). Called NFETCH, the subroutine can be used to store data in either an in-core array or an external file. This routine, developed before NASTRAN's OUTPUT2 module became available, is essentially superceded by OUTPUT2, although the user must explicitly list in advance each NASTRAN data block to be written by OUTPUT2, instead of checkpointing the run.

TRANSIENT RESPONSE TO INPUT ACCELERATIONS

Wingate et al.[56] have developed a transient analysis postprocessor which allows the user to prescribe (1) input acceleration forcing functions, and (2) nonzero initial conditions when a modal formulation is used. The standard NASTRAN release provides neither of these capabilities.

Not having the first capability, however, is merely an inconvenience to a user since it can be overcome by placing a large mass at the appropriate point and applying an input force equal to the product of the total mass and the desired input acceleration. In contrast, the lack of the second is truly a program deficiency (although NASTRAN's direct approach has no such restriction).

The procedure developed requires a NASTRAN normal mode analysis, in which the modes and other data blocks are written onto a user tape. This tape is then the input to the postprocessor. The time integration is performed with a fourth-order Runge-Kutta procedure.

RIGID LINKS

Occasionally the finite element analyst must include rigid links in his model. Typical applications include offset plates, connections between beam and plate elements, or any situation in which a very stiff member is desired. Since the inclusion of a high-stiffness member might cause matrix ill-conditioning problems, the NASTRAN user generally defines rigid links using multipoint constraint (MPC) equations. To automate the generation of the necessary MPC cards, Anderson [57] has written a NASTRAN preprocessor called RIGID. For each rigid link, RIGID generates the six constraint equations required to fix the distance between any two grid points.

A similar capability is also available to clients of the Structural Dynamics Research Corporation as part of SDRC's package of processors [58].

PROCESSORS FOR VEHICLE DYNAMICS

Here we describe a package of SDRC-developed processors which were intended primarily to aid in the dynamic analysis of vehicle systems (e.g., automobiles) [58].

Several preprocessor modules are included in the package. One generates rigid links and is described in the preceding section. Another module was written to generate NASTRAN concentrated mass elements (CONM2) to account for rotatory inertia in the beam element (BAR). A third module generates constraint relationships (MPCs) to transform the effects of isolation elements (such as mounts and bushings) to the vehicle coordinate axes (fore-aft, side to side, verticle).

To facilitate the description of subsystem properties for a NASTRAN analysis, two interface modules were written: (1) a substructuring interface module to generate symmetric substructures, to reorder substructure matrices if component and system sequencing differs, and to insert substructures into NASTRAN via binary buffered files (INPUTT4); and (2) a modal modeling interface module which accepts modal information either from experiment or from finite element computer programs.

Although the overall vehicle dynamics capability developed by SDRC goes beyond that mentioned here, we have emphasized only those modules considered to be NASTRAN pre- or postprocessors.

ANTENNA RADIATION

The program described in this section is of such scope and size that it probably ought not be considered a NASTRAN postprocessor. Rather, NASTRAN might be considered a preprocessor to it. In any case, a general purpose program for the analysis of reflecting antenna systems has been developed by Cook [59].

Called the General Antenna Package, the program determines the effects of structural deformation on the radiation properties of reflecting surfaces such as antennas. The mathematical model chosen for the radiation problem is analogous to that used for the structural problem, providing compatibility between the two parts of the analysis. The role played by NASTRAN is the calculation of the structural deformation, which in turn is used as input to the radiation step.

Although the General Antenna Package possesses wide-ranging capability, a more complete discussion of it here would be beyond the scope of this paper.

STRUCTURAL MODIFICATION REANALYSIS

Frequently the engineering designer is interested in assessing the effect on a structure of changes in the various member components of the structure. Rather than reanalyzing the entire structure following some change, it is often more economical to determine only the change in the solution relative to that of some reference configuration whose solution is known. Various procedures have been proposed for this computation, some exact and some approximate.

Levy [60] has proposed an exact method of solution based on a parallel element approach. Since the implementation was designed to accept input specifically in NASTRAN format, Levy's program is treated here as a NASTRAN postprocessor. Examples presented demonstrated substantial computational advantages to the post-processing approach over complete reanalysis when at least two changes are considered.

CONCLUDING REMARKS

We have discussed some 35 preprocessors and postprocessors which have been developed to interface with the NASTRAN structural analysis computer program. The number, diversity, and high quality of these processors attest to the wide acceptance that NASTRAN is achieving throughout government and industry. This in turn helps to justify the development of nontrivial pre- and postprocessors. Moreover, the availability of a number of useful processors should also influence NASTRAN's popularity. All these interrelationships thus promote a higher-quality, more efficient approach to structural analysis and provide a common ground for increased communication throughout the user community.

ACKNOWLEDGMENTS

The authors are indebted to all the people who contributed material for inclusion herein. Without their cooperation, this survey would not have been possible. This work was funded by the Navy NASTRAN Systems Office at the Naval Ship Research and Development Center, whose support is gratefully acknowledged.

REFERENCES

1 MacNeal, R.H., ed., The NASTRAN Theoretical Manual, NASA SP-221 (01), Washington, D.C., 1972.

2 McCormick, C.W., ed., The NASTRAN User's Manual, NASA SP-222 (01), Washington, D.C., 1973.

3 The NASTRAN Programmer's Manual, NASA SP-223 (01), Washington, D.C., 1972.

4 MacNeal, R.H., and McCormick, C.W., "The NASTRAN Computer Program for Structural Analysis," Society of Automotive Engineers, Inc., National Aeronautics and Space Engineering and Manufacturing Meeting, Los Angeles, California, Oct. 6-10, 1969.

5 Butler, T.G., and Michel, D., NASTRAN: A Summary of the Functions and Capabilities of the NASA Structural Analysis Computer System, NASA SP-260, Washington, D.C., 1971.

6 Raney, J.P., and Weidman, D.J., "NASTRAN Overview: Development, Dynamics Application, Maintenance, Acceptance," The Shock and Vibration Bulletin, Bulletin 42, Part 5, January 1972, pp. 109-127.

7 Butler, T.G., "Technical and Social Impact of NASTRAN," NASTRAN: Users' Experiences, NASA TM X-2637, 1972, pp. 15-27.

8 Kamel, H.A., and Eisenstein, H.K., "Automatic Mesh Generation in Two and Three Dimensional Interconnected Domains," High Speed Computing of Elastic Structures, Tome 2, edited by B. Fraeijs de Veubeke, Universite de Liege, Belgium, 1971.

9 Eppink, R.T., "A Complete Data Generation Program for Large Complex Ship Structures," Proceedings of the Third Navy-NASTRAN Colloquium, Naval Ship Research and Development Center, Washington, D.C., March 1972, pp. 93-106.

10 Brophy, J.M., "Computer-Aided Cartridge Case Design Using Finite Element Stress Analysis: The Automation of Finite Element Configuration," Frankford Arsenal Report R2054, September 1972.

11 Burk, R.C., and Held, F.H., "An Interactive Graphics System to Facilitate Finite Element Structural Analysis," NASTRAN: Users' Experiences, NASA TM X-2893, 1973, pp. 679-695.

12 Cook, W.L., "Automated Input Preparation for NASTRAN," Goddard Space Flight Center Report X-321-69-237, April 1969.

13 Huang, P. C., and Matra, J.P., Jr., "Planform Input Generator (PING), A NASTRAN Pre-Processor for Lifting Surfaces - Theoretical Development, User's Manual, and Program Listing," Naval Ordnance Laboratory Technical Report 73-199, December 1973.

14 Ice, M.W., "SAIL II User's Manual," Boeing Computer Services, Inc., Seattle, Washington.

15 Ice, M.W., "NASTRAN User Interfaces - Automated Input Innovations," NASTRAN: Users' Experiences, NASA TM X-2378, 1971, pp. 669-678.

16 Katow, M.S., and Cooper, B.M., "NASTRAN Data Generation and Management Using Interactive Graphics," NASTRAN: Users' Experiences, NASA TM X-2637, 1972, pp. 399-406.

17 Lorensen, W., "IGFES - An Interactive Graphics Finite Element System," paper in these proceedings.

18 Martens, M.A., Grooms, H.R., Kitagawa, M., and Bentley, H.C., "NARFEM User's Manual," North American Rockwell Space Division Report SD 71-302, December 1970.

19 Furuike, T., and Yahata, S., "Vehicle Structural Analysis - Pre and Postprocessor Data Handling," NASTRAN: Users' Experiences, NASA TM X-2378, 1971, pp. 697-707.

20 Meyer, P.M., "Data Generator for the Idealization for Finite Element Structural Analysis of Naval Ship Flat Plated Grillages with Multiple Openings," Naval Ship Research and Development Center Report 3807, May 1972.

21 McKee, J.M., and Marcus, E.T., "A General Purpose Data Generator for Finite Element Analysis," Naval Ship Research and Development Center Report 4066, April 1973.

22 McKee, J.M., "Automated Structural Modeling for Use with NASTRAN," Proceedings of the Third Navy-NASTRAN Colloquium, Naval Ship Research and Development Center, Washington, D.C., March 1972, pp. 83-92.

23 Rockwell, R.D., and Pincus, D.S., "Computer-Aided Input/Output for Use with the Finite Element Method of Structural Analysis," Naval Ship Research and Development Center Report 3402, August 1970.

24 Rockwell, R.D., "Computer Aided Input/Output for Use with Finite Element Analyses," Proceedings of the Third Navy-NASTRAN Colloquium, Naval Ship Research and Development Center, Washington, D.C., March 1972, pp. 58-77.

25 Rockwell, R.D., "Computer-Aided Input/Output for Use with Finite Element Structural Analyses," Naval Ship Research and Development Center Report 3844, February 1973.

26 "NASTRAN.LINKØ-2, A NASTRAN Preprocessor," Technical Bulletin STB 7310, SCI-TEK, Inc., Wilmington, Del., February 1973.

27 Chestnutt, D.L., "SWATH Ship Pre- and Post-Processors for NASTRAN," Proceedings of the Fourth Navy-NASTRAN Colloquium, Naval Ship Research and Development Center, Bethesda, Md., March 1973, pp. 15-17.

28 O'Hara, G.P., "CONN4: A NASTRAN Threaded Connection Aid," Proceedings of the Fourth Navy-NASTRAN Colloquium, Naval Ship Research and Development Center, Bethesda, Md., March 1973, pp. 18-37.

29 Giles, G.L., and Blackburn, C.L., "Procedure for Efficiently Generating, Checking, and Displaying NASTRAN Input and Output Data for Analysis of Aerospace Vehicle Structures," NASTRAN: Users' Experiences, NASA TM X-2378, 1971, pp. 679-696.

30 Smith, W.W., "A Special NASTRAN Program for Input Checking and Undeformed Structure Plotting," NASTRAN: Users' Experiences, NASA TM X-2378, 1971, pp. 559-568.

31 Smith, W.W., "An Interactive NASTRAN Preprocessor," NASTRAN: Users' Experiences, NASA TM X-2893, 1973, pp. 641-659.

32 Shomrat, J., Schweid, E., and Newman, M., "The Sage System," paper in this volume.

33 Kelly, B.M., "IDEAL - An Interactive Graphics Aid in the Idealization of a Structural Model," Naval Ship Research and Development Center Report 4014, in preparation.

34 McKee, J.M., "Verifying NASTRAN Data," Proceedings of the Third Navy-NASTRAN Colloquium, Naval Ship Research and Development Center, March 1972, pp. 109-114.

35 FASTDRAW Interactive Graphics System, McDonnell-Douglas Automation Company, Document B1298083, 1973.

36 Cronk, M., "An Interactive Computer Graphics Program for NASTRAN," NASTRAN: Users' Experiences, NASA TM X-2378, 1971, pp. 659-667.

37 Anderson, P. L., and Buell, W.R., "NASTIE, A Program Converting SAMIS Element Data to NASTRAN Bulk Data," Ford Motor Company Report 1612D, March 15, 1973.

38 Giles, G.L., and Dutton, J.H., "Application of NASTRAN to the Analysis of a Space Shuttle Orbiter Structure," NASTRAN: Users' Experiences, NASA TM X-2378, 1971, pp. 451-464.

39 Eshleman, A.L., Jr., and Anderson, G.E., "User's Manual, Automated Structural Design Computer Program," McDonnell-Douglas Corporation Document DAC-33447, June 1970.

40 Everstine, G.C., "The BANDIT Computer Program for the Reduction of Matrix Bandwidth for NASTRAN," Naval Ship Research and Development Center Report 3827, March 1972.

41 Everstine, G.C., "The BANDIT Computer Program for the Reduction of Matrix Bandwidth for NASTRAN," NASTRAN: Users' Experiences, NASA TM X-2637, 1972, pp. 407-414.

42 Cuthill, E.H., and McKee, J.M., "Reducing the Bandwidth of Sparse Symmetric Matrices," Proceedings of the 24th National Conference ACM 1969, pp. 157-172.

43 Levy, R., "Resequencing of the Structural Stiffness Matrix to Improve Computational Efficiency," JPL Quarterly Review, vol. 1, no. 2, July 1971, pp. 61-70.

44 Levy, R., "Savings in NASTRAN Decomposition Time by Sequencing to Reduce Active Columns," NASTRAN: Users' Experiences, NASA TM X-2378, 1971, pp. 627-631.

45 Levy, R., "Structural Stiffness Matrix Wavefront Resequencing Program (WAVEFRONT)," JPL Technical Report 32-1526, vol. XIV, 1972, pp. 50-55.

46 Cuthill, E.H., "Several Strategies for Reducing the Bandwidth of Matrices," Sparse Matrices and Their Applications, edited by D.J. Rose and R.A. Willoughby, Plenum Press, New York, 1972, pp. 157-166.

47 McCormick, C.W., "Review of NASTRAN Development Relative to Efficiency of Execution," NASTRAN: Users' Experiences, NASA TM X-2893, 1973, pp. 7-28.

48 "PVEC, A Computer Program for Generating Partition Vectors that Merge Structural Matrices from Several NASTRAN Substructures," Memo G-2538-76, Boeing Computer Services, Inc., Seattle, Washington, 1971.

49 Puccinelli, E.F., and Jackson, C.E., Jr., "VIEW - A Modification of the RAVFAC View Factor Program for Use with the NASTRAN Thermal Analyzer," NASTRAN: Users' Experiences, NASA TM X-2637, 1972, pp. 455-463.

50 Jackson, C.E., Jr., "Programmer's Manual for VIEW," Goddard Space Flight Center Report X-322-73-120, March 1973.

51 Lovin, J.K., and Lubkowitz, A.W., "User's Manual for 'RAVFAC', A Radiation View Factor Digital Computer Program," Contract #NAS 8-30154, Huntsville Research and Engineering Center, Huntsville, Alabama, 1969.

52 Kelly, B.M., "The NASTRAN Contour Plotter," NASTRAN: Users' Experiences, NASA TM X-2637, 1972, pp. 385-397.

53 Rogers, J.L., Jr., "A Method for Transferring NASTRAN Data Between Dissimilar Computers," NASTRAN: Users' Experiences, NASA TM X-2893, 1973, pp. 633-639.

54 Rogers, J.L., Jr., "Intercomputer Transfer in Full Precision of Arbitrary Data on Magnetic Tape Employing NASTRAN User Tape Format," NASA TM X-2901, 1973.

55 Ice, M.W., "Reading Data from a NASTRAN Checkpoint/Restart Tape," Memo G-2538-090, Boeing Computer Services, Inc., Seattle, Washington, Oct. 27, 1971.

56 Wingate, R.T., Jones, T.C., and Stephens, M.V., "NASTRAN Postprocessor Program for Transient Response to Input Accelerations," NASTRAN: Users' Experiences, NASA TM X-2893, 1973, pp. 707-734.

57 Anderson, P. L., "RIGID - Automatic Generation of Rigid Links for the NASTRAN Program," Ford Motor Company Report 1205-1, September 12, 1972.

58 McClelland, W. A., and Klosterman, A.L., "Using NASTRAN for Dynamic Analysis of Vehicle Systems," Proceedings of the International Conference on Vehicle Structural Mechanics, Society of Automotive Engineers, Inc., Detroit, Michigan, March 26-28, 1974; also, SDRC Report, Structural Dynamics Research Corporation, Cincinnati, Ohio, 1974.

59 Cook, W.L., "A General-Purpose Conversational Software System for the Analysis of Distorted Multi-Surface Antennas," Ph.D. Dissertation, School of Engineering and Applied Science, The George Washington University, February 19, 1973.

60 Levy, R., "A NASTRAN Postprocessor for Structural Modification Reanalysis," NASTRAN: Users' Experiences, NASA TM X-2378, 1971, pp. 737-747.

APPENDIX

SOURCES AND AVAILABILITY

The following information is provided for each program:

1. Person to contact for information
2. Availability
3. Special arrangements

Quite often programs are available from COSMIC

COSMIC
Barrow Hall
University of Geogia
Athen, Georgia 30601

SAILII, PAVEC

1. M. W. Ice or J. Tocher
   Boeing Computer Services, P. O. Box 24346
   Seattle, Washington 98124
2. Contacts
3. May buy or use program on Boeing system

AXIS, SHELBY, COONS, MOVE, BANDAID

1. W. L. Cook
   COMSAT
   Clarksburg, Maryland 20734
   (301) 428-4416
2. Contact or COSMIC
3. Buy from COSMIC

IGFES

1. W. Lorensen
   Watervliet Arsenal
   Watervliet, New York 12189
   (301) 273-4610
2. Contact
3. Paragraph 1-8 (7) of AR 18-1 states: The Freedom of Information Act, 5 US Code 552 (AR345-20) has been interpreted as not requiring the release of Government-owned programs to the general public. However, requests for such programs may be favorably considered by HQDA on a case-by-case basis (e.g. an educational institution or other nonprofit activity). Requests for Government-owned ADP programs, with explanatory documentation and appropriate recommendation will be forwarded to HQDA.

NASTRAN LINKØ-2

1. Bill Rapley
   SCI-TEK, Inc.
   1707 Gilpin Ave.
   Willmington, Del. 19806
   (302) 658-2431

2. Contact
3. Use on SCI-TEK system

SAGE

1. J. Shomrat
   Israel Aircraft Ind., Ltd.
   Lod Airport
   Israel
2. Contact
3. Buy

WAVEFRONT, Structural Modification Reanalysis

1. Roy Levy
   Jet Propulsion Laboratory
   4800 Oak Grove Drive
   Pasadena, Calif. 91103
2. Contact or COSMIC
3. Buy from COSMIC

GRIDXY

1. John Higgins
   Frankford Arsenal
   Philadelphia, Pa. 19137
   (215) 831-5273
2. Contact
3. Paragraph 1-8 (7) of AR 18-1 states: The Freedom of Information Act, 5 US Code 552 (AR345-20) has been interpreted as not requiring the release of Government-owned programs to the general public. However, requests for such programs may be favorably considered by HQDA on a case-by-case basis (e.g. an educational institution or other nonproft activity). Requests for Government-owned ADP programs, with explanatory documentation and appropriate recommendation will be forwarded to HQDA.

FEM

1. R. C. Burk or Fred Held
   McDonnell Douglas Astronautics Co. - EAST
   P. O. Box 516
   St. Louis, Mo. 63166
   (314) 232-6108
2. Contact
3. Will sell to qualified individuals

DATGEN, CUTUP, HIDE, BANDIT, GENDA, IDEAL, CONPLT

1. Gordon Everstine or James McKee
   Dept. of the Navy
   NSRDC
   Carderock Labs
   Bethesda, Md. 20034
2. Contacts
3. Limited distribution to qualified individuals

CRONK

1. Michael Cronk
   General Dynamics-CONVAIR, P. O. Box 80844
   San Diego, Calif.
   (714) 277-8900 ext. 2435
2. Contact
3. Will sell to qualified individuals

Transient Response Program

1. R. T. Wingate
   NASA-Langley Research Center
   Hampton, Va. 23665
   (804) 827-2675
2. Contact
3. Available to qualified individuals

VIEW

1. C. E. Jackson, Jr.
   NASA/Goddard
   Greenbelt, Md. 20771
   (301) 982-5275
2. Contact or Goddard Program Library (#B00173)
3. Write or call: Mrs. Pat Barnes
   Code 531.2
   Goddard Space Flight Center
   Greenbelt, Md. 20771
   (301) 982-6796
   Available to qualified individuals

FAST DRAW

1. John Flanders or Gerry Folk
   McDonnell Douglas Automation Co.
   Box 516
   St. Louis, Mo. 63166
   (314) 232-7876
2. Contacts
3. Use on McDonnell Douglas network

Jet Propulsion Laboratory Preprocessor

1. M. S. Katow or B. M. Cooper
   Jet Propulsion Laboratory
   4800 Oak Grove Drive
   Pasedena, Calif. 91103
   (213) 354-4726
2. Contact
3. Available to qualified individuals

NASTY

1. Edward Carl
   Structural Dynamics Research Corp (SDRC)
   5729 Dragon Way
   Cincinnati, Ohio 45227
2. Contact
3. Use on SDRC's system

PING, BING

1. Dr. Pao C. Huang and John P. Matra, Jr.
   U. S. Naval Ordnance Lab.
   White Oak Silver Spring, Md. 20910
2. Contact
3. Available to qualified individuals

CONSIDERATIONS FOR DEVELOPING A GENERAL FINITE ELEMENT PRE- AND POSTPROCESSING SYSTEM

V. A. Tischler and L. J. D. Bernier
Air Force Flight Dynamics Laboratory
Wright-Patterson Air Force Base, Ohio

ABSTRACT

After the need for a general pre- and postprocessing system is established, the system is described in terms of its five major modules. Additional consideration is given to the required hardware and software, and the method of operation is presented in general terms. Finally, the system is described in terms of its current capability.

INTRODUCTION

For over a decade, the Air Force has been involved in finite element methods. In developing this methodology, however, emphasis has been justifiably centered on developing efficient analysis algorithms and complete element libraries with less attention given to optimizing the overall analysis capability. Consequently, as fast as potential user organizations become attracted to the versatility and power of available finite element programs, they also become discouraged and discontent due to the intolerable effort required to prepare data for, and reduce data from, these same programs. Also, due to the considerable amount of time required to learn to use the more general and powerful programs, users tend to adopt one program. This, of course, is undesirable because it forces the user to accept the weaknesses and limitations, as well as the strengths, of that particular program. It also discourages users from exploring the capabilities of smaller special purpose programs, which in many cases could satisfy their needs more efficiently.

Ideally, a user should feel comfortable using any of the analysis programs available to him. A major step in this direction is being made by standardizing and automating, to as great a degree as possible, the input and output requirements of all available programs.

Although the current development of better pre- and postprocessors is enhancing a user's ability to apply individual programs to more complicated problems, there still remains a lack of effective communication between structures organizations.

Pre- and postprocessors improve a user's ability to exercise a given program and develop a better understanding of its versatility and limitations. But the user is still more "capability bound" than he should be because most users simply do not have the time required to learn to use the total available capability.

For the development engineer, responsible for advancing the state-of-the-art in finite elements and providing users with an effective capability, a need exists to evaluate efficiently and to streamline the total available capability. By using a standardized pre- and postprocessing system, individual programs can be more easily compared. Consequently, unnecessary and unwanted redundancies can be eliminated and the relative strengths and weaknesses of the individual programs better understood.

Also, by standardizing user requirements, the prospects of developing structural data libraries becomes more attractive. Using these libraries as a common data bank would greatly improve communication among Air Force organizations and between Air Force and industry organizations. Consequently, the Air Force would increase its ability to monitor effectively industrial efforts at all levels of hardware development and support. Equally important, by increasing its ability to conduct complicated structural analyses quickly, the Air Force would be able to make more positive contributions in the areas of analysis and design and, further, would decrease its dependency on industrial support after the hardware is placed in the Air Force inventory.

SYSTEM DESCRIPTION

The system being considered will provide a standard, efficient approach for using any finite element program, without sacrificing the inherent flexibility of the individual programs. Emphasis is being placed on making the system interactive and user oriented. To accomplish this, extensive use will be made of modular programming, standardization, and interactive graphics.

The system will have three major subsystems: the preprocessor, the analysis module, and the postprocessor. Each of these subsystems will be modularized to take advantage of all available software and to provide a convenient means for updating the system. The key to the success of the total system, however, lies in the ability of the conversion routines to standardize and minimize user requirements to interface completely with the analysis programs. Figure 1 illustrates the major system modules and indicates how they are intended to interrelate.

Executive Monitor

The executive monitor will act as the total system coordinator and will provide the required interface between the user and the system or the remaining system modules. In addition to guiding the user efficiently through an analysis, the executive monitor will have tutorial responsibilities. Brief, simple menus will be available for the experienced user, while the inexperienced user will have equally simple but more descriptive menus to assist him in choosing between alternatives. Also, by requiring major system modules to interrelate through the executive monitor, the modularity of the entire system will be enhanced and major system changes will be more easily effected.

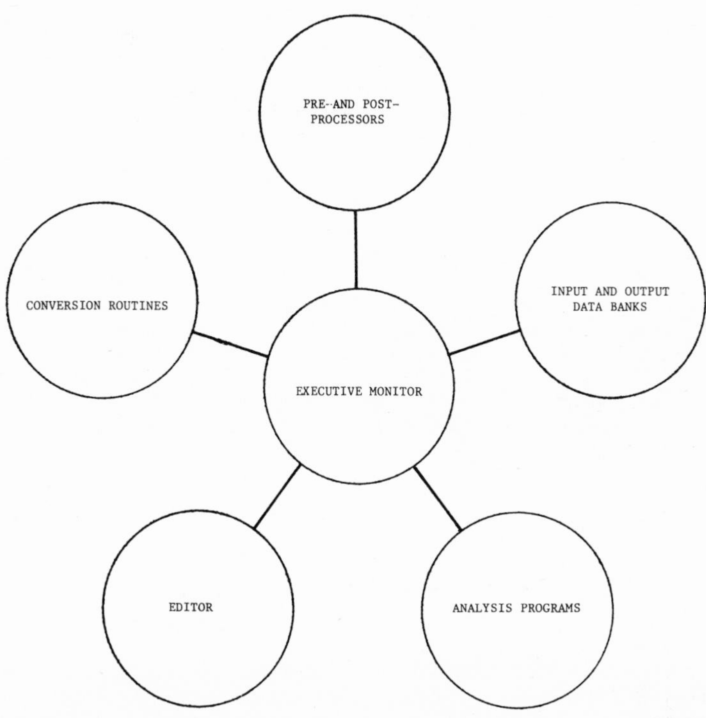

Fig. 1 Major system modules

Pre- and Postprocessors

The pre- and postprocessors[1] will be used to automate data synthesis and reduction, thereby releasing the user from tedious and time-consuming data manipulation and, consequently, improving the overall efficiency of the analysis. What follows is a synopsis of features necessary to make the proposed system functionally attractive.

---

[1] Appendix A is a compendium of pre- and postprocessors known to the authors, either through the literature or personal contacts.

In the development of preprocessors, grid and element generators have received by far the most attention. Figure 2 illustrates the various classes of structures generally encountered by a structural analyst. Data for most 2-D and 3-D truss networks are currently generated either by hand or using a digitizer, but for structures exhibiting some degree of regularity, a structural input data language can be used [1, 2]. The generation of data for 2-D structures, such as plates, is easily the most automated, and algorithms, as well as programs, are quite abundant in the literature [3, 4, 5, 6, 7, 8, 9, 10, 11]. In spite of the progress made in this area, more work is required to facilitate modeling around discontinuities and transitioning from high-density to low-density modeling areas and to guard against generation of undesirable high-aspect ratio elements.

The 3-D thin-walled structures are most representative of the class of structures generally encountered in aerospace structures. Mesh generation for shells of revolution are treated in the literature [12], and programs are available. In addition, there are special purpose programs, written to generate data for a specific type of structure, such as a wing or fuselage [13, 14]. More general 3-D shell-modeling programs are also available [3, 4, 5, 10, 12, 15, 16], although few, if any, exhibit the degree of generality required to deal effectively with this class of structures. For fairly complicated 3-D shell structures having some degree of regularity, it is possible to use a structural data language. However, when geometries are completely irregular, the user must, once again, resort to hand calculations and, possibly, the digitizer.

Finally, little was found in the literature concerning programs specifically written to generate thick-walled structures [7, 16]. For many cases, however, it seems reasonable to assume that the shell generators and the input structural data languages could be easily extended to this class of structures.

Other necessary ingredients needed to improve the efficiency of any preprocessing capability are segment generators [12, 17], to take advantage of repeating or symmetric structures; an automatic substructuring capability; load generators, especially for thermal, pressure, and inertial loads; a material properties table; a geometric properties generator; and a bandwidth optimizer [12, 18, 19]. A repeat option should also be available for assigning identical constraints, loads, material, and geometric properties to various elements or nodes.

The postprocessors should provide complete descriptions of force, displacement, stress, strain, and time histories when required [17]. Also, ordering schemes should be available to isolate areas of interest quickly in the tabular data. Force data should include nodal forces and shear flows, either in a tabular form or superimposed on the geometric model. Displacement output should feature tabular data as well as the display of the deformed model, either alone or superimposed on the undeformed model [20]. Provisions should also be made to allow scaling of the displacements [17, 20]. Stress-strain data, including effective and principal stresses and strains, should be available in a tabular form or superimposed on the geometric model. When solid elements are used in an

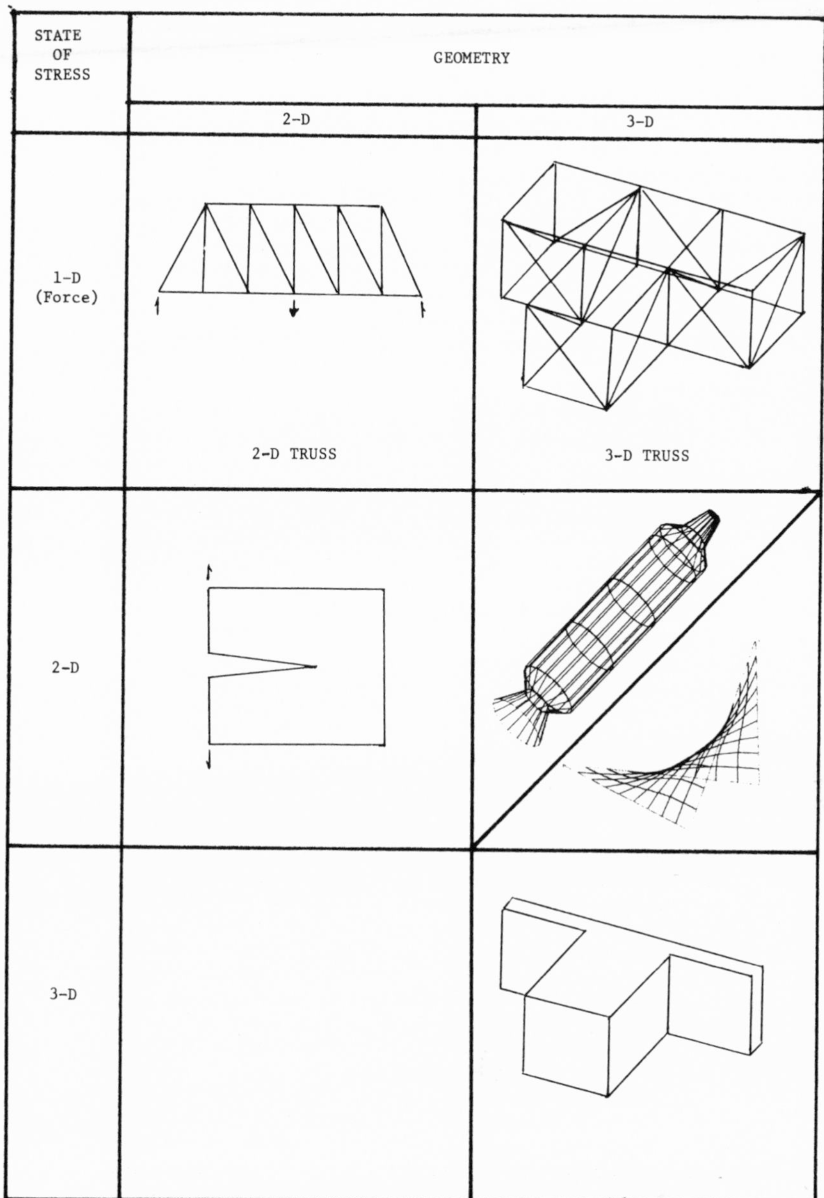

STATE OF STRESS
GEOMETRY
2-D
3-D
1-D (Force)
2-D TRUSS
3-D TRUSS
2-D
3-D

Fig. 2 General classification of structures

analysis, arbitrary cuts through the structure should be provided so that stress-strain data can be viewed on any cut plane [17, 20]. Contour plots of all these data should be made available to the user [10, 21, 22]. Finally, equilibrium checks should be available for any arbitrary cut in the model for purposes of ensuring a well-conditioned stiffness matrix.

To make the pre- and postprocessing system truly flexible, a variety of display options should be available. These include the ability to view the structure at any angle [23, 24, 25] (this feature takes on greater significance if it can be done dynamically), hidden line algorithms [10], a zooming capability [26], an exploded view capability [26], individual display of various elements [17], and elements or nodes with tabular data superimposed [26]. Also, shading or the use of colors can be used quite advantageously to identify element types, material, and geometric properties or ranges of stress-strain values.

Input and Output Data Banks

These data banks will be used to store data in both standard and special formats. A standard format will be system dependent and will consist of data common to all programs. The special format will depend on the particular program used and will consist of data peculiar to the requirements of that program, be it a preprocessor, analysis program, or postprocessor. These data banks will be accessible either on- or off-line, on-line being defined as available through the resident computing system and off-line being defined as available to the vector graphics unit from a local external storage device.

Editor

The primary function of the editor will be to manipulate and change data in the input and output data banks. Although a special editing capability will be included in the pre- and postprocessing system for purposes of providing on-line support and making the system more machine independent, the editing functions of the resident system can certainly be used independently of the pre- and postprocessing system. For example, on the CDC 6600 files can be edited from the alphanumeric displays via the editing capability of INTERCOM. Finally, a special editor will be written for use with the vector graphics system in an off-line mode.

Conversion Routines

In using the proposed system, data will be prepared in a standard format, either by hand or using available preprocessors, without regard to the particular analysis program to be used. This is made possible by the fact that all finite element analysis programs have essentially the same data requirements, although their formats may vary. These data include geometric data, such as grid point coordinate data, element connection data, thicknesses, areas, inertias, and the like, as well as boundary condition data, load data, and material properties data. Once these data are generated, the conversion routines will be used to transform the data to a card image data stream consistent with the requirements of the analysis program used.

Conversion routines will also be used to transform output from the analysis programs to a standard format. In addition to providing the user with standardized tabular data, this capability will also facilitate the data manipulations required to use the available postprocessors.

And finally, although conversion routines will be used primarily to interface the analysis portion of the system with the pre- and postprocessing parts of the system, they will also minimize user requirements while using the available system pre- and postprocessors.

Hardware and Software Considerations

The entire pre- and postprocessing system will reside in the user's resident computing system and will be accessible in both an off-line and on-line mode. Figure 3 shows the suggested interaction of the total capability. Alphanumeric and vector graphic display units will be used in the on-line mode, while the off-line mode will consist of batch jobs.

Batch jobs will be used primarily to take advantage of system modules that exceed the on-line limitations of the resident computing system and to enter data received from external sources, although provisions will be made to allow the user to access the total system capability in this mode. Alphanumeric displays will interact directly with the system and have the ability to utilize the entire system capability.

The vector graphics displays will be used to extend the capability of the currently less expensive and more abundant alphanumeric displays. In addition to being able to communicate directly with the pre- and postprocessing system, the vector graphics unit will be supported by its own minicomputer. This minicomputer will be charged with a dual responsibility. First, it will provide the necessary medium for the user to take full advantage of the display characteristics of his particular display scope. And second, it will support an editing capability independent of that found in the pre- and postprocessing system. Consequently, the user will be able to display and edit the geometric model, tabular data, or results in an off-line mode. To accomplish this, it will be necessary to make the input and output data banks accessible either on- or off-line.

Additional features anticipated for the proposed system are a digitizing capability to be used with the vector graphics scope, either on- or off-line; a high-speed reproduction capability for hard copies of images produced on the vector graphics scope; a restart capability to allow the user to conduct an analysis in more than one sitting and provide a convenient means of mixing operational modes; and an off-line hard-copy capability for both tabular listings and plots.

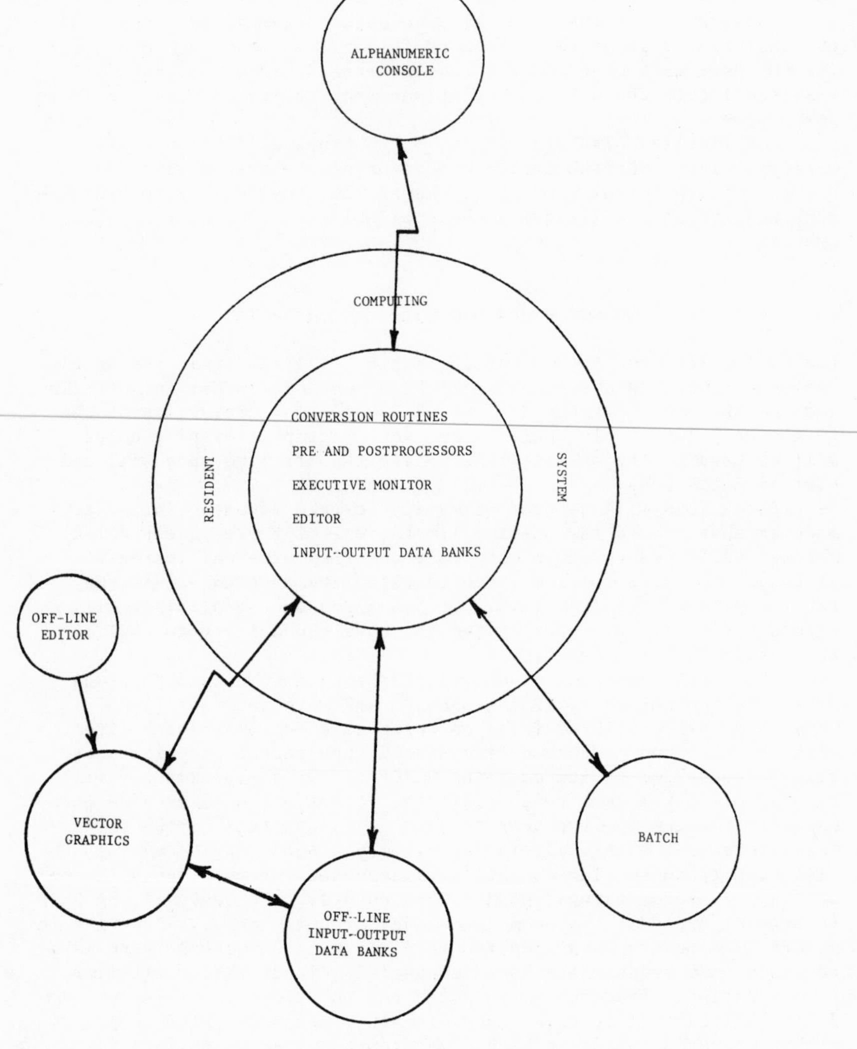

Fig. 3 Suggested interaction of the total capability

Method of Operation

Figure 4 is a simplified schematic of the proposed system. It is included here to provide a clearer understanding of the nature of the system and its intended operation. For purposes of understanding the system shown in Fig. 4, assume that the user has made contact with the system from an interactive vector graphics terminal. First, a display of the available analyses will appear along with a list of programs capable of conducting these analyses. After choosing the type of analysis to be conducted, the user will generate the data in a standard format, either directly or using the available preprocessors. All data generated in this manner will be stored in the input data bank. At any time during or after data generation, the user will have the ability to edit the data.

After the data have been generated and reviewed for correctness, the user will choose the analysis program he intends to use. At this point, it may be necessary to generate additional data, in the special format, to satisfy any peculiarities of the chosen analysis program. With this accomplished, the system conversion routines will transform all input data to a card image stream compatible with the chosen analysis program.

The analysis will then be conducted either on- or off-line depending on the size of the analysis and the chosen analysis program. At this point, a number of output options will be made available to the user, including a hard copy of the analysis results in the special format.

With the analysis completed, the conversion routines will transform the output data to a standard format and store the data in the output data bank. The user will then review a portion, or all, of the data, and, further, have the option of continuing on or rerunning the analysis. If the user chooses to rerun the analysis the system will regress to the preprocessing phase. If, on the other hand, the user is satisfied with the output at this point, he will have a number of postprocessors at his disposal to assist him in taking a closer look at the results.

After reviewing the results to his satisfaction, the user will be able to obtain hard copies, including tabular data and plots of any, or all, of the results generated by the analysis program and the postprocessors.

Having completed the analysis the user will have the option of using the system again or stopping. If he chooses to use the system again, the system will loop back to the beginning. Because the user may want to conduct a different analysis on the same model or use a different program to conduct the same analysis, the data in the input data bank will not be destroyed until the user indicates he wishes to initiate a completely new analysis. Restart options will be scattered throughout the system to provide the user the flexibility of conducting an analysis in more than one sitting. For a more detailed schematic of the proposed system refer to Appendix B.

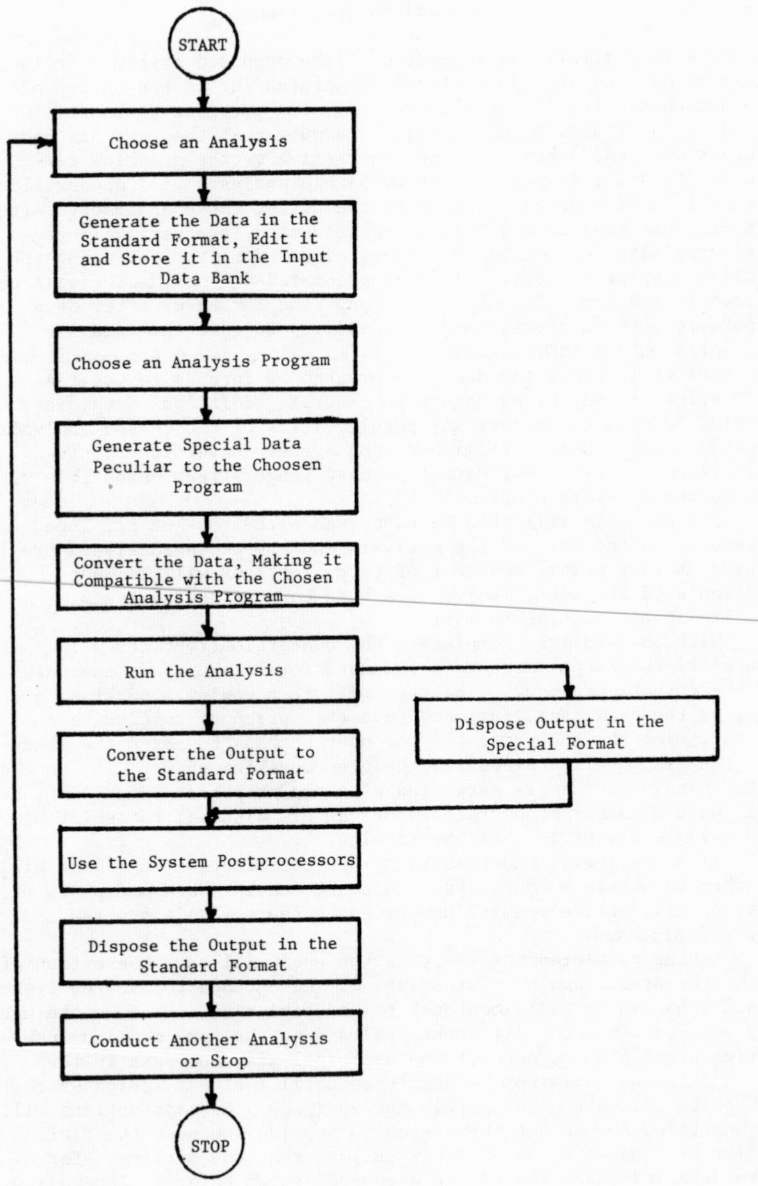

Fig. 4 General flow chart of the proposed system

EVOLUTION OF THE PROTOTYPE SYSTEM

Once it was decided to develop a general pre- and postprocessing system, two short-term goals were established. One was to provide the user with a workable system in the shortest possible time, and the other was to assess the value of a more general system by developing a less sophisticated, but representative, prototype system.

The CLAAS Interactive Graphics System developed by M. J. Cronk was available for our use [27, 28]. The CLAAS system consisted of a pre- and postprocessing interactive graphics program plus an off-line structural analysis program based on the finite element method. First, data for the analysis program was preprocessed by the graphics program, then an off-line analysis was made, and finally the results, plus the original data, were postprocessed by the graphics program. A closed system existed in the sense that the original data could be altered at any time by the graphics program and then run off-line by the analysis program.

Unfortunately, as a structural analysis tool the analysis program was limited in scope. Not only was the element library of bars and shear panels inadequate for solving many aircraft structural problems, but also the finite element model had to be constructed so that the stiffness matrix was tri-diagonal. This latter restriction necessitated dividing the finite element model up into "lines" of nodes where each line contained the same number of nodes. Nodes were connected by elements such that the nodes in line I were connected only to nodes in line I, line I - 1, and line I + 1. The line concept necessarily leads to a requirement for dummy nodes.

However, the graphics program was well thought out. Although it primarily processed data defined within the line concept, it could also process data in the form of substructures where a substructure was defined as any arbitrary collection of nodes. Since substructures were usually functional collections of nodes such as ribs, bulkheads, spars, etc., they gave the user the flexibility he needed and wanted in working with his finite element model. Thus, since the graphics program performed most of the desirable functions of a pre- and postprocessing system, it was adopted as the basis for our system.

In order to complete the definition of our system, our goals became:

1. To select representative structural analysis programs to be used either on- or off-line with the graphics program
2. To extend the graphics program's pre- and postprocessing capability to handle the expanded finite element libraries of the analysis programs
3. To enhance the preprocessing capability of the graphics program by attaching finite element model generators AXIS [12], SHELBY [12], and COONS [12], and the plot program EZPLOT [29]
4. To write adequate conversion programs which not only converted data written within the line concept to the input formats of the selected analysis programs but also converted results from these same programs to the input format of the postprocessing part of the graphics program

SAVFEG: A Prototype System

We gave our system the acronym SAVFEG - Structural Analysis Via the Finite Element Method with Graphics. SAVFEG was developed on the CDC 6600 under the Scope 3.3 operating system with a CDC 274 digigraphic console.

To accomplish our goals four structural analysis programs were selected:

1. ANALYZE, developed in-house by Dr. V. B. Venkayya. Since this program can run in $50{,}000_8$ core memory or less for an intermediate size problem (180 degrees of freedom, 200 elements), it was selected to run on-line in the SAVFEG system.
2. ASOP - A Structural Optimization Program
3. MAGIC - An automated general purpose system for structural analysis.
4. NASTRAN - NASA STRuctural ANalysis Program

Programs 2, 3, and 4 are run off-line in the SAVFEG system.

The element library, at present, includes only bars, shear panels, and quadrilateral and triangular membrane elements. However, by choosing ASOP, MAGIC, and NASTRAN, the scope of the element library can be expanded as needed in the future. Thus, SAVFEG can be used either as a closed on-line system or a closed off-line system.

Basically the SAVFEG system operates in the following manner. The SAVFEG preprocessor either generates the finite element model data internally or accepts data from batch. All data must be defined within the line concept, except those which are in the form of substructures. Depending on the analysis program chosen, a conversion routine rewrites the data in the format expected by that program. After analysis, either on- or off-line, a conversion routine rewrites the output results in the format expected by the SAVFEG postprocessor. After reviewing the results of the analysis, the user can again make changes to the model via SAVFEG, store the updated model for later use, or rerun the analysis with the same or a different analysis program.

The SAVFEG preprocessor allows the user to:

1. Generate, if desired, his finite element model via AXIS, SHELBY, or COONS. Data for these programs can be input through either batch or the console. EZPLOT can be used to view the generated model, partial or full, and a hard-copy capability is provided from the CALCOMP plotter.
2. Check the input data for format errors and make corrections if necessary at the console.
3. Check visually the finite element model to locate errors. Hard-copy output of any view, partial or full, of the model is available via the CALCOMP plotter.
4. Display bars, shear panels, quadrilateral and triangular membranes, loads, constraints, and nodes. All of these can be added to the model or deleted from the model at the console. In addition, any geometric or material properties associated with any of the above can be changed.
5. Process the updated data via the punch, printer, permanent file, or tape.

The SAVFEG postprocessor allows the user to:

1. Develop static and deflected views of the model and overlay these views. Hard-copy output of any view, partial or full, of the model is available via the CALCOMP plotter.
2. Develop plots of bar loads or stresses directly on a view of the model for any loading condition.
3. Display a readout of the loads in an individual bar along with the bar area for any loading condition.
4. Display the average shear flow in each panel or membrane for each loading condition or the maximum of all loading conditions.
5. Display a readout of the shear flows on each side of a panel or a membrane for any loading condition.
6. Display all corner forces for a panel or a membrane for any loading condition.
7. Display stresses σ_x, σ_y, σ_{xy} for a membrane for any loading condition.

SUMMARY AND CONCLUDING REMARKS

After establishing the need for a general pre- and postprocessing system, the system was described in terms of its five major modules. Additional consideration was given to the required hardware and software, and the method of operation was presented in general terms. Finally, the prototype system was described in terms of its current capability.

Although the prototype system is filling an immediate need, by the very nature of its development it is unacceptable in terms of flexibility and efficiency. Consequently, a new, longer-term effort has been initiated to develop a more sophisticated system that will eventually replace SAVFEG. Also, the feasibility of developing a structures data library will be explored and its impact on the proposed finite element system evaluated.

ACKNOWLEDGMENT

The development of SAVFEG is being supported by Dr. P. Poirier and Messrs. J. Hudson, T. Rowland, G. Grimes, and J. Smith of the Computer Services Branch at Wright-Patterson AFB.

REFERENCES

1 Boeing Document D2-125179-8, "The ASTRA System," May 1972.

2 Bjaaland, H. K., and Nelson, M. F., "The Input of Structural Topology to a Computer," R71-38, Oct. 1971, Massachusetts Institute of Technology, Cambridge, Mass.

3 Kamel, H. A., et. al., "Some Developments in the Analysis of Complex Ship Structures," EES Series Report 36, Aug. 1972, Tucson: College of Engineering, University of Arizona.

4 Kamel, H. A., and Eisenstein, H. K., "Automatic Mesh Generation of Two and Three Dimensional Inter-Connected Domains," a paper

presented at a symposium on "High Speed Computing of Elastic Structures," sponsored by the International Union of Theoretical and Applied Mechanics; Liege, Belgium, Aug. 23-28, 1970.

5 Akyuz, Fevzican A., "Natural Coordinate System: An Automatic Input Data Generation Scheme for a Finite Element Method," Nuclear Engineering and Design, Vol. II, No. 2, Mar. 1970, pp. 197-205.

6 Brophy, Jo Anne M., "Computer-Aided Cartridge Design Using Finite Element Stress Analysis: The Automation of Finite Element Configuration," R-2054, Sept. 1972, Frankford Arsenal, Philadelphia, Pa.

7 Buell, W. R., and Bush, B. A., "Mesh Generation - A Survey," Journal of Engineering for Industry, Transactions of the ASME, Feb. 1973.

8 Frederick, C. O., et. al., "Two-Dimensional Automatic Mesh Generation for Structural Analysis," International Journal for Numerical Methods in Engineering, Vol. 2, No. 1, 1970, pp. 133-144.

9 Meyer, P. M., "Data Generator for the Idealization for Finite Element Structural Analysis of Naval Ship Flat Plated Grillages with Multiple Openings," Report 3807, May 1972, Naval Ship Research and Development Center, Bethesda, Maryland.

10 Rockwell, R. D., "Computer-Aided Input/Output for Use with Finite Element Structural Analyses," Report 3844, Feb. 1973, Naval Ship Research and Development Center, Bethesda, Maryland.

11 Crisp, R. J., "FEMG - Finite element Mesh Generation Program," RD/C/N309, May 1969, Central Electricity Generating Board, London, S.E.1.

12 Cook, W. L., "Automated Input Data Preparation for NASTRAN," GSC-11039, Apr. 1969, Goddard Space Flight Center, Greenbelt, Maryland.

13 Giles, G. L., "Procedure for Automating Aircraft Wing Structural Design," Proceedings of the American Society of Civil Engineers, Journal of the Structural Division, Vol. 97, ST1, 1970, pp. 99-113.

14 Giles, G. L., and Blackburn, C. L., "Procedure for Efficiently Generating, Checking and Displaying NASTRAN Input and Output Data for Analysis of Aerospace Vehicle Structures," presented at the NASTRAN User's Colloquium, Langley Research Center, Sept. 13-15, 1971.

15 Coons, S. A., "Surfaces for Computer-Aided Design of Space Forms," MAC-TR-41, June 1967, Project MAC, Massachusetts Institute of Technology, Cambridge, Mass.

16 Zienkiewicz, O. C., and Phillips, D. V., "An Automatic Mesh Generation Scheme for Plane and Curved Surfaces by 'Isoparametric' Coordinates," International Journal for Numerical Methods in Engineering, Vol. 3, 1971, pp. 519-528.

17 Bousquet, R. D., et. al., "The Development of Computer Graphics for Large Scale Finite Element Codes," presented at the Third Conference on Matrix Methods in Structural Mechanics, Wright-Patterson Air Force Base, Ohio, Oct. 19-21, 1971.

18 Everstine, Gordon C., "The Bandit Computer Program for the Reduction of Matrix Bandwidth for NASTRAN," Report 3827, Mar. 1972, Naval Ship Research and Development Center, Bethesda, Maryland.

19 Cuthill, E. H., and McKee, J. M., "Reducing the Bandwidth of Sparse Symmetric Matrices," Proceedings of the 24th National ACM Conference, 1969, pp. 157-172.

20 Yates, D. N., et. al., "The DAISY Code. Lockheed's Development and Experience," presented at the ONR International Symposium on Numerical and Computer Methods in Structural Mechanics, University of Illinois, Champaign, Illinois, Sept. 8-10, 1971.

21 McCue, G. A., et. al., "FORTRAN IV Stereographic Function Representation and Contouring Program," SID 65-1182, Sept. 1965, North American Aviation Inc.

22 Smith, R. R., "An Algorithm for Plotting Contours in Arbitrary Planar Regions," NUC TP 267, Oct. 1971, Naval Undersea Research and Development Center, San Diego, Calif.

23 Canright, R., and Swigert, P., "PLOT3D - A Package of FORTRAN Subprograms to Draw Three-Dimensional Surfaces," NASA TM X-1598, June 1968, Lewis Research Center, Cleveland, Ohio.

24 Craidon, C. B., "Description of a Digital Computer Program for Airplane Configuration Plots," NASA TM X-2074, Sept. 1970, Langley Research Center, Hampton, Virginia.

25 Matz, D. G., "EZPLOT: A NASTRAN Aid for Plotting Undeformed Structures," ENJE-TM-73-20, Aeronautical Systems Division, W-PAFB, Ohio, Aug. 1973.

26 Batdorf, W. J., et. al., "Lockheed-Georgia Company 3D Structural Analysis Display Program," Report ER11269, Jan. 1972, Lockheed-Georgia Co., Marietta, Georgia.

27 Cronk, M. J., "User's Manual for the CLAAS Interactive Graphics Program - P5420," Contract No. F33G15-72-C-2088, Jan. 1971, General Dynamics, Convair Aerospace Division.

28 Cronk, M. J., "Program Manual for the CLAAS Interactive Graphics Program," Contract No. F33G15-72-C-2088, Jan. 1971, General Dynamics, Convair Aerospace Division.

APPENDIX A

This section is a compendium of pre- and postprocessors. It is not intended to be all inclusive but rather to serve as a convenient reference to those programs the authors are either aware of or have in their possession.

The programs are listed in alphabetical order by program name, and the source is given in terms of a reference where the program was found. A brief description is provided, and where known, system peculiarities and program availability are given.

AXIS

Source: Kamel, H. A., AME Dept., University of Arizona, Tucson, Arizona.

Description: This program accepts output from MESH2 and features mesh generation and modification capabilities, triangular and quadrilateral elements, concentrated and distributed loading, cyclic execution with possibility of model, load, and boundary condition modification. Model, loads, and boundary conditions may be preserved on paper tape or written on disk and read again in any sequence. The deflected shape is plotted in dotted lines over the original model, and stresses are displayed as symbols of varying intensity. The intensity of a vector represents the scaled von Mises criterion and denotes the severity of the loading. Selected deflection and stress values may be printed on teletype or the printer. A local analysis procedure is incorporated, where results from one step can be transferred internally to serve as boundary conditions during the next step. The process may be carried on indefinitely. Thermal stress analysis and mixed boundary conditions are incorporated. Present capacity for a 16K system is 350 nodes and 700 elements.

Specifics: Developed for the PDP 15 computer.

Availability: Inquiries should be made through Dr. Hussein A. Kamel.

AXIS

Source: Cook, W. L., "Automated Input Data Preparation for NASTRAN," GSC-11039, April 1969, Goddard Space Flight Center, Greenbelt, Maryland.

Description: The program generates NASTRAN input data for shells described by the rotation of a plane curve about an axis. In addition to providing standard NASTRAN control cards along with the usual material and geometric data, AXIS also generates PLOAD cards from a mathematical definition of a pressure load, and SEQGP cards when resequencing is necessary to reduce the bandwidth of the stiffness matrix.

Specifics: This program is coded in FORTRAN and requires approximately 72K octal words memory on the CDC 6600 computer.

Availability: For all government and GSFC-sponsored contractor personnel, the program is available through:

Goddard Space Flight Center
Computer Program Library
Greenbelt, Maryland 20771

For all others, the program is available through:

COSMIC
University of Georgia
Athens, Georgia 30601

BANDAID

Source: Cook, W. L., "Automated Input Data Preparation for NASTRAN," GSC-11039, April 1969, Goddard Space Flight Center, Greenbelt, Maryland.

Description: Given a NASTRAN data deck BANDAID will automatically resequence the grid points of a structural idealization to achieve a reduced bandwidth in the stiffness matrix. Output includes SEQGP bulk data cards.

Specifics: This program is coded in PL1.

Availability: For all government and GFSC-sponsored contractor personnel, the program is available through:

Goddard Space Flight Center
Computer Program Library
Greenbelt, Maryland 20771

For all others, the program is available through:

COSMIC
University of Georgia
Athens, Georgia 30601

BANDIT

Source: Everstein, G. C., "The BANDIT Computer Program for the Reduction of Matrix Bandwidth for NASTRAN," Report 3827, March 1963, Naval Ship Research and Development Center, Bethesda, Maryland.

Description: Given a NASTRAN data deck BANDIT will automatically resequence the grid points of a structural idealization to achieve a reduced bandwidth in the stiffness matrix. Output includes SEQGP bulk data cards.

Specifics: BANDIT will run on all NASTRAN computers.

Availability: Inquiries should be made through the Computation and Mathematics Dept., Naval Ship Research and Development Center, Bethesda, Maryland.

BLADE

Source: Kielb, R., Aeronautical Systems Division, Analysis Branch, Wright-Patterson Air Force Base, Ohio.

Description: BLADE generates 3-D idealizations for turbine engine blade type structures. Output is in a NASTRAN format and includes grid, element, property, temperature, and pressure cards for solid or hollow blades.

Specifics: This program is coded in FORTRAN extended and used on the CDC 6600 computer.

Availability: Inquiries should be made through Mr. Robert Kielb.

CLAAS

Source: Cronk, M. J., "User's Manual for the CLAAS Interactive Graphics Program - P5420," Contract No. F33G15-72-C-2088, Jan. 1971, General Dynamics, Convair Aerospace Division.

Description: An interactive computer graphics program used to review and modify a proposed finite element model, generate a tape containing the results of the CLAAS analysis, and review the results of the analysis.

Specifics: The program is compatible with CDC 6000 computers that use the CDC 274 CRT and CALCOMP plotter.

Availability: Inquiries should be made through Mr. M. Cronk, Convair Aerospace Division, General Dynamics, San Diego, California.

CONPLT

Source: Rockwell, R. D., "Computer-Aided Input/Output for Use with Finite Element Structural Analyses," Report 3844, Feb. 1973, Naval Ship Research and Development Center, Bethesda, Maryland.

Description: Given the finite element idealization and necessary analysis results, CONPLT plots contours and graphs of user-specified "lines" of nodes or elements.

Specifics: This program is coded in FORTRAN, operates on the CDC 6000 computers, and is used in conjunction with the SC4020 plotter.

Availability: Inquiries should be made through the Structures Dept., Naval Ship Research and Development Center, Bethesda, Maryland.

COONS

Source: Cook, W. L., "Automated Input Data Preparation for NASTRAN," GSC-11039, April 1969, Goddard Space Flight Center, Greenbelt, Maryland.

Description: Given a description of four bounding curves of a free-form shell structure, the COONS program will create a smooth surface passing through those curves and generate the NASTRAN input data describing the surface. In addition to providing standard NASTRAN control cards along with the usual material and geometric data, COONS also generates PLOAD cards from a mathematical definition of a pressure load and SEQGP cards when resequencing is necessary to reduce the bandwidth of the stiffness matrix.

Specifics: This program is coded in FORTRAN and requires approximately 54K octal words of memory on the CDC 6600 computer.

Availability: For all government and GFSC-sponsored contractor personnel, the program is available through:

Goddard Space Flight Center
Computer Program Library
Greenbelt, Maryland 20771

For all others, the program is available through:

COSMIC
University of Georgia
Athens, Georgia 30601

DATGEN

Source: Meyer, P. M., "Data Generator for the Idealization for Finite Element Structural Analysis of Naval Ship Flat Plated Grillages with Multiple Openings," Report 3807, May 1972, Naval Ship Research and Development Center, Bethesda, Maryland.

Description: This program generates finite element data for flat-plated ship structural grillages with up to five openings and uniform pressure loads normal to the planar surface. The output consists of a tape that can be read directly by NASTRAN to produce a plot of the idealization.

Specifics: The program is coded in FORTRAN and operates on a CDC 6700 computer in conjunction with the SC4020 plotter.

Availability: Inquiries should be made through the Structures Dept., Naval Ship Research and Development Center, Bethesda, Maryland.

EZPLOT

Source: Matz, D. G., "EZPLOT: A NASTRAN Aid for Plotting Undeformed Structures," ENJE-TM-73-20, Aug. 1973, Aeronautical Systems Division, Wright-Patterson Air Force Base, Ohio.

Description: EZPLOT accepts a standard NASTRAN data deck and is used to provide "hard-copy" orthographic projections of the undeformed structure.

Specifics: This program is coded in FORTRAN extended and is used in conjunction with the CALCOMP plotter. The program requires approximately 32K octal words of memory on the CDC 6600 computer.

Availability: Inquiries should be made through Mr. D. G. Matz.

FEDGE

Source: Akyuz, Fevzican A., "Natural Coordinate System: An Automatic Input Data Generation Scheme for a Finite-Element Method," Nuclear Engineering and Design, Vol. II, No. 2, Mar. 1970, pp 197-205.

Description: The program provides for an automatic means of discretization for continuous domains of any form by using the concept of natural coordinate systems, a concept which is introduced in the source. The basic algorithm is based on a suitable classification of the topological properties of complex geometrical configurations in one-, two-, or three-dimensional space.

Specifics: The program was developed for a 32K IBM 7094-7044 direct coupled system, but can be used on other systems having a FORTRAN II compiler and FAP assembler.

Availability: COSMIC
University of Georgia
Athens, Georgia 30601

FEMG

Source: Crisp, R. J., "FEMG - Finite Element Mesh Generation Program," RD/C/N309, May 1969, Central Electricity Generating Board, London, S.E.1.

Description: A program that generates, correlates, and checks data required to define a 2-D triangular mesh for a finite element stress analysis programs. This program is currently compatible with analysis programs FEP, FEX, SAFE, and TESS.

Specifics: The program is coded in FORTRAN and was developed on an IBM 360. It is capable of accepting data from a DMAC coordinate digitizer and uses the CALCOMP plotter to produce hard copies.

Availability: Unknown.

FLOW 2

Source: Kamel, H. A., AME Dept., University of Arizona, Tucson, Arizona.

Description: This program accepts output from the MESH2 program and includes model generation, modification capabilities, prescribed temperatures (potential), source intensities, and plots of potential (temperatures) and velocities using different intensity levels and scaled vectors. Model, loads, and boundary conditions may be punched on paper tape and read again at a later time in any sequence. Local analysis capabilities are included. Results from one step may be transferred internally as boundary conditions in the next step. The process may be repeated indefinitely, and results may be transferred to MEMB or AXIS for subsequent thermal stress computations. Capacity, with a 16K system, is over 350 nodes and 700 elements.

Specifics: Developed for the PDP 15 computer.

Availability: Inquiries should be made through Dr. Hussein A. Kamel.

Fuselage Input Data Generator

Source: Giles, G. L., and Blackburn, C. L., "Procedure for Efficiently Generating, Checking and Displaying NASTRAN Input and Output Data for Analysis of Aerospace Vehicle Structures," presented at the NASTRAN User's Colloquium, Langley Research Center, Sept. 13-15, 1971.

Description: The program was developed as an input data generator for fuselage type structures but is applicable to virtually any type of stiffened shell structure.

Specifics: The program is coded in FORTRAN and requires approximately 70K octal words of memory on the CDC 6600 computer.

Availability: Unknown.

GENDA

Source: Rockwell, R. D., "Computer-Aided Input/Output for Use with Finite Element Structural Analyses," Report 3844, Feb. 1973, Naval Ship Research and Development Center, Bethesda, Maryland.

Description: GENDA allows the user to segment a 3-D shell structure and treat each segment as a 2-D shell structure for purposes of generating the necessary finite element data. The segments are idealized by the methods of Rockwell and Pincus,[2] after which point GENDA maps the pseudostructure back to its actual domain. Finally the bandwidth is reduced using the Cuthill-McKee algorithm.[3] Output consists of plots and the desired geometric and bookkeeping data.

---

[2] Rockwell, R. D., and Pincus, D. S., "Computer-Aided Input/Output for Use with the Finite Element Method of Structural Analysis," NSRDC Report 3402, Aug. 1970.

[3] Cuthill, E., and McKee, J., "Reducing the Bandwidth of Sparse Symmetric Matrices," paper presented at the 1969 National ACM Conference, San Francisco, California, Aug. 1969.

Specifics: This program is coded in FORTRAN, operates on the CDC 6000 computers, and uses the SC4020 plotter.

Availability: Inquiries should be made through the Structures Dept., Naval Ship Research and Development Center, Bethesda, Maryland.

GEOMETRY PROCESSOR

Source: Kitagawa, M., "Geometry Processor," Space Division of Rockwell International, Calif.

Description: The program calculates the Cartesian coordinates from basic values such as the radius and angles for cylindrical or cone type structures. The coordinates and elements are graphically displayed by CRT plots and punched in BCD format. The program also contains a "Table of Equivalents" which relabels the coordinates or element identification.

Specifics: This program is coded in 65% FORTRAN and 35% assembly language (SC4020).

Availability: COSMIC
University of Georgia
Athens, Georgia 30601

GRAMA

Source: Kielb, R., Aeronautical Systems Division, Analysis Branch, Wright-Patterson Air Force Base, Ohio.

Description: GRAMA is a 2-D mesh generator capable of generating grid points and quadrilateral elements. The program also uses a finite difference relaxation scheme to determine the final grid point coordinates.

Specifics: This program is coded in FORTRAN extended and used on the CDC 6600 computer.

Availability: Inquiries should be made through Mr. Robert Kielb.

GRIDXY

Source: Brophy, Jo Anne M., "Computer-Aided Cartridge Case Design Using Finite Element Stress Analysis: The Automation of Finite Element Configuration," Report R-2054, Sept. 1972, Frankford Arsenal, Philadelphia.

Description: This program generates finite element input data for axisymmetric structures modeled using triangular or quadrilateral ring elements. Specifically, it generates a nodal point network and a complete description of the physical and mathematical properties of the structure. The output format is flexible.

Specifics: This program is coded in FORTRAN extended, operates on CDC 6000 computers, and requires approximately 51K octal words memory.

HIDE

Source: Rockwell, R. D., "Computer-Aided Input/Output for Use with Finite Element Structural Analyses," Report 3844, Feb. 1973, Naval Ship Research and Development Center, Bethesda, Maryland.

Description: Given the finite element idealization, HIDE produces a plot tape with and/or without hidden lines. Each element, or part, is plotted only after it is examined regarding whether it can be all, or partly, seen from the observer's viewpoint.

Specifics: This program is coded in FORTRAN, operates on the CDC 6000 computers, and is used in conjunction with the SC4020 plotter.

Availability: Inquiries should be made through the Structures Dept., Naval Ship Research and Development Center, Bethesda, Maryland.

IDLZ

Source: Rockwell, R. D., and Pincus, D. S., "Computer Aided Input/Output for Use with the Finite Element Method of Structural Analysis," Report 3402, Aug. 1970, Naval Ship Research and Development Center, Bethesda, Maryland.

Description: IDLZ uses the Rockwell-Pincus method to generate 2-D finite element meshes.

Specifics: This program is coded in FORTRAN and used on the CDC 6000 computers.

Availability: Inquiries should be made through the Structures Dept., Naval Ship Research and Development Center, Bethesda, Maryland.

MEMB

Source: Kamel, H. A., AME Dept., University of Arizona, Tucson, Arizona.

Description: This program accepts output from MESH2 and features mesh generation and modification, one- and two-dimensional arrays of nodes, elements, concentrated and distributed loads, and triangular and quadrilateral elements as well as load and boundary condition alterations. Model loads and boundary conditions may be punched out on paper tape or written on disk and read again in any sequence. The deflected shape is plotted as a dotted shape superimposed on the original model. For stress representation, symbols of varying intensity are plotted to represent the von Mises criterion. The intensity of the symbol denotes the level of stress. Selected displacement and stress values may also be printed on the teletype or the printer. Local analysis capabilities (successive zooming) is allowed. During each step results from one step are automatically injected in the next solution. Thermal stresses and mixed boundary conditions are incorporated. Present capacity for a 16K system is over 350 nodes and 700 elements.

Specifics: Developed for the PDP 15.

Availability: Inquiries should be made through Dr. Hussein A. Kamel.

MESH

Source: Zienkiewicz, O. C., "Triangular Mesh Generation Program," University of Wales, Scotland.

Description: MESH is a 2-D triangulation scheme used to generate finite element data for a series of analysis programs.

Specifics: This program is coded in FORTRAN and requires approximately 36K octal words of memory on the CDC 6600 computer.

Availability: Inquiries should be made through Dr. O. C. Zienkiewicz.

MESH2

Source: Kamel, H. A., AME Dept., University of Arizona, Tucson, Arizona.

Description: This program includes point-by-point generation, automatic generation of series points on straight lines and circular arcs, automatic generation of two-dimensional node groups bound by straight lines and circular arcs, element generation (triangles and quadrilaterals), automatic generation of one- and two-dimensional arrangements of triangles and quadrilaterals, elimination of single elements, graphical display of complete model or special areas of interest (automatically magnified), automatic bandwidth optimization, optional teletype output (complete or selective), results output in binary or ASCII format on paper punch or dec-tape and disk output to be read at a later point in time to enable subsequent modification, and serves as input generator for later PDP 15 analysis packages or other general purpose analysis programs.

Specifics: Developed for the PDP 15 computer.

Availability: Inquiries should be made through Dr. Hussein A. Kamel.

MESH3

Source: Kamel, H. A., AME Dept., University of Arizona, Tucson, Arizona.

Description: This program features an automatic internal numbering scheme, point-by-point generation for key points only, generation of straight lines and circular arcs in space, line and arc identification by labels, automatic generation of triangular, quadrilateral elements and line stiffeners on surfaces bound by lines and arcs, graphical isometric display of complete or partial model, model rotation around three axes, optional teletype output (complete or selective). Results are output on punched paper tape or on disk in binary or ASCII format. The program also serves as an input generator to later analysis packages on the PDP 15 or other general purpose programs.

Specifics: Developed for the PDP 15 computer.

Availability: Inquiries should be made through Dr. Hussein A. Kamel.

MOVE

Source: Cook, W. L., "Automated Input Data Preparation for

NASTRAN," GSC-11039, April 1969, Goddard Space Flight Center, Greenbelt, Maryland.

Description: For a structural model consisting of two or more identical segments, this program generates the complete structure given the grid points and elements of a single segment in a NASTRAN format.

Specifics: This program is coded in PL1.

Availability: For all government and GFSC-sponsored contractor personnel, the program is available through:

Goddard Space Flight Center
Computer Program Library
Greenbelt, Maryland 20771

For all others, the program is available through:

COSMIC
University of Georgia
Athens, Georgia 30601

NARFEM

Source: Furuike, T., and Yahata, S., "Vehicle Structural Analysis - Pre and Post Processor Data Handling," presented at the NASTRAN User's Colloquium, Langley Research Center, Sept. 13-15, 1971.

Description: A general preprocessor that is used to generate, visually check, and modify a finite element model. The program has a variable output format that is compatible with several structural analysis programs.

Specifics: Unknown.

Availability: Unknown.

NARSLAGS

Source: Furuike, T., and Yahata, S., "Vehicle Structural Analysis - Pre and Post Processor Data Handling," presented at the NASTRAN User's Colloquium, Langley Research Center, Sept. 13-15, 1971.

Description: This program generates loads for a stress model from input data of selected points. The program also generates loads due to structural weight. Both distributed mass and concentrated mass can be handled to formulate the mass matrix.

Specifics: Unknown.

Availability: Unknown.

PLOT 2D

Source: "A Computer Program to Display the Dynamic Stress Analysis of Structures," a program developed by the Visual Computing Corporation, Culver City, California, for the Naval Civil Engineering Laboratory, Port Hueneme, California.

Description: This program supercedes the computer program DYSAND and is used to process a display data file and produce pictorial displays. The displays graphically represent the physical environment of the structure and provide a means of visually analyzing its dynamic behavior.

Specifics: The program is coded in FORTRAN and is used in conjunction with the SC4020 plotter.

Availability: Inquiries can be made through the Structures Division, Naval Civil Engineering Laboratory, Port Hueneme, California.

PLOT 3D

Source: Canright, R., and Swigert, P., "PLOT 3D - A Package of FORTRAN Subprograms to Draw Three-Dimensional Surfaces," NASA TM X-1598, June 1958, Lewis Research Center, Cleveland, Ohio.

Description: A package of FORTRAN subprograms to draw 3-D surfaces of the form Z = f (x, y). The function f and the bounding values of x and y are the input to PLOT 3D. The surface defined may be drawn after arbitrary rotations and the output is by off-line incremental plotter or on-line microfilm recorder.

Specifics: PLOT 3D is implemented on both the IBM 7044/7094 Direct Couple System - California Products Digital Incremental Plotter, controller model 780, printer model 765; and the IBM 360/67 - Control Data Corporation microfilm system model 280. The subprograms are coded in FORTRAN and readily adaptable to other hardware.

Availability: The subroutines are available in Appendix A of the source.

PLOTS

Source: Zimmerman, K. L., "PLOTS - The Plotting Subroutine Incorporated in the BRLESC Finite Computer Program," AD 749 786, Aug. 1972, U.S. Army, Aberdeen Research and Development Center, Aberdeen Proving Ground, Maryland.

Description: PLOTS is a subroutine that calls standard BRLESC plotting subroutines which plot finite element grids, deformed and undeformed, and contours of any component of stress, strain, or temperature. Finite element grids, whether deformed or undeformed, are plotted one element at a time. In deformed finite element grids, the actual nodal point displacements are magnified and then added to the original nodal point coordinates. The resultant deformed grid is plotted over the outline of the undeformed solid. The option to plot an enlarged portion of a grid is also available.

Specifics: This subroutine is coded in FORTRAN and calls standard BRLESC plotting subroutines.

Availability: A listing of the subroutine is available from Appendix B of the source.

SAGPAC

Source: Serpanos, J. E., "SAGPAC - Structural Analysis Graphics Package: A Fortran IV Computer Program," Technical Note 4062-47, June 1970, Weapons Development Department, Naval Weapons Center, China Lake, California.

Description: SAGPAC is a general postprocessor that will plot (1) basic and deflected structures, (2) contour lines of selected stresses and strains to include zooming capabilities on critical areas of the structure, and (3) stresses or strains versus radial or axial distances.

Specifics: This program is coded in FORTRAN and used in conjunction with the SC4060 plotter.

Availability: Unknown.

SAIL

Source: Boeing Document D2-125179-8, "The ASTRA System," May 1972.

Description: A structural input data language that uses FORTRAN-like statements to generate input for ASTRA (Advanced Structural Analyzer). The bulk of SAIL's capability is contained in the following three features: defining and manipulating substructures, computational loops, and external data generators (similar to a FORTRAN subroutine).

Specifics: The program is coded in FORTRAN and is currently being used on the IBM 370-165 at the Boeing Company. It can be used in conjunction with the SC4020 and CALCOMP plotters and has been successfully adapted to other finite element programs, notably NASTRAN and SAMECS.

Availability: Inquiries can be made through Boeing Computer Services, Inc.

SHELBY

Source: Cook, W. L., "Automated Input Data Preparation for NASTRAN," GSC-11039, Goddard Space Flight Center, Greenbelt, Maryland, April 1969.

Description: The program generates NASTRAN input data for shells described by translation of a plane curve along an arbitrary axis in space. In addition to providing standard NASTRAN control cards along with the usual material and geometric data, SHELBY also generates PLOAD cards from a mathematical definition of a pressure load and SEQGP cards when resequencing is necessary to reduce the bandwidth of the stiffness matrix.

Specifics: This program is coded in FORTRAN and requires approximately 71K octal words memory on the CDC 6600 computer.

Availability: For all government and GFSC-sponsored contractor personnel, the program is available through:

Goddard Space Flight Center
Computer Program Library
Greenbelt, Maryland 20771

For all others, the program is available through:

COSMIC
University of Georgia
Athens, Georgia 30601

SHELL

Source: Kielb, R., Aeronautical Systems Division, Analysis Branch, Wright-Patterson Air Force Base, Ohio.

Description: SHELL is an axisymmetric finite element mesh generator that is capable of generating quadrilateral or triangular plate elements and grid cards for any angular portion of an axisymmetric structure. The output of this program is compatible with NASTRAN.

Specifics: This program is coded in FORTRAN extended and used on the CDC 6600.

Availability: Inquiries should be made through Mr. Robert Kielb.

Wing and Empennage Input Data Generator

Source: Giles, G. L., and Blackburn, C. L., "Procedure for Efficiently Generating, Checking and Displaying NASTRAN Input and Output Data for Analysis of Aerospace Vehicle Structures," presented at the NASTRAN User's Colloquium, Langley Research Center, Sept. 13-15, 1971.

Description: This program is used to generate finite element data for wing and empennage structures defined within a given aerodynamic shape.

Specifics: The program is coded in FORTRAN and requires approximately 70K octal words of memory on the CDC 6600 computer.

Source: McCue, G. A., et al., "FORTRAN IV Stereographic Function Representation and Contouring Program," SID 65-1182, Sept. 1965, North American Aviation Inc.

Description: A package of FORTRAN subroutines that will generate stereographic drawings, stereographic contour maps, and perspective projections.

Specifics: This package of subroutines use the SC4020 as an output device.

Availability: A listing of subroutines is available in Appendix II of the source.

Source: McCue, G. A., and Duprie, H. J., "Improved FORTRAN IV Function Contouring Program," SID 65-672, April 1965, North American Aviation Inc.

Description: A numerical technique for generating contour maps having two or three independent variables. The algorithm has been programmed as a group of subroutines which generate output for subsequent plotting.

Specifics: This program is coded in FORTRAN and is used in conjunction with the SC4020 plotter.

Availability: The subroutines are available in Appendix 2 of the source.

Source: Dailey, G., "Computer Programs for Plotting Finite Element Patterns of Two- and Three-Dimensional Structures," TG 1140, Oct. 1970, John Hopkins University, Applied Physics Lab., Maryland.

Description: Two computer programs for plotting finite element patterns of two- and three-dimensional structures are described. One program draws two-dimensional structures to scale, and the other draws perspective views of three-dimensional structures. Both programs calculate the degrees of freedom and semibandwidth of the stiffness matrix.

Specifics: The programs are written in PL1 for the IBM 360-91 computer. The program should run on any IBM 360 computer with more than 200K bytes main storage and seven or nine track tape drives and a CALCOMP plotter with the proper software support.

Availability: Unknown.

Source: Smith, R. R., "An Algorithm for Plotting Contours in Arbitrary Planar Regions," AD 731 987, Oct. 1971, Naval Undersea Research and Development Center, San Diego, California.

Description: An algorithm for generating contour plots of a function defined on an arbitrary planar region. The three main features of the algorithm are: (1) any planar region which can be enclosed by a single boundary curve that does not cross itself can be treated; (2) function values can be specified for any arbitrary distribution of points in the region; (3) the region specified by the data points is subdivided into a collection of adjacent triangles whose union is the entire region. This technique is useful for problems other than contour plotting, in particular for finite element calculations.

Specifics: This program is coded in FORTRAN.

Availability: A program listing is available from Appendix B of the source.

APPENDIX B

Schematic of the General Pre- and Postprocessing System

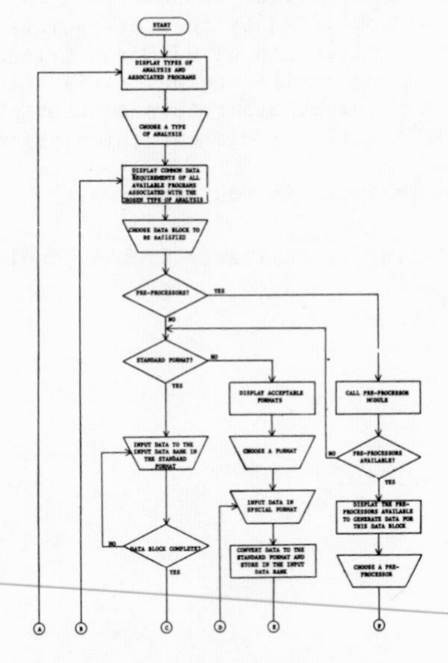

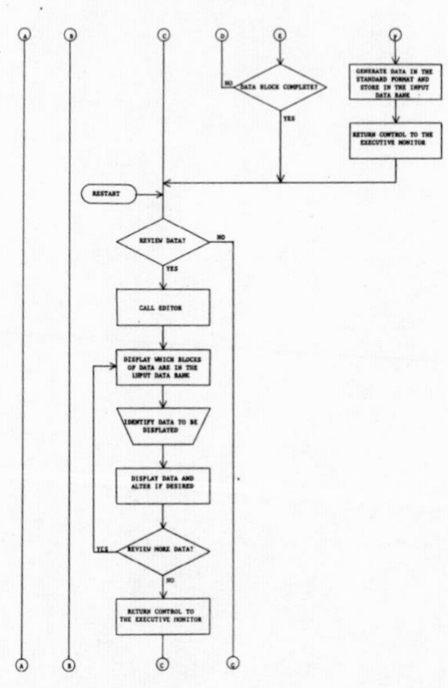

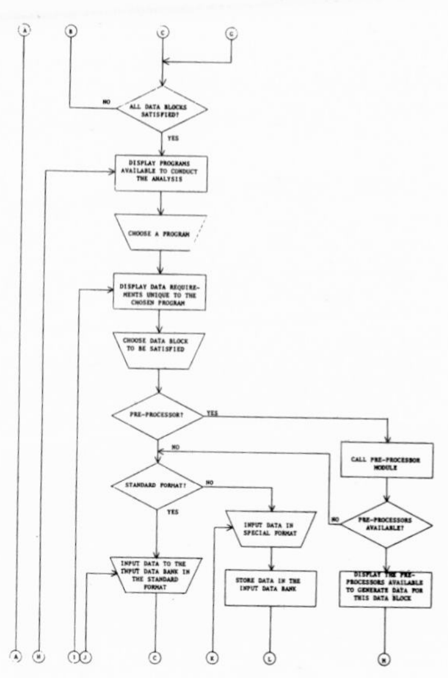
ALL DATA BLOCKS SATISFIED?
NO
YES
DISPLAY PROGRAMS AVAILABLE TO CONDUCT THE ANALYSIS
CHOOSE A PROGRAM
DISPLAY DATA REQUIREMENTS UNIQUE TO THE CHOSEN PROGRAM
CHOOSE DATA BLOCK TO BE SATISFIED
PRE-PROCESSOR?
YES
NO
CALL PRE-PROCESSOR MODULE
STANDARD FORMAT?
NO
YES
INPUT DATA IN SPECIAL FORMAT
PRE-PROCESSORS AVAILABLE?
NO
INPUT DATA TO THE INPUT DATA BANK IN THE STANDARD FORMAT
STORE DATA IN THE INPUT DATA BANK
DISPLAY THE PRE-PROCESSORS AVAILABLE TO GENERATE DATA FOR THIS DATA BLOCK

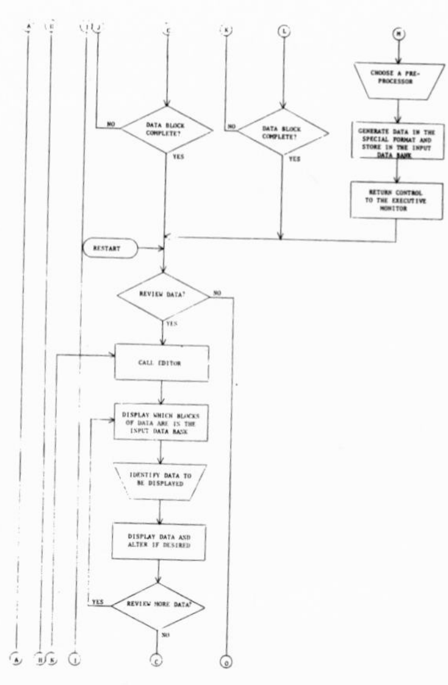
CHOOSE A PRE-PROCESSOR
DATA BLOCK COMPLETE?
NO
YES
DATA BLOCK COMPLETE?
NO
YES
GENERATE DATA IN THE SPECIAL FORMAT AND STORE IN THE INPUT DATA BANK
RETURN CONTROL TO THE EXECUTIVE MONITOR
RESTART
REVIEW DATA?
NO
YES
CALL EDITOR
DISPLAY WHICH BLOCKS OF DATA ARE IN THE INPUT DATA BANK
IDENTIFY DATA TO BE DISPLAYED
DISPLAY DATA AND ALTER IF DESIRED
REVIEW MORE DATA?
YES
NO

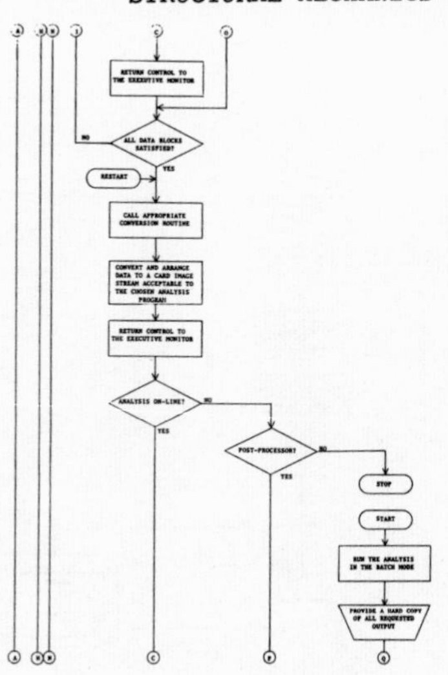
A
H
H
I
C
G
RETURN CONTROL TO THE EXECUTIVE MONITOR
NO
ALL DATA BLOCKS SATISFIED?
YES
RESTART
CALL APPROPRIATE CONVERSION ROUTINE
CONVERT AND ARRANGE DATA TO A CARD IMAGE STREAM ACCEPTABLE TO THE CHOSEN ANALYSIS PROGRAM
RETURN CONTROL TO THE EXECUTIVE MONITOR
ANALYSIS ON-LINE?
NO
YES
POST-PROCESSOR?
NO
YES
STOP
START
RUN THE ANALYSIS IN THE BATCH MODE
PROVIDE A HARD COPY OF ALL REQUESTED OUTPUT
A
H
H
C
F
Q

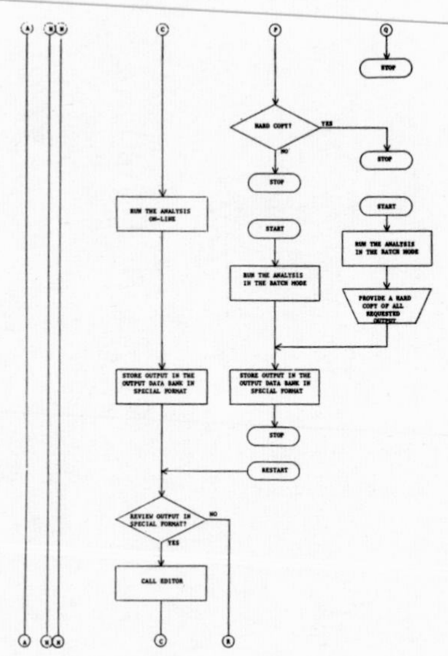
A
H
H
C
F
Q
STOP
HARD COPY?
YES
NO
STOP
STOP
START
RUN THE ANALYSIS ON-LINE
START
RUN THE ANALYSIS IN THE BATCH MODE
RUN THE ANALYSIS IN THE BATCH MODE
PROVIDE A HARD COPY OF ALL REQUESTED OUTPUT
STORE OUTPUT IN THE OUTPUT DATA BANK IN SPECIAL FORMAT
STORE OUTPUT IN THE OUTPUT DATA BANK IN SPECIAL FORMAT
STOP
RESTART
REVIEW OUTPUT IN SPECIAL FORMAT?
NO
YES
CALL EDITOR
A
H
H
C
B

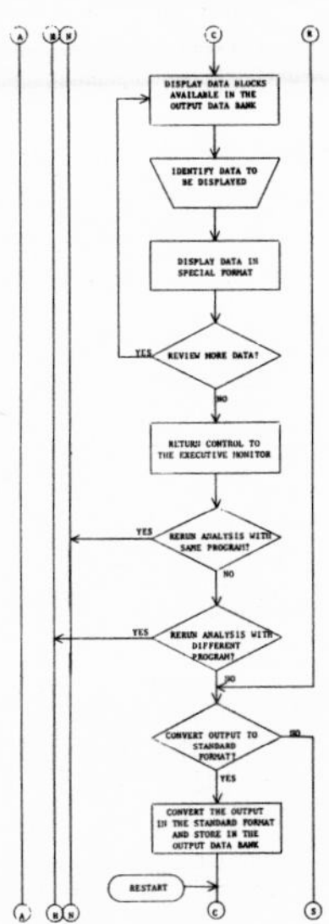
A
N
N
C
R
DISPLAY DATA BLOCKS AVAILABLE IN THE OUTPUT DATA BANK
IDENTIFY DATA TO BE DISPLAYED
DISPLAY DATA IN SPECIAL FORMAT
REVIEW MORE DATA?
YES
NO
RETURN CONTROL TO THE EXECUTIVE MONITOR
RERUN ANALYSIS WITH SAME PROGRAM?
YES
NO
RERUN ANALYSIS WITH DIFFERENT PROGRAM?
YES
NO
CONVERT OUTPUT TO STANDARD FORMAT?
NO
YES
CONVERT THE OUTPUT IN THE STANDARD FORMAT AND STORE IN THE OUTPUT DATA BANK
RESTART
A
N
N
C
S

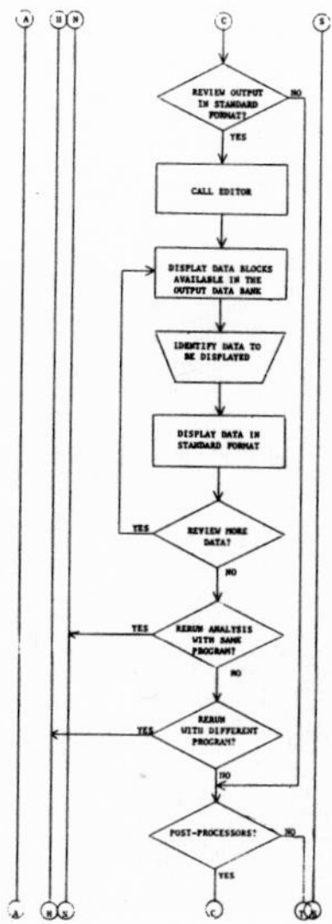
A
N
C
S
REVIEW OUTPUT IN STANDARD FORMAT?
NO
YES
CALL EDITOR
DISPLAY DATA BLOCKS AVAILABLE IN THE OUTPUT DATA BANK
IDENTIFY DATA TO BE DISPLAYED
DISPLAY DATA IN STANDARD FORMAT
REVIEW MORE DATA?
YES
NO
RERUN ANALYSIS WITH SAME PROGRAM?
YES
NO
RERUN WITH DIFFERENT PROGRAM?
YES
NO
POST-PROCESSORS?
NO
YES
A
N
S
C

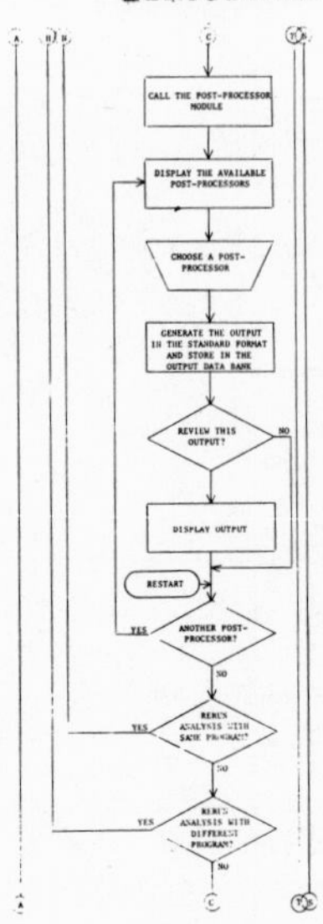
A
C
T
S
CALL THE POST-PROCESSOR MODULE
DISPLAY THE AVAILABLE POST-PROCESSORS
CHOOSE A POST-PROCESSOR
GENERATE THE OUTPUT IN THE STANDARD FORMAT AND STORE IN THE OUTPUT DATA BANK
REVIEW THIS OUTPUT?
NO
DISPLAY OUTPUT
RESTART
YES
ANOTHER POST-PROCESSOR?
NO
YES
YES
NO
A
C
T
S

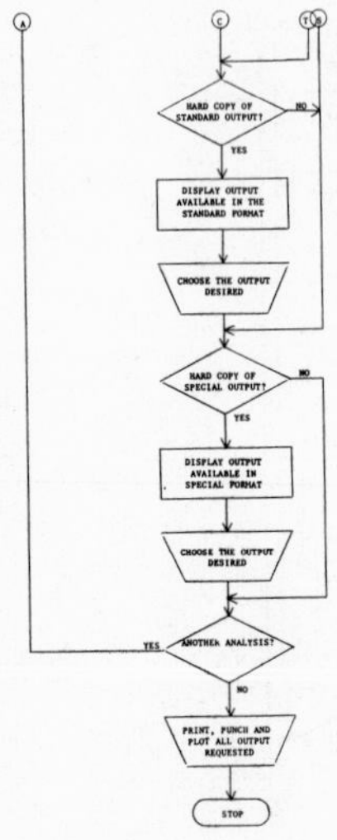
A
C
T
S
HARD COPY OF STANDARD OUTPUT?
NO
YES
DISPLAY OUTPUT AVAILABLE IN THE STANDARD FORMAT
CHOOSE THE OUTPUT DESIRED
HARD COPY OF SPECIAL OUTPUT?
NO
YES
DISPLAY OUTPUT AVAILABLE IN SPECIAL FORMAT
CHOOSE THE OUTPUT DESIRED
YES
ANOTHER ANALYSIS?
NO
PRINT, PUNCH AND PLOT ALL OUTPUT REQUESTED
STOP

DESIGN OF PRE- AND POSTPROCESSORS

Ervin D. Herness and James L. Tocher
Boeing Computer Services, Inc.
Seattle, Washington

ABSTRACT

Pre- and postprocessors are rapidly becoming the key components of all general purpose structural analysis programs. They can perform a wide variety of data generation, data checking, and data display tasks which, if done by hand, are both tedious and time consuming. Pre- and postprocessor technology is being enhanced by new computer operating systems, improved alphanumeric terminals, low cost interactive graphics terminals, and minicomputers. Most present day pre- and postprocessors, however, are not well designed. They are seldom readily portable, are too dependent upon a particular operating system, and are designed for a particular display device. In this paper pre- and postprocessor design guidelines are given which provide for portability, flexibility, and relative independence from operating systems, graphics devices, and even the central structural analysis program.

NOMENCLATURE

cps	=	Characters per second
CRT	=	Cathode ray tube
I/O	=	Input/output
PPPs	=	Pre- and postprocessors
RJE	=	Remote job entry
TSO	=	IBM's Time Share Option operating system (or similar systems on other computers)

INTRODUCTION

Pre- and postprocessors (PPPs) are rapidly becoming required features of all general purpose structural analysis programs. With the unit cost of computing going down and the cost of labor going up, PPPs provide a way to reduce large labor costs significantly. In 1970 Tocher and Felippa [1] estimated that typically 80 percent of all finite element development effort went into the central analysis program, while only 20 percent went into data generators and graphics displays. This is contrasted with a production environment in which the labor costs for most analyses are two to ten times the computer costs. Since PPPs are generally designed to save labor, it should be obvious that effort put into building good PPPs can pay off handsomely in reducing future structural analysis costs.

ADVANTAGES OF PRE- AND POSTPROCESSORS

PPPs offer substantial advantages to users of large structural analysis systems. They obviate most of the drudgery ordinarily associated with data preparation, debugging, and output interpretation, thereby reducing structural analysis flow time by impressive amounts. In addition to automation of time-consuming but computationally trivial data generation and reduction tasks, PPPs can also automate the bookkeeping and transcription processes associated with plotting, editing, error correction, and interfacing with other programs.

PPPs have the additional advantage of being cheap to develop and inexpensive to use (unlike the central analysis programs which utilize them). Because of their mathematical and logical simplicity, they can be easily and rapidly coded and checked out, and they execute rapidly and inexpensively.

APPLICATIONS OF PRE- AND POSTPROCESSORS TO STRUCTURAL COMPUTING

PPPs are used for a variety of tasks surrounding the central structural analysis program. Other papers in this volume cover this area in depth. However, a summary of common applications will be given in order to illustrate the breadth of PPP development and use.

Preprocessors

Preprocessors are programs which are used to prepare, check, and alter data before the central structural analysis is performed. Preprocessors have been written to:

1. Generate nodal point coordinates and boundary conditions along lines, meshes, and curved surfaces
2. Generate element connectivity for linear, planar, and solid elements in one-, two-, and three-dimensional mesh patterns
3. Generate nodal loads and element pressures for distributed loadings on arbitrarily oriented surfaces
4. Check data for syntax and reasonableness
5. Reorder nodal point numbers to reduce the bandwidth of the equations in order to reduce solution time and storage
6. Graphically display the configuration and element properties of the structure specified by the input data
7. Prepare connectivity data for the coupling of two or more substructures
8. Transform input data prepared for one structural analysis program into a format suitable for some other analysis program

Postprocessors

Postprocessors are programs which are used to manipulate and display results computed by the central structural analysis program. Postprocessors have been written to:

1. Plot contours of stresses, strains, bending moments, and deflected shapes

2. Compute principal stresses
3. Plot the deflected shape of a structure
4. Plot the natural vibration mode shapes
5. Generate time history plots of deflections and stresses
6. Print stress values adjacent to members on plots of structural idealizations composed of rods, beams, stringers, and plates
7. Prepare plots and tables suitable for inclusion in formal reports summarizing an analysis
8. Combine solution cases to produce results for various loading combinations
9. Extract maximum stress values and stress exceedances
10. Extract data from output files for use by subsequent processing programs
11. Convert output data into a format suitable for another analysis program

These lists are not intended to be exhaustive; rather, they are designed to show the variety of tasks for which PPPs are used.

NEW SUPPORTING HARDWARE

At least five types of peripheral hardware can be exploited effectively in PPP development and use. These devices, some of which are relatively new while others are considered old war horses, are

1. Quiet, low-speed alphanumeric terminals
2. Storage tube vector graphics scopes
3. Refresh tube vector graphics scopes
4. Plotters
5. Minicomputers

The most significant features of the new alphanumeric terminals are that they are quiet, they can operate at 30 cps instead of 10 cps, and many of them are portable enough to take home for evening and weekend work. Typical rental costs range from $80.00 to $150.00 per month.

The new storage tube terminals (e.g., the Tektronix 4010) are similar to the alphanumeric terminals in that they operate by ordinary telephone dial-up to the computer. However, they are also capable of drawing lines on a screen. Although normally driven at 30 cps by the computer, they are capable of being driven over 30 times as fast (up to 960 cps) and can fill the screen with line drawings and graphs at a dizzying rate. The line resolution is excellent, and one can rent such a graphics terminal for about $250.00 per month. Two disadvantages are that lines and characters cannot be selectively removed from a display on a storage tube and, at the 30 cps rate, the picture drawing speed may seem somewhat slow to persons used to refresh-type screens.

The refresh CRT scopes such as the IBM 2250 and the CDC 243 are very capable but also an order of magnitude more expensive (several thousand dollars per month rental) than the storage tube scopes. With the refresh capability the display screen can detect commands from a light-pen, and numbers or lines can be changed individually without redrawing the complete display. A disadvantage of a refresh scope for displaying large structures is that the number of lines and characters that can be drawn is limited by the size of the refresh buffer for the scope. Also, a considerable cost in telephone lines and computer resources is incurred by this class of scope.

Plotters have been around for many years and are, of course, a required item of hardware for many PPPs. Plotters such as the CALCOMP and SC4020 are most commonly used, with large high-accuracy, bed-type plotters such as the Gerber getting much less usage.

Minicomputers such as DEC PDP, Hewlett-Packard, NOVA, Varian, Honeywell, IBM System 3, etc. are rapidly becoming widely available and will have a significant impact on engineering analysis. They can be used as an interface with a large computer to handle a large number of alphanumeric and graphics terminals. They can also stand alone in either a batch or a mini-time-sharing mode handling many significant analyses and are especially effective for problems requiring high central processor activity and relatively low I/O. The cost of arithmetic operations on a "mini" is frequently much lower than on a large scale computer, even considering the fact that the mini's short word length requires that many operations be done in double precision.

Many of the tasks done by PPPs can be handled by a minicomputer. A major design factor which must be considered in developing PPPs on a mini is the structural data base. Storage limitations of the mini and the difficulties of transfer of the data base to and from the large-scale computer (possibly by magnetic tape) must be investigated before money is spent for developing PPPs on a mini. Since experience with this kind of application on a mini is very limited, one must proceed with care and not be overcome by the enthusiasm of a salesman or another user whose experiences have been quite successful in a different type of application.

LIVING WITH NEW OPERATING SYSTEMS AND LIKING IT

In the past few years there have been significant changes in operating systems which have radically altered the engineer's computing environment. The new operating systems can support remote job entry (RJE), batch job submission, and interactive computing. Systems such as TSO on the IBM 370 and KRONOS on the CDC 6600 present powerful computing options that are especially useful for pre- and postprocessing work. This evolution in computing environment is shown in Fig. 1.

Batch computing requires the repeated correction and submission of card decks. There is always the risk that the deck will be lost or dropped. A day (or more) after submission the deck and listing will be returned. After his wait, the user frequently finds that the job has failed, producing little or no useful data. The batch mode minimizes computer expense, but the cost of lost time is often ignored because that substantial expense is not readily visible.

The next step above batch is RJE, which is provided by teleprocessing. A high-speed terminal remote from the central computer is used to read cards and print the output. RJE eliminates most mail/messenger service problems and effectively puts the computer closer to the user. This is a desirable arrangement for an organization which has a large enough computing volume to afford such a terminal or for a remote service bureau office which services many local customers.

IN THE BEGINNING WAS BATCH:

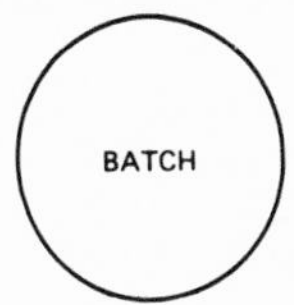

THEN EMERGED TELEPROCESSING . . . .

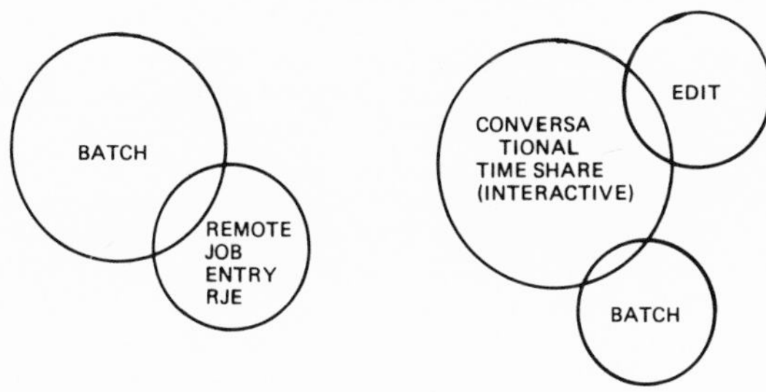

AND THEN EVOLVED TSO

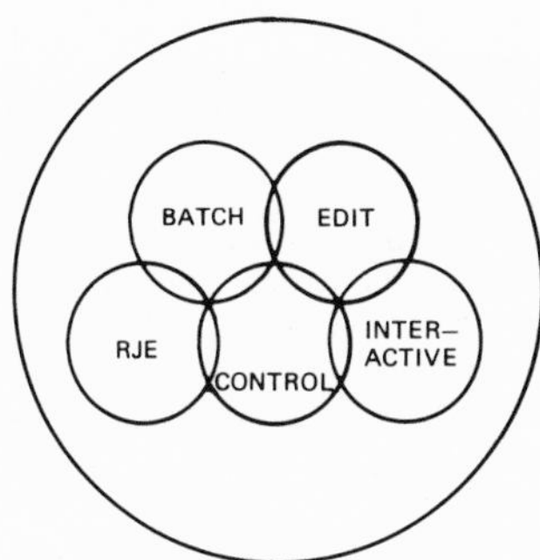

Fig. 1 Evolution of the computing environment

Teleprocessing also brought with it interactive computing. The user dials a number on his telephone and couples his teletype (or one of the newer quiet alphanumeric terminals) directly to the computer. The computer's response to editing commands is almost instantaneous (if the computer is not servicing too many other users at the same time), and small computational tasks are handled rapidly. The major drawback of interactive systems for large-scale structural analysis is simply that the system cannot tolerate the degradation caused by the execution of a large problem. In addition, the user waiting time and the cost of such an execution in the interactive mode would be prohibitive. These drawbacks are sometimes circumvented by deferring execution to an overnight batch operation.

The operating systems which have now emerged combine the best of the three worlds of batch, RJE, and time sharing. These multiple function operating systems are of the IBM TSO type. For the sake of simplicity, all such systems (regardless of computer) will be designated as TSO throughout the remainder of this paper. The major features of TSO are job control and editing from a low-speed terminal from a desk in the engineer's own work area, interactive computing, remote job entry, and batch computing.

The job control and editing capability provide for data and control card editing, launching of a job into the batch stream, spying on a job to ascertain its status in the queue, and selective printing of the output.

The computing environment provided by TSO is ideal for using PPPs. If the I/O and resource requirements for a PPP are small, it can run efficiently in the interactive environment of TSO. Furthermore, TSO can effectively support large PPPs through fast turnaround batch. After the preprocessor has generated and certified the input data (the data base for the structural analysis), the large-scale structural analysis program may be launched as an overnight batch job. The resulting output file (the data base for the postprocessors) can then be saved as a permanent file for subsequent interrogation by the postprocessors in the interactive mode.

DESIGN WEAKNESSES IN PRESENT DAY PPPs

Most PPPs in use today share one glaring weakness: they are designed for one type of computer and one graphics device and will interface with only one structural analysis program. This inflexibility means that PPPs cannot be readily shared among users. This inability to share PPPs affects an organization deeply. The organization is hobbled by PPPs with computer-dependent code, dependence on a specific graphics device, operating system dependence, and a lack of a well-organized data base. A near-crisis situation develops every time there is a minor change in the operating system, a replacement of one graphics device with another, or a change in the central structural analysis program. The addition of a new structural analysis program to the repertoire is usually met in one of two ways: either the new program is ignored (partly because its PPPs do not fit the system), or new versions of the established PPPs are developed (at no small cost) to match the new program. It is no wonder that there is little exchange of PPPs; they simply are too device dependent to be portable.

There are, of course, exceptions to this narrow design syndrome. NASTRAN, although certainly not a PPP, is able to operate on IBM, CDC, and Univac computers and drive a wide variety of plotters. The authors consider it a minor computing wonder that the 6600 and 360 versions of NASTRAN can produce plots on either the CALCOMP or SC4020 plotters without requiring modifications to installation system routines.

Two experiences at Boeing Computer Services (BCS) have demonstrated the value of PPP portability. The SAIL (Structural Analysis Input Language) program [2] has been modified to generate data for three different analysis programs (ASTRA [3], SAMECS [4], and NASTRAN) running on either the CDC 6600 or the IBM 360/370's. User response has been very favorable since engineers are no longer restricted to a particular structural program with whose input they are familiar. With one common input generation language the user can prepare data which is automatically transcribed to the correct format for the designated analysis program.

Another example of portability is the BCS Interactive Graphics System [5]. This library of interactive graphics routines has been used to drive the following devices:

1. IBM 2250 scope connected to an IBM 360/65 running under O/S
2. CDC 243 scope connected to a CDC 6600 running under KRONOS
3. Tektronix 4010 scope dialed into a CDC 6600 running under KRONOS
4. CALCOMP and SC4020 plotters utilizing hardcopy files produced by programs using any of the above three interactive graphics devices

To the user, the library appears to be independent of graphics device, operating system, and computer. User acceptance has been enthusiastic, and proper design has minimized conversion problems.

PRE- AND POSTPROCESSOR DESIGN

In this section some design guidelines are given for portable and flexible PPPs. The relation of a set of PPPs to the data base and to the main structural analysis program is shown in Fig. 2. At the top of the figure a series of preprocessors is shown which are used in an interactive mode to generate and validate the input data base. Once the input data is generated, corrected, and verified, the large-scale (batch) structural analysis package is run to produce the output data base. The postprocessors, shown at the bottom of the figure, are operated in the interactive mode, accessing the output data base to perform additional calculations to help the analyst interpret his results. The TSO environment permits the analyst to be very much in the loop with the PPPs. He can make immediate decisions as to the validity of the data and correct errors as they are identified during preprocessing. Furthermore, in the postprocessing phase, the analyst can quickly view and analyze his results and take further action accordingly (request plots, printout, etc.). The display scope is very much a part of this environment. In developing PPPs along these lines, two factors should be considered: the data base and PPP portability.

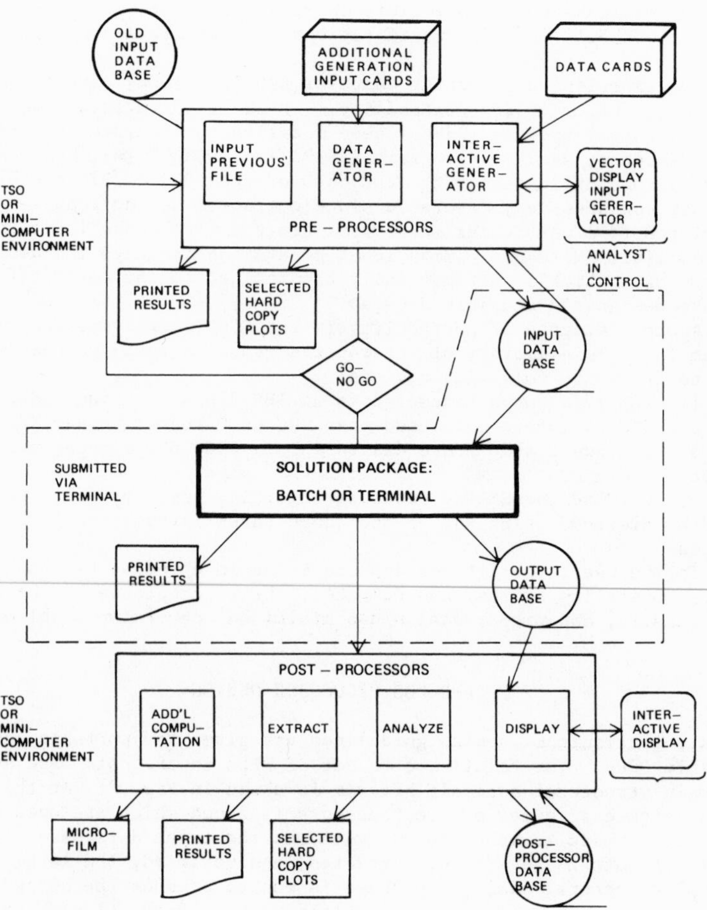

Fig. 2 Pre- and postprocessor design

The Data Base

The design of the data base is the key to the PPP system. The data base should be self-contained, easily read and updated from many sources, and easily checked. The user should have the option of preserving a copy of the updated data base whenever he so chooses. These concepts are developed next.

A data base should be self-contained and logically complete. All pointers which describe the length of each array are contained within the data base. Key information about the type of data (e.g., integer, real, double precision, or alphanumeric) should be included. The first record should describe the content and format of the rest of the data base. Data should not be stored in program data statements or elsewhere in the code. Because many PPPs may contribute to the data base and additional ones may be added at any time, the data structure should be designed to be easily read and updated. Updating implies not only data replacement but also expansion of existing data arrays or addition of new ones (with attendant changes in the first, or pointer, record).

Array sizes should be variable in length. Extremely long arrays should be broken up into a number of records so that they can be read into a small program as well as a large program. Whenever there is a possibility of the data base's being accessed and updated by two programs simultaneously, status words should be maintained which can lock out data base access until updates have been completed. In other words, it should be permissible for two processors to simultaneously read the data base, but if one processor is writing into the data base, other processors must be locked out.

Checking the accuracy and consistency of the data base is an important function of PPPs. A check to ensure that all numbers are within an expected range is often a simple way to identify input errors. Areas of plates, lengths of beams, and other physical quantities can all be checked to be real and positive. A processor which performs such checks frequently uncovers numerous errors.

The ability to save (preserve) the data at a given point in time must be a part of the data base design. It should be possible to initialize the current data base from one or more previous data bases. Such action should generate appropriate record information to show the data base history.

Providing for Portability

A portable program is one which can run on several different computers with only minor modifications. PPPs should not only be computer independent but should also be relatively independent of graphics devices. The design rules for portability of PPPs are:

1. Keep it simple
2. Use higher-level languages
3. Stay away from special compiler features
4. Isolate I/O, machine language, and word/byte dependent sections as separate subroutines
5. Use existing mathematical utility and graphics libraries as basic building blocks

These concepts are developed next.

If there is a general rule that can be applied to PPP design, it would be "keep it simple." The immediate gains that arise from complicated logic and coding very seldom pay off in the long run. The extra time needed to check out complicated logic often results in development delays. When the delay becomes too long, the requirements for developing the processor may go away because the engineer has found another way to solve his problem.

The use of a higher-level language such as FORTRAN IV is essential for PPP portability. Because there are variations in FORTRAN on different computers, special features of a compiler should be avoided. For example, the UNIVAC 1108 compiler allows the DO loop index to start at zero (e.g., DO 1 I=0,N) or to be decremented (e.g., DO 1 I=1,N,-M). This sort of FORTRAN code is not permitted by most compilers. Nonstandard returns are not available on all compilers. A not-so-obvious part of program design that can become very compiler dependent is the length of a subroutine. A 500 statement subroutine is too large for some compilers whereas a 4000 statement subroutine may not be too large for another. As a rule, most subroutines should be less than 500 statements long.

Machine word length varies significantly from machine to machine, and thus the way words are used plays an important part in portability. On the UNIVAC 1108 and IBM 7094 the single precision FORTRAN word is 36 bits long and each word may contain 6 characters. On the IBM 360/370 the basic FORTRAN word is 32 bits with 4 characters per word. On the CDC 6600 the FORTRAN word is 60 bits long with 10 characters per word. In all of the above the floating point word has the sign bit and the exponent on the left end of the word. On the FACOM 230-60 computer (made by FUJITSU Ltd., Tokyo) the FORTRAN word is 36 bits with 4 characters per word. However, the exponent of the floating point word is on the right end of the word. This means that the exponent for a double precision word is in the second word instead of the first. Obviously any program that directly accesses the most significant part of a word (without using the SINGLE function of the compiler) is not easily portable to the FACOM computer. PPPs designed for portability should have special functions isolated as separate subroutines. All of the I/O should be handled by a small number of subroutines. The fact that I/O may be written on several physical devices means that it is inherently machine dependent. Even FORTRAN I/O is often limited and device dependent. What works on a disk may not work on a tape. The reading and writing of records in nonsequential order on a previously written sequential file will cause many FORTRAN I/O packages to fail.

If possible, variable length records should be blocked into fixed length buffers and written or read as fixed length records. If variable length records are to be written, it is more efficient to write them as dimensioned arrays. For example, the statement

```
WRITE (6) (A(I), I=1,N)
```

should be replaced by the subroutine call

```
CALL WRT(6,A,N)
```

where subroutine WRT is coded as follows:

```
SUBROUTINE WRT(IU,A,N)
DIMENSION A(N)
WRITE(IU) A
RETURN
END
```

Of course the use of machine language should be isolated into small, lowest-level subroutines and used sparingly. FORTRAN equivalent routines should also be coded as backup. Any byte or bit dependent logic should also be so isolated.

The use of well-tested mathematical, utility, and graphics subroutine libraries as basic building modules of PPPs enables the development of portable programs quickly. Figure 3 shows a typical PPP using such libraries. Notice that the actual code the developer must write is a small part of the total. Well-tested libraries containing frequently used mathematical and utility subroutines can save the PPP designer many days of development and checkout. A graphics subroutine library which appears to be device independent to the user and yet is able to interface with a variety of graphics devices and operating systems is highly desirable. Careful separation of the basic functions can make conversion to another system straightforward.

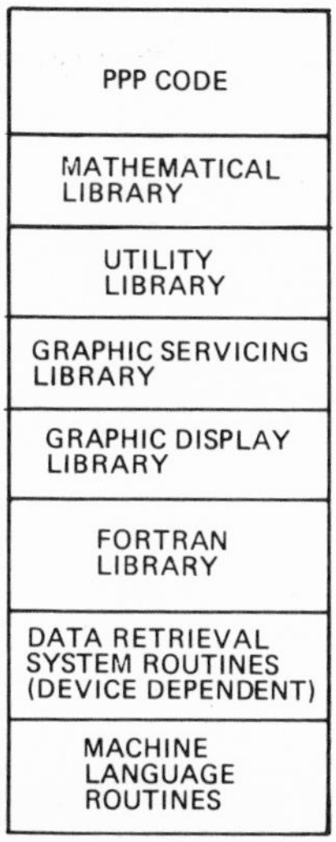

PPP CODE
MATHEMATICAL LIBRARY
UTILITY LIBRARY
GRAPHIC SERVICING LIBRARY
GRAPHIC DISPLAY LIBRARY
FORTRAN LIBRARY
DATA RETRIEVAL SYSTEM ROUTINES (DEVICE DEPENDENT)
MACHINE LANGUAGE ROUTINES

Fig. 3 Basic modules of a PPP

CONCLUSIONS

The variety of PPP applications is growing rapidly, and their production usage is becoming widespread. New operating systems of the TSO type are well suited for these applications. New terminals, graphics devices, and minicomputers offer great potential but also require careful PPP system design. Consideration of the design concepts described in this paper can increase portability and flexibility, which will pay off handsomely in reducing future costs.

ACKNOWLEDGMENT

The authors would like to thank Richard N. Karnes for his careful review and helpful suggestions on this paper.

REFERENCES

1 Tocher, J. L. and Felippa, C. A., "Computer Graphics Applied to Production Structural Analysis," Proceedings of the Symposium of the International Union of Theoretical and Applied Mechanics, Liege, Belgium, August 1970, pp. 521-545.

2 Ice, M. W. and Herness, E. D., "ASTRA User's Manual," Boeing Document D2-125579-5, April 1968.

3 Ice, M. W. and Herness, E. D., "Introduction to the ASTRA System," Boeing Document D2-125579-8, October 1967 (Revised May 1972).

4 Connacher, N., McElroy, M., and Hansen, S. D., "SAMECS Structural Analysis System -- User's Document," Boeing Document D6-23757-ITN, February 1969.

5 Quenneville, C. E., "BCS Interactive Graphics System," BCS Document G2081, January 1974.

A PROTOTYPE MACHINE-INDEPENDENT PREPROCESSOR

S. J. Fenves, D. R. Rehak, and D. Gouirand
Carnegie-Mellon University
Pittsburgh, Pennsylvania

ABSTRACT

The major difficulty in developing application programs for time-shared use or converting batch programs to time sharing is to provide appropriate modes of interaction between the user and the program. It is inefficient to require each application programmer to provide facilities for various modes of interaction and data manipulations needed by the user. In this paper, the needs of users are described, and a prototype preprocessor, easily interfaced with application program, is presented and exemplified. The preprocessor is written entirely in FORTRAN and can be implemented on most computers by changing a few internal tables.

INTRODUCTION

With the advent of widely available commercial time-sharing systems and networks, the computer has become a direct, practical problem-solving tool for a much larger class of engineers and designers than ever before. Interactive computing, where the engineer makes decisions while connected to the computer, is becoming increasingly practical. These decisions typically involve examination of the output generated by the program and modification of one or more input parameters, followed by a command to reexecute the program. Thus, the majority of interactive processing in engineering may be characterized as computer-aided design, in the sense that the engineer makes the design decisions and the computer performs the processing necessary to predict the consequences of these decisions.

In such a symbiotic, intimate man-machine interaction situation, the mode or level of communication becomes of paramount importance. The engineer must be able to communicate with the program in his own terminology and at his own level of competence. Formating and coding rules for preparing input data in a batch environment are simply not acceptable in an interacting environment, because the frequent references to the program's user's manual for the appropriate formats and codes detract too much from the immediacy of solving the problem at hand.

A number of engineering problem-solving systems have been designed from start to operate in an interactive mode, and several others have been successfully converted to such a mode [1]. However, this expenditure of effort has been warranted to date only for fairly substantial systems. The authors know of no systematic software aids available to developers of small-to-medium-scale programs providing the requisite facilities to the users of their programs.

In this paper, the present state of man-machine communication generally available to interactive users is first described. Next, the needed features are discussed from both the user's and the programmer's point of view. A prototype system incorporating most of the desired features is described and exemplified with two applications.

PRESENT STATUS

As mentioned above, a small fraction of programs are written directly for interactive use. There is a general reluctance about writing such programs, because of both the extra programming required to handle the interactive communication and the dependence on the I/O (input-output) facilities of the available system, which tend to make the programs less portable and generally available. Also, conversions made from batch to time-sharing environments generally have been successful only when the original systems or programs had reasonable sophisticated I/O facilities of their own, notably systems with problem-oriented language (POL) inputs.

However, the vast majority of programs developed by engineers for use by other engineers in an interactive environment are written in some version of a higher-level language such as FORTRAN or BASIC supported by the time-sharing system provided by the vendor. These languages typically have processing and output capabilities identical to those of the batch versions. Thus, the output formating for meaningful display of program results poses no great problem.

On the other hand, the input facilities of the available time-sharing languages fall in two classes. First, the standard formated input, including alphanumeric fields, is allowed. This method is patently impractical in a time-shared environment, for counting of spaces for proper alignment of fields is tedious and highly error prone. For this reason, time-sharing systems allow for free-field input of data items, delimited by separators such as blanks, commas, or semicolons. The penalty paid for this convenience is that alphanumeric input is not allowed, so that all options, choices, etc. must be specified by numeric codes. A typical input sequence for a time-shared mechanics application program is shown below.

```
1
1,1,3,0,0,2,2,0,2,0,0,3
2,1,0.,0.,0.,100.,0.,3.E7
2,10.,0.,0.,0.,0.,0.,0.
1000.,0.,0.
2,1,0.,0.,0.,100.,0.,3.E7
2,10.,0.,0.,0.,0.,0.,0.
1000.,0.,0.
2,1,0.,0.,0.,100.,0.,3.E7
2,10.,0.,0.,0.,0.,0.,0.
1000.,0.,0.
2,4
3,4
```

Thus, the so-called free-field input has made the use of interactive programs even less legible and harder to use than the worst batch programs with their complicated input forms.

What is needed, therefore, is an input mechanism which more closely satisfies the user's needs for ease of use, legibility, and ease of memorization yet places no significantly greater burden on the programmer than the conventional READ or INPUT statement. These desired facilities are further discussed in the next section.

REQUIREMENTS

The requirements for an input facility to serve engineering time-shared use properly can be conveniently discussed in terms of the users of the system and the programmers writing the application programs.

User Requirements

The user requirements can be subdivided into two classes: conversational facilities and data management facilities.

Conversational requirements

Users of engineering application programs can be thought of in three categories, each of which requires a different mode of input.

1. A user who has never used a program previously, or has completely forgotten its use, needs a great deal of assistance from the program to enter the required data. The best method for providing this assistance is by means of a "Socratic dialogue," in which the program outputs a question or request for each data item needed,

together with the mnemonic label of the variable and some descriptive text describing the data item (options available, units, etc.) typically found in the Remarks column of a user's manual. The user needs to input only the actual value for each variable. This mode is called "prompting mode" here.

2. A casual or infrequent user of the program would know most (if not all) of the labels associated with the variables. Such a user would be best served by a facility such as the labeled data input in ICES and other POLs or the NAMELIST available in some versions of FORTRAN, where the input consists of pairs of variable labels (names) and their values. The learning of the labels is facilitated by the fact that they are output in the prompting mode when the user first learns the use of the program. This mode will be referred to as the "labeled mode".

3. Finally, there is the experienced user who may use the program on a daily basis and who will know both the meaning and the order of the entire set of variables. Such a user would be well served if he could input just the numerical values for the variables, in what will be referred to as the "direct mode".

Thus, a well-designed input system for interactive engineering use should allow each user to pick the mode best suited to his level of familiarity with the program concerned.

Data management facilities

There are two types of desirable data management facilities for convenient use of engineering computer-aided design programs.

First, in most programs of any complexity, certain data elements may be assigned a "default" value, to be used unless overwritten by specific values pertaining to the problem at hand. The facility for using the predefined default values should be conveniently available to the user, at least in the prompting and labeled modes.

Second, as indicated in the Introduction, the user typically wants to rerun a program with certain data items modified. Such modifications either may occur in one session at the terminal, where the user is immediately ready to modify certain data items as a consequence of the results obtained on the previous run, or may be spread out over several sessions, where the data are first saved and then later reloaded and modified as desired.

Programmer's Requirements

The satisfaction of the user requirements imposes severe demands on the programmer if every application program is to contain all of the facilities enumerated.

Most of the application programs of the class considered are typically written by applications-oriented people (engineers, designers, etc.) whose major contribution is the engineering model and the corresponding computational algorithm. These application programmers would prefer to handle user communication by means of conventional FORTRAN READ (list) statements, which completely describe the program's input requirements, regardless of the mode of interaction (batch or time shared), and certainly without having to concern themselves with the various types of user requirements discussed in the previous section.

PREP: A PROTOTYPE INTERACTION SYSTEM

Motivation

It is obvious from the preceding discussion that a major gap exists between the users' requirements for truly meaningful interactive use and the application programmers' reluctance and/or inability to provide the facilities for satisfying these requirements.

It is further apparent that this mismatch is not restricted to a particular program but occurs in all engineering application programs. It therefore appears feasible to develop a single, program-independent facility for bridging the gap. In the following, a prototype system called PREP is described which attempts to provide a simple facility for converting batch input-oriented programs to interactive use.

PREP as Seen by the User

A session at a terminal using an application program which incorporates PREP consists of three parts, as follows:

1. An initial dialog , in which the user specifies, in response to queries from PREP:

    a. Whether the prompting mode is to be used; if not, the user is free to use either the labeled or direct mode and can freely mix the two.

    b. Whether a file containing previous data is to be loaded; if so, the file name is requested, and, if the named file exists, it will be loaded.

    c. If a file has been loaded, whether the contents of the data file is to be printed out before each input request.

2. The actual input , during which the user supplies the input data to the application program. Data are grouped into sets. Each set is processed as follows:

    a. An identifying header message is output.

    b. Previous values are displayed, if requested.

    c. The user enters the data or modifications in the mode selected.

3. A terminal dialog , in which the user again specifies in response to queries:

    a. Whether the program is to be rerun, allowing the user to modify previously entered data values.

    b. Whether the current data are to be saved on a data file.

    c. If a rerun is requested, whether to display previous values and whether the user wishes to repeat the initial dialog.

During the initial input, the user may specify default values, if provided for, by a carriage return in the prompting mode or by omitting the label-value pair in the labeled mode. During subsequent runs, the user may retain the previously entered values by the same convention.

PREP as Seen by the Programmer

In order to convert an existing application program for use through PREP, the following modifications need to be made:

1. All data appearing in the READ statements must be placed into a COMMON block/PREPCM/. This can be readily accomplished by using EQUIVALENCE statements.
2. All READ statements must be changed to calls to the SUBROUTINE PREP. Each such call defines a set of input data. Upon return from PREP, the values will be placed in the COMMON block regardless of mode of entry or source.
3. An initial call to PREP, as the first executable statement, so as to establish the initial dialog must be provided.
4. A terminal call to PREP to perform the terminal dialog must be provided; in addition, a transfer statement to the beginning of the program must be provided.

In addition to the program modification, the programmer must prepare the data tables used by PREP. A table generation program is available to assist the programmer in this task. Data are entered at two levels:

1. For each set of data, corresponding to distinct calls to PREP, the header message to be printed out is input.
2. For each variable within the set, the following information is entered:
   a. The name or label to be used in the labeled mode
   b. The type of variable expected (integer, real, double precision, or alphanumeric)
   c. An indication of whether the variable is a scalar or a vector
   d. If a vector, the actual size (which may be a fixed integer or a reference to another variable) and the maximum, or DIMENSIONed size
   e. Default value, if the variable is defaultable
   f. The prompting message to be printed out in the prompting mode

The output from the table generation program contains almost all the information needed to produce a user's manual automatically.

Operation of PREP

As implied, PREP operates essentially as a table-driven interpreter. Each call to PREP, other than the initial and terminal calls, has one integer argument which designates the set of data involved. An internal table contains pointers to the first through the last variable in the set. In the prompting mode, the subroutine simply loops over these variables, outputs the prompting message and label, and then stores the variable entered. In the labeled mode, the subroutine reads the label-value pair, matches the labels against those in the table, and stores the corresponding variable in its assigned location.

EXAMPLES

Two examples of the use of PREP will be illustrated. For the sake of legibility, the output produced at the terminal has been retyped.

Beam Program

The PREP program has been incorporated into a general beam analysis program BEAM supplied by W. D. Pilkey, University of Virginia [2]. This program is typical of a small program where the entire computation may be performed in an interactive mode. A segment of the original user's manual is reproduced below

Line 7 (Inspan support line,
repeat NSUP times)

MSUP(I),MTYP(I)

Symbols in proper order	Definition and explanation
MSUP(I)	Station no. of inspan support
MTYP(I)	1 shear release, V = 0 2 moment release, M = 0 3 guided support, θ = 0 4 rigid support, W = 0

The output from the table preparation program is as follows:

```
INPUT FOR READ  7

CHARACTERISTICS OF THE INSPAN SUPPORTS

DATA ITEM    VARIABLE   TYPE   LOC    WORDS    CONTROL BY    OPTION
    41       STATSUP    INTE    996     10     NBINSPAN       RQRD
PROMPTING MESSAGE
STATION NUMBER OF INSPAN SUPPORT

DATA ITEM    VARIABLE   TYPE   LOC    WORDS    CONTROL BY    OPTION
    42       SUPTYPE    INTE   1006     10     NBINSPAN       RQRD
PROMPTING MESSAGE
TYPE OF SUPPORT:
SHEAR RELEASE, ENTER 1
MOMENT RELEASE, ENTER 2
GUIDED SUPPORT, ENTER 3
RIGID SUPPORT, ENTER 4
```

In an on-line session, the modified BEAM program was started with the following initial dialog:

```
INPUT PROMPTING DESIRED?
? Y
FILE TO BE LOADED?
? N
```

The data set corresponding to the segment introduced above was entered as follows:

```
CHARACTERISTICS OF THE INSPAN SUPPORTS

ENTER STATSUP
STATION NUMBER OF INSPAN SUPPORT
? 2 3

ENTER SUPTYPE
TYPE OF SUPPORT:
SHEAR RELEASE, ENTER 1
MOMENT RELEASE, ENTER 2
GUIDED SUPPORT, ENTER 3
RIGID SUPPORT, ENTER 4
? 4 4
```

After the results were obtained, the following terminal dialog took place:

```
RERUN PROGRAM?
? Y
SAVE DATA FILE?
? Y
FILE WRITTEN
ENTER FILE NAME
? BBB
FILE SAVED
DISPLAY OLD VALUES?
? Y
TABLE RELOAD CANCELLED
REINITIALIZE?
? N
```

When the program reached the segment in question, the following dialog took place:

```
CHARACTERISTICS OF THE INSPAN SUPPORTS

CURRENT VALUES ARE
STATSUP    (  1) =              2
STATSUP    (  2) =              3
SUPTYPE    (  1) =              4
SUPTYPE    (  2) =              4
?
```

The blank following the question mark indicates that the previous values are to be retained. On a subsequent run, the following initial dialog took place:

```
INPUT PROMPTING DESIRED?
? N
FILE TO BE LOADED?
? Y
ENTER FILE NAME
? BBB
FILE ACCESSED
```

Upon encountering the same segment, the support types were modified through the following dialog:

```
CHARACTERISTICS OF THE INSPAN SUPPORTS

CURRENT VALUES ARE
STATSUP    ( 1) =              2
STATSUP    ( 2) =              3
SUPTYPE    ( 1) =              4
SUPTYPE    ( 2) =              4
? SUPTYPE 2 3
```

The last line is an example of labeled input. The complete conversion of the original program to the version presented here took one of the authors (Gouirand) approximately 12 man-days.

Preprocessor for SAMMSOR III

SAMMSOR III [3] is a major program which is normally run in the batch or remote-job entry mode. However, the preparation of the data can significantly benefit from the facilities of PREP. A separate preprocessor program was written to accept its input data by means of PREP and to produce a formated output file ready for input to SAMMSOR III. The variable labels (names) and the prompting messages were copied essentially verbatim from the SAMMSOR III user's manual. A typical interactive session for preparing one set of input data is shown in Appendix A. The development of the complete preprocessor program required approximately 6 man-days.

SUMMARY AND CONCLUSIONS

Engineering application programs can be considerably enhanced and made more responsive to a time-shared environment by providing interaction facilities which are independent of both the programs and the hardware environment and which can be used by application programmers essentially in the same way they use the FORTRAN-supplied READ facilities.

A prototype for such an interactive preprocessor of data has been presented and exemplified. While it lacks many desirable refinements, it does provide for most of the interaction facilities needed for meaningful, conversational use of engineering application programs.

It is hoped that the need for such preprocessors is more readily recognized and that suppliers of time-shared computer services will eventually make similar facilities directly available to their customers.

ACKNOWLEDGMENTS

This work was performed as part of the project "Computer Aids for Structural Analysis and Design," sponsored by the Office of Naval Research under contract N00014-67-A-0314-0014. The basic input routine, SCAN, used by PREP was supplied by John W. Melin, University of Illinois. This routine contains the only machine-dependent features of the system, in the form of data tables, and has been successfully implemented by Dr. Melin and one of the authors (Rehak) on over 10 different machines and operating systems.

REFERENCES

1 "Interactive ICES Under TSO", Report No. R73-10, Department of Civil Engineering, MIT, 1973.

2 Pilkey, W. D. and Jay, A., "Structural Members" in this volume.

3 Haisler, W. E., Stricklin, J. A. and Van Riesemann, W. A., "SAMMSOR III - A Finite Element Program to Determine Stiffness and Mass Matrices of Ring-Stiffened Shells of Revolution", Report No. TEES-RPT-72-26 and SLA-73-0126, Aerospace Engineering Department, Texax A & M University, 1972.

APPENDIX A

The following represents an actual session for preparing a set of input data for the SAMMSOR III program using the labeled mode. Two features should be noted:

1. The equal sign is used as a convenient separator between the label and the corresponding data.
2. Whenever required, input data are omitted in the labeled mode; PREP automatically reverts to the prompting mode for the remaining variables.

```
TABLES ACCESSED
TABLES READ
INPUT PROMPTING DESIRED?
? N
FILE TO BE LOADED?
? N
DUE TO THE SPECIAL STRUCTURE OF THE PREP PACKAGE
IT IS ONLY POSSIBLE TO ENTER ONE CASE AT A TIME
NO COMMENTS ARE PERMITTED
SOME DATA NECESSARY TO HELP PREP WORK CORRECTLY
IS REQUIRED ON THREE OCCASIONS

RUN CONTROL DATA

? DSCRATCH=32 SSCRATCH=94

CASE IDENTIFICATION DATA

? TAPE1=13

CASE CONTROL DATA

? NELEMS=4 NPROP=1 NTHICK=1 NHARMNHARMO=3 NPRINT=0 SHELL=1
NSETRING=1
? NSETOON=1

SHELL MATERIAL PROPERTIES

? ELEM1=1 ELEM2=4,YOUNG1=.1234 YOUNG2=.1423 SHEARMOD=6.E7
POISSON=.327
? DENSITY=.084763

SHELL THICKNESS

? TELEM1=1 TELEM2=4 THICKNESS=.25

SHELL SEGMENTATION

? NBSEGMENT=4
```

```
SEGMENT DATA

? NODE1=1 2 3 4 AXIAL1=0 1 2 3 RADIAL1=0 1 4 9 SLOPE1=0 2 4 6
? NODE2=2 3 4 5 AXIAL2=1 2 3 4 RADIAL2=1 4 9 16 SLOPE2=2 4 6 8
? CLASS=3 3 3 3 NERAT1=1 1 1 1 RA1=1 1 1 1 NERAT2=0 0 0 0
RA2=0 0 0 0
? NERAT3=0 0 0 0 RA3=0 0 0 0 NERAT4=0 0 0 0 RA4=0 0 0 0
NERAT5=0 0 0 0
? RA5=0 0 0 0

RING STIFFENERS

? RELEM1=1 RELEM2=4 NRING1=1

ENTER NFLANGE1
NUMBER OF FLANGES PER RING STIFFENER IN CURRENT GROUP
? 2

ENTER DENS1
MATERIAL DENSITY
? 0.0734

ENTER YOUNG3
YOUNG'S MODULUS
? .4526

FLANGES

? 2

CHARACTERISTICS OF THE FLANGES

? DISTANCE=0.24 0.374

ENTER HEIGHT
HEIGHT OF FLANGE IN DIRECTION NORMAL TO SHELL
? 0.987 1.008

ENTER LENGTH
'LENGTH OF FLANGE IN MERIDIONAL DIRECTION
? 231.95 345.67
```

```
THE FOLLOWING IS THE LISTING OF THE DECK OF
CARDS EQUIVALENT TO THE INPUT READ
    1   32   94
    0   13
    4    3    0    1    1    1
    1    4 1.234E-01 1.423E-01 6.000E+07 3.270E-01 8.476E-02
    1    4 2.500E-01
    1     0.        0,        0.
    2 1.000E+00 1.000E+00 2.000E+00
    3    1 1.000E+00    0     0.        0     0.        0     0.         0     0.
    2 1.000E+00 1.000E+00 2.000E+00
    3 2.000E+00 4.000E+00 4.000E+00
    3    1 1.000E+00    0     0.        0     0.        0     0.         0     0.
    3 2.000E+00 4.000E+00 4.000E+00
    4 3.000E+00 9.000E+00 6.000E+00
    3    1 1.000E+00    0     0.        0     0.        0     0.         0     0.
    4 3.000E+00 9.000E+00 6.000E+00
    5 4.000E+00 1.600E+01 8.000E+00
    3    1 1.000E+00    1     0.        0     0.        0     0.         0     0.
    1    4    1    2 7.340E-02 4.526E-01
2.400E-01 9.870E-01 2.319E+02
3.740E-01 1.008E+00 3.457E+02
END OF CASE
 DATA TRANSFER COMPLETED
 INTRODUCE NAME OF FILE UNDER WHICH DATA
 IS TO BE SAVED XXXXXX
 ? DATA
 ? Y
 FILE SAVED
 RERUN PROGRAM?
 ? N
 SAVE DATA FILE?
 ? N
 STOP.
```

FOR THE COMPUTER GOURMET - GRAPHICS

Jackie Potts
Naval Ship Research and Development Center
Bethesda, Maryland

ABSTRACT

Computer graphics is the pièce de résistance of today's sophisticated computer endeavors. It can be divided into two main divisions--passive and interactive. Two of the most popular subdivisions of passive graphics are microfilm recorders and plotters. Plotters can be further divided into flatbed, drum, and electronic. Under the glamorous interactive division are grouped computer animation and computer-generated movies. This is a survey paper which considers the history of all the types of computer graphics and a discussion of the typical hardware of each type including its main attributes as well as its advantages and disadvantages. This paper also contains in the appendix a comprehensive glossary of computer graphics terminology.

INTRODUCTION

Just as Camarones à la Espanola, Escargots à la Cablisienne, or Escargots à la Mode de l'Abbaye are deemed necessary to an epicurean; just as, according to the senses of a wine connoisseur, Lobster Absintha or Supremes of Pheasant Berchoux should be accompanied by a Chablis, Bordeaux Blanc, or emerald dry wine and Tournedos à la Bearnaise, Entrecôte à la Mirabeau, or chateaubriand should be accompanied by a Cabernet Sauvigon or Pinot Noir wine; just as an art cognoscente will select only a Mona Lisa or a Picasso; just as an automobile dilettante will only consider a Jaguar; just as a musical savant will accept no less than a Mackintosh pre-amp, amplifier, and tuner with a Tanberg tape deck and Dolby unit playing through Bose 901 speakers--in the same way, the computer gourmet will insist that computer graphics be part of his computer endeavors.

The sophisticated computer graphics that we use today is a far cry from the graphics of our remote ancestors. In the Stone Age our ancestors drew simple lines on the walls of their caves with crude sticks. Pictures enabled Stone Age men and women to communicate with each other and with all the ages that have followed. These early forms of graphics were the first recorded attempts of one person to interface with another. In his turn, Archimedes drew pictures in the sand. In fact, in all ages pictures of various degrees of sophistication have contributed to man's communication with his human world.

Ever since Archimedes, countless scientists have been grateful for the ability to use drawings, diagrams, and physical analogies to solve intractable problems. Leonardo da Vinci, Galileo, Lord Kelvin, and others through the centuries have felt that knowledge had to have good mathematical symbolism to be scientific. In 1882 Babbage, wishing to ensure the accuracy of the output from his calculating engine, had stereotyped plates made. Today digital symbolism is added to alphanumeric and numeric symbolism to create accurate, real-time, pictorial, and interpretable communication which aids a decision-making environment.

In the present era men and women using computer graphics teach, contribute to the knowledge and improvement of our environment, solve highly technical problems, and draw complicated engineering, architectural, medical, artistic, and constructural designs and simulations on picture tubes with an elaborate joystick. How true that computer graphics forces the computer machine both to understand man's natural language and to extend and expand the human being's capability for creative thinking! How true that computer graphics provides the Rolls Royce for computer users!

OBJECTIVES AND MAJOR DIVISIONS

Since computer graphics has become a very vital link in Homo sapiens's interaction with his machine world, it deserves a closer investigation. Just what is it? Whence did it materialize? Are all computer graphics alike? What are the kinds of computer graphics? What are the essential elements of each type? What are some applications of each type? What groups are contributing to its growth? What are its future benefits for mankind? This overview will attempt to answer these questions. All the major types of graphics will be discussed. First will come the microfilm recorders. Plotters will follow and then--the most publicized--interactive graphics. This will be followed by a discussion of four different applications of computer-generated animation and movies. The last division of this overview concerns itself with some of the drawbacks of computer graphics. In the author's opinion the drawbacks must be corrected for computer graphics to live up to its future potential.

In all areas a brief description and history will be presented along with several applications. In some instances manufacturers will be mentioned. No attempt has been made to include all manufacturers or even the most powerful and well known. Nor does mention of a manufacturer or product imply the author's endorsement. Due to limitations of space and time this overview must limit itself to being as general and as representative as possible rather than covering in minute detail any special aspects of the entire field. Thus only the highlights in the history and current state-of-the-art can be mentioned, and only the major types can be discussed in detail.

The material in this paper was gathered from manuals, specification sheets, books, journals, proceedings, magazines, newspapers, courses, attendance and participation in conferences and symposia, and individual conversations. The industrial and technical details presented are true to the best of the author's knowledge, but they are subject to rapid change.

GLOSSARY OF TERMS

Of first consideration should be the definition of computer graphics. Computer graphics is "the art of image generation and manipulation." It "usually applies to computer-generated displays which contain lines and points. However the term has been used to indicate displays containing only alphanumeric data." This explanation is provided by the SHARE, Inc.[1] Glossary.

In an attempt to communicate effectively and to agree on terminology, the author has included at the end of this overview the glossary of computer graphic terms which was used as reference. In the Appendix the words "display elements," "coordinate," and "display" are in italics because they are defined elsewhere in the glossary. This glossary was prepared under the author's supervision in association with SHARE, Inc., and is reprinted here with the approval of the SHARE executive board. It has been published in the Proceedings of SHARE XXXVI and in Computer Graphics, the quarterly magazine of SIGGRAPH, the special interest group on Graphics of the Association of Computer Machinery (ACM). The glossary has been considered by the American National Standards Institute (ANSI) Technical Committee X3K5 for use in the American Standard Vocabulary for Information and Processing and as a basis of graphic terms and graphic definitions in the International Business Machine (IBM) Data Processing Glossary GC 20-1699-4, published in December 1972.

PASSIVE COMPUTER GRAPHICS

Computer graphics has two main divisions. They are interactive and passive. Interactive computer graphics is associated with the glamorous graphics as illustrated by commercials on television. This graphics uses a display console while the console is in the "interactive mode." Passive graphics includes plotters, microfilm recorders, and other devices which do not allow immediate active participation.

Microfilm Recorders

The charactron which is part of some microfilm computer graphics appeared early in the life of computers. The charactron is a special purpose cathode-ray tube which was developed by Joseph T. McNaney [1] and engineers of Consolidated Vultee Aircraft Corporation. It was described in the January 1952 issue of the Digital Computer Newsletter as a matrix which contains character-shaped openings that are located between the electron gun and the flourescent screen. The stream of electrons that is directed through the matrix openings forms a shaped beam which produces a display of characters on the

---

[1] SHARE Inc., is composed of users of large scale scientific computer machines. Its by-laws,which were revised in August 1973, state that its principal purposes "shall be to foster research and development, and the exchange and public dissemination of data pertaining to computer science in the best scientific tradition."

screen of the tube. These characters can be either read or photographed. The character selection in the matrix is designated by an electrostatic deflection system, while the character images on the screen are positioned by either electrostatic or electromagnetic deflection. In this way the proper sequence of applied deflection voltages selects and thus positions the matrix characters and translates input signals into visual intelligence [2]. Within the year 1952-1953 dry printing techniques and other refinements had been made to the charactron. It was suggested that one of its uses might be in computer display, but there was no sign at that time that anyone anticipated its use as a microfilm input or output peripheral and thus a major contributor to the advancement of computer graphics.

It remained until the Western Joint Computer Conference of 1956 for the charactron to be considered as an aid in debugging, in the editing of input data, in the plotting of graphs, and in real-time simulation. In 1956 multiple graphs could be plotted on the same frame.

Early in the fifties Convair, a division of General Dynamics, San Diego, California, installed a charactron on its ERA 1103 Computer. It had a type C7A cathode-ray tube (CRT) which could display alphanumeric characters at a rate of 10,000 characters per second. The equipment included a CRT with a 7-inch diameter screen, a test generator for noncomputer alignment adjustments, a Beattie 35 millimeter magazine camera, and a Kenyon camera which was camera and photo laboratory combined. A 6-bit code selected the proper alphanumerical character from a 6 x 6 matrix; a 20-bit code positioned the characters on the face of the tube.

Stromberg Carlson (SC)[2] entered the field in the late fifties. David Taylor Model Basin (DTMB)[3] obtained one of the initial SC 4020s and published their first programmer's manual in June 1961. For a short while DTMB used a high-speed automatic processing camera which permitted visual observation of a selected frame of film eight seconds after exposure. The frame to be viewed was selected by an instruction in the program. This camera was called the Kelvin-Hughes [3]. North American[4] installed their first SC 4020 in November 1961 and produced a plot package which was modified by SC and republished in 1965 under the name SCORES.

The SC 4020 is a peripheral system which processes a specially coded magnetic tape obtained as output from a computer program to produce graphical, pictorial, or alphanumeric output. As the tape is read, the desired lines and characters are displayed on a specialized CRT. Sensitized paper and/or film is exposed to the display, and the results are developed and returned to the programmer. The raster orientation of most of the SC 4020 routines are comparable to the origin of the Cartesian coordinate system. The (0,0) point is on the lower left-hand corner, and the (1023,1023) point is on the upper right-hand corner [4].

---

[2] In the late sixties Stromberg Carlson became Stromberg Datagraphics (SD). Thus the use of the letters SD in the discussion of the 4060 microfilm recorder and SC for the 4020.

[3] Now Naval Ship Research and Development Center (NSRDC).

[4] Now Rockwell International.

Fig. 1 shows an SC 4020 as viewed by the user. Its hardware consists of a 7-inch, shaped-beam CHARACTRON[5] tube as a generator for characters and lines. Two of its best features are an F-30

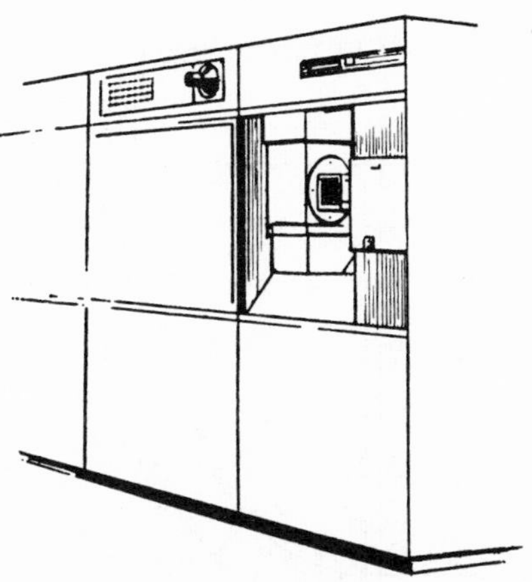

Fig. 1 SC 4020

vector generator and an F-20 variable length generator. The vector generator means that two addressable points can be connected with a straight line, thus obtaining a continuous graph. The variable-length axis generator gives the capability of drawing a continuous horizontal or vertical line the full frame width, starting at any plotting position. The position of a character to be plotted is specified by selecting one of 1,024 horizontal positions with any one of 1,024 vertical positions. For normal printing, up to 64 lines with up to 128 characters per line may be recorded on a frame of film, which, when developed, yields a page of output. For negative output the information appears as clear (transparent) areas on a totally dark (black) background. A Fulton Reversal Processor provides a film output at 20 feet per minute. From this 18 x 24 inch paper copy is produced on a Zerox 1824 printer. This is a 114.5 magnification. Besides the tube, F-30 vector generator, and F-20 axis generator already mentioned, the recorder system consists of the necessary electronic controls and logical circuitry; an automatic, electrically operated 35 mm (Richardson) camera for high-quality permanent photographic recordings;an F-10 Typewriter Simulator which positions each character automatically along a printing line, returning to a margin upon completion of a line or receipt of a carriage return signal; an F-50 tape adaptor which accepts input directly from a tape unit and processes it into a form acceptable to the recorder: and an F-55 DC power supply. Figure 2 is a sample of the Sc 4020 output from a naval architecture computer program showing the longitudinal design of a shell of a ship.

---

[5] CHARACTRON is a trade mark of the Stromberg Carlson Corporation (1964).

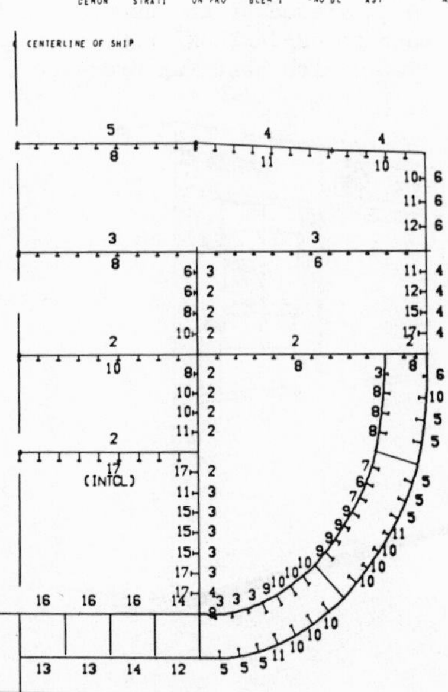

Fig. 2 Sample of SC 4020 output from ship design program of Natale Nappi

In the late sixties the Stromberg Datagraphics (SD) 4060 was produced. It contains an individual computer, called the Product Control Unit, as well as a CRT, camera, film developer, and a hard-copy unit. The resolution has a matrix of (4096, 3072) and the beam in the CRT is shaped by a 119-character stencil. Lines may be drawn in four different widths. A special hardware feature is the drawing of characters with the stroke generator [5].

There are other microfilm recorders but Stromberg Datagraphics is the most well-known company in this field. About 12 years ago the users of the SC 4020 formed an organization named UAIDE.[6] In the Fall of 1973 UAIDE formally joined the National Microfilm Association.

Plotters

The next form of passive graphics to be considered in this article is the plotter. As early as 1953 when computers were first beginning to show their importance to science and industry, two companies were displaying computer plotting devices. They were the Benson Lehner Corporation and the Logistic Research, Inc.

The Benson-Lehner Corporation announced in 1953 the production of an incremental plotter for use by the Computer Research Corporation Model 105 (CRC-105) Digital Computer. This plotter permitted

[6] UAIDE - Users Association of Information Display Equipment.

preparation of punched tapes and the plotting and reading of curves. From a single tape input it could plot a curve on an area 11 x 17 inches. The plotter would advance one increment in the abscissa and the tape reader would advance one step at each reading of the plotter. Then the increment in ordinate was prepared by feeding the tape reader to the plotter. The slope of the curve could not be more than one.

The Logistic Research, Inc. (LOGRINC) developed two automatic graph followers for use on the Magnetic Drum Digital Differential Analyzer (MADDIDA) - 44A, developed by the Northrop Aircraft Corporation, and the CRC-105. As an input or output device either model could be used without additional equipment. As an output medium the two-dimensional followers plotted any two variables against one another. The three-dimensional device could handle functions of two variables. Families of curves on a single sheet could be placed on a drum, and a separately externally control input would cause the follower to switch automatically from one curve to another. The separate input or "z" axis control was also incremental and bi-directional. Both the two-dimensional and three-dimensional models were built on the principle that an electric impulse when received caused an increment movement of the pen. They were a "combination of the Digital Plotter and the necessary photo-electric and electronic components for automatically following curves and converting the data into electrical impulses" [2].

At the present time we have much more advanced peripheral devices which employ the use of flatbed, drum, or electronic digital plotters and which can operate either off-line or on-line. Figure 3 shows a schematic of both off-line and on-line plotting systems.

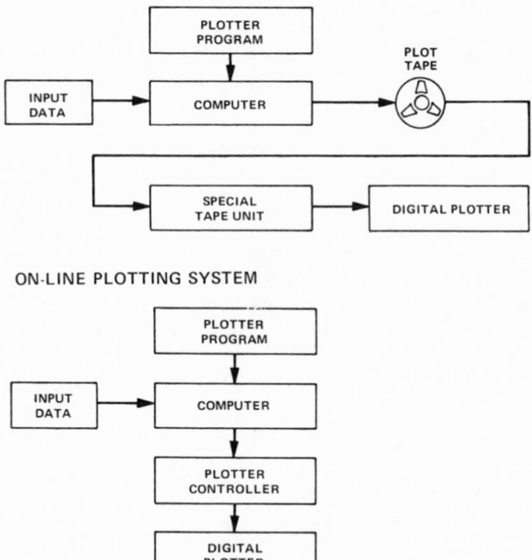

Fig. 3 Schematic of off-line and on-line plotting systems

Plotters are manufactured by many companies--among them, Gerber Scientific Instruments, Hewlett-Packard, and Electronic Products, Inc. One company which produces all three types mentioned in the previous paragraph is the California Computer Products, Inc. (CALCOMP). Their drum plotters, such as shown in Fig. 4, present computer output

Fig. 4 CALCOMP drum plotter

data in an uninterrupted manner. Continuous plots up to 120 feet in length are produced on a 21-inch or 30-inch drum by the rotary motion of the drum and lateral motion of the pen carriage. CALCOMP claims that these plotters are drift-free and the operation is fully incremental. Digital commands activate step motors to produce a recording of the pen movement relative to the surface of the paper. The pictorial representation may include any desired combination of letters, symbols, lines, and axes with unlimited choice of scale factors, letter and symbol sizes, and printing angles. When the digital plotter is used off-line, the plot data and required control commands are recorded on magnetic tape. For plotting purposes the tape is then played on a CALCOMP magnetic tape unit. This unit, which includes the control logic circuits that locate the plot data on the tape, decodes the circuits that supply the X and Y axis drive signals to the plotter and the commands to the pen. When the digital plotter is used on-line, the plot data and control commands are supplied either directly to it by a plotter adapter or through interface electrons supplied by the manufacturer.

Flatbed plotters are used for applications where total visibility is a prime concern. They can also handle a variety of preprinted forms and special materials not practically handled by a drum type plotter. The plot is produced by a motion of a beam and pen carriage. Z-axis commands are used to raise or lower the pen. Either ball point or liquid ink pens may be used.

The third type of plotter is the electronic digital. These plotters are employed for ultra-high-speed plotting and microfilm recording of computer output data. They operate on the same basic incremental principle as the ink-on-paper plotter. The plot is generated electronically on the screen of an 8 x 10 inch CRT and automatically recorded on microfilm or paper. It is advertised that one model can accept input commands at rates up to 100,000 characters per second. When this model is used off-line with a special CALCOMP tape unit, all data on a complete 2,400-foot roll of magnetic tape can be plotted in eight minutes [6].

Three-dimensional work on a plotter was sponsored in part by the United States Atomic Energy Commission and reported on by David L. Nelson as early as March 1966 [7]. The program plotted a three-dimensional perspective of the data on an off-line plotter. It plotted any two-dimensional array interpreting the value of the array as the z component and the address of the array as the x and y components. Invisible data points were deleted.

Applications of digital plotters are growing constantly. In science, plotters have been used in atomic structure analysis, nuclear explosion, seismograms, weather data mapping, radio telescope studies, satellite flight tracking, and submarine hull designs. In medical and psychiatric diagnosis, plotters have been used in such studies as the biomedical research on the graph of sound vibrations in the inner ear. In industry, applications include automatic drafting, Critical Path Method (CPM), and Planning Evaluation Review Techniques (PERT). In engineering, plotters have aided in air traffic recordings, traffic pattern analysis, oil survey measurements, test data graphics, and contour mapping. In business, the studies of inventory and budget control, advertising and market research, and profit and loss trends as well as financial analysis have been aided. Several examples of various outputs of plotters are presented in Fig. 5-9. Figure 5 shows how a plotter contributed to decision making. An interactive computer program was employed to arrive at

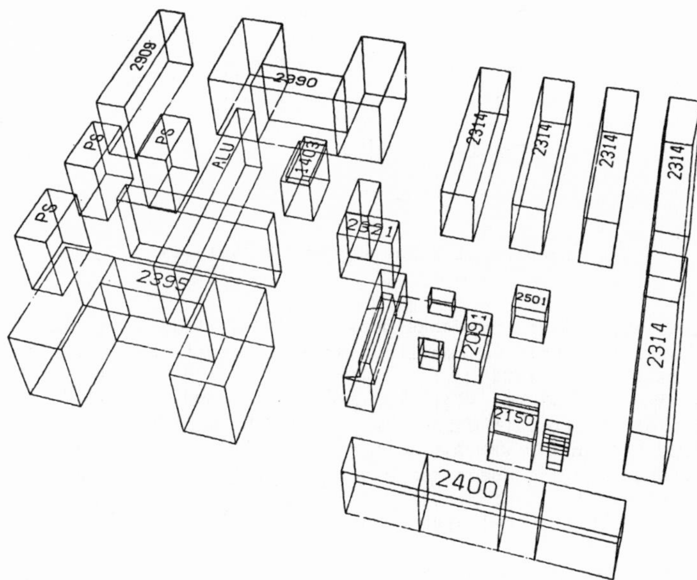

Fig. 5 APL 360/91 layout by Woody Anderson

the proper layout of the computer equipment. Then a plotter was activated to record the decision permanently. Figure 6 shows the

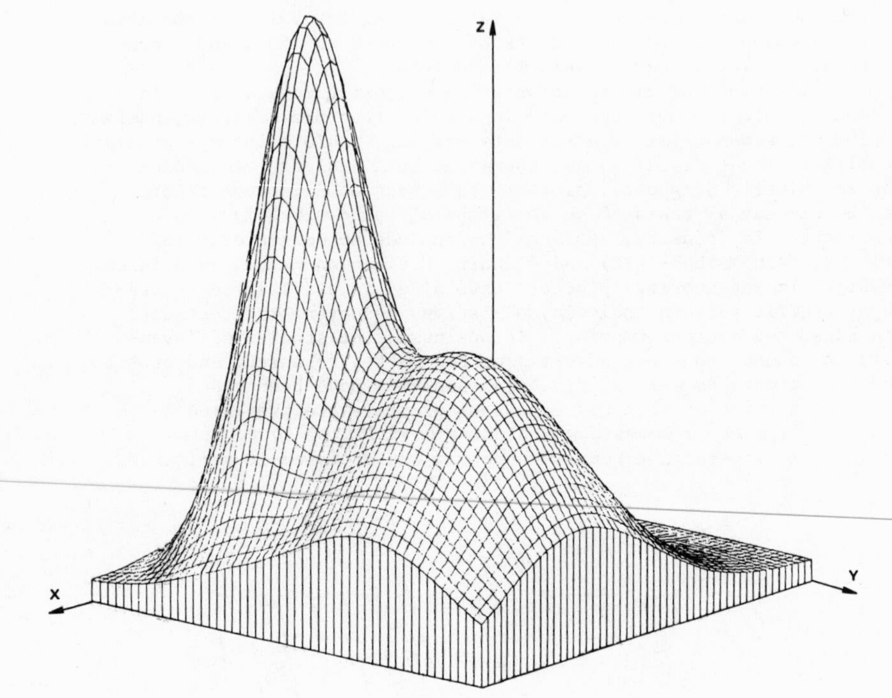

Fig. 6 Results of test problem at Johns Hopkins University

results of a test problem at the Johns Hopkins University, Applied Physics Laboratory (APL) using a complicated mathematical equation. Figure 7 presents another drawing of a solution of a mathematical equation, this time from NSRDC. Figure 8 shows a reduced drawing of the compartments of an aircraft which was produced by Robert H. Thompson of NSRDC. Figure 9 was produced on a flatbed plotter at the National Aeronautics and Space Administration (NASA), Langley, Virginia, from information sent by satellite. It is a view of the Chesapeake Bay Bridge Tunnel area of Hampton, Norfolk, and Portsmouth, Virginia.

GRID THETA=135 GAMMA=90 Q=QS=2.5 F=.125

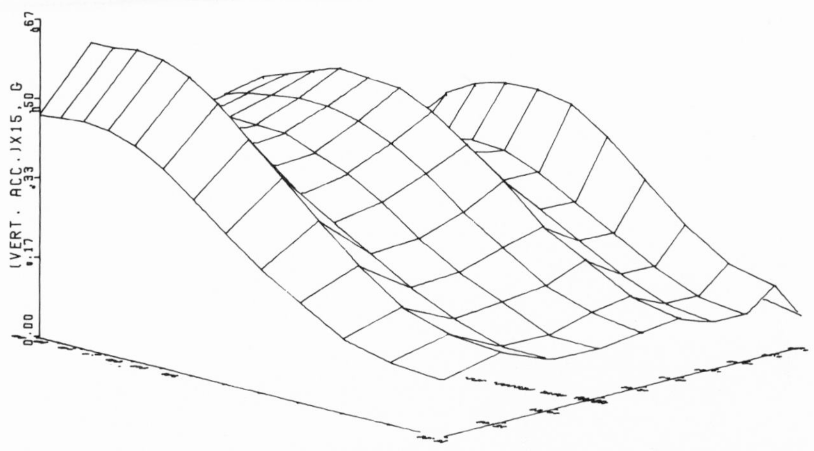

Fig. 7 Computer drawing of a mathematical solution

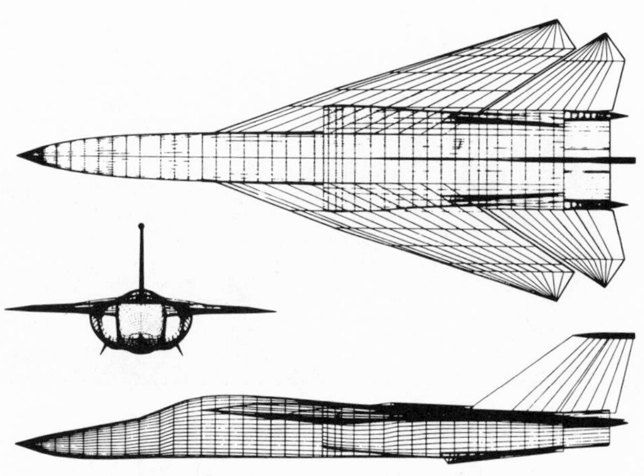

Fig. 8 Reduced drawing of compartments of an aircraft produced by Bob Thompson

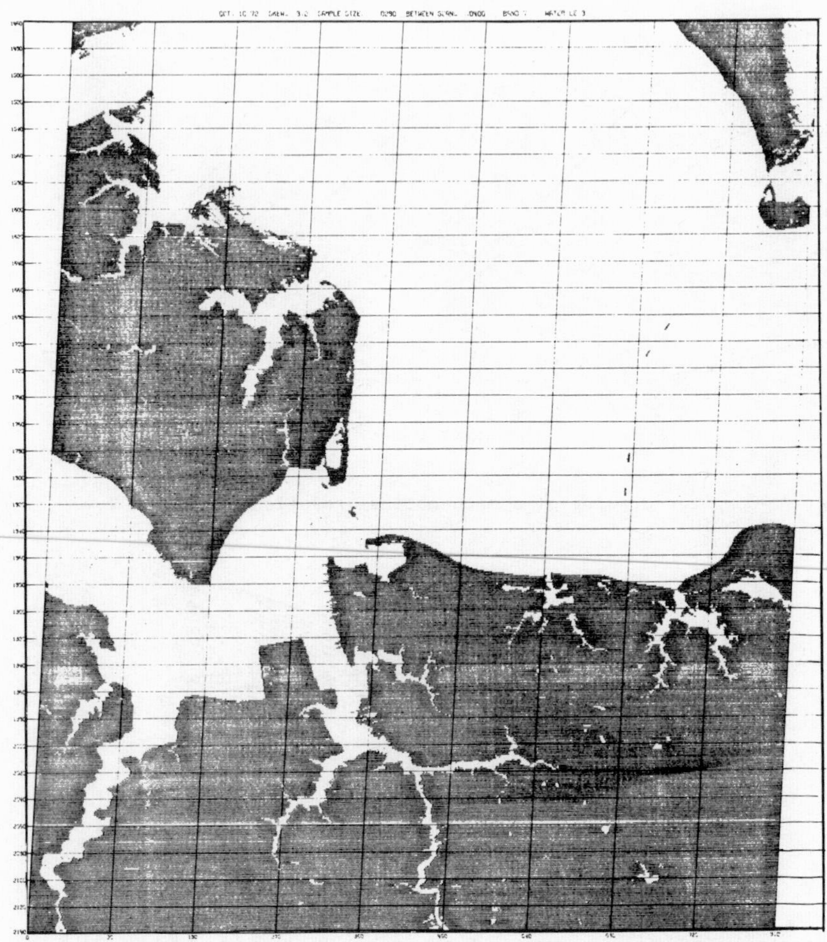

Fig. 9 View of Chesapeake Bay Tunnel area of Hampton, Norfolk, and Portsmouth, Va., produced on a flatbed plotter from information sent by satellite

Miscellaneous

One of the first passive forms of computer graphics output was on a computer printer along with other output of the program. Even though the printer was slow and expensive, it was effective. Today--with the advent of high-speed printers and with the costs of the installation of off-line graphics--interest in printer graphics has been revived. Figure 10 shows a picture of a printer which was

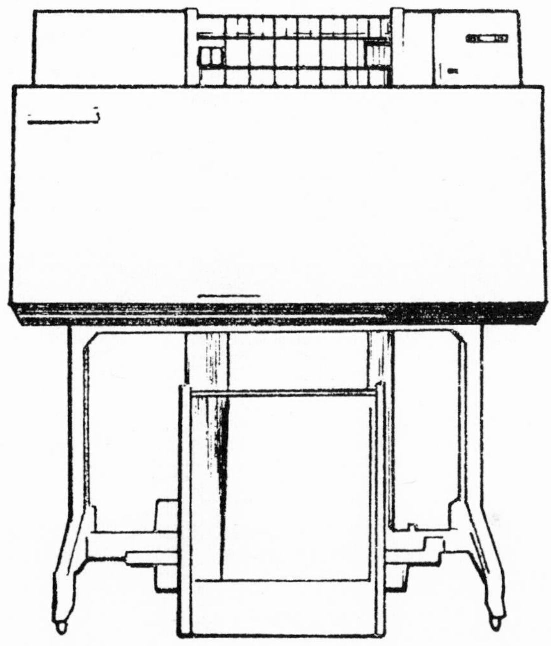

Fig. 10 Picture of printer which was used as a plotter

used as a plotter. Figure 11 shows a more artistic endeavor of the printer.

Fig. 11 Artistic endeavor of the printer

The last type of passive graphics to be mentioned in this overview is the ink pen recorder, which gives simultaneous graphs of one or several variables depending upon the number of channels. A typical example of this type of recorder is the Brush Recorder System. In Fig. 12, eight different displacement histories for finite

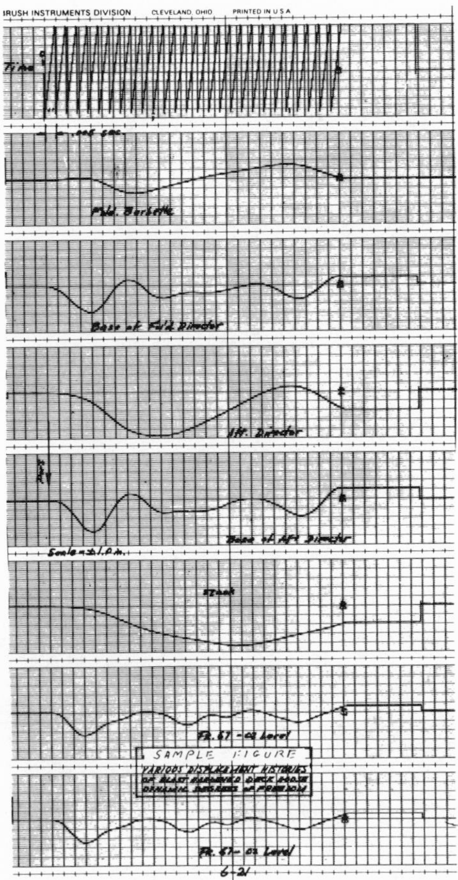

Fig. 12 Eight different displacement histories for finite analysis

element analysis are shown. Figure 13 shows a typical Brush Recorder system. This system uses as input the output tape of a computer pro-

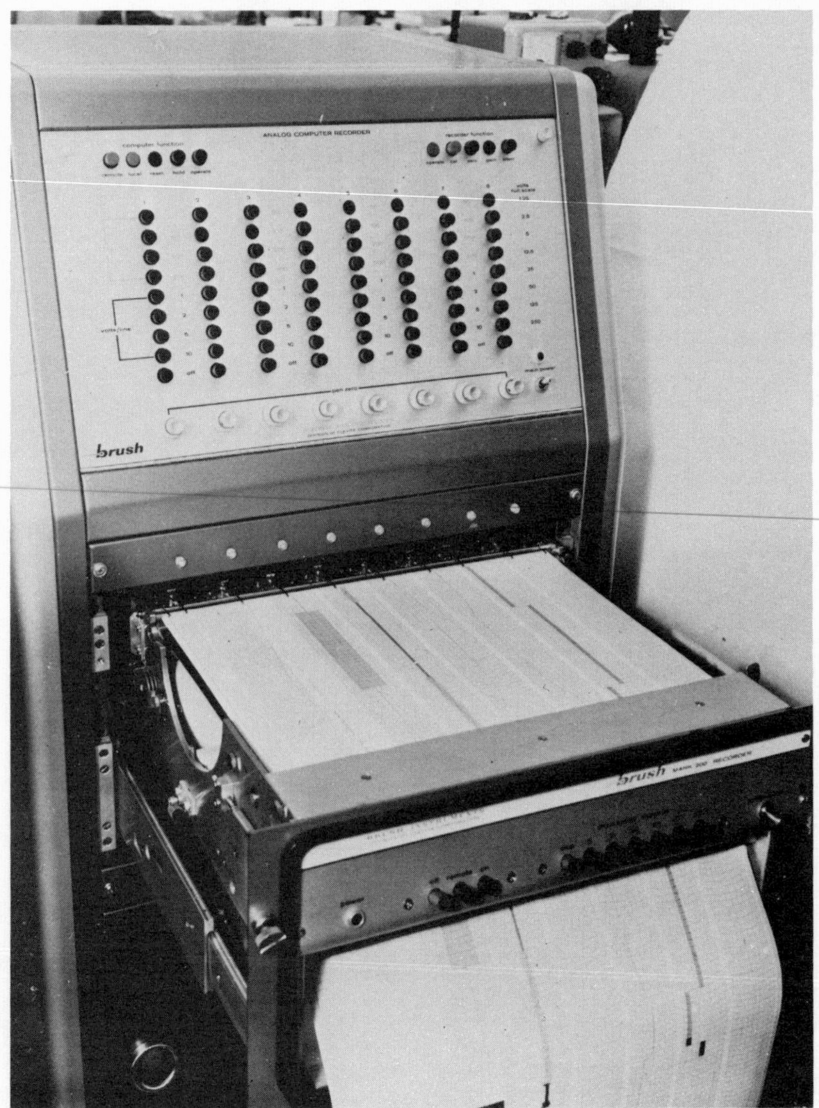

Fig. 13 Brush Recorder System

gram and converts the information on the tape to a form usable for plotting the graphs. The position of each pen along the x axis is controlled by a time base signal. The position of each pen along the y axis is controlled by signals representing the values of the variable.

INTERACTIVE

History

If plotters and the charactron can be traced back to the early fifties, so can interactive graphics, which is the construction, storage, retrieval, manipulation, alteration, and analysis of pictorial data, using an on-line display console with manual input (interactive devices) capability. More than twenty years ago the sophisticated programmers for the ORDVAC--an Ordnance Department eight-digital, shifting-register, breadboard type computer at the Aberdeen Proving Ground--utilized the cathode-ray tube (CRT) as a graphic debugging device in addition to making the most of the CRT in its original design role as an address counter. As early as in the Semi-Automatic Ground Environment (SAGE) Air Defense System Computer, the CRT display console was given moderate input capabilities [2]. The Whirlwind I, which became operational in 1950, used the CRT as an output device [10].

It was in the SAGE project that group displays were introduced and a light gun or light cannon was added as an input device. The light gun was the forefather of the current light pen; it permitted the operator (now also the user) to select desired displayed objects as targets and thus to direct the operation of the computer program.

One of the drawbacks of the early cathode-ray tubes was the inability to maintain a picture on the screen. The display image appeared once for an instant and then quickly faded out of sight. The refresh rate was so low that regeneration became quite apparent. This phenomenon is called flicker. It was corrected for the most part by refreshing the CRT from data stored in the memory of the computer. Even today there is a percentage of flicker in many cathode-ray displays.

Quite a few years had to intervene before a display console became really popular with computer users and programmers and was widely used as an interactive computer input/output device for applications. The breakthrough occurred in 1963 and was accomplished by I. E. Sutherland and T. E. Johnson with the SKETCHPAD programs. These programs proved that an interactive display console could be used as an input/output device to accept or exhibit data in both pictorial and alphanumeric characters and to provide a more efficient means for controlling the sequence of a program than any other device such as an on-line typewriter [11, 12]. Licklider is quite enthusiastic over the advances made in interactive graphics because of the SKETCHPAD. He reports that Ivan Sutherland mentioned his SKETCHPAD project in 1962 during a discussion period of the session on Man-Computer Communication at the Spring Joint Computer Conference of the American Federation of Information Processing

Societies (AFIPS). At the end of the session, Sutherland showed to those who lingered their first glimpse of his very dramatic on-line graphical compositions [20]. Since this is quite the most significant advancement in interactive computer graphics, so far more details are worthwhile.

The basic elements of the SKETCHPAD--which revolutionized interactive graphics--are points, lines, and arcs. Its lines are ideally straight, and its arcs are ideally circular. The lines used can be constrained to be parallel or perpendicular to each other or to be parallel with the axes of the coordinate frames. Two points can be constrained to be parallel or perpendicular to each other or to be parallel with the axes of the coordinate frames. Two points can be declared identical, and their identity can be enforced by the computer and its programs.

By means of the SKETCHPAD, a user could employ a CRT display and a light pen to construct pictures while a computer monitored the user's motions and also constructed a set of data representing the picture which was being drawn. The data structure thus created represented the topological properties of the drawing. These properties were displayed on the CRT and served as names or labels in the data structure, which were chosen as easily as pointing the light pen at them. By use of this mechanism the user could extend his data structure to include numerical and other nonpictorial attributes simply by pointing his light pen at the desired element and then typing in the additional information desired. In the same way the different elements of the system which the drawing represented and their relationship could be designed with the aid of the light pen, which selected the elements and then indicated their relationship. By allowing the user to make this selection even if the relationships were as simple as the elementary one of the definition of a line in terms of its end points or as complicated as that of geometrical constraints with limitations of size, parallelism of two lines, or perpendicularity, accurate displays could be drawn in spite of precision inaccuracies of the light pen.

From 1962 on use of the CRT display console mushroomed. In the fall of 1962 the first GM DAC-1/7090 was shipped. It was known as the Windshield Design and was in production in 1965. The prototype of the earliest IBM graphic console was available on the IBM 7044 and used in Alpine. The IBM 2250 appeared at the 1964 Fall Joint Computer Conference. The third and fourth versions of the IBM 2250 were announced in 1966 and 1967 respectively. The IBM 2285 was presented at the 1968 Fall Joint Computer Conference. In IBM, interest was so pronounced that interactive graphics was the subject of an entire issue of the IBM Systems Journal in 1968.

In 1965 Bell Telephone installed a computer system which was built around display consoles, Centralized Records Business Office (CRBO). This system consisted of 28 Raytheon 401 display terminals divided between two Raytheon 425 control units.

In the medical field the IBM System 360 Model 50 MFT (multiprogramming with a fixed number of tasks) with 256,000 bytes of core was placed in operation in 1967 for Hobbs-Baylor Medical School. The system used an IBM 2260 Display Station and in 1968 served both scientific users at Baylor University College of Medicine and the hospital data management system at the Texas Institute for Rehabilitation and Research (TIRR) [13].

The Graphics Data Processor, as described by E. J. Smura in 1968, was capable of producing a half-tone picture of a 16th-intensity-level picture or a four-bit-intensity-level picture. The equipment consisted of Scientific Data Systems (SDS) 930 digital computer into which graphical data were entered and stored in digital form, a graphical recorder which used several cathode-ray tubes to present the graphical data for recording or direct view, and an interface between the computer and recorder for interpreting data from the computer and converting them to analog form so that they were suitable for presentation to an output device [14].

By 1968 CRTs were being used in interactive graphics systems for data display, geographical display associated with command and control systems, and structural data files. Computer animation sprung off in the late sixties. As early as 1969 SHARE, Inc. started a film festival under the sponsorship of the Graphics Division.[7]

At the present time interactive graphics is used in almost all fields including commercial aviation. To help take care of a large number of passengers--it flew more than six million passengers in California in 1972--Pacific Southwest Airlines (PSA) installed an on-line terminal system. It consists of a National Cash Register (NCR) 101 computer, 140 identical Bunker Ramo 2210 CRT data terminals, and four Bunker Ramo multistation control units. The size of the 2210 screen is only about three inches. But, in normal working position with the eyes at a distance of about 15 inches from the screen, the alphanumeric display can be read like a book. The keyboard is designed for use by nontypists. In the future if the need arises, the 2210s can be unplugged and replaced with larger CRT terminals capable of displaying up to 960 characters and fitted with typist's keyboards. While a customer is at the counter or on the telephone, the agent uses the CRT terminal to key in flight numbers, date, and number of seats desired. There is an instantaneous display. The NCR 101 computer automatically deducts the seats from the inventory. If the desired flight is sold out, the computer displays up to six alternate flights. The display can also be used to cancel reservations and to check on the customer's PSA travel card status [15].

Input

In the late sixties several interactive devices were available. These were the alphanumeric and function keyboards, the light pen, the cursor, the joystick, the mouse, and the tablet. The Rand tablet, originated at RAND Corporation, was one of the first tablets to be developed. It was developed further and marketed by Bolt, Beranek, and Newman. All the input devices are defined in the glossary.

---

7 The Graphics Division is now the Graphics Group in the Integrated Systems Division.

Types of Interactive Graphics CRTs

Interactive graphics CRTs may be classified into two main groups. They are the storage tube and the refreshed scope. The storage tube is a CRT with special long-persistence phosphor which retains any displayed image without refreshing until the image is erased. It usually has a small screen. It can be located quite far from a computer and is relatively inexpensive, as low as $3,800. It cannot display dynamic phenomena,and there can be no changes of a picture part without the picture being erased. Even so the picture may still have new information added. The generation of display is slow. A cursor is used to make additions. Because of its characteristics it does not occupy the core space of a large computer. Usually it has a minicomputer attached to it. Two manufacturers of storage tubes are Computek and Tektronix. The main host or connected minicomputer for both companies is the PDP, manufactured by Digital Equipment Company. Storage tubes can also be connected with other computers such as those produced by IBM, NOVA, and VARIAN.

The refreshed scope has a large screen and can demand a computer fairly close. These scopes are expensive and the computer is extra,but this cost is overshadowed by the ability to display dynamic phenomena. Another advantage is that almost any part of the picture can be changed without erasing the entire picture. As in the storage tube, new information may be added to the display. The amount of display is limited by the sign of the display buffer. Even at the present time the flicker develops if the information is excessive. Thus sophisticated users should look instantly for overcrowding of their input when flicker occurs. The generation of the display is fast, but it does need adjusting from time to time. Different intensities may be displayed. The light pen as well as the cursor may be used. Windowing and hardware rotation are available. In this case of the refreshed scope, the display occupies core space. Just a few of the manufacturers of refreshed scopes are IBM, ADAGE, and IDIOM.

Description of Several Systems

In the following paragraphs are detailed descriptions of the systems of several manufacturers. An attempt was made to select typical examples rather than every system.

Adage

The Adage Graphics Terminals, Fig. 14, are refresh type or stroke drawing CRT display stations. The terminals contain a DPR2 digital processor, a Graphics Coordinate Transformation Array, and both a character and vector generator. The DPR2 is a general purpose digital computer. It performs all processing and control functions including that of the conventional input/output devices. In fact, it is the only programmable element of the entire system. There are from 4,096 to 32,768 30-bit words of core. The number of priority

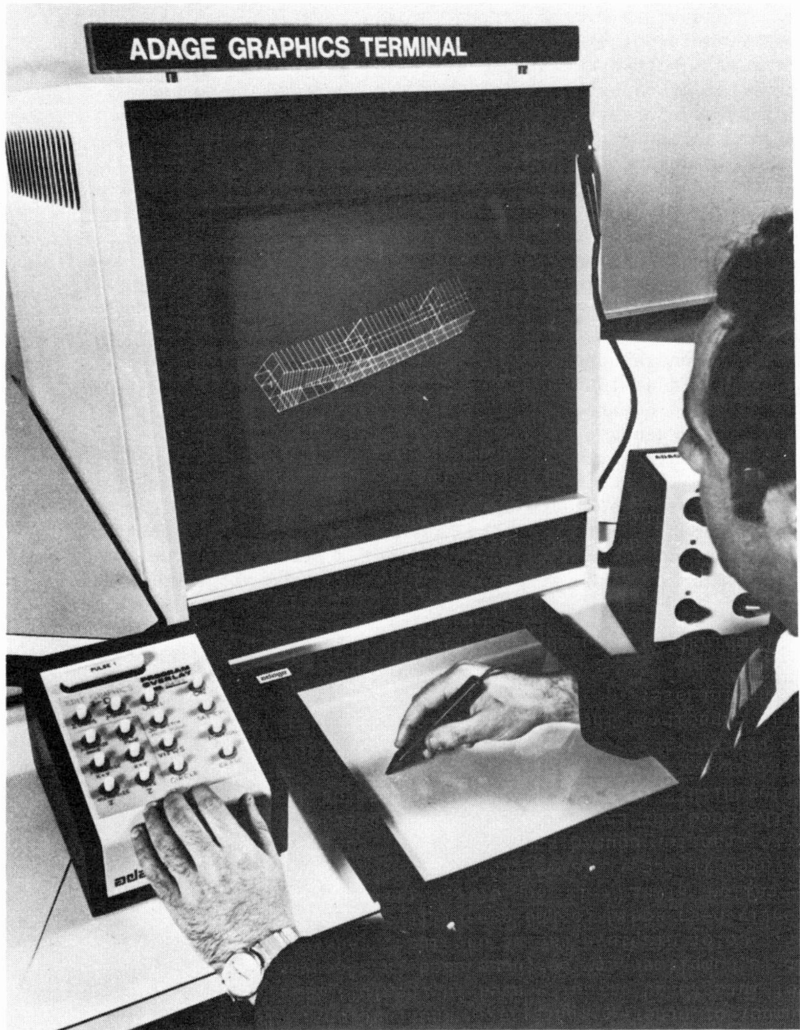

Fig. 14 Adage graphics terminals

interrupt channels can vary from five to twenty-five. The Graphics Coordinate Transformation Array is a hybrid unit, which transforms "all vector endpoints by the current scale, rotation, and displacement factors before they are displayed" [17]. The character generator serves as a high-speed stroke writer. Its vocabulary has sixty-four characters, but another thirty-two can be added. The characters can be displayed in several ways, upright or italicized, in one of three sizes, and at one of three intensities. With the analog output of the Transformation Array the vector generator drives the CRT beam. This is controlled by a processor register which specifies the desired display mode. A light pen, sixteen keyboard function switches, and two pedal function switches are the interactive devices that are provided with the system. Other peripheral items such as the Adage Data Tablet, joystick, bowling ball, six variable control dials, and auxiliary keyboards may be obtained. Each piece of hardware contains its own register. These registers are addressable from the central processor. A three-dimensional hardware windowing is also available as an option. With this the programmer can designate the desired upper and lower bounds for the x, y, and z coordinates. Whenever the CRT beam goes beyond any of the designated values, the vector generally will blank the beams. Perhaps the outstanding characteristic of the Adage hardware is its ability to represent the z coordinate by a varying intensity of the electron beam. Because of this feature the graphic displays can be described so that they appear to be three-dimensional. Still depth cueing by apparent change in the size of the image must be programmed [17].

Computek

The Computek terminal, Fig. 15, consists of a display controller and

Fig. 15 Computek computer terminal

Type 611 direct view storage tube. The display controller can utilize either a single-byte, a two-byte, or a four-byte mode. The single-byte mode is the one used for commands and character transmissions. The two-byte and four-byte modes are used for graphics. In the two-byte mode, vectors of coarse definition are drawn. In the four-byte mode, vectors can be drawn between any two of the 1,024 x 1,024 addressable points. The type 611 storage display unit is built by Tektronix. It has a 6 1/2 x 8 1/4 inch screen with a revolution of approximately 300 x 400 line pairs and an erase time of half a second. Also there is a stroke generator available. This generator allows a continuous curve to be drawn when four bytes of data are transmitted. Other features of the terminal are that it can accept data at the rate of 2,000 characters per second and that it is provided with several "back panel" outputs which can be switched between ground and five volts under computer control. This latter feature permits a camera to be actuated, a capping shutter to be triggered, and other external devices to be synchronized with the flow of data. There are four cursor control keys. If one of the keys is depressed briefly, the cursor is stepped one character in the indicated direction. If the key is held down, the cursor slews at three inches per second until released. The joystick when used in this system completely controls the beam position and the circular cursor; in other words, it takes control over the four cursor action keys. There are two versions of the Computek tablet. One has a resolution of eight bits. The other has a resolution of ten bits. This tablet has the free-running, stroked, and demand modes. In the free-running mode, data are updated every 300 microseconds. In the stroked mode, data are updated when the stylus is moved. In the demand mode, data are updated upon request. There is an interface provided for attaching the tablet to its own modem for general use. Another interface is provided for use with Computek terminal [18].

Control Data Corporation

NSRDC has onboard Control Data Corporation (CDC) equipment for its interactive graphics. The system is composed of a CDC 6700, a CDC 1700, and the Digigraphics 274 Display Console. The host computer, the CDC 6700, has dual central processors (6600 and 6400) with a memory of 131,000 (131K) 60-bit words. The small computer, the CDC 1700, serves two special functions. First it is a broad band communication interface between the terminal and the host computer so that there is no limit to the distance between the two computers. Second, it relieves the main frame of many of its terminal control and maintenance functions. The Digigraphics 274 Display Console is connected to the CDC 1700. It allows the user to enter data directly, and create, display, store, retrieve, and modify any graphics forms desired.

IDIIOM

The Information Displays, Inc. Input/Output machine (IDIIOM), Fig. 16, is characteristic of the sophistication of a refreshed scope

Fig. 16 IDIIOM graphic terminal

display processing unit. It has seventeen registers including triple-rank input registers which significantly reduce the load on the driving minicomputer. It has several different kinds of jump subroutines, a variety of register to register transfer instructions and a number of other computerlike instructions. In fact, the number of integrated circuits used in the IDIIOM II display processing unit is complimentary to those used in a conventional minicomputer. Color displays can be obtained. A new operating system has recently been added to the capability of the system. The new system --called HIGHER --is one of the first FORTRAN-type operating systems which in itself is written in FORTRAN. Because of its ability to transfer unused programs out of the minicomputer core without the need for the programmer to keep track of subroutine utilization, HIGHER, is unusually well suited for a mini-based system. It is estimated that the HIGHER operating system can work with about 25 percent less core than current operating systems of comparable capacity. The IDIIOM's unique capabilities from a hardware standpoint are:

1. Combination of high resolution TV with computer generated data
2. Programmable character set which operates at speeds equal to a hard-wired set
3. Ability to drive up to six independent displays and support complete operator interaction on each one

IDIgraf

The IDIgraf produced by Information Displays, Inc. is a cheaper refreshed display terminal than the IDIIOM. One of its main characteristics is that it can do a great deal of both alphanumeric and graphics hardware editing that does not require connection to any computer. Another feature is a frew slew cursor mode such that when the cursor matches any eight raster units of any graphical element, like a line, it captures the line by causing the line to blink. Perhaps the best feature economically is that it is one of the few--if not the only one--that offers pention color stroke writing. It is significant that this display has up to twice the resolution of any low-cost TV display. The IDIgraf contains a 17-inch rectangular CRT with bonded safety face plate. The display areas is nominally 10 inches square, as shown in Fig. 17.

Fig. 17 IDIgraf graphic terminal

Tektronix

Tektronix, Inc. was founded in 1946 and has only lately started competing in interactive graphic terminal systems. True it produced glass--lately ceramic--CRT tubes for many other computer companies for years, but it was known more for its oscilloscopes and other electronic equipment until about 1970 when it acquired control of a great percentage of Corning's graphics capability. Now, Tektronix offers a line of interactive graphic computer terminals featuring CRT display from a direct view storage tube. A special feature of Tektronix terminals, Fig. 18, is the stand-alone design with control and interface circuitry in the mounting pedestal. Space is provided in the pedestal minibus for computer interface cards, plus one or two peripheral instrument interfaces. The 11-inch diagonal storage

Fig. 18 Tektronix terminals

CRT in the present Tektronix terminals handles graphic displays up to 5.6 inches high by 7.5 inches wide or alphanumeric displays of 35 lines of 72 characters each. A new terminal is scheduled for release sometime in 1974. It features a storage tube with 19-inch diagonal measure which makes it the largest storage tube on the market. Normally storage scopes have been smaller in diagonal size than refresh scopes.

Vector General

The basic Vector General system, which is about the same in price as the IDIIOM, is one of the oldest systems. For its minicomputer it employs a Digital Equipment Company (DEC) PDP 11-05 with screen refresh and control information. A unique quality about the Vector General terminal is that its three-dimensional capabilities and perspective are both incorporated into the hardware. The terminal has a 21-inch interactive screen with light pen, table, joystick, and ten control knobs. The control knobs are similar to those in the Adage terminal.

Color

As early as 1966 color cathode-ray tubes were available but because of their comparatively limited resolution were not used extensively. The early color systems were usually projection display systems which used either single or multiple channel additives. The light valve, scribing, and film-based projection display systems used the multiple-channel additive for their multicolor accomplishments. This process formed one clear and opaque monochromatic image for each of the primary colors. Intermediate colors were then obtained by mixing the images of the primary colors on the screen.

Two different methods of the single-channel additive were experimented with. One was known as the Dufay color mosaic and involved the dot-interface. The attempts to apply this method to the pioneer display systems were apparently unsuccessful. The other method was known as the lenticular and involved the line interface. General Precision Laboratories (GPL) was successful in the production of the lenticular process and produced "Lenticolor." In both of these approaches to the solution of the color problem the film structure itself was more important than the optical-mechanical elements.

The transparent film base of the lenticular film was "embossed" with cylindrical ribs during manufacture. The curvature of these cylindrical lenses and the index of refraction of the film base were such that a bundle of parallel rays arriving at the film base were focused into a set of parallel linear elements at the opposite side of the base, where the emulsion was coated. By varying the direction of arrival of the parallel bundle, different linear elements were exposed [20]. The lenticolor process used three directions which coincided with three bands across the face of the lens. Any standard projector could be used simply by placing the three parallel red, blue, and green filter bands across the projection lens. The CRT display was photographed with three different sets of symbols or graphics--each time one-third of the lens was photographed while the other two-thirds was covered. Thus the CRT display was photographed three times. It took approximately ten seconds for the reverse processing of the film.

Animation

In the years since 1964 when Knowlton first discussed his BEFLIX language for producing computer-animated movies, film making under computer direction has been employed in a variety of disciplines. Computer-generated movies and animation have become a new communication interface between computers and human beings, scientists and other scientists, scientist and laymen, and educators and students. Computer picture animation seems to be especially suited for the physical sciences, mathematics, art, and advertising. Perhaps its most powerful and valuable use can be its role of convincing laymen that the computer is not a powerful and undisciplined robot controlled giant but a helpful ally.

Because of its potential power several examples of computer movies and animation will be discussed in this paper. One report is taken from work done at a university. The second method discussed was developed both at a university and research laboratory. The other two methods are from commercial companies employing two different processes.

The first method of computer animation to be discussed was reported on in 1970. At that time Philips--at the University of Michigan--used as a display medium a bi-stable storage tube terminal made by Computek. He claimed that this tube, Model 400/20, retained complex, flicker-free frames for approximately fifteen minutes with undiminished brightness. The host computer was an International Business Machine (IBM) 360/67 with a Digital Equipment Corporation PDP/8 computer interfaced to act as a data concentrator. The computer-animated films were made by photographing static displays a frame at a time. When a complete frame had been drawn the last character that was transmitted caused one of the "back panel" outputs to go high. This, in turn, actuated an Airiflex 16-mm pin-registered camera. The animation motor of the camera permitted the selection of two shutter speeds or the equivalent of bulb operation. A screen erase signal was sent to the display, and another frame was started. Then the information was sent to a postprocessor. In this case the postprocessor used a SC 4020, which employed a subroutine package developed by Polytechnic Institute of Brooklyn, POLYGRAPHICS. The postprocessor consisted of about 200 lines of FORTRAN IV Code. It read its input from a tape or file that contained SC 4020 hardware code [19]. The SC 4020 is described elsewhere in this paper.

In the on-line system developed by S. E. Anderson at the Johns Hopkins University Applied Physics Laboratory an IBM 360/91 is used to drive a 2250/3 cathode-ray tube in a time-shared environment. Picture language commands can be typed in from the alphanumeric keyboard. A light pen can be employed to scan and edit the code. Then the dynamic sequence can be displayed. In a "movie editor" mode the user can depress the programmed function keyboard switches and thus advance a selected number of frames. The acceptable picture sequence can be made into a permanent record in one of three ways, one of which is on-line. A Polaroid camera can be used to obtain a snapshot directly from the CRT screen. The off-line methods obtain a plot by means of a CALCOMP x-y plotter or a 16 or 35 mm movie by means of the SC 4020 microfilm recorder. Two major programs, adapted from Anderson's work while at Syracuse University, were written to accept this picture language. They are Hopkins Implementation of Aided Motion Pictures (HICAMP) and Hopkins Implementation of Computer Aided Movie Perspectives (HICAMPER). HICAMP produces movies of planar, two-dimensional objects. HICAMPER produces movies of three-dimensional objects in perspective. Stereoscopic animations can be prepared by simply creating two slightly divergent views of a three-dimensional object and choosing two viewing points at approximately the interpupillary distance apart. If the viewer uses an image splitter when the perspective views corresponding to both right and left eye images are displayed on the CRT screen, an illusion of three-dimensional sight is obtained [21].

Even "Walt Disney's World" has been transformed by computer graphics. About 1967 commercial computer-generated animation was developed when special purpose computers were designed specifically for animation. These systems are hybrids of both digital and analog computers. The analog techniques are employed for structures, basic shapes, and animation while the digital techniques supply control, storage, and the timing of the total animation. Video techniques are used for detailed imagery and the surface characterization of images [22]. Two of the leading animation firms are Computer Image Corporation and Mathematical Applications Group, Inc. (MAGI). Computer Image has developed two animation systems. They are SCANIMATE and CAESAR. SCANIMATE scans and animates artwork by television pickup and input of real artwork created by an artist. The first step in the SCANIMATE system is the preparation of "Kodaliths." Kodaliths is the preparation of black and white artwork on transparent sheets. A Kodalith is equivalent to a dozen or more of the conventional cels[8] produced in conventional animation. After its preparation the Kodalith is then placed in front of a special TV camera with a high-quality scanning format pickup device which allows up to seven separate photos to be set up. At this point the art-work conversion unit converts the images that have been received into electrical signals which serve as input to the computer and which are restructured as an image on a cathode-ray tube. There are two modes of manipulating the image, (1) direct--as a joystick is moved, so moves the image, (2) parametric --the user determines the motion parameters, and the image moves accordingly. Color animation is added at the completion of the animation. Figure 19 gives a schematic outline of the SCANIMATE Computer System developed by Computer Image Corporation. The cameras used are the Airi for

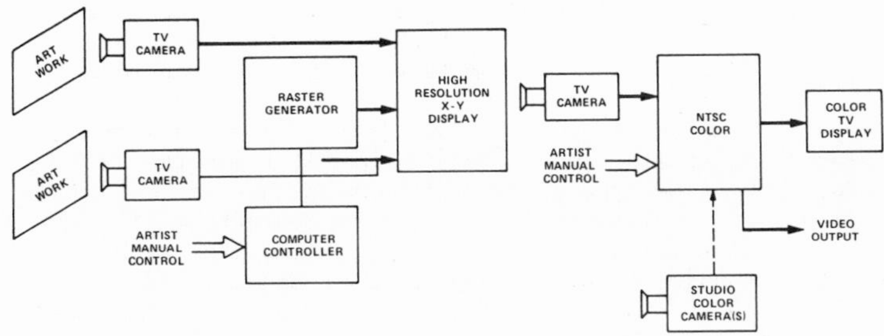

Fig. 19 Schematic of SCANIMATE animation system

the 16-mm work and the Mitchell or ACIME for 35-mm work. The camera and computer are linked with a synchronous motor. The speed of the camera is 24 frames per second, but the computer calculates, updates, and draws the image at a rate of 48 frames per second. CAESAR (Computer Animated Episodes by Single Axis Rotation) is a more advanced system than SCANIMATE. It is claimed that CAESAR provides flexibility and to-the-frame control of animation and color.

---

8 A cel is one of the transparent sheets of celluloid on which objects or sections of objects are drawn or painted in the making of animated cartoons for motion pictures and television.

The MAGI system is a visual simulation technique which produces fully computer-generated perspective views of three-dimensional objects. Whereas computer graphics systems usually display an object as a collection of line segments which seem to distinguish the boundaries between surface, the simulation approach treats an object as a set of three-dimensional surfaces that reflect light. It is this reflected light impinging on photographic film that forms an image of the object. Once the three basic components of a camera, a light source, and object or objects have been defined, geometric ray tracing computes a "picture" of the object as it is shown in the simulated camera. Geometric ray tracing tracks individual rays of light from the light source to the object to the camera. Each reflected ray provides the intensity at a single point on the picture. After a sufficient number of points have been computed, the entire area of intensity data may be displayed on a cathode-ray tube. The heart of the MAGI system is the three-dimensional modeling technique, known as the combinatorial geometry method [23].

CONCLUSIONS

The life ahead seems quite rosy for computer graphics. Surely there must be some pitfalls. Does it have any problems? If so, what are the major ones?

First, it might be called disjointed. It certainly seems to go in all directions at once without any guidance. There appears to be great interest by many people. But where do "youngsters" in the graphics field go for help? Sometimes they go to manufacturer's classes. Here it is obvious that they learn about only one manufacturer's type of computer graphics. Sometimes they go to workshops, conferences, symposia, and universities. At these they all too often receive the same type of instructions as they would in a manufacturer's course. How can this drawback be corrected? All who teach computer graphics can and must become truly familiar with all areas of computer graphics before attempting to teach it. If some expert is called upon to discuss his specialty he should continuously emphasize that he is talking about only a part of the field of computer graphics. Is this too much to ask? As a former teacher, the author feels that such frankness will put the teaching in proper perspective and lead to enhancement of both the teacher's knowledge and ethical reputation.

A second pitfall might be that many personnel working on computer graphics put too much emphasis on interactive graphics. True, this is the glamorous area, but there are other areas. No one area should be neglected. After all there are plotters and microfilm recorders in many interactive systems. The author looks forward to a future system which include a highly developed scanner or digitizer that can prepare input from all types of written drawings and designs without even requiring the user's presence most of the time. Then the system will, for debugging purposes, record the input on a display scope and if requested produce a permanent record. The input will also be sent to a computer program. The output will be produced on the scope and recorded permanently, if desired. Graphics in the desired format will be obtained. A permanent and separate record will be made of the graphics work so that single or multiple copies may be produced later. The system will include a scope, scanner, and plotter or microfilm. Does this sound farfetched? Such systems are almost in production today.

The graf/pen sonic digitizer, Fig. 20, which is produced by

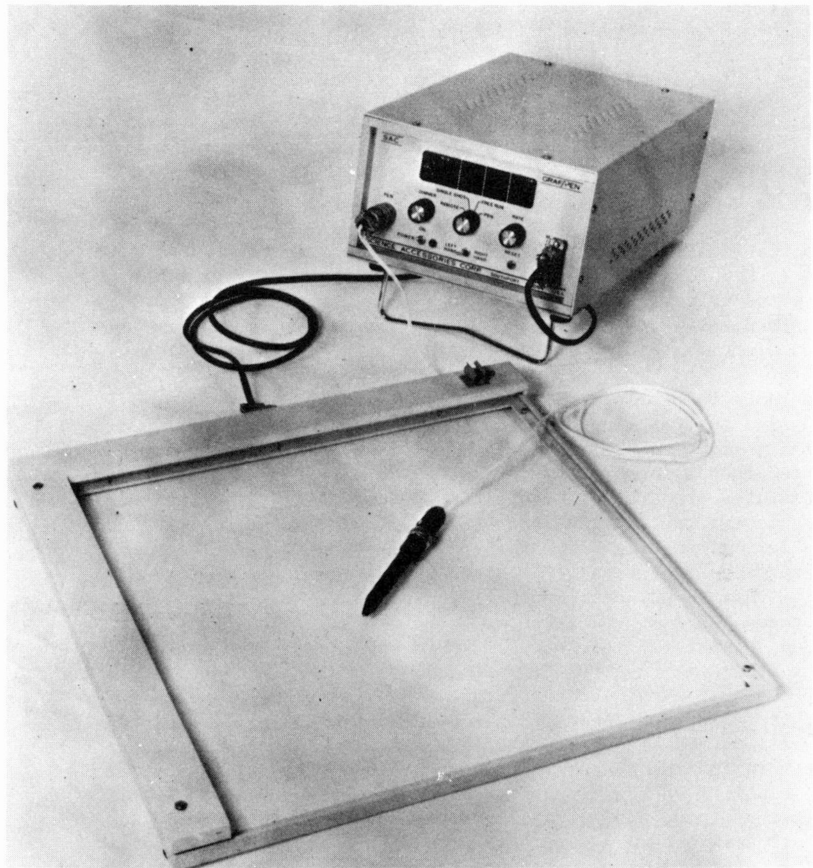

Fig. 20 Graf/pen sonic digitizer

the Science Accessories Corporation, is a big step in this direction. Its basic components are a pen, a control unit, and two-linear microphone sensors. The pen emits a hypersonic sound wave for each point to be digitized; the time required for the sound wave to reach a sensor mounted on the left side of the field represents the X coordinate; similarly, the time to reach a sensor mounted at the top of the field is the Y coordinate. Among its functions, the control unit converts the times into digits in the binary or binary-coded decimal codes for computer consumption.

A third pitfall is that computer graphics speaks with many tongues. In fact by 1967 more than thirty languages and dialects had been developed for programming conversational instruction. This lack of standardization was mentioned recently by General C. W. Meyers, Director of the Defense Systems Management School, and it has been bemoaned by others through the years. It is true that there are several types of users of interactive graphics systems requiring different capabilities and convenience factors. For instance, the computer system programmer specializing in computer graphics may need to be thoroughly familiar with a basic machine language. But each manufacturer should not have a different machine language for each model, nor should each manufacturer have his own assembly language intelligent to only a few personnel. In addition, the program and interface language should be universal enough that the applications or user programmer is familiar with this language and can talk intelligently with the system programmer. In other words, standardization should present common grounds of terminology. Also, an ordinary user should be able to obtain results in computer graphics without adding a knowledge of computer language to his specialty.

The American National Standards Institute (ANSI) has a Committee for Drawings and Drafting Practice (Y14). This committee has a subcommittee (Y14.26) whose purpose is "to investigate and develop justified recommendations for appropriate national standards for computer assisted preparation of engineering drawings." The chairman of the subcommittee is R. W. Rau. Also with the support of the Society for Information Display (SID), The Institute of Electrical and Electronics Engineers (IEEE), and the Electronic Institute of America (EIA), an Ad Hoc subcommittee on Display Parameters has been formed. This subcommittee is under the ANSI/X3 Computer Standards Committee. Sol Sherr is chairman of this committee. In addition, the Department of Defense has assigned Wallace Dietrich of the Naval Ship Engineering Center to be the principal contact for its Computer Aided Design/Numerical Control (CDNC) Standardization Area. In a fourth attempt at standardizing, a tri-service ad hoc committee on Interactive Graphics (TRICOG) has been organized. The TRICOG Technical Working Group is chaired by Robert M. Dunn. Will any or all of these efforts really develop a standardization? Can those several groups get together for common benefit? Some prominent areas are not covered by any of these groups. Will they be covered by one of these groups or another group in the future? Whether computer graphics lives up to its potential or falls by the wayside when a new glamorous development is introduced depends in great part upon these committees and future ones formed by concerned parties. Computer graphics must encourage and adhere to standardization!

A fourth pitfall of computer graphics might be its expense. An estimate of the cost of computer graphics should include not only the dollars required for lease or purchase of all physical inventories which include all remote interconnecting and local hardware but also the cost of all software and the education of programmers and users. In addition, wait or turnaround time, capability, and productivity should be considered. Counting all the overhead, the total cost of a console with interaction devices and host computer interface is still in the $100,000 bracket. CRT terminals have taken great strides in the past few years with the aid of minicomputers. "Minis" such as the CDC 1700 and the PDP 8 and 10 have become valuable work horses. In the future they will contribute even more to the advancement of computer graphics, for they can take care of the critical interface and core problems. More and more graphics CRTs, microfilm recorders, and plotters employ "minis" to do their mundane tasks. While minicomputers have reduced the cost of computer graphics, minicomputers as expected, have several drawbacks. A digital buffer is needed. The software is still inadequate and tends to be exclusively for the standalone system. Most minis have too few general purpose registers and too little core, inadequate bit and byte manipulation instructions, and poor operating systems. One possible solution to this might be a powerful intermediate-priced microprogrammed graphics. Another solution might be an improvement of the mini itself. The manufacturers need to listen to the views of their customers and satisfy their customers' requirements. In their turn, users need to be reasonable, definite, realistic, and cooperative.

A fifth pitfall might be the failure to satisfy the impatience of its users. No matter who the user--system or application programmer, scientist or manager, student or teacher, under 30 or over, occasional user or computer professional, employed or homemaker, educated or underprivileged--he or she expects quick turnaround time. This speed of response has always been requested in batch work. Even quicker response is expected of terminals and scopes. We need to turn to advances in technology for this.

Many past advancements for humanity could be cited. Many future achievements could be predicted for the computer graphics user who is a true gourmet. Hopefully his taste will always demand the finest and best contributions from computers and computer professionals.

ACKNOWLEDGMENTS

There is great indebtedness to a multitude of people only some of whom can be mentioned here. Robert P. Rich of the Applied Physics Laboratory, Johns Hopkins University, and Charles Bitterli and William Chapman, members of his staff, were especially helpful in the tracing of the early history of computer graphics. Also Rod Allen, COMPAID, Max Armour, Bell Helicopter Company, and Nora Taylor and Philip Battey, Naval Ship Research and Development Center, went to great efforts in order to help the author obtain source material. A special indebtedness is due Robert Herring, Naval Research Laboratory, who not only contributed source material but also contrubuted valuable reviewing comments, and to O. Dale Smith, Rockwell International, who fostered the author's initial graphics endeavors and has always given encouragement. The author's genuine thanks goes to Edward Habib, Code 1745, and Robert D. Short, Code 174, Naval Ship Research and Development Center, for their tolerance and helpfulness, and to the secretaries for their many skills and wonderful cooperation.

REFERENCES

1 McNaney, Joseph T., "The Type C19K Charactron Tube and Its Application to Air Surveillance Systems," 1955 National Convention IRE Convention Record, Part 5, 1955 p. 31.

2 Digital Computer Newsletter, Office of Naval Research Mathematical Science Division, Vol. 1, No. 1, April 1949, through Vol. 5, No. 4, Oct 1953.

3 Mejia, R., Battery, P., Baynes, P., Hairston, E., Hardy, D., and Cuthill, E., "SC 4020 Microfilm Recorder Programming Manual 1," David Taylor Model Basin Report 1469, Jun 1961.

4 Cheriwinski, Lois, "Illustrations of UAIDE SCORES Package Routines for the SC Microfilm Recorder," David Taylor Model Basin AML-29-68 Technical Note, Apr 1968.

5 Brown, G.D., Bish, C.H., Berman, R.A., "The Integrated Graphics System for the SC 4060: II. System Programmer's Guide," Rand Corporation Memorandum RM-5661-PR, Jan 1969.

6 "CALCOMP Drum-Type Digital Plotter Incremental and Zip Mode," California Computer Products, Inc., 221B 25M 1068, 1968.

7 Nelson, David L., "Perspective Plotting of Two Dimensional Arrays-Plot 3D, a Computer Program for a Digital Plotter," Technical Report 553, University of Maryland, Department of Physics and Astronomy, Mar 1966.

8 Karasek, F.W., "GC/MS Data System," R/D Research Development, Vol 24, No. 10, Oct 1973, p. 40.

9 "Finnigan Model 6000 GS/MS Interactive Data System," Brochure, Finnigan Corporation, Sep 1973.

10 Everett, R.R., "The Whirlwind I Computer," Review of Electronic Digital Computers, Joint AIEE-IRE Converence, 70, Feb 1952.

11 Sutherland, "SKETCHPAD - A Man-Machine Graphical Communication System," AFIPS Conference Proceedings, Spring Joint Computer Conference, Vol. 23, 1963, pp. 329-346.

12 Johnson, T.E., "SKETCHPAD": A Computer Program for Drawing in Three-Dimensions," AFIPS Conference Proceedings, Spring Joint Computer Conference, Vol. 23, 1963, pp. 347-353.

13 Hobbs, William F., Levey, Allan H., and McBride, Jane, "The Baylor Medical School Teleprocessing System - Operational Time-Sharing on a System/360 Computer," AFIPS Spring Joint Computer Conference, 1968, p. 31.

14 Smura, E.J., "Graphical Data Processing," AFIPS Conference Proceedings, Spring Joint Computer Conference, 1968, p. 111.

15 "Reservations via CRT at Pacific Southwest Airlines," CRT Notes, No. 21, November 1973.

16 Newman, W.M., and Sproull, Robert F., "Principles of Interactive Computer Graphics," McGraw-Hill Book Company, 1973.

17 van Dam, A., and Bergeron, R. Daniel, "Software Capabilities of the Adage Graphics Terminals," International Symposium Computer Graphics 70, Brunel University, Uxbridge, Middlesex, England, Vol. 2.

18 van Dam, A., and Michener, James C., "Storage Tube Graphics: Comparison of Terminals," International Symposium Computer Graphics 70, Brunel University, Uxbridge, Middlesex, England, Vol. 3.

19 Philips, Richard L., "Production of Computer Animated Films From a Remote Storage Tube Terminal," International Symposium Computer Graphics 70, Brunel University, Uxbridge, Middlesex, England, Vol. 1.

20 Gruenberger, F., (Editor), Computer Graphics Utility/Production/Art, Academic Press, 1967.

21 Anderson, S.E., "Generating Computer Animated Movies from a Graphic Console," Computer Graphics Quarterly SIGGRAPH-ACM, Vol. 4, No. 2, Fall 1970, p. 35.

22 Davis, Hatfield, "Computer Animation," Filmmakers Newsletter, Dec 1970, pp. 24-26.

23 Goldstein, Robert A., and Nagel, Roger, "Three-D Visual Simulation," Simulation Councils, Inc., Jan 1971, pp. 25-31.

24 Machover, C., "Computer Graphics for MIS," Proceedings Symposium Interactive Computer Graphics for Project Managers, Defense Systems Management School, Oct 1973.

APPENDIX - GLOSSARY

This Computer Graphics Glossary records the application of terms and concepts used within the computer graphics discipline. It is under continual review and revision to maintain it as a current reference in the field. In the text of this Glossary italicized terms are defined elsewhere in the Glossary and within a set of synonyms the definition follows the preferred term.

The Glossary was prepared under the auspices of SHARE, Inc., which owns the copyright. It has been published in an issue of SHARE's Secretarial Distribution (SSD), and may not be reproduced without permission of SHARE's Board of Directors. With the sole idea of advancing technical knowledge, SHARE, Inc. (which is not responsible for the contents) has given the author written permission to include the Glossary in this paper.

ABSOLUTE VECTOR: A directed *line segment* whose end points are measured in absolute units from a point designated as the *origin*.

ADDRESSABLE POINT: Any place on the *display surface* to which the *display writer* may be directed. These positions are specified by *coordinates*. Such addressable positions are finite in number and form a discrete grid over the *display surface*.

AIMING CIRCLE: A circle or other pattern of light projected by a *light pen* onto the surface of the *display* to guide the accurate positioning of the pen and/or to describe the *light pen's* field of view.

AUDIBLE ALARM: A short tone sounded to get the attention of the *display console* operator.

BLANKED ELEMENT: *Display data* that produce no visible output but change the position of the *display writer*.

BLINKING CONTROL: A programming technique or hardware option on a *display device* in which a *display element* is repeatedly displayed and erased. Usually used to attract the attention of the user.

BUFFER: (See DISPLAY BUFFER).

CAD: Computer Aided Design.

CADM: Computed Aided Design and Manufacturing.

CAI: Computer Aided Instruction.

CAM: Computed Aided Manufacturing.

CATHODE-RAY TUBE (CRT or crt): An electron tube whose face is covered with a *phosphor* that emits light when energized by its electron beam.

CHARACTER GENERATOR: Hardware which will generate a finite set of characters onto a *display surface*.

CLIPPING: (See SCISSORING).

COM: Computer Output Microfilm.

CONTRAST: The relationship of the brightest to the darkest portions of a *display image*.

CONTROL BALL: A ball that can be moved in at least two degrees of freedom and which moves one or more *display elements* by providing *coordinate* input to the *display device* (syn.: ROLLING BALL; see JOY-STICK).

COORDINATE: An ordered set of data values, either absolute or relative, which specifies a location.

CRT: Cathode-Ray Tube.

CRT DISPLAYS: Displays utilizing *cathode-ray tube* as the viewing element. (See RASTER SCAN, STORAGE TUBE, and DIRECTED BEAM.)

CURSOR: A movable marker visible on a *CRT display* used to indicate the position at which the next operation (e.g., insertion, replacement, or erasure of a character) is to take place.

DETECTABLE ELEMENT: A *display element* which has the characteristics which permit it to be identified by a *light pen detect*.

DIGITIZER: A device that codes images or shapes into digital computer usable data.

DIRECT VIEW STORAGE TUBE (DVST): (See STORAGE TUBE).

DIRECTED BEAM: The *CRT* method of tracing the elements of a *display image* in any sequence given by the computer program, where the beam motion is analogous to the pen movements of a *flatbed plotter*. This contrasts with the *raster scan* method which requires the *display elements* to be sorted in the order of their appearance (usually top to bottom, left to right).

DISPLAY: 1. A visual presentation of data. 2. Sometimes used as a synonym for a *display device*.

DISPLAY BACKGROUND: That portion of a *display image* which cannot be altered by the user. Sometimes called the static portion of the *display*.

DISPLAY BUFFER: A storage device or memory area that holds all orders and *coordinate* data required to generate a *display image*. This could include a portion of computer memory, direct access storage, or a special purpose storage device.

DISPLAY COMMAND: A display I/O instruction, such as those used to start or stop the *display device*.

DISPLAY CONSOLE: A hardware complex consisting of a *display device* plus one or more computer input devices. The types of input devices commonly employed are alphanumeric keyboards, *function keys*, *tablets*, *joysticks*, *control balls*, and *light pens*.

DISPLAY CONSOLE OPERATOR: A person using a *display device* and interacting with a computer at a *display console*. "User" is often used in place of "display operator,"

DISPLAY DATA: 1. Any collection of data intended for *display*. 2. Information in a *display buffer* which directly produces the *display image*.

DISPLAY DEVICE: A device capable of presenting information on a viewing surface or image area which uses *display elements* such as points, *line segments*, and/or alphanumeric characters to construct the display. This term usually refers to a *CRT* but also includes such devices as *plotters*, microfilm records, and page printers.

DISPLAY DEVICE COORDINATE SYSTEM: The set of numerical values assigned to the addressable points on the *display surface*.

DISPLAY ELEMENTS: The basic hardware-generated functions, such as points, *line segments*, characters, etc., produced by the *display writer*, used to construct a *display image*. *Display elements* can be combined to form *display entities*; e.g., *line segments* combined to form a square.

DISPLAY FILE: (See DISPLAY DATA, definition 2).

DISPLAY FOREGROUND: The collection of *display elements, entities*, and/or *groups* of a *display image* that are subject to change by the program or by the use in *interactive mode*.

DISPLAY FRAMES: Analogous to the successive frames in a motion picture film.

DISPLAY GROUP: A collection of *display entities* which can be manipulated as a unit and which can be further combined to form larger groups.

DISPLAY IMAGE: The collection of *display elements, entities*, and/or *groups* that are visually represented together on the viewing surface of a *display device*. Only one *display image* can be presented on a device's viewing surface at any one time.

DISPLAY MENU: (See MENU).

DISPLAY PANEL: A *display image* for use in the *interactive mode*. Panels usually include a *menu*.

DISPLAY SURFACE: That medium (paper, film, *CRT* screen, etc.) upon which the *display writer* produces a *display image*.

DISPLAY SPACE: The area defined by the *display device coordinate system*.

DISPLAY WRITER: The part of a *display device* used to create a visible mark on the display (e.g., ballpoint pen, liquid ink stylus, cutting stylus, laser beam, or electron beam).

DRUM PLOTTER: A *plotter* which draws an image on a recording medium (paper, film, etc.) supported by a drum; the *plotting head* moves parallel to the line of rotation for one axis while the drum rotation provides the other axis.

FLATBED PLOTTER: A *plotter* which draws an image on a recording medium (paper, film, etc.) affixed to a flat table.

FLICKER: A blinking or pulsation of a *display image* on a *CRT*. Flicker occurs when the *refresh rate* is so low that *regeneration* becomes noticeable.

FLYING SPOT SCANNER: A system that encodes a picture by *raster scanning* and recording the brightness at each addressable point.

FUNCTION BUTTON: (See FUNCTION KEY).

FUNCTION KEY: A push button or switch which may be pressed to send an identifiable interrupt to the display control program.

FUNCTION KEY BOARD: An input device for an interactive display console consisting of a number of *function keys*.

FUNCTION MENU: (See MENU).

GAS PANEL: (See PLASMA PANEL).

GRAPHIC (adj.): The adjective "graphic" can be used in place of the adjective "display" in the terms in this glossary. "Graphic" usually refers to those devices which draw lines and points.

GRAPHIC DATA: (See DISPLAY DATA).

GRAPHIC LANGUAGE: The software interface between the programmer and the *display device*.

GRAPHICS: The art of image generation and manipulation. "Graphics" usually applies to computer-generated displays which contain lines and points. However, the term has been used to indicate displays containing only alphanumeric data.

Interactive graphics is a technique of using a *display console* in the *interactive mode*.

Passive graphics is a technique of using a *display device* in the *passive mode*, usually associated with *plotters* and microfilm recorders.

HARD COPY: A permanent copy of a *display image*.

HELP MODE: (See PROMPT MODE).

HIDDEN LINES: *Line segments* which are obscured from view in a projected image of a three-dimensional object.

IMAGE: (See DISPLAY IMAGE).

IMAGE SPACE: (See DISPLAY SPACE).

INCREMENT SIZE: (See PLOTTER STEP SIZE).

INCREMENTAL VECTOR: (See RELATIVE VECTOR).

INTENSITY LEVEL: One of the discrete levels of brightness of the light emitted by a *CRT*, usually under program control.

INTERACTIVE GRAPHICS: (See GRAPHICS).

INTERACTIVE MODE: A method of operation that allows on-line man-machine communication. Commonly used to enter data and to direct the course of a program.

JOYSTICK: A level that can be moved in at least two degrees of freedom and which moves one or more *display elements* by providing *coordinate* input to the *display*.

LIGHT BUTTON: A *display element* which may be selected by an input device and is programmed to operate as a *function key* (also called VIRTUAL PUSHBUTTON).

LIGHT GUN: (See LIGHT PEN).

LIGHT PEN; A *stylus* which detects within a limited area (e.g., *aiming circle*) light generation on a *CRT*. Also see LIGHT PEN DETECT.

LIGHT PEN DETECT: The sensing by a *light pen* of light generated on a *CRT*. It can provide an interrupt which may be interpreted by a *display* control program to determine either positional or *display element* identifying information.

LINE SEGMENT: A finite section of a line, usually described by its end point.

LINEARITY: A measurement of straightness of a plotted *line segment*.

MAPPING FUNCTION: A transformation which converts all the points of one coordinate system into another.

MENU: A list of options on a *display* allowing an operator to select his next action by indicating one or more choices with an input device.

MODEL: The representation in computer storage of all relevant application data.

MODEL SPACE: The coordinate system of the *model's* geometric data.

MOUSE: A hand-held device, with two perpendicular wheels, which is rolled around on a flat surface to provide coordinate input to the *display device*. (See CONTROL BALL.)

ORIGIN: A reference point whose *coordinates* are all zero.

PAGE: In *graphics*, a single *display image* of a set of *display images*. These *display images* are usually alphanumeric text.

PAGING: In *graphics*, the process of replacing one *page* with another *page*. Usually the *pages* are displayed in consecutive order, either forward or backward.

PARALLAX: The apparent displacement of an object as seen from two different points; e.g., the difference between where the eye perceives an object and where the *light pen* can perceive it.

PASSIVE GRAPHICS: (See GRAPHICS).

PASSIVE MODE: A method of operation that does not allow any on-line interaction or alteration.

PFK: Program Function Key.

PFKB: Program Function Key Board.

PHOSPHOR: The chemical coating on the inside face on a *CRT* which glows with visible light when energized by an electron beam.

PLASMA PANEL: A *display device* consisting of a flat, gas-filled panel containing a grid of wires. Energizing grid intersection points ionize the gas, thus emitting light.

PLOTTER: A device for making a permanent copy of a *display image*.

PLOTTER STEP SIZE: Length between two adjacent addressable points in the horizontal or vertical direction. This is analogous to a *raster unit* on a *CRT*. Also called *increment size*.

PLOTTING HEAD: A *display writer* used on mechanical plotters.

PROGRAM FUNCTION KEY: (See FUNCTION KEY).

PROGRAM FUNCTION KEY BOARD: (See FUNCTION KEY BOARD).

PROMPT MODE: Prompting using separate *display panels*.

PROMPTING; The method of informing the user of possible actions.

RASTER SCAN: A technique for generating or recording an image with an intensity controlled, line-by-line sweep across the entire *display surface*. (This technique is used to generate a picture on a TV set and to digitize an image with a flying spot scanner.)

RASTER UNIT: The distance between two adjacent addressable points in the horizontal or vertical direction on a *CRT display*. Analogous to *plotter step size*.

REFRESH RATE: The rate at which a display is *regenerated*.

REGENERATION: The process of repeatedly displaying an image on a *CRT display* device. Since the image is retained by the phosphor for only a short period of time, the image must be regenerated in order to remain visible. (See REFRESH RATE and FLICKER.)

RELATIVE VECTOR: A *vector* whose starting point is the end point of the preceding *display element;* and whose end point is specified as a displacement from the starting point.

REPEATABILITY: A measure of the hardware accuracy of the retrace of a *display element*.

RESOLUTION: 1. Plotter step size or *raster unit*. 2. The smallest distance between two display elements which can be visually detected as two distinct elements.

ROLLING BALL: (See CONTROL BALL).

RUBBER-BANDING: A technique for displaying a straight line which has one end fixed and the other end following a *stylus* or other input device.

SCALING: Transforming one or more *display elements* by multiplying all dimensions by a constant value, thus magnifying or reducing these elements in a *display image*.

SCISSORING: Removing parts of *display elements* which lie outside defined bounds.

SCROLLING: The continuous vertical or horizontal movement of *display elements* within a *window*. As new data is moved into the *window* at one edge, the old data is moved out at the opposite edge. The *window* may include the entire *display*.

SKIATRON TUBE: A *CRT* whose electron beam causes a phosphor surface of a tube to darken rather than brighten. The image is viewed by illumination from the rear in a dark trace against the otherwise transparent or translucent face of the tube.

STORAGE TUBE: A *CRT* which retains an image for an extended period of time without *regeneration*.

STYLUS: A hand-held object which provides *coordinate* input to the *display device*.

TABLET: An input device which digitizes coordinate data indicated by *stylus* position.

TRACKING: Following or determining the position of a moving input device (e.g., the writing tip of a *stylus)*.

<u>TRACKING SYMBOL</u>: A cross or other symbol in a display used for indicating the position of a *stylus*.

<u>UNBLANKED ELEMENT</u>: A visible *display element*. (Cf. BLANKED ELEMENT.)

<u>VECTOR</u>: A directed *line segment* (i.e., one in which one end point is identified as the initial point, and the other end point as the terminal point). (See RELATIVE VECTOR and ABSOLUTE VECTOR.)

<u>VIRTUAL PUSHBUTTON</u>: (See LIGHT BUTTON).

<u>VOLTAGE PENCIL</u>: (See STYLUS).

<u>WINDOW</u>: A bounded area within a *display image* that contains a *scissored* subset of the displayable data of the *model*.

<u>ZOOMING</u>: Continuous *scaling* of all elements of a *window* to give the appearance of moving towards or away from the object of interest.

THE SAGE SYSTEM

J. Shomrat, E. Schweid, and M. Newman
Israel Aircraft Industries Ltd.
Lod Airport, Israel

ABSTRACT

A computer graphics preprocessor, SAGE, used in connection with large finite element structural models, is described. The SAGE system is designed to operate in a time-sharing environment and is compatible with a TEKTRONIX 4002 or 4010 graphic display terminal of the storage CRT type, connected to an XDS SIGMA-7 computer. The software consists essentially of a preprocessor and a display program, written mostly in FORTRAN. Display capabilities include any combinations of spatial rotations, translations, sectioning by two planes, and blow-ups of selected model regions. In addition, the user may correct or modify his model on line through a series of update commands. It has been found that the SAGE system is highly cost effective from the standpoint of savings in engineering man-hours, computer run time reductions, and minimization of schedule slippages.

INTRODUCTION

Finite element structural analysis programs such as NASTRAN are used to analyze complex structures consisting of a large number of elements and nodes. A large amount of data, usually in the form of punched cards, must therefore be prepared. The introduction of errors while preparing this data is almost unavoidable, and error elimination through manual checking of each data card is a very inefficient and time-consuming process. The SAGE system was designed to make the process of verifying the geometry and connectivity of a model more efficient by enabling the user to display, manipulate, and correct his model using a graphic display terminal connected on-line to a central computer and operating in a time-sharing environment. The system has been in productive use since mid-1972, mainly for verifying and correcting NASTRAN data, and has proven to be an invaluable preprocessing tool for the structural analyst.

HARDWARE CONFIGURATION

The SAGE system has been designed to operate with a TEKTRONIX 4002 or 4010 graphic display terminal of the storage CRT type, connected to a XDS SIGMA-7 computer via a 1200 bits/second full-duplex line. It is used in a time-sharing environment under the UTS operating system. The choice of this graphic terminal was based mainly on economic considerations. In addition, this terminal has the advantage that a large amount of information may be displayed with no flicker problems as opposed to refreshed CRTs. The major disadvantages of the terminal are found to be its small screen size (14" x 16"), the relatively long time required to display a model, and the lack of selective erasure capability.

SOFTWARE DESCRIPTION

The SAGE system consists of essentially two separate programs: a preprocessor and a display program. The basic specifications for the design of a program required that it be flexible and expandable in the sense that data for several finite element programs might be processed and also that new element types could be added as needed. For this reason, as well as for programming efficiency, the program was built in a modular fashion. Except for a small number of subroutines, which are programmed in assembly language, the system is programmed in FORTRAN.

PREPROCESSOR

The preprocessor is a program used in a batch mode. Its inputs are the same data cards which are to be used by the analysis program. The preprocessor performs the following major operations:

1. Interpretation of data cards and creation of files for nodal data, element data, and coordinate system data. These files have a format independent of the analysis program so that the only module which is analysis-program dependent is the interpreter.
2. Transformation of all nodal coordinates to a basic rectangular system.
3. Merging of element and nodal data to create a display file. The display file is organized by elements so that all data pertaining to an element, such as element number and type, node numbers of the element and coordinates of each node, are sequential.

Each display file generated is defined by name and may be accessed from the terminal by that name.

DISPLAY PROGRAM

The display program is constructed of several modules as follows:

a. Command interpreter
b. Element display generator
c. Display manipulation routines
d. Display file updating routines
e. Terminal communication routines

The functions of these modules are explained below.

Command Interpreter

The user, sitting at the terminal, controls the operation of the program by commands, each of which consists of a command name with up to four characters, followed in most cases by parameters. The commands are entered through the keyboard and in addition, certain parameters are entered with a cursor device (joystick or thumb-wheels). The command interpreter reads the command name, processes the parameters, and then calls the appropriate module to act on the command.

Element Display Generator

The element display generator reads the display file one record at a time, scans the file element by element, and accesses an appropriate routine, according to the element type, which generates the sequence of coordinates (in the model coordinate system) that describe the element and its special symbology. This sequence of coordinates is transferred to the display manipulation module. In addition, node and/or element numbers may be displayed.

The user can ask to display elements by element type and/or element number. Each element is displayed by its edges and nodes plus special symbols which serve to distinguish between element types (the display of these symbols is optional). Examples of several element symbols are shown in Fig. 1.

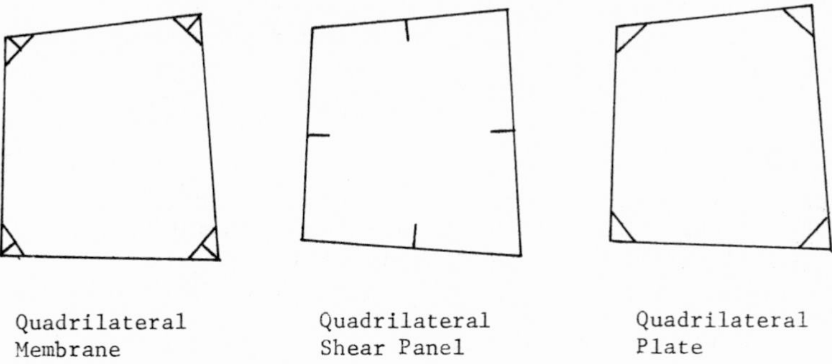

Fig. 1 Typical element display symbology

Display Manipulation Module

The display manipulation module causes the element data to generate various model displays by means of straight lines in the screen coordinate system. The required manipulations may be specified directly by user commands or may be performed automatically as a consequence of other user operations. They consist of any combination of spatial rotations, translations, sectioning by two planes, and blowing-up any area of the display. Specification of the cutting plane or area to be blown up is input by the joystick or thumbwheel cursor device.

Display File Updating Module

Several commands exist which enable the user to correct and modify his model on line. These commands allow the user to define new nodes, change coordinates of existing nodes, define new elements, change the node definition of existing elements, and delete elements. When such a command is processed, the display file is updated and the correction is immediately displayed.

Terminal Communication Routines

The low-level terminal communication routines perform formating of coordinates as required by the terminal and handle the input and output to and from the terminal.

EXAMPLE OF SYSTEM USAGE

In order to demonstrate the display, checking, and error detection capabilities of the SAGE system, a representative case sequence is illustrated below. Clearly, many alternate mode of operation are possible, and the steps described may be changed extensively at the option of the user.

Step 1. Display the entire model

A side view of the model, consisting of an aircraft fuselage section, is shown in Fig. 2. The model is automatically scaled so as to be completely within the limits of the screen.

Step 2. Examine display and detect errors

Some obvious modeling errors are evident in the display, but their sources are not entirely clear. Therefore, the structure is rotated into other orientations as shown, for example, in Fig. 3, revealing the previous errors more clearly and some additional ones besides. The preprocessor provides a data printout which can further assist the user in detecting the actual input data errors.

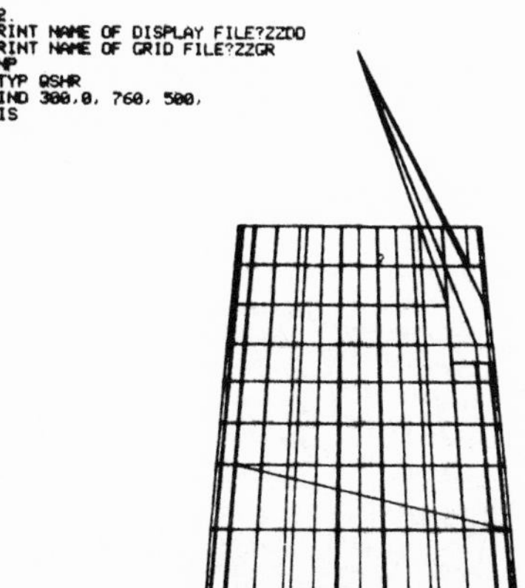
SL2.
PRINT NAME OF DISPLAY FILE?ZZDO
PRINT NAME OF GRID FILE?ZZGR
?INP
?DTYP QSHR
?WIND 300.0, 760, 500,
?DIS

Fig. 2 Display of an aircraft fuselage section showing some outstanding errors

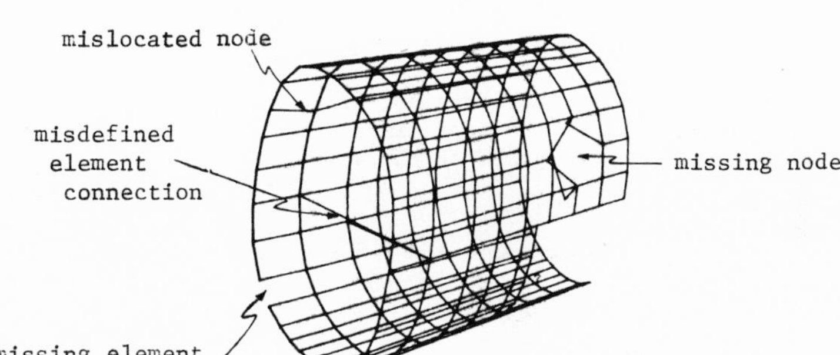
ROTO X-45 Y-45 Z-45
?DIS
mislocated node
misdefined element connection
missing node
missing element

Fig. 3 Rotated view of structure revealing additional modeling errors

The missing node error implies that the coordinates of the node in question have not been defined (e.g., a missing grid card). In the SAGE system, such nodes are given a coordinate location of (0.0, 0.0, 0.0).

Step 3. Isolate selected substructural regions for subsequent error correction activities

In Fig. 4, the structure has been sliced longitudinally, retaining the left section for subsequent error correction activities. In addition, node numbers have been added to the display. This substructure can now be rotated, translated, sectioned, and blown up at the request of the user, as in the case of the complete structural model.

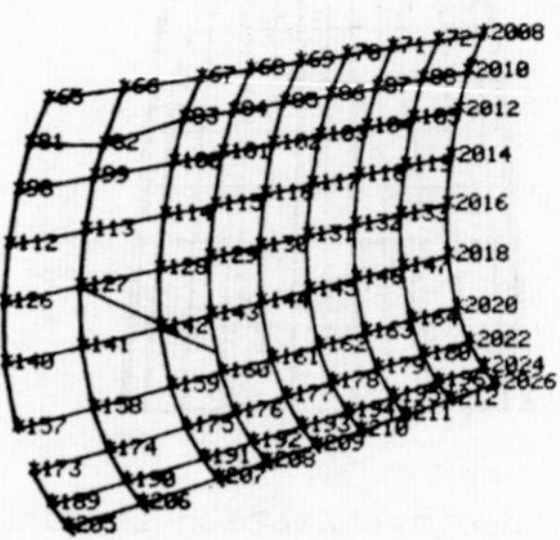

Fig. 4 Left substructure region showing node numbers

Step 4. Correct misdefined nodes and display corrections

It is evident from the display that node no. 82 has been incorrectly defined. Figure 5 shows the typed-in correction and a display of this correction superposed on the structural model.

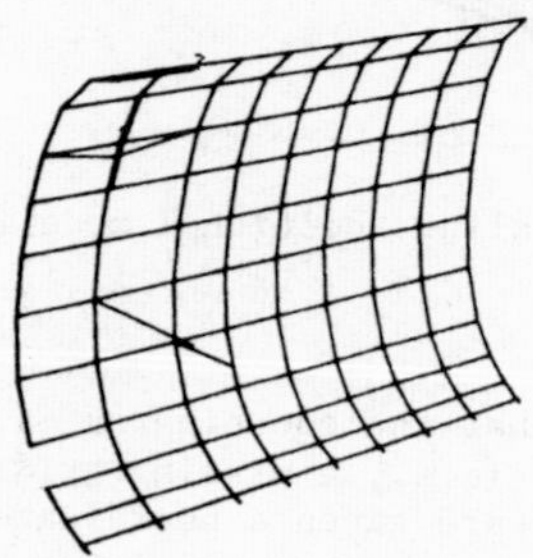

CHGR
INPUT FOR CGRD IS ID,CP,X1,X2,X3,CD,PS?82. 0. -27.18 328.3 50.372 0. 0.

Fig. 5 Correction of misdefined node no. 82

Step 5. Add missing elements

First, the element identification numbers are requested and displayed as shown in Fig. 6. In this particular case, a shear panel element, no. 3025, immediately to the left of panel no. 3026 is found to be missing. Figure 7 shows the typed-in element data and the addition of this element to the model.

DIS E

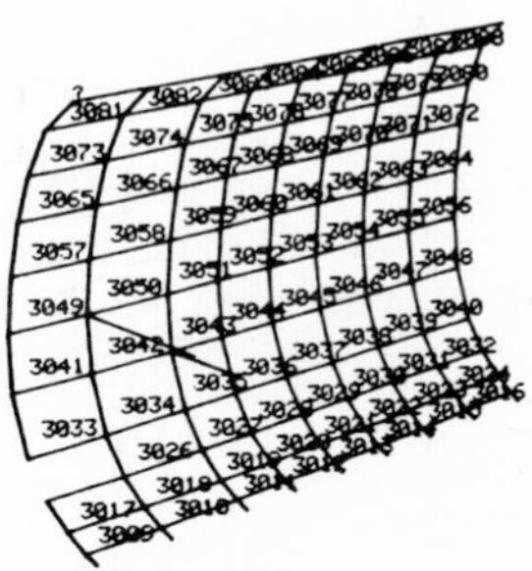

Fig. 6 Left substructure region showing element identification numbers

ADDE
PRINT ELEMENT NAME AND ID ?QSHR 3025.
PRINT GRID NUMBERS ?157. 158. 174. 173.

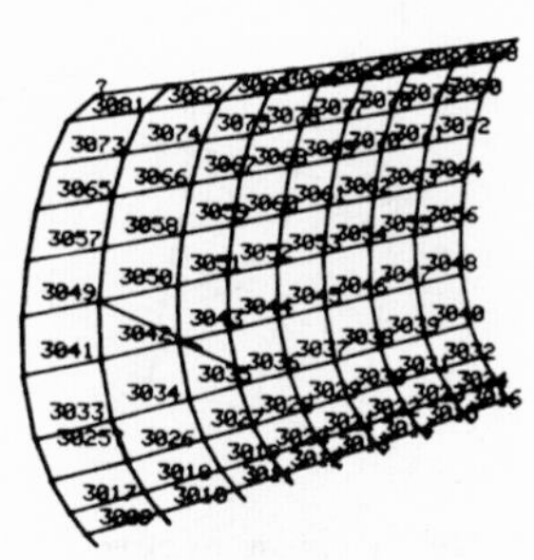

Fig. 7 Addition of shear panel element no. 3025

Step 6 Correct misdefined elements

Figure 8 shows a blow-up (zoom) in the region of element no. 3042. Note that the existence of shear panels within quadrilateral boundaries are indicated by midside tick lines perpendicular to the sides and extending into the panel regions. This display symbology makes it clear that one of the corner nodes of the panel in question has been misdefined. Figure 9 indicates the correction instruction (node no. 90 replaced by node no. 128) and shows the correct element superposed on the invalid one.

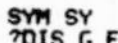

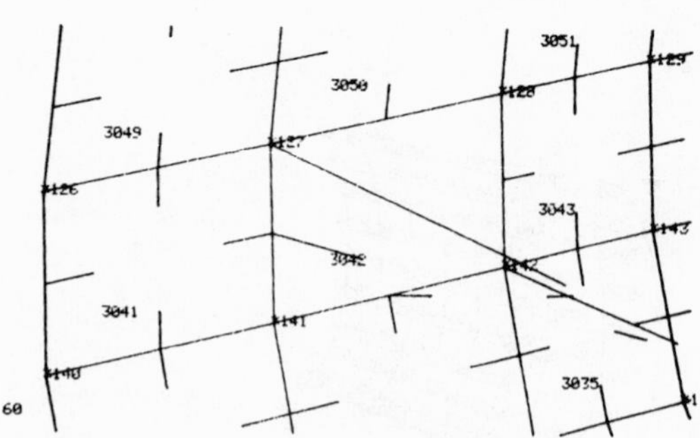

Fig. 8 Blow-up (zoom) in region of misdefined panel no. 3042

?CHGE 3042 90 128

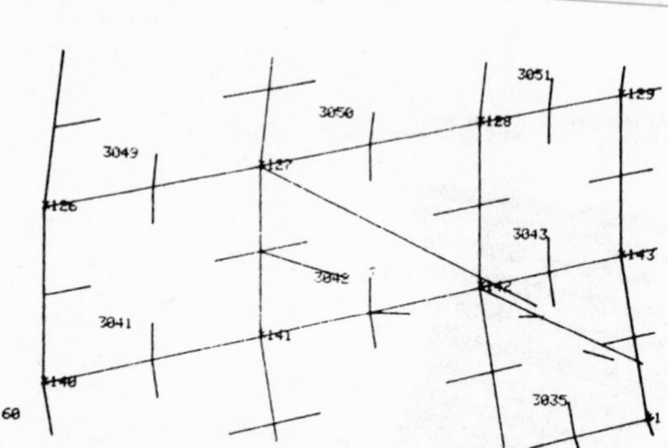

Fig. 9 Correction of panel no. 3042 definition error

The above procedures are repeated for other model regions until all indicated errors have been corrected. In Fig. 10, for example, a longitudinal sectioning of the right side of the structure, containing a missing node, has been performed, and node numbers have been indicated. It is found that node no. 63, used to define panel corner points, has not been defined via coordinates on a grid card. As mentioned earlier, the SAGE system assigns the coordinates (0.0, 0.0, 0.0) to all such missing nodes. Figure 11 shows the input of the node definition data and the modeling change resulting from the correction.

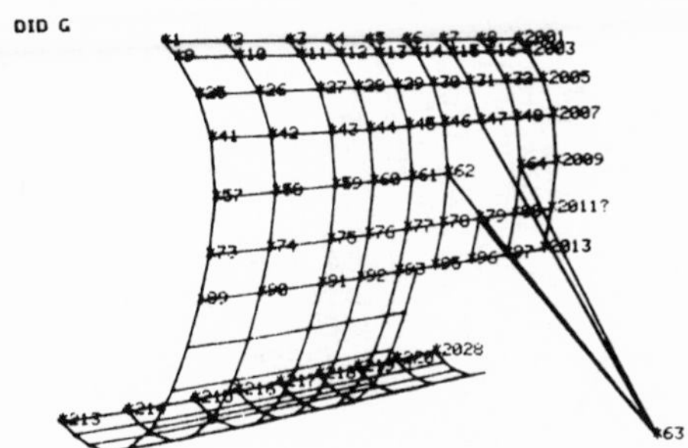

Fig. 10 Right section showing undefined (missing) node no. 63

```
ADDG
INPUT= ID CP X1 X2 X3 CD PS?63. 0. 20.988 369. 60.625 0. 0.
```

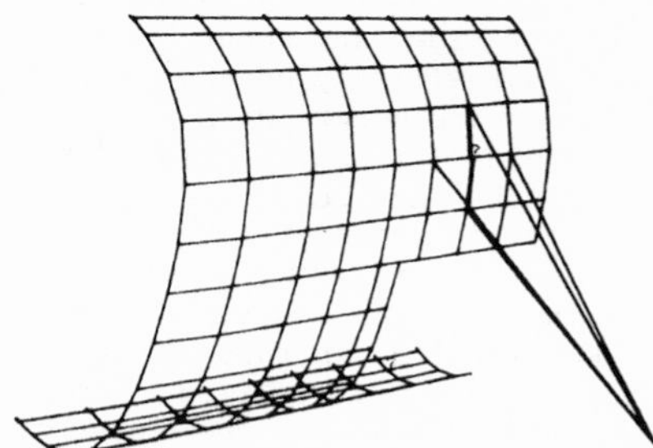

Fig. 11 Addition of missing node

Step 7. Redisplay entire model to verify its final validity

The result of this operation is shown in Fig. 12. It is worthwhile mentioning that the user may terminate this or any other display operation at any time by pressing the BREAK key on the terminal.

EFFECTIVENESS EVALUATION

Considering a typical case of a finite element model consisting of approximately 1,000 nodes and as many elements, our previous experiences indicate that the conventional process of verifying and correcting all errors by conventional, nongraphical examination methods would require on the order of 80-100 man-hours and perhaps 3-5 reruns on a large-scale computer, amounting to roughly 2 hours of computer time. In addition, the calendar time consumed in the process might be on the level of 2 to 3 weeks.

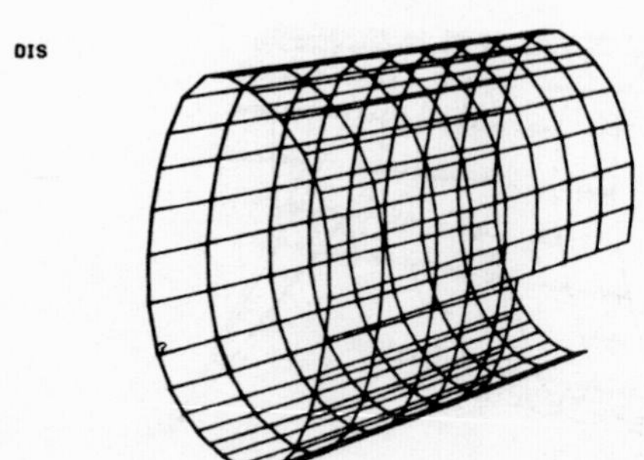

Fig. 12 Final, corrected model display

Using the SAGE system, it has been found that the same job can typically be completed with an expenditure of 8-10 man-hours, 15 to 20 minutes of computer time, and about 2 days of elapsed calendar time, i.e., for approximately one-tenth the time and man-hour expenditure involved in the nongraphics process. Further, the cost of using the graphics terminal is almost insignificant compared to the other expenses incurred.

Aside from the savings in computer costs, manpower, and schedule time, the SAGE system has given our structural engineers increased confidence in the accuracy and reliability of their computerized applications work. Although this advantage cannot be directly measured in material terms, its importance from the standpoint of improved work quality cannot be overemphasized.

DEDICATION

This paper is dedicated to the memory of Gideon Kain.

THE GIFTS SYSTEM

H. A. Kamel and M. W. McCabe
Aerospace & Mechanical Engineering Dept.
University of Arizona
Tucson, Arizona

ABSTRACT

The paper describes a graphics-oriented general purpose analysis package aimed particularly at ship structure design using the finite element method on a time-sharing system.

INTRODUCTION

The acronym given the system, GIFTS, stands for a Graphics-oriented Interactive Finite element analysis package for Time Sharing systems. Its design goals are:

1. Continuity of operation, so as to be suitable for structural design applications.
2. Full accessibility to all data pertaining to the problem at all times through a time-sharing graphics terminal. The use of graphical display is preferred whenever possible.
3. Low-cost graphics terminals, to allow access to a large number of users with minimum capital outlay.
4. Minimum core utilization, to ensure high priority on a large time-sharing computer system and to enable execution on minicomputers.
5. A high degree of machine independence.
6. Inherent flexibility so that the program may be expanded and maintained in response to user demands and hardware innovations. Some of the changes may be introduced by the user to help him solve special problems or may be shared with other users within the frame of a user's club.

In designing the program system, the concept of a general purpose program is dropped in favor of a general purpose Unified Data Base (UDB), which contains data describing the problem at hand and is permanently stored on disk as a set of random access files of fixed format. Program modules are then written to operate on the UDB.

The same concept enables the generation and display modules of the system to be used in conjunction with other analysis programs such as NASTRAN and DAISY. Although the current system addresses itself mainly to ship structure design, care has been taken that the UDB caters to all types of finite element and related computations, such as heat transfer and the analysis of solids.

The need for minimum core allocation has prompted a study in sparsely populated matrix representation and the mechanics of submatrix propagation during the solution procedure. An interesting result is that the core requirement is practically independent of problem size. Buffering, however, may be needed to speed up the solution of large problems.

Iterative solution schemes are an attractive alternative in certain cases, and different schemes are being tested for their suitability in an interactive environment. Some preliminary results are available. Another potentially useful technique is that of substructuring. Work is under way to implement the method, and results are expected within the near future.

The creation of the UDB has shown a need for the classification of the display models according to topology and geometry rather than according to the engineering discipline involved. In this fashion, the same display routines may be used with different model types.

BASIC MOTIVATION, SURVEY, AND CRITIQUE OF OTHER AVAILABLE SYSTEMS

If the finite element method is to replace classical techniques in the daily analysis of complex problems, the communication between the user and the computer has first to be improved. Since communication is a two-way process, attention has to be given to both channels of information flow. The speeding up of the information flow from the engineer to the computer is primarily achieved through automatic mesh generation, which is a form of information packing. To speed up the flow of data from the computer to the user, graphic display devices are the obvious answer. If the engineer is regarded as a processor quite different in characteristics from the computer, particularly with respect to its peripheral devices, then an efficient system should utilize the most suitable combination of input/output devices--the human hand is at best a slow and clumsy peripheral device.

In addition to the above, economic considerations play an important part. The user should not be expected to invest unreasonable sums for initial equipment not to pay unnecessarily high rental charges.

To achieve this goal, one of two alternatives is feasible minicomputer with graphics capabilities or a time-sharing system in conjunction with inexpensive storage tube terminals. Refreshed scope systems, superior in certain aspects such as speed of picture plotting, seem to be economically feasible only in conjunction with minicomputer systems.

Time-sharing systems may be nationwide or, for large companies, a local in-house facility. Increasing attention has been paid lately to the problem of program transferability. Nationwide networks seem to provide an attractive medium through which both programs and data may be shared. This step, in many instances successfully applied in business activities, should be considered more seriously by the scientific user.

Full control of the entire modeling, analysis, and display process should be exercised from the terminal. For a time-sharing user to be dependent on the input of a card deck in a computer center 2000 miles away contradicts the very nature of the time-sharing concept. One exception is the use of magnetic tapes to dump a large disk area in order to reduce storage charges. The concept of "terminal independence" is an important criterion by which to judge such a system.

The design of a graphics finite element system should tackle all stages of the analysis, namely, model generation, model editing, model display, load generation, load editing, load display, boundary condition generation editing and display, analysis, and result display. An overoccupation with the mesh generation and display, for example, and an almost total neglect of load and boundary condition manipulation is typical of many systems and represents an incomplete solution to the problem.

In handling complex models, it should be possible to mask most of the information and only display certain portions at a time. This criterion is most important for practical size problems.

The ability to generate automatically large amounts of data should always be accompanied by the capability of editing such data in order to provide for local detail and occasional irregularities.

In general, it seems that systems available today do not fulfill all the requirements to the degree that the GIFTS system does. Some were designed on sound system principles but handle only small problems. Other systems, designed to handle large complex models, grew mostly in an organic fashion, as the need arose, with no clear design goal in mind. They typically suffer from overplotting, or the tendency to show too much information on the screen.

In order to evaluate an interactive graphics system, the users feel that the following items should provide at least a partial basis for evaluation and possible comparisons.

1. Well-defined data base
2. Data base accessible in a simple manner
3. Suitability to minicomputer application
4. Mode of operation (batch, remote batch, dedicated interactive, interactive time sharing)
5. Percentage of coding in FORTRAN IV (or other universal high-level language)
6. Simplicity of plotting via FORTRAN callable routines
7. Disk file types (sequential, random access)
8. Can new files be generated from the program?
9. Number of different computers implemented on
10. Can mesh generation routines be called from the terminal?
11. Can generation, computation, and data editing be taken in any sequence?
12. Can the analysis be performed by programs other than the package, such as NASTRAN or DAISY?
13. Can the system merge data bases to form a new larger model?
14. Can the system extract one data base from another?
15. Can the system handle more than one data base (more than one problem) within the same disk space?
16. Does the system have a dictionary relating internal and external node numbering schemes, so that node renumbering does not

prevent the user from using the original numbering scheme in communicating with the program?

17. Is the system bandwidth limited?
18. What is the recommended maximum number of unknowns for the interactive analysis program?
19. What is the maximum number of loading cases?
20. Availability of 2-D model display and edit programs
21. Availability of 3-D shell model display and edit programs
22. Availability of 3-D solid model display and edit programs
23. Does the system have load display and editing capabilities?
24. Does the system have boundary condition display and editing capabilities?
25. Can one display the deformed shape?
26. Can one control the displacement magnification factor?
27. How does the system display stresses?
    a. contours
    b. vectors
    c. coded symbols
28. Can the system display
    a. individual stress components
    b. principle stresses
    c. von Mises generalized stress criterion, or similar quantities
29. Can the system display displacements and stresses
    a. simultaneously
    b. individually
30. Can the system select special areas for display, and how
    a. by topological considerations
    b. by boxing in three-dimensional space
    c. by windowing two-dimensionally on the screen
    d. by node numbers
    e. by element numbers
    f. by element types
31. Can the display be rotated into
    a. a number of fixed positions--how many
    b. to any position
32. Is perspective depth included?
33. Can the system display node numbers?
34. Can the system display element numbers?
35. Can the system display material codes on element plot?
36. Can the system display element thickness codes on element plot?
37. Can loading cases be superimposed, with weighting factors, to produce composite displays?
38. Availability of 2-D mesh generators
39. Availability of 3-D shell mesh generators
40. Availability of 3-D solids mesh generators
41. Does the system have automated load generation?
42. Does the system have automated boundary condition generation?
43. Does the system have its own analysis package?

44. What basic element types does the analysis package have?
    a. rods
    b. beams with shear and torsion capabilities
    c. triangular membrane elements
    d. quadrilateral membrane elements
    e. triangular axisymmetric elements
    f. quadratic axisymmetric elements
    g. triangular plate elements
    h. quadrilateral plate elements
    i. solid tetrahedron elements
    j. solid brick (8 nodes) elements
    k. others
45. Does the system design allow for substructuring?
46. Does the system design allow for local analyses performed subsequent to a major analysis with fully automated data transfer?
47. Does the system have dynamics capabilities (vibrational modes)?
48. Does the system have transient response capabilities?
49. Does the system have nonlinear capabilities?
50. Core requirement on a large computer system
51. Minimum and maximum terminal configuration (with purchase or lease prices)

DESCRIPTION OF THE GIFTS SYSTEM

The package being designed is not a single program. While a general purpose program is perhaps necessary and desirable for large batch oriented applications, it would be the wrong configuration for a time-sharing graphics package. The GIFTS system is designed, therefore, as a comprehensive collection of programs, subprograms, and subroutines, as well as a standardized data base. Thus, the programs are only a part of the system and may be modified, expanded, or replaced as the need arises. Each program or subprogram operates on the data base and results in data modification. In some instances it may add new items to the data base. Programs are thus data manipulators, and the central concept is that of the Unified Data Base. Although the GIFTS system is meant to be self-contained, it is possible, thanks to its modularity, to substitute any large analysis package, such as DAISY or NASTRAN, using a suitable interface, for one or more of its components. It is also possible for GIFTS, or any of its modules, to serve as a subprogram for a more comprehensive system.

The UDB contains information describing the model nodes, the elements, the freedom patterns, the loads being applied, the boundary conditions, the stiffness matrix before or after decomposition, stresses, material properties, etc. Additionally the data base maintains a status list which keeps track of the progress of the solution and prevents the inexperienced user as much as possible from committing flagrant errors. It also provides the experienced user with automatic selection of operations in order to avoid excessive decision making on an uninteresting level. Data storage, in the form of the UDB, is as transparent as program coding. Information is stored mostly as random access files with fixed record length. All data are accessible through FORTRAN coding. It is possible to preserve all files of a particular job on tape through a "save" routine and to recall the job at a later time in order to continue the analysis. In this manner disk storage costs may be reduced.

One of the basic assumptions of the system is that the analysis of a problem is conducted in a number of steps. In between steps the data are preserved in the UDB. A certain amount of disk software is required, particularly random file data access. Fortunately, all large systems, and most small systems, now have such a capability. In case of nonavailability of true random access one may simulate it through relatively simple subroutines, although use of such inadequate computer systems is discouraged for reasons of efficiency.

The user of the system is highly conscious of the program progress in a physical sense. He knows, for example, whether the stiffness matrix has been generated or whether the loads have already been applied. He may retrieve the status of the computation through the use of an information package. The user also has complete access to data on the disk. The files are all described in sufficient detail, and the user's manual contains a guide on how to read and write records from the particular files, using the standard systems software as much as possible. In this fashion the user is encouraged to write his own supplementary routines in order to suit his particular application and his current needs.

Since the data base contains all the information required in order to perform most of the standard finite element operations, one may be able to access the same data base with several different programs. For example, one may use a heat conduction program to calculate the temperature distribution within a model and follow up with the application of an axisymmetric solids program to calculate resulting thermal stresses. It is also possible, due to a special file protection feature, that a user has access to several different problems within the same disk space. It is equally possible to provide special packages that may merge several data banks and produce a new UDB under a different name. Several programmers may access the same data bank in order to operate as a team during complex model generation or result examination.

The most economical graphics terminal which offers satisfactory performance for engineering purposes appears to be a storage tube device coupled with a hard-copy attachment, a high-speed data transmission interface, and, possibly, a writing tablet.

THE UNIFIED DATA BASE (UDB)

The GIFTS system is designed to operate in core with minimum disk requirements. It is therefore imperative that all data pertaining to the problem be disk resident. Although the package is aimed at ship design using the finite element method, it is desirable that the data base be sufficiently general to allow, for example, heat transfer computation, axisymmetrical solids, and so on. For this reason, a standard disk file data storage format that is both efficient and general had to be designed. Allocation of file space is dynamic in the sense that a default problem size is originally assumed. As the model is finalized, the file sizes are cut down to the actual problem size. Should the model exceed the default size parameters, files are automatically increased to accommodate the additional data. All this is performed in a fully automatic fashion, without inconveniencing the user in any way. The proposed UDB has the following advantages.

1. It is possible under the new system to save and restore the job conveniently so that the analysis may be performed in stages.
2. Through the common data base it is possible to interface different programs with relative ease, including large batch programs.
3. A medium is provided for interchange of finite element oriented programs, such as mesh generators and solution schemes.
4. It is suited for operation on a computer network where both data and programs may be accessed via high speed data links.
5. The approach is equally feasible for use on mini-computers.

The following is a brief description of the various data files and their contents.

The parameter file. This file is a collection of lists containing various parameters relating to the problem. It has various switches indicating problem type and mode of operation, as well as a status list that registers the progress of the computation. All subprograms begin by loading the parameter file in core, and then check to see if all prerequisites for the requested computation are satisfied. If the operation results in a change of status, it is registered in the status list, and the file is written back on disk upon successful termination. The file also contains a directory of additional non-standard files defined by the user. It includes, as well, weighting factors that may be applied to the various loading cases for the purpose of display. In eigenvalue and transient response analysis, these values may be used to store frequencies or time values.

The point file. This is a random access file in which each record has a number of parameters pertaining to an individual node. In each record there is a dictionary relating the system and user numbering schemes. This allows possible renumbering of nodes, for example, by an automatic bandwidth optimizer, without causing the user to lose track of his original node numbers. Each record also contains the coordinates of the node, the value of an associated lumped mass, prescribed constraints and allowed freedoms, a pointer to possible inclined boundary conditions and kinematic dependency on a number of nodes, thereby allowing future implementation of reduced substructures. Kinematically consistent masses may be introduced as special files in a manner similar to the stiffness matrix.

The element file. This is a random access file in which each record describing an element gives its type, the total number of associated points, the number of active structural nodes, the connectivity, material pointer, and the thickness group pointer. The material pointer allows the use of several materials within the same problem.

Instead of having each element contain its own thickness parameters, it was decided that it should contain a thickness pointer to a group of thickness parameters. It is assumed that only a limited number of scantling data will be used even in a problem of large size.

The thickness file. This file has a number of records, each containing one or more thickness parameters. For elements such as beams with shear and torsional rigidity, several cross-sectional parameters are required. In the case of substructuring, the file may contain an alphanumeric reference to the file containing the condensed stiffness matrix of the substructure (superelement).

The material file. This file has parameters describing material type and properties. Each isotropic material is allowed five parameters (Young's modulus, Poisson's ratio, shear modulus, density, and thermal expansion coefficient). In case of more complex properties, such as in temperature-dependent, orthotropic materials or multilayered composites, one may use a succession of records in order to register all the pertinent parameters. In each case a standard material type code controls the interpretation of the stored values.

The stiffness directory. This program has information used to keep track of disk addresses of submatrices. A sparse matrix representation in which the individual submatrices are addressable by pointers is adopted. Each record of the directory file contains information pertaining to one row of the master stiffness matrix, indicating the number of submatrices, their position and their storage address numbers. A backwards reference parameter, which indicates the lowest node number linked to the present node, is also used. In cases where iterative solution techniques are preferred, a variation on the basic scheme is adopted.

If kinematically consistent mass matrices are stored, the same stiffness directory may be used to track the submatrices of the new matrix, which will be stored, however, in a special file distinct from the stiffness file. The same also holds true for elastic instability problems, where one or more geometric stiffness matrices may be required.

The stiffness file. This file contains the stiffness submatrices. After triangularization, the same file will contain the triangularized matrix. Similar files may be built for kinematically consistent mass or geometric stiffness matrices.

The load file. This contains load values for all loading cases handled. In transient response analysis, the load vectors represent load values at certain instances of time.

The deflection file. This file contains a full set of deflection vectors for all loading cases. In eigenvector computation the file is used for eigenvector storage.

The stress file. This file has stress values for all loading cases.

The save/restore file. This is a sequential file created by a special routine in order to collect and save all data pertinent to the problem in question on tape. This is performed by a standard program that may be called by the user at any stage.

MACHINE INDEPENDENCE

Coding Language

The system has been coded using a sufficiently nondemanding subset of FORTRAN IV, which can be handled by large systems and by most minicomputers, without much need for change. Machine-dependent functions, such as integer packing routines used to preserve core storage, are isolated. These are usually written in assembly language. In packing integers into single words, the basic word length of the machine plays an important part. As a rule, a number of integers are packed into one word so that for the smallest word length anticipated a reasonable range of integers may be represented. Numerical output statements, created by FORTRAN commands, are designed to fit on a standard teletype or its CRT equivalent.

Mass Storage Handling Routines

In most commerical systems, disk files may be stored in sequential or random access formats. It is assumed that both types of files are accessible through a five-character name, and a four-character extension, of which the first is a blank. In the random access option a file is defined which consists of a number of records all of equal length. The number of records, or the size of each record, may be adjusted later in the computation. A read or write statement may be initiated for any of the records in any order. If only a part of the record is written back, the rest of the record remains unchanged. Where such random access file capabilities do not exist, special routines are provided to simulate the process. In the case of systems that use a different number of words for representing integer and floating point variables, automatic record size adjustment is provided.

Graphics Coding

Only a small basic set of commands is used for simplicity. FORTRAN callable routines are used to draw vectors and characters and to move the beam to a new position.

OUTLINE OF THE BASIC THREE-DIMENSIONAL SHELL PACKAGE

Figure 1 is a diagram showing the flow of operations between the various subprograms. The heavy flow lines indicate the normal path.

Lighter flow lines denote possible changes of sequence. Program that operate in an interactive mode are represented by ordinary boxes. Those operating in a remote batch mode are enclosed in doubly framed baxes. In this section we give the program mnemonics and brief descriptions of their basic functions.

EDITP (edit parameter program). The program is designed to initiate new jobs or change the nature of old ones. It may also be used to delete old jobs, rename old jobs, or override current status lists. In this fashion, the user has veto power over any of the program modules.

BULKM (bulk model generator). The BULKM program is a sophisticated three-dimensional model generation package designed to produce large amounts of data with a minimum number of parameters. By the nature of the operation, the program is most efficient where structural regularity can be assumed. It can generate straight lines and arcs using a minimum of key points. Intermediate nodes on these line boundaries are generated at equal spaces or crowded near either end, near the center, or near both ends.

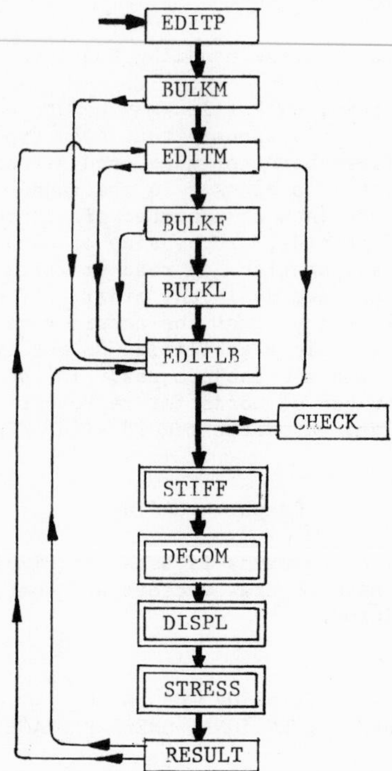

Fig. 1 Overview of the GIFTS package

Each line is given an alphanumeric identifier by which it may be referenced later (see Fig. 2). The program can generate different

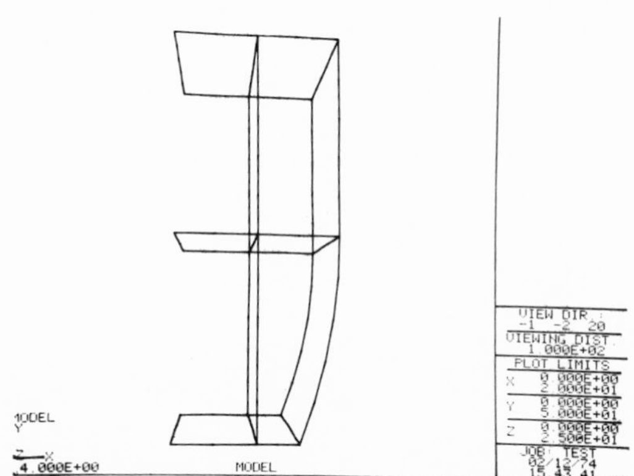

Fig. 2 Model outlines as generated by BULKM

types of grids, using lines and arcs as boundaries. It automatically chooses the node numbering scheme. The program generates plate and stiffener elements automatically (Fig. 3).

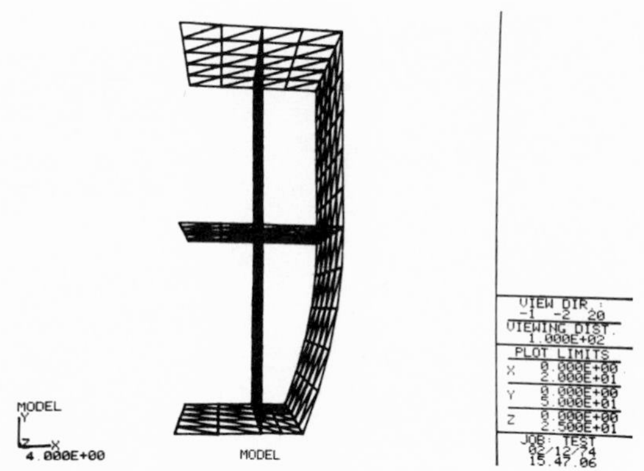

Fig. 3 Automatic element generation using BULKM

BULKM can be used to generate models of different types. For example, triangular and quadrilateral elements may be interpreted as being of the membrane or plate variety depending on the job parameters, which may also be influenced by the BULKM program. The program is also used to specify thicknesses, moments of inertia, etc. Special instructions help define material properties.

EDITM (model editor and display program for three-dimensional shells). The editor is a program designed to modify models after a BULKM pass. Its primary aim is to introduce local detail in an otherwise regular sturcture. Nodes may be shifted, added, or deleted. Elements may be added, deleted, or modified. Thickness parameters and material properties may be redefined. The display capabilities of EDITM are more sophisticated than those of BULKM, and include perspective, boxing, and node and element number display (Fig. 4).

BULKF (automatic freedom generator). This program scans the element list and assigns freedoms to nodal points based on the class of problem, the type of elements connected to the nodes, and their geometric orientation. The program will, for example, only allow freedoms in the plane of a grid of membrane elements, as the elements are not capable of supporting loads out of their plane [1].

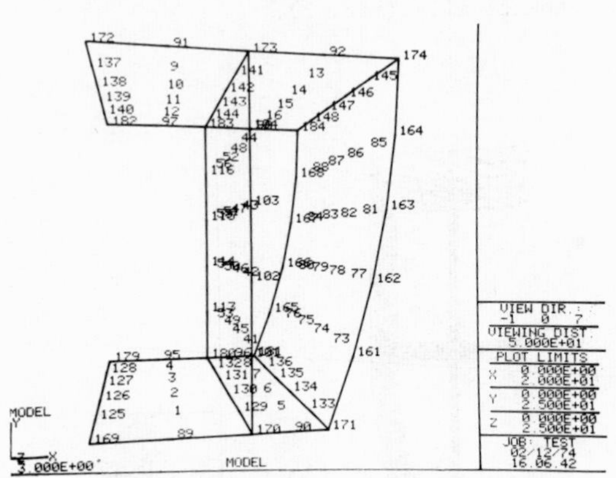

Fig. 4 Display of node numbers with windowing

BULKLB (automatic load and boundary condition generator). The BULKLB program is designed to produce loading on a structure generated by BULKM and to introduce line and surface boundary conditions. It can be used to introduce uniform pressure loads, static liquid heads, weight, and inertia, as well as line, surface, and concentrated loads.

EDITLB (edit and display routine for loads and boundary conditions for three-dimensional shells). This is a program designed to introduce special loads and boundary conditions after BULKLB has been called. It may be used directly instead of BULKLB, if the loading is simple enough or if the model was not generated by BULKM. The program has display capabilities and will represent the load components or resultants as vectors (Figs.5 and 6). Instructions controlling the display are similar to those of the usual editor EDITM.

CHECK (final model check program). This program conducts final checks on model detail. It ensures, for example, that all rods and beams have finite length and that triangles and quadrilaterals have a finite area and an acceptable aspect ratio. The program flags out warnings to the user whenever possible.

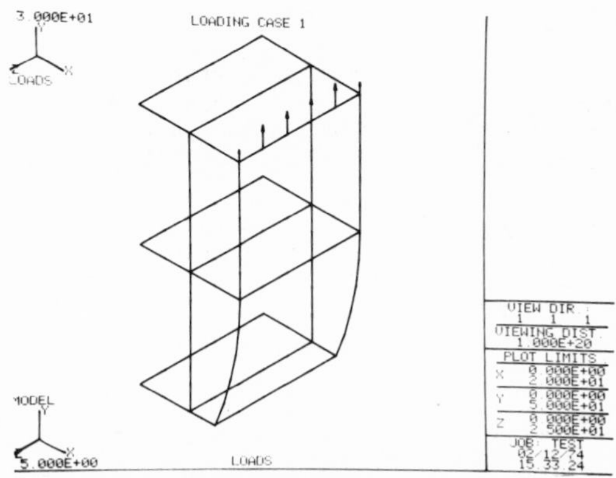

Fig. 5 Edge load applied by the BULKLB routine

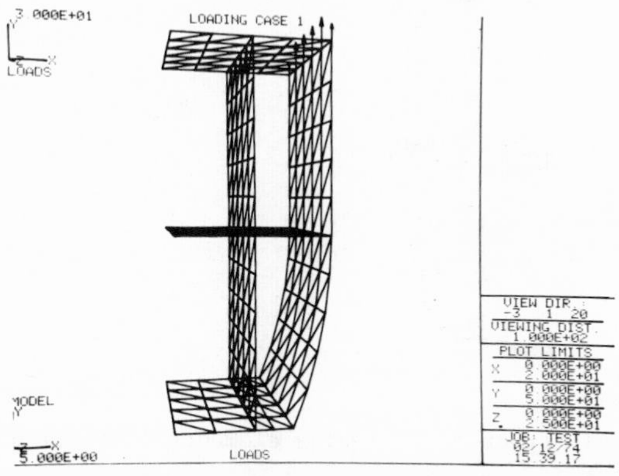

Fig. 6 Simultaneous display of elements and loads

STIFF (stiffness assembler). Program STIFF computes element stiffness matrices and assembles them into the appropriate file. It starts by predicting and allocating the storage to the various submatrices and establishes the stiffness directory through simulating but not actually performing the assembly and decomposition operations.

STIFFI (stiffness assembler for iterative solution). This program is similar to STIFF, except that it prepares for an iterative rather than a direct solution procedure.

DECOM (stiffness decomposition program). This program is based on a generalized Cholesky scheme using submatrices instead of scalar entries. It will operate only if a stiffness matrix is present. It will result in the destruction of the $\underline{K}$ matrix, and therefore resets the appropriate status word.

DEFL (deflection solution procedure). This procedure obtains the final deflection values using the decomposed matrix and load files.

GAUSR (Gaus-Seidal iterative scheme with relaxation factor). This is an alternative to DECOM and DEFL. It adjusts the value of the relaxation factor automatically to give a high convergence rate. The process is controlled by the user.

RELAX2 (group relaxation scheme using two nodes simultaneously). This is an alternative iterative scheme under consideration.

STRESS (stress routine). This routine computes stress for all elements from the already available displacements and stores the results in the appropriate file.

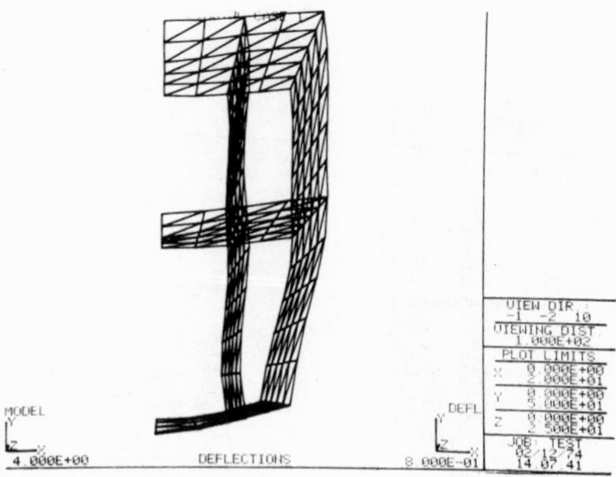

Fig. 7 Deflection of structure including elements (magnified)

RESULT (result display). This program is similar in nature to the model display program. Additional functions are available pertaining to the selection of results. It is designed to display one loading configuration at a time. This loading configuration may be a combination of all available loading cases, using certain weighting factors. Examples of deflection displays are given in Figs. 7 and 8. Stress representation is shown in Figs. 9 and 10. A stress level key is included in the plot.

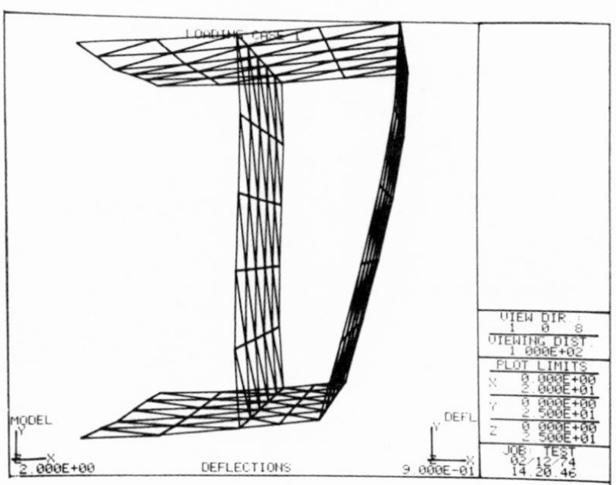

Fig. 8 Deflection of lower portion of the structure obtained by windowing

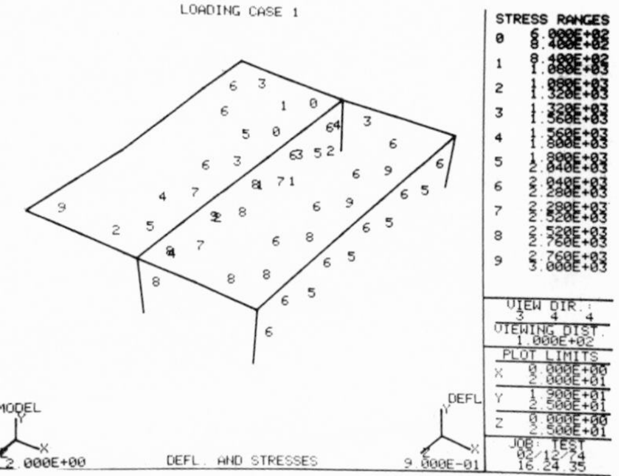

Fig. 9 Stress representation in intermediate horizontal shell

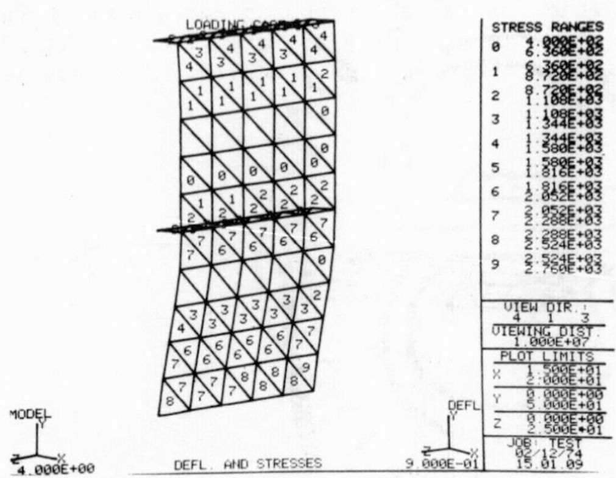

Fig. 10 Stress representation in outer shell with elements included

SAVER (save/restore capability). This package performs the function of saving a job on secondary storage (DECtape, MAGtape) or restoring a job from such a secondary file.

DUMP (data dump routine). This routine is designed to allow the user to inspect all files on disk in order to satisfy himself of their correctness. The program can dump any part or, if requested, all of any file.

THE SPARSE SUBMATRIX REPRESENTATION

Banded versus Partitioned Approaches

The banded approach of stiffness matrix storage is only justified in two-dimensional problems. The reason for this is related not only to storage and computational efficiency but also to the convenience in model generation. In the modeling of complex configurations the user tries to solve two problems simultaneously--the ease of modeling and description and that of efficient computation. The banded approach results in a conflict between the two objectives which may only be solved by a process of node renumbering or bandwidth optimization. Solution methods, such as the wavefront technique [2], essentially merge the two processes so that equation renumbering is done at the same time as the stiffness matrix is generated and decomposed. The resulting program is bound to be complicated. Furthermore, it suffers from the major disadvantage, which is also present in the classical banded approach, that problem size limitations are highly dependent on the amount of core available.

Previous remarks regarding the conflict between convenient model description and an efficient solution scheme is present to a greater extent in the case of automatic mesh generators. Since it is the purpose of this system to provide the user with automatic data generators whenever possible, it is necessary to adopt a scheme based on sparse submatrix representation (see also [3]). Such a scheme has the following advantages:

1. Core requirements are practically independent of problem size. This is of importance in time-sharing activities where I/O is relatively inexpensive and core storage is at a premium. It also is applicable to minicomputers with limited core, if disk space is available.

2. The processes of model generation and solution of equations do not present a conflict. On the contrary, the model generation scheme may utilize the logic inherent in the model connectivity in order to provide a near optimal node arrangement. This is accomplished with practically no additional computational effort.

The process described relies on the availability of random disk access. Should the random access operation prove to be inefficient, it may be speeded up by creating some user-controlled buffers or by replacing the system handler altogether with a special one. In both cases the changes are modular and do not alter the mainframe of the program. The resulting package is callable in the same manner as before. The user has, therefore, access to the data.

In order to explain the sparse scheme and to outline the principles involved in choosing the node numbers, some examples were investigated. It is impossible to discuss in detail all cases examined. The following simple problem and discussion should illustrate the resulting concept.

Example: Structure with Several Substructures

Let us consider a structure with five substructures S1 to S5 interconnected by interfaces I12 to I45 as shown in Fig. 11a.

As a basic plan of operation, we list the substructures first, then the interfaces. The stiffness matrix is divided into four regions, SS, SI, IS, and II (Fig. 11b). Each area of interaction between two items is called a block. Each block is divided, in itself, into submatrices, some of which are nonzero. In Fig. 11c the pattern of the $\underline{K}$ is shown for the case where the interfaces are listed in the order they occur in the structure. The X symbols denote the original nonzero blocks. The block propagation occurs in two stages, primary and secondary. The primary propagation wave occurs during the decomposition process as the diagonals corresponding to the substructures, S1 to S5, act as pivots. The secondary propagation, in which the interfaces act as pivots, results, in this case, in no further full blocks. Blocks generated during the primary propagation process are symbolized by the letter P and those during the secondary propagation (none in this case) by S.

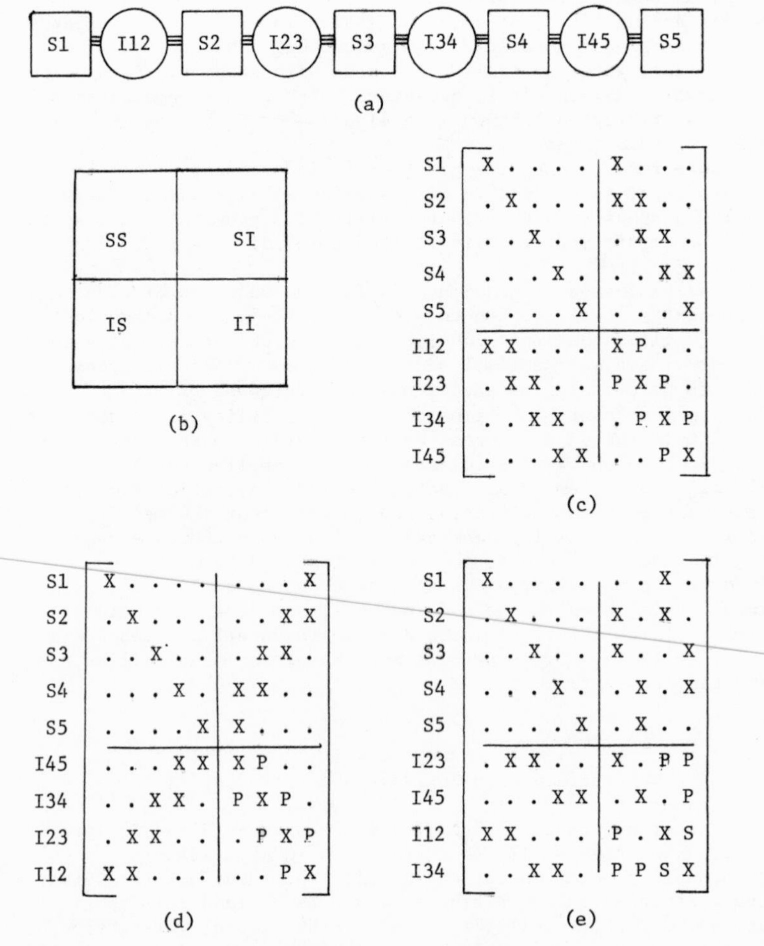

Fig. 11 Structure with substructures and interfaces in series

It must be noted here that the primary propagation phase resulted in the creation of a pattern within the II area. This pattern may be predicted if we consider each interface as a supernode, with each connected substructure as a superelement. The three-band pattern that evolves may be triangularized directly without further block propagation.

Should one reverse the order of the interfaces, as in (d), the result in unchanged. Rearrangement, as in (e), results in some secondary propagation.

Further investigation of detailed submatrix propagation within the nonzero blocks leads to the following additional conclusions.

1. Although no primary block propagation occurs in the SI part of the stiffness matrix, submatrix propagation does occur within each of the subregions. This propagation, however, is restricted within the individual blocks, and normally extends from the first nonzero submatrix down to the bottom row of submatrices within the block. The order in which the submatrices propagate within an SI block is influenced by the order in which the nodes are numbered within the substructure and the interface it is connected to. It is advisable, whenever there is a choice, to arrange the numbers so that the submatrices occurring in the interaction blocks occur as near the bottom of the block as possible.

2. A sparse matrix representation, if properly handled, seems to require approximately the same storage space as a renumbered system.

CLASSIFICATION OF FINITE ELEMENT DISPLAYS

In devising a general purpose graphics-oriented element package, it is obvious that the display functions must be, to a certain extent, shared by the different programs. In order to benefit from this fact, it is necessary to classify the model display process and techniques. In this section we attempt to establish such a classification.

In general, a graphics routine displays the geometry of the model, modified by certain functions, whose magnitude, direction, and variation over the model are of interest. Some of the functions may modify the geometry, as is the case with displacement functions. Most other functions, such as loads and stresses, are superimposed on the current geometry. It is possible to plot several functions simultaneously, such as deformations and stresses (see Fig. 9).

Classification of Finite Element Models

Finite element models may be classified according to the geometric space they occupy and the class of finite elements utilized. The following possibilities come to mind.

E1S2 models. These are models involving one-dimensional elements in two-dimensional space. Examples are two-dimensional pin-jointed trusses and frames and two-dimensional ducting problems. Elements are uniaxial rods, beams, or tubes.

E1S3 models. These are models involving one-dimensional elements in three-dimensional space such as three-dimensional trusses and frames and three-dimensional ducting.

E2S2 models. These are two-dimensional elements in two-dimensional space such as in two-dimensional elasticity, flat plates, axisymmetric solids problems, and two-dimensional flow and heat conduction problems. E2S2 models may also contain one-dimensional elements.

E2S3 models. These are two-dimensional elements in three-dimensional space such as complex shell structures. Such models may also include one-dimensional elements.

E3S3 models. These are three-dimensional elements in three-dimensional space such as in three-dimensional solids, and fluid flow and heat conduction problems.

Classification of Functions to be Displayed

There are basically three types of functions: scalar, vector, and tensor functions. Scalar functions include temperatures, head, and von Mises yield criterion. Vector functions are represented by deflections, loads velocities, and temperature gradients. Tensor functions are represented by stresses and strains. It is always possible to reduce tensor functions to vectors and scalars for ease of representation.

Functions may be assigned to nodes (deflections, loads) or to elements (stresses). It is sometimes possible to translate element functions into node functions by a process of averaging, interpolation, extrapolation, and/or smoothing.

Data Reduction for Plotting

In handling the graphical display of three-dimensional complex structures, it becomes quickly apparent that selective output is necessary in order for the user to be able to interpret the display effectively. Selection may be based on topological and geometric basis or reduction through computer processing.

Selection of display items according to topology and geometry. If the logic employed in the construction of the model is preserved in the files, which is the case in the GIFTS system, only the boundaries (edges) of the areas need be plotted (Figs.2, 5 and 9). This reduces the amount of information and speeds up the plotting process. An alternative may be provided by computer elimination of all internal boundaries [4].

Another method based on geometry is that of "successive windowing," where windowing is interpreted in the most general sense. A series of windowing operators is applied to a model. Each operator reduces the amount of data to be plotted by dividing the space into two portions, one of which is invisible. Such operators may incorporate a Cartesian box, a cylinder, a sphere, or a plane.

Selection based on range. In the case of scalar functions, only values between certain limits are plotted. This concept is extended in the GIFTS system so that several ranges may be displayed simultaneously, each designated by a special symbol chosen by the user (Figs.9 and 10).

Display of scalar functions. Plots using different intensities [4] or using the technique described above seem to be the most feasible. It is also possible, in two-dimensional cases, to create a contour plot.

Display of vector quantities. These are best drawn as line segments pointing in the direction of the vector, whose length is proportional to the value of the function (Fig. 5). Other examples may be found in reference [4].

Display of tensor quantities. In displaying stresses, for example, it is found most suitable to reduce the stress tensor to a scalar function (such as the von Mises criterion) or to a set of vectors (such as the principal stresses). The GIFTS plot routine offers both possibilities.

Display of solid models. Here data reduction is imperative. It is recommended that only the edges of the model be plotted. Upon request, sections through the solid are generated and the appropriate functions are plotted on these sections using two-dimensional display techniques. In addition, facilities to change the viewing line and to introduce perspective and windowing should be sufficient to produce a powerful display tool.

One comment worth including is that sections are best taken along topologically meaningful node and element groups. Experiments with computed intersections of an arbitrary plane with a complex solid (Fig. 12) have proved convenient but rather expensive.

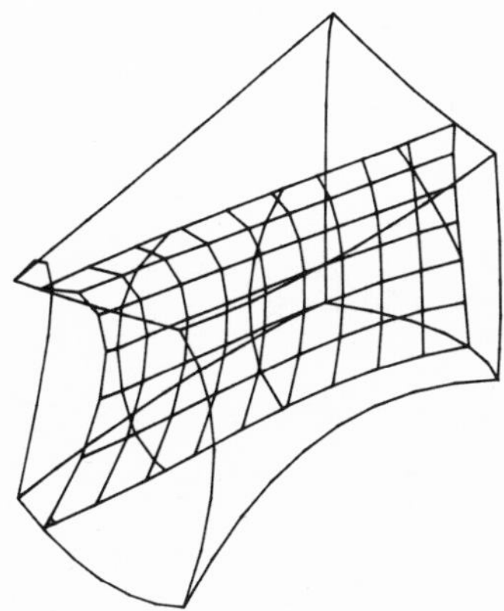

Fig. 12 Intersection of a plane with an isoparametric curved solid region

ACKNOWLEDGMENTS

This research has been supported by the Office of Naval Research under Contract Number N00014-67-A-0209-0016. The University of Arizona Computer Center and the Engineering Experiment Station have given valuable assistance and advice on many occasions.

A great amount of help has been rendered by Paul Shanta. The manuscript was typed by Carol Geisler and Betty Holmstrom.

REFERENCES

1 Kamel, H. A., Lui, D., "Application of the Finite Element Method to Ship Structures," Journal of Computers and Structures, Vol. 1, 1971, pp. 103-130.

2 Irons, B. M., Kan, D. K., "Equation Solving Algorithms for the Finite Element Method," Numerical and Computer Methods in Structural Mechanics, Academic Press, New York, 1973, pp. 497-511.

3 McCormick, C. W., "The NASTRAN Program for Structural Analysis", Advances in Computational Methods in Structural Mechanics and Design, UAH Press, 1972, pp. 551-572.

4 Kamel, H. A., McCabe, M. W., "A Two Dimensional Package for Finite Element Analysis Utilizing Interactive Mini-Computer Graphics", presented at the ICCAS Conference, August 28-30, 1973, Tokyo, Japan.

AN INTERACTIVE GRAPHICS FINITE ELEMENT SYSTEM

W. E. Lorensen
Watervliet Arsenal
Watervliet, New York

ABSTRACT

IGFES, an Interactive Graphics Finite Element System, is described. This general purpose pre- and postprocessor for finite element analysis provides two-dimensional grid generation and a variety of output display techniques including contours, perspectives of surfaces, and two-dimensional plots of dependent variables. IGFES currently supports NASTRAN and executes in 80K bytes on an IBM 360/44. At this writing, supported graphics devices are the Tektronix 4002A storage display and Calcomp 563 incremental plotter.

INTRODUCTION

Justification for development of graphics pre- and postprocessors has been given wide exposure in recent literature [1,2,3,4]. Structural analysts generally agree that alternatives to conventional input preparation and output analysis for finite element codes are necessary if the full capabilities of existing codes are to be exploited. IGFES, an Interactive Grapnics Finite Element System, has been designed to meet these requirements.

The following description of the design considerations and functional capabilities of IGFES is meant to give users and graphics system designers an overview of a general purpose graphics package to support finite element analyses. Throughout the discussion, the modularity of the system and the number of user options are stressed. Future developments in the field will presumably make use of IGFES.

NASTRAN [5,6] was chosen as the testbed finite element code because of its widespread acceptance in the military and aerospace structural analysis communities. IGFES is written in FORTRAN IV, and each of its functional modules runs in 80,000 bytes on an IBM 360/44 under the OS/MFT operating system. Graphics devices are supported through a general purpose package of graphics programs similar to TCS [7]. Currently, the Tektronix 4002A storage display and the Calcomp 563 incremental plotter are supported by TCS-like packages.

DESIGN PHILOSOPHY

A number of design criteria are required prior to the coding of the interactive system. The objectives of generality and transportability should be given the highest priority. Interactiveness can be obtained using either of two approaches. One calls for the modification of finite element codes to include an interactive capability; the other generates input before the analysis and analyzes output after the analysis. The primary advantages of the former technique are that intermediate feedback from the analysis program can be obtained while it is executing, parametric studies can be made easily, and the complete finite element analysis cycle can take place in one session. However, the disadvantages of such an approach make it infeasible. First, the system's generality is lost. Each finite element code must be modified to provide the graphics capability. As updates and revisions are made to the analysis program, changes to the graphics package are necessary. Furthermore, most finite element analysis programs cannot meet the response constraints of an interactive system.

Consequently, the latter pre- and postprocessor approach is deemed the most general and practical. If the system is designed modularly, analysis code-dependent routines can be sufficiently isolated so as to facilitate support for additional codes. The IGFES system meets these criteria by interfacing with the finite element code through two files. The input file for the program is generated from the internal data structure of the IGFES input module. The output file which IGFES accepts is generated by a program which reduces the finite element output into a standard interface file. Using this philosophy, only two subprograms must be written to support a new analysis program. The importance of modularity will become more apparent when the output module is discussed in detail, for IGFES permits the use of any of the output module programs with data other than those produced by a finite element analysis. The other design criteria are more obvious and easily justified. IGFES is easy to use, a consideration which motivates engineers to use and experiment with finite element programs; IGFES supports all NASTRAN triangular and quadrilateral elements; IGFES is written in ANSI FORTRAN; all graphics display dependent routines are isolated; IGFES provides the user with a variety of output display techniques; and the design allows other techniques to be added easily.

INPUT MODULE

The input module design calls for a grid generation program, a boundary condition generator, and a structural editor. IGFES will eventually contain a program to lead the user through the entire input generation, which in the case of NASTRAN will include prompting for the executive control cards, case control, and those bulk data cards not produced by the grid and boundary condition generator. At this writing, only the grid generator has been completed.

The grid generator is comprised of three phases. All three phases run under control of a monitor which invokes appropriate routines as requested by the user. Phase one is the boundary input phase. Once the boundaries are described, the grid region definition phase, phase two, is entered, and the user denotes a four "cornered" subset of the boundaries which defines a mapping from a rectangular region onto this arbitrary four-sided region. Based on the generation algorithm chosen, the number of grid points the user requests in this region, and the type of element chosen, a finite element idealization is displayed. Phase three permits regeneration of regions, generation of new regions, and various manipulations of the generated regions.

The boundary input phase accepts as input points defining line segments which describe the boundary. Either straight lines or arbitrary curves can be used to depict the boundaries since a cubic spline interpolation procedure is used to represent the boundary internally [8]. As each point terminating a segment is entered, its position is interpolated as a function of chord length. The method, which results in two cyclic functions describing a boundary, provides a convenient method of handling multivalued boundaries analytically. Also, by using chord length as a parameter, node distribution algorithms for the sides of grid regions are simplified. Multiple boundaries are allowed, and all input can be entered through cards, interactively with a graphics input device such as a joystick, or with a combination of cards and graphic input.

Once the user informs the program as to the nodal density in a grid region, he enters phase two and graphically denotes the sides defining the region. Figures 1 through 4 illustrate grid regions.

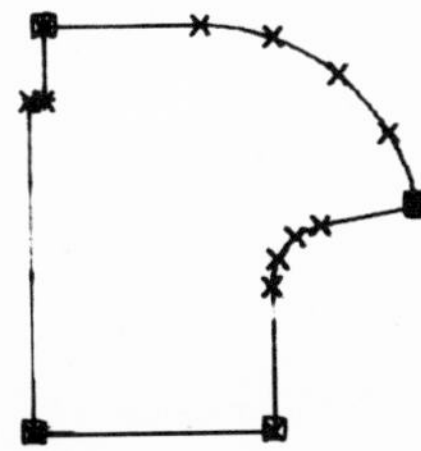

Fig. 1 Input boundary denoted by ✕ , grid region corners by □ , and enforced points by a ✱

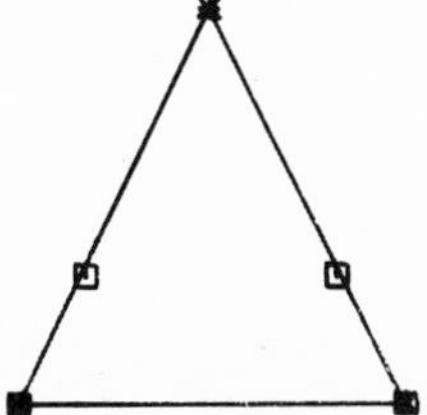

Fig. 2 Definition of a grid region on a triangular boundary

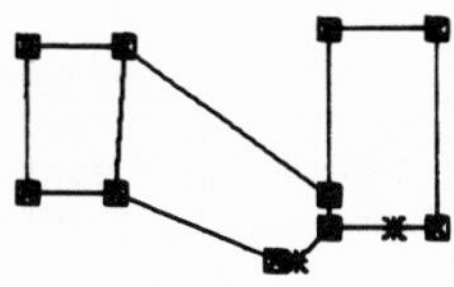

Fig. 3 Multiple grid regions

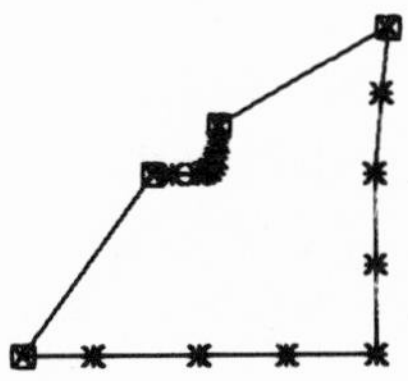

Fig. 4 All input points are enforced.

When defining a grid region side, several user options are available. Generated nodes on a side can be forced to coincide with specific input points; sides can be generated on the interior of a boundary; and the previous grid region definition can be recalled. As each side is defined, the length of that side is calculated from the parametric splines, and this length is used as a parameter in a nodal distribution algorithm. The distribution algorithm is defined by the user prior to the grid region definition. A default algorithm can be used which results in equal nodal spacing along a side. By varying two parameters between zero and one, the user can effectively specify an infinity of algorithms ranging from equal spacing to a geometric progression of spacing. A separate algorithm can be specified for each side and may be applied in a counterclockwise direction on the side.

With the nodal density and grid region defined, the program prompts the user for an element type and proceeds to generate and display the grid according to one of two generation methods. Method one is iterative, calculating interior grid point locations based on the boundary grid points. Method two is direct and connects grid points on opposite sides with straight lines, distributing nodes on these lines according to the appropriate side distribution algorithms. Figures 5 through 10 illustrate examples of the side and grid generation algorithms.

Phase three of the generation process is the monitor manipulation phase. Options available allow the user to perform multiple grid region generations. If this is the case, the system automatically eliminates duplicate grid points at grid generation region boundaries. This option can be suppressed to allow modeling of contact surfaces between structures. Other monitor options allow selecting the grid generation method, defining the node distribution algorithms for the grid region sides, displaying all generated regions, numbering the nodes on the sides of the grid regions, punching connection and grid cards for the regions, windowing a portion of the generated regions, and entering node and element label starting values and increments.

A more detailed description of grid generator usage is available upon request from the author.

OUTPUT MODULE

The IGFES output module executes under control of a machine-dependent monitor, although its associated display routines can be executed independently in either an interactive or batch mode. The output module currently contains a NASTRAN-IGFES interface routine, a data selection program, a surface approximation program, a surface fit analysis program, a contour plotter, a perspective surface plotter, and a two-dimensional plotting program. Detailed user instructions for the output module are available from the author.

NASIGIO: NASTRAN-IGFES Interface. This routine, running in a noninteractive mode, produces a file to be accessed by the output module application routines. NASIGIO accepts tape files produced by the

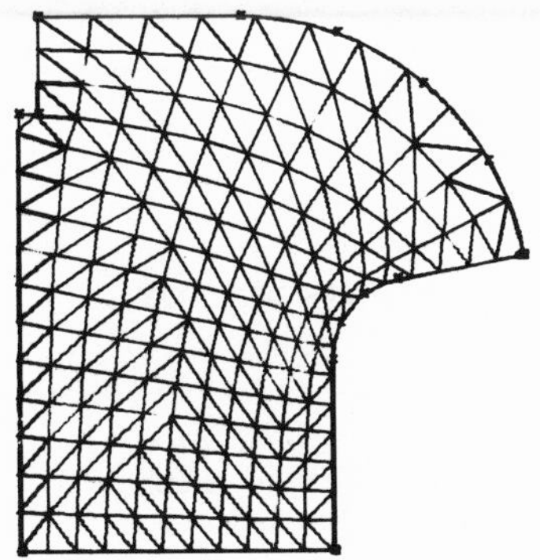

Fig. 5 Generation displaying smaller elements in concave area

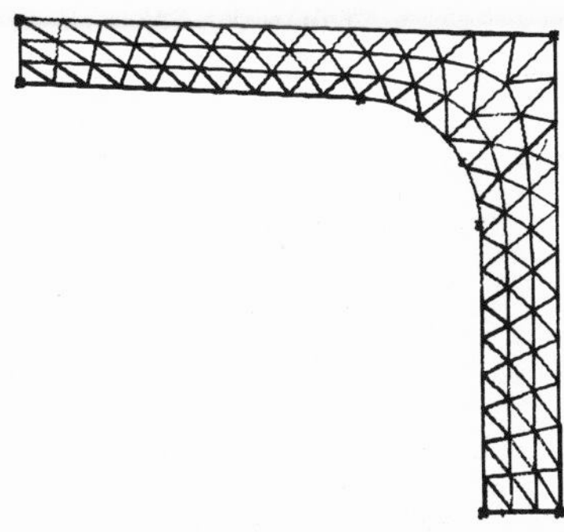

Fig. 6 Enforced corner on a region side

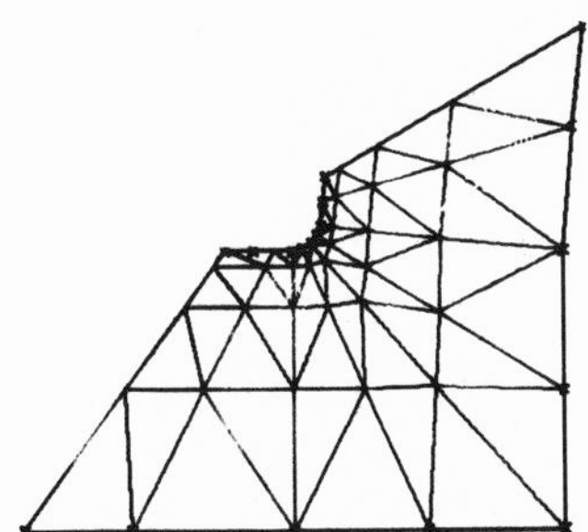

Fig. 7 Method 2, geometric distribution on nodes on two sides

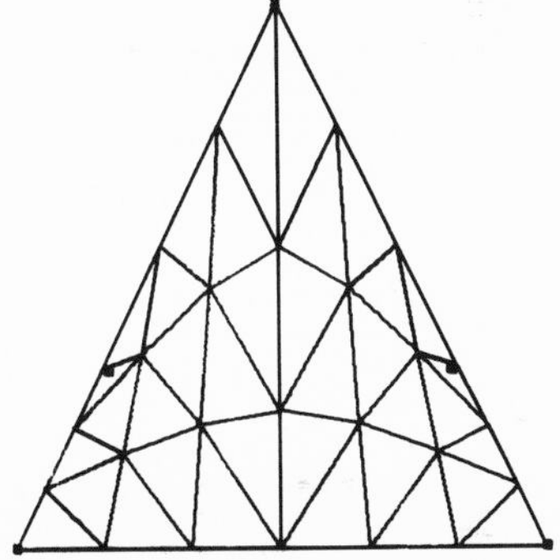

Fig. 8 Generation applied to a triangular boundary

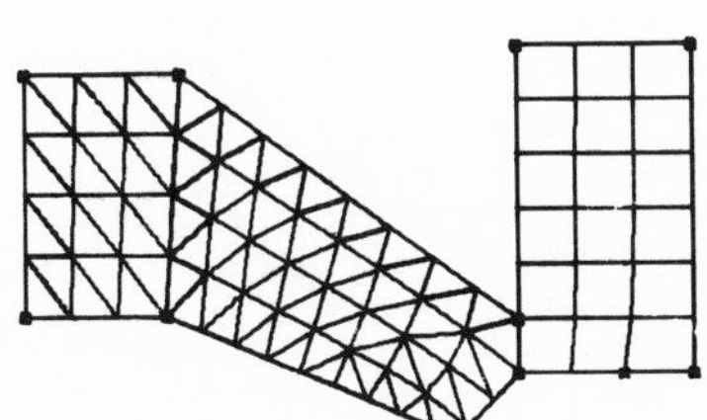

Fig. 9 Multiple generations, mixed elements

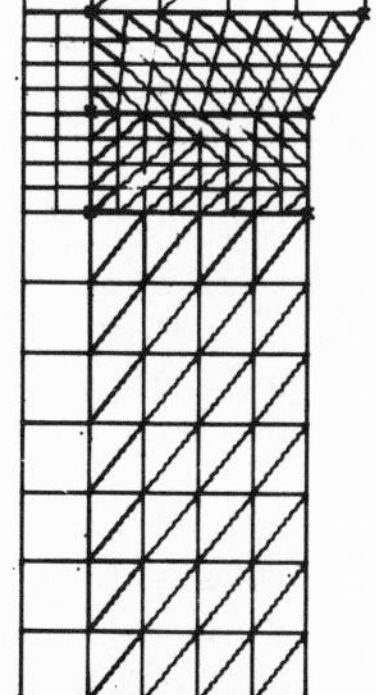

Fig. 10 Graphic windowing and multiple generations

Table 1 NASIGIO Interface File

Record number	Description
1	Number of files on tape, number of data sets per file
2	Number of words per control record
3	Control record: number of title records, number of words per title record, number of distinct boundaries, number of boundary nodes, number of dependent variables, number of data records
4	Pointers to boundary node vector, defining distinct boundaries
5	Boundary node vectors
6	Dependent variable titles
7	Independent and dependent variable maxima and minima
8	ID, independent variables, dependent variables
.	
.	
.	
7 + Number of data points	

All records except record 1 are repeated for each subcase, element type, eigenvalue, and so on.

NASTRAN OUTPUT2 module. NASTRAN ALTER statements [5] are used to dump the required files onto the tape. The files currently used are BGPDT, the basic grid point definition table; EQEXIN, equivalence table between external and internal grid points; SIL, the scalar index list; EST, the element summary table; and an appropriate stress and displacement file, e.g. OES1 and OUG1 for static analyses. Given this information, NASIGIO collects all data associated with each subcase, transforms stresses from the elemental to the global coordinate system, computes maxima and minima for each independent and dependent variable, determines which grid points lie on the boundaries of a structure, and produces the standard interface file shown in Table 1. The dependent variables include all the components of stress, principal stresses, displacements, velocities, and eigenvectors. Note that NASIGIO compresses all data into an X,Y,Z format, the general form of data required by the display routines. NASIGIO can be used independent of IGFES as a general NASTRAN data reduction program.

SELECT: Data Selection Program. SELECT works directly from the interface file and allows the user to interactively select the subcase and dependent variable he wishes to display. Once the data have been selected, this routine produces files for the other output module routines. SELECT must be executed prior to the other programs in the system if the user is analyzing NASTRAN data.

SURFACE: Surface Approximation Program. SURFACE, working with X,Y,Z data values, performs a piecewise doubly cubic spline approximation of the data. The user has the option of specifying a smoothing parameter which can reflect the user's confidence in the finite element data. Since the splines are doubly cubic, second order partial derivatives are continuous, resulting in smooth looking surfaces.

FIT: Surface Fit Analysis Program. FIT works in conjunction with SURFACE and graphically displays the deviation of the analysis data from the approximating surface.

CONTOUR: Contour Display Program. CONTOUR, using the doubly cubic spline produced by SURFACE, draws curves of constant dependent variables superimposed on a picture of the structural boundaries. The user interactively controls the region he wishes to view and the contour levels to be tracked. Figures 11 and 12 illustrate the versatility of CONTOUR. A detailed description of a rectangular region version of the program and associated analysis can be found in [9].

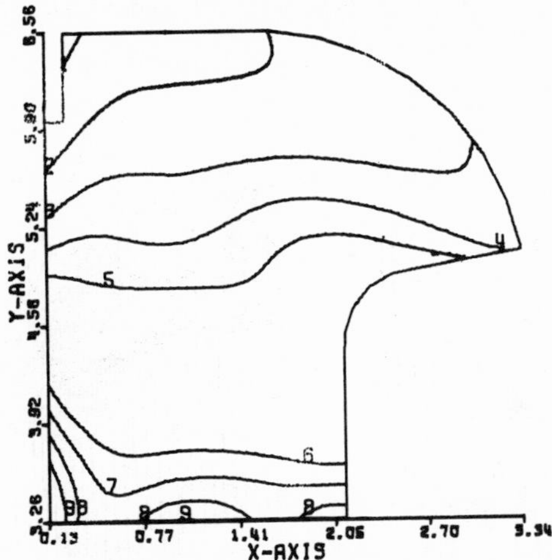

Fig. 11 Sample isostress output from CONTOUR

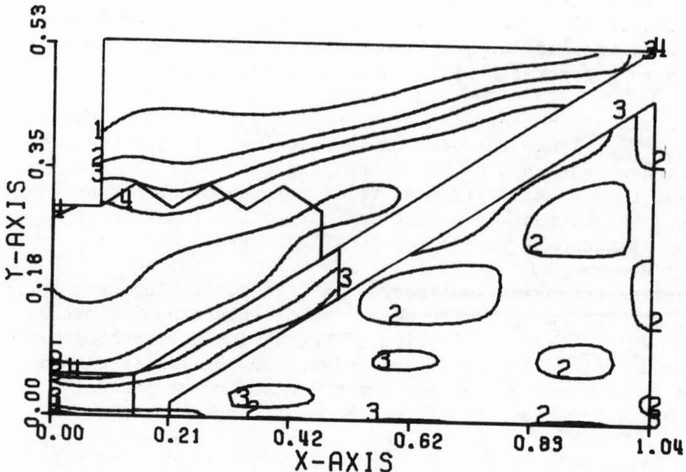

Fig. 12 CONTOUR display on a multiple region structure

PERSPV: Perspective Surface Display Program. PERSPV is a modified implementation of Kubert's algorithm described in [10]. The current version, which only displays rectangular regions of a surface, allows the user to view the displacement and stress surfaces approximated by SURFACE. Interactive options permit user control of view, region to be viewed, removal of hidden lines, and display of sidebars on the surface. Figure 13 shows stress surfaces for two NASTRAN analyses.

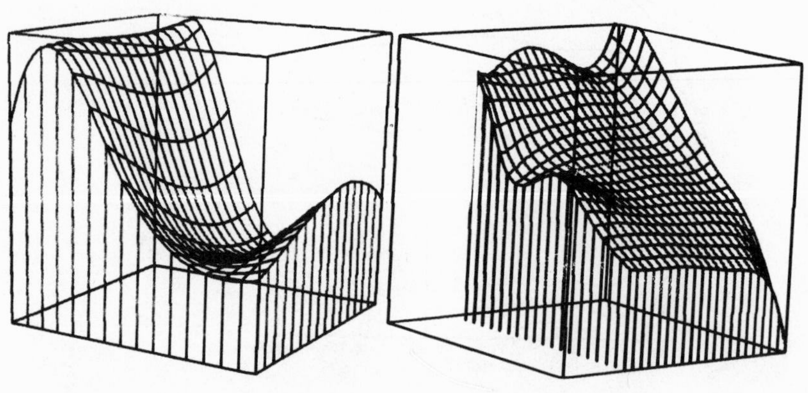

Fig. 13 Sample output from PERSPV

XYPLOT: Two-Dimensional Plotting Program. Displacements or stresses can be plotted versus one of the independent variables of position while the other is fixed. XYPLOT allows multiple plots to be superimposed and is useful in displaying the independent variable along boundaries or other lines of interest. Sample output from XYPLOT appears in Fig. 14.

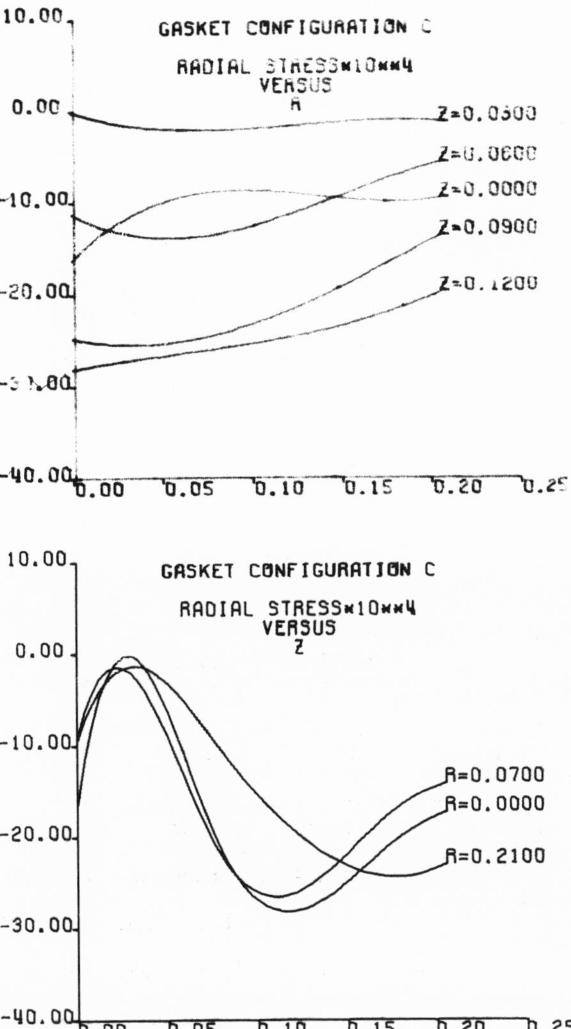

Fig. 14 Sample output from XYPLOT

CONCLUSIONS

IGFES offers a low cost, modular alternative to conventional finite element input and output processing. While most finite element graphics systems have been designed for specific application programs, the generality of this system lends itself to support a number of programs as well as to execute a variety of display devices. Also, recognizing the importance of output display as well as input generation, IGFES has been designed to allow alternative display techniques to be added in the future.

Future plans call for extension of the grid generation process to three dimensions, completion of the remaining input module routines, and implementation of the system on an interactive, standalone, refreshed graphics system.

ACKNOWLEDGMENTS

The author is indebted to R. Bair of the Computer Science Office at Watervliet Arsenal for his contribution of the contour program and for overall design and implementation responsibility of the output module. Acknowledgment is also due to R. Scanlon of the Applied Mathematics and Mechanics Division at Watervliet for his contributions of the two-and three-dimensional spline analyses and programs.

REFERENCES

1 Cronk, M., "An Interactive Computer Graphics Program for NASTRAN," NASTRAN: Users' Experience, NASA, TM-X-2378, 1971, pp. 659-667.

2 Young, H., "Gun Barrel Design by Interactive Computer Graphics," Proceedings of the Joint Annual Meeting of the American Ordnance Association, May 1972, American Ordnance Association, Washington, D.C., pp. 12D1-12D22.

3 Katow, S., and Cooper, B., "NASTRAN Data Generation and Management Using Interactive Graphics," NASTRAN: Users' Experience, NASA, TM-X-2637, 1972, pp. 399-406.

4 Bousquet, R. D., and Yates, D. N., "A Low Cost Interactive Graphics System for Large Scale Finite Element Analyses," Computers and Structures, Vol. 3, 1973, pp. 1321-1330.

5 McCormick, C. W., "The NASTRAN User's Manual," SP-222(01), June 1972, NASA, Washington, D.C.

6 McNeal, R. H., "The NASTRAN Theoretical Manual," SP-221(01), April 1972, NASA, Washington, D.C.

7 "Terminal Control System Users Manual," 1971, Tektronix, Inc., Beaverton, Ore.

8 Scanlon, R. D., "An Interpolating Cubic Spline Fortran Subroutine," WVT-7010, Jan 1970, Watervliet Arsenal, Watervliet, N.Y.

9 Bair, R., "Computations of Smooth Contours from Non-Uniform Data," WVT-7265, Dec 1972, Watervliet Arsenal, Watervliet, N.Y.

10 Kubert, B., Szabo, J., and Guilieri, S., "The Perspective Representation of Functions of Two Variables," Journal of the Association for Computing Machinery, Vol. 15, No. 2, April 1968, pp. 193-204.

APPLICATIONS OF CONTINUOUS TONE COMPUTER-GENERATED IMAGES IN STRUCTURAL MECHANICS

H. N. Christiansen
Brigham Young University, Provo, Utah
University of Utah, Salt Lake City, Utah

ABSTRACT

The recent development of continuous tone computer-generated images (which refers to the rendering of shaded objects by means of a raster-driven cathode ray tube) has created the opportunity to display pattern type information derived from mathematical models in a new and often realistic form. The paper describes the process by which these pictures are created and the application of the technique to the display of kinematic and elastic systems which have been modeled using the finite element approach.

INTRODUCTION

The display of kinematic and elastic systems using the medium of continuous tone computer-generated images has been under investigation by the author at the University of Utah since the fall of 1969. During this time, attention has been focused upon the display of kinematic systems and static and dynamic elastic analysis results produced by finite element procedures.

Initial effort was applied to the creation of a simulator for foldable plate and truss systems and their display in black and white pictures. This was followed by the use of the simulator to create solutions for two-dimensional elasticity problems. These results were initially displayed by shading the elements according to the magnitude of the stress or strain component selected for presentation. Later these same functions were displayed by use of color variation.

The next project was the creation of a simulator for the generation of pictures of distorted structural frameworks. Given the output of a framework analysis routine, this display system creates a sequence of panels which represent the distorted framework, its loading system, and support constraints.

Two years ago, programming was started on a general display routine which would facilitate the production of still and motion pictures for three-dimensional models whose surfaces could be represented as mosaics of triangular and quadrilateral panels. This routine has been extensively used to display models of sonar devices, automotive parts, thin arch dams, and delta wing airplanes. Recently this routine was modified to accept complex mathematical solutions for both displacements and special scalar functions like temperature and pressure.

Following sections describe the process by which these pictures are produced and present a review (including display examples) of each project.

COMPUTER-GENERATED DISPLAYS

Following a modified adage that "a picture is worth a thousand dumps," the past dozen (or so) years has seen the unsteady rise of the field of computer graphics. Initially the research work centered around problems associated with the reduction of three-dimensional data onto a two-dimensional surface. Problems of perspective were treated by Smith [1] and Johnson [2] and efficient hidden-line removal was discussed by Roberts [3]. Building upon this base, line-drawing technology has developed into a full-fledged discipline. The development of shaded pictures was addressed at the University of Utah (Wylie, Romney, Evans, and Erdahl [4], Warnock [5], Romney [6], and Watkins [7]), General Electric [8], and MAGI [9]. Difficult problems of shadows and movable lights were discussed in 1970 by Bouknight and Kelley [10]. More recently, pseudocolor image enhancement techniques have been developed at Rand [11], and halftone mapping technology advanced at IBM [12].

Computer Generation of Continuous Tone Pictures

Computer generation of continuous tone pictures refers to the rendering of shaded objects by means of a raster-driven cathode ray tube. The image produced is recorded on film by a camera mounted in front and pointing at the screen. As the picture is being computed and displayed one "scan" line at a time, the camera, with the shutter open, performs an integration function.

Just as the complete image is a sequence of lines, so an individual line is a sequence of dots or groups of dots. The intensity of each dot (or group) must be computed and converted to an analog signal which, in turn, controls the scope intensity. As the relationship between the signal and scope intensity is usually nonlinear, a correction table is utilized. The displays shown in this paper each contain 1024 lines with each line containing 1024 groups of dots. On the system available to the author, the option of color requires the computation and display of three images. These images are the red, blue, and green component intensities for the prescribed scene and background colors. Prior to each pass, the appropriate color filter must be inserted between the scope and the camera.

All scenes, on the University of Utah system, are composed of polygonal elements, and the hidden surface problem is solved as a result of comparisons of elements in a scene in order to establish depth relationships. This computation is aided by accepting the previous scan line definition as the basis for the initial estimate of the new scan line.

Some systems have been devised (MAGI [9], Comba [13], Weiss [14], and Mahl [15]) which have the capability to remove hidden parts for curved surfaces by restricting the class of possible surfaces and/or accepting relatively long execution times. In 1971, Gouraud [16] suggested a method which accepts a finite element approximation of the surface and shades the elements according to a linear variation of light intensity so that visual discontinuities between adjacent polygons mostly disappear. Gouraud shading suffers in that the derivatives of the intensity functions are not continuous; however, the method is both general and efficient from a computational standpoint.

In 1972 Watkins reduced the hidden surface problem to hardware including the Gouraud shading feature. Most of the computer-generated displays contained in this paper were made using this equipment. The computation time required by the "Visible Surface Processor" is negligible, and as other components of the system are developed from software to hardware form, the time required to generate an image can be dramatically reduced. For example, the University of Utah has a movie production system in operation which utilizes the original equipment developed by Watkins but otherwise is largely dependent upon software. This system produces a single 512 scan line image in from eight to sixteen seconds (even slower for software that produces an improved image quality). By comparison, Case Western Reserve University has purchased a system which has almost every required function reduced to hardware form (including rotation and translation). The Case system can make thirty new 525 scan line pictures each second.

Recently, Phong [17] has devised a shading procedure based upon a linear variation, over each element, of the orientation of the normal to the surface. In this way, the light intensity function is continuous and its derivatives are approximately continuous. Results using Phong shading are very impressive for curved surfaces having a modest number of elements. This process is currently being reduced to hardware and will be available as part of future "Visible Surface Processors."

The complication of shadows is avoided by modeling the location of the light source at the position of observation. The intensity of the reflected light is then made a linear function of the cosine squared of the angle between the normal to the surface and the direction to the observer (i.e., light source). The constant term in the formula represents the minimum level of diffused light intensity. This is helpful in preserving a definition of the boundary of a curved surface against a dark background.

Two shading formats are available. The first, called "flat" element shading, preserves the identification of the edges of the elements by utilizing a single normal for the entire element (even if the element is a warped quadrilateral). This normal is the average of the normals calculated at the nodes according to the cross product of the two vectors formed along the element edges associated with the node. Even with the approximation of an average normal, the light intensity will still change over the element surface due to variations in the vector from the surface to the observer. This effect may be deleted by the approximate modeling of the light source at infinity behind the observer, a modification which simplifies the computations but, in the view of the author, tends to reduce the three-dimensional appearance of the pictures.

The second format, called "smooth" element shading, involves curved surface simulation according to the Gouraud method. This requires approximations for the normal to the surface at each node, which, in turn, requires either the averaging of normals based upon the individual elements associated with the node or the local least-square fitting of the surface by an appropriate mathematical function. The averaging scheme has been utilized by the author in an effort to reduce computation time.

SIMULATION OF KINEMATIC FOLDED-PLATE SYSTEMS

The requirement for simulation was generated by Resch [18], the creator of "Kinematic Folded Plate Systems". The common property of these systems is that, by allowing only folding along the edges of a continuous line pattern, a flat sheet may be transformed into a variety of shapes. Figure 1 illustrates a paper model of one such system which is a repetition of only two nonidentical elements.

Resch's initial investigations involved the manual folding of sheets of paper upon which the pattern had been scribed. The model could then be moved by hand to produce a variety of shell forms, which could be stabilized by providing sufficient supports or the addition of truss and/or plate elements. This approach had three major drawbacks. First, the process required repetition (the model shown in Fig. 1 took 30 hours to fold) to investigate any parameter change in the "star pattern" or "wrinkle" definition. Second, the dimensions of elements added to stabilize the shell were unique to the chosen shape. Finally, the paper models were not useful in an analysis of stress and strain.

The functions of kinematic and elastic analysis were accomplished as a result of an iterative finite element stiffness approach to the linear elasticity problem. Developments which allowed this approach include the modeling of the folds as elastic hinges and the generation of applied force systems which tend to restore the original dimensions (but not orientation) of the elements in the following iteration. These procedures were documented by the author in 1971 [19].

Starting with a very regular geometry, the simulator has been used to create a variety of three-dimensional shapes. Probably of greatest potential interest are the domelike extensions, an example of which is shown in the computer generated picture of Fig. 2.

COMPUTER-GENERATED PHOTOELASTIC DISPLAYS

Following the development of the kinematic simulator, attention was turned to the display of stress and strain using continuous tone images. It was hoped that the advantage of the photoelastic method to "see the stresses" might be passed on to finite element and finite difference users. The initial work, in black and white, centered upon variation of the light intensity and geometry modifications to indicate the stress or strain level. More recent work has introduced

Fig. 1

Fig. 2

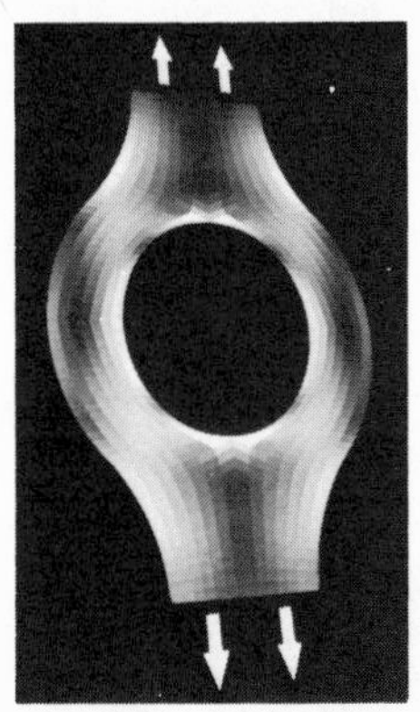

Fig. 3

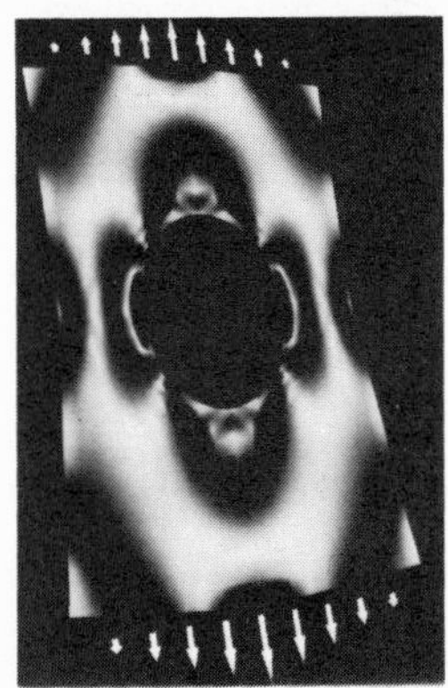

Fig. 4

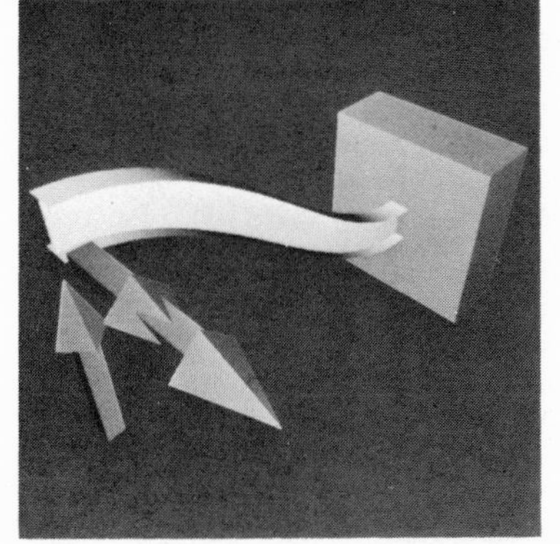

Fig. 5

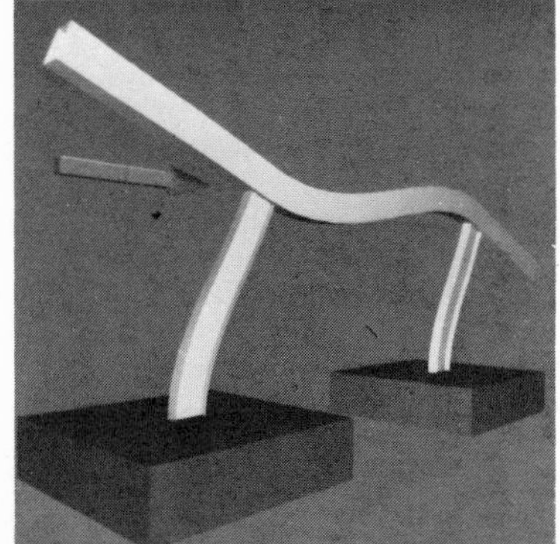

Fig. 6

color stress fringes to portray the same functions.

The most primitive scheme involves the shading of distorted grids according to the magnitude of the selected stress or strain function. Elements with high values appear bright, and those with low values appear relatively dark. Figure 3 illustrates a "flat" format treatment for a strip with a circular ring section.

A slight change in the black to white scheme results in the display of fringe patterns. The effect is similar to stress optic patterns but provides complete choice in the function to be displayed. The modification is to display a new function, ϕ, instead of the stress, σ, where

$$\phi = \cos^2(\alpha\sigma+\beta)$$

The constants α and β affect the number of fringes and shifts in the pattern. Figure 4 presents a "smooth" version of this technique applied to a rectangular plate with a circular hole. "Smooth" versions use results which were calculated directly for the nodes (by a least-square fit of an incomplete cubic to a local set of nodal displacements), while "flat" versions represent the average of the four nodal values for each element. The function displayed in Fig. 3 and Fig. 4 is maximum principal stress.

The major difficulty with this method is the lack of identification of stress level associated with any particular fringe. That is, if one scans from black to white, he may not know if the stress level has increased or decreased. Later illustrations suggest that color stress fringes are an answer to this problem.

Color Stress Fringes

In the color stress fringe procedure, the elements are colored according to the level of the selected function in either a "flat" or a "smooth" format. This is accomplished by specifying the components of a sequence of color fringes and the range of the function which is to correspond to a complete range of color fringes. For function values below or above the specified range, the element or node (depending upon format) is assigned the color of the lowest or highest fringe. For a value within the fringe range, the selected color is a combination of the appropriate adjacent fringe colors. Two functional forms of color mixing have been investigated. These are cosine-squared and linear variations. The cosine-squared variation gives the appearance, as might be expected, of one fringe color or the next with narrow bands of intermediate colors, while linear variation gives a wider region of mixed color. While the differences have been slight, the author prefers the effect given by the linear or "sawtooth" variation scheme.

Figures 7 and 8 show examples of this type of display. The color code for these two pictures (in order of increasing stress) is blue, turquoise, green, yellow, red, and magenta. The normal spectral order from cold to warm colors (excluding magenta) seems a natural choice.

It seems appropriate at this point to admit what is obvious;

that is, the difficulty in generating pictures which have continuously smooth variation in color from one fringe to the next. While this is mostly due to limitations of the author in the field of photography, the problem is made more difficult by some "rounding" of intensity values by the visible surface processor and differences in film characteristics between Polaroid (the medium used for test pictures) and the sheet film used to obtain color negatives. The quality of the pictures is also degraded by the number and type of photographic processes applied to the original picture by the time it appears in published form. The pictures shown in this paper are photographs of photographs of prints from the original negatives.

An alternate procedure for functions of two variables is to introduce geometry modifications in the form of computed out-of-plane coordinates which are proportional to the selected scalar function. This warped surface display technique is illustrated in Fig. 11 in connection with a discussion of the display of complex solutions.

SIMULATION OF DISTORTED STRUCTURAL FRAMEWORKS

The structural framework simulator is a postprocessor which accepts data files generated by standard three-dimensional framework analysis routines and produces images of distorted models. The resulting pictures consist of the structural elements, the loads (represented by computer generated arrows),and supports (models of rigid surfaces). Each structural element is represented by a large number of warped quadrilateral surfaces which define the cross section of the element and trace the distorted shape. Beam segment lengths are controlled by the specification of the maximum allowable segment length and the minimum number of segments for any particular element. The arrows and supports are mosaics of planar elements.

The procedures by which the distorted shapes may be developed from the basic geometry and nodal displacements were recently presented by the author [20] and consist of an extended exercise in linear transformations involving five distinct Cartesian coordinate systems.

Member force and moment resultants may be computed from the nodal displacement and rotation information and indicated on the display through the use of color. That is, each beam segment may be colored according to the level of a particular function of the axial coordinate in the same manner as described for color stress fringes.

Examples of Distorted Frameworks

The simulator was initially applied (for check-out purposes) to a single beam. Figure 5 illustrates the simulation of an I section subjected to a lateral displacement at the near end. This displacement requires both a shear force (shown by a single-headed arrow) and an applied moment (indicated by a double-headed arrow). The program was next used to display two-dimensional systems such as the bent shown subject to a side-sway loading in Fig. 6. In this structure the columns are I sections and the horizontal beam (in an undistorted

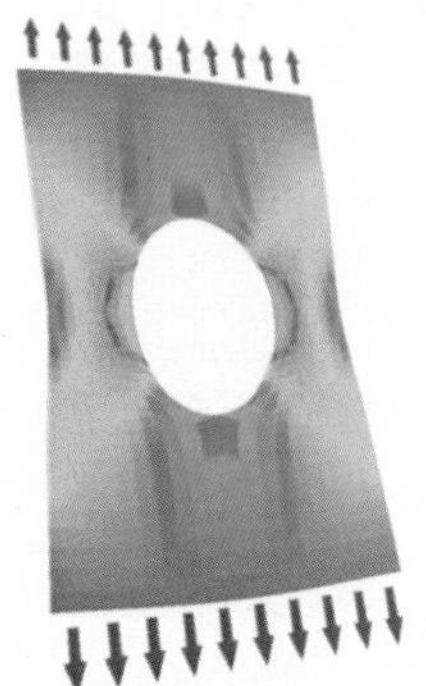

Fig. 7

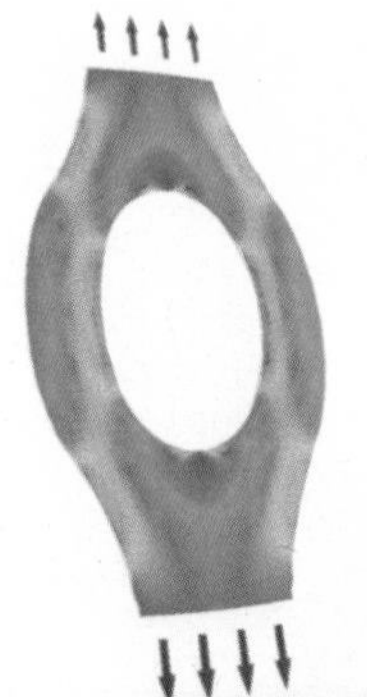

Fig. 8

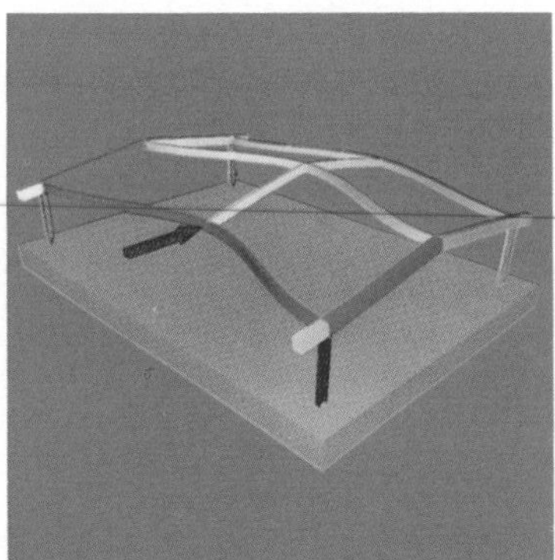

Fig. 9

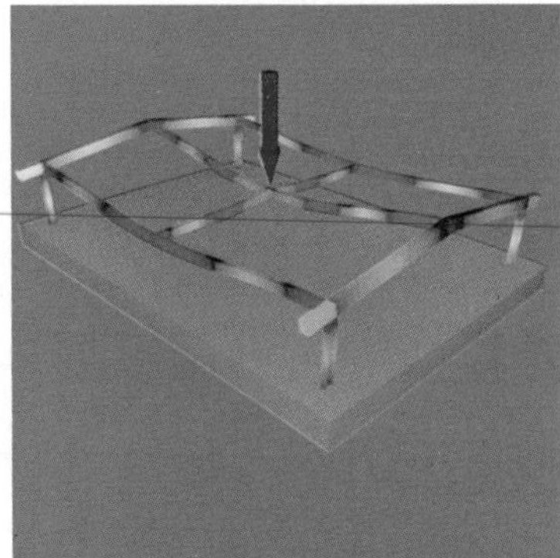

Fig. 10

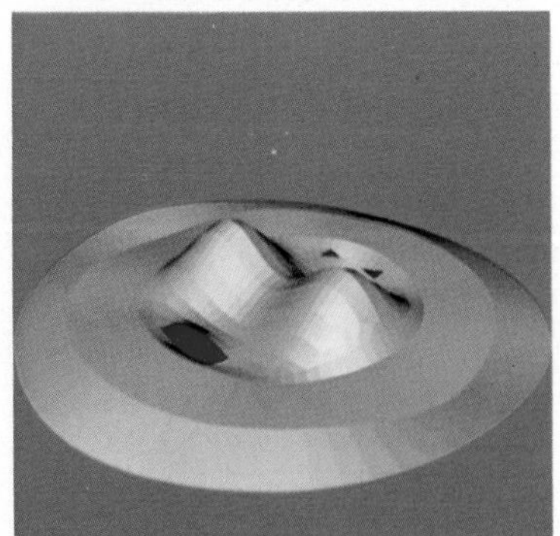

Fig. 11

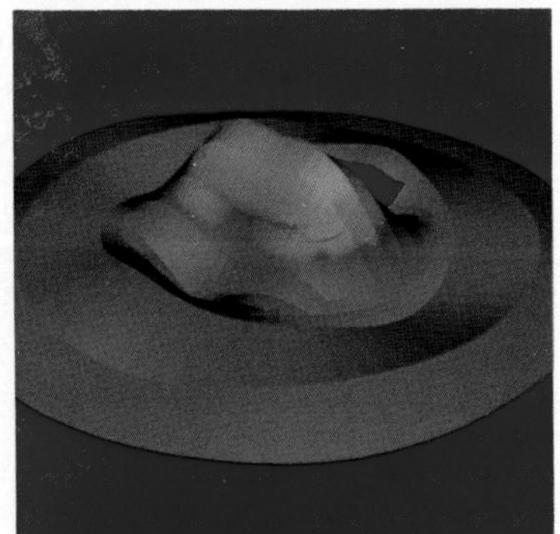

Fig. 12

sense) is a hollow rectangular section. In this picture, the effect of modeling the light source at the observer results in the web of the nearest column being brighter than the flanges. The reverse is true of the far column. Figures 9 and 10 illustrate the application of the simulator to the display of a three—dimensional framework for a small building. The structure has been distorted according to a lateral load in Fig. 9,and color has been used to indicate the absolute value of the torsion function. Figure 10 shows the structure distorted by the application of a vertical force with color used to describe the absolute value of the bending moment function for the major principal axis of the crossection. The color code for Figs. 9 and 10 (going away from zero) is white, blue, turquoise, green, yellow, and red.

DISPLAY OF THREE-DIMENSIONAL FINITE ELEMENT MODELS

There is no difficulty in accepting a one-to-one relationship between structural elements and display elements for two-dimensional problems. However, in three dimensions each structural element may have several surfaces, and each surface will result in one or more display elements. Thus, as a matter of economy and computer storage limitation, it is useful to supply only those surfaces which will appear on the exterior in the display.

The elements provided are grouped according to the "smooth" surface to which they belong, and much of the user control takes place at this level. Such control includes the option to withhold particular surfaces from a given display, the specification of the surface color components, the definition of a local translation vector for exploded views, and susceptibility of the surface to fringing and/or "peel away" functions. The "peel away" capability allows the viewing of interior or otherwise hidden surfaces by eliminating elements from the display as they penetrate a user-specified plane. Controls at the system level include rotation, translations of the global coordinate axes, distance from the coordinate origin to the observer, selection of format, scaling of the local translation vectors, and the selection of new displacement vectors and scalar function files.

The computer coding which controls the display of the displacements and scalar functions has several options. To obtain views of the model as distorted by static loadings or vibration, the user supplies an amplification factor to be applied to the displacements before they are added to the nodal coordinates. The geometry may also be modified according to the scalar function in the manner of a display of a function of two variables as a surface above a reference plane. This requires scaling factors for each coordinate direction. Fringe systems may also be defined to represent variations in the scalar function or components of the displacements in a specified direction. Practice is usually required to find reasonable scaling factors, and restraint is advised in calling combinations of options into play.

Examples of Three-Dimensional Models

Displays are presented for two finite element models of unclassified United States Navy electromechanical sonar transducers. The first device, shown in Figs. 13 and 14, is a T-shaped sonar element. A "smooth" format version of the undistorted structure is shown in Fig. 13 while Fig. 14 illustrates a "flat" version of the model distorted according to a "free-free" normal mode of vibration. The second device, shown in Fig. 15, distorted according to an N = 4 harmonic mode of free vibration, is a segmented piezoelectric cylinder. The data for these two models were supplied by personnel at the Naval Undersea Center (NUC), San Diego. Figure 16 is a "flat" format version of a finite element model of an automotive lower suspension arm. These data were supplied by Ford Motor Company based upon a NASTRAN finite element model. A thin arch concrete dam and some surrounding countryside are shown in Fig. 17, based upon data provided by the University of California, Berkeley. A model of a delta wing airplane vibrating in its lowest frequency normal mode is illustrated in Fig. 18. This model was supplied by Stanford University.

Display of Complex Solutions

The display routine for three-dimensional models has recently been modified to accept complex mathematical solutions (in the form of amplitude and phase shift values) for both structural displacements and special scalar functions. This program has been used to display pressure patterns in a fluid medium which surrounds a segmented piezoelectric cylinder. The axisymmetric vibration of the cylinder is induced by a sinusoidal variation of voltage. The dominant resulting effect is a pressure wave which moves from the interior surface of the cylinder toward the axis where it splits into two axially directed pressure waves. Since the solution is axisymmetric, it may be described as a function of the radial and axial coordinates and displayed as a warping of that plane. Figure 11 shows such a display with the elevation above the plane being an expression of the presure function. The intersection of the cylinder and the warped plane is shown in red. In this "flat" version, the pressure wave has already arrived at the axis and split into two axially directed waves. Figure 12 shows the pressure function at a previous instant when the wave is in the process of splitting. In this display, color fringes have been added to reinforce the impression of the pressure function.

Generation of Color Movies

Both the conventional and complex solution routines have been modified to allow the generation of black and white and color movies. The conventional routine has been used to generate a color movie which describes the details of the display model of the T-shaped sonar transducer shown in Figs. 13 and 14 and three normal modes of vibration. It has also been used to describe five normal modes of free vibration and a flutter mode for the delta wing model. The

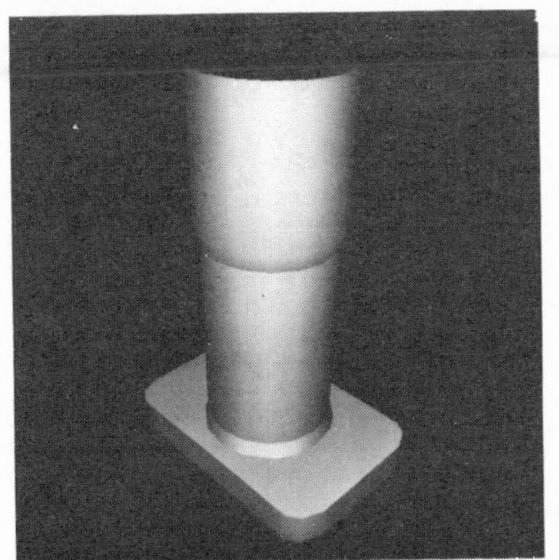

Fig. 13

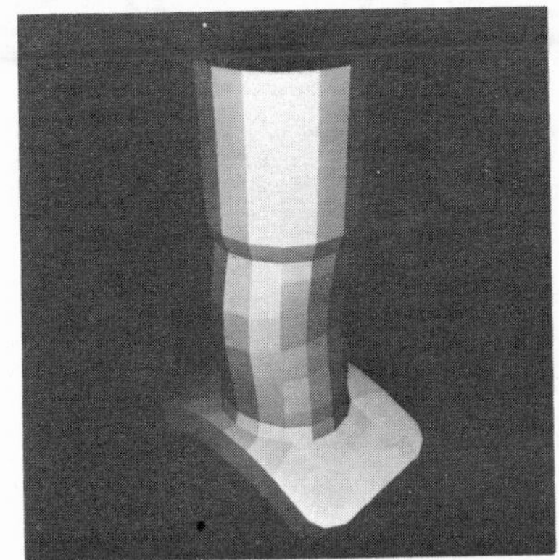

Fig. 14

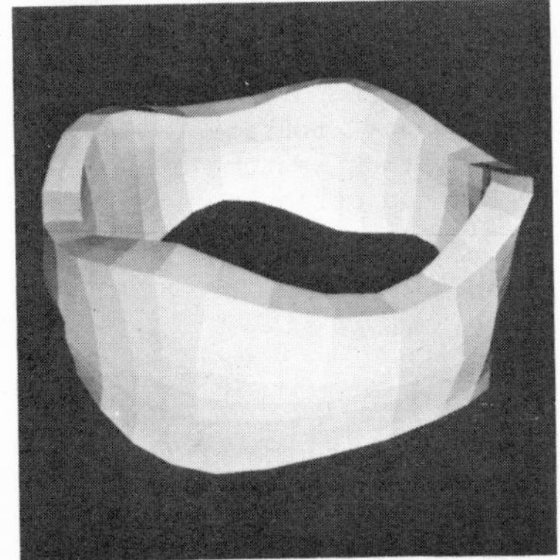

Fig. 15

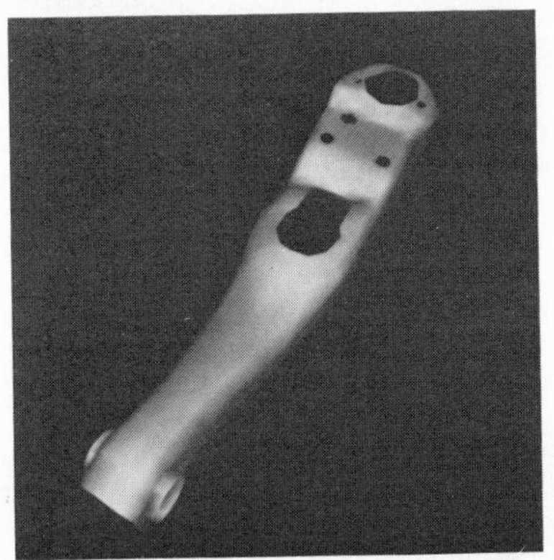

Fig. 16

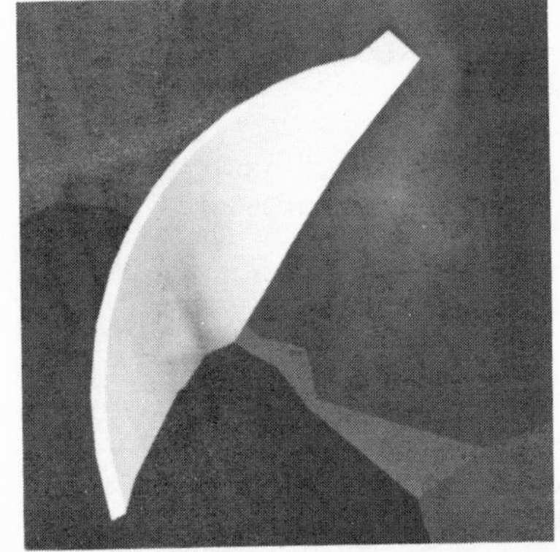

Fig. 17

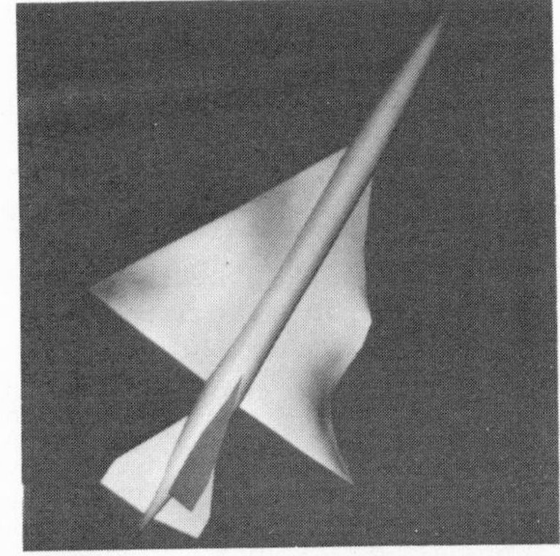

Fig. 18

complex routine has been used to generate a color movie of the pressure patterns shown in Figs. 11 and 12. The distorted framework simulator has also been modified for the generation of movies but has, so far, only been used to display key frames.

When using the movie production options, one may specify (in addition to the options available in the production of single frames):

1. The number of frames to be made under automatic control
2. Frequency of vibration
3. Smoothly accelerated motion for rigid body rotation and translation, zooming, and explosion of selected surfaces
4. The advancement of the "peel away" plane in space

When using the movie production system, the advancement of the film and the insertion of the color filters is under computer control.

CURRENT WORK

Current work seeks to reduce computation times and to extend the capabilities previously described by increasing the menu of command functions and generating more sophisticated "peel away" routines. In the latter category is an effort which will allow brick systems of finite elements to be penetrated by a plane, with the option to create a new grid automatically on the cutting plane and to warp and/or to color this new surface.

ACKNOWLEDGMENT

This work is part of a continuing research effort in computer graphics at the University of Utah, under the sponsorship of the Advanced Research Projects Agency and the United States Navy. The author appreciates the facilities provided by these organizations and the cooperation and assistance given by the faculty and staff of the Computer Science Division, especially the photographic processing services of Mike Milochik.

REFERENCES

1 Smith, A. F., "Method for Computer Visualization," AMC Technical Report No. 8436-TM-2, Sept. 1960, Electronic Systems Laboratory, M.I.T.

2 Johnson, T. E., "Sketchpad III: Three Dimensional Graphical Communication with a Digital Computer," Report ESL-TM-173, June 1963, Electronic Systems Laboratory, M.I.T.

3 Roberts, L. G. "Machine Perception of Three-Dimensional Solids," Technical Report No. 315, May 1963, Lincoln Laboratory, M.I.T.

4 Wylie, C., Romney, G., Evans, D. C., and Erdahl, A., "Half-Tone Perspective Drawings by Computer," Proceedings Fall Joint Computer Conference, American Federation for Information Processing

Systems, Vol. 31, Nov. 1967

5 Warnock, J. E., "A Hidden Surface Algorithm for Computer Generated Halftone Pictures," Tech. Rep. TR 4-15, 1969, Div. Computer Science, University of Utah, Salt Lake City.

6 Romney, G. W., "Computer Assisted Assembly and Rendering of Solids," Tech. Rep. TR 4-20, 1970, Div. Computer Science, University of Utah, Salt Lake City.

7 Watkins, G. S., "A Real Time Visible Surface Algorithm" Tech. Rep. UTEC-CSc-70-101, July 1970, Div. Computer Science, Univeristy of Utah, Salt Lake City.

8 Rougelot, R. S. and Shoemaker, R., "G.E. Real Time Display," NASA Rep. NAS 9-3916, General Electric Co., Syracuse, N. Y.

9 MAGI, Mathematical Applications Group Inc., "3-D Simulated Graphics," Datamation, Vol. 14, Feb. 1968, p. 69.

10 Bouknight, W. J., and Kelley, K., "An Algorithm for Producing Half-Tone Computer Graphics Presentations with Shadows and Moveable Light Sources," Proceedings Spring Joint Computer Conference, American Federation for Information Processing Systems, Vol. 36, 1970, pp. 1-10.

11 Lamar, J. V., Stratton, R. H., and Simac, J. J., "Computer Techniques for Pseudocolor Image Enhancement," Proceedings First USA-Japan Computer Conference, 1972, pp. 316-319.

12 June-Min, P., "Computer-Aided Mapping Technology for Geographic Data," Proceedings First USA-Japan Computer Conference, 1972, pp. 325-333.

13 Comba, P. G., "A Procedure for Detecting Intersections of Three Dimensional Objects," Rep. 39.020, Jan. 1967. IBM New York Scientific Center, New York, N. Y.

14 Weiss, R. A., "Be Vision, A Package of IBM 7090 Fortran Programs to Draw Orthographic Views of Combinations of Planes and Quadric Surfaces," Jour. Ass. Comput. Mach., Vol. 13, April 1966, pp. 194-204.

15 Mahl, R., "Visible Surface Algorithms for Quadric Patches," Tech. Rep. UTEC-CSc-70-111, Dec. 1970, Div. Computer Science, University of Utah, Salt Lake City.

16 Gouraud, H., "Computer Display of Curved Surfaces," June 1971, Unpublished Ph.D. Thesis, Div. Computer Science, University of Utah, Salt Lake City.

17 Phong, Bui-Tuong, "Illumination for Computer Generated Images," Aug. 1973, Unpublished Ph.D. Thesis, Div. Computer Science, University of Utah, Salt Lake City.

18 Resch, R. D., and Christiansen, H. N., "The Design and Analysis of Kinematic Folded Plate Systems," Proceedings Symposium for Folded Plates and Prismatic Structures, International Association for Shell Structures, Oct. 1970, Vienna, Austria

19 Christiansen, H. N. "Displays of Kinematic and Elastic Systems," Proceedings of the Third Conference on Matrix Methods in Structural Mechanics, Oct. 1971, Wright-Patterson Air Force Base, Dayton, Ohio.

20 Christiansen, H. N., "Computer Simulation of Distorted Structural Frameworks," Preprint 2080, Oct. 1973, American Society of Civil Engineers Annual and National Environmental Engineering Meeting, New York.

IV. TRENDS IN COMPUTER-RELATED DEVELOPMENTS

SOFTWARE DEVELOPMENT UTILIZING PARALLEL PROCESSING

E. I. Field, S. E. Johnson, and H. Stralberg
Universal Analytics, Inc.
Los Angeles, California

ABSTRACT

The concepts of parallel processing are described in terms of the ILLIAC IV computer system design and the ILSA (ILLIAC IV Structural Analysis) finite element program which was developed as an extension of NASTRAN for installation on the ILLIAC. As a result, from any IBM, CDC, or UNIVAC installation of NASTRAN on the ARPA network of computers the user will have access to ILSA for performing the functions of structural matrix generation, assembly, and solution on the ILLIAC. Also, the concept of pipeline processing is briefly described to illustrate the multiprocessing design of the CDC STAR and Texas Instruments ASC hardware systems. A summary of critical issues is presented to assist in planning the development of structural mechanics software for these fourth-generation computers.

INTRODUCTION

The recent advances in computer technology of the fourth-generation computers (ILLIAC IV, CDC STAR, and Texas Instruments ASC) is opening up a new era for engineering mechanics research and development. Specifically, the finite element technology, developed to define constructs for structural, fluid, and heat transfer problems which are highly amenable to computerization, already has been limited by the capacities and speeds of current third-generation computers. The consequence has been to focus attention on perfecting limited approaches to solve critical problems by reducing model sizes and/or applying two-dimensional approximations to three-dimensional phenomena. Now, with the quantum jump in capability realized by multiprocessors which employ the new parallel or pipeline technology, these limitations no longer need to plague the analyst. Unfortunately, much of the requisite technology, along with the necessary empirical data for arbitrary three-dimensional analyses, has not been developed.

Recognizing the need for a generalized structural analysis tool to serve the community as a test bed for such development, Universal Analytics, Inc. (UAI), in cooperation with the Advanced Research Projects Agency (ARPA) and the Defense Nuclear Agency (DNA), developed the ILSA, ILLIAC IV Structural Analysis program. This program is designed to incorporate the latest developments in finite element technology and to exploit the parallel processing capabilities of the ILLIAC IV, the first of several multiprocessors to be available for such development work.

In order to visualize the considerable advantages of parallel processing for structural analyses, a detailed view of the ILLIAC IV is presented. Because the ILLIAC IV can be accessed via a network of host computers located throughout the United States and abroad, the ARPANET resources are also described. With this background, the design of the ILSA program is presented with special emphasis on how finite element analysis has been adapted to parallel processing. To further illustrate how multiprocessing techniques can be used, an example of pipeline matrix processing is also described.

Along with the excitement of the new opportunities for development in the field of engineering mechanics come the more serious questions of how to go about exploiting these new multiprocessing capabilities. Not only must considerable effort be spent on implementing algorithms appropriate to the architecture of these fourth-generation computers, but there are also complex management considerations involved in undertaking such an effort. This paper, therefore, concludes with a summary of these critical issues.

ARPANET AND THE ILLIAC IV

The system[1] that includes the ILLIAC IV processor is an integrated system of data processing, information storage, and communications equipment located at NASA Ames Research Center, Moffett Field, California. The system, through a high-speed communications network (ARPANET) provides large-scale computational and file management services to a growing community of users located throughout the nation and abroad. Through ARPANET, specialized analysis and research groups seeking solutions to a range of contemporary problems have access to required computational power not otherwise available to them, and not economically feasible on an individual or regional basis.

The system is being developed at Ames Research Center by the NASA Institute for Advanced Computation (IAC). The Advanced Research Projects Agency, ARPA, of the Department of Defense underwrote the research and production that were required in the early phases of development. Since 1971, the system at Ames has been sponsored and managed by a multiagency board of owners, and today IAC receives its technical direction from this board. The board currently includes representatives of NASA, ARPA, and IAC.

The system being developed by IAC is the largest of several systems of computing resources available to users through ARPANET. Other resources vary from small interactive systems to large conventional processors. This network approach to the distribution of computer system services is one of the principal economic justifications for the development and operation of large systems--on the scale of the IAC system--for users whose applications require such systems but who cannot economically justify a dedicated system. The

---

[1] Additional summary information is available in a Press Seminar Handout prepared by the Institute for Advanced Computations, Ames Research Center, Moffett Field, California, August 22, 1973.

small systems on the network are used for communications and for program and data preparation tasks, allowing larger resources to be applied to tasks commensurate with their special capabilities.

The IAC system includes two one-of-a-kind devices, the ILLIAC IV Processor and the UNICON Laser Memory. The ILLIAC IV, which is the principal processing resource in the system, is capable of executing at the rate of 150 million operations per second with an average instruction execution time of between 5 and 10 nanoseconds (that is, an average instruction executes in 7 billionths of a second). Mass storage in the system is provided by the UNICON memory which has a total on-line capacity of over 700 billion bits or 85 billion characters of storage. This combination of processing speed and data storage capacity represents the single most powerful computational facility available in the world today.

The potential value of the system at Ames is greatly enhanced by its accessibility to a large number of disparate and geographically remote users through the ARPA communications network. All users, wherever located, access the system remotely via ARPANET for both data transfer and interactive submission of processing requests.

The ARPA Network

ARPANET is a wide-bank (currently 50,000 bits per second) communications network linking together computing centers and terminal access points throughout the country. The network has been constantly growing, both in the number of computing facilities available through it and in its geographic extent, since its inception in 1969. It should be noted that the participating institutions are primarily universities, with a sprinkling of public and private research organizations.

A primary purpose underlying ARPA's sponsorship of the network is to establish the feasibility and to develop the technical and management basis for computer system networks on a large scale. In the future, data rates in excess of one million bits per second are expected in ARPANET, and worldwide expansion has already begun.

ARPANET not only represents a linking together of equipment but also includes a standard set of management procedures and communications protocol. For example, network procedures include a file transfer facility that allows the easy transfer of files to and from virtually any computing center in the network. There is also a high degree of standardization in the control languages of the various network systems.

Network users of any of the available computing and data storage services interact with the resources of their choice, either directly through data terminal devices tied into the network or indirectly through interaction with local computer centers which are in turn tied into the network. A set of geographically dispersed network resources can be readily combined, and each applied to an appropriate part of the solution of a single problem. Regardless of how the resources of the system are intended to be used, the procedures for access and interaction are generally the same. The potential of the network as a resource for servicing and making available the multitude of current and yet-to-be-developed structural mechanics programs is obvious.

The IAC System

The IAC system is a remotely accessible large-scale computing facility available to a variety of user groups. Figure 1 is a high-level block diagram showing the major functional elements of the system. Looking at these elements first from a user's point of view, the communications processors tie the system to ARPANET. They provide data error checking and perform the necessary communications protocol.

The central processors, which are off-the-shelf Digital Equipment Corporation PDP-10 processors, control the user job sequences and execute all of the utility programs provided by the system to users. These processors interpret service requests from users and either execute the necessary utility programs or pass on the requests when they are for services provided by other resources in the system. The resource management processors control and allocate the major resources in the system; they do not directly execute user programs. These resources include the data storage devices and the ILLIAC IV processor.

All of the system processors communicate with each other through central memory. Central memory also functions both as a program and data store for the processors and as an intermediate buffer for data transfers between any two storage devices in the system and between the system and ARPANET. Central memory is thus pictured as the central device in the system in Fig. 1.

The major devices in the system are the UNICON laser memory and the ILLIAC IV processor. The UNICON laser memory, with an on-line capacity of 700 billion bits, is the largest storage device in a central file system which also includes a large buffer disk and magnetic tapes. Users do not directly address or specify the UNICON memory for data storage; rather, the location of all data files in the system is controlled by the central file system.

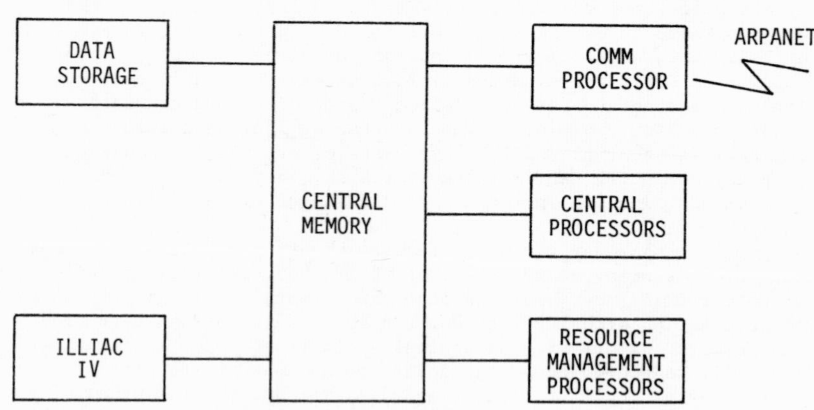

Fig. 1 Major functional elements of the system

The block labeled "ILLIAC IV" in Fig. 1 represents both the ILLIAC IV processor and the large ILLIAC IV disk memory system, which has a capacity of over 32 million 32-bit words. The ILLIAC IV processor, which is the major processor in the system, is totally dedicated to the execution of user code. It is seen by the central system as a large peripheral device providing a special user service, namely, very high-speed parallel processing on large data volumes. These services can be used independently or in combination in the solution of a single problem. The separate nature of these services is a result of the architecture of the system. Although fully integrated, it can be functionally utilized as three independent subsystems, namely, the ILLIAC IV, the information storage subsystem, and the central system.

Once the program and data have been stored on the system at Ames, the user defines the general processing tasks to be performed, including those steps to be done on the ILLIAC IV. This definition can be done interactively or in a batch sequence. The central system then initiates the movement of the data files from the UNICON memory to the ILLIAC IV memory system. The transfer takes place in a series of steps under control of resource management processors. Once a sufficient quantity of data has been transferred to the ILLIAC IV disk memory, the program itself is moved from the UNICON to the processor memory, which functions as working storage for the ILLIAC IV. Execution then begins. Refer to Fig. 2 for an illustration of data movement within the ILLIAC IV subsystem.

The ILLIAC IV subsystem, which is distinct from the central system but controlled by it, includes the ILLIAC IV processor, a 256K (32-bit word) processor memory, and a 32-million-word main disk memory device (see Fig. 2). The high execution rate of the ILLIAC IV is achieved principally through the parallel structure of its processor, presented here in general terms.

The ILLIAC IV processor executes a common instruction sequence simultaneously on a large number of otherwise independent sets of data. This simple description contains the three key phrases in understanding the parallelism implemented in the ILLIAC IV: (1) a common instruction sequence, (2) simultaneously, and (3) independent sets of data.

The instruction sequence is similar to that of any modern large-scale processor. "Simultaneously" means that each instruction in the program sequence operates on every one of the independent data sets at the same time. The number of independent data sets can vary from one up to 512, depending on the required word size of each data item and the skill of the implementing programmer. With a single data stream, the ILLIAC IV is functionally identical to a conventional, nonparallel processor. The machine architecture, as seen below, facilitates 64, 128, or 512 parallel data streams which are 64, 32, or 8 bits wide respectively. Both 64-bit and 32-bit words are standard ILLIAC IV word sizes and are fully implemented in parallel in the hardware and in the programming languages. Eight-bit wide parallelism may be thought of as a byte mode with more limited implementation in the hardware and software.

To achieve this high degree of parallelism, the processor structure consists of a single Control Unit (CU) that performs instruction decoding and program control and 64 arithmetic and logic units referred to as Processing Elements (PEs). The CU reads the instruction

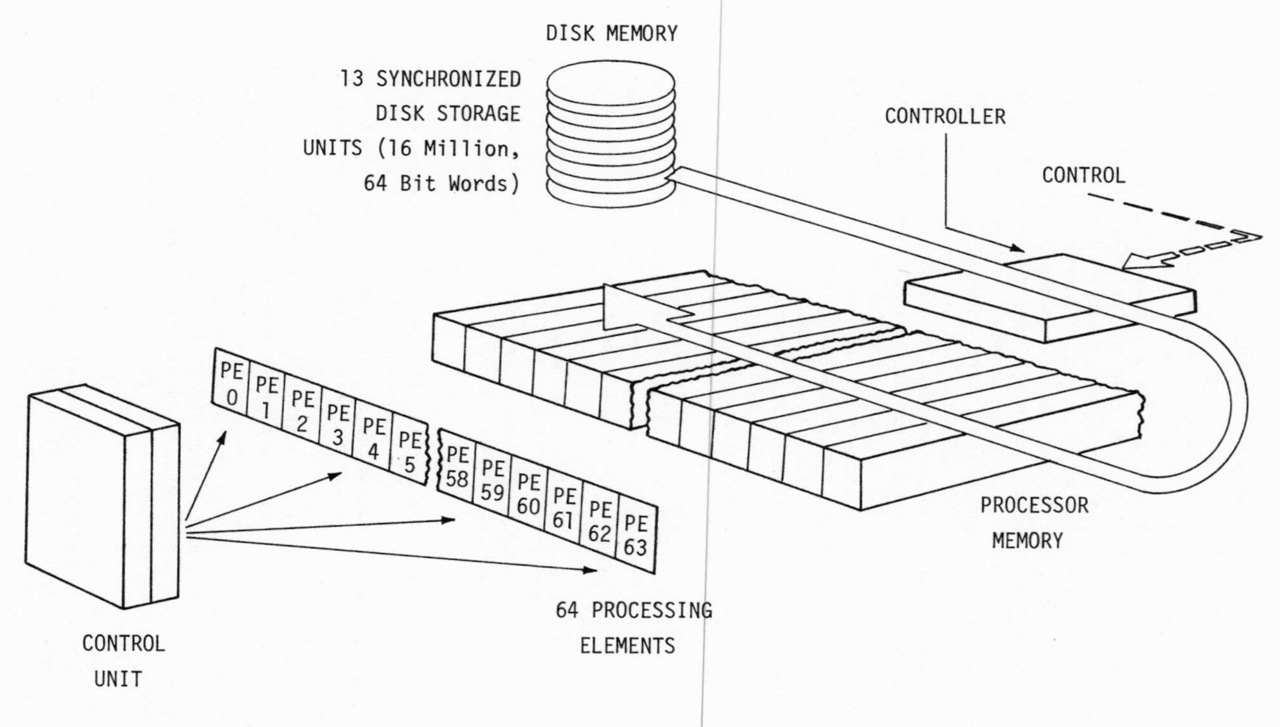

Fig. 2 Data movement within the ILLIAC IV subsystem

sequence stored in processor memory, decodes each instruction, and generates identical control signals for each of the PEs. The entire set of PEs (or any subset within it) executes the same instruction simultaneously, under CU control, each on different sets of data. Under program control, any subset of PEs can be selected not to execute the current instruction.

The instruction set and instruction execution times of each individual PE in the ILLIAC IV processor equal or exceed those of existing conventional processors. For example, a full 64-bit, floating point, normalized ADD takes approximately 300 nanoseconds. A MULTIPLY under the same conditions executes in about 600 nanoseconds.

There is a single instruction sequence, stored in processor memory. Interleaved in this instruction sequence, normally, are computational steps executed by the PEs , and program control instructions executed by the CU itself. Thus the CU, in addition to controlling the PEs , executes instructions such as branching, loop counting, and the generation of external system calls. In order to provide a sophisticated control capacity (for example, complex loop indexing), the CU has a complete instruction set including arithmetic and logic, byte and bit, and special control instructions. The CU is, if viewed as a stand-alone device, a full-scale processor.

Another vital function of the CU is to direct the transfer of data from one PE to another. Each PE memory segment is hard-wired to four neighboring PEs to provide high-speed, core-to-core transfer rates. Two methods of transfer are available to the program user. One is broadcasting , which takes a single word from one PE and broadcasts it to all PEs. The second provides for routing of one word from each PE(i) the same distance j to PE(i + j) for all PEs simultaneously.

The high execution rate of the ILLIAC IV is achieved not only through the parallel structure of the PEs but also through execution overlap. The execution of instructions within the CU is overlapped in time with the execution of instructions by the PEs. The accessing of instructions from processor memory is overlapped with the execution of these instructions. Finally, PE operand fetches are overlapped with PE instruction execution.

The processor memory is working storage for both instructions and data for the ILLIAC IV processor. This memory may be thought of as an array of 64 columns and 2048 rows. Each column (which is either two 32-bit words or one 64-bit word in width) is associated with an individual PE. The CU can access the entire processor memory, while each PE accesses only its associated column. As described earlier, however, data can also be moved from one PE to another under control of the CU. The main memory storage for the ILLIAC IV is the disk memory. This device is a fixed-head rotating disk system with a capacity of about 32 million 32-bit words. The system is composed physically of 13 disks which rotate synchronously with a 40-millisecond rotation period and which provide a maximum data transfer rate of about 10^9 bits per second. The entire processor memory can be written out to disk or loaded from disk in just one revolution. That is, 256K 32-bit words can be transferred in 40 milliseconds.

However, it is this column of data in each PE and the concept of the parallel processing on different data sets in each PE that provide the central themes from which the design originated for the matrix manipulative operations of ILSA.

ILSA DEVELOPMENT

Two major problems were faced in the design of ILSA [1]. The first was the selection of analytical capability and the method for its implementation. The second was the problem of implementation itself.

Many finite element structural programs were already in existence (e.g., NASTRAN, SAMIS, STARDYNE, SAP, FORMAT), each with its own unique approach to problem formulation and solution. Some programs were designed to treat a general set of structural problems; others were highly specialized. Of all of the available programs, NASTRAN (NASA STRuctural ANalysis) offered the broadest range of modeling and analytical capability. It was the most advanced, best organized, and most widely accepted program available to the general public. Hence, NASTRAN was selected to be the primary model for the development of ILSA. This solved the problem of selecting suitable analytical capabilities, and it provided a methodology for their implementation.

The second problem, the more difficult and on-going problem, was that of actually implementing these capabilities. This problem was partially solved by exploiting the sophisticated modular design of NASTRAN. Each basic function of ILSA was assigned to a separate nearly independent module. These modules were constrained to intercommunicate solely by way of a carefully designed set of data files. Though these design decisions were reflected in difficult technical problems, they were amenable to rational solution. The more complex aspects of implementation revolved around the status of the ILLIAC hardware and software development.

At the outset, the documentation for the ILLIAC was scant, and in some cases out of date. Significant classes of software functions had not even been specified, such as for input and output, for data format conversion, and for status displays of memory and control registers. All initial development had to be performed with the SSK3 ILLIAC simulator program installed on the Burroughs B6700. It was estimated to be 50,000 times slower than the ILLIAC itself. Consequently, program testing was severely limited. Finally, as the ILLIAC was being installed and tested, many of its supporting software functions were being developed. As these capabilities became available many programming changes had to be made to the ILSA modules already developed.

Now that the ILLIAC system has stabilized, in terms of both its hardware and software, installation and testing of applications software can proceed. The processing modules of ILSA are also now ready for installation. What remains is to validate the design, test its operation, and complete the development of the executive control and intercomputer communications components of ILSA. The following section describes the general methodology used in the development of ILSA for the ILLIAC system.

General Methodology

The analytical approach to the specification of a structural model and to the determination of its response to static and dynamic loadings for ILSA was patterned after that used by NASTRAN. In fact, the NASTRAN program itself has been incorporated into ILSA to serve as the user/program interface on a host computer, remote from the ILLIAC complex. NASTRAN will communicate with the ILLIAC via data files to be transmitted over the ARPANET. Also, this design retains for the user his full access to all the NASTRAN capabilities on the host computer.

The user, operating through the host computer, prepares his input using NASTRAN. In order to limit the number of ILSA submodules to be developed initially and at the same time to provide the most advanced modeling capability, the multitude of elements available with NASTRAN were supplanted by three of the most advanced finite elements available: the beam, the isoparametric shell/plate, and the isoparametric solid elements. The other input options of NASTRAN are retained for specification of the model geometry, element connectivities, support conditions and loadings, and so on.

The reporting of results to the user are also performed by NASTRAN. These results are derived from solution data files returned to the host computer from the ILLIAC via the ARPANET. The user has the option of selectively requesting any or all of these results. He also will have the option of plotting the structural model and its deformed shape.

Paramount in the design of ILSA is the full retention, independent of the ILLIAC, of all the NASTRAN capabilities on the host computer. In fact, the ILSA program operating on the ILLIAC could simply be conceived of as a modular extension of NASTRAN. Eventually the remaining abstraction capabilities of NASTRAN could be added to ILSA to provide even further exploitation of the ILLIAC computing power.

User control of all processing steps is exercised through the options provided by the NASTRAN executive and case control decks as input at the host computer. The NASTRAN executive includes a specialized language compiler which takes matrix-processing instructions, which could come from user/input, and generates a table—driven sequence of module calls appropriate to the processing steps requested. This method of process control is retained in the ILSA design for executive control on the ILLIAC. That is, a table-driven sequence monitor will be implemented as part of ILSA during installation on the ILLIAC. The tabular listing of operational sequences will be generated in NASTRAN and passed over the ARPANET to the ILSA executive on the ILLIAC. This table is used not only to direct module calls but also to maintain a completions status for check pointing and restarting in the case of planned or unplanned interruptions.

In order to implement this design, the first three modules developed were those required to implement a full static analysis capability. This experience then was used to develop the two modules needed for implementing a direct integration approach for transient dynamic analysis in ILSA. The following section highlights the key design concepts of each module used to exploit the parallel processing features of the ILLIAC.

Implementation

Each of the static and dynamic analysis modules of ILSA presented its own unique design problems. Overlaying each of these, however, was the basic problem of organization so that each would be compatible with the other. Compounding this one issue was the obvious newness of the parallel processing concepts. After careful study, the central theme emerged focusing attention on matrix storage schemes.

The ILLIAC architecture presented two controlling limitations. First, data that is created in one PE and then stored out on the disk can only be read back into that same PE. Second, core memory for each PE is severely limited to only 2000 64-bit words each. The second limitation suggested matrix data had to be stored efficiently; that is, with a minimum of indexing variables. This, of course, led to the use of a monotonically ascending internal numbering scheme and storage of the matrix in 64 x 64 word blocks. The ILLIAC word length of 64 also allowed for a very efficient packing of binary indexing pointers required for operating on the matrix.

The first limitation, however, was not overcome quite so easily. Only after several design iterations and carefully examining the interaction among each of the modular functions did the solution become clear. The equation solver for static (and also dynamic) analysis dictated the entire organization of the element matrix generator and the matrix assembler. Because this was the case, the design of the equation solver is described first. It should be noted that this module, with some modification, is used for both static and dynamic transient analysis.

Linear Equation Solver

The linear equation solver (LEQS) module solves matrix equations of the form:

$$[A] \left\{ \begin{array}{c} C \\ --- \\ X \end{array} \right\} = \left\{ \begin{array}{c} B1+Y \\ ---- \\ B2 \end{array} \right\}$$

where [A] is the structural stiffness matrix, {Y} the reaction forces of boundary conditions, {X} the displacements at all other degrees of freedom, {B1} the load matrix of boundary conditions, {B2} the load matrix of all other degrees of freedom, and {C} the user-specified displacements at boundary conditions. The degrees of freedom with boundary conditions are actually spread among other degrees of freedom, not isolated as might be inferred from the partitioned equation above. Input to LEQS consists of matrices [A], {B}, and {C} where {B} = [B1 ⋮ B2]. The matrices are read from ILLIAC disk files prepared by the ASSEM modules. [A] is a symmetric matrix. Only its upper triangular partition is required by LEQS. Boundary conditions may be user specified or program generated. LEQS generated a boundary condition, with zero displacement, whenever a zero or almost zero diagonal entry in [A] is encountered during Gaussian elimination.

Output of LEQS consists of vectors {X} and {Y}. The reaction vector {Y} is separated on two output files; one contains reactions at program-defined boundary conditions, the other at user-specified boundary conditions.

The present version of LEQS is designed to solve the maximum size problem that requires each entry in [A] to be read only once during each of the LEQS calls to the decomposition, forward, and backward substitution routines. The [A] matrix is represented in the form shown in Fig. 3. Most nonzero entries are inside the ragged band whose bandwidth MB is defined in the figure. The steps of the ragged band contain 64 rows each. Bandwidth MB is defined by ASSEM and is always a multiple of 64. In addition to the entries inside the band a limited number of nonzero entries are allowed outside the band. More precisely, the number of nonzero columns outside the band resulting from Gaussian elimination must not exceed 64 in any one row. The maximum allowed size of MB is 5 x 64 = 320. To solve larger problems, it would be required to include a scheme for multiple reads of the [A] matrix partitions during Gaussian elimination, which can be implemented efficiently. Such a scheme would tend to make the solution time I-0 dependent, rather than computationally dependent.

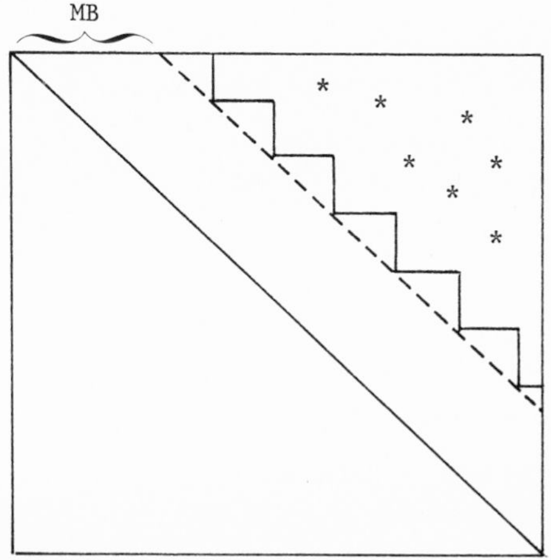

MB: Bandwidth
*: Special Points

Fig. 3 Banded [A] matrix with scattered special points

The [A] matrix is read in blocks of 64 x 64, stored in PE rows, such that all entries in the same matrix column are stored in the same PE. This storage scheme allows full utilization of parallel processing capability during Gaussian elimination. Since each column is entirely within one PE, it becomes possible to operate simultaneously on 64 columns at a time.

Submodule DECØM reads [A] into memory in 64 x 64 blocks. When one block has been decomposed and is no longer needed, the results are written to disk and stored row-wise. Then a new block of [A] is read into that same location. The row-wise storage of the 64 x 64 blocks of the decomposed matrix is the most desirable arrangement for application to forward and backward substitution.

The load vectors {B}, as well as the output vectors {X} and {Y} and the vectors of specified displacments {C}, are all stored such that each row is within one PE. This arrangement allows the processes of forward substitution and backward substitution to operate simultaneously on 64 rows at a time.

This arrangement provides nearly complete utilization of parallel processing. It will be so efficient that, for the largest matrix the ILLIAC can handle, the processing will be nearly I/O bound. Storage schemes which reduce the number of computations therefore could only be justified were they also to reduce the volume of data to be transferred to and from disk. One possible candidate that shows promise is the full wavefront approach. But, because of the extra routing that would be required, the gains made would most likely be offset by the time required to move the data between PEs.

Matrix Assembler

Stepping now to the module which assembles the matrices operated on by LEQS, the design requirements of this module come into focus. This matrix assembler module (ASSEM) also has the responsibility for preparing the integer and logical data for control of LEQS processing. The objectives of ASSEM are:

1. To determine the exact layout in ILLIAC memory of the structural stiffness matrix. This includes finding the matrix bandwidth MB (see LEQS above), identifying all special points (that is, nonzero coefficients outside the band defined by MB), and assigning a PE for processing of special columns (columns outside the band with nonzero entries). This task requires scanning of the file of element connectivity data created during the generation of each element matrix. It is accomplished by an iterative scheme to determine MB. Each iteration involves a scan of the entire element connectivity file. The assumption that MB is a multiple of 64 ensures that only a few iterations are required. After MB is computed, the element connectivity file is overlayed with information on PE allocation for special columns.

2. To prepare logical information relating to special columns and boundary conditions, required by LEQS in a compact form so that it will require minimum memory space during LEQS execution. This information is stored in binary form.

3. To assemble the element structural mass, damping, stiffness, and load matrices and to incorporate the user-defined load data. Each entry in the structural matrices must be allocated to a specific PE (as determined by its required position in the assembled matrix). Each entry in the assembled matrix may contain contributions from many element matrices, with each contribution coming from a different PE, the PE where it was created during matrix generation. In general, matrix assembly requires moving of data from each PE to every other PE. Each time an entry is routed from the PE in which it was created to its assigned PE, entries should be found in all other PEs that need to be routed the same distance. When this is feasible, the routing process achieves optimum efficiency.

To meet this design objective, a system of 63 queues is established within each PE, where each queue k contains the element matrix entries to be routed a distance k. For each k, a routing algorithm selects the first entry in queue k in each PE and routes up to 64 matrix entries at a time.

Due to limited core space, only a small portion of the full structural matrices may be created at one time. For each such small portion, a scan is required through all element-generated matrices. This requires extensive rereading of element data from the ILLIAC disk and could easily lead to excessive I/O requirements.

The following two techniques were incorporated to minimize the dependence on I/O during matrix assembly:

1. Read only element matrix data required by the section of the final matrix being assembled. The element connectivity data are used to determine which pages (a page is 16 PE rows [1024 words] of data) from the element matrix file contain data to be inserted in the current section of the assembled structural matrix. If the elements are organized in some systematic fashion (for instance, according to the lowest degree of freedom in each element), this will usually enable the skipping of large sections of the element matrix file.

2. Perform I/O and computations in parallel. This requires issuing requests for input of additional element matrix data as far in advance as possible of when the data is due to be assembled. This feature minimizes the time spent waiting for input to be completed.

These processing requirements helped dictate the design of the next module to be described, the element matrix generator.

Element Matrix Generator

The element matrix generator module (EMATG) consists of three independent submodules, BEAM, SHELL, and SØLID. This separation provided the maximum core memory for generation and assembly of each element matrix. This separation also dictates that all elements of the same type be processed together. The architecture of the 64 separate PEs suggests immediately the possibility of processing one element in each PE to achieve full parallelism. This works well for beam elements. But, the isoparametric element requires much more core storage space than is available in just one PE.

The organization of SHELL and SØLID emerged from the integration process used for building isoparametric element matrices. One PE is assigned to each integration point. For example, a shell element with nine integration points specified (a 3 x 3 mesh) will be provided nine PEs for building that element's matrices. Therefore, seven elements, each with nine integration points, can be processed simultaneously, leaving only one PE inactive.

For the most efficient use of core storage, the contributions to the element matrices at each integration point are evaluated entry by entry, summed, and stored in the sorted sequence ready for final output to disk. Prior to output, these matrices are spread out over as many PEs as required depending on their size. If M is the number of PEs required for storage of one element's matrices, 64/M sets of element matrices are computed to fill all of core before initiating a write to disk.

In order to implement the schemes indicated above, it was required to develop routines that would efficiently:

1. Allocate the PEs used to integrate and store each element matrix

2. Copy required element data from the input PE of each element to all PEs used to integrate the element

3. Sum the matrix entries contributed from each integration point over the PEs used for integration

4. Insert each matrix entry in the PE where it is to be stored prior to output

The above tasks have no parallel in the design of matrix generation programs for sequential computers; hence, they contributed significantly to the total design effort of EMATG.

In addition to these tasks, considerable design effort was required to avoid unnecessary data transfers between PEs during integration itself. The generalized stiffness matrix can be defined by the equation

$$[K] = [T]^T \left(\int_V [C]^T [D][C] \, dV \right) [T]$$

where [C] is the strain matrix, [D] is the material matrix, and [T] is the product of all required transformations. By decomposing [D] to the product $[U]^T[U]$ and bringing the transformations inside the integral, the above equation can be written simply as

$$[K] = \int_V [B]^T [B] \, dV$$

where

$$[B]^T = [T]^T [C]^T [U]^T$$

This approach reduces the total number of multiplications by applying the transformations only on one side of a smaller matrix $[C]^T \; [U]^T$ rather than to both sides of the larger matrix $[C^T \cdot D \cdot C]$. Of particular value, a single row of [K] can be computed at a time and routed to its assigned PE, thus minimizing core storage for intermediate results.

During the development phases of programming to meet these objectives, a specialized mapping of core storage was required. This map helped visualize the sequence in which the various segments of data would overlay each other. In developing these overlay maps, it was soon discovered that the designer had to face potential maximum conditions. The usual attempts to economize as one would on sequential processors failed. Instead, trade-offs were made in an effort to minimize the maximums. In reality, this simplified the programming effort by eliminating a lot of unnecessary sophistication.

The above three modules constitute the static analysis capability of ILSA. They provide, also, the data required by the dynamic analysis module of ILSA, to be described next.

Dynamic Response Equation Solver

The Dynamic Response Equation Solver (DYRES) module of ILSA performs a direct integration of the finite difference equation:

$$[A0]\left\{\frac{x_n}{y_n}\right\} = \frac{1}{3}\left\{\frac{P_n + P_{n-1} + P_{n-2}}{R_n + R_{n-1} + R_{n-2}}\right\} +$$

$$[A1]\left\{\frac{x_{n-1}}{y_{n-1}}\right\} + [A2]\left\{\frac{x_{n-2}}{y_{n-2}}\right\}$$

where

$$[A0] \equiv \left[\frac{M}{\Delta t^2} + \frac{B}{2\Delta t} + \frac{K}{3}\right]$$

$$[A1] \equiv \left[\frac{2M}{\Delta t^2} - \frac{K}{3}\right]$$

$$[A2] \equiv \left[\frac{-M}{\Delta t^2} + \frac{B}{2\Delta t} - \frac{K}{3}\right]$$

and [M], [B], and [K] are the customary mass, damping, and stiffness matrices, respectively. These equations are those solved now in NASTRAN.

The iterative nature of the dynamic solution algorithm naturally organizes DYRES into logical loops. The outer loop algorithm is named the Dynamic Response INterval INitialization (DRININ). DRININ is executed at the beginning of the solution and at the beginning of each time the value of Δt changes. The inner loop is named the Dynamic Response ITERation (DRITER). DRITER computes $\{x_n\}$ and $\{R_n\}$. DRININ uses a modified version of the static analysis matrix decomposition routine DECØM. DRITER uses modified versions of the static analysis routines for both the forward and backward substitution.

The major problem encountered in designing the initialization routine DRININ centered around the limited storage space on the ILLIAC disk. Seven matrices [M], [B], [K], [A0], [A1], and [A2] plus the decomposed [A0] compete for space on the disk. For the largest problem (8192 degrees of freedom and an average semibandwidth of 384), only three of these matrices will fit on disk. Obviously, considerable planning is required to be assured that the data is available when needed.

To best accommodate this restriction on space, only the upper half of the symmetric matrices is stored. This concession, however, requires special programming to postmultiply the full matrix by a vector while generating the right-hand side at each iteration. First the upper-half matrix is loaded into memory in row blocks, each row is multiplied into the vector entries, and the individual products are routed and summed. Then, these blocks must be transposed by routing to obtain the remaining products which in turn are routed and added to the previous partial sum.

Coordination of all these steps, which must also include building the user-specified time-dependent load vectors, requires that the driver module DYRES be carefully tuned to achieve reasonable efficiency for both small and large problems. This tuning will require incorporating performance parameters to be obtained from actual execution experience.

In order to better understand the level of design effort required to utilize multiprocessing, the following section contrasts parallel with pipeline processing for the simple task of multiplying a matrix by a vector.

COMPARING PARALLEL WITH PIPELINE PROCESSING

The ILLIAC IV and the STAR multiprocessors employ completely different processing concepts. Both achieve their efficiency when large numbers of identical operations are grouped together. The ILLIAC performs a sequence of parallel operations on 64 groups of data simultaneously, thereby reducing processing time by a factor of nearly 64. The STAR, in contrast, pipelines these operations, passing a stream of data through a sequence of operations. As long as data continues to enter the pipeline, keeping it full, the time it takes to process any one piece of data is the time required for the single longest operation in the pipeline. Of course, more than one pipeline may be operating at once.

To illustrate the differences, consider the in-core premultiplication of a vector by a matrix:

$$[A]\{b\} = \{c\}$$

Figure 4 illustrates how this operation would utilize a parallel processor with four PEs. The [A] and {b} data would be stored across PE memories as shown. In the first step, {b} is multiplied by the circled terms of [A], the diagonal entries. In step 2, {b} is now multiplied by the first off-diagonal terms of [A]. However, the results are not aligned in the same memory block with the preceding intermediate results. Thus, the results of this second (and subsequent) set of multiplies must be routed across array memory before being added to the running sum. Upon completion of step 4, as shown, the {c} vector entries are aligned in the same PE as the corresponding entries of {b}.

In contrast, the pipeline processing of the STAR is illustrated in Fig. 5 for the identical problem. The pipe, in this case, is divided into five stages: three fetches from memory to the arithmetic

PE_1	PE_2	PE_3	PE_4
b_1	b_2	b_3	b_4

STEP 1

a_{11}	a_{12}	a_{13}	a_{14}
a_{21}	a_{22}	a_{23}	a_{24}
a_{31}	a_{32}	a_{33}	a_{34}
a_{41}	a_{42}	a_{43}	a_{44}

Multiply

b_1a_{11}	b_2a_{22}	b_3a_{33}	b_4a_{44}

Step 1 intermediate result

STEP 2

			a_{14}
a_{21}			
	a_{32}		
		a_{43}	

Multiply

b_1a_{21}	b_2a_{32}	b_3a_{43}	b_4a_{14}

Result of multiply

Route 1 to right and add

$b_1a_{11}+$ b_4a_{14}	$b_2a_{22}+$ b_1a_{21}	$b_3a_{33}+$ b_2a_{32}	$b_4a_{44}+$ b_3a_{43}

Step 2 intermediate result

Fig. 4 Sequence of parallel operations using four PEs

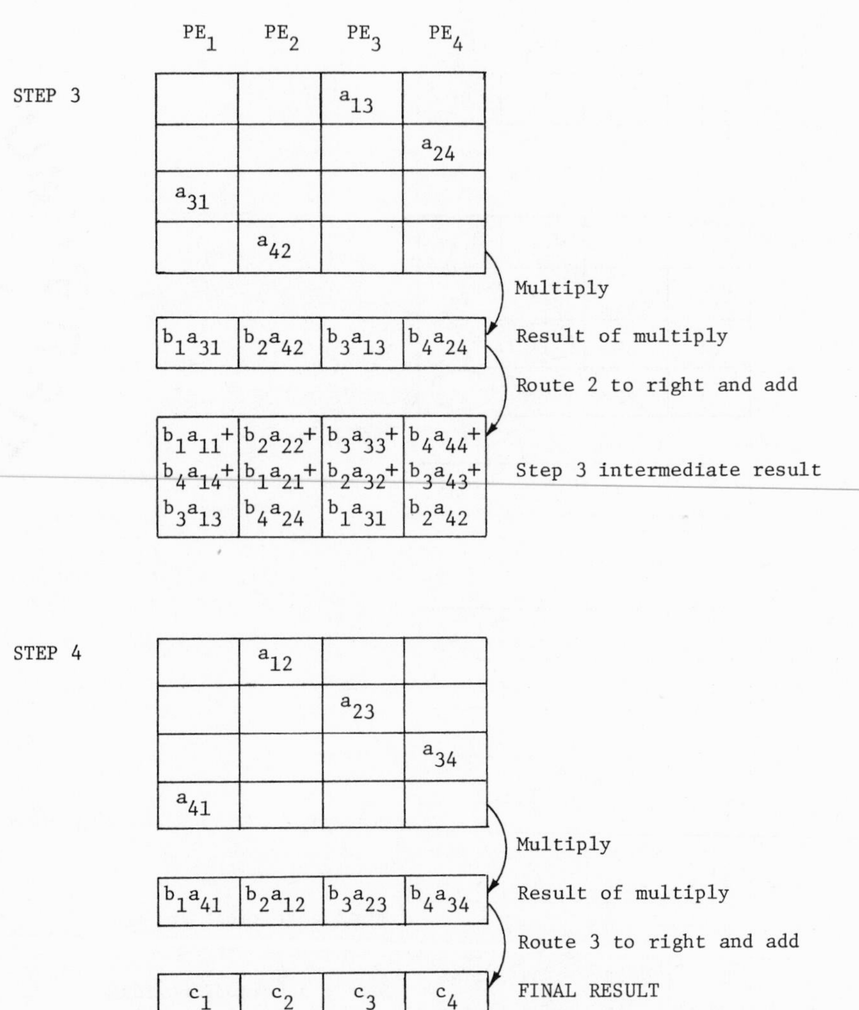
PE1 PE2 PE3 PE4
STEP 3
a13
a24
a31
a42
Multiply
b1a31 b2a42 b3a13 b4a24
Result of multiply
Route 2 to right and add
b1a11+ b4a14+ b3a13
b2a22+ b1a21+ b4a24
b3a33+ b2a32+ b1a31
b4a44+ b3a43+ b2a42
Step 3 intermediate result
STEP 4
a12
a23
a34
a41
Multiply
b1a41 b2a12 b3a23 b4a34
Result of multiply
Route 3 to right and add
c1 c2 c3 c4
FINAL RESULT

Fig. 4 (cont.)

t_1	t_2	t_3	t_4	t_5	t_6	t_7	t_8
Fetch a_{11}	Fetch a_{21}	Fetch a_{31}	Fetch a_{41}	Fetch a_{12}	Fetch a_{22}	Fetch a_{32}	Fetch a_{42}
	Fetch b_1	Fetch b_1	Fetch b_1	Fetch b_1	Fetch b_2	Fetch b_2	Fetch b_2
		Fetch c_1	Fetch c_2	Fetch c_3	Fetch c_4	Fetch c_1	Fetch c_2
			$c_1=c_1+a_{11}*b_1$	$c_2=c_2+a_{21}*b_1$	$c_3=c_3+a_{31}*b_1$	$c_4=c_4+a_{41}+b_1$	$c_1=c_1+a_{12}*b_2$
				Store c_1	Store c_2	Store c_3	Store c_4

t_9	t_{10}	t_{11}	t_{12}	t_{13}	t_{14}	t_{15}	t_{16}
Fetch a_{13}	Fetch a_{23}	Fetch a_{33}	Fetch a_{43}	Fetch a_{14}	Fetch a_{24}	Fetch a_{34}	Fetch a_{44}
Fetch b_2	Fetch b_3	Fetch b_3	Fetch b_3	Fetch b_3	Fetch b_4	Fetch b_4	Fetch b_4
Fetch c_3	Fetch c_4	Fetch c_1	Fetch c_2	Fetch c_3	Fetch c_4	Fetch c_1	Fetch c_2
$c_2=c_2+a_{22}*b_2$	$c_3=c_3+a_{32}*b_2$	$c_4=c_4+a_{42}*b_2$	$c_1=c_1+a_{13}*b_3$	$c_2=c_2+a_{23}*b_3$	$c_3=c_3+a_{33}*b_3$	$c_4=c_4+a_{43}*b_3$	$c_1=c_1+a_{14}*b_4$
Store c_1	Store c_2	Store c_3	Store c_4	Store c_1	Store c_2	Store c_3	Store c_4

t_{17}	t_{18}	t_{19}	t_{20}
Fetch b_4			
Fetch c_3	Fetch c_4		
$c_2=c_2+a_{24}*b_4$	$c_3=c_3+a_{34}*b_4$	$c_4=c_4+a_{44}*b_4$	
Store c_1	Store c_2	Store c_3	Store c_4

Fig. 5 Sequence of pipeline operations in five stages

unit, a multiply and add, and a store from the arithmetic unit to memory. Note that in this example, the multiplication by [A] proceeds column-wise. If it were to proceed row-wise, the running sum for each term of {c} would not have been stored before the next fetch required it. Of course, a {c} term could be fetched before entering the pipe, and then the result could be stored in completion of processing that row of [A]. This would cause the pipe to empty for each row of [A], thereby effectively destroying pipeline efficiency. For this reason, the same term of {b} is fetched repeatedly just to keep the pipe full.

This simple example helps illustrate the unusual considerations required for exploiting the advantages of multiprocessing computers. Exciting new opportunities arise for developing new algorithms, designing new higher level languages, and testing new mathematical constructs. The many problems of program development for exploiting these capabilities need to be examined. The next section attempts to identify some of the major considerations and presents some guidelines for planning a development effort.

IMPACT OF FOURTH-GENERATION COMPUTERS

The rapidly accelerating technology which is producing the new fourth-generation computer systems is compounding the management problems of selecting suitable hardware, planning for future software development, and allocating the resources required to do the job. Much more must be known about these problems before adequate planning can take place. For example, the factors effecting new code development, the problems of adapting existing codes, and the potential benefit/cost parameters are only some of the issues to be studied before good management decisions can be made. The following section outlines some of the major issues affecting the decision-making process.

Summary of Key New Capabilities

The most recent developments in new computer technology incorporate a variety of capabilities. The systems now available, or soon to be made available, can be summarized as follows:

1. IBM 360/195 - Provides a serial/overlap ability which combines efficient input/output buffering that can be overlapped with computations. Uses hardware for queuing computations to minimize memory fetches within the arithmetic units. These features provide a semipipeline capability with a minimum of programming effort required.

2. CDC 7600 - Provides a serial/overlap ability combining efficient buffered memory utilization with an instruction stack to minimize memory fetches and to overlap arithmetic operations. Thus, a semipipeline capability is provided requiring modest reprogramming to achieve full optimization.

3. Texas Instruments ASC - Provides a parallel pipeline ability which completely overlaps distinct hardware functions, reducing computation time for the sequence of operations to that time required

for the longest single operation in the sequence. Initial hardware to be released soon will provide two such pipelines. Subsequent releases will provide four pipelines with the ability to process in parallel. Reprogramming will be required. However, familiar higher-level language constructs will be available.

4. CDC STAR - Provides a multiple pipeline ability much like the ASC processor. Offers parallelism via overlap of the separate pipeline processing functions. Initial delivery of limited versions will be available in 1974 with plans for full system delivery late that year or early the next year. Modest reprogramming effort required to adapt existing software.

5. BURROUGHS ILLIAC IV - Provides a one-of-a-kind parallel processing ability as described earlier. Requires considerable redesign and complete applications redevelopment utilizing specialized assembly and new higher-level languages. The ILLIAC IV may be accessed from conventional hardware systems via the ARPANET and has available to it the UNICON for massive on-line data storage.

Each of these systems has its own special features which must be carefully considered in light of the overall design of a program to be developed for a particular analysis. As was noted, a prime concern for future development work is the level of systems software to be made available. Several questions should be asked prior to initiating such a development effort on any of these systems.

1. Can existing code be adapted to operate on the new system?
2. Will existing code efficiently utilize the features of the system?
3. Is there an interim capability available for development and testing of these programs (e.g., the SSK3 simulator on the Burroughs 6700 for the ILLIAC IV) prior to delivery and installation of the hardware?
4. Is there adequate systems software planned to assist the user in the tasks of debugging his programs?
5. Will the planned system software specifications be in a state of flux prior to and during installation of the hardware?
6. If new languages are to be defined (as is the case with the ILLIAC), how much special training effort is required in order to utilize them?
7. Once the hardware is delivered, how long will it take to stabilize sufficiently for effective utilization by the applications programmer?
8. Will systems support personnel be available to answer questions and assist in the actual installation of programs on the new system?

Each of these questions emanated from the specific experience of working on the development of ILSA for the ILLIAC IV. They all deal with the problems a user will have while attempting to interface with the system. However, these are not the only problems to be encountered.

Special attention must be given to preplanning the actual design of any applications program which is expected to exploit the multiprocessing capabilities of these new computers. To make this preplanning successful, the designer should experience the process of coding at least a typical section of his program. This experience is of paramount importance on at least two counts. First, he must learn to think parallel or pipeline. The standard conventions commonly accepted for serial processing often no longer apply. Second, the user must familiarize himself with the details of execution timings for all steps in the processing sequence. This requirement cannot be overemphasized. Complete planning of each step in the analysis is essential if the inevitable cascading effects of decision errors or omissions are to be avoided.

Execution Times

Of utmost importance to any management decision is the question of cost effectiveness. Part of this question can be resolved by measuring relative performance parameters. Table 1 summarizes one set of comparisons. The table entries indicate the computation speeds of 64-bit precision operations in millions of operations per second, memory to memory. For each operation, two limiting conditions are assumed, full overlap with $N = \infty$ and single step operations with $N = 1$.

Table 1. ILLIAC IV timing comparisons

Operation	Steps per Stage	Millions of Operations per Second				
		IBM 360/75	IBM 360/195	CDC 7600	CDC STAR	ILLIAC* IV
Addition	$N = \infty$	0.24	4.6	5.2	50	50
	$N = 1$	0.24	0.55	1.6	0.57	0.78
Multiplication	$N = \infty$	0.14	4.6	5.2	25	44
	$N = 1$	0.14	0.53	1.5	0.57	0.69
Division	$N = \infty$	0.096	1.7	2.0	12.5	17
	$N = 1$	0.096	0.43	0.93	0.56	0.27

Presented at an ILLIAC IV users' conference held October 31, 1973 and sponsored by R & D Associates, Santa Monica, Calif.

*Single quadrant of the ILLIAC system (64 PEs)

These numbers reveal the significant increases in power obtained by the STAR and the ILLIAC even over the current CDC 7600 and IBM hardware. A word of caution must be introduced, however. These are theoretical timings based on optimum conditions. No computer ever operates under these conditions all the time. For example, the ILLIAC may be reduced from 50 million additions per second to 37 million simply by operating at a slower clock speed, and to 9.1 million by operating in a nonoverlap mode. One or both of these conditions may occur as hardware design problems are worked out.

Real problems seldom lend themselves to ideal program sequencing. Even when segments of code provide near optimum efficiency, the difficulties of routing data through the system, accessing mass storage for reading new data, or writing results may severely degrade the overall system performance. For large matrix manipulative processing on the ILLIAC, the input/output features effectively control program design. The extra programming effort to schedule input/output operations so as to minimize wait time on reading from or writing to disk yields much greater rewards than saving a few steps in the internal computations. If storage is optimized on disk, the total execution time for the decomposition of a matrix of order 8000 with 380 semibandwidth is estimated at 1.5 seconds. If the input/output operations are not optimized, the decomposition could take as much as ten times that estimate.

This recognition that such dramatic losses of efficiency can virtually wipe out any potential gains has given strong impetus to the development of new system software. The conventional boundaries between job control, assembly language, and algorithmic languages are disappearing. Access to and control by the programmer of the basic hardware functions is required in order to achieve the full potential of these new systems. The resulting demands for professionalism on the part of the programmer are all too obvious.

Software

The advent of these new computer capabilities is already bringing about dramatic shifts in the attention being given to structural mechanics software. Analyses are being contemplated of a size that would require the dedication of a conventional system for days. New hybrid schemes are being developed which combine the finite element and the finite difference approaches [2]. New structural modeling concepts are also being adopted to utilize efficiently the architecture of these new machines [3]. The mathematical models are being expressed in different forms to facilitate parallel and/or pipeline processing.

Programmers must become intimately familiar with the architecture and timing consequences. Even operations research techniques [1] are being applied to evaluate the probabilities of data being available when needed. New documentation requirements are being imposed. New flowcharting techniques are required to depict the multidimensional sequences of operations for both parallel and pipeline processing. A key problem to be solved is the display of the time dimension for such processing.

The impact of fourth—generation computers on structural mechanics will be significant and envigorating. The availability of these new resources will facilitate the technological advances needed to protect our environment. The consequences of these advances will be to enhance the safety of shipping, of ground and air transportation, of oil pipelines, and of nuclear reactors. Most of these structures exemplify some of the following problem areas:

1. Nonlinear material properties, including temperature dependence
2. Plastic deformation and component failure
3. Nonlinear dynamic response
4. Temperature shock loadings
5. Nonlinear soil/structure interaction including slippage
6. Fluid/structure interaction

Finally, the focus of attention can now be shifted away from perfecting only limited approaches for solving these critical problems. The requisite technology must be developed and the necessary empirical data must be collected in order to mount a substantial attack on these arbitrary three-dimensional problems that face the engineering profession.

REFERENCES

1 "Interim Systems Manual, ILSA," DNA001-72-C-0108, Feb. 1973, Universal Analytics, Inc., Los Angeles, Calif.

2 Frazier, G. A., Alexander, J. H., Petersen, C. M., "3-D Seismic Code for ILLIAC IV, Interim Report," SSS-R-73-1506, Feb. 1973, Systems, Science and Software, La Jolla, Calif.

3 Frazier, G. A., Private Communication, Systems, Science and Software, La Jolla, Calif., 1973.

4 ILLIAC IV Systems Characteristics and Programming Manual, 66000D IL4-PM1, May 1972, Burroughs Corporation, Defense, Space and Special Systems Group, Detroit, Michigan.

5 Systems Guide for the ILLIAC IV User, SG-I1000-0000-C, July 1973, Institute for Advanced Computation, Ames Research Center, Moffett Field, Calif.

6 "Symposium on Complexity of Sequential and Parallel Numerical Algorithms, Program and Abstracts," Department of Computer Science, Carnegie-Mellon University, Pittsburgh, Pa., May 1973.

7 "Press Seminar Handout," Prepared by the Institute for Advanced Computations, Ames Research Center, Moffett Field, Calif., Aug. 1973.

8 Graham, W. R., "A Review of Capabilities and Limitations of Parallel and Pipeline Computers," Numerical and Computer Methods in Structural Mechanics, Academic Press, Inc., New York and London, 1973, pp. 479-495.

9 Gozalez, M. J., and Ramamoorthy, C. V., "Program Suitability for Parallel Processing," IEEE Transactions and Computer, C-20 (6), 1971, pp. 647-654.

10 McIntyre, D. E., "An Introduction to the ILLIAC IV Computers," Datamation, April 1970, pp. 60-67.

11 McIntyre, D. E., "ILLIAC IV Language Evaluation - A Preliminary Report," ILLIAC IV Document No. 213, May 1970, University of Illinois.

12 Wirsching, J. E., and Alberts, A. A., "Application of the STAR Computer to Problems in the Numerical Calculation of Electromagnetic Fields," AFWL-TR-69-165, April 1970, Air Force Weapons Laboratory, Kirtland AFB, New Mexico.

13 Wirsching, J. E., Alberts, A. A., McIntyre, D. E., and Carroll, A. B., "Application of the ILLIAC IV Computer to Problems in the Numerical Calculation of Electromagnetic Fields," AFWL-TR-69-91, March 1970, Air Force Weapons Laboratory, Kirtland AFB, New Mexico.

COMPUTER NETWORKS: CAPABILITIES AND LIMITATIONS

Ira W. Cotton
Institute for Computer Sciences and Technology
National Bureau of Standards Washington, D.C. 20234

ABSTRACT

This paper is intended to provide an introduction to the capabilities and limitations involved with the use of today's computer networks. While the treatment is at times brisk, this is intentional, in order to emphasize the applicability of networks for a wide range of applications--and the equally wide range of problems.

INTRODUCTION

This paper was composed with the assistance of a computer 3,000 miles away. A CRT terminal in my office connected to a computer network provided convenient access to the distant computer and to a special set of text preparation programs not available anywhere else in the world [1]. Paragraphs were added over a period of weeks, not sequentially, but as they occurred to me. Occasionally, short paragraphs and especially bibliographic citations were copied from a data base of articles which I had written previously using the same system. Each time, I had the option of obtaining a fresh version of the manuscript without aggravating my secretary. In addition, copies of the manuscript could be transmitted--automatically, through the network--to friends and colleagues who were also connected to the network, for their comments and opinion. I have grown so used to the power and flexibility of preparing manuscripts in this way that I would be hard put indeed to go back to the old manual methods.

It's still not a bed of roses, though. I did have to ask my secretary to type this last draft because the proper font wasn't available on my printer. System availability is also a problem. I have very little trouble getting on to my computer in the morning, because it's on the West Coast and users there are still asleep, but in the afternoon congestion becomes a problem. When the system goes down for maintenance, scheduled or otherwise, I'm out of luck--no editing that day. (I suspect this particular machine of both ESP and malevolence--it always goes down when I have a deadline). Finally, getting assistance over the phone is sometimes awkward, even though the staff at the other end couldn't be nicer.

From what I've just told you, I think you can see that my own little application of preparing this paper is a good example of both the benefits and the problems of using computer networks. Before considering these issues in somewhat greater detail, however, we should attend to some matters of vocabulary.

NETWORK TAXONOMY

The first networks were basically time-sharing systems to which remote terminals were connected via communications lines. These networks have been called "star" networks because their topology, consisting of a number of communications lines converging on a single node, resembles a starburst. The circuit configurations in a star network may become very complex, with clusters of terminals sharing lines by means of multiplexers and concentrators, but the principle remains the same: terminal access to a time-sharing system. For this reason, some people prefer to exclude star networks from consideration as computer networks, or at least to require the designation "terminal-oriented" whenever such networks are discussed. Our approach will be the latter. Star networks are perhaps less interesting technically than other types of true computer-to-computer networks, but their profusion demands coverage here.

True computer-to-computer networks provide for the interconnection of many, possibly dissimilar, computer systems. Such networks may be classified on a number of bases. The topology may be centralized or distributed, depending on the degree to which all nodes are connected to one or more focal nodes. If the topology is distributed, the network may or may not be fully connected. (The connectivity of a network is the average number of internodal links.) The network may be homogeneous or heterogeneous, depending on whether or not all the host systems are roughly identical (or at least software compatible).

Another classification of networks relates to the means by which communication paths, or links, are established for the transmission of data. So-called circuit-switched networks establish an open link between two communicating processes for the duration of the dialogue. The circuit remains open whether or not the processes are communicating. Message-switched systems, on the other hand, establish a link for each message to be exchanged which remains open only for the duration of that particular message exchange.

Recent developments in network technology have clouded this distinction somewhat. It is necessary to make a distinction between physical and virtual links. A network such as Tymnet [2] may establish virtual circuit switching over what is essentially a message-switched subsystem, while networks such as ARPANET [3] utilize functionally identical subsystems to accomplish packet switching, where pieces of the same message may not even follow the same route from source to destination. The distinction between the two examples cited may reduce to one of routing: is it fixed or dynamic? Fixed routing is analogous to circuit switching, however the routing is actually implemented; dynamic routing is analogous to message switching, with the same proviso.

The real point of all the vocabulary is (1) to be able to understand the jargon in the literature on networks or in salesmen's presentations and (2) to be able to relate these terms to their effects on network performance. Two excellent references in this regard are [4, 5]. We have fortunately passed from the first stage of network development and experimentation to the stage of network introspection and refinement [6, 7]. Despite this, no single network type has yet been demonstrated to be superior under all circumstances.

For example, centralization may be most efficient for a computer utility under a single management such as General Electric [8] for reasons of economies of scale and reliability. Networks such as Tymnet and ARPANET, however, must of necessity be distributed since they undertake to link existing facilities. This simple contrast suggests that the applications intended for the network have an impact on the best network design. Let us examine some of these applications.

NETWORK APPLICATIONS

Network applications vary widely in the ease with which they may be implemented and the benefits which they will yield. Some relatively simple applications may offer large paybacks, but some others, which will require greater effort and longer periods of time to effectively implement, may offer even greater paybacks. These latter applications require a reorientation of the way people view computers and a change in their habits.

The following catalog of network applications is arranged in roughly ascending order of complexity and potential payoff.

Access

The most immediate and obvious use of a network is for the wide and relatively inexpensive access to resources not available locally. Such resources may include "raw" computer power on a variety of machines, software, and data bases. Networks make possible a great reduction in the cost of accessing these resources by permitting users to share communications facilities. Costs go down as utilization of the lines goes up; the use of higher capacity lines also offers economies of scale.

If you don't have access to a local computer, use of a network can be a lifesaver. Networks really do make it possible to implement most applications without an on-site computer. For example, all the systems programming for the ILLIAC IV (located on the West Coast) by the staff at the University of Illinois is now done remotely through the ARPANET. About the only applications which would be very awkward to attempt through a network are those requiring manual intervention through the main computer console. Few programs are written this way anymore.

Backup

The availability of a backup machine may be a very important benefit from a network, regardless of whether the processing to be backed up is normally done on the network or not. A network can serve as backup for a local installation provided that some advance precautions are taken, such as locating duplicate copies of essential files at sites in the network. Then, if the local machine fails, the network can be utilized. The same principle applies to the regular use of machines on the network. By storing duplicate copies of files at other locations, the failure of a particular node will not prevent essential work from being done elsewhere. Unfortunately, the present level of reliability achieved on some networks forces this to be a regular procedure.

Shared Resources

The access and backup benefits derive from the ability of one user to access many resources. Another primary benefit derives from the ability of a single resource to serve many users. For resources with high fixed costs but moderate variable costs (such as the costs associated with developing and using a software package), benefits are realized when utilization increases, since the fixed costs can be shared by more users. This results in lower unit cost for all users and may make feasible the development and use of specialized resources which could not otherwise be cost justified.

Examples of such resources include larger processors and mass storage devices as well as specialized software and data bases. Larger processors permit more sophisticated or lengthy applications to be undertaken. The use of specialized software through a network allows it to be maintained more easily by the originators. The ability to maintain large data bases at a single location and to provide access through a network is a capability whose importance has only recently been recognized.

Geographic dispersion of users may be a real factor in raising utilization rates of network resources. Time zone differentials between widely separated groups of users permit each group to utilize resources during what seems to them to be "prime time" without interfering with other groups. Intercontinental networks such as G.E., Tymnet, and ARPANET are already taking advantage of this feature [8, 9].

Distributed Multiprocessing

A computer network which supports computer-to-computer communications (as opposed to just terminal-to-computer communications) may permit programs to be written which operate on more than one computer system simultaneously. I call this "distributed multiprocessing" in contrast to centralized multiprocessing (the garden variety), where different parts of the same program may run simultaneously on the several CPUs of a single computer system.

An example of this distributed multiprocessing is provided by the McRoss system that coordinates the operation of two or more cooperating air traffic control simulation programs [10]. With the use of special protocols which have been developed on the DEC PDP-10 for the ARPANET, these programs may run in a single system or in many different systems simultaneously. Each simulation program, called Route Oriented Simulation System (ROSS), models the airspace of one air traffic control center in detail [11]. A flight between any two points may be simulated by running programs for each of the intervening air traffic control centers.

Running the several programs required for a flight on several systems has the advantage that the failure of one of the systems need not kill the simulation. When a system fails, the other systems can adjust to the change in configuration. Running on several systems should speed up the simulation, and there is a particular elegence in modeling a distributed air traffic control situation on distributed computer systems.

Teleconferencing

Teleconferencing refers to the use of computer networks for personal communications among widely dispersed groups of people. Aspects of this application were discussed at a recent conference session [12], and systems have been implemented to support multiparty conferences via computer [13]. My own experience is that a computer network can indeed greatly facilitate personal interactions among dispersed communities. Lost in the publicity on the technical accomplishments of the ARPANET has been the great social accomplishment of integrating the contributions of researchers all around the country. The network is regularly used for the distribution of documentation, requests for comments, comments, gripes, and the like, and on-line collaboration is not uncommon.

Most regular participants have "mailboxes" or directories at particular sites to which text files may be sent. Users regularly link to one another for assistance or simply to chat. Users have sorted themselves out into a number of special interest groups, and a kind of distributed network organization has evolved. This sort of facilitation of personal interaction which networks can achieve is perhaps their most powerful capability, and one which has only just begun to be exploited. Full exploitation will only come with time, as more and more users adjust their work patterns to take advantage of the capabilities.

NETWORK SERVICES

The services which attract users to networks--access to computer power, programs, and data bases--are primarily functions of the host systems on the network. One view of networks is that they ought to be completely transparent with respect to the delivery of these services. This is a shortsighted approach, however, since there are supporting services which a network can and must provide if the primary services are to be used effectively. Indeed, lack of such supporting services may result in the primary services not be useable at all.

The special administrative problems presented by computer networks have only recently been recognized [14]. The problem of network user documentation and user assistance in general is of great importance. The difficulties encountered in providing user assistance are compounded in a network environment by the remoteness of the user from user consultants and reference documents. Present network service documentation runs the gamut from very sparce (but sometimes adequate for simple systems) to voluminous (yet frequently inadequate for complex systems). If a user must refer to a number of different documents that are not well-coordinated and at the same time suffers a lack of sympathetic consultant services, life may be very difficult. Woe piles upon woe when the user is a customer of several different network services, each having separate and very much different documentation.

The use of multiple network services is not at all unusual and may be desirable or essential for many customers. The problems facing the multiple service user are just beginning when he attempts to log-in to the somewhat different services. Even systems connected to the same computer network (e.g., ARPANET) may have totally different and incompatible log-in sequences and command language syntax and semantics. This gross lack of compatibility leads to frustration and the waste of considerable effort and expense. The need for standards at the interface between users and networks to facilitate multiple network use has been recognized, and techniques are being investigated to accomplish this [15].

Some other services which can be provided include on-line directories to services, on-line status reports of available facilities, tutorial files giving instruction in the use of resources, and the ability to send and receive "network mail". The ARPANET has taken this concept a step further by implementing a "Resource Sharing Executive" which permits any of a number of different systems to provide these same services to network customers [16]. A user asking for such a service doesn't even know in advance which system will provide it.

NETWORK PROBLEMS

Obviously, not all these benefits will be realized without problems. Computer networks are expensive to set up and to operate, system incompatibilities still stand as obstructions to resource sharing, and a variety of nontechnical difficulties frequently interfere. Additional detail on the problems which are described here may be found in [17].

Network Costs

A wide variety of services are currently available in the commercial sector. What is billed as "network services" may range from dial-up access to a local service bureau to use of a national telecommunications network to link the geographically distributed computers of the many divisions of a large corporation. Naturally, the more specialized the service and the larger the investment required, the fewer the firms offering the particular service. While there are many local service bureaus, there are only a handful of companies offering both access to computer services and the use of a national, and in some cases international, communications network. It is encouraging to note that this handful is growing, as new carriers such as Packet Communications, Inc. and Telenet, Inc. are receiving approval from the FCC to set up networks based on the packet-switched technology developed for the ARPANET. For any particular network service required, you should be able to locate at least several companies offering it.

Determining what networking will cost may be as simple as consulting a vendor's price schedule or as difficult as computing the installation cost of a private network. Actually, even consulting a price schedule is not as simple or straightforward as it may seem. The costs of using a network are some combination of equipment charges (e.g., terminal rental), line charges (leased or dial-up toll calls), connect charges to the network and/or computer, and computer utilization charges (including CPU utilization, peripheral utilization, and on- and off-line storage charges). The relative weight of these different factors varies from user to user, from application to application, and from vendor to vendor. You will need to specify your computing and traffic requirements very carefully before you will be able to determine what any particular service is likely to cost.

If your service requirements are sufficiently great, it may be most economical for you to set up your own private network. A number of programs and services are now available to help users design their own networks [18, 19]. The use of such services can provide you with the near-minimum cost to implement your own network, and thus enable you to compare this cost with that of using commercial services. The limitations of these programs are that most are directed at terminal networks which, although they may permit multiple computers to be included, do not account for intercomputer traffic. Substantial development work is in progress on the automated design of true intercomputer networks [20], but the results are not generally available outside the research community.

Technical Difficulties

Reliability

The most serious technical problem facing computer network developers is to provide service with sufficient high reliability that customers may safely come to rely on it. The standards for such reliability are very high--in my own estimation, even 90 percent reliability for the network as a whole is not good enough. Users demand 99 percent reliability for the network as a whole, which implies that the individual components must possess even higher reliability, or the degree of redundancy must be increased from that found in most networks today.

Integral to reliability is the ability to diagnose rapidly and respond to failures in the network. The network control center developed for the ARPANET is the most sophisticated approach to this problem [21]. Such a control center serves the multiple function of continually monitoring the network and diagnosing problem areas, coordinating corrective measures,and providing a central point to which users may direct inquiries and complaints.

Flexibility

Networks, as they presently exist, still do not permit all the sorts of devices to be connected to them or all the types of interconnections to be made which may be desired. The problem is partly one of hardware or systems software and partly one of protocols.

For example, few intercomputer networks permit synchronous terminals to be directly connected; yet such terminals are widely used in star networks for remote job submission. There may also be a severe speed restriction on the types of asynchronous terminals which can be connected. Protocol problems are generally responsible for the inability to interconnect networks or to accomplish meaningful data exchanges between computers of different manufacture.

We recently encountered both problems at NBS in an attempt to connect a sponsor to the ARPANET [22]. We wished to provide a way for a remote minicomputer (which, incidentally, was driving a number of graphic displays) to submit programs to a large host on the network. Neither hardware not software existed to do the job We finally were able to connect them physically to the network through another minicomputer at our end which could communicate with them via a synchronous interface and with the network via an asynchronous interface. We then had to ship their data to an intermediate host on the network (in Boston) for reformating before an actual job could be submitted to the desired host (in California). There is some satisfaction in implementing makeshift solutions like this because they do work, but this is obviously not the preferred way.

Security

Security is an issue which has been conveniently ignored by most networks, but which will have to be addressed by them eventually. Security may seem to be of little importance in an academic environment, but even there may be found instances of sensitive files which require adequate protection from unauthorized examination and tampering (e.g., personnel files and files of student grades). Networks which offer service commercially have the responsibility to develop measures to protect their customers' sensitive information.

Network Standards

With computer networks developing at such a rapid pace and showing signs of becoming an important segment of the computer/communications industry, the question of standards is becoming more important. Standards are needed to permit both the access of multiple networks by users with a single set of terminal facilities and the eventual interconnection of networks.

The American National Standards Institute (ANSI) is the focus of voluntary standards activities in the United States. ANSI Committee X3 on Computers and Information Processing is the primary vehicle for developing these standards. A task group is now at work to determine the areas for possible standardization in the area of networks and to develop recommendations. These recommendations have the opportunity to become American National Standards after a review and balloting procedure, and they may also be considered as the American position for submission to the International Standards Organization (ISO).

ISO is the primary organization in the international sphere for the development of voluntary standards. ISO considers recommendations from the various national organizations, of which ANSI is but one. Other special organizations may also submit recommendations to ISO. One such organization is the International Federation for Information Processing (IFIP), which has commissioned a task group to investigate issues in network standards. This group is already planning a series of experiments in network interconnection.

A final organization of importance in the area of standards is the National Bureau of Standards, empowered by Congress to coordinate the development of data-processing standards in the federal government. The vehicle for this program is the Federal Information Processing Standards (FIPS), some two dozen of which have already been issued. Where they exist, ANSI standards are generally adopted as the FIPS, and NBS staff members such as myself participate on many ANSI committees.

The importance of standards to the user is that they facilitate portability of software, hardware, and people. Software standards permit programs to be run on many different computers. Hardware standards permit terminals and other peripherals to be connected to many different systems. Procedural standards make it easier for people to move from one network to another as their needs dictate. Potential users of network services will want to inquire as to which standards the candidate network conforms. Experienced users and those with special requirements may wish to participate in future standardization efforts. All of the organizations mentioned above encourage such participation.

Nontechnical Difficulties

Regulation

Up to the present time, most networks have been able to avoid entanglements in tariff questions through one means or another. However, as networks continue to grow in size and importance, there are likely to be tariff decisions made which will affect networks.

There have been two major issues before the Federal Communications Commission that have direct bearing on computer networks and their required data communications. These investigations covered specialized common carriers and the interdependence of computers and communications, and the resulting rulings are often confused and considered together. However, as Enslow has pointed out [23], it is important to realize that these issues are separate and distinct in their effects on both computer networks and communications.

The major question addressed in the Specialized Common Carrier inquiry, FCC Docket No. 18920, was whether or not carriers other than the presently established ones would be permitted to offer competitive services. The FCC decision, released in June 1971, came almost eight years after MCI first filed for authority to construct a Chicago to St. Louis microwave system; however, the ruling covered all of the applications pending before it. The Commissioner's position strongly supported free and competitive entry into the market.

In November 1966, the FCC initiated Docket No. 16979 to examine the "Regulatory and Policy Problems Presented by Interdependence of Computer and Communication Services and Facilities." Another lengthy study was required before the Commission issued its final order in March 1971. Although all of the items raised in the initial inquiry were not ruled on, there were important decisions made on the regulatory status of publicly offered teleprocessing services. Enslow characterized the decision in terms of a spectrum of service offerings between pure computing and pure communications.

"Pure" remote computing utilizes communications services, but that use is only incidental to the primary function of the service. It was ruled that this service would be unregulated. The other end of the spectrum is circuit switching, which requires some computation and logical decisions to be made by the switching processor. However, this is incidental to the primary service, which is "essentially communications" and therefore fully subject to regulation.

The Commission's ruling also covered message-switched service, which, though it requires more computation, is still a "communications" service and regulated. What the ruling did not settle was the status of services where the "incidental" test fails. These questions have been handled on a case-by-case basis, although, as in the case of packet switching, the trend has been to permit new services to be offered.

Financial Considerations

Financial problems include the costs associated with the various network components and raising the necessary capital to develop and operate a network. Networks are not inexpensive to develop or operate, with substantial costs arising both from hardware and software components. Capitalization requirements present a barrier to entry for potential networks, be they academic or commercial.

Furthermore, most costs are fixed over a given operating period, resulting in a high sensitivity to variations in demand on the part of network financiers. A major challenge to network developers is to reduce the overhead and thereby the costs associated with such components as host software specific to the network. The communications components of networks represent opportunities for savings resulting from economies of scale; so another challenge is to raise network loading to the point where these benefits are realized.

Sociological Problems

These problems arise from the large and diverse organization which typifies computer networks. A large, distributed network establishes a community of developers, maintainers, and users who must interact with each other. The technology of the network requires that each of these people operate in a way in which they may not be accustomed. Users in particular must adapt to certain constraints of a network in order to accomplish their desired tasks. For example, less user assistance through personal interaction is characteristic of networks. User assistance is provided instead through better documentation and on-line tutorials in automated form. To the extent that users are unable to adapt to this mode of operation, serious impediments to network success may arise.

Computer networks have also interfered with the bureaucratic prerogatives of organizations. The reluctance of organizations to yield control of their own facilities and begin to rely on other facilities (even when cost savings can be demonstrated) has been noted [24, 25]. Vested interests and local empire building may be serious impediments to the acceptance of networks in some cases.

CONCLUSIONS

Computer networks as they exist today and are planned for the near future offer growing capabilities with ever diminishing limitations. However, we are not likely soon to see a well-integrated (inter) national system develop, as, for example, the telephone system. Rather, a number of private networks and specialized communities will develop; indeed, are developing. We may hope that, over time and through some combination of consumer pressure and government regulation, order and compatibility will prevail.

In the meantime, users will have to live with what is available. Those who are unwilling or unable to adapt to some of the peculiarities of networks may find them unusable. Knowledgeable users (or those who are willing to become knowledgeable) will find many advantages from the use of networks.

ACKNOWLEDGMENTS

The contributions of my colleagues at the National Bureau of Standards, Thomas N. Pyke, Jr., Robert P. Blanc, and Dr. Marshall D. Abrams, to the preparation of this article are gratefully acknowledged.

This work was supported in part by the National Science Foundation under Grant AG-350.

REFERENCES

1 Englebart, Douglas C., Watson, Richard W. and Norton, James C., "The augmented knowledge workshop." National Computer Conference, 1973, pp. 9-21.

2 Tymes, L. R., "Tymnet, a terminal oriented communications network," Spring Joint Computer Conference, 1971, pp. 211-216.

3 Heart, Frank E., Kahn, Robert E., Crother, William R., Walden, David C., "The interface message processor for the ARPA computer network." Spring Joint Computer Conference, 1970, pp. 511-567.

4 Blanc, Robert P., "Review of computer networking technology." National Bureau of Standards Technical Note 804, February 1974.

5 Pyke, Thomas N., Jr. and Blanc, Robert P., "Computer networking technology -- a state of the art review." Computer, Vol. 6, No. 8, August, 1973, pp 12-19.

6 Aupperle, Eric M., "Merit network re-examined." Compcon 73, pp. 25-30.

7 Kahn, Robert E. and Crother, William R., "A study of the ARPA network design and performance." Bolt Beranek and Newman, Report No. 2161, August 1971.

8 Feeney, George J., "The future of computer utilities." Proc. First International Conference on Computer Communications, Washington, D.C., October 1972, pp 237-239.

9 Combs, Bill, "Tymnet: a distributed network." Datamation, Vol. 19, No. 7, July 1973, pp. 40-43.

10 Thomas, Robert H. and Henderson, D. Austin, "McRoss -- a multi-computer programming system." Spring Joint Computer Conference, 1972, pp. 281-293.

11 Sutherland, W., Myer, T., Henderson, D., and Thomas, E., "Ross, a route oriented simulation system." In Proc. 5th Conf. Applications of Simulation, New York, December 1971.

12 Conrath, David W., "Teleconferencing: The computer, communication and organization." Proc., First International Conference on Computer Communication, October 1972, pp. 143-144.

13 Turoff, Murray, "Human communication via data networks." Computer Decisions, January 1973, pp. 25-29

14 Abrams, Marshall D. "Remote computing: the administrative side." Computer Decisions, October 1973, pp. 42-46.

15 Neumann, Albrecht J. "User procedures standardization for network access." National Bureau of Standards Technical Note 799, October 1973.

16 Thomas, Robert H. "A resource sharing executive for the ARPANET." National Computer Conference, 1973, pp. 155-163.

17 Cotton, Ira W. "Network management survey." National Bureau of Standards Technical Note 805, February 1974.

18 Raymond, R.C. and McKee, D.J., "A design model for teleprocessing systems." DATACOMM 73 Symposium, November 1973, pp. 131-140.

19 Whitney, V. Kevin M., Doll, Dixon R., "A database system for the management and design of telecommunication networks." DATACOMM 73 Symposium, November 1973, pp. 141-147.

20 Frank, Howard, Kahn, Robert E. and Kleinrock, Leonard, "Computer communications network design -- experience with theory and practice." Spring Joint Computer Conference 1972, pp. 255-270.

21 McKenzie, Alexander A., Cosell, Bernard P., McQuillan, John M., Thrope, Martin J., "The network control center for the ARPA network." Computer Communication - Impacts and Implications, Proc. First International Conference on Computer Communication, 1972, pp. 185-191.

22 Abrams, Marshall D., Hudson, J.A., Meissner, P., Pyke, T.N., Rosenthal, R., Ulmer, F.H., "Use of computer networks in support of interactive graphics for computer-aided design and engineering." National Bureau of Standards, NBSIR 73-217, June 1972.

23 Enslow, Philip H., Jr., "Non-technical issues in network design -- economic, legal, social, and other considerations." Computer, Vol, 6, No. 8 August 1973, pp. 20-30.

24 Herzog, Bertram, "Organizational issues and the computing network market." Compcon 73, pp. 11-14.

25 Brooks, Frederick, P., Jr., Ferrell, James K., and Gallie, Thomas M., "Organizational, financial and political aspects of a three-university computing center." Proceedings IFIP Congress, 1968, pp. 923-927.

TRENDS IN COMPUTER SOFTWARE

Rona B. Stillman
National Bureau of Standards
Washington, D.C.

ABSTRACT

Promising approaches for producing more reliable software are surveyed. Potential improvements in programming languages, automated testing tools, and proof of correctness techniques (validation condition generation and theorem proving) are discussed.

INTRODUCTION

The software industry, like other spheres of modern endeavor, has entered a period of self-evaluation and redirection. In the past, when software was regarded as an esoteric product, a novelty, or a toy for the technically inclined, a great many excesses were tolerated and forgiven. Software was expensive, unreliable, difficult to understand, debug, and test, and so brittle that even relatively minor perturbations in environment could not be handled satisfactorily. Computers have now become an exceedingly important and ubiquitous part of our daily lives, e.g., computers are used by business in accounting, credit, and inventory systems, by health care institutions to diagnose disease and perform blood and tissue typing, by engineers in designing mechanical parts, by the military in deploying nuclear weapons, and by the mass transit industry in handling air traffic control, seat reservations, and routing.

Nonetheless, software quality is still uneven at best. Programs thought to be correct (and released for general use) will suddenly produce wrong results, no results, or behave otherwise erratically, because some special condition in the data or in the environment was not accounted for in the logic of the program. Current practice is to design and implement a software system, the paramount considerations usually being production speed, running efficiency, and/or minimal use of storage, and then to test it for some arbitrary subset of possible input values and environmental conditions. The system is accepted as correct when it executes the test cases correctly or when time/money runs out, whichever occurs first. It is, therefore, not surprising that software products are usually unreliable and that confidence in new software is generally nil.

These facts have become increasingly clear to software producers:

1. Software (system software and application software) which is reliable, robust, understandable, testable, and maintainable must

be built. Society cannot and will not tolerate repeated software failures in systems affecting public health, welfare, and safety.

2. It is worthwhile to sacrifice something in the way of production time, running efficiency, and/or storage required to obtain more reliable code. While the details of this tradeoff are, as yet, unknown, it is apparent that computer costs (time, space) are falling, while software errors are becoming potentially more dangerous and expensive.

3. Whereas software has become increasingly more sophisticated and complex, techniques for producing and ensuring quality in software have not kept pace. Such techniques must now be developed, improved, and integrated into the software production process.

This paper will describe techniques for improving software reliability, some of which have already been used successfully (although, perhaps, on a limited scale), some of which are now ready for use or will be in the near future, and some of which are still objects of research.

IMPROVING PROGRAMMING LANGUAGES

Most useful computer programs (i.e., those having appreciable length and structure as opposed to simple tutorial exercises) tend to be very difficult to understand, debug, test, and modify. Poor programming language design--or poor choice of a particular programming language for a particular task--is sometimes the cause. A good programming language must be more than just computationally complete, i.e., must do more than merely permit the programmer to write his algorithms. Rather, it should facilitate good program design and encourage a clear, readable, natural style of coding and data representation. It should be nearly self-documenting. Well-written programs should be easily tested and amenable to verification (even, eventually, to proof of correctness). Moreover, it must be possible to write a fast compiler for the language which produces efficient object code (to discourage either writing quirky code or reverting to assembly language). The characteristics of such a language would include the following items.

Rich Set of Data Representations

It is unnatural to define all data objects either as single elements or as n-dimensional arrays of single elements. The code required to manipulate data which have been forced into these restricted formats is usually complex, opaque, and error-prone. A better route is for the programming language to provide for tables (or files) consisting of entries (or records) of considerable structural complexity, for stacks, for lists, for queues, for character strings, etc., and for convenient operators to manipulate them.

Stringent Variable/Data Specification Requirements

It is dangerous to permit a variable to be used which has not been previously and explicitly defined. Naming conventions (as in FORTRAN) which are used in the interest of terseness and convenience cause insidious errors and encourage sloppy programming. For example, consider the following error sequence which can occur if variables can be referenced without having been explicitly defined:

```
IVAR ← 1
   .
   .
   .
IVOR ← IVAR + 1
```

In this case, the programmer meant to increment IVAR by 1 but misspelled the name as "IVOR". Instead of generating an error message at compile time, when it would be cheap and convenient to correct the mistake, the compiler "does you the favor" of creating a new variable, IVOR, and giving it the value of IVAR + 1. The programmer is left to his own resources to ferret out the bug which shows up during execution.

It can be argued, perhaps, that the problem is caused not because implicit variable definitions are permitted but rather because the compiler doesn't flag all instances where undeclared variables are used. I think, however, that it is disconcerting to see an undeclared variable suddenly appear in a program which I am trying to understand, and that it is undesireable not to have, anywhere in the program, a complete description of the data used.

In fact, I think that in the future, programming languages will encourage more rather than less in the way of explicit data definition. That is, various attributes could be assigned to variables as a means of giving some semantic content to operations involving these variables. For example, the following variables could be assigned "units of measure":

```
VSPEED   FEET/SEC

VTIME    SEC

VDIST    FEET
```

Then any attempt to set

```
VSPEED ← VDISTANCE * VTIME
```

or

```
VDIST ← VTIME
```

would be caught as a semantic error by the compiler.

Similarly, if the legal range of values of a variable is defined by the programmer (perhaps as part of the variable declaration), then dynamic data-bound checks can be made as well. Whenever a new value is assigned to the variable during execution, a data value tolerance test is made, and out-of-tolerance values are reported. An explicit definition of the legal range of values of a variable is helpful also in constructing relevant test cases, maintaining the program, and as a means of documentation.

Sophisticated Approaches to Controlling Access to Shared Data

A serious shortcoming in most programming languages today is that there is no satisfactory capability for controlling access to data objects. That is, there is no mechanism by which a programmer can decide which portions of his program will be permitted to access a given data object, and thereby be assured that the data is inaccessible elsewhere. Usually, the programmer has the following limited choice: he may declare a data object as global, in which case it can be accessed from anywhere in the program, or he can declare it local, in which case it exists only inside the block in which it was declared. In order to permit just three subroutines out of 200 in his program to access the same data object (without passing it as a parameter), the programmer is forced to declare the data global and thereby open it to (the slings and arrows of) the other 197 subroutines in his program. Furthermore, there is no way for the programmer to determine what level of access a program block has to a data object; i.e., there is no facility for allowing two of the subroutines read/write access, while the third has only read access.

Along these lines, if the programmer could decide in every subroutine how each parameter would be passed, it would eliminate some spurious side effects. Consider the following main program, with one function FUN.

```
MAIN PROG
x ← 2                          FUN (z)

y ← FUN(x)                     z ← z+3

PRINT x , y                    FUN(z) ← z*z

END
```

By passing x as an argument (in, say, FORTRAN), the programmer has unwittingly given the function FUN the right to change x and, indeed, will find that in the main program x is 5, not 2. It would be clearer and safer to insist that parameters be specified either as INPUT (or read only parameters) or as OUTPUT (read/write parameters). From the implementer's point of view, this amounts to specifying whether a parameter will be called by value or called by reference, respectively.

Many interface errors involve passing bad parameters. Since it is easier and cheaper to check parameter passing at compile time than at run time, since compile time checks are possible only if the program is compiled as a unit, and since link-editing can be more expensive than compiling, the practice of independent compilation should be reexamined.

Validation-Oriented Syntax

The syntax of a language should be designed to minimize the scope of an error, to permit errors to be uncovered as early as possible, and to reduce the chance that an error will result in a syntactically correct statement, i.e., to reduce the chance that an error will pass undetected through the compiler. For example, consider the following statement for incrementing a variable:

$$x \leftarrow x+3$$

By mistyping the operator +, the following syntactically correct statements could result:

$$x \leftarrow x-3$$

$$x \leftarrow x*3$$

$$x \leftarrow x/3 \text{ , etc.}$$

Furthermore, requiring that the variable name x appear twice just serves to increase the potential for error, e.g., $x \leftarrow y+3$.

A less error-prone syntax for incrementing a variable would be

INC x 3

Misspelling the operator INC would most likely result in a syntax error, e.g., ONC x 3 , which would be flagged by the compiler. Moreover, the variable name occurs only once [1].

Encourage Structured Programming

Software which is built from a very limited and well-defined set of control structures is thought to be more reliable than conventionally written (i.e., unstructured) code [2, 3]. The argument is that programmers use their unrestrictured GO-TO rights to construct an intricate maze of arbitrary transfers, directing control helter-skelter through the program and thereby obscuring the underlying logic. The execution characteristics of such a program are extremely difficult to analyze, and the programmer is unlikely to know exactly what is going on. On the other hand, if the program is built as a hierarchy of modules, and if strict rules are enforced governing the transfer of control within and between modules, the logic will be much more explicit, and the program should be simpler to understand, document, debug, test, and maintain. The computational completeness of certain restricted classes of control structures has been proven by Böhm and Jacopini [4], Kosaraju [5], and others.

The ability to block or group statements (by means of brackets or control delimiters) is essential so that all structuring methods can be applied recursively. It also facilitates a natural, readable style of coding. Without blocking, a programmer who wants to perform two or more successive actions on each branch of a conditional statement is forced to distort his logic and disrupt the flow of his program. Consider, for example, the following program which reads an integer n and, if n is zero, increments COUNTER1 by 2 and reads another input; if n is positive it increments COUNTER1 by 1, decrements COUNTER2 by 1, and reads another input; otherwise, it decrements COUNTER1 by 1, increments COUNTER2 by 1, and halts:

```
        INTEGER n, COUNTER1 , COUNTER2
        COUNTER1=0
        COUNTER2=0
1       READ n
        IF n=0      BEGIN
                    COUNTER1=COUNTER1+2
                    GOTO 1
                    END

        IF n>0      BEGIN
                    COUNTER1=COUNTER1+1
                    COUNTER2=COUNTER2-1
                    GOTO 1
                    END
        COUNTER1=COUNTER1-1
        COUNTER2=COUNTER2+1
        HALT
```

Without the ability to block statements, this program becomes:

```
        INTEGER n, COUNTER1, COUNTER2
        COUNTER1=0
        COUNTER2=0
1       READ n
        IF n=0      GOTO 2
        IF n>0      GOTO 3
        COUNTER1=COUNTER1-1
        COUNTER2=COUNTER2+1
        HALT
2       COUNTER1=COUNTER1+2
        GOTO 1
3       COUNTER1=COUNTER1+1
        COUNTER2=COUNTER2-1
        GOTO 1
```

Optionally Compilable Assertions

Among the most fundamental and difficult tasks in programming is that of program design, i.e., specifying precisely what is to be done, deciding upon a data representation, subdividing complex tasks into simpler subtasks, and--most important--making all assumptions explicit at every interface. Fuzzy, poorly defined interfaces between program modules and implicit assumptions made within modules are a major source of subtle, insidious errors and cause difficulty in modifying and understanding a program.

Optionally compilable assertions are test conditions that the programmer, at his own discretion, can insert into his code; i.e., they are logical statements over the program variables which are similar in syntax to the predicate portion of an IF-statement. The assertions can be handled either by the compiler (as another statement form) or by a preprocessor. For example, before the first executable statement of SUBROUTINE A (PARAM1, PARAM2, PARAM3), the programmer may assert that (PARAM1≥0 and PARAM2≥0 and PARAM3≥0), that is, that all input parameters must be positive. Whenever SUBROUTINE A is entered, this assertion will be evaluated, and if true, the execution will proceed. Otherwise, a message will be produced indicating that a negative value was passed to SUBROUTINE A in violation of the assertion. Since they indicate where, in the execution, certain error conditions have occurred, assertions can be very helpful in debugging and in validating a program. They are valuable also in formalizing the interfaces between software modules which have been written by different programmers, and hence localizing the effects of an error.

Because the assertions (which are really error traps) introduce additional overhead, it seems prudent to make them optionally compilable. That is, by setting a switch, the assertions will either be compiled with the code (for use in testing and validation) or will be treated as comments, i.e., ignored so as not to degrade the performance of the program in normal operation.

AUTOMATED SOFTWARE TESTING TOOLS

The tasks of debugging, modifying, testing, documenting, and, in general, understanding the logical structure of a program are greatly facilitated by the use of software testing tools. There are two main categories of analysis: static analysis, which is performed without executing the software, and dynamic analysis, which is dependent upon information collected while the software is in execution.

Static Analyzers

These software tools accept a subject program as input and produce the following type of information as output:

1. Display of the program structure and logic flow

2. Description of the global data, i.e., the data which is shared among the subroutines

3. Subroutine/global variables referenced listing

4. Global variable/subroutines where referenced listing

5. Subroutine/subroutines referenced listing

6. Subroutine/subroutines where referenced listing

7. Entry point/subroutine listing

8. Subroutine/entry points listing

9. Description of the disconnected portions of code, i.e. code which cannot be reached from the start state

10. Description of the blocked portions of code, i.e., code from which an exit state cannot be reached

Research is being conducted to develop other tools which analyze the possible execution paths of a program and output a (hopefully minimal) subset of paths which exercise every statement and/or branch option in the program [6, 7] or, better still, generate the input data which causes these paths to be executed. These potential path analyzers could also be used to identify execution paths which include a particular instruction or sequence of instructions. This information is extremely valuable to a programmer in constructing a set of test cases that will thoroughly exercise his code. The major challenge in developing potential path analyzers is finding some appropriate way to deal with the enormous number of possible execution paths of even relatively simple programs. Perhaps modularly designed structured programs offer some promise in this regard: if each module is analyzed independently of the others, and <u>then</u> the flow from module to module is considered, the combinatorial problem will be eased.

Dynamic Analyzers

Dynamic analyzers are software tools which, by inserting traps in the subject program, cause the following type of information to be produced in addition to the program's normal output:

1. Number of times each statement in the program has been executed in a single run or series of runs

2. Number of times each transfer in the program has been executed in a single run or series of runs

3. Number of times each subroutine in the program has been entered during a single run or series of runs

4. Amount of time spent in each subroutine during a single run or series of runs

5. For each statement assigning a new value to a specified variable, the maximum, minimum, first, and last value assigned during the computation

The operation of a dynamic analyzer is shown schematically in Fig. 1.

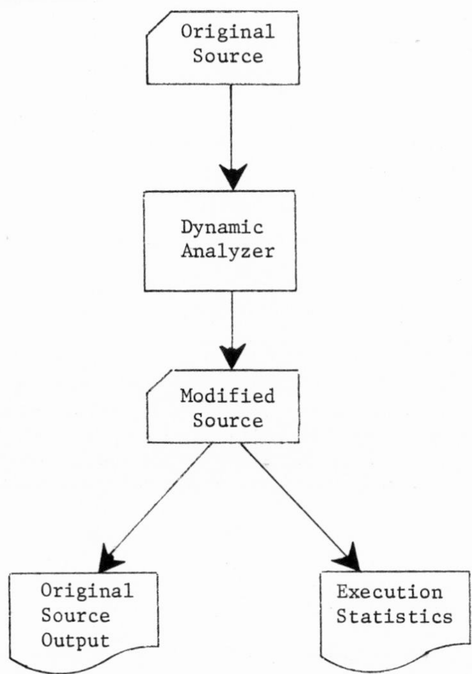

Fig. 1 Dynamic analyzer

Because the programmer can now accurately analyze the effectiveness of his test cases, i.e., he knows how many times each statement (transfer, subroutine) has been exercised, and, in particular, he knows which statements (transfers, subroutines) have never been exercised, he can construct a set of test cases that is both thorough and minimally redundant. Regression testing, required during validation and maintenance phases, is simplified as well. When a portion of code has been altered (corrected, improved, etc.), those test runs involving the changed code, i.e., the set of tests that must be reevaluated, is readily identified.

Although the traps inserted by the dynamic analyzer will usually be removed before the program begins normal operation (the traps introduce considerable overhead in both space and time), it may sometimes be desirable to leave them intact for awhile. For example, if a program is to be optimized, it is extremely important to know which portions of code are repeatedly executed during normal operation. Small improvements in these will result in a significantly more efficient program. Conversely, if a portion of code is executed only rarely, it might not be worthwhile to bother optimizing it at all. In a similar vein, a precise description of the normally running program in terms of the types of instructions executed, number of calls made to specific system routines, time spent performing certain functions, average running time, etc. is essential if an accurate model of the program is to be built.

It should be noted that the utility of the tools discussed above depends strongly on the format and the quantity of the output they produce. Programmers will ignore tools that produce too much output for their purposes or produce poorly organized output that is difficult to interpret. Therefore, the programmer should be given a wide range of options in specifying the degree of detail, the amount, and the format of the output produced by the tools.

MACHINE-GENERATED PROOFS OF CORRECTNESS

Testing a program thoroughly serves to increase confidence in its reliability. However, no set of test cases (short of an exhaustive list of all possible inputs, which is generally impractical and often impossible) will ever guarantee its correctness in any mathematical sense. A rigorous proof of correctness consists of two separate but related subtasks:

1. Given the subject program together with certain additional information provided by the programmer, generate (automatically) a set of potential theorems, the proof of which ensures the correctness of the program. The potential theorems are called validation conditions.

2. Prove (automatically) each of the validation conditions.

The overall process of proving that a program is correct is depicted in Fig. 2.

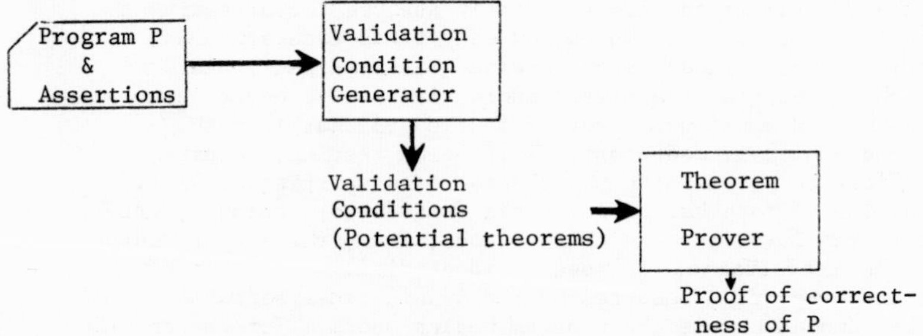

Fig. 2 Proof of program correctness

Clearly, the property of being correct is a relative one. That is, a program is said to be correct if it executes as the programmer intended. In addition to his program, then, the programmer must supply some definition of his intentions. These take the form of assertions (i.e., sentences in the first-order predicate calculus) over the program variables. Each assertion $A(V_1,\ldots,V_n)$ is associated with some statement S_i in the program, with the understanding that when execution reaches statement S_i, then $A(V_1,\ldots,V_n)$ is true of the program variables. In particular, a characteristic assertion called the initial assertion, $I(V_1,\ldots,V_n)$, is associated with the START statement, and a characteristic assertion called the final assertion, $F(V_1,\ldots,V_n)$, is associated with the HALT statement. By means of these, the programmer has defined correctness as follows: If the initial values of the program variables $(V_1,\ldots,V_n)$ satisfy condition $I(V_1,\ldots,V_n)$, and the program terminates, then the final values of $(V_1,\ldots,V_n)$ must satisfy $F(V_1,\ldots,V_n)$.

Generating the Validation Conditions

Like every system of deductive reasoning, generating the validation conditions involves the application of valid rules of inference to a set of valid axioms. In this case, the axioms are the semantic description of the statements in a programming language and have the general format

$$A_0 \ \{\text{STATEMENT TYPE}\} \ A_1$$

where A_0 is an assertion on the program variables before executing the statement, and A_1 is an assertion on the program variables after executing the statement. Thus, the axioms say the following: If A_0 is true of the program variables when control reaches the statement, and if the statement is executed, then A_1 is (the strongest assertion that is) true of the program variables after execution.

Consider a single statement program with initial and final assertions specified:

$I(V_1,\ldots,V_n)$	Start
	Statement
$F(V_1,\ldots,V_n)$	Halt

This program is said to be correct if

$$\forall V_1,\ldots,V_n(A_1(V_1,\ldots,V_n) \supset F(V_1,\ldots,V_n))$$

where A_1 is the consequent of initial condition I, as prescribed by the statement axiom. $\forall V_1,\ldots,V_n(A_1(V_1,\ldots,V_n) \supset F(V_1,\ldots,V_n))$, a potential theorem of the first-order predicate calculus, is the validation condition of the program. A logical proof that the validation condition is true, then, is a rigorous proof of correctness of the program.

It is not difficult to extend this line of reasoning inductively to programs of any finite length. The consequent of any statement, as prescribed by the appropriate axiom, simply serves as the premise assertion for the next statement. Thus, generating a validation condition consists of repeatedly transforming the assertion at hand, each transformation being determined by the semantic definition, i.e., the axiom associated with that statement type in the programming language.

A few very simple definitions are now required to establish a meaningful working vocabulary.

Program. A program $P = (S_1, S_2, \ldots, S_n)$, $n \geq 1$, is a finite sequence of statements.

Tagged Statement. A tagged statement is one that has an assertion associated with it. For instance, the START and HALT statements are always tagged statements.

Control Path. A control path $CP = (S_{i_1}, S_{i_2}, \ldots, S_{i_m})$, $m \geq 1$, is a finite sequence of statements such that control can pass from statement S_{i_j} to statement $S_{i_{j+1}}$ for $1 \leq j \leq m-1$.

Tagged Path. A tagged path $T = (S_{i_1}, S_{i_2}, \ldots, S_{i_m})$ is a control path such that the first statement, S_{i_1}, and the last statement, S_{i_m} are tagged, and none between are tagged.

It should be apparent that a program P is verified IFF every tagged path in P is verified. That is, a program as a whole is said to execute correctly IFF every possible path in the program executes correctly. Hence, the proof of correctness of a program consists of proving the validation condition for each of the tagged paths in the program. To ensure that the length of each tagged path and the total number of tagged paths remain finite, it is necessary to demand that at least one statement in every loop in the program be tagged. Thus no loop can ever be enclosed within the confines of a tagged path. Since loops are iterative by their nature, the assertions required to "cut" these loops are called iterative assertions [8, 9].

Proving the Validation Conditions

It is usually thought that proving theorems requires a certain amount of ingenuity or human intelligence. In fact, however, any theorem that can be proven at all can be proven by executing some purely clerical algorithm [10].

Most deductive algorithms are based upon the notions of unsatisfiability and refutation, rather than upon the notions of satisfiability and proof. To prove a theorem T, (the validation conditions are theorems), it is sufficient to exhibit just one counter example ~ T. In practice, the relevant mathematical axioms, the hypothesis of T and the negation of the conclusion of T are expressed as a finite set S of sentences in the first-order predicate calculus. By using specified rules of inference, the deductive algorithm seeks to produce a logical contradition, that is, a pair of mutually exclusive inferences, from S. Early systems employed exhaustive instantiation as their only rule of inference. They made repeated substitutions of constants for variables in the sentences of S, producing successively larger sets of instances until a complementary pair of instances was found. The substitutions were made in a systematic and exhaustive manner, guaranteeing that if a contradiction existed, it would be included, eventually, in the steadily expanding sets of instances. For problems of reasonable mathematic interest, however, this expansion is a virtual explosion. The irony of the situation is that almost all of the instances generated are irrelevant to the problem, and the number which are relevant is quite small.

In 1965, J. A. Robinson formulated the resolution principle, a sound and effective rule of inference that reduced the combinatorial explosion by a factor in excess of 10^{50}. Whereas an instantiation-based system substitutes constants for variables in the sentences of S, a resolution-based system deals with the sentences of S in their more general form. Rather than blindly instantiating every sentence in S and waiting for a contradiction to surface, resolution produces a new sentence from two existing sentences only if these sentences, or instances of them, contain a pair of complementary literals [11]. Nonetheless, the major shortcoming of the basic resolution procedure is that it is wasteful of both time and space. Many of the sentences produced turn out to be irrelevant to the refutation which is eventually found; many turn out to provide less information than a sentence already included in the problem. Since, in the basic procedure, a sentence once generated is never discarded, for many problems the available resources are exhausted before any proof can be found.

Various proof strategies have been devised in an attempt to make resolution-based systems more efficient. These proof strategies are logical adjuncts to resolution and fall into three main categories:

1. Editing strategies, which are procedures embodying a criterion for detecting and deleting redundant and/or irrelevant sentences. One difficulty with editing strategies (e.g., subsumption [11], purity [11]) is that as more sentences are added to the problem, proportionately more time is required to determine if a given sentence can be deleted. That is, the efficiency of editing strategies is, generally, inversely proportional to the size of the problem. This is a very serious drawback since interesting problems are often very large. Some interesting hardware/software techniques

involving the use of associative processors have been suggested to alleviate this difficulty [12].

2. Search strategies, which specify the order in which sentences are to be considered for resolution or define certain criteria which limit the candidates for resolution. For example, Wos's unit preference strategy [13] resolves shorter sentences before longer ones, with unit sentences (i.e., sentences composed of only one literal) being resolved first. J. A. Robinson's P1-Resolution [14] prohibits the resolution of two negative sentences (i.e., sentences which contain at least one complemented literal). The success of search strategies, however, is erratic and unpredictable. They perform well on some problems and poorly on others, and it is difficult, if not impossible, to predict if a given problem will run well under a given search strategy.

3. Augmenting resolution as the rule of inference, to increase its power and efficiency. For example, attempts have been made to devise a system that will recognize the special characteristics of the equality relation and handle it accordingly, rather than treating it as just another predicate (e.g., paramodulation [15]). Systems using higher-order logics (first-order logic permits quantification over variable symbols; second-order logic permits quantification over function and predicate symbols; by introducing new function and predicate symbols which admit the old functions and predicates in their domain, we obtain higher-order logics) are being investigated since they allow a more compact and natural expression of problems. However, it is considerably more difficult to automate higher-order logics, and, to date, there have not been any significant advances in this area.

None of these approaches, alone or in combination, has been powerful enough to make automatic theorem proving practical. Perhaps human intelligence should intrude at key decision points in the proof process, providing a sense of purpose and direction that is notably missing from completely mechanized proof procedures, while the bulk of the computation is still performed by the computer. Hand-generated proofs have been suggested by R. L. London [16, 17] and C. A. R. Hoare [18, 19] among others, but these can be tedious and are subject to error in much the same way as the original program was. Since it is a simpler task to check the correctness of a proof than it is to generate it, proofs produced by hand but verified by machine are another interesting possibility.

CONCLUSION

Because of its importance and generality (software reliability has technical, economic, and political implications), this paper was devoted to the trend toward greater concern for software reliability. Other trends have been treated, directly and indirectly, in other papers in the symposium. As new hardware and novel configurations are developed, software will be built to exploit them. System software of greater complexity and sophistication will be required to manage effectively minicomputers, multiprocessors, parallel processors, pipeline processors, microprocessors, associative memories, and hybrid systems of assorted flavors. Networks (in their many variations) present problems in developing a standard user/network communications interface and deciding how best to share and protect available resources. There is a need to develop tools which are available to users of different languages running on different machines in a network, i.e., to develop a language-independent interface between a user and his programming language which makes a variety of standard tools available to him. As the conversational style of program construction becomes the norm, more and better programmer support features will be provided; e.g., a "programming environment" will be offered which will include compatible interpreters and compilers, intelligent text editors, on-line documentation and interactive help routines, automated testing tools, verification condition generators, and theorem provers. Finally, provided reliable software can be built, we will see a continuation and acceleration of the trend toward ever more ubiquitous and sophisticated software applications.

REFERENCES

1 Elspas, B., Green, N., and Levitt, K., "Software Reliability," Computer, Jan./Feb. 1972.

2 Dijkstra, E. W., "Notes on Structured Programming," Technische Hogeschool, Eindhoven, Aug. 1969.

3 Mills, H. D., "Mathematical Foundations for Structured Programming," FSD 72-6012, Federal Systems Division, IBM, Feb. 1972.

4 Böhn, C., and Jacopini, G., "Flow Diagrams, Turing Machines and Languages with only Two Formation Rules," Communications of the ACM, Vol. 9, No. 5, May 1966.

5 Kosaraju, S. R., "Analysis of Structured Programs," Quality Software Tech. Report, Electrical Engineering Department, Johns Hopkins University, Nov. 1972.

6 Brown, J. R., DeSalvió, A. J., Heine, D. E., and Purdy, J. G., "Automated Software Quality Assurance: A Case Study of Three Systems," ACM SIGPLAN Symposium on Computer Program Test Methods, June 1972.

7 Hoffman, R. H., "Automated Verification System User's Guide," TRW Note No. 72-FMT-891, Project Apollo, Task MSC/TRW A-527, Jan. 1972.

8 Floyd, R. W., "Assigning Meanings to Programs," Proceedings of Symposium in Applied Mathematics, Vol XIX, ed. J. T. Schwartz, American Mathematical Society, Providence, Rhode Island, 1967.

9 King, J. C., "A Program Verifier," Ph.D. Thesis, Carnegie-Mellon University, Pittsburgh, Pa., 1969.

10 Robinson, J. A., "Review of Automatic Theorem-Proving," Proceedings of Symposium in Applied Mathematics, Vol XIX, ed. J. T. Schwartz, American Mathematical Society, Providence, Rhode Island, 1967.

11 Robinson, J. A., "A Machine-Oriented Logic Based on the Resolution Principle," Journal of the ACM, Vol. 12, No. 1, Jan. 1965.

12 Stillman, R. B., "The Concept of Weak Substitution in Theorem-Proving," Journal of the ACM, Vol. 20, No. 4, Oct. 1973.

13 Wos, L., Carson, D., and Robinson, G., "The Unit Preference Strategy in Theorem-Proving," Proceedings of the FJCC, 1964.

14 Robinson, J. A., "Automatic Deduction with Hyper-Resolution," International Journal of Computer Mathematics, Vol. 1, 1965.

15 Robinson, G., and Wos, L., "Paramodulation and Theorem-Proving in First-Order Theories with Equality," Machine Intelligence 4, eds. B. Meltzer and D. Michie, Edinburgh University Press, Edinburgh, 1969.

16 London, R. L., "Computer Programs Can be Proved Correct," Proceedings of the Fourth Systems Symposium--Formal Systems and Non-Numerical Problem Solving by Computers, Case Western Reserve University, Nov. 1968.

17 London, R. L. "Software Reliability Through Proving Programs Correct," Publ. 71C 6-C, IEEE Computer Society, New York, Mar. 1971.

18 Hoare, C. A. R., "An Axiomatic Basis for Computer Programming," Communications of the ACM, Vol. 12, No. 10, Oct. 1969.

19 Hoare, C. A. R., "Proof of a Program: FIND," Communications of the ACM, Vol. 14, No. 1, Jan. 1971.

GENESYS: MACHINE-INDEPENDENT SOFTWARE SHARING

R. J. Allwood
The GENESYS Centre
University of Technology
Loughborough, Leicestershire, U.K.

ABSTRACT

Software for the future, in order to be enduring and profitable, must be flexible enough to meet changing design criteria and varying computers. The GENESYS Centre was sponsored by the U.K. Government as a master program to be used on a wide range of computers to solve engineering problems. The commercial and operational procedures adopted to fully exploit the advantages of GENESYS are discussed in this paper.

INTRODUCTION

Computers continue to get bigger, faster, and cheaper to use. While most are aware of this, the rate of change is often not appreciated. Amongst several investigators, Sharpe [1] has shown that during 16 years of the commercial sale of computers, the effectiveness/cost of hardware has improved so as to reduce the cost of a piece of scientific computation by approximately 24 percent per annum. Development at or near this rate may be expected for many years. Paralleling this fall in the cost of computing is the steady rise of programmers' salaries, currently in the United Kingdom 10 percent. These are two of the factors which combine to make the investment in software at any computer installation roughly equivalent now to the investment in hardware. A third factor, however, propels us forward to an even greater imbalance in favor of the software investment. The ever-increasing available computer power opens up further applications generally when the run-time economics compete with that of traditional methods. However, since computing languages make very little progress, development of the necessary software for new applications inevitably results in larger and larger programs. Technical application programs now range from 10,000 FORTRAN statements upwards with NASTRAN at 300,000, possibly the largest to date. These three factors are pushing us steadily forward from the free software era of the late 1950s through to unbundling of the late 1960s to perhaps an era of free hardware given away with software in the late 1970s. Identifying this trend clarifies a rational view for the development of software for the future. Resources will be found for the development of these major technological programs only if the investment will bring in a proper return. To secure this the software must be durable. In a world of rapidly changing technology, permanence can only be bought by flexibility, flexibility to meet changing design criteria exemplified by Codes of Practice, changes in material properties and availability, but above all flexibility to meet that most rapidly changing component, the computer upon which the software is to run. The first twenty years of programming effort in technical applications has been dominated by the need to rewrite or at least substantially modify working software to transfer it to each succeeding new generation of machine. Perhaps it has not mattered until now, with the software industry learning its trade and playing a relatively insignificant part economically. For the future there can be no doubt that software must be machine-independent. We can no longer afford to continue re-inventing the wheel.

GENESYS

In 1969 the U.K. Government made the decision that it should sponsor a system under which a library of integrated program suites could be made readily available on a wide range of computers. A master program called GENESYS (GENeral Engineering SYStem) was developed to solve the technical problems involved. The GENESYS Centre was established in 1970 to coordinate the development of the master program and an initial library of integrated suites. In September 1972, the programs were made commercially available, starting with a library of four integrated suites (known as subsystems). Eighteen subsystems are now available or being developed. The programs run on twelve combinations of computer/operating systems, and they are mounted on forty installations in the United Kingdom and three other countries. The Centre has twenty full time staff members promoting and supporting the use of the software, and is rapidly reaching a self-supporting commercial situation. GENESYS, which has been well described by its authors, Alcock & Shearing [2, 3], will be described only briefly in this paper. Attention will be focused rather more on the commercial and operational procedures adopted to exploit the powerful solution GENESYS provides to the problems of sharing software.

The subsystems in the library are developed and run under the control of the master program GENESYS. In a number of external ways GENESYS is similar to ICES [4], but it differs internally in design and implementation. The most important difference is that GENESYS was designed to be as machine independent as possible. Written in a low level version of FORTRAN (even ANSI FORTRAN is not machine independent) GENESYS has a few machine dependent routines amounting to approximately five percent of the total number of statements. These handle the reading and writing of characters, the transfer of records to and from peripheral devices, and the loading and calling of program modules. These routines have now been written for the twelve implementations already referred to (see Appendix I), and each implementation took approximately two to four man months effort by programmers familiar with the operating system and FORTRAN compilers of each particular machine. The total size of GENESYS is approximately 11,000 FORTRAN statements.

GENESYS performs two types of operation. For the programmer developing a subsystem it provides facilities through which a set

user and programmer manuals. The author determines the leasing charges and the royalty rate for bureau users, and the Centre adds an appropriate commission to these rates to cover the costs of handling the subsystem. The author enters into a contract with the Centre to maintain the subsystem, that is, to correct any errors which may be found during use. Where necessary the author also agrees to provide a technical service to answer the in-depth questions which users or prospective users may have about the program.

Distributing Software

We turn now to the task of distributing and maintaining the software. The majority of the software in the GENESYS library has been written by programmers working outside the GENESYS Centre. We do not take on any software unless it is documented to our standards, evidence of tests is produced for us to examine, and a benchmark test is supplied. For larger subsystems we arrange field trials before we are prepared to distribute such a subsystem. Programs satisfying these requirements are then distributed to users requesting them and automatically to all computer service bureaus. Generally they are distributed in source form, although compiled forms in binary are distributed to the common types of computers.

The one facet of the distribution operation which has so far not yielded to a satisfactory solution is that of incompatible magnetic tapes format between machines of different manufacture. We feel that this is the only remaining machine dependence deliberately fostered by computer manufacturers which we have not yet been able to circumvent. Currently we are forced to transport boxes of cards between key base machines upon which a suitable number of magnetic tapes may be produced.

Distribution is one thing; maintenance is entirely another. It must always be assumed that software contains errors and a formal procedure is necessary for rectifying them when detected. We have established a strict error recording procedure whereby when a complaint is made from either a user or a computer center, the symptoms of the fault are recorded immediately on preprinted forms with several copies. Our first action is to inform users of all errors, and this is done through a bimonthly error notice sent to all members which includes where possible an instruction to the user of ways to avoid the error. Unless the error is catastrophic, the correction is accumulated with other corrections for each subsystem anticipating updates only two or three times per year. An update, which is a set of editing statements to the source form of the current version, is prepared either by our own staff or by the original author of the subsystem. When an update is agreed upon, the editing statements are passed to the Centre's distribution staff, who go through the standard procedure of editing our own source form copy, running a series of benchmark tests, and then initiating the copying and distributing process of the update. Two weeks after the update has been distributed to all centers, a software notice is sent to all members summarizing the errors which have been cleared and any still outstanding. The updated subsystem is given a new version number which is printed out at the head of each set of results.

Use Promotion

Software needs to be sold together with the services required by users. In the promotion of GENESYS we use all the standard marketing devices of mailing shots, brochures, and leaflets, and we also have a field sales staff calling on prospective users. We find that the most convincing sales staff are engineers able to talk at the same level as the users, rather than programmers or systems experts. Success in this task depends not only upon selling ability but also on the provision of the backup services which users, whether sophisticated or not, expect and demand. The most obvious service is that of providing courses or short seminars to train engineers in the use of existing subsystems. More important to users, however, is the simple provision of an individual consultant service. Users expect a dynamic response to their own questions, particularly to the simple questions. All Centre staff members are geared to treating this service as top priority, and approximately 15 percent of our resources are used on the task. A charge is made to the user at a fixed hourly rate if more than an hour is spent providing advice, and users generally expect to be charged. Since, as already described, most of the programs on the GENESYS library were developed outside the Centre, it is essential for the programs to be documented to our standards for us to be able to provide a fast response. To each subsystem we allocate one or more Centre staff member as first-line contact men with reserves for backup during holidays or sickness. Each contact man has the responsibility of being familiar with the user manual of the subsystems he looks after. In some instances this familiarity extends to the programmer's manual. By this means, and by providing reserved telephones for "hot-line" questions from members, we are able to provide an immediate response to the majority of questions put to us by users. To some users this is the most important feature of GENESYS.

INDUSTRY REACTION

Judged on a purely commercial basis, the construction industry's reaction to GENESYS has been most favourable. Income within the first year of commercial operation exceeded half the Centre's expenses, and we expect to be self-supporting in two to three years time even with the planned expansion to a staff of 30 by then. The number of organizations contracted to the Centre has reached 100, and in the process of renewing the contracts which have finished their first year's term only 1 out of 30 has failed to renew. It is clear that industry is prepared to pay for a service, particularly for a service of support and advice given from a basis of sound documentation and comparable engineering experience. The construction industry market in the United Kingdom for this service is based on a total turnover of £4,000 m. per year. Design fees for this work total some £100 m., of which perhaps 10 to 20 percent may be transferred to computers.

Resistance within the industry appears to be mainly from the middle management, where substantial software investment has already been incurred. Senior management see the waste of resources likely if no change is made in programming methods, whilst new, untrained staff see the value of standard data preparation procedures and a centralized source of advice. A large part of the market is still relatively untapped, and GENESYS receives a very ready acceptance in this area. This is particularly so if an existing subsystem already matches the apparent needs of the user.

Many of the questions posed by prospective users from industry are answered by the services already described in this paper. One question which is commonly raised has not yet been discussed. This is the question of responsibility for errors in the software. In the contracts we have entered into, our position is stated quite simply that although we accept the responsibility of correcting errors in the software as quickly as possible we do not accept any consequential liabilities caused by that error. We believe that this is the only basis upon which a software service can be supplied. Unknown and extraneous factors can create errors in computer results. These range from faulty operation of the hardware through to misuse of the software outside the range of problems it was developed for. We believe too that it would be counter productive to assume for the software the responsibility now carried by the designer for the artifact he produces. Engineers frequently engage other people's services to discharge their function but never dissemble their responsibilities when using these services.

Four immediate effects can already be seen from the existence of the GENESYS program library:

1. Substantial investments are being made by national organizations in GENESYS subsystems as already described. The motivations for this are machine-independence and a ready-made marketing outlet

2. The facilities within GENESYS which allow many program modules to access a common data-bank are causing more programmers and users to think in terms of truly integrated program suites: examples are RC building design linked with quantities, bridge deck design linked with abutments and foundations

3. The existence of the GENESYS Centre focuses the views of engineers using a common set of programs. These user groups will become large and influential bodies allowing an otherwise unrepresented and fragmented viewpoint to be mobilized

4. The services offered by computer service bureaus can be compared on an equal basis as never before

THE FUTURE

The immediate future for GENESYS looks assured with a steady growth of the library of up to 50 or so subsystems and a penetration of 50 percent of the market. With three or four hundred user organizations this could profitably support a center with 40 staff members. The technical developments necessary to achieve this growth are basically straightforward. Generalized graphical input/output is already being included, and a more effective version for interactive working is needed. A substantial part of the growth will come almost automatically as hardware developments lead to minicomputers capable of running GENESYS at a price of £10,000.

There are technical problems that will be more difficult to solve, that is, to solve in the same machine independent spirit as is embodied in GENESYS now. Examples are in the handling of files. The present GENESYS files are not structured; what form should be introduced? Or, should GENESYS allow specialized routines to be introduced to access data files of unique forms? One problem which seems to recede satisfactorily is the dependence GENESYS has on FORTRAN. The demise of the language predicted so confidently seems now unlikely. Any improvements in FORTRAN can be easily accomodated by simply updating the compiler sections of GENESYS.

The most intractable problems we see ahead are not computer problems. The formulation of Codes of Practice in terms appropriate to computer programs would seem to be a generation or two away. The development of satisfactory code modules allowing a new Code of Practice to be plugged in easily seems equally far off despite all the work done on decision tables. Away from technical problems entirely, the social effects on introducing CAD techniques on a large scale are not negligible. The detailer/draughtsman receiving a bonus to his salary which depends upon the number of drawings he produces per week will not welcome a program that takes away all the tedious (but simple) details and leaves him only the challenging (but time-consuming) jobs. On the other side of the coin one of the advantages of using CAD programs is to reduce the time lag between the work of the draughtsman and the site workers. Reducing this time increases the effectiveness of any withdrawal of labor by the draughtsmen.

Professionally, the full exploitation of integrated program suites may backfire too. Once suitable file structures are devised the whole team of architect, structural engineer, services engineer, and contractor become more closely linked. Perhaps too closely. Will all the team be necessary, and if so, who stays? At least the professions will have to work more closely than now to stay in competition with the integrated building firm properly exploiting CAD to examine design alternatives as never before.

ACKNOWLEDGMENT

GENESYS and the GENESYS Centre have been funded by the Directorate of Research Requirements, the Department of the Environment, and we acknowledge this invaluable support without which no start could have been made. Many engineers have contributed also by guidance through many committees. We particularly acknowledge the contribution of Brian Scruby, Senior Partner, Sir Frederick Snow and Partners, Chairman of the GENESYS Board.

REFERENCES

1 Sharpe, W. F., The Economics of Computers, Columbia University Press, 1969, p. 342.

2 Alcock, D. G., and Shearing, B. H., "GENESYS - An Attempt to Rationalize the Use of Computers in Structural Engineering," The Structural Engineer, Vol. 48, No. 4, April 1970, pp. 143-152.

3 "GENESYS Reference Manual," The GENESYS Centre, University of Technology, Loughborough, Leics., 1972.

4 Roos, D., ICES System: General Description, R67-49 M.I.T. Press, 1967.

APPENDIX I

GENESYS Computer Range

Computer Series	Range Type	Operating System	Mark of O.S.
ICL 1900	1902S	Manual	-
	1902A	Exec	
	1904A, 1905E	George 2	9B
	1905F, 1903A	George 3	6.6, 7.6
UNIVAC 1100	1107	Exec 8	27
	1108	Exec 2	
ICL System 4	4/50	J Level	1600
	4/75, 4/72	MULTIJOB	
CDC 6000	6500, 6600, 7600	SCOPE	3.4, 3.2, 2.0
IBM 360	360/30, 360/40	DOS	25
	360/50, 360/65, 360/75	OS	MFT, MVT, 20.6, 21.6
IBM 370	370/135, 370/145	OS/VS	VS1.1
	370/153	DOS/VS	

APPENDIX II

Library

Subsystem	Status
<u>RC-Building/1</u> - for the design and detailing of the components of reinforced concrete buildings, producing moment envelopes, stresses, column loads, and bar-bending and bar-fixing schedules for beams, columns, and slabs. Sponsor - Department of the Environment, DRR	Available
<u>Bridge/1</u> - for the analysis of straight bridges which may be considered as continuous beams,with complex sections varying in depth supported in many ways. HA and HB loads can be distributed automatically with minimum data preparation requirements. Sponsor - Department of the Environment, DRR	Available
<u>Slab-Bridge/1</u> - for the analysis of skewed or curved decks with or without beams using the finite element method. Handles point, line, or distributed loads; settlements may also be analyzed. Sponsor - Department of the Environment, HECB	Available
<u>Frame-Analysis/1</u> - for the calculation of deflections, forces, bending moments and reactions in any plane frame, grillage, or space-frame. A wide variety of loads, settlement, and temperature effects can be analyzed. Sponsor - Department of the Environment, DRR	Available
<u>Slip-Circle/1</u> - for the calculation of the factor of safety of slopes against failure along a circle. Complex strata of soils, phreatic surface, and point loads at the top of the slope can be allowed for. Author - Geocomp UK Ltd.	Available
<u>UBM/1</u> - for the calculation of the ultimate bending moment and ultimate axial load of any concrete cross section with or without steel, using the stress-strain relationship assumed in CP110.	March 1974

Subsystem	Status
Author - Building Design Partnership	
Highways/1 - for the calculation of horizontal and vertical alignments of a road and the calculation of earthwork quantities from specified cross section templates. The original terrain is specified by levels on a regular grid.	Mid 1974
Drainage/1 - for the design of a drainage scheme, particularly for roads. This subsystem can be used directly from the results of Highways/1.	Late 1974
Sponsor - Department of the Environment, HECB	
Flat-Slab/1 - for the design and detailing of reinforced concrete buildings with flush slabs supported on columns. Beams are included and all components designed and detailed to CP110.	1975
Sponsor - Property Services Agency	
Portal-Frames/1 - for the design of single story multibay steel frames. All members will be sized and joint details calculated.	1974
Sponsor - Constrado, British Steel Corporation	
Pile-2D - for the analysis of a group of piles assuming 2-D behaviour. The user determines the point of support for each pile.	1974
Author - Stirling Maynard and Partners	
Backwater - for the calculation of the longitudinal profile of a river or water channel.	1974
Author - General & Engineering Computer Serv. Ltd.	
Storm-Water - for the sizing of the pipes of a drainage network. The method used is the RRL Hydrograph method.	Mid 1974
Author - Sir Frederick Snow & Partners	
Water-Distribution - for the analysis of fluid flow through a network of pipes. A comprehensive range of pumps, valves, and reservoirs may be included in the system	Mid 1974
Author - Water Research Association	

Subsystem	Status
Composite-Construction - for the analysis and design of a range of sections composed of RC slabs and steel beams. The program determines the cost of each section from user specified factors. Author - Fitzroy Computers Ltd.	Early 1974
Subframe - for the analysis of a continuous beam including the stiffness of supporting columns. Redistribution of moments to CP110 is allowed from factors submitted in input data. Author - Fitzroy Computers Ltd.	Early 1974
General-Slip - for the calculation of the factor of safety of a soil embankment failing along a general failure surface. This subsystem is similar to SLIP-CIRCLE but allows noncircular failure surfaces to be tried. Author -Geocomp Ltd.	Early 1974
DAMWAND - for the calculation of stresses and displacements in sheet-piles. Several construction phases can be handled. Author - Rijkswaterstaat, Holland	Mid 1974

INDEXES

SUBJECT INDEX OF STRUCTURAL MECHANICS COMPUTER PROGRAMS

(See the Alphabetical Index for page numbers)

ALPHABETICAL INDEX OF STRUCTURAL MECHANICS COMPUTER PROGRAMS